STUDENT'S SOLUTIONS MANUAL

Cindy Trimble & Associates

Kevin Bodden
Lewis & Clark Community College

Randy Gallaher
Lewis & Clark Community College

ELEMENTARY ALGEBRA

SECOND EDITION

Michael Sullivan, III
Joliet Junior College

Katherine R. Struve
Columbus State Community College

Janet Mazzarella
Southwestern College

Prentice Hall
is an imprint of

Reproduced by Pearson Prentice Hall from electronic files supplied by the author.

Publishing as Prentice Hall, Upper Saddle River, NJ 07458.

1 2 3 4 5 6 BRR 10 09 08

ISBN-13: 978-0-321-58930-9 Standalone
ISBN-10: 0-321-58930-0 Standalone

ISBN-13: 978-0-321-58956-9 Component
ISBN-10: 0-321-58956-4 Component

Prentice Hall
is an imprint of

www.pearsonhighered.com

Contents

Chapter 1

Section 1.2

1.2 Quick Checks

1. In the statement $6 \cdot 8 = 48$, 6 and 8 are called <u>factors</u> and 48 is called the <u>product</u>.

2. 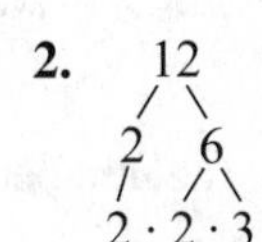
The prime factorization of 12 is $2 \cdot 2 \cdot 3$.

3. 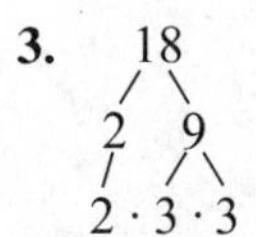
The prime factorization of 18 is $2 \cdot 3 \cdot 3$.

4. 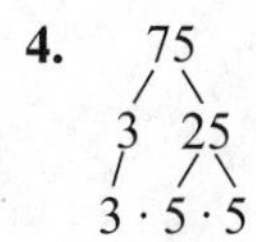
The prime factorization of 75 is $3 \cdot 5 \cdot 5$.

5. 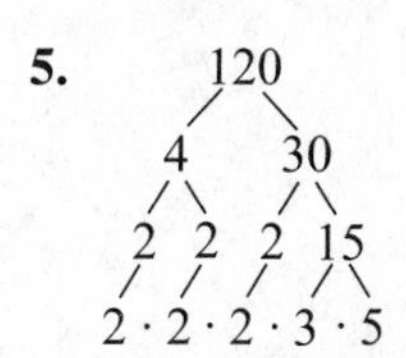
The prime factorization of 120 is $2 \cdot 2 \cdot 2 \cdot 3 \cdot 5$.

6. 131 is a prime number.

7. 459
   9 51
   $3 \cdot 3 \cdot 3 \cdot 17$
The prime factorization of 459 is $3 \cdot 3 \cdot 3 \cdot 17$.

8. $6 = 2 \cdot 3$
   $8 = 2 \cdot \quad \cdot 2 \cdot 2$
   $2 \cdot 3 \cdot 2 \cdot 2$
The LCM is $2 \cdot 3 \cdot 2 \cdot 2 = 24$.

9. $5 = 5$
   $10 = 5 \cdot 2$
   $5 \cdot 2$
The LCM is $2 \cdot 5 = 10$.

10. $45 = 3 \cdot 3 \cdot 5$
    $72 = 3 \cdot 3 \cdot \quad 2 \cdot 2 \cdot 2$
    $3 \cdot 3 \cdot 5 \cdot 2 \cdot 2 \cdot 2$
The LCM is $2 \cdot 2 \cdot 2 \cdot 3 \cdot 3 \cdot 5 = 360$.

11. $7 = 7$
    $3 = \quad 3$
    $7 \cdot 3$
The LCM is $3 \cdot 7 = 21$.

12. $12 = 2 \cdot 2 \cdot 3$
    $18 = 2 \cdot \quad 3 \cdot 3$
    $30 = 2 \cdot \quad 3 \cdot \quad 5$
    $2 \cdot 2 \cdot 3 \cdot 3 \cdot 5$
The LCM is $2 \cdot 2 \cdot 3 \cdot 3 \cdot 5 = 180$.

13. Fractions which represent the same portion of a whole are called equivalent <u>fractions</u>.

14. Multiply the numerator and denominator of $\frac{1}{2}$ by 5.
$\frac{1}{2} = \frac{1 \cdot 5}{2 \cdot 5} = \frac{5}{10}$

15. Multiply the numerator and denominator of $\frac{5}{8}$ by 6.
$\frac{5}{8} = \frac{5 \cdot 6}{8 \cdot 6} = \frac{30}{48}$

16. The denominators are 4 and 6.
    $4 = 2 \cdot 2$
    $6 = 2 \cdot \quad 3$
LCD $= 2 \cdot 2 \cdot 3 = 12$
$\frac{1}{4} = \frac{1 \cdot 3}{4 \cdot 3} = \frac{3}{12}$
$\frac{5}{6} = \frac{5 \cdot 2}{6 \cdot 2} = \frac{10}{12}$

17. The denominators are 12 and 15.
    $12 = 2 \cdot 2 \cdot 3$
    $15 = \quad 3 \cdot 5$
LCD $= 2 \cdot 2 \cdot 3 \cdot 5 = 60$

$$\frac{5}{12}=\frac{5\cdot 5}{12\cdot 5}=\frac{25}{60}$$
$$\frac{4}{15}=\frac{4\cdot 4}{15\cdot 4}=\frac{16}{60}$$

18. The denominators are 20 and 16.
$$20=2\cdot 2\cdot 5$$
$$16=2\cdot 2\cdot \quad 2\cdot 2$$
$$\text{LCD}=2\cdot 2\cdot 5\cdot 2\cdot 2=80$$
$$\frac{9}{20}=\frac{9\cdot 4}{20\cdot 4}=\frac{36}{80}$$
$$\frac{11}{16}=\frac{11\cdot 5}{16\cdot 5}=\frac{55}{80}$$

19. $\frac{45}{80}=\frac{3\cdot 3\cdot 5}{2\cdot 2\cdot 2\cdot 2\cdot 5}=\frac{3\cdot 3\cdot \not{5}}{2\cdot 2\cdot 2\cdot 2\cdot \not{5}}=\frac{9}{16}$

20. $\frac{4}{9}=\frac{2\cdot 2}{3\cdot 3}=\frac{4}{9}$

21. $\frac{20}{50}=\frac{2\cdot 2\cdot 5}{2\cdot 5\cdot 5}=\frac{\not{2}\cdot 2\cdot \not{5}}{\not{2}\cdot 5\cdot \not{5}}=\frac{2}{5}$

22. $\frac{30}{105}=\frac{2\cdot 3\cdot 5}{3\cdot 5\cdot 7}=\frac{2\cdot \not{3}\cdot \not{5}}{\not{3}\cdot \not{5}\cdot 7}=\frac{2}{7}$

23. The 1 is two places to the right of the decimal; this is the hundredths place.

24. The 2 is one place to the right of the decimal; this the tenths place.

25. The 8 is four places to the left of the decimal; this is the thousands place.

26. The 9 is three places to the right of the decimal; this is the thousandths place.

27. The 3 is one place to the left of the decimal; this is the ones place.

28. The 2 is five places to the left of the decimal; this is the ten thousands place.

29. The number 1 is in the tenths place. The number to its right is 7. Since 7 is greater than 5, we round to 0.2.

30. The number 3 is in the hundredths place. The number to its right is 2. Since 2 is less than 5, we round to 0.93.

31. The number 9 is in the hundredths place. The number to its right is 6. Since 6 is greater than 5, we round to 1.40.

32. The number 8 is in the thousandths place. The number to its right is 3. Since 3 is less than 5, we round to 14.398.

33. The number 0 is in the hundredths place. The number to its right is 4. Since 4 is less than 5, we round to 690.00.

34. The number 9 is in the tenths place. The number to its right is 8. Since 8 is greater than 5, we round to 60.0.

35.
```
   0.4
5)2.0
  2 0
  ---
    0
```
$\frac{2}{5}=0.4$

36.
```
   0.428571
7)3.000000
  2 8
  ---
    20
    14
    --
     60
     56
     --
      40
      35
      --
       50
       49
       --
        10
         7
        --
         3
```
$\frac{3}{7}=0.\overline{428571}$

37.
```
   1.375
8)11.000
   8
   --
   3 0
   2 4
   ---
     60
     56
     --
      40
      40
      --
       0
```
$\frac{11}{8}=1.375$

38. $$\begin{array}{r} 0.833 \\ 6\overline{)5.000} \\ \underline{4\,8} \\ 20 \\ \underline{18} \\ 20 \\ \underline{18} \\ 2 \end{array}$$

$\frac{5}{6} = 0.8\overline{3}$

39. $$\begin{array}{r} 0.55 \\ 9\overline{)5.00} \\ \underline{4\,5} \\ 50 \\ \underline{45} \\ 5 \end{array}$$

$\frac{5}{9} = 0.\overline{5}$

40. $0.65 = \frac{65}{100} = \frac{\cancel{5} \cdot 13}{\cancel{5} \cdot 20} = \frac{13}{20}$

41. $0.2 = \frac{2}{10} = \frac{\cancel{2}}{\cancel{2} \cdot 5} = \frac{1}{5}$

42. $0.625 = \frac{625}{1000} = \frac{5 \cdot \cancel{125}}{8 \cdot \cancel{125}} = \frac{5}{8}$

43. The word percent means parts per hundred, so 35% means 35 parts out of 100 parts or $\frac{35}{100}$.

44. $23\% = 23\% \cdot \frac{1}{100\%} = \frac{23}{100} = 0.23$

45. $1\% = 1\% \cdot \frac{1}{100\%} = \frac{1}{100} = 0.01$

46. $72.4\% = 72.4\% \cdot \frac{1}{100\%} = \frac{72.4}{100} = 0.724$

47. $127\% = 127\% \cdot \frac{1}{100\%} = \frac{127}{100} = 1.27$

48. $89.26\% = 89.26\% \cdot \frac{1}{100\%} = \frac{89.26}{100} = 0.8926$

49. To convert a decimal to a percent, multiply the decimal by 100%.

50. $0.15 = 0.15 \cdot \frac{100\%}{1} = 15\%$

51. $0.8 = 0.8 \cdot \frac{100\%}{1} = 80\%$

52. $1.3 = 1.3 \cdot \frac{100\%}{1} = 130\%$

53. $0.398 = 0.398 \cdot \frac{100\%}{1} = 39.8\%$

54. $0.004 = 0.004 \cdot \frac{100\%}{1} = 0.4\%$

1.2 Exercises

55.
```
 25
 / \
5 · 5
```
The prime factorization of 25 is $5 \cdot 5$.

57.
```
   28
   / \
  4   7
 / \   \
2 · 2 · 7
```
The prime factorization of 28 is $2 \cdot 2 \cdot 7$.

59.
```
 21
 / \
3 · 7
```
The prime factorization of 21 is $3 \cdot 7$.

61.
```
     36
    /  \
   4    9
  / \  / \
 2 · 2 · 3 · 3
```
The prime factorization of 36 is $2 \cdot 2 \cdot 3 \cdot 3$.

63.
```
   20
   / \
  4   5
 / \   \
2 · 2 · 5
```
The prime factorization of 20 is $2 \cdot 2 \cdot 5$.

65.
```
  30
  / \
 5   6
 |  / \
 5 · 2 · 3
```
The prime factorization of 30 is $2 \cdot 3 \cdot 5$.

67. 50
/ \
2 25
/ / \
$2 \cdot 5 \cdot 5$
The prime factorization of 50 is $2 \cdot 5 \cdot 5$.

69. 53 is a prime number.

71.
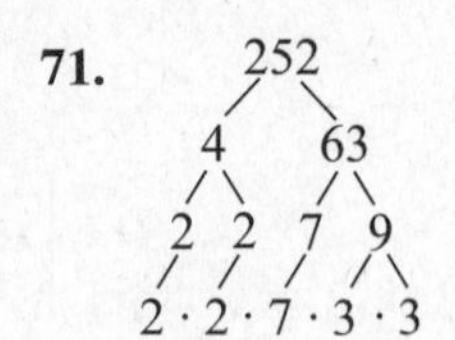

The prime factorization of 252 is $2 \cdot 2 \cdot 3 \cdot 3 \cdot 7$.

73.
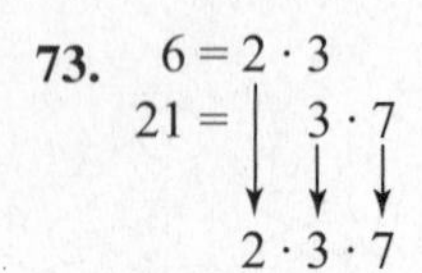

The LCM is $2 \cdot 3 \cdot 7 = 42$.

75. $12 = 2 \cdot 2 \cdot 3$
$10 = 2 \cdot \quad 5$
$2 \cdot 2 \cdot 3 \cdot 5$
The LCM is $2 \cdot 2 \cdot 3 \cdot 5 = 60$.

77. $15 = 3 \cdot 5$
$14 = \quad 2 \cdot 7$
$3 \cdot 5 \cdot 2 \cdot 7$
The LCM is $2 \cdot 3 \cdot 5 \cdot 7 = 210$.

79. $30 = 2 \cdot 3 \cdot 5$
$45 = \quad 3 \cdot 5 \cdot 3$
$2 \cdot 3 \cdot 5 \cdot 3$
The LCM is $2 \cdot 3 \cdot 3 \cdot 5 = 90$.

81. $5 = 5$
$6 = \quad 2 \cdot 3$
$12 = \quad 2 \cdot 3 \cdot 2$
$5 \cdot 2 \cdot 3 \cdot 2$
The LCM is $2 \cdot 2 \cdot 3 \cdot 5 = 60$.

83. $3 = 3$
$8 = \quad 2 \cdot 2 \cdot 2$
$9 = 3 \cdot \quad 3$
$3 \cdot 2 \cdot 2 \cdot 2 \cdot 3$
The LCM is $2 \cdot 2 \cdot 2 \cdot 3 \cdot 3 = 72$.

85. $\frac{2}{3} = \frac{2 \cdot 4}{3 \cdot 4} = \frac{8}{12}$

87. $\frac{3}{4} = \frac{3 \cdot 6}{4 \cdot 6} = \frac{18}{24}$

89. $7 = \frac{7 \cdot 3}{1 \cdot 3} = \frac{21}{3}$

91. $2 = 2$
$8 = 2 \cdot 2 \cdot 2$
$\text{LCD} = 2 \cdot 2 \cdot 2 = 8$
$\frac{1}{2} = \frac{1 \cdot 4}{2 \cdot 4} = \frac{4}{8}$
The equivalent fractions are $\frac{4}{8}$ and $\frac{3}{8}$.

93. $5 = \quad 5$
$3 = 3$
$\text{LCD} = 3 \cdot 5 = 15$
$\frac{3}{5} = \frac{3 \cdot 3}{5 \cdot 3} = \frac{9}{15}$
$\frac{2}{3} = \frac{2 \cdot 5}{3 \cdot 5} = \frac{10}{15}$
The equivalent fractions are $\frac{9}{15}$ and $\frac{10}{15}$.

95. $6 = 2 \cdot \quad 3$
$8 = 2 \cdot 2 \cdot 2$
$\text{LCD} = 2 \cdot 2 \cdot 2 \cdot 3 = 24$
$\frac{5}{6} = \frac{5 \cdot 4}{6 \cdot 4} = \frac{20}{24}$
$\frac{5}{8} = \frac{5 \cdot 3}{8 \cdot 3} = \frac{15}{24}$
The equivalent fractions are $\frac{20}{24}$ and $\frac{15}{24}$.

97. $15 = \quad 3 \cdot 5$
$20 = 2 \cdot 2 \cdot \quad 5$
$\text{LCD} = 2 \cdot 2 \cdot 3 \cdot 5 = 60$
$\frac{11}{15} = \frac{11 \cdot 4}{15 \cdot 4} = \frac{44}{60}$
$\frac{7}{20} = \frac{7 \cdot 3}{20 \cdot 3} = \frac{21}{60}$
The equivalent fractions are $\frac{44}{60}$ and $\frac{21}{60}$.

99. $9 = 3 \cdot 3$
$18 = 2 \cdot 3 \cdot 3$
$30 = 2 \cdot 3 \cdot 5$

$\text{LCD} = 2 \cdot 3 \cdot 3 \cdot 5 = 90$

$\frac{2}{9} = \frac{2 \cdot 10}{9 \cdot 10} = \frac{20}{90}$

$\frac{7}{18} = \frac{7 \cdot 5}{18 \cdot 5} = \frac{35}{90}$

$\frac{7}{30} = \frac{7 \cdot 3}{30 \cdot 3} = \frac{21}{90}$

The equivalent fractions are $\frac{20}{90}$, $\frac{35}{90}$, and $\frac{21}{90}$.

101. $\frac{14}{21} = \frac{2 \cdot 7}{3 \cdot 7} = \frac{2 \cdot \cancel{7}}{3 \cdot \cancel{7}} = \frac{2}{3}$

103. $\frac{38}{18} = \frac{2 \cdot 19}{2 \cdot 9} = \frac{\cancel{2} \cdot 19}{\cancel{2} \cdot 9} = \frac{19}{9}$

105. $\frac{22}{44} = \frac{2 \cdot 11}{2 \cdot 2 \cdot 11} = \frac{\cancel{2} \cdot \cancel{11}}{\cancel{2} \cdot 2 \cdot \cancel{11}} = \frac{1}{2}$

107. $\frac{32}{40} = \frac{2 \cdot 2 \cdot 2 \cdot 2 \cdot 2}{2 \cdot 2 \cdot 2 \cdot 5} = \frac{\cancel{2} \cdot \cancel{2} \cdot \cancel{2} \cdot 2 \cdot 2}{\cancel{2} \cdot \cancel{2} \cdot \cancel{2} \cdot 5} = \frac{4}{5}$

109. The 0 is two places to the right of the decimal; this is the hundredths place.

111. The 5 is two places to the left of the decimal; this is the tens place.

113. The 6 is three places to the right of the decimal; this is the thousandths place.

115. The number 2 is in the tenths place. The number to the right of 2 is 0. Since 0 is less than 5, we round to 578.2.

117. The number 5 is in the tens place. The number to the right of 5 is 4. Since 4 is less than 5, we round to 350.

119. The number 9 is in the thousandths place. The number to the right of 9 is 8. Since 8 is greater than 5, we round to 3682.010.

121. The number 9 is in the ones place. The number to its right is also 9. Since 9 is greater than 5, we round to 30.

123.
```
    0.625
8 ) 5.000
    4 8
      20
      16
       40
       40
        0
```
$\frac{5}{8} = 0.625$

125.
```
    0.285714
7 ) 2.000000
    1 4
      60
      56
       40
       35
        50
        49
         10
          7
          30
          28
           2
```
$\frac{2}{7} = 0.\overline{285714}$

127.
```
     0.3125
16 ) 5.0000
     4 8
       20
       16
        40
        32
         80
         80
          0
```
$\frac{5}{16} = 0.3125$

129. $$\begin{array}{r} 0.230769 \\ 13\overline{)3.000000} \\ \underline{2\,6} \\ 40 \\ \underline{39} \\ 100 \\ \underline{91} \\ 90 \\ \underline{78} \\ 120 \\ \underline{117} \\ 3 \end{array}$$

$\frac{3}{13} = 0.\overline{230769}$

131. $$\begin{array}{r} 1.16 \\ 25\overline{)29.00} \\ \underline{25} \\ 4\,0 \\ \underline{2\,5} \\ 1\,50 \\ \underline{1\,50} \\ 0 \end{array}$$

$\frac{29}{25} = 1.16$

133. $$\begin{array}{r} 2.16 \\ 6\overline{)13.0} \\ \underline{12} \\ 1\,0 \\ \underline{6} \\ 40 \\ \underline{36} \\ 4 \end{array}$$

$\frac{13}{6}$ rounded to the nearest tenth is 2.2.

135. $$\begin{array}{r} 2.666 \\ 3\overline{)8.000} \\ \underline{6} \\ 2\,0 \\ \underline{1\,8} \\ 20 \\ \underline{18} \\ 20 \\ \underline{18} \\ 2 \end{array}$$

$\frac{8}{3}$ rounded to the nearest hundredth is 2.67.

137. $$\begin{array}{r} 0.5185 \\ 27\overline{)14.0000} \\ \underline{13\,5} \\ 50 \\ \underline{27} \\ 230 \\ \underline{216} \\ 140 \\ \underline{135} \\ 5 \end{array}$$

$\frac{14}{27}$ rounded to the nearest thousandth is 0.519.

139. $0.75 = \frac{75}{100} = \frac{3 \cdot \cancel{25}}{4 \cdot \cancel{25}} = \frac{3}{4}$

141. $0.9 = \frac{9}{10}$

143. $0.982 = \frac{982}{1000} = \frac{\cancel{2} \cdot 491}{\cancel{2} \cdot 500} = \frac{491}{500}$

145. $0.2525 = \frac{2525}{10{,}000} = \frac{\cancel{25} \cdot 101}{\cancel{25} \cdot 400} = \frac{101}{400}$

147. $37\% = 37\% \cdot \frac{1}{100\%} = \frac{37}{100} = 0.37$

149. $6.02\% = 6.02\% \cdot \frac{1}{100\%} = \frac{6.02}{100} = 0.0602$

151. $0.1\% = 0.1\% \cdot \frac{1}{100\%} = \frac{0.1}{100} = 0.001$

153. $0.2 = 0.2 \cdot \frac{100\%}{1} = 20\%$

155. $0.275 = 0.275 \cdot \frac{100\%}{1} = 27.5\%$

157. $2 = 2 \cdot \frac{100\%}{1} = 200\%$

159. The least number of months is the LCM of 3, 7, and 12.

$$\begin{aligned} 3 &= 3 \\ 7 &= \phantom{3 \cdot{}} 7 \\ 12 &= 3 \cdot \phantom{7 \cdot{}} 2 \cdot 2 \\ \text{LCM} &= 3 \cdot 7 \cdot 2 \cdot 2 \end{aligned}$$

The least number of months is $2 \cdot 2 \cdot 3 \cdot 7 = 84$.

161. The number of days is the LCM of 4 and 10.

$4 = 2 \cdot 2$

$10 = 2 \cdot \quad 5$

$\text{LCM} = 2 \cdot 2 \cdot 5$

Bob gives Sam both types of medication on the same day every $2 \cdot 2 \cdot 5 = 20$ days.

163. 325 of 500

$$\frac{325}{500} = \frac{5 \cdot 5 \cdot 13}{2 \cdot 2 \cdot 5 \cdot 5 \cdot 5} = \frac{\cancel{5} \cdot \cancel{5} \cdot 13}{2 \cdot 2 \cdot \cancel{5} \cdot \cancel{5} \cdot 5} = \frac{13}{20}$$

The fraction of students that work at least 25 hours per week is $\frac{13}{20}$.

165. 840 of 1200

$$\begin{array}{r} 0.7 \\ 1200\overline{)840.0} \\ \underline{840\ 0} \\ 0 \end{array}$$

$$\frac{840}{1200} = 0.7 = 0.7 \cdot \frac{100\%}{1} = 70\%$$

70% of adult Americans believe that they eat healthily.

167. 85 of 110

$$\begin{array}{r} 0.772 \\ 110\overline{)85.000} \\ \underline{77\ 0} \\ 8\ 00 \\ \underline{7\ 70} \\ 300 \\ \underline{220} \\ 80 \end{array}$$

$$\frac{85}{110} = 0.7\overline{72} \approx 0.7727 \cdot \frac{100\%}{1} = 77.27\%$$

The score is 77.27%.

169. **(a)** $\frac{8}{24} = 0.\overline{3}$

$$0.3333 \cdot \frac{100\%}{1} = 33.33\%$$

Jackson sleeps for 33.33% of the day.

(b) $\frac{4}{24} = 0.1\overline{6}$

$$0.1667 \cdot \frac{100\%}{1} = 16.67\%$$

Jackson works for 16.67% of the day.

(c) $\frac{6}{24} = 0.25$

$$0.25 \cdot \frac{100\%}{1} = 25\%$$

Jackson goes to school and studies 25% of the day.

171. 3 of 14

$$\begin{array}{r} 0.21428 \\ 14\overline{)3.00000} \\ \underline{2\ 8} \\ 20 \\ \underline{14} \\ 60 \\ \underline{56} \\ 40 \\ \underline{28} \\ 120 \\ \underline{112} \\ 8 \end{array}$$

$$\frac{3}{14} \approx 0.2143$$

$$0.2143 \cdot \frac{100\%}{1} = 21.43\%$$

21.43% is saturated fat.

173. The process results in the following 25 prime numbers: 2, 3, 5, 7, 11, 13, 17, 19, 23, 29, 31, 37, 41, 43, 47, 53, 59, 61, 67, 71, 73, 79, 83, 89, 97

Section 1.3

Preparing for the Number Systems and the Real Number Line

P1.

$$\begin{array}{r} 0.625 \\ 8\overline{)5.000} \\ \underline{4\ 8} \\ 20 \\ \underline{16} \\ 40 \\ \underline{40} \\ 0 \end{array}$$

$$\frac{5}{8} = 0.625$$

P2.

$$\begin{array}{r} 0.8181 \\ 11\overline{)9.0000} \\ \underline{8\,8} \\ 20 \\ \underline{11} \\ 90 \\ \underline{88} \\ 20 \\ \underline{11} \\ 9 \end{array}$$

$\frac{9}{11} = 0.8181\ldots$ or $0.\overline{81}$

1.3 Quick Checks

1. The first 4 positive odd numbers are 1, 3, 5, and 7. If we let O represent this set, then $O = \{1, 3, 5, 7\}$.

2. The states whose names begin with the letter *A* are Alabama, Alaska, Arizona, and Arkansas. If we let *A* represent this set, then $A = \{$Alabama, Alaska, Arizona, Arkansas$\}$.

3. There are no states whose names begin with the letter *Z*. If we let *Z* represent this set, then $Z = \{\ \}$ or $\varnothing$.

4. Every integer is a rational number. True

5. Real numbers that can be represented with a terminating decimal are called rational.

6. 12 is the only natural number.

7. 0 and 12 are the whole numbers.

8. -5, 12, and 0 are the integers.

9. $\frac{11}{5}$, -5, 12, $2.\overline{76}$, 0, and $\frac{18}{4}$ are the rational numbers.

10. π is the only irrational number.

11. All the numbers listed are real numbers.

12. Number line from -3 to 4 with points plotted at -2, 0, $\frac{1}{2}$, 3, 3.5.

13. The symbols $<, >, \le, \ge$ are called inequality symbols.

14. $2 < 9$ because 2 lies to the left of 9 on the real number line.

15. $-5 < -3$ because -5 lies to the left of -3 on the real number line.

16. $\frac{4}{5} > \frac{1}{2}$ because $\frac{4}{5} = \frac{8}{10}$ and $\frac{1}{2} = \frac{5}{10}$. Having 8 parts out of 10 is more than having 5 parts out of 10. Also, $\frac{4}{5} = 0.8$ and $\frac{1}{2} = 0.5$ and 0.8 lies to the right of 0.5 on the real number line.

17. $\frac{4}{7} > 0.5$ because $\frac{4}{7} = 0.\overline{571428}$ and $0.\overline{571428}$ lies to the right of 0.5 on the real number line.

18. $\frac{4}{3} = \frac{20}{15}$

19. $-\frac{4}{3} < -\frac{5}{4}$

20. The distance from zero to a point on a real number line whose coordinate is *a* is called the absolute value of *a*.

21. $|-15| = 15$ because the distance from 0 to -15 on the real number line is 15.

22. $\left|-\frac{3}{4}\right| = \frac{3}{4}$ because the distance from 0 to $-\frac{3}{4}$ on the real number line is $\frac{3}{4}$.

1.3 Exercises

23. The set of whole numbers less than 5 is $A = \{0, 1, 2, 3, 4\}$.

25. The set of natural numbers less than 5 is $D = \{1, 2, 3, 4\}$.

27. The set of even natural numbers between 4 and 15 is $E = \{6, 8, 10, 12, 14\}$.

29. 3 is the only natural number.

31. -4, 3, and 0 are the integers.

33. 2.303003000... is the only irrational number.

35. All the numbers listed are real numbers.

37. π is the only irrational number.

39. $\frac{5}{5}=1$, so $\frac{5}{5}$ is the only whole number.

41. Number line with points at -2, -1.5, 0, $\frac{3}{3}$, $\frac{4}{3}$; axis labels -3, -2, -1, 0, 1, 2, 3.

43. -2 lies to the right of -3 on the number line. The statement is true.

45. Since $-6=-6$, the statement is true.

47. Since the decimal equivalent of $\frac{3}{2}$ is 1.5, the statement is true.

49. Since $\pi = 3.14159...$, $\pi > 3.14$. The statement is false.

51. Since -1 lies to the left of 0 on the number line, $-1 < 0$.

53. Since $\frac{5}{8}=0.625$ and $\frac{6}{11}=0.\overline{54}$, $\frac{5}{8}>\frac{6}{11}$.

55. Since $\frac{6}{13}=0.\overline{461538}$, $\frac{6}{13}>0.46$.

57. Since $\frac{42}{6}$ simplifies to 7, $\frac{42}{6}=7$.

59. $|-12|=12$ because the distance from 0 to -12 on the real number line is 12.

61. $|4|=4$ because the distance from 0 to 4 on the real number line is 4.

63. $\left|-\frac{3}{8}\right|=\frac{3}{8}$ because the distance from 0 to $-\frac{3}{8}$ on the real number line is $\frac{3}{8}$.

65. $|-2.1|=2.1$ because the distance from 0 to -2.1 on the real number line is 2.1.

67. (a) Number line with points at -4.5, -1, $-\frac{1}{2}$, $\frac{3}{5}$, 1, 3.5, $|-7|$; axis labels -7 -6 -5 -4 -3 -2 -1 0 1 2 3 4 5 6 7.

(b) From left to right on the number line, the order of the numbers is -4.5, -1, $-\frac{1}{2}$, $\frac{3}{5}$, 1, 3.5, $|-7|$.

(c) **(i)** -1, 1, and $|-7|=7$ are integers.

(ii) All numbers listed are rational numbers.

		Natural	Whole	Integers	Rational	Irrational	Real
69.	−100			√	√		√
71.	−10.5				√		√
73.	$\frac{75}{25}$	√	√	√	√		√
75.	7.56556555...					√	√

77. The whole numbers are a subset of the integers. The statement is true.

79. The set of rational numbers and the set of irrational numbers have no elements in common. The statement is false.

81. The natural numbers are a subset of the whole numbers. The statement is true.

83. Every terminating decimal can be expressed as the ratio of two integers. The statement is true.

85. 0 is an integer that is neither negative or positive. The statement is true.

87. Non-terminating and non-repeating decimals are irrational numbers.

89. The set of rational numbers combined with the irrational numbers comprises the set of real numbers.

91. The only number that is both nonnegative and non-positive is 0.

93. Both elements of *Y* are also elements of *X*, so $Y \subseteq X$. The statement is true.

95. Both elements of *Y* are also elements of *Z*, so $Y \subseteq Z$. The statement is true.

97. $A \cup B$ consists of all elements that are in either *A* or *B*, so $A \cup B = \{7, 8, 9, 10, 11, 12, 13, 14, 15\}$.

99. $B \cup C$ consists of all elements that are in either *B* or *C*, so $B \cup C = \{10, 11, 12, 13, 14, 15\}$.

101. $A \cap C$ is the set of all elements common to both *A* and *C*, so $A \cap C = \{11, 12\}$.

103. $A \cap B$ is the set of even whole numbers less than 11, so $A \cap B = \{2, 4, 6, 8, 10\}$.

105. **(a)** Answers may vary. One possibility: There are subsets with no elements, subsets with one element, subsets with two elements, subsets with three elements, and subsets with four elements.
0: { }
1: {1}, {2}, {3}, {4}
2: {1, 2}, {1, 3}, {1, 4}, {2, 3}, {2, 4}, {3, 4}
3: {1, 2, 3}, {1, 2, 4}, {1, 3, 4}, {2, 3 4}
4: {1, 2, 3, 4}

(b) There are a total of 16 subsets.

107. A rational number is any number that can be written as the quotient of two integers, denominator not equal to zero. Natural numbers, whole numbers and integers are rational numbers. Terminating and repeating decimals are also rational numbers.

Section 1.4

Preparing for Adding, Subtracting, Multiplying, and Dividing Integers

P1. $\frac{16}{36} = \frac{2\cdot2\cdot2\cdot2}{2\cdot2\cdot3\cdot3} = \frac{\not{2}\cdot\not{2}\cdot2\cdot2}{\not{2}\cdot\not{2}\cdot3\cdot3} = \frac{2\cdot2}{3\cdot3} = \frac{4}{9}$

1.4 Quick Checks

1. The answer to an addition problem is called the sum.

2.

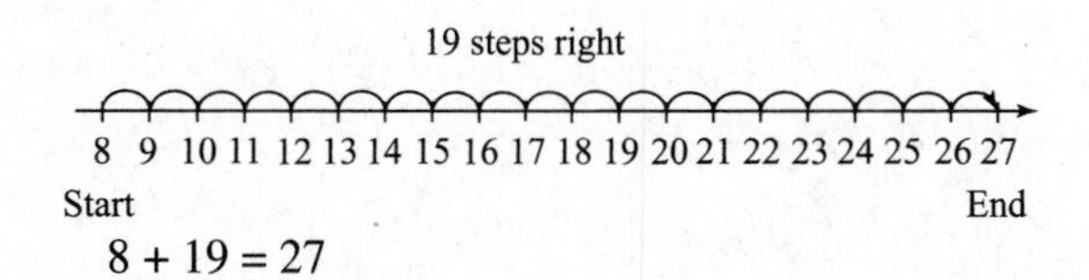

$8 + 19 = 27$

3.

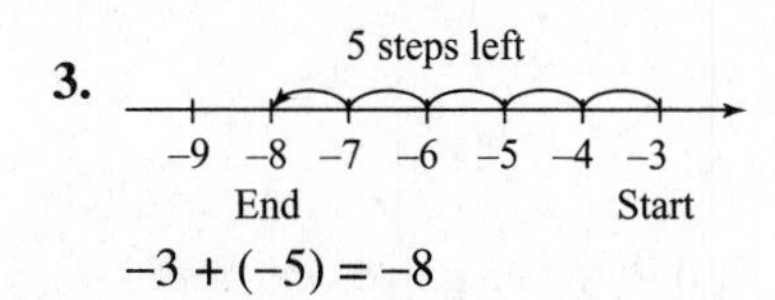

$-3 + (-5) = -8$

4.

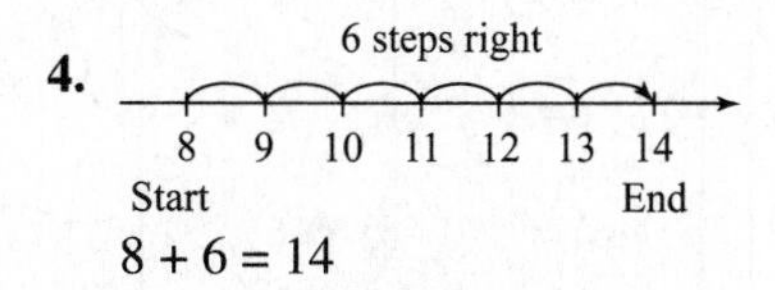

$8 + 6 = 14$

5.

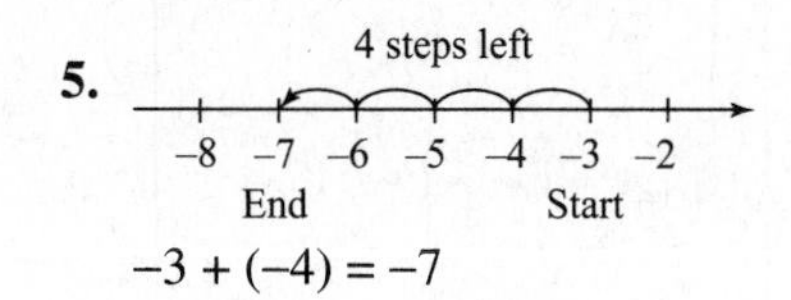

$-3 + (-4) = -7$

6.

12 steps left

−18 −16 −14 −12 −10 −8 −6 −4

End Start

$-5 + (-12) = -17$

7.

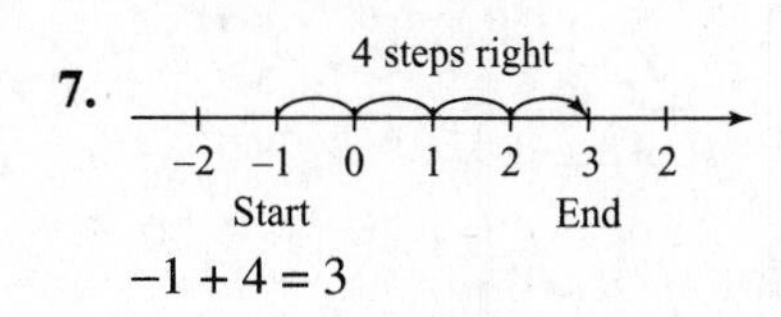

$-1 + 4 = 3$

8.

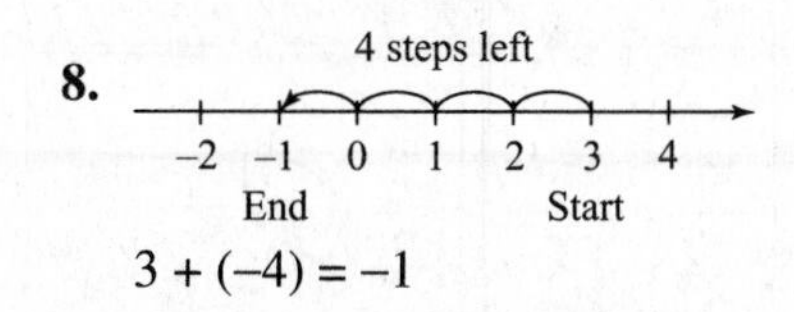

$3 + (-4) = -1$

9.

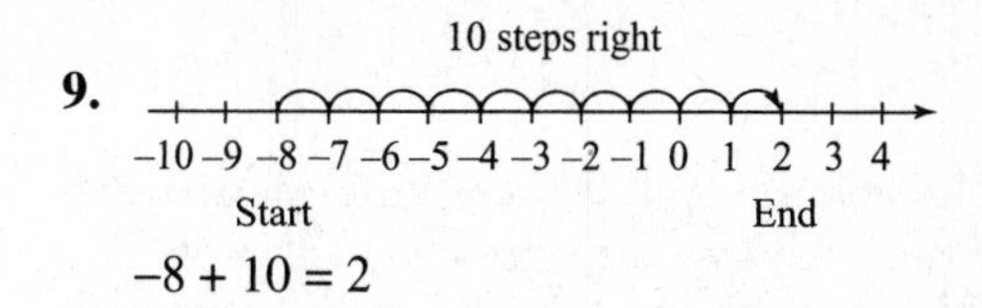

$-8 + 10 = 2$

10.

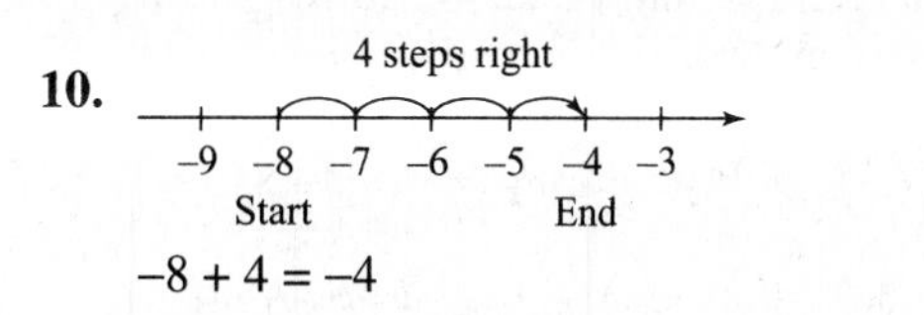

$-8 + 4 = -4$

11.

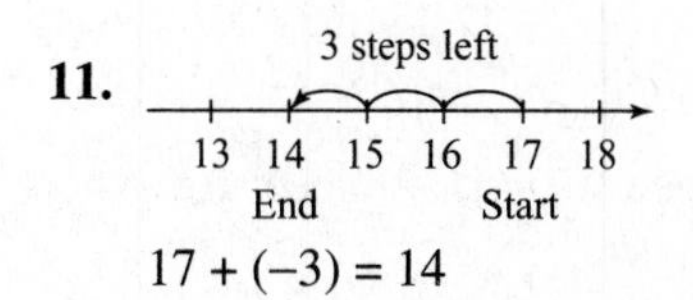

$17 + (-3) = 14$

12.

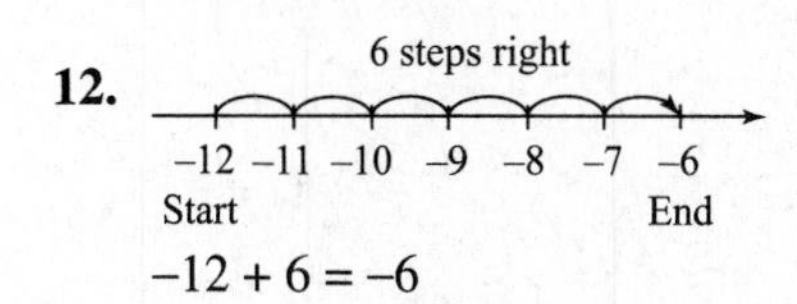

$-12 + 6 = -6$

13.

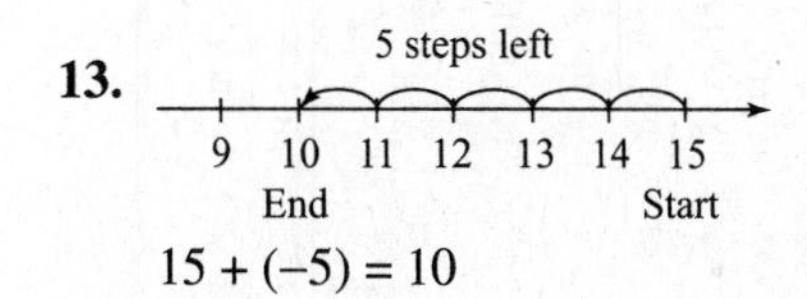

$15 + (-5) = 10$

14. The sum of two negative integers will be negative.

15. $|-11| = 11$
$|7| = 7$
The signs of −11 and 7 are different, so we subtract the absolute values: $11 - 7 = 4$.
The larger absolute value, 11, corresponds to a negative number in the original problem, so the sum is negative.
$-11 + 7 = -4$

16. $|5| = 5$
$|-8| = 8$
The signs of 5 and −8 are different, so we subtract the absolute values: $8 - 5 = 3$.
The larger absolute value, 8, corresponds to a negative number in the original problem, so the sum is negative.
$5 + (-8) = -3$

17. $|-8| = 8$
$|-16| = 16$
The signs of −8 and −16 are the same, so we add the absolute values: $8 + 16 = 24$.
Both numbers in the original problem are negative, so the sum is negative.
$-8 + (-16) = -24$

18. $|-94| = 94$
$|38| = 38$
The signs of −94 and 38 are different, so we subtract the absolute values: 94 − 38 = 56. The larger absolute value, 94, corresponds to a negative number in the original problem, so the sum is negative.
$-94 + 38 = -56$

19. The additive inverse of 7 is −7 because $7 + (-7) = 0$.

20. The additive inverse of $\frac{3}{7}$ is $-\frac{3}{7}$ because $\frac{3}{7}+\left(-\frac{3}{7}\right)=0$.

21. The additive inverse of −21 is $-(-21) = 21$ because $-21 + 21 = 0$.

22. The additive inverse of $-\frac{8}{5}$ is $-\left(-\frac{8}{5}\right)=\frac{8}{5}$ because $-\frac{8}{5}+\frac{8}{5}=0$.

23. The additive inverse of −5.75 is $-(-5.75) = 5.75$ because $-5.75 + 5.75 = 0$.

24. The answer to a subtraction problem is called the <u>difference</u>.

25. The subtraction problem $-3 - 10$ is equivalent to $-3 + \underline{(-10)}$.

26. $59 - (-21) = 59 + 21 = 80$

27. $-32 - 146 = -32 + (-146) = -178$

28. 17 minus 35 is 17 − 35.
$17 - 35 = 17 + (-35) = -18$

29. −382 subtracted from −2954 is $-2954 - (-382)$.
$-2954 - (-382) = -2954 + 382 = -2572$

30.
$$\begin{aligned} 8-13+5-21 &= 8+(-13)+5+(-21) \\ &= -5+5+(-21) \\ &= 0+(-21) \\ &= -21 \end{aligned}$$

31.
$$\begin{aligned} -27-49+18 &= -27+(-49)+18 \\ &= -76+18 \\ &= -58 \end{aligned}$$

32.
$$\begin{aligned} 3-(-14)-8+3 &= 3+14+(-8)+3 \\ &= 17+(-8)+3 \\ &= 9+3 \\ &= 12 \end{aligned}$$

33.
$$\begin{aligned} &-825 + 375 - (-735) + 265 \\ &= -825 + 375 + 735 + 265 \\ &= -450 + 735 + 265 \\ &= 285 + 265 \\ &= 550 \end{aligned}$$

34. The product of two integers with the same sign is <u>positive</u>.

35. $-3(7) = -21$

36. $13(-4) = -52$

37. $5 \cdot 16 = 80$

38. $-9(-12) = 108$

39. $(-13)(-25) = 325$

40. The product of thirteen negative factors is negative. True

41. $-3 \cdot 9 \cdot (-4) = -27 \cdot (-4) = 108$

42.
$$\begin{aligned} (-3)\cdot(-4)\cdot(-5)\cdot(-6) &= 12\cdot(-5)\cdot(-6) \\ &= -60\cdot(-6) \\ &= 360 \end{aligned}$$

43. The reciprocal of 6 is $\frac{1}{6}$.

44. The reciprocal of −2 is $-\frac{1}{2}$.

45. The quotient of two negative numbers is positive. True

46. $\frac{20}{-4}=\frac{5\cdot 4}{-1\cdot 4}=\frac{5\cdot \cancel{4}}{-1\cdot \cancel{4}}=\frac{5}{-1}=-5$

47. $\frac{707}{-101}=\frac{7\cdot 101}{-1\cdot 101}=\frac{7\cdot \cancel{101}}{-1\cdot \cancel{101}}=\frac{7}{-1}=-7$

48. $-63\div(-7)=\frac{-63}{-7}=\frac{9\cdot(-7)}{1\cdot(-7)}=\frac{9\cdot \cancel{(-7)}}{1\cdot \cancel{(-7)}}=9$

1.4 Exercises

49. $8+7=15$

51. $-5+9=4$

53. $9+(-5)=4$

55. $-11+(-8)=-19$

57. $-16+37=21$

59. $-119+(-209)=-328$

61. $-14+21+(-18)=7+(-18)=-11$

63. $$\begin{aligned}74+(-13)+(-23)+5&=61+(-23)+5\\&=38+5\\&=43\end{aligned}$$

65. The additive inverse of -325 is $-(-325)=325$ because $-325+325=0$.

67. The additive inverse of 125 is -125 because $125+(-125)=0$.

69. $23-12=23+(-12)=11$

71. $9-17=9+(-17)=-8$

73. $-20-8=-20+(-8)=-28$

75. $13-(-41)=13+41=54$

77. $-36-(-36)=-36+36=0$

79. $0-41=0+(-41)=-41$

81. $-93-(-62)=-93+62=-31$

83. $86-(-86)=86+86=172$

85. $5\cdot 8=40$

87. $8(-7)=-56$

89. $0\cdot(-21)=0$

91. $(-48)(-3)=144$

93. $(-42)3=-126$

95. $-5\cdot 6\cdot 3=-30\cdot 3=-90$

97. $-10(3)(-7)=-30(-7)=210$

99. $$\begin{aligned}(-2)(4)(-1)(3)(5)&=-8(-1)(3)(5)\\&=8(3)(5)\\&=24(5)\\&=120\end{aligned}$$

101. The reciprocal of 8 is $\frac{1}{8}$.

103. The reciprocal of -4 is $-\frac{1}{4}$.

105. The reciprocal of 1 is $\frac{1}{1}$ or 1.

107. $10\div 2=\frac{10}{2}=\frac{5\cdot 2}{1\cdot 2}=\frac{5\cdot \cancel{2}}{1\cdot \cancel{2}}=5$

109. $\frac{-56}{-8}=\frac{7\cdot(-8)}{1\cdot(-8)}=\frac{7\cdot \cancel{(-8)}}{1\cdot \cancel{(-8)}}=7$

111. $\frac{-45}{3}=\frac{-15\cdot 3}{1\cdot 3}=\frac{-15\cdot \cancel{3}}{1\cdot \cancel{3}}=-15$

113. $\frac{35}{10}=\frac{7\cdot 5}{2\cdot 5}=\frac{7\cdot \cancel{5}}{2\cdot \cancel{5}}=\frac{7}{2}$

115. $\frac{60}{-42}=\frac{10\cdot 6}{-7\cdot 6}=\frac{10\cdot \cancel{6}}{-7\cdot \cancel{6}}=\frac{10}{-7}=-\frac{10}{7}$

117. $\frac{-105}{-12}=\frac{35\cdot(-3)}{4\cdot(-3)}=\frac{35\cdot \cancel{(-3)}}{4\cdot \cancel{(-3)}}=\frac{35}{4}$

119. $-4\cdot 18=-72$

121. $-16-(-76)=-16+76=60$

123. $-9\cdot(-19)=171$

125. $\frac{120}{-8}=\frac{15\cdot 8}{-1\cdot 8}=\frac{15\cdot \cancel{8}}{-1\cdot \cancel{8}}=-15$

127. $-98+56=-42$

129. $\frac{75}{|-20|}=\frac{75}{20}=\frac{15\cdot 5}{4\cdot 5}=\frac{15\cdot \cancel{5}}{4\cdot \cancel{5}}=\frac{15}{4}$

131. $|-14|+|-26|=14+26=40$

133. $|-389|-627 = 389-627$
$= 389+(-627)$
$= -238$

135. The sum of 28 and −21 is written as $28 + (-21) = 7$.

137. −21 minus 47 is written as $-21 - 47 = -21 + (-47) = -68$.

139. −12 multiplied by 18 is written as $-12 \cdot 18 = -216$.

141. −36 divided by −108 is written as

$$-36 \div (-108) = \frac{-36}{-108} = \frac{1 \cdot (-36)}{3 \cdot (-36)} = \frac{1 \cdot \cancel{(-36)}}{3 \cdot \cancel{(-36)}} = \frac{1}{3}.$$

143. −3.25 points; the number is negative because the price fell.

145. −6 yards; the number is negative because the team lost yards.

147. −\$48; the number is negative because the account is overdrawn.

149. $8 + 3 + 3 = 11 + 3 = 14$
Loren and Richard walked 14 miles.

151. $563 + (-46) + 233 + (-63) + (-32) = 655$
Martha's new balance is \$655.

153. $725 + (-120) + (-590) + 310 = 325$
The company has 325 cases on hand. There is not enough stock to fill an order for 450 cases. The difference is
$325 + (-450) = -125$ cases.

155. $25{,}350 - (-375) = 25{,}350 + 375 = 25{,}725$
The distance between them is 25,725 feet.

157. $-3 + (-5) = -8$
$-3(-5) = 15$
The integers are −3 and −5.

159. $-12 + 2 = -10$
$-12(2) = -24$
The integers are −12 and 2.

161. **(a)** $\frac{1}{1} = 1, \frac{2}{1} = 2, \frac{3}{2} = 1.5, \frac{5}{3} \approx 1.667, \frac{8}{5} = 1.6,$
$\frac{13}{8} = 1.625, \frac{21}{13} \approx 1.615,$
$\frac{34}{21} \approx 1.619, \frac{55}{34} \approx 1.618, \ldots$

(b) The golden ratio is about 1.618.

(c) Answers may vary.

163. The problem $42 \div 4$ may be written equivalently as $42 \cdot \frac{1}{4}$.

Section 1.5

Preparing for Adding, Subtracting, Multiplying, and Dividing Rational Numbers Expressed as Fractions and Decimals

P1. $12 = 2 \cdot 2 \cdot 3$
$16 = 2 \cdot 2 \cdot 2 \cdot 2$
$\text{LCD} = 2 \cdot 2 \cdot 2 \cdot 2 \cdot 3 = 48$
The LCD of $\frac{5}{12}$ and $\frac{3}{16}$ is 48.

P2. $30 = 5 \cdot 6$
$\frac{4}{5} = \frac{4}{5} \cdot \frac{6}{6} = \frac{4 \cdot 6}{5 \cdot 6} = \frac{24}{30}$

1.5 Quick Checks

1. $\frac{-4}{14} = \frac{-2 \cdot 2}{7 \cdot 2} = \frac{-2 \cdot \cancel{2}}{7 \cdot \cancel{2}} = -\frac{2}{7}$

2. $-\frac{18}{30} = -\frac{3 \cdot 6}{5 \cdot 6} = -\frac{3 \cdot \cancel{6}}{5 \cdot \cancel{6}} = -\frac{3}{5}$

3. $\frac{24}{-4} = \frac{6 \cdot 4}{-1 \cdot 4} = \frac{6 \cdot \cancel{4}}{-1 \cdot \cancel{4}} = \frac{6}{-1} = -6$

4. $\frac{3}{4} \cdot \frac{9}{8} = \frac{3 \cdot 9}{4 \cdot 8} = \frac{27}{32}$

5. $$\begin{aligned}\frac{-5}{7}\cdot\frac{56}{15}&=\frac{-5\cdot 56}{7\cdot 15}\\&=\frac{-1\cdot 5\cdot 7\cdot 8}{7\cdot 3\cdot 5}\\&=\frac{-1\cdot \cancel{5}\cdot \cancel{7}\cdot 8}{\cancel{7}\cdot 3\cdot \cancel{5}}\\&=-\frac{8}{3}\end{aligned}$$

6. $$\begin{aligned}\frac{12}{45}\cdot\left(-\frac{18}{20}\right)&=\frac{12}{45}\cdot\frac{-18}{20}\\&=\frac{12\cdot(-18)}{45\cdot 20}\\&=\frac{2\cdot 2\cdot 3\cdot 2\cdot 3\cdot(-3)}{3\cdot 3\cdot 5\cdot 2\cdot 2\cdot 5}\\&=\frac{\cancel{2}\cdot \cancel{2}\cdot \cancel{3}\cdot 2\cdot \cancel{3}\cdot(-3)}{\cancel{3}\cdot \cancel{3}\cdot 5\cdot \cancel{2}\cdot \cancel{2}\cdot 5}\\&=\frac{2\cdot(-3)}{5\cdot 5}\\&=-\frac{6}{25}\end{aligned}$$

7. $$\begin{aligned}-\frac{25}{75}\cdot\left(-\frac{9}{4}\right)&=\frac{-25}{75}\cdot\frac{-9}{4}\\&=\frac{-25\cdot(-9)}{75\cdot 4}\\&=\frac{-1\cdot 5\cdot 5\cdot(-1)\cdot 3\cdot 3}{3\cdot 5\cdot 5\cdot 2\cdot 2}\\&=\frac{\cancel{5}\cdot \cancel{5}\cdot \cancel{3}\cdot 3}{\cancel{3}\cdot \cancel{5}\cdot \cancel{5}\cdot 2\cdot 2}\\&=\frac{3}{2\cdot 2}\\&=\frac{3}{4}\end{aligned}$$

8. $$\begin{aligned}\frac{7}{3}\cdot\frac{1}{14}\cdot\left(-\frac{9}{11}\right)&=\frac{7}{3}\cdot\frac{1}{14}\cdot\frac{-9}{11}\\&=\frac{7\cdot 1\cdot(-9)}{3\cdot 14\cdot 11}\\&=\frac{7\cdot 3\cdot(-3)}{3\cdot 2\cdot 7\cdot 11}\\&=\frac{\cancel{7}\cdot \cancel{3}\cdot(-3)}{\cancel{3}\cdot 2\cdot \cancel{7}\cdot 11}\\&=\frac{-3}{2\cdot 11}\\&=-\frac{3}{22}\end{aligned}$$

9. Two numbers are called multiplicative inverses, or reciprocals, if their product is equal to <u>one</u>.

10. The reciprocal of 12 is $\frac{1}{12}$ because $12\cdot\frac{1}{12}=1$.

11. The reciprocal of $\frac{7}{5}$ is $\frac{5}{7}$ because $\frac{7}{5}\cdot\frac{5}{7}=1$.

12. The reciprocal of $-\frac{1}{4}$ is –4 because

$-\frac{1}{4}\cdot(-4)=1$.

13. The reciprocal of $-\frac{31}{20}$ is $-\frac{20}{31}$ because

$-\frac{31}{20}\cdot\left(-\frac{20}{31}\right)=1$.

14. $$\frac{5}{7}\div\frac{7}{10}=\frac{5}{7}\cdot\frac{10}{7}=\frac{5\cdot 10}{7\cdot 7}=\frac{50}{49}$$

15. $$\begin{aligned}-\frac{9}{12}\div\frac{14}{7}&=-\frac{9}{12}\cdot\frac{7}{14}\\&=-\frac{3\cdot 3\cdot 7}{2\cdot 2\cdot 3\cdot 2\cdot 7}\\&=-\frac{\cancel{3}\cdot 3\cdot \cancel{7}}{2\cdot 2\cdot \cancel{3}\cdot 2\cdot \cancel{7}}\\&=-\frac{3}{2\cdot 2\cdot 2}\\&=-\frac{3}{8}\end{aligned}$$

16. $$\begin{aligned}\frac{8}{35}\div\left(\frac{-1}{10}\right)&=\frac{8}{35}\cdot\left(\frac{10}{-1}\right)\\&=\frac{2\cdot 2\cdot 2\cdot 2\cdot 5}{5\cdot 7\cdot(-1)}\\&=\frac{2\cdot 2\cdot 2\cdot 2\cdot \cancel{5}}{\cancel{5}\cdot 7\cdot(-1)}\\&=\frac{2\cdot 2\cdot 2\cdot 2}{7\cdot(-1)}\\&=-\frac{16}{7}\end{aligned}$$

17. $-\frac{18}{63} \div \left(-\frac{54}{35}\right) = \frac{-18}{63} \cdot \left(\frac{-35}{54}\right)$
$= \frac{-1 \cdot 2 \cdot 3 \cdot 3 \cdot (-1) \cdot 5 \cdot 7}{3 \cdot 3 \cdot 7 \cdot 2 \cdot 3 \cdot 3 \cdot 3}$
$= \frac{\not{2} \cdot \not{3} \cdot \not{3} \cdot 5 \cdot \not{7}}{\not{3} \cdot \not{3} \cdot \not{7} \cdot \not{2} \cdot 3 \cdot 3 \cdot 3}$
$= \frac{5}{3 \cdot 3 \cdot 3}$
$= \frac{5}{27}$

18. $\frac{-5}{7} + \frac{3}{7} = \frac{-5+3}{7}$

19. $-\frac{9}{10} - \frac{3}{10} = \frac{-9}{10} - \frac{3}{10}$
$= \frac{-9-3}{10}$
$= \frac{-12}{10}$
$= \frac{2 \cdot (-6)}{2 \cdot 5}$
$= \frac{\not{2} \cdot (-6)}{\not{2} \cdot 5}$
$= -\frac{6}{5}$

20. $\frac{8}{11} + \frac{2}{11} = \frac{8+2}{11} = \frac{10}{11}$

21. $-\frac{18}{35} + \frac{3}{35} = \frac{-18}{35} + \frac{3}{35}$
$= \frac{-18+3}{35}$
$= \frac{-15}{35}$
$= \frac{-3 \cdot 5}{5 \cdot 7}$
$= \frac{-3 \cdot \not{5}}{\not{5} \cdot 7}$
$= -\frac{3}{7}$

22. $\frac{19}{63} - \frac{10}{63} = \frac{19-10}{63} = \frac{9}{63} = \frac{1 \cdot 9}{7 \cdot 9} = \frac{1 \cdot \not{9}}{7 \cdot \not{9}} = \frac{1}{7}$

23. $12 = 2 \cdot 2 \cdot 3$
$18 = 2 \cdot 3 \cdot 3$
$\text{LCD} = 2 \cdot 2 \cdot 3 \cdot 3 = 36$
$\frac{5}{12} - \frac{5}{18} = \frac{5}{12} \cdot \frac{3}{3} - \frac{5}{18} \cdot \frac{2}{2}$
$= \frac{15}{36} - \frac{10}{36}$
$= \frac{15-10}{36}$
$= \frac{5}{36}$

24. $14 = 2 \cdot 7$
$21 = 3 \cdot 7$
$\text{LCD} = 2 \cdot 3 \cdot 7 = 42$
$\frac{3}{14} + \frac{10}{21} = \frac{3}{14} \cdot \frac{3}{3} + \frac{10}{21} \cdot \frac{2}{2}$
$= \frac{9}{42} + \frac{20}{42}$
$= \frac{9+20}{42}$
$= \frac{29}{42}$

25. $6 = 2 \cdot 3$
$12 = 2 \cdot 2 \cdot 3$
$\text{LCD} = 2 \cdot 2 \cdot 3 = 12$
$-\frac{23}{6} + \frac{7}{12} = \frac{-23}{6} \cdot \frac{2}{2} + \frac{7}{12}$
$= \frac{-46}{12} + \frac{7}{12}$
$= \frac{-46+7}{12}$
$= \frac{-39}{12}$
$= \frac{-1 \cdot 3 \cdot 13}{2 \cdot 2 \cdot 3}$
$= \frac{-1 \cdot \not{3} \cdot 13}{2 \cdot 2 \cdot \not{3}}$
$= \frac{-13}{2 \cdot 2}$
$= -\frac{13}{4}$

26. $5 = 5$
$11 = 11$
$\text{LCD} = 5 \cdot 11 = 55$

$$\begin{aligned}\frac{3}{5}+\left(-\frac{4}{11}\right) &= \frac{3}{5}\cdot\frac{11}{11}+\frac{-4}{11}\cdot\frac{5}{5}\\ &= \frac{33}{55}+\frac{-20}{55}\\ &= \frac{33+(-20)}{55}\\ &= \frac{13}{55}\end{aligned}$$

27.
$$\begin{aligned}-2+\frac{7}{16} &= \frac{-2}{1}+\frac{7}{16}\\ &= \frac{-2}{1}\cdot\frac{16}{16}+\frac{7}{16}\\ &= \frac{-32}{16}+\frac{7}{16}\\ &= \frac{-32+7}{16}\\ &= \frac{-25}{16}\\ &= -\frac{25}{16}\end{aligned}$$

28.
$$\begin{aligned}6+\left(\frac{-9}{4}\right) &= \frac{6}{1}+\left(\frac{-9}{4}\right)\\ &= \frac{6}{1}\cdot\frac{4}{4}+\left(\frac{-9}{4}\right)\\ &= \frac{24}{4}+\left(\frac{-9}{4}\right)\\ &= \frac{24+(-9)}{4}\\ &= \frac{15}{4}\end{aligned}$$

29.
$$\begin{array}{r} 9.670\\ +\,11.344\\ \hline 21.014\end{array}$$
So $9.67 + 11.344 = 21.014$.

30.
$$\begin{array}{r} 81.96\\ -\,17.39\\ \hline 64.57\end{array}$$
So $81.96 - 17.39 = 64.57$.

31.
$$\begin{array}{r} 14.950\\ 7.118\\ +\,0.300\\ \hline 22.368\end{array}$$
So $14.95 + 7.118 + 0.3 = 22.368$.

32. $\begin{array}{r} 345.6700 \\ -\ 8.0912 \\ \hline 337.5788 \end{array}$

So 345.67 − 8.0912 = 337.5788.

33. $\begin{array}{r} 180.782 \\ -\ 100.300 \\ \hline 80.482 \end{array}$

$-180.782 + 100.3 + 9.07 = -80.482 + 9.07$

$\begin{array}{r} 80.482 \\ -\ 9.070 \\ \hline 71.412 \end{array}$

So −180.782 + 100.3 + 9.07 = −71.412.

34. $\begin{array}{r} 74.280 \\ +\ 14.832 \\ \hline 89.112 \end{array}$

So $-74.28 - 14.832 = -74.28 + (-14.832)$
$= -89.112.$

35.

$\begin{array}{rl} 23.9 & \text{one digit to the right of the decimal point} \\ \times\ 0.2 & \text{one digit to the right of the decimal point} \\ \hline 4.78 & \text{two digits to the right of the decimal point} \end{array}$

36.

$\begin{array}{rl} 9.1 & \text{one digit to the right of the decimal point} \\ \times\ 7.24 & \text{two digits to the right of the decimal point} \\ \hline 364 & \\ 182 & \\ 637 & \\ \hline 65.884 & \text{three digits to the right of the decimal point} \end{array}$

37.

$\begin{array}{rl} -3.45 & \text{two digits to the right of the decimal point} \\ \times\ 0.03 & \text{two digits to the right of the decimal point} \\ \hline -0.1035 & \text{four digits to the right of the decimal point} \end{array}$

38.

$\begin{array}{rl} 257 & \text{no digits to the right of the decimal point} \\ \times -3.5 & \text{one digit to the right of the decimal point} \\ \hline 1285 & \\ 771 & \\ \hline -899.5 & \text{one digit to the right of the decimal point} \end{array}$

39.

$\begin{array}{rl} -0.03 & \text{two digits to the right of the decimal point} \\ \times\ -0.45 & \text{two digits to the right of the decimal point} \\ \hline 0.0135 & \text{four digits to the right of the decimal point} \end{array}$

40.

$\begin{array}{rl} 9.9 & \text{one digit to the right of the decimal point} \\ \times\ 0.002 & \text{three digits to the right of the decimal point} \\ \hline 0.0198 & \text{four digits to the right of the decimal point} \end{array}$

41. $$\begin{array}{r} 0.25 \\ 73\overline{)18.25} \\ \underline{14\,6} \\ 3\,65 \\ \underline{3\,65} \\ 0 \end{array}$$

So, $\frac{18.25}{73} = 0.25.$

42. $\frac{1.0032}{0.12} = \frac{1.0032}{0.12} \cdot \frac{100}{100} = \frac{100.32}{12}$

$$\begin{array}{r} 8.36 \\ 12\overline{)100.32} \\ \underline{96} \\ 4\,3 \\ \underline{3\,6} \\ 72 \\ \underline{72} \\ 0 \end{array}$$

So $\frac{1.0032}{0.12} = 8.36.$

43. $\frac{-4.2958}{45.7} = \frac{-4.2958}{45.7} \cdot \frac{10}{10} = \frac{-42.958}{457}$

$$\begin{array}{r} 0.094 \\ 457\overline{)42.958} \\ \underline{41\,13} \\ 1\,828 \\ \underline{1\,828} \\ 0 \end{array}$$

So $\frac{-4.2958}{45.7} = -0.094.$

44. $\frac{0.1515}{-5.05} = \frac{0.1515}{-5.05} \cdot \frac{100}{100} = \frac{15.15}{-505}$

$$\begin{array}{r} 0.03 \\ 505\overline{)15.15} \\ \underline{15\,15} \\ 0 \end{array}$$

So $\frac{0.1515}{-5.05} = -0.03.$

1.5 Exercises

45. $\frac{14}{21} = \frac{2 \cdot 7}{3 \cdot 7} = \frac{2 \cdot \cancel{7}}{3 \cdot \cancel{7}} = \frac{2}{3}$

47. $\frac{38}{-18} = \frac{2 \cdot 19}{2 \cdot (-9)} = \frac{\cancel{2} \cdot 19}{\cancel{2} \cdot (-9)} = \frac{19}{-9} = -\frac{19}{9}$

49. $-\frac{22}{44} = -\frac{1 \cdot 22}{2 \cdot 22} = -\frac{1 \cdot \cancel{22}}{2 \cdot \cancel{22}} = -\frac{1}{2}$

51. $\frac{32}{40} = \frac{4 \cdot 8}{5 \cdot 8} = \frac{4 \cdot \cancel{8}}{5 \cdot \cancel{8}} = \frac{4}{5}$

53. $\frac{6}{5} \cdot \frac{2}{5} = \frac{6 \cdot 2}{5 \cdot 5} = \frac{12}{25}$

55. $$\begin{aligned} \frac{5}{-2} \cdot 10 &= \frac{5}{-2} \cdot \frac{10}{1} \\ &= \frac{5 \cdot 10}{-2 \cdot 1} \\ &= \frac{5 \cdot 2 \cdot 5}{-1 \cdot 2} \\ &= \frac{5 \cdot \cancel{2} \cdot 5}{-1 \cdot \cancel{2}} \\ &= \frac{5 \cdot 5}{-1} \\ &= -25 \end{aligned}$$

57. $$\begin{aligned} -\frac{3}{2} \cdot \frac{4}{9} &= -\frac{3 \cdot 4}{2 \cdot 9} \\ &= -\frac{3 \cdot 2 \cdot 2}{2 \cdot 3 \cdot 3} \\ &= -\frac{\cancel{3} \cdot \cancel{2} \cdot 2}{\cancel{2} \cdot \cancel{3} \cdot 3} \\ &= -\frac{2}{3} \end{aligned}$$

59. $$\begin{aligned} -\frac{22}{3} \cdot \left(-\frac{12}{11}\right) &= \frac{-22 \cdot (-12)}{3 \cdot 11} \\ &= \frac{-2 \cdot 11 \cdot 3 \cdot (-4)}{3 \cdot 11} \\ &= \frac{-2 \cdot \cancel{11} \cdot \cancel{3} \cdot (-4)}{\cancel{3} \cdot \cancel{11}} \\ &= \frac{-2 \cdot (-4)}{1} \\ &= 8 \end{aligned}$$

61. $5 \cdot \frac{31}{15} = \frac{5}{1} \cdot \frac{31}{15}$
$= \frac{5 \cdot 31}{1 \cdot 15}$
$= \frac{5 \cdot 31}{1 \cdot 5 \cdot 3}$
$= \frac{\not{5} \cdot 31}{1 \cdot \not{5} \cdot 3}$
$= \frac{31}{1 \cdot 3}$
$= \frac{31}{3}$

63. $\frac{3}{4} \cdot \frac{8}{11} = \frac{3 \cdot 8}{4 \cdot 11}$
$= \frac{3 \cdot 4 \cdot 2}{4 \cdot 11}$
$= \frac{3 \cdot \not{4} \cdot 2}{\not{4} \cdot 11}$
$= \frac{3 \cdot 2}{11}$
$= \frac{6}{11}$

65. The reciprocal of $\frac{3}{5}$ is $\frac{5}{3}$.

67. The reciprocal of -5 or $-\frac{5}{1}$ is $-\frac{1}{5}$.

69. $\frac{4}{9} \div \frac{8}{15} = \frac{4}{9} \cdot \frac{15}{8}$
$= \frac{4 \cdot 15}{9 \cdot 8}$
$= \frac{4 \cdot 3 \cdot 5}{3 \cdot 3 \cdot 4 \cdot 2}$
$= \frac{\not{4} \cdot \not{3} \cdot 5}{\not{3} \cdot 3 \cdot \not{4} \cdot 2}$
$= \frac{5}{3 \cdot 2}$
$= \frac{5}{6}$

71. $-\frac{1}{3} \div 3 = -\frac{1}{3} \div \frac{3}{1} = -\frac{1}{3} \cdot \frac{1}{3} = -\frac{1 \cdot 1}{3 \cdot 3} = -\frac{1}{9}$

73. $\frac{5}{6} \div \left(-\frac{5}{4}\right) = \frac{5}{6} \cdot \left(-\frac{4}{5}\right)$
$= \frac{5 \cdot (-4)}{6 \cdot 5}$
$= \frac{5 \cdot (-2) \cdot 2}{2 \cdot 3 \cdot 5}$
$= \frac{\not{5} \cdot (-2) \cdot \not{2}}{\not{2} \cdot 3 \cdot \not{5}}$
$= -\frac{2}{3}$

75. $\frac{36}{28} \div \frac{22}{14} = \frac{36}{28} \cdot \frac{14}{22}$
$= \frac{36 \cdot 14}{28 \cdot 22}$
$= \frac{9 \cdot 4 \cdot 2 \cdot 7}{7 \cdot 4 \cdot 2 \cdot 11}$
$= \frac{9 \cdot \not{4} \cdot \not{2} \cdot \not{7}}{\not{7} \cdot \not{4} \cdot \not{2} \cdot 11}$
$= \frac{9}{11}$

77. $-8 \div \frac{2}{3} = -\frac{8}{1} \cdot \frac{3}{2}$
$= \frac{-8 \cdot -3}{1 \cdot 2}$
$= \frac{-4 \cdot 2 \cdot 3}{1 \cdot 2}$
$= \frac{-4 \cdot \not{2} \cdot 3}{1 \cdot \not{2}}$
$= \frac{-4 \cdot 3}{1}$
$= -12$

79. $-8 \div \left(-\frac{1}{4}\right) = -\frac{8}{1} \cdot \left(-\frac{4}{1}\right) = \frac{-8 \cdot (-4)}{1 \cdot 1} = 32$

81. $\frac{3}{4} + \frac{3}{4} = \frac{3+3}{4} = \frac{6}{4} = \frac{2 \cdot 3}{2 \cdot 2} = \frac{3}{2}$

83. $\frac{9}{8} - \frac{5}{8} = \frac{9-5}{8} = \frac{4}{8} = \frac{1 \cdot 4}{2 \cdot 4} = \frac{1}{2}$

85. $\frac{6}{7} - \left(-\frac{8}{7}\right) = \frac{6}{7} + \frac{8}{7} = \frac{6+8}{7} = \frac{14}{7} = \frac{2 \cdot 7}{7} = 2$

87. $-\frac{5}{3} + 2 = -\frac{5}{3} + \frac{2}{1} = -\frac{5}{3} + \frac{6}{3} = \frac{-5+6}{3} = \frac{1}{3}$

89. $6-\frac{7}{2}=\frac{6}{1}-\frac{7}{2}=\frac{12}{2}-\frac{7}{2}=\frac{12-7}{2}=\frac{5}{2}$

91. $$\begin{aligned}-\frac{4}{3}+\frac{1}{4}&=-\frac{4}{3}\cdot\frac{4}{4}+\frac{1}{4}\cdot\frac{3}{3}\\&=-\frac{16}{12}+\frac{3}{12}\\&=\frac{-16+3}{12}\\&=-\frac{13}{12}\end{aligned}$$

93. $$\begin{aligned}\frac{7}{5}+\left(-\frac{23}{20}\right)&=\frac{7}{5}\cdot\frac{4}{4}+\left(-\frac{23}{20}\right)\\&=\frac{28}{20}+\left(-\frac{23}{20}\right)\\&=\frac{28+(-23)}{20}\\&=\frac{5}{20}\\&=\frac{1\cdot 5}{4\cdot 5}\\&=\frac{1}{4}\end{aligned}$$

95. $$\begin{aligned}\frac{7}{15}-\left(-\frac{4}{3}\right)&=\frac{7}{15}+\frac{4}{3}\\&=\frac{7}{15}+\frac{4}{3}\cdot\frac{5}{5}\\&=\frac{7}{15}+\frac{20}{15}\\&=\frac{7+20}{15}\\&=\frac{27}{15}\\&=\frac{9\cdot 3}{5\cdot 3}\\&=\frac{9}{5}\end{aligned}$$

97. $15=3\cdot 5$
$10=2\cdot 5$
$\text{LCD}=2\cdot 3\cdot 5=30$
$$\begin{aligned}\frac{8}{15}-\frac{7}{10}&=\frac{8}{15}\cdot\frac{2}{2}-\frac{7}{10}\cdot\frac{3}{3}\\&=\frac{16}{30}-\frac{21}{30}\\&=\frac{16-21}{30}\\&=\frac{-5}{30}\\&=\frac{-1\cdot 5}{6\cdot 5}\\&=-\frac{1}{6}\end{aligned}$$

99. $10=2\cdot 5$
$8=2\cdot 2\cdot 2$
$\text{LCD}=2\cdot 2\cdot 2\cdot 5=40$
$$\begin{aligned}-\frac{33}{10}-\left(-\frac{33}{8}\right)&=-\frac{33}{10}+\frac{33}{8}\\&=-\frac{33}{10}\cdot\frac{4}{4}+\frac{33}{8}\cdot\frac{5}{5}\\&=-\frac{132}{40}+\frac{165}{40}\\&=\frac{-132+165}{40}\\&=\frac{33}{40}\end{aligned}$$

101. $12=2\cdot 2\cdot 3$
$18=2\cdot 3\cdot 3$
$\text{LCD}=2\cdot 2\cdot 3\cdot 3=36$
$$\begin{aligned}\frac{19}{12}-\left(-\frac{41}{18}\right)&=\frac{19}{12}+\frac{41}{18}\\&=\frac{19}{12}\cdot\frac{3}{3}+\frac{41}{18}\cdot\frac{2}{2}\\&=\frac{57}{36}+\frac{82}{36}\\&=\frac{57+82}{36}\\&=\frac{139}{36}\end{aligned}$$

103. $3 = 3$
$9 = 3 \cdot 3$
$6 = 2 \cdot 3$
$\text{LCD} = 2 \cdot 3 \cdot 3 = 18$

$$\begin{aligned}-\frac{2}{3}+\left(-\frac{5}{9}\right)+\frac{5}{6} &= -\frac{2}{3}\cdot\frac{6}{6}+\left(-\frac{5}{9}\cdot\frac{2}{2}\right)+\frac{5}{6}\cdot\frac{3}{3}\\ &= -\frac{12}{18}+\left(-\frac{10}{18}\right)+\frac{15}{18}\\ &= \frac{-12+(-10)+15}{18}\\ &= -\frac{7}{18}\end{aligned}$$

105.
$$\begin{array}{r}10.5\\ -\,4.0\\ \hline 6.5\end{array}$$
So, $-10.5 + 4 = -6.5$.

107.
$$\begin{array}{r}3.5\\ +\,4.9\\ \hline 8.4\end{array}$$
So, $-(-3.5) + 4.9 = 3.5 + 4.9 = 8.4$.

109.
$$\begin{array}{r}39.10\\ +\,16.82\\ \hline 55.92\end{array}$$
So, $39.1 - (-16.82) = 39.1 + 16.82 = 55.92$.

111.
$$\begin{array}{r}6.70\\ -\,5.21\\ \hline 1.49\end{array}$$
So, $-5.21 - (-6.7) = -5.21 + 6.7 = 1.49$.

113.
$$\begin{array}{r}45.00\\ -\,2.45\\ \hline 42.55\end{array}$$
So, $45 - 2.45 = 42.55$.

115.
$$\begin{array}{r}4.3\\ \times\,5.8\\ \hline 344\\ 215\\ \hline 24.94\end{array}$$
So, $4.3 \times 5.8 = 24.94$.

117.
$$\begin{array}{r}120\\ \times\,0.075\\ \hline 600\\ 840\\ \hline 9.000\end{array}$$
So, $0.075 \times 120 = 9$.

119. $\frac{136.08}{5.6} = \frac{1360.8}{56} = 24.3$

$$\begin{array}{r}24.3\\ 56\overline{)1360.8}\\ \underline{112}\\ 240\\ \underline{224}\\ 16\,8\\ \underline{16\,8}\\ 0\end{array}$$

121. $\frac{25.48}{0.052} = \frac{25{,}480}{52} = 490$

$$\begin{array}{r}490\\ 52\overline{)25480}\\ \underline{208}\\ 468\\ \underline{468}\\ 0\end{array}$$

123.
$$\begin{aligned}-1.25-(-0.6)+1.6 &= -1.25+0.6+1.6\\ &= -0.65+1.6\\ &= 0.95\end{aligned}$$

125. $6 = 2 \cdot 3$
$15 = 3 \cdot 5$
$\text{LCD} = 2 \cdot 3 \cdot 5 = 30$

$$\begin{aligned}-\frac{5}{6}+\frac{7}{15} &= -\frac{5}{6}\cdot\frac{5}{5}+\frac{7}{15}\cdot\frac{2}{2}\\ &= -\frac{25}{30}+\frac{14}{30}\\ &= \frac{-25+14}{30}\\ &= -\frac{11}{30}\end{aligned}$$

127.
$$\begin{aligned}-\frac{10}{21}\cdot\frac{14}{5} &= -\frac{10\cdot 14}{21\cdot 5}\\ &= -\frac{2\cdot 5\cdot 2\cdot 7}{3\cdot 7\cdot 5}\\ &= -\frac{2\cdot \cancel{5}\cdot 2\cdot \cancel{7}}{3\cdot \cancel{7}\cdot \cancel{5}}\\ &= -\frac{2\cdot 2}{3}\\ &= -\frac{4}{3}\end{aligned}$$

129. $\frac{3}{8}\div\left(-\frac{9}{16}\right)=\frac{3}{8}\cdot\left(-\frac{16}{9}\right)$
$=\frac{3\cdot(-16)}{8\cdot 9}$
$=\frac{3\cdot(-2)\cdot 8}{8\cdot 3\cdot 3}$
$=\frac{\cancel{3}\cdot(-2)\cdot\cancel{8}}{\cancel{8}\cdot\cancel{3}\cdot 3}$
$=-\frac{2}{3}$

131. $-\frac{5}{12}+\frac{2}{12}=\frac{-5+2}{12}=\frac{-3}{12}=\frac{-1\cdot 3}{3\cdot 4}=-\frac{1}{4}$

133. $-\frac{2}{7}-\frac{17}{5}=-\frac{2}{7}\cdot\frac{5}{5}-\frac{17}{5}\cdot\frac{7}{7}$
$=-\frac{10}{35}-\frac{119}{35}$
$=\frac{-10-119}{35}$
$=-\frac{129}{35}$

135.
```
 10.3
- 8.7
  1.6
```
So, –8.7 – (–10.3) = –8.7 + 10.3 = 1.6.

137. $12 = 2\cdot 2\cdot 3$
$28 = 2\cdot 2\cdot 7$
LCD $= 2\cdot 2\cdot 3\cdot 7 = 84$
$\frac{1}{12}+\left(-\frac{5}{28}\right)=\frac{1}{12}\cdot\frac{7}{7}+\left(-\frac{5}{28}\right)\cdot\frac{3}{3}$
$=\frac{7}{84}+\left(-\frac{15}{84}\right)$
$=\frac{7+(-15)}{84}$
$=\frac{-8}{84}$
$=\frac{-2\cdot 4}{21\cdot 4}$
$=-\frac{2}{21}$

139.
```
 -12.03
×   4.2
   2406
  4812
-50.526
```
So, –12.03 × 4.2 = –50.526.

141. $36\cdot\left(-\frac{4}{9}\right)=\frac{36}{1}\cdot\left(-\frac{4}{9}\right)$
$=\frac{36\cdot(-4)}{1\cdot 9}$
$=\frac{4\cdot 9\cdot(-4)}{1\cdot 9}$
$=\frac{4\cdot\cancel{9}\cdot(-4)}{1\cdot\cancel{9}}$
$=\frac{4\cdot(-4)}{1}$
$=-16$

143. $-27\div\frac{9}{5}=-\frac{27}{1}\div\frac{9}{5}$
$=-\frac{27}{1}\cdot\frac{5}{9}$
$=\frac{-27\cdot 5}{1\cdot 9}$
$=\frac{-3\cdot 9\cdot 5}{1\cdot 9}$
$=\frac{-3\cdot\cancel{9}\cdot 5}{1\cdot\cancel{9}}$
$=\frac{-3\cdot 5}{1}$
$=-15$

145.
```
 10.20
- 3.62
  6.58
```
So, 3.62 – 10.2 = –6.58.

147. $\frac{-145.518}{18.42}=\frac{-14{,}551.8}{1842}=-7.9$
```
          7.9
1842)14551.8
     12894
      1657 8
      1657 8
           0
```

149. $7 = 7$
$14 = 2\cdot 7$
$21 = 3\cdot 7$
LCD $= 2\cdot 3\cdot 7 = 42$

$$\frac{12}{7}-\frac{17}{14}-\frac{48}{21}=\frac{12}{7}\cdot\frac{6}{6}-\frac{17}{14}\cdot\frac{3}{3}-\frac{48}{21}\cdot\frac{2}{2}$$
$$=\frac{72}{42}-\frac{51}{42}-\frac{96}{42}$$
$$=\frac{72-51-96}{42}$$
$$=\frac{-75}{42}$$
$$=\frac{-3\cdot 25}{3\cdot 14}$$
$$=-\frac{25}{14}$$

151. $54.2-18.78-(-2.5)+20.47$
$=54.2-18.78+2.5+20.47$
$=58.39$

153. $400\times 25.8\times 0.003=10{,}320\times 0.003$
$=30.96$

155. $12 = 2 \cdot 2 \cdot 3$
$6 = 2 \cdot 3$
$8 = 2 \cdot 2 \cdot 2$
$\text{LCD} = 2 \cdot 2 \cdot 2 \cdot 3 = 24$

$$-\frac{11}{12}-\left(-\frac{1}{6}\right)+\frac{7}{8}=-\frac{11}{12}+\frac{1}{6}+\frac{7}{8}$$
$$=-\frac{11}{12}\cdot\frac{2}{2}+\frac{1}{6}\cdot\frac{4}{4}+\frac{7}{8}\cdot\frac{3}{3}$$
$$=-\frac{22}{24}+\frac{4}{24}+\frac{21}{24}$$
$$=\frac{-22+4+21}{24}$$
$$=\frac{3}{24}$$
$$=\frac{1\cdot 3}{8\cdot 3}$$
$$=\frac{1}{8}$$

157. $24 \cdot 7 = 168$ hours in one week

$$168\cdot\frac{1}{8}=\frac{168}{1}\cdot\frac{1}{8}$$
$$=\frac{168\cdot 1}{1\cdot 8}$$
$$=\frac{21\cdot 8\cdot 1}{1\cdot 8}$$
$$=\frac{21\cdot \not{8}\cdot 1}{1\cdot \not{8}}$$
$$=21$$

She spends 21 hours per week watching TV.

159. $$36\cdot\frac{2}{3}\cdot\frac{3}{4}=\frac{36}{1}\cdot\frac{2}{3}\cdot\frac{3}{4}$$
$$=\frac{36\cdot 2\cdot 3}{3\cdot 4}$$
$$=\frac{9\cdot 4\cdot 2\cdot 3}{3\cdot 4}$$
$$=\frac{9\cdot \not{4}\cdot 2\cdot \not{3}}{\not{3}\cdot \not{4}}$$
$$=\frac{9\cdot 2}{1}$$
$$=18$$

This term, 18 students will pass the class.

161. $-43.29 + (-25.50) = -68.79$
Maria has a balance of $-\$68.79$.

163. $2.75 + 0.87 + (-1.12) + 0.52 + (-0.62)$
$= 2.4$
The net change was \$2.40.

165. $$d(P, Q)=|3.5-(-9.7)|$$
$$=|3.5+9.7|$$
$$=|13.2|$$
$$=13.2$$

167. $$d(P, Q)=\left|\frac{7}{5}-\left(-\frac{13}{3}\right)\right|$$
$$=\left|\frac{7}{5}+\frac{13}{3}\right|$$
$$=\left|\frac{7}{5}\cdot\frac{3}{3}+\frac{13}{3}\cdot\frac{5}{5}\right|$$
$$=\left|\frac{21}{15}+\frac{65}{15}\right|$$
$$=\left|\frac{21+65}{15}\right|$$
$$=\left|\frac{86}{15}\right|$$
$$=\frac{86}{15}$$

Putting the Concepts Together (Sections 1.2–1.5)

1. $8 = 2 \cdot 2 \cdot 2$
$20 = 2 \cdot 2 \cdot \quad 5$
$\downarrow\ \downarrow\ \downarrow\ \downarrow$
$\text{LCD} = 2 \cdot 2 \cdot 2 \cdot 5 = 40$

$\frac{7}{8}\cdot\frac{5}{5}=\frac{35}{40}$

$\frac{9}{20}\cdot\frac{2}{2}=\frac{18}{40}$

2. $\frac{21}{63}=\frac{7\cdot3\cdot1}{7\cdot3\cdot3}=\frac{1}{3}$

3. $\frac{2}{7}=7\overline{)2.000000}$ with quotient 0.285714

$$\begin{array}{r} 14 \\ \hline 60 \\ 56 \\ \hline 40 \\ 35 \\ \hline 50 \\ 49 \\ \hline 10 \\ 7 \\ \hline 30 \\ 28 \\ \hline \end{array}$$

$\frac{2}{7}=0.\overline{285714}$

4. $0.375=\frac{375}{1000}=\frac{3\cdot5\cdot5\cdot5}{2\cdot2\cdot2\cdot5\cdot5\cdot5}=\frac{3}{2\cdot2\cdot2}=\frac{3}{8}$

5. $12.3\%=12.3\%\frac{1}{100\%}=\frac{12.3}{100}=0.123$

6. $0.0625=0.0625\cdot\frac{100\%}{1}=6.25\%$

7. **(a)** -12, $-\frac{14}{7}=-2$, 0, and 3 are the integers.

(b) -12, $-\frac{14}{7}$, -1.25, 0, 3, and 11.2 are the rational numbers.

(c) $\sqrt{2}$ is the only irrational number.

(d) All the numbers listed are real numbers.

8. $\frac{1}{8}<0.5$ because $0.5=\frac{1}{2}=\frac{4}{8}$ and $\frac{1}{8}<\frac{4}{8}$.

9. $17+(-28)=-11$

10. $-23+(-42)=-65$

11. $18-45=18+(-45)=-27$

12. $3-(-24)=3+24=27$

13. $-18-(-12.5)=-18+12.5=-5.5$

14. $(-5)(2)=-10$

15. $25(-4)=-100$

16. $(-8)(-9)=72$

17. $\frac{-35}{7}=\frac{-5\cdot7}{7}=\frac{-5\cdot\cancel{7}}{\cancel{7}}=-5$

18. $\frac{-32}{-2}=\frac{-2\cdot16}{-2}=\frac{\cancel{-2}\cdot16}{\cancel{-2}}=16$

19. $27\div-3=\frac{27}{-3}=\frac{9\cdot3}{-1\cdot3}=\frac{9\cdot\cancel{3}}{-1\cdot\cancel{3}}=-9$

20. $-\frac{4}{5}-\frac{11}{5}=\frac{-4-11}{5}$

$=\frac{-15}{5}$

$=\frac{-3\cdot5}{5}$

$=\frac{-3\cdot\cancel{5}}{\cancel{5}}$

$=-3$

21. $7-\frac{4}{5}=\frac{7}{1}-\frac{4}{5}$

$=\frac{7}{1}\cdot\frac{5}{5}-\frac{4}{5}$

$=\frac{35}{5}-\frac{4}{5}$

$=\frac{35-4}{5}$

$=\frac{31}{5}$

22. $12=2\cdot2\cdot3$

$18=2\cdot3\cdot3$

$\text{LCD}=2\cdot2\cdot3\cdot3=36$

$$\frac{7}{12}+\frac{5}{18}=\frac{7}{12}\cdot\frac{3}{3}+\frac{5}{18}\cdot\frac{2}{2}$$
$$=\frac{21}{36}+\frac{10}{36}$$
$$=\frac{21+10}{36}$$
$$=\frac{31}{36}$$

23. $12 = 2 \cdot 2 \cdot 3$
$18 = 2 \cdot 3 \cdot 3$
$\text{LCD} = 2 \cdot 2 \cdot 3 \cdot 3 = 36$

$$-\frac{5}{12}-\frac{1}{18}=-\frac{5}{12}\cdot\frac{3}{3}-\frac{1}{18}\cdot\frac{2}{2}$$
$$=-\frac{15}{36}-\frac{2}{36}$$
$$=\frac{-15-2}{36}$$
$$=\frac{-17}{36}$$
$$=-\frac{17}{36}$$

24. $$\frac{6}{25}\cdot 15\cdot\frac{1}{2}=\frac{6}{25}\cdot\frac{15}{1}\cdot\frac{1}{2}$$
$$=\frac{6\cdot 15\cdot 1}{25\cdot 1\cdot 2}$$
$$=\frac{2\cdot 3\cdot 3\cdot 5\cdot 1}{5\cdot 5\cdot 1\cdot 2}$$
$$=\frac{\cancel{2}\cdot 3\cdot 3\cdot \cancel{5}}{\cancel{5}\cdot 5\cdot \cancel{2}}$$
$$=\frac{3\cdot 3}{5}$$
$$=\frac{9}{5}$$

25. $$\frac{2}{7}\div(-8)=\frac{2}{7}\div\left(-\frac{8}{1}\right)$$
$$=\frac{2}{7}\cdot\left(-\frac{1}{8}\right)$$
$$=\frac{2\cdot(-1)}{7\cdot 8}$$
$$=\frac{2\cdot(-1)}{7\cdot 2\cdot 4}$$
$$=\frac{\cancel{2}\cdot(-1)}{7\cdot\cancel{2}\cdot 4}$$
$$=\frac{-1}{7\cdot 4}$$
$$=-\frac{1}{28}$$

26. $\frac{0}{-8}=0$

27.
```
   3.56
 + 7.20
  10.76
```
So $3.56 - (-7.2) = 3.56 + 7.2 = 10.76$.

28.
```
  18.946
 - 11.300
   7.646
```
So $18.946 - 11.3 = 7.646$.

29. $$62.488\div 42.8=\frac{62.488}{42.8}$$
$$=\frac{62.488}{42.8}\cdot\frac{10}{10}$$
$$=\frac{624.88}{428}$$

```
        1.46
428)624.88
    428
    196 8
    171 2
     25 68
     25 68
         0
```
So $62.488 \div 42.8 = 1.46$.

30.
```
   7.94
 ×  2.8
  6 352
 15 88
 22.232
```

Section 1.6

Preparing for Properties of Real Numbers

P1. $12 + 3 + (-12) = 15 + (-12) = 3$

P2. $$\frac{3}{4}\cdot 11\cdot\frac{4}{3}=\frac{3}{4}\cdot\frac{11}{1}\cdot\frac{4}{3}$$
$$=\frac{3\cdot 11}{4\cdot 1}\cdot\frac{4}{3}$$
$$=\frac{3\cdot 11\cdot 4}{4\cdot 1\cdot 3}$$
$$=\frac{\cancel{3}\cdot 11\cdot\cancel{4}}{\cancel{4}\cdot 1\cdot\cancel{3}}$$
$$=\frac{11}{1}$$
$$=11$$

1.6 Quick Checks

1. The product of any real number and the number 1 is that number.

2. $96 \text{ inches} = 96 \text{ inches} \cdot \frac{1 \text{ foot}}{12 \text{ inches}}$
$= \frac{96}{12} \text{ feet}$
$= \frac{2 \cdot 2 \cdot 2 \cdot 2 \cdot 2 \cdot 3}{2 \cdot 2 \cdot 3} \text{ feet}$
$= \frac{\cancel{2} \cdot \cancel{2} \cdot 2 \cdot 2 \cdot 2 \cdot \cancel{3}}{\cancel{2} \cdot \cancel{2} \cdot \cancel{3}} \text{ feet}$
$= 8 \text{ feet}$

3. $500 \text{ minutes} = 500 \text{ minutes} \cdot \frac{1 \text{ hour}}{60 \text{ minutes}}$
$= \frac{500}{60} \text{ hours}$
$= \frac{2 \cdot 2 \cdot 5 \cdot 5 \cdot 5}{2 \cdot 2 \cdot 3 \cdot 5} \text{ hours}$
$= \frac{\cancel{2} \cdot \cancel{2} \cdot \cancel{5} \cdot 5 \cdot 5}{\cancel{2} \cdot \cancel{2} \cdot 3 \cdot \cancel{5}} \text{ hours}$
$= \frac{25}{3} \text{ hours}$
$= 8\frac{1}{3}$ hours or 8 hours, 20 minutes

4. $88 \text{ ounces} = 88 \text{ ounces} \cdot \frac{1 \text{ pound}}{16 \text{ ounces}}$
$= \frac{88}{16} \text{ pounds}$
$= \frac{2 \cdot 2 \cdot 2 \cdot 11}{2 \cdot 2 \cdot 2 \cdot 2} \text{ pounds}$
$= \frac{\cancel{2} \cdot \cancel{2} \cdot \cancel{2} \cdot 11}{\cancel{2} \cdot \cancel{2} \cdot \cancel{2} \cdot 2} \text{ pounds}$
$= \frac{11}{2} \text{ pounds}$
$= 5\frac{1}{2}$ pounds or 5 pounds, 8 ounces

5. The Commutative Property of Addition states that for any real numbers a and b, $a + b = \underline{b + a}$.

6. The sum of any real number and its opposite is equal to 0.

7. $(-8) + 22 + 8 = (-8) + 8 + 22 = 0 + 22 = 22$

8. $\frac{8}{15} + \frac{3}{20} + \left(-\frac{8}{15}\right) = \frac{8}{15} + \left(-\frac{8}{15}\right) + \frac{3}{20}$
$= 0 + \frac{3}{20}$
$= \frac{3}{20}$

9. $2.1 + 11.98 + (-2.1) = 2.1 + (-2.1) + 11.98$
$= 0 + 11.98$
$= 11.98$

10. $-8 \cdot (-13) \cdot \left(-\frac{3}{4}\right) = -8 \cdot \left(-\frac{3}{4}\right) \cdot (-13)$
$= -\overset{2}{\cancel{8}} \cdot \left(-\frac{3}{\underset{1}{\cancel{4}}}\right) \cdot (-13)$
$= -2 \cdot (-3) \cdot (-13)$
$= 6 \cdot (-13)$
$= -78$

11. $\frac{5}{22} \cdot \frac{18}{331} \cdot \left(-\frac{44}{5}\right) = \frac{5}{22} \cdot \left(-\frac{44}{5}\right) \cdot \frac{18}{331}$
$= \frac{\overset{1}{\cancel{5}}}{\underset{1}{\cancel{22}}} \cdot \left(-\frac{\overset{2}{\cancel{44}}}{\underset{1}{\cancel{5}}}\right) \cdot \frac{18}{331}$
$= \frac{1}{1} \cdot \left(-\frac{2}{1}\right) \cdot \frac{18}{331}$
$= -\frac{2}{1} \cdot \frac{18}{331}$
$= -\frac{36}{331}$

12. $100{,}000 \cdot 349 \cdot 0.00001$
$= 100{,}000 \cdot 0.00001 \cdot 349$
$= 1 \cdot 349$
$= 349$

13. $14 + 101 + (-101) = 14 + (101 + (-101))$
$= 14 + 0$
$= 14$

14. $14 \cdot \frac{1}{5} \cdot 5 = 14 \cdot \left(\frac{1}{5} \cdot 5\right) = 14 \cdot 1 = 14$

15. $-34.2 + 12.6 + (-2.6)$
$= -34.2 + (12.6 + (-2.6))$
$= -34.2 + 10$
$= -24.2$

16. $\frac{19}{2}\cdot\frac{4}{38}\cdot\frac{50}{13}=\left(\frac{19}{2}\cdot\frac{4}{38}\right)\cdot\frac{50}{13}$

$=\left(\frac{\cancel{19}^{1}}{\cancel{2}_{1}}\cdot\frac{\cancel{4}^{2}}{\cancel{38}_{2}}\right)\cdot\frac{50}{13}$

$=\frac{2}{2}\cdot\frac{50}{13}$

$=1\cdot\frac{50}{13}$

$=\frac{50}{13}$

17. $\frac{0}{22}=0$ because 0 is the dividend.

18. $\frac{-11}{0}$ is undefined because 0 is the divisor.

19. $-\frac{0}{5}=0$ because 0 is the dividend.

20. $\frac{5678}{0}$ is undefined because 0 is the divisor.

1.6 Exercises

21. $13 \text{ feet} = 13 \text{ feet}\cdot\frac{12 \text{ inches}}{1 \text{ foot}}$

$=13\cdot 12 \text{ inches}$

$=156 \text{ inches}$

23. 4500 centimeters

$= 4500 \text{ centimeters}\cdot\frac{1 \text{ meter}}{100 \text{ centimeters}}$

$=\frac{4500}{100} \text{ meters}$

$= 45 \text{ meters}$

25. $42 \text{ quarts} = 42 \text{ quarts}\cdot\frac{1 \text{ gallon}}{4 \text{ quarts}}$

$=\frac{42}{4} \text{ gallons}$

$=10\frac{1}{2} \text{ gallons}$

$=10 \text{ gallons, } 2 \text{ quarts}$

27. $180 \text{ ounces} = 180 \text{ ounces}\cdot\frac{1 \text{ pound}}{16 \text{ ounces}}$

$=\frac{180}{16} \text{ pounds}$

$=11\frac{1}{4} \text{ pounds}$

$=11 \text{ pounds, } 4 \text{ ounces}$

29. 16,200 seconds

$=16{,}200 \text{ seconds}\cdot\frac{1 \text{ hour}}{3600 \text{ seconds}}$

$=\frac{16{,}200}{3600} \text{ hours}$

$=4\frac{1}{2} \text{ hours}$

$= 4 \text{ hours, } 30 \text{ minutes}$

31. $16 + (-16) = 0$ illustrates the Additive Inverse Property.

33. $\frac{3}{4}\cdot\frac{5}{5}=\frac{3}{4}\cdot 1=\frac{3}{4}$ illustrates the Multiplicative Identity Property.

35. $12\cdot\frac{1}{12}=1$ illustrates the Multiplicative Inverse Property.

37. $34.2 + (-34.2) = 0$ illustrates the Additive Inverse Property.

39. $\frac{0}{a}=0$

41. $\frac{2}{3}\cdot\left(-\frac{12}{43}\right)\cdot\frac{3}{2}=\frac{2}{3}\cdot\frac{3}{2}\cdot\left(-\frac{12}{43}\right)$ illustrates the Commutative Property of Multiplication since the order in which the numbers are multiplied changes.

43. $5.23 + 4.98 + (-5.23)$
$= 5.23 + (-5.23) + 4.98$ illustrates the Commutative Property of Addition since the order in which the numbers are added changes.

45. $\frac{a}{0}$ is undefined.

47. $54+29+(-54)=54+(-54)+29$

$=0+29$

$=29$

49. $\frac{9}{5}\cdot\frac{5}{9}\cdot 18 = 1\cdot 18 = 18$

51. $-25\cdot 13\cdot\frac{1}{5} = -25\cdot\frac{1}{5}\cdot 13 = -5\cdot 13 = -65$

53. $$\begin{aligned}347+456+(-456) &= 347+(456+(-456))\\ &= 347+0\\ &= 347\end{aligned}$$

55. $\frac{9}{2}\cdot\left(-\frac{10}{3}\right)\cdot 6 = \frac{\overset{3}{\cancel{9}}}{\underset{1}{\cancel{2}}}\cdot\left(-\frac{\overset{5}{\cancel{10}}}{\underset{1}{\cancel{3}}}\right)\cdot 6 = 3\cdot(-5)\cdot 6 = -15\cdot 6 = -90$

57. $\frac{7}{0}$ is undefined because 0 is the divisor.

59. $100(-34)(0.01) = 100(0.01)(-34) = 1(-34) = -34$

61. $569.003\cdot 0 = 0$

63. $\frac{45}{3902}+\left(-\frac{45}{3902}\right) = 0$

65. $$\begin{aligned}-\frac{5}{44}\cdot\frac{80}{3}\cdot\frac{11}{5} &= -\frac{5}{44}\cdot\frac{11}{5}\cdot\frac{80}{3}\\ &= -\frac{\overset{1}{\cancel{5}}}{\underset{4}{\cancel{44}}}\cdot\frac{\overset{1}{\cancel{11}}}{\underset{1}{\cancel{5}}}\cdot\frac{80}{3}\\ &= -\frac{1}{4}\cdot\frac{80}{3}\\ &= -\frac{1}{\underset{1}{\cancel{4}}}\cdot\frac{\overset{20}{\cancel{80}}}{3}\\ &= -\frac{20}{3}\end{aligned}$$

67. $$\begin{aligned}321.03+(-32.84)+(-85.03)+(-120.56)+120.56 &= 321.03+(-32.84)+(-85.03)+[(-120.56)+120.56]\\ &= 321.03+(-32.84)+(-85.03)+0\\ &= 288.19+(-85.03)\\ &= 203.16\end{aligned}$$

Alberto's balance is $203.16.

69. $$\begin{aligned}-3-(4-10) &= -3-[4+(-10)]\\ &= -3-(-6)\\ &= -3+6\\ &= 3\end{aligned}$$

71. $-15+10-(4-8) = -15+10-[4+(-8)]$
$= -15+10-(-4)$
$= -15+10+4$
$= -5+4$
$= -1$

73. $\frac{30 \text{ miles}}{1 \text{ hour}}$
$= \frac{30 \text{ miles}}{1 \text{ hour}} \cdot \frac{1 \text{ hour}}{3600 \text{ seconds}} \cdot \frac{5280 \text{ feet}}{1 \text{ mile}}$
$= \frac{30 \cdot 5280 \text{ feet}}{3600 \text{ seconds}}$
$= 44$ feet per second

75. Zero does not have a multiplicative inverse because two numbers are multiplicative inverses if their product is one. There is no real number such that the product of that number and 0 is equal to 1.

77. The quotient $\frac{0}{4} = 0$ because $4 \cdot 0 = 0$. The quotient $\frac{4}{0}$ is undefined because we should be able to determine a real number □ such that 0 · □ = 4. But because the product of 0 and every real number is 0, there is no replacement value for □.

79. The product of a nonzero real number and its multiplicative inverse (reciprocal) equals 1, the multiplicative identity.

Section 1.7

Preparing for Exponents and the Order of Operations

P1. $9 + (-19) = -10$

P2. $28 - (-7) = 28 + 7 = 35$

P3.
$$-7 \cdot \frac{8}{3} \cdot 36 = -7 \cdot \left(\frac{8}{3} \cdot \frac{36}{1}\right) = -7 \cdot \left(\frac{8 \cdot 36}{3 \cdot 1}\right) = -7 \cdot \left(\frac{8 \cdot 3 \cdot 12}{3 \cdot 1}\right) = -7 \cdot \left(\frac{8 \cdot \cancel{3} \cdot 12}{\cancel{3} \cdot 1}\right) = -7 \cdot \left(\frac{96}{1}\right) = -672$$

P4. $\frac{100}{-15} = \frac{5 \cdot 20}{-3 \cdot 5} = \frac{\cancel{5} \cdot 20}{-3 \cdot \cancel{5}} = \frac{20}{-3} = -\frac{20}{3}$

1.7 Quick Checks

1. The expression 11 · 11 · 11 · 11 · 11 contains five factors of 11, so $11 \cdot 11 \cdot 11 \cdot 11 \cdot 11 = 11^5$.

2. The expression (–7)(–7)(–7)(–7) contains four factors of –7, so $(-7)^4$.

3. The expression (–2) · (–2) · (–2) contains three factors of –2, so $(-2) \cdot (-2) \cdot (-2) = (-2)^3$.

4. $2^4 = 2 \cdot 2 \cdot 2 \cdot 2 = 16$

5. $(-7)^2 = (-7) \cdot (-7) = 49$

6. $\left(-\frac{1}{6}\right)^3 = \left(-\frac{1}{6}\right) \cdot \left(-\frac{1}{6}\right) \cdot \left(-\frac{1}{6}\right) = -\frac{1}{216}$

7. $(0.9)^2 = (0.9) \cdot (0.9) = 0.81$

8. $-2^4 = -(2 \cdot 2 \cdot 2 \cdot 2) = -16$

9. $(-2)^4 = (-2) \cdot (-2) \cdot (-2) \cdot (-2) = 16$

10. $1 + 7 \cdot 2 = 1 + 14 = 15$

11. $-11 \cdot 3 + 2 = -33 + 2 = -31$

12. $18 + 3 \div \left(-\frac{1}{2}\right) = 18 + 3 \cdot \left(-\frac{2}{1}\right)$
$= 18 + (-6)$
$= 12$

13. $9 \cdot 4 - 5 = 36 - 5 = 31$

14. $$\begin{aligned}\frac{15}{2} \div (-5) - \frac{3}{2} &= \frac{15}{2} \cdot \left(-\frac{1}{5}\right) - \frac{3}{2}\\ &= -\frac{3}{2} - \frac{3}{2}\\ &= -\frac{6}{2}\\ &= -3\end{aligned}$$

15. $8(2 + 3) = 8(5) = 40$

16. $(2 - 9) \cdot (5 + 4) = (-7) \cdot (9) = -63$

17. $\left(\frac{6}{7}+\frac{8}{7}\right)\cdot\left(\frac{11}{8}+\frac{5}{8}\right)=\left(\frac{14}{7}\right)\cdot\left(\frac{16}{8}\right)=2\cdot 2=4$

18. $\frac{2+5\cdot 6}{-3\cdot 8-4}=\frac{2+30}{-24-4}=\frac{32}{-28}=\frac{8\cdot \cancel{4}}{-7\cdot \cancel{4}}=-\frac{8}{7}$

19. $$\begin{aligned}\frac{(12+14)\cdot 2}{13\cdot 2+13\cdot 5} &= \frac{26\cdot 2}{13\cdot 2+13\cdot 5}\\ &= \frac{52}{26+65}\\ &= \frac{52}{91}\\ &= \frac{4\cdot \cancel{13}}{7\cdot \cancel{13}}\\ &= \frac{4}{7}\end{aligned}$$

20. $$\begin{aligned}\frac{4+3\div \frac{1}{7}}{2\cdot 9-3} &= \frac{4+3\cdot 7}{2\cdot 9-3}\\ &= \frac{4+21}{18-3}\\ &= \frac{25}{15}\\ &= \frac{5\cdot \cancel{5}}{3\cdot \cancel{5}}\\ &= \frac{5}{3}\end{aligned}$$

21. $$\begin{aligned}4\cdot[2\cdot(3+7)-15] &= 4\cdot[2\cdot 10-15]\\ &= 4\cdot[20-15]\\ &= 4\cdot[5]\\ &= 20\end{aligned}$$

22. $$\begin{aligned}&2\cdot\{4\cdot[26-(9+7)]-15\}-10\\ &= 2\cdot\{4[26-16]-15\}-10\\ &= 2\cdot\{4[10]-15\}-10\\ &= 2\cdot\{40-15\}-10\\ &= 2\cdot\{25\}-10\\ &= 50-10\\ &= 40\end{aligned}$$

23. $\frac{7-5^2}{2}=\frac{7-25}{2}=\frac{-18}{2}=\frac{-9\cdot \cancel{2}}{\cancel{2}}=-9$

24. $3(7-3)^2 = 3(4)^2 = 3\cdot 16 = 48$

25. $$\begin{aligned}\frac{(-3)^2+7(1-3)}{3\cdot 2+5} &= \frac{9+7(1-3)}{6+5}\\ &= \frac{9+7(-2)}{11}\\ &= \frac{9+(-14)}{11}\\ &= \frac{-5}{11}\\ &= -\frac{5}{11}\end{aligned}$$

26. $$\begin{aligned}2+5\cdot 3^2-\frac{3}{2}\cdot 2^2 &= 2+5\cdot 9-\frac{3}{2}\cdot 4\\ &= 2+45-6\\ &= 47-6\\ &= 41\end{aligned}$$

27. $\frac{(4-10)^2}{2^3-5}=\frac{(-6)^2}{2^3-5}=\frac{36}{8-5}=\frac{36}{3}=\frac{12\cdot \cancel{3}}{\cancel{3}}=12$

28. $$\begin{aligned}-3[(-4)^2-5(8-6)]^2 &= -3[(-4)^2-5(2)]^2\\ &= -3[16-5(2)]^2\\ &= -3[16-10]^2\\ &= -3[6]^2\\ &= -3[36]\\ &= -108\end{aligned}$$

29. $\frac{(2.9+7.1)^2}{5^2-15}=\frac{(10)^2}{25-15}=\frac{100}{10}=\frac{10\cdot \cancel{10}}{\cancel{10}}=10$

30. $\left(\dfrac{4^2-4(-3)(1)}{7\cdot 2}\right)^2=\left(\dfrac{16-4(-3)}{14}\right)^2$
$=\left(\dfrac{16+12}{14}\right)^2$
$=\left(\dfrac{28}{14}\right)^2$
$=(2)^2$
$=2^2$
$=4$

1.7 Exercises

31. The expression $5\cdot 5$ contains two factors of 5, so $5\cdot 5=5^2$.

33. The expression $\frac{3}{5}\cdot\frac{3}{5}\cdot\frac{3}{5}$ contains three factors of $\frac{3}{5}$, so $\frac{3}{5}\cdot\frac{3}{5}\cdot\frac{3}{5}=\left(\frac{3}{5}\right)^3$.

35. $8^2=8\cdot 8=64$

37. $(-8)^2=(-8)(-8)=64$

39. $10^3=10\cdot 10\cdot 10=1000$

41. $\left(\dfrac{3}{4}\right)^3=\left(\dfrac{3}{4}\right)\left(\dfrac{3}{4}\right)\left(\dfrac{3}{4}\right)=\dfrac{27}{64}$

43. $(1.5)^2=(1.5)(1.5)=2.25$

45. $-3^2=-(3\cdot 3)=-9$

47. $-1^{20}=-\underbrace{(1\cdot 1\cdot 1\cdot\cdots\cdot 1)}_{\text{20 factors}}=-1$

49. $0^4=0\cdot 0\cdot 0\cdot 0=0$

51. $\left(-\dfrac{1}{2}\right)^6$
$=\left(-\dfrac{1}{2}\right)\left(-\dfrac{1}{2}\right)\left(-\dfrac{1}{2}\right)\left(-\dfrac{1}{2}\right)\left(-\dfrac{1}{2}\right)\left(-\dfrac{1}{2}\right)$
$=\dfrac{1}{64}$

53. $\left(-\dfrac{1}{3}\right)^3=\left(-\dfrac{1}{3}\right)\left(-\dfrac{1}{3}\right)\left(-\dfrac{1}{3}\right)=-\dfrac{1}{27}$

55. $2+3\cdot 4=2+12=14$

57. $-5\cdot 3+12=-15+12=-3$

59. $100\div 2\cdot 50=50\cdot 50=2500$

61. $156-3\cdot 2+10=156-6+10$
$=150+10$
$=160$

63. $(2+3)\cdot 4=5\cdot 4=20$

65. $8\div 4\cdot 2=2\cdot 2=4$

67. $\dfrac{4+2}{2+8}=\dfrac{6}{10}=\dfrac{2\cdot 3}{2\cdot 5}=\dfrac{3}{5}$

69. $\dfrac{14-6}{6-14}=\dfrac{8}{-8}=\dfrac{1\cdot 8}{-1\cdot 8}=-1$

71. $13-[3+(-8)4]=13-[3+(-32)]$
$=13-[-29]$
$=13+29$
$=42$

73. $(-8.75-1.25)\div(-2)=-10\div(-2)=5$

75. $4-2^3=4-8=-4$

77. $15+4\cdot 5^2=15+4\cdot 25=15+100=115$

79. $-2^3+3^2\div(2^2-1)=-8+9\div(4-1)$
$=-8+9\div 3$
$=-8+3$
$=-5$

81. $\left(\dfrac{4^2-3}{12-2\cdot 5}\right)^2=\left(\dfrac{16-3}{12-10}\right)^2=\left(\dfrac{13}{2}\right)^2=\dfrac{169}{4}$

83. $-2\cdot[5\cdot(9-3)-3\cdot 6]=-2\cdot[5\cdot 6-3\cdot 6]$
$=-2\cdot[30-18]$
$=-2\cdot 12$
$=-24$

85. $\left(\frac{4}{3}+\frac{5}{6}\right)\left(\frac{2}{5}-\frac{9}{10}\right)=\left(\frac{8}{6}+\frac{5}{6}\right)\left(\frac{4}{10}-\frac{9}{10}\right)$
$=\left(\frac{13}{6}\right)\left(\frac{-5}{10}\right)$
$=\left(\frac{13}{6}\right)\left(\frac{-1}{2}\right)$
$=\frac{13\cdot(-1)}{6\cdot 2}$
$=-\frac{13}{12}$

87. $-2.5+4.5\div 1.5=-2.5+3=0.5$

89. $4+2\cdot(6-2)=4+2\cdot 4=4+8=12$

91. $\frac{12-16\div 4+(-24)}{16\cdot 2-4\cdot 0}=\frac{12-4+(-24)}{32-0}$
$=\frac{8+(-24)}{32}$
$=\frac{-16}{32}$
$=\frac{-1\cdot 16}{2\cdot 16}$
$=-\frac{1}{2}$

93. $\left(\frac{2-(-4)^3}{5^2-7\cdot 2}\right)^2=\left(\frac{2-(-64)}{25-14}\right)^2$
$=\left(\frac{2+64}{11}\right)^2$
$=\left(\frac{66}{11}\right)^2$
$=6^2$
$=36$

95. $\frac{5^2-10}{3^2+6}=\frac{25-10}{9+6}=\frac{15}{15}=1$

97. $\left|6\cdot(5-3^2)\right|=\left|6\cdot(5-9)\right|$
$=\left|6\cdot(-4)\right|$
$=\left|-24\right|$
$=24$

99. $\frac{81}{8}+\frac{13}{4}\div\frac{1}{2}=\frac{81}{8}+\frac{13}{4}\cdot\frac{2}{1}$
$=\frac{81}{8}+\frac{13}{2}$
$=\frac{81}{8}+\frac{52}{8}$
$=\frac{133}{8}$

101. $\frac{-7}{20}+\frac{3}{8}\div\frac{1}{2}=\frac{-7}{20}+\frac{3}{8}\cdot\frac{2}{1}$
$=\frac{-7}{20}+\frac{6}{8}$
$=\frac{-7}{20}+\frac{3}{4}$
$=\frac{-7}{20}+\frac{15}{20}$
$=\frac{8}{20}$
$=\frac{2\cdot 4}{5\cdot 4}$
$=\frac{2}{5}$

103. $\frac{21-3^2}{1+3}=\frac{21-9}{1+3}=\frac{12}{4}=\frac{3\cdot 4}{1\cdot 4}=3$

105. $\frac{3}{4}\cdot\left[\frac{5}{4}\div\left(\frac{3}{8}-\frac{1}{8}\right)-3\right]=\frac{3}{4}\cdot\left[\frac{5}{4}\div\frac{2}{8}-3\right]$
$=\frac{3}{4}\cdot\left[\frac{5}{4}\div\frac{1}{4}-3\right]$
$=\frac{3}{4}\cdot\left[\frac{5}{4}\cdot\frac{4}{1}-3\right]$
$=\frac{3}{4}\cdot[5-3]$
$=\frac{3}{4}\cdot 2$
$=\frac{6}{4}$
$=\frac{3\cdot 2}{2\cdot 2}$
$=\frac{3}{2}$

107. $\left(\frac{4}{3}\right)^3 - \left(\frac{1}{2}\right)^2 \cdot \left(\frac{8}{3}\right) + 2 \div 3$
$= \frac{64}{27} - \frac{1}{4} \cdot \left(\frac{8}{3}\right) + 2 \div 3$
$= \frac{64}{27} - \frac{8}{12} + 2 \div 3$
$= \frac{64}{27} - \frac{2}{3} + \frac{2}{3}$
$= \frac{64}{27} + \left(-\frac{2}{3} + \frac{2}{3}\right)$
$= \frac{64}{27} + 0$
$= \frac{64}{27}$

109. $\frac{5^2 - 3^3}{|4 - 4^2|} = \frac{25 - 27}{|4 - 16|}$
$= \frac{-2}{|-12|}$
$= \frac{-2}{12}$
$= \frac{-1 \cdot 2}{6 \cdot 2}$
$= -\frac{1}{6}$

111. $72 = 8 \cdot 9 = (2 \cdot 2 \cdot 2) \cdot (3 \cdot 3) = 2^3 \cdot 3^2$

113. $48 = 16 \cdot 3 = (2 \cdot 2 \cdot 2 \cdot 2) \cdot 3 = 2^4 \cdot 3$

115. $(4 \cdot 3 + 6) \cdot 2 = (12 + 6) \cdot 2 = 18 \cdot 2 = 36$

117. $(4 + 3) \cdot (4 + 2) = 7 \cdot 6 = 42$

119. $(6 - 4) + (3 - 1) = 2 + 2 = 4$

121. $479 + 0.075(479) = 479 + 35.925 = 514.925$
The total amount is \$514.93.

123. $2 \cdot 3.1416 \cdot 6^2 + 2 \cdot 3.1416 \cdot 6 \cdot 10$
$= 2 \cdot 3.1416 \cdot 36 + 2 \cdot 3.1416 \cdot 6 \cdot 10$
$= 226.1952 + 376.992$
$= 603.1872$
The surface area is about 603.19 square inches.

125. $1000(1 + 0.03)^2 = 1000(1.03)^2$
$= 1000(1.0609)$
$= 1060.9$
The amount of money is \$1060.90.

127. $\angle XYZ = \angle XYQ + \angle QYZ$
$= 46.5° + 69.25°$
$= 115.75°$

129. The sum $2x^2 + 4x^2$ is not equal to $6x^4$ because when we combine like terms, we add the coefficients of the like terms and keep the variables and exponents the same. Put another way, $2x^2 + 4x^2 = (2 + 4)x^2 = 6x^2$.

Section 1.8

Preparing for Simplifying Algebraic Expressions

P1. $-3 + 8 = 5$

P2. $-7 - 8 = -7 + (-8) = -15$

P3. $-\frac{4}{3}(27) = -\frac{4}{3}\left(\frac{27}{1}\right)$
$= -\frac{4 \cdot 27}{3 \cdot 1}$
$= -\frac{4 \cdot 3 \cdot 9}{3 \cdot 1}$
$= -\frac{4 \cdot \not{3} \cdot 9}{\not{3} \cdot 1}$
$= -\frac{4 \cdot 9}{1}$
$= -36$

1.8 Quick Checks

1. To evaluate an algebraic expression means to substitute the numerical value for each variable into the expression and simplify.

2. Substitute 4 for k.
$-3k + 5 = -3(4) + 5 = -12 + 5 = -7$

3. Substitute 12 for t.
$\frac{5}{4}t - 6 = \frac{5}{4}(12) - 6 = \frac{60}{4} - 6 = 15 - 6 = 9$

4. Substitute −2 for y.
$$\begin{aligned}-2y^2-y+8&=-2(-2)^2-(-2)+8\\&=-2(4)-(-2)+8\\&=-8-(-2)+8\\&=-8+2+8\\&=-8+8+2\\&=0+2\\&=2\end{aligned}$$

5. Substitute 8 for x and 16 for y.
$$\begin{aligned}7.50x+10y&=7.50(8)+10(16)\\&=60+160\\&=220\end{aligned}$$
The value is $220.

6. The algebraic expression $5x^2+3xy$ has two terms: $5x^2$ and $3xy$.

7. The algebraic expression $9ab-3bc+5ac-ac^2$ has four terms: $9ab$, $-3bc$, $5ac$, and $-ac^2$.

8. The algebraic expression $\frac{2mn}{5}-\frac{3n}{7}$ has two terms: $\frac{2mn}{5}$ and $-\frac{3n}{7}$.

9. The algebraic expression $\frac{m^2}{3}-8$ has two terms: $\frac{m^2}{3}$ and −8.

10. The coefficient of $2z^2$ is 2.

11. The coefficient of $xy=1\cdot xy$ is 1.

12. The coefficient of $-b=-1\cdot b$ is −1.

13. The coefficient of 5 is 5.

14. The coefficient of $-\frac{2}{3}z$ is $-\frac{2}{3}$.

15. The coefficient of $\frac{x}{6}=\frac{1}{6}\cdot x$ is $\frac{1}{6}$.

16. $-\frac{2}{3}p^2$ and $\frac{4}{5}p^2$ are like terms. They have the same variable raised to the same power.

17. $\frac{m}{6}=\frac{1}{6}m$ and $4m$ are like terms. They have the same variable raised to the same power.

18. $3a^2b$ and $-2ab^2$ are unlike terms. The variable a is raised to the second power in $3a^2b$ and to the first power in $-2ab^2$.

19. $8a$ and 11 are unlike terms. $8a$ has a variable and 11 does not.

20. $6(x+2)=6\cdot x+6\cdot 2=6x+12$

21.
$$\begin{aligned}-5(x+2)&=-5\cdot x+(-5)\cdot 2\\&=-5x+(-10)\\&=-5x-10\end{aligned}$$

22.
$$\begin{aligned}-2(k-7)&=-2\cdot k-(-2)\cdot 7\\&=-2k-(-14)\\&=-2k+14\end{aligned}$$

23.
$$\begin{aligned}(8x+12)\frac{3}{4}&=8x\cdot\frac{3}{4}+12\cdot\frac{3}{4}\\&=\overset{2}{\cancel{8}}x\cdot\frac{3}{\underset{1}{\cancel{4}}}+\overset{3}{\cancel{12}}\cdot\frac{3}{\underset{1}{\cancel{4}}}\\&=2x\cdot 3+3\cdot 3\\&=6x+9\end{aligned}$$

24. $3x-8x=(3-8)x=-5x$

25. $-5x^2+x^2=-5x^2+1x^2=(-5+1)x^2=-4x^2$

26.
$$\begin{aligned}-7x-x+6-3&=-7x-1x+6-3\\&=(-7-1)x+(6-3)\\&=-8x+3\end{aligned}$$

27.
$$\begin{aligned}4x-12x-3+17&=(4-12)x+(-3+17)\\&=-8x+14\end{aligned}$$

28.
$$\begin{aligned}3a+2b-5a+7b-4&=3a-5a+2b+7b-4\\&=(3-5)a+(2+7)b-4\\&=-2a+9b-4\end{aligned}$$

29.
$$\begin{aligned}&(5ac+2b)+(7ac-5a)+(-b)\\&=5ac+2b+7ac-5a+(-1)b\\&=5ac+7ac+2b+(-1)b-5a\\&=(5+7)ac+(2+(-1))b-5a\\&=12ac+1b-5a\\&=12ac+b-5a\end{aligned}$$

30. $5ab^2+7a^2b+3ab^2-8a^2b$
$=5ab^2+3ab^2+7a^2b-8a^2b$
$=(5+3)ab^2+(7-8)a^2b$
$=8ab^2-1a^2b$
$=8ab^2-a^2b$

31. $\frac{4}{3}rs-\frac{3}{2}r^2+\frac{2}{3}rs-5$
$=\frac{4}{3}rs+\frac{2}{3}rs-\frac{3}{2}r^2-5$
$=\left(\frac{4}{3}+\frac{2}{3}\right)rs-\frac{3}{2}r^2-5$
$=\frac{6}{3}rs-\frac{3}{2}r^2-5$
$=2rs-\frac{3}{2}r^2-5$

32. To simplify an algebraic expression means to remove all parentheses and combine like terms.

33. $3x+2(x-1)-7x+1=3x+2x-2-7x+1$
$=3x+2x-7x-2+1$
$=-2x-1$

34. $m+2n-3(m+2n)-(7-3n)$
$=m+2n-3m-6n-7+3n$
$=m-3m+2n-6n+3n-7$
$=-2m-1n-7$
$=-2m-n-7$

35. $2(a-4b)-(a+4b)+b$
$=2a-8b-a-4b+b$
$=2a-a-8b-4b+b$
$=2a-1a-8b-4b+1b$
$=(2-1)a+(-8-4+1)b$
$=1a+(-11)b$
$=a-11b$

36. $\frac{1}{2}(6x+4)-\frac{1}{3}(12-9x)=3x+2-4+3x$
$=3x+3x+2-4$
$=6x-2$

1.8 Exercises

37. Substitute 4 for x.
$2x+5=2(4)+5=8+5=13$

39. Substitute 3 for x.
$x^2+3x-1=3^2+3(3)-1$
$=9+3(3)-1$
$=9+9-1$
$=18-1$
$=17$

41. Substitute –5 for k.
$4-k^2=4-(-5)^2$
$=4-25$
$=-21$

43. Substitute 8 for x and 10 for y.
$\frac{5x}{y}+y^2=\frac{5(8)}{10}+10^2$
$=\frac{5(8)}{10}+100$
$=\frac{40}{10}+100$
$=4+100$
$=104$

45. Substitute 3 for x and 2 for y.
$\frac{9x-5y}{x+y}=\frac{9(3)-5(2)}{3+2}=\frac{27-10}{5}=\frac{17}{5}$

47. Substitute 3 for x and 4 for y.
$(x+3y)^2=(3+3\cdot 4)^2$
$=(3+12)^2$
$=15^2$
$=225$

49. Substitute 3 for x and 4 for y.
$x^2+9y^2=3^2+9(4^2)$
$=9+9(16)$
$=9+144$
$=153$

51. $2x^3+3x^2-x+6$ can be written as $2x^3+3x^2+(-1\cdot x)+6$. The terms are $2x^3$, $3x^2$, $-x$, and 6. The coefficient of $2x^3$ is 2. The coefficient of $3x^2$ is 3. The coefficient of $-x$ is –1. The coefficient of 6 is 6.

53. $z^2+\frac{2y}{3}$ can be written as $1\cdot z^2+\frac{2}{3}y$. The terms are z^2 and $\frac{2y}{3}$. The coefficient of z^2 is 1. The coefficient of $\frac{2y}{3}$ is $\frac{2}{3}$.

55. $8x$ and 8 are unlike terms. $8x$ has a variable and 8 does not.

57. 54 and −21 are like terms. They are both constants.

59. $12b$ and $-b$ are like terms. They have the same variable raised to the same power.

61. r^2s and rs^2 are unlike terms. The variable r is raised to the second power in r^2s and the first power in rs^2.

63. $3(m+2)=3\cdot m+3\cdot 2=3m+6$

65. $$\begin{aligned}(3n^2+2n-1)6&=3n^2\cdot 6+2n\cdot 6-1\cdot 6\\&=18n^2+12n-6\end{aligned}$$

67. $$\begin{aligned}-(x-y)&=-1\cdot(x-y)\\&=-1\cdot x-(-1)\cdot y\\&=-x-(-y)\\&=-x+y\end{aligned}$$

69. $$\begin{aligned}(8x-6y)(-0.5)&=8x\cdot(-0.5)-6y(-0.5)\\&=-4x-(-3y)\\&=-4x+3y\end{aligned}$$

71. $5x-2x=(5-2)x=3x$

73. $4z-6z+8z=(4-6+8)z=6z$

75. $$\begin{aligned}2m+3n+8m+7n&=2m+8m+3n+7n\\&=(2+8)m+(3+7)n\\&=10m+10n\end{aligned}$$

77. $$\begin{aligned}0.3x^7+x^7+0.9x^7&=(0.3+1+0.9)x^7\\&=2.2x^7\end{aligned}$$

79. $-3y^6+13y^6=(-3+13)y^6=10y^6$

81. $$\begin{aligned}&-(6w-12y-13z)\\&=-1\cdot 6w-(-1)\cdot 12y-(-1)\cdot 13z\\&=-6w-(-12y)-(-13z)\\&=-6w+12y+13z\end{aligned}$$

83. $$\begin{aligned}5(k+3)-8k&=5k+15-8k\\&=5k-8k+15\\&=-3k+15\end{aligned}$$

85. $7n-(3n+8)=7n-3n-8=4n-8$

87. $$\begin{aligned}(7-2x)-(x+4)&=7-2x-x-4\\&=7-3x-4\\&=-3x+3\end{aligned}$$

89. $$\begin{aligned}(7n-8)-(3n-6)&=7n-8-3n+6\\&=4n-8+6\\&=4n-2\end{aligned}$$

91. $$\begin{aligned}-6(n-3)+2(n+1)&=-6n+18+2n+2\\&=-6n+2n+18+2\\&=-4n+20\end{aligned}$$

93. $\frac{2}{3}x+\frac{1}{6}x=\frac{4}{6}x+\frac{1}{6}x=\frac{5}{6}x$

95. $$\begin{aligned}\frac{1}{2}(8x+5)-\frac{2}{3}(6x+12)&=4x+\frac{5}{2}-4x-8\\&=4x-4x+\frac{5}{2}-8\\&=\frac{5}{2}-\frac{16}{2}\\&=-\frac{11}{2}\end{aligned}$$

97. $$\begin{aligned}2(0.5x+9)-3(1.5x+8)&=x+18-4.5x-24\\&=x-4.5x+18-24\\&=-3.5x-6\end{aligned}$$

99. $$\begin{aligned}&3.2(x+1.6)+1.4(2x-3.7)\\&=3.2x+5.12+2.8x-5.18\\&=3.2x+2.8x+5.12-5.18\\&=6x-0.06\end{aligned}$$

101. **(a)** $5x+3x=5(4)+3(4)=20+12=32$

(b) $5x+3x=8x=8(4)=32$

103. (a) $-2a^2+5a^2=-2(-3)^2+5(-3)^2$
$=-2(9)+5(9)$
$=-18+45$
$=27$

(b) $-2a^2+5a^2=3a^2=3(-3)^2=3(9)=27$

105. (a) $4z-3(z+2)=4(6)-3(6+2)$
$=4(6)-3(8)$
$=24-24$
$=0$

(b) $4z-3(z+2)=4z-3z-6$
$=z-6$
$=6-6$
$=0$

107. (a) $5y^2+6y-2y^2+5y-3$
$=5(-2)^2+6(-2)-2(-2)^2+5(-2)-3$
$=5(4)+6(-2)-2(4)+5(-2)-3$
$=20-12-8-10-3$
$=-13$

(b) $5y^2+6y-2y^2+5y-3$
$=5y^2-2y^2+6y+5y-3$
$=3y^2+11y-3$
$=3(-2)^2+11(-2)-3$
$=3(4)+11(-2)-3$
$=12-22-3$
$=-13$

109. (a) $\frac{1}{2}(4x-2)-\frac{2}{3}(3x+9)$
$=\frac{1}{2}(4\cdot 3-2)-\frac{2}{3}(3\cdot 3+9)$
$=\frac{1}{2}(12-2)-\frac{2}{3}(9+9)$
$=\frac{1}{2}(10)-\frac{2}{3}(18)$
$=5-12$
$=-7$

(b) $\frac{1}{2}(4x-2)-\frac{2}{3}(3x+9)=2x-1-2x-6$
$=2x-2x-1-6$
$=-7$

111. (a) $3a+4b-7a+3(a-2b)$
$=3(2)+4(5)-7(2)+3(2-2\cdot 5)$
$=3(2)+4(5)-7(2)+3(2-10)$
$=3(2)+4(5)-7(2)+3(-8)$
$=6+20-14-24$
$=-12$

(b) $3a+4b-7a+3(a-2b)$
$=3a+4b-7a+3a-6b$
$=3a-7a+3a+4b-6b$
$=-a-2b$
$=-2-2(5)$
$=-2-10$
$=-12$

113. Let $h=4$, $b=5$, $B=17$.
$\frac{1}{2}h(b+B)=\frac{1}{2}(4)(5+17)$
$=\frac{1}{2}(4)(22)$
$=2(22)$
$=44$

115. Let $a=6$, $b=3$, $c=-4$, $d=-2$.
$\frac{a-b}{c-d}=\frac{6-3}{-4-(-2)}=\frac{6-3}{-4+2}=\frac{3}{-2}=-\frac{3}{2}$

117. Let $a=7$, $b=8$, $c=1$.
$b^2-4ac=8^2-4(7)(1)=64-28=36$

119. Let $m=125$.
$59.95+0.15m=59.95+0.15(125)$
$=59.95+18.75$
$=78.70$
The cost of renting the truck is \$78.70.

121. Let $a=156$ and $c=421$.
$12a+7c=12(156)+7(421)$
$=1872+2947$
$=4819$
The revenue for 156 adult tickets and 421 children's tickets is \$4819.

123. (a) $2w+2(3w-4)=2w+6w-8$
$=8w-8$

(b) Let $w=5$.
$8w-8=8(5)-8=40-8=32$
The perimeter is 32 yards.

125. Let $s = 2950$ and $b = 2050$.
$0.055s + 0.0325b$
$= 0.055(2950) + 0.0325(2050)$
$= 162.25 + 66.625$
$= 228.875$
The annual interest is about \$228.88.

127. Answers may vary. One possibility:
$2.75(-3x^2 + 7x - 3) - 1.75(-3x^2 + 7x - 3)$
$= (2.75 - 1.75)(-3x^2 + 7x - 3)$
$= (1)(-3x^2 + 7x - 3)$
$= -3x^2 + 7x - 3$

129. The sum $2x^2 + 4x^2$ is not equal to $6x^4$ because when we combine like terms, we add the coefficients of the like terms and keep the variables and exponents the same. Put another way, $2x^2 + 4x^2 = (2+4)x^2 = 6x^2$.

Chapter 1 Review

1.
```
    24
   /  \
  4    6
 /\    /\
2 · 2 · 2 · 3
```
$24 = 2 \cdot 2 \cdot 2 \cdot 3$

2.
```
 87
 /\
3 · 29
```
$87 = 3 \cdot 29$

3.
```
    81
   /  \
  9    9
 /\    /\
3 · 3 · 3 · 3
```
$81 = 3 \cdot 3 \cdot 3 \cdot 3$

4.
```
  124
  /\
 4  31
 /\   \
2 · 2 · 31
```
$124 = 2 \cdot 2 \cdot 31$

5. 17 is prime.

6. $18 = 3 \cdot 3 \cdot 2$
$24 = 3 \quad\;\; \cdot 2 \cdot 2 \cdot 2$
$3 \cdot 3 \cdot 2 \cdot 2 \cdot 2 = 72$

7. $4 = 2 \cdot 2$
$8 = 2 \cdot 2 \cdot 2$
$18 = 2 \cdot \quad\;\; 3 \cdot 3$
$2 \cdot 2 \cdot 2 \cdot 3 \cdot 3 = 72$

8. $\dfrac{7}{15} = \dfrac{7}{15} \cdot \dfrac{2}{2} = \dfrac{14}{30}$

9. $3 = \dfrac{3}{1} = \dfrac{3}{1} \cdot \dfrac{4}{4} = \dfrac{12}{4}$

10. $6 = 2 \cdot 3$
$8 = 2 \quad\; \cdot 2 \cdot 2$
$\text{LCD} = 2 \cdot 3 \cdot 2 \cdot 2 = 24$
$\dfrac{1}{6} = \dfrac{1}{6} \cdot \dfrac{4}{4} = \dfrac{4}{24}$
$\dfrac{3}{8} = \dfrac{3}{8} \cdot \dfrac{3}{3} = \dfrac{9}{24}$

11. $16 = 2 \cdot 2 \cdot 2 \quad\; \cdot 2$
$24 = 2 \cdot 2 \cdot 2 \cdot 3$
$\text{LCD} = 2 \cdot 2 \cdot 2 \cdot 3 \cdot 2 = 48$
$\dfrac{9}{16} = \dfrac{9}{16} \cdot \dfrac{3}{3} = \dfrac{27}{48}$
$\dfrac{7}{24} = \dfrac{7}{24} \cdot \dfrac{2}{2} = \dfrac{14}{48}$

12. $\dfrac{25}{60} = \dfrac{5 \cdot 5}{5 \cdot 3 \cdot 2 \cdot 2} = \dfrac{5}{3 \cdot 2 \cdot 2} = \dfrac{5}{12}$

13. $\dfrac{125}{250} = \dfrac{5 \cdot 5 \cdot 5}{5 \cdot 5 \cdot 5 \cdot 2} = \dfrac{1}{2}$

14. $\dfrac{96}{120} = \dfrac{3 \cdot 2 \cdot 2 \cdot 2 \cdot 2 \cdot 2}{3 \cdot 2 \cdot 2 \cdot 2 \cdot 5} = \dfrac{2 \cdot 2}{5} = \dfrac{4}{5}$

15. 21.76

16. 15

17. $\frac{8}{9} = 9\overline{)8.00}$ → 0.88

$$\begin{array}{r} 0.88 \\ 9\overline{)8.00} \\ \underline{7\,2} \\ 80 \end{array}$$

$\frac{8}{9} = 0.88... = 0.\overline{8}$

18. $\frac{9}{32} = 32\overline{)9.00000}$

$$\begin{array}{r} 0.28125 \\ 32\overline{)9.00000} \\ \underline{6\,4} \\ 2\,60 \\ \underline{2\,56} \\ 40 \\ \underline{32} \\ 80 \\ \underline{64} \\ 160 \\ \underline{160} \\ 0 \end{array}$$

$\frac{9}{32} = 0.28125$

19.

$$\begin{array}{r} 1.833 \\ 6\overline{)11.000} \\ \underline{6} \\ 5\,0 \\ \underline{4\,8} \\ 20 \\ \underline{18} \\ 20 \end{array}$$

$\frac{11}{6}$ rounded to the nearest hundredth is 1.83.

20.

$$\begin{array}{r} 2.37 \\ 8\overline{)19.00} \\ \underline{16} \\ 3\,0 \\ \underline{2\,4} \\ 60 \\ \underline{56} \\ 40 \end{array}$$

$\frac{19}{8}$ rounded to the nearest tenth is 2.4.

21. $0.6 = \frac{6}{10} = \frac{3 \cdot 2}{5 \cdot 2} = \frac{3}{5}$

22. $0.375 = \frac{375}{1000} = \frac{5 \cdot 5 \cdot 5 \cdot 3}{5 \cdot 5 \cdot 5 \cdot 2 \cdot 2 \cdot 2} = \frac{3}{2 \cdot 2 \cdot 2} = \frac{3}{8}$

23. $0.864 = \frac{864}{1000}$
$= \frac{3 \cdot 3 \cdot 3 \cdot 2 \cdot 2 \cdot 2 \cdot 2 \cdot 2}{5 \cdot 5 \cdot 5 \cdot 2 \cdot 2 \cdot 2}$
$= \frac{3 \cdot 3 \cdot 3 \cdot 2 \cdot 2}{5 \cdot 5 \cdot 5}$
$= \frac{108}{125}$

24. $41\% = 41\% \cdot \frac{1}{100\%} = \frac{41}{100} = 0.41$

25. $760\% = 760\% \cdot \frac{1}{100\%} = \frac{760}{100} = 7.60$

26. $9.03\% = 9.03\% \cdot \frac{1}{100\%} = \frac{9.03}{100} = 0.0903$

27. $0.35\% = 0.35\% \cdot \frac{1}{100\%} = \frac{0.35}{100} = 0.0035$

28. $0.23 = 0.23 \cdot \frac{100\%}{1} = 23\%$

29. $1.17 = 1.17 \cdot \frac{100\%}{1} = 117\%$

30. $0.045 = 0.045 \cdot \frac{100\%}{1} = 4.5\%$

31. $3 = 3 \cdot \frac{100\%}{1} = 300\%$

32. **(a)** $\frac{12}{20} = \frac{3 \cdot 2 \cdot 2}{5 \cdot 2 \cdot 2} = \frac{3}{5}$

The student earned $\frac{3}{5}$ of the point.

(b) $\frac{3}{5} = 0.6$

$0.6 \cdot \frac{100\%}{1} = 60\%$
The student earned 60% of the points.

33. The set of whole numbers less than 7 is $A = \{0, 1, 2, 3, 4, 5, 6\}$.

34. The set of natural numbers less than or equal to 3 is $B = \{1, 2, 3\}$.

35. The set of integers greater than −3 and less than or equal to 5 is $C = \{-2, -1, 0, 1, 2, 3, 4, 5\}$.

36. The set of integers greater than or equal to −2 and less than 4 is $D = \{-2, -1, 0, 1, 2, 3\}$.

37. $\frac{9}{3} = 3$ and 11 are the natural numbers.

38. 0, $\frac{9}{3} = 3$, and 11 are the whole numbers.

39. −6, 0, $\frac{9}{3} = 3$, and 11 are the integers.

40. −6, −3.25, 0, $\frac{9}{3}$, 11, and $\frac{5}{7}$ are the rational numbers.

41. 5.030030003... is the only irrational number.

42. All the numbers listed are real numbers.

43.

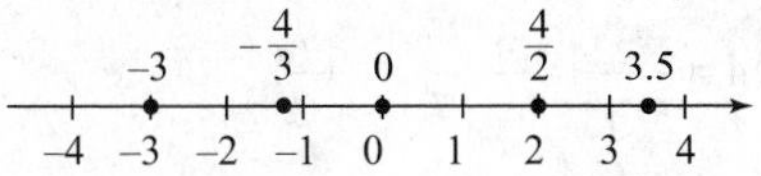

44.

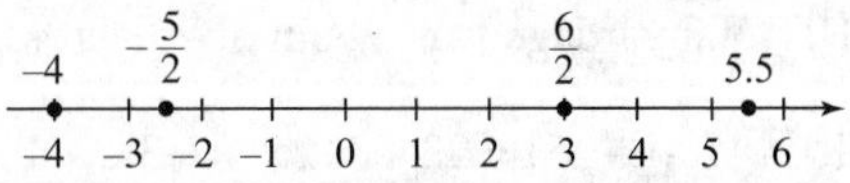

45. Since −3 lies to the left of −1 on the real number line, $-3 < -1$. The statement is false.

46. Since $5 = 5$, the statement $5 \le 5$ is true.

47. Since −5 lies to the left of −3 on the real number line, $-5 \le -3$ is a true statement.

48. Since $\frac{1}{2} = 0.5$, the statement is true.

49. $-\left|\frac{1}{2}\right| = -\frac{1}{2}$

50. $|-7| = 7$

51. $-|-6| = -6$

52. $-|-8.2| = -8.2$

53. $\frac{1}{4} = 0.25$

54. Since −6 lies to the left of 0 on the real number line, $-6 < 0$.

55. Since $\frac{3}{4} = 0.75$ and $0.83 > 0.75$, then $0.83 > \frac{3}{4}$.

56. Since −2 lies to the right of −10 on the real number line, $-2 > -10$.

57. $|-4| = 4$
$|-3| = 3$
Since 4 lies to the right of 3 on the real number line $4 > 3$ and $|-4| > |-3|$.

58. $\frac{4}{5} = \frac{4 \cdot 6}{5 \cdot 6} = \frac{24}{30}$
$\left|-\frac{5}{6}\right| = \frac{5}{6} = \frac{5 \cdot 5}{6 \cdot 5} = \frac{25}{30}$
Since $\frac{24}{30} < \frac{25}{30}$, then $\frac{4}{5} < \left|-\frac{5}{6}\right|$.

59. A rational number is any number that may be written as the quotient of two integers where the denominator does not equal zero. Both terminating decimals and repeating decimals are rational numbers. An irrational number is a non-repeating, non-terminating decimal.

60. The set of positive integers is called the natural numbers.

61. $-2 + 9 = 7$

62. $6 + (-10) = -4$

63. $-23 + (-11) = -34$

64. $-120 + 25 = -95$

65. $-|-2 + 6| = -|4| = -4$

66. $-|-15| + |-62| = -15 + 62 = 47$

67. $$\begin{aligned} -110 + 50 + (-18) + 25 &= -60 + (-18) + 25 \\ &= -78 + 25 \\ &= -53 \end{aligned}$$

68. $-28 + (-35) + (-52) = -63 + (-52) = -115$

69. $-10 - 12 = -10 + (-12) = -22$

70. $18 - 25 = 18 + (-25) = -7$

71. $-11-(-32)=-11+32=21$

72. $0-(-67)=0+67=67$

73. $34-18+10=34+(-18)+10=16+10=26$

74. $-49-8+21=-49+(-8)+21$
$=-57+21$
$=-36$

75. $-6(-2)=12$

76. $4(-10)=-40$

77. $13(-86)=-1118$

78. $-19\times 423=-8037$

79. $(11)(13)(-5)=143(-5)=-715$

80. $(-53)(-21)(-10)=1113(-10)=-11{,}130$

81. $\frac{-20}{-4}=\frac{-4\cdot 5}{-4\cdot 1}=\frac{\cancel{-4}\cdot 5}{\cancel{-4}\cdot 1}=\frac{5}{1}=5$

82. $\frac{60}{-5}=\frac{5\cdot 12}{5\cdot(-1)}=\frac{\cancel{5}\cdot 12}{\cancel{5}\cdot(-1)}=\frac{12}{-1}=-12$

83. $\frac{|-55|}{11}=\frac{55}{11}=\frac{11\cdot 5}{11\cdot 1}=\frac{\cancel{11}\cdot 5}{\cancel{11}\cdot 1}=\frac{5}{1}=5$

84. $-\left|\frac{-100}{4}\right|=-\left|\frac{4\cdot(-25)}{4}\right|$
$=-\left|\frac{\cancel{4}\cdot(-25)}{\cancel{4}}\right|$
$=-|-25|$
$=-25$

85. $\frac{120}{-15}=\frac{15\cdot 8}{15\cdot(-1)}=\frac{\cancel{15}\cdot 8}{\cancel{15}\cdot(-1)}=\frac{8}{-1}=-8$

86. $\frac{64}{-20}=\frac{4\cdot 16}{4\cdot(-5)}=\frac{\cancel{4}\cdot 16}{\cancel{4}\cdot(-5)}=-\frac{16}{5}$

87. $\frac{-180}{54}=\frac{-10\cdot 18}{3\cdot 18}=\frac{-10\cdot\cancel{18}}{3\cdot\cancel{18}}=-\frac{10}{3}$

88. $\frac{-450}{105}=\frac{-30\cdot 15}{7\cdot 15}=\frac{-30\cdot\cancel{15}}{7\cdot\cancel{15}}=-\frac{30}{7}$

89. The additive inverse of 13 is −13 since $13+(-13)=0$.

90. The additive inverse of −45 is 45 since $-45+45=0$.

91. −43 plus 101 is written as $-43+101=58$.

92. 45 plus −28 is written as $45+(-28)=17$.

93. −10 minus −116 is written as $-10-(-116)=-10+116=106$.

94. 74 minus 56 is written as $74-56=74+(-56)=18$.

95. The sum of 13 and −8 is written as $13+(-8)=5$.

96. The difference between −60 and −10 is written as $-60-(-10)=-60+10=-50$.

97. −21 multiplied by −3 is written as $-21\cdot(-3)=63$.

98. 54 multiplied by −18 is written as $54\cdot(-18)=-972$.

99. −34 divided by −2 is written as
$-34\div(-2)=\frac{-34}{-2}=\frac{-2\cdot 17}{-2\cdot 1}=\frac{17}{1}=17.$

100. −49 divided by 14 is written as
$-49\div 14=\frac{-49}{14}=\frac{-7\cdot 7}{2\cdot 7}=-\frac{7}{2}.$

101. $20+(-6)+12=14+12=26$
His total yardage was a gain of 26 yards.

102. $10+12+(-25)=22+(-25)=-3$
The temperature at midnight was −3°F.

103. $6-(-18)=6+18=24$
The difference in temperature was 24°F.

104. $11\cdot 5=55$
$8\cdot 4=32$
$55+32=87$
Sarah's test score was 87 points.

105. $\frac{32}{64}=\frac{1\cdot 32}{2\cdot 32}=\frac{1\cdot\cancel{32}}{2\cdot\cancel{32}}=\frac{1}{2}$

106. $-\frac{27}{81}=-\frac{1\cdot 27}{3\cdot 27}=-\frac{1\cdot\cancel{27}}{3\cdot\cancel{27}}=-\frac{1}{3}$

107. $\frac{-100}{150}=\frac{-2\cdot 50}{3\cdot 50}=\frac{-2\cdot \cancel{50}}{3\cdot \cancel{50}}=\frac{-2}{3}=-\frac{2}{3}$

108. $\frac{35}{-25}=\frac{5\cdot 7}{5\cdot(-5)}=\frac{\cancel{5}\cdot 7}{\cancel{5}\cdot(-5)}=\frac{7}{-5}=-\frac{7}{5}$

109. $\frac{2}{3}\cdot\frac{15}{8}=\frac{2\cdot 15}{3\cdot 8}=\frac{2\cdot 3\cdot 5}{3\cdot 2\cdot 4}=\frac{\cancel{2}\cdot\cancel{3}\cdot 5}{\cancel{3}\cdot\cancel{2}\cdot 4}=\frac{5}{4}$

110.
$$\begin{aligned}-\frac{3}{8}\cdot\frac{10}{21}&=-\frac{3\cdot 10}{8\cdot 21}\\&=-\frac{3\cdot 2\cdot 5}{2\cdot 4\cdot 3\cdot 7}\\&=-\frac{\cancel{3}\cdot\cancel{2}\cdot 5}{\cancel{2}\cdot 4\cdot\cancel{3}\cdot 7}\\&=-\frac{5}{4\cdot 7}\\&=-\frac{5}{28}\end{aligned}$$

111.
$$\begin{aligned}\frac{5}{8}\cdot\left(-\frac{2}{25}\right)&=\frac{5\cdot(-2)}{8\cdot 25}\\&=\frac{5\cdot 2\cdot(-1)}{2\cdot 4\cdot 5\cdot 5}\\&=\frac{\cancel{5}\cdot\cancel{2}\cdot(-1)}{\cancel{2}\cdot 4\cdot\cancel{5}\cdot 5}\\&=\frac{-1}{4\cdot 5}\\&=-\frac{1}{20}\end{aligned}$$

112.
$$\begin{aligned}5\cdot\left(-\frac{3}{10}\right)&=\frac{5}{1}\cdot\left(-\frac{3}{10}\right)\\&=\frac{5\cdot(-3)}{1\cdot 10}\\&=\frac{5\cdot(-3)}{1\cdot 2\cdot 5}\\&=\frac{\cancel{5}\cdot(-3)}{1\cdot 2\cdot\cancel{5}}\\&=-\frac{3}{2}\end{aligned}$$

113.
$$\begin{aligned}\frac{24}{17}\div\frac{18}{3}&=\frac{24}{17}\cdot\frac{3}{18}\\&=\frac{24\cdot 3}{17\cdot 18}\\&=\frac{6\cdot 4\cdot 3}{17\cdot 6\cdot 3}\\&=\frac{\cancel{6}\cdot 4\cdot\cancel{3}}{17\cdot\cancel{6}\cdot\cancel{3}}\\&=\frac{4}{17}\end{aligned}$$

114.
$$\begin{aligned}-\frac{5}{12}\div\frac{10}{16}&=-\frac{5}{12}\cdot\frac{16}{10}\\&=-\frac{5\cdot 16}{12\cdot 10}\\&=-\frac{5\cdot 4\cdot 2\cdot 2}{4\cdot 3\cdot 5\cdot 2}\\&=-\frac{\cancel{5}\cdot\cancel{4}\cdot 2\cdot\cancel{2}}{\cancel{4}\cdot 3\cdot\cancel{5}\cdot\cancel{2}}\\&=-\frac{2}{3}\end{aligned}$$

115.
$$\begin{aligned}-\frac{27}{10}\div 9&=-\frac{27}{10}\div\frac{9}{1}\\&=-\frac{27}{10}\cdot\frac{1}{9}\\&=-\frac{27\cdot 1}{10\cdot 9}\\&=-\frac{3\cdot 9\cdot 1}{10\cdot 9}\\&=-\frac{3\cdot\cancel{9}\cdot 1}{10\cdot\cancel{9}}\\&=-\frac{3}{10}\end{aligned}$$

116.
$$\begin{aligned}20\div\left(-\frac{5}{8}\right)&=\frac{20}{1}\div\left(-\frac{5}{8}\right)\\&=\frac{20}{1}\cdot\left(-\frac{8}{5}\right)\\&=\frac{20\cdot(-8)}{1\cdot 5}\\&=\frac{4\cdot 5\cdot(-8)}{1\cdot 5}\\&=\frac{4\cdot\cancel{5}\cdot(-8)}{1\cdot\cancel{5}}\\&=\frac{4\cdot(-8)}{1}\\&=-32\end{aligned}$$

117. $\frac{2}{9}+\frac{1}{9}=\frac{2+1}{9}=\frac{3}{9}=\frac{3\cdot 1}{3\cdot 3}=\frac{\cancel{3}\cdot 1}{\cancel{3}\cdot 3}=\frac{1}{3}$

118. $-\frac{6}{5}+\frac{4}{5}=\frac{-6+4}{5}=\frac{-2}{5}=-\frac{2}{5}$

119. $\frac{5}{7}-\frac{2}{7}=\frac{5-2}{7}=\frac{3}{7}$

120.
$$\begin{aligned}\frac{7}{5}-\left(-\frac{8}{5}\right)&=\frac{7}{5}+\frac{8}{5}\\&=\frac{7+8}{5}\\&=\frac{15}{5}\\&=\frac{5\cdot 3}{5}\\&=\frac{\cancel{5}\cdot 3}{\cancel{5}}\\&=3\end{aligned}$$

121.
$$\begin{aligned}\frac{3}{10}+\frac{1}{20}&=\frac{3}{10}\cdot\frac{2}{2}+\frac{1}{20}\\&=\frac{6}{20}+\frac{1}{20}\\&=\frac{6+1}{20}\\&=\frac{7}{20}\end{aligned}$$

122. $12 = 4\cdot 3$
$9 = 3\cdot 3$
$\text{LCD} = 4\cdot 3\cdot 3 = 36$
$$\begin{aligned}\frac{5}{12}+\frac{4}{9}&=\frac{5}{12}\cdot\frac{3}{3}+\frac{4}{9}\cdot\frac{4}{4}\\&=\frac{15}{36}+\frac{16}{36}\\&=\frac{15+16}{36}\\&=\frac{31}{36}\end{aligned}$$

123. $35 = 5\cdot 7$
$49 = 7\cdot 7$
$\text{LCD} = 5\cdot 7\cdot 7 = 245$
$$\begin{aligned}-\frac{7}{35}-\frac{2}{49}&=-\frac{7}{35}\cdot\frac{7}{7}-\frac{2}{49}\cdot\frac{5}{5}\\&=-\frac{49}{245}-\frac{10}{245}\\&=\frac{-49-10}{245}\\&=-\frac{59}{245}\end{aligned}$$

124. $6 = 2\cdot 3$
$4 = 2\cdot 2$
$\text{LCD} = 2\cdot 2\cdot 3 = 12$
$$\begin{aligned}\frac{5}{6}-\left(-\frac{1}{4}\right)&=\frac{5}{6}+\frac{1}{4}\\&=\frac{5}{6}\cdot\frac{2}{2}+\frac{1}{4}\cdot\frac{3}{3}\\&=\frac{10}{12}+\frac{3}{12}\\&=\frac{10+3}{12}\\&=\frac{13}{12}\end{aligned}$$

125.
$$\begin{aligned}-2-\left(-\frac{5}{12}\right)&=-2+\frac{5}{12}\\&=-\frac{2}{1}\cdot\frac{12}{12}+\frac{5}{12}\\&=-\frac{24}{12}+\frac{5}{12}\\&=\frac{-24+5}{12}\\&=-\frac{19}{12}\end{aligned}$$

126.
$$\begin{aligned}-5+\frac{9}{4}&=-\frac{5}{1}\cdot\frac{4}{4}+\frac{9}{4}\\&=-\frac{20}{4}+\frac{9}{4}\\&=\frac{-20+9}{4}\\&=-\frac{11}{4}\end{aligned}$$

127. $10 = 2 \cdot 5$
$5 = 5$
$2 = 2$
$\text{LCD} = 2 \cdot 5 = 10$

$$\begin{aligned}-\frac{1}{10}+\left(-\frac{2}{5}\right)+\frac{1}{2} &= -\frac{1}{10}+\left(-\frac{2}{5}\cdot\frac{2}{2}\right)+\frac{1}{2}\cdot\frac{5}{5} \\ &= -\frac{1}{10}+\left(-\frac{4}{10}\right)+\frac{5}{10} \\ &= \frac{-1+(-4)+5}{10} \\ &= \frac{0}{10} \\ &= 0\end{aligned}$$

128. $6 = 2 \cdot 3$
$4 = 2 \cdot 2$
$24 = 2 \cdot 2 \cdot 2 \cdot 3$
$\text{LCD} = 2 \cdot 2 \cdot 2 \cdot 3 = 24$

$$\begin{aligned}-\frac{5}{6}-\frac{1}{4}+\frac{3}{24} &= -\frac{5}{6}\cdot\frac{4}{4}-\frac{1}{4}\cdot\frac{6}{6}+\frac{3}{24} \\ &= -\frac{20}{24}-\frac{6}{24}+\frac{3}{24} \\ &= \frac{-20-6+3}{24} \\ &= -\frac{23}{24}\end{aligned}$$

129.
$$\begin{array}{r} 30.3 \\ +\,18.2 \\ \hline 48.5 \end{array}$$
So, $30.3 + 18.2 = 48.5$.

130.
$$\begin{array}{r} 43.02 \\ -\,18.36 \\ \hline 24.66 \end{array}$$
So, $-43.02 + 18.36 = -24.66$.

131.
$$\begin{array}{r} 201.37 \\ -\,118.39 \\ \hline 82.98 \end{array}$$
So, $201.37 - 118.39 = 82.98$.

132.
$$\begin{array}{r} 35.10 \\ +\,18.64 \\ \hline 53.74 \end{array}$$
So, $-35.1-18.64 = -35.1+(-18.64) = -53.74$.

133.
$$\begin{array}{r} 2.01 \\ \times\,0.04 \\ \hline 0.0804 \end{array}$$
So, $(-0.04)(-2.01) = 0.0804$.

134.
$$\begin{array}{r} 87.3 \\ \times\,2.98 \\ \hline 6984 \\ 7857 \\ 1746 \\ \hline 260.154 \end{array}$$
So, $(87.3)(-2.98) = -260.154$.

135. $\dfrac{69.92}{3.8} = \dfrac{699.2}{38} = 18.4$

$$\begin{array}{r} 18.4 \\ 38\overline{)699.2} \\ 38 \\ \hline 319 \\ 304 \\ \hline 15\,2 \\ 15\,2 \\ \hline 0 \end{array}$$

136. $-\dfrac{1.08318}{0.042} = -\dfrac{1083.18}{42} = -25.79$

$$\begin{array}{r} 25.79 \\ 42\overline{)1083.18} \\ 84 \\ \hline 243 \\ 210 \\ \hline 33\,1 \\ 29\,4 \\ \hline 3\,78 \\ 3\,78 \\ \hline 0 \end{array}$$

137.
$$\begin{aligned}12.5-18.6+8.4 &= 12.5+(-18.6)+8.4 \\ &= -6.1+8.4 \\ &= 2.3\end{aligned}$$

138.
$$\begin{aligned}-13.5+10.8-20.2 &= -13.5+10.8+(-20.2) \\ &= -2.7+(-20.2) \\ &= -22.9\end{aligned}$$

139. $12.9\times1.4\times(-0.3) = 18.06\times(-0.3) = -5.418$

140.
$$\begin{aligned}2.4\times6.1\times(-0.05) &= 14.64\times(-0.05) \\ &= -0.732\end{aligned}$$

141.
$$\begin{aligned}&256.75+(-175.68)+(-180.00) \\ &= 81.07+(-180.00) \\ &= -98.93\end{aligned}$$
Lee's checking account balance is –$98.93. The account is overdrawn.

142. $36\cdot\frac{2}{3}=\frac{36}{1}\cdot\frac{2}{3}$
$=\frac{36\cdot 2}{1\cdot 3}$
$=\frac{3\cdot 12\cdot 2}{1\cdot 3}$
$=\frac{\not{3}\cdot 12\cdot 2}{1\cdot \not{3}}$
$=\frac{12\cdot 2}{1}$
$=24$

Jarred had 24 friends that wanted the Panthers to win.

143. $15-3\frac{1}{2}=15-\frac{7}{2}$
$=\frac{15}{1}-\frac{7}{2}$
$=\frac{15}{1}\cdot\frac{2}{2}-\frac{7}{2}$
$=\frac{30}{2}-\frac{7}{2}$
$=\frac{30-7}{2}$
$=\frac{23}{2}$ or $11\frac{1}{2}$

The length of the remaining piece is $11\frac{1}{2}$ inches.

144. net price = price $\times$ quantity
$=\$35\times 5$
$=\$175$
sales tax $=\$175\times 6.75\%$
$=\$175\times 0.0675$
$=\$11.81$

Sierra spent \$175 + \$11.81 = \$186.81 on the clothes.

145. $(5\cdot 12)\cdot 10=5\cdot(12\cdot 10)$ illustrates the associative property of multiplication since the grouping of multiplication changes.

146. $20\cdot\frac{1}{20}=1$ illustrates the multiplicative inverse property.

147. $\frac{8}{3}\cdot\frac{3}{8}=1$ illustrates the multiplicative inverse property.

148. $\frac{5}{3}\cdot\left(-\frac{18}{61}\right)\cdot\frac{3}{5}=\frac{5}{3}\cdot\frac{3}{5}\cdot\left(-\frac{18}{61}\right)$ illustrates the commutative property of multiplication since the order in which the numbers are multiplied changes.

149. $9\cdot 73\cdot\frac{1}{9}=9\cdot\frac{1}{9}\cdot 73$ illustrates the commutative property of multiplication since the order in which the numbers are multiplied changes.

150. $23.9+(-23.9)=0$ illustrates the additive inverse property.

151. $36+0=36$ illustrates the identity property of addition.

152. $-49+0=-49$ illustrates the identity property of addition.

153. $23+5+(-23)=23+(-23)+5$ illustrates the commutative property of addition since the order of the addition changes.

154. $\frac{7}{8}=\frac{7}{8}\cdot\frac{3}{3}$ illustrates the multiplicative identity property since $\frac{3}{3}=1$.

155. $14\cdot 0=0$ illustrates the multiplication property of zero.

156. $-5.3+(5.3+2.8)=(-5.3+5.3)+2.8$ illustrates the associative property of addition since the grouping of the addition changes.

157. $144+29+(-144)=144+(-144)+29$
$=0+29$
$=29$

158. $76+99+(-76)=76+(-76)+99$
$=0+99$
$=99$

159. $\frac{19}{3}\cdot 18\cdot\frac{3}{19}=\frac{19}{3}\cdot\frac{3}{19}\cdot 18=1\cdot 18=18$

160. $\frac{14}{9}\cdot 121\cdot\frac{9}{14}=\frac{14}{9}\cdot\frac{9}{14}\cdot 121=1\cdot 121=121$

161. $3.4+42.56+(-42.56)$
$=3.4+[42.56+(-42.56)]$
$=3.4+0$
$=3.4$

162. $5.3+3.6+(-3.6)=5.3+[3.6+(-3.6)]$
$=5.3+0$
$=5.3$

163. $\frac{9}{7}\cdot\left(-\frac{11}{3}\right)\cdot 7=\frac{9}{7}\cdot 7\cdot\left(-\frac{11}{3}\right)$
$=9\cdot\left(-\frac{11}{3}\right)$
$=\overset{3}{\cancel{9}}\cdot\left(-\frac{11}{\cancel{3}}\right)$
$=3\cdot(-11)$
$=-33$

164. $\frac{13}{5}\cdot\frac{18}{39}\cdot 5=\frac{13}{5}\cdot 5\cdot\frac{18}{39}$
$=13\cdot\frac{18}{39}$
$=\cancel{13}\cdot\frac{18}{\underset{3}{\cancel{39}}}$
$=\frac{18}{3}$
$=\frac{6\cdot 3}{1\cdot 3}$
$=6$

165. $\frac{7}{0}$ is undefined because 0 is the divisor.

166. $\frac{0}{100}=0$ because 0 is the dividend.

167. $1000(-334)(0.001)=1000(0.001)(-334)$
$=1(-334)$
$=-334$

168. $400(0.5)(0.01)=400(0.01)(0.5)=4(0.5)=2$

169. $43,569,003\cdot 0=0$

170. $154\cdot\frac{1}{154}=1$

171. $\frac{3445}{302}+\left(-\frac{3445}{302}\right)=0$

172. $130\cdot\frac{42}{42}=130\cdot 1=130$

173. $-\frac{7}{48}\cdot\frac{20}{3}\cdot\frac{12}{7}=-\frac{7}{48}\cdot\frac{12}{7}\cdot\frac{20}{3}$
$=-\frac{\overset{1}{\cancel{7}}}{\underset{4}{\cancel{48}}}\cdot\frac{\overset{1}{\cancel{12}}}{\underset{1}{\cancel{7}}}\cdot\frac{20}{3}$
$=-\frac{1}{4}\cdot\frac{20}{3}$
$=-\frac{1}{\underset{1}{\cancel{4}}}\cdot\frac{\overset{5}{\cancel{20}}}{3}$
$=-\frac{5}{3}$

174. $\frac{9}{8}\cdot\left(-\frac{25}{13}\right)\cdot\frac{48}{9}=\frac{9}{8}\cdot\frac{48}{9}\cdot\left(-\frac{25}{13}\right)$
$=\frac{\overset{1}{\cancel{9}}}{\underset{1}{\cancel{8}}}\cdot\frac{\overset{6}{\cancel{48}}}{\underset{1}{\cancel{9}}}\cdot\left(-\frac{25}{13}\right)$
$=6\cdot\left(-\frac{25}{13}\right)$
$=-\frac{150}{13}$

175. The expression $3\cdot 3\cdot 3\cdot 3$ contains four factors of 3, so $3\cdot 3\cdot 3\cdot 3=3^4$.

176. The expression $\frac{2}{3}\cdot\frac{2}{3}\cdot\frac{2}{3}$ contains three factors of $\frac{2}{3}$, so $\frac{2}{3}\cdot\frac{2}{3}\cdot\frac{2}{3}=\left(\frac{2}{3}\right)^3$.

177. The expression $(-4)(-4)$ contains two factors of -4, so $(-4)(-4)=(-4)^2$.

178. The expression $(-3)(-3)(-3)$ contains three factors of -3, so $(-3)(-3)(-3)=(-3)^3$.

179. $5^3=5\cdot 5\cdot 5=125$

180. $2^5 = 2 \cdot 2 \cdot 2 \cdot 2 \cdot 2 = 32$

181. $(-3)^4 = (-3) \cdot (-3) \cdot (-3) \cdot (-3) = 81$

182. $(-4)^3 = (-4) \cdot (-4) \cdot (-4) = -64$

183. $-3^4 = -(3 \cdot 3 \cdot 3 \cdot 3) = -81$

184. $\left(\frac{1}{2}\right)^6 = \frac{1}{2} \cdot \frac{1}{2} \cdot \frac{1}{2} \cdot \frac{1}{2} \cdot \frac{1}{2} \cdot \frac{1}{2} = \frac{1}{64}$

185.
$$\begin{aligned} -2+16 \div 4 \cdot 2-10 &= -2+4 \cdot 2-10 \\ &= -2+8-10 \\ &= 6-10 \\ &= -4 \end{aligned}$$

186.
$$\begin{aligned} -4+3[2^3+4(2-10)] &= -4+3[2^3+4(-8)] \\ &= -4+3[8+4(-8)] \\ &= -4+3[8-32] \\ &= -4+3[-24] \\ &= -4-72 \\ &= -76 \end{aligned}$$

187.
$$\begin{aligned} (12-7)^3+(19-10)^2 &= 5^3+9^2 \\ &= 125+81 \\ &= 206 \end{aligned}$$

188.
$$\begin{aligned} 5-(-12 \div 2 \cdot 3)+(-3)^2 &= 5-(-6 \cdot 3)+(-3)^2 \\ &= 5-(-18)+(-3)^2 \\ &= 5-(-18)+9 \\ &= 5+18+9 \\ &= 32 \end{aligned}$$

189. $\frac{2 \cdot (4+8)}{3+3^2} = \frac{2 \cdot (12)}{3+9} = \frac{24}{12} = \frac{2 \cdot 12}{1 \cdot 12} = 2$

190.
$$\begin{aligned} \frac{3 \cdot (5+2^2)}{2 \cdot 3^3} &= \frac{3 \cdot (5+4)}{2 \cdot 27} \\ &= \frac{3 \cdot 9}{2 \cdot 27} \\ &= \frac{27}{2 \cdot 27} \\ &= \frac{1}{2} \end{aligned}$$

191.
$$\begin{aligned} \frac{6 \cdot [12-3 \cdot (5-2)]}{5 \cdot [21-2 \cdot (4+5)]} &= \frac{6 \cdot [12-3 \cdot 3]}{5 \cdot [21-2 \cdot 9]} \\ &= \frac{6 \cdot [12-9]}{5 \cdot [21-18]} \\ &= \frac{6 \cdot 3}{5 \cdot 3} \\ &= \frac{6}{5} \end{aligned}$$

192.
$$\begin{aligned} \frac{4 \cdot [3+2 \cdot (8-6)]}{5 \cdot [14-2 \cdot (2+3)]} &= \frac{4 \cdot [3+2 \cdot 2]}{5 \cdot [14-2 \cdot 5]} \\ &= \frac{4 \cdot [3+4]}{5 \cdot [14-10]} \\ &= \frac{4 \cdot 7}{5 \cdot 4} \\ &= \frac{7}{5} \end{aligned}$$

193. Let $x = 5$ and $y = -2$.
$x^2 - y^2 = 5^2 - (-2)^2 = 25 - 4 = 21$

194. Let $x = 3$ and $y = -3$.
$$\begin{aligned} x^2 - 3y^2 &= 3^2 - 3(-3)^2 \\ &= 9-3(9) \\ &= 9-27 \\ &= -18 \end{aligned}$$

195. Let $x = -1$ and $y = -4$.
$$\begin{aligned} (x+2y)^3 &= (-1+2(-4))^3 \\ &= (-1-8)^3 \\ &= (-9)^3 \\ &= -729 \end{aligned}$$

196. Let $a = 5$, $b = -10$, $x = -3$, $y = 2$.
$$\begin{aligned} \frac{a-b}{x-y} &= \frac{5-(-10)}{-3-2} \\ &= \frac{5+10}{-3-2} \\ &= \frac{15}{-5} \\ &= \frac{3 \cdot 5}{-1 \cdot 5} \\ &= -3 \end{aligned}$$

197. $3x^2 - x + 6$ can be written as $3x^2 + (-1) \cdot x + 6$. The terms are $3x^2$, $-x$, and 6. The coefficient of $3x^2$ is 3. The coefficient of $-x$ is -1. The coefficient of 6 is 6.

198. $2x^2y^3 - \frac{y}{5}$ can be written as $2x^2y^3 + \left(-\frac{1}{5}\right)\cdot y$.

The terms are $2x^2y^3$ and $-\frac{y}{5}$. The coefficient of $2x^2y^3$ is 2. The coefficient of $-\frac{y}{5}$ is $-\frac{1}{5}$.

199. $4xy^2$ and $-6xy^2$ are like terms. They have the same variables raised to the same power.

200. $-3x$ and $4x^2$ are unlike terms. They have the same variable, but it is raised to different powers.

201. $-6y$ and -6 are unlike terms. $-6y$ has a variable and -6 does not.

202. -10 and 4 are like terms. They are both constants.

203. $4x - 6x - x = (4 - 6 - 1)x = -3x$

204.
$$\begin{aligned} 6x-10-10x-5 &= 6x-10x-10-5 \\ &= (6-10)x+(-10-5) \\ &= -4x-15 \end{aligned}$$

205.
$$\begin{aligned} &0.2x^4+0.3x^3-4.3x^4 \\ &= 0.2x^4-4.3x^4+0.3x^3 \\ &= (0.2-4.3)x^4+0.3x^3 \\ &= -4.1x^4+0.3x^3 \end{aligned}$$

206.
$$\begin{aligned} &-3(x^4-2x^2-4) \\ &= -3\cdot x^4-(-3)\cdot 2x^2-(-3)\cdot 4 \\ &= -3x^4-(-6x^2)-(-12) \\ &= -3x^4+6x^2+12 \end{aligned}$$

207. $20-(x+2) = 20-x-2 = 20-2-x = 18-x$

208.
$$\begin{aligned} &-6(2x+5)+4(4x+3) \\ &= -6\cdot 2x+(-6)\cdot 5+4\cdot 4x+4\cdot 3 \\ &= -12x-30+16x+12 \\ &= -12x+16x-30+12 \\ &= 4x-18 \end{aligned}$$

209.
$$\begin{aligned} &5-(3x-1)+2(6x-5) \\ &= 5-3x+1+2\cdot 6x-2\cdot 5 \\ &= 5-3x+1+12x-10 \\ &= -3x+12x+5+1-10 \\ &= 9x-4 \end{aligned}$$

210.
$$\begin{aligned} &\frac{1}{6}(12x+18)-\frac{2}{5}(5x+10) \\ &= \frac{1}{6}\cdot 12x+\frac{1}{6}\cdot 18+\left(-\frac{2}{5}\right)\cdot 5x+\left(-\frac{2}{5}\right)\cdot 10 \\ &= 2x+3-2x-4 \\ &= 2x-2x+3-4 \\ &= -1 \end{aligned}$$

211. Let $m = 315$.
$$\begin{aligned} 19.95+0.25m &= 19.95+0.25(315) \\ &= 19.95+78.75 \\ &= 98.7 \end{aligned}$$
The total daily cost is $98.70.

Chapter 1 Test

1.
$$\begin{aligned} 2 &= 2 \\ 6 &= 2\cdot 3 \\ 14 &= 2 \qquad \cdot 7 \\ & \downarrow \;\; \downarrow \;\; \downarrow \\ & 2\cdot 3\cdot 7 \end{aligned}$$
The LCM is $2 \cdot 3 \cdot 7 = 42$.

2. $\frac{21}{66} = \frac{3\cdot 7}{2\cdot 3\cdot 11} = \frac{\cancel{3}\cdot 7}{2\cdot\cancel{3}\cdot 11} = \frac{7}{22}$

3.
$$\begin{array}{r} 1.444 \\ 9\overline{)13.000} \\ \underline{9} \\ 4\,0 \\ \underline{3\,6} \\ 40 \\ \underline{36} \\ 40 \end{array}$$

$\frac{13}{9}$ rounded to the nearest hundredth is 1.44.

4. $0.425 = \frac{425}{1000} = \frac{17\cdot\cancel{25}}{40\cdot\cancel{25}} = \frac{17}{40}$

5. $0.6\% = 0.6\%\cdot\frac{1}{100\%} = \frac{0.6}{100} = 0.006$

6. $0.183 = 0.183\cdot\frac{100\%}{1} = 18.3\%$

7. $\frac{4}{15}-\left(-\frac{2}{30}\right)=\frac{4}{15}\cdot\frac{2}{2}-\left(-\frac{2}{30}\right)$
$=\frac{8}{30}-\left(-\frac{2}{30}\right)$
$=\frac{8}{30}+\frac{2}{30}$
$=\frac{8+2}{30}$
$=\frac{10}{30}$
$=\frac{1\cdot 10}{3\cdot 10}$
$=\frac{1}{3}$

8. $\frac{21}{4}\cdot\frac{3}{7}=\frac{21\cdot 3}{4\cdot 7}$
$=\frac{3\cdot 7\cdot 3}{4\cdot 7}$
$=\frac{3\cdot \cancel{7}\cdot 3}{4\cdot \cancel{7}}$
$=\frac{3\cdot 3}{4}$
$=\frac{9}{4}$

9. $-16\div\frac{3}{20}=-16\cdot\frac{20}{3}$
$=\frac{-16}{1}\cdot\frac{20}{3}$
$=\frac{-16\cdot 20}{1\cdot 3}$
$=-\frac{320}{3}$

10. $14-110-(-15)+(-21)$
$=14+(-110)+15+(-21)$
$=-96+15+(-21)$
$=-81+(-21)$
$=-102$

11. $\begin{array}{r} 14.50 \\ -\ 2.34 \\ \hline 12.16 \end{array}$

So, $-14.5 + 2.34 = -12.16$.

12. $(-4)(-1)(-5) = 4(-5) = -20$

13. $16\div 0=\frac{16}{0}$ is undefined because 0 is the divisor.

14. $-20 - (-6) = -20 + 6 = -14$

15. $-110\div(-2)=\frac{-110}{-2}=\frac{-2\cdot 55}{-2\cdot 1}=55$

16. **(a)** 6 is the only natural number.

(b) 0 and 6 are the whole numbers

(c) -2, 0, and 6 are the integers.

(d) All those listed, -2, $-\frac{1}{2}$, 0, 2.5, and 6, are the rational numbers.

(e) There are no irrational numbers.

(f) All those listed are real numbers.

17. $-|-14|=-14$
Since $-14<-12$, $-|-14|<-12$.

18. $\left|-\frac{2}{5}\right|=|-0.4|=0.4$

So, $\left|-\frac{2}{5}\right|=0.4$.

19. $-16\div 2^2\cdot 4+(-3)^2=-16\div 4\cdot 4+9$
$=-4\cdot 4+9$
$=-16+9$
$=-7$

20. $\frac{4(-9)-3^2}{25+4(-6-1)}=\frac{4(-9)-9}{25+4(-7)}$
$=\frac{-36-9}{25-28}$
$=\frac{-45}{-3}$
$=\frac{-3\cdot 15}{-3\cdot 1}$
$=15$

21. $8-10[6^2-5(2+3)]=8-10[6^2-5(5)]$
$=8-10[36-25]$
$=8-10[11]$
$=8-110$
$=-102$

22. Let $x = -1$ and $y = 3$.

$$\begin{aligned}(x-2y)^3 &= (-1-2\cdot 3)^3\\ &= (-1-6)^3\\ &= (-7)^3\\ &= (-7)(-7)(-7)\\ &= -343\end{aligned}$$

23. $-6(2x+5)-(4x-2)$

$$\begin{aligned}&= -6\cdot 2x - 6\cdot 5 - 4x - (-2)\\ &= -12x - 30 - 4x + 2\\ &= -12x - 4x - 30 + 2\\ &= -16x - 28\end{aligned}$$

24. $\frac{1}{2}(4x^2+8)-6x^2+5x$

$$\begin{aligned}&= \frac{1}{2}\cdot 4x^2 + \frac{1}{2}\cdot 8 - 6x^2 + 5x\\ &= 2x^2 + 4 - 6x^2 + 5x\\ &= 2x^2 - 6x^2 + 5x + 4\\ &= -4x^2 + 5x + 4\end{aligned}$$

25. $675.15+(-175.50)+(-78)+110.20$

$$\begin{aligned}&= 499.65 + (-78) + 110.20\\ &= 421.65 + 110.20\\ &= 531.85\end{aligned}$$

Latoya has $531.85 in her bank account.

26.

$$\begin{aligned}2(x+5)+2x &= 2\cdot x + 2\cdot 5 + 2x\\ &= 2x + 10 + 2x\\ &= 2x + 2x + 10\\ &= 4x + 10\end{aligned}$$

Chapter 2

Section 2.1

Preparing for Linear Equations: The Addition and Multiplication Properties of Equality

P1. The additive inverse of 3 is –3 because $3 + (-3) = 0$.

P2. The multiplicative inverse of $-\frac{4}{3}$ is $-\frac{3}{4}$ because $-\frac{4}{3}\left(-\frac{3}{4}\right)=1$.

P3. $\frac{2}{3}\left(\frac{3}{2}\right)=\frac{2\cdot 3}{3\cdot 2}=\frac{\overset{1}{\cancel{2}}\cdot\overset{1}{\cancel{3}}}{\underset{1}{\cancel{3}}\cdot\underset{1}{\cancel{2}}}=\frac{1\cdot 1}{1\cdot 1}=1$

P4. $-4(2x+3)=-4\cdot 2x+(-4)\cdot 3$
$=-8x+(-12)$
$=-8x-12$

P5. $11-(x+6)=11-x-6=5-x$ or $-x+5$

2.1 Quick Checks

1. The values of the variable that result in a true statement are called <u>solutions</u>.

2. $a-4=-7$
$-3-4\overset{?}{=}-7$
$-7=-7$ True
Since the left side equals the right side when we replace *a* by –3, $a=-3$ is a solution of the equation.

3. $\frac{1}{2}+x=10$
$\frac{1}{2}+\frac{21}{2}\overset{?}{=}10$
$\frac{22}{2}\overset{?}{=}10$
$11=10$ False
Since the left side does not equal the right side when we replace *x* by $\frac{21}{2}$, $x=\frac{21}{2}$ is *not* a solution of the equation.

4. $3x-(x+4)=8$
$3(6)-(6+4)\overset{?}{=}8$
$3(6)-10\overset{?}{=}8$
$18-10\overset{?}{=}8$
$8=8$ True
Since the left side equals the right side when we replace *x* by 6, $x = 6$ is a solution of the equation.

5. $-9b+3+7b=-3b+8$
$-9(-3)+3+7(-3)\overset{?}{=}-3(-3)+8$
$27+3+(-21)\overset{?}{=}9+8$
$9=17$ False
Since the left side does not equal the right side when we replace *b* by –3, $b = -3$ is *not* a solution of the equation.

6. $x-11=21$
$x-11+11=21+11$
$x=32$
Check: $x-11=21$
$32-11\overset{?}{=}21$
$21=21$ True
Because $x = 32$ satisfies the original equation, the solution is 32, or the solution set is {32}.

7. $y+7=21$
$y+7-7=21-7$
$y=14$
Check: $y+7=21$
$14+7\overset{?}{=}21$
$21=21$ True
Because $y = 14$ satisfies the original equation, the solution is 14, or the solution set is {14}.

8. $-8+a=4$
$-8+a+8=4+8$
$a=12$
Check: $-8+a=4$
$-8+12\overset{?}{=}4$
$4=4$ True
Because $a = 12$ satisfies the original equation, the solution is 12, or the solution set is {12}.

9. $-3=12+c$
$-3-12=12+c-12$
$-15=c$
Check: $-3=12+c$
$-3\overset{?}{=}12+(-15)$
$-3=-3$ True
Because $c = -15$ satisfies the original equation, the solution is –15, or the solution set is {–15}.

10. $z - \frac{2}{3} = \frac{5}{3}$

$z - \frac{2}{3} + \frac{2}{3} = \frac{5}{3} + \frac{2}{3}$

$z = \frac{7}{3}$

Check: $z - \frac{2}{3} = \frac{5}{3}$

$\frac{7}{3} - \frac{2}{3} \stackrel{?}{=} \frac{5}{3}$

$\frac{5}{3} = \frac{5}{3}$ True

Because $z = \frac{7}{3}$ satisfies the original equation, the solution is $\frac{7}{3}$, or the solution set is $\left\{\frac{7}{3}\right\}$.

11. $p + \frac{5}{4} = \frac{1}{4}$

$p + \frac{5}{4} - \frac{5}{4} = \frac{1}{4} - \frac{5}{4}$

$p = \frac{-4}{4}$

$p = -1$

Check: $p + \frac{5}{4} = \frac{1}{4}$

$-1 + \frac{5}{4} \stackrel{?}{=} \frac{1}{4}$

$-\frac{4}{4} + \frac{5}{4} \stackrel{?}{=} \frac{1}{4}$

$\frac{1}{4} = \frac{1}{4}$ True

Because $p = -1$ satisfies the original equation, the solution is -1, or the solution set is $\{-1\}$.

12. $\frac{3}{8} = w - \frac{1}{4}$

$\frac{3}{8} + \frac{1}{4} = w - \frac{1}{4} + \frac{1}{4}$

$\frac{3}{8} + \frac{2}{8} = w$

$\frac{5}{8} = w$

Check: $\frac{3}{8} = w - \frac{1}{4}$

$\frac{3}{8} \stackrel{?}{=} \frac{5}{8} - \frac{1}{4}$

$\frac{3}{8} \stackrel{?}{=} \frac{5}{8} - \frac{2}{8}$

$\frac{3}{8} = \frac{3}{8}$ True

Because $w = \frac{5}{8}$ satisfies the original equation, the solution is $\frac{5}{8}$, or the solution set is $\left\{\frac{5}{8}\right\}$.

13. $\frac{5}{4} + x = \frac{1}{6}$

$\frac{5}{4} + x - \frac{5}{4} = \frac{1}{6} - \frac{5}{4}$

$x = \frac{2}{12} - \frac{15}{12}$

$x = -\frac{13}{12}$

Check: $\frac{5}{4} + x = \frac{1}{6}$

$\frac{5}{4} + \left(-\frac{13}{12}\right) \stackrel{?}{=} \frac{1}{6}$

$\frac{15}{12} + \left(-\frac{13}{12}\right) \stackrel{?}{=} \frac{1}{6}$

$\frac{2}{12} \stackrel{?}{=} \frac{1}{6}$

$\frac{1}{6} = \frac{1}{6}$ True

Because $x = -\frac{13}{12}$ satisfies the original equation, the solution is $-\frac{13}{12}$, or the solution set is $\left\{-\frac{13}{12}\right\}$.

14. $p + 1472.25 = 13{,}927.25$

$p + 1472.25 - 1472.25 = 13{,}927.25 - 1472.25$

$p = 12{,}455$

Since $p = 12{,}455$, the price of the car was \$12,455 before the extra charges.

15. Dividing both sides of an equation by 3 is the same as multiplying both sides by $\underline{\frac{1}{3}}$.

16.
$$8p = 16$$
$$\frac{1}{8}(8p) = \frac{1}{8}(16)$$
$$\left(\frac{1}{8}\cdot 8\right)p = \frac{1}{8}(16)$$
$$p = 2$$
Check: $8p = 16$
$$8(2) \stackrel{?}{=} 16$$
$$16 = 16 \quad \text{True}$$
Because $p = 2$ satisfies the original equation, the solution is 2, or the solution set is {2}.

17.
$$-7n = 14$$
$$-\frac{1}{7}(-7n) = -\frac{1}{7}(14)$$
$$\left(-\frac{1}{7}\cdot -7\right)n = -\frac{1}{7}(14)$$
$$n = -2$$
Check: $-7n = 14$
$$-7(-2) \stackrel{?}{=} 14$$
$$14 = 14 \quad \text{True}$$
Because $n = -2$ satisfies the original equation, the solution is −2, or the solution set is {−2}.

18.
$$6z = 15$$
$$\frac{1}{6}(6z) = \frac{1}{6}(15)$$
$$\left(\frac{1}{6}\cdot 6\right)z = \frac{15}{6}$$
$$z = \frac{5}{2}$$
Check: $6z = 15$
$$6\left(\frac{5}{2}\right) \stackrel{?}{=} 15$$
$$\overset{3}{\cancel{6}}\left(\frac{5}{\underset{1}{\cancel{2}}}\right) \stackrel{?}{=} 15$$
$$15 = 15 \quad \text{True}$$
Because $z = \frac{5}{2}$ satisfies the original equation, the solution is $\frac{5}{2}$, or the solution set is $\left\{\frac{5}{2}\right\}$.

19.
$$-12b = 28$$
$$-\frac{1}{12}(-12b) = -\frac{1}{12}(28)$$
$$\left(-\frac{1}{12}\cdot -12\right)b = -\frac{28}{12}$$
$$b = -\frac{7}{3}$$
Check: $-12b = 28$
$$-12\left(-\frac{7}{3}\right) \stackrel{?}{=} 28$$
$$-\overset{4}{\cancel{12}}\left(-\frac{7}{\underset{1}{\cancel{3}}}\right) \stackrel{?}{=} 28$$
$$28 = 28 \quad \text{True}$$
Because $b = -\frac{7}{3}$ satisfies the original equation, the solution is $-\frac{7}{3}$, or the solution set is $\left\{-\frac{7}{3}\right\}$.

20. False: To solve $-\frac{4}{3}z = 16$, multiply each side of the equation by $-\frac{3}{4}$.

21.
$$\frac{4}{3}n = 12$$
$$\frac{3}{4}\left(\frac{4}{3}n\right) = \frac{3}{4}(12)$$
$$\left(\frac{3}{4}\cdot\frac{4}{3}\right)n = \frac{3}{4}(12)$$
$$n = 9$$
Check: $\frac{4}{3}n = 12$
$$\frac{4}{3}(9) \stackrel{?}{=} 12$$
$$12 = 12 \quad \text{True}$$
Because $n = 9$ satisfies the original equation, the solution is 9, or the solution set is {9}.

22.
$$-21 = \frac{7}{3}k$$
$$\frac{3}{7}(-21) = \frac{3}{7}\left(\frac{7}{3}k\right)$$
$$\frac{3}{7}(-21) = \left(\frac{3}{7}\cdot\frac{7}{3}\right)k$$
$$-9 = k$$

Check: $-21 = \frac{7}{3}k$

$-21 \stackrel{?}{=} \frac{7}{3}(-9)$

$-21 = -21$ True

Because $k = -9$ satisfies the original equation, the solution is -9, or the solution set is $\{-9\}$.

23. $15 = -\frac{z}{2}$

$-2(15) = -2\left(-\frac{z}{2}\right)$

$-30 = z$

Check: $15 = -\frac{z}{2}$

$15 \stackrel{?}{=} -\frac{-30}{2}$

$15 = 15$ True

Because $z = -30$ satisfies the original equation, the solution is -30 or the solution set is $\{-30\}$.

24. $\frac{3}{8}b = \frac{9}{4}$

$\frac{8}{3}\left(\frac{3}{8}b\right) = \frac{8}{3}\cdot\frac{9}{4}$

$\left(\frac{8}{3}\cdot\frac{3}{8}\right)b = \frac{\overset{2}{\cancel{8}}}{\underset{1}{\cancel{3}}}\cdot\frac{\overset{3}{\cancel{9}}}{\underset{1}{\cancel{4}}}$

$b = 6$

Check: $\frac{3}{8}b = \frac{9}{4}$

$\frac{3}{8}(6) \stackrel{?}{=} \frac{9}{4}$

$\frac{3}{\underset{4}{\cancel{8}}}(\overset{3}{\cancel{6}}) \stackrel{?}{=} \frac{9}{4}$

$\frac{9}{4} = \frac{9}{4}$ True

The solution is 6, or the solution set is $\{6\}$.

25. $-\frac{4}{9} = \frac{-t}{6}$

$-\frac{4}{9} = -\frac{1}{6}t$

$-6\left(-\frac{4}{9}\right) = -6\left(-\frac{1}{6}t\right)$

$\overset{2}{\cancel{-6}}\left(-\frac{4}{\underset{3}{\cancel{9}}}\right) = \left(-6\cdot-\frac{1}{6}\right)t$

$\frac{8}{3} = t$

Check: $-\frac{4}{9} = \frac{-t}{6}$

$-\frac{4}{9} \stackrel{?}{=} -\frac{1}{6}\left(\frac{8}{3}\right)$

$-\frac{4}{9} \stackrel{?}{=} -\frac{1}{\underset{3}{\cancel{6}}}\left(\frac{\overset{4}{\cancel{8}}}{3}\right)$

$-\frac{4}{9} = -\frac{4}{9}$ True

The solution is $\frac{8}{3}$, or the solution set is $\left\{\frac{8}{3}\right\}$.

26. $\frac{1}{4} = -\frac{7}{10}m$

$-\frac{10}{7}\left(\frac{1}{4}\right) = -\frac{10}{7}\left(-\frac{7}{10}m\right)$

$-\frac{\overset{5}{\cancel{10}}}{7}\left(\frac{1}{\underset{2}{\cancel{4}}}\right) = \left(-\frac{10}{7}\cdot-\frac{7}{10}\right)m$

$-\frac{5}{14} = m$

Check: $\frac{1}{4} = -\frac{7}{10}m$

$\frac{1}{4} \stackrel{?}{=} -\frac{7}{10}\left(-\frac{5}{14}\right)$

$\frac{1}{4} \stackrel{?}{=} -\frac{\overset{1}{\cancel{7}}}{\underset{2}{\cancel{10}}}\left(-\frac{\overset{1}{\cancel{5}}}{\underset{2}{\cancel{14}}}\right)$

$\frac{1}{4} = \frac{1}{4}$ True

The solution is $-\frac{5}{14}$, or the solution set is $\left\{-\frac{5}{14}\right\}$.

2.1 Exercises

27. $3x-1=5;\ x=2$
$$3(2)-1 \stackrel{?}{=} 5$$
$$6-1 \stackrel{?}{=} 5$$
$$5=5 \quad \text{True}$$
Yes, $x = 2$ is a solution of the equation.

29. $4-(m+2)=3(2m-1);\ m=1$
$$4-(1+2) \stackrel{?}{=} 3(2(1)-1)$$
$$4-3 \stackrel{?}{=} 3(2-1)$$
$$1 \stackrel{?}{=} 3(1)$$
$$1=3 \quad \text{False}$$
No, $m = 1$ is not a solution of the equation.

31. $8k-2=4;\ k=\frac{3}{4}$
$$8\left(\frac{3}{4}\right)-2 \stackrel{?}{=} 4$$
$$6-2 \stackrel{?}{=} 4$$
$$4=4 \quad \text{True}$$
Yes, $k=\frac{3}{4}$ is a solution of the equation.

33. $r+1.6=2r+1;\ r=0.6$
$$0.6+1.6 \stackrel{?}{=} 2(0.6)+1$$
$$2.2 \stackrel{?}{=} 1.2+1$$
$$2.2=2.2 \quad \text{True}$$
Yes, $r = 0.6$ is a solution of the equation.

35.
$$x-9=11$$
$$x-9+9=11+9$$
$$x=20$$
Check: $x-9=11$
$$20-9 \stackrel{?}{=} 11$$
$$11=11 \quad \text{True}$$
The solution is 20, or the solution set is $\{20\}$.

37.
$$x+4=-8$$
$$x+4-4=-8-4$$
$$x=-12$$
Check: $x+4=-8$
$$-12+4 \stackrel{?}{=} -8$$
$$-8=-8 \quad \text{True}$$
The solution is -12, or the solution set is $\{-12\}$.

39.
$$12=n-7$$
$$12+7=n-7+7$$
$$19=n$$
Check: $12=n-7$
$$12 \stackrel{?}{=} 19-7$$
$$12=12 \quad \text{True}$$
The solution is 19 or the solution set is $\{19\}$.

41.
$$-8=x+5$$
$$-8-5=x+5-5$$
$$-13=x$$
Check: $-8=x+5$
$$-8 \stackrel{?}{=} -13+5$$
$$-8=-8 \quad \text{True}$$
The solution is -13 or the solution set is $\{-13\}$.

43.
$$x-\frac{2}{3}=\frac{4}{3}$$
$$x-\frac{2}{3}+\frac{2}{3}=\frac{4}{3}+\frac{2}{3}$$
$$x=\frac{6}{3}$$
$$x=2$$
Check: $x-\frac{2}{3}=\frac{4}{3}$
$$2-\frac{2}{3} \stackrel{?}{=} \frac{4}{3}$$
$$\frac{6}{3}-\frac{2}{3} \stackrel{?}{=} \frac{4}{3}$$
$$\frac{4}{3}=\frac{4}{3} \quad \text{True}$$
The solution is 2 or the solution set is $\{2\}$.

45.
$$z+\frac{1}{2}=\frac{3}{4}$$
$$z+\frac{1}{2}-\frac{1}{2}=\frac{3}{4}-\frac{1}{2}$$
$$z=\frac{3}{4}-\frac{2}{4}$$
$$z=\frac{1}{4}$$

Check: $z+\frac{1}{2}=\frac{3}{4}$

$\frac{1}{4}+\frac{1}{2}\stackrel{?}{=}\frac{3}{4}$

$\frac{1}{4}+\frac{2}{4}\stackrel{?}{=}\frac{3}{4}$

$\frac{3}{4}=\frac{3}{4}$ True

The solution is $\frac{1}{4}$ or the solution set is $\left\{\frac{1}{4}\right\}$.

47. $\frac{5}{12}=x-\frac{3}{8}$

$\frac{5}{12}+\frac{3}{8}=x-\frac{3}{8}+\frac{3}{8}$

$\frac{10}{24}+\frac{9}{24}=x$

$\frac{19}{24}=x$

Check: $\frac{5}{12}=x-\frac{3}{8}$

$\frac{5}{12}\stackrel{?}{=}\frac{19}{24}-\frac{3}{8}$

$\frac{5}{12}\stackrel{?}{=}\frac{19}{24}-\frac{9}{24}$

$\frac{5}{12}\stackrel{?}{=}\frac{10}{24}$

$\frac{5}{12}=\frac{5}{12}$ True

The solution is $\frac{19}{24}$ or the solution set is $\left\{\frac{19}{24}\right\}$.

49. $w+3.5=-2.6$

$w+3.5-3.5=-2.6-3.5$

$w=-6.1$

Check: $w+3.5=-2.6$

$-6.1+3.5\stackrel{?}{=}-2.6$

$-2.6=-2.6$ True

The solution is –6.1 or the solution set is {–6.1}.

51. $5c=25$

$\frac{5c}{5}=\frac{25}{5}$

$c=5$

Check: $5c=25$

$5(5)\stackrel{?}{=}25$

$25=25$ True

The solution is 5 or the solution set is {5}.

53. $-7n=28$

$\frac{-7n}{-7}=\frac{28}{-7}$

$n=-4$

Check: $-7n=28$

$-7(-4)\stackrel{?}{=}28$

$28=28$ True

The solution is –4 or the solution set is {–4}.

55. $4k=14$

$\frac{4k}{4}=\frac{14}{4}$

$k=\frac{7}{2}$

Check: $4k=14$

$4\cdot\frac{7}{2}\stackrel{?}{=}14$

$2\cdot 7\stackrel{?}{=}14$

$14=14$ True

The solution is $\frac{7}{2}$ or the solution set is $\left\{\frac{7}{2}\right\}$.

57. $-6w=15$

$\frac{-6w}{-6}=\frac{15}{-6}$

$w=-\frac{5}{2}$

Check: $-6w=15$

$-6\cdot\left(-\frac{5}{2}\right)\stackrel{?}{=}15$

$3\cdot 5\stackrel{?}{=}15$

$15=15$ True

The solution is $-\frac{5}{2}$ or the solution set is $\left\{-\frac{5}{2}\right\}$.

59. $\frac{5}{3}a=35$

$\frac{3}{5}\left(\frac{5}{3}a\right)=\frac{3}{5}(35)$

$a=3(7)$

$a=21$

Check: $\frac{5}{3}a=35$

$\frac{5}{3}(21)\stackrel{?}{=}35$

$5(7)\stackrel{?}{=}35$

$35=35$ True

The solution is 21 or the solution set is {21}.

61. $-\frac{3}{11}p = -33$

$-\frac{11}{3}\left(-\frac{3}{11}p\right) = -\frac{11}{3}(-33)$

$p = -11(-11)$

$p = 121$

Check: $-\frac{3}{11}p = -33$

$-\frac{3}{11}(121) \stackrel{?}{=} -33$

$-3(11) \stackrel{?}{=} -33$

$-33 = -33$ True

The solution is 121 or the solution set is {121}.

63. $\frac{n}{5} = 8$

$\frac{5}{1} \cdot \frac{n}{5} = \frac{5}{1} \cdot 8$

$n = 40$

Check: $\frac{n}{5} = 8$

$\frac{40}{5} \stackrel{?}{=} 8$

$8 = 8$ True

The solution is 40, or solution set is {40}.

65. $\frac{6}{5} = 2x$

$\frac{1}{2} \cdot \frac{6}{5} = \frac{1}{2} \cdot 2x$

$\frac{3}{5} = x$

Check: $\frac{6}{5} = 2x$

$\frac{6}{5} \stackrel{?}{=} 2 \cdot \frac{3}{5}$

$\frac{6}{5} = \frac{6}{5}$ True

The solution is $\frac{3}{5}$ or the solution set is $\left\{\frac{3}{5}\right\}$.

67. $5y = -\frac{5}{3}$

$\frac{1}{5} \cdot 5y = \frac{1}{5} \cdot \left(-\frac{5}{3}\right)$

$y = -\frac{1}{3}$

Check: $5y = -\frac{5}{3}$

$5 \cdot \left(-\frac{1}{3}\right) \stackrel{?}{=} -\frac{5}{3}$

$-\frac{5}{3} = -\frac{5}{3}$ True

The solution is $-\frac{1}{3}$ or the solution set is $\left\{-\frac{1}{3}\right\}$.

69. $\frac{1}{2}m = \frac{9}{2}$

$2 \cdot \frac{1}{2}m = 2 \cdot \frac{9}{2}$

$m = 9$

Check: $\frac{1}{2}m = \frac{9}{2}$

$\frac{1}{2} \cdot 9 \stackrel{?}{=} \frac{9}{2}$

$\frac{9}{2} = \frac{9}{2}$ True

The solution is 9 or the solution set is {9}.

71. $-\frac{3}{8}t = \frac{1}{6}$

$-\frac{8}{3} \cdot \left(-\frac{3}{8}t\right) = -\frac{8}{3} \cdot \frac{1}{6}$

$t = -\frac{8}{18}$

$t = -\frac{4}{9}$

Check: $-\frac{3}{8}t = \frac{1}{6}$

$-\frac{3}{8} \cdot \left(-\frac{4}{9}\right) \stackrel{?}{=} \frac{1}{6}$

$-\frac{1}{2} \cdot \left(-\frac{1}{3}\right) \stackrel{?}{=} \frac{1}{6}$

$\frac{1}{6} = \frac{1}{6}$ True

The solution is $-\frac{4}{9}$ or the solution set is $\left\{-\frac{4}{9}\right\}$.

73. $\frac{5}{24} = \frac{-y}{8}$

$-8 \cdot \frac{5}{24} = -8 \cdot \frac{-y}{8}$

$-\frac{5}{3} = y$

Check: $\frac{5}{24}=\frac{-y}{8}$

$\frac{5}{24}\stackrel{?}{=}\frac{-\left(-\frac{5}{3}\right)}{8}$

$\frac{5}{24}\stackrel{?}{=}\frac{5}{3}\cdot\frac{1}{8}$

$\frac{5}{24}=\frac{5}{24}$ True

The solution is $-\frac{5}{3}$ or the solution set is $\left\{-\frac{5}{3}\right\}$.

75. $n-4=-2$

$n-4+4=-2+4$

$n=2$

Check: $n-4=-2$

$2-4\stackrel{?}{=}-4$

$-2=-2$ True

The solution is 2 or the solution set is $\{2\}$.

77. $b+12=9$

$b+12-12=9-12$

$b=-3$

Check: $b+12=9$

$-3+12\stackrel{?}{=}9$

$9=9$ True

The solution is -3 or the solution set is $\{-3\}$.

79. $2=3x$

$\frac{2}{3}=\frac{3x}{3}$

$\frac{2}{3}=x$

Check: $2=3x$

$2\stackrel{?}{=}3\cdot\frac{2}{3}$

$2=2$ True

The solution is $\frac{2}{3}$ or the solution set is $\left\{\frac{2}{3}\right\}$.

81. $-4q=24$

$\frac{-4q}{-4}=\frac{24}{-4}$

$q=-6$

Check: $-4q=24$

$-4(-6)\stackrel{?}{=}24$

$24=24$ True

The solution is -6 or the solution set is $\{-6\}$.

83. $-39=x-58$

$-39+58=x-58+58$

$19=x$

Check: $-39=x-58$

$-39\stackrel{?}{=}19-58$

$-39=-39$ True

The solution is 19 or the solution set is $\{19\}$.

85. $-18=-301+x$

$301-18=301-301+x$

$283=x$

Check: $-18=-301+x$

$-18\stackrel{?}{=}-301+283$

$-18=-18$ True

The solution is 283 or the solution set is $\{283\}$.

87. $\frac{x}{5}=-10$

$5\cdot\frac{x}{5}=5\cdot(-10)$

$x=-50$

Check: $\frac{x}{5}=-10$

$\frac{-50}{5}\stackrel{?}{=}-10$

$-10=-10$ True

The solution is -50 or the solution set is $\{-50\}$.

89. $m-56.3=-15.2$

$m-56.3+56.3=-15.2+56.3$

$m=41.1$

Check: $m-56.3=-15.2$

$41.1-56.3\stackrel{?}{=}-15.2$

$-15.2=-15.2$ True

The solution is 41.1 or the solution set is $\{41.1\}$.

91. $-40=-6c$

$\frac{-40}{-6}=\frac{-6c}{-6}$

$\frac{20}{3}=c$

Check: $-40=-6c$

$-40\stackrel{?}{=}-6\cdot\frac{20}{3}$

$-40\stackrel{?}{=}-2\cdot 20$

$-40=-40$ True

The solution is $\frac{20}{3}$ or the solution set is $\left\{\frac{20}{3}\right\}$.

93.
$$14 = -\frac{7}{2}c$$
$$-\frac{2}{7}\cdot 14 = -\frac{2}{7}\cdot\left(-\frac{7}{2}c\right)$$
$$-2\cdot 2 = c$$
$$-4 = c$$

Check: $14 = -\frac{7}{2}c$
$$14 \stackrel{?}{=} -\frac{7}{2}\cdot(-4)$$
$$14 \stackrel{?}{=} -7\cdot(-2)$$
$$14 = 14 \quad \text{True}$$

The solution is –4 or the solution set is {–4}.

95.
$$\frac{3}{4} = -\frac{x}{16}$$
$$-16\cdot\frac{3}{4} = -16\cdot\left(-\frac{x}{16}\right)$$
$$-4\cdot 3 = x$$
$$-12 = x$$

Check: $\frac{3}{4} = -\frac{x}{16}$
$$\frac{3}{4} \stackrel{?}{=} -\frac{-12}{16}$$
$$\frac{3}{4} = \frac{3}{4} \quad \text{True}$$

The solution is –12 or the solution set is {–12}.

97.
$$x - \frac{5}{16} = \frac{3}{16}$$
$$x - \frac{5}{16} + \frac{5}{16} = \frac{3}{16} + \frac{5}{16}$$
$$x = \frac{8}{16}$$
$$x = \frac{1}{2}$$

Check: $x - \frac{5}{16} = \frac{3}{16}$
$$\frac{1}{2} - \frac{5}{16} \stackrel{?}{=} \frac{3}{16}$$
$$\frac{8}{16} - \frac{5}{16} \stackrel{?}{=} \frac{3}{16}$$
$$\frac{3}{16} = \frac{3}{16} \quad \text{True}$$

The solution is $\frac{1}{2}$ or the solution set is $\left\{\frac{1}{2}\right\}$.

99.
$$-\frac{3}{16} = -\frac{3}{8} + z$$
$$\frac{3}{8} - \frac{3}{16} = \frac{3}{8} - \frac{3}{8} + z$$
$$\frac{6}{16} - \frac{3}{16} = z$$
$$\frac{3}{16} = z$$

Check: $-\frac{3}{16} = -\frac{3}{8} + z$
$$-\frac{3}{16} \stackrel{?}{=} -\frac{3}{8} + \frac{3}{16}$$
$$-\frac{3}{16} \stackrel{?}{=} -\frac{6}{16} + \frac{3}{16}$$
$$-\frac{3}{16} = -\frac{3}{16} \quad \text{True}$$

The solution is $\frac{3}{16}$ or the solution set is $\left\{\frac{3}{16}\right\}$.

101.
$$\frac{5}{6} = -\frac{2}{3}z$$
$$-\frac{3}{2}\cdot\frac{5}{6} = -\frac{3}{2}\cdot\left(-\frac{2}{3}z\right)$$
$$-\frac{1}{2}\cdot\frac{5}{2} = z$$
$$-\frac{5}{4} = z$$

Check: $\frac{5}{6} = -\frac{2}{3}z$
$$\frac{5}{6} \stackrel{?}{=} -\frac{2}{3}\left(-\frac{5}{4}\right)$$
$$\frac{5}{6} \stackrel{?}{=} \frac{10}{12}$$
$$\frac{5}{6} = \frac{5}{6} \quad \text{True}$$

The solution is $-\frac{5}{4}$ or the solution set is $\left\{-\frac{5}{4}\right\}$.

103.
$$y + 1562.35 = 20{,}062.15$$
$$y + 1562.35 - 1562.35 = 20{,}062.15 - 1562.35$$
$$y = 18{,}499.80$$

The price of the car before the extra charges was $18,499.80.

105.
$$p - 17 = 51$$
$$p - 17 + 17 = 51 + 17$$
$$p = 68$$

The original price of the sleeping bag was $68.

107. $4h = 48$

$$\frac{4h}{4} = \frac{48}{4}$$

$$h = 12$$

Rebecca purchased 12 Happy Meals.

109.

$$45 = \frac{3000}{12} \cdot r$$

$$\frac{12}{3000} \cdot 45 = \frac{12}{3000} \cdot \frac{3000}{12} \cdot r$$

$$\frac{540}{3000} = r$$

$$0.18 = r$$

The annual interest rate is 0.18 or 18%.

111.

$$x + \lambda = 48$$

$$x + \lambda - \lambda = 48 - \lambda$$

$$x = 48 - \lambda$$

113. $14 = \theta x$

$$\frac{14}{\theta} = \frac{\theta x}{\theta}$$

$$\frac{14}{\theta} = x$$

115. Let $x = -\frac{2}{9}$.

$$x + \lambda = \frac{16}{3}$$

$$-\frac{2}{9} + \lambda = \frac{16}{3}$$

$$-\frac{2}{9} + \lambda + \frac{2}{9} = \frac{16}{3} + \frac{2}{9}$$

$$\lambda = \frac{48}{9} + \frac{2}{9}$$

$$\lambda = \frac{50}{9}$$

117. Let $x = \frac{7}{8}$.

$$-\frac{3}{4} = \theta x$$

$$-\frac{3}{4} = \theta\left(\frac{7}{8}\right)$$

$$\frac{8}{7}\left(-\frac{3}{4}\right) = \frac{8}{7}\left(\frac{7}{8}\theta\right)$$

$$-\frac{2(3)}{7} = \theta$$

$$-\frac{6}{7} = \theta$$

119. The solution of an equation is the value of the variable that satisfies the equation.

121. An algebraic expression differs from an equation in that the expression does not contain an equals sign and an equation does. An equation using the expression $x - 10$ is $x - 10 = 22$ solving for x, $x = 32$.

Section 2.2

Preparing for Linear Equations: Using the Properties Together

P1.

$$6 - (4 + 3x) + 8 = 6 - 4 - 3x + 8$$
$$= 6 - 4 + 8 - 3x$$
$$= 2 + 8 - 3x$$
$$= 10 - 3x$$

P2. $2(3x + 4) - 5$ for $x = -1$:

$$2[3(-1) + 4] - 5 = 2[-3 + 4] - 5$$
$$= 2[1] - 5$$
$$= 2 - 5$$
$$= -3$$

2.2 Quick Checks

1. To solve the equation $2x - 11 = 40$, the first step is to add 11 to each side of the equation.

2.

$$5x - 4 = 11$$
$$5x - 4 + 4 = 11 + 4$$
$$5x = 15$$
$$\frac{5x}{5} = \frac{15}{5}$$
$$x = 3$$

Check: $5x - 4 = 11$

$$5(3) - 4 \stackrel{?}{=} 11$$
$$15 - 4 \stackrel{?}{=} 11$$
$$11 = 11 \quad \text{True}$$

Because $x = 3$ satisfies the equation, the solution is $x = 3$, or the solution set is $\{3\}$.

3.

$$8 - 5r = -2$$
$$-8 + 8 - 5r = -8 + (-2)$$
$$-5r = -10$$
$$\frac{-5r}{-5} = \frac{-10}{-5}$$
$$r = 2$$

Check: $8 - 5r = -2$

$$8 - 5(2) \stackrel{?}{=} -2$$
$$8 - 10 \stackrel{?}{=} -2$$
$$-2 = -2 \quad \text{True}$$

Because $r = 2$ satisfies the equation, the solution is $r = 2$, or the solution set is $\{2\}$.

4.
$$8 = \frac{2}{3}k - 4$$
$$8 + 4 = \frac{2}{3}k - 4 + 4$$
$$12 = \frac{2}{3}k$$
$$\frac{3}{2}(12) = \frac{3}{2}\left(\frac{2}{3}k\right)$$
$$18 = k$$

Check: $8 = \frac{2}{3}k - 4$
$$8 \stackrel{?}{=} \frac{2}{3}(18) - 4$$
$$8 \stackrel{?}{=} 12 - 4$$
$$8 = 8 \text{ True}$$

Because $k = 18$ satisfies the equation, the solution is $k = 18$, or the solution set is $\{18\}$.

5.
$$-\frac{3}{2}n + 2 = -\frac{1}{4}$$
$$-\frac{3}{2}n + 2 - 2 = -\frac{1}{4} - 2$$
$$-\frac{3}{2}n = -\frac{1}{4} - \frac{8}{4}$$
$$-\frac{3}{2}n = -\frac{9}{4}$$
$$-\frac{2}{3}\left(-\frac{3}{2}n\right) = -\frac{2}{3}\left(-\frac{9}{4}\right)$$
$$n = \frac{3}{2}$$

Check: $-\frac{3}{2}n + 2 = -\frac{1}{4}$
$$-\frac{3}{2}\left(\frac{3}{2}\right) + 2 \stackrel{?}{=} -\frac{1}{4}$$
$$-\frac{9}{4} + 2 \stackrel{?}{=} -\frac{1}{4}$$
$$-\frac{9}{4} + \frac{8}{4} \stackrel{?}{=} -\frac{1}{4}$$
$$-\frac{1}{4} = -\frac{1}{4} \text{ True}$$

Because $n = \frac{3}{2}$ satisfies the equation, the solution is $n = \frac{3}{2}$, or the solution set is $\left\{\frac{3}{2}\right\}$.

6.
$$7b - 3b + 3 = 11$$
$$4b + 3 = 11$$
$$4b + 3 - 3 = 11 - 3$$
$$4b = 8$$
$$\frac{4b}{4} = \frac{8}{4}$$
$$b = 2$$

Check: $7b - 3b + 3 = 11$
$$7(2) - 3(2) + 3 \stackrel{?}{=} 11$$
$$14 - 6 + 3 \stackrel{?}{=} 11$$
$$11 = 11 \text{ True}$$

Since $b = 2$ results in a true statement, the solution of the equation is 2, or the solution set is $\{2\}$.

7.
$$-3a + 4 + 4a = 13 - 27$$
$$4 + a = -14$$
$$-4 + 4 + a = -4 + (-14)$$
$$a = -18$$

Check: $-3a + 4 + 4a = 13 - 27$
$$-3(-18) + 4 + 4(-18) \stackrel{?}{=} 13 - 27$$
$$54 + 4 - 72 \stackrel{?}{=} 13 - 27$$
$$-14 = -14 \text{ True}$$

Since $a = -18$ results in a true statement, the solution of the equation is -18, or the solution set is $\{-18\}$.

8.
$$6c - 2 + 2c = 18$$
$$8c - 2 = 18$$
$$8c - 2 + 2 = 18 + 2$$
$$8c = 20$$
$$\frac{8c}{8} = \frac{20}{8}$$
$$c = \frac{5}{2}$$

Check: $6c - 2 + 2c = 18$
$$6\left(\frac{5}{2}\right) - 2 + 2\left(\frac{5}{2}\right) \stackrel{?}{=} 18$$
$$15 - 2 + 5 \stackrel{?}{=} 18$$
$$18 = 18 \text{ True}$$

Since $c = \frac{5}{2}$ results in a true statement, the solution of the equation is $\frac{5}{2}$, or the solution set is $\left\{\frac{5}{2}\right\}$.

9.
$$-12 = 5x - 3x + 4$$
$$-12 = 2x + 4$$
$$-12 - 4 = 2x + 4 - 4$$
$$-16 = 2x$$
$$\frac{-16}{2} = \frac{2x}{2}$$
$$-8 = x$$

Check: $-12 = 5x - 3x + 4$
$$-12 \stackrel{?}{=} 5(-8) - 3(-8) + 4$$
$$-12 \stackrel{?}{=} -40 + 24 + 4$$
$$-12 = -12 \quad \text{True}$$

Since $x = -8$ results in a true statement, the solution of the equation is -8, or the solution set is $\{-8\}$.

10. $2(y+5) - 3 = 11$
$$2y + 10 - 3 = 11$$
$$2y + 7 = 11$$
$$2y + 7 - 7 = 11 - 7$$
$$2y = 4$$
$$\frac{2y}{2} = \frac{4}{2}$$
$$y = 2$$

Check: $2(y+5) - 3 = 11$
$$2(2+5) - 3 \stackrel{?}{=} 11$$
$$2(7) - 3 \stackrel{?}{=} 11$$
$$14 - 3 \stackrel{?}{=} 11$$
$$11 = 11 \quad \text{True}$$

Since $y = 2$ results in a true statement, the solution of the equation is 2, or the solution set is $\{2\}$.

11. $\frac{1}{2}(4 - 6x) + 5 = 3$
$$2 - 3x + 5 = 3$$
$$-3x + 7 = 3$$
$$-3x + 7 - 7 = 3 - 7$$
$$-3x = -4$$
$$\frac{-3x}{-3} = \frac{-4}{-3}$$
$$x = \frac{4}{3}$$

Check: $\frac{1}{2}(4 - 6x) + 5 = 3$
$$\frac{1}{2}\left(4 - 6 \cdot \frac{4}{3}\right) + 5 \stackrel{?}{=} 3$$
$$\frac{1}{2}(4 - 8) + 5 \stackrel{?}{=} 3$$
$$\frac{1}{2}(-4) + 5 \stackrel{?}{=} 3$$
$$-2 + 5 \stackrel{?}{=} 3$$
$$3 = 3 \quad \text{True}$$

Since $x = \frac{4}{3}$ results in a true statement, the solution of the equation is $\frac{4}{3}$, or the solution set is $\left\{\frac{4}{3}\right\}$.

12. $4 - (6 - x) = 11$
$$4 - 6 + x = 11$$
$$-2 + x = 11$$
$$2 - 2 + x = 2 + 11$$
$$x = 13$$

Check: $4 - (6 - x) = 11$
$$4 - (6 - 13) \stackrel{?}{=} 11$$
$$4 - (-7) \stackrel{?}{=} 11$$
$$4 + 7 \stackrel{?}{=} 11$$
$$11 = 11 \quad \text{True}$$

Since $x = 13$ results in a true statement, the solution of the equation is 13, or the solution set is $\{13\}$.

13. $8 + \frac{2}{3}(2n - 9) = 10$
$$8 + \frac{4}{3}n - 6 = 10$$
$$\frac{4}{3}n + 2 = 10$$
$$\frac{4}{3}n + 2 - 2 = 10 - 2$$
$$\frac{4}{3}n = 8$$
$$\frac{3}{4}\left(\frac{4}{3}n\right) = \frac{3}{4}(8)$$
$$n = 6$$

$$\text{Check: } 8+\frac{2}{3}(2n-9)=10$$
$$8+\frac{2}{3}(2\cdot 6-9) \stackrel{?}{=} 10$$
$$8+\frac{2}{3}(12-9) \stackrel{?}{=} 10$$
$$8+\frac{2}{3}(3) \stackrel{?}{=} 10$$
$$8+2 \stackrel{?}{=} 10$$
$$10=10 \quad \text{True}$$

Since $n = 6$ results in a true statement, the solution of the equation is 6, or the solution set is $\{6\}$.

14. $\frac{1}{3}(2x+9)+\frac{x}{3}=5$
$$\frac{2x}{3}+\frac{9}{3}+\frac{x}{3}=5$$
$$3+\frac{3x}{3}=5$$
$$3+\frac{x}{1}=5$$
$$3+x=5$$
$$3-3+x=5-3$$
$$x=2$$

$$\text{Check: } \frac{1}{3}(2x+9)+\frac{x}{3}=5$$
$$\frac{1}{3}[2(2)+9]+\frac{2}{3} \stackrel{?}{=} 5$$
$$\frac{1}{3}(4+9)+\frac{2}{3} \stackrel{?}{=} 5$$
$$\frac{1}{3}(13)+\frac{2}{3} \stackrel{?}{=} 5$$
$$\frac{13}{3}+\frac{2}{3} \stackrel{?}{=} 5$$
$$\frac{15}{3} \stackrel{?}{=} 5$$
$$5=5 \text{ True}$$

Since $x = 2$ results in a true statement, the solution of the equation is 2, or the solution set is $\{2\}$.

15. $3x+4=5x-8$
$$3x+4-3x=5x-8-3x$$
$$4=2x-8$$
$$4+8=2x-8+8$$
$$12=2x$$
$$\frac{12}{2}=\frac{2x}{2}$$
$$6=x$$

$$\text{Check: } 3x+4=5x-8$$
$$3(6)+4 \stackrel{?}{=} 5(6)-8$$
$$18+4 \stackrel{?}{=} 30-8$$
$$22=22 \quad \text{True}$$

Since $x = 6$ results in a true statement, the solution of the equation is 6, or the solution set is $\{6\}$.

16. $10m+3=6m-11$
$$10m+3-6m=6m-11-6m$$
$$4m+3=-11$$
$$4m+3-3=-11-3$$
$$4m=-14$$
$$\frac{4m}{4}=\frac{-14}{4}$$
$$m=-\frac{7}{2}$$

$$\text{Check: } 10m+3=6m-11$$
$$10\left(-\frac{7}{2}\right)+3 \stackrel{?}{=} 6\left(-\frac{7}{2}\right)-11$$
$$-35+3 \stackrel{?}{=} -21-11$$
$$-32=-32 \quad \text{True}$$

Since $m=-\frac{7}{2}$ results in a true statement, the solution of the equation is $-\frac{7}{2}$, or the solution set is $\left\{-\frac{7}{2}\right\}$.

17. False: To solve the equation $13 - 2(7x + 1) + 8x = 12$, the first step is to remove the parentheses using the Distributive Property.

18. $-9x+3(2x-3)=-10-2x$
$$-9x+6x-9=-10-2x$$
$$-3x-9=-10-2x$$
$$3x-3x-9=3x-10-2x$$
$$-9=x-10$$
$$-9+10=x-10+10$$
$$1=x$$

The solution to the equation is $x = 1$, or the solution set is $\{1\}$.

19. $3-4(p+5)=5(p+2)-12$
$3-4p-20=5p+10-12$
$-4p-17=5p-2$
$4p-4p-17=4p+5p-2$
$-17=9p-2$
$-17+2=9p-2+2$
$-15=9p$
$\frac{-15}{9}=\frac{9p}{9}$
$-\frac{5}{3}=p$

The solution to the equation is $p=-\frac{5}{3}$, or the solution set is $\left\{-\frac{5}{3}\right\}$.

20. $400+20(h-40)=640$
$400+20h-800=640$
$20h-400=640$
$20h-400+400=640+400$
$20h=1040$
$\frac{20h}{20}=\frac{1040}{20}$
$h=52$

Marcella worked 52 hours that week.

2.2 Exercises

21. $3x+4=7$
$3x+4-4=7-4$
$3x=3$
$\frac{3x}{3}=\frac{3}{3}$
$x=1$

Check: $3x+4=7$
$3(1)+4 \stackrel{?}{=} 7$
$3+4) \stackrel{?}{=} 7$
$7=7$ True

The solution is 1 or the solution set is {1}.

23. $2y-1=-5$
$2y-1+1=-5+1$
$2y=-4$
$\frac{2y}{2}=\frac{-4}{2}$
$y=-2$

Check: $2y-1=-5$
$2(-2)-1 \stackrel{?}{=} -5$
$-4-1 \stackrel{?}{=} -5$
$-5=-5$ True

The solution is -2 or the solution set is $\{-2\}$.

25. $-3p+1=10$
$-3p+1-1=10-1$
$-3p=9$
$\frac{-3p}{-3}=\frac{9}{-3}$
$p=-3$

Check: $-3p+1=10$
$-3(-3)+1 \stackrel{?}{=} 10$
$9+1 \stackrel{?}{=} 10$
$10=10$ True

The solution is -3, or the solution set is $\{-3\}$.

27. $8y+3=15$
$8y+3-3=15-3$
$8y=12$
$\frac{8y}{8}=\frac{12}{8}$
$y=\frac{3}{2}$

Check: $8y+3=15$
$8\left(\frac{3}{2}\right)+3 \stackrel{?}{=} 15$
$4\cdot 3+3 \stackrel{?}{=} 15$
$12+3 \stackrel{?}{=} 15$
$15=15$ True

The solution is $\frac{3}{2}$, or the solution set is $\left\{\frac{3}{2}\right\}$.

29. $5-2z=11$
$-5+5-2z=-5+11$
$-2z=6$
$\frac{-2z}{-2}=\frac{6}{-2}$
$z=-3$

Check: $5-2z=11$
$5-2(-3) \stackrel{?}{=} 11$
$5+6 \stackrel{?}{=} 11$
$11=11$ True

The solution is -3 or the solution set is $\{-3\}$.

31. $\frac{2}{3}x+1=9$

$\frac{2}{3}x+1-1=9-1$

$\frac{2}{3}x=8$

$\frac{3}{2}\cdot\frac{2}{3}x=\frac{3}{2}\cdot 8$

$x=12$

Check: $\frac{2}{3}x+1=9$

$\frac{2}{3}\cdot 12+1 \stackrel{?}{=} 9$

$8+1 \stackrel{?}{=} 9$

$9=9$ True

The solution is 12 or the solution set is {12}.

33. $\frac{7}{2}y-1=13$

$\frac{7}{2}y-1+1=13+1$

$\frac{7}{2}y=14$

$\frac{2}{7}\cdot\frac{7}{2}y=\frac{2}{7}\cdot 14$

$y=4$

Check: $\frac{7}{2}y-1=13$

$\frac{7}{2}(4)-1 \stackrel{?}{=} 13$

$14-1 \stackrel{?}{=} 13$

$13=13$ True

The solution is 4, or the solution set is {4}.

35. $3x-7+2x=-17$

$5x-7=-17$

$5x-7+7=-17+7$

$5x=-10$

$\frac{5x}{5}=\frac{-10}{5}$

$x=-2$

Check: $3x-7+2x=-17$

$3(-2)-7+2(-2) \stackrel{?}{=} -17$

$-6-7+(-4) \stackrel{?}{=} -17$

$-13-4 \stackrel{?}{=} -17$

$-17=-17$ True

The solution is –2, or the solution set is {–2}.

37. $2k-7k-8=17$

$-5k-8=17$

$-5k-8+8=17+8$

$-5k=25$

$\frac{-5k}{-5}=\frac{25}{-5}$

$k=-5$

Check: $2k-7k-8=17$

$2(-5)-7(-5)-8 \stackrel{?}{=} 17$

$-10+35-8 \stackrel{?}{=} 17$

$17=17$ True

The solution is –5, or the solution set is {–5}.

39. $2(x+1)=-14$

$2x+2=-14$

$2x+2-2=-14-2$

$2x=-16$

$\frac{2x}{2}=\frac{-16}{2}$

$x=-8$

Check: $2(x+1)=-14$

$2(-8+1) \stackrel{?}{=} -14$

$2(-7) \stackrel{?}{=} -14$

$-14=-14$ True

The solution is –8 or the solution set is {–8}.

41. $-3(2+r)=9$

$-6-3r=9$

$-6-3r+6=9+6$

$-3r=15$

$\frac{-3r}{-3}=\frac{15}{-3}$

$r=-5$

Check: $-3(2+r)=9$

$-3[2+(-5)] \stackrel{?}{=} 9$

$-3(-3) \stackrel{?}{=} 9$

$9=9$ True

The solution is –5, or the solution set is {–5}.

43. $17=2-(n+6)$

$17=2-n-6$

$17=-n-4$

$4+17=-n-4+4$

$21=-n$

$-1\cdot 21=-n\cdot -1$

$-21=n$

Check: $17 = 2-(n+6)$
$17 \stackrel{?}{=} 2-(-21+6)$
$17 \stackrel{?}{=} 2-(-15)$
$17 = 2+15$
$17 = 17$ True
The solution is −21, or the solution set is {−21}.

45. $-8 = 5-(7-z)$
$-8 = 5-7+z$
$-8 = -2+z$
$2-8 = -2+2+z$
$-6 = z$
Check: $-8 = 5-(7-z)$
$-8 \stackrel{?}{=} 5-[7-(-6)]$
$-8 \stackrel{?}{=} 5-(13)$
$-8 = -8$ True
The solution is −6, or the solution set is {−6}.

47. $2x+9 = x+1$
$2x+9-x = x+1-x$
$x+9 = 1$
$x+9-9 = 1-9$
$x = -8$
Check: $2x+9 = x+1$
$2(-8)+9 \stackrel{?}{=} -8+1$
$-16+9 \stackrel{?}{=} -7$
$-7 = -7$ True
The solution is −8, or the solution set is {−8}.

49. $2t-6 = 3-t$
$2t-6+t = 3-t+t$
$3t-6 = 3$
$3t-6+6 = 3+6$
$3t = 9$
$\frac{3t}{3} = \frac{9}{3}$
$t = 3$
Check: $2t-6 = 3-t$
$2(3)-6 \stackrel{?}{=} 3-3$
$6-6 \stackrel{?}{=} 0$
$0 = 0$ True
The solution is 3, or the solution set is {3}.

51. $14-2n = -4n+7$
$14-2n+4n = -4n+7+4n$
$14+2n = 7$
$14+2n-14 = 7-14$
$2n = -7$
$\frac{2n}{2} = -\frac{7}{2}$
$n = -\frac{7}{2}$
Check: $14-2n = -4n+7$
$14-2\left(-\frac{7}{2}\right) \stackrel{?}{=} -4\left(-\frac{7}{2}\right)+7$
$14+7 \stackrel{?}{=} 2(7)+7$
$21 \stackrel{?}{=} 14+7$
$21 = 21$ True
The solution is $-\frac{7}{2}$, or the solution set is $\left\{-\frac{7}{2}\right\}$.

53. $-3(5-3k) = 6k+6$
$-15+9k = 6k+6$
$-15+9k-6k = 6k+6-6k$
$-15+3k = 6$
$15-15+3k = 15+6$
$3k = 21$
$\frac{3k}{3} = \frac{21}{3}$
$k = 7$
Check: $-3(5-3k) = 6k+6$
$-3(5-3\cdot 7) \stackrel{?}{=} 6\cdot 7+6$
$-3(5-21) \stackrel{?}{=} 42+6$
$-3(-16) \stackrel{?}{=} 48$
$48 = 48$ True
The solution is 7, or the solution set is {7}.

55. $2(2x+3) = 3(x-4)$
$4x+6 = 3x-12$
$4x+6-3x = 3x-12-3x$
$x+6 = -12$
$x+6-6 = -12-6$
$x = -18$
Check: $2(2x+3) = 3(x-4)$
$2[2(-18)+3] \stackrel{?}{=} 3(-18-4)$
$2(-36+3) \stackrel{?}{=} 3(-22)$
$2(-33) \stackrel{?}{=} -66$
$-66 = -66$ True
The solution is −18, or the solution set is {−18}.

57.
$$\begin{aligned} 3+2(x-1)&=5x\\ 3+2x-2&=5x\\ 2x+1&=5x\\ -2x+2x+1&=5x-2x\\ 1&=3x\\ \frac{1}{3}&=\frac{3x}{3}\\ \frac{1}{3}&=x \end{aligned}$$

Check:
$$\begin{aligned} 3+2(x-1)&=5x\\ 3+2x-2&=5x\\ 2x+1&=5x\\ 2\left(\frac{1}{3}\right)+1&\stackrel{?}{=}5\left(\frac{1}{3}\right)\\ \frac{2}{3}+1&\stackrel{?}{=}\frac{5}{3}\\ \frac{2}{3}+\frac{3}{3}&\stackrel{?}{=}\frac{5}{3}\\ \frac{5}{3}&=\frac{5}{3} \quad \text{True} \end{aligned}$$

The solution is $\frac{1}{3}$, or the solution set is $\left\{\frac{1}{3}\right\}$.

59.
$$\begin{aligned} 9(6+a)+33a&=10a\\ 54+9a+33a&=10a\\ 54+42a&=10a\\ -42a+54+42a&=-42a+10a\\ 54&=-32a\\ \frac{54}{-32}&=\frac{-32a}{-32}\\ -\frac{27}{16}&=a \end{aligned}$$

Check:
$$\begin{aligned} 9(6+a)+33a&=10a\\ 9\left[6\left(-\frac{27}{16}\right)\right]+33\left(-\frac{27}{16}\right)&\stackrel{?}{=}10\left(-\frac{27}{16}\right)\\ 9\left(\frac{69}{16}\right)+33\left(-\frac{27}{16}\right)&\stackrel{?}{=}10\left(-\frac{27}{16}\right)\\ \frac{621}{16}-\frac{891}{16}&\stackrel{?}{=}\frac{-270}{16}\\ \frac{-270}{16}&=\frac{-270}{16} \quad \text{True} \end{aligned}$$

The solution is $-\frac{27}{16}$, or the solution set is $\left\{-\frac{27}{16}\right\}$.

61.
$$\begin{aligned} -5x+11&=1\\ -5x+11-11&=1-11\\ -5x&=-10\\ \frac{-5x}{-5}&=\frac{-10}{-5}\\ x&=2 \end{aligned}$$

Check:
$$\begin{aligned} -5x+11&=1\\ -5(2)+11&\stackrel{?}{=}1\\ -10+11&\stackrel{?}{=}1\\ 1&=1 \quad \text{True} \end{aligned}$$

The solution is 2, or the solution set is {2}.

63.
$$\begin{aligned} 4m+5&=2\\ 4m+5-5&=2-5\\ 4m&=-3\\ \frac{4m}{4}&=\frac{-3}{4}\\ m&=-\frac{3}{4} \end{aligned}$$

Check:
$$\begin{aligned} 4m+5&=2\\ 4\cdot\left(-\frac{3}{4}\right)+5&\stackrel{?}{=}2\\ -3+5&\stackrel{?}{=}2\\ 2&=2 \quad \text{True} \end{aligned}$$

The solution is $-\frac{3}{4}$, or the solution set is $\left\{-\frac{3}{4}\right\}$.

65.
$$\begin{aligned} -2(3n-2)&=2\\ -6n+4&=2\\ -6n+4-4&=2-4\\ -6n&=-2\\ \frac{-6n}{-6}&=\frac{-2}{-6}\\ n&=\frac{1}{3} \end{aligned}$$

Check:
$$\begin{aligned} -2(3n-2)&=2\\ -2\left(3\cdot\frac{1}{3}-2\right)&\stackrel{?}{=}2\\ -2(1-2)&\stackrel{?}{=}2\\ -2(-1)&\stackrel{?}{=}2\\ 2&=2 \quad \text{True} \end{aligned}$$

The solution is $\frac{1}{3}$, or the solution set is $\left\{\frac{1}{3}\right\}$.

67. $4k-(3+k)=-(2k+3)$
$$\begin{aligned}4k-3-k&=-2k-3\\-3+3k&=-2k-3\\3-3+3k&=-2k-3+3\\3k&=-2k\\2k+3k&=0\\5k&=0\\\frac{5k}{5}&=\frac{0}{5}\\k&=0\end{aligned}$$
Check: $4k-(3+k)=-(2k+3)$
$$\begin{aligned}4(0)-(3+0)&\stackrel{?}{=}-(2\cdot 0+3)\\0-3&\stackrel{?}{=}-3\\-3&=-3\quad\text{True}\end{aligned}$$
The solution is 0, or the solution set is $\{0\}$.

69.
$$\begin{aligned}2y+36&=6+6y\\2y+36-2y&=6+6y-2y\\36&=6+4y\\36-6&=6+4y-6\\30&=4y\\\frac{30}{4}&=\frac{4y}{4}\\\frac{15}{2}&=y\end{aligned}$$
Check: $2y+36=6+6y$
$$\begin{aligned}2\cdot\frac{15}{2}+36&\stackrel{?}{=}6+6\cdot\frac{15}{2}\\15+36&\stackrel{?}{=}6+45\\51&=51\quad\text{True}\end{aligned}$$
The solution is $\frac{15}{2}$, or the solution set is $\left\{\frac{15}{2}\right\}$.

71. $\frac{1}{2}(-4k+28)=6+14k$
$$\begin{aligned}-2k+14&=6+14k\\-2k+14+2k&=6+14k+2k\\14&=6+16k\\14-6&=6+16k-6\\8&=16k\\\frac{8}{16}&=\frac{16k}{16}\\\frac{1}{2}&=k\end{aligned}$$
Check: $\frac{1}{2}(-4k+28)=6+14k$
$$\begin{aligned}\frac{1}{2}\left[-4\left(\frac{1}{2}\right)+28\right]&\stackrel{?}{=}6+14\left(\frac{1}{2}\right)\\\frac{1}{2}(-2+28)&\stackrel{?}{=}6+7\\\frac{1}{2}(26)&\stackrel{?}{=}13\\13&=13\quad\text{True}\end{aligned}$$
The solution is $\frac{1}{2}$, or the solution set is $\left\{\frac{1}{2}\right\}$.

73. $-\frac{5}{2}(x+6)+\frac{3}{2}x=-8$
$$\begin{aligned}-\frac{5}{2}x-\frac{30}{2}+\frac{3}{2}x&=-8\\-15-x&=-8\\15-15-x&=15-8\\-x&=7\\\frac{-x}{-1}&=\frac{7}{-1}\\x&=-7\end{aligned}$$
Check: $-\frac{5}{2}(x+6)+\frac{3}{2}x=-8$
$$\begin{aligned}-\frac{5}{2}(-7+6)+\frac{3}{2}(-7)&\stackrel{?}{=}-8\\-\frac{5}{2}(-1)-\frac{21}{2}&\stackrel{?}{=}-8\\\frac{5}{2}-\frac{21}{2}&\stackrel{?}{=}-8\\-\frac{16}{2}&\stackrel{?}{=}-8\\-8&=-8\quad\text{True}\end{aligned}$$
The solution is -7, or the solution set is $\{-7\}$.

75. $-3(2y+3)-1=-4(y+6)+2y$
$$\begin{aligned}-6y-9-1&=-4y-24+2y\\-6y-10&=-2y-24\\2y-6y-10&=2y-2y-24\\-4y-10&=-24\\10-4y-10&=10-24\\-4y&=-14\\\frac{-4y}{-4}&=\frac{-14}{-4}\\y&=\frac{7}{2}\end{aligned}$$

Check: $-3(2y+3)-1=-4(y+6)+2y$

$$-3\left(2\cdot\frac{7}{2}+3\right)-1\stackrel{?}{=}-4\left(\frac{7}{2}+6\right)+2\cdot\frac{7}{2}$$
$$-3(7+3)-1\stackrel{?}{=}-4\left(\frac{19}{2}\right)+7$$
$$-3(10)-1\stackrel{?}{=}-38+7$$
$$-30-1\stackrel{?}{=}-31$$
$$-31=-31 \quad \text{True}$$

The solution is $\frac{7}{2}$, or the solution set is $\left\{\frac{7}{2}\right\}$.

77. $x+(x+6)=38$
$$x+x+6=38$$
$$2x+6=38$$
$$-6+2x+6=38-6$$
$$2x=32$$
$$\frac{2x}{2}=\frac{32}{2}$$
$$x=16$$
$x+6=16+6=22$

There are 16 grams of fat in McDonald's Southwestern salad and 22 grams of fat in a Burger King Tendercrisp Chicken garden salad.

79. $2w+2(2w+2)=30$
$$2w+4w+4=30$$
$$6w+4=30$$
$$6w+4-4=30-4$$
$$6w=26$$
$$\frac{6w}{6}=\frac{26}{6}$$
$$w=\frac{13}{3} \text{ or } 4\frac{1}{3}$$
$$2w+2=2\left(\frac{13}{3}\right)+2=\frac{26}{3}+\frac{6}{3}=\frac{32}{3} \text{ or } 10\frac{2}{3}$$

The width is $\frac{13}{3}$ or $4\frac{1}{3}$ feet and the length is $\frac{32}{3}$ or $10\frac{2}{3}$ feet.

81. $40x+4(1.5x)=368$
$$40x+6x=368$$
$$46x=368$$
$$\frac{46x}{46}=\frac{368}{46}$$
$$x=8$$

Jennifer's regular pay rate is \$8 per hour.

83.
$$\begin{aligned} 2x+2(x+5) &= 42 \\ 2x+2x+10 &= 42 \\ 4x+10 &= 42 \\ 4x+10-10 &= 42-10 \\ 4x &= 32 \\ \frac{4x}{4} &= \frac{32}{4} \\ x &= 8 \end{aligned}$$

$x + 5 = 8 + 5 = 13$

Yes, Becky has enough wallpaper.

85.
$$\begin{aligned} 8[4-6(x-1)]+5[(2x+3)-5] &= 18x-338 \\ 8[4-6x+6]+5[2x+3-5] &= 18x-338 \\ 8(10-6x)+5(2x-2) &= 18x-338 \\ 80-48x+10x-10 &= 18x-338 \\ 70-38x &= 18x-338 \\ 70-38x+38x &= 18x-338+38x \\ 70 &= 56x-338 \\ 70+338 &= 56x-338+338 \\ 408 &= 56x \\ \frac{408}{56} &= \frac{56x}{56} \\ \frac{51}{7} &= x \end{aligned}$$

Check: $8[4-6(x-1)]+5[(2x+3)-5]=18x-338$

$$\begin{aligned} 8\left[4-6\left(\frac{51}{7}-1\right)\right]+5\left[\left(2\cdot\frac{51}{7}+3\right)-5\right] &\stackrel{?}{=} 18\cdot\frac{51}{7}-338 \\ 8\left[4-6\left(\frac{44}{7}\right)\right]+5\left[\left(\frac{102}{7}+\frac{21}{7}\right)-5\right] &\stackrel{?}{=} \frac{918}{7}-338 \\ 8\left[\frac{28}{7}-\frac{264}{7}\right]+5\left[\frac{123}{7}-\frac{35}{7}\right] &\stackrel{?}{=} \frac{918}{7}-338 \\ 8\left(-\frac{236}{7}\right)+5\left(\frac{88}{7}\right) &\stackrel{?}{=} \frac{918}{7}-\frac{2366}{7} \\ -\frac{1888}{7}+\frac{440}{7} &\stackrel{?}{=} -\frac{1448}{7} \\ -\frac{1448}{7} &= -\frac{1448}{7} \end{aligned}$$

The solution is $\frac{51}{7}$, or the solution set is $\left\{\frac{51}{7}\right\}$.

87.
$$\begin{aligned} 3(36.7-4.3x)-10 &= 4(10-2.5x)-8(3.5-4.1x) \\ 110.1-12.9x-10 &= 40-10x-28+32.8x \\ -12.9x+100.1 &= 22.8x+12 \\ -22.8x-12.9x+100.1 &= -22.8x+22.8x+12 \\ -35.7x+100.1 &= 12 \\ -35.7x+100.1-100.1 &= 12-100.1 \\ -35.7x &= -88.1 \\ \frac{-35.7x}{-35.7} &= \frac{-88.1}{-35.7} \\ x &\approx 2.47 \end{aligned}$$

89.
$$\begin{aligned} 3.5\{4-[6-(2x+3)]+5\} &= -18.4 \\ 3.5\{4-[6-2x-3]+5\} &= -18.4 \\ 3.5\{4-[3-2x]+5\} &= -18.4 \\ 3.5\{4-3+2x+5\} &= -18.4 \\ 3.5\{6+2x\} &= -18.4 \\ 21+7x &= -18.4 \\ 7x &= -39.4 \\ x &= \frac{-39.4}{7} \\ x &\approx -5.6 \end{aligned}$$

91. $3d+2x=12;\ x=-4$
$$\begin{aligned} 3d+2(-4) &= 12 \\ 3d-8 &= 12 \\ 3d-8+8 &= 12+8 \\ 3d &= 20 \\ \frac{3d}{3} &= \frac{20}{3} \\ d &= \frac{20}{3} \end{aligned}$$

93. $\frac{2}{3}x-d=1;\ x=-\frac{3}{8}$
$$\begin{aligned} \frac{2}{3}\left(-\frac{3}{8}\right)-d &= 1 \\ -\frac{1}{4}-d &= 1 \\ -\frac{1}{4}-d+\frac{1}{4} &= 1+\frac{1}{4} \\ -d &= \frac{5}{4} \\ d &= -\frac{5}{4} \end{aligned}$$

95. Answers may vary. Possible answer: $6x-2(x+1)$ is an expression, while $6x-2(x+1)=6$ is an equation. An algebraic equation involves an equals sign, while an algebraic expression does not.

97. Answers may vary. Possible answer: This can lead to a correct answer if the next step is to add $-7x$ to both sides of the equation. However, the first step should be to combine like terms.

Section 2.3

Preparing for Solving Linear Equations Involving Fractions and Decimals; Classifying Equations

P1. $5=5$
$4=2\cdot 2$
$\text{LCD}=2\cdot 2\cdot 5=20$
The LCD of $\frac{3}{5}$ and $\frac{3}{4}$ is 20.

P2. $8=2\cdot 2\cdot 2$
$12=2\cdot 2\cdot 3$
$\text{LCD}=2\cdot 2\cdot 2\cdot 3=24$
The LCD of $\frac{3}{8}$ and $-\frac{7}{12}$ is 24.

2.3 Quick Checks

1. The least common denominator is the smallest number that each denominator has as a common multiple.

2. The LCD of 5, 4, and 2 is 20.
$$\begin{aligned} \frac{2x}{5}-\frac{x}{4} &= \frac{3}{2} \\ 20\left(\frac{2x}{5}-\frac{x}{4}\right) &= 20\left(\frac{3}{2}\right) \\ 20\left(\frac{2x}{5}\right)-20\left(\frac{x}{4}\right) &= 20\left(\frac{3}{2}\right) \\ 8x-5x &= 30 \\ 3x &= 30 \\ \frac{3x}{3} &= \frac{30}{3} \\ x &= 10 \end{aligned}$$
Check:
$$\begin{aligned} \frac{2x}{5}-\frac{x}{4} &= \frac{3}{2} \\ \frac{2(10)}{5}-\frac{10}{4} &\stackrel{?}{=} \frac{3}{2} \\ 4-\frac{5}{2} &\stackrel{?}{=} \frac{3}{2} \\ \frac{8}{2}-\frac{5}{2} &\stackrel{?}{=} \frac{3}{2} \\ \frac{3}{2} &= \frac{3}{2} \quad \text{True} \end{aligned}$$
The solution of the equation is 10, or the solution set is $\{10\}$.

3. The LCD of 6 and 9 is 18.

$$\frac{5}{6}x+\frac{1}{9}=-\frac{1}{6}x-\frac{1}{6}$$
$$18\left(\frac{5}{6}x+\frac{1}{9}\right)=18\left(-\frac{1}{6}x-\frac{1}{6}\right)$$
$$18\left(\frac{5}{6}x\right)+18\left(\frac{1}{9}\right)=18\left(-\frac{1}{6}x\right)-18\left(\frac{1}{6}\right)$$
$$15x+2=-3x-3$$
$$3x+15x+2=3x-3x-3$$
$$18x+2=-3$$
$$18x+2-2=-3-2$$
$$18x=-5$$
$$\frac{18x}{18}=\frac{-5}{18}$$
$$x=-\frac{5}{18}$$

Check:
$$\frac{5}{6}x+\frac{1}{9}=-\frac{1}{6}x-\frac{1}{6}$$
$$\frac{5}{6}\left(-\frac{5}{18}\right)+\frac{1}{9}\stackrel{?}{=}-\frac{1}{6}\left(-\frac{5}{18}\right)-\frac{1}{6}$$
$$-\frac{25}{108}+\frac{1}{9}\stackrel{?}{=}\frac{5}{108}-\frac{1}{6}$$
$$-\frac{25}{108}+\frac{12}{108}\stackrel{?}{=}\frac{5}{108}-\frac{18}{108}$$
$$-\frac{13}{108}=-\frac{13}{108}\quad\text{True}$$

The solution of the equation is $-\frac{5}{18}$, or the solution set is $\left\{-\frac{5}{18}\right\}$.

4. The result of multiplying the equation $\frac{1}{5}x+7=\frac{3}{10}$ by 10 is $2x + \underline{70} = 3$.

5.
$$\frac{a}{3}-\frac{1}{3}=-5$$
$$3\left(\frac{a}{3}-\frac{1}{3}\right)=3(-5)$$
$$3\left(\frac{a}{3}\right)-3\left(\frac{1}{3}\right)=3(-5)$$
$$a-1=-15$$
$$a-1+1=-15+1$$
$$a=-14$$

Check:
$$\frac{a}{3}-\frac{1}{3}=-5$$
$$\frac{-14}{3}-\frac{1}{3}\stackrel{?}{=}-5$$
$$\frac{-15}{3}\stackrel{?}{=}-5$$
$$-5=-5\quad\text{True}$$

The solution is -14, or the solution set is $\{-14\}$.

6. The LCD of 4 and 5 is 20.

$$\frac{3x-3}{4}-1=\frac{3}{5}x$$
$$20\left(\frac{3x-3}{4}-1\right)=20\left(\frac{3}{5}x\right)$$
$$20\left(\frac{3x-3}{4}\right)-20(1)=20\left(\frac{3}{5}x\right)$$
$$5(3x-3)-20=4(3x)$$
$$15x-15-20=12x$$
$$15x-35=12x$$
$$15x-35-15x=12x-15x$$
$$-35=-3x$$
$$\frac{-35}{-3}=\frac{-3x}{-3}$$
$$\frac{35}{3}=x$$

Check:
$$\frac{3x-3}{4}-1=\frac{3}{5}x$$
$$\frac{3\left(\frac{35}{3}\right)-3}{4}-1\stackrel{?}{=}\frac{3}{5}\left(\frac{35}{3}\right)$$
$$\frac{35-3}{4}-1\stackrel{?}{=}7$$
$$\frac{32}{4}-1\stackrel{?}{=}7$$
$$8-1\stackrel{?}{=}7$$
$$7=7\quad\text{True}$$

The solution is $\frac{35}{3}$, or the solution set is $\left\{\frac{35}{3}\right\}$.

7. To clear the decimals in the equation $0.25x + 5 = 7 - 0.3x$, multiply both sides of the equation by $\underline{100}$.

8.
$$0.2z=20$$
$$10\cdot 0.2z=10\cdot 20$$
$$2z=200$$
$$\frac{2z}{2}=\frac{200}{2}$$
$$z=100$$

Check: $0.2z = 20$
$0.2(100) \stackrel{?}{=} 20$
$20 = 20$ True
The solution is 100, or the solution set is $\{100\}$.

9. $0.15p - 2.5 = 5$
$100(0.15p - 2.5) = 100(5)$
$100 \cdot 0.15p - 100 \cdot 2.5 = 100 \cdot 5$
$15p - 250 = 500$
$15p - 250 + 250 = 500 + 250$
$15p = 750$
$\frac{15p}{15} = \frac{750}{15}$
$p = 50$
Check: $0.15p - 2.5 = 5$
$0.15(50) - 2.5 \stackrel{?}{=} 5$
$7.5 - 2.5 \stackrel{?}{=} 5$
$5 = 5$ True
The solution is 50, or the solution set is $\{50\}$.

10. The coefficient of the first term of the equation $n + 0.25n = 50$ is 1.

11. $p + 0.05p = 52.5$
$1p + 0.05p = 52.5$
$1.05p = 52.5$
$100 \cdot 1.05p = 100 \cdot 52.5$
$105p = 5250$
$\frac{105p}{105} = \frac{5250}{105}$
$p = 50$
The solution of the equation is 50, or the solution set is $\{50\}$.

12. $c - 0.25c = 120$
$1c - 0.25c = 120$
$0.75c = 120$
$100 \cdot 0.75c = 100 \cdot 120$
$75c = 12{,}000$
$\frac{75c}{75} = \frac{12{,}000}{75}$
$c = 160$
The solution of the equation is 160, or the solution set is $\{160\}$.

13. $0.36y - 0.5 = 0.16y + 0.3$
$0.20y - 0.5 = 0.3$
$0.2y = 0.8$
$10(0.2y) = 10(0.8)$
$2y = 8$
$y = 4$
The solution to the equation is 4, or the solution set is $\{4\}$.

14. $0.12x + 0.05(5000 - x) = 460$
$0.12x + 250 - 0.05x = 460$
$0.07x + 250 = 460$
$0.07x = 210$
$100(0.07x) = 100(210)$
$7x = 21{,}000$
$x = 3000$
The solution to the equation is 3000, or the solution set is $\{3000\}$.

15. True

16. $3(x+4) = 4 + 3x + 18$
$3x + 12 = 3x + 22$
$3x + 12 - 3x = 3x + 22 - 3x$
$12 = 22$
The statement 12 = 22 is false, so the equation is a contradiction. The solution set is $\varnothing$ or $\{\ \}$.

17. $\frac{1}{3}(6x - 9) - 1 = 6x - [4x - (-4)]$
$2x - 3 - 1 = 6x - [4x + 4]$
$2x - 4 = 6x - 4x - 4$
$2x - 4 = 2x - 4$
$2x - 4 - 2x = 2x - 4 - 2x$
$-4 = -4$
The statement $-4 = -4$ is true for all real numbers x. The solution set is the set of all real numbers.

18. $-5 - (9x + 8) + 23 = 7 + x - (10x - 3)$
$-5 - 9x - 8 + 23 = 7 + x - 10x + 3$
$-9x + 10 = -9x + 10$
$-9x + 10 + 9x = -9x + 10 + 9x$
$10 = 10$
The statement 10 = 10 is true for all real numbers x. The solution set is the set of all real numbers.

19. $\frac{3}{2}x-8=x+7+\frac{1}{2}x$

$\frac{3}{2}x-8=\frac{2}{2}x+7+\frac{1}{2}x$

$\frac{3}{2}x-8=\frac{3}{2}x+7$

$\frac{3}{2}x-8-\frac{3}{2}x=\frac{3}{2}x+7-\frac{3}{2}x$

$-8=7$

The statement $-8 = 7$ is false, so the equation is a contradiction. The solution set is $\varnothing$ or $\{\ \}$.

20. When the variable is eliminated from a linear equation and a true statement results, the solution set is all real numbers.

21. When the variable is eliminated from a linear equation and a false statement results, the solution set is the empty set ($\varnothing$).

22. $2(x-7)+8=6x-(4x+2)-4$

$2x-14+8=6x-4x-2-4$

$2x-6=2x-6$

$2x-6-2x=2x-6-2x$

$-6=-6$

The statement $-6 = -6$ is true for all values of x, so the equation is an identity. The solution set is the set of all real numbers.

23. $\frac{4(7-x)}{3}=x$

$3\left(\frac{4(7-x)}{3}\right)=3x$

$4(7-x)=3x$

$28-4x=3x$

$28-4x+4x=3x+4x$

$28=7x$

$4=x$

The equation has solution $x = 4$, so it is a conditional equation. The solution set is $\{4\}$.

24. $\frac{1}{2}(4x-6)=6\left(\frac{1}{3}x-\frac{1}{2}\right)+4$

$2x-3=2x-3+4$

$2x-3=2x+1$

$2x-3-2x=2x+1-2x$

$-3=1$

The statement $-3 = 1$ is false, so the equation is a contradiction. The solution set is $\varnothing$ or $\{\ \}$.

25. $4(5x-4)+1=-2+20x$

$20x-16+1=-2+20x$

$20x-15=-2+20x$

$20x-15-20x=-2+20x-20x$

$-15=-2$

The statement $-15 = -2$ is false, so the equation is a contradiction. The solution set is $\varnothing$ or $\{\ \}$.

26. $0.04x+0.06(x+250)=65$

$0.04x+0.06x+15=65$

$0.10x+15=65$

$0.1x=50$

$10(0.1x)=10(50)$

$x=500$

She invested $500 in the savings account.

2.3 Exercises

27. $\frac{2k-1}{4}=2$

$4\left(\frac{2k-1}{4}\right)=4(2)$

$2k-1=8$

$2k-1+1=8+1$

$2k=9$

$\frac{2k}{2}=\frac{9}{2}$

$k=\frac{9}{2}$

Check: $\frac{2k-1}{4}=2$

$\frac{2\left(\frac{9}{2}\right)-1}{4}\overset{?}{=}2$

$\frac{9-1}{4}\overset{?}{=}2$

$\frac{8}{4}\overset{?}{=}2$

$2=2$ True

The solution is $\frac{9}{2}$, or the solution set is $\left\{\frac{9}{2}\right\}$.

29. $\frac{3x+2}{4}=\frac{x}{2}$

$4\left(\frac{3x+2}{4}\right)=4\left(\frac{x}{2}\right)$

$3x+2=2x$

$3x+2-3x=2x-3x$

$2=-x$

$-1(2)=-1(-x)$

$-2=x$

Check: $$\frac{3x+2}{4}=\frac{x}{2}$$
$$\frac{3(-2)+2}{4}=\frac{-2}{2}$$
$$\frac{-6+2}{4}=-1$$
$$\frac{-4}{4}=-1$$
$$-1=-1 \text{ True}$$

The solution is –2, or the solution set is {–2}.

31. $$\frac{1}{5}x+\frac{3}{2}=\frac{3}{10}$$
$$10\left(\frac{1}{5}x+\frac{3}{2}\right)=10\left(\frac{3}{10}\right)$$
$$10\left(\frac{1}{5}x\right)+10\left(\frac{3}{2}\right)=3$$
$$2x+15=3$$
$$2x+15-15=3-15$$
$$2x=-12$$
$$\frac{2x}{2}=\frac{-12}{2}$$
$$x=-6$$

Check: $$\frac{1}{5}x+\frac{3}{2}=\frac{3}{10}$$
$$\frac{1}{5}(-6)+\frac{3}{2}\stackrel{?}{=}\frac{3}{10}$$
$$-\frac{12}{10}+\frac{15}{10}\stackrel{?}{=}\frac{3}{10}$$
$$\frac{3}{10}=\frac{3}{10} \text{ True}$$

The solution is –6, or the solution set is {–6}.

33. $$\frac{-2x}{3}+1=\frac{5}{9}$$
$$9\left(\frac{-2x}{3}+1\right)=9\left(\frac{5}{9}\right)$$
$$9\left(\frac{-2x}{3}\right)+9(1)=5$$
$$-6x+9=5$$
$$-6x+9-9=5-9$$
$$-6x=-4$$
$$\frac{-6x}{-6}=\frac{-4}{-6}$$
$$x=\frac{2}{3}$$

Check: $$\frac{-2x}{3}+1=\frac{5}{9}$$
$$\frac{-2\cdot\frac{2}{3}}{3}+1\stackrel{?}{=}\frac{5}{9}$$
$$-\frac{4}{9}+1\stackrel{?}{=}\frac{5}{9}$$
$$-\frac{4}{9}+\frac{9}{9}\stackrel{?}{=}\frac{5}{9}$$
$$\frac{5}{9}=\frac{5}{9} \text{ True}$$

The solution is $\frac{2}{3}$, or the solution set is $\left\{\frac{2}{3}\right\}$.

35. $$\frac{a}{4}-\frac{a}{3}=-\frac{1}{2}$$
$$12\left(\frac{a}{4}-\frac{a}{3}\right)=12\left(-\frac{1}{2}\right)$$
$$12\left(\frac{a}{4}\right)-12\left(\frac{a}{3}\right)=-6$$
$$3a-4a=-6$$
$$-a=-6$$
$$\frac{-a}{-1}=\frac{-6}{-1}$$
$$a=6$$

Check: $$\frac{a}{4}-\frac{a}{3}=-\frac{1}{2}$$
$$\frac{6}{4}-\frac{6}{3}\stackrel{?}{=}-\frac{1}{2}$$
$$\frac{18}{12}-\frac{24}{12}\stackrel{?}{=}-\frac{6}{12}$$
$$-\frac{6}{12}=-\frac{6}{12} \text{ True}$$

The solution is 6, or the solution set is {6}.

37. $$\frac{5}{4}(2a-10)=-\frac{3}{2}a$$
$$4\cdot\frac{5}{4}(2a-10)=4\cdot\left(-\frac{3}{2}a\right)$$
$$5(2a-10)=2(-3a)$$
$$10a-50=-6a$$
$$10a-50-10a=-6a-10a$$
$$-50=-16a$$
$$\frac{-50}{-16}=\frac{-16a}{-16}$$
$$\frac{25}{8}=a$$

Check: $\frac{5}{4}(2a-10)=-\frac{3}{2}a$

$$\frac{5}{4}\left[2\left(\frac{25}{8}\right)-10\right] \stackrel{?}{=} -\frac{3}{2}\cdot\frac{25}{8}$$
$$\frac{5}{4}\left(\frac{25}{4}-10\right) \stackrel{?}{=} -\frac{75}{16}$$
$$\frac{125}{16}-\frac{200}{16} \stackrel{?}{=} -\frac{75}{16}$$
$$-\frac{75}{16}=-\frac{75}{16} \quad \text{True}$$

The solution is $\frac{25}{8}$, or the solution set is $\left\{\frac{25}{8}\right\}$.

39.
$$\frac{y}{10}+3=\frac{y}{4}+6$$
$$20\left(\frac{y}{10}+3\right)=20\left(\frac{y}{4}+6\right)$$
$$20\left(\frac{y}{10}\right)+20(3)=20\left(\frac{y}{4}\right)+20(6)$$
$$2y+60=5y+120$$
$$-2y+2y+60=-2y+5y+120$$
$$60=3y+120$$
$$60-120=3y+120-120$$
$$-60=3y$$
$$\frac{-60}{3}=\frac{3y}{3}$$
$$-20=y$$

Check: $\frac{y}{10}+3=\frac{y}{4}+6$
$$\frac{-20}{10}+3 \stackrel{?}{=} \frac{-20}{4}+6$$
$$-2+3 \stackrel{?}{=} -5+6$$
$$1=1 \quad \text{True}$$

The solution is −20, or the solution set is {−20}.

41.
$$\frac{4x-9}{3}+\frac{x}{6}=\frac{x}{2}-2$$
$$6\left(\frac{4x-9}{3}+\frac{x}{6}\right)=6\left(\frac{x}{2}-2\right)$$
$$6\left(\frac{4x-9}{3}\right)+6\left(\frac{x}{6}\right)=6\left(\frac{x}{2}\right)-6(2)$$
$$2(4x-9)+x=3x-12$$
$$8x-18+x=3x-12$$
$$9x-18=3x-12$$
$$9x-18-3x=3x-12-3x$$
$$6x-18=-12$$
$$6x-18+18=-12+18$$
$$6x=6$$
$$\frac{6x}{6}=\frac{6}{6}$$
$$x=1$$

Check: $\frac{4x-9}{3}+\frac{x}{6}=\frac{x}{2}-2$
$$\frac{4\cdot 1-9}{3}+\frac{1}{6} \stackrel{?}{=} \frac{1}{2}-2$$
$$-\frac{10}{6}+\frac{1}{6} \stackrel{?}{=} \frac{3}{6}-\frac{12}{6}$$
$$-\frac{9}{6}=-\frac{9}{6} \quad \text{True}$$

The solution is 1, or the solution set is {1}.

43.
$$0.4w=12$$
$$10(0.4w)=10(12)$$
$$4w=120$$
$$\frac{4w}{4}=\frac{120}{4}$$
$$w=30$$

Check: $0.4w=12$
$$0.4(30) \stackrel{?}{=} 12$$
$$12=12 \quad \text{True}$$

The solution is 30, or the solution set is {30}.

45.
$$-1.3c=5.2$$
$$10(-1.3c)=10(5.2)$$
$$-13c=52$$
$$\frac{-13c}{-13}=\frac{52}{-13}$$
$$c=-4$$

Check: $-1.3c=5.2$
$$-1.3(-4) \stackrel{?}{=} 5.2$$
$$5.2=5.2 \quad \text{True}$$

The solution is −4, or the solution set is {−4}.

47.
$$\begin{aligned} 1.05p &= 52.5 \\ 100(1.05p) &= 100(52.5) \\ 105p &= 5250 \\ \frac{105p}{105} &= \frac{5250}{105} \\ p &= 50 \end{aligned}$$

Check: $1.05p = 52.5$
$$\begin{aligned} 1.05(50) &\stackrel{?}{=} 52.5 \\ 52.5 &= 52.5 \quad \text{True} \end{aligned}$$

The solution is 50, or the solution set is {50}.

49.
$$\begin{aligned} p + 1.5p &= 12 \\ 2.5p &= 12 \\ 10(2.5p) &= 10(12) \\ 25p &= 120 \\ \frac{25p}{25} &= \frac{120}{25} \\ p &= 4.8 \end{aligned}$$

Check: $p + 1.5p = 12$
$$\begin{aligned} 4.8 + 1.5(4.8) &\stackrel{?}{=} 12 \\ 4.8 + 7.2 &\stackrel{?}{=} 12 \\ 12 &= 12 \quad \text{True} \end{aligned}$$

The solution is 4.8, or the solution set is {4.8}.

51.
$$\begin{aligned} p + 0.05p &= 157.5 \\ 1.05p &= 157.5 \\ 100(1.05p) &= 100(157.5) \\ 105p &= 15,750 \\ \frac{105p}{105} &= \frac{15,750}{105} \\ p &= 150 \end{aligned}$$

Check: $p + 0.05p = 157.5$
$$\begin{aligned} 150 + 0.05(150) &\stackrel{?}{=} 157.5 \\ 150 + 7.5 &\stackrel{?}{=} 157.5 \\ 157.5 &= 157.5 \quad \text{True} \end{aligned}$$

The solution is 150, or the solution set is {150}.

53.
$$\begin{aligned} 0.3x + 2.3 &= 0.2x + 1.1 \\ 10(0.3x + 2.3) &= 10(0.2x + 1.1) \\ 10(0.3x) + 10(2.3) &= 10(0.2x) + 10(1.1) \\ 3x + 23 &= 2x + 11 \\ -2x + 3x + 23 &= -2x + 2x + 11 \\ x + 23 &= 11 \\ x + 23 - 23 &= 11 - 23 \\ x &= -12 \end{aligned}$$

Check: $0.3x + 2.3 = 0.2x + 1.1$
$$\begin{aligned} 0.3(-12) + 2.3 &\stackrel{?}{=} 0.2(-12) + 1.1 \\ -3.6 + 2.3 &\stackrel{?}{=} -2.4 + 1.1 \\ -1.3 &= -1.3 \quad \text{True} \end{aligned}$$

The solution is −12, or the solution set is {−12}.

55.
$$\begin{aligned} 0.65x + 0.3x &= x - 3 \\ 0.95x &= x - 3 \\ 100(0.95x) &= 100(x - 3) \\ 95x &= 100x - 300 \\ 95x - 100x &= 100x - 300 - 100x \\ -5x &= -300 \\ \frac{-5x}{-5} &= \frac{-300}{-5} \\ x &= 60 \end{aligned}$$

Check: $0.65x + 0.3x = x - 3$
$$\begin{aligned} 0.65(60) + 0.3(60) &\stackrel{?}{=} 60 - 3 \\ 39 + 18 &\stackrel{?}{=} 57 \\ 57 &= 57 \quad \text{True} \end{aligned}$$

The solution is 60, or the solution set is {60}.

57.
$$\begin{aligned} 3 + 1.5(z + 2) &= 3.5z - 4 \\ 3 + 1.5z + 3 &= 3.5z - 4 \\ 1.5z + 6 &= 3.5z - 4 \\ -1.5z + 1.5z + 6 &= -1.5z + 3.5z - 4 \\ 6 &= 2z - 4 \\ 6 + 4 &= 2z - 4 + 4 \\ 10 &= 2z \\ \frac{10}{2} &= \frac{2z}{2} \\ 5 &= z \end{aligned}$$

Check: $3 + 1.5(z + 2) = 3.5z - 4$
$$\begin{aligned} 3 + 1.5(5 + 2) &\stackrel{?}{=} 3.5(5) - 4 \\ 3 + 1.5(7) &\stackrel{?}{=} 17.5 - 4 \\ 3 + 10.5 &\stackrel{?}{=} 13.5 \\ 13.5 &= 13.5 \quad \text{True} \end{aligned}$$

The solution is 5, or the solution set is {5}.

59.
$$\begin{aligned} 0.02(2c - 24) &= -0.4(c - 1) \\ 100[0.02(2c - 24)] &= 100[-0.4(c - 1)] \\ 2(2c - 24) &= -40(c - 1) \\ 4c - 48 &= -40c + 40 \\ 4c - 48 + 40c &= -40c + 40 + 40c \\ 44c - 48 &= 40 \\ 44c - 48 + 48 &= 40 + 48 \\ 44c &= 88 \\ \frac{44c}{44} &= \frac{88}{44} \\ c &= 2 \end{aligned}$$

Check: $0.02(2c - 24) = -0.4(c - 1)$
$$\begin{aligned} 0.02[2(2) - 24] &\stackrel{?}{=} -0.4(2 - 1) \\ 0.02(4 - 24) &\stackrel{?}{=} -0.4(1) \\ 0.02(-20) &\stackrel{?}{=} -0.4 \\ -0.4 &= -0.4 \quad \text{True} \end{aligned}$$

The solution is 2, or the solution set is {2}.

61. $0.15x+0.10(250-x)=28.75$
$$0.15x+25-0.1x=28.75$$
$$25+0.05x=28.75$$
$$-25+25+0.05x=-25+28.75$$
$$0.05x=3.75$$
$$100(0.05x)=100(3.75)$$
$$5x=375$$
$$\frac{5x}{5}=\frac{375}{5}$$
$$x=75$$
Check: $0.15x+0.10(250-x)=28.75$
$$0.15(75)+0.10(250-75)\stackrel{?}{=}28.75$$
$$11.25+0.10(175)\stackrel{?}{=}28.75$$
$$11.25+17.5\stackrel{?}{=}28.75$$
$$28.75=28.75 \quad \text{True}$$
The solution is 75, or the solution set is {75}.

63. $4z-3(z+1)=2(z-3)-z$
$$4z-3z-3=2z-6-z$$
$$z-3=z-6$$
$$-z+z-3=-z+z-6$$
$$-3=-6$$
This is a false statement, so the equation is a contraction. The solution set is $\varnothing$ or { }.

65. $6q-(q-3)=2q+3(q+1)$
$$6q-q+3=2q+3q+3$$
$$5q+3=5q+3$$
$$-5q+5q+3=-5q+5q+3$$
$$3=3$$
This is a true statement. The equation is an identity. The solution set is the set of all real numbers.

67. $9a-5(a+1)=2(a-3)$
$$9a-5a-5=2a-6$$
$$4a-5=2a-6$$
$$-2a+4a-5=-2a+2a-6$$
$$2a-5=-6$$
$$2a-5+5=-6+5$$
$$2a=-1$$
$$\frac{2a}{2}=\frac{-1}{2}$$
$$a=-\frac{1}{2}$$
This is a conditional equation. The solution set is $\left\{-\frac{1}{2}\right\}$.

69. $\frac{4x-9}{6}-\frac{x}{2}=\frac{x}{6}+3$
$$6\left(\frac{4x-9}{6}-\frac{x}{2}\right)=6\left(\frac{x}{6}+3\right)$$
$$6\left(\frac{4x-9}{6}\right)-6\left(\frac{x}{2}\right)=6\left(\frac{x}{6}\right)+6(3)$$
$$4x-9-3x=x+18$$
$$x-9=x+18$$
$$-x+x-9=-x+x+18$$
$$-9=18$$
This is a false statement, so the equation is a contradiction. The solution set is $\varnothing$ or { }.

71. $\frac{5z+1}{5}=\frac{2z-3}{2}$
$$10\left(\frac{5z+1}{5}\right)=10\left(\frac{2z-3}{2}\right)$$
$$2(5z+1)=5(2z-3)$$
$$10z+2=10z-15$$
$$-10z+10z+2=-10z+10z-15$$
$$2=-15$$
This is a false statement, so the equation is a contradiction. The solution set is $\varnothing$ or { }.

73. $\frac{q}{3}+\frac{4}{5}=\frac{5q+12}{15}$
$$15\left(\frac{q}{3}+\frac{4}{5}\right)=15\left(\frac{5q+12}{15}\right)$$
$$15\left(\frac{q}{3}\right)+15\left(\frac{4}{5}\right)=5q+12$$
$$5q+12=5q+12$$
$$-5q+5q+12=-5q+5q+12$$
$$12=12$$
This is a true statement. The equation is an identity. The solution set is the set of all real numbers.

75. $-3(2n+4)=10n$
$$-6n-12=10n$$
$$6n-6n-12=6n+10n$$
$$-12=16n$$
$$\frac{-12}{16}=\frac{16n}{16}$$
$$-\frac{3}{4}=n$$
The solution is $-\frac{3}{4}$ or the solution set is $\left\{-\frac{3}{4}\right\}$.

77. $-2x+5x=4(x+2)-(x+8)$
$$3x=4x+8-x-8$$
$$3x=3x$$
$$-3x+3x=-3x+3x$$
$$0=0$$
This is a true statement. The equation is an identity. The solution set is the set of all real numbers.

79. $-6(x-2)+8x=-x+10-3x$
$$-6x+12+8x=-x+10-3x$$
$$12+2x=10-4x$$
$$12+2x+4x=10-4x+4x$$
$$12+6x=10$$
$$-12+12+6x=-12+10$$
$$6x=-2$$
$$\frac{6x}{6}=\frac{-2}{6}$$
$$x=-\frac{1}{3}$$
The solution is $-\frac{1}{3}$ or the solution set is $\left\{-\frac{1}{3}\right\}$.

81. $\frac{3}{4}x=\frac{1}{2}x-5$
$$4\left(\frac{3}{4}x\right)=4\left(\frac{1}{2}x-5\right)$$
$$3x=2x-20$$
$$-2x+3x=-2x+2x-20$$
$$x=-20$$
The solution is –20, or the solution set is {–20}.

83. $\frac{1}{2}x+2=\frac{4x+1}{4}$
$$4\left(\frac{1}{2}x+2\right)=4\left(\frac{4x+1}{4}\right)$$
$$2x+8=4x+1$$
$$-2x+2x+8=-2x+4x+1$$
$$8=2x+1$$
$$8-1=2x+1-1$$
$$7=2x$$
$$\frac{7}{2}=\frac{2x}{2}$$
$$\frac{7}{2}=x$$
The solution is $\frac{7}{2}$, or the solution set is $\left\{\frac{7}{2}\right\}$.

85. $0.3p+2=0.1(p+5)+0.2(p+1)$
$$0.3p+2=0.1p+0.5+0.2p+0.2$$
$$0.3p+2=0.3p+0.7$$
$$-0.3p+0.3p+2=-0.3p+0.3p+0.7$$
$$2=0.7$$
This is a false statement, so the equation is a contradiction. The solution set is ∅ or { }.

87. $-0.7x=1.4$
$$10(-0.7x)=10(1.4)$$
$$-7x=14$$
$$\frac{-7x}{-7}=\frac{14}{-7}$$
$$x=-2$$
The solution is –2, or the solution set is {–2}.

89. $\frac{3(2y-1)}{5}=2y-3$
$$5\left[\frac{3(2y-1)}{5}\right]=5(2y-3)$$
$$3(2y-1)=5(2y-3)$$
$$6y-3=10y-15$$
$$-6y+6y-3=-6y+10y-15$$
$$-3=4y-15$$
$$-3+15=4y-15+15$$
$$12=4y$$
$$\frac{12}{4}=\frac{4y}{4}$$
$$3=y$$
The solution is 3, or the solution set is {3}.

91. $0.6x-0.2(x-4)=0.4(x-2)$
$$0.6x-0.2x+0.8=0.4x-0.8$$
$$0.4x+0.8=0.4x-0.8$$
$$-0.4x+0.4x+0.8=-0.4x+0.4x-0.8$$
$$0.8=-0.8$$
This is a false statement, so the equation is a contradiction. The solution set is ∅ or { }.

93. $\frac{3x-2}{4}=\frac{5x-1}{6}$
$$12\left(\frac{3x-2}{4}\right)=12\left(\frac{5x-1}{6}\right)$$
$$3(3x-2)=2(5x-1)$$
$$9x-6=10x-2$$
$$-9x+9x-6=-9x+10x-2$$
$$-6=x-2$$
$$-6+2=x-2+2$$
$$-4=x$$
The solution is –4, or the solution set is {–4}.

95.
$$\begin{aligned}0.3x+2.6x&=5.7-1.8+2.8x\\2.9x&=3.9+2.8x\\2.9x-2.8x&=3.9+2.8x-2.8x\\0.1x&=3.9\\10(0.1x)&=10(3.9)\\x&=39\end{aligned}$$
The solution is 39, or the solution set is {39}.

97.
$$\begin{aligned}\frac{3}{2}x-6&=\frac{2(x-9)}{3}+\frac{1}{6}x\\6\left(\frac{3}{2}x-6\right)&=6\left[\frac{2(x-9)}{3}+\frac{1}{6}x\right]\\9x-36&=4(x-9)+x\\9x-36&=4x-36+x\\9x-36&=5x-36\\-5x+9x-36&=-5x+5x-36\\4x-36&=-36\\4x-36+36&=-36+36\\4x&=0\\\frac{4x}{4}&=\frac{0}{4}\\x&=0\end{aligned}$$
The solution is 0, or the solution set is {0}.

99.
$$\begin{aligned}\frac{2}{3}\left[4-\left(\frac{x}{2}+6\right)-2x\right]+3&=\frac{5x}{6}\\6\left\{\frac{2}{3}\left[4-\left(\frac{x}{2}+6\right)-2x\right]+3\right\}&=6\left(\frac{5x}{6}\right)\\4\left[4-\left(\frac{x}{2}+6\right)-2x\right]+18&=5x\\16-4\left(\frac{x}{2}+6\right)-8x+18&=5x\\16-2x-24-8x+18&=5x\\10-10x&=5x\\10&=15x\\\frac{2}{3}&=x\end{aligned}$$
The solution is $\frac{2}{3}$, or the solution set is $\left\{\frac{2}{3}\right\}$.

101.
$$\begin{aligned}2.8x+13.754&=4-2.95x\\5.75x+13.754&=4\\5.75x&=-9.754\\x&=\frac{-9.754}{5.75}\\x&\approx-1.70\end{aligned}$$

103.
$$\begin{aligned}x-\{1.5x-2[x-3.1(x+10)]\}&=0\\x-\{1.5x-2[x-3.1x-31]\}&=0\\x-\{1.5x-2[-2.1x-31]\}&=0\\x-\{1.5x+4.2x+62\}&=0\\x-\{5.7x+62\}&=0\\x-5.7x-62&=0\\-4.7x-62&=0\\-4.7x&=62\\x&=\frac{62}{-4.7}\\x&\approx-13.2\end{aligned}$$

105.
$$\begin{aligned}1.06p&=53\\100(1.06p)&=100(53)\\106p&=5300\\\frac{106p}{106}&=\frac{5300}{106}\\p&=50\end{aligned}$$
The price of the jeans is \$50.

107.
$$\begin{aligned}x+0.06x&=19,080\\1.06x&=19,080\\100(1.06x)&=100(19,080)\\106x&=1,908,000\\\frac{106x}{106}&=\frac{1,908,000}{106}\\x&=18,000\end{aligned}$$
The cost of the car before taxes was \$18,000.

109.
$$\begin{aligned}w+0.04w&=8.84\\1.04w&=8.84\\w&=\frac{8.84}{1.04}\\w&=8.5\end{aligned}$$
Bob's hourly wage was \$8.50 before the raise.

111.
$$\begin{aligned}p-0.25p&=60\\0.75p&=60\\p&=\frac{60}{0.75}\\p&=80\end{aligned}$$
The original price of the MP3 player was \$80.

113.
$$\begin{aligned}0.25q+0.10(2q+3)&=7.05\\100[0.25q+0.10(2q+3)]&=100(7.05)\\25q+10(2q+3)&=705\\25q+20q+30&=705\\45q+30&=705\\45q&=675\\q&=15\end{aligned}$$
There were 15 quarters in the piggy bank.

115. $2x+2(x+3)=2\left(\frac{1}{2}x\right)+2(x+6)$
$2x+2x+6=x+2x+12$
$4x+6=3x+12$
$x+6=12$
$x=6$
The first rectangle has width 6 units.

117. $1442.50=0.15(x-7300)+730$
$100(1442.50)=100[0.15(x-7300)+730]$
$144,250=15(x-7300)+73,000$
$144,250=15x-109,500+73,000$
$144,250=15x-36,500$
$180,750=15x$
$12,050=x$
Your adjusted gross income was $12,050.

119. A linear equation with one solution is $2x + 5 = 11$. A linear equation with no solution is $2x + 5 = 6 + 2x - 9$. A linear equation that is an identity is $2(x + 5) - 3 = 4x - (2x - 7)$. To form an identity or a contradiction, the variable expressions must be eliminated, leaving either a true (identity) or a false (contradiction) statement.

121. The student didn't multiply each term of the equation by 6. Solving using that (incorrect) method gives the second step $4x - 5 = 3x$, and solving for x gives $x = 5$. The correct method to solve the equation is to multiply ALL terms by 6. The correct second step is $4x - 30 = 3x$, producing the correct solution $x = 30$.

Section 2.4

Preparing for Evaluating Formulas and Solving Formulas for a Variable

P1. $2L + 2W$ for $L = 7$ and $W = 5$:
$2(7) + 2(5) = 14 + 10 = 24$

P2. In 0.5873, the number 8 is in the hundredths place. The number to the right of 8 is 7. Since 7 is greater than or equal to 5, we round 0.5873 to 0.59.

2.4 Quick Checks

1. A formula is an equation that describes how two or more variables are related.

2. $F=\frac{9}{5}C+32$
$F=\frac{9}{5}(15)+32$
$F=27+32$
$F=59$
The temperature is 59° Fahrenheit.

3. $c = a + 30$
$c = 10 + 30$
$c = 40$
A size 10 dress in the United States is a Continental dress size 40.

4. $E = 250 + 0.05S$
$E = 250 + 0.05(1250)$
$E = 250 + 62.5$
$E = 312.5$
The earnings of the salesman were $312.50.

5. $N = p + 0.06p$
$N = 5600 + 0.06(5600)$
$N = 5600 + 336$
$N = 5936$
The new population is 5936 persons.

6. The total amount borrowed in a loan is called principal. Interest is the money paid for the use of the money.

7. The amount Bill invested, P, is $2500. The interest rate, r, is 3% = 0.03. Because 8 months is $\frac{2}{3}$ of a year, $t=\frac{2}{3}$.
$I = Prt$
$I=2500\cdot 0.03\cdot\frac{2}{3}$
$I = 50$
Bill earned $50 on his investment. At the end of 8 months he had $2500 + $50 = $2550.

8. $A=\frac{1}{2}h(B+b)$
$A=\frac{1}{2}\cdot 4.5(9+7)$
$A=\frac{1}{2}\cdot 4.5\cdot 16$
$A=36$
The area of the trapezoid is 36 square inches.

9. The area of a circle is found using the formula $A=\pi r^2$.

10. **(a)** The radius of the circle is

$\frac{1}{2}(4 \text{ feet}) = 2 \text{ feet.}$

Area of remaining garden

$$\begin{aligned}&= \text{Area of rectangle} - \text{Area of circle}\\&= lw - \pi r^2\\&= (20 \text{ feet})(10 \text{ feet}) - \pi(2 \text{ feet})^2\\&= 200 \text{ feet}^2 - 4\pi \text{ feet}^2\\&\approx 187 \text{ feet}^2\end{aligned}$$

Approximately 187 square feet of garden will receive grass.

(b) Cost for sod

$$\begin{aligned}&= 187 \text{ square feet} \cdot \frac{\$0.25}{1 \text{ square foot}}\\&= \$46.75\end{aligned}$$

The sod will cost $46.75.

11. The radius of the pad is $\frac{1}{2}(6 \text{ feet}) = 3 \text{ feet.}$

$$\begin{aligned}\text{Area} &= \pi r^2\\&= \pi(3 \text{ feet})^2\\&= 9\pi \text{ feet}^2\\&\approx 28.27 \text{ feet}^2\end{aligned}$$

The area of the pad is about 28.27 square feet.

12.

$$\begin{aligned}\text{Area of 18" pizza} &= \pi r^2\\&= \pi(9 \text{ inches})^2\\&= 81\pi \text{ inches}^2\\&\approx 254.47 \text{ inches}^2\end{aligned}$$

$$\begin{aligned}\text{Area of 9" pizza} &= \pi r^2\\&= \pi(4.5 \text{ inches})^2\\&= 20.25\pi \text{ inches}^2\\&\approx 63.62 \text{ inches}^2\end{aligned}$$

Cost per square inch of 18" pizza:

$\frac{\$16.99}{254.47} \approx \$0.07$

Cost per square inch of 9" pizza:

$\frac{\$8.99}{63.62} \approx \$0.14$

The 18" pizza is the better buy.

13.

$$\begin{aligned}F &= \frac{9}{5}C + 32\\F - 32 &= \frac{9}{5}C + 32 - 32\\F - 32 &= \frac{9}{5}C\\\frac{5}{9}(F - 32) &= \frac{5}{9}\left(\frac{9}{5}C\right)\\\frac{5}{9}(F - 32) &= C \text{ or } C = \frac{5F - 160}{9}\end{aligned}$$

14.

$$\begin{aligned}S &= 2\pi rh + 2\pi r^2\\S - 2\pi r^2 &= 2\pi rh + 2\pi r^2 - 2\pi r^2\\S - 2\pi r^2 &= 2\pi rh\\\frac{S - 2\pi r^2}{2\pi r} &= \frac{2\pi rh}{2\pi r}\\\frac{S - 2\pi r^2}{2\pi r} &= h\end{aligned}$$

15. False: Solving $x - y = 6$ for y results in $y = x - 6$.

16.

$$\begin{aligned}x + 2y &= 7\\x + 2y - x &= 7 - x\\2y &= 7 - x\\\frac{2y}{2} &= \frac{7 - x}{2}\\y &= \frac{7 - x}{2}\\y &= \frac{7}{2} - \frac{x}{2}\end{aligned}$$

17.

$$\begin{aligned}5x - 3y &= 15\\5x - 3y - 5x &= 15 - 5x\\-3y &= 15 - 5x\\\frac{-3y}{-3} &= \frac{15 - 5x}{-3}\\y &= \frac{15 - 5x}{-3}\\y &= -5 + \frac{5}{3}x\end{aligned}$$

18. $\frac{3}{4}a+2b=7$

$\frac{3}{4}a+2b-\frac{3}{4}a=7-\frac{3}{4}a$

$2b=7-\frac{3}{4}a$

$\frac{1}{2}(2b)=\frac{1}{2}\left(7-\frac{3}{4}a\right)$

$b=\frac{7}{2}-\frac{3}{8}a$ or $b=\frac{28-3a}{8}$

19. $3rs+\frac{1}{2}t=12$

$3rs+\frac{1}{2}t-3rs=12-3rs$

$\frac{1}{2}t=12-3rs$

$2\left(\frac{1}{2}t\right)=2(12-3rs)$

$t=24-6rs$

20. (a) $d=rt$

$\frac{d}{r}=\frac{rt}{r}$

$\frac{d}{r}=t$

(b) Use $t=\frac{d}{r}$ with $d = 550$ and $r = 60$.

$t=\frac{d}{r}$

$t=\frac{550}{60}$

$t=\frac{55}{6}=9\frac{1}{6}$

It will take them $9\frac{1}{6}$ hours, or 9 hours and 10 minutes.

21. (a) $I=Prt$

$\frac{I}{Pr}=\frac{Prt}{Pr}$

$\frac{I}{Pr}=t$

(b) Use $t=\frac{I}{Pr}$ with $I = 35$, $P = 1000$, and $r=7\%=0.07$.

$t=\frac{I}{Pr}$

$t=\frac{35}{1000(0.07)}$

$t=\frac{35}{70}$

$t=\frac{1}{2}$

The $1000 must be invested for $\frac{1}{2}$ year, or 6 months, at 7% interest to earn $35 in interest.

22. (a) $R=qp$

$\frac{R}{q}=\frac{qp}{q}$

$\frac{R}{q}=p$

(b) Use $p=\frac{R}{q}$ with $R = 5000$ and $q = 125$.

$p=\frac{R}{q}$

$p=\frac{5000}{125}$

$p=40$

The price is $40.

2.4 Exercises

23. $f = 3.281m$
$m = 335$:
$f = 3.281(335) \approx 1099.14$
$m = 300$:
$f = 3.281(300) = 984.3$
The Rogun dam is about 1099.14 feet high, the Nurek dam is 984.3 feet high.

25. $S = P - 0.20P$; $P = 130$
$S = 130 - 0.20(130) = 130 - 26 = 104$
The sale price is $104.00.

27. $E = 500 + 0.15S$; $S = 1000$
$E = 500 + 0.15(1000) = 500 + 150 = 650$
The earnings are $650.

29. $C=\frac{5}{9}(F-32)$; $F=68$

$C=\frac{5}{9}(68-32)=\frac{5}{9}(36)=20$

The temperature is 20°C.

31. $I = Prt$; $P = 200$, $r = 3\% = 0.03$, $t = \frac{6}{12} = 0.5$

$I = 200(0.03)(0.5)$
$I = 3$
Therese's investment will earn \$3.

33. **(a)** $P = 2l + 2w$; $l = 16$, $w = 9$
$P = 2(16) + 2(9) = 32 + 18 = 50$
The perimeter is 50 units.

(b) $A = lw = 16 \cdot 9 = 144$
The area is 144 square units.

35. **(a)** $P = 2l + 2w$; $l = 12.5$, $w = 5.6$
$P = 2(12.5) + 2(5.6) = 25 + 11.2 = 36.2$
The perimeter is 36.2 meters.

(b) $A = lw = 12.5(5.6) = 70$
The area is 70 square meters.

37. **(a)** $P = 4s$; $s = 9$
$P = 4 \cdot 9 = 36$
The perimeter is 36 units.

(b) $A = s^2 = 9^2 = 81$
The area is 81 square units.

39. **(a)** $C = 2\pi r$; $r = 5$
$C = 2\pi(5) = 10\pi = 10(3.14) = 31.4$
The circumference is 31.4 centimeters.

(b) $A = \pi r^2 = 3.14(5)^2 = 3.14(25) = 78.5$
The area is 78.5 square centimeters.

41. $A = \pi r^2$; $r = \frac{14}{3}$

$$A = \pi\left(\frac{14}{3}\right)^2 = \frac{22}{7}\left(\frac{196}{9}\right) = \frac{616}{9} \approx 68.44$$

The area is 68.44 square inches.

43.
$$d = rt$$
$$\frac{d}{t} = \frac{rt}{t}$$
$$\frac{d}{t} = r$$

45.
$$C = \pi d$$
$$\frac{C}{\pi} = \frac{\pi d}{\pi}$$
$$\frac{C}{\pi} = d$$

47.
$$I = Prt$$
$$\frac{I}{Pr} = \frac{Prt}{Pr}$$
$$\frac{I}{Pr} = t$$

49.
$$A = \frac{1}{2}bh$$
$$\frac{2}{h}(A) = \frac{2}{h}\left(\frac{1}{2}bh\right)$$
$$\frac{2A}{h} = b$$

51.
$$P = a + b + c$$
$$P - b - c = a + b + c - b - c$$
$$P - b - c = a$$

53.
$$A = P + Prt$$
$$A - P = Prt$$
$$\frac{A - P}{Pt} = \frac{Prt}{Pt}$$
$$\frac{A - P}{Pt} = r$$

55.
$$A = \frac{1}{2}h(B + b)$$
$$2A = h(B + b)$$
$$\frac{2A}{h} = B + b$$
$$\frac{2A}{h} - B = b$$

57.
$$3x + y = 12$$
$$-3x + 3x + y = -3x + 12$$
$$y = -3x + 12$$

59.
$$10x - 5y = 25$$
$$-10x + 10x - 5y = -10x + 25$$
$$-5y = -10x + 25$$
$$\frac{-5y}{-5} = \frac{-10x + 25}{-5}$$
$$y = 2x - 5$$

61.
$$4x + 3y = 13$$
$$-4x + 4x + 3y = -4x + 13$$
$$3y = -4x + 13$$
$$\frac{3y}{3} = \frac{-4x + 13}{3}$$
$$y = \frac{-4x + 13}{3}$$

63.
$$\frac{1}{2}x-\frac{1}{6}y=2$$
$$6\left(\frac{1}{2}x-\frac{1}{6}y\right)=6\cdot 2$$
$$6\left(\frac{1}{2}x\right)-6\left(\frac{1}{6}y\right)=12$$
$$3x-y=12$$
$$-3x+3x-y=-3x+12$$
$$-y=-3x+12$$
$$-1\cdot(-y)=-1(-3x+12)$$
$$y=3x-12$$

65. (a)
$$P=R-C$$
$$-R+P=-R+R-C$$
$$-R+P=-C$$
$$\frac{-R+P}{-1}=\frac{-C}{-1}$$
$$R-P=C$$

(b) $C=R-P$; $P = 1200$, $R = 1650$
$C = 1650 - 1200 = 450$
The cost is \$450.

67. (a)
$$I = Prt$$
$$\frac{I}{Pt}=\frac{Prt}{Pt}$$
$$\frac{I}{Pt}=r$$

(b) $r=\frac{I}{Pt}$; $I = 225$, $P = 5000$, $t = 1.5$
$$r=\frac{225}{5000(1.5)}=\frac{225}{7500}=0.03$$
The rate is 0.03 or 3%.

69. (a)
$$Z=\frac{x-\mu}{\sigma}$$
$$Z\cdot\sigma=\left(\frac{x-\mu}{\sigma}\right)\cdot\sigma$$
$$Z\sigma=x-\mu$$
$$Z\sigma+\mu=x-\mu+\mu$$
$$Z\sigma+\mu=x$$

(b) $x=Z\sigma+\mu$; $Z = 2$, $\mu = 100$, $\sigma = 15$
$x = 2(15) + 100 = 30 + 100 = 130$

71. (a)
$$y=mx+5$$
$$y-5=mx+5-5$$
$$y-5=mx$$
$$\frac{y-5}{x}=\frac{mx}{x}$$
$$\frac{y-5}{x}=m$$

(b) $m=\frac{y-5}{x}$; $x = 3$, $y = -1$
$$m=\frac{-1-5}{3}=\frac{-6}{3}=-2$$

73. (a)
$$A=P+Prt$$
$$A-P=P+Prt-P$$
$$A-P=Prt$$
$$\frac{A-P}{Pt}=\frac{Prt}{Pt}$$
$$\frac{A-P}{Pt}=r$$

(b) $r=\frac{A-P}{Pt}$; $A = 540$, $P = 500$, $t = 2$
$$r=\frac{540-500}{500(2)}=\frac{40}{1000}=0.04$$
The rate is 0.04 or 4%.

75. (a)
$$V=\pi r^2h$$
$$\frac{V}{\pi r^2}=\frac{\pi r^2h}{\pi r^2}$$
$$\frac{V}{\pi r^2}=h$$

(b) $h=\frac{V}{\pi r^2}$; $V = 320\pi$, $r = 8$
$$h=\frac{320\pi}{\pi(8)^2}=\frac{320\pi}{64\pi}=5$$
The height is 5 mm.

77. (a)
$$A=\frac{1}{2}bh$$
$$\frac{2}{h}\cdot A=\frac{2}{h}\cdot\frac{1}{2}bh$$
$$\frac{2A}{h}=b$$

(b) $b = \dfrac{2A}{h}$; $A = 45, h = 5$

$b = \dfrac{2(45)}{5} = \dfrac{90}{5} = 18$

The base is 18 feet.

79. $A = 37, H = 178, W = 82$

$$\begin{aligned} E &= 66.67 + 13.75W + 5H - 6.76A \\ &= 66.67 + 13.75(82) + 5(178) - 6.76(37) \\ &= 66.67 + 1127.5 + 890 - 250.12 \\ &= 1834.05 \end{aligned}$$

The Basal Energy Expenditure is 1834.05.

81. (a) $V = \pi r^2 h$

$\dfrac{V}{\pi r^2} = \dfrac{\pi r^2 h}{\pi r^2}$

$\dfrac{V}{\pi r^2} = h$

(b) $h = \dfrac{V}{\pi r^2}$; $V = 90\pi, r = 3$

$h = \dfrac{90\pi}{\pi(3)^2} = \dfrac{90}{9} = 10$

The height is 10 inches.

83.

Size	Area: $A = \pi r^2$	Price per square inch
12"	$A = \pi \cdot 6^2 \approx 113.097$	$\frac{\$9.99}{113.097} = \$0.088$
8"	$A = \pi \cdot 4^2 \approx 50.265$	$\frac{\$4.49}{50.265} = \$0.089$

Since $0.088 < $0.089, the medium pizza is a better deal.

85. (a) $d = rt$

$\dfrac{d}{r} = \dfrac{rt}{r}$

$\dfrac{d}{r} = t$

Let $d = 600$, $r = 50$.

$t = \dfrac{600}{50} = 12$

Jason expects the trip to take 12 hours.

(b) $12(28) = 336$

Jason can expect $336.

87. Region = Rectangle + Trapezoid

$$\begin{aligned}\text{Area} &= lw+\frac{1}{2}h(B+b)\\ &= 8(5)+\frac{1}{2}(3)(5+2)\\ &= 40+\frac{21}{2}\\ &= 50.5\end{aligned}$$

The area is 50.5 square inches.

89. $\text{figure} = \text{cone}+\frac{1}{2}\cdot\text{sphere}$

$$\begin{aligned}\text{Volume} &= \frac{1}{3}\pi r^2 h+\frac{1}{2}\cdot\frac{4}{3}\pi r^3\\ &= \frac{1}{3}\pi(4)^2(10)+\frac{2}{3}\pi(4)^3\\ &= \frac{160}{3}\pi+\frac{128}{3}\pi\\ &= \frac{288}{3}\pi\\ &= 96\pi\end{aligned}$$

The amount of ice cream is $96\pi \approx 301.59$ cubic cm.

91. (a)

$$\begin{aligned}P &= D-0.02(I-234{,}600)\\ -D+P &= -D+D-0.02(I-234{,}600)\\ -D+P &= -0.02(I-234{,}600)\\ \frac{-D+P}{-0.02} &= \frac{-0.02(I-234{,}600)}{-0.02}\\ \frac{D-P}{0.02} &= I-234{,}600\\ \frac{D-P}{0.02}+234{,}600 &= I-234{,}600+234{,}600\\ \frac{D-P}{0.02}+234{,}600 &= I\end{aligned}$$

(b)

$$\begin{aligned}I &= \frac{D-P}{0.02}+234{,}600\\ &= \frac{15{,}821-15{,}500}{0.02}+234{,}600\\ &= \frac{321}{0.02}+234{,}600\\ &= 16{,}050+234{,}600\\ &= 250{,}650\end{aligned}$$

The adjusted gross income is \$250,650.

93. (a) 7 feet 6 inches $= 7\frac{1}{2}$ feet

8 feet 2 inches $= 8\frac{1}{6}$ feet

$$A = lw = \left(7\frac{1}{2}\right)\left(8\frac{1}{6}\right) = \left(\frac{15}{2}\right)\left(\frac{49}{6}\right) = \frac{245}{4} = 61\frac{1}{4}$$

You need 62 tiles to cover the floor.

(b) $\$6(62) = \$372$
It will cost \$372.

(c) Yes; $\$372 > \$350$

95. (a) Region = Rectangle − Circle
$\text{Area} = lw - \pi r^2 \approx (90)(60) - 3.14159(12)^2 = 5400 - 452.38896 = 4947.61104$
The grass area is 4948 square feet.

(b) $4948(0.25) = 1237$
It will cost \$1237 to sod the lawn.

97. (a) $l = 5 \text{ ft} = 5 \times 12 \text{ in.} = 60 \text{ in.}$; $w = 18$ in.
$A = lw = 60(18) = 1080$
The area is 1080 square inches.

(b) $l = 5$ ft; $w = 18 \text{ in.} = \frac{18}{12} \text{ ft} = 1.5 \text{ ft}$
$A = lw = 5(1.5) = 7.5$
The area is 7.5 square feet.

99. To convert from square inches to square feet, multiply square inches by $\frac{1 \text{ ft}^2}{144 \text{ in.}^2}$.

101. Answers may vary. One possibility: Both are correct. The first answer can be expanded into the second answer.

Putting the Concepts Together (Sections 2.1–2.4)

1. (a) $4 - (6 - x) = 5x - 8$; $x = \frac{3}{2}$

$$4 - \left(6 - \frac{3}{2}\right) \stackrel{?}{=} 5\left(\frac{3}{2}\right) - 8$$
$$4 - \left(\frac{12}{2} - \frac{3}{2}\right) \stackrel{?}{=} \frac{15}{2} - 8$$
$$\frac{8}{2} - \frac{9}{2} \stackrel{?}{=} \frac{15}{2} - \frac{16}{2}$$
$$-\frac{1}{2} = -\frac{1}{2} \quad \text{True}$$

Yes, $x = \frac{3}{2}$ is a solution of the equation.

(b) $4-(6-x)=5x-8;\ x=-\frac{5}{2}$

$$4-\left[6-\left(-\frac{5}{2}\right)\right] \stackrel{?}{=} 5\left(-\frac{5}{2}\right)-8$$
$$4-\left(\frac{12}{2}+\frac{5}{2}\right) \stackrel{?}{=} \frac{-25}{2}-8$$
$$\frac{8}{2}-\frac{17}{2} \stackrel{?}{=} -\frac{25}{2}-\frac{16}{2}$$
$$-\frac{9}{2}=-\frac{41}{2} \quad \text{False}$$

No, $x=-\frac{5}{2}$ is *not* a solution of the equation.

2. (a) $\frac{1}{2}(x-4)+3x=x+\frac{1}{2};\ x=-4$

$$\frac{1}{2}(-4-4)+3(-4) \stackrel{?}{=} -4+\frac{1}{2}$$
$$\frac{1}{2}(-8)+(-12) \stackrel{?}{=} -\frac{8}{2}+\frac{1}{2}$$
$$-4+(-12) \stackrel{?}{=} -\frac{7}{2}$$
$$-16=-\frac{7}{2} \quad \text{False}$$

No, $x=-4$ is *not* a solution of the equation.

(b) $\frac{1}{2}(x-4)+3x=x+\frac{1}{2};\ x=1$

$$\frac{1}{2}(1-4)+3(1) \stackrel{?}{=} 1+\frac{1}{2}$$
$$\frac{1}{2}(-3)+3 \stackrel{?}{=} \frac{2}{2}+\frac{1}{2}$$
$$-\frac{3}{2}+\frac{6}{2} \stackrel{?}{=} \frac{3}{2}$$
$$\frac{3}{2}=\frac{3}{2} \quad \text{True}$$

Yes, $x=1$ is a solution of the equation.

3.
$$x+\frac{1}{2}=-\frac{1}{6}$$
$$6\left(x+\frac{1}{2}\right)=6\left(-\frac{1}{6}\right)$$
$$6x+3=-1$$
$$6x+3-3=-1-3$$
$$6x=-4$$
$$\frac{6x}{6}=\frac{-4}{6}$$
$$x=-\frac{2}{3}$$

Check:
$$x+\frac{1}{2}=-\frac{1}{6}$$
$$-\frac{2}{3}+\frac{1}{2} \stackrel{?}{=} -\frac{1}{6}$$
$$-\frac{4}{6}+\frac{3}{6} \stackrel{?}{=} -\frac{1}{6}$$
$$-\frac{1}{6}=-\frac{1}{6} \quad \text{True}$$

The solution is $-\frac{2}{3}$ or the solution set is $\left\{-\frac{2}{3}\right\}$.

4.
$$-0.4m=16$$
$$10(-0.4m)=10(16)$$
$$-4m=160$$
$$\frac{-4m}{-4}=\frac{160}{-4}$$
$$m=-40$$

Check:
$$-0.4m=16$$
$$-0.4(-40) \stackrel{?}{=} 16$$
$$16=16 \quad \text{True}$$

The solution is −40, or the solution set is {−40}.

5.
$$14=-\frac{7}{3}p$$
$$-\frac{3}{7}\cdot 14=-\frac{3}{7}\cdot\left(-\frac{7}{3}p\right)$$
$$-3\cdot 2=p$$
$$-6=p$$

Check:
$$14=-\frac{7}{3}p$$
$$14 \stackrel{?}{=} -\frac{7}{3}(-6)$$
$$14=14 \quad \text{True}$$

The solution is −6, or the solution set is {−6}.

6.
$$8n-11=13$$
$$8n-11+11=13+11$$
$$8n=24$$
$$\frac{8n}{8}=\frac{24}{8}$$
$$n=3$$

Check:
$$8n-11=13$$
$$8(3)-11 \stackrel{?}{=} 13$$
$$24-11 \stackrel{?}{=} 13$$
$$13=13 \quad \text{True}$$

The solution is 3, or the solution set is {3}.

7.
$$\frac{5}{2}n-4=-19$$
$$\frac{5}{2}n-4+4=-19+4$$
$$\frac{5}{2}n=-15$$
$$\frac{2}{5}\cdot\frac{5}{2}n=\frac{2}{5}\cdot(-15)$$
$$n=-6$$

Check: $\frac{5}{2}n-4=-19$
$$\frac{5}{2}\cdot(-6)-4\stackrel{?}{=}-19$$
$$-15-4\stackrel{?}{=}-19$$
$$-19=-19 \quad \text{True}$$

The solution is –6, or the solution set is {–6}.

8.
$$-(5-x)=2(5x+8)$$
$$-5+x=10x+16$$
$$-16-5+x=10x+16-16$$
$$-21+x=10x$$
$$-21+x-x=10x-x$$
$$-21=9x$$
$$\frac{-21}{9}=\frac{9x}{9}$$
$$-\frac{7}{3}=x$$

Check: $-(5-x)=2(5x+8)$
$$-\left[5-\left(-\frac{7}{3}\right)\right]\stackrel{?}{=}2\left[5\left(-\frac{7}{3}\right)+8\right]$$
$$-\left(\frac{15}{3}+\frac{7}{3}\right)\stackrel{?}{=}2\left(-\frac{35}{3}+\frac{24}{3}\right)$$
$$-\frac{22}{3}\stackrel{?}{=}2\left(-\frac{11}{3}\right)$$
$$-\frac{22}{3}=-\frac{22}{3} \quad \text{True}$$

The solution is $-\frac{7}{3}$, or the solution set is $\left\{-\frac{7}{3}\right\}$.

9.
$$7(x+6)=2x+3x-15$$
$$7x+42=5x-15$$
$$7x+42-42=5x-15-42$$
$$7x=5x-57$$
$$-5x+7x=-5x+5x-57$$
$$2x=-57$$
$$\frac{2x}{2}=\frac{-57}{2}$$
$$x=-\frac{57}{2}$$

Check: $7(x+6)=2x+3x-15$
$$7\left(-\frac{57}{2}+6\right)\stackrel{?}{=}2\left(-\frac{57}{2}\right)+3\left(-\frac{57}{2}\right)-15$$
$$7\left(-\frac{45}{2}\right)\stackrel{?}{=}-\frac{114}{2}-\frac{171}{2}-\frac{30}{2}$$
$$-\frac{315}{2}=-\frac{315}{2} \quad \text{True}$$

The solution is $-\frac{57}{2}$, or the solution set is $\left\{-\frac{57}{2}\right\}$.

10.
$$-7a+5+8a=2a+8-28$$
$$5+a=2a-20$$
$$5+a+20=2a-20+20$$
$$25+a=2a$$
$$25+a-a=2a-a$$
$$25=a$$

Check: $-7a+5+8a=2a+8-28$
$$-7(25)+5+8(25)\stackrel{?}{=}2(25)+8-28$$
$$-175+5+200\stackrel{?}{=}50+8-28$$
$$30=30 \quad \text{True}$$

The solution is 25, or the solution set is {25}.

11.
$$-\frac{1}{2}(x-6)+\frac{1}{6}(x+6)=2$$
$$6\left[-\frac{1}{2}(x-6)+\frac{1}{6}(x+6)\right]=6\cdot 2$$
$$-3(x-6)+1(x+6)=12$$
$$-3x+18+x+6=12$$
$$-2x+24=12$$
$$-2x+24-24=12-24$$
$$-2x=-12$$
$$\frac{-2x}{-2}=\frac{-12}{-2}$$
$$x=6$$

Check: $-\frac{1}{2}(x-6)+\frac{1}{6}(x+6)=2$

$$-\frac{1}{2}(6-6)+\frac{1}{6}(6+6) \stackrel{?}{=} 2$$
$$-\frac{1}{2}(0)+\frac{1}{6}(12) \stackrel{?}{=} 2$$
$$0+2 \stackrel{?}{=} 2$$
$$2=2 \text{ True}$$

The solution is 6, or the solution set is {6}.

12.
$$0.3x-1.4=-0.2x+6$$
$$10(0.3x-1.4)=10(-0.2x+6)$$
$$3x-14=-2x+60$$
$$3x-14+14=-2x+60+14$$
$$3x=-2x+74$$
$$2x+3x=2x-2x+74$$
$$5x=74$$
$$\frac{5x}{5}=\frac{74}{5}$$
$$x=14.8$$

Check: $0.3x-1.4=-0.2x+6$
$$0.3(14.8)-1.4 \stackrel{?}{=} -0.2(14.8)+6$$
$$4.44-1.4 \stackrel{?}{=} -2.96+6$$
$$3.04=3.04 \text{ True}$$

The solution is $\frac{74}{5}=14.8$, or the solution set is {14.8}.

13. $5+3(2x+1)=5x+x-10$
$$5+6x+3=5x+x-10$$
$$8+6x=6x-10$$
$$8+6x-6x=6x-10-6x$$
$$8=-10$$

This is a false statement, so the equation is a contradiction. The solution set is ∅ or { }.

14. $3-2(x+5)=-2(x+2)-3$
$$3-2x-10=-2x-4-3$$
$$-2x-7=-2x-7$$
$$2x-2x-7=2x-2x-7$$
$$-7=-7$$

This is a true statement. The equation is an identity. The solution set is the set of all real numbers.

15. $0.024x+0.04(7500-x)=220$
$$0.024x+300-0.04x=220$$
$$300-0.016x=220$$
$$-300+300-0.016x=-300+220$$
$$-0.016x=-80$$
$$1000(-0.016x)=1000(-80)$$
$$-16x=-80,000$$
$$x=5000$$

You should invest $5000 in the CD.

16. (a)
$$A=\frac{1}{2}h(B+b)$$
$$\frac{2}{h}\cdot A=\frac{2}{h}\cdot\frac{1}{2}h(B+b)$$
$$\frac{2A}{h}=B+b$$
$$\frac{2A}{h}-B=B+b-B$$
$$\frac{2A}{h}-B=b$$

(b) $b=\frac{2A}{h}-B$; $A=76$, $h=8$, $B=13$

$$b=\frac{2(76)}{8}-13=\frac{152}{8}-13=19-13=6$$

The length of base b is 6 inches.

17. (a)
$$V=\pi r^2h$$
$$\frac{V}{\pi r^2}=\frac{\pi r^2h}{\pi r^2}$$
$$\frac{V}{\pi r^2}=h$$

(b) $h=\frac{V}{\pi r^2}$; $V=117\pi$, $r=3$

$$h=\frac{117\pi}{\pi(3)^2}=\frac{117\pi}{9\pi}=13$$

The height is 13 inches.

18.
$$3x+2y=14$$
$$-3x+3x+2y=-3x+14$$
$$2y=-3x+14$$
$$\frac{2y}{2}=\frac{-3x+14}{2}$$
$$y=-\frac{3}{2}x+7$$

Section 2.5

Preparing for Introduction to Problem Solving: Direct Translation Problems

P1.
$$x + 34.95 = 60.03$$
$$x + 34.95 - 34.95 = 60.03 - 34.95$$
$$x = 25.08$$
The solution set is $\{25.08\}$.

P2.
$$x + 0.25x = 60$$
$$1x + 0.25x = 60$$
$$1.25x = 60$$
$$100 \cdot 1.25x = 100 \cdot 60$$
$$125x = 6000$$
$$\frac{125x}{125} = \frac{6000}{125}$$
$$x = 48$$
The solution set is $\{48\}$.

2.5 Quick Checks

1. The sum of 5 and 17 is represented mathematically as $5 + 17$.

2. The product of -2 and 6 is represented mathematically as $-2 \cdot 6$.

3. The quotient of 25 and 3 is represented mathematically as $\frac{25}{3}$.

4. The difference of 7 and 4 is represented mathematically as $7 - 4$.

5. Twice a less 2 is represented mathematically as $2a - 2$.

6. Three plus the quotient of z and 4 is represented mathematically as $3 + \frac{z}{4}$.

7. Anne earned $z + 50$ dollars.

8. Melissa paid $x - 15$ dollars for her sociology book.

9. Tim has $75 - d$ quarters.

10. The width of the platform is $3l - 2$ feet.

11. The number of dimes is $2q + 3$.

12. The number of red M&Ms in the bowl is $3b - 5$.

13. To translate an English sentence into a mathematical statement we use equations.

14. "The product of 3 and y is equal to 21" is represented mathematically as $3y = 21$.

15. "The sum of 3 and x is equivalent to the product of 5 and x" is represented mathematically as $3 + x = 5x$.

16. "The difference of x and 10 equals the quotient of x and 2" is represented mathematically as $x - 10 = \frac{x}{2}$.

17. "Three less than a number y is five times y" is represented mathematically as $y - 3 = 5y$.

18. Letting variables represent unknown quantities and then expressing relationships among the variables in the form of equations is called mathematical modeling.

19. We want to know how much each person paid for the pizza. Let s be the amount that Sean pays. Then Connor pays $\frac{2}{3}s$. The total amount they pay is \$15, so $s + \frac{2}{3}s = 15$.
$$s + \frac{2}{3}s = 15$$
$$\frac{5}{3}s = 15$$
$$s = \frac{3}{5}(15)$$
$$s = 9$$
$$\frac{2}{3}s = \frac{2}{3}(9) = 6$$
Sean pays \$9 for the pizza, and Connor pays \$6.

20. False: If n represents the first of three consecutive odd integers, then $n + 2$ and $n + 4$ represent the next two odd integers.

21. We want to find three consecutive even integers that sum to 270. Let n be the first even integer. Then the next even integer is $n + 2$, and the even integer after that is $n + 4$.

$$\begin{aligned} n+(n+2)+(n+4) &= 270 \\ n+n+2+n+4 &= 270 \\ 3n+6 &= 270 \\ 3n &= 264 \\ n &= 88 \end{aligned}$$

$n + 2 = 88 + 2 = 90$
$n + 4 = 88 + 4 = 92$
The integers are 88, 90, and 92.

22. We want to find four consecutive odd integers that sum to 72. Let n be the first odd integers. Then the next three odd integers are $n + 2$, $n + 4$, and $n + 6$.

$$\begin{aligned} n+(n+2)+(n+4)+(n+6) &= 72 \\ 4n+12 &= 72 \\ 4n+12-12 &= 72-12 \\ 4n &= 60 \\ \frac{4n}{4} &= \frac{60}{4} \\ n &= 15 \end{aligned}$$

The integers are 15, 17, 19, and 21.

23. We want to find the lengths of the three pieces of ribbon. Let x be the length of the shortest piece. Then the length of the longest piece is $x + 24$, and the length of the third piece is $\frac{1}{2}(x+24)$.
The sum of the lengths of the pieces is the length of the original ribbon, 76 inches.

$$\begin{aligned} x+(x+24)+\frac{1}{2}(x+24) &= 76 \\ x+x+24+\frac{1}{2}x+12 &= 76 \\ 2\left(x+x+24+\frac{1}{2}x+12\right) &= 2(76) \\ 2x+2x+48+x+24 &= 152 \\ 5x+72 &= 152 \\ 5x &= 80 \\ x &= 16 \end{aligned}$$

$x + 24 = 16 + 24 = 40$

$\frac{1}{2}(x+24) = \frac{1}{2}(16+24) = \frac{1}{2}(40) = 20$

The lengths of the pieces are 16 inches, 20 inches, and 40 inches.

24. We want to know the amount invested in each type of investment. Let s be the amount invested in stocks. Then the amount invested in bonds is $2s$. The total amount invested is $18,000.

$$\begin{aligned} s+2s &= 18,000 \\ 3s &= 18,000 \\ s &= 6000 \end{aligned}$$

$2s = 2(6000) = 12,000$
The amount invested in stocks is $6000, and the amount invested in bonds is $12,000.

25. We are looking for the number of miles for which the rental costs are the same. Let m be the number of miles driven. Renting from E-Z Rental would cost $30 + 0.15m$. Renting from Do It Yourself Rental would cost $15 + 0.25m$.

$$\begin{aligned} 30+0.15m &= 15+0.25m \\ 30 &= 15+0.10m \\ 15 &= 0.1m \\ 10\cdot 15 &= 10\cdot 0.1m \\ 150 &= m \end{aligned}$$

The costs are the same if 150 miles are driven.

26. We are looking for the number of minutes for which the monthly costs of the plans will be the same. Let m be the number of minutes. The monthly cost of plan A is $15 + 0.05m$, and the monthly cost of plan B is $0.20m$.

$$\begin{aligned} 15+0.05m &= 0.20m \\ 15 &= 0.15m \\ 100 &= m \end{aligned}$$

The monthly costs are the same for 100 minutes.

2.5 Exercises

27. The sum of –5 and a number: $-5 + x$

29. The product of a number and $\frac{2}{3}$: $x\left(\frac{2}{3}\right)$ or $\frac{2}{3}x$

31. Half of a number: $\frac{1}{2}x$

33. A number less –25: $x - (-25)$

35. The quotient of a number and 3: $\frac{x}{3}$

37. $\frac{1}{2}$ more than a number: $x+\frac{1}{2}$

39. 9 more than 6 times a number: $6x + 9$

41. Twice the sum of 13.7 and a number: $2(13.7 + x)$

43. The sum of twice a number and 31: $2x + 31$

45. Let r be the number of runs scored by the Richmond Braves. Then $r + 5$ is the number of runs scored by the Columbus Clippers.

47. Let b be the amount Bill has in his bank. Then $b + 0.55$ is the amount Jan has in her bank.

49. Let j be the amount Janet will get. Then $200 - j$ is the amount Kathy will get.

51. Let a be the number of adults who visited the show. Then $1433 - a$ is the number of children who visited the show.

53. $x + 15 = -34$

55. $35 = 3x - 7$

57. $\frac{x}{-4} + 5 = 36$

59. $2(x + 6) = x + 3$

61. We want to find the number. Let n be the number.
$$n + (-12) = 71$$
$$n - 12 = 71$$
$$n = 83$$
Is the sum of 83 and −12 equal to 71? Yes. The number is 83.

63. We want to find the number. Let n be the number.
$$2n - 25 = -53$$
$$2n = -28$$
$$n = -14$$
Is 25 less than twice −14 equal to −53? Yes. The number is −14.

65. We want to find three consecutive integers that sum to 165. Let n be the first integer. Then the other integers are $n + 1$ and $n + 2$, respectively.
$$n + (n+1) + (n+2) = 165$$
$$n + n + 1 + n + 2 = 165$$
$$3n + 3 = 165$$
$$3n = 162$$
$$n = 54$$
If $n = 54$, then $n + 1 = 55$, and $n + 2 = 56$. Are 54, 55, and 56 consecutive integers? Yes. Do they sum to 165? Yes. The numbers are 54, 55 and 56.

67. We want to find the lengths of the bridges. If x is the length of the Verrazano-Narrows Bridge, then the length of the Golden Gate Bridge is $x - 60$.
$$x + (x - 60) = 8460$$
$$2x - 60 = 8460$$
$$2x = 8520$$
$$x = 4260$$
If $x = 4260$, then $x - 60 = 4200$. Is 4200 equal to 60 less than 4260? Yes. Do 4260 and 4200 sum to 8460? Yes.
The Verrazano-Narrows Bridge is 4260 feet long and the Golden Gate Bridge is 4200 feet long.

69. We want to find the price of the motorcycle before the extra charges. Let m be the price of the motorcycle before the extra charges. Then the total price is $m + 679.79$, which is also 11,894.79.
$$m + 679.79 = 11{,}894.79$$
$$m = 11{,}215$$
Is 11,215 plus 679.79 equal to 11,894.79? Yes. The price of the motorcycle before the extra charges is $11,215.

71. We want to find the amount invested in each type of investment. Let x be the amount invested in CDs. Then $x + 3000$ is the amount invested in bonds. The total invested is 20,000.
$$x + (x + 3000) = 20{,}000$$
$$2x + 3000 = 20{,}000$$
$$2x = 17{,}000$$
$$x = 8500$$
If $x = 8500$, $x + 3000 = 11{,}500$. Is 11,500 3000 greater than 8500? Yes. Is the sum of 8500 and 11,500 equal to 20,000? Yes. The amount invested in CDs is $8500, the amount invested in bonds is $11,500.

73. We want to find the amount invested in each type of investment. Let x be the amount in stocks. Then $\frac{3}{5}x$ is the amount invested in bonds. The total invested in 32,000.
$$x + \frac{3}{5}x = 32{,}000$$
$$\frac{5}{5}x + \frac{3}{5}x = 32{,}000$$
$$\frac{8}{5}x = 32{,}000$$
$$\frac{5}{8} \cdot \frac{8}{5}x = \frac{5}{8} \cdot 32{,}000$$
$$x = 20{,}000$$
$$\frac{3}{5}x = 12{,}000$$

Is 12,000 $\frac{3}{5}$ of 20,000. Yes. Is the sum of 20,000 and 12,000 equal to 32,000? Yes. You should invest $20,000 in stocks and $12,000 in bonds.

75. We want to find the amount of dietary fiber in each cereal. Let x be the amount of dietary fiber in Kellogg's Smart Start Cereal. Then $4x$ is the amount of dietary fiber in Kashi Go Lean Crunch.

$$\begin{aligned} x+4x &= 10 \\ 5x &= 10 \\ x &= 2 \end{aligned}$$

If $x = 2$, then $4x = 8$. Is 8 four times 2? Yes. Is the sum of 8 and 2 equal to 10? Yes.
There are 2 grams of dietary fiber in Kellogg's Smart Start Cereal, and 8 grams in Kashi Go Lean Crunch.

77. We want to find Elizabeth Morrell's adjusted gross income (AGI). Let x be her AGI. Then her husband's AGI was
$x - 2549$.

$$\begin{aligned} x+(x-2549) &= 55,731 \\ 2x-2549 &= 55,731 \\ 2x &= 58,280 \\ x &= 29,140 \end{aligned}$$

If $x = 29,140$, then $x - 2549 = 26,591$. Do 29,140 and 26,591 sum to 55,731? Yes. Elizabeth Morrell's adjusted gross income was $29,140.

79. We want to find the number of miles for which the companies charge the same amount. Let x be the number of miles. Then EZ-Rental charges $35 + 0.15x$ and Do It Yourself Rental charges $20 + 0.25x$.

$$\begin{aligned} 35+0.15x &= 20+0.25x \\ 35 &= 20+0.10x \\ 15 &= 0.10x \\ 150 &= x \end{aligned}$$

If 150 miles are driven, EZ-Rental charges $35 + 0.15(150) = 35 + 22.5 = \$57.50$ and Do It Yourself Rental charges
$20 + 0.25(150) = 20 + 37.5 = \$57.50$, so the charges are the same. The cost of renting is the same when 150 miles are driven.

81. We want to find the number of pages for which the cost is the same. Let x be the number of pages. Then Hewlett-Packard costs $200 + 0.03x$ and the Brother costs $240 + 0.01x$.

$$\begin{aligned} 200+0.03x &= 240+0.01x \\ 200+0.02x &= 240 \\ 0.02x &= 40 \\ x &= 2000 \end{aligned}$$

If 2000 pages are printed, the Hewlett-Packard costs $200 + 0.03(2000) = 200 + 60 = \$260$ and the Brother costs
$240 + 0.01(2000) = 240 + 20 = \$260$, so the costs are the same. The cost is the same for 2000 pages.

83. We want to find the adjusted gross incomes (AGIs). Let j be Jenson Beck's AGI. Then $j - 249$ is Maureen Beck's AGI, since Jensen Beck's AGI is 249 more than Maureen's.

$$\begin{aligned} j+(j-249) &= 72,193 \\ 2j-249 &= 72,193 \\ 2j &= 72,442 \\ j &= 36,221 \end{aligned}$$

If $j = 36,221$, then $j - 249 = 35,972$. Is 36,221 less 249 equal to 35,972? Yes. Do 36,221 and 35,972 sum to 72,193? Yes. Jensen Beck's adjusted gross income was $36,221, Maureen Beck's was $35,972.

85. We want to find the number of runs scored in each game. There are 4 numbers since there is a losing number of runs and a winning number of runs for each game. Let n be the losing number of runs in the first game. Since the scores were consecutive integers, the numbers of runs scored were n,
$n + 1$, $n + 2$, and $n + 3$.

$$\begin{aligned} n+(n+1)+(n+2)+(n+3) &= 26 \\ 4n+6 &= 26 \\ 4n &= 20 \\ n &= 5 \end{aligned}$$

If $n = 5$, then $n + 1 = 6$, $n + 2 = 7$, and $n + 3 = 8$.
Are 5, 6, 7, and 8 consecutive integers that sum to 26? Yes. The number of runs scored were 5, 6, 7, and 8.

87. Answers may vary.

89. Answers may vary.

91. We want to find the measure of each angle. Let x be the measure of the second angle. Then the smallest angle has measure $\frac{1}{2}x$ and the third angle has measure $4\left(\frac{1}{2}x\right)+40$. Their measures sum to 180.

$$x+\left(\frac{1}{2}x\right)+\left[4\left(\frac{1}{2}x\right)+40\right]=180$$
$$x+\frac{1}{2}x+2x+40=180$$
$$\frac{2}{2}x+\frac{1}{2}x+\frac{4}{2}x=140$$
$$\frac{7}{2}x=140$$
$$x=40$$

If $x = 40$, then $\frac{1}{2}x = 20$ and $4\left(\frac{1}{2}x\right)+40=120.$

Is 20 $\frac{1}{2}$ of 40? Yes. Is 120 40 more than 4 times 20? Yes. Do 40, 20, and 120 sum to 180? Yes. The angles measure 20°, 40°, and 120°.

93. The process of taking a verbal description of a problem and developing a mathematical equation that can be used to solve the problem is mathematical modeling. We often make assumptions to make the mathematics more manageable in the mathematical models.

95. Both students are correct, and the value for n will be the same for both students. Answers may vary.

Section 2.6

Preparing for Problem Solving: Direct Translation Problems Involving Percent

P1. $45\% = 45\% \cdot \frac{1}{100\%} = \frac{45}{100} = 0.45$

P2. $0.2875 = 0.2875 \cdot \frac{100\%}{1} = 28.75\%$

2.6 Quick Checks

1. Percent means "divided by <u>100</u>."

2. The word "of" translates into <u>multiplication</u> in mathematics, so 40% of 120 means 0.40 <u>times</u> 120.

3. We want to know the unknown number. Let n represent the number.
$n = 0.89 \cdot 900$
$n = 801$
801 is 89% of 900.

4. We want to know the unknown number. Let n represent the number.
$n = 0.035 \cdot 72$
$n = 2.52$
2.52 is 3.5% of 72.

5. We want to know the unknown number. Let n represent the number.
$n = 1.50 \cdot 24$
$n = 36$
36 is 150% of 24.

6. We want to know the unknown number. Let n represent the number.
$8\frac{3}{4}\% = 8.75\% = 0.0875$
$n = 0.0875 \cdot 40$
$n = 3.5$
3.5 is $8\frac{3}{4}\%$ of 40.

7. We want to know the percentage. Let x represent the percent.
$8 = x \cdot 20$
$\frac{8}{20} = \frac{20x}{20}$
$0.4 = x$
$40\% = x$
The number 8 is 40% of 20.

8. We want to know the percentage. Let x represent the percent.
$15 = x \cdot 40$
$\frac{15}{40} = \frac{40x}{40}$
$0.375 = x$
$37.5\% = x$
The number 15 is 37.5% of 40.

9. We want to know the percentage. Let x represent the percent.
$12.3 = x \cdot 60$
$\frac{12.3}{60} = \frac{60x}{60}$
$0.205 = x$
$20.5\% = x$
The number 12.3 is 20.5% of 60.

10. We want to know the percentage. Let x represent the percent.

$$44 = x \cdot 40$$
$$\frac{44}{40} = \frac{40x}{40}$$
$$1.10 = x$$
$$110\% = x$$

The number 44 is 110% of 40.

11. We want to know a number. Let x represent the number.

$$14 = 0.28 \cdot x$$
$$\frac{14}{0.28} = \frac{0.28x}{0.28}$$
$$50 = x$$

14 is 28% of 50.

12. We want to know a number. Let x represent the number.

$$111 = 0.74 \cdot x$$
$$\frac{111}{0.74} = \frac{0.74x}{0.74}$$
$$150 = x$$

111 is 74% of 150.

13. We want to know a number. Let x represent the number.

$$14.8 = 0.185 \cdot x$$
$$\frac{14.8}{0.185} = \frac{0.185x}{0.185x}$$
$$80 = x$$

14.8 is 18.5% of 80.

14. We want to know a number. Let x represent the number.

$$102 = 1.36 \cdot x$$
$$\frac{102}{1.36} = \frac{1.36x}{1.36}$$
$$75 = x$$

102 is 136% of 75.

15. We want to know the number of U.S. residents 25 years of age or older in 2007 who have bachelor's degrees. Let x represent the number of people 25 years of age or older in 2007 who have bachelor's degrees.

$$0.17 \cdot 195{,}000{,}000 = x$$
$$33{,}150{,}000 = x$$

The number of residents 25 years of age or older in 2007 who have bachelor's degrees is 33,150,000.

16. We want to know Janet's new salary. Let n represent her new salary, which is her old salary plus the 2.5% raise.

$$n = 39{,}000 + 0.025 \cdot 39{,}000$$
$$n = 39{,}000 + 975$$
$$n = 39{,}975$$

Janet's new salary is $39,975.

17. We want to know the price of the car before the sales tax. Let p represent the price of the car. The sales tax is 7% of the price of the car before sales tax.

$$p + 0.7p = 7811$$
$$1.07p = 7811$$
$$\frac{1.07p}{1.07} = \frac{7811}{1.07}$$
$$p = 7300$$

The price of the car before sales tax was $7300.

18. We want to know the wholesale price of the gasoline. Let p represent the wholesale price of the gasoline. The selling price is the sum of the wholesale price and the markup.

$$p + 0.80p = 4.50$$
$$1.80p = 4.50$$
$$\frac{1.80p}{1.80} = \frac{4.50}{1.80}$$
$$p = 2.50$$

The gas station pays $2.50 per gallon for the gasoline.

19. We want to know the original price of the recliners. Let p represent the original price. The sale price is the original price less the 25% discount.

$$p - 0.25p = 494.25$$
$$0.75p = 494.25$$
$$\frac{0.75p}{0.75} = \frac{494.25}{0.75}$$
$$p = 659$$

The original price of the recliners was $659.

20. We want to know the value of Albert's house one year ago. Let x represent the value one year ago. The value now is the value a year ago minus the 2% loss.

$$x - 0.02x = 148{,}000$$
$$0.98x = 148{,}000$$
$$\frac{0.98x}{0.98} = \frac{148{,}000}{0.98}$$
$$x \approx 151{,}020$$

Albert's house was worth $151,020 one year ago.

2.6 Exercises

21. $n = 0.50(160)$
$n = 80$
80 is 50% of 160.

23. $0.07(200) = n$
$14 = n$
7% of 200 is 14.

25. $n = 0.16(30)$
$n = 4.8$
4.8 is 16% of 30.

27. $31.5 = 0.15x$
$\frac{31.5}{0.15} = \frac{0.15x}{0.15}$
$210 = x$
31.5 is 15% of 210.

29. $0.60(120) = x$
$72 = x$
60% of 120 is 72.

31. $24 = 1.20x$
$\frac{24}{1.20} = \frac{1.20x}{1.20}$
$20 = x$
24 is 120% of 20.

33. $p \cdot 60 = 24$
$\frac{60p}{60} = \frac{24}{60}$
$p = 0.4$
$p = 40\%$
24 is 40% of 60.

35. $1.5 = p \cdot 20$
$\frac{1.5}{20} = \frac{20p}{20}$
$0.075 = p$
$7.5\% = p$
1.5 is 7.5% of 20.

37. $p \cdot 300 = 600$
$\frac{300p}{300} = \frac{600}{300}$
$p = 2$
$p = 200\%$
600 is 200% of 300.

39. Let x be the price of the tennis racket before tax. Then the sales tax is $0.06x$. The total cost is \$57.24.
$x + 0.06x = 57.24$
$1.06x = 57.24$
$x = 54$
The tennis racket cost \$54 before tax.

41. Let x be Todd's salary after the pay cut.
$x = 120,000 - 0.15(120,000)$
$x = 120,000 - 18,000$
$x = 102,000$
Todd's salary is \$102,000 after the pay cut.

43. Let x be Mrs. Fisher's original investment. Then the amount she lost is $0.09x$.
$x - 0.09x = 22,750$
$0.91x = 22,750$
$x = 25,000$
Mrs. Fisher's original investment was \$25,000.

45. Let x be the price before the discount. The amount of the discount is $0.25x$.
$x - 0.25x = 51$
$0.75x = 51$
$x = 68$
The price of the merchandise was \$68 before the discount.

47. Let x be the original price of the suit. Then the discount is $0.30x$.
$x - 0.30x = 399$
$0.70x = 399$
$x = 570$
The original price of the suit was \$570.

49. Let x be the original price of the table. Then the discount is $0.4x$.
$x - 0.4x = 240$
$0.6x = 240$
$x = 400$
The original price of the table was \$400.

51. Let v be the number of votes that the winner received. Then the loser received $0.60v$ votes.
$v + 0.60v = 848$
$1.60v = 848$
$v = 530$
If $v = 530$, then $0.60v = 318$.
The winner received 530 votes and the loser received 318 votes.

53. Let h be the value of the house. Then $0.03h$ is the amount of Melanie's commission.

$$0.03h = 8571$$
$$h = 285,700$$

The value of the house was $285,700.

55. Let x = number of votes that separated the two candidates.
220,369 = total number of votes cast.
0.48%, or 0.0048 is the fraction of votes separating the two candidates.
$x = (0.0048)220,369$
$x = 1057.7$ rounded up
$x = 1058$ votes

57. Let x be the number of males that have never married.
$x = 0.29(106)$
$x = 30.74$
30.74 million males aged 18 years or older have never married.

59. Let x = percent of population that held an Associate's degree in 2004.

$$x = \frac{13,243,927}{186,534,177} \cdot 100\% = 7.1\%$$

61. Let x = percent of population that held a Bachelor's degree in 2004.

$$x = \frac{32,083,878}{186,534,177} \cdot 100\% = 17.2\%$$

63. Let x be the initial selling price. Then the sale price is $x - 0.25x$. This needs to be $6 more than the purchase price of $12.

$$x - 0.25x = 12 + 6$$
$$0.75x = 18$$
$$x = 24$$

The shirts should be priced at $24.

65. The amount of change is 21 – 15 = 6.

$$\frac{6}{21} \approx 0.286$$

The gas mileage decreases by 28.6% due to the extra weight.

67. (a) value after 1st year $= 55,670 - 0.25(55,670)$
$= 55,670 - 13,917.50$
$= 41,752.50$

value after 2nd year
$= 41,752.50 - 0.25(41,752.50)$
$= 41,752.50 - 10,438.125$
$= 31,314.375$

The car will be worth $31,314.38 after two years.

(b) The amount of change is
55,670 – 31,314.38 = 24,355.62.

$$\frac{24,355.62}{55,670} \approx 0.437$$

The overall decrease is 43.7%.

69. The amount of change is 4.39 – 3.89 = 0.50.

$$\frac{0.50}{3.89} \approx 0.1285$$

The percent increase is 12.9%.

71. The equation should be $x + 0.05x = 12.81$. Jack's new hourly wage is a percentage of his current hourly wage, so multiplying 5% to his original wage gives his hourly raise.

Section 2.7

Preparing for Problem Solving: Geometry and Uniform Motion

P1.

$$q + 2q - 30 = 180$$
$$3q - 30 = 180$$
$$3q = 210$$
$$q = 70$$

The solution set is {70}.

P2.

$$30w + 20(w + 5) = 300$$
$$30w + 20w + 100 = 300$$
$$50w + 100 = 300$$
$$50w = 200$$
$$w = 4$$

The solution set is {4}.

2.7 Quick Checks

1. Complementary angles are angles whose measures sum to <u>90</u> degrees.

2. This is a complementary angle problem. We are looking for the measures of two angles whose sum is 90°. Let x represent the measure of the smaller angle. Then $x + 12$ represents the measure of the larger angle.

$$x + (x + 12) = 90$$
$$2x + 12 = 90$$
$$2x = 78$$
$$x = 39$$

$x + 12 = 39 + 12 = 51$
The two complementary angles measure 39° and 51°.

3. This is a supplementary angle problem. We are looking for the measures of two angles whose sum is 180°. Let x represent the measure of the smaller angle. Then $2x - 30$ represents the measure of the larger angle.

$$x+(2x-30)=180$$
$$3x-30=180$$
$$3x=210$$
$$x=70$$

$2x - 30 = 2(70) - 30 = 140 - 30 = 110$

The two supplementary angles measure 70° and 110°.

4. The sum of the measures of the angles of a triangle is 180 angles.

5. This is an "angles of a triangle" problem. We know that the sum of the measures of the interior angles of a triangle is 180°. Let x represent the measure of the largest angle. Then $\frac{1}{3}x$ represents the measure of the smallest angle, and $x - 65$ represents the measure of the middle angle.

$$x+\left(\frac{1}{3}x\right)+(x-65)=180$$
$$2\frac{1}{3}x-65=180$$
$$\frac{7}{3}x=245$$
$$x=105$$

$$\frac{1}{3}x=\frac{1}{3}(105)=35$$

$x - 65 = 105 - 65 = 40$

The measures of the angles of the triangle are 35°, 40°, and 105°.

6. True

7. False; the perimeter of a rectangle can be found by adding twice the length of the rectangle to twice the width of the rectangle.

8. This is a perimeter problem. We want to find the width and length of a garden. Let w represent the width of the garden. Then $2w$ represents the length of the garden. We know that the perimeter is 9 feet, and that the formula for the perimeter of a rectangle is $P = 2l + 2w$.

$$2(2w)+2w=9$$
$$4w+2w=9$$
$$6w=9$$
$$w=\frac{9}{6}$$
$$w=\frac{3}{2}$$

$$2w=2\left(\frac{3}{2}\right)=3$$

The width of the garden is $\frac{3}{2}=1.5$ feet and the length is 3 feet.

9. This problem is about the surface area of a rectangular box. The formula for the surface area of a rectangular box is $SA = 2lw + 2lh + 2hw$, where l is the length of the box, w is the width of the box, and h is the height of the box. We are given the surface area, the length, and the width of the rectangular box.

$$SA = 2lw+2lh+2hw$$
$$62=2\cdot3\cdot2+2\cdot3\cdot h+2\cdot h\cdot 2$$
$$62=12+6h+4h$$
$$62=12+10h$$
$$50=10h$$
$$5=h$$

The height of the box is 5 feet.

10. False; when using $d = rt$ to calculate the distance traveled, it is necessary to travel at a constant speed.

11. This is a uniform motion problem. We want to know the average speed of each biker. Let r represent the average speed of Luis. Then $r + 5$ represents José's average speed. Each biker rides for 3 hours.

	Rate	· **Time**	= **Distance**
Luis	r	3	$3r$
José	$r+5$	3	$3(r+5)$

Since the bikers are 63 miles apart after 3 hours, the sum of the distances they biked is 63.

$$3r+3(r+5)=63$$
$$3r+3r+15=63$$
$$6r+15=63$$
$$6r=48$$
$$r=8$$

$r + 5 = 8 + 5 = 13$
Luis' average speed was 8 miles per hour and José's average speed was 13 miles per hour.

12. This is a uniform motion problem. We want to know how long it takes to catch up with Tanya, and how far you are from your house when you catch her. Let x represent the amount of time you drive before catching Tanya. Then $x + 2$ represents the number of hours that Tanya runs before you catch her.

	Rate	· Time	= Distance
Tanya	8	$x + 2$	$8(x + 2)$
You	40	x	$40x$

When you catch up with Tanya, the distances the two of you have traveled is the same.

$$\begin{aligned} 8(x+2) &= 40x \\ 8x+16 &= 40x \\ 16 &= 32x \\ \frac{1}{2} &= x \end{aligned}$$

You catch up to Tanya after driving for $\frac{1}{2}$ hour.

$$40x = 40\left(\frac{1}{2}\right) = 20$$

You and Tanya are 20 miles from your house when you catch up to her.

2.7 Exercises

13. This is a supplementary angle problem. We are looking for the measures of two angles whose sum is 180°. Let x represent the measure of the second angle. Then $3x - 10$ represents the measure of the first angle.

$$\begin{aligned} (3x-10)+x &= 180 \\ 4x-10 &= 180 \\ 4x &= 190 \\ x &= 47.5 \end{aligned}$$

$3x - 10 = 3(47.5) - 10 = 142.5 - 10 = 132.5$
The angles measure 47.5° and 132.5°.

15. We want to find complementary angles. The measures of complementary angles sum to 90°. Let x be the measure of the smaller angle. Then the measure of the larger angle is $x + 2$.

$$\begin{aligned} x+(x+2) &= 90 \\ 2x+2 &= 90 \\ 2x &= 88 \\ x &= 44 \end{aligned}$$

If $x = 44$, then $x + 2 = 46$.
Are 44 and 46 consecutive even integers that sum to 90? Yes. The measures of the angles are 44° and 46°.

17.
$$\begin{aligned} x+(x+2)+(2x+10) &= 180 \\ 4x+12 &= 180 \\ 4x &= 168 \\ x &= 42 \end{aligned}$$

The angles measure 42°, $(42 + 2)° = 44°$, and $(2 \cdot 42 + 10)° = (84 + 10)° = 94°$.

19. We want to find the measures of the angles of a triangle. The measures of the angels of a triangle sum to 180°. Let x be the measure of the first angle. Then $x + 2$ and $x + 4$ are the measures of the next two angles, respectively.

$$\begin{aligned} x+(x+2)+(x+4) &= 180 \\ 3x+6 &= 180 \\ 3x &= 174 \\ x &= 58 \end{aligned}$$

If $x = 58$, then $x + 2 = 60$, and $x + 4 = 62$. Do 58, 60, and 62 sum to 180? Yes. The measures of the angles are 58°, 60°, and 62°.

21. We want the dimensions of a rectangle. The perimeter of a rectangle is twice the length plus twice the width. Let w be the width of the rectangle. Then the length is $2w + 8$. The perimeter is 88 feet.

$$\begin{aligned} 2w+2(2w+8) &= 88 \\ 2w+4w+16 &= 88 \\ 6w+16 &= 88 \\ 6w &= 72 \\ w &= 12 \end{aligned}$$

If $w = 12$, then $2w + 8 = 32$. Is the sum of twice 12 and twice 32 equal to 88? Yes. The length of the rectangle is 32 feet, and the width is 12 feet.

23. We want to find the dimensions of the rectangular field. The perimeter of a rectangle is twice the length plus twice the width.
Let w be the width (shorter dimension) of the original field. then the length of the field is $2w$ and the sides of the smaller squares are all w. There are 7 sides that need fencing and 294 feet of fencing were used.

$7w = 294$
$w = 42$
If $w = 42$, then $2w = 84$. Would it require 294 feet of fencing to enclose a field that is 42 feet by 84 feet and divide it into two equal squares? Yes. The field is 42 feet by 84 feet.

25. We want to find the sides of a triangle. The perimeter of a triangle is the sum of the lengths of the legs. Let x be the length of each of the congruent legs. The base is 45 inches and the perimeter is 98 inches.
$x + x + 45 = 98$
$2x = 53$
$x = 26.5$
Is the sum of 26.5, 26.5, and 45 equal to 98? Yes. Each leg is 26.5 inches.

27. **(a)** Let t be the time since the cars left Chicago. The distance traveled by the car going east is $62t$.

(b) The distance traveled by the car going west is $68t$.

(c) The total distance is $62t + 68t$.

(d) The equation is $62t + 68t = 585$.

29.

	Rate	· Time	= Distance
Martha	528	$t + 10$	$528(t + 10)$
Mom	880	t	$880t$

$528(t + 10) = 880t$

31. Let l be the length of the rectangle. Then the width is $\frac{1}{2}l - 3$. The perimeter is 36 inches.
$2l + 2\left(\frac{1}{2}l - 3\right) = 36$
$2l + l - 6 = 36$
$3l - 6 = 36$
$3l = 42$
$l = 14$
If $l = 14$, then $\frac{1}{2}l - 3 = 4$.
The length is 14 inches, the width is 4 inches.

33. Let l be the length of the billboard. The height is 15 feet and the perimeter is 110 feet.
$2l + 2(15) = 110$
$2l + 30 = 110$
$2l = 80$
$l = 40$
The length of the billboard is 40 feet.

35. The area of a trapezoid is $A = \frac{1}{2}h(B + b)$, where h is the height and the bases are B and b. Let B be the longer base. Then the shorter base is $B - 8$. The height is 60 feet and the area is 2160 square feet.
$\frac{1}{2}(60)[B + (B - 8)] = 2160$
$30[2B - 8] = 2160$
$60B - 240 = 2160$
$60B = 2400$
$B = 40$
If $B = 40$, then $B - 8 = 32$. The bases are 40 feet and 32 feet.

37. **(a)** Let l be the length of the garden. Then the width is $2l - 3$. The perimeter is 60 yards.
$2l + 2(2l - 3) = 60$
$2l + 4l - 6 = 60$
$6l - 6 = 60$
$6l = 66$
$l = 11$
If $l = 11$, then $2l - 3 = 19$. The length of the garden is 11 feet and the width is 19 feet.

(b) $A = lw = 11(19) = 209$
The area of the garden is 209 square feet.

39. Since the southbound boat is traveling at 47 mph, the northbound boat is traveling at $47 + 16 = 63$ mph.
After t hours, the northbound boat will have gone $63t$ miles and the southbound boat will have gone $47t$ miles.
$63t + 47t = 1430$
$110t = 1430$
$t = 13$
The boats will be 1430 miles apart after 13 hours.

41. Let x be the speed of the slower car. Then $x + 12$ is the speed of the faster car. The slower car stops after 4 hours and 30 minutes (4.5 hours) and has traveled $4.5x$ miles. The faster car stops after 4 hours and has traveled $4(x + 12)$ miles. The difference between their stopping points is

24 miles.

$$\begin{aligned} 4(x+12)-4.5x &= 24 \\ 4x+48-4.5x &= 24 \\ -0.5x &= -24 \\ x &= 48 \end{aligned}$$

If $x = 48$, then $x + 12 = 60$. The slower car is traveling at 48 mph, and the faster car is traveling at 60 mph.

43. Let t be the amount of time spent traveling at 62 mph. Then $6 - t$ is the amount of time spent traveling at 54 mph. The distance traveled at 62 mph is $62t$, and the distance at 54 mph is $54(6 - t)$.

$$\begin{aligned} 62t+54(6-t) &= 360 \\ 62t+324-54t &= 360 \\ 8t+324 &= 360 \\ 8t &= 36 \\ t &= 4.5 \end{aligned}$$

If $t = 4.5$, then $6 - t = 1.5$.
The trip consisted of 4.5 hours on the freeway and 1.5 hours on the 2-lane highway.

45. Let r be the rate at which she jogs.

	Rate	· Time	= Distance
jog	r	$\frac{10}{60}=\frac{1}{6}$	$\frac{1}{6}r$
walk	$r-4$	$\frac{30}{60}=\frac{1}{2}$	$\frac{1}{2}(r-4)$

The distances are equal.

$$\begin{aligned} \frac{1}{6}r &= \frac{1}{2}(r-4) \\ 6\left(\frac{1}{6}r\right) &= 6\left[\frac{1}{2}(r-4)\right] \\ r &= 3(r-4) \\ r &= 3r-12 \\ 12 &= 2r \\ 6 &= r \end{aligned}$$

Carol jogs at 6 mph.

47. Let x = measure of base angles.
Let $2x - 16$ = measure of vertex angle.
180 = sum of all angles

$$\begin{aligned} x+x+2x-16 &= 180 \\ 4x-16 &= 180 \\ 16+4x-16 &= 180+16 \\ 4x &= 196 \\ \frac{4x}{4} &= \frac{196}{4} \\ x &= 49° \end{aligned}$$

Substitute x into $2x - 16$ for vertex angle of 82°.
49°, 49°, 82°

49. The marked angles are equal in measure because they are alternate interior angles.

$$\begin{aligned} 20-2x &= 3x+5 \\ 20 &= 5x+5 \\ 15 &= 5x \\ 3 &= x \end{aligned}$$

51. The marked angles are supplementary because they are interior angles on the same side of the transversal.

$$\begin{aligned} (8x+12)+(3x+3) &= 180 \\ 11x+15 &= 180 \\ 11x &= 165 \\ x &= 15 \end{aligned}$$

53. The marked angles are equal in measure because they are corresponding angles.

$$\begin{aligned} 4x+2 &= \frac{3}{2}x+32 \\ 2(4x+2) &= 2\left(\frac{3}{2}x+32\right) \\ 8x+4 &= 3x+64 \\ 5x &= 60 \\ x &= 12 \end{aligned}$$

55. Complementary angles are those whose sum 90° and supplementary angles sum to 180°.

57. Answers may vary. The equation $65t + 40t = 115$ is describing the sum of distances whereas the equation $65t - 40t = 115$ represents the difference of distances.

Section 2.8

Preparing for Solving Linear Inequalities in One Variable

P1. Since 4 is to the left of 19 on the number line, $4 < 19$.

P2. Since −11 is to the right of −24 on the number line, $-11 > -24$.

P3. $\frac{1}{4} = 0.25$

P4. Since $\frac{5}{6}=\frac{25}{30}$, $\frac{4}{5}=\frac{24}{30}$, and 25 > 24, then $\frac{5}{6}>\frac{4}{5}$.

2.8 Quick Checks

1. True: When graphing an inequality that contains a > or a < symbol, we use parentheses.

2. $n \ge 8$

3. $a < -6$

4. $x > -1$

5. $p \le 0$

6. True

7. False; the inequality $x < -4$ is written in interval notation as $(-\infty, -4)$.

8. $[-3, \infty)$

9. $(-\infty, 12)$

10. $(-\infty, 2.5]$

11. $(125, \infty)$

12. To solve an inequality means to find the set of all replacement values of the variable for which the statement is true.

13. The Addition Property of Inequality states that the direction, or sense, of each inequality remains the same when the same quantity is added to each side of the inequality.

14.
$$\begin{aligned} 5n-4 &> 11 \\ 5n-4+4 &> 11+4 \\ 5n &> 15 \\ \frac{5n}{5} &> \frac{15}{5} \\ n &> 3 \end{aligned}$$
The solution set is $\{n|n > 3\}$ or $(3, \infty)$.

15.
$$\begin{aligned} -2x+3 &< 7-3x \\ -2x+3+3x &< 7-3x+3x \\ x+3 &< 7 \\ x+3-3 &< 7-3 \\ x &< 4 \end{aligned}$$
The solution set is $\{x|x < 4\}$ or $(-\infty, 4)$.

16.
$$\begin{aligned} 5n+8 &\le 4n+4 \\ 5n+8-4n &\le 4n+4-4n \\ n+8 &\le 4 \\ n+8-8 &\le 4-8 \\ n &\le -4 \end{aligned}$$
The solution set is $\{n|n \le -4\}$ or $(-\infty, -4]$.

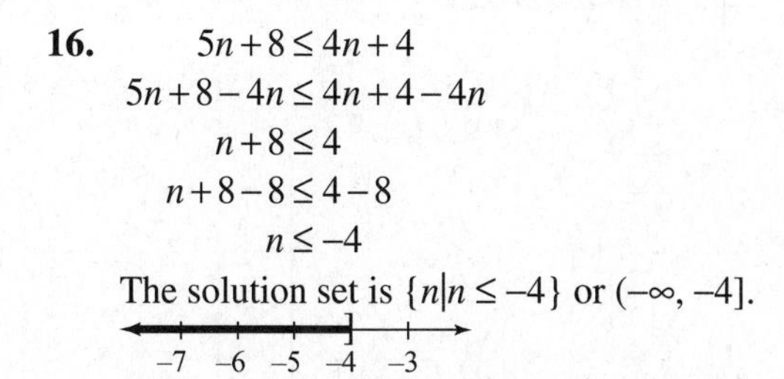

17.
$$\begin{aligned} 3(4x-8)+12 &> 11x-13 \\ 12x-24+12 &> 11x-13 \\ 12x-12 &> 11x-13 \\ 12x-12-11x &> 11x-13-11x \\ x-12 &> -13 \\ x-12+12 &> -13+12 \\ x &> -1 \end{aligned}$$
The solution set is $\{x|x > -1\}$ or $(-1, \infty)$.

18. When solving an inequality, we reverse the direction of the inequality symbol when we multiply or divide by a negative number.

19. False; the solution to the inequality $-\frac{1}{2}x > 9$ is $\{x|x < -18\}$. This solution is written in interval notation as $(-\infty, -18)$.

20.
$$\begin{aligned} 6k &< -36 \\ \frac{6k}{6} &< \frac{-36}{6} \\ k &< -6 \end{aligned}$$
The solution set is $\{k|k < -6\}$ or $(-\infty, -6)$.

21. $2n \geq -5$

$\frac{2n}{2} \geq \frac{-5}{2}$

$n \geq -\frac{5}{2}$

The solution set is $\left\{n \middle| n \geq -\frac{5}{2}\right\}$ or $\left[-\frac{5}{2}, \infty\right)$.

−4 −3 −2 −1 0

22. $-\frac{3}{2}k > 12$

$-\frac{2}{3}\left(-\frac{3}{2}k\right) < -\frac{2}{3}(12)$

$k < -8$

The solution set is $\{k|k < -8\}$ or $(-\infty, -8)$.

−11 −10 −9 −8 −7

23. $-\frac{4}{3}p \leq -\frac{4}{5}$

$-\frac{3}{4}\left(-\frac{4}{3}p\right) \geq -\frac{3}{4}\left(-\frac{4}{5}\right)$

$p \geq \frac{3}{5}$

The solution set is $\left\{p \middle| p \geq \frac{3}{5}\right\}$ or $\left[\frac{3}{5}, \infty\right)$.

−1 0 1 2 3

24. $3x - 7 > 14$

$3x - 7 + 7 > 14 + 7$

$3x > 21$

$\frac{3x}{3} > \frac{21}{3}$

$x > 7$

The solution set is $\{x|x > 7\}$ or $(7, \infty)$.

6 7 8 9 10

25. $-4n - 3 < 9$

$-4n - 3 + 3 < 9 + 3$

$-4n < 12$

$\frac{-4n}{-4} > \frac{12}{-4}$

$n > -3$

The solution set is $\{n|n > -3\}$ or $(-3, \infty)$.

−4 −3 −2 −1 0

26. $2x - 6 < 3(x + 1) - 5$

$2x - 6 < 3x + 3 - 5$

$2x - 6 < 3x - 2$

$2x - 6 - 2x < 3x - 2 - 2x$

$-6 < x - 2$

$-6 + 2 < x - 2 + 2$

$-4 < x$

The solution set is $\{x|x > -4\}$ or $(-4, \infty)$.

−5 −4 −3 −2 −1

27. $-4(x + 6) + 18 \geq -2x + 6$

$-4x - 24 + 18 \geq -2x + 6$

$-4x - 6 \geq -2x + 6$

$-4x - 6 + 2x \geq -2x + 6 + 2x$

$-2x - 6 \geq 6$

$-2x - 6 + 6 \geq 6 + 6$

$-2x \geq 12$

$\frac{-2x}{-2} \leq \frac{12}{-2}$

$x \leq -6$

The solution set is $\{x|x \leq -6\}$ or $(-\infty, -6]$.

−9 −8 −7 −6 −5

28. $\frac{1}{2}(x + 2) > \frac{1}{5}(x + 17)$

$10\left[\frac{1}{2}(x + 2)\right] > 10\left[\frac{1}{5}(x + 17)\right]$

$5(x + 2) > 2(x + 17)$

$5x + 10 > 2x + 34$

$5x + 10 - 2x > 2x + 34 - 2x$

$3x + 10 > 34$

$3x + 10 - 10 > 34 - 10$

$3x > 24$

$\frac{3x}{3} > \frac{24}{3}$

$x > 8$

The solution set is $\{x|x > 8\}$ or $(8, \infty)$.

7 8 9 10 11

29. $\frac{4}{3}x-\frac{2}{3}\le\frac{4}{5}x+\frac{3}{5}$

$15\left(\frac{4}{3}x-\frac{2}{3}\right)\le15\left(\frac{4}{5}x+\frac{3}{5}\right)$

$20x-10\le12x+9$

$20x-10-12x\le12x+9-12x$

$8x-10\le9$

$8x-10+10\le9+10$

$8x\le19$

$\frac{8x}{8}\le\frac{19}{8}$

$x\le\frac{19}{8}$

The solution set is $\left\{x \middle| x\le\frac{19}{8}\right\}$ or $\left(-\infty, \frac{19}{8}\right]$.

0 1 2 3 4

30. When solving an inequality, if the variable is eliminated and the result is a true statement, the solution is all real numbers or $(-\infty, \infty)$.

31. When solving an inequality, if the variable is eliminated and the result is a false statement, the solution is the empty set or $\varnothing$.

32. $-2x+7(x-5)\le6x+32$

$-2x+7x-35\le6x+32$

$5x-35\le6x+32$

$5x-35-5x\le6x+32-5x$

$-35\le x+32$

$-35-32\le x+32-32$

$-67\le x$

The solution set is $\{x|x\ge-67\}$ or $[-67, \infty)$.

–65 –66 –67 –68 –69

33. $-x+7-8x\ge2(8-5x)+x$

$-9x+7\ge16-10x+x$

$-9x+7\ge16-9x$

$-9x+7+9x\ge16-9x+9x$

$7\ge16$

The statement $7 \ge 16$ is a false statement, so this inequality has no solution. The solution set is $\varnothing$ or { }.

–2 –1 0 1 2

34. $\frac{3}{2}x+5-\frac{5}{2}x<4x-3(x+1)$

$-\frac{2}{2}x+5<4x-3x-3$

$-x+5<x-3$

$-x+5+x<x-3+x$

$5<2x-3$

$5+3<2x-3+3$

$8<2x$

$\frac{8}{2}<\frac{2x}{2}$

$4<x$

The solution set is $\{x|x>4\}$ or $(4, \infty)$.

3 4 5 6 7

35. $0.8x+3.2(x+4)\ge2x+12.8+3x-x$

$0.8x+3.2x+12.8\ge4x+12.8$

$4x+12.8\ge4x+12.8$

$4x+12.8-4x\ge4x+12.8-4x$

$12.8\ge12.8$

Since 12.8 is always greater than or equal to 12.8, the solution set for this inequality is $\{x|x$ is any real number$\}$ or $(-\infty, \infty)$.

–2 –1 0 1 2

36. We want to know the maximum number of boxes of supplies that the worker can move on the elevator. Let b represent the number of boxes. Then the weight of the boxes is $91b$, and the weight of the worker and the boxes is $180 + 91b$. This weight cannot be more than 2000 pounds.

$180+91b\le2000$

$91b\le1820$

$\frac{91b}{91}\le\frac{1820}{91}$

$b\le20$

The number of boxes must be less than or equal to 20, so the maximum number of boxes the worker can move in the elevator is 20.

2.8 Exercises

37. $x>2$

0 2

$(2, \infty)$

39. $x\le-1$

–1 0

$(-\infty, -1]$

41. $z \geq -3$

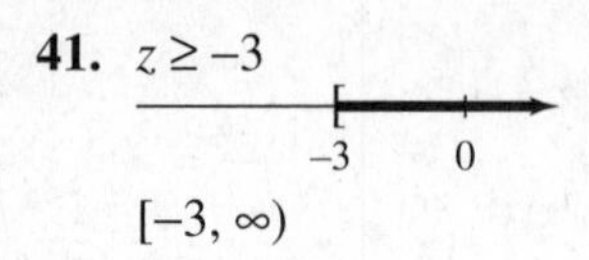

$[-3, \infty)$

43. $x < 4$

0 4

$(-\infty, 4)$

45. $(-\infty, 2)$

47. $\varnothing$ or $\{\ \}$

49. $(-\infty, \infty)$

51. Adding 7 to each side does not change the direction of the inequality symbol. The symbol remains <. We used the Addition Principle of Inequality.

53. Multiplying each side by 3, a positive number, does not change the direction of the inequality symbol. The symbol remains >. We used the Multiplication Principle of Inequality.

55. Subtracting 2 or adding –2 to each side does not change the direction of the inequality symbol. The symbol remains ≤. We used the Addition Principle of Inequality.

57. Dividing each side by –3 or multiplying each side by $-\frac{1}{3}$, a negative number, reverses the direction of the inequality symbol. The symbol becomes ≤. We used the Multiplication Principle of Inequality.

59. $x + 1 < 5$

$x < 4$

$\{x | x < 4\}$

$(-\infty, 4)$

0 4

61. $x - 6 \geq -4$

$x \geq 2$

$\{x | x \geq 2\}$

$[2, \infty)$

0 2

63. $3x \leq 15$

$x \leq 5$

$\{x | x \leq 5\}$

$(-\infty, 5]$

0 5

65. $-5x < 35$

$x > -7$

$\{x | x > -7\}$

$(-7, \infty)$

–7 0

67. $3x - 7 > 2$

$3x > 9$

$x > 3$

$\{x | x > 3\}$

$(3, \infty)$

0 3

69. $3x - 1 \geq 3 + x$

$2x - 1 \geq 3$

$2x \geq 4$

$x \geq 2$

$\{x | x \geq 2\}$

$[2, \infty)$

0 2

71. $1 - 2x \leq 3$

$-2x \leq 2$

$x \geq -1$

$\{x | x \geq -1\}$

$[-1, \infty)$

–1 0

73. $-2(x + 3) < 8$

$-2x - 6 < 8$

$-2x < 14$

$x > -7$

$\{x | x > -7\}$

$(-7, \infty)$

–7 0

75. $4-3(1-x)\le 3$
$4-3+3x\le 3$
$1+3x\le 3$
$3x\le 2$
$x\le \frac{2}{3}$

$\left\{x \middle| x\le \frac{2}{3}\right\}$

$\left(-\infty, \frac{2}{3}\right]$

77. $\frac{1}{2}(x-4)>x+8$
$\frac{1}{2}x-2>x+8$
$-2>\frac{1}{2}x+8$
$-10>\frac{1}{2}x$
$-20>x$
$\{x|x<-20\}$
$(-\infty, -20)$

79. $4(x-1)>3(x-1)+x$
$4x-4>3x-3+x$
$4x-4>4x-3$
$-4>-3$
This is a false statement. Therefore, there is no solution. The solution set is $\varnothing$ or { }.

81. $5(n+2)-2n\le 3(n+4)$
$5n+10-2n\le 3n+12$
$3n+10\le 3n+12$
$10\le 12$
The solution to this inequality is all real numbers, since 10 is always less than 12.
$\{n|n \text{ is any real number}\}$
$(-\infty, \infty)$

83. $2n-3(n-2)<n-4$
$2n-3n+6<n-4$
$-n+6<n-4$
$-2n+6<-4$
$-2n<-10$
$n>5$
$\{n|n>5\}$
$(5, \infty)$

85. $4(2w-1)\ge 3(w+2)+5(w-2)$
$8w-4\ge 3w+6+5w-10$
$8w-4\ge 8w-4$
$-4\ge -4$
The solution to this inequality is all real numbers, since –4 is always equal to –4.
$\{w|w \text{ is any real number}\}$
$(-\infty, \infty)$

87. $3y-(5y+2)>4(y+1)-2y$
$3y-5y-2>4y+4-2y$
$-2y-2>2y+4$
$-4y-2>4$
$-4y>6$
$y<-\frac{3}{2}$

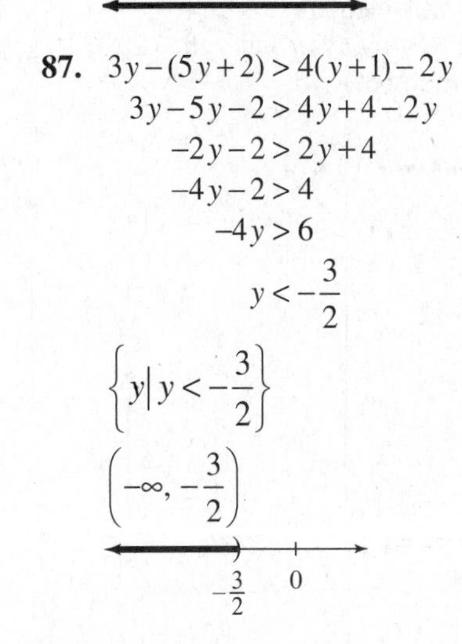

$\left\{y \middle| y<-\frac{3}{2}\right\}$

$\left(-\infty, -\frac{3}{2}\right)$

89. $x\ge 16{,}000$

91. $x\le 20{,}000$

93. $x>12{,}000$

95. $x>0$

97. $x\le 0$

99. $-1<x-5$
$4<x$
$\{x|x>4\}$
$(4, \infty)$

101. $-\frac{3}{4}x > -\frac{9}{16}$

$-\frac{4}{3}\cdot\left(-\frac{3}{4}\right)x < -\frac{4}{3}\cdot\left(-\frac{9}{16}\right)$

$x < \frac{3}{4}$

$\left\{x \middle| x < \frac{3}{4}\right\}$

$\left(-\infty, \frac{3}{4}\right)$

0 $\frac{3}{4}$

103. $3(x+1) > 2(x+1) + x$

$3x+3 > 2x+2+x$

$3x+3 > 3x+2$

$3 > 2$

The solution to this inequality is all real numbers, since 3 is always greater than 2.

$\{x|x \text{ is any real number}\}$

$(-\infty, \infty)$

105. $-4a+1 > 9+3(2a+1)+a$

$-4a+1 > 9+6a+3+a$

$-4a+1 > 7a+12$

$-11a+1 > 12$

$-11a > 11$

$a < -1$

$\{a|a < -1\}$

$(-\infty, -1)$

−1 0

107. $n+3(2n+3) > 7n-3$

$n+6n+9 > 7n-3$

$7n+9 > 7n-3$

$9 > -3$

The solution to this inequality is all real numbers, since 9 is always greater than −3.

$\{n|n \text{ is any real number}\}$

$(-\infty, \infty)$

109. $\frac{x}{2} \geq 1-\frac{x}{4}$

$4\left(\frac{x}{2}\right) \geq 4\left(1-\frac{x}{4}\right)$

$2x \geq 4-x$

$3x \geq 4$

$x \geq \frac{4}{3}$

$\left\{x \middle| x \geq \frac{4}{3}\right\}$

$\left[\frac{4}{3}, \infty\right)$

0 $\frac{4}{3}$

111. $\frac{x+5}{2}+4 > \frac{2x+1}{3}+2$

$6\left(\frac{x+5}{2}+4\right) > 6\left(\frac{2x+1}{3}+2\right)$

$3(x+5)+24 > 2(2x+1)+12$

$3x+15+24 > 4x+2+12$

$3x+39 > 4x+14$

$39 > x+14$

$25 > x$

$\{x|x < 25\}$

$(-\infty, 25)$

0 25

113. $-5z-(3+2z) > 3-7z$

$-5z-3-2z > 3-7z$

$-3-7z > 3-7z$

$-3 > 3$

This inequality is false. Therefore, there is no solution. The solution set is $\varnothing$ or { }.

0

115. $1.3x+3.1 < 4.5x-15.9$

$3.1 < 3.2x-15.9$

$19 < 3.2x$

$5.9375 < x$

$\{x|x > 5.9375\}$

$(5.9375, \infty)$

0 5.9375

117. Let x be the number of miles. Then the charge for 1 week is $(55 + 0.18x)$ dollars.

$$55+0.18x \le 280$$
$$0.18x \le 225$$
$$x \le 1250$$

You can drive at most 1250 miles.

119. Let f be Yvette's score on the final exam. Then she has six scores total (72, 78, 66, 81, f, and f), since the final exam counts as two tests.

$$72+78+66+81+f+f \ge 360$$
$$297+2f \ge 360$$
$$2f \ge 63$$
$$f \ge 31.5$$

Assuming that only whole-number scores are possible, the minimum score on the final is 32.

121. Let m be the number of minutes. Then Imperial Telephone charges $(10 + 0.03m)$ dollars and Mayflower Communications charges $(6 + 0.04m)$ dollars.

$$10+0.03m < 6+0.04m$$
$$4 < 0.01m$$
$$400 < m$$

Imperial Telephone is cheaper for more than 400 minutes.

123.

$$150,000 \le 2.98I - 76.11$$
$$150,076.11 \le 2.98I$$
$$50,361.11 \le I$$

The bank will lend at least $150,000 for an annual income of $50,361.11 or greater.

125. Let x be the score on the fifth test. The average is then $\dfrac{68+82+87+89+x}{5}$.

$$80 \le \frac{68+82+87+89+x}{5}$$
$$80 \le \frac{326+x}{5}$$
$$400 \le 326+x$$
$$74 \le x$$

You need a score of at least 74 to get a B.

127.

$$-3 < x+30 < 16$$
$$-3-30 < x+30-30 < 16-30$$
$$-33 < x < -14$$

The solution set is $\{x|-33 < x < -14\}$.

129.

$$-6 \le \frac{3x}{2} \le 9$$
$$2(-6) \le 2\left(\frac{3x}{2}\right) \le 2(9)$$
$$-12 \le 3x \le 18$$
$$\frac{-12}{3} \le \frac{3x}{3} \le \frac{18}{3}$$
$$-4 \le x \le 6$$

The solution set is $\{x|-4 \le x \le 6\}$.

131.

$$-7 \le 2x-3 < 15$$
$$-7+3 \le 2x-3+3 < 15+3$$
$$-4 \le 2x < 18$$
$$-\frac{4}{2} \le \frac{2x}{2} < \frac{18}{2}$$
$$-2 \le x < 9$$

The solution set is $\{x|-2 \le x < 9\}$.

133.

$$4 < 6-\frac{x}{2} \le 10$$
$$4-6 < 6-\frac{x}{2}-6 \le 10-6$$
$$-2 < -\frac{x}{2} \le 4$$
$$-2(-2) > -2\left(-\frac{x}{2}\right) \ge -2(4)$$
$$4 > x \ge -8$$

This is the same as $-8 \le x < 4$. The solution set is $\{x|-8 \le x < 4\}$.

135. A left parenthesis is used to indicate that the solution is greater than a number. A left bracket is used to show that the solution is greater than or equal to a given number.

137. When solving an inequality and the variables are eliminated and a true statement results, the solution is all real numbers. When solving an inequality and the variables are eliminated and a false statement results, the solution is the empty set.

Chapter 2 Review

1. $3x + 2 = 7; x = 5$

$$3(5)+2 \stackrel{?}{=} 7$$
$$15+2 \stackrel{?}{=} 7$$
$$17 = 7 \quad \text{False}$$

No, $x = 5$ is *not* a solution to the equation.

2. $5m - 1 = 17$; $m = 4$

$$\begin{aligned} 5(4)-1 &\stackrel{?}{=} 17 \\ 20-1 &\stackrel{?}{=} 17 \\ 19 &= 17 \quad \text{False} \end{aligned}$$

No, $m = 4$ is *not* a solution to the equation.

3. $6x + 6 = 12$; $x = \dfrac{1}{2}$

$$\begin{aligned} 6\left(\frac{1}{2}\right)+6 &\stackrel{?}{=} 12 \\ 3+6 &\stackrel{?}{=} 12 \\ 9 &= 12 \quad \text{False} \end{aligned}$$

No, $x = \dfrac{1}{2}$ is *not* a solution to the equation.

4. $9k + 3 = 9$; $k = \dfrac{2}{3}$

$$\begin{aligned} 9\left(\frac{2}{3}\right)+3 &\stackrel{?}{=} 9 \\ 6+3 &\stackrel{?}{=} 9 \\ 9 &= 9 \quad \text{True} \end{aligned}$$

Yes, $k = \dfrac{2}{3}$ is a solution to the equation.

5.

$$\begin{aligned} n-6 &= 10 \\ n-6+6 &= 10+6 \\ n &= 16 \end{aligned}$$

Check:

$$\begin{aligned} n-6 &= 10 \\ 16-16 &\stackrel{?}{=} 10 \\ 10 &= 10 \quad \text{True} \end{aligned}$$

The solution is 16, or the solution set is {16}.

6.

$$\begin{aligned} n-8 &= 12 \\ n-8+8 &= 12+8 \\ n &= 20 \end{aligned}$$

Check:

$$\begin{aligned} n-8 &= 12 \\ 20-8 &\stackrel{?}{=} 12 \\ 12 &= 12 \quad \text{True} \end{aligned}$$

The solution is 20, or the solution set is {20}.

7.

$$\begin{aligned} x+6 &= -10 \\ x+6-6 &= -10-6 \\ x &= -16 \end{aligned}$$

Check:

$$\begin{aligned} x+6 &= -10 \\ -16+6 &\stackrel{?}{=} -10 \\ -10 &= -10 \quad \text{True} \end{aligned}$$

The solution is −16, or the solution set is {−16}.

8.

$$\begin{aligned} x+2 &= -5 \\ x+2-2 &= -5-2 \\ x &= -7 \end{aligned}$$

Check:

$$\begin{aligned} x+2 &= -5 \\ -7+2 &\stackrel{?}{=} -5 \\ -5 &= -5 \quad \text{True} \end{aligned}$$

The solution is −7, or the solution set is {−7}.

9.

$$\begin{aligned} -100 &= m-5 \\ -100+5 &= m-5+5 \\ -95 &= m \end{aligned}$$

Check:

$$\begin{aligned} -100 &= m-5 \\ -100 &\stackrel{?}{=} -95-5 \\ -100 &= -100 \quad \text{True} \end{aligned}$$

The solution is −95, or the solution set is {−95}.

10.

$$\begin{aligned} -26 &= m-76 \\ -26+76 &= m-76+76 \\ 50 &= m \end{aligned}$$

Check:

$$\begin{aligned} -26 &= m-76 \\ -26 &\stackrel{?}{=} 50-76 \\ -26 &= -26 \quad \text{True} \end{aligned}$$

The solution is 50, or the solution set is {50}.

11.

$$\begin{aligned} \frac{2}{3}y &= 16 \\ \frac{3}{2}\cdot\frac{2}{3}y &= \frac{3}{2}\cdot 16 \\ y &= 3\cdot 8 \\ y &= 24 \end{aligned}$$

Check:

$$\begin{aligned} \frac{2}{3}y &= 16 \\ \frac{2}{3}\cdot 24 &\stackrel{?}{=} 16 \\ 2\cdot 8 &\stackrel{?}{=} 16 \\ 16 &= 16 \quad \text{True} \end{aligned}$$

The solution is 24, or the solution set is {24}.

12.

$$\begin{aligned} \frac{x}{4} &= 20 \\ 4\cdot\frac{x}{4} &= 4\cdot 20 \\ x &= 80 \end{aligned}$$

Check:

$$\begin{aligned} \frac{x}{4} &= 20 \\ \frac{80}{4} &\stackrel{?}{=} 20 \\ 20 &= 20 \quad \text{True} \end{aligned}$$

The solution is 80, or the solution set is {80}.

13. $-6x = 36$

$\frac{-6x}{-6} = \frac{36}{-6}$

$x = -6$

Check: $-6x = 36$

$-6(-6) \stackrel{?}{=} 36$

$36 = 36$ True

The solution is –6 or the solution set is {–6}.

14. $-4x = -20$

$\frac{-4x}{-4} = \frac{-20}{-4}$

$x = 5$

Check: $-4x = -20$

$-4 \cdot 5 \stackrel{?}{=} -20$

$-20 = -20$ True

The solution is 5, or the solution set is {5}.

15. $z + \frac{5}{6} = \frac{1}{2}$

$z + \frac{5}{6} - \frac{5}{6} = \frac{1}{2} - \frac{5}{6}$

$z = \frac{3}{6} - \frac{5}{6}$

$z = -\frac{2}{6}$

$z = -\frac{1}{3}$

Check: $z + \frac{5}{6} = \frac{1}{2}$

$-\frac{1}{3} + \frac{5}{6} \stackrel{?}{=} \frac{1}{2}$

$-\frac{2}{6} + \frac{5}{6} \stackrel{?}{=} \frac{1}{2}$

$\frac{3}{6} \stackrel{?}{=} \frac{1}{2}$

$\frac{1}{2} = \frac{1}{2}$ True

The solution is $-\frac{1}{3}$ or the solution set is $\left\{-\frac{1}{3}\right\}$.

16. $m - \frac{1}{8} = \frac{1}{4}$

$m - \frac{1}{8} + \frac{1}{8} = \frac{1}{4} + \frac{1}{8}$

$m = \frac{2}{8} + \frac{1}{8}$

$m = \frac{3}{8}$

Check: $m - \frac{1}{8} = \frac{1}{4}$

$\frac{3}{8} - \frac{1}{8} \stackrel{?}{=} \frac{1}{4}$

$\frac{2}{8} \stackrel{?}{=} \frac{1}{4}$

$\frac{1}{4} = \frac{1}{4}$ True

The solution is $\frac{3}{8}$ or the solution set is $\left\{\frac{3}{8}\right\}$.

17. $1.6x = 6.4$

$\frac{1.6x}{1.6} = \frac{6.4}{1.6}$

$x = 4$

Check: $1.6x = 6.4$

$1.6(4) \stackrel{?}{=} 6.4$

$6.4 = 6.4$ True

The solution is 4, or the solution set is {4}.

18. $1.8m = 9$

$\frac{1.8m}{1.8} = \frac{9}{1.8}$

$m = 5$

Check: $1.8m = 9$

$1.8(5) \stackrel{?}{=} 9$

$9 = 9$ True

The solution is 5, or the solution set is {5}.

19. $p - 1200 = 18{,}900$

$p - 1200 + 1200 = 18{,}900 + 1200$

$p = 20{,}100$

The original price was $20,100.

20. $3c = 7.65$

$\frac{3c}{3} = \frac{7.65}{3}$

$c = 2.55$

Each cup of coffee cost $2.55.

21. $5x - 1 = -21$

$5x - 1 + 1 = -21 + 1$

$5x = -20$

$\frac{5x}{5} = \frac{-20}{5}$

$x = -4$

Check: $5x - 1 = -21$

$5(-4) - 1 \stackrel{?}{=} -21$

$-20 - 1 \stackrel{?}{=} -21$

$-21 = -21$ True

The solution is –4, or the solution set is {–4}.

22.
$$\begin{aligned} -3x+7&=-5\\ -3x+7-7&=-5-7\\ -3x&=-12\\ \frac{-3x}{-3}&=\frac{-12}{-3}\\ x&=4 \end{aligned}$$
Check:
$$\begin{aligned} -3x+7&=-5\\ -3(4)+7&\stackrel{?}{=}-5\\ -12+7&\stackrel{?}{=}-5\\ -5&=-5 \quad \text{True} \end{aligned}$$
The solution is 4, or the solution set is {4}.

23.
$$\begin{aligned} \frac{2}{3}x+5&=11\\ \frac{2}{3}x+5-5&=11-5\\ \frac{2}{3}x&=6\\ \frac{3}{2}\cdot\frac{2}{3}x&=\frac{3}{2}\cdot 6\\ x&=9 \end{aligned}$$
Check:
$$\begin{aligned} \frac{2}{3}x+5&=11\\ \frac{2}{3}\cdot 9+5&\stackrel{?}{=}11\\ 6+5&\stackrel{?}{=}11\\ 11&=11 \quad \text{True} \end{aligned}$$
The solution is 9, or the solution set is {9}.

24.
$$\begin{aligned} \frac{5}{7}x-2&=-17\\ \frac{5}{7}x-2+2&=-17+2\\ \frac{5}{7}x&=-15\\ \frac{7}{5}\cdot\frac{5}{7}x&=\frac{7}{5}\cdot(-15)\\ x&=-21 \end{aligned}$$
Check:
$$\begin{aligned} \frac{5}{7}x-2&=-17\\ \frac{5}{7}(-21)-2&\stackrel{?}{=}-17\\ -15-2&\stackrel{?}{=}-17\\ -17&=-17 \quad \text{True} \end{aligned}$$
The solution is –21, or the solution set is {–21}.

25.
$$\begin{aligned} -2x+5+6x&=-11\\ 4x+5&=-11\\ 4x+5-5&=-11-5\\ 4x&=-16\\ \frac{4x}{4}&=\frac{-16}{4}\\ x&=-4 \end{aligned}$$
Check:
$$\begin{aligned} -2x+5+6x&=-11\\ -2(-4)+5+6(-4)&\stackrel{?}{=}-11\\ 8+5-24&\stackrel{?}{=}-11\\ -11&=-11 \quad \text{True} \end{aligned}$$
The solution is –4, or the solution set is {–4}.

26.
$$\begin{aligned} 3x-5x+6&=18\\ -2x+6&=18\\ -2x+6-6&=18-6\\ -2x&=12\\ \frac{-2x}{-2}&=\frac{12}{-2}\\ x&=-6 \end{aligned}$$
Check:
$$\begin{aligned} 3x+5x+6&=18\\ 3(-6)-5(-6)+6&\stackrel{?}{=}18\\ -18+30+6&\stackrel{?}{=}18\\ 18&=18 \quad \text{True} \end{aligned}$$
The solution is –6, or the solution set is {–6}.

27.
$$\begin{aligned} 2m+0.5m&=10\\ 2.5m&=10\\ 10\cdot 2.5m&=10\cdot 10\\ 25m&=100\\ \frac{25m}{25}&=\frac{100}{25}\\ m&=4 \end{aligned}$$
Check:
$$\begin{aligned} 2m+0.5m&=10\\ 2(4)+0.5(4)&\stackrel{?}{=}10\\ 8+2&\stackrel{?}{=}10\\ 10&=10 \quad \text{True} \end{aligned}$$
The solution is 4, or the solution set is {4}.

28.
$$\begin{aligned} 1.4m+m&=-12\\ 2.4m&=-12\\ \frac{2.4m}{2.4}&=\frac{-12}{2.4}\\ m&=-5 \end{aligned}$$
Check:
$$\begin{aligned} 1.4m+m&=-12\\ 1.4(-5)+(-5)&\stackrel{?}{=}-12\\ -7+(-5)&\stackrel{?}{=}-12\\ -12&=-12 \quad \text{True} \end{aligned}$$
The solution is –5, or the solution set is {–5}.

29.
$$\begin{aligned}-2(x+5)&=-22\\ -2x-10&=-22\\ -2x-10+10&=-22+10\\ -2x&=-12\\ \frac{-2x}{-2}&=\frac{-12}{-2}\\ x&=6\end{aligned}$$
Check:
$$\begin{aligned}-2(x+5)&=-22\\ -2(6+5)&\stackrel{?}{=}-22\\ -2(11)&\stackrel{?}{=}-22\\ -22&=-22 \quad \text{True}\end{aligned}$$
The solution is 6, or the solution set is {6}.

30.
$$\begin{aligned}3(2x+5)&=-21\\ 6x+15&=-21\\ 6x+15-15&=-21-15\\ 6x&=-36\\ \frac{6x}{6}&=\frac{-36}{6}\\ x&=-6\end{aligned}$$
Check:
$$\begin{aligned}3(2x+5)&=-21\\ 3[2(-6)+5]&\stackrel{?}{=}-21\\ 3(-12+5)&\stackrel{?}{=}-21\\ 3(-7)&\stackrel{?}{=}-21\\ -21&=-21 \quad \text{True}\end{aligned}$$
The solution is –6, or the solution set is {–6}.

31.
$$\begin{aligned}5x+4&=-7x+20\\ 5x+4-4&=-7x+20-4\\ 5x&=-7x+16\\ 7x+5x&=7x-7x+16\\ 12x&=16\\ \frac{12x}{12}&=\frac{16}{12}\\ x&=\frac{4}{3}\end{aligned}$$
Check:
$$\begin{aligned}5x+4&=-7x+20\\ 5\left(\frac{4}{3}\right)+4&\stackrel{?}{=}-7\left(\frac{4}{3}\right)+20\\ \frac{20}{3}+\frac{12}{3}&\stackrel{?}{=}\frac{-28}{3}+\frac{60}{3}\\ \frac{32}{3}&=\frac{32}{3} \quad \text{True}\end{aligned}$$
The solution is $\frac{4}{3}$, or the solution set is $\left\{\frac{4}{3}\right\}$.

32.
$$\begin{aligned}-3x+5&=x-15\\ -3x+5-5&=x-15-5\\ -3x&=x-20\\ -x-3x&=-x+x-20\\ -4x&=-20\\ \frac{-4x}{-4}&=\frac{-20}{-4}\\ x&=5\end{aligned}$$
Check:
$$\begin{aligned}-3x+5&=x-15\\ -3(5)+5&\stackrel{?}{=}5-15\\ -15+5&\stackrel{?}{=}-10\\ -10&=-10 \quad \text{True}\end{aligned}$$
The solution is 5, or the solution set is {5}.

33.
$$\begin{aligned}4(x-5)&=-3x+5x-16\\ 4x-20&=2x-16\\ -2x+4x-20&=-2x+2x-16\\ 2x-20&=-16\\ 2x-20+20&=-16+20\\ 2x&=4\\ \frac{2x}{2}&=\frac{4}{2}\\ x&=2\end{aligned}$$
Check:
$$\begin{aligned}4(x-5)&=-3x+5x-16\\ 4(2-5)&\stackrel{?}{=}-3(2)+5(2)-16\\ 4(-3)&\stackrel{?}{=}-6+10-16\\ -12&=-12 \quad \text{True}\end{aligned}$$
The solution is 2, or the solution set is {2}.

34.
$$\begin{aligned}4(m+1)&=m+5m-10\\ 4m+4&=6m-10\\ -4m+4m+4&=-4m+6m-10\\ 4&=2m-10\\ 4+10&=2m-10+10\\ 14&=2m\\ \frac{14}{2}&=\frac{2m}{2}\\ 7&=m\end{aligned}$$
Check:
$$\begin{aligned}4(m+1)&=m+5m-10\\ 4(7+1)&\stackrel{?}{=}7+5(7)-10\\ 4(8)&\stackrel{?}{=}7+35-10\\ 32&=32 \quad \text{True}\end{aligned}$$
The solution is 7, or the solution set is {7}.

35. $x+x+4=24$
$$2x+4=24$$
$$2x+4-4=24-4$$
$$2x=20$$
$$\frac{2x}{2}=\frac{20}{2}$$
$$x=10$$
$x+4=14$
Skye is 14 years old.

36. $2w+2(w+10)=96$
$$2w+2w+20=96$$
$$4w+20=96$$
$$4w+20-20=96-20$$
$$4w=76$$
$$\frac{4w}{4}=\frac{76}{4}$$
$$w=19$$
$w+10=29$
The width is 19 yards and the length is 29 yards.

37. $\frac{6}{7}x+3=\frac{1}{2}$
$$14\left(\frac{6}{7}x+3\right)=14\left(\frac{1}{2}\right)$$
$$14\left(\frac{6}{7}x\right)+14(3)=7$$
$$12x+42=7$$
$$12x+42-42=7-42$$
$$12x=-35$$
$$\frac{12x}{12}=\frac{-35}{12}$$
$$x=-\frac{35}{12}$$

Check: $\frac{6}{7}x+3=\frac{1}{2}$
$$\frac{6}{7}\left(-\frac{35}{12}\right)+3\stackrel{?}{=}\frac{1}{2}$$
$$-\frac{5}{2}+\frac{6}{2}\stackrel{?}{=}\frac{1}{2}$$
$$\frac{1}{2}=\frac{1}{2} \quad \text{True}$$

The solution is $-\frac{35}{12}$, or the solution set is $\left\{-\frac{35}{12}\right\}$.

38. $\frac{1}{4}x+6=\frac{5}{6}$
$$12\left(\frac{1}{4}x+6\right)=12\left(\frac{5}{6}\right)$$
$$12\left(\frac{1}{4}x\right)+12(6)=2\cdot 5$$
$$3x+72=10$$
$$3x+72-72=10-72$$
$$3x=-62$$
$$\frac{3x}{3}=\frac{-62}{3}$$
$$x=-\frac{62}{3}$$

Check: $\frac{1}{4}x+6=\frac{5}{6}$
$$\frac{1}{4}\left(-\frac{62}{3}\right)+6\stackrel{?}{=}\frac{5}{6}$$
$$-\frac{31}{6}+\frac{36}{6}\stackrel{?}{=}\frac{5}{6}$$
$$\frac{5}{6}=\frac{5}{6} \quad \text{True}$$

The solution is $-\frac{62}{3}$, or the solution set is $\left\{-\frac{62}{3}\right\}$.

39. $\frac{n}{2}+\frac{2}{3}=\frac{n}{6}$
$$6\left(\frac{n}{2}+\frac{2}{3}\right)=6\left(\frac{n}{6}\right)$$
$$6\left(\frac{n}{2}\right)+6\left(\frac{2}{3}\right)=n$$
$$3n+4=n$$
$$-3n+3n+4=-3n+n$$
$$4=-2n$$
$$\frac{4}{-2}=\frac{-2n}{-2}$$
$$-2=n$$

Check: $\frac{n}{2}+\frac{2}{3}=\frac{n}{6}$
$$\frac{-2}{2}+\frac{2}{3}\stackrel{?}{=}\frac{-2}{6}$$
$$-\frac{3}{3}+\frac{2}{3}\stackrel{?}{=}-\frac{1}{3}$$
$$-\frac{1}{3}=-\frac{1}{3} \quad \text{True}$$

The solution is –2, or the solution set is {–2}.

40.
$$\frac{m}{8}+\frac{m}{2}=\frac{3}{4}$$
$$8\left(\frac{m}{8}+\frac{m}{2}\right)=8\left(\frac{3}{4}\right)$$
$$8\left(\frac{m}{8}\right)+8\left(\frac{m}{2}\right)=2\cdot 3$$
$$m+4m=6$$
$$5m=6$$
$$\frac{5m}{5}=\frac{6}{5}$$
$$m=\frac{6}{5}$$

Check: $\frac{m}{8}+\frac{m}{2}=\frac{3}{4}$
$$\frac{\frac{6}{5}}{8}+\frac{\frac{6}{5}}{2}\stackrel{?}{=}\frac{3}{4}$$
$$\frac{3}{20}+\frac{6}{10}\stackrel{?}{=}\frac{3}{4}$$
$$\frac{15}{20}=\frac{15}{20}\quad\text{True}$$

The solution is $\frac{6}{5}$, or the solution set is $\left\{\frac{6}{5}\right\}$.

41.
$$1.2r=-1+2.8$$
$$1.2r=1.8$$
$$10(1.2r)=10(1.8)$$
$$12r=18$$
$$\frac{12r}{12}=\frac{18}{12}$$
$$r=\frac{3}{2}$$

Check: $1.2r=-1+2.8$
$$1.2\left(\frac{3}{2}\right)\stackrel{?}{=}1.8$$
$$1.8=1.8\quad\text{True}$$

The solution is $\frac{3}{2}$, or the solution set is $\left\{\frac{3}{2}\right\}$.

42.
$$0.2x+0.5x=2.1$$
$$0.7x=2.1$$
$$10(0.7x)=10(2.1)$$
$$7x=21$$
$$\frac{7x}{7}=\frac{21}{7}$$
$$x=3$$

Check: $0.2x+0.5x=2.1$
$$0.2(3)+0.5(3)\stackrel{?}{=}2.1$$
$$0.6+1.5\stackrel{?}{=}2.1$$
$$2.1=2.1\quad\text{True}$$

The solution is 3, or the solution set is {3}.

43.
$$1.2m-3.2=0.8m-1.6$$
$$10(1.2m-3.2)=10(0.8m-1.6)$$
$$12m-32=8m-16$$
$$-8m+12m-32=-8m+8m-16$$
$$4m-32=-16$$
$$4m-32+32=-16+32$$
$$4m=16$$
$$\frac{4m}{4}=\frac{16}{4}$$
$$m=4$$

Check: $1.2m-3.2=0.8m-1.6$
$$1.2(4)-3.2\stackrel{?}{=}0.8(4)-1.6$$
$$4.8-3.2\stackrel{?}{=}3.2-1.6$$
$$1.6=1.6\quad\text{True}$$

The solution is 4, or the solution set is {4}.

44.
$$0.3m+0.8=0.5m+1$$
$$10(0.3m+0.8)=10(0.5m+1)$$
$$3m+8=5m+10$$
$$-3m+3m+8=-3m+5m+10$$
$$8=2m+10$$
$$8-10=2m+10-10$$
$$-2=2m$$
$$\frac{-2}{2}=\frac{2m}{2}$$
$$-1=m$$

Check: $0.3m+0.8=0.5m+1$
$$0.3(-1)+0.8\stackrel{?}{=}0.5(-1)+1$$
$$-0.3+0.8\stackrel{?}{=}-0.5+1$$
$$0.5=0.5\quad\text{True}$$

The solution is −1, or the solution set is {−1}.

45.
$$\frac{1}{2}(x+5)=\frac{3}{4}$$
$$4\cdot\frac{1}{2}(x+5)=4\cdot\frac{3}{4}$$
$$2(x+5)=3$$
$$2x+10=3$$
$$2x+10-10=3-10$$
$$2x=-7$$
$$\frac{2x}{2}=\frac{-7}{2}$$
$$x=-\frac{7}{2}$$

Check:
$$\begin{aligned}\frac{1}{2}(x+5)&=\frac{3}{4}\\ \frac{1}{2}\left(-\frac{7}{2}+5\right)&\stackrel{?}{=}\frac{3}{4}\\ \frac{1}{2}\left(-\frac{7}{2}+\frac{10}{2}\right)&\stackrel{?}{=}\frac{3}{4}\\ \frac{1}{2}\left(\frac{3}{2}\right)&\stackrel{?}{=}\frac{3}{4}\\ \frac{3}{4}&=\frac{3}{4}\quad\text{True}\end{aligned}$$

The solution is $-\frac{7}{2}$, or the solution set is $\left\{-\frac{7}{2}\right\}$.

46.
$$\begin{aligned}-\frac{1}{6}(x-1)&=\frac{2}{3}\\ -6\cdot\left[-\frac{1}{6}(x-1)\right]&=-6\cdot\frac{2}{3}\\ x-1&=-4\\ x-1+1&=-4+1\\ x&=-3\end{aligned}$$

Check:
$$\begin{aligned}-\frac{1}{6}(x-1)&=\frac{2}{3}\\ -\frac{1}{6}(-3-1)&\stackrel{?}{=}\frac{2}{3}\\ -\frac{1}{6}(-4)&\stackrel{?}{=}\frac{2}{3}\\ \frac{2}{3}&=\frac{2}{3}\end{aligned}$$

The solution is –3, or the solution set is {–3}.

47.
$$\begin{aligned}0.1(x+80)&=-0.2+14\\ 10[0.1(x+80)]&=10(-0.2+14)\\ 1(x+80)&=-2+140\\ x+80&=138\\ x+80-80&=138-80\\ x&=58\end{aligned}$$

Check:
$$\begin{aligned}0.1(x+80)&=-0.2+14\\ 0.1(58+80)&\stackrel{?}{=}-0.2+14\\ 0.1(138)&\stackrel{?}{=}13.8\\ 13.8&=13.8\quad\text{True}\end{aligned}$$

The solution is 58, or the solution set is {58}.

48.
$$\begin{aligned}0.35(x+6)&=0.45(x+7)\\ 100[0.35(x+6)]&=100[0.45(x+7)]\\ 35(x+6)&=45(x+7)\\ 35x+210&=45x+315\\ -35x+35x+210&=-35x+45x+315\\ 210&=10x+315\\ 210-315&=10x+315-315\\ -105&=10x\\ \frac{-105}{10}&=\frac{10x}{10}\\ -10.5&=x\end{aligned}$$

Check:
$$\begin{aligned}0.35(x+6)&=0.45(x+7)\\ 0.35(-10.5+6)&\stackrel{?}{=}0.45(-10.5+7)\\ 0.35(-4.5)&\stackrel{?}{=}0.45(-3.5)\\ -1.575&=-1.575\quad\text{True}\end{aligned}$$

The solution is –10.5, or the solution set is {–10.5}.

49.
$$\begin{aligned}4x+2x-10&=6x+5\\ 6x-10&=6x+5\\ -6x+6x-10&=-6x+6x+5\\ -10&=5\end{aligned}$$

This is a false statement, so the equation is a contradiction. The solution set is ∅ or { }.

50.
$$\begin{aligned}-2(x+5)&=-5x+3x+2\\ -2x-10&=-2x+2\\ 2x-2x-10&=2x-2x+2\\ -10&=2\end{aligned}$$

This is a false statement, so the equation is a contradiction. The solution set is ∅ or { }.

51.
$$\begin{aligned}-5(2n+10)&=6n-50\\ -10n-50&=6n-50\\ 10n-10n-50&=10n+6n-50\\ -50&=16n-50\\ -50+50&=16n-50+50\\ 0&=16n\\ \frac{0}{16}&=\frac{16n}{16}\\ 0&=n\end{aligned}$$

This is a conditional equation. The solution set is {0}.

52.
$$\begin{aligned} 8m+10&=-2(7m-5)\\ 8m+10&=-14m+10\\ 14m+8m+10&=14m-14m+10\\ 22m+10&=10\\ 22m+10-10&=10-10\\ 22m&=0\\ \frac{22m}{22}&=\frac{0}{22}\\ m&=0 \end{aligned}$$
This is a conditional equation. The solution set is {0}.

53.
$$\begin{aligned} 10x-2x+18&=2(4x+9)\\ 8x+18&=8x+18\\ -8x+8x+18&=-8x+8x+18\\ 18&=18 \end{aligned}$$
This is a true statement. The equation is an identity. The solution set is the set of all real numbers.

54.
$$\begin{aligned} -3(2x-8)&=-3x-3x+24\\ -6x+24&=-6x+24\\ 6x-6x+24&=6x-6x+24\\ 24&=24 \end{aligned}$$
This is a true statement. The equation is an identity. The solution set is the set of all real numbers.

55.
$$\begin{aligned} p-0.20p&=12.60\\ 0.8p&=12.60\\ 10(0.8p)&=10(12.60)\\ 8p&=126\\ \frac{8p}{8}&=\frac{126}{8}\\ p&=15.75 \end{aligned}$$
The shirt's original price was $15.75.

56.
$$\begin{aligned} 0.10x+0.05(2x-1)&=0.55\\ 0.10x+0.10x-0.05&=0.55\\ 0.2x-0.05&=0.55\\ 0.2x&=0.6\\ 10\cdot 0.2x&=10\cdot 0.6\\ 2x&=6\\ \frac{2x}{2}&=\frac{6}{2}\\ x&=3 \end{aligned}$$
Juanita found 3 dimes.

57. $A = lw$; $l = 8$, $w = 6$
$A = 8(6) = 48$
The area is 48 square inches.

58. $P = 4s$; $s = 16$
$P = 4(16) = 64$
The perimeter is 64 centimeters.

59. $P = 2l + 2w$; $P = 16$, $l = \frac{13}{2}$
$$\begin{aligned} 16&=2\left(\frac{13}{2}\right)+2w\\ 16&=13+2w\\ -13+16&=-13+13+2w\\ 3&=2w\\ \frac{3}{2}&=\frac{2w}{2}\\ \frac{3}{2}&=w \end{aligned}$$
The width is $\frac{3}{2}$ yards.

60. $C = \pi d$; $d = \frac{15}{\pi}$
$$C=\pi\left(\frac{15}{\pi}\right)=15$$
The circumference is 15 millimeters.

61.
$$\begin{aligned} V&=LWH\\ \frac{V}{LW}&=\frac{LWH}{LW}\\ \frac{V}{LW}&=H \end{aligned}$$

62.
$$\begin{aligned} I&=Prt\\ \frac{I}{rt}&=\frac{Prt}{rt}\\ \frac{I}{rt}&=P \end{aligned}$$

63.
$$\begin{aligned} S&=2LW+2LH+2WH\\ S-2LH&=2LW+2LH+2WH-2LH\\ S-2LH&=2LW+2WH\\ S-2LH&=W(2L+2H)\\ \frac{S-2LH}{2L+2H}&=\frac{W(2L+2H)}{2L+2H}\\ \frac{S-2LH}{2L+2H}&=W \end{aligned}$$

64.
$$\rho = mv + MV$$
$$\rho - mv = mv + MV - mv$$
$$\rho - mv = MV$$
$$\frac{\rho - mv}{V} = \frac{MV}{V}$$
$$\frac{\rho - mv}{V} = M$$

65.
$$2x + 3y = 10$$
$$-2x + 2x + 3y = -2x + 10$$
$$3y = -2x + 10$$
$$\frac{3y}{3} = \frac{-2x + 10}{3}$$
$$y = \frac{-2x + 10}{3}$$

66.
$$6x - 7y = 14$$
$$6x - 7y + 7y = 14 + 7y$$
$$6x = 14 + 7y$$
$$\frac{6x}{6} = \frac{14 + 7y}{6}$$
$$x = \frac{14 + 7y}{6}$$

67. (a)
$$A = P(1 + r)^t$$
$$\frac{A}{(1 + r)^t} = \frac{P(1 + r)^t}{(1 + r)^t}$$
$$\frac{A}{(1 + r)^t} = P$$

(b) $P = \frac{A}{(1 + r)^t}$; $A = 3000$, $t = 6$,

$r = 5\% = 0.05$

$$P = \frac{3000}{(1 + 0.05)^6} = \frac{3000}{1.05^6} \approx 2238.65$$

You would have to deposit \$2238.65.

68. (a)
$$A = 2\pi rh + 2\pi r^2$$
$$A - 2\pi r^2 = 2\pi rh + 2\pi r^2 - 2\pi r^2$$
$$A - 2\pi r^2 = 2\pi rh$$
$$\frac{A - 2\pi r^2}{2\pi r} = \frac{2\pi rh}{2\pi r}$$
$$\frac{A - 2\pi r^2}{2\pi r} = h$$

(b) $h = \frac{A - 2\pi r^2}{2\pi r}$; $A = 72\pi$, $r = 4$

$$h = \frac{72\pi - 2\pi(4^2)}{2\pi(4)}$$
$$h = \frac{\pi(72 - 32)}{\pi(8)}$$
$$h = \frac{40}{8}$$
$$h = 5$$

The height is 5 centimeters.

69. $I = Prt$; $P = 500$, $r = 3\% = 0.03$, $t = \frac{9}{12} = \frac{3}{4}$

$$I = 500(0.03)\left(\frac{3}{4}\right) = 11.25$$

Samuel's investment will earn \$11.25.

70. $d = 3$, $r = \frac{3}{2}$

$$A = \pi r^2 = \pi\left(\frac{3}{2}\right)^2 = \frac{9}{4}\pi \approx 7.069$$

The area is about 7.1 square feet.

71. the difference between a number and 6: $x - 6$

72. eight subtracted from a number: $x - 8$

73. the product of –8 and a number: $-8x$

74. the quotient of a number and 10: $\frac{x}{10}$

75. twice the sum of 6 and a number: $2(6 + x)$

76. four times the difference of 5 and a number: $4(5 - x)$

77. $6 + x = 2x + 5$

78. $6x - 10 = 2x + 1$

79. $x - 8 = \frac{1}{2}x$

80. $\frac{6}{x} = 10 + x$

81. $4(2x + 8) = 16$

82. $5(2x - 8) = -24$

83. Let s be Sarah's age. Then $s + 7$ is Jacob's age.

84. Let c be Consuelo's speed. Then $2c$ is José's speed.

85. Let m be Max's amount. Then $m - 6$ is Irene's amount.

86. Let v be Victor's amount. Then $350 - v$ is Larry's amount.

87. We want to find Lee Lai's weight one year ago. Let n be the weight.

$$n - 28 = 125$$
$$n = 153$$

Is the difference between 153 and 125 28? Yes. Lee Lai's weight was 153 pounds one year ago.

88. We want to find three consecutive integers. Let n be the first integer. Then $n + 1$ and $n + 2$ are the next two integers.

$$n + (n + 1) + (n + 2) = 39$$
$$3n + 3 = 39$$
$$3n = 36$$
$$n = 12$$

If $n = 12$, then $n + 1 = 13$ and $n + 2 = 14$. Are the numbers 12, 13, 14 consecutive integers? Yes. Do they sum to 39? Yes. The integers are 12, 13, and 14.

89. We want to find how much each will receive. Let j be the amount received by Juan. Then $j - 2000$ is the amount received by Roberto.

$$j + (j - 2000) = 20,000$$
$$2j - 2000 = 20,000$$
$$2j = 22,000$$
$$j = 11,000$$

If $j = 11,000$, then $j - 2000 = 9000$. Do 11,000 and 9000 differ by 2000? Yes. Do 11,000 and 9000 sum to 20,000? Yes. Juan will receive $11,000 and Roberto will receive $9000.

90. We want to find the number of miles for which the cost will be the same. Let x be the number of miles driven. ABC-Rental charges $30 + 0.15x$ and U-Do-It Rental charges $15 + 0.3x$.

$$30 + 0.15x = 15 + 0.3x$$
$$30 = 15 + 0.15x$$
$$15 = 0.15x$$
$$100 = x$$

ABC-Rental's cost will be $30 + 0.15(100) = 30 + 15 = \$45$ and U-Do-It-Rental's cost will be $15 + 0.3(100) = 15 + 30 = \$45$, and they are the same. The cost will be the same for 100 miles.

91. $x = 0.065(80)$

$x = 5.2$

5.2 is 6.5% of 80.

92.
$$18 = 0.3x$$
$$\frac{18}{0.3} = \frac{0.3x}{0.3}$$
$$60 = x$$

18 is 30% of 60.

93.
$$15.6 = p \cdot 120$$
$$\frac{15.6}{120} = \frac{120p}{120}$$
$$0.13 = p$$
$$13\% = p$$

15.6 is 13% of 120.

94.
$$1.1 \cdot x = 55$$
$$\frac{1.1x}{1.1} = \frac{55}{1.1}$$
$$x = 50$$

110% of 50 is 55.

95. Let x be the cost before tax. Then $0.06x$ is the tax amount.

$$x + 0.06x = 19.61$$
$$1.06x = 19.61$$
$$x = 18.5$$

The leotard cost $18.50 before sales tax.

96. Let x be the previous hourly fee. Then $0.085x$ is the amount of the increase.

$$x + 0.085x = 32.55$$
$$1.085x = 32.55$$
$$x = 30$$

Mei Ling's previous hourly fee was $30.

97. Let x be the sweater's original price. Then $0.70x$ is the discounted amount.

$$x - 0.70x = 12$$
$$0.3x = 12$$
$$x = 40$$

The sweater's original price was $40.

98. Let x be the store's price. Then $0.80x$ is the markup amount.

$$x + 0.80x = 360$$
$$1.8x = 360$$
$$x = 200$$

The store paid $200 for the suit.

99. Let x be the total value of the computers. Then Tanya earns $500 plus $0.02x$.

$$500 + 0.02x = 3000$$
$$0.02x = 2500$$
$$x = 125{,}000$$

Tanya must sell computers worth a total of $125,000.

100. Let x be the winner's amount. then $0.80x$ is the loser's amount.

$$x + 0.80x = 900$$
$$1.8x = 900$$
$$x = 500$$

If $x = 500$, then $0.8(500) = 400$. The winner received 500 votes, whereas the loser received 400 votes.

101. We want to find complementary angles. The measures of complementary angles sum to 90°. Let x be the measure of the second angle. Then $6x + 20$ is the measure of the first angle.

$$x + (6x + 20) = 90$$
$$7x + 20 = 90$$
$$7x = 70$$
$$x = 10$$

If $x = 10$, then
$6x + 20 = 6(10) + 20 = 60 + 20 = 80$. The measures of the angles are 10° and 80°.

102. We want to find supplementary angles. The measures of supplementary angles sum to 180°. Let x be the measure of the second angle. Then $2x - 60$ is the measure of the first angle.

$$x + (2x - 60) = 180$$
$$3x - 60 = 180$$
$$3x = 240$$
$$x = 80$$

If $x = 80$, then
$2x - 60 = 2(80) - 60 = 160 - 60 = 100$. The measures of the angles are 80° and 100°.

103. We want to find the measures of the angles of the triangle. The measures of the angles of a triangle sum to 180°. Let x be the measure of the first angle. Then $2x$ is the measure of the second and $2x + 30$ is the measure of the third.

$$x + (2x) + (2x + 30) = 180$$
$$5x + 30 = 180$$
$$5x = 150$$
$$x = 30$$

If $x = 30$, then $2x = 60$, and
$2x + 30 = 60 + 30 = 90$. The measures of the angles are 30°, 60°, and 90°.

104. We want to find the measures of the angles of the triangle. The measures of the angles of a triangle sum to 180°. Let x be the measure of the first angle. Then $x - 5$ is the measure of the second angle and $2(x - 5) - 5$ is the measure of the third angle.

$$x + (x - 5) + 2(x - 5) - 5 = 180$$
$$x + x - 5 + 2x - 10 - 5 = 180$$
$$4x - 20 = 180$$
$$4x = 200$$
$$x = 50$$

If $x = 50$, then $x - 5 = 45$ and
$2(x - 5) - 5 = 2(45) - 5 = 85$. The measures of the angles are 50°, 45°, and 85°.

105. We want to find the dimensions of the rectangle. Let w be the width of the rectangle. Then $2w + 15$ is the length of the rectangle. The perimeter of a rectangle is the sum of twice the length and twice the width. The perimeter is 78 inches.

$$2(2w + 15) + 2w = 78$$
$$4w + 30 + 2w = 78$$
$$6w + 30 = 78$$
$$6w = 48$$
$$w = 8$$

If $w = 8$, then $2w + 15 = 2(8) + 15 = 31$. The length is 31 inches and the width is 8 inches.

106. We want to find the dimensions of the rectangle. Let w be the width of the rectangle. then $4w$ is the length of the rectangle. The perimeter of a rectangle is the sum of twice the length and twice the width. The perimeter is 70 cm.

$$2(4w) + 2w = 70$$
$$8w + 2w = 70$$
$$10w = 70$$
$$w = 7$$

If $w = 7$, then $4w = 28$. The width of the rectangle is 7 cm and the length is 28 cm.

107. **(a)** We want to find the dimensions of the rectangular garden. Let l be the length. Then the width is $2l$. The perimeter is twice the length plus twice the width and is 120 feet.

$$2l + 2(2l) = 120$$
$$2l + 4l = 120$$
$$6l = 120$$
$$l = 20$$

If $l = 20$, then $2l = 40$. The garden's length is 20 feet and the width is 40 feet.

(b) $A = lw = 20(40) = 800$
The area of the garden is 800 square feet.

108. The area of a trapezoid is $A=\frac{1}{2}h(B+b)$, where h is the height and the bases are B and b. Let B be the longer base. Then $B - 10$ is the shorter base. The height is 80 feet and the area is 3600 square feet.

$$\frac{1}{2}(80)[B+(B-10)]=3600$$
$$40(2B-10)=3600$$
$$80B-400=3600$$
$$80B=4000$$
$$B=50$$

If $B = 50$, then $B - 10 = 40$. The bases are 50 feet and 40 feet.

109. Let t be the time at which they are 35 miles apart.

	Rate	· Time	= Distance
slow	18	t	$18t$
fast	25	t	$25t$

The difference of the distances is 35, since they are traveling in the same direction.

$$25t-18t=35$$
$$7t=35$$
$$t=5$$

They will be 35 miles apart after 5 hours.

110. Let r be the speed of the faster train.

	Rate	· Time	= Distance
East	r	6	$6r$
West	$r - 10$	6	$6(r - 10)$

The sum of their distances is 720, since they are traveling in opposite directions.

$$6r+6(r-10)=720$$
$$6r+6r-60=720$$
$$12r=780$$
$$r=65$$

The faster train is traveling at 65 mph.

111. $x\le -3$

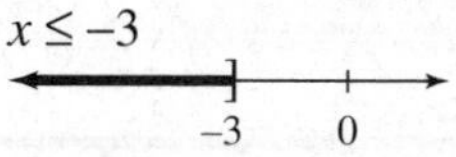

−3 0

112. $x>4$

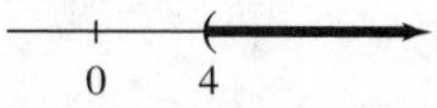

0 4

113. $m<2$

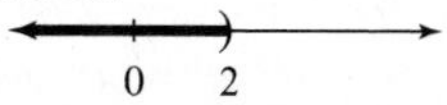

0 2

114. $m\ge -5$

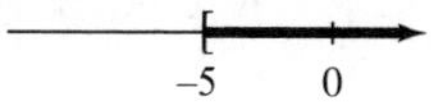

−5 0

115. $0<n$

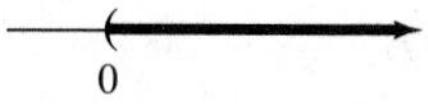

0

116. $-3\le n$

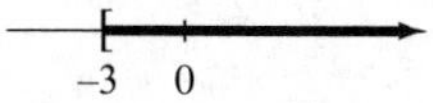

−3 0

117. $x<-4$

$(-\infty, -4)$

118. $x\ge 7$

$[7, \infty)$

119. $[2, \infty)$

120. $(-\infty, 3)$

121.

$$4x+3<2x-10$$
$$2x+3<-10$$
$$2x<-13$$
$$x<-\frac{13}{2}$$

$$\left\{x \middle| x<-\frac{13}{2}\right\}$$

$$\left(-\infty, -\frac{13}{2}\right)$$

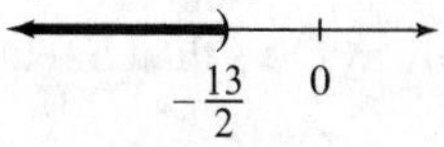

$-\frac{13}{2}$ 0

122.

$$3x-5\ge -12$$
$$3x\ge -7$$
$$x\ge -\frac{7}{3}$$

$$\left\{x \middle| x\ge -\frac{7}{3}\right\}$$

$$\left[-\frac{7}{3}, \infty\right)$$

$-\frac{7}{3}$ 0

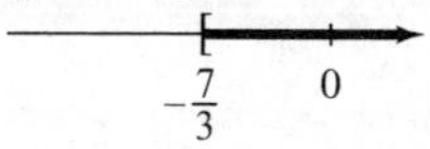

123. $-4(x-1) \le x+8$
$-4x+4 \le x+8$
$-5x+4 \le 8$
$-5x \le 4$
$x \ge -\frac{4}{5}$

$\left\{x \middle| x \ge -\frac{4}{5}\right\}$

$\left[-\frac{4}{5}, \infty\right)$

$-\frac{4}{5}$ 0

124. $6x-10 < 7x+2$
$-10 < x+2$
$-12 < x$
$\{x|x > -12\}$
$(-12, \infty)$

−12 0

125. $-3(x+7) > -x-2x$
$-3x-21 > -3x$
$-21 > 0$
This is a false statement. Therefore, there is no solution. The solution set is $\varnothing$ or { }.

0

126. $4x+10 \le 2(2x+7)$
$4x+10 \le 4x+14$
$10 \le 14$
The solution to this inequality is all real numbers, since 10 is always less than or equal to 14.
$\{x|x \text{ is any real number}\}$

127. $\frac{1}{2}(3x-1) > \frac{2}{3}(x+3)$
$6 \cdot \frac{1}{2}(3x-1) > 6 \cdot \frac{2}{3}(x+3)$
$3(3x-1) > 4(x+3)$
$9x-3 > 4x+12$
$5x-3 > 12$
$5x > 15$
$x > 3$
$\{x|x > 3\}$
$(3, \infty)$

0 3

128. $\frac{5}{4}x+2 < \frac{5}{6}x-\frac{7}{6}$
$12\left(\frac{5}{4}x+2\right) < 12\left(\frac{5}{6}x-\frac{7}{6}\right)$
$15x+24 < 10x-14$
$5x+24 < -14$
$5x < -38$
$x < -\frac{38}{5}$

$\left\{x \middle| x < -\frac{38}{5}\right\}$

$\left(-\infty, -\frac{38}{5}\right)$

$-\frac{38}{5}$ 0

129. Let m be the number of miles driven.
$19.95+0.2m \le 32.95$
$0.2m \le 13$
$m \le 65$
A customer can drive at most 65 miles.

130. Let s be the score on his third game.
$\frac{148+155+s}{3} > 151$
$\frac{303+s}{3} > 151$
$303+s > 453$
$s > 150$
Travis must score more than 150.

Chapter 2 Test

1. $x+3 = -14$
$x+3-3 = -14-3$
$x = -17$
Check:
$x+3 = -14$
$-17+3 \stackrel{?}{=} -14$
$-14 = -14$ True
The solution is −17, or the solution set is {−17}.

2. $-\frac{2}{3}m = \frac{8}{27}$
$-\frac{3}{2} \cdot \left(-\frac{2}{3}m\right) = -\frac{3}{2} \cdot \frac{8}{27}$
$m = -\frac{4}{9}$

Check: $$-\frac{2}{3}m=\frac{8}{27}$$
$$-\frac{2}{3}\cdot\left(-\frac{4}{9}\right)\stackrel{?}{=}\frac{8}{27}$$
$$\frac{8}{27}=\frac{8}{27}\quad\text{True}$$

The solution is $-\frac{4}{9}$, or the solution set is $\left\{-\frac{4}{9}\right\}$.

3.
$$5(2x-4)=5x$$
$$10x-20=5x$$
$$-10x+10x-20=-10x+5x$$
$$-20=-5x$$
$$\frac{-20}{-5}=\frac{-5x}{-5}$$
$$4=x$$

Check: $$5(2x-4)=5x$$
$$5(2\cdot 4-4)\stackrel{?}{=}5\cdot 4$$
$$5(8-4)\stackrel{?}{=}20$$
$$5(4)\stackrel{?}{=}20$$
$$20=20\quad\text{True}$$

The solution is 4, or the solution set is {4}.

4.
$$-2(x-5)=5(-3x+4)$$
$$-2x+10=-15x+20$$
$$2x-2x+10=2x-15x+20$$
$$10=-13x+20$$
$$10-20=-13x+20-20$$
$$-10=-13x$$
$$\frac{-10}{-13}=\frac{-13x}{-13}$$
$$\frac{10}{13}=x$$

Check: $$-2(x-5)=5(-3x+4)$$
$$-2\left(\frac{10}{13}-5\right)\stackrel{?}{=}5\left(-3\cdot\frac{10}{13}+4\right)$$
$$-2\left(\frac{10}{13}-\frac{65}{13}\right)\stackrel{?}{=}5\left(-\frac{30}{13}+\frac{52}{13}\right)$$
$$-2\left(-\frac{55}{13}\right)\stackrel{?}{=}5\left(\frac{22}{13}\right)$$
$$\frac{110}{13}=\frac{110}{13}\quad\text{True}$$

The solution is $\frac{10}{13}$ or the solution set is $\left\{\frac{10}{13}\right\}$.

5.
$$-\frac{2}{3}x+\frac{3}{4}=\frac{1}{3}$$
$$12\left(-\frac{2}{3}x+\frac{3}{4}\right)=12\left(\frac{1}{3}\right)$$
$$12\left(-\frac{2}{3}x\right)+12\left(\frac{3}{4}\right)=12\left(\frac{1}{3}\right)$$
$$-8x+9=4$$
$$-8x+9-9=4-9$$
$$-8x=-5$$
$$\frac{-8x}{8}=\frac{-5}{-8}$$
$$x=\frac{5}{8}$$

Check: $$-\frac{2}{3}x+\frac{3}{4}=\frac{1}{3}$$
$$-\frac{2}{3}\left(\frac{5}{8}\right)+\frac{3}{4}\stackrel{?}{=}\frac{1}{3}$$
$$-\frac{5}{12}+\frac{9}{12}\stackrel{?}{=}\frac{1}{3}$$
$$\frac{4}{12}\stackrel{?}{=}\frac{1}{3}$$
$$\frac{1}{3}=\frac{1}{3}\quad\text{True}$$

The solution is $\frac{5}{8}$ or the solution set is $\left\{\frac{5}{8}\right\}$.

6.
$$-0.6+0.4y=1.4$$
$$10(-0.6+0.4y)=10(1.4)$$
$$-6+4y=14$$
$$6-6+4y=6+14$$
$$4y=20$$
$$\frac{4y}{4}=\frac{20}{4}$$
$$y=5$$

Check: $$-0.6+0.4y=1.4$$
$$-0.6+0.4(5)\stackrel{?}{=}1.4$$
$$-0.6+2\stackrel{?}{=}1.4$$
$$1.4=1.4\quad\text{True}$$

The solution is 5, or the solution set is {5}.

7.
$$8x+3(2-x)=5(x+2)$$
$$8x+6-3x=5x+10$$
$$6+5x=5x+10$$
$$6+5x-5x=5x+10-5x$$
$$6=10$$

This is a false statement. The equation is a contradiction. The solution set is ∅ or { }.

8.
$$\begin{aligned} 2(x+7) &= 2x-2+16 \\ 2x+14 &= 2x+14 \\ -2x+2x+14 &= -2x+2x+14 \\ 14 &= 14 \end{aligned}$$
This is a true statement. The equation is an identity. The solution set is the set of all real numbers.

9. (a)
$$\begin{aligned} V &= lwh \\ \frac{V}{wh} &= \frac{lwh}{wh} \\ \frac{V}{wh} &= l \end{aligned}$$

(b) $l = \frac{V}{wh}$; $V = 540,\ w = 6,\ h = 10$

$$l = \frac{540}{6(10)} = \frac{540}{60} = 9$$
The length is 9 inches.

10. (a)
$$\begin{aligned} 2x+3y &= 12 \\ -2x+2x+3y &= -2x+12 \\ 3y &= -2x+12 \\ \frac{3y}{3} &= \frac{-2x+12}{3} \\ y &= -\frac{2}{3}x+4 \end{aligned}$$

(b) $y = -\frac{2}{3}x+4;\ x = 8$
$$\begin{aligned} y &= -\frac{2}{3}(8)+4 \\ y &= -\frac{16}{3}+\frac{12}{3} \\ y &= -\frac{4}{3} \end{aligned}$$

11. Let x be the number.
$6(x - 8) = 2x - 5$

12.
$$\begin{aligned} 18 &= 0.30x \\ \frac{18}{0.30} &= \frac{0.30x}{0.30} \\ 60 &= x \end{aligned}$$
18 is 30% of 60.

13. We want to find three consecutive integers. Let n be the first integer. Then $n + 1$ and $n + 2$ are the next two integers, respectively. They sum to 48.
$$\begin{aligned} n+(n+1)+(n+2) &= 48 \\ 3n+3 &= 48 \\ 3n &= 45 \\ n &= 15 \end{aligned}$$
If $n = 15$, then $n + 1 = 16$ and $n + 2 = 17$. Do 15, 16, and 17 sum to 48? Yes. Are 15, 16, and 17 consecutive integers? Yes. The integers are 15, 16, and 17.

14. We need to find the lengths of the three sides. Let m be the length of the middle side. Then the length of the longest side is $m + 2$, and the length of the shortest side is $m - 14$. The perimeter, or the sum of the three sides, is 60.
$$\begin{aligned} m+(m+2)+(m-14) &= 60 \\ 3m-12 &= 60 \\ 3m &= 72 \\ m &= 24 \end{aligned}$$
If $m = 24$, then $m + 2 = 26$, and $m - 14 = 10$. Do 24, 26, and 10 sum to 60? Yes. The lengths of the sides are 10 inches, 24 inches, and 26 inches.

15. Let t be the time at which they are 350 miles apart.

	Rate	· Time	= Distance
Kimberly	40	t	$40t$
Clay	60	t	$60t$

The sum of their distances is 350 since they are traveling in opposite directions.
$$\begin{aligned} 40t+60t &= 350 \\ 100t &= 350 \\ t &= 3.5 \end{aligned}$$
They will be 350 miles apart in 3.5 hours.

16. Let x be the length of the shorter piece. Then $3x + 1$ is the length of the longer piece. The sum of the lengths of the pieces is 21.
$$\begin{aligned} x+(3x+1) &= 21 \\ 4x+1 &= 21 \\ 4x &= 20 \\ x &= 5 \end{aligned}$$
$3x + 1 = 3(5) + 1 = 15 + 1 = 16$
The shorter piece is 5 feet and the longer piece is 16 feet.

17. Let x be the original price of the backpack. Then the discount amount was $0.20x$

$$x - 0.20x = 28.80$$
$$0.80x = 28.80$$
$$10 \cdot 0.8x = 10 \cdot 28.8$$
$$8x = 288$$
$$x = 36$$

The original price of the backpack was $36.

18. $3(2x-5) \le x+15$

$$6x - 15 \le x + 15$$
$$5x - 15 \le 15$$
$$5x \le 30$$
$$x \le 6$$

$\{x | x \le 6\}$

$(-\infty, 6]$

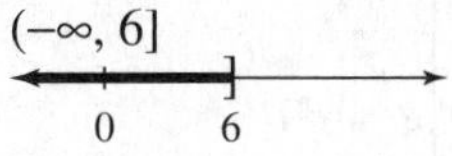

19. $-6x - 4 < 2(x-7)$

$$-6x - 4 < 2x - 14$$
$$-4 < 8x - 14$$
$$10 < 8x$$
$$\frac{5}{4} < x$$

$\left\{x \middle| x > \frac{5}{4}\right\}$

$\left(\frac{5}{4}, \infty\right)$

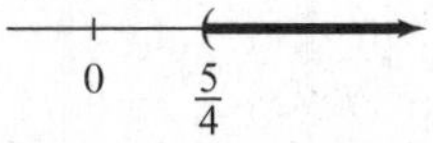

20. Let m be the number of minutes Danielle can use.

$$30 + 0.35m \le 100$$
$$0.35m \le 70$$
$$m \le 200$$

She can use her cell phone at most 200 minutes.

Chapter 3

Section 3.1

Preparing for the Rectangular Coordinate System and Equations in Two Variables

P1.

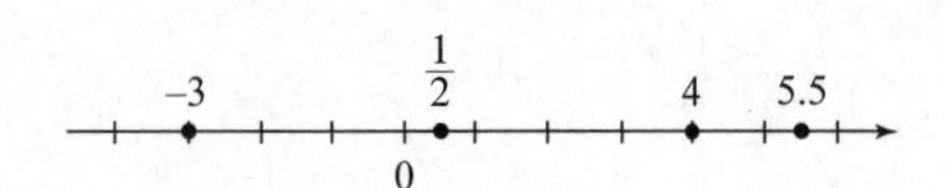

P2. **(a)** $3x + 5$ for $x = 4$:
$3(4) + 5 = 12 + 5 = 17$

(b) $3x + 5$ for $x = -1$:
$3(-1) + 5 = -3 + 5 = 2$

P3. **(a)** $2x - 5y$ for $x = 3$, $y = 2$:
$2(3) - 5(2) = 6 - 10 = -4$

(b) $2x - 5y$ for $x = 1$, $y = -4$:
$2(1) - 5(-4) = 2 + 20 = 22$

P4.
$$3x+5=14$$
$$3x+5-5=14-5$$
$$3x=9$$
$$\frac{3x}{3}=\frac{9}{3}$$
$$x=3$$
The solution set is $\{3\}$.

P5.
$$5(x-3)-2x=3x+12$$
$$5x-15-2x=3x+12$$
$$3x-15=3x+12$$
$$3x-15-3x=3x+12-3x$$
$$-15=12$$
The statement $-15 = 12$ is false, so the equation is a contradiction. The solution set is $\varnothing$.

3.1 Quick Checks

1. In the rectangular coordinate system, we call the horizontal number line the x-axis and we call the vertical real number line the y-axis. The point where these two axes intersect is called the origin.

2. If (x, y) are the coordinates of a point P, then x is called the x-coordinate of P and y is called the y-coordinate of P.

3. False; the point whose ordered pair is $(-2, 4)$ is located in Quadrant II.

4. False; the ordered pairs $(3, 2)$ and $(2, 3)$ do not represent the same point in the Cartesian plane.

5.

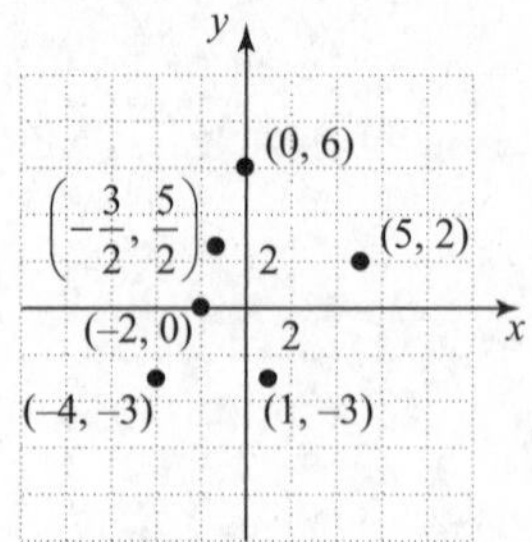

(a) $(5, 2)$ lies in quadrant I.

(b) $(-4, -3)$ lies in quadrant III.

(c) $(1, -3)$ lies in quadrant IV.

(d) $(-2, 0)$ lies on the x-axis.

(e) $(0, 6)$ lies on the y-axis.

(f) $\left(-\frac{3}{2}, \frac{5}{2}\right) = (-1.5, 2.5)$ lies in quadrant II.

6.

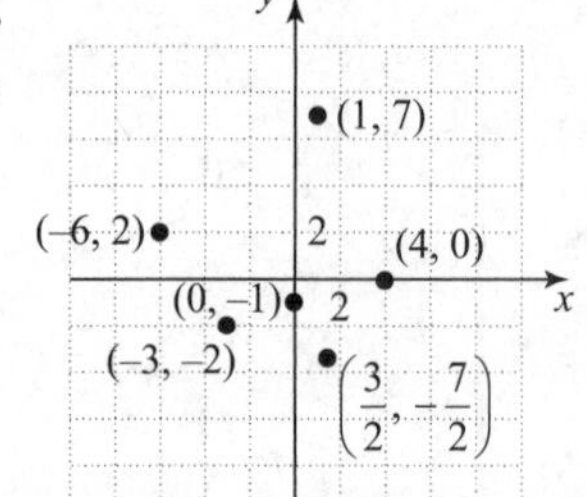

(a) $(-6, 2)$ lies in quadrant II.

(b) $(1, 7)$ lies in quadrant I.

(c) $(-3, -2)$ lies in quadrant III.

(d) $(4, 0)$ lies on the x-axis.

(e) $(0, -1)$ lies on the y-axis.

(f) $\left(\frac{3}{2}, -\frac{7}{2}\right) = (1.5, -3.5)$ lies in quadrant IV.

7.

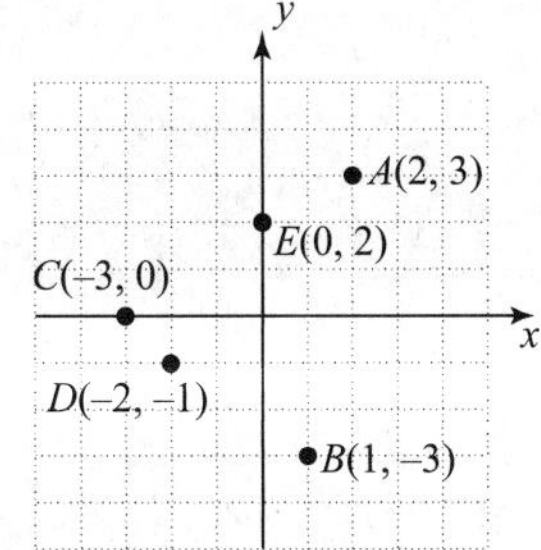

(a) $A(2, 3)$

(b) $B(1, -3)$

(c) $C(-3, 0)$

(d) $D(-2, -1)$

(e) $E(0, 2)$

8. True

9. (a)

$$\begin{aligned} x+4y &= 12 \\ 4+4(2) &\stackrel{?}{=} 12 \\ 4+8 &\stackrel{?}{=} 12 \\ 12 &= 12 \quad \text{True} \end{aligned}$$

The statement is true, so (4, 2) satisfies the equation $x + 4y = 12$.

(b)

$$\begin{aligned} x+4y &= 12 \\ -2+4(4) &\stackrel{?}{=} 12 \\ -2+16 &\stackrel{?}{=} 12 \\ 14 &= 12 \quad \text{False} \end{aligned}$$

The statement 14 = 12 is false, so (–2, 4) does not satisfy the equation $x + 4y = 12$.

(c)

$$\begin{aligned} x+4y &= 12 \\ 1+4(8) &\stackrel{?}{=} 12 \\ 1+32 &\stackrel{?}{=} 12 \\ 33 &= 12 \quad \text{False} \end{aligned}$$

The statement 33 = 12 is false, so (1, 8) does not satisfy the equation $x + 4y = 12$.

10. (a)

$$\begin{aligned} y &= 4x+3 \\ 3 &\stackrel{?}{=} 4(1)+3 \\ 3 &\stackrel{?}{=} 4+3 \\ 3 &= 7 \quad \text{False} \end{aligned}$$

The statement 3 = 7 is false, so (1, 3) does not satisfy the equation $y = 4x + 3$.

(b)

$$\begin{aligned} y &= 4x+3 \\ -5 &\stackrel{?}{=} 4(-2)+3 \\ -5 &\stackrel{?}{=} -8+3 \\ -5 &= -5 \quad \text{True} \end{aligned}$$

The statement is true, so (–2, –5) satisfies the equation $y = 4x + 3$.

(c)

$$\begin{aligned} y &= 4x+3 \\ -3 &\stackrel{?}{=} 4\left(-\frac{3}{2}\right)+3 \\ -3 &\stackrel{?}{=} -6+3 \\ -3 &= -3 \quad \text{True} \end{aligned}$$

The statement is true, so $\left(-\frac{3}{2}, -3\right)$ satisfies the equation $y = 4x + 3$.

11. Substitute 3 for x and solve for y.

$$\begin{aligned} 2x+y &= 10 \\ 2(3)+y &= 10 \\ 6+y &= 10 \\ 6-6+y &= 10-6 \\ y &= 4 \end{aligned}$$

The ordered pair that satisfies the equation is (3, 4).

12. Substitute 1 for y and solve for x.

$$\begin{aligned} -3x+2y &= 11 \\ -3x+2(1) &= 11 \\ -3x+2 &= 11 \\ -3x+2-2 &= 11-2 \\ -3x &= 9 \\ \frac{-3x}{-3} &= \frac{9}{-3} \\ x &= -3 \end{aligned}$$

The ordered pair that satisfies the equation is (–3, 1).

13. Substitute $\frac{1}{2}$ for x and solve for y.

$$\begin{aligned} 4x+3y &= 0 \\ 4\left(\frac{1}{2}\right)+3y &= 0 \\ 2+3y &= 0 \\ -2+2+3y &= -2 \\ 3y &= -2 \\ \frac{3y}{3} &= \frac{-2}{3} \\ y &= -\frac{2}{3} \end{aligned}$$

The ordered pair that satisfies the equation is $\left(\frac{1}{2}, -\frac{2}{3}\right)$.

14. $y = 5x - 2$

$x = -2$: $y = 5(-2) - 2$
$y = -10 - 2$
$y = -12$

$x = 0$: $y = 5(0) - 2$
$y = 0 - 2$
$y = -2$

$x = 1$: $y = 5(1) - 2$
$y = 5 - 2$
$y = 3$

x	y	(x, y)
−2	−12	(−2, −12)
0	−2	(0, −2)
1	3	(1, 3)

15. $y = -3x + 4$

$x = -1$: $y = -3(-1) + 4$
$y = 3 + 4$
$y = 7$

$x = 2$: $y = -3(2) + 4$
$y = -6 + 4$
$y = -2$

$x = 5$: $y = -3(5) + 4$
$y = -15 + 4$
$y = -11$

x	y	(x, y)
−1	7	(−1, 7)
2	−2	(2, −2)
5	−11	(5, −11)

16. $2x + y = -8$

$x = -5$: $2(-5) + y = -8$
$-10 + y = -8$
$y = 2$

$y = -4$: $2x + (-4) = -8$
$2x - 4 = -8$
$2x = -4$
$x = -2$

$x = 2$: $2(2) + y = -8$
$4 + y = -8$
$y = -12$

x	y	(x, y)
−5	2	(−5, 2)
−2	−4	(−2, −4)
2	−12	(2, −12)

17. $2x - 5y = 18$

$x = -6$: $2(-6) - 5y = 18$
$-12 - 5y = 18$
$-5y = 30$
$y = -6$

$y = -4$: $2x - 5(-4) = 18$
$2x + 20 = 18$
$2x = -2$
$x = -1$

$x = 2$: $2(2) - 5y = 18$
$4 - 5y = 18$
$-5y = 14$
$y = -\frac{14}{5}$

x	y	(x, y)
−6	−6	(−6, −6)
−1	−4	(−1, −4)
2	$-\frac{14}{5}$	$\left(2, -\frac{14}{5}\right)$

18. (a) $C = 10 + 1.32607x$

$x = 50$: $C = 10 + 1.32607(50)$
$C = 76.30$

$x = 100$: $C = 10 + 1.32607(100)$
$C = 142.61$

$x = 150$: $C = 10 + 1.32607(150)$
$C = 208.91$

x therms	50 therms	100 therms	150 therms
C ($)	$76.30	$142.61	$208.91

(b)

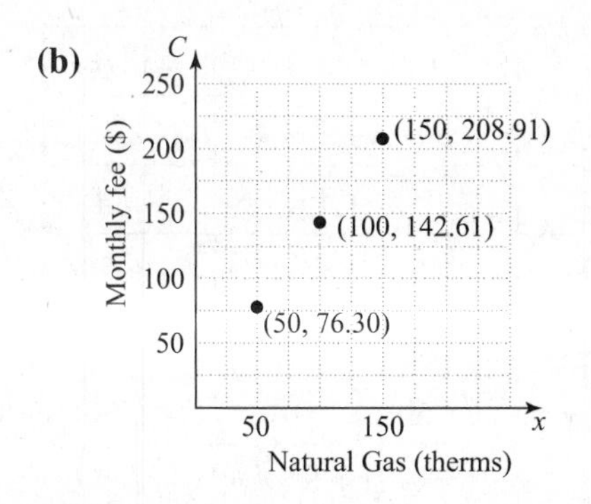

3.1 Exercises

19. Quadrant I: B
Quadrant II: A, E
Quadrant III: C
Quadrant IV: D, F

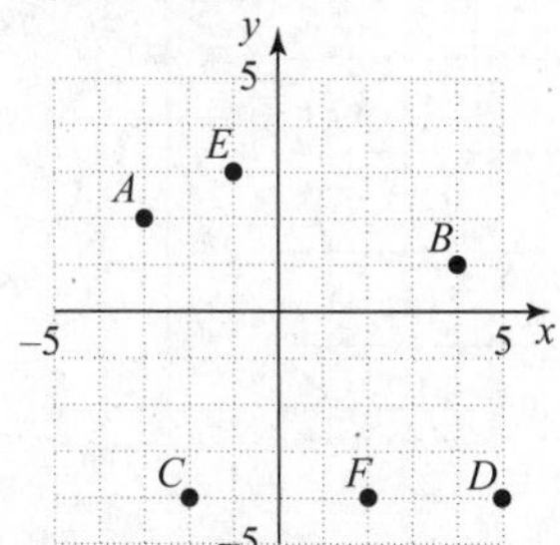

21. Quadrant I: C, E
Quadrant III: F
Quadrant IV: B
x-axis: A, G
y-axis: D, G

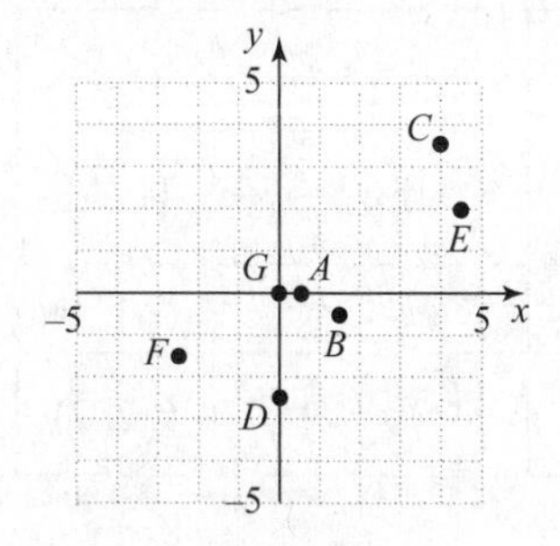

23. Positive x-axis: A
Negative x-axis: D
Positive y-axis: C
Negative y-axis: B

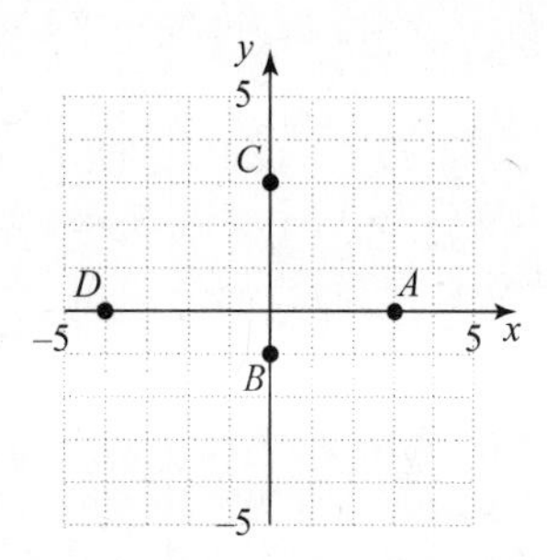

25. $A(4, 0)$; $B(-3, 2)$; $C(1, -4)$; $D(-2, -4)$; $E(3, 5)$; $F(0, -3)$;
Quadrant I: E; Quadrant II: B;
Quadrant III: D; Quadrant IV: C;
Positive x-axis: A; Negative y-axis: F

27. $y = -3x + 5$
$A(-2, -1)$
$-1 \stackrel{?}{=} -3(-2) + 5$
$-1 \stackrel{?}{=} 6 + 5$
$-1 \neq 11$
No
$B(2, -1)$
$-1 \stackrel{?}{=} -3(2) + 5$
$-1 \stackrel{?}{=} -6 + 5$
$-1 = -1$
Yes
$C\left(\frac{1}{3}, 4\right)$
$4 \stackrel{?}{=} -3\left(\frac{1}{3}\right) + 5$
$4 \stackrel{?}{=} -1 + 5$
$4 = 4$
Yes

29. $3x + 2y = 4$
$A(0, 2)$
$3(0) + 2(2) \stackrel{?}{=} 4$
$0 + 4 \stackrel{?}{=} 4$
$4 = 4$
Yes
$B(1, 0)$
$3(1) + 2(0) \stackrel{?}{=} 4$
$3 + 0 \stackrel{?}{=} 4$
$3 \neq 4$
No
$C(4, -4)$
$3(4) + 2(-4) \stackrel{?}{=} 4$
$12 - 8 \stackrel{?}{=} 4$
$4 = 4$
Yes

31. $\frac{4}{3}x + y - 1 = 0$

$A(3, -3)$

$$\frac{4}{3}(3) + (-3) - 1 \stackrel{?}{=} 0$$
$$4 - 3 - 1 \stackrel{?}{=} 0$$
$$0 = 0$$

Yes

$B(-6, -9)$

$$\frac{4}{3}(-6) + (-9) - 1 \stackrel{?}{=} 0$$
$$-8 - 9 - 1 \stackrel{?}{=} 0$$
$$-18 \neq 0$$

No

$C\left(\frac{3}{4}, 0\right)$

$$\frac{4}{3}\left(\frac{3}{4}\right) + 0 - 1 \stackrel{?}{=} 0$$
$$1 + 0 - 1 \stackrel{?}{=} 0$$
$$0 = 0$$

Yes

33. Let $x = 4$.

$$x + y = 5$$
$$4 + y = 5$$
$$y = 1$$

The ordered pair is (4, 1).

35. Let $y = -1$.

$$2x + y = 9$$
$$2x + (-1) = 9$$
$$2x = 10$$
$$x = 5$$

The ordered pair is (5, −1).

37. Let $x = -3$.

$$-3x + 2y = 15$$
$$-3(-3) + 2y = 15$$
$$9 + 2y = 15$$
$$2y = 6$$
$$y = 3$$

The ordered pair is (−3, 3).

39.

x	$y = -x$	(x, y)
−3	$-(-3) = 3$	(−3, 3)
0	$-0 = 0$	(0, 0)
1	−1	(1, −1)

41.

x	$y = -3x + 1$	(x, y)
−2	$-3(-2) + 1 = 7$	(−2, 7)
−1	$-3(-1) + 1 = 4$	(−1, 4)
4	$-3(4) + 1 = -11$	(4, −11)

43. $2x + y = 6$

$$y = -2x + 6$$

x	$y = -2x + 6$	(x, y)
−1	$-2(-1) + 6 = 8$	(−1, 8)
2	$-2(2) + 6 = 2$	(2, 2)
3	$-2(3) + 6 = 0$	(3, 0)

45. $y = 6$

x	$y = 6$	(x, y)
−4	6	(−4, 6)
1	6	(1, 6)
12	6	(12, 6)

47. $x - 2y + 6 = 0$

$$x = 2y - 6$$

or

$$x - 2y + 6 = 0$$
$$-2y = -x - 6$$
$$y = \frac{1}{2}x + 3$$

$x = 2y - 6$	$y = \frac{1}{2}x + 3$	(x, y)
1	$\frac{1}{2}(1) + 3 = \frac{7}{2}$	$\left(1, \frac{7}{2}\right)$
$2(1) - 6 = -4$	1	(−4, 1)
−2	$\frac{1}{2}(-2) + 3 = 2$	(−2, 2)

49.

$$y = 5 + \frac{1}{2}x$$
$$y - 5 = \frac{1}{2}x$$
$$2(y - 5) = x$$

$x=2(y-5)$	$y=5+\frac{1}{2}x$	(x, y)
$2(7-5)=4$	7	(4, 7)
−4	$5+\frac{1}{2}(-4)=3$	(−4, 3)
$2(2-5)=-6$	2	(−6, 2)

51. $\frac{x}{2}+\frac{y}{3}=-1$

$\frac{x}{2}=-\frac{y}{3}-1$

$x=-\frac{2y}{3}-2$

or

$\frac{x}{2}+\frac{y}{3}=-1$

$\frac{y}{3}=-\frac{x}{2}-1$

$y=-\frac{3x}{2}-3$

$x=-\frac{2}{3}y-2$	$y=-\frac{3}{2}x-3$	(x, y)
0	$-\frac{3}{2}(0)-3=-3$	(0, −3)
$-\frac{2}{3}(0)-2=-2$	0	(−2, 0)
$-\frac{2}{3}(-6)-2=2$	−6	(2, −6)

53. $y=-3x-10$

In $A(\_\_, -16)$, $y=-16$.

$-16=-3x-10$

$-6=-3x$

$2=x$

$A(2, -16)$

In $B(-3, \_)$, $x=-3$.

$y=-3(-3)-10=9-10=-1$

$B(-3, -1)$

In $C(\_, -9)$, $y=-9$.

$-9=-3x-10$

$1=-3x$

$-\frac{1}{3}=x$

$C\left(-\frac{1}{3}, -9\right)$

55. $x=-\frac{1}{3}y$

In $A(2, \_)$, $x=2$

$2=-\frac{1}{3}y$

$-6=y$

$A(2, -6)$

In $B(\_, 0)$, $y=0$

$x=-\frac{1}{3}(0)=0$

$B(0, 0)$

In $C\left(\_, -\frac{1}{2}\right)$, $y=-\frac{1}{2}$

$x=-\frac{1}{3}\left(-\frac{1}{2}\right)=\frac{1}{6}$

$C\left(\frac{1}{6}, -\frac{1}{2}\right)$

57. $x=4$

Here the x-coordinate will be 4, regardless of the value of the y-coordinate.

$A(4, -8)$

$B(4, -19)$

$C(4, 5)$

59. $y=\frac{2}{3}x+2$

In $A(\_, 4)$, $y=4$.

$4=\frac{2}{3}x+2$

$2=\frac{2}{3}x$

$3=x$

$A(3, 4)$

In $B(-6, \_)$, $x=-6$.

$y=\frac{2}{3}(-6)+2=-4+2=-2$

$B(-6, -2)$

In $C\left(\frac{1}{2}, \_\right)$, $x=\frac{1}{2}$

$y=\frac{2}{3}\left(\frac{1}{2}\right)+2=\frac{1}{3}+2=\frac{7}{3}$

$C\left(\frac{1}{2}, \frac{7}{3}\right)$

61. $\frac{1}{2}x - 3y = 2$

In $A(-4, \_)$, $x = -4$

$\frac{1}{2}(-4) - 3y = 2$

$-2 - 3y = 2$

$-3y = 4$

$y = -\frac{4}{3}$

$A\left(-4, -\frac{4}{3}\right)$

In $B(\_, -1)$, $y = -1$

$\frac{1}{2}x - 3(-1) = 2$

$\frac{1}{2}x + 3 = 2$

$\frac{1}{2}x = -1$

$x = -2$

$B = (-2, -1)$

In $C\left(-\frac{2}{3}, \_\right)$, $x = -\frac{2}{3}$

$\frac{1}{2}\left(-\frac{2}{3}\right) - 3y = 2$

$-\frac{1}{3} - 3y = 2$

$-3y = \frac{7}{3}$

$y = -\frac{7}{9}$

$C\left(-\frac{2}{3}, -\frac{7}{9}\right)$

63. $0.5x - 0.3y = 3.1$

In $A(20, \_)$, $x = 20$

$0.5(20) - 0.3y = 3.1$

$10 - 0.3y = 3.1$

$-0.3y = -6.9$

$y = 23$

$A(20, 23)$

In $B(\_, -17)$, $y = -17$

$0.5x - 0.3(-17) = 3.1$

$0.5x + 5.1 = 3.1$

$0.5x = -2$

$x = -4$

$B(-4, -17)$

In $C(2.6, \_)$, $x = 2.6$

$0.5(2.6) - 0.3y = 3.1$

$1.3 - 0.3y = 3.1$

$-0.3y = 1.8$

$y = -6$

$C(2.6, -6)$

65. $C = 9.95n + 4.95$

(a) $n = 2$

$C = 9.95(2) + 4.95$

$= 19.9 + 4.95$

$= 24.85$

It will cost \$24.85 to order 2 CDs.

(b) $n = 5$

$C = 9.95(5) + 4.95$

$= 49.75 + 4.95$

$= 54.7$

It will cost \$54.70 to order 5 CDs.

(c) Let $C = 64.65$

$64.65 = 9.95n + 4.95$

$59.70 = 9.95n$

$6 = n$

If you have \$64.65, you can order 6 CDs.

(d) It costs \$34.80 to order 3 CDs.

67. $P = 0.444n + 21.14$

(a) $n = 0$

$P = 0.444(0) + 21.14 = 0.21.14 = 21.14$

21.14% of U.S. population 25 years or older had a bachelor's degree in 1990.

(b) $n = 10$

$P = 0.444(10) + 21.14$

$= 4.44 + 21.14$

$= 25.58$

25.58% of U.S. population 25 years or older had a bachelor's degree in 2000.

(c) $n = 30$

$P = 0.444(30) + 21.14$

$= 13.32 + 21.14$

$= 34.46$

34.46% of U.S. population 25 years or older will have a bachelor's degree in 2020.

(d) Let $P = 50$.

$$50 = 0.444n + 21.14$$
$$28.86 = 0.444n$$
$$65 = n$$

65 years after 1990 = 2055.

(e) Answers will vary.

69. $4a + 2b = -8$

$$4a = -2b - 8$$
$$a = -\frac{1}{2}b - 2$$

or

$$4a + 2b = -8$$
$$2b = -4a - 8$$
$$b = -2a - 4$$

$a = -\frac{1}{2}b - 2$	$b = -2a - 4$	(a, b)
2	$-2(2) - 4 = -8$	$(2, -8)$
$-\frac{1}{2}(-4) - 2 = 0$	-4	$(0, -4)$
$-\frac{1}{2}(6) - 2 = -5$	6	$(-5, 6)$

71. $\frac{2p}{5} + \frac{3q}{10} = 1$

$$\frac{2p}{5} = -\frac{3q}{10} + 1$$
$$p = -\frac{3}{4}q + \frac{5}{2}$$

or

$$\frac{2p}{5} + \frac{3q}{10} = 1$$
$$\frac{3q}{10} = -\frac{2p}{5} + 1$$
$$q = -\frac{4}{3}p + \frac{10}{3}$$

$p = -\frac{3}{4}q + \frac{5}{2}$	$q = -\frac{4}{3}p + \frac{10}{3}$	(p, q)
0	$-\frac{4}{3}(0) + \frac{10}{3} = \frac{10}{3}$	$\left(0, \frac{10}{3}\right)$
$-\frac{3}{4}(0) + \frac{5}{2} = \frac{5}{2}$	0	$\left(\frac{5}{2}, 0\right)$
-10	$-\frac{4}{3}(-10) + \frac{10}{3} = \frac{50}{3}$	$\left(-10, \frac{50}{3}\right)$

73. $y = -2x + k$; (1, 2) is a solution.
$2 = -2(1) + k$
$2 = -2 + k$
$4 = k$

75. $7x - ky = -4$; (2, 9) is a solution.
$7(2) - k(9) = -4$
$14 - 9k = -4$
$-9k = -18$
$k = 2$

77. $kx - 4y = 6$; $\left(-8, -\frac{5}{2}\right)$ is a solution.
$k(-8) - 4\left(-\frac{5}{2}\right) = 6$
$-8k + 10 = 6$
$-8k = -4$
$k = \frac{1}{2}$

79. $3x - 2y = -6$
Points may vary.
$x = 0$: $3(0) - 2y = -6$
$-2y = -6$
$y = 3$
(0, 3)
$y = 0$: $3x - 2(0) = -6$
$3x = -6$
$x = -2$
(−2, 0)
$x = -4$: $3(-4) - 2y = -6$
$-12 - 2y = -6$
$-2y = 6$
$y = -3$
(−4, −3)

x	y	(x, y)
0	3	(0, 3)
−2	0	(−2, 0)
−4	−3	(−4, −3)

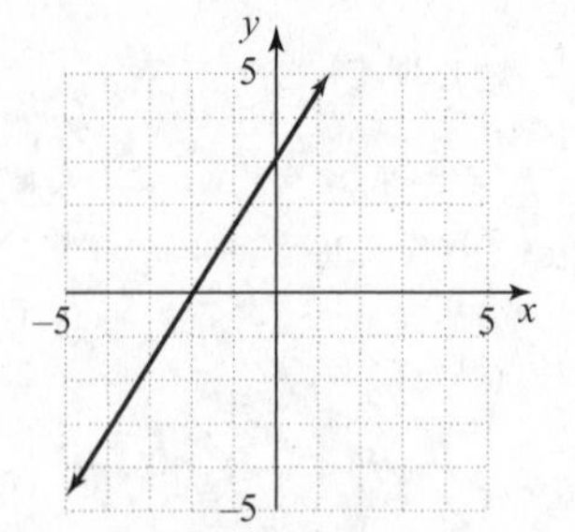

The figure is a line.

81.

x	$y = x^2 - 4$	(x, y)
−2	$(-2)^2 - 4 = 0$	(−2, 0)
−1	$(-1)^2 - 4 = -3$	(−1, −3)
0	$0^2 - 4 = -4$	(0, −4)
1	$1^2 - 4 = -3$	(1, −3)
2	$2^2 - 4 = 0$	(2, 0)

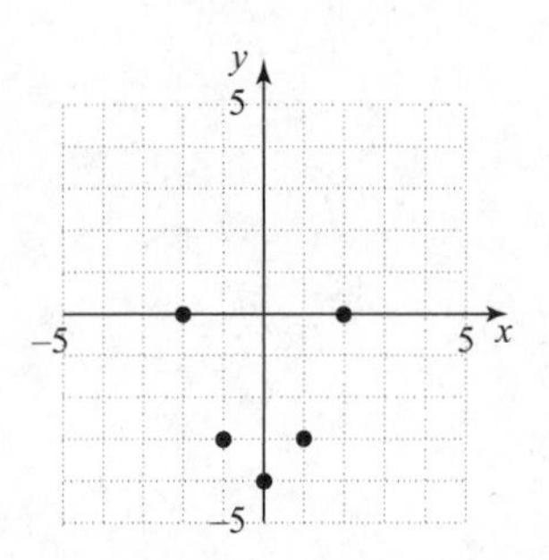

83.

x	$y = -x^3 + 2$	(x, y)
−2	$-(-2)^3 + 2 = 10$	(−2, 10)
−1	$-(-1)^3 + 2 = 3$	(−1, 3)
0	$-(0)^3 + 2 = 2$	(0, 2)
1	$-(1)^3 + 2 = 1$	(1, 1)
2	$-(2)^3 + 2 = -6$	(2, −6)

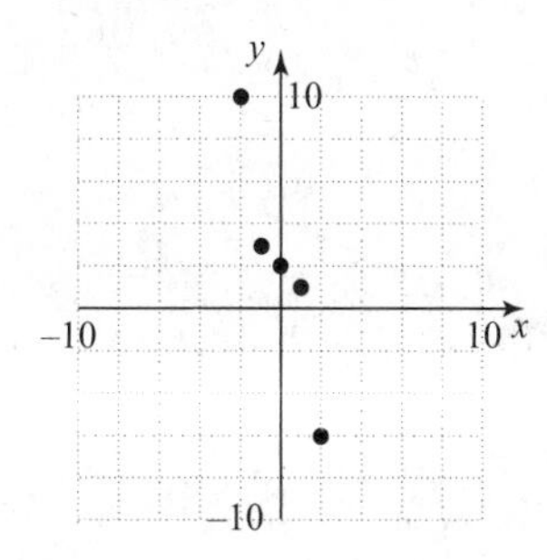

85. Answers may vary. One possibility: The quadrants are numbered I, II, III, and IV, counterclockwise starting from the upper right. The signs of the coordinates of a point determine the quadrant.
Quadrant I: $x > 0$, $y > 0$
Quadrant II: $x < 0$, $y > 0$
Quadrant III: $x < 0$, $y < 0$
Quadrant IV: $x > 0$, $y < 0$
A point with one coordinate of 0 lies on either the x-axis (second coordinate 0) or the y-axis (first coordinate 0).

87.

X	Y1
-3	-15
-2	-13
-1	-11
0	-9
1	-7
2	-5
3	-3

Y1=2X-9

89.

X	Y1
-3	11
-2	10
-1	9
0	8
1	7
2	6
3	5

Y1=-X+8

91. $y + 2x = 13$
$y = 13 - 2x$

X	Y1
-3	19
-2	17
-1	15
0	13
1	11
2	9
3	7

Y1=13-2X

93.

X	Y1
-3	-53
-2	-23
-1	-5
0	1
1	-5
2	-23
3	-53

Y1=-6X²+1

Section 3.2

Preparing for Graphing Equations in Two Variables

P1. $4x = 24$
$$\frac{4x}{4} = \frac{24}{4}$$
$$x = 6$$
The solution set is {6}.

P2. $-3y = 18$
$$\frac{-3y}{-3} = \frac{18}{-3}$$
$$y = -6$$
The solution set is {−6}.

P3. $2x + 5 = 13$
$$2x + 5 - 5 = 13 - 5$$
$$2x = 8$$
$$\frac{2x}{2} = \frac{8}{2}$$
$$x = 4$$
The solution set is {4}.

3.2 Quick Checks

1. Points may vary.

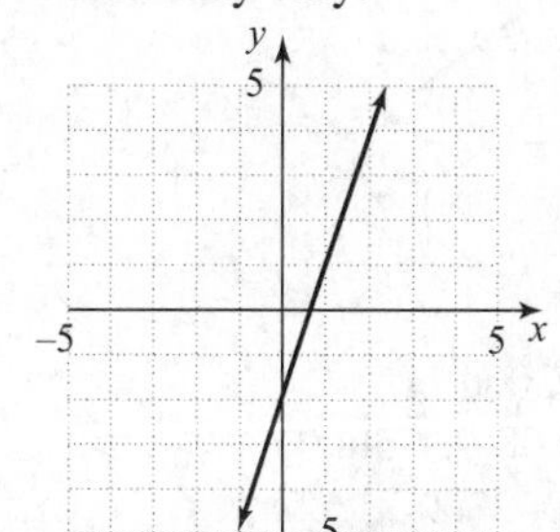

2. Points may vary.

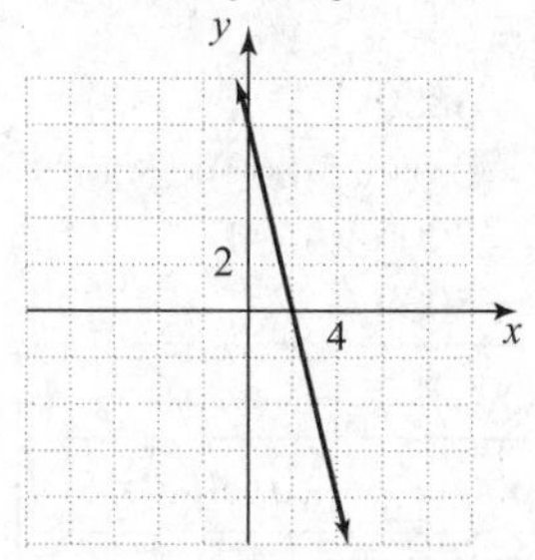

3. A <u>linear</u> equation is an equation of the form $Ax + By = C$, where A, B, and C are real numbers, and A and B are not both zero. Equations written in this form are said to be in <u>standard form</u>.

4. The equation $4x - y = 12$ is a linear equation in two variables because it is written in the form $Ax + By = C$ with $A = 4$, $B = -1$, and $C = 12$.

5. The equation $5x - y^2 = 10$ is not a linear equation because y is squared.

6. The equation $5x = 20$ is a linear equation in two variables because it is written in the form $Ax + By = C$ with $A = 5$, $B = 0$, and $C = 20$.

7. Points may vary.

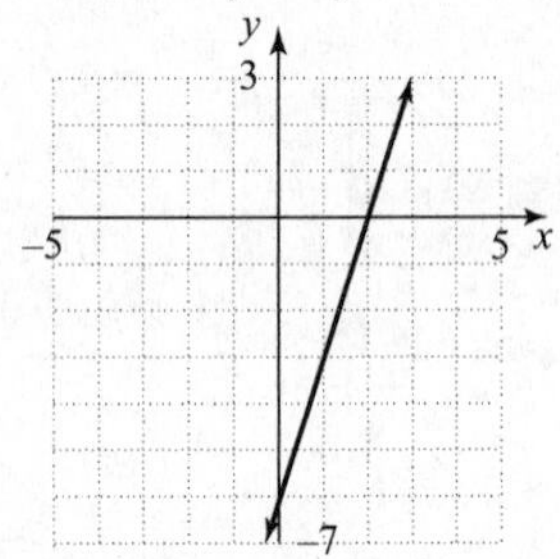

8. Points may vary.

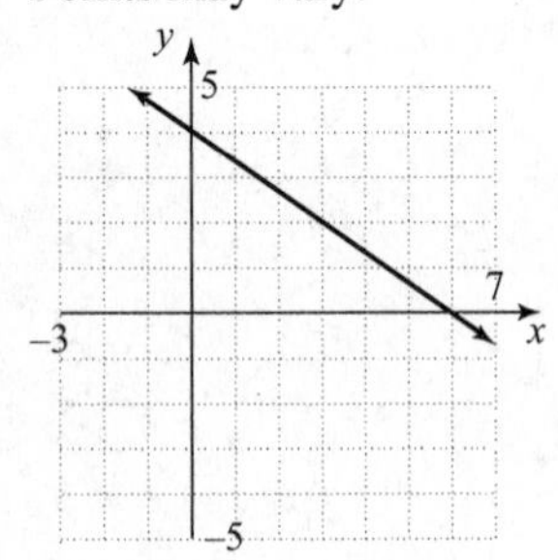

9. **(a)** $S = 0.08x + 3000$

$x = 0$: $S = 0.08(0) + 3000$
$= 0 + 3000$
$= 3000$

$x = 10{,}000$: $S = 0.08(10{,}000) + 3000$
$= 800 + 3000$
$= 3800$

$x = 25{,}000$: $S = 0.08(25{,}000) + 3000$
$= 2000 + 3000$
$= 5000$

x	S	(x, S)
0	3000	(0, 3000)
10,000	3800	(10,000, 3800)
25,000	5000	(25,000, 5000)

(b)

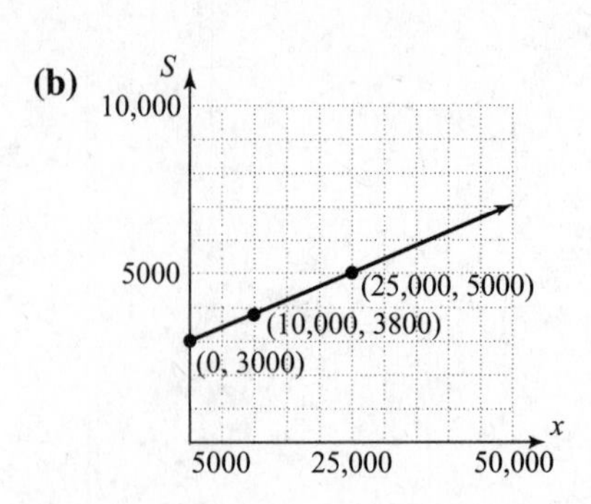

10. The <u>intercepts</u> are the points, if any, where a graph crosses or touches the coordinate axes.

11. The intercepts of the graph are the points (0, 3) and (4, 0). The x-intercept is 4. The y-intercept is 3.

12. The only intercept of the graph is the point (0, −2). There is no x-intercept. The y-intercept is −2.

13. False; to find the y-intercept(s), if any, of the graph of an equation, let $x = 0$ in the equation and solve for y.

For 14–18, additional points may vary.

14. $x + y = 3$
x-intercept, let $y = 0$:
$x + 0 = 3$
$x = 3$
y-intercept, let $x = 0$:
$0 + y = 3$
$y = 3$
Plot the points (3, 0), (0, 3), and an additional point.

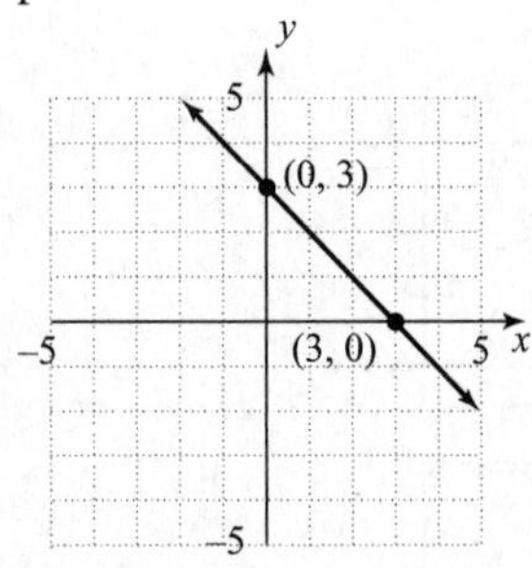

15. $2x - 5y = 20$
x-intercept, let $y = 0$:
$2x - 5(0) = 20$
$2x = 20$
$x = 10$
y-intercept, let $x = 0$:

$$2(0) - 5y = 20$$
$$-5y = 20$$
$$y = -4$$

Plot the points (10, 0), (0, –4), and an additional point.

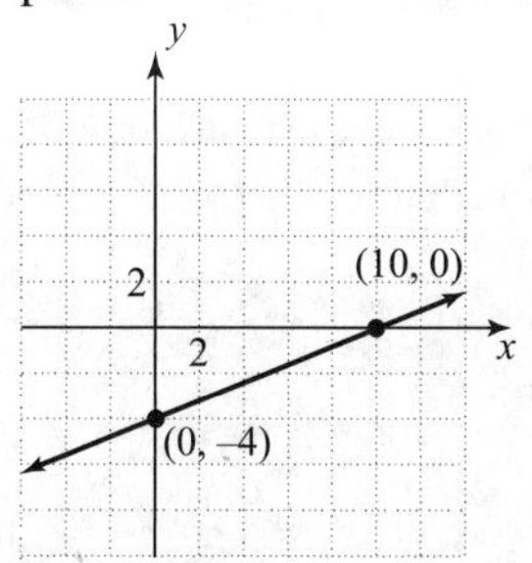

16. $\frac{3}{2}x - 2y = 9$

x-intercept, let $y = 0$:

$$\frac{3}{2}x - 2(0) = 9$$
$$\frac{3}{2}x = 9$$
$$x = \frac{2}{3} \cdot 9$$
$$x = 6$$

y-intercept, let $x = 0$:

$$\frac{3}{2}(0) - 2y = 9$$
$$-2y = 9$$
$$y = -\frac{9}{2}$$

Plot the points (6, 0), $\left(0, -\frac{9}{2}\right)$, and an additional point.

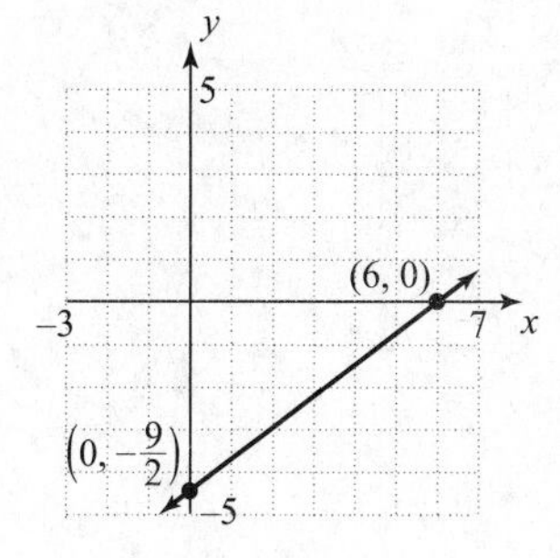

17. $y = \frac{1}{2}x$

x-intercept, let $y = 0$:

$$0 = \frac{1}{2}x$$
$$0 = x$$

y-intercept, let $x = 0$:

$$y = \frac{1}{2}(0)$$
$$y = 0$$

The intercepts are the same point, (0, 0), so let $x = 2$ to get a second point:

$$y = \frac{1}{2}(2)$$
$$y = 1$$

Plot the points (0, 0), (2, 1), and an additional point.

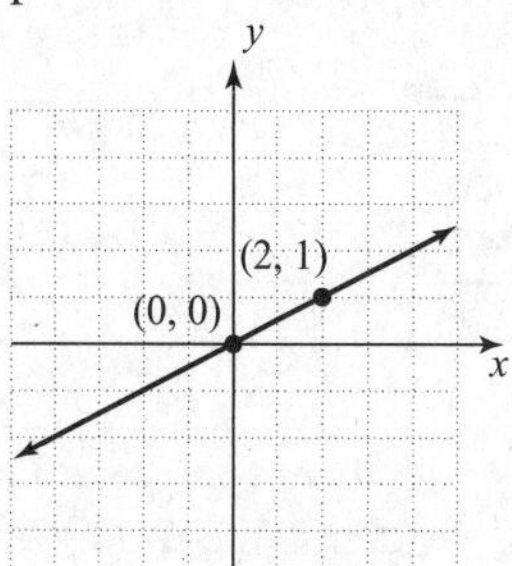

18. $4x + y = 0$

x-intercept, let $y = 0$:

$$4x + 0 = 0$$
$$4x = 0$$
$$x = 0$$

y-intercept, let $x = 0$:

$$4(0) + y = 0$$
$$y = 0$$

The intercepts are the same point, (0, 0), so let $x = 1$ to get a second point:

$$4(1) + y = 0$$
$$4 + y = 0$$
$$y = -4$$

Plot the points (0, 0), (1, –4), and an additional point.

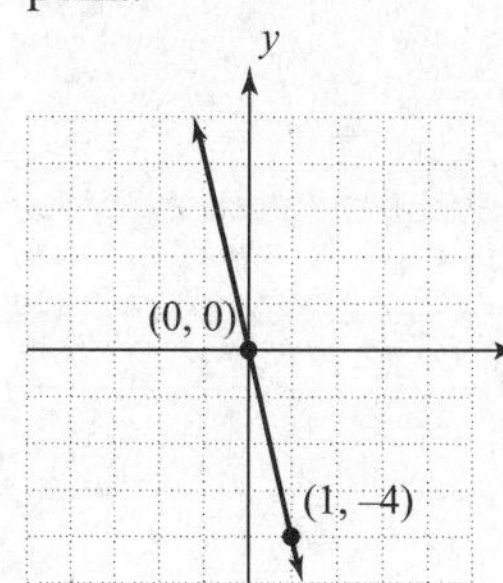

19. The graph of $x = -5$ is a vertical line whose x-intercept is -5.

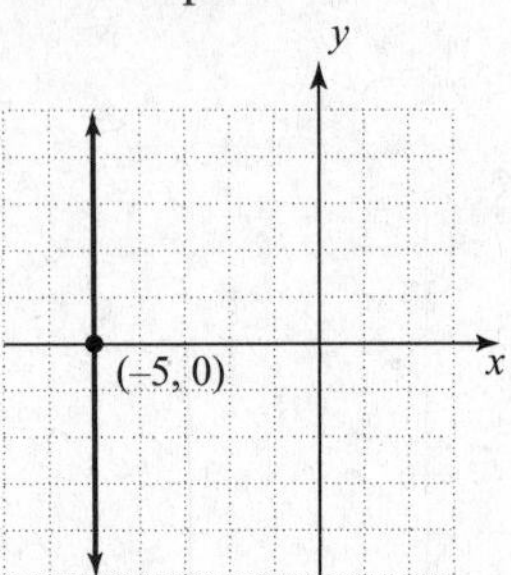

20. The graph of $y = -4$ is a horizontal line whose y-intercept is -4.

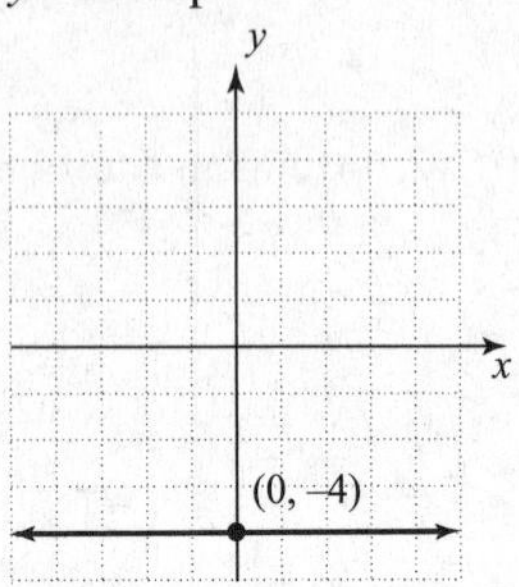

3.2 Exercises

21. Yes; the equation is written in the form $Ax + By = C$ with $A = 2$, $B = -5$, and $C = 10$.

23. No; variables cannot be squared in a linear equation.

25. No; variables cannot be in the denominator of a fraction in a linear equation.

27. Yes; the equation can be rewritten as $y = 1$, which is in the form $Ax + By = C$ with $A = 0$, $B = 1$, and $C = 1$.

In Problems 29–45, points may vary.

29. $y = 2x$

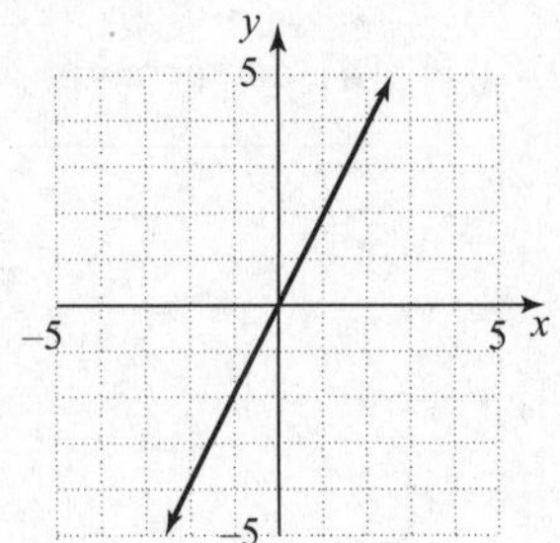

31. $y = 4x - 2$

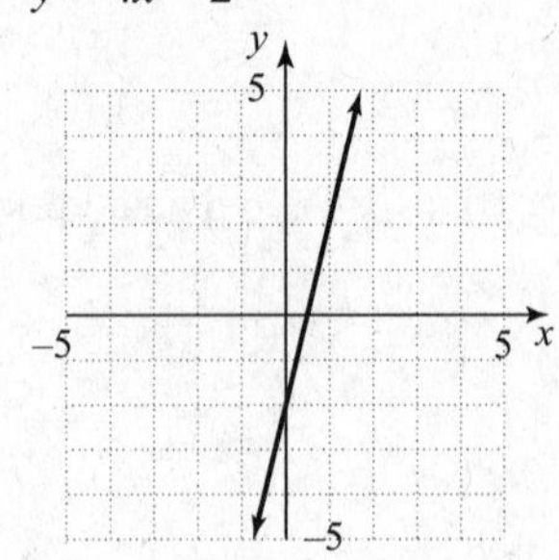

33. $y = -2x + 5$

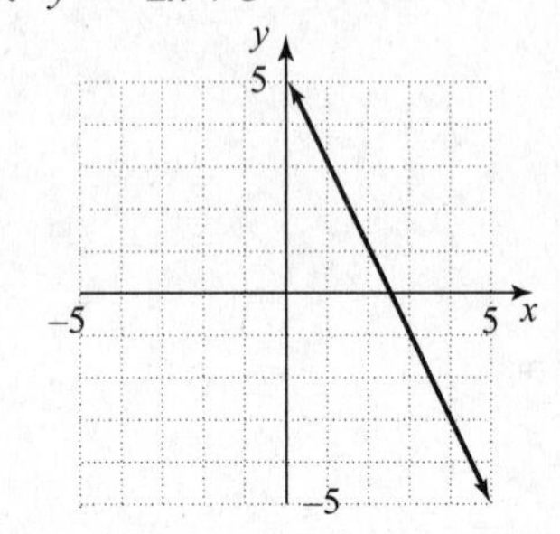

35. $x + y = 5$

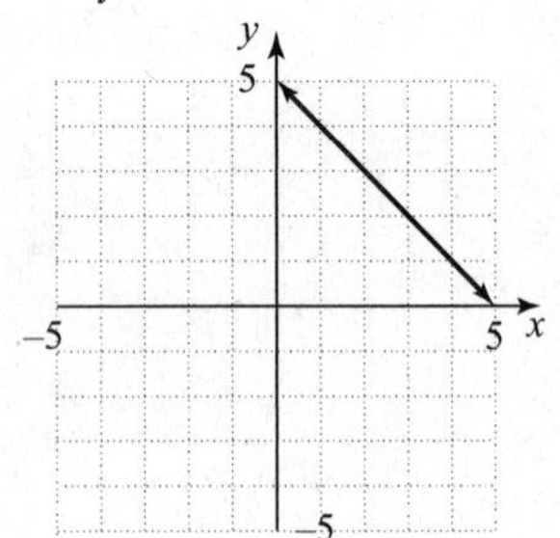

37. $-2x + y = 6$

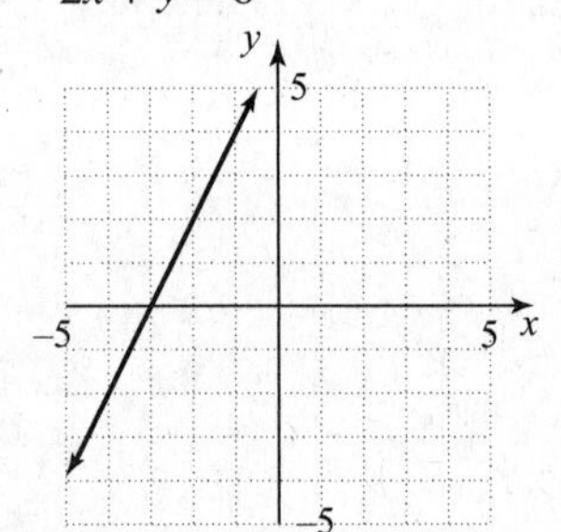

39. $4x - 2y = -8$

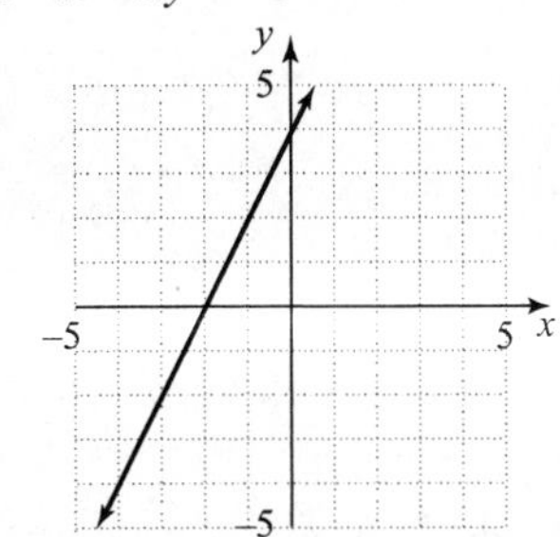

41. $x = -4y$

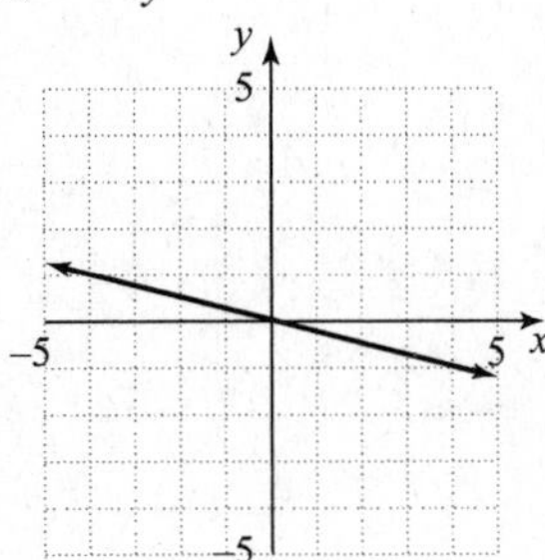

43. $y + 7 = 0$

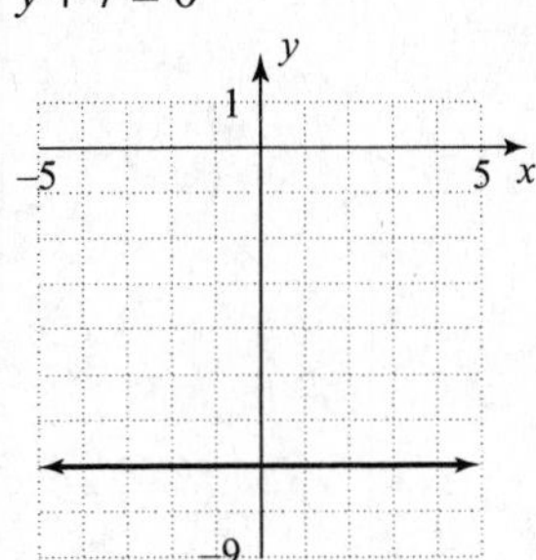

45. $y - 2 = 3(x + 1)$

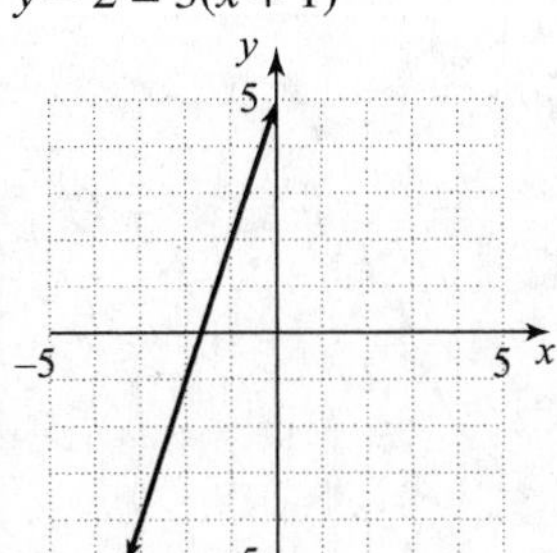

47. The graph crosses the y-axis at (0, −5) and the x-axis at (5, 0). The intercepts are (0, −5) and (5, 0).

49. The graph crosses the y-axis at (0, 4) and the x-axis at (2, 0). The intercepts are (0, 4) and (2, 0).

51. The graph crosses the y-axis at (0, −3) and does not cross the x-axis (no x-intercept). The only intercept is (0, −3).

53. The graph does not cross the y-axis (no y-intercept) and crosses the x-axis at (−5, 0). The only intercept is (−5, 0).

55. $2x + 3y = -12$
y-intercept, let $x = 0$:
$$2(0) + 3y = -12$$
$$3y = -12$$
$$y = -4$$
x-intercept, let $y = 0$:
$$2x + 3(0) = -12$$
$$2x = -12$$
$$x = -6$$
The intercepts are (0, −4) and (−6, 0).

57. $x = -6y$
y-intercept, let $x = 0$:
$$0 = -6y$$
$$0 = y$$
x-intercept, let $y = 0$:
$$x = -6(0)$$
$$x = 0$$
The only intercept is (0, 0).

59. $y = x - 5$
y-intercept, let $x = 0$:
$$y = 0 - 5$$
$$y = -5$$
x-intercept, let $y = 0$:
$$0 = x - 5$$
$$5 = x$$
The intercepts are (0, −5) and (5, 0).

61. $\frac{x}{6} + \frac{y}{8} = 1$
y-intercept, let $x = 0$:
$$\frac{0}{6} + \frac{y}{8} = 1$$
$$\frac{y}{8} = 1$$
$$y = 8$$
x-intercept, let $y = 0$:
$$\frac{x}{6} + \frac{0}{8} = 1$$
$$\frac{x}{6} = 1$$
$$x = 6$$
The intercepts are (0, 8) and (6, 0).

63. $x = 4$ is the equation of a vertical line which has x-intercept (4, 0) and no y-intercept.

65. $y + 2 = 0$, or $y = -2$, is the equation of a horizontal line which has y-intercept (0, −2) and no x-intercept.

In Problems 67–81, additional points may vary.

67. $3x + 6y = 18$
y-intercept, $x = 0$:
$3(0)+6y=18$
$6y=18$
$y=3$
(0, 3)
x-intercept, $y = 0$:
$3x+6(0)=18$
$3x=18$
$x=6$
(6, 0)

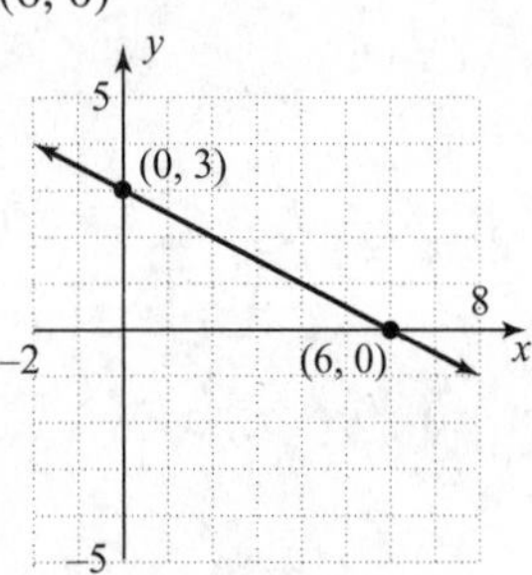

69. $-x + 5y = 15$
y-intercept, $x = 0$:
$-0+5y=15$
$5y=15$
$y=3$
(0, 3)
x-intercept, $y = 0$:
$-x+5(0)=15$
$-x=15$
$x=-15$
(−15, 0)

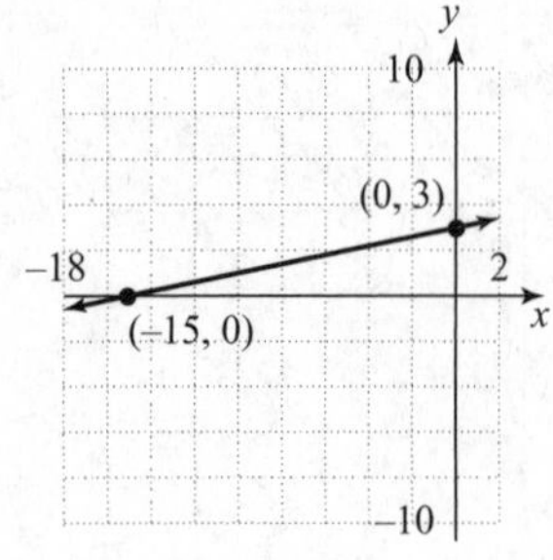

71. $\frac{1}{2}x = y+3$
y-intercept, $x = 0$:
$\frac{1}{2}(0) = y+3$
$0 = y+3$
$-3 = y$
(0, −3)
x-intercept, $y = 0$:
$\frac{1}{2}x = 0+3$
$\frac{1}{2}x = 3$
$x = 6$
(6, 0)

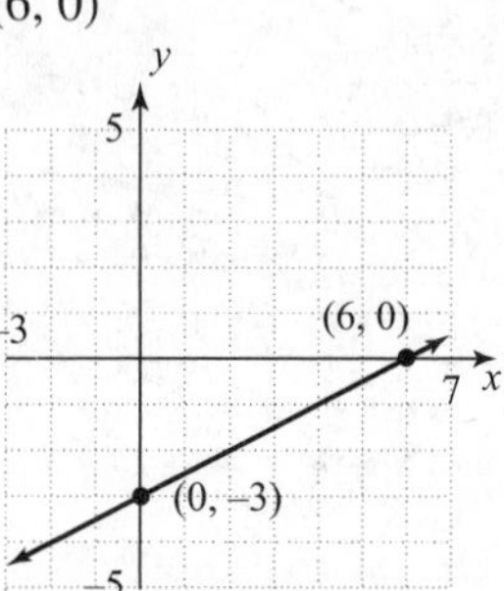

73. $9x - 2y = 0$
y-intercept, $x = 0$:
$9(0)-2y=0$
$-2y=0$
$y=0$
(0, 0)
x-intercept, $y = 0$:
$9x-2(0)=0$
$9x=0$
$x=0$
(0, 0)

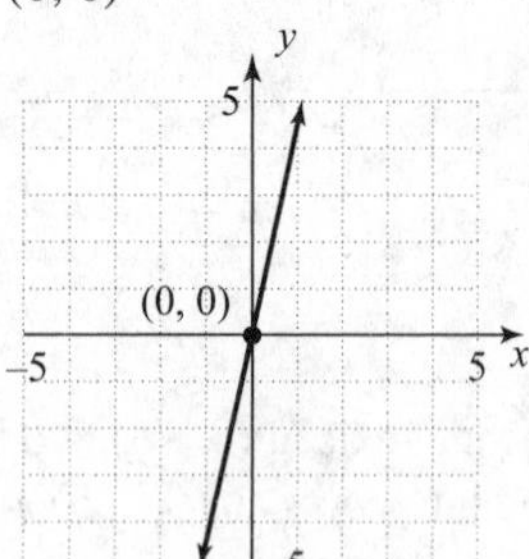

75. $y = -\frac{1}{2}x + 3$

y-intercept, $x = 0$

$y = -\frac{1}{2}(0) + 3$

$y = 3$

$(0, 3)$

x-intercept, $y = 0$

$0 = -\frac{1}{2}x + 3$

$-3 = -\frac{1}{2}x$

$6 = x$

$(6, 0)$

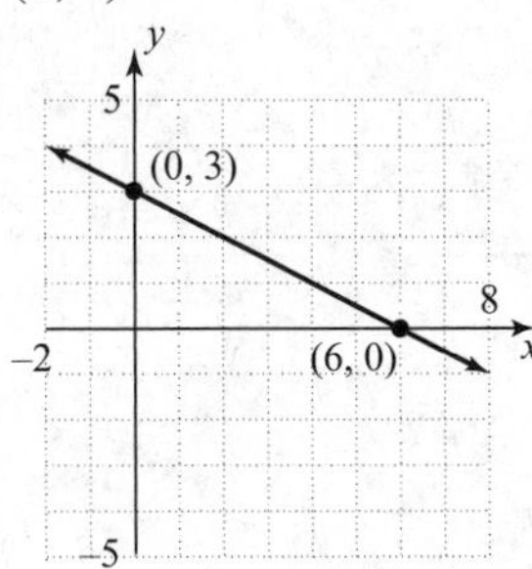

77. $\frac{1}{3}y + 2 = 2x$

y-intercept, $x = 0$:

$\frac{1}{3}y + 2 = 2(0)$

$\frac{1}{3}y = -2$

$y = -6$

$(0, -6)$

x-intercept, $y = 0$:

$\frac{1}{3}(0) + 2 = 2x$

$2 = 2x$

$1 = x$

$(1, 0)$

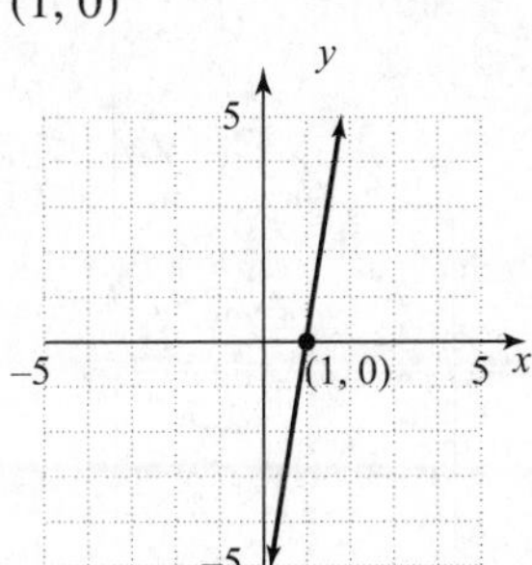

79. $\frac{x}{2} + \frac{y}{3} = 1$

y-intercept, $x = 0$:

$\frac{0}{2} + \frac{y}{3} = 1$

$\frac{y}{3} = 1$

$y = 3$

$(0, 3)$

x-intercept, $y = 0$

$\frac{x}{2} + \frac{0}{3} = 1$

$\frac{x}{2} = 1$

$x = 2$

$(2, 0)$

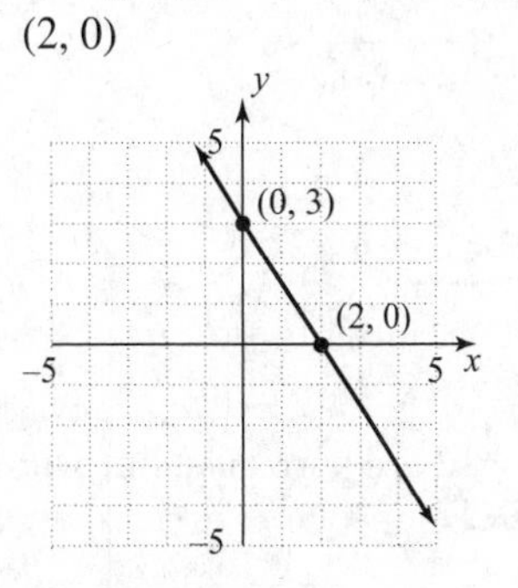

81. $4y - 2x + 1 = 0$

y-intercept, $x = 0$:

$4y - 2(0) + 1 = 0$

$4y + 1 = 0$

$4y = -1$

$y = -\frac{1}{4}$

$\left(0, -\frac{1}{4}\right)$

x-intercept, $y = 0$

$4(0) - 2x + 1 = 0$

$-2x + 1 = 0$

$-2x = -1$

$x = \frac{1}{2}$

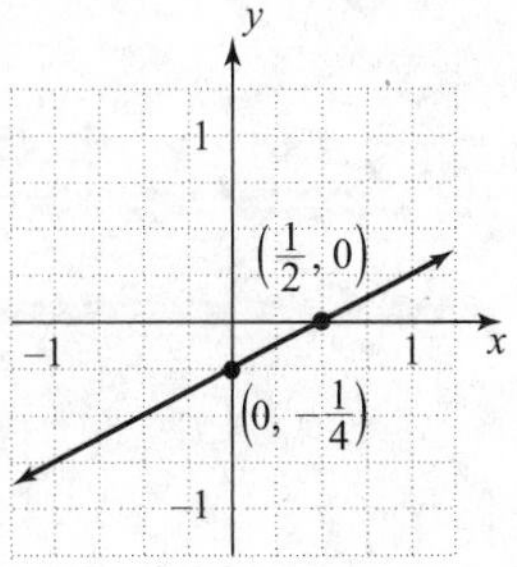

83. $x = 5$ is a vertical line with an x-intercept of $(5, 0)$.

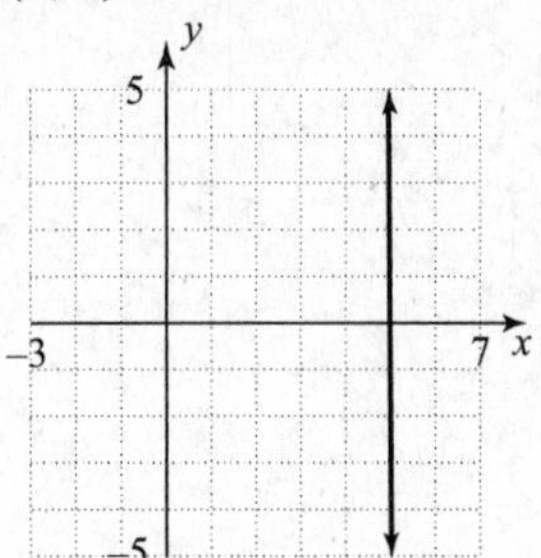

85. $y = -6$ is a horizontal line with a y-intercept of $(0, -6)$.

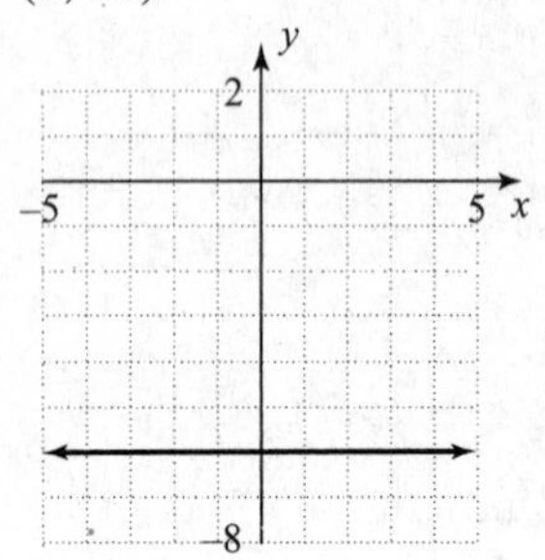

87. $y - 12 = 0$ or $y = 12$ is a horizontal line with a y-intercept of $(0, 12)$.

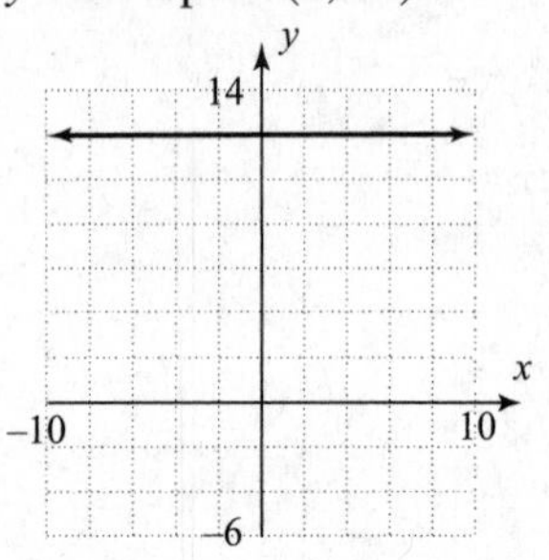

89. $3x - 5 = 0$ or $x = \frac{5}{3}$ is a vertical line with an x-intercept of $\left(\frac{5}{3}, 0\right)$.

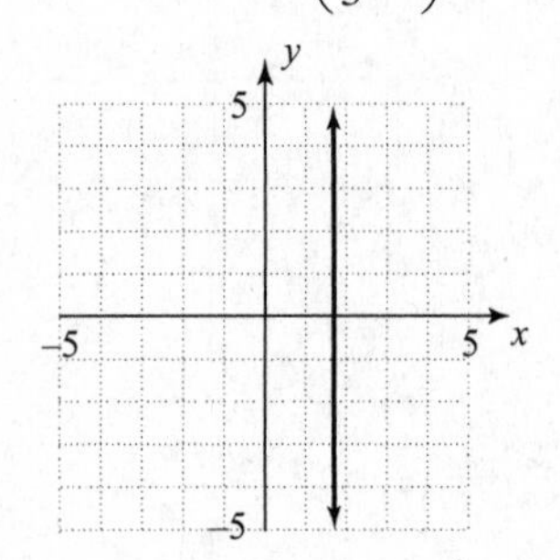

91. $y = 2x - 5$

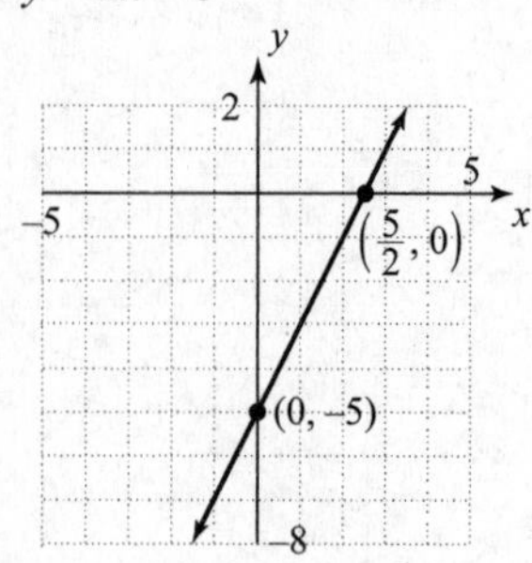

93. $y = -5$

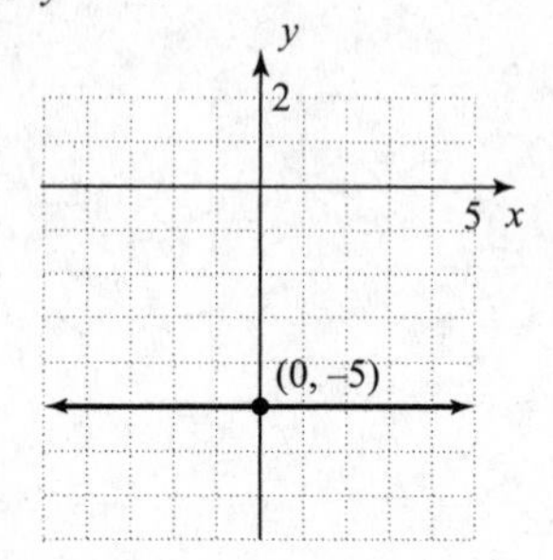

95. $2x + 5y = -20$

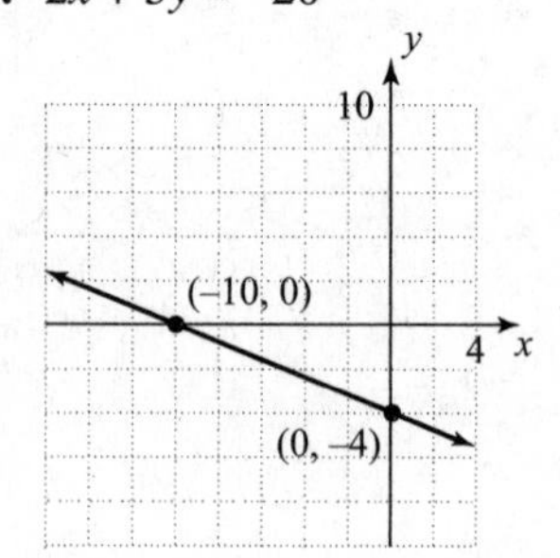

97. $2x = -6y + 4$

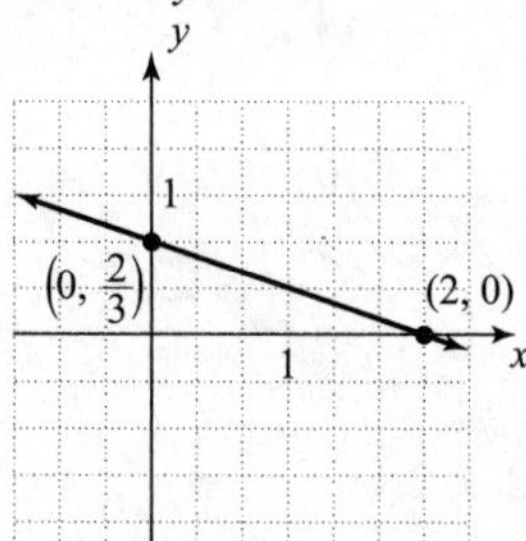

99. $x - 3 = 0$

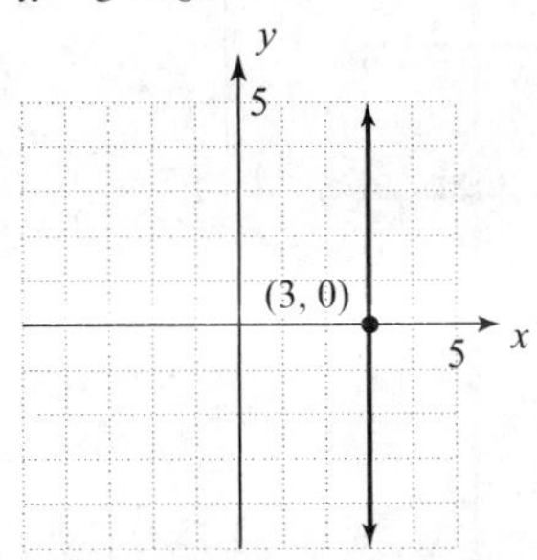

101. $3y - 12 = 0$

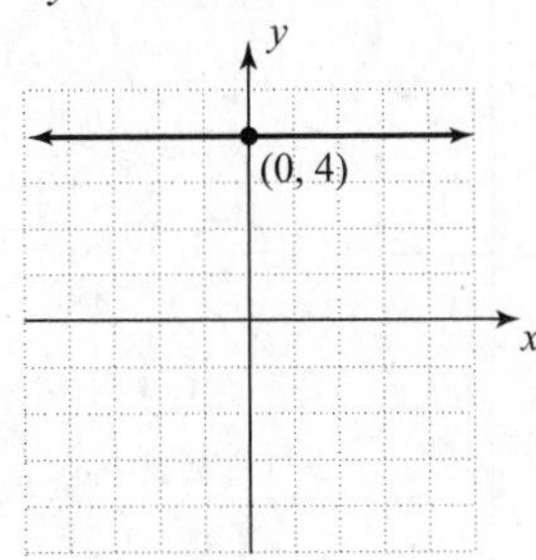

103. $4x + 3y = 18$; $(3, y)$

$$\begin{aligned} 4(3)+3y &= 18 \\ 12+3y &= 18 \\ 3y &= 6 \\ y &= 2 \end{aligned}$$

105. $3x + 5y = 11$; $(x, -2)$

$$\begin{aligned} 3x+5(-2) &= 11 \\ 3x-10 &= 11 \\ 3x &= 21 \\ x &= 7 \end{aligned}$$

107.

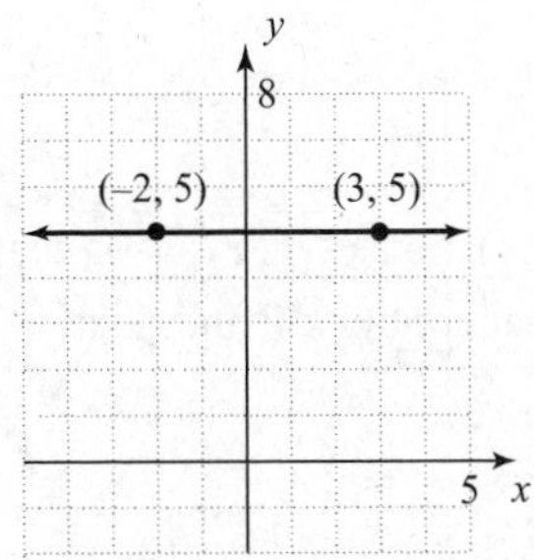

The line is horizontal. The equation is $y = 5$.

109.

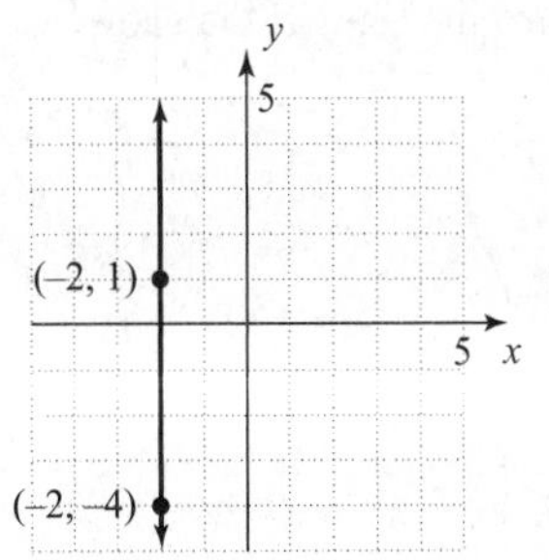

The line is vertical. The equation is $x = -2$.

111. The line is horizontal and has y-intercept $(0, 4)$. The equation is $y = 4$.

113. The line is vertical and has x-intercept $(-9, 0)$. The equation is $x = -9$.

115. If the x-coordinate is twice the y-coordinate, then $x = 2y$.

117. If the y-coordinate is two more than the x-coordinate, then $y = x + 2$.

119. **(a)** $E = 100n + 500$

$$\begin{aligned} n = 0:\ E &= 100(0)+500 \\ &= 0+500 \\ &= 500 \\ n = 4:\ E &= 100(4)+500 \\ &= 400+500 \\ &= 900 \\ n = 10:\ E &= 100(10)+500 \\ &= 1000+500 \\ &= 1500 \end{aligned}$$

The ordered pairs are $(0, 500)$, $(4, 900)$, and $(10, 1500)$.

(b)

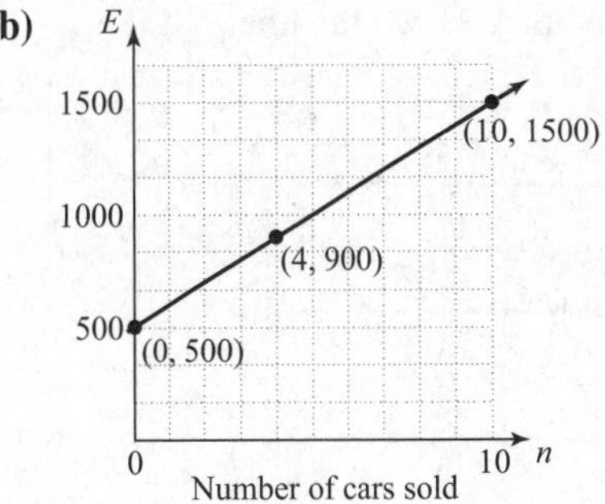

(c) If she sells 0 cars, her earnings are \$500.

121. The "steepness" of the lines is the same.

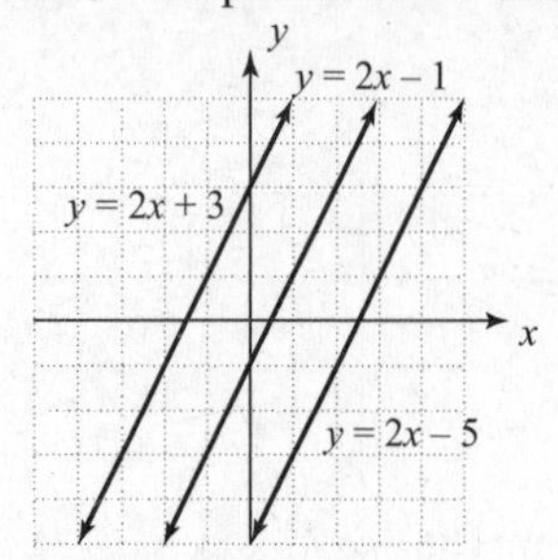

123. The lines get steeper as the coefficient of x gets larger.

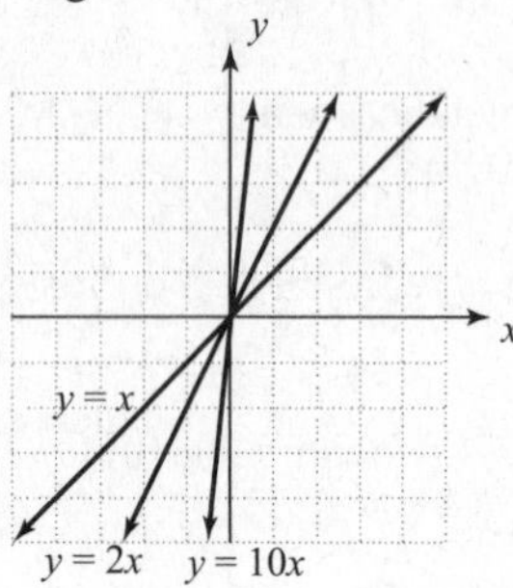

125. The graph crosses the y-axis at (0, −6) and the x-axis at (−2, 0) and (3, 0). The intercepts are (0, −6), (−2, 0), and (3, 0).

127. The graph crosses the y-axis at (0, 14) and the x-axis at (−3, 0), (2, 0), and (5, 0). The intercepts are (0, 14), (−3, 0), (2, 0), and (5, 0).

129. The graph of an equation is the set of all ordered pairs (x, y) that make the equation a true statement.

131. Two points are required, but it is good to plot a third point to check the line.

133. $y = 2x - 9$

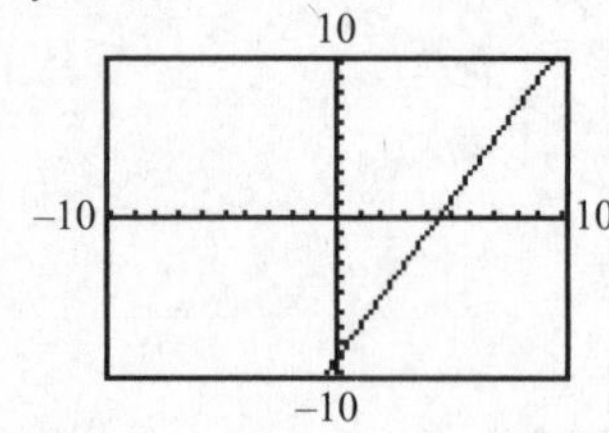

135. $y + 2x = 13$ or $y = 13 - 2x$

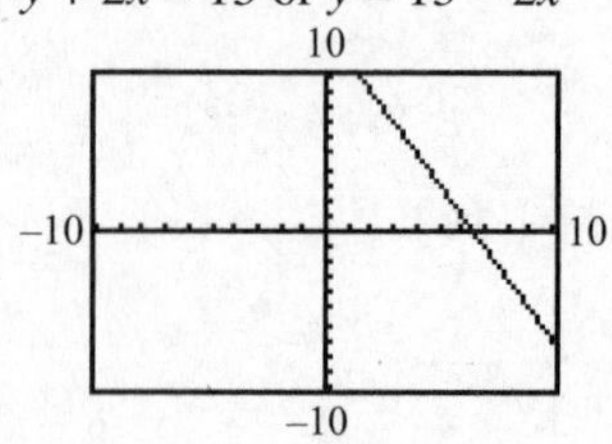

137. $y = -6x^2 + 1$

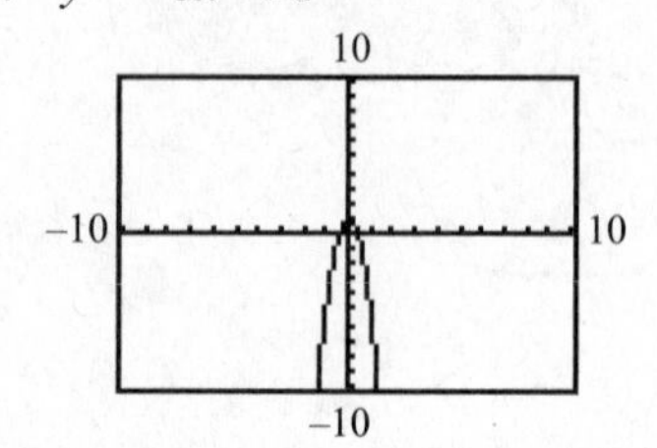

Section 3.3

Preparing for Slope

P1. $\dfrac{5-2}{8-7} = \dfrac{3}{1} = 3$

P2. $\dfrac{3-7}{9-3} = \dfrac{-4}{6} = \dfrac{-2 \cdot 2}{2 \cdot 3} = -\dfrac{2}{3}$

P3. $\dfrac{-3-4}{6-(-1)} = \dfrac{-7}{6+1} = \dfrac{-7}{7} = -1$

3.3 Quick Checks

1. Let run = 10, rise = 6
$\text{slope} = \dfrac{\text{rise}}{\text{run}} = \dfrac{6}{10} = \dfrac{3}{5}$

2. False; if $P = (x_1, y_1)$ and $Q = (x_2, y_2)$, then the slope of the line that contains P and Q is $\dfrac{y_2 - y_1}{x_2 - x_1}$, not $\dfrac{x_2 - x_1}{y_2 - y_1}$.

3. True

4. If the graph of a line goes up as you move to the right, then the slope of this line must be positive.

5.

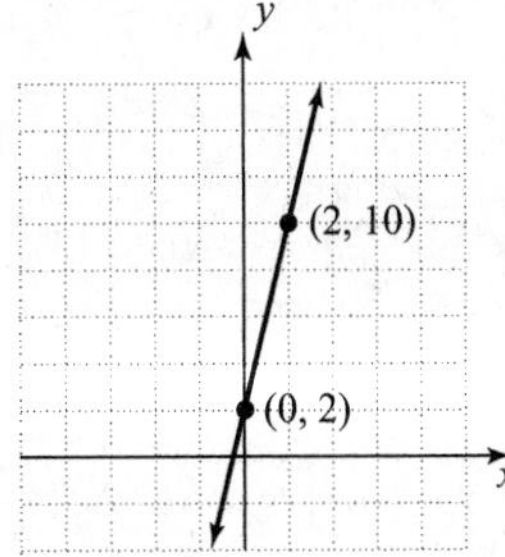

Let $(0, 2) = (x_1, y_1)$ and $(2, 10) = (x_2, y_2)$.

$$m = \frac{y_2 - y_1}{x_2 - x_1} = \frac{10-2}{2-0} = \frac{8}{2} = \frac{4}{1} = 4$$

The value of y increases by 4 when x increases by 1.

6.

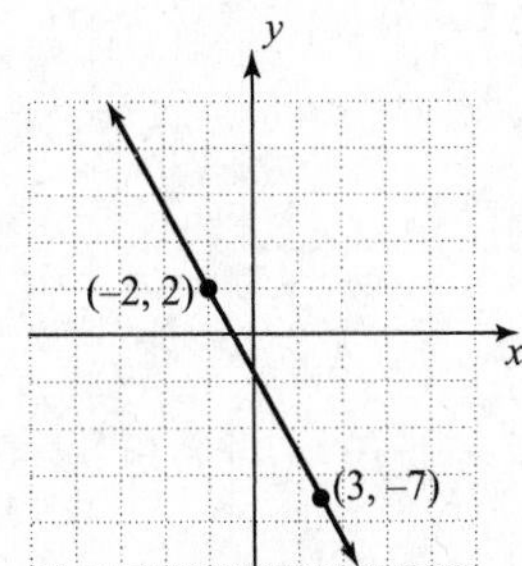

Let $(-2, 2) = (x_1, y_1)$ and $(3, -7) = (x_2, y_2)$.

$$m = \frac{y_2 - y_1}{x_2 - x_1} = \frac{-7-2}{3-(-2)} = \frac{-9}{5} = -\frac{9}{5}$$

The value of y decreases by 9 when x increases by 5.

7. The slope of a horizontal line is <u>0</u>, while the slope of a vertical line is <u>undefined</u>.

8.

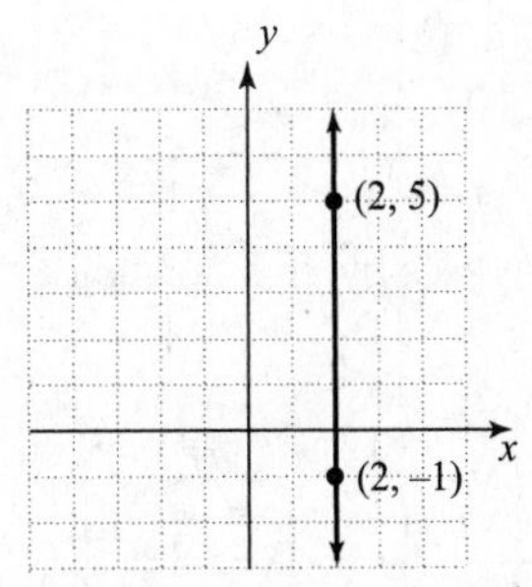

Let $(2, 5) = (x_1, y_1)$ and $(2, -1) = (x_2, y_2)$.

$$m = \frac{y_2 - y_1}{x_2 - x_1} = \frac{-1-5}{2-2} = \frac{-6}{0}$$

The slope of the line is undefined. When y increases by 1, there is no change in x.

9.

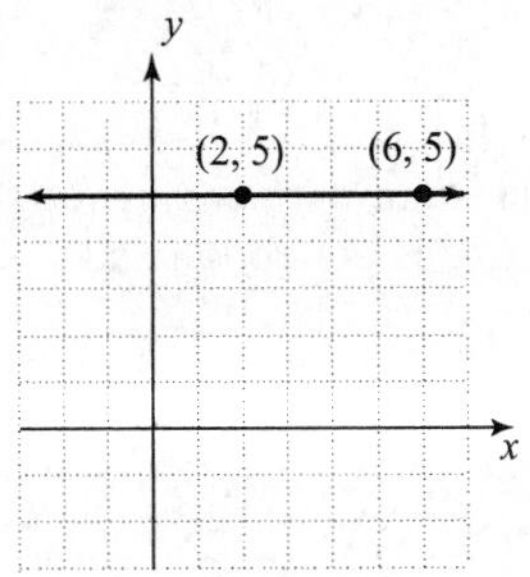

Let $(2, 5) = (x_1, y_1)$ and $(6, 5) = (x_2, y_2)$.

$$m = \frac{y_2 - y_1}{x_2 - x_1} = \frac{5-5}{6-2} = \frac{0}{4} = 0$$

There is no change in y when x increases by 1.

10. **(a)** $m = \frac{\text{rise}}{\text{run}} = \frac{1}{2}$

y will increase by 1 unit when x increases by 2 units. Starting at (1, 2) and moving 1 unit up and 2 units to the right, we end up at (3, 3).

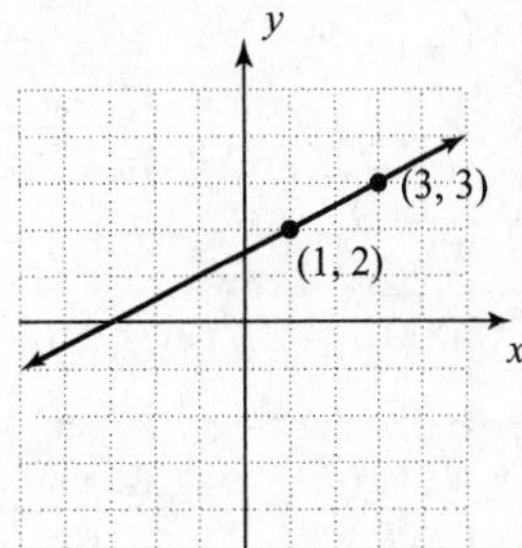

(b) $m = \frac{\text{rise}}{\text{run}} = -3 = \frac{-3}{1}$

y will decrease by 3 units when x increases by 1 unit. Starting at (1, 2) and moving 3 units down and 1 unit to the right, we end up at (2, −1).

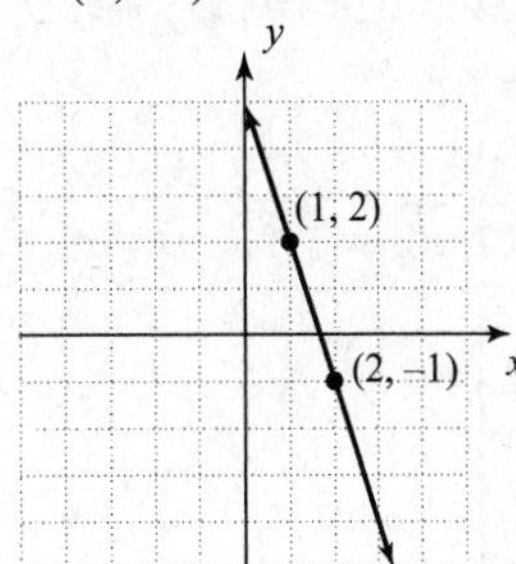

(c) $m = \frac{\text{rise}}{\text{run}} = 0 = \frac{0}{1}$

There is no change in y when x increases or decreases; the line is horizontal and passes through (1, 2).

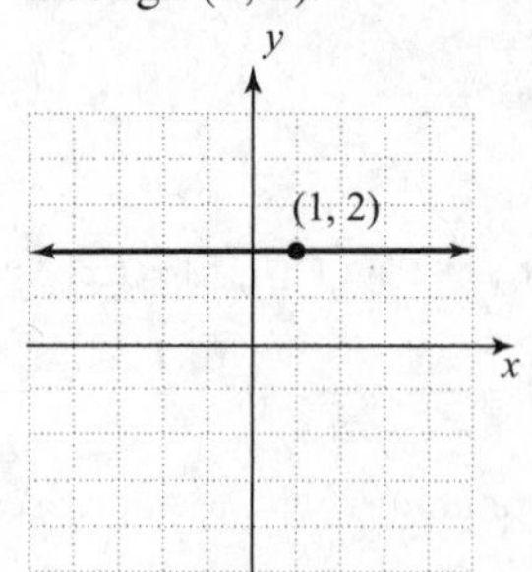

11. $\text{grade} = \frac{\text{rise}}{\text{run}} = \frac{4 \text{ feet}}{50 \text{ feet}} = 0.08 = 8\%$

The grade of the road is 8%.

12. Let $(x_1, y_1) = (10{,}000, 1370)$ and $(x_2, y_2) = (14{,}000, 1850)$.

$$m = \frac{y_2 - y_1}{x_2 - x_1} = \frac{1850 - 1370}{14{,}000 - 10{,}000} = \frac{480}{4000} = 0.12$$

The unit of measure of y is dollars while the unit of measure of x is miles driven. Between 10,000 and 14,000 miles driven, the annual cost of operating a Cobalt is $0.12 per mile, on average.

3.3 Exercises

13. The line passes through $(0, 3) = (x_1, y_1)$ and $(2, 0) = (x_2, y_2)$.

$$m = \frac{y_2 - y_1}{x_2 - x_1} = \frac{0-3}{2-0} = \frac{-3}{2} = -\frac{3}{2}$$

15. The line passes through $(-4, -4) = (x_1, y_1)$ and $(8, 2) = (x_2, y_2)$.

$$m = \frac{y_2 - y_1}{x_2 - x_1} = \frac{2-(-4)}{8-(-4)} = \frac{6}{12} = \frac{1}{2}$$

17. The line passes through $(-3, 3) = (x_1, y_1)$ and $(3, -1) = (x_2, y_2)$.

$$m = \frac{y_2 - y_1}{x_2 - x_1} = \frac{-1-3}{3-(-3)} = \frac{-4}{6} = -\frac{2}{3}$$

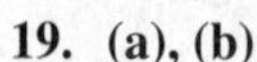

19. (a), (b)

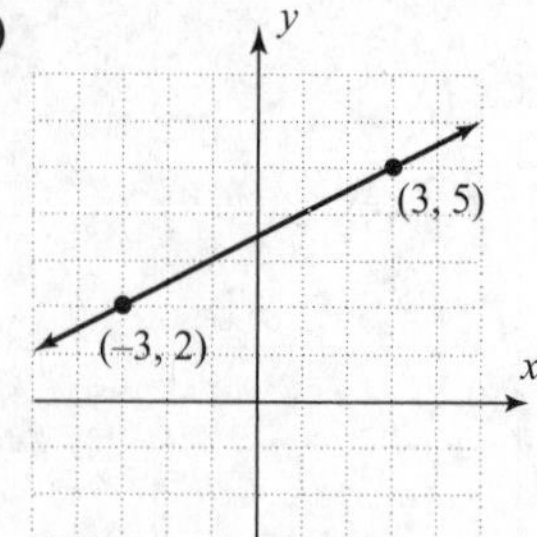

(c) Let $(-3, 2) = (x_1, y_1)$ and $(3, 5) = (x_2, y_2)$.

$$m = \frac{y_2 - y_1}{x_2 - x_1} = \frac{5-2}{3-(-3)} = \frac{3}{6} = \frac{1}{2}$$

The value of y increases by 1 unit when x increases by 2 units.

21. (a), (b)

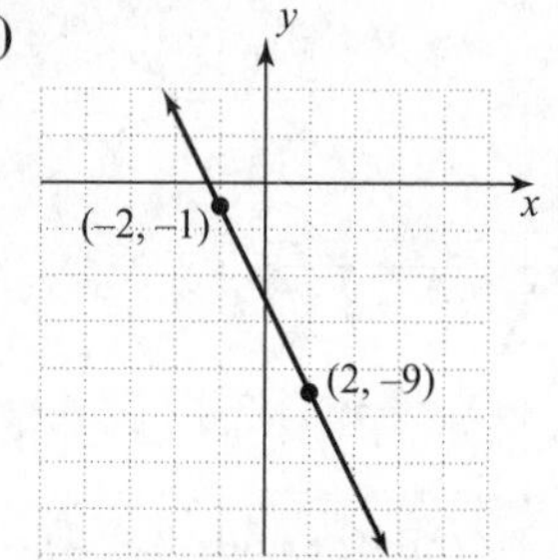

(c) Let $(2, -9) = (x_1, y_1)$ and $(-2, -1) = (x_2, y_2)$.

$$m = \frac{-1-(-9)}{-2-2} = \frac{8}{-4} = -2$$

The value of y decreases by 2 units when x increases by 1 unit.

23. $(10, 4) = (x_1, y_1)$; $(6, 12) = (x_2, y_2)$

$$m = \frac{y_2 - y_1}{x_2 - x_1} = \frac{12-4}{6-10} = \frac{8}{-4} = -2 = \frac{-2}{1}$$

The value of y decreases by 2 units when x increases by 1 unit.

25. $(4, -4) = (x_1, y_1)$; $(12, -12) = (x_2, y_2)$

$$m = \frac{y_2 - y_1}{x_2 - x_1} = \frac{-12-(-4)}{12-4} = \frac{-8}{8} = -1 = \frac{-1}{1}$$

The value of y decreases by 1 unit when x increases by 1 unit.

27. $(7, -2) = (x_1, y_1)$; $(4, 3) = (x_2, y_2)$

$$m = \frac{y_2 - y_1}{x_2 - x_1} = \frac{3-(-2)}{4-7} = \frac{5}{-3} = \frac{-5}{3}$$

The value of y decreases by 5 units when x increases by 3 units.

29. $(0, 6) = (x_1, y_1); (-4, 0) = (x_2, y_2)$

$$m = \frac{y_2 - y_1}{x_2 - x_1} = \frac{0-6}{-4-0} = \frac{-6}{-4} = \frac{3}{2}$$

The value of y increases by 3 units when x increases by 2 units.

31. $(-4, -1) = (x_1, y_1); (2, 3) = (x_2, y_2)$

$$m = \frac{y_2 - y_1}{x_2 - x_1} = \frac{3-(-1)}{2-(-4)} = \frac{4}{6} = \frac{2}{3}$$

The value of y increases by 2 units when x increases by 3 units.

33. $\left(\frac{1}{2}, \frac{3}{4}\right) = (x_1, y_1); \left(-\frac{5}{2}, -\frac{1}{4}\right) = (x_2, y_2)$

$$m = \frac{y_2 - y_1}{x_2 - x_1} = \frac{-\frac{1}{4} - \frac{3}{4}}{-\frac{5}{2} - \frac{1}{2}} = \frac{-1}{-3} = \frac{1}{3}$$

The value of y increases by 1 unit when x increases by 3 units.

35. $\left(\frac{1}{2}, \frac{1}{3}\right) = (x_1, y_1); \left(\frac{3}{4}, \frac{5}{6}\right) = (x_2, y_2)$

$$\begin{aligned} m &= \frac{y_2 - y_1}{x_2 - x_1} \\ &= \frac{\frac{5}{6} - \frac{1}{3}}{\frac{3}{4} - \frac{1}{2}} \\ &= \frac{\frac{5}{6} - \frac{2}{6}}{\frac{3}{4} - \frac{2}{4}} \\ &= \frac{\frac{3}{6}}{\frac{1}{4}} \\ &= \frac{\frac{1}{2}}{\frac{1}{4}} \\ &= \frac{1}{2} \div \frac{1}{4} \\ &= \frac{1}{2} \cdot \frac{4}{1} \\ &= 2 \\ &= \frac{2}{1} \end{aligned}$$

The value of y increases by 2 units when x increases by 1 unit.

37. The line is vertical, so the slope is undefined.

39. The line is horizontal, so the slope is 0.

41. $(4, -6) = (x_1, y_1); (-1, -6) = (x_2, y_2)$

$$m = \frac{y_2 - y_1}{x_2 - x_1} = \frac{-6-(-6)}{-1-4} = \frac{0}{-5} = 0$$

There is no change in the y values, the line is horizontal.

43. $(3, 9) = (x_1, y_1); (3, -2) = (x_2, y_2)$

$$m = \frac{y_2 - y_1}{x_2 - x_1} = \frac{-2-9}{3-3} = \frac{-11}{0}$$

The slope is undefined. The line is vertical.

45. $(4, 2); m = 1$

Plot (4, 2). Since $m = 1 = \frac{1}{1}$, there is a 1-unit increase in y for every 1-unit increase in x.

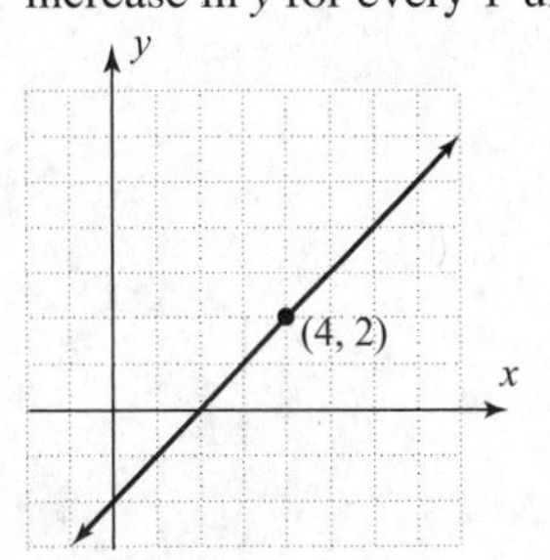

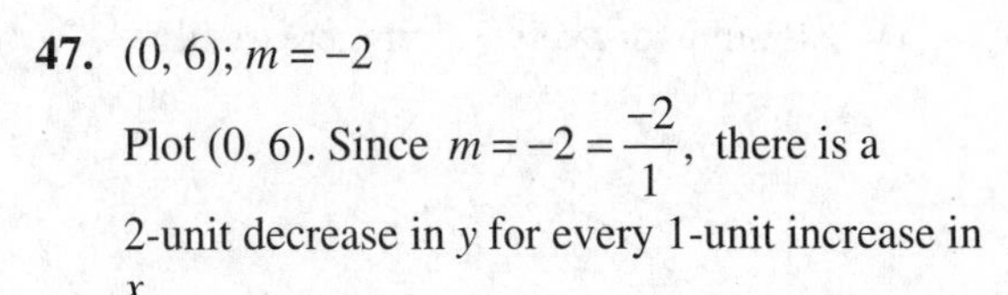

47. $(0, 6); m = -2$

Plot (0, 6). Since $m = -2 = \frac{-2}{1}$, there is a 2-unit decrease in y for every 1-unit increase in x.

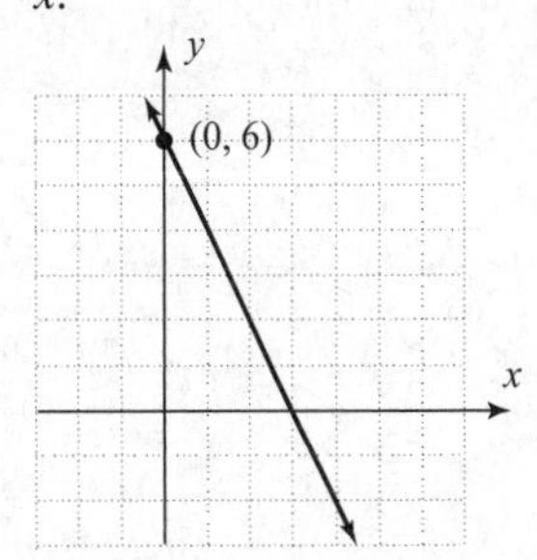

49. $(-1, 0);\ m = \frac{1}{4}$

Plot (−1, 0). Since $m = \frac{1}{4}$, there is a 1-unit increase in y for every 4-unit increase in x.

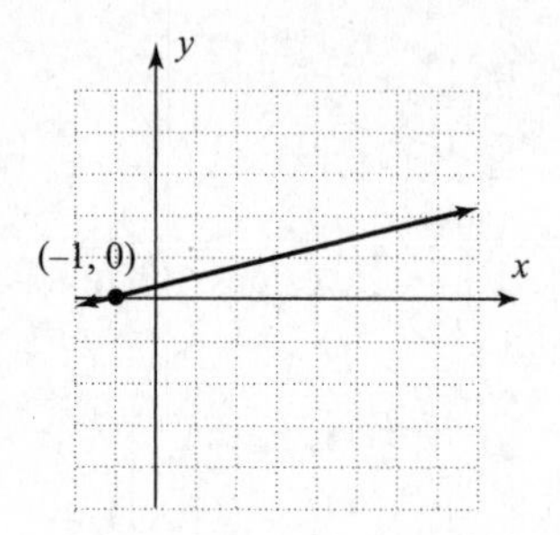

51. (2, −3); $m = 0$
Plot (2, −3). Since $m = 0$, the line is horizontal.

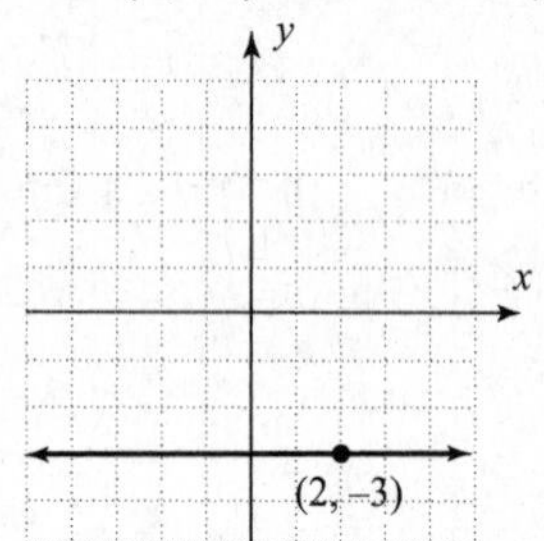

53. (2, 1); $m = \frac{2}{3}$

Plot (2, 1). Since $m = \frac{2}{3}$, there is a 2-unit increase in y for every 3-unit increase in x.

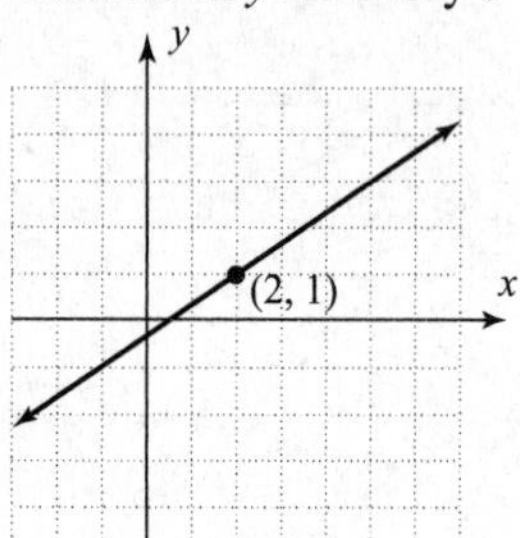

55. (−1, 4); $m = -\frac{5}{3}$

Plot (−1, 4). Since $m = -\frac{5}{3} = -\frac{5}{3}$, there is a 5-unit decrease in y for every 3-unit increase in x.

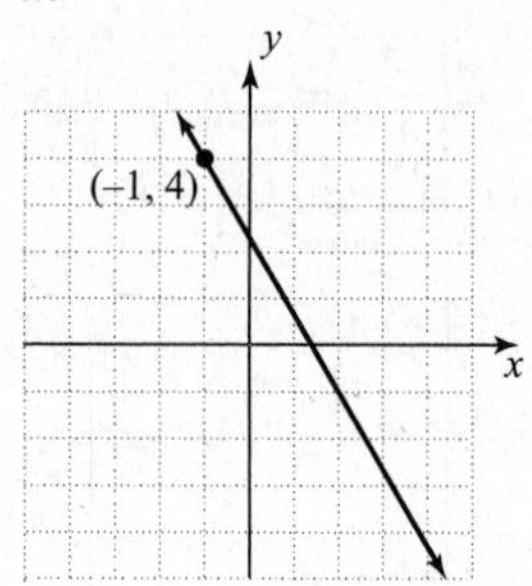

57. (0, 0); *m* is undefined.
Plot (0, 0), which is the origin. Since *m* is undefined, the line is vertical—it is the y-axis.

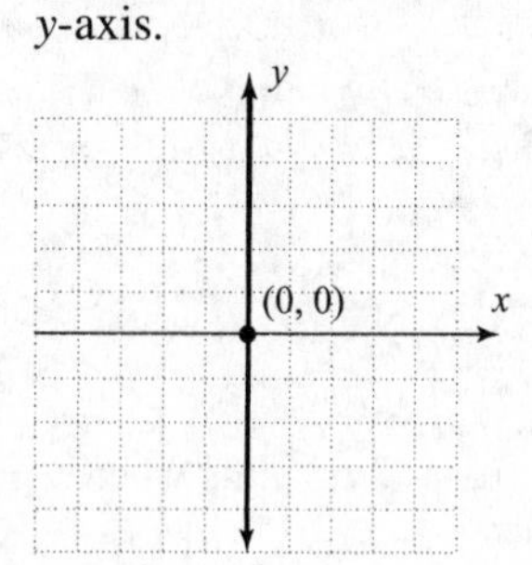

59. (0, 2); $m = -4$

Plot (0, 2). Since $m = -4 = \frac{-4}{1}$, there is a 4-unit decrease in y for every 1-unit increase in x.

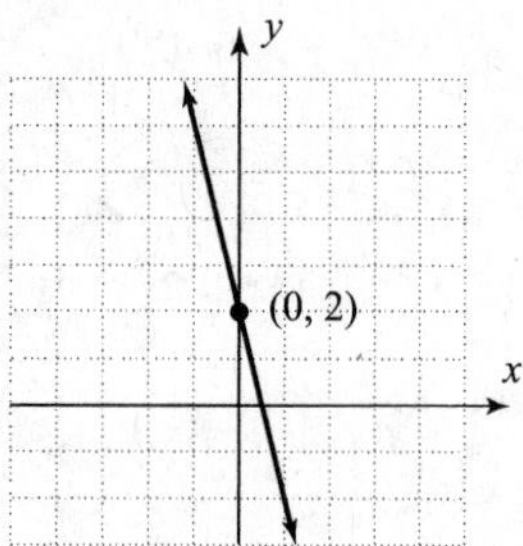

61. (2, −3); $m = \frac{3}{4}$

Plot (2, −3). Since $m = \frac{3}{4}$, there is a 3-unit increase in y for every 4-unit increase in x.

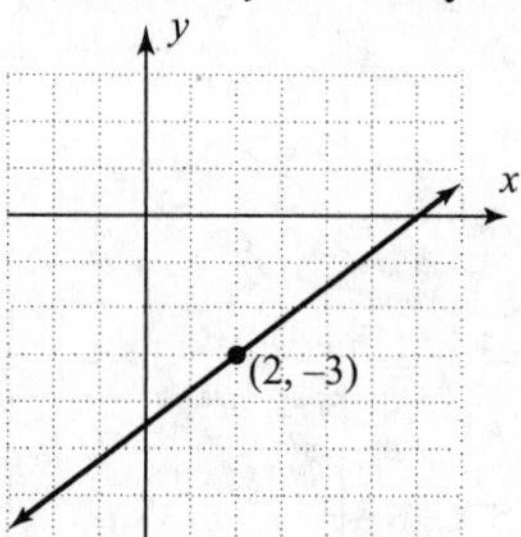

63. (2, −1)

$m_1 = 2 = \frac{2}{1}$

$m_2 = -\frac{1}{2} = \frac{-1}{2}$

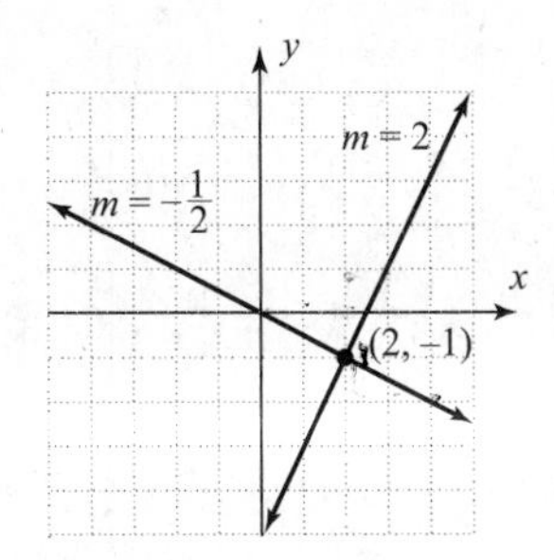

65. $m = \frac{3}{4}$
(−1, −2)
(2, 1)

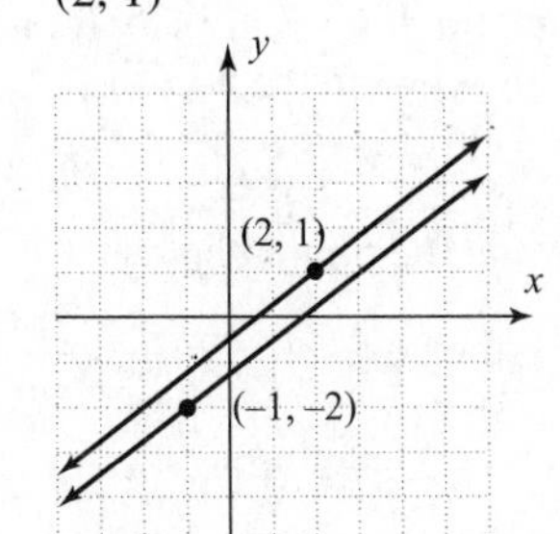

67. For every 1 foot = 12 inches horizontally (run), there is a 4-inch elevation (rise). The pitch (slope) is $\frac{4}{12} = \frac{1}{3}$.

69. The pitch (slope) is $\frac{2}{5}$ with a horizontal distance (run) of 30 in. Let x be the height.

$$\frac{x}{30} = \frac{2}{5}$$

$$x = \frac{2}{5}(30) = 12$$

He should add 12 in. = 1 ft to the height.

71. $\frac{\text{rise}}{\text{run}} = \frac{200}{1250} = 0.16 = 16\%$
The grade is 16%.

73. (0, 123 million); (70, 281 million)

$$m = \frac{(281 \text{ million}) - (123 \text{ million})}{70 - 0}$$
$$= 158 \text{ million}$$
$$\approx 2.26 \text{ million}$$

The population is increasing at an average rate of about 2.26 million people per year.

In 75–77 points may vary.

75. (−2, 1), (0, −5) lie on the line.

$$m = \frac{-5-1}{0-(-2)} = \frac{-6}{2} = -3$$

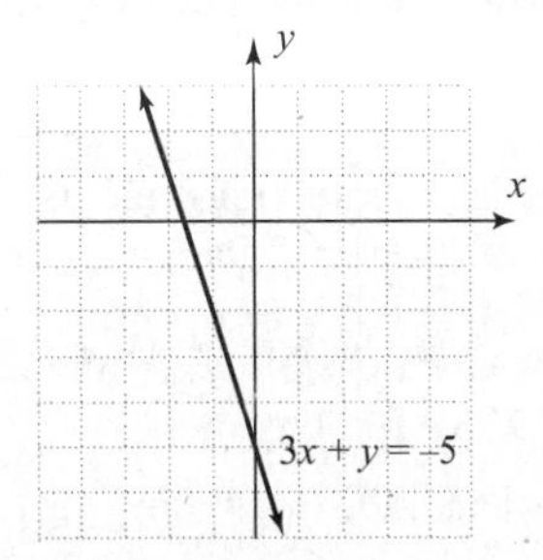

77. (−2, −2) and (0, 4) lie on the line.

$$m = \frac{4-(-2)}{0-(-2)} = \frac{6}{2} = 3$$

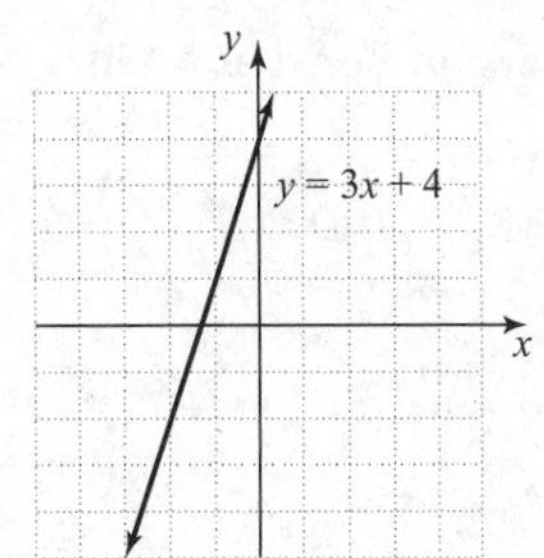

79. $(2a, a) = (x_1, y_1);\ (3a, -a) = (x_2, y_2)$

$$m = \frac{y_2 - y_1}{x_2 - x_1} = \frac{-a - a}{3a - 2a} = \frac{-2a}{a} = -2$$

81. $(2p+1, q-4) = (x_1, y_1);$
$(3p+1, 2q-4) = (x_2, y_2)$

$$m = \frac{y_2 - y_1}{x_2 - x_1} = \frac{2q-4-(q-4)}{3p+1-(2p+1)}$$
$$= \frac{2q-4-q+4}{3p+1-2p-1}$$
$$= \frac{q}{p}$$

83. $(a+1, b-1) = (x_1, y_1);$
$(2a-5, b+5) = (x_2, y_2)$

$$m = \frac{y_2 - y_1}{x_2 - x_1} = \frac{b+5-(b-1)}{2a-5-(a+1)}$$
$$= \frac{b+5-b+1}{2a-5-a-1}$$
$$= \frac{6}{a-6}$$

85. Let $R_1 = 1000$ and $Q_1 = 400$, so $R_2 = 1200$ and $Q_2 = 500$.

$$MR = \frac{R_2 - R_1}{Q_2 - Q_1} = \frac{1200 - 1000}{500 - 400} = \frac{200}{100} = 2$$

The marginal revenue, or rate of change, is 2. For every hot dog sold, revenue increases by $2.

87. Answers may vary. One possibility: A line with one *x*-intercept, say $(x_1, 0)$, but no *y*-intercept is a vertical line. $(x_1, 1)$ and $(x_1, 3)$ could lie on the line. Since the line is vertical, the slope is undefined.

Section 3.4

Preparing for Slope-Intercept Form of a Line

P1.
$$\begin{aligned} 4x + 2y &= 10 \\ -4x + 4x + 2y &= -4x + 10 \\ 2y &= -4x + 10 \\ \frac{2y}{2} &= \frac{-4x + 10}{2} \\ y &= -2x + 5 \end{aligned}$$

P2.
$$\begin{aligned} 10 &= 2x - 8 \\ 10 + 8 &= 2x - 8 + 8 \\ 18 &= 2x \\ \frac{18}{2} &= \frac{2x}{2} \\ 9 &= x \end{aligned}$$

The solution set is $\{9\}$.

3.4 Quick Checks

1. Comparing $y = 4x - 3$ to $y = mx + b$, we see that $m = 4$ and $b = -3$. The slope is 4 and the *y*-intercept is -3.

2.
$$\begin{aligned} 3x + y &= 7 \\ y &= -3x + 7 \end{aligned}$$

Comparing $y = -3x + 7$ to $y = mx + b$, we see that $m = -3$ and $b = 7$. The slope is -3 and the *y*-intercept is 7.

3.
$$\begin{aligned} 2x + 5y &= 15 \\ 5y &= -2x + 15 \\ y &= -\frac{2}{5}x + 3 \end{aligned}$$

Comparing $y = -\frac{2}{5}x + 3$ to $y = mx + b$, we see that $m = -\frac{2}{5}$ and $b = 3$. The slope is $-\frac{2}{5}$ and the *y*-intercept is 3.

4. Comparing $y = 8$ to $y = mx + b$, we see that $m = 0$ and $b = 8$. The slope is 0 and the *y*-intercept is 8.

5. $x = 3$ cannot be compared to $y = mx + b$. It has an undefined slope, and no *y*-intercept.

6. $y = 2x - 5$
The slope is $m = 2$ and the *y*-intercept is $b = -5$. Plot $(0, -5)$, then use $m = 2 = \frac{2}{1} = \frac{\text{rise}}{\text{run}}$ to find a second point on the graph.

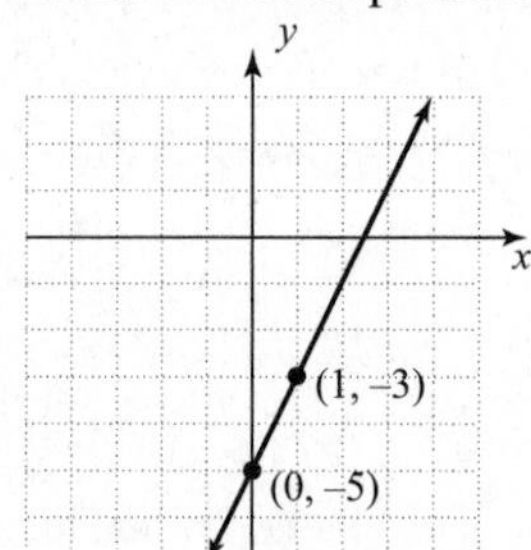

7. $y = \frac{1}{2}x - 5$

The slope is $m = \frac{1}{2}$ and the *y*-intercept is $b = -5$. Plot $(0, -5)$, then use $m = \frac{1}{2} = \frac{\text{rise}}{\text{run}}$ to find a second point on the graph.

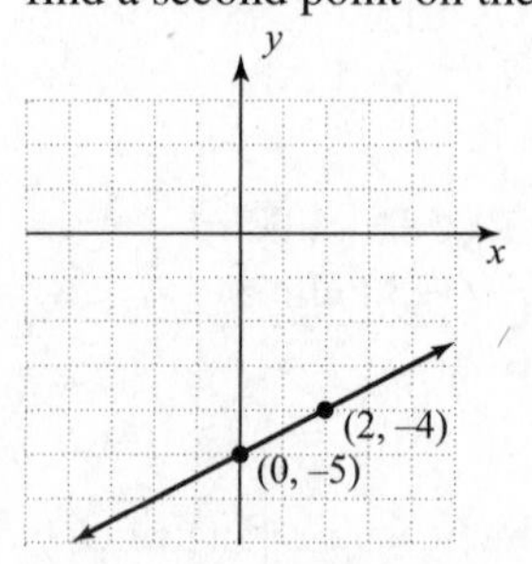

8. $y = -3x + 1$
The slope is $m = -3$ and the y-intercept is $b = 1$.
Plot (0, 1), then use $m = -3 = \frac{-3}{1} = \frac{\text{rise}}{\text{run}}$ to find a second point on the graph.

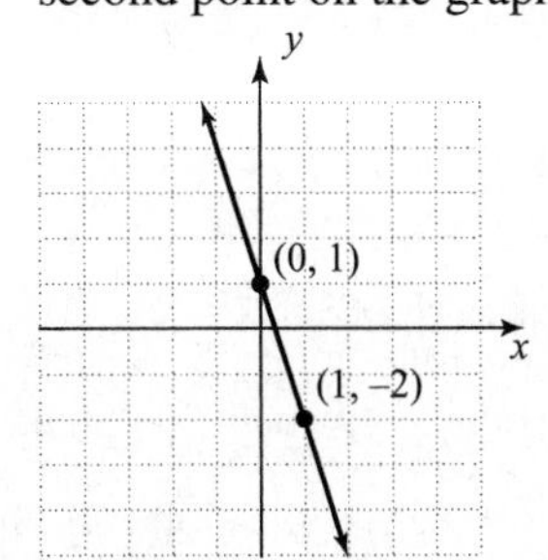

9. $y = -\frac{3}{2}x + 4$

The slope is $m = -\frac{3}{2}$ and the y-intercept is $b = 4$. Plot (0, 4), then use $m = -\frac{3}{2} = \frac{-3}{2} = \frac{\text{rise}}{\text{run}}$ to find a second point on the graph.

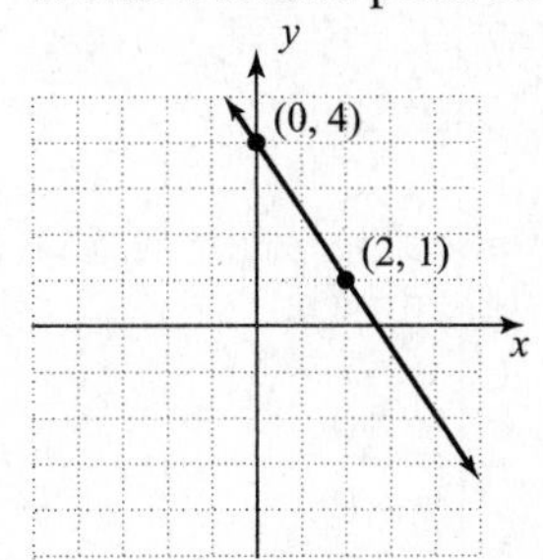

10. $-2x + y = -3$
$y = 2x - 3$
The slope is $m = 2$ and the y-intercept is $b = -3$. Plot (0, –3), then use $m = 2 = \frac{2}{1} = \frac{\text{rise}}{\text{run}}$ to find a second point on the graph.

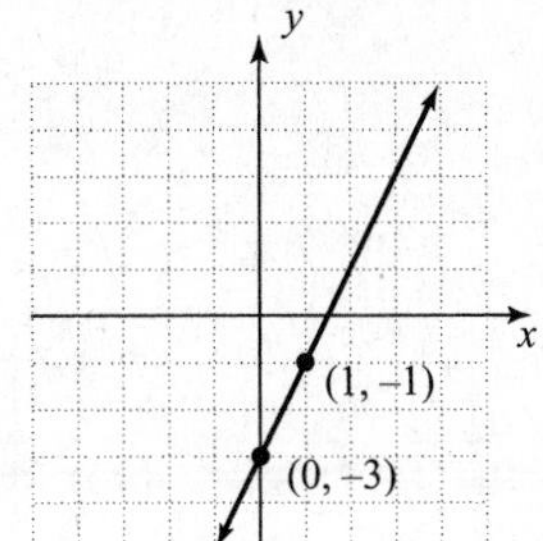

11. $6x - 2y = 2$
$-2y = -6x + 2$
$y = 3x - 1$
The slope is $m = 3$ and the y-intercept is $b = -1$. Plot (0, –1), then use $m = 3 = \frac{3}{1} = \frac{\text{rise}}{\text{run}}$ to find a second point on the graph.

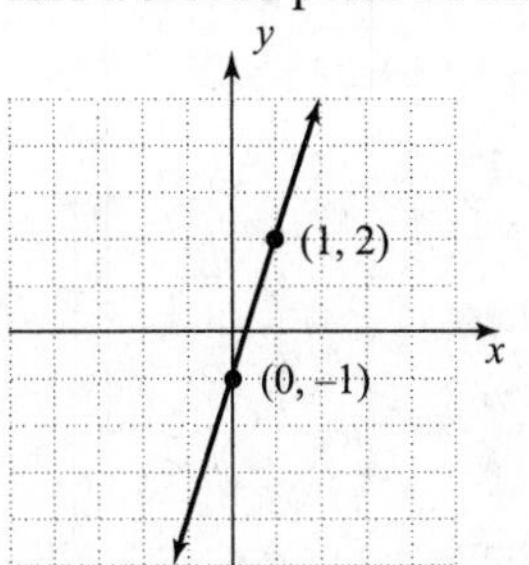

12. $3x + 5y = 0$
$5y = -3x$
$y = -\frac{3}{5}x$

The slope is $m = -\frac{3}{5}$ and the y-intercept is $b = 0$. Plot (0, 0), then use $m = -\frac{3}{5} = \frac{-3}{5} = \frac{\text{rise}}{\text{run}}$ to find a second point on the graph.

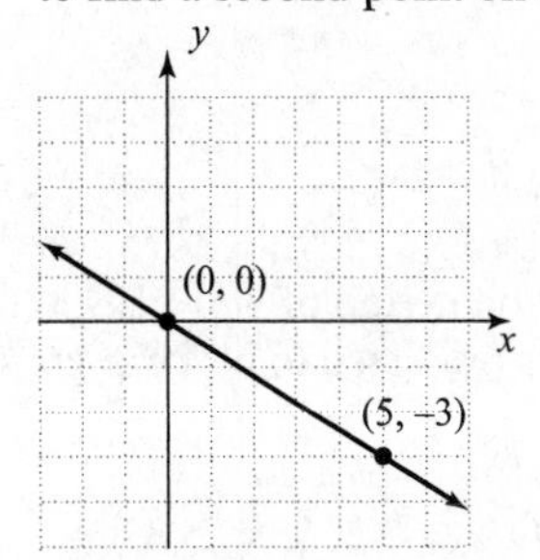

13. List three techniques that can be used to graph a line: point plotting, using intercepts, using slope and a point.

14. Substitute 3 for m and –2 for b in $y = mx + b$.
$y = 3x - 2$

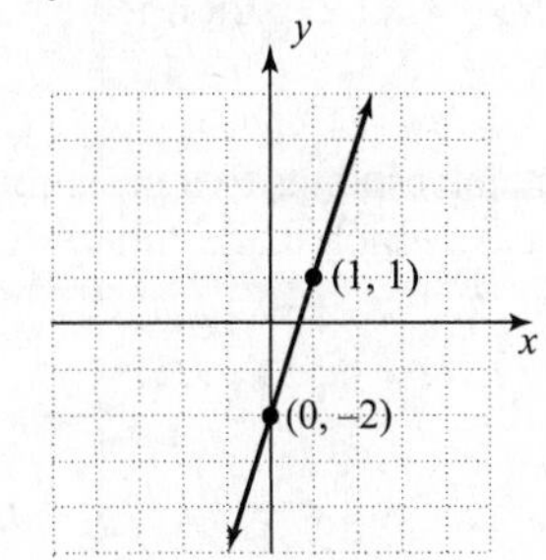

15. Substitute $-\frac{1}{4}$ for m and 3 for b in $y = mx + b$.

$y = -\frac{1}{4}x + 3$

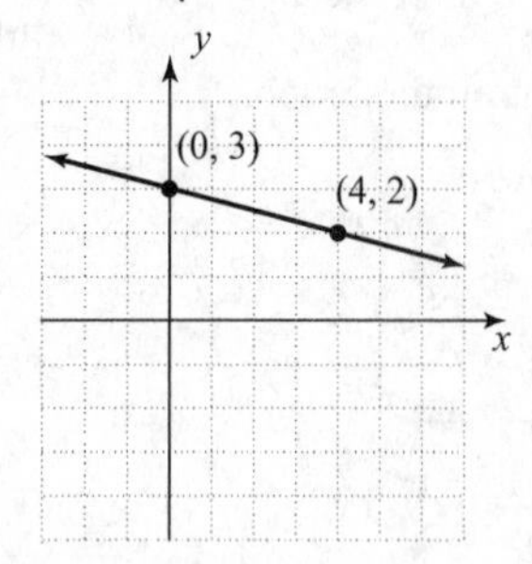

16. Substitute 0 for m and −1 for b in $y = mx + b$.

$y = 0x - 1$

$y = -1$

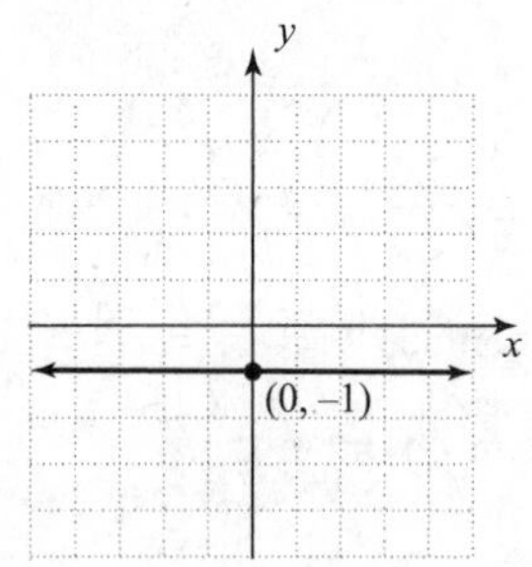

17. **(a)** $y = 143x - 2215$

$x = 30$: $y = 143(30) - 2215 = 2075$

After a gestation period of 30 weeks, the birth weight is predicted to be 2075 grams.

(b) $y = 143x - 2215$

$x = 36$: $y = 143(36) - 2215 = 2933$

After a gestation period of 36 weeks, the birth weight is predicted to be 2933 grams.

(c) The slope is $m = 143 = \frac{\text{rise}}{\text{run}} = \frac{143 \text{ grams}}{1 \text{ week}}$.

The birth weight increases by 143 grams as the length of the gestation period increases by 1 week.

(d) The y-intercept corresponds to a gestation period of 0 weeks, which makes no sense.

(e)

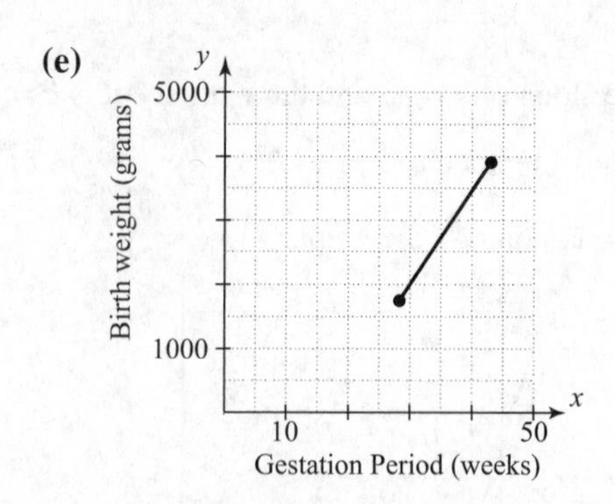

18. **(a)** The rate of change is given as \$0.38 per mile or $\frac{\$0.38}{1 \text{ mile}}$, which is the slope m. The cost of \$50 does not change with the number of miles driven, so this is the y-intercept, b.

$y = 0.38x + 50$

(b) $x = 75$: $y = 0.38(75) + 50 = 78.5$

If the truck is driven for 75 miles, the cost of the rental is \$78.50.

(c) $y = 84.20$:

$$\begin{aligned} 84.20 &= 0.38x + 50 \\ 34.20 &= 0.38x \\ 90 &= x \end{aligned}$$

If the cost of renting the truck was \$84.20, then 90 miles were driven.

(d)

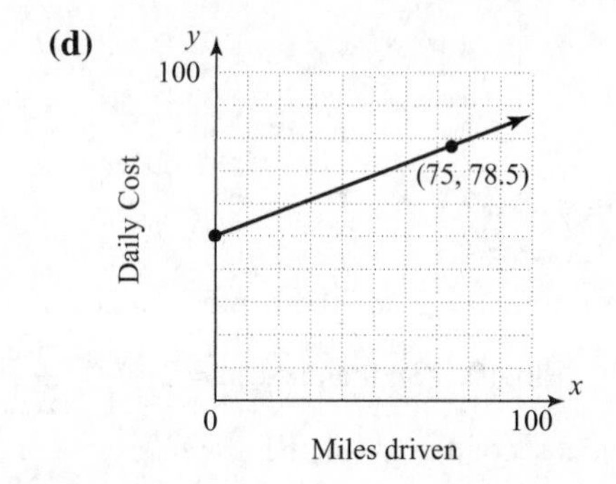

3.4 Exercises

19. $y = 5x + 2$

Slope: $m = 5$

y-intercept: $b = 2$

21. $y = x - 9$

Slope: $m = 1$

y-intercept: $b = -9$

23. $y = -10x + 7$

Slope: $m = -10$

y-intercept: $b = 7$

25. $y = -x - 9$
Slope: $m = -1$
y-intercept: $b = -9$

27. $2x + y = 4$
$y = -2x + 4$
Slope: $m = -2$
y-intercept: $b = 4$

29. $2x + 3y = 24$
$3y = -2x + 24$
$y = -\frac{2}{3}x + 8$
Slope: $m = -\frac{2}{3}$
y-intercept: $b = 8$

31. $5x - 3y = 9$
$-3y = -5x + 9$
$y = \frac{5}{3}x - 3$
Slope: $m = \frac{5}{3}$
y-intercept: $b = -3$

33. $x - 2y = 5$
$-2y = -x + 5$
$y = \frac{1}{2}x - \frac{5}{2}$
Slope: $m = \frac{1}{2}$
y-intercept: $b = -\frac{5}{2}$

35. $y = -5$
This is a horizontal line.
slope: $m = 0$
y-intercept: -5

37. $x = 6$
This is a vertical line.
Slope: m = undefined
y-intercept: none

39. $y = x + 3$
$m = 1$
$b = 3$

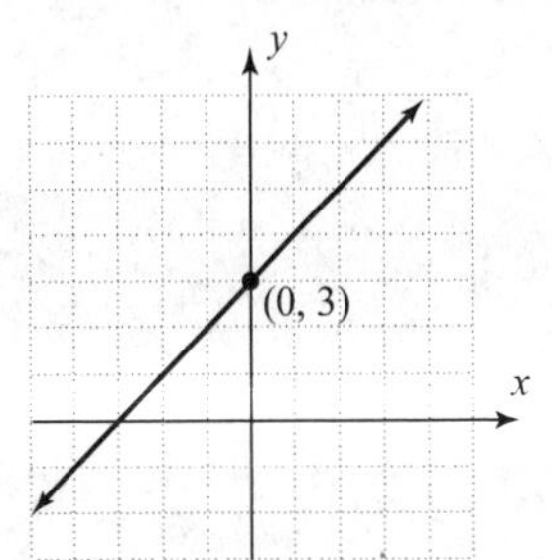

41. $y = -2x - 3$
$m = -2$
$b = -3$

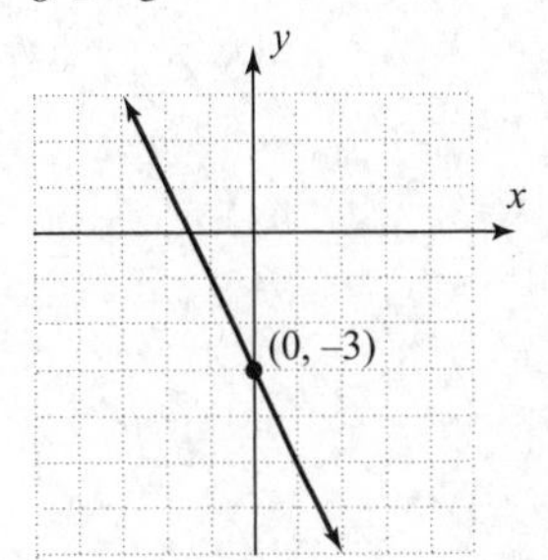

43. $y = -\frac{2}{3}x + 2$
$m = -\frac{2}{3}$
$b = 2$

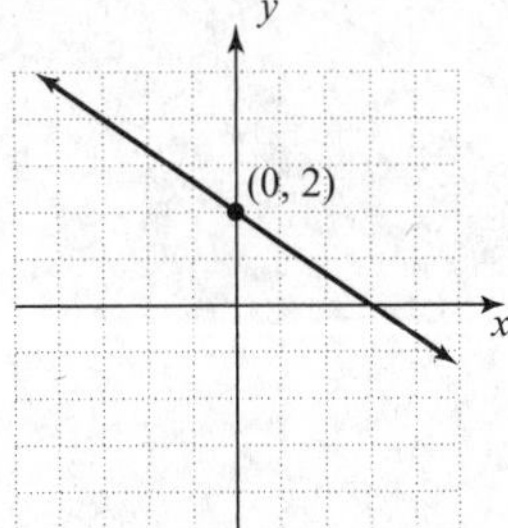

45. $y = -\frac{5}{2}x - 2$
$m = -\frac{5}{2}$
$b = -2$

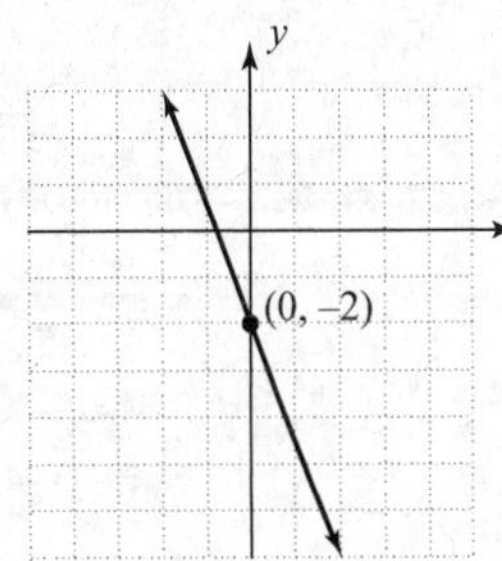

47. $4x+y=5$
$$y=-4x+5$$
$$m=-4=\frac{-4}{1}$$
$$b=5$$

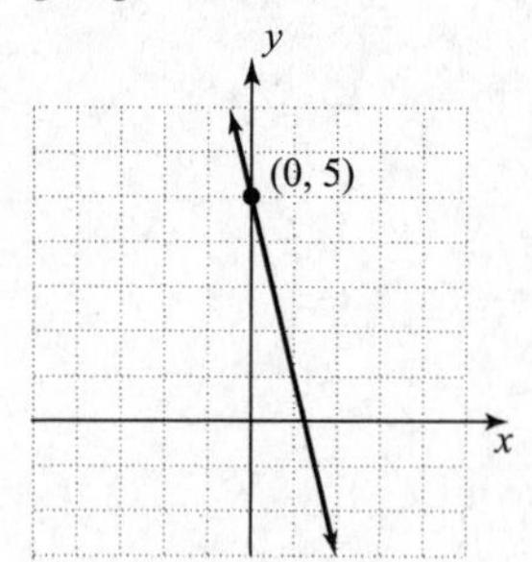

49. $x+2y=-6$
$$2y=-x-6$$
$$y=-\frac{1}{2}x-3$$
$$m=-\frac{1}{2}$$
$$b=-3$$

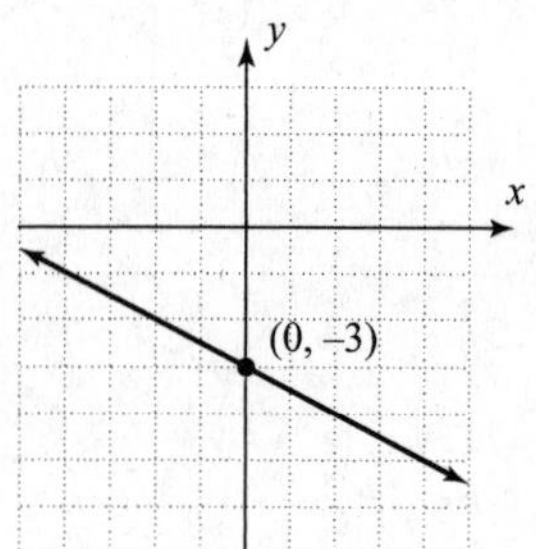

51. $3x-2y=10$
$$-2y=-3x+10$$
$$y=\frac{3}{2}x-5$$
$$m=\frac{3}{2}$$
$$b=-5$$

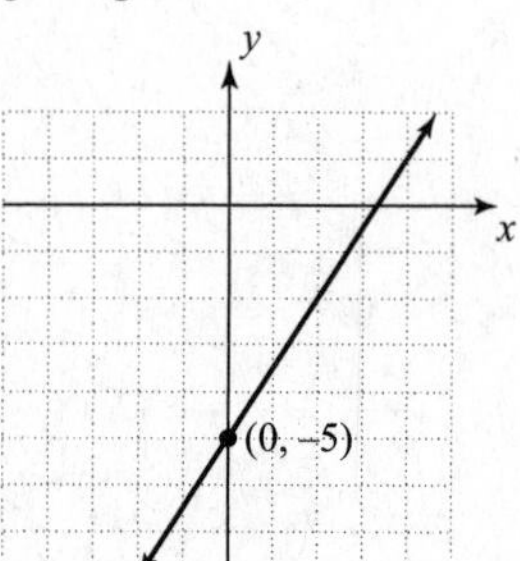

53. $6x+3y=-15$
$$3y=-6x-15$$
$$y=-2x-5$$
$$m=-2=\frac{-2}{1}$$
$$b=-5$$

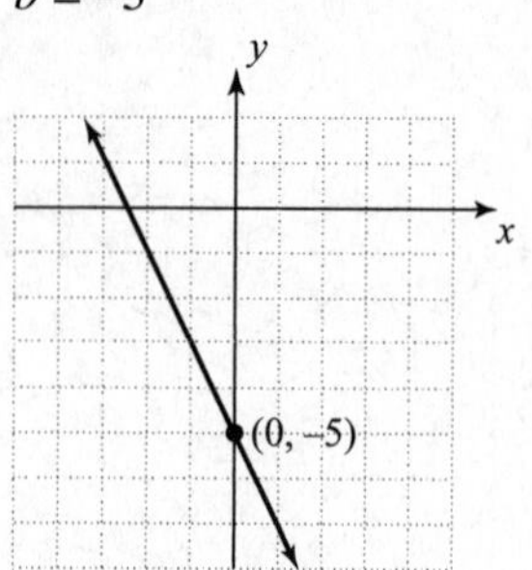

55. $m=-1$; $b=8$
$y=-1x+8$
$y=-x+8$

57. $m=\frac{6}{7}$; $b=-6$
$$y=\frac{6}{7}x-6$$

59. $m=-\frac{1}{3}$; $b=\frac{2}{3}$
$$y=-\frac{1}{3}x+\frac{2}{3}$$

61. m is undefined; x-intercept -5
This is a vertical line.
$x=-5$

63. $m=0$; $b=3$
$y=0x+3$
$y=3$

65. $m=5$; $b=0$
$y=5x+0$
$y=5x$

67. $y = 2x - 7$
$m = 2, b = -7$
x-intercept $\left(\frac{7}{2}, 0\right)$
y-intercept $(0, -7)$

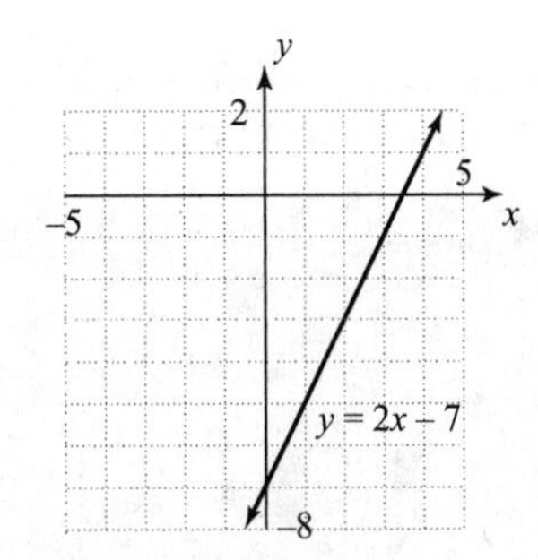

69. $3x - 2y = 24$
$$-2y = -3x + 24$$
$$y = \frac{3}{2}x - 12$$
$m = \frac{3}{2},\ b = -12$
x-intercept $(8, 0)$
y-intercept $(0, -12)$

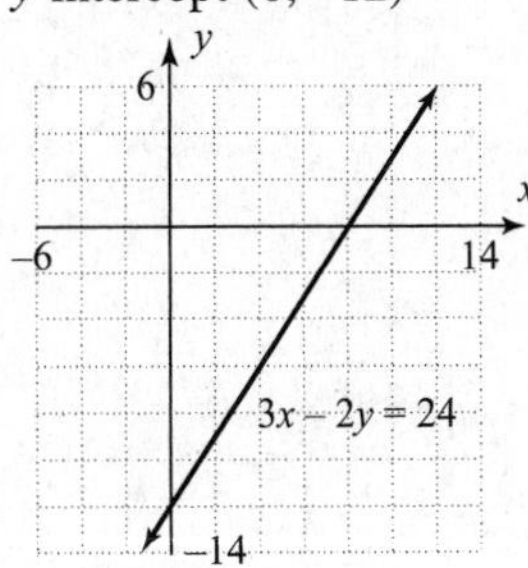

71. $y = -5$ or $y = 0x - 5$
$m = 0, b = -5$
Horizontal line

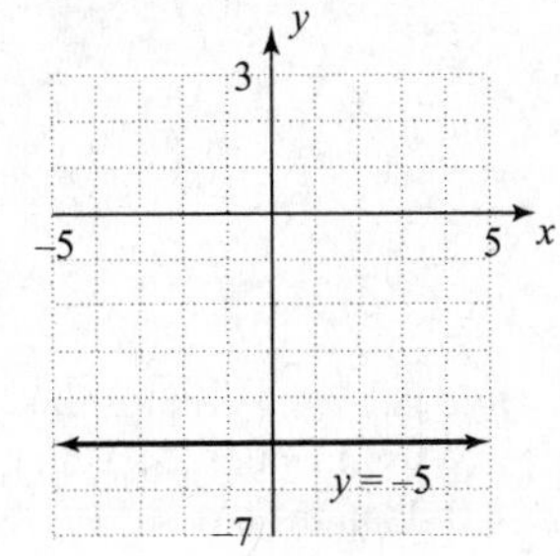

73. $x = -6$
Vertical line
Undefined slope
no y-intercept

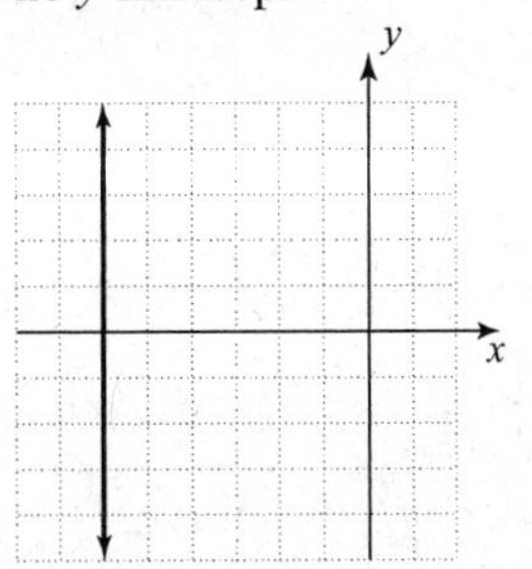

75. $6x - 4y = 0$
$$-4y = -6x + 0$$
$$y = \frac{6}{4}x + 0$$
$$y = \frac{3}{2}x + 0$$
$m = \frac{3}{2},\ b = 0$
x- and y-intercept $(0, 0)$

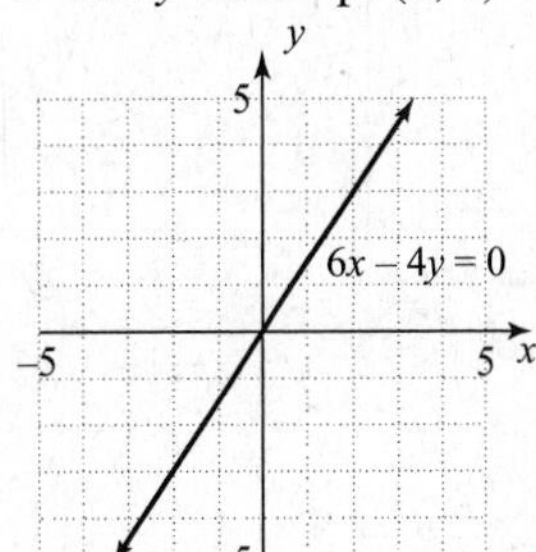

77. $y = -\frac{5}{3}x + 6$
$m = -\frac{5}{3}, b = 6$
x-intercept $\left(\frac{18}{5}, 0\right)$
y-intercept $(0, 6)$

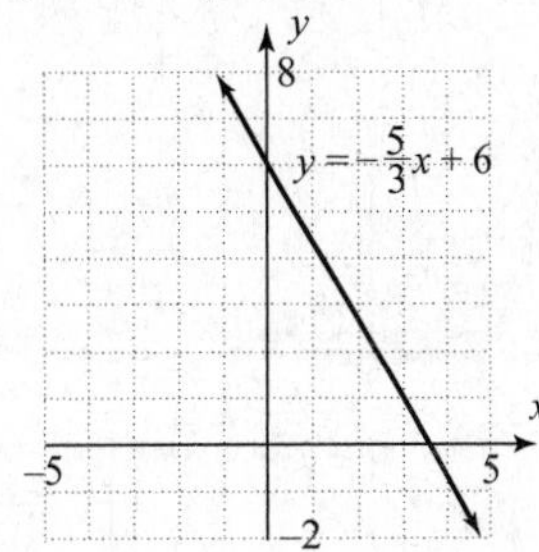

79. $2y = x + 4$ or $y = \frac{1}{2}x + 2$

$m = \frac{1}{2},\ b = 2$

x-intercept $(-4, 0)$
y-intercept $(0, 2)$

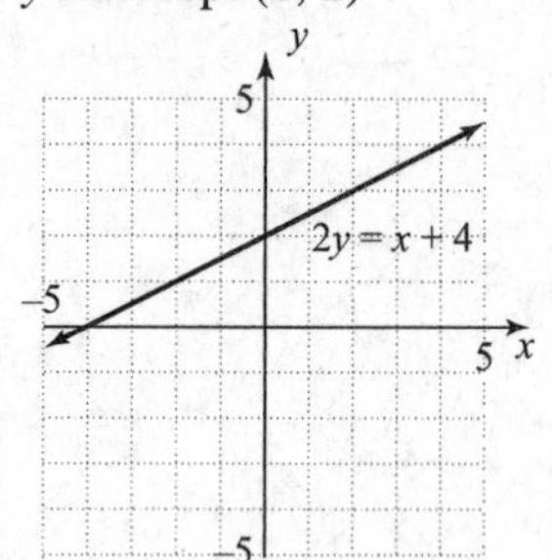

81. $y = \frac{x}{3} = \frac{1}{3}x + 0$

$m = \frac{1}{3}$

$b = 0$

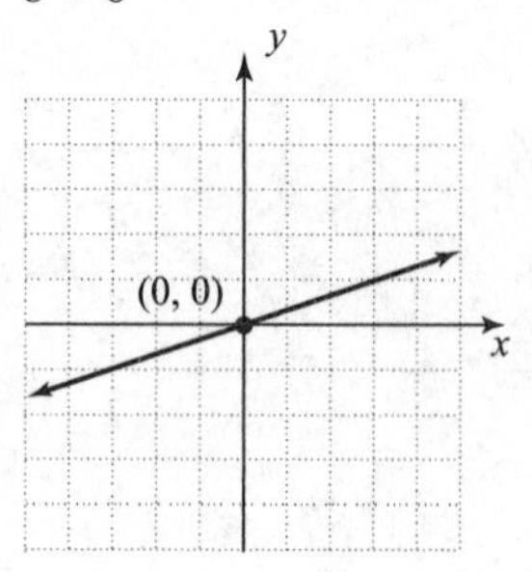

83. $2x = -8y$

$8y = -2x$

$y = -\frac{2}{8}x = -\frac{1}{4}x + 0$

$m = -\frac{1}{4}$

$b = 0$

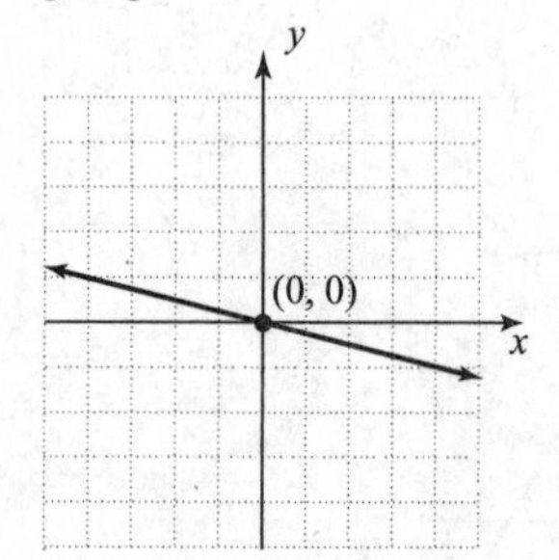

85. $y = -\frac{2x}{3} + 1 = -\frac{2}{3}x + 1$

$m = -\frac{2}{3}$

$b = 1$

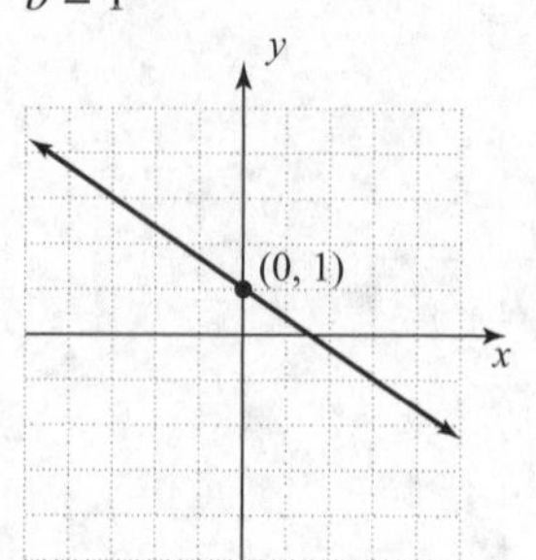

87. $x + 2 = -7$

$x = -9$

Vertical line
Undefined slope
No y-intercept

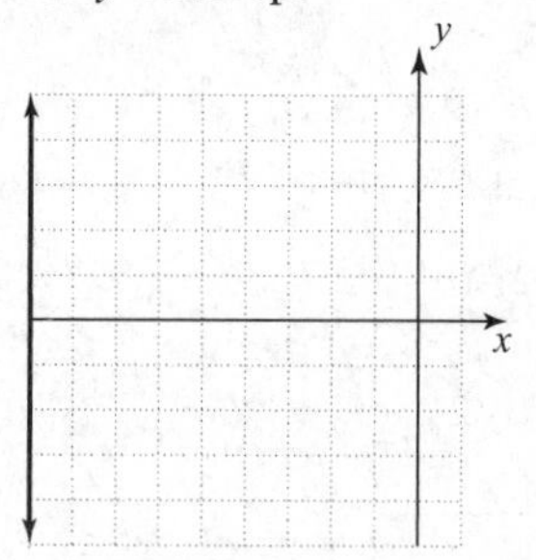

89. $5x + y + 1 = 0$

$y = -5x - 1$

$m = -5$

$b = -1$

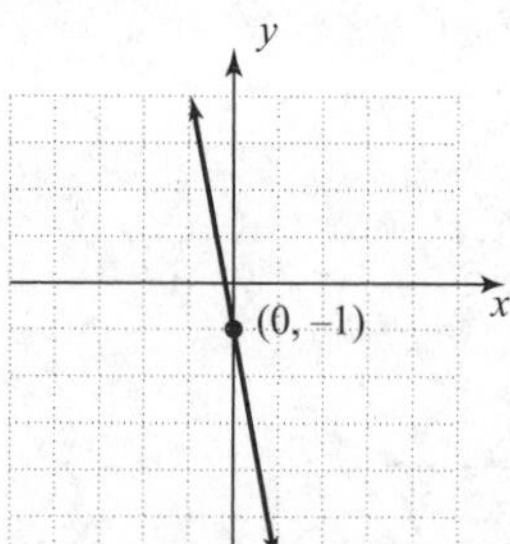

91. **(a)** The rate of change is 8% or 0.08 of the amount of sales. With no sales, the salary is \$400. Thus, $m = 0.08$ and $b = 400$.
$y = 0.08x + 400$

(b) For sales of \$1200, $x = 1200$.
$y = 0.08(1200) + 400 = 96 + 400 = 496$
Dien's income is \$496 if he sells \$1200 worth of merchandise.

(c)

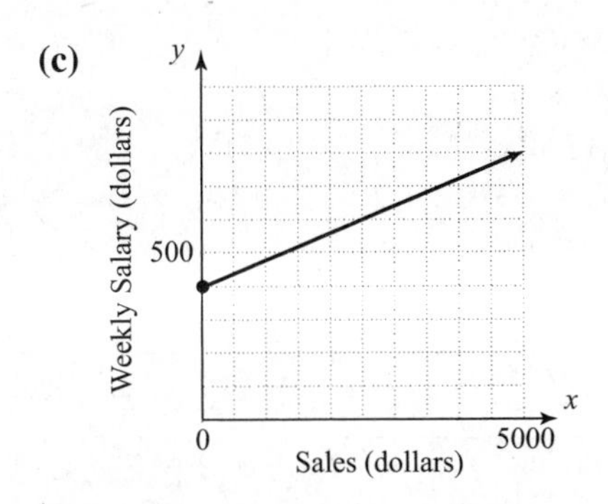

93. $y = -3.583x + 56$

(a) In 1999, $x = 4$.
$y = -3.583(4) + 56 = -14.332 + 56 = 41.668$
In 1999, the cost per minute was 47¢.

(b) Find x when $y = 20.17$.
$$20.17 = -3.583x + 56$$
$$-35.83 = -3.583x$$
$$10 = x$$
$x = 10$ corresponds to 10 years after 1995, so the cost was 20.17¢ per minute in 2005.

(c) The cost per minute is decreasing by 3.583¢ per year.

(d) No, the cost will never be 0 or negative.

(e)

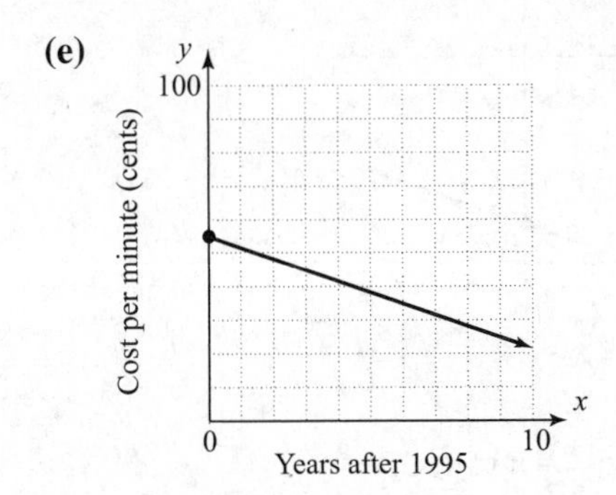

95. $2x + By = 12$
$$By = -2x + 12$$
$$y = \frac{-2}{B}x + \frac{12}{B}$$
$$m = -\frac{2}{B}$$
$$\frac{1}{2} = -\frac{2}{B}$$
$$\frac{B}{2} = -2$$
$$B = -4$$

97. $Ax - 2y = 10$
$$-2y = -Ax + 10$$
$$y = \frac{A}{2}x - 5$$
$$m = \frac{A}{2}$$
$$-2 = \frac{A}{2}$$
$$-4 = A$$

99. $x + By = \frac{1}{2}$
$$By = -x + \frac{1}{2}$$
$$y = -\frac{1}{B}x + \frac{1}{2B}$$
$$b = \frac{1}{2B}$$
$$-\frac{1}{6} = \frac{1}{2B}$$
$$-\frac{B}{6} = \frac{1}{2}$$
$$B = -3$$

101. (a) Variable cost = $m = 40$
Fixed cost = $b = 4000$
$y = 40x + 4000$

(b) $x = 500$
$$y = 40(500) + 4000$$
$$= 20{,}000 + 4000$$
$$= 24{,}000$$
The daily cost of manufacturing 500 calculators is $24,000.

(c) $y = 19{,}000$
$$19{,}000 = 40x + 4000$$
$$15{,}000 = 40x$$
$$375 = x$$
375 calculators were manufactured.

(d)

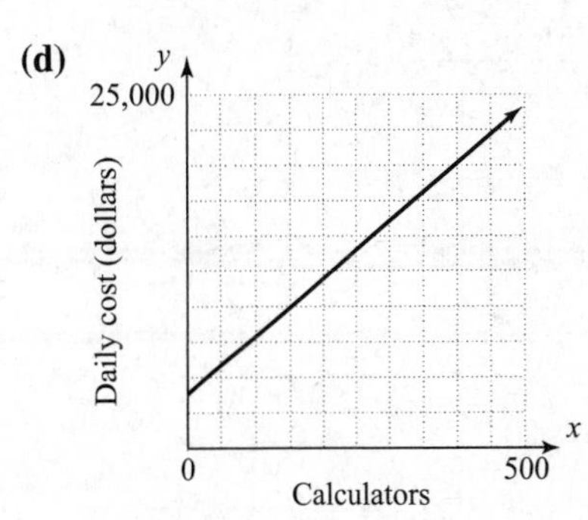

103. The slope of the line shown is positive and the y-intercept is negative.

(a) $y = 3x - 2$

(b) $y = -2x + 5$

(c) $y = 3$

(d) $2x + 3y = 6$ or $3y = -2x + 6$ or $y = -\frac{2}{3}x + 2$

(e) $3x - 2y = 8$ or $-2y = -3x + 8$ or $y = \frac{3}{2}x - 4$

(f) $4x - y = -4$ or $y = 4x + 4$

(g) $-5x + 2y = 12$ or $2y = 5x + 12$ or $y = \frac{5}{2}x + 6$

(h) $x - y = -3$ or $y = x + 3$

Using the slope-intercept forms of the lines, the equations (a) or (e) could have the graph.

Section 3.5

Preparing for Point-Slope Form of a Line

P1.
$$\begin{aligned} y-3 &= 2(x+1) \\ y-3 &= 2x+2 \\ y-3+3 &= 2x+2+3 \\ y &= 2x+5 \end{aligned}$$

P2. $\frac{7-3}{4-2} = \frac{4}{2} = 2$

3.5 Quick Checks

1. The point-slope form of a nonvertical line whose slope is m that contains the point (x_1, y_1) is $\underline{y - y_1 = m(x - x_1)}$.

2. True

3. $m = 3$; $(x_1, y_1) = (2, 1)$
$$\begin{aligned} y - y_1 &= m(x - x_1) \\ y-1 &= 3(x-2) \\ y-1 &= 3x-6 \\ y &= 3x-5 \end{aligned}$$

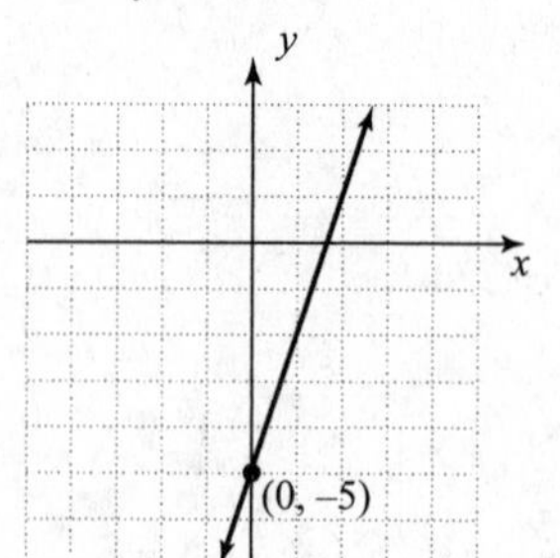

4. $m = \frac{1}{3}$; $(x_1, y_1) = (3, -4)$
$$\begin{aligned} y - y_1 &= m(x - x_1) \\ y-(-4) &= \frac{1}{3}(x-3) \\ y+4 &= \frac{1}{3}x-1 \\ y &= \frac{1}{3}x-5 \end{aligned}$$

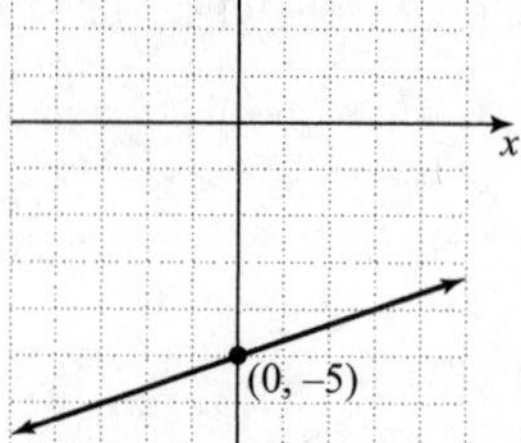

5. $m = -4$; $(x_1, y_1) = (-2, 5)$
$$\begin{aligned} y - y_1 &= m(x - x_1) \\ y-5 &= -4(x-(-2)) \\ y-5 &= -4(x+2) \\ y-5 &= -4x-8 \\ y &= -4x-3 \end{aligned}$$

6. $m = -\frac{5}{2}$; $(x_1, y_1) = (-4, 5)$

$$\begin{aligned} y - y_1 &= m(x - x_1) \\ y - 5 &= -\frac{5}{2}(x - (-4)) \\ y - 5 &= -\frac{5}{2}(x + 4) \\ y - 5 &= -\frac{5}{2}x - 10 \\ y &= -\frac{5}{2}x - 5 \end{aligned}$$

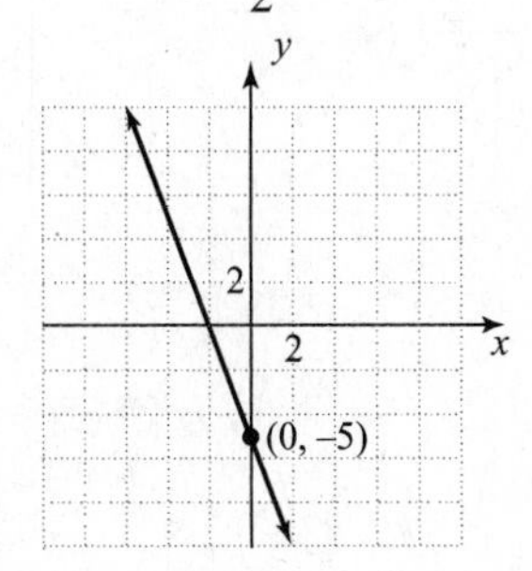

7. Since the line is horizontal, the slope, m, is 0.
$m = 0$; $(x_1, y_1) = (-2, 3)$

$$\begin{aligned} y - y_1 &= m(x - x_1) \\ y - 3 &= 0(x - (-2)) \\ y - 3 &= 0 \\ y &= 3 \end{aligned}$$

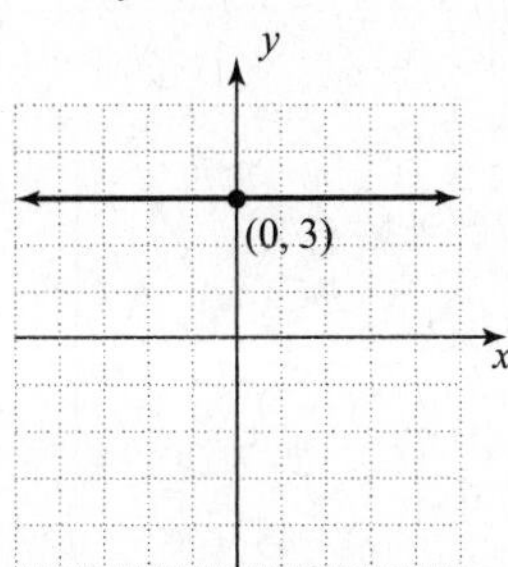

8. Let $(x_1, y_1) = (0, 2)$ and $(x_2, y_2) = (3, 5)$.

$$m = \frac{y_2 - y_1}{x_2 - x_1} = \frac{5 - 2}{3 - 0} = \frac{3}{3} = 1$$

$$\begin{aligned} y - y_1 &= m(x - x_1) \\ y - 2 &= 1(x - 0) \\ y - 2 &= x \\ y &= x + 2 \end{aligned}$$

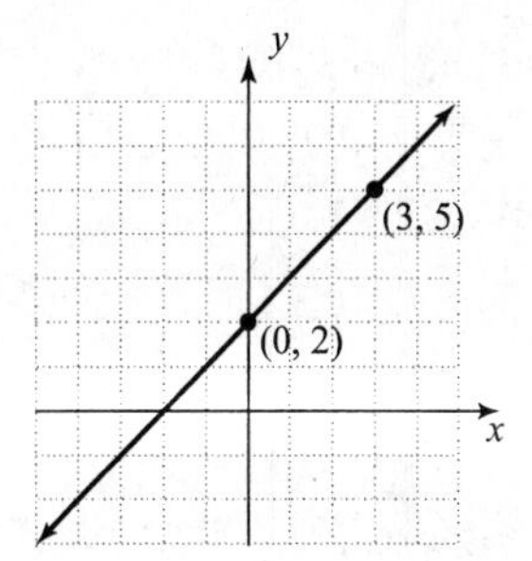

9. Let $(x_1, y_1) = (-1, 4)$ and $(x_2, y_2) = (1, -2)$.

$$m = \frac{y_2 - y_1}{x_2 - x_1} = \frac{-2 - 4}{1 - (-1)} = \frac{-6}{1 + 1} = \frac{-6}{2} = -3$$

$$\begin{aligned} y - y_1 &= m(x - x_1) \\ y - 4 &= -3(x - (-1)) \\ y - 4 &= -3(x + 1) \\ y - 4 &= -3x - 3 \\ y &= -3x + 1 \end{aligned}$$

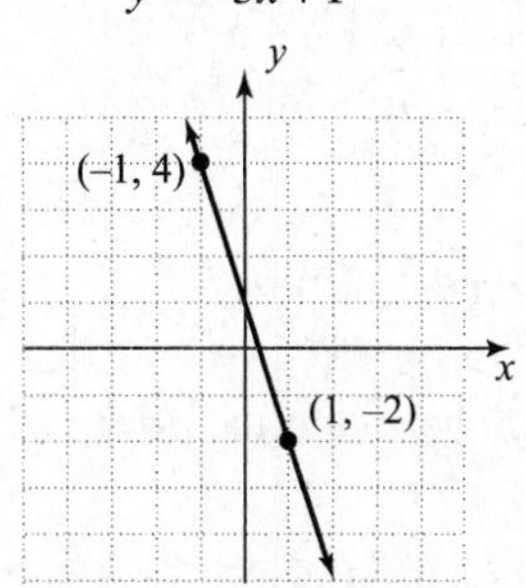

10. Let $(x_1, y_1) = (3, 2)$ and $(x_2, y_2) = (3, -4)$.

$$m = \frac{y_2 - y_1}{x_2 - x_1} = \frac{-4 - 2}{3 - 3} = \frac{-6}{0}$$

The slope is undefined, so the line is vertical. The equation of the line is $x = 3$.

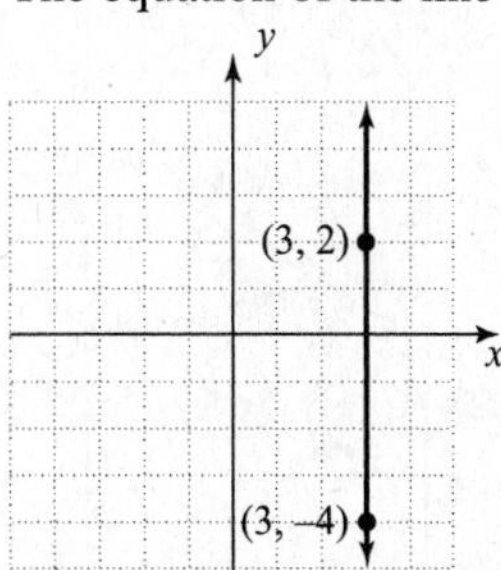

11. List the five forms that are used when writing the equation of a line: Horizontal line, $y = b$; Vertical line, $x = a$; Point-slope, $y - y_1 = m(x - x_1)$; Slope-intercept, $y = mx + b$; Standard form, $Ax + By = C$.

12. (a) When $x = 3.90$, then $y = 400$, so $(x_1, y_1) = (3.9, 400)$. When $x = 4.10$, then $y = 380$, so $(x_2, y_2) = (4.1, 380)$.

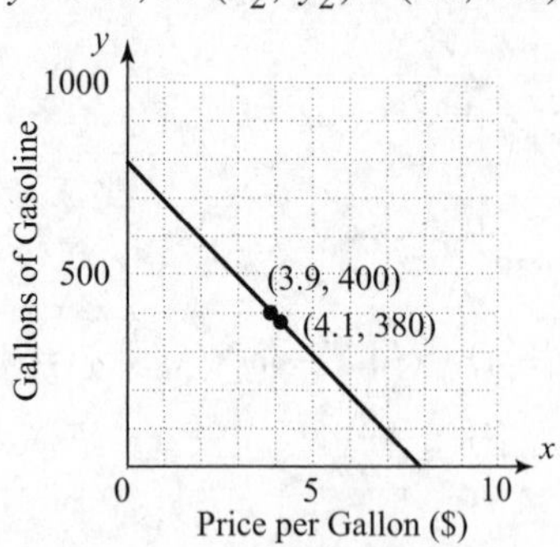

$$m = \frac{y_2 - y_1}{x_2 - x_1} = \frac{380 - 400}{4.1 - 3.9} = \frac{-20}{0.2} = -100$$

$$y - y_1 = m(x - x_1)$$
$$y - 400 = -100(x - 3.9)$$
$$y - 400 = -100x + 390$$
$$y = -100x + 790$$

(b) Let $x = 4.00$.
$y = -100x + 790$
$y = -100(4) + 790 = 390$
390 gallons will be sold if the price is $4.00 per gallon.

(c) The slope is −100. The number of gallons of gasoline sold will decrease by 100 if the price per gallon increases by $1.

3.5 Exercises

13. (2, 5); slope = 3

$$y - y_1 = m(x - x_1)$$
$$y - 5 = 3(x - 2)$$
$$y - 5 = 3x - 6$$
$$y = 3x - 1$$

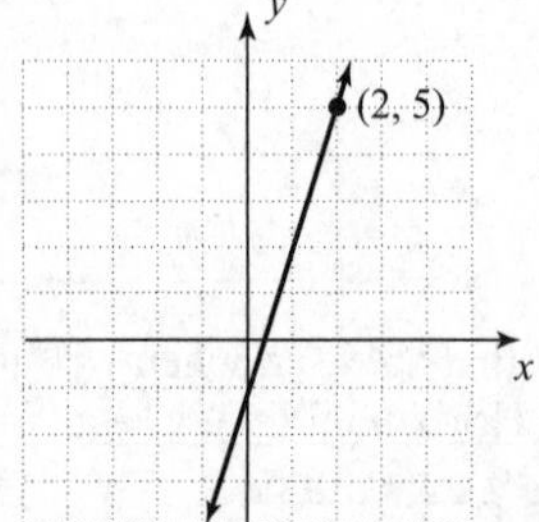

15. (−1, 2); slope = −2

$$y - y_1 = m(x - x_1)$$
$$y - 2 = -2[x - (-1)]$$
$$y - 2 = -2(x + 1)$$
$$y - 2 = -2x - 2$$
$$y = -2x$$

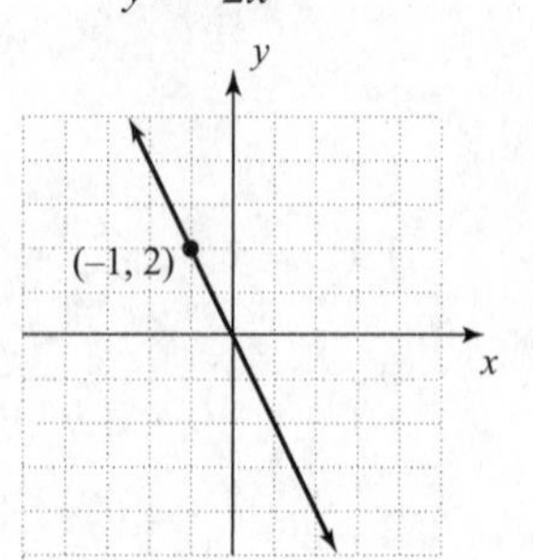

17. (8, −1); slope = $\frac{1}{4}$

$$y - y_1 = m(x - x_1)$$
$$y - (-1) = \frac{1}{4}(x - 8)$$
$$y + 1 = \frac{1}{4}x - 2$$
$$y = \frac{1}{4}x - 3$$

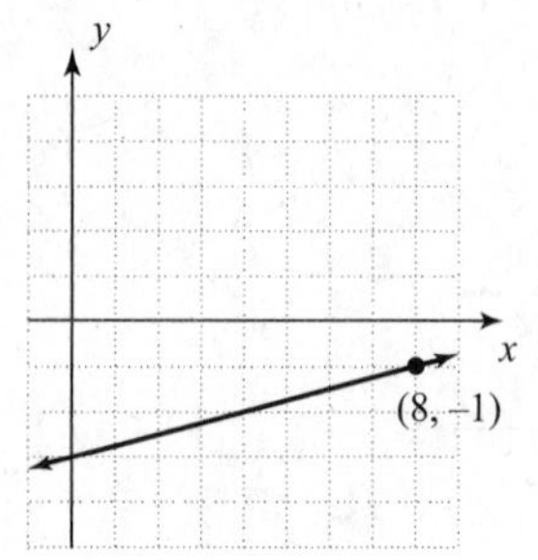

19. (0, 13); slope = −6

$$y - y_1 = m(x - x_1)$$
$$y - 13 = -6(x - 0)$$
$$y - 13 = -6x$$
$$y = -6x + 13$$

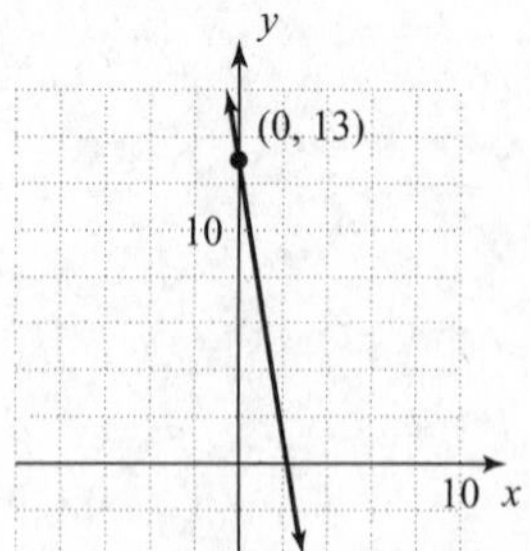

21. (5, –7); slope = 0

$$y-y_1=m(x-x_1)$$
$$y-(-7)=0(x-5)$$
$$y+7=0$$
$$y=-7$$

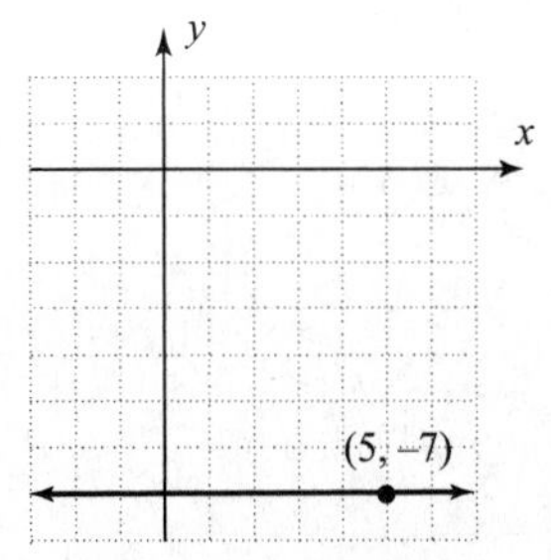

23. (–4, 5); undefined slope
Since the slope is undefined, the line is vertical.
The equation is $x = -4$.

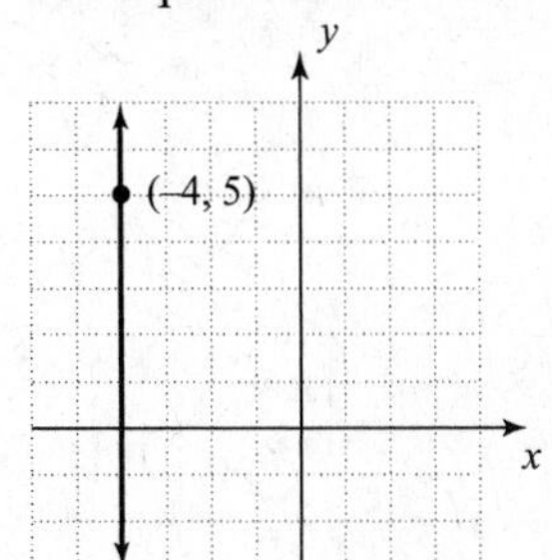

25. (–3, 0); slope $=\frac{2}{3}$

$$y-y_1=m(x-x_1)$$
$$y-0=\frac{2}{3}[x-(-3)]$$
$$y=\frac{2}{3}(x+3)$$
$$y=\frac{2}{3}x+2$$

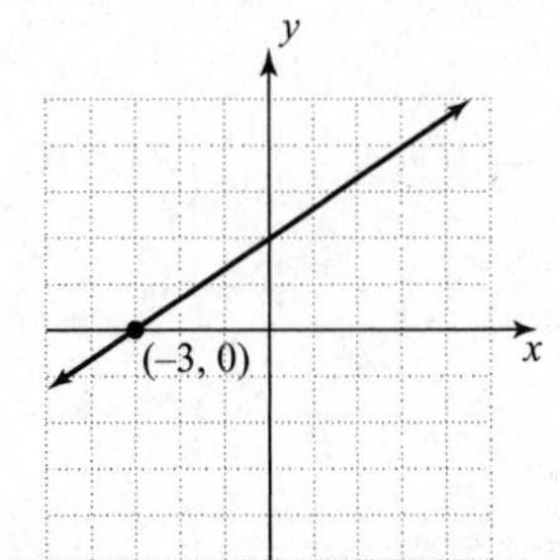

27. (–8, 6); slope $=-\frac{3}{4}$

$$y-y_1=m(x-x_1)$$
$$y-6=-\frac{3}{4}[x-(-8)]$$
$$y-6=-\frac{3}{4}(x+8)$$

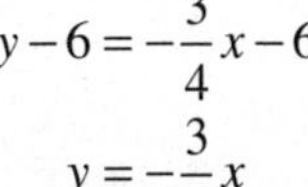

$$y-6=-\frac{3}{4}x-6$$
$$y=-\frac{3}{4}x$$

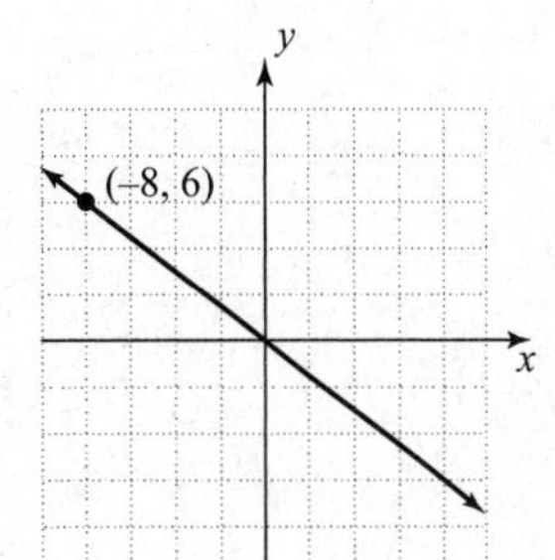

29. (–3, 10); vertical
$x=-3$

31. (–1, –5); horizontal
$y=-5$

33. (0.2, –4.3); horizontal
$y=-4.3$

35. $\left(\frac{1}{2}, \frac{7}{4}\right)$; vertical

$$x=\frac{1}{2}$$

37. (0, 4), (–2, 0)

$$m=\frac{y_2-y_1}{x_2-x_1}=\frac{0-4}{-2-0}=\frac{-4}{-2}=2$$
$$y-y_1=m(x-x_1)$$
$$y-4=2(x-0)$$
$$y-4=2x$$
$$y=2x+4$$

39. (1, 2), (0, 6)

$$m=\frac{y_2-y_1}{x_2-x_1}=\frac{6-2}{0-1}=\frac{4}{-1}=-4$$
$$y-y_1=m(x-x_1)$$
$$y-2=-4(x-1)$$
$$y-2=-4x+4$$
$$y=-4x+6$$

41. $(-3, 2), (1, -4)$

$$m = \frac{y_2 - y_1}{x_2 - x_1} = \frac{-4-2}{1-(-3)} = \frac{-6}{4} = -\frac{3}{2}$$

$$\begin{aligned} y - y_1 &= m(x - x_1) \\ y - 2 &= -\frac{3}{2}[x-(-3)] \\ y - 2 &= -\frac{3}{2}(x+3) \\ y - 2 &= -\frac{3}{2}x - \frac{9}{2} \\ y &= -\frac{3}{2}x - \frac{9}{2} + \frac{4}{2} \\ y &= -\frac{3}{2}x - \frac{5}{2} \end{aligned}$$

43. $(-3, -11), (2, -1)$

$$m = \frac{y_2 - y_1}{x_2 - x_1} = \frac{-1-(-11)}{2-(-3)} = \frac{-1+11}{2+3} = \frac{10}{5} = 2$$

$$\begin{aligned} y - y_1 &= m(x - x_1) \\ y - (-11) &= 2[x-(-3)] \\ y + 11 &= 2(x+3) \\ y + 11 &= 2x + 6 \\ y &= 2x - 5 \end{aligned}$$

45. $(4, -3), (-3, -3)$

$$m = \frac{y_2 - y_1}{x_2 - x_1} = \frac{-3-(-3)}{-3-4} = \frac{-3+3}{-7} = \frac{0}{-7} = 0$$

$$\begin{aligned} y - y_1 &= m(x - x_1) \\ y - (-3) &= 0(x-4) \\ y + 3 &= 0 \\ y &= -3 \end{aligned}$$

47. $(2, -1), (2, -9)$

$$m = \frac{y_2 - y_1}{x_2 - x_1} = \frac{-9-(-1)}{2-2} = \frac{-9+1}{0}$$

Since the slope is undefined, the line is vertical.
$x = 2$

49. $(0.1, 0.6), (0.5, 0.7)$

$$m = \frac{y_2 - y_1}{x_2 - x_1} = \frac{0.7-0.6}{0.5-0.1} = \frac{0.1}{0.4} = 0.25$$

$$\begin{aligned} y - y_1 &= m(x - x_1) \\ y - 0.6 &= 0.25(x - 0.1) \\ y - 0.6 &= 0.25x - 0.025 \\ y &= 0.25x + 0.575 \end{aligned}$$

51. $\left(\frac{1}{2}, -\frac{9}{4}\right), \left(\frac{5}{2}, -\frac{1}{4}\right)$

$$m = \frac{y_2 - y_1}{x_2 - x_1} = \frac{-\frac{1}{4} - \left(-\frac{9}{4}\right)}{\frac{5}{2} - \frac{1}{2}} = \frac{-\frac{1}{4} + \frac{9}{4}}{\frac{4}{2}} = \frac{2}{2} = 1$$

$$\begin{aligned} y - y_1 &= m(x - x_1) \\ y - \left(-\frac{9}{4}\right) &= 1\left(x - \frac{1}{2}\right) \\ y + \frac{9}{4} &= x - \frac{1}{2} \\ y &= x - \frac{2}{4} - \frac{9}{4} \\ y &= x - \frac{11}{4} \end{aligned}$$

53. $(4, -2), m = 5$

$$\begin{aligned} y - y_1 &= m(x - x_1) \\ y - (-2) &= 5(x-4) \\ y + 2 &= 5x - 20 \\ y &= 5x - 22 \end{aligned}$$

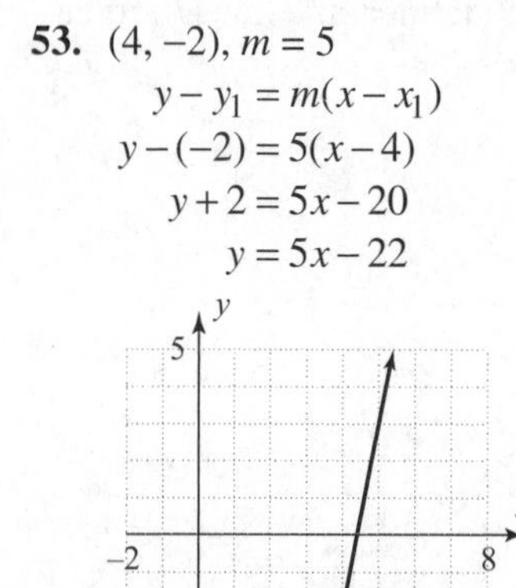

55. $(-3, 5)$, horizontal
$y = 5$

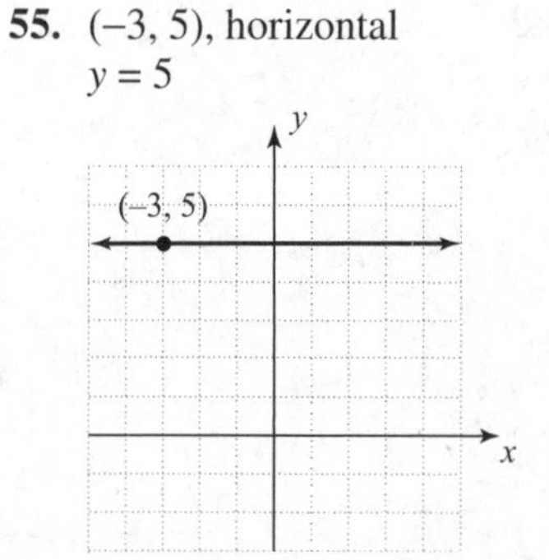

57. $(1, 3), (-4, -2)$

$$m = \frac{y_2 - y_1}{x_2 - x_1} = \frac{-2-3}{-4-1} = \frac{-5}{-5} = 1$$

$$\begin{aligned} y - y_1 &= m(x - x_1) \\ y - 3 &= 1(x-1) \\ y - 3 &= x - 1 \\ y &= x + 2 \end{aligned}$$

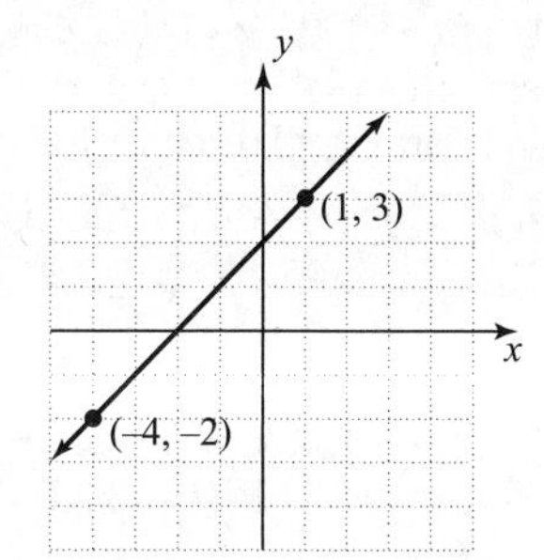

59. $(-2, 3),\ m = \frac{1}{2}$

$$y - y_1 = m(x - x_1)$$
$$y - 3 = \frac{1}{2}[x - (-2)]$$
$$y - 3 = \frac{1}{2}(x + 2)$$
$$y - 3 = \frac{1}{2}x + 1$$
$$y = \frac{1}{2}x + 4$$

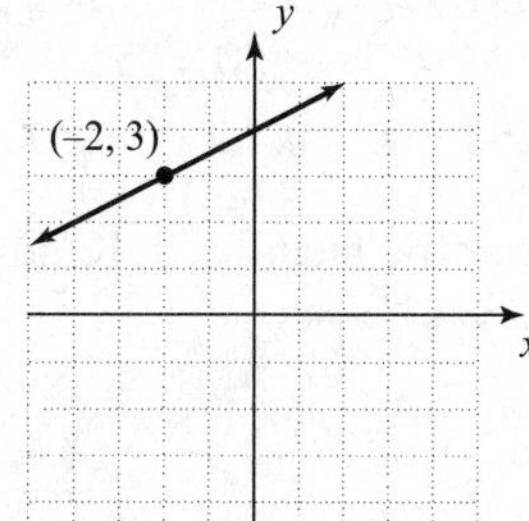

61. (5, 2), vertical line
$x = 5$

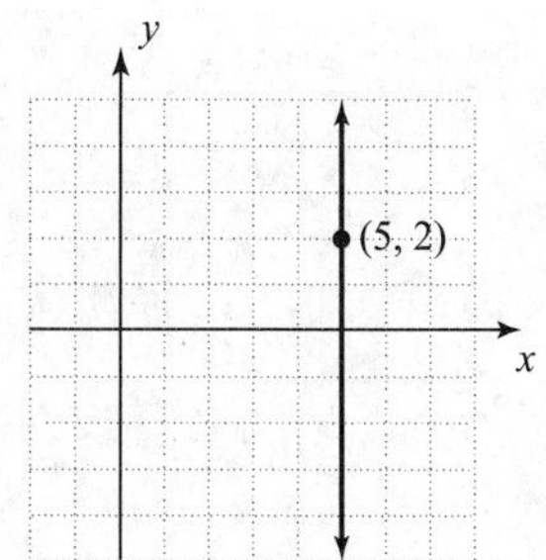

63. $(3, -19), (-1, 9)$

$$m = \frac{y_2 - y_1}{x_2 - x_1} = \frac{9 - (-19)}{-1 - 3} = \frac{28}{-4} = -7$$
$$y - y_1 = m(x - x_1)$$
$$y - (-19) = -7(x - 3)$$
$$y + 19 = -7x + 21$$
$$y = -7x + 2$$

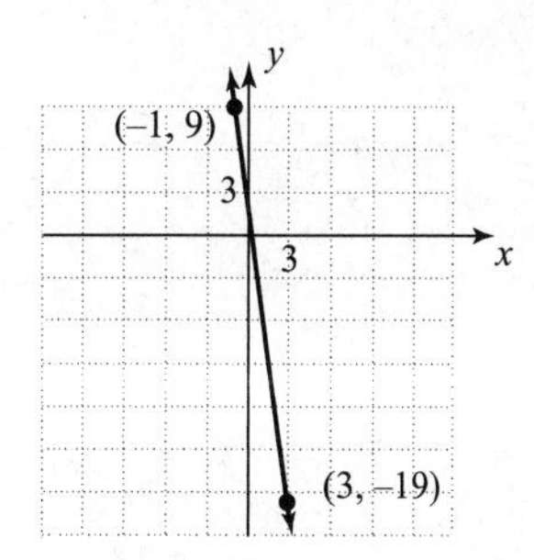

65. $(6, 3),\ m = -\frac{2}{3}$

$$y - y_1 = m(x - x_1)$$
$$y - 3 = -\frac{2}{3}(x - 6)$$
$$y - 3 = -\frac{2}{3}x + 4$$
$$y = -\frac{2}{3}x + 7$$

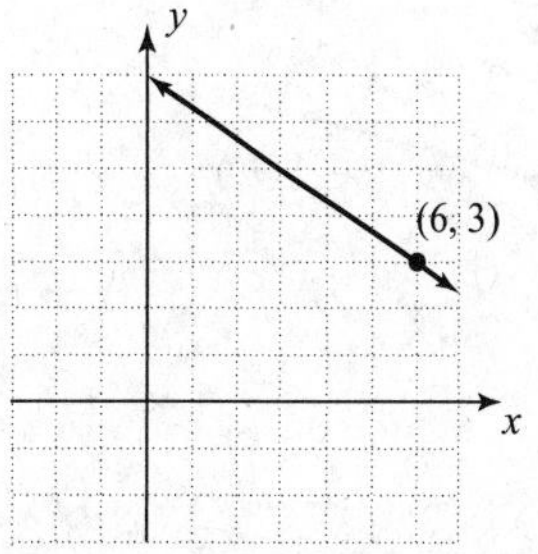

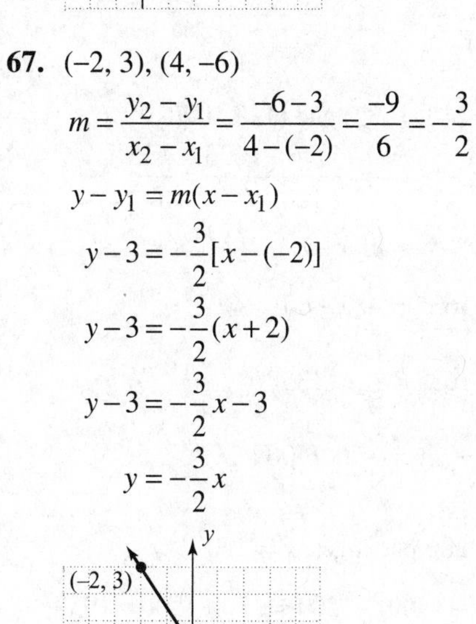

67. $(-2, 3), (4, -6)$

$$m = \frac{y_2 - y_1}{x_2 - x_1} = \frac{-6 - 3}{4 - (-2)} = \frac{-9}{6} = -\frac{3}{2}$$
$$y - y_1 = m(x - x_1)$$
$$y - 3 = -\frac{3}{2}[x - (-2)]$$
$$y - 3 = -\frac{3}{2}(x + 2)$$
$$y - 3 = -\frac{3}{2}x - 3$$
$$y = -\frac{3}{2}x$$

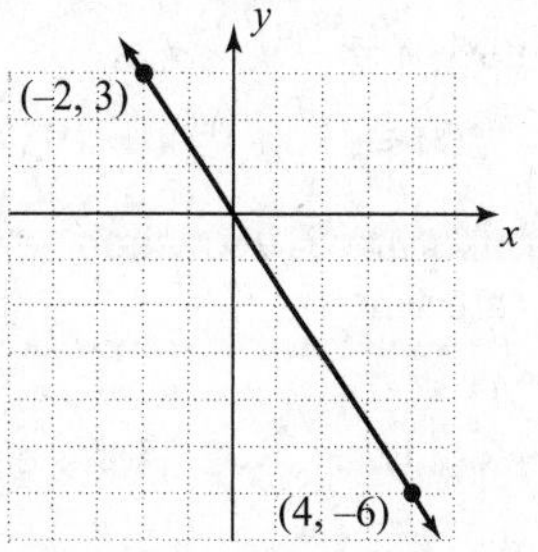

y
x
(–2, 3)
(4, –6)

69. x-intercept: 5 (5, 0)
y-intercept: −2 (0, −2)

$$m = \frac{y_2 - y_1}{x_2 - x_1} = \frac{-2-0}{0-5} = \frac{-2}{-5} = \frac{2}{5}$$

$$y = \frac{2}{5}x - 2$$

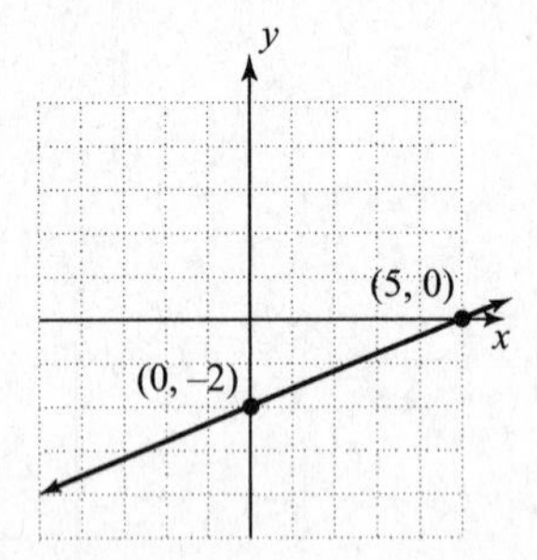

71. **(a)** (60, 1635) indicates that when 60 packages are shipped, the expenses for the department are $1635.

(b)

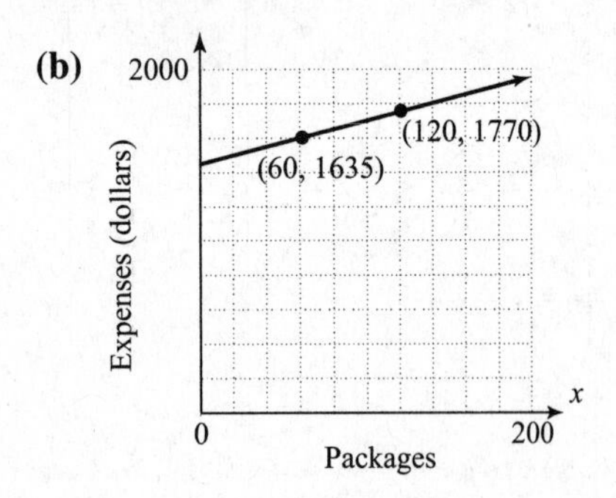

(c) Use (60, 1635) and (120, 1770).

$$m = \frac{y_2 - y_1}{x_2 - x_1} = \frac{1770-1635}{120-60} = \frac{135}{60} = \frac{9}{4}$$

$$y - y_1 = m(x - x_1)$$
$$y - 1635 = \frac{9}{4}(x - 60)$$
$$y - 1635 = \frac{9}{4}x - 135$$
$$y = \frac{9}{4}x + 1500$$

(d) For 200 packages, $x = 200$.

$$y = \frac{9}{4}(200) + 1500 = 450 + 1500 = 1950$$

The total expenses are $1950 when 200 packages are sent.

(e) Expenses increase by $\$\frac{9}{4} = 2.25$ for each additional package sent.

73. **(a)** 1980 is 0 years after 1980.
2005 is 25 years after 1980.
The ordered pairs are (0, 820) and (25, 985).

(b)

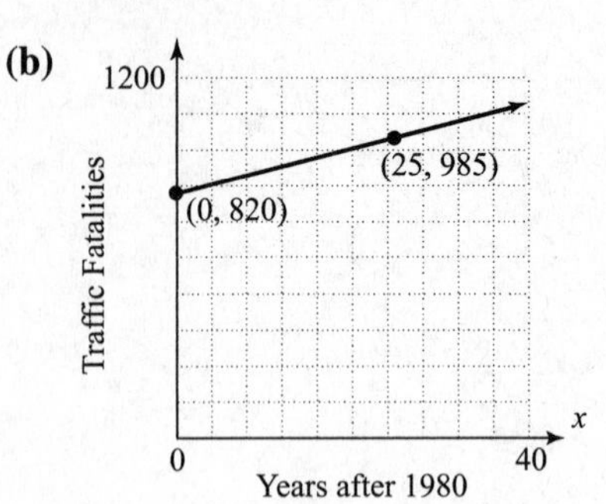

(c) Use (0, 820) and (25, 985).

$$m = \frac{y_2 - y_1}{x_2 - x_1} = \frac{985-820}{25-0} = \frac{165}{25} = 6.6$$

$y = mx + b$; $b = 820$
$$y = 6.6x + 820$$

(d) 2000 is 20 years after 1980, so $x = 20$.

$$y = 6.6(20) + 820 = 132 + 820 = 952$$

There were 952 traffic fatalities in 2000.

(e) The number of traffic fatalities in Kentucky increases by 6.6 each year.

75. (−4, 2); slope = 3

$$y = mx + b$$
$$y = 3x + b$$
$$2 = 3(-4) + b$$
$$2 = -12 + b$$
$$14 = b$$
$$y = 3x + 14$$

77. (3, −8); slope = −2

$$y = mx + b$$
$$y = -2x + b$$
$$-8 = -2(3) + b$$
$$-8 = -6 + b$$
$$-2 = b$$
$$y = -2x - 2$$

79. $\left(\frac{2}{3}, \frac{1}{2}\right)$; slope = 6

$$\begin{aligned} y &= mx + b \\ y &= 6x + b \\ \frac{1}{2} &= 6\left(\frac{2}{3}\right) + b \\ \frac{1}{2} &= 4 + b \\ -\frac{7}{2} &= b \\ y &= 6x - \frac{7}{2} \end{aligned}$$

81. (6, −13) and (−2, −5)

$$\begin{aligned} m &= \frac{y_2 - y_1}{x_2 - x_1} \\ &= \frac{-5-(-13)}{-2-6} \\ &= \frac{-5+13}{-8} \\ &= \frac{8}{-8} \\ &= -1 \end{aligned}$$

$$\begin{aligned} y &= mx + b \\ y &= -x + b \\ -13 &= -(6) + b \\ -13 &= -6 + b \\ -7 &= b \\ y &= -x - 7 \end{aligned}$$

83. (5, −1) and (−10, −4)

$$m = \frac{y_2 - y_1}{x_2 - x_1} = \frac{-4-(-1)}{-10-5} = \frac{-4+1}{-15} = \frac{-3}{-15} = \frac{1}{5}$$

$$\begin{aligned} y &= mx + b \\ y &= \frac{1}{5}x + b \\ -1 &= \frac{1}{5}(5) + b \\ -1 &= 1 + b \\ -2 &= b \\ y &= \frac{1}{5}x - 2 \end{aligned}$$

85. (−4, 8) and (2, −1)

$$m = \frac{y_2 - y_1}{x_2 - x_1} = \frac{-1-8}{2-(-4)} = \frac{-9}{2+4} = \frac{-9}{6} = -\frac{3}{2}$$

$$\begin{aligned} y &= mx + b \\ y &= -\frac{3}{2}x + b \\ 8 &= -\frac{3}{2}(-4) + b \\ 8 &= 6 + b \\ 2 &= b \\ y &= -\frac{3}{2}x + 2 \end{aligned}$$

87. Answers may vary. One possibility: Yes, although the equations (in point-slope form) will be different, they will simplify to the same slope-intercept form.

Section 3.6

Preparing for Parallel and Perpendicular Lines

P1. The reciprocal of 3 is $\frac{1}{3}$ since $3\left(\frac{1}{3}\right) = 1$.

P2. The reciprocal of $-\frac{3}{5}$ is $-\frac{5}{3}$ since $-\frac{3}{5}\left(-\frac{5}{3}\right) = 1$.

3.6 Quick Checks

1. Two nonvertical lines are parallel if and only if their <u>slopes</u> are equal and they have different <u>y-intercepts</u>. Vertical lines are parallel if they have different <u>x-intercepts</u>.

2. The slope of $y = 2x + 1$ is 2 and the slope of $y = -2x - 3$ is −2. Since the slopes are different, the lines are not parallel.

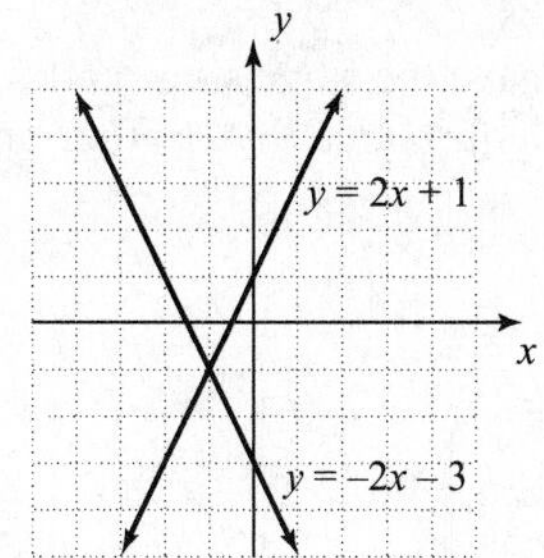

3. $6x+3y=3$
$$3y=-6x+3$$
$$y=-2x+1$$
The slope of the line is –2 and the y-intercept is 1.
$$10x+5y=10$$
$$5y=-10x+10$$
$$y=-2x+2$$
The slope of the line is –2 and the y-intercept is 2. Because the lines have the same slope, but different y-intercepts, the lines are parallel.

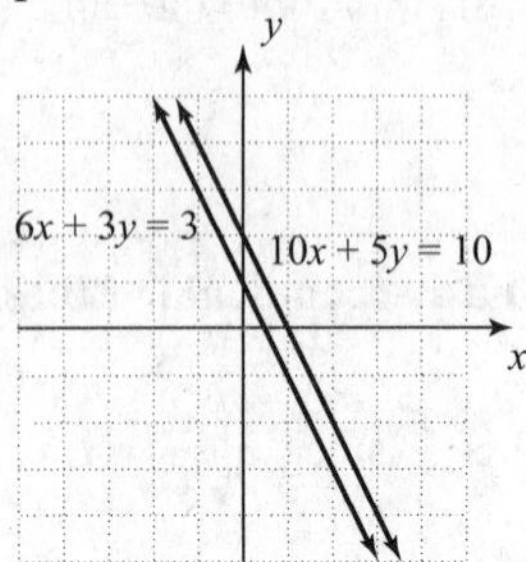

4. $4x+5y=10$
$$5y=-4x+10$$
$$y=-\frac{4}{5}x+2$$
The slope of the line is $-\frac{4}{5}$ and the y-intercept is 2.
$$8x+10y=20$$
$$10y=-8x+20$$
$$y=-\frac{4}{5}x+2$$
The slope of the line is $-\frac{4}{5}$ and the y-intercept is 2.
Because the lines have the same slope and y-intercept, they are the same line, so they are not parallel.

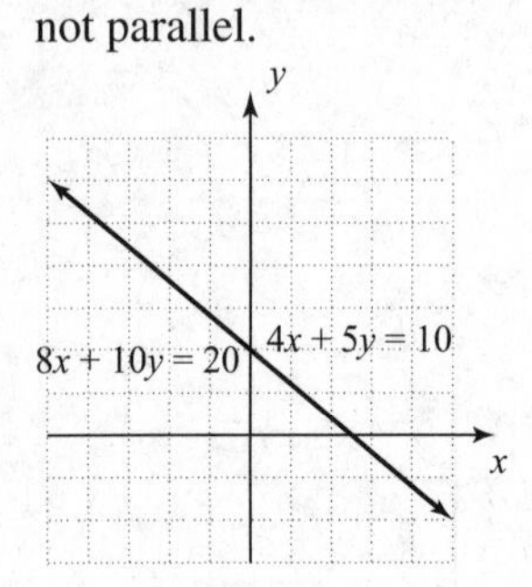

5. The slope of $y = 2x + 1$ is 2.
$m = 2$; (2, 3)
$$y-y_1=m(x-x_1)$$
$$y-3=2(x-2)$$
$$y-3=2x-4$$
$$y=2x-1$$

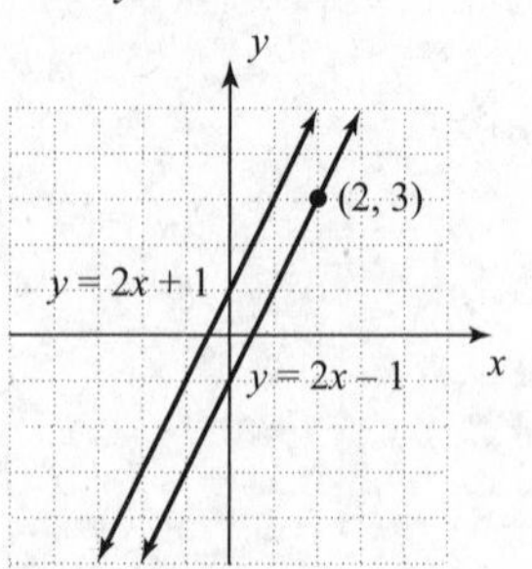

6. $3x+2y=4$
$$2y=-3x+4$$
$$y=-\frac{3}{2}x+2$$
The slope is $-\frac{3}{2}$.
$m=-\frac{3}{2}$; (–2, 3)
$$y-y_1=m(x-x_1)$$
$$y-3=-\frac{3}{2}(x-(-2))$$
$$y-3=-\frac{3}{2}(x+2)$$
$$y-3=-\frac{3}{2}x-3$$
$$y=-\frac{3}{2}x$$

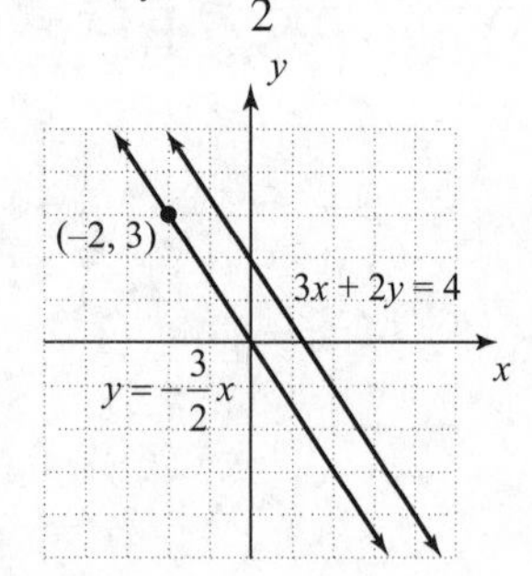

7. $x = -2$ is a vertical line, so any line parallel to $x = -2$ will also be vertical. The vertical line through (3, 1) has equation $x = 3$.

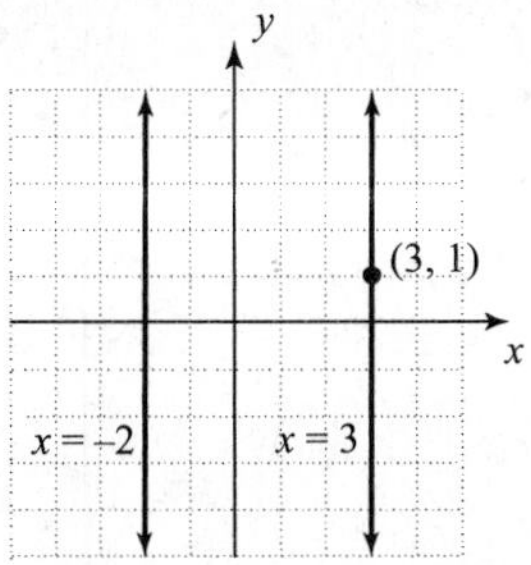

8. $y+3=0$
$y=-3$
$y = -3$ is a horizontal line, so any line parallel to $y = -3$ will also be horizontal. The horizontal line through (–2, 5) has equation $y = 5$.

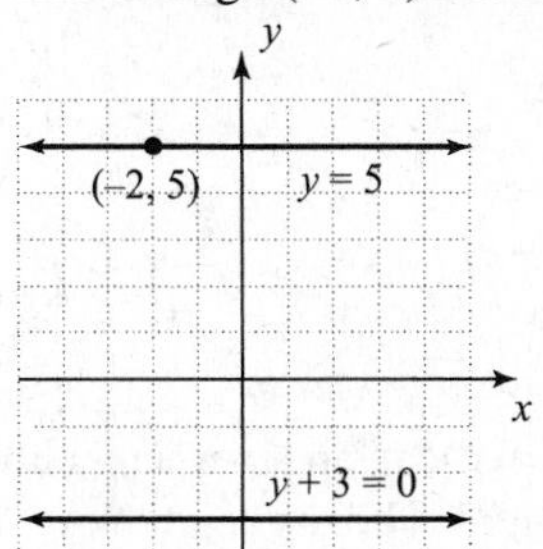

9. Given any two nonvertical lines, if the product of their slopes is –1, then the lines are perpendicular.

10. False; L_1 and L_2 are not perpendicular because their slopes are not negative reciprocals.

11. The negative reciprocal of –4 is $\frac{-1}{-4}=\frac{1}{4}$. Any line whose slope is $\frac{1}{4}$ will be perpendicular to a line whose slope is –4.

12. The negative reciprocal of $\frac{5}{4}$ is $\frac{-1}{\frac{5}{4}}=-\frac{4}{5}$. Any line whose slope is $-\frac{4}{5}$ will be perpendicular to a line whose slope is $\frac{5}{4}$.

13. The negative reciprocal of $-\frac{1}{5}$ is $\frac{-1}{-\frac{1}{5}}=\frac{5}{1}=5$. Any line whose slope is 5 will be perpendicular to a line whose slope is $-\frac{1}{5}$.

14. The slope of $y = 4x - 3$ is 4. The slope of $y=-\frac{1}{4}x-4$ is $-\frac{1}{4}$. Since $4\cdot\left(-\frac{1}{4}\right)=-1$, the lines are perpendicular.

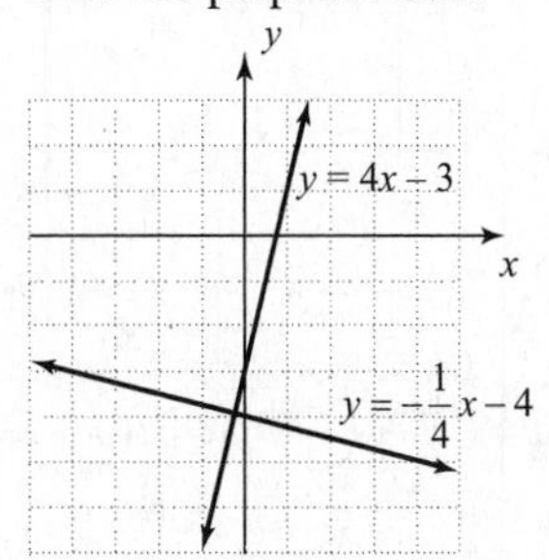

15. $2x-y=3$
$-y=-2x+3$
$y=2x-3$
The slope of the line is 2.
$x-2y=2$
$-2y=-x+2$
$y=\frac{1}{2}x-1$
The slope of the line is $\frac{1}{2}$. Since $2\cdot\frac{1}{2}=1\neq-1$, the lines are not perpendicular.

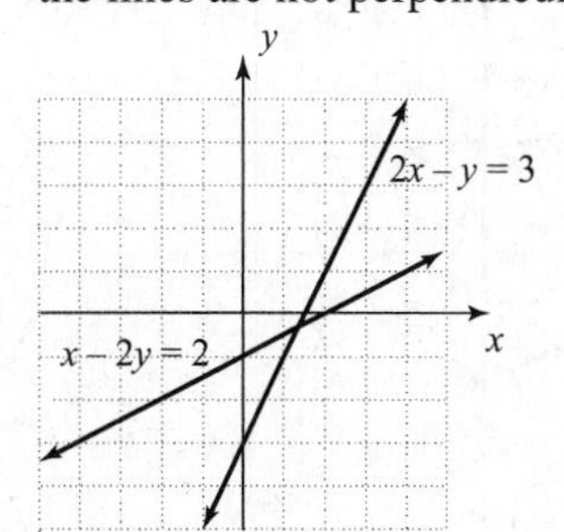

16. $5x+2y=8$
$2y=-5x+8$
$y=-\frac{5}{2}x+4$
The slope of the line is $-\frac{5}{2}$.

$$2x-5y=10$$
$$-5y=-2x+10$$
$$y=\frac{2}{5}x-2$$

The slope of the line is $\frac{2}{5}$. Since $-\frac{5}{2}\cdot\frac{2}{5}=-1$, the lines are perpendicular.

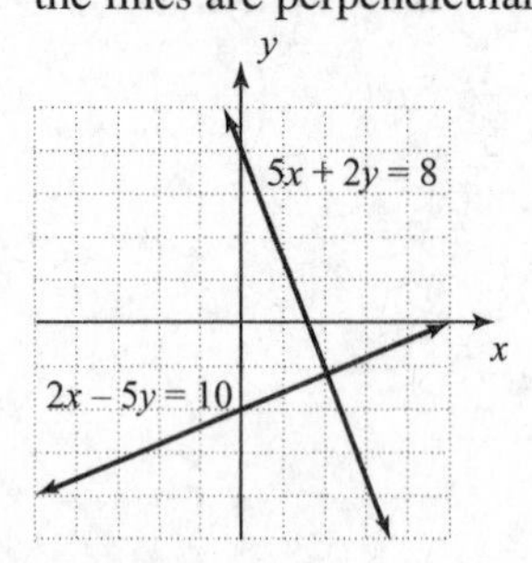

17. The slope of $y = 2x + 1$ is 2. A line perpendicular to $y = 2x + 1$ will have slope $\frac{-1}{2}=-\frac{1}{2}$.

$$m=-\frac{1}{2};\ (-4,\ 2)$$
$$y-y_1=m(x-x_1)$$
$$y-2=-\frac{1}{2}(x-(-4))$$
$$y-2=-\frac{1}{2}(x+4)$$
$$y-2=-\frac{1}{2}x-2$$
$$y=-\frac{1}{2}x$$

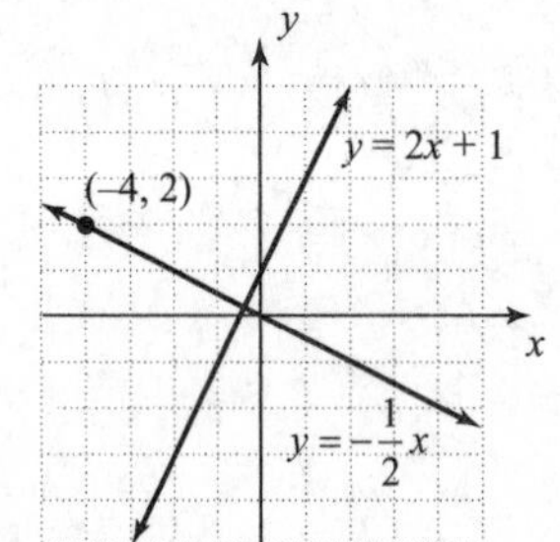

18. $2x+3y=3$
$$3y=-2x+3$$
$$y=-\frac{2}{3}x+1$$

The slope of $y=-\frac{2}{3}x+1$ is $-\frac{2}{3}$. A line perpendicular to $y=-\frac{2}{3}x+1$ will have slope $\frac{-1}{-\frac{2}{3}}=\frac{3}{2}$.

$$m=\frac{3}{2};\ (-2,\ -1)$$
$$y-y_1=m(x-x_1)$$
$$y-(-1)=\frac{3}{2}(x-(-2))$$

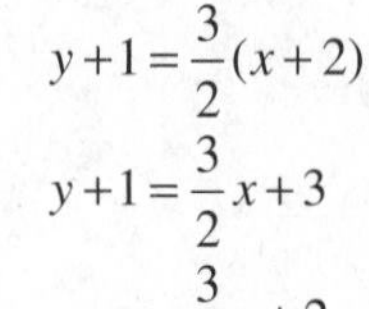

$$y+1=\frac{3}{2}(x+2)$$
$$y+1=\frac{3}{2}x+3$$
$$y=\frac{3}{2}x+2$$

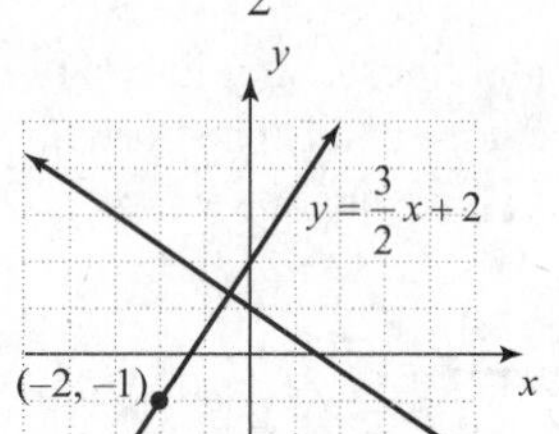

19. The line $x = -4$ is vertical, so a perpendicular line will be horizontal. The horizontal line through $(-1, -5)$ has equation $y = -5$.

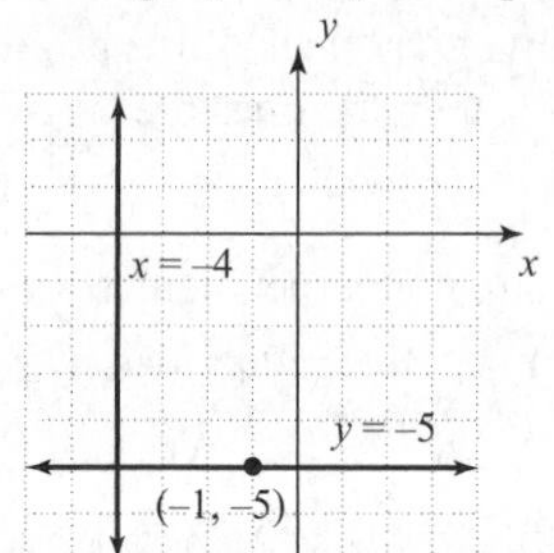

20. $y+2=0$
$$y=-2$$

The line $y = -2$ is horizontal, so a perpendicular line will be vertical. The vertical line through $(3, -2)$ has equation $x = 3$.

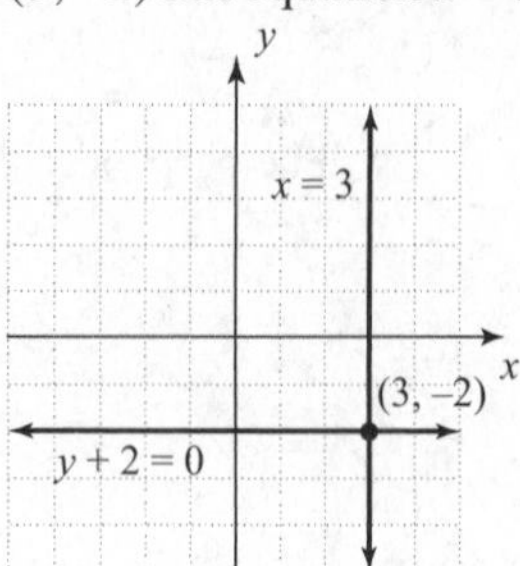

3.6 Exercises

	Slope of the Given Line	Slope of a Line Parallel to the Given Line	Slope of a Line Perpendicular to the Given Line
21.	$m=-3$	$m_1=m=-3$	$m_2=\frac{-1}{m}=\frac{-1}{-3}=\frac{1}{3}$
23.	$m=\frac{1}{2}$	$m_1=m=\frac{1}{2}$	$m_2=\frac{-1}{m}=\frac{-1}{\frac{1}{2}}=-1\left(\frac{2}{1}\right)=-2$
25.	$m=-\frac{4}{9}$	$m_1=m=-\frac{4}{9}$	$m_2=\frac{-1}{m}=\frac{-1}{-\frac{4}{9}}=-1\left(-\frac{9}{4}\right)=\frac{9}{4}$
27.	$m=0$	$m_1=m=0$	$m_2=\frac{-1}{m}=\frac{-1}{0}=$ undefined

29. $L_1: y=x-3;\ m_1=1$
$L_2: y=1-x=-x+1;\ m_2=-1$
Since $m_1m_2=(1)(-1)=-1,$ the lines are perpendicular.

31. $L_1:\ y=\frac{3}{4}x+2;\ m_1=\frac{3}{4},\ b_1=2$
$L_2: y=0.75x-1=\frac{3}{4}x-1;\ m_2=\frac{3}{4},\ b_2=1$
Since $m_1=m_2$ and $b_1\neq b_2$, the lines are parallel.

33. $L_1: y=-\frac{5}{3}x-6;\ m_1=-\frac{5}{3}$
$L_2: y=\frac{3}{5}x-1;\ m_2=\frac{3}{5}$
Since $m_1m_2=\left(-\frac{5}{3}\right)\left(\frac{3}{5}\right)=-1,$ the lines are perpendicular.

35. $L_1: x+y=-3$ or $y=-x-3;\ m_1=-1$
$L_2: y-x=1$ or $y=x+1;\ m_2=1$
Since $m_1m_2=(-1)(1)=-1,$ the lines are perpendicular.

37. $L_1: 2x-5y=5$ or $y=\frac{2}{5}x-1;\ m_1=\frac{2}{5}$
$L_2: 5x+2y=4$ or $y=-\frac{5}{2}x+2;\ m_2=-\frac{5}{2}$
Since $m_1m_2=\left(\frac{2}{5}\right)\left(-\frac{5}{2}\right)=-1,$ the lines are perpendicular.

39. $L_1: 4x - 5y - 15 = 0$ or $y = \frac{4}{5}x - 3$; $m_1 = \frac{4}{5}$, $b_1 = -3$

$L_2: 8x - 10y + 5 = 0$ or $y = \frac{4}{5}x + \frac{1}{2}$;

$m_2 = \frac{4}{5}$, $b_2 = \frac{1}{2}$

Since $m_1 = m_2$ and $b_1 \neq b_2$, the lines are parallel.

41. $L_1: 4x = 3y + 3$ or $y = \frac{4}{3}x - 1$; $m_1 = \frac{4}{3}$, $b_1 = -1$

$L_2: 6y = 8x + 36$ or $y = \frac{4}{3}x + 6$; $m_2 = \frac{4}{3}$, $b_2 = 6$

Since $m_1 = m_2$ and $b_1 \neq b_2$, the lines are parallel.

43. $(4, -2)$; $y = 3x - 1$

$y = 3x - 1$ has slope 3. A parallel line also has slope 3.

$$y - y_1 = m(x - x_1)$$
$$y - (-2) = 3(x - 4)$$
$$y + 2 = 3x - 12$$
$$y = 3x - 14$$

45. $(-3, 8)$; $y = -4x + 5$

$y = -4x + 5$ has slope -4. A parallel line also has slope -4.

$$y - y_1 = m(x - x_1)$$
$$y - 8 = -4[x - (-3)]$$
$$y - 8 = -4(x + 3)$$
$$y - 8 = -4x - 12$$
$$y = -4x - 4$$

47. $(3, -7)$; $y = 4$

$y = 4$ is a horizontal line and has slope 0. A parallel line has equation $y = b$.

$y = -7$

49. $(-1, 10)$; $x = 10$

$x = 10$ is a vertical line and the slope is undefined. A parallel line has equation $x = c$.

$x = -1$

51. $(10, 2)$; $3x - 2y = 5$

$$3x - 2y = 5$$
$$-2y = -3x + 5$$
$$y = \frac{3}{2}x - \frac{5}{2}$$

The line has slope $\frac{3}{2}$. A parallel line also has slope $\frac{3}{2}$.

$$y - y_1 = m(x - x_1)$$
$$y - 2 = \frac{3}{2}(x - 10)$$
$$y - 2 = \frac{3}{2}x - 15$$
$$y = \frac{3}{2}x - 13$$

53. $(-1, -10)$; $x + 2y = 4$

$$x + 2y = 4$$
$$2y = -x + 4$$
$$y = -\frac{1}{2}x + 2$$

The line has slope $-\frac{1}{2}$. A parallel line also has slope $-\frac{1}{2}$.

$$y - y_1 = m(x - x_1)$$
$$y - (-10) = -\frac{1}{2}[x - (-1)]$$
$$y + 10 = -\frac{1}{2}(x + 1)$$
$$y + 10 = -\frac{1}{2}x - \frac{1}{2}$$
$$y = -\frac{1}{2}x - \frac{21}{2}$$

55. $(3, 5)$; $y = \frac{1}{2}x - 2$

$y = \frac{1}{2}x - 2$ has slope $\frac{1}{2}$. A perpendicular line has slope $\frac{-1}{\frac{1}{2}} = -2$.

$$y - y_1 = m(x - x_1)$$
$$y - 5 = -2(x - 3)$$
$$y - 5 = -2x + 6$$
$$y = -2x + 11$$

57. $(-4, -1)$; $y = -4x + 1$
$y = -4x + 1$ has slope -4. A perpendicular line has slope $\frac{-1}{-4} = \frac{1}{4}$.

$$y - y_1 = m(x - x_1)$$
$$y - (-1) = \frac{1}{4}[x - (-4)]$$
$$y + 1 = \frac{1}{4}(x + 4)$$
$$y + 1 = \frac{1}{4}x + 1$$
$$y = \frac{1}{4}x$$

59. $(-2, 1)$; x-axis
The x-axis is horizontal. A perpendicular line is vertical. The vertical line through $(-2, 1)$ is $x = -2$.

61. $(7, 5)$; y-axis
The y-axis is vertical. A perpendicular line is horizontal. The horizontal line through $(7, 5)$ is $y = 5$.

63. $(0, 0)$; $2x + 5y = 7$

$$2x + 5y = 7$$
$$5y = -2x + 7$$
$$y = -\frac{2}{5}x + \frac{7}{5}$$

The line has slope $-\frac{2}{5}$. A perpendicular line has slope $\frac{-1}{-\frac{2}{5}} = (-1)\left(-\frac{5}{2}\right) = \frac{5}{2}$.

$$y - y_1 = m(x - x_1)$$
$$y - 0 = \frac{5}{2}(x - 0)$$
$$y = \frac{5}{2}x$$

65. $(-10, -3)$; $5x - 3y = 4$

$$5x - 3y = 4$$
$$-3y = -5x + 4$$
$$y = \frac{5}{3}x - \frac{4}{3}$$

The line has slope $\frac{5}{3}$. A perpendicular line has slope $\frac{-1}{\frac{5}{3}} = -\frac{3}{5}$.

$$y - y_1 = m(x - x_1)$$
$$y - (-3) = -\frac{3}{5}[x - (-10)]$$
$$y + 3 = -\frac{3}{5}(x + 10)$$
$$y + 3 = -\frac{3}{5}x - 6$$
$$y = -\frac{3}{5}x - 9$$

67. $(3, -5) = (x_1, y_1)$; $m = 7$

$$y - y_1 = m(x - x_1)$$
$$y - (-5) = 7(x - 3)$$
$$y + 5 = 7x - 21$$
$$y = 7x - 26$$

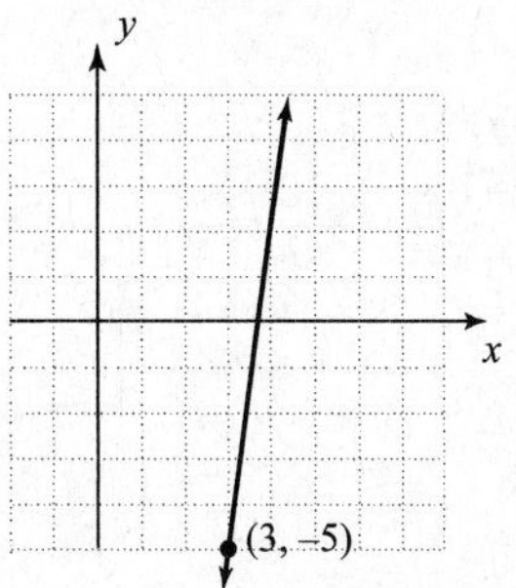

69. The slope of $y = -5x + 3$ is -5. A perpendicular line has slope $\frac{-1}{-5} = \frac{1}{5}$.

$$(2, 9) = (x_1, y_1)$$
$$y - y_1 = m(x - x_1)$$
$$y - 9 = \frac{1}{5}(x - 2)$$
$$y - 9 = \frac{1}{5}x - \frac{2}{5}$$
$$y = \frac{1}{5}x + \frac{43}{5}$$

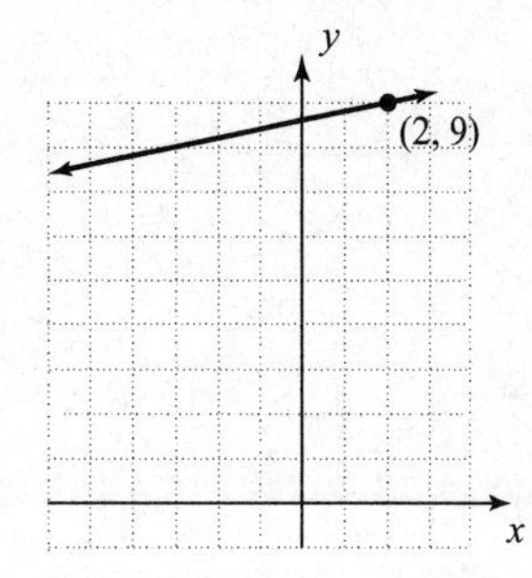

71. The slope of $y = -7x + 2$ is -7. A parallel line will also have slope -7.

$$(6, -1) = (x_1, y_1)$$
$$y - y_1 = m(x - x_1)$$
$$y - (-1) = -7(x - 6)$$
$$y + 1 = -7x + 42$$
$$y = -7x + 41$$

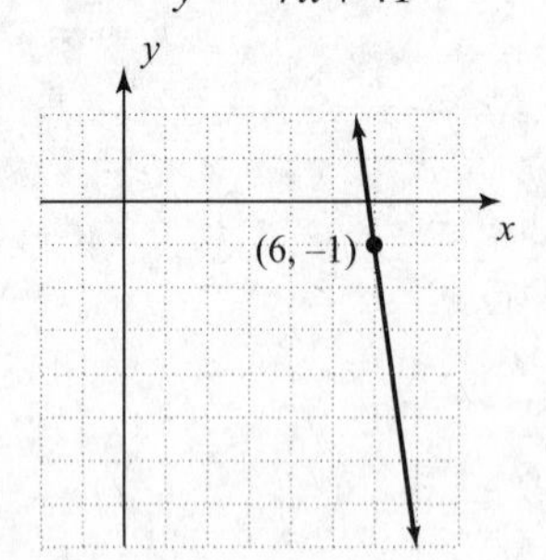

73. $(-6, 2)$ and $(-1, -8)$

$$m = \frac{-8-2}{-1-(-6)} = \frac{-10}{-1+6} = \frac{-10}{5} = -2$$
$$y - y_1 = m(x - x_1)$$
$$y - 2 = -2[x - (-6)]$$
$$y - 2 = -2(x + 6)$$
$$y - 2 = -2x - 12$$
$$y = -2x - 10$$

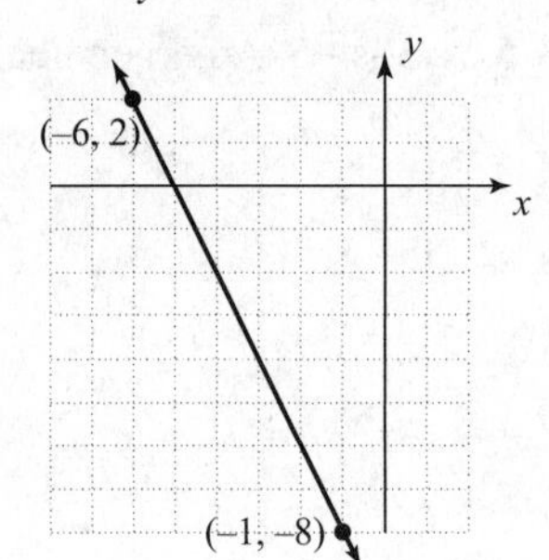

75. $m = 3, b = -2$
$y = mx + b$
$y = 3x - 2$

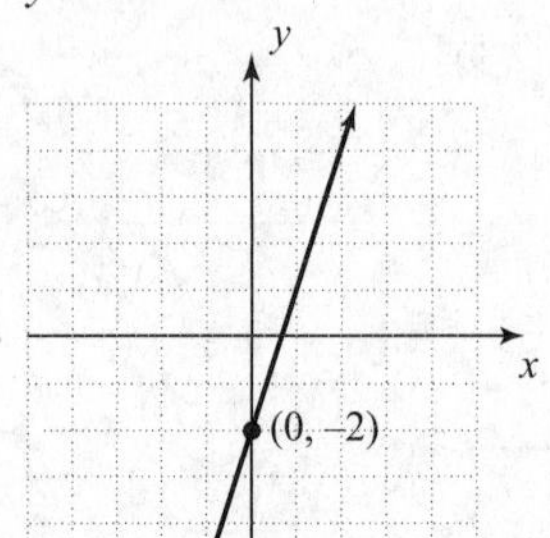

77. The line $x = -6$ is vertical. A parallel line will also be vertical. The vertical line through $(5, 1)$ is $x = 5$.

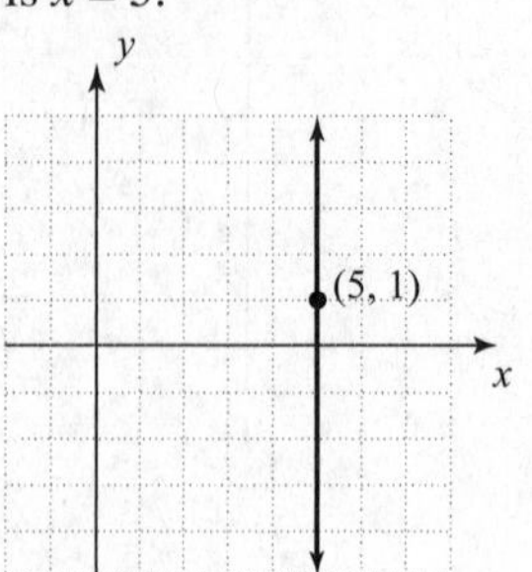

79. $4x + 3y = 9$

$$3y = -4x + 9$$
$$y = -\frac{4}{3}x + 3$$

The slope of the line is $-\frac{4}{3}$. A parallel line will also have slope $-\frac{4}{3}$.

$$(3, -2) = (x_1, y_1)$$
$$y - y_1 = m(x - x_1)$$
$$y - (-2) = -\frac{4}{3}(x - 3)$$
$$y + 2 = -\frac{4}{3}x + 4$$
$$y = -\frac{4}{3}x + 2$$

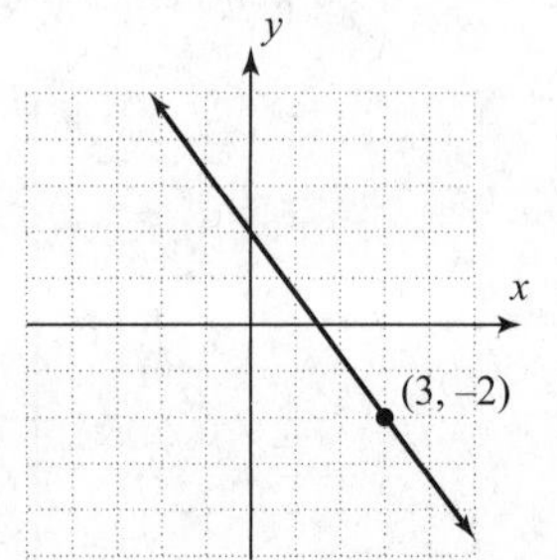

81. $x - 2y = -10$

$$-2y = -x - 10$$
$$y = \frac{1}{2}x + 5$$

The line has slope $\frac{1}{2}$. A perpendicular line has slope $\frac{-1}{\frac{1}{2}} = -1\left(\frac{2}{1}\right) = -2$.

$$\begin{aligned}(-1, -3) &= (x_1, y_1)\\ y - y_1 &= m(x - x_1)\\ y - (-3) &= -2[x - (-1)]\\ y + 3 &= -2(x + 1)\\ y + 3 &= -2x - 2\\ y &= -2x - 5\end{aligned}$$

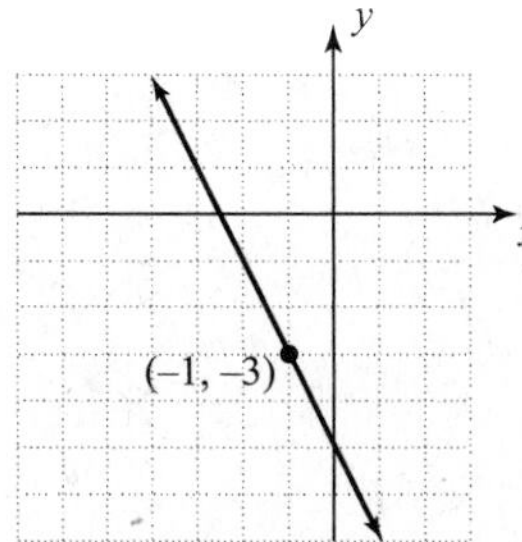

83. (a) $L_1 : (0, -1)$ and $(-2, -7)$

$$m_1 = \frac{-7-(-1)}{-2-0} = \frac{-7+1}{-2} = \frac{-6}{-2} = 3$$

$L_2 : (-1, 5)$ and $(2, -4)$

$$m_2 = \frac{-4-5}{2-(-1)} = \frac{-9}{2+1} = \frac{-9}{3} = -3$$

(b) Since $m_1 \neq m_2$ and $m_1 \neq \frac{-1}{m_2}$, the lines are neither parallel nor perpendicular.

85. (a) $L_1 : (2, 8)$ and $(7, 18)$

$$m_1 = \frac{18-8}{7-2} = \frac{10}{5} = 2$$

$L_2 : (-2, -3)$ and $(6, 13)$

$$m_2 = \frac{13-(-3)}{6-(-2)} = \frac{13+3}{6+2} = \frac{16}{8} = 2$$

(b) Since $m_1 = m_2$, the lines are parallel.

87. (a) $L_1 : (-2, -5)$ and $(4, -2)$

$$m = \frac{-2-(-5)}{4-(-2)} = \frac{-2+5}{4+2} = \frac{3}{6} = \frac{1}{2}$$

$L_2 : (-8, -5)$ and $(0, -1)$

$$m_2 = \frac{-1-(-5)}{0-(-8)} = \frac{-1+5}{0+8} = \frac{4}{8} = \frac{1}{2}$$

(b) Since $m_1 = m_2$, the lines are parallel.

89. (a) $L_1 : (-6, -9)$ and $(3, 6)$

$$m_1 = \frac{6-(-9)}{3-(-6)} = \frac{6+9}{3+6} = \frac{15}{9} = \frac{5}{3}$$

$L_2 : (10, -8)$ and $(-5, 1)$

$$m_2 = \frac{1-(-8)}{-5-10} = \frac{1+8}{-15} = \frac{9}{-15} = -\frac{3}{5}$$

(b) Since $m_1 m_2 = \left(\frac{5}{3}\right)\left(-\frac{3}{5}\right) = -1$, the lines are perpendicular.

91.

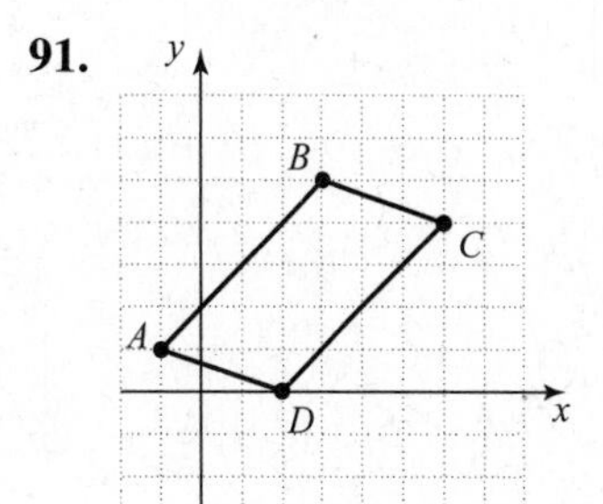

$$\text{Slope } AB = m_1 = \frac{5-1}{3-(-1)} = \frac{4}{3+1} = \frac{4}{4} = 1$$

$$\text{Slope } BC = m_2 = \frac{4-5}{6-3} = \frac{-1}{3} = -\frac{1}{3}$$

$$\text{Slope } CD = m_3 = \frac{0-4}{2-6} = \frac{-4}{-4} = 1 = m_1$$

$$\text{Slope } DA = m_4 = \frac{1-0}{-1-2} = \frac{1}{-3} = -\frac{1}{3} = m_2$$

Since the slopes of the opposite sides are the same, the figure is a parallelogram.

93.

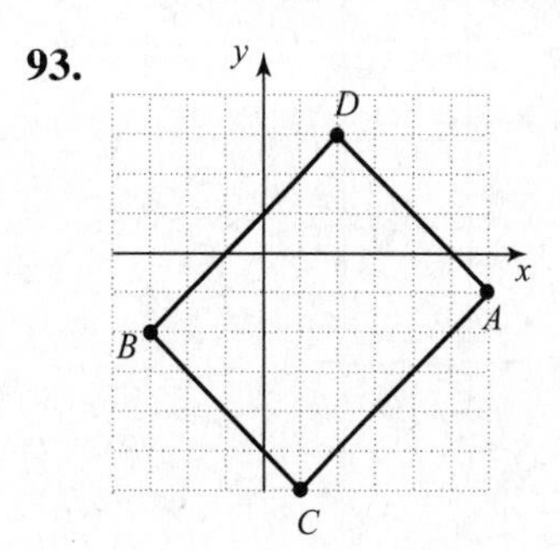

$$\begin{aligned}\text{Slope } AC = m_1 &= \frac{-6-(-1)}{1-6}\\ &= \frac{-6+1}{-5}\\ &= \frac{-5}{-5}\\ &= 1\end{aligned}$$

$$\text{Slope } CB = m_2 = \frac{-2-(-6)}{-3-1} = \frac{-2+6}{-4} = \frac{4}{-4} = -1 = \frac{-1}{m_1}$$

$$\text{Slope } BD = m_3 = \frac{3-(-2)}{2-(-3)} = \frac{3+2}{2+3} = \frac{5}{5} = 1 = m_1$$

$$\text{Slope } DA = m_4 = \frac{-1-3}{6-2} = \frac{-4}{4} = -1 = m_2$$

Since the slopes of the opposite sides are the same, the figure is a parallelogram.
Since the slopes of the adjacent sides AC and CB are negative reciprocals, the sides are perpendicular so the figure is a rectangle.

95.

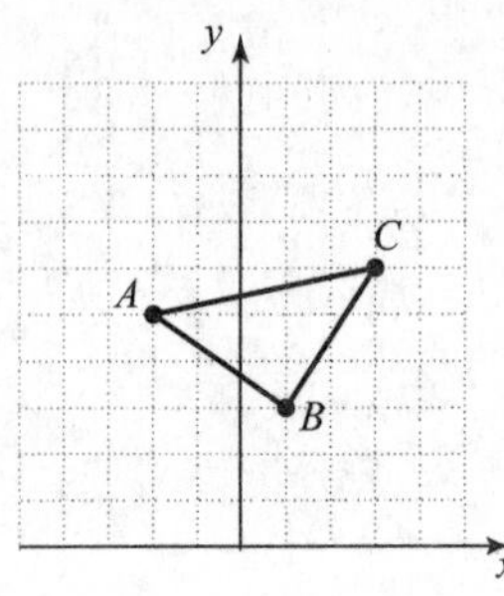

$$\text{Slope } AB = m_1 = \frac{3-5}{1-(-2)} = \frac{-2}{1+2} = -\frac{2}{3}$$

$$\text{Slope } BC = m_2 = \frac{6-3}{3-1} = \frac{3}{2} = \frac{-1}{m},$$

$$\text{Slope } CA = m_3 = \frac{5-6}{-2-3} = \frac{-1}{-5} = \frac{1}{5}$$

Since the slopes of AC and BC are negative reciprocals, the triangle is a right triangle.

97.

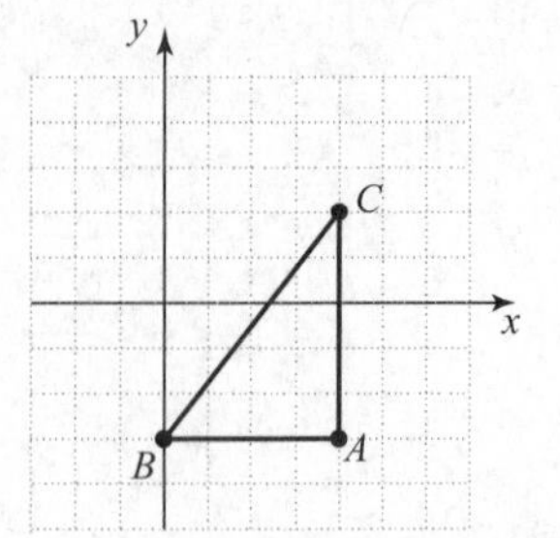

$$\text{Slope } AB = m_1 = \frac{-3-(-3)}{0-4} = \frac{-3+3}{-4} = \frac{0}{-4} = 0$$

$$\text{Slope } BC = m_2 = \frac{2-(-3)}{4-0} = \frac{2+3}{4} = \frac{5}{4}$$

$$\text{Slope } CA = m_3 = \frac{-3-2}{4-4} = \frac{-5}{0} = \text{undefined}$$

A vertical line (with undefined slope) is perpendicular to a line with 0 slope, thus the triangle is a right triangle.

99. $-3y = 6x-12$
$y = -2x+4$
Slope $= -2$
$4x + By = -2$
$By = -4x-2$
$y = -\frac{4}{B}x - \frac{2}{B}$

Slope $= -\frac{4}{B}$

Parallel lines have the same slope.

$-2 = -\frac{4}{B}$
$-2B = -4$
$B = 2$

101. $Ax + 6y = -6$
$6y = -Ax - 6$
$y = -\frac{A}{6}x - 1$

Slope $= -\frac{A}{6}$

$12 - 6y = -9x$
$-6y = -9x - 12$
$y = \frac{3}{2}x + 2$

$\text{Slope} = \frac{3}{2}$

The slopes of perpendicular lines are negative reciprocals.

$-\frac{A}{6} = \frac{-1}{\frac{3}{2}}$

$-\frac{A}{6} = -\frac{2}{3}$

$3A = 12$

$A = 4$

103.

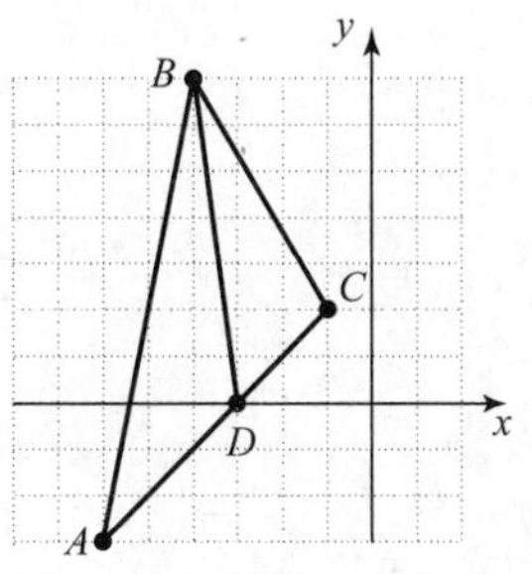

$\text{Slope } AC = m_1 = \frac{2-(-3)}{-1-(-6)} = \frac{2+3}{-1+6} = \frac{5}{5} = 1$

$\text{Slope } BD = m_2 = \frac{0-7}{-3-(-4)}$

$= \frac{-7}{-3+4}$

$= \frac{-7}{1}$

$= -7$

Since $m_1 m_2 = 1(-7) = -7 \neq 1$, $\overline{BD}$ is not a perpendicular to $\overline{AC}$, so it is not an altitude of triangle ABC.

105. Answers may vary. One possibility:
If $m_1 = m_2$ and $b_1 = b_2$, the lines are identical.
If $m_1 = m_2$ and $b_1 \neq b_2$, the lines are parallel.
If $m_1 \neq m_2$ and $m_1 \neq \frac{-1}{m_2}$ the lines intersect in one point, but they are not perpendicular.
If $m_1 = \frac{-1}{m_2}$ the lines intersect in one point and are perpendicular.

Putting the Concepts Together (Sections 3.1–3.6)

1. $(1, -2)$; $4x - 3y = 10$

$4x - 3y = 10$

$4(1) - 3(-2) \stackrel{?}{=} 10$

$4 + 6 \stackrel{?}{=} 10$

$10 = 10$ True

$(1, -2)$ is a solution to the equation $4x - 3y = 10$.

In Problems 2 and 3, points may vary.

2. $y = \frac{2}{3}x - 1$

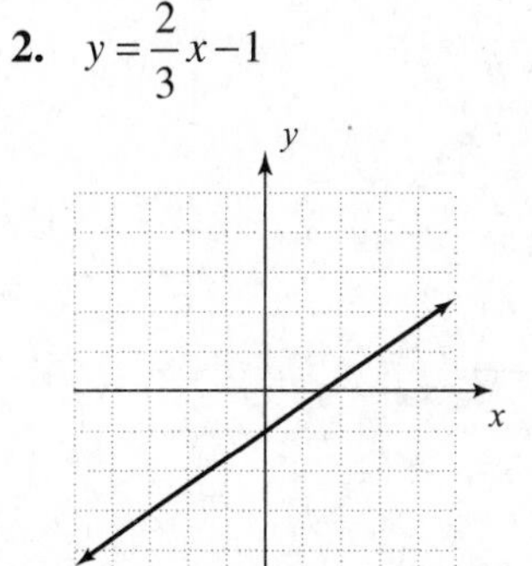

3. $-5x + 2y = 10$

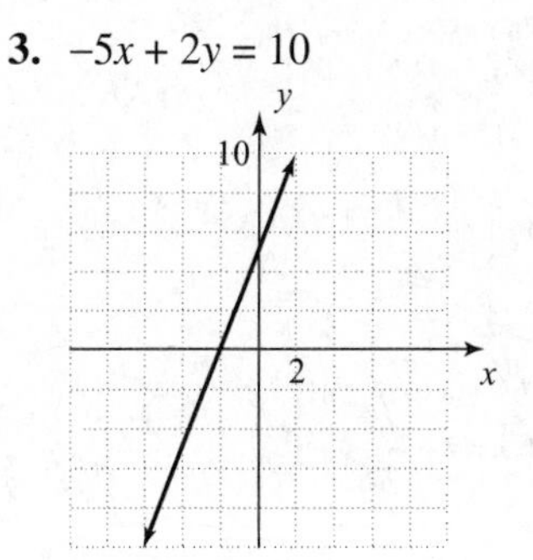

4. $-8x + 2y = 6$

(a) For the x-intercept, let $y = 0$.

$-8x + 2(0) = 6$

$-8x = 6$

$x = -\frac{6}{8}$

$x = -\frac{3}{4}$

The x-intercept is $-\frac{3}{4}$.

(b) For the y-intercept, let $x = 0$.

$-8(0) + 2y = 6$

$2y = 6$

$y = 3$

The y-intercept is 3.

5. $4x + 3y = 6$

$y = 0$: $4x + 3(0) = 6$

$4x = 6$

$x = \frac{6}{4}$

$x = \frac{3}{2}$

The x-intercept is $\frac{3}{2}$.

$x = 0$: $4(0) + 3y = 6$

$3y = 6$

$y = 2$

The y-intercept is 2.

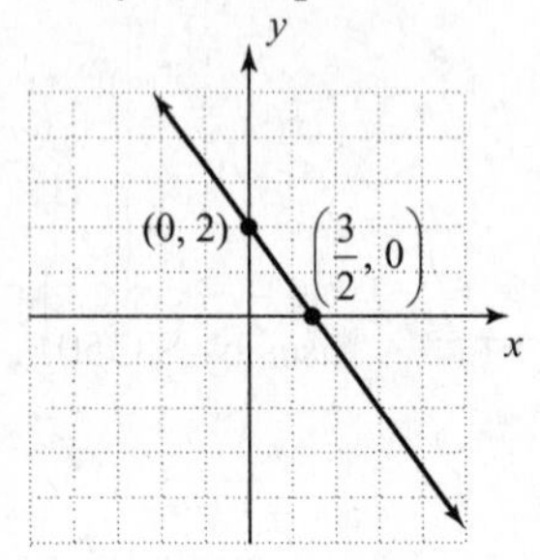

6. $6x + 9y = -12$

$9y = -6x - 12$

$y = -\frac{6}{9}x - \frac{12}{9}$

$y = -\frac{2}{3}x - \frac{4}{3}$

(a) The slope is $m = -\frac{2}{3}$.

(b) The y-intercept is $b = -\frac{4}{3}$.

7. $(3, -5) = (x_1, y_1)$; $(-6, -2) = (x_2, y_2)$

$$m = \frac{y_2 - y_1}{x_2 - x_1} = \frac{-2-(-5)}{-6-3} = \frac{-2+5}{-9} = \frac{3}{-9} = -\frac{1}{3}$$

The slope is $-\frac{1}{3}$.

8. $2y = -5x - 4$

$y = -\frac{5}{2}x - 2$

(a) The given line has slope $-\frac{5}{2}$. A perpendicular line will have slope $\frac{-1}{-\frac{5}{2}} = \frac{2}{5}$.

(b) The given line has slope $-\frac{5}{2}$. A parallel line will also have slope $-\frac{5}{2}$.

9. L_1: $10x + 5y = 2$

$5y = -10x + 2$

$y = -2x + \frac{2}{5}$

L_1 has slope -2 and y-intercept $\frac{2}{5}$.

L_2: $y = -2x + 3$

L_2 has slope -2 and y-intercept 3.

Since the lines have the same slope but different y-intercepts, they are parallel.

10. $m = 3$; $b = 1$

$y = mx + b$

$y = 3x + 1$

11. $m = -6$; $(x_1, y_1) = (-1, 4)$

$y - y_1 = m(x - x_1)$

$y - 4 = -6[x - (-1)]$

$y - 4 = -6(x + 1)$

$y - 4 = -6x - 6$

$y = -6x - 2$

12. $(x_1, y_1) = (4, -1)$; $(x_2, y_2) = (-2, 11)$

$$m = \frac{y_2 - y_1}{x_2 - x_1} = \frac{11-(-1)}{-2-4} = \frac{11+1}{-6} = \frac{12}{-6} = -2$$

$y - y_1 = m(x - x_1)$

$y - (-1) = -2(x - 4)$

$y + 1 = -2x + 8$

$y = -2x + 7$

13. The slope of $y = \frac{2}{5}x - 5$ is $\frac{2}{5}$. A perpendicular line will have slope $\frac{-1}{\frac{2}{5}} = -\frac{5}{2}$.

$m = -\frac{5}{2}$; $(x_1, y_1) = (-8, 0)$

$$\begin{aligned} y - y_1 &= m(x - x_1) \\ y - 0 &= -\frac{5}{2}[x - (-8)] \\ y &= -\frac{5}{2}(x + 8) \\ y &= -\frac{5}{2}x - 20 \end{aligned}$$

14.

$$\begin{aligned} -8y + 2x &= -1 \\ -8y &= -2x - 1 \\ y &= \frac{1}{4}x + \frac{1}{8} \end{aligned}$$

The slope of the line is $\frac{1}{4}$. A parallel line will also have slope $\frac{1}{4}$.

$m = \frac{1}{4}$; $(x_1, y_1) = (-8, 3)$

$$\begin{aligned} y - y_1 &= m(x - x_1) \\ y - 3 &= \frac{1}{4}[x - (-8)] \\ y - 3 &= \frac{1}{4}(x + 8) \\ y - 3 &= \frac{1}{4}x + 2 \\ y &= \frac{1}{4}x + 5 \end{aligned}$$

15. The horizontal line through (–6, –8) has equation $y = -8$.

16. A line with undefined slope is a vertical line. The vertical line through (2, 6) has equation $x = 2$.

17. Let $(80, 1180) = (x_1, y_1)$ and $(50, 850) = (x_2, y_2)$.

$$m = \frac{y_2 - y_1}{x_2 - x_1} = \frac{850 - 1180}{50 - 80} = \frac{-330}{-30} = 11$$

The average rate to ship an additional package is $11 per package.

18. (a) Let $(0.7, 3543) = (x_1, y_1)$ and $(0.8, 4378) = (x_2, y_2)$.

$$\begin{aligned} m &= \frac{y_2 - y_1}{x_2 - x_1} \\ &= \frac{4378 - 3543}{0.8 - 0.7} \\ &= \frac{835}{0.1} \\ &= 8350 \end{aligned}$$

$$\begin{aligned} y - y_1 &= m(x - x_1) \\ y - 3543 &= 8350(x - 0.7) \\ y - 3543 &= 8350x - 5845 \\ y &= 8350x - 2302 \end{aligned}$$

The equation is $y = 8350x - 2302$ where x is the weight, in carats, of the diamond and y is the price.

(b) For every 1-carat increase in the weight of a diamond, the cost increases by $8350.

(c) $y = 8350x - 2302$; $x = 0.76$
$y = 8350(0.76) - 2302 = 4044$
A 0.76-carat diamond costs $4044.

Section 3.7

Preparing for Linear Inequalities in Two Variables

P1.

$$\begin{aligned} x - 4 &> 5 \\ x - 4 + 4 &> 5 + 4 \\ x &> 9 \end{aligned}$$

The solution set is $\{x | x > 9\}$.

P2.

$$\begin{aligned} 3x + 1 &\le 10 \\ 3x + 1 - 1 &\le 10 - 1 \\ 3x &\le 9 \\ \frac{3x}{3} &\le \frac{9}{3} \\ x &\le 3 \end{aligned}$$

The solution set is $\{x | x \le 3\}$.

P3.

$$\begin{aligned} 2(x + 1) - 6x &> 18 \\ 2x + 2 - 6x &> 18 \\ -4x + 2 &> 18 \\ -4x + 2 - 2 &> 18 - 2 \\ -4x &> 16 \\ \frac{-4x}{-4} &< \frac{16}{-4} \\ x &< -4 \end{aligned}$$

The solution set is $\{x | x < -4\}$.

3.7 Quick Checks

1. (a) Let $x = 2$ and $y = 1$ in the inequality.
$$2x + y > 7$$
$$2(2) + 1 > 7?$$
$$4 + 1 > 7?$$
$$5 > 7 \quad \text{False}$$
The statement $5 > 7$ is false, so (2, 1) is not a solution to the inequality.

(b) Let $x = 3$ and $y = 4$ in the inequality.
$$2x + y > 7$$
$$2(3) + 4 > 7?$$
$$6 + 4 > 7?$$
$$10 > 7 \quad \text{True}$$
The statement $10 > 7$ is true, so (3, 4) is a solution to the inequality.

(c) Let $x = -1$ and $y = 10$ in the inequality.
$$2x + y > 7$$
$$2(-1) + 10 > 7?$$
$$-2 + 10 > 7?$$
$$8 > 7 \quad \text{True}$$
The statement $8 > 7$ is true, so (−1, 10) is a solution to the inequality.

2. (a) Let $x = 2$ and $y = 1$ in the inequality.
$$-3x + 2y \le 8$$
$$-3(2) + 2(1) \le 8?$$
$$-6 + 2 \le 8?$$
$$-4 \le 8 \quad \text{True}$$
The statement $-4 \le 8$ is true, so (2, 1) is a solution to the inequality.

(b) Let $x = 3$ and $y = 4$ in the inequality.
$$-3x + 2y \le 8$$
$$-3(3) + 2(4) \le 8?$$
$$-9 + 8 \le 8?$$
$$-1 \le 8 \quad \text{True}$$
The statement $-1 \le 8$ is true, so (3, 4) is a solution to the inequality.

(c) Let $x = -1$ and $y = 10$ in the inequality.
$$-3x + 2y \le 8$$
$$-3(-1) + 2(10) \le 8?$$
$$3 + 20 \le 8?$$
$$23 \le 8 \quad \text{False}$$
The statement $23 \le 8$ is false, so (−1, 10) is not a solution to the inequality.

3. When drawing the boundary line for the graph of $Ax + By \ge C$, we use a solid line. When drawing the boundary line for the graph of $Ax + By < C$, we use a dashed line.

4. The boundary line separates the xy-plane into two regions, called half-planes.

5. Graph the line $y = -2x + 1$ with a dashed line since the inequality is strict.
Use (0, 0) as a test point.
$$y < -2x + 1$$
$$0 < -2(0) + 1?$$
$$0 < 1 \quad \text{True}$$
Since $0 < 1$ is a true statement, shade the half-plane containing (0, 0).

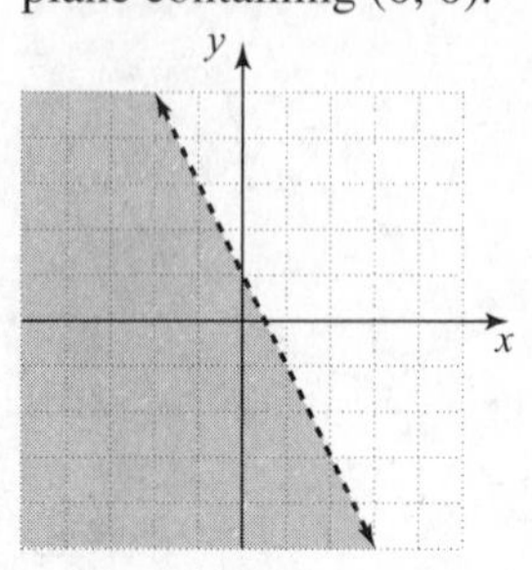

6. Graph the line $y = 3x + 2$ with a solid line since the inequality is nonstrict.
Use (0, 0) as a test point.
$$y \ge 3x + 2$$
$$0 \ge 3(0) + 2?$$
$$0 \ge 2 \quad \text{False}$$
Since $0 \ge 2$ is a false statement, shade the half-plane that does not contain (0, 0).

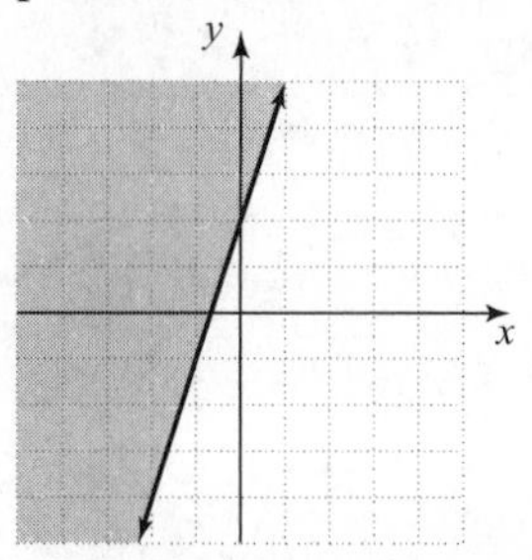

7. Graph the line $2x + 3y = 6$ with a solid line since the inequality is nonstrict.
x-intercept, $y = 0$:
$$2x + 3(0) = 6$$
$$2x = 6$$
$$x = 3$$
y-intercept, $x = 0$:
$$2(0) + 3y = 6$$
$$3y = 6$$
$$y = 2$$
Use (0, 0) as a test point.
$$2x + 3y \le 6$$
$$2(0) + 3(0) \le 6?$$
$$0 \le 6 \quad \text{True}$$
Since $0 \le 6$ is a true statement, shade the half-

plane containing (0, 0).

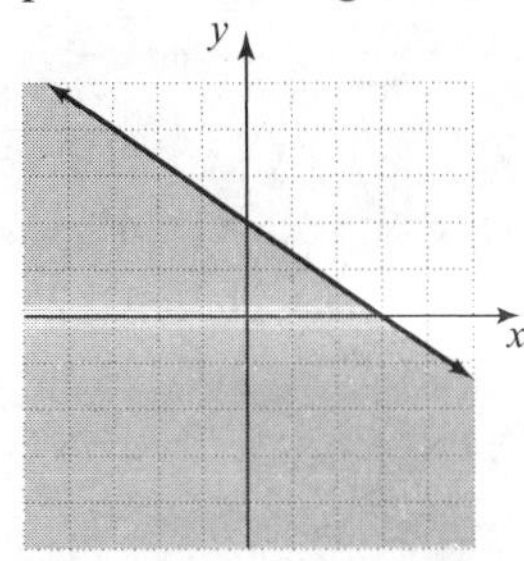

8. Graph the line $4x - 6y = 12$ with a dashed line since the inequality is strict.

x-intercept, $y = 0$:

$$4x - 6(0) = 12$$
$$4x = 12$$
$$x = 3$$

y-intercept, $x = 0$:

$$4(0) - 6y = 12$$
$$-6y = 12$$
$$y = -2$$

Use (0, 0) as a test point.

$$4x - 6y > 12$$
$$4(0) - 6(0) > 12?$$
$$0 > 12 \quad \text{False}$$

Since $0 > 12$ is a false statement, shade the half-plane that does not contain (0, 0).

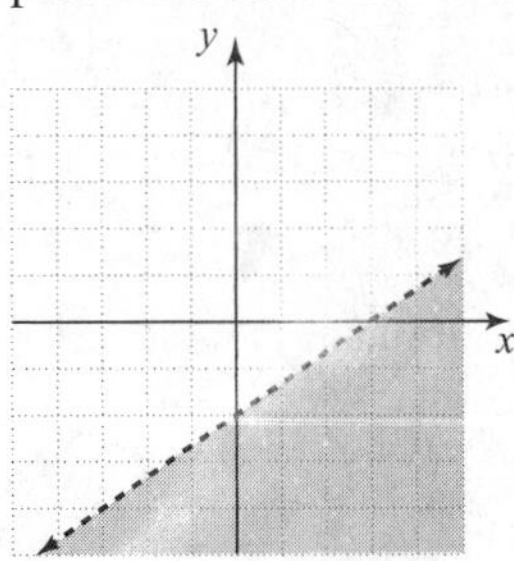

9. Graph the line $3x + y = 0$ with a dashed line since the inequality is strict.

$$3x + y = 0$$
$$y = -3x$$

Slope -3, y-intercept 0.

Use (1, 1) as a test point.

$$3x + y < 0$$
$$3(1) + 1 < 0?$$
$$4 < 0 \quad \text{False}$$

Since $4 < 0$ is a false statement, shade the half-plane that does not contain (1, 1).

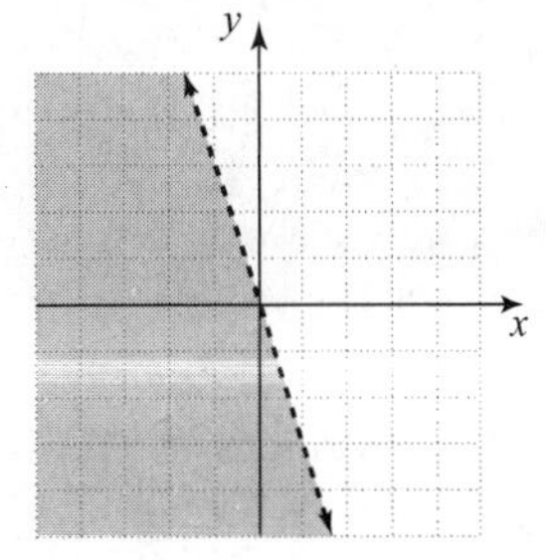

10. Graph the line $2x - 5y = 0$ with a solid line since the inequality is nonstrict.

$$2x - 5y = 0$$
$$-5y = -2x$$
$$y = \frac{2}{5}x$$

Slope $\frac{2}{5}$, y-intercept 0.

Use (1, 1) as a test point.

$$2x - 5y \le 0$$
$$2(1) - 5(1) \le 0?$$
$$2 - 5 \le 0?$$
$$-3 \le 0 \quad \text{True}$$

Since $-3 \le 0$ is a true statement, shade the half-plane containing (1, 1).

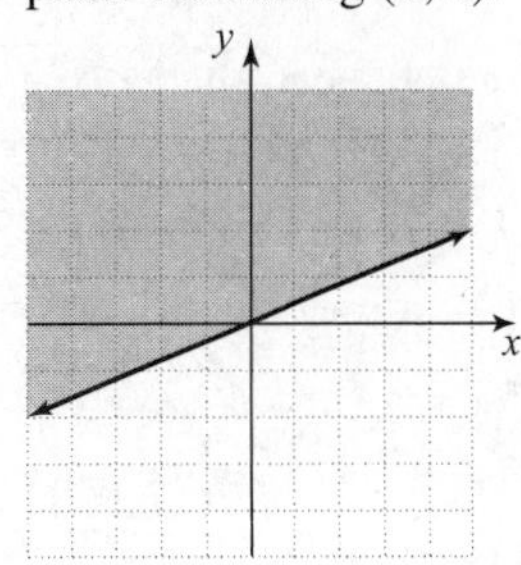

11. Graph the horizontal line $y = -1$ with a dashed line, since the inequality is strict. Use (0, 0) as a test point.

$$y < -1$$
$$0 < -1 \quad \text{False}$$

Since $0 < -1$ is a false statement, shade the half-plane that does not contain (0, 0).

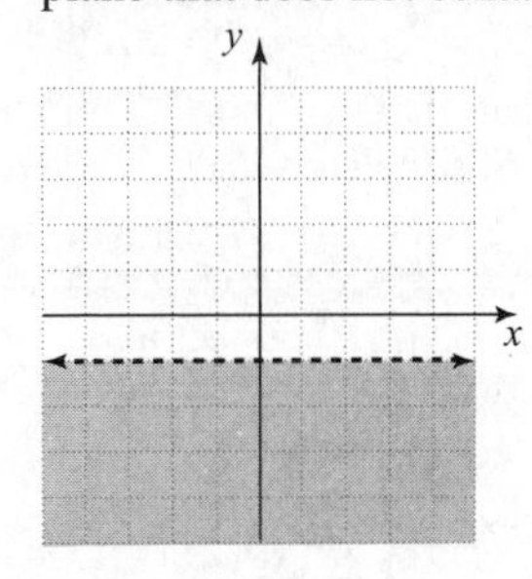

12. $3y - 9 = 0$
$3y = 9$
$y = 3$
Graph the horizontal line $y = 3$ with a solid line since the inequality is nonstrict.
Use (0, 0) as a test point.
$3y - 9 \geq 0$
$3(0) - 9 \geq 0?$
$-9 \geq 0$ False
Since $-9 \geq 0$ is a false statement, shade the half-plane that does not contain (0, 0).

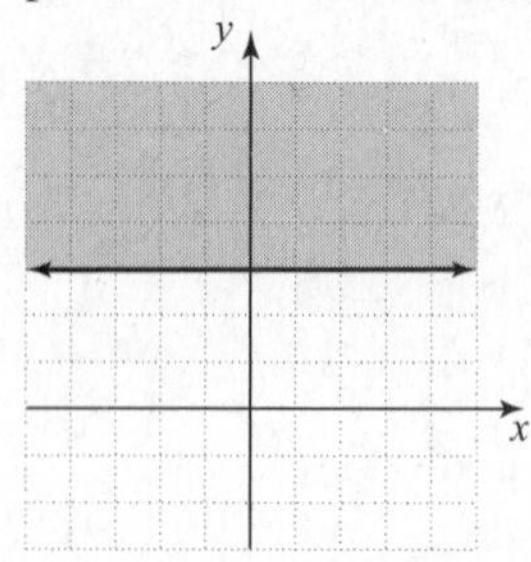

13. Graph the vertical line $x = 6$ with a dashed line since the inequality is strict.
Use (0, 0) as a test point.
$x > 0$
$0 > 6$ False
Since $0 > 6$ is a false statement, shade the half-plane that does not contain (0, 0).

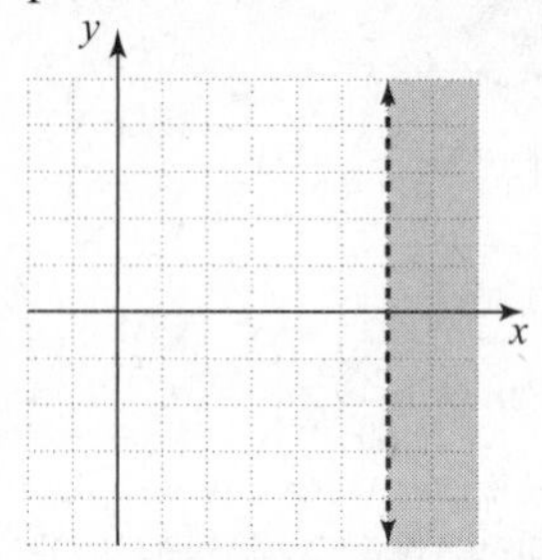

14. **(a)** We want to determine the number of suckers and taffy sticks Kevin can buy. Since each sucker costs $0.20, *s* suckers will cost 0.20*s* dollars. Since each taffy stick costs $0.25, *t* taffy sticks will cost 0.25*t* dollars.
Since Kevin can spend no more than $2 we use a nonstrict inequality.
$0.2s + 0.25t \leq 2$

(b) Let $s = 6$ and $t = 3$.
$0.2(6) + 0.25(3) \leq 2?$
$1.2 + 0.75 \leq 2?$
$1.95 \leq 2$ True
Since $1.95 \leq 2$ is a true statement, Kevin can buy 6 suckers and 3 taffy sticks.

(c) Let $s = 5$ and $t = 5$.
$0.2(5) + 0.25(5) \leq 2?$
$1 + 1.25 \leq 2?$
$2.25 \leq 2$ False
Since $2.25 \leq 2$ is a false statement, Kevin cannot buy 5 suckers and 5 taffy sticks.

3.7 Exercises

15. $y > -x + 2$
$A(2, 4)$: $4 > -2 + 2?$
$4 > 0$ True
$B(3, -6)$: $-6 > 3 + 2?$
$-6 > 5$ False
$C(0, 0)$: $0 > -0 + 2?$
$0 > 2$ False
$A(2, 4)$ is a solution.

17. $y \leq 3x - 1$
$A(-6, -15)$: $-15 \leq 3(-6) - 1?$
$-15 \leq -18 - 1?$
$-15 \leq -19$ False
$B(0, 0)$: $0 \leq 3(0) - 1?$
$0 \leq 0 - 1?$
$0 \leq -1$ False
$C(-1, -6)$: $-6 \leq 3(-1) - 1?$
$-6 \leq -3 - 1?$
$-6 \leq -4$ True
$C(-1, -6)$ is a solution.

19. $3x \geq 2y$
$A(-8, -12)$: $3(-8) \geq 2(-12)?$
$-24 \geq -24$ True
$B(3, 5)$: $3(3) \geq 2(5)?$
$9 \geq 10$ False
$C(-5, -8)$: $3(-5) \geq 2(-8)?$
$-15 \geq -16$ True
$A(-8, -12)$ and $C(-5, -8)$ are solutions.

21. $2x - 3y < -6$
$A(2, -1)$: $2(2) - 3(-1) < -6?$
$4 + 3 < -6?$
$7 < -6$ False
$B(4, 8)$: $2(4) - 3(8) < -6?$
$8 - 24 < -6?$
$-16 < -6$ True
$C(-3, 0)$: $2(-3) - 3(0) < -6?$
$-6 - 0 < -6?$
$-6 < -6$ False
$B(4, 8)$ is a solution.

23. $x \le 2$
$A(7, 2)$: $7 \le 2$ False
$B(2, 5)$: $2 \le 2$ True
$C(4, 2)$: $4 \le 2$ False
$B(2, 5)$ is a solution.

25. $y > -1$
$A(-1, 1)$: $1 > -1$ True
$B(3, -1)$: $-1 > -1$ False
$C(4, -2)$: $-2 > -1$ False
$A(-1, 1)$ is a solution.

27. $y > 3x - 2$
Graph $y = 3x - 2$ with a dashed line.
Test (0, 0): $0 > 3(0) - 2$?
$0 > -2$ True
Shade the half-plane containing (0, 0).

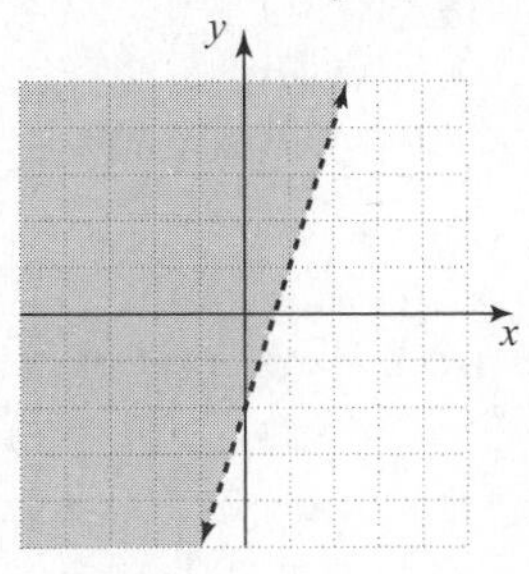

29. $y \le -x + 1$
Graph $y = -x + 1$ with a solid line.
Test (0, 0): $0 \le -0 + 1$?
$0 \le 1$ True
Shade the half-plane containing (0, 0).

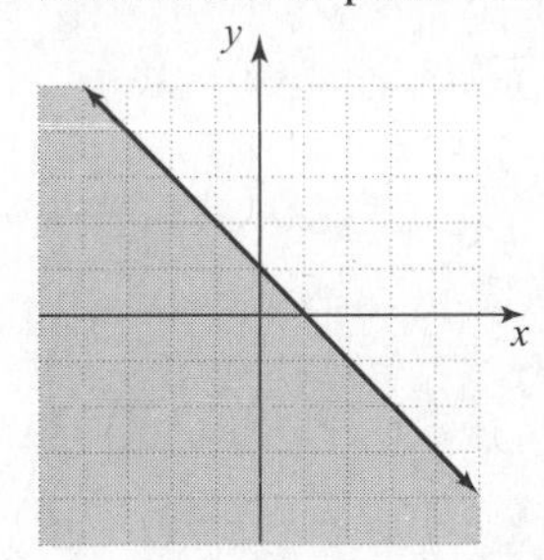

31. $y < \frac{x}{2}$

Graph $y = \frac{x}{2}$ with a dashed line.

Test (1, 1): $1 < \frac{1}{2}$ False
Shade the half-plane *not* containing (1, 1).

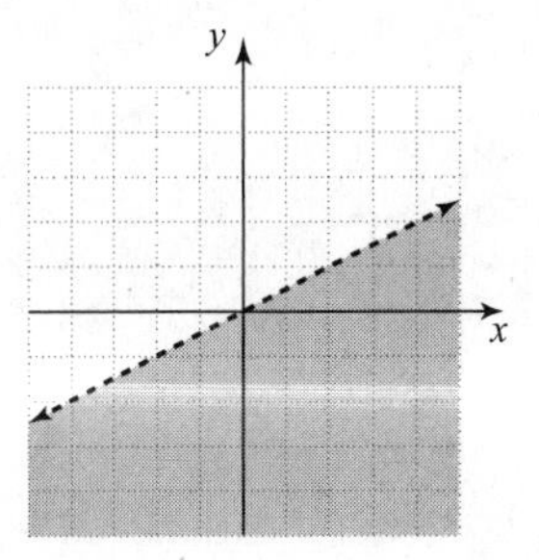

33. $y > 5$
Graph $y = 5$ with a dashed line.
Test (0, 0): $0 > 5$ False
Shade the half-plane *not* containing (0, 0).

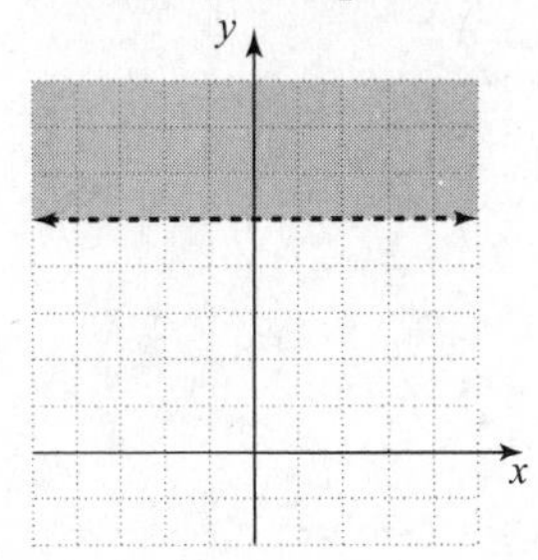

35. $y \le \frac{2}{5}x + 3$

Graph $y = \frac{2}{5}x + 3$ with a solid line.

Test (0, 0): $0 \le \frac{2}{5}(0) + 3$?
$0 \le 3$ True
Shade the half-plane containing (0, 0).

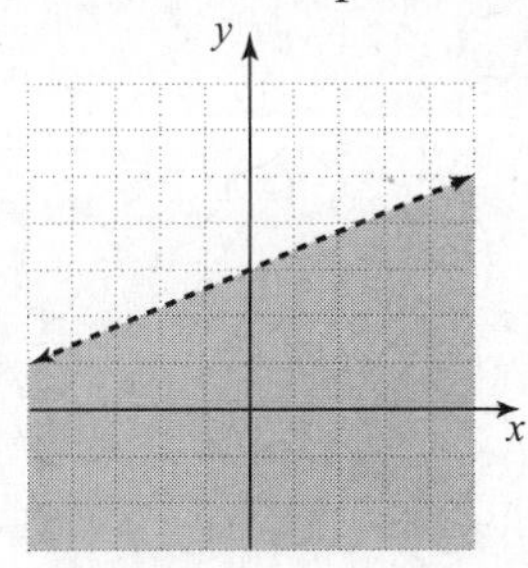

37. $y \ge -\frac{4}{3}x + 2$

Graph $y = -\frac{4}{3}x + 2$ with a solid line.

Test (0, 0): $0 \ge -\frac{4}{3}(0) + 2$?
$0 \ge 2$ False
Shade the half-plane *not* containing (0, 0).

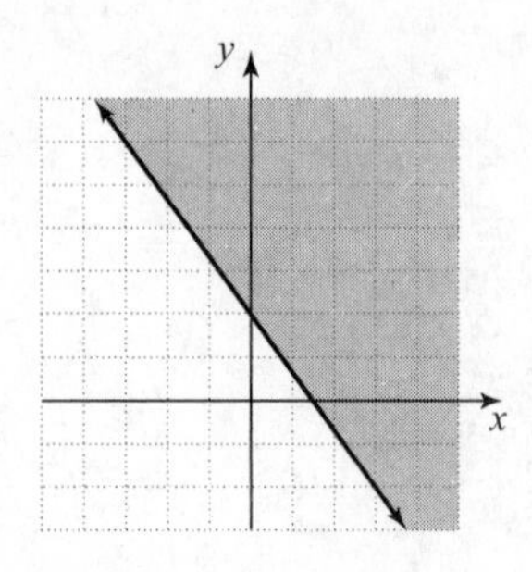

39. $x < 2$
Graph $x = 2$ with a dashed line.
Test (0, 0): $0 < 2$ True
Shade the half-plane containing (0, 0).

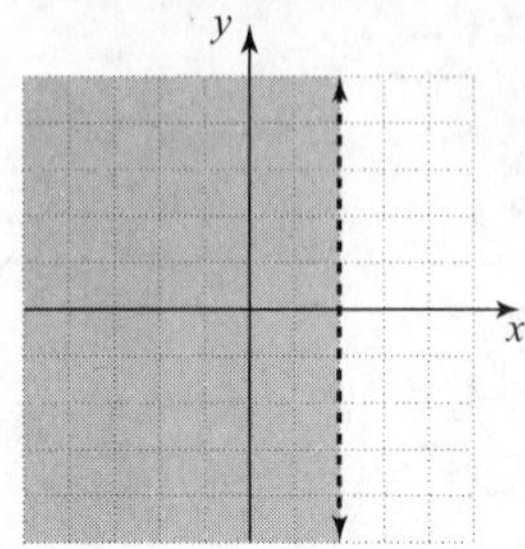

41. $3x - 4y < 12$
Graph $3x - 4y = 12$ with a dashed line.
Test (0, 0): $3(0) - 4(0) < 12?$
$0 < 12$ True
Shade the half-plane containing (0, 0).

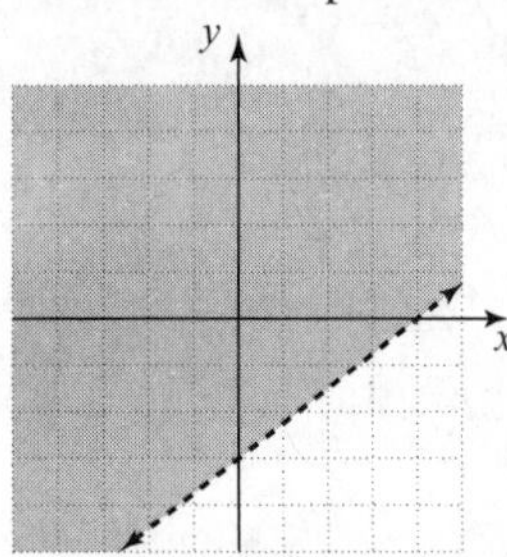

43. $2x + y \geq -4$
Graph $2x + y = 4$ with a solid line.
Test (0, 0): $2(0) + 0 \geq -4?$
$0 \geq -4$ True
Shade the half-plane containing (0, 0).

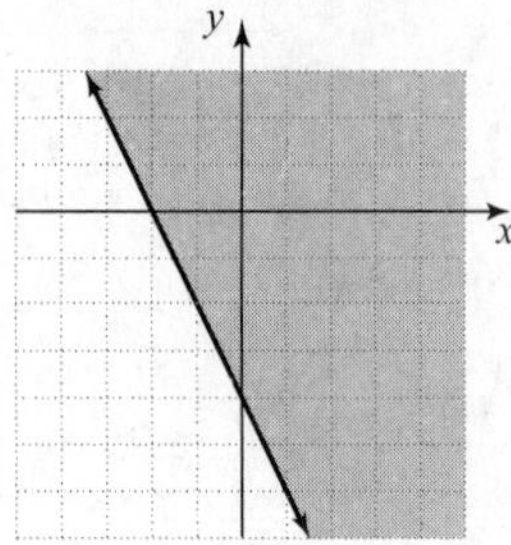

45. $x + y > 0$
Graph $x + y = 0$ with a dashed line.
Test (1, 1): $1 + 1 > 0?$
$2 > 0$ True
Shade the half-plane containing (1, 1).

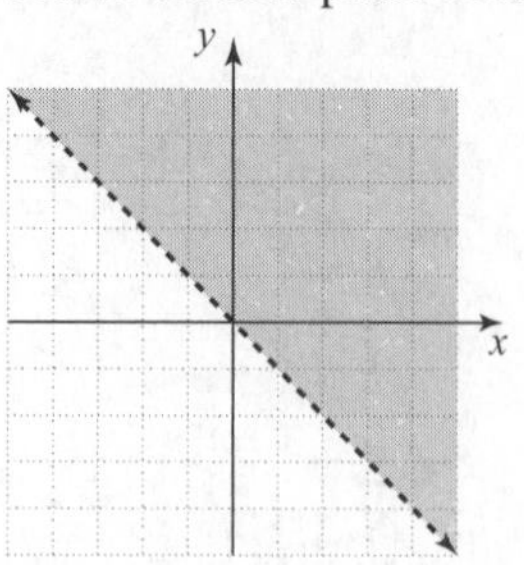

47. $5x - 2y < -8$
Graph $5x - 2y = -8$ with a dashed line.
Test (0, 0): $5(0) - 2(0) < -8?$
$0 < -8$ False
Shade the half-plane *not* containing (0, 0).

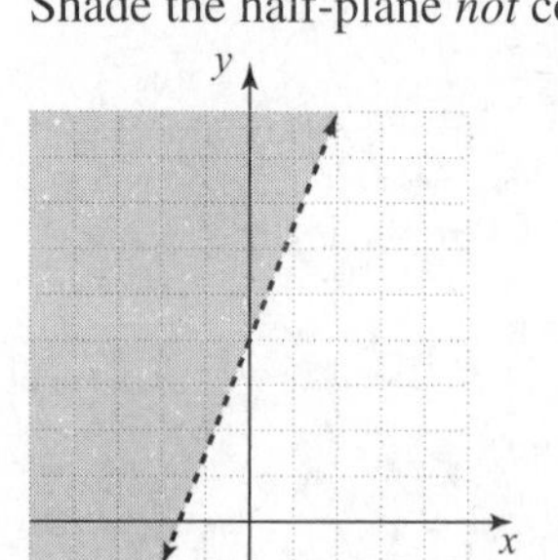

49. $x > -1$
Graph $x = -1$ with a dashed line.
Test (0, 0): $0 > -1$ True
Shade the half-plane containing (0, 0).

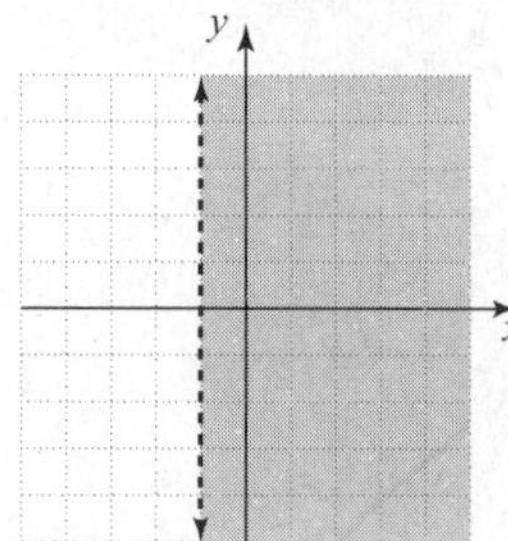

51. $y \le 4$
Graph $y = 4$ with a solid line.
Test (0, 0): $0 \le 4$ True
Shade the half-plane containing (0, 0).

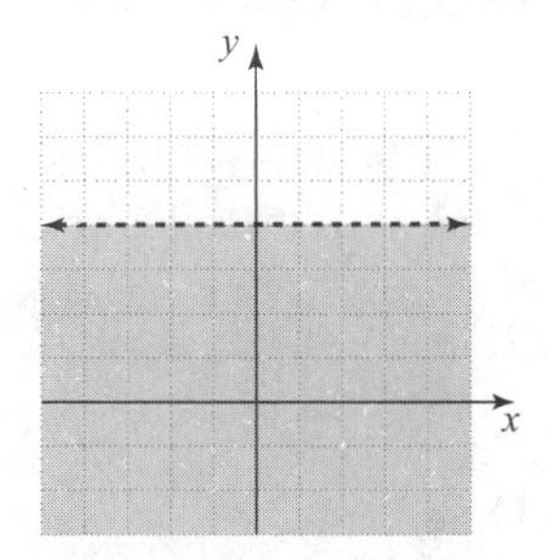

53. $\frac{x}{3} - \frac{y}{5} \ge 1$

Graph $\frac{x}{3} - \frac{y}{5} = 1$ with a solid line.

Test (0, 0): $\frac{0}{3} - \frac{0}{5} \ge 1?$
$0 \ge 1$ False
Shade the half-plane *not* containing (0, 0).

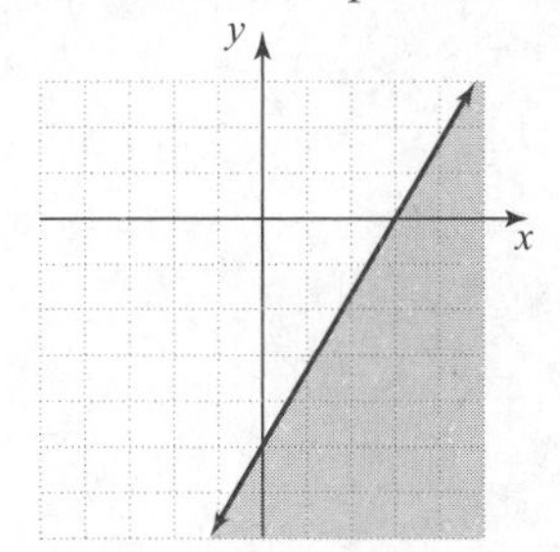

55. $-3 \ge x - y$
Graph $-3 = x - y$ with a solid line.
Test (0, 0): $-3 \ge 0 - 0?$
$-3 \ge 0$ False
Shade the half-plane *not* containing (0, 0).

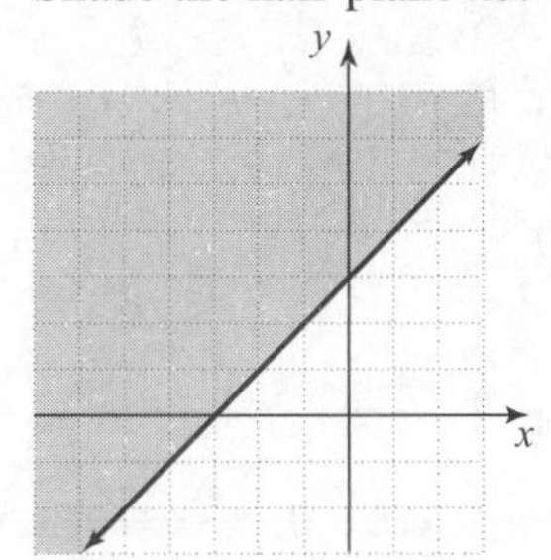

57. $x + y \ge 26$
Graph $x + y = 26$ with a solid line.
Test (0, 0): $0 + 0 \ge 26?$
$0 \ge 26$ False
Shade the half-plane *not* containing (0, 0).

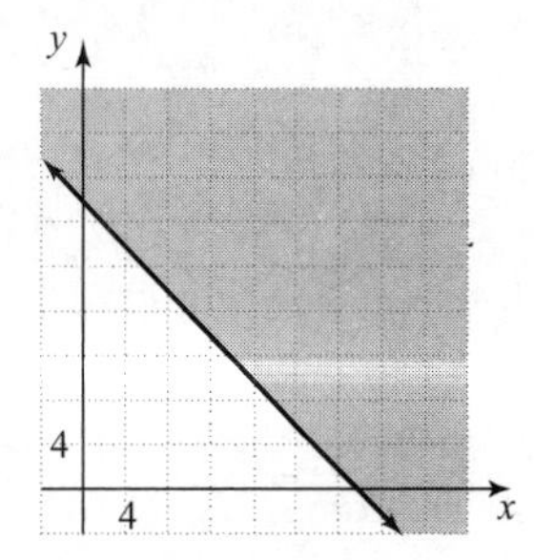

59. $\frac{y}{-2} \le 4$ or $y \ge -8$

Graph $y = -8$ with a solid line.
Test (0, 0): $0 \ge -8$ True
Shade the half-plane containing (0, 0).

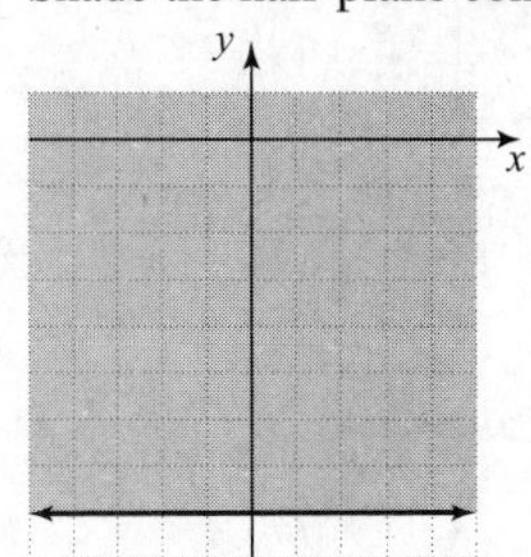

61. $x \le y - 3$
Graph $x = y - 3$ with a solid line.
Test (0, 0): $0 \le 0 - 3?$
$0 \le -3$ False
Shade the half-plane *not* containing (0, 0).

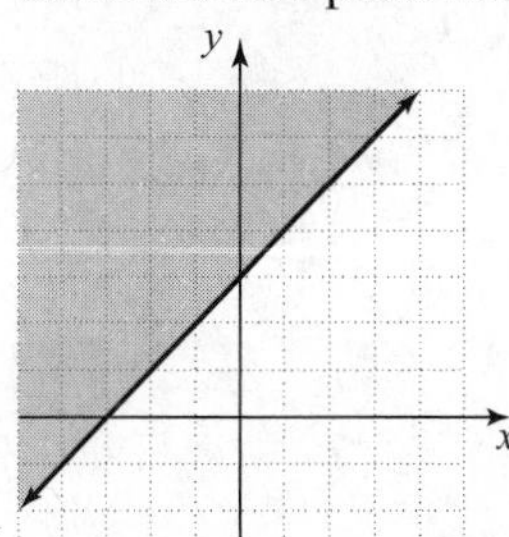

63. $x + 3y < 0$
Graph $x + 3y = 0$ with a dashed line.
Test (1, 1): $1 + 3(1) < 0?$
$4 < 0$ False
Shade the half-plane *not* containing (1, 1).

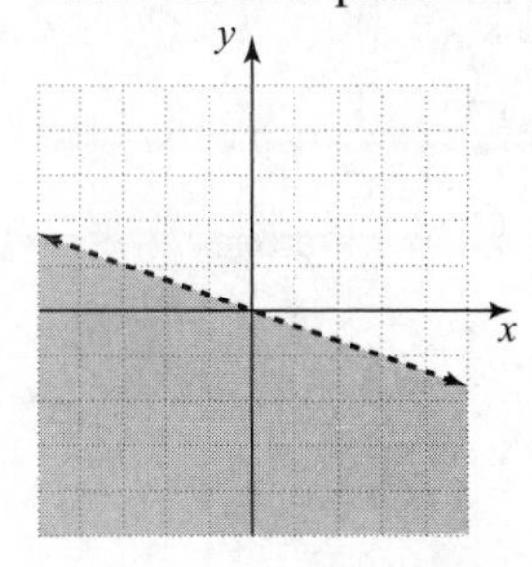

65. $2x - \frac{1}{2}y \geq 5$

Graph $2x - \frac{1}{2}y = 5$ with a solid line.

Test (0, 0): $2(0) - \frac{1}{2}(0) \geq 5?$

$0 \geq 5$ False

Shade the half-plane *not* containing (0, 0).

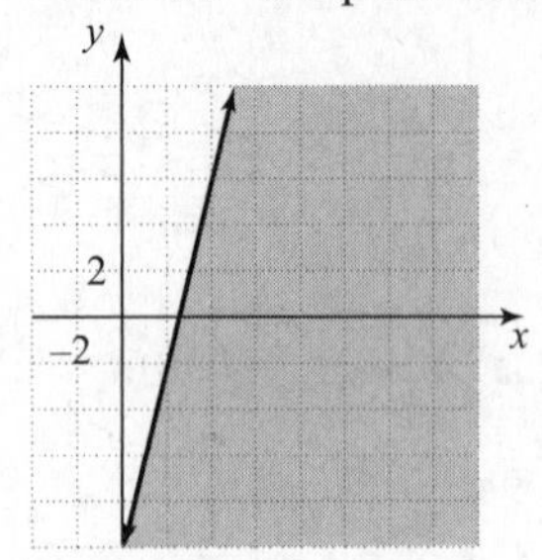

67. $x(-2) > -1$

$-2x > -1$

$x < \frac{1}{2}$

Graph $x = \frac{1}{2}$ with a dashed line.

Test (0, 0): $0 < \frac{1}{2}$ True

Shade the half-plane containing (0, 0).

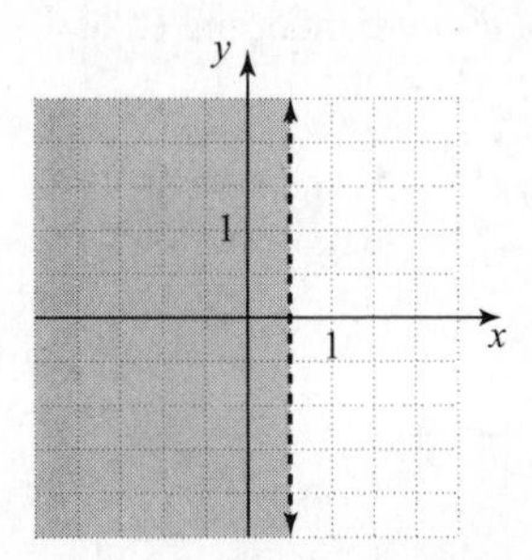

69. (a) Let s be the number of students and a the number of adults. The cost of s students is $3s$; the cost for a adults is $5a$.
$3s + 5a \leq 120$

(b) Test $s = 32$, $a = 6$:
$3(32) + 5(6) \leq 120?$
$96 + 30 \leq 120?$
$126 \leq 120$ False
No, there is not enough money.

(c) Test $s = 29$, $a = 4$:
$3(29) + 5(4) \leq 120?$
$87 + 20 \leq 120?$
$107 \leq 120$ True
Yes, there is enough money.

71. (a) Let s be the number of camping stoves and b be the number of sleeping bags he carries. The weight of s stoves is $3.1s$; the weight of b bags is $5.7b$.
$3.1s + 5.7b \leq 22$

(b) Test $s = 3$, $b = 2$:
$3.1(3) + 5.7(2) \leq 22?$
$9.3 + 11.4 \leq 22?$
$20.7 \leq 22$ True
No, it will not be too heavy.

(c) Test $s = 2$, $b = 3$:
$3.1(2) + 5.7(3) \leq 22?$
$6.2 + 17.1 \leq 22?$
$23.3 \leq 22$ False
Yes, it will be too heavy.

73. $3x - 2y > 6$ and $x + y < 2$

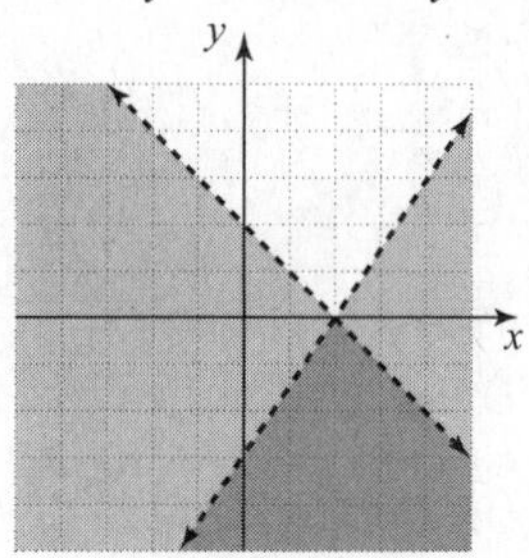

75. $y > \frac{3}{4}x - 1$ and $x \geq 0$

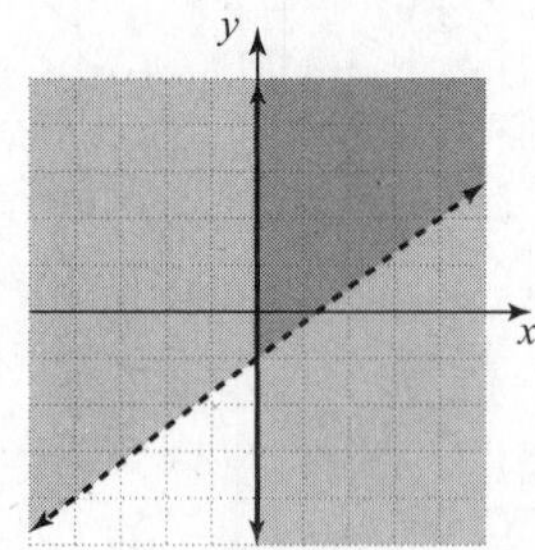

77. $x < -3$ and $y \le 4$

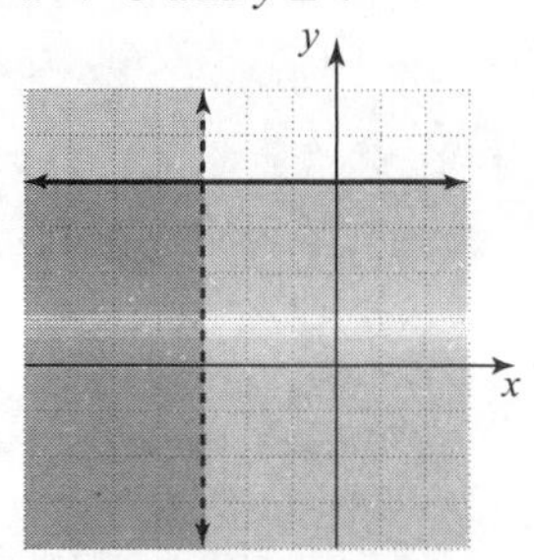

79. Answers may vary.

81. $y > 3$

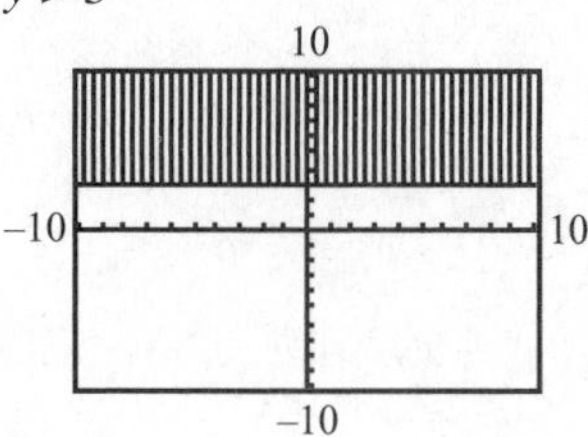

83. $y < 5x$

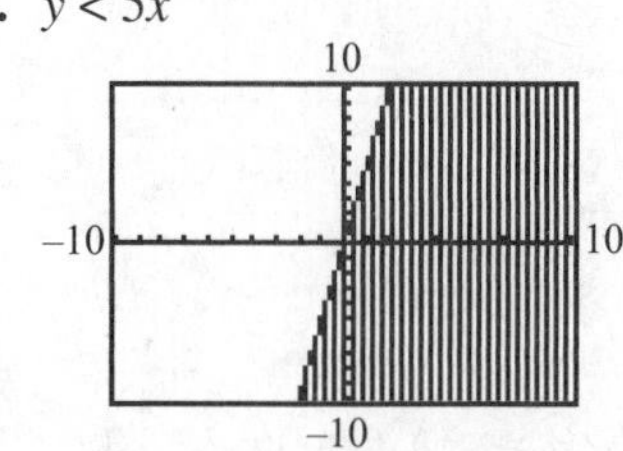

85. $y > 2x + 3$

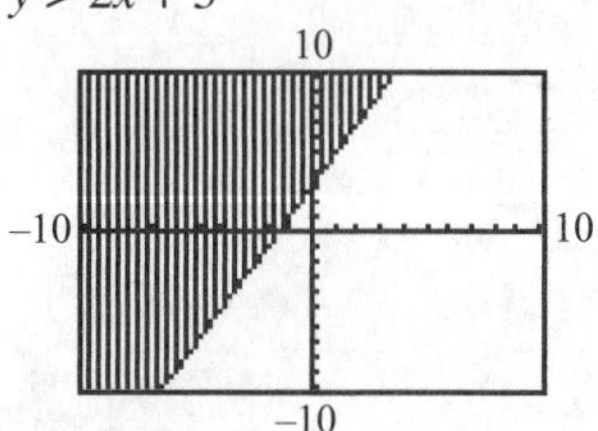

87. $y \le \frac{1}{2}x - 5$

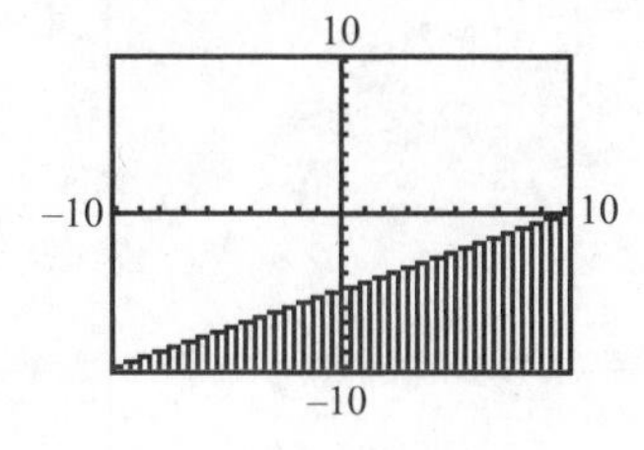

89. $3x + y \le 4$

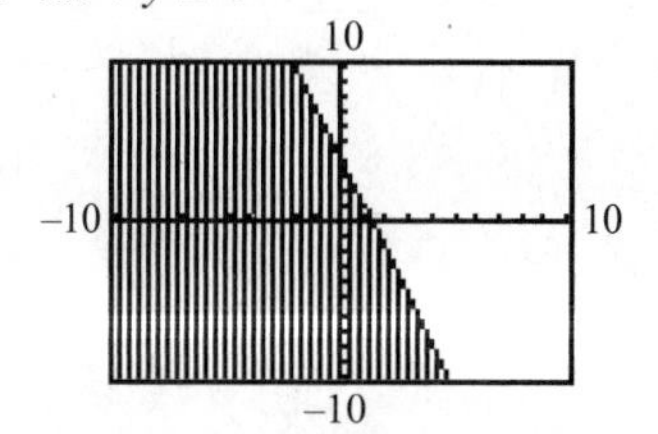

91. $2x + 5y \le -10$

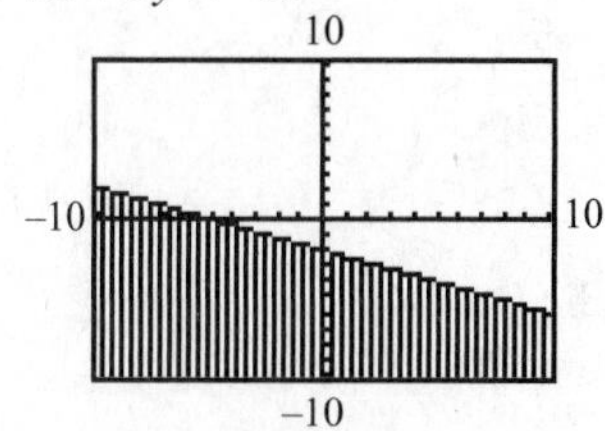

Chapter 3 Review

1. Plot $A(3, -2)$, which lies in quadrant IV.

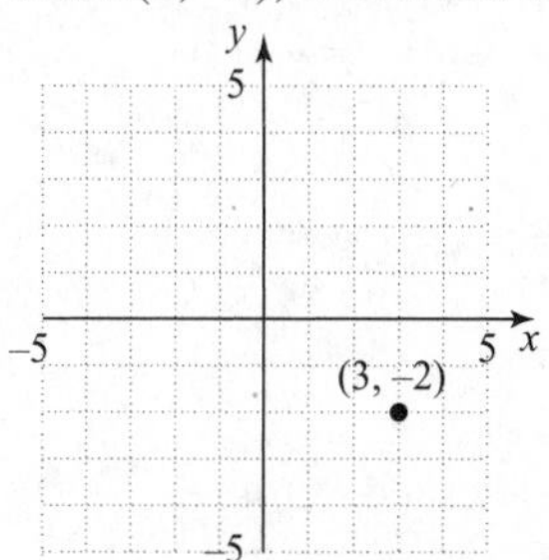

2. Plot $B(-1, -3)$, which lies in quadrant III.

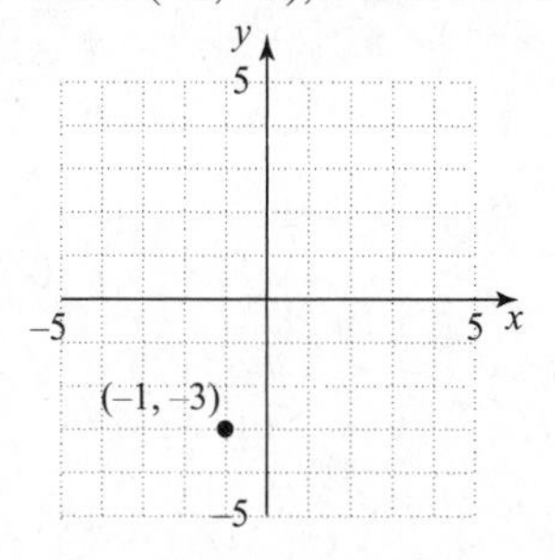

3. Plot $C(-4, 0)$, which lies on the x-axis.

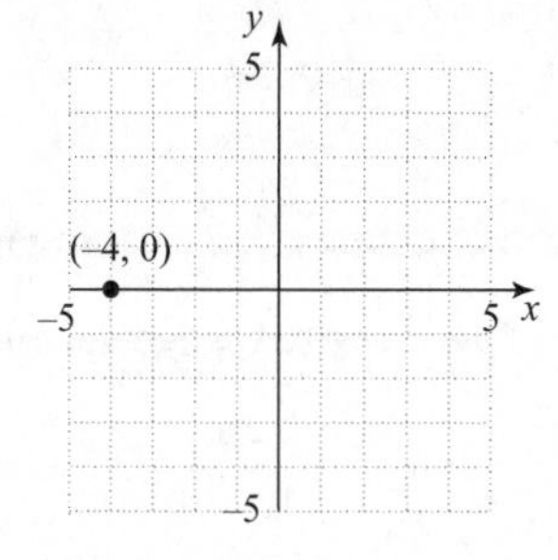

4. Plot $D(0, 2)$ which lies on the y-axis.

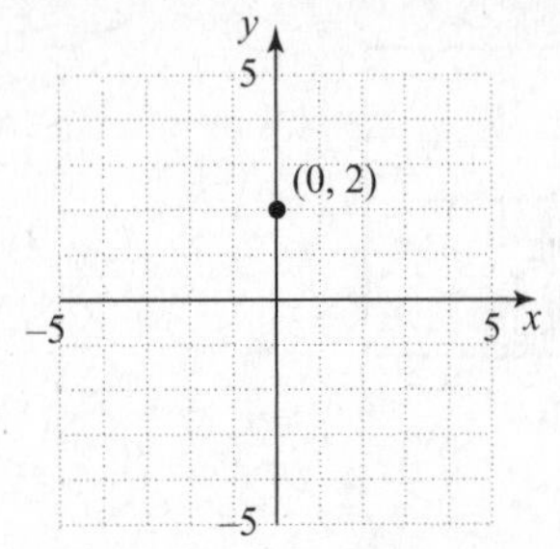

5. $A(1, 4)$ lies in quadrant I.
$B(-3, 0)$ lies on the x-axis.

6. $A(0, 1)$ lies on the y-axis.
$B(-2, 2)$ lies in quadrant II.

7. $y = 3x - 7$
$A(-1, -10)$
$-10 \stackrel{?}{=} 3(-1)-7$
$-10 \stackrel{?}{=} -3-7$
$-10 = -10$
Yes
$B(-7, 0)$
$0 \stackrel{?}{=} 3(-7)-7$
$0 \stackrel{?}{=} -21-7$
$0 \neq -28$
No

8. $4x - 3y = 2$
$A(2, 2)$
$4(2)-3(2) \stackrel{?}{=} 2$
$8-6 \stackrel{?}{=} 2$
$2 = 2$
Yes
$B(6, 5)$
$4(6)-3(5) \stackrel{?}{=} 2$
$24-15 \stackrel{?}{=} 2$
$9 \neq 2$
No

9. $-x = 3y$

(a) Let $x = 4$.
$-4 = 3y$
$-\frac{4}{3} = y$

Therefore, $\left(4, -\frac{4}{3}\right)$ satisfies the equation.

(b) Let $y = 2$.
$-x = 3(2)$
$-x = 6$
$x = -6$
Therefore, $(-6, 2)$ satisfies the equation.

10. $x - y = 0$

(a) Let $x = 4$.
$4 - y = 0$
$-y = -4$
$y = 4$
Therefore, $(4, 4)$ satisfies the equation.

(b) Let $y = -2$.
$x-(-2) = 0$
$x+2 = 0$
$x = -2$
Therefore, $(-2, -2)$ satisfies the equation.

11. $3x - 2y = 10$ or $-2y = -3x + 10$ or $y = \frac{3}{2}x - 5$

x	$y = \frac{3}{2}x - 5$	(x, y)
-2	$\frac{3}{2}(-2)-5 = -8$	$(-2, -8)$
0	$\frac{3}{2}(0)-5 = -5$	$(0, -5)$
4	$\frac{3}{2}(4)-5 = 1$	$(4, 1)$

12.

x	$y = -x + 2$	(x, y)
-3	$-(-3) + 2 = 5$	$(-3, 5)$
2	$-2 + 2 = 0$	$(2, 0)$
4	$-4 + 2 = -2$	$(4, -2)$

13. $y=-\frac{1}{3}x-4$ or $\frac{1}{3}x=-y-4$ or

$x=-3y-12$

$x=-3y-12$	$y=-\frac{1}{3}x-4$	(x, y)
$-3(2)-12=-18$	2	$(-18, 2)$
-6	$-\frac{1}{3}(-6)-4=-2$	$(-6, -2)$
$-3(-12)-12=24$	-12	$(24, -12)$

14.

$$\begin{aligned}3x-2y&=7\\-2y&=-3x+7\\y&=\frac{3}{2}x-\frac{7}{2}\end{aligned}$$

$$\begin{aligned}3x-2y&=7\\3x&=2y+7\\x&=\frac{2}{3}y+\frac{7}{3}\end{aligned}$$

$x=\frac{2}{3}y+\frac{7}{3}$	$y=\frac{3}{2}x-\frac{7}{2}$	(x, y)
$\frac{2}{3}(-8)+\frac{7}{3}=-3$	-8	$(-3, -8)$
3	$\frac{3}{2}(3)-\frac{7}{2}=1$	$(3, 1)$
$\frac{2}{3}(4)+\frac{7}{3}=5$	4	$(5, 4)$

15.

x	$C=5+2x$	(x, C)
1	$5+2(1)=7$	$(1, 7)$
2	$5+2(2)=9$	$(2, 9)$
3	$5+2(3)=11$	$(3, 11)$

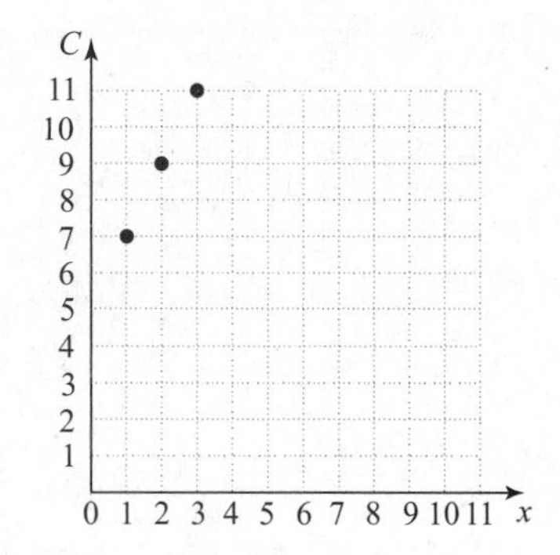

16.

x	$E = 1000 + 0.10$	(x, E)
500	$1000 + 0.10(500) = 1050$	(500, 1050)
1000	$1000 + 0.10(1000) = 1100$	(1000, 1100)
2000	$1000 + 0.10(2000) = 1200$	(2000, 1200)

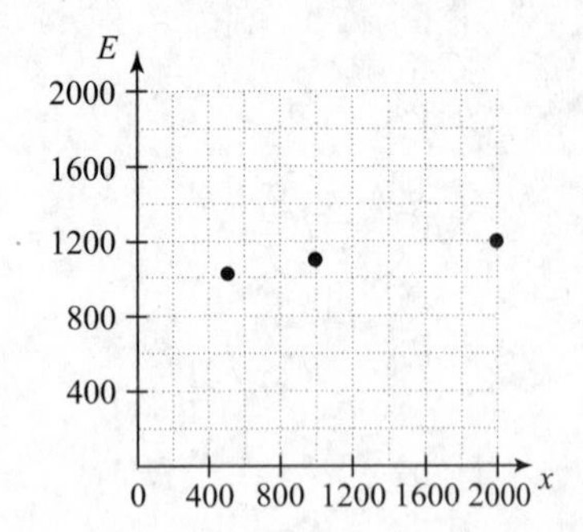

In Problems 17–20, points may vary.

17. $y = -2x$

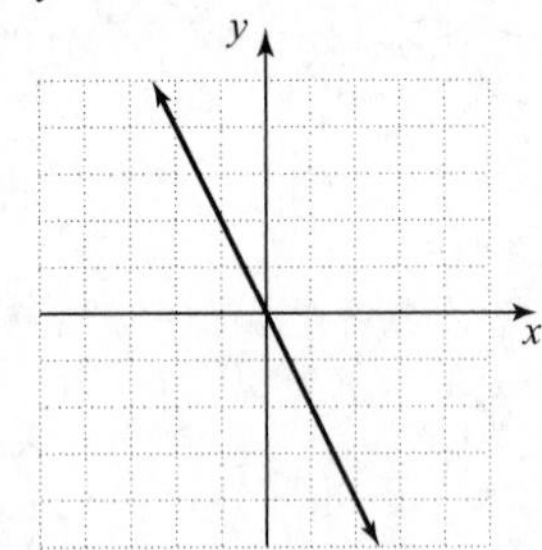

18. $y = x$

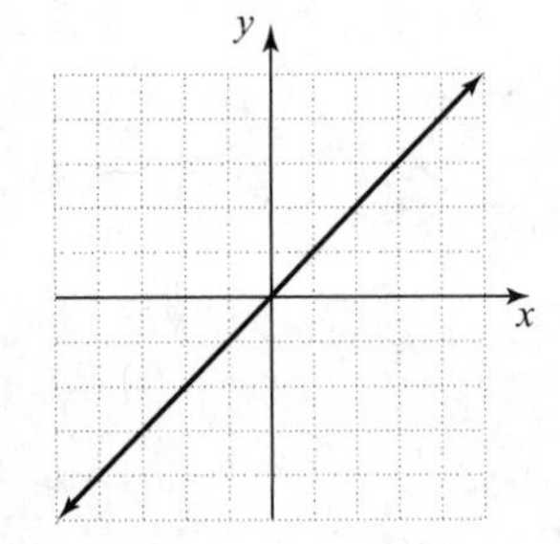

19. $4x + y = -2$

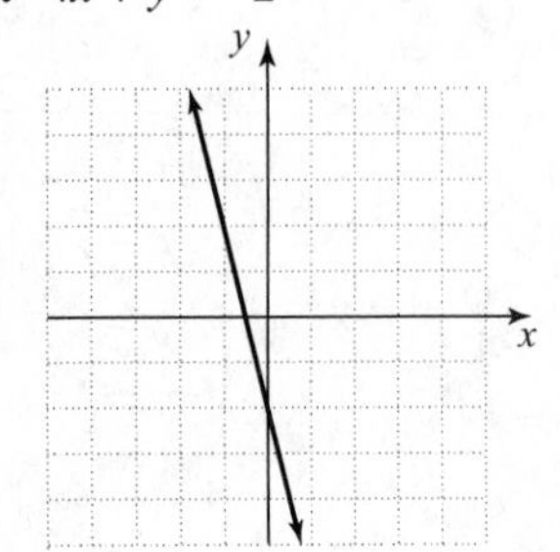

20. $3x - y = -1$

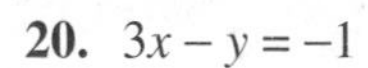

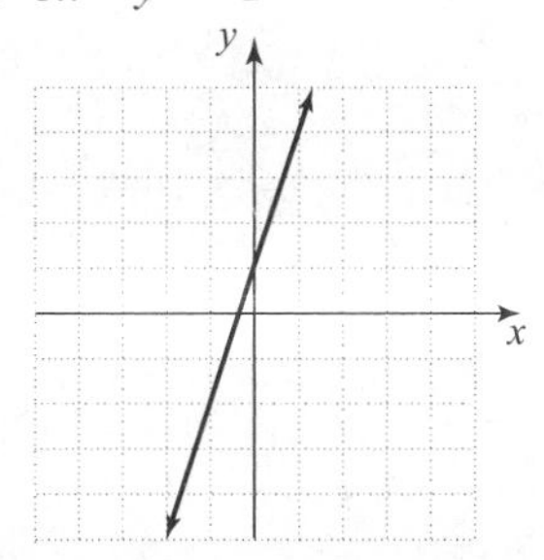

21.

p	$C = 40 + 2p$	(p, C)
20	$40 + 2(20) = 80$	(20, 80)
50	$40 + 2(50) = 140$	(50, 140)
80	$40 + 2(80) = 200$	(80, 200)

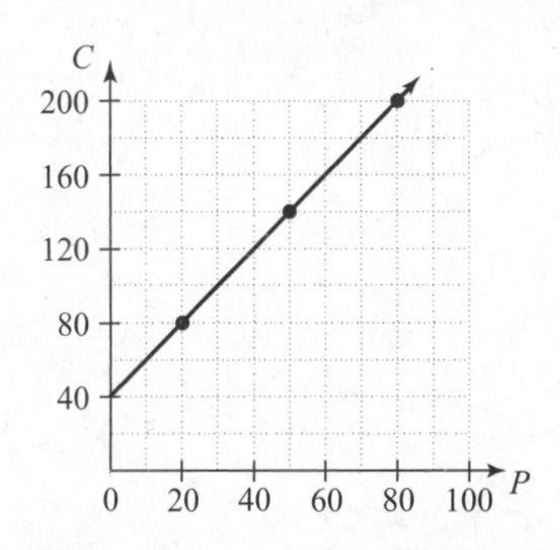

22.

p	$F = 500 + 3p$	(p, F)
100	$500 + 3(100) = 800$	(100, 800)
200	$500 + 3(200) = 1100$	(200, 1100)
500	$500 + 3(500) = 2000$	(500, 2000)

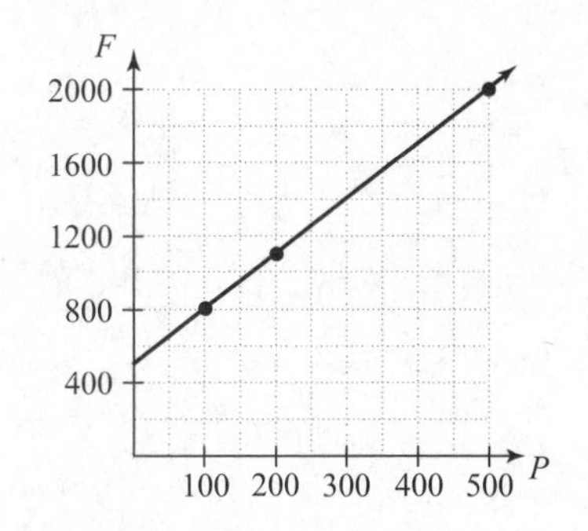

23. The graph crosses the x-axis at (–2, 0) and the y-axis at (0, –4). The intercepts are (–2, 0) and (0, –4).

24. The graph crosses the y-axis at (0, 1) and does not cross the x-axis. The only intercept is (0, 1).

25. $-3x + y = 9$
y-intercept: let $x = 0$.
$$-3(0) + y = 9$$
$$y = 9$$
x-intercept: let $y = 0$.
$$-3x + 0 = 9$$
$$-3x = 9$$
$$x = -3$$
The intercepts are (0, 9) and (−3, 0).

26. $y = 2x - 6$
y-intercept: let $x = 0$.
$$y = 2(0) - 6$$
$$y = -6$$
x-intercept: let $y = 0$.
$$0 = 2x - 6$$
$$6 = 2x$$
$$3 = x$$
The intercepts are (0, −6) and (3, 0).

27. $x = 3$ is the equation of a vertical line which has x-intercept (3, 0) and no y-intercept.

28. $2x - 5y = 2$
y-intercept: let $x = 0$.

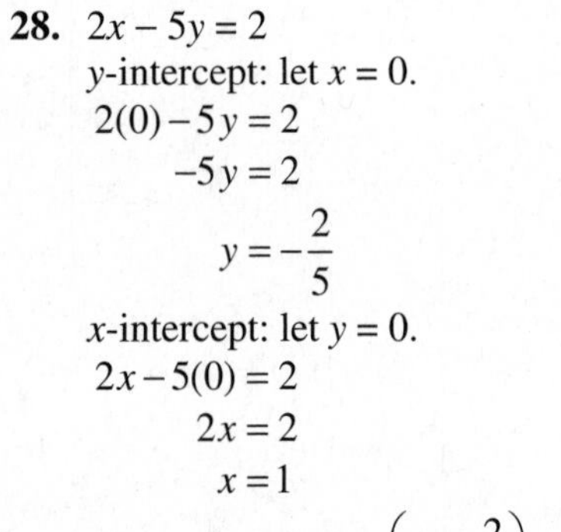

$$2(0) - 5y = 2$$
$$-5y = 2$$
$$y = -\frac{2}{5}$$
x-intercept: let $y = 0$.
$$2x - 5(0) = 2$$
$$2x = 2$$
$$x = 1$$
The intercepts are $\left(0, -\frac{2}{5}\right)$ and (1, 0).

29. $y - 3x = 3$
y-intercept: let $x = 0$.

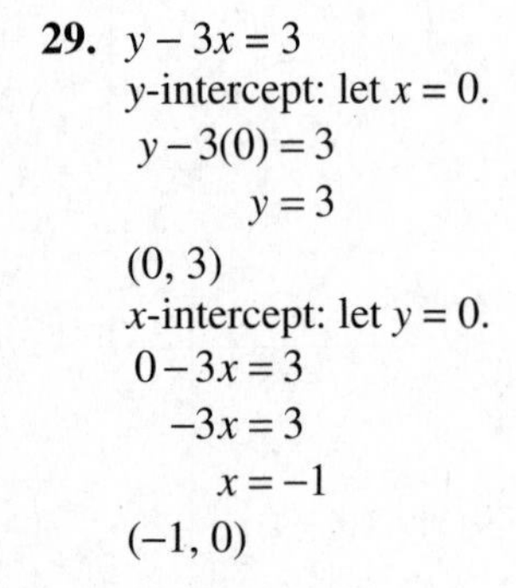

$$y - 3(0) = 3$$
$$y = 3$$
(0, 3)
x-intercept: let $y = 0$.
$$0 - 3x = 3$$
$$-3x = 3$$
$$x = -1$$
(−1, 0)

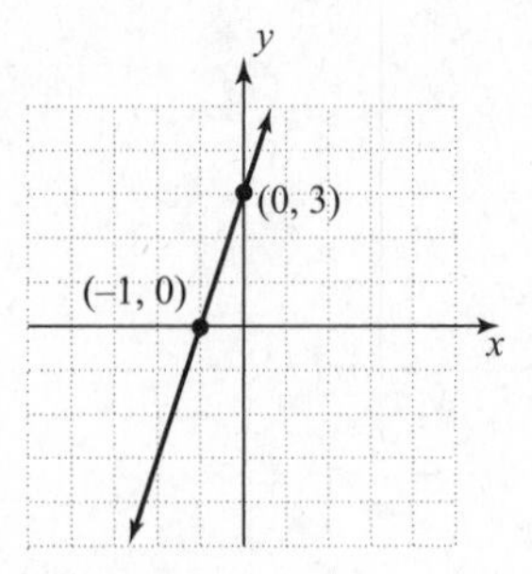

30. $2x + 5y = 0$
y-intercept: let $x = 0$.
$$2(0) + 5y = 0$$
$$5y = 0$$
$$y = 0$$
(0, 0)
x-intercept: let $y = 0$.
$$2x + 5(0) = 0$$
$$2x = 0$$
$$x = 0$$
(0, 0)

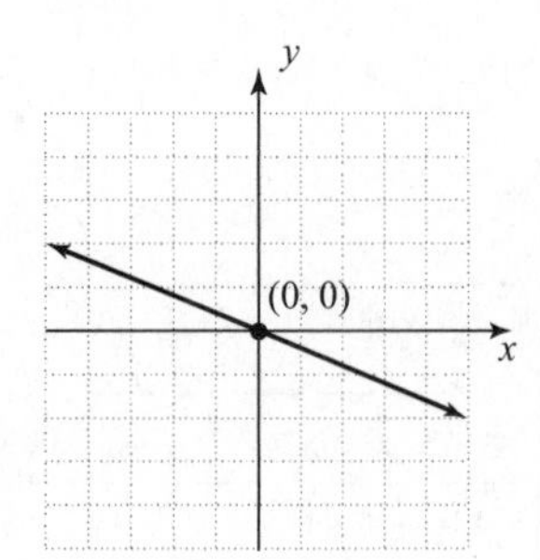

31. $\frac{x}{3} + \frac{y}{2} = 1$
y-intercept: let $x = 0$.
$$\frac{0}{3} + \frac{y}{2} = 1$$
$$\frac{y}{2} = 1$$
$$y = 2$$
(0, 2)
x-intercept: let $y = 0$.
$$\frac{x}{3} + \frac{0}{2} = 1$$
$$\frac{x}{3} = 1$$
$$x = 3$$
(3, 0)

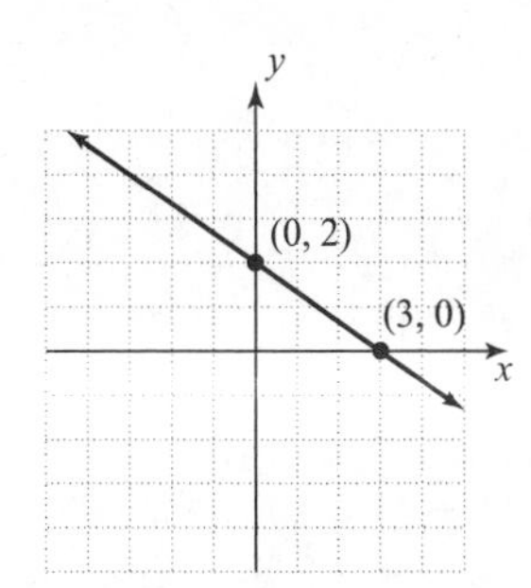

32. $y = -\frac{3}{4}x + 3$
y-intercept: let $x = 0$.
$y = -\frac{3}{4}(0) + 3$
$y = 3$
(0, 3)
x-intercept: let $y = 0$.
$0 = -\frac{3}{4}x + 3$
$-3 = -\frac{3}{4}x$
$4 = x$
(4, 0)

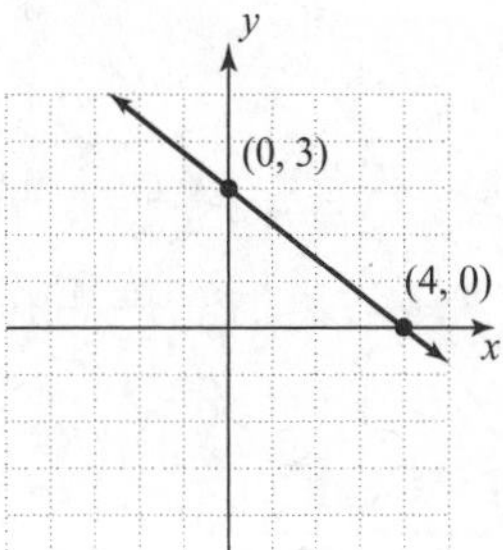

33. $x = -2$ is a vertical line with an x-intercept of (–2, 0).

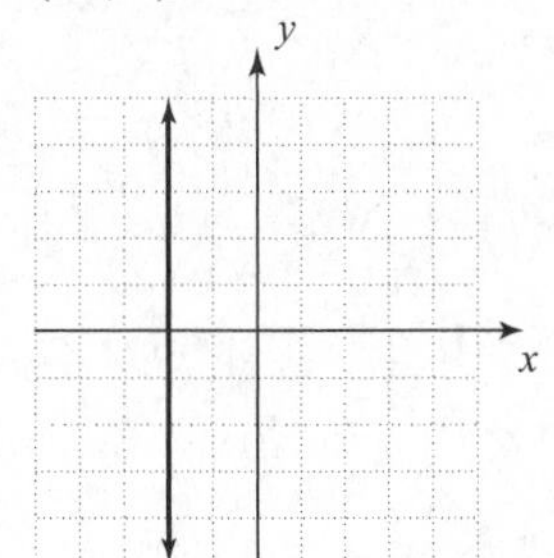

34. $y = 3$ is a horizontal line with a y-intercept of (0, 3).

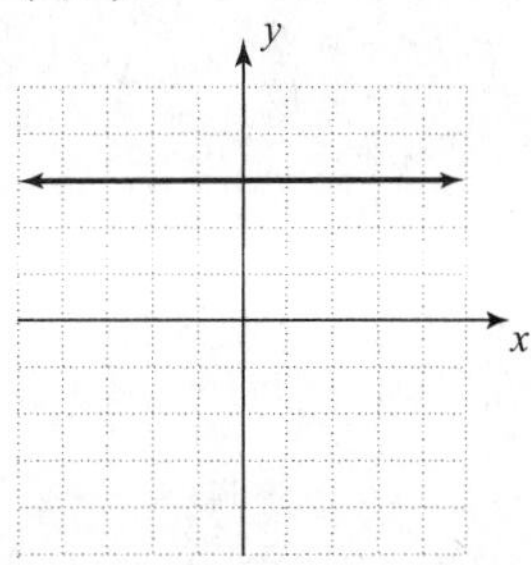

35. $y = -4$ is a horizontal line with a y-intercept of (0, –4).

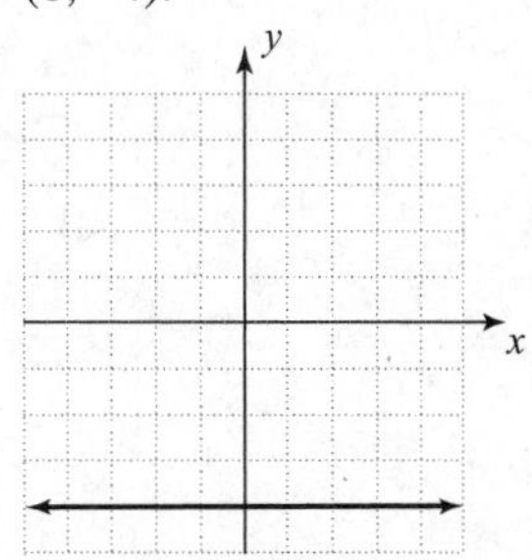

36. $x = 1$ is a vertical line with an x-intercept of (1, 0).

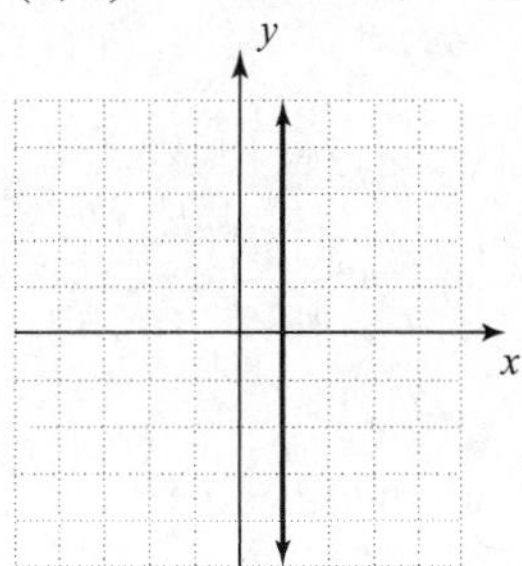

37. The line passes through the points (–3, 0) and (0, 4). Let $(-3, 0) = (x_1, y_1)$ and $(0, 4) = (x_2, y_2)$.

$$m = \frac{y_2 - y_1}{x_2 - x_1} = \frac{4 - 0}{0 - (-3)} = \frac{4}{3}$$

38. The line passes through the points (–4, 4) and (–1, –2). Let $(-4, 4) = (x_1, y_1)$ and $(-1, -2) = (x_2, y_2)$.

$$m = \frac{y_2 - y_1}{x_2 - x_1} = \frac{-2 - 4}{-1 - (-4)} = \frac{-6}{3} = -2$$

39. Let $(-4, 6) = (x_1, y_1)$ and $(-3, -2) = (x_2, y_2)$.

$$m = \frac{y_2 - y_1}{x_2 - x_1} = \frac{-2 - 6}{-3 - (-4)} = \frac{-8}{1} = -8$$

40. Let $(4, 1) = (x_1, y_1)$ and $(0, -7) = (x_2, y_2)$.

$$m = \frac{y_2 - y_1}{x_2 - x_1} = \frac{-7-1}{0-4} = \frac{-8}{-4} = 2$$

41. Let $\left(\frac{1}{2}, -\frac{3}{4}\right) = (x_1, y_1)$ and $\left(\frac{5}{2}, -\frac{1}{4}\right) = (x_2, y_2)$.

$$m = \frac{y_2 - y_1}{x_2 - x_1} = \frac{-\frac{1}{4} - \left(-\frac{3}{4}\right)}{\frac{5}{2} - \frac{1}{2}} = \frac{-\frac{1}{4} + \frac{3}{4}}{\frac{4}{2}} = \frac{\frac{1}{2}}{2} = \frac{1}{4}$$

42. Let $\left(-\frac{1}{2}, \frac{2}{3}\right) = (x_1, y_1)$ and $\left(\frac{3}{2}, \frac{1}{3}\right) = (x_2, y_2)$.

$$m = \frac{y_2 - y_1}{x_2 - x_1} = \frac{\frac{1}{3} - \frac{2}{3}}{\frac{3}{2} - \left(-\frac{1}{2}\right)} = \frac{-\frac{1}{3}}{\frac{4}{2}} = \frac{-\frac{1}{3}}{2} = -\frac{1}{6}$$

43. Let $(-3, -6) = (x_1, y_1)$ and $(-3, -10) = (x_2, y_2)$.

$$m = \frac{y_2 - y_1}{x_2 - x_1} = \frac{-10-(-6)}{-3-(-3)} = \frac{-10+6}{-3+3} = \frac{-4}{0}$$ is undefined.

44. Let $(-5, -1) = (x_1, y_1)$ and $(-1, -1) = (x_2, y_2)$.

$$m = \frac{y_2 - y_1}{x_2 - x_1} = \frac{-1-(-1)}{-1-(-5)} = \frac{-1+1}{-1+5} = \frac{0}{4} = 0$$

45. Let $\left(\frac{3}{4}, \frac{1}{2}\right) = (x_1, y_1)$ and $\left(-\frac{1}{4}, \frac{1}{2}\right) = (x_2, y_2)$.

$$m = \frac{y_2 - y_1}{x_2 - x_1} = \frac{\frac{1}{2} - \frac{1}{2}}{-\frac{1}{4} - \frac{3}{4}} = \frac{0}{-1} = 0$$

46. Let $\left(\frac{1}{3}, -\frac{3}{5}\right) = (x_1, y_1)$ and $\left(\frac{3}{9}, -\frac{1}{5}\right) = (x_2, y_2)$.

$$m = \frac{y_2 - y_1}{x_2 - x_1} = \frac{-\frac{1}{5} - \left(-\frac{3}{5}\right)}{\frac{3}{9} - \frac{1}{3}} = \frac{-\frac{1}{5} + \frac{3}{5}}{\frac{1}{3} - \frac{1}{3}} = \frac{\frac{2}{5}}{0}$$ is undefined.

47. $(-2, -3)$; $m = 4$

Plot $(-2, -3)$. Since $m = 4 = \frac{4}{1}$, there is a 4-unit increase in y for every 1-unit increase in x.

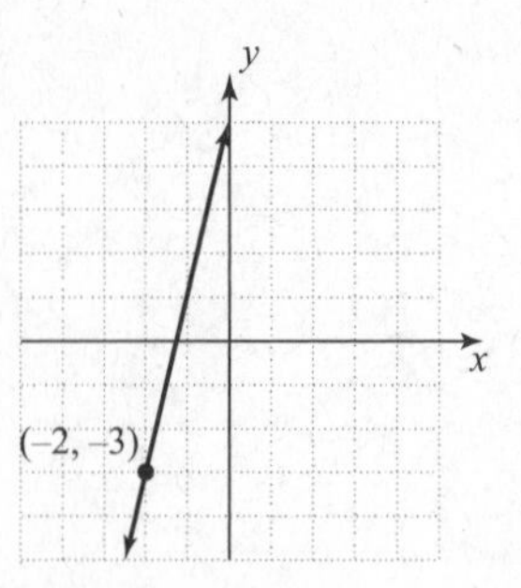

48. $(1, -3)$; $m = -2$

Plot $(1, -3)$. Since $m = -2 = \frac{-2}{1}$, there is a 2-unit decrease in y for every 1-unit increase in x.

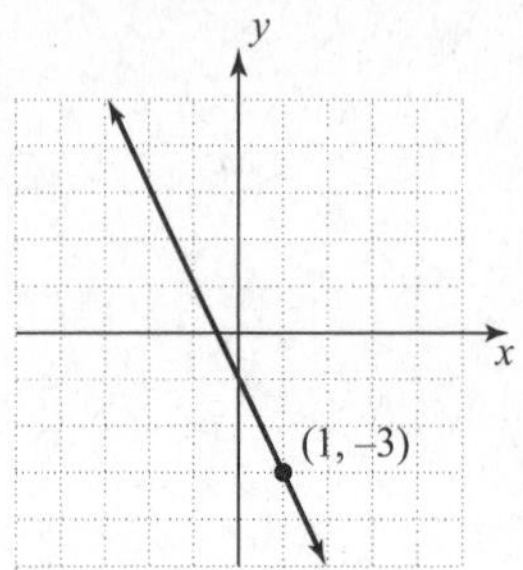

49. $(0, 1)$; $m = -\frac{2}{3}$

Plot $(0, 1)$. Since $m = -\frac{2}{3} = \frac{-2}{3}$, there is a 2-unit decrease in y for every 3-unit increase in x.

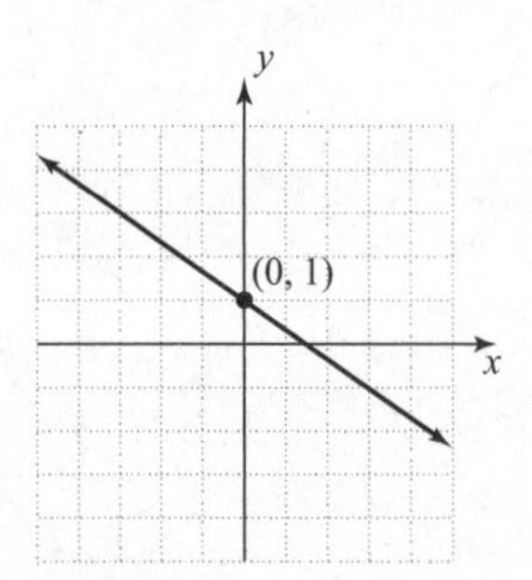

50. $(2, 3)$; $m = 0$

Plot $(2, 3)$. Since $m = 0$, the line is horizontal.

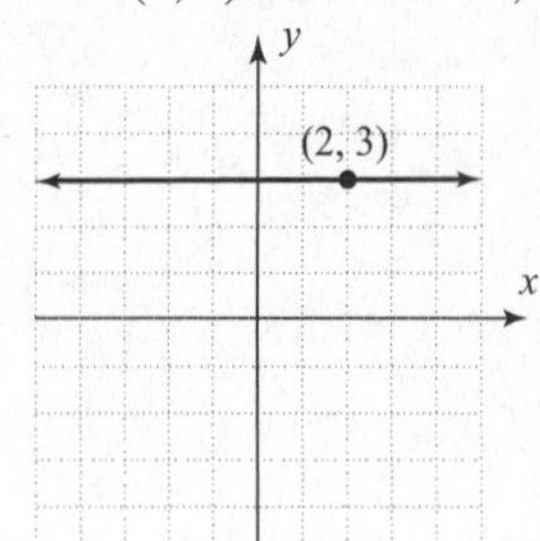

51. Let $(x_1, y_1) = (20, 1400)$ and $(x_2, y_2) = (50, 2750)$.

$$m = \frac{y_2 - y_1}{x_2 - x_1} = \frac{2750 - 1400}{50 - 20} = \frac{1350}{30} = 45$$

The cost to produce 1 additional bicycle is $45.

52. $\text{grade} = \text{slope} = \frac{\text{rise}}{\text{run}} = \frac{-5}{100} = -\frac{1}{20} = -0.05$

or 5%

53. $y = -x + \frac{1}{2}$

slope: $m = -1$

y-intercept: $b = \frac{1}{2}$

54. $y = x - \frac{3}{2}$

slope: $m = 1$

y-intercept: $b = -\frac{3}{2}$

55. $3x - 4y = -4$

$$-4y = -3x - 4$$

$$y = \frac{3}{4}x + 1$$

slope: $m = \frac{3}{4}$

y-intercept: $b = 1$

56. $2x + 5y = 8$

$$5y = -2x + 8$$

$$y = -\frac{2}{5}x + \frac{8}{5}$$

slope: $m = -\frac{2}{5}$

y-intercept: $b = \frac{8}{5}$

57. $y = \frac{1}{3}x + 1$

slope: $m = \frac{1}{3}$

y-intercept: $b = 1$

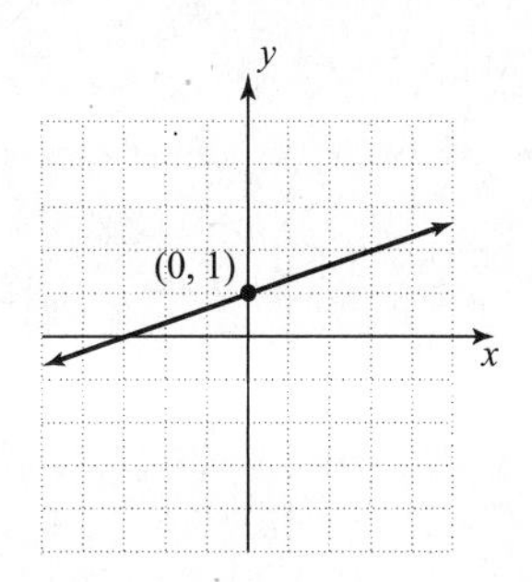

58. $y = -\frac{x}{2} - 1$

slope: $m = -\frac{1}{2}$

y-intercept: $b = -1$

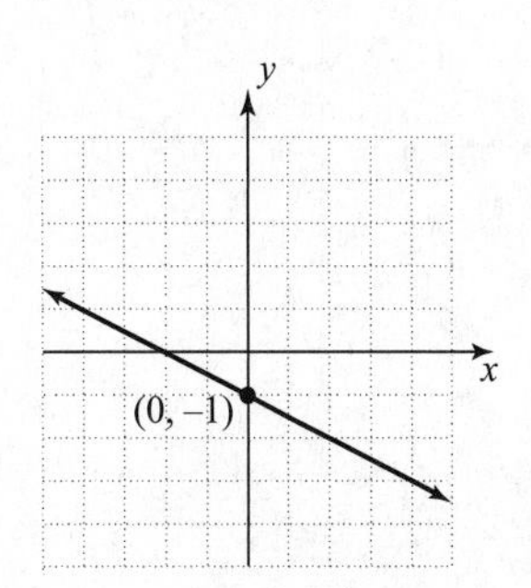

59. $y = -\frac{2x}{3} - 2$

slope: $m = -\frac{2}{3}$

y-intercept: $b = -2$

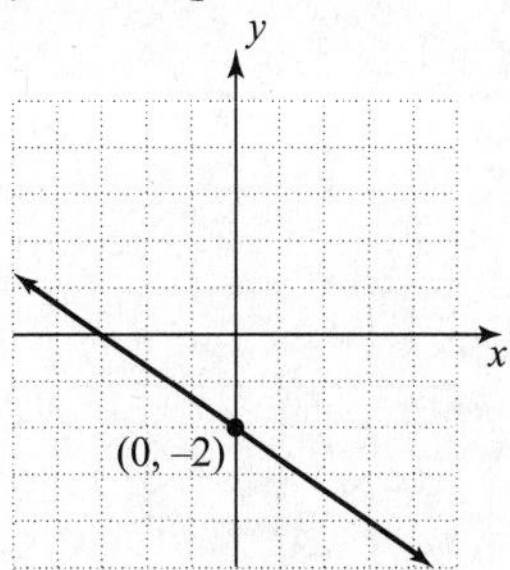

60. $y = \frac{3x}{4} + 3$

slope: $m = \frac{3}{4}$

y-intercept: $b = 3$

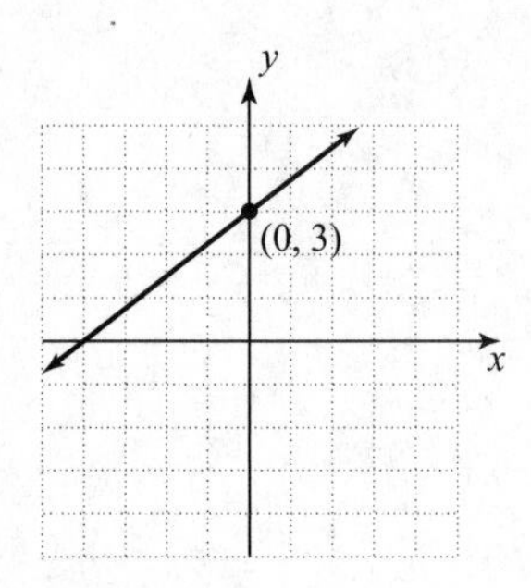

61. $y = x$
slope: $m = 1$
y-intercept: $b = 0$

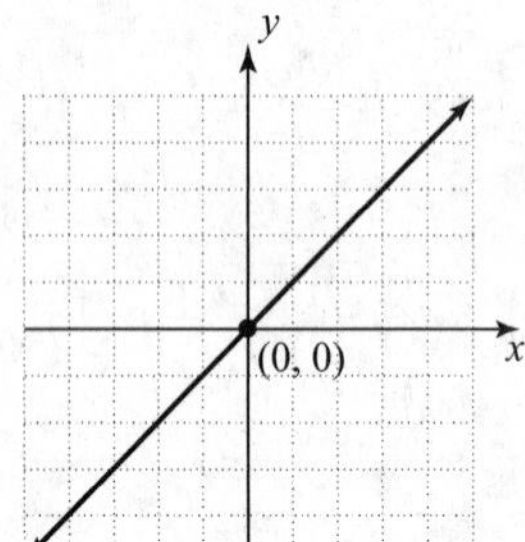

62. $y = -2x$
slope: $m = -2$
y-intercept: $b = 0$

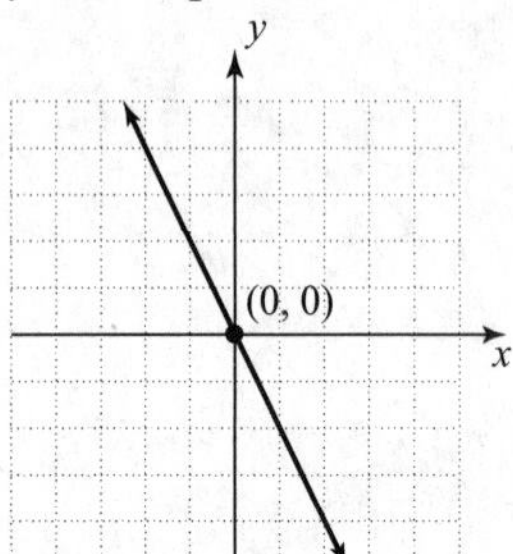

63.
$$2x - y = -4$$
$$-y = -2x - 4$$
$$y = 2x + 4$$
slope: $m = 2$
y-intercept: $b = 4$

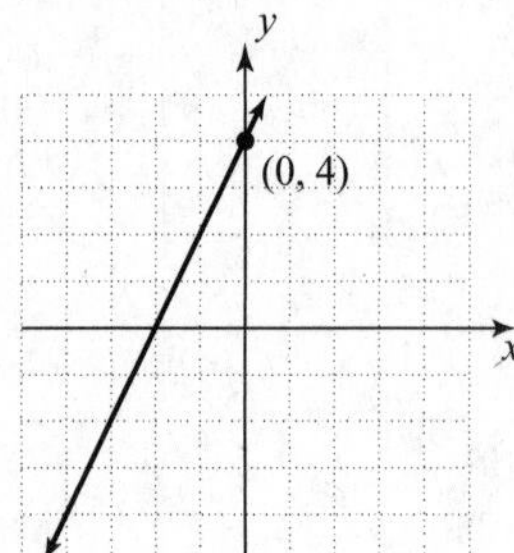

64.
$$-4x + 2y = 2$$
$$2y = 4x + 2$$
$$y = 2x + 1$$
slope: $m = 2$
y-intercept: $b = 1$

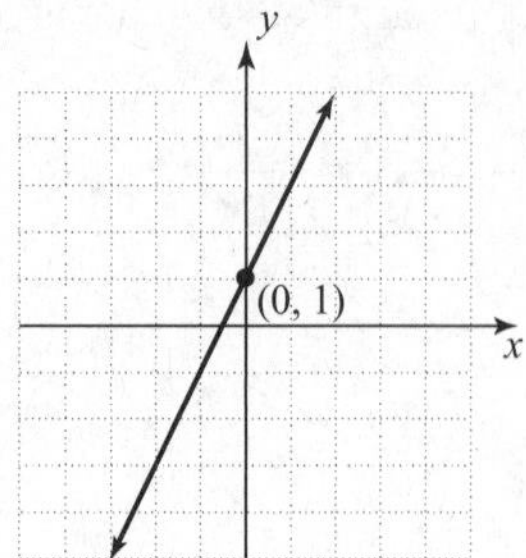

65. $m = -\frac{3}{4}$; $b = \frac{2}{3}$
$$y = mx + b$$
$$y = -\frac{3}{4}x + \frac{2}{3}$$

66. $m = \frac{1}{5}$; $b = 10$
$$y = mx + b$$
$$y = \frac{1}{5}x + 10$$

67. m undefined; x-intercept $= -12$
This is a vertical line.
$x = -12$

68. $m = 0$; $b = -4$
$y = mx + b$
$y = 0x - 4$
$y = -4$

69. $m = 1$; $b = -20$
$y = mx + b$
$y = 1x - 20$
$y = x - 20$

70. $m = -1$; $b = -8$
$y = mx + b$
$y = -1x - 8$
$y = -x - 8$

71. $C = 120 + 80d$

(a) Let $d = 3$.
$C = 120 + 80(3) = 120 + 240 = 360$
It will cost $360 to rent for 3 days.

(b) Let $C = 680$.
$$680 = 120 + 80d$$
$$560 = 80d$$
$$7 = d$$
The car was rented for 7 days.

(c)

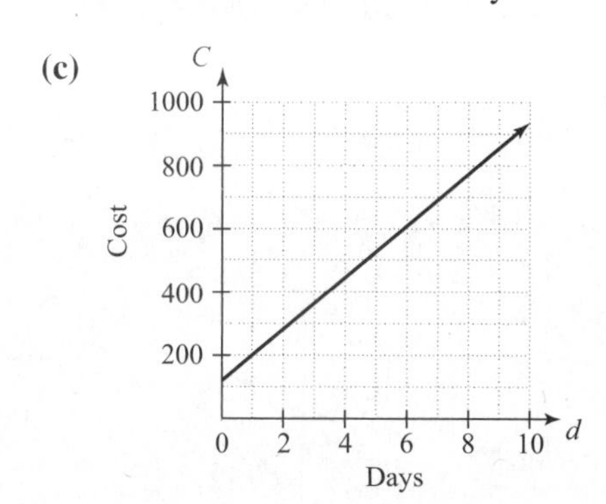

72. (a) $d = 22$ days, $C = \$418$: (22, 418)
$d = 35$ days, $C = \$665$: (35, 665)
$$m = \frac{665-418}{35-22} = \frac{247}{13} = \frac{19}{1} \text{ or } 19$$

(b) It costs $19 more for each additional day.

(c) $C = 19d$

(d) Let $d = 8$.
$C = 19(8) = 152$
It will cost $152 to rent for 8 days.

73. (0, −3); slope = 6
$$y - y_1 = m(x - x_1)$$
$$y - (-3) = 6(x - 0)$$
$$y + 3 = 6x$$
$$y = 6x - 3$$

74. (4, 0); slope = −2
$$y - y_1 = m(x - x_1)$$
$$y - 0 = -2(x - 4)$$
$$y = -2x + 8$$

75. (3, −1); slope $= -\frac{1}{2}$
$$y - y_1 = m(x - x_1)$$
$$y - (-1) = -\frac{1}{2}(x - 3)$$
$$y + 1 = -\frac{1}{2}x + \frac{3}{2}$$
$$y = -\frac{1}{2}x + \frac{1}{2}$$

76. (−1, −3); slope $= \frac{2}{3}$
$$y - y_1 = m(x - x_1)$$
$$y - (-3) = \frac{2}{3}[x - (-1)]$$
$$y + 3 = \frac{2}{3}(x + 1)$$
$$y + 3 = \frac{2}{3}x + \frac{2}{3}$$
$$y = \frac{2}{3}x - \frac{7}{3}$$

77. $\left(-\frac{4}{3}, -\frac{1}{2}\right)$; horizontal
A horizontal line has slope 0. The horizontal line through $\left(-\frac{4}{3}, -\frac{1}{2}\right)$ has equation $y = -\frac{1}{2}$.

78. $\left(-\frac{4}{7}, \frac{8}{5}\right)$; vertical
A vertical line has undefined slope. The vertical line through $\left(-\frac{4}{7}, \frac{8}{5}\right)$ has equation $x = -\frac{4}{7}$.

79. (−5, 2); slope is undefined.
A line with undefined slope is vertical. The vertical line through (−5, 2) has equation $x = -5$.

80. (6, 0); slope = 0
A line with 0 slope is horizontal. The horizontal line through (6, 0) has equation $y = 0$.

81. $(x_1, y_1) = (-7, 0)$ and $(x_2, y_2) = (0, 8)$
$$m = \frac{y_2 - y_1}{x_2 - x_1} = \frac{8-0}{0-(-7)} = \frac{8}{7}$$
$$y - y_1 = m(x - x_1)$$
$$y - 0 = \frac{8}{7}[x - (-7)]$$
$$y = \frac{8}{7}(x + 7)$$
$$y = \frac{8}{7}x + 8$$

82. $(x_1, y_1) = (0, -6)$ and $(x_2, y_2) = (4, 0)$

$$m = \frac{y_2 - y_1}{x_2 - x_1} = \frac{0-(-6)}{4-0} = \frac{6}{4} = \frac{3}{2}$$

$$y - y_1 = m(x - x_1)$$
$$y-(-6) = \frac{3}{2}(x-0)$$
$$y+6 = \frac{3}{2}x$$
$$y = \frac{3}{2}x - 6$$

83. $(x_1, y_1) = (3, 5)$ and $(x_2, y_2) = (-2, -10)$

$$m = \frac{y_2 - y_1}{x_2 - x_1} = \frac{-10-5}{-2-3} = \frac{-15}{-5} = 3$$

$$y - y_1 = m(x - x_1)$$
$$y-5 = 3(x-3)$$
$$y-5 = 3x-9$$
$$y = 3x-4$$

84. $(x_1, y_1) = (-15, 1)$ and $(x_2, y_2) = (-5, -3)$

$$m = \frac{y_2 - y_1}{x_2 - x_1} = \frac{-3-1}{-5-(-15)} = \frac{-4}{-5+15} = \frac{-4}{10} = -\frac{2}{5}$$

$$y - y_1 = m(x - x_1)$$
$$y-1 = -\frac{2}{5}[x-(-15)]$$
$$y-1 = -\frac{2}{5}(x+15)$$
$$y-1 = -\frac{2}{5}x-6$$
$$y = -\frac{2}{5}x-5$$

85. $(x_1, y_1) = (3, 12)$ and $(x_2, y_2) = (5, 4)$

$$m = \frac{y_2 - y_1}{x_2 - x_1} = \frac{4-12}{5-3} = \frac{-8}{2} = -4$$

$$y - y_1 = m(x - x_1)$$
$$y-12 = -4(x-3)$$
$$y-12 = -4x+12$$
$$y = -4x+24$$

$A = -4d + 24$

86. $(x_1, y_1) = (0, 15)$ and $(x_2, y_2) = (6, 13)$

$$m = \frac{y_2 - y_1}{x_2 - x_1} = \frac{13-15}{6-0} = \frac{-2}{6} = -\frac{1}{3}$$

$$y - y_1 = m(x - x_1)$$
$$y-15 = -\frac{1}{3}(x-0)$$
$$y-15 = -\frac{1}{3}x$$
$$y = -\frac{1}{3}x+15$$

$$F = -\frac{1}{3}m+15$$

87. $y = -\frac{1}{3}x+2$

$m_1 = -\frac{1}{3}$

$x - 3y = 3$ or $y = \frac{1}{3}x-1$

$m_2 = \frac{1}{3}$

Since $m_1 \neq m_2$, the lines are not parallel.

88. $y = \frac{1}{2}x-4$

$m_1 = \frac{1}{2}$

$x - 2y = 6$ or $y = \frac{1}{2}x-3$

$m_2 = \frac{1}{2}$

Since $m_1 = m_2$, the lines are parallel.

89. $(3, -1)$; $y = -x + 5$
The slope of the line is -1. A parallel line also has slope -1.

$$y - y_1 = m(x - x_1)$$
$$y-(-1) = -1(x-3)$$
$$y+1 = -x+3$$
$$y = -x+2$$

90. $(-2, 4)$; $y = 2x - 1$
The slope of the line is 2. A parallel line also has slope 2.
$$\begin{aligned} y - y_1 &= m(x - x_1) \\ y - 4 &= 2[x - (-2)] \\ y - 4 &= 2(x + 2) \\ y - 4 &= 2x + 4 \\ y &= 2x + 8 \end{aligned}$$

91. $(-1, 10)$; $3x + y = -7$ or $y = -3x - 7$
The slope of the line is -3. A parallel line also has slope -3.
$$\begin{aligned} y - y_1 &= m(x - x_1) \\ y - 10 &= -3[x - (-1)] \\ y - 10 &= -3(x + 1) \\ y - 10 &= -3x - 3 \\ y &= -3x + 7 \end{aligned}$$

92. $(4, -5)$; $6x + 2y = 5$ or $y = -3x + \frac{5}{2}$
The slope of the line is -3. A parallel line also has slope -3.
$$\begin{aligned} y - y_1 &= m(x - x_1) \\ y - (-5) &= -3(x - 4) \\ y + 5 &= -3x + 12 \\ y &= -3x + 7 \end{aligned}$$

93. $(5, 19)$; y-axis
The y-axis is vertical. A parallel line will also be vertical. The vertical line through $(5, 19)$ has equation $x = 5$.

94. $(-1, -12)$; x-axis
The x-axis is horizontal. A parallel line will also be horizontal. The horizontal line through $(-1, -12)$ has equation $y = -12$.

95.
$$\begin{aligned} 3x - 2y &= 5 \\ -2y &= -3x + 5 \\ y &= \frac{3}{2}x - \frac{5}{2} \end{aligned}$$
The slope of the line is $\frac{3}{2}$. The slope of a perpendicular line is $\frac{-1}{\frac{3}{2}} = -\frac{2}{3}$.

96.
$$\begin{aligned} 4x - 9y &= 1 \\ -9y &= -4x + 1 \\ y &= \frac{4}{9}x - \frac{1}{9} \end{aligned}$$
The slope of the line is $\frac{4}{9}$. The slope of a perpendicular line is $\frac{-1}{\frac{4}{9}} = -\frac{9}{4}$.

97.
$$\begin{aligned} x + 3y &= 3 \\ 3y &= -x + 3 \\ y &= -\frac{1}{3}x + 1 \end{aligned}$$
$m_1 = -\frac{1}{3}$
$y = 3x + 1$
$m_2 = 3$
Since $m_1 m_2 = \left(-\frac{1}{3}\right)(3) = -1$, the lines are perpendicular.

98.
$$\begin{aligned} 5x - 2y &= 2 \\ -2y &= -5x + 2 \\ y &= \frac{5}{2}x - 1 \end{aligned}$$
$m_1 = \frac{5}{2}$
$y = \frac{2}{5}x + 12$
$m_2 = \frac{2}{5}$
Since $m_1 m_2 = \left(\frac{5}{2}\right)\left(\frac{2}{5}\right) = 1 \neq -1$, the lines are not perpendicular.

99. $(-3, 4)$; $y = -3x + 1$
The slope of the line is -3. The slope of a perpendicular line is $\frac{-1}{-3} = \frac{1}{3}$.
$$\begin{aligned} y - y_1 &= m(x - x_1) \\ y - 4 &= \frac{1}{3}[x - (-3)] \\ y - 4 &= \frac{1}{3}(x + 3) \\ y - 4 &= \frac{1}{3}x + 1 \\ y &= \frac{1}{3}x + 5 \end{aligned}$$

100. (4, −1); $y = 2x - 1$
The slope of the line is 2. The slope of a perpendicular line is $\frac{-1}{2}$.

$$y - y_1 = m(x - x_1)$$
$$y - (-1) = -\frac{1}{2}(x - 4)$$
$$y + 1 = -\frac{1}{2}x + 2$$
$$y = -\frac{1}{2}x + 1$$

101. (1, −3); $2x - 3y = 6$
$$-3y = -2x + 6$$
$$y = \frac{2}{3}x - 2$$

The slope of the line is $\frac{2}{3}$. The slope of a perpendicular line is $\frac{-1}{\frac{2}{3}} = -\frac{3}{2}$.

$$y - y_1 = m(x - x_1)$$
$$y - (-3) = -\frac{3}{2}(x - 1)$$
$$y + 3 = -\frac{3}{2}x + \frac{3}{2}$$
$$y = -\frac{3}{2}x - \frac{3}{2}$$

102. $\left(-\frac{3}{5}, \frac{2}{5}\right)$; $x + y = -7$
$$y = -x - 7$$
The slope of the line is −1. The slope of a perpendicular line is $\frac{-1}{-1} = 1$.

$$y - y_1 = m(x - x_1)$$
$$y - \frac{2}{5} = 1\left[x - \left(-\frac{3}{5}\right)\right]$$
$$y - \frac{2}{5} = x + \frac{3}{5}$$
$$y = x + 1$$

103. $y \le 3x + 4$
$A(2, 0)$: $0 \le 3(2) + 4?$
$0 \le 6 + 4?$
$0 \le 10$ True
$B(-4, -8)$: $-8 \le 3(-4) + 4$
$-8 \le -12 + 4$
$-8 \le -8$ True
$C(7, 26)$: $26 \le 3(7) + 4?$
$26 \le 21 + 4?$
$26 \le 25$ False
$A(2, 0)$ and $B(-4, -8)$ are solutions.

104. $y > \frac{1}{3}x + 4$

$A(6, -2)$: $-2 > \frac{1}{3}(6) + 4?$
$-2 > 2 + 4?$
$-2 > 6$ False

$B(0, 4)$: $4 > \frac{1}{3}(0) + 4?$
$4 > 4$ False

$C(-18, -1)$: $-1 > \frac{1}{3}(-18) + 4?$
$-1 > -6 + 4?$
$-1 > -2$ True
$C(-18, -1)$ is a solution.

105. $y < -\frac{1}{4}x + 2$

Graph $y = -\frac{1}{4}x + 2$ with a dashed line.

Test (0, 0): $0 < -\frac{1}{4}(0) + 2?$
$0 < 2$ True
Shade the half-plane containing (0, 0).

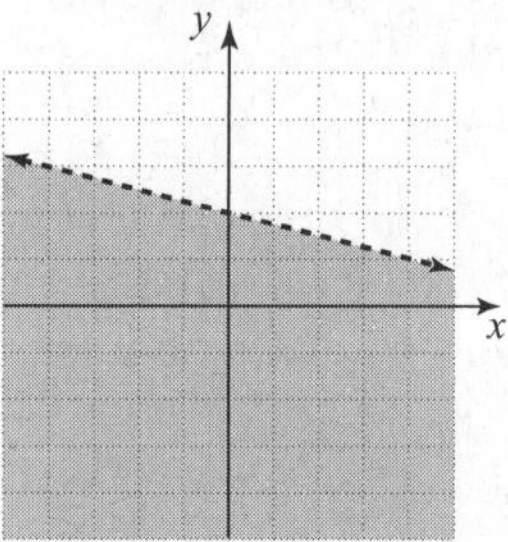

106. $y > 2x - 1$
Graph the line $y = 2x - 1$ with a dashed line.
Test (0, 0): $0 > 2(0) - 1?$
$0 > -1$ True
Shade the half-plane containing (0, 0).

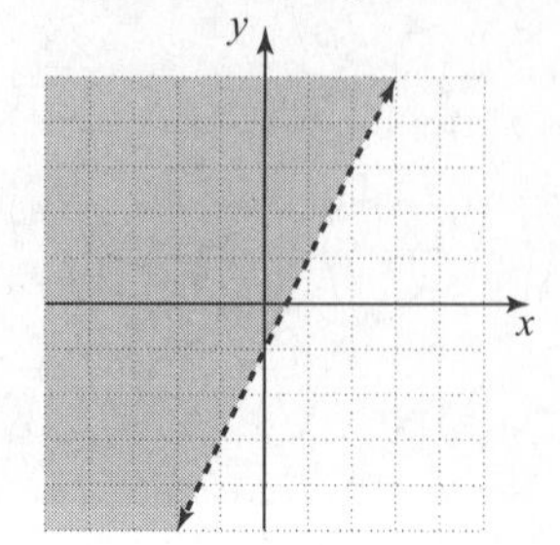

107. $3x + 2y \geq -6$
Graph the line $3x + 2y = -6$ with a solid line.
Test (0, 0): $3(0) + 2(0) \geq -6$
$0 \geq -6$ True
Shade the half-plane containing (0, 0).

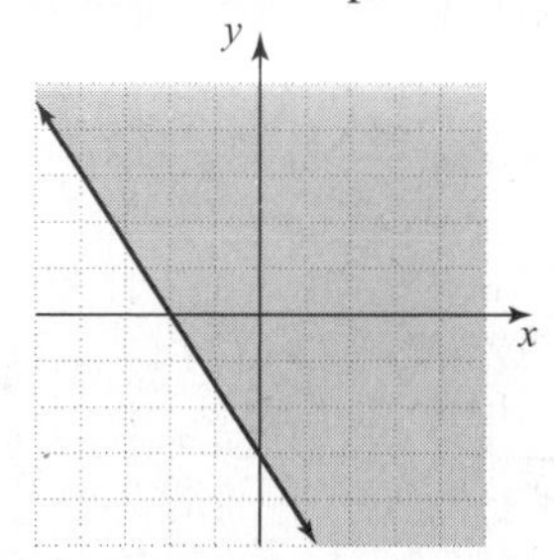

108. $-2x + y \geq 4$
Graph the line $-2x + y = 4$ with a solid line.
Test (0, 0): $-2(0) + 0 \geq 4?$
$0 \geq 4$ False
Shade the half-plane *not* containing (0, 0).

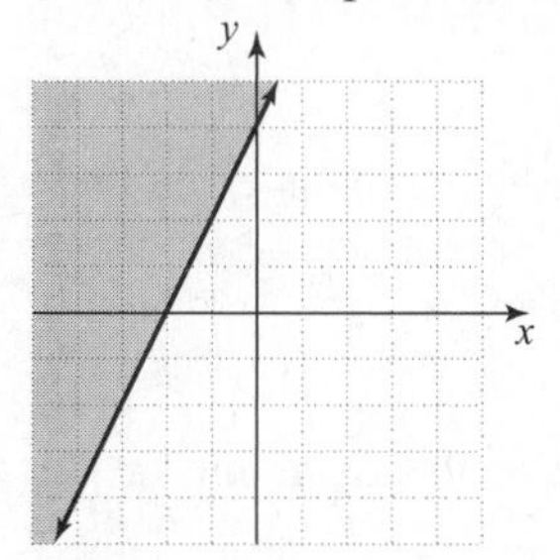

109. $x - 3y \leq 0$
Graph the line $x - 3y = 0$ with a solid line.
Test (0, 1): $0 - 3(1) \leq 0?$
$-3 \leq 0$ True
Shade the half-plane containing (0, 1).

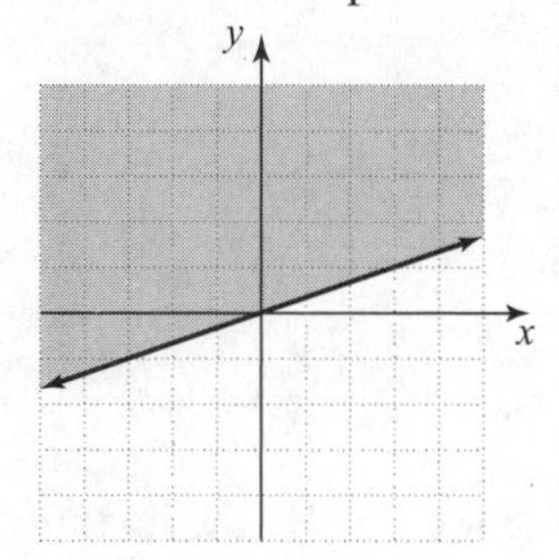

110. $x - 4y \geq 4$
Graph $x - 4y = 4$ with a solid line.
Test (0, 0): $0 - 4(0) \geq 4?$
$0 \geq 4$ False
Shade the half-plane *not* containing (0, 0).

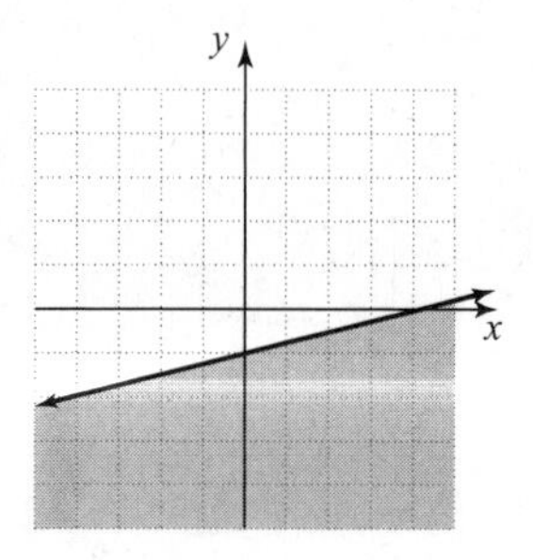

111. $x < -3$
Graph $x = -3$ with a dashed line.
Test (0, 0): $0 < -3$ False
Shade the half-plane *not* containing (0, 0).

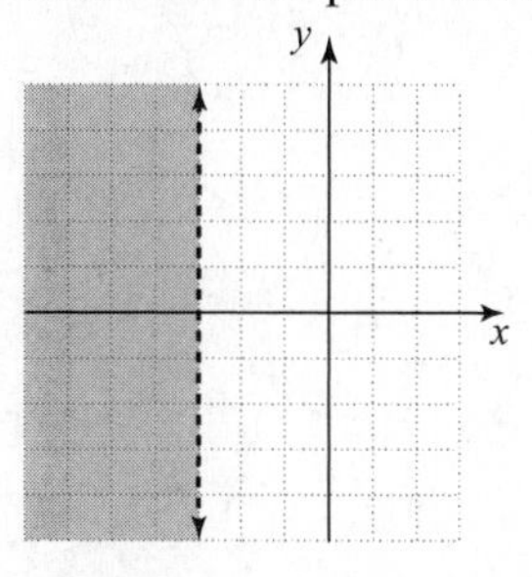

112. $y > 2$
Graph $y = 2$ with a dashed line.
Test (0, 0): $0 > 2$ False
Shade the half-plane *not* containing (0, 0).

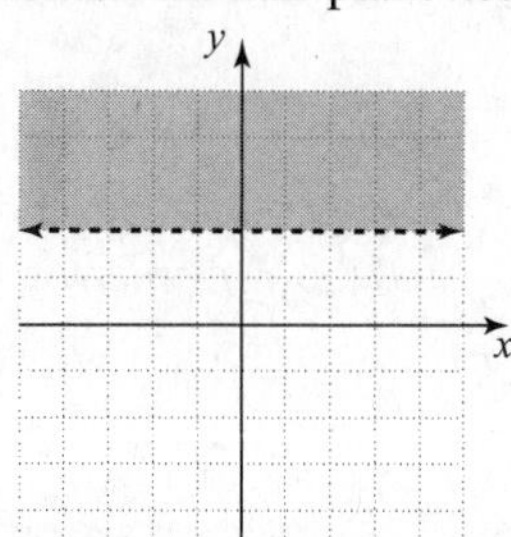

113. Each quarter is worth \$0.25, so x quarters are worth $0.25x$. Each dime is worth \$0.10, so y dimes are worth $0.1y$.
$0.25x + 0.1y \geq 12$

114. $2x - \frac{1}{2}y \leq 10$

Chapter 3 Test

1. $3x - 4y = -17$; (−3, −2)
$3(-3) - 4(-2) \stackrel{?}{=} -17$
$-9 + 8 \stackrel{?}{=} -17$
$-1 = -17$ False
(−3, −2) is not a solution to the equation.

2. $3x - 9y = 12$

(a) Let $y = 0$.
$$3x - 9(0) = 12$$
$$3x = 12$$
$$x = 4$$
The x-intercept is (4, 0).

(b) Let $x = 0$.
$$3(0) - 9y = 12$$
$$-9y = 12$$
$$y = -\frac{4}{3}$$
The y-intercept is $\left(0, -\frac{4}{3}\right)$.

3. $4x - 3y = -24$
$$-3y = -4x - 24$$
$$y = \frac{4}{3}x + 8$$
$$y = mx + b$$

(a) The slope is $m = \frac{4}{3}$.

(b) The y-intercept is $b = 8$, or (0, 8).

4. $y = -\frac{3}{4}x + 2$

The slope is $m = -\frac{3}{4}$ and the y-intercept is $b = 2$.

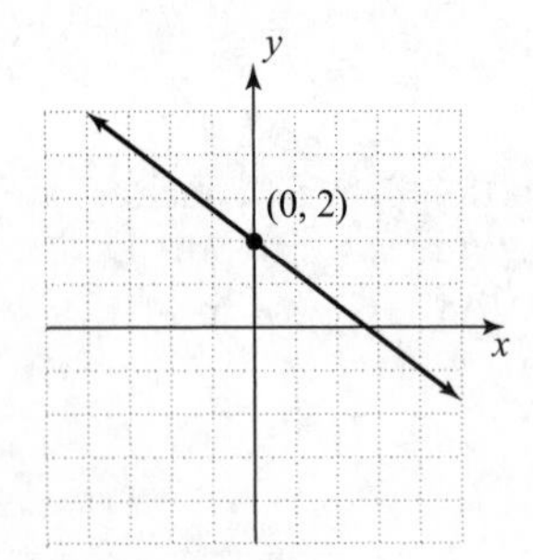

5. $3x - 6y = -12$
$$-6y = -3x - 12$$
$$y = \frac{1}{2}x + 2$$

The slope is $m = \frac{1}{2}$ and the y-intercept is $b = 2$.

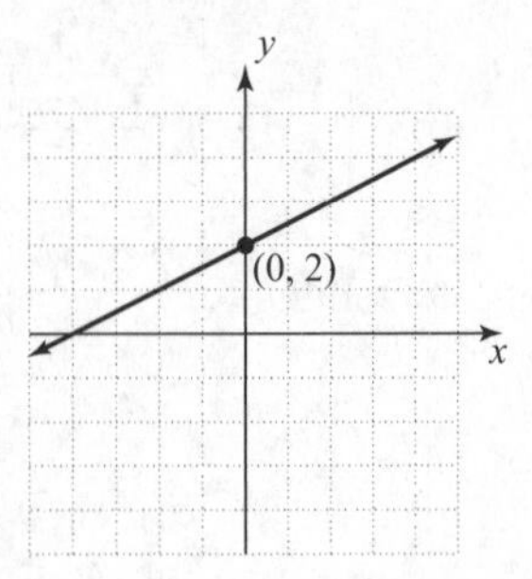

6. Let $(2, -2) = (x_1, y_1)$ and $(-4, -1) = (x_2, y_2)$.
$$m = \frac{y_2 - y_1}{x_2 - x_1} = \frac{-1-(-2)}{-4-2} = \frac{-1+2}{-6} = -\frac{1}{6}$$

7. $3y = 2x - 1$
$$y = \frac{2}{3}x - \frac{1}{3}$$
$$y = mx + b$$

The slope of the line is $m = \frac{2}{3}$.

(a) A line perpendicular to the given line has slope $\frac{-1}{\frac{2}{3}} = -\frac{3}{2}$.

(b) A line parallel to the given line has the same slope, $\frac{2}{3}$.

8. $L_1:\ 3x - 7y = 2$
$$-7y = -3x + 2$$
$$y = \frac{3}{7}x - \frac{2}{7}$$

The line has slope $m_1 = \frac{3}{7}$.

$L_2:\ y = \frac{7}{3}x + 4$

The line has slope $m_2 = \frac{7}{3}$.

Since $m_1 \neq m_2$ and $m_1 m_2 \neq -1$, the lines are neither parallel nor perpendicular.

9. slope = −4 and y-intercept is −15
$m = -4$ and $b = -15$
$y = mx + b$
$y = -4x - 15$

10. slope = 2 and contains (–3, 8)

$$y-y_1=m(x-x_1)$$
$$y-8=2[x-(-3)]$$
$$y-8=2(x+3)$$
$$y-8=2x+6$$
$$y=2x+14$$

11. $(x_1, y_1)=(-3, -2)$ and $(x_2, y_2)=(-4, 1)$

$$m=\frac{y_2-y_1}{x_2-x_1}=\frac{1-(-2)}{-4-(-3)}=\frac{1+2}{-4+3}=\frac{3}{-1}=-3$$

$$y-y_1=m(x-x_1)$$
$$y-(-2)=-3[x-(-3)]$$
$$y+2=-3(x+3)$$
$$y+2=-3x-9$$
$$y=-3x-11$$

12. $y=\frac{1}{2}x+2$ has slope $m=\frac{1}{2}$.

A parallel line also has slope $m=\frac{1}{2}$.

Let $(x_1, y_1)=(4, 0)$.

$$y-y_1=m(x-x_1)$$
$$y-0=\frac{1}{2}(x-4)$$
$$y=\frac{1}{2}x-2$$

13. $4x-6y=5$

$$-6y=-4x+5$$
$$y=\frac{2}{3}x-\frac{5}{6}$$

This line has slope $m=\frac{2}{3}$. A perpendicular line has slope $m=\frac{-1}{\frac{2}{3}}=-\frac{3}{2}$.

Let $(x_1, y_1)=(4, 2)$.

$$y-y_1=m(x-x_1)$$
$$y-2=-\frac{3}{2}(x-4)$$
$$y-2=-\frac{3}{2}x+6$$
$$y=-\frac{3}{2}x+8$$

14. A horizontal line has slope $m = 0$. The horizontal line through (3, 5) has equation $y = 5$.

15. A line with undefined slope is vertical. The vertical line through (–2, –1) has equation $x = -2$.

16. Let $(x_1, y_1)=(20, 560)$ and $(x_2, y_2)=(30, 640)$.

$$m=\frac{y_2-y_1}{x_2-x_1}=\frac{640-560}{30-20}=\frac{80}{10}=\frac{8}{1} \text{ or } 8$$

The average rate to ship a package is $8.

17. $y \geq x-3$

Graph $y = x - 3$ with a solid line.

Test (0, 0): $0 \geq 0-3$

$0 \geq -3$ True

Shade the half-plane containing (0, 0).

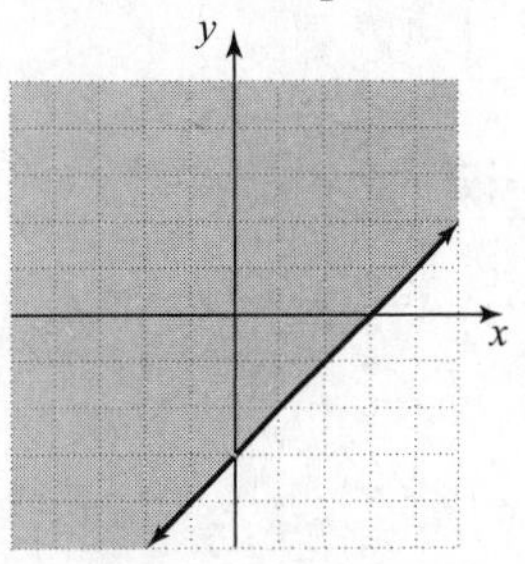

18. $-2x - 4y < 8$

Graph $-2x - 4y = 8$ with a dashed line.

Test (0, 0): $-2(0)-4(0)<8$?

$0<8$ True

Shade the half-plane containing (0, 0).

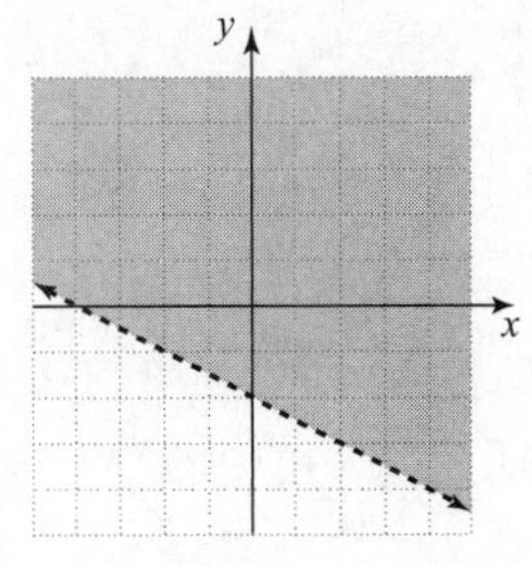

19. $x \leq -4$

Graph $x = -4$ with a solid line.

Test (0, 0): $0 \leq -4$ False

Shade the half-plane *not* containing (0, 0).

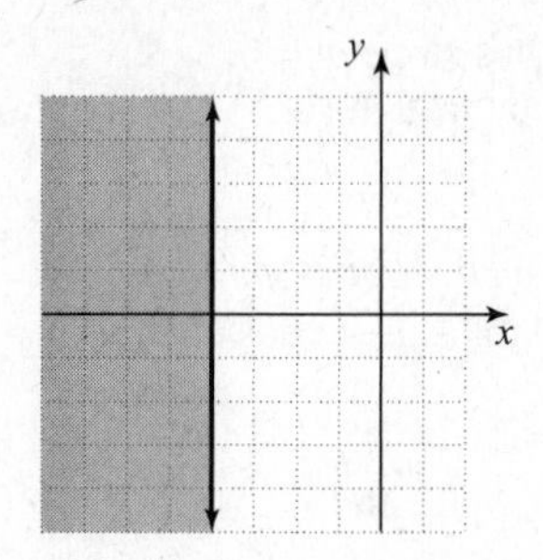

Cumulative Review Chapters 1–3

1. $200 \div 25 \cdot (-2) = 8 \cdot (-2) = -16$

2. $\frac{3}{4} + \frac{1}{6} - \frac{2}{3} = \frac{9}{12} + \frac{2}{12} - \frac{8}{12} = \frac{9+2-8}{12} = \frac{3}{12} = \frac{1}{4}$

3. $$\frac{8-3(5-3^2)}{7-2\cdot 6} = \frac{8-3(5-9)}{7-12} = \frac{8-3(-4)}{-5} = \frac{8+12}{-5} = \frac{20}{-5} = -4$$

4. $(-3)^3 + 3(-3)^2 - 5(-3) - 7$
$= -27 + 3(9) - 5(-3) - 7$
$= -27 + 27 + 15 - 7$
$= 0 + 15 - 7$
$= 15 - 7$
$= 8$

5. $8m - 5m^2 - 3 + 9m^2 - 3m - 6$
$= -5m^2 + 9m^2 + 8m - 3m - 3 - 6$
$= 4m^2 + 5m - 9$

6. $8(n+2) - 7 = 6n - 5$
$8n + 16 - 7 = 6n - 5$
$8n + 9 = 6n - 5$
$8n = 6n - 14$
$2n = -14$
$n = -7$
The solution set is $\{-7\}$.

7. $\frac{2}{5}x + \frac{1}{6} = -\frac{2}{3}$
$30\left(\frac{2}{5}x + \frac{1}{6}\right) = 30\left(-\frac{2}{3}\right)$
$12x + 5 = -20$
$12x = -25$
$x = \frac{-25}{12}$
The solution set is $\left\{-\frac{25}{12}\right\}$.

8. $A = \frac{1}{2}h(b+B)$
$2(A) = 2\left(\frac{1}{2}h(b+B)\right)$
$2A = h(b+B)$
$2A = hb + hB$
$2A - hb = hB$
$\frac{2A - hb}{h} = B$
$B = \frac{2A - hb}{h}$ or $B = \frac{2A}{h} - b$

9. $6x - 7 > -31$
$6x > -24$
$x > -4$
$\{x \mid x > -4\}$ or $(-4, \infty)$

−8 −6 −4 −2 0 2 4 6 8

10. $5(x-3) \ge 7(x-4) + 3$
$5x - 15 \ge 7x - 28 + 3$
$5x - 15 \ge 7x - 25$
$5x \ge 7x - 10$
$-2x \ge -10$
$x \le 5$
$\{x \mid x \le 5\}$ or $(-\infty, 5]$

−8 −6 −4 −2 0 2 4 5 6 8

11.

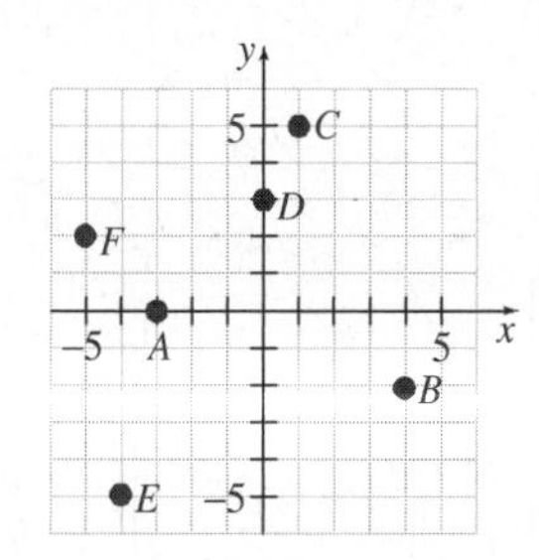

12. For $y=-\frac{1}{2}x+4$, the slope is $-\frac{1}{2}$ and the y-intercept is 4. Begin at (0, 4) and move to the right 2 units and down 1 units to find the point (2, 3). We can also move 2 units to the left and 1 units up to find the point (–2, 5).

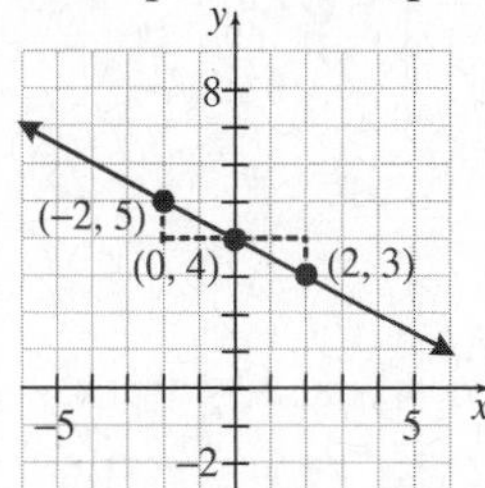

13. $4x-5y=15$

$$-5y=-4x+15$$
$$y=\frac{-4x+15}{-5}$$
$$y=\frac{4}{5}x-3$$

The slope is $\frac{4}{5}$ and the y-intercept is –3. Begin at the point (0, –3) and move to the right 5 units and up 4 units to find the point (5, 1).

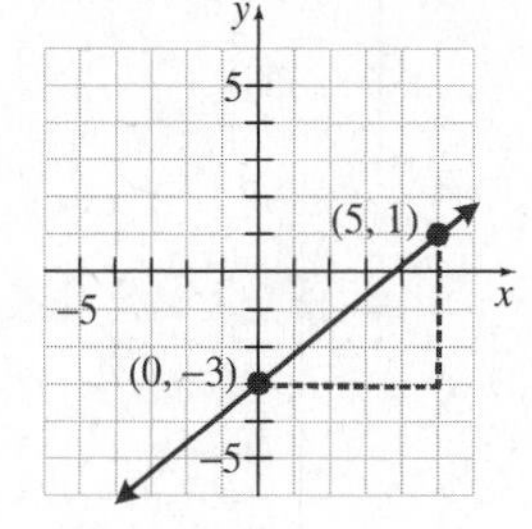

14. $m=\frac{y_2-y_1}{x_2-x_1}=\frac{10-(-2)}{-6-3}=\frac{12}{-9}=-\frac{4}{3}$

$$y-y_1=m(x-x_1)$$
$$y-10=-\frac{4}{3}(x-(-6))$$
$$y-10=-\frac{4}{3}(x+6)$$
$$y-10=-\frac{4}{3}x-8$$
$$y=-\frac{4}{3}x+2 \text{ or } 4x+3y=6$$

15. The slope of the line we seek is $m=-3$, the same as the slope of the line $y=-3x+10$. Thus, the equation of the line we seek is:

$$y-7=-3(x-(-5))$$
$$y-7=-3(x+5)$$
$$y-7=-3x-15$$
$$y=-3x-8 \text{ or } 3x+y=-8$$

16. Replace the inequality symbol with an equal sign to obtain $x-3y=12$. Because the inequality is strict, graph $x-3y=12$ $\left(y=\frac{1}{3}x-4\right)$ using a dashed line.

Test Point: $(0, 0)$: $0-3(0)\overset{?}{>}12$

$$0\overset{?}{>}12 \quad \text{False}$$

Therefore, $(0, 0)$ is a not a solution to $x-3y>12$. Shade the half-plane that does not contain $(0, 0)$.

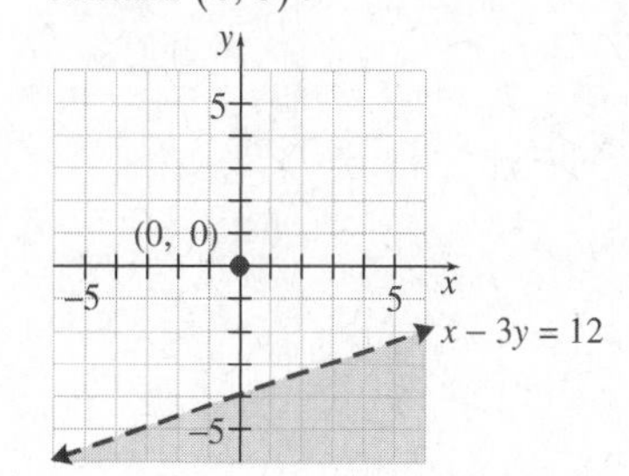

17. Let x = score on final exam.

$$93 \le \frac{94+95+90+97+2x}{6} \le 100$$
$$93 \le \frac{376+2x}{6} \le 100$$
$$6(93) \le 6\left(\frac{376+2x}{6}\right) \le 6(100)$$
$$558 \le 376+2x \le 600$$
$$558-376 \le 376+2x-376 \le 600-376$$
$$182 \le 2x \le 224$$
$$\frac{182}{2} \le \frac{2x}{2} \le \frac{224}{2}$$
$$91 \le x \le 112$$

Shawn needs to score at least 91 on the final exam to earn an A (assuming the maximum score on the exam is 100).

18. Let x = weight in pounds.

$$0.2x-2 \ge 30$$
$$0.2x-2+2 \ge 30+2$$
$$0.2x \ge 32$$
$$\frac{0.2x}{0.2} \ge \frac{32}{0.2}$$
$$x \ge 160$$

A person 62 inches tall would be considered obese if they weighed 160 pounds or more.

19. Let x = measure of the smaller angle.
Larger angle: $15+2x$

$$x+(15+2x)=180$$
$$3x+15=180$$
$$3x+15-15=180-15$$
$$3x=165$$
$$\frac{3x}{3}=\frac{165}{3}$$
$$x=55$$

The angles measure $55°$ and $125°$.

20. Let h = height of cylinder in inches.

$$S=2\pi r^2+2\pi rh$$
$$100=2\pi(2)^2+2\pi(2)h$$
$$100=8\pi+4\pi h$$
$$100-8\pi=4\pi h$$
$$h=\frac{100-8\pi}{4\pi}\approx 5.96$$

The cylinder should be about 5.96 inches tall.

21. Let x = the first even integer.
$x+2$ = the second even integer:
$x+4$ = the third even integer:

$$x+(x+2)=22+(x+4)$$
$$2x+2=x+26$$
$$2x+2-x=x+26-x$$
$$x+2=26$$
$$x+2-2=26-2$$
$$x=24$$

The three consecutive even integers are 24, 26, and 28.

Chapter 4

Section 4.1

Preparing for Solving Systems of Linear Equations by Graphing

P1. $y = 2x - 3$
The slope is $m = 2$ and the y-intercept is $b = -3$. Plot the y-intercept $(0, -3)$. Use the slope $m = \frac{2}{1} = \frac{\text{rise}}{\text{run}}$ to find a second point on the graph.

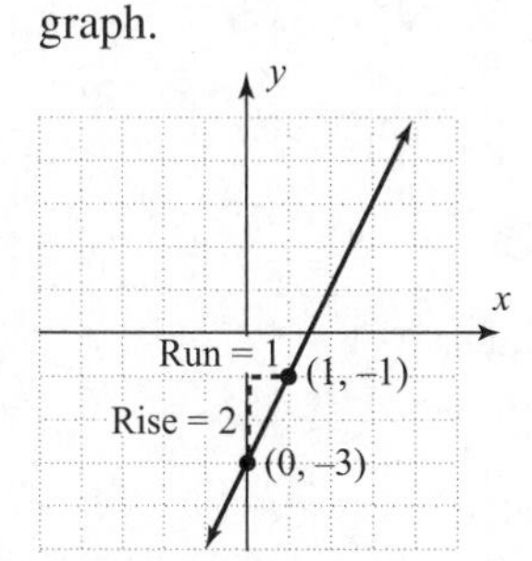

P2. $3x + 4y = 12$
x-intercept, let $y = 0$:

$$3x + 4(0) = 12$$
$$3x = 12$$
$$x = 4$$

y-intercept, let $x = 0$:

$$3(0) + 4y = 12$$
$$4y = 12$$
$$y = 3$$

Additional point: Let $x = 2$.

$$3(2) + 4y = 12$$
$$6 + 4y = 12$$
$$4y = 6$$
$$y = \frac{6}{4} = \frac{3}{2}$$

Plot the points $(4, 0)$, $(0, 3)$, and $\left(2, \frac{3}{2}\right)$.

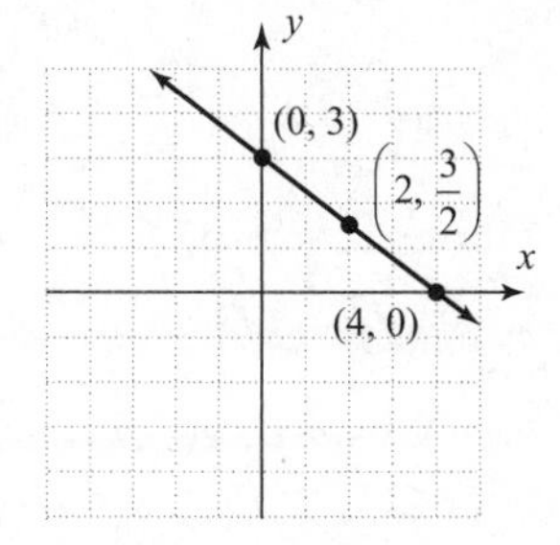

P3. $2x + 6y = 12$

$$6y = -2x + 12$$
$$y = -\frac{1}{3}x + 2$$

The slope is $-\frac{1}{3}$ and the y-intercept is 2.

$$-3x - 9y = 18$$
$$-9y = 3x + 18$$
$$y = -\frac{1}{3}x - 2$$

The slope is $-\frac{1}{3}$ and the y-intercept is -2.

Because the lines have the same slope, $-\frac{1}{3}$, but different y-intercepts, they are parallel.

4.1 Quick Checks

1. A system of linear equations is a grouping of two or more linear equations, each of which contains one or more variables.

2. A solution of a system of equations consists of values of the variables that satisfy each equation of the system.

3. $\begin{cases} 2x + 3y = 7 & (1) \\ 3x + y = -7 & (2) \end{cases}$

(a) $(3, 1)$

$$(1):\ 2(3) + 3(1) \stackrel{?}{=} 7$$
$$6 + 3 \stackrel{?}{=} 7$$
$$9 = 7 \quad \text{False}$$

$(3, 1)$ is not a solution.

(b) $(-4, 5)$

$$(1):\ 2(-4) + 3(5) \stackrel{?}{=} 7$$
$$-8 + 15 \stackrel{?}{=} 7$$
$$7 = 7 \quad \text{True}$$

$$(2):\ 3(-4) + 5 \stackrel{?}{=} -7$$
$$-12 + 5 \stackrel{?}{=} -7$$
$$-7 = -7 \quad \text{True}$$

$(-4, 5)$ is a solution.

(c) $(-2, -1)$

(1): $2(-2)+3(-1) \stackrel{?}{=} 7$

$-4-3 \stackrel{?}{=} 7$

$-7=7$ False

$(-2, -1)$ is not a solution.

4. $\begin{cases} 3x-6y=6 & (1) \\ -2x+4y=-4 & (2) \end{cases}$

(a) $(2, 0)$

(1): $3(2)-6(0) \stackrel{?}{=} 6$

$6-0 \stackrel{?}{=} 6$

$6=6$ True

(2): $-2(2)+4(0) \stackrel{?}{=} -4$

$-4+0 \stackrel{?}{=} -4$

$-4=-4$ True

$(2, 0)$ is a solution.

(b) $(0, -1)$

(1): $3(0)-6(-1) \stackrel{?}{=} 6$

$0+6 \stackrel{?}{=} 6$

$6=6$ True

(2): $-2(0)+4(-1) \stackrel{?}{=} -4$

$0-4 \stackrel{?}{=} -4$

$-4=-4$ True

$(0, -1)$ is a solution.

(c) $(4, 1)$

(1): $3(4)-6(1) \stackrel{?}{=} 6$

$12-6 \stackrel{?}{=} 6$

$6=6$ True

(2): $-2(4)+4(1) \stackrel{?}{=} -4$

$-8+4 \stackrel{?}{=} -4$

$-4=-4$ True

$(4, 1)$ is a solution.

5. $\begin{cases} y=-2x+9 & (1) \\ y=3x-11 & (2) \end{cases}$

We graph $y = -2x + 9$ using slope -2 and y-intercept 9. We graph $y = 3x - 11$ using slope 3 and y-intercept -11.

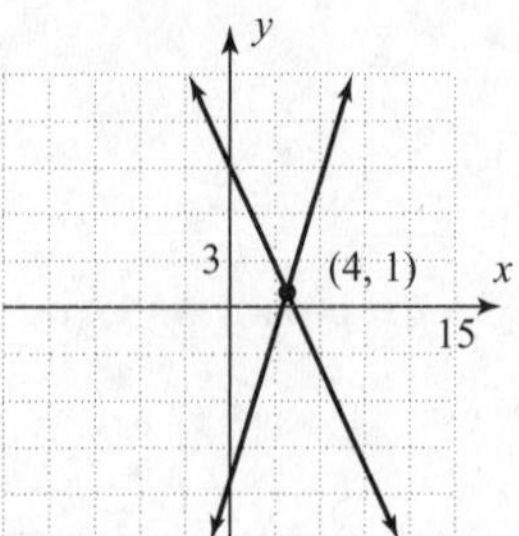

The point of intersection is $(4, 1)$.

Check:

(1): $1 \stackrel{?}{=} -2(4)+9$

$1 \stackrel{?}{=} -8+9$

$1=1$ True

(2): $1 \stackrel{?}{=} 3(4)-11$

$1 \stackrel{?}{=} 12-11$

$1=1$ True

The solution is $(4, 1)$.

6. $\begin{cases} 4x+y=-3 & (1) \\ 3x-y=-11 & (2) \end{cases}$

(1): $4x+y=-3$

$y=-4x-3$

Graph using slope -4 and y-intercept -3.

(2): $3x-y=-11$

$-y=-3x-11$

$y=3x+11$

Graph using slope 3 and y-intercept 11.

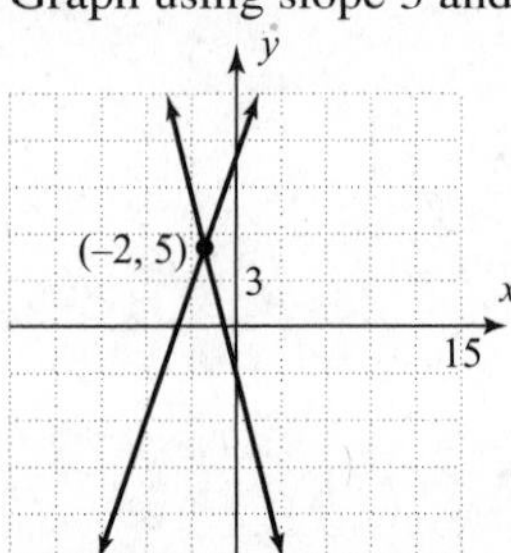

The intersection point is $(-2, 5)$.

Check:

(1): $4(-2)+5 \stackrel{?}{=} -3$

$-8+5 \stackrel{?}{=} -3$

$-3=-3$ True

(2): $3(-2)-5 \stackrel{?}{=} -11$

$-6-11 \stackrel{?}{=} -11$

$-11 \stackrel{?}{=} -11$ True

The solution is $(-2, 5)$.

7. $\begin{cases} y = 2x + 4 & (1) \\ 2x - y = 1 & (2) \end{cases}$

Graph (1) using slope 2 and y-intercept 4. Graph (2) using intercepts $\left(\frac{1}{2}, 0\right)$ and (0, –1).

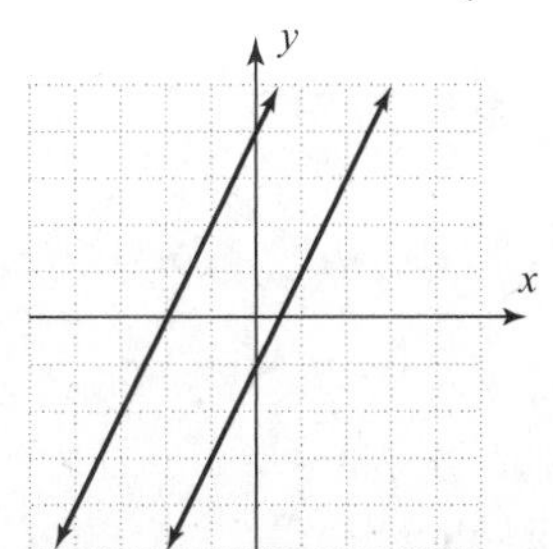

The lines are parallel and do not intersect. The system of equations has no solution or the solution set is ∅.

8. $\begin{cases} y = 3x + 2 & (1) \\ -6x + 2y = 4 & (2) \end{cases}$

Graph (1) using slope 3 and y-intercept 2.

(2): $-6x + 2y = 4$

$$2y = 6x + 4$$
$$y = 3x + 2$$

Graph using slope 3 and y-intercept 2.

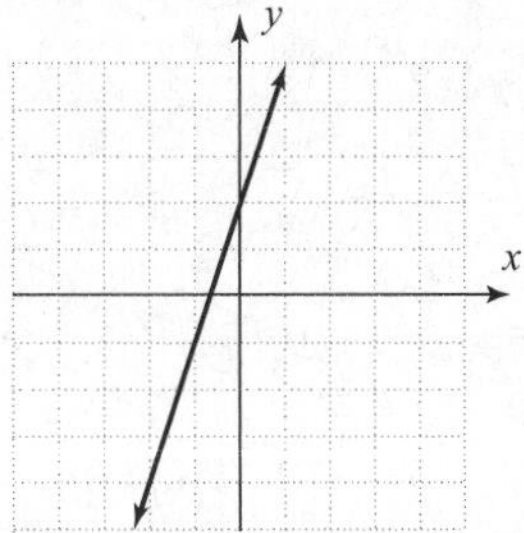

The equations are the same, so the lines coincide. The system of equations has infinitely many solutions.

9. If a system of linear equations in two variables has at least one solution, the system is consistent.

10. If a system of linear equations in two variables has exactly one solution, the lines are independent.

11. A system of linear equations in two variables that has no solutions is inconsistent.

12. True

13. $\begin{cases} 7x - 2y = 4 & (1) \\ 2x + 7y = 7 & (2) \end{cases}$

(1): $7x - 2y = 4$

$$-2y = -7x + 4$$
$$y = \frac{7}{2}x - 2$$

(2): $2x + 7y = 7$

$$7y = -2x + 7$$
$$y = -\frac{2}{7}x + 1$$

The equations have different slopes, so the lines intersect in exactly one point. The system is consistent and the equations are independent.

14. $\begin{cases} 6x + 4y = 4 & (1) \\ -12x - 8y = -8 & (2) \end{cases}$

(1): $6x + 4y = 4$

$$4y = -6x + 4$$
$$y = -\frac{3}{2}x + 1$$

(2): $-12x - 8y = -8$

$$-8y = 12x - 8$$
$$y = -\frac{3}{2}x + 1$$

The equations have the same slope, $-\frac{3}{2}$, and the same y-intercept, 1, so the lines coincide. There are infinitely many solutions. The system is consistent and the equations are dependent.

15. $\begin{cases} 3x - 4y = 8 & (1) \\ -6x + 8y = 8 & (2) \end{cases}$

(1): $3x - 4y = 8$

$$-4y = -3x + 8$$
$$y = \frac{3}{4}x - 2$$

(2): $-6x + 8y = 8$

$$8y = 6x + 8$$
$$y = \frac{3}{4}x + 1$$

The equations have the same slope, $\frac{3}{4}$, but different y-intercepts, so the lines are parallel. Therefore, the system has no solution and is inconsistent.

16. We want to know the number of miles for which the cost of both trucks is the same. Let m represent the number of miles and let C represent the cost.
EZ-Rental: $C = 0.40m + 20$
U-Move-It: $C = 0.25m + 35$

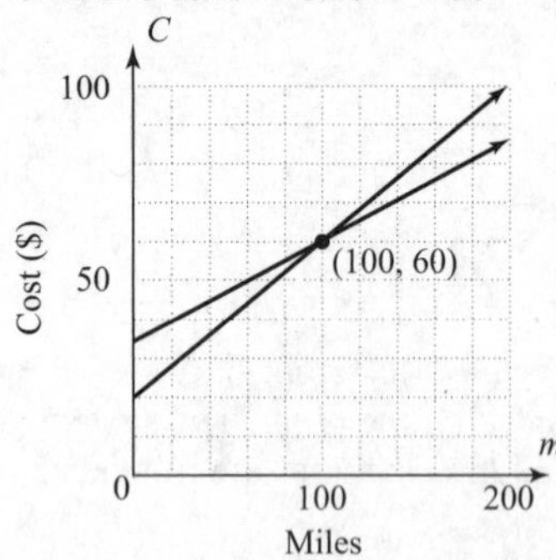

The charge will be the same for 100 miles.

EZ-Rental: $C = 0.40(100) + 20$
$C = 40 + 20$
$C = 60$

U-Move-It: $C = 0.25(100) + 35$
$C = 25 + 35$
$C = 60$

For 100 miles, both costs are the same, \$60.

4.1 Exercises

17. $\begin{cases} x - y = -4 & (1) \\ 3x + y = -4 & (2) \end{cases}$

(a) (2, 6)
(1): $2 - 6 \stackrel{?}{=} -4$
$-4 = -4$ True

(2): $3(2) + 6 \stackrel{?}{=} -4$
$6 + 6 \stackrel{?}{=} -4$
$12 = -4$ False
(2, 6) is not a solution.

(b) (−2, 2)
(1): $-2 - 2 \stackrel{?}{=} -4$
$-4 = -4$ True

(2): $3(-2) + 2 \stackrel{?}{=} -4$
$-6 + 2 \stackrel{?}{=} -4$
$-4 = -4$ True
(−2, 2) is a solution.

(c) (2, −2)
(1): $2 - (-2) \stackrel{?}{=} -4$
$2 + 2 \stackrel{?}{=} -4$
$4 = -4$ False
(2, −2) is not a solution.

19. $\begin{cases} 3x - y = 2 & (1) \\ -15x + 5y = -10 & (2) \end{cases}$

(a) (1, −1)
(1): $3(1) - (-1) \stackrel{?}{=} 2$
$3 + 1 \stackrel{?}{=} 2$
$4 = 2$ False
(1, −1) is not a solution.

(b) (−2, −8)
(1): $3(-2) - (-8) \stackrel{?}{=} 2$
$-6 + 8 \stackrel{?}{=} 2$
$2 = 2$ True

(2): $-15(-2) + 5(-8) \stackrel{?}{=} -10$
$30 - 40 \stackrel{?}{=} -10$
$-10 = -10$ True
(−2, −8) is a solution.

(c) (0, −2)
(1): $3(0) - (-2) \stackrel{?}{=} 2$
$0 + 2 \stackrel{?}{=} 2$
$2 = 2$ True

(2): $-15(0) + 5(-2) \stackrel{?}{=} -10$
$0 - 10 \stackrel{?}{=} -10$
$-10 = -10$ True
(0, −2) is a solution.

21. $\begin{cases} 6x - 2y = 1 & (1) \\ y = 3x + 2 & (2) \end{cases}$

(a) $\left(0, -\frac{1}{2}\right)$

(1): $6(0) - 2\left(-\frac{1}{2}\right) \stackrel{?}{=} 1$
$0 + 1 \stackrel{?}{=} 1$
$1 = 1$ True

(2): $-\frac{1}{2} \stackrel{?}{=} 3(0) + 2$
$-\frac{1}{2} = 2$ False

$\left(0, -\frac{1}{2}\right)$ is not a solution.

(b) (−2, −4)

(1): $6(-2)-2(-4) \stackrel{?}{=} 1$
$-12+8 \stackrel{?}{=} 1$
$-4=1$ False

(−2, −4) is not a solution.

(c) (0, 2)

(1): $6(0)-2(2) \stackrel{?}{=} 1$
$0-4 \stackrel{?}{=} 1$
$-4=1$ False

(0, 2) is not a solution.

23. $\begin{cases} 2x-y=-1 & (1) \\ 3x+2y=-5 & (2) \end{cases}$

(1): $2x-y=-1$
$-y=-2x-1$
$y=2x+1$

(2): $3x+2y=-5$
$2y=-3x-5$
$y=-\frac{3}{2}x-\frac{5}{2}$

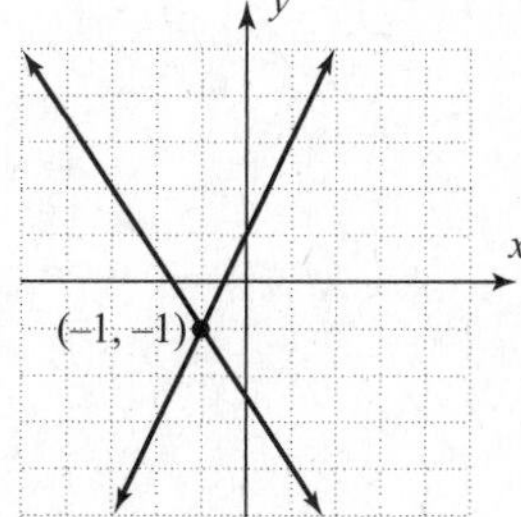

Check (−1, −1):

(1): $2(-1)-(-1) \stackrel{?}{=} -1$
$-2+1 \stackrel{?}{=} -1$
$-1=-1$ True

(2): $3(-1)+2(-1) \stackrel{?}{=} -5$
$-3-2 \stackrel{?}{=} -5$
$-5=-5$ True

The solution is (−1, −1).

25. $\begin{cases} y=x+5 & (1) \\ y=-\frac{1}{5}x-1 & (2) \end{cases}$

Both equations are in slope-intercept form.

Check (−5, 0):

(1): $0 \stackrel{?}{=} -5+5$
$0=0$ True

(2): $0 \stackrel{?}{=} -\frac{1}{5}(-5)-1$
$0 \stackrel{?}{=} 1-1$
$0=0$ True

The solution is (−5, 0).

27. $\begin{cases} y=\frac{3}{4}x-4 & (1) \\ y=-\frac{1}{2}x+1 & (2) \end{cases}$

Both equations are in slope-intercept form.

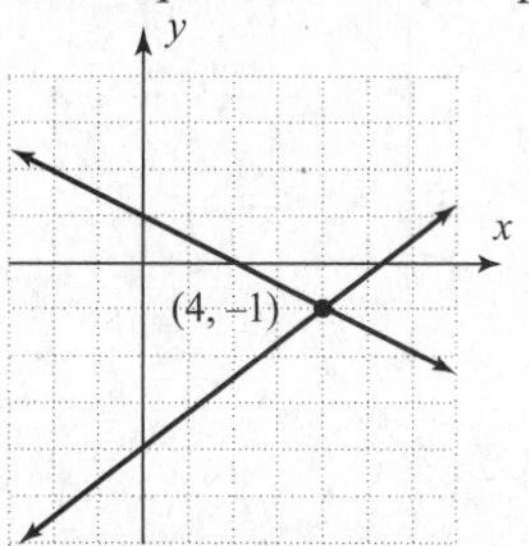

Check (4, −1):

(1): $-1 \stackrel{?}{=} \frac{3}{4}(4)-4$
$-1 \stackrel{?}{=} 3-4$
$-1=-1$ True

(2): $-1 \stackrel{?}{=} -\frac{1}{2}(4)+1$
$-1 \stackrel{?}{=} -2+1$
$-1=-1$ True

The solution is (4, −1).

29. $\begin{cases} x-4=0 & (1) \\ 3x-5y=22 & (2) \end{cases}$

(1): $x-4=0$
$x=4$

(2): $3x-5y=22$
$-5y=-3x+22$
$y=\frac{3}{5}x-\frac{22}{5}$

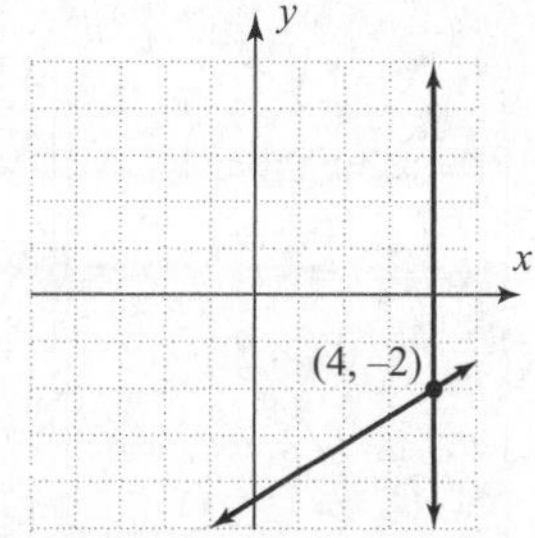

Check (4, −2):

(1): $4-4 \stackrel{?}{=} 0$

$0=0$ True

(2): $3(4)-5(-2) \stackrel{?}{=} 22$

$12+10 \stackrel{?}{=} 22$

$22=22$ True

The solution is (4, −2).

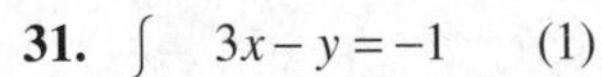

31. $\begin{cases} 3x-y=-1 & (1) \\ -6x+2y=-4 & (2) \end{cases}$

(1): $3x-y=-1$

$-y=-3x-1$

$y=3x+1$

(2): $-6x+2y=-4$

$2y=6x-4$

$y=3x-2$

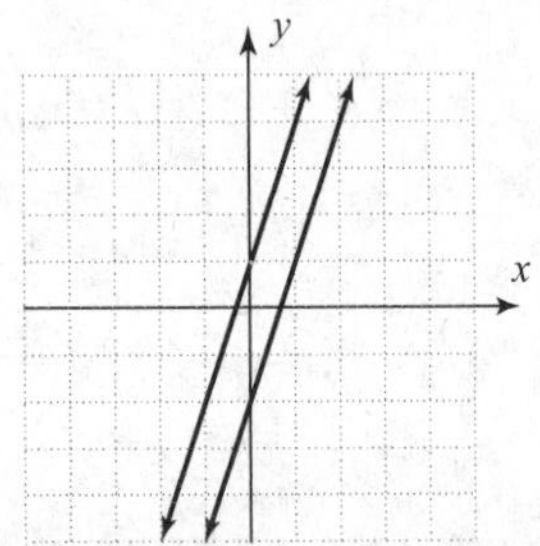

The lines are parallel, so there is no solution.

33. $\begin{cases} x+y=-2 & (1) \\ 3x-4y=8 & (2) \end{cases}$

(1): $x+y=-2$

$y=-x-2$

(2): $3x-4y=8$

$-4y=-3x+8$

$y=\frac{3}{4}x-2$

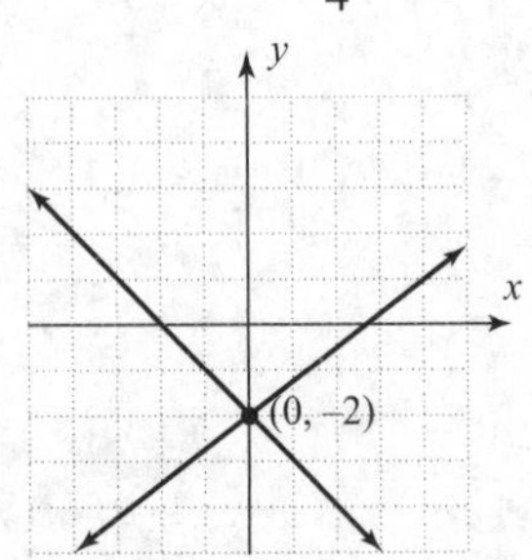

Check (0, −2):

(1): $0+(-2) \stackrel{?}{=} -2$

$-2=-2$ True

(2): $3(0)-4(-2) \stackrel{?}{=} 8$

$0+8 \stackrel{?}{=} 8$

$8=8$ True

The solution is (0, −2).

35. $\begin{cases} 2y=6-4x & (1) \\ 6x=9-3y & (2) \end{cases}$

(1): $2y=6-4x$

$2y=-4x+6$

$y=-2x+3$

(2): $6x=9-3y$

$3y=-6x+9$

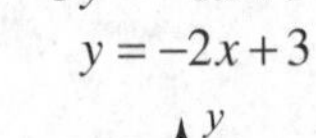

$y=-2x+3$

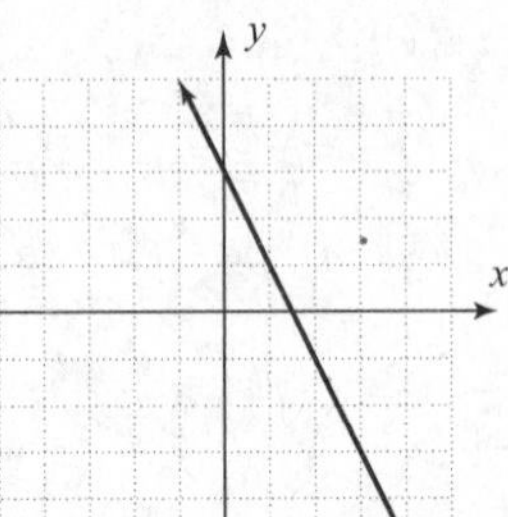

The lines coincide, so there are infinitely many solutions.

37. $\begin{cases} y=-x+3 & (1) \\ 3y=2x+9 & (2) \end{cases}$

(1): $y=-x+3$

(2): $3y=2x+9$

$y=\frac{2}{3}x+3$

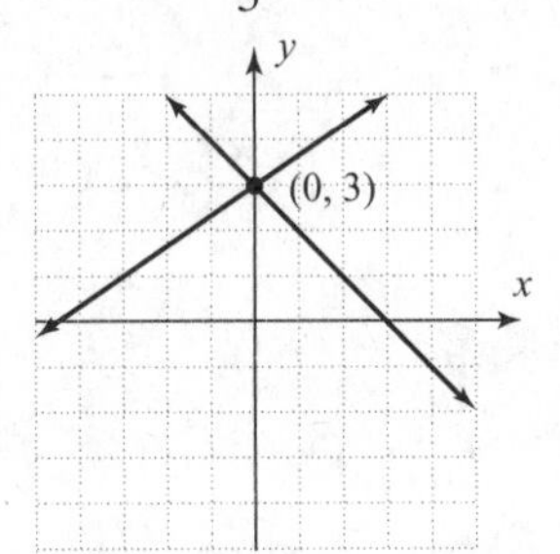

Check (0, 3):

(1): $3 \stackrel{?}{=} -0+3$

$3=3$ True

(2): $3(3) \stackrel{?}{=} 2(0)+9$

$9 \stackrel{?}{=} 0+9$

$9=9$ True

The solution is (0, 3).

39. The system has one solution. It is consistent and the equations are independent. The solution is (3, 2).

41. The system has no solution. It is inconsistent.

43. The system has infinitely many solutions. It is consistent and the equations are dependent.

45. The system has one solution. It is consistent and the equations are independent. The solution is $(-1, -2)$.

47. Since the lines have different slopes, they will intersect. The system has one solution. It is consistent and the equations are independent.

49. $\begin{cases} 6x+2y=12 & (1) \\ 3x+y=12 & (2) \end{cases}$

(1): $6x+2y=12$
$$2y=-6x+12$$
$$y=-3x+6$$
(2): $3x+y=12$
$$y=-3x+12$$
The lines have the same slope and different y-intercepts, so the lines are parallel. The system has no solution. It is inconsistent.

51. $\begin{cases} x+2y=2 & (1) \\ 2x+4y=4 & (2) \end{cases}$

(1): $x+2y=2$
$$2y=-x+2$$
$$y=-\frac{1}{2}x+1$$
(2): $2x+4y=4$
$$4y=-2x+4$$
$$y=-\frac{1}{2}x+1$$
The lines have the same slope and y-intercept, so the lines coincide. The system has infinitely many solutions. It is consistent and the equations are dependent.

53. $\begin{cases} y=4 & (1) \\ x=4 & (2) \end{cases}$

Equation (1) is a horizontal line and equation (2) is a vertical line. They intersect. There is one solution. The system is consistent and the equations are independent.

55. $\begin{cases} x-2y=-4 & (1) \\ -x+2y=-4 & (2) \end{cases}$

(1): $x-2y=-4$
$$-2y=-x-4$$
$$y=\frac{1}{2}x+2$$
(2): $-x+2y=-4$
$$2y=x-4$$
$$y=\frac{1}{2}x-2$$
The lines have the same slope but different y-intercepts, so the lines are parallel. The system has no solution. It is inconsistent.

57. $\begin{cases} x+2y=4 & (1) \\ x+1=5-2y & (2) \end{cases}$

(1): $x+2y=4$
$$2y=-x+4$$
$$y=-\frac{1}{2}x+2$$
(2): $x+1=5-2y$
$$2y=-x+4$$
$$y=-\frac{1}{2}x+2$$
The lines have the same slope and y-intercept, so the lines coincide. The system has infinitely many solutions. It is consistent and the equations are dependent.

59. $\begin{cases} y=2x+9 \\ y=-3x-6 \end{cases}$

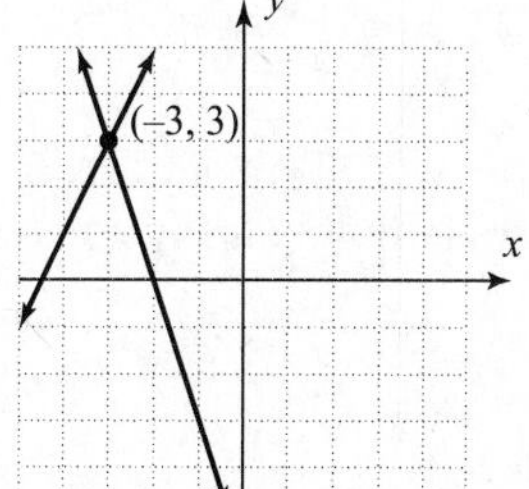

The system is consistent. The equations are independent. The solution is $(-3, 3)$.

61. $\begin{cases} x-2y=6 \\ 2x-4y=0 \end{cases}$

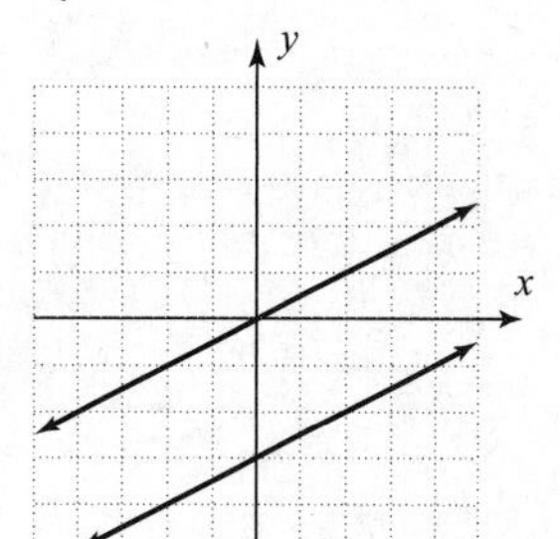

The system is inconsistent. There is no solution.

63. $\begin{cases} 3x-2y=-2 \\ 2x+y=8 \end{cases}$

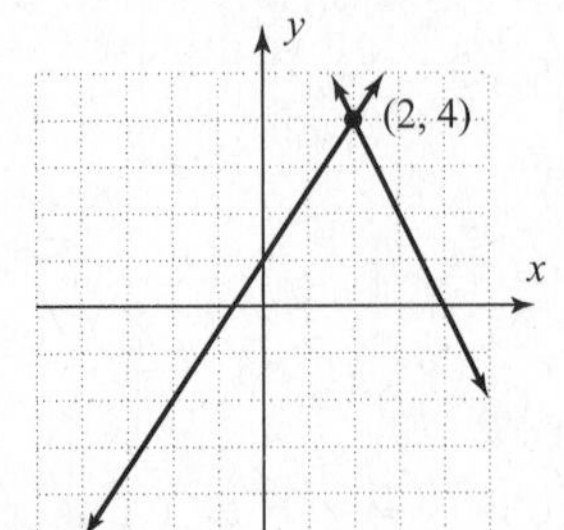

The system is consistent. The equations are independent. The solution is (2, 4).

65. $\begin{cases} y=x+2 \\ 3x-3y=-6 \end{cases}$

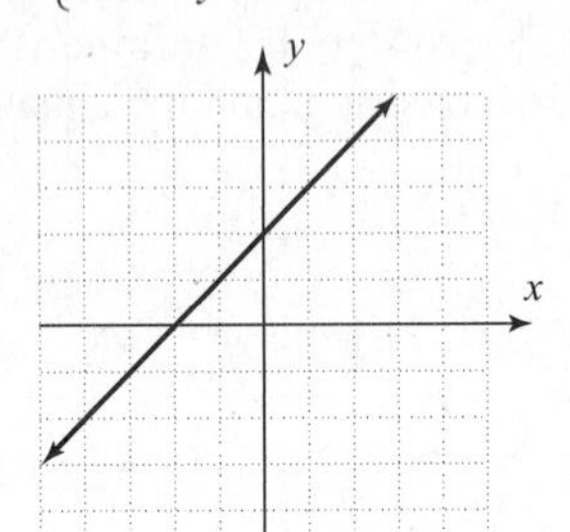

The system is consistent. The equations are dependent. There are infinitely many solutions.

67. $\begin{cases} -5x-2y=-2 \\ 10x+4y=4 \end{cases}$

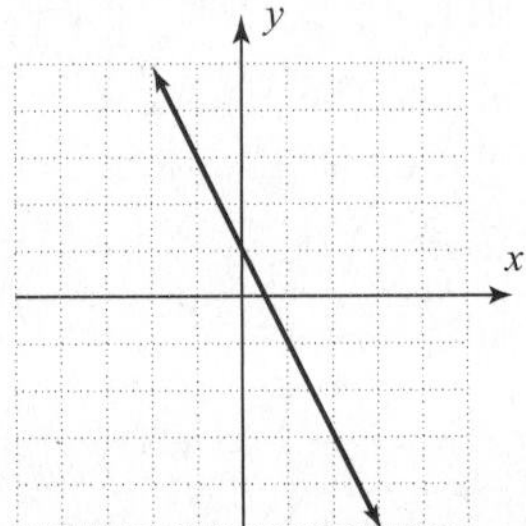

The system is consistent. The equations are dependent. There are infinitely many solutions.

69. $\begin{cases} 3x=4y-12 \\ 2x=-4y-8 \end{cases}$

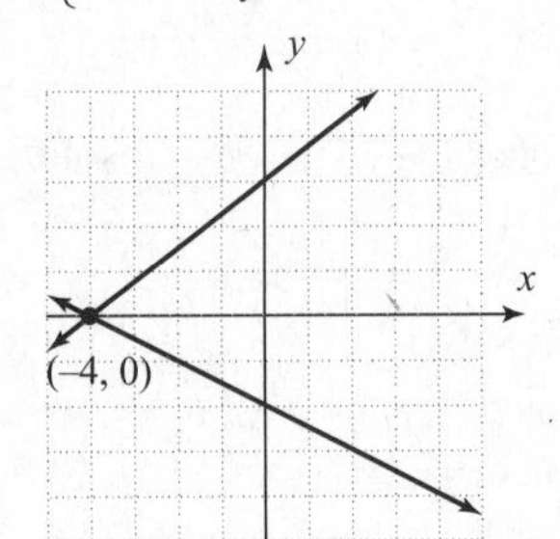

The system is consistent. The equations are independent. The solution is (–4, 0).

71. $\begin{cases} y=15x+1000 \\ y=25x \end{cases}$

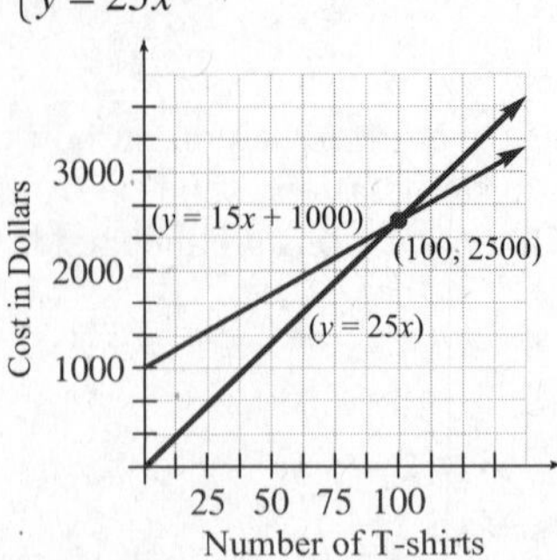

The break-even point is (100, 2500). The company needs to sell 100 T-shirts to break even at a cost/revenue of $2500.

73. $\begin{cases} y=0.8x+3.5 \\ y=1.5x \end{cases}$

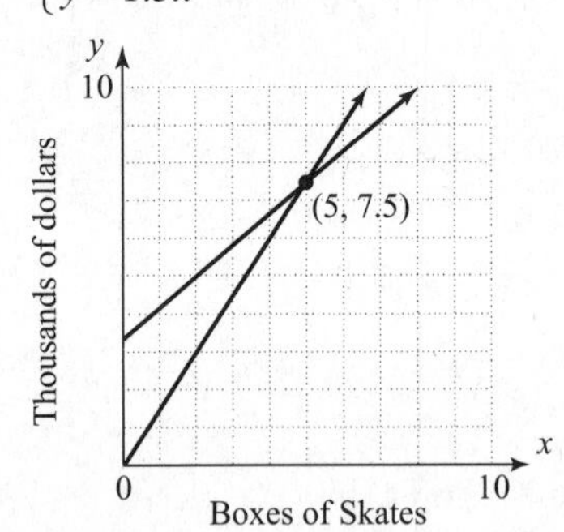

The break-even point is (5, 7.5). The company needs to sell 5 boxes of skates to break even at a cost/revenue of $7500.

75. Let x be the number of miles Liza drives, and let y be the cost. The equation for Wheels-to-Go is $y = 0.20x + 50$. The equation for Acme is $y = 0.12x + 62$. The system is:

$\begin{cases} y=0.2x+50 \\ y=0.12x+62 \end{cases}$

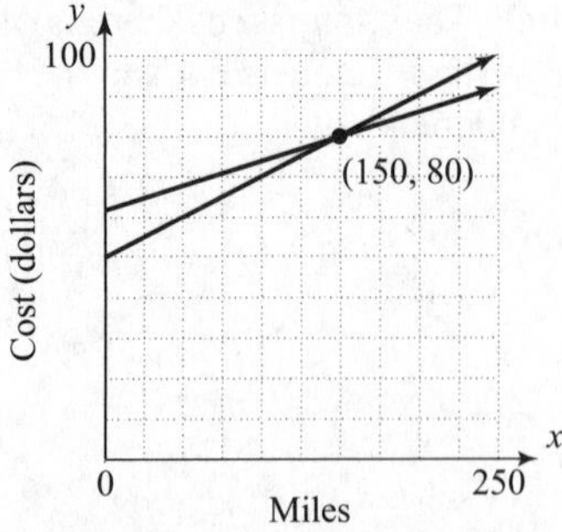

The solution is (150, 80), so the cost for driving 150 miles is the same ($80) for both companies. The per-mile charge for Acme is less, so she should use Acme if she will drive more than 150 miles.

77. Let x be the number of minutes and y be the cost per month. The equation for Plan A is $y = 0.05x + 8.95$. The equation for Plan B is $y = 0.07x + 5.95$.

$$\begin{cases} y = 0.05x + 8.95 \\ y = 0.07x + 5.95 \end{cases}$$

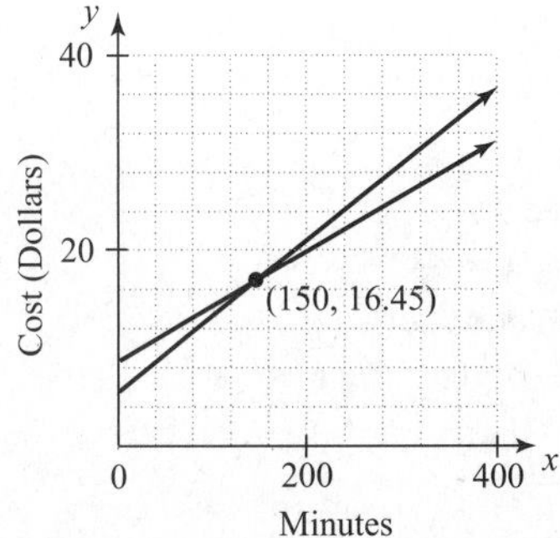

The solution is (150, 16.45). The monthly cost is the same ($16.45) when 150 minutes are used. The monthly fee for Plan B is less when fewer than 150 minutes are used, so choose Plan B if you typically use 100 minutes per month.

79. The system is dependent when the equations represent the same line.

$$3x - y = -4$$
$$-y = -3x - 4$$
$$y = 3x + 4$$

The slope is 3, the y-intercept is 4.

$y = cx + 4$

The slope is c, the y-intercept is 4.

The system is dependent when $c = 3$.

81. The system is dependent when the equations represent the same line.

$$x + 3 = 3(x - y)$$
$$x + 3 = 3x - 3y$$
$$3y = 2x - 3$$
$$y = \frac{2}{3}x - 1$$

The slope is $\frac{2}{3}$, the y-intercept is -1.

$$2y + 3 = 2cx - y$$
$$3y = 2cx - 3$$
$$y = \frac{2c}{3}x - 1$$

The slope is $\frac{2c}{3}$, the y-intercept is -1. The system is dependent when $\frac{2c}{3} = \frac{2}{3}$ or when $c = 1$.

83. $\begin{cases} y = 3x - 1 & (1) \\ y = -2x + 5 & (2) \end{cases}$

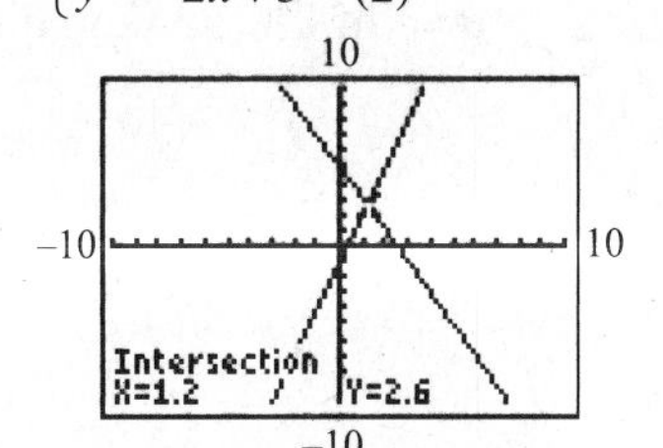

The solution is (1.2, 2.6).

85. $\begin{cases} 3x - y = -1 & (1) \\ -4x + y = -3 & (2) \end{cases}$

(1) $3x - y = -1$
$$y = 3x + 1$$

(2) $-4x + y = -3$
$$y = 4x - 3$$

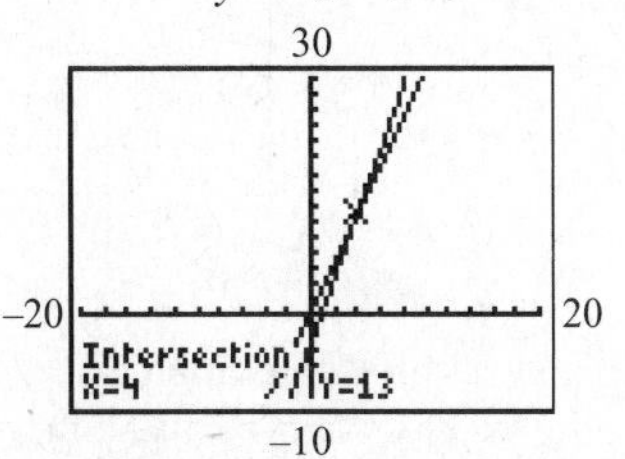

The solution is (4, 13).

87. $\begin{cases} 4x - 3y = 1 & (1) \\ -8x + 6y = -2 & (2) \end{cases}$

(1) $4x - 3y = 1$
$$-3y = -4x + 1$$
$$y = \frac{4}{3}x - \frac{1}{3}$$

(2) $-8x + 6y = -2$
$$6y = 8x - 2$$
$$y = \frac{4}{3}x - \frac{1}{3}$$

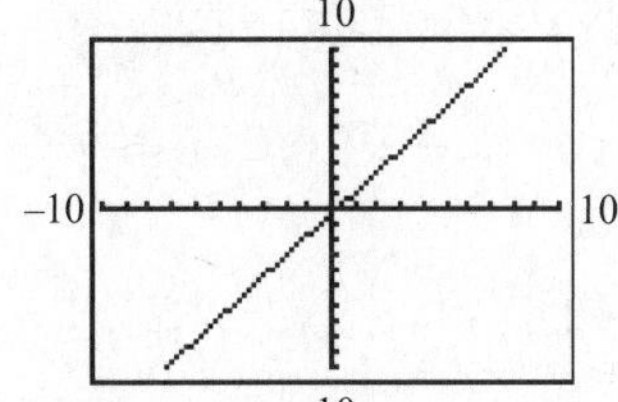

The lines coincide. There are infinitely many solutions.

89. The three possibilities for a solution of a system of two linear equation containing two variables are (1) lines intersecting at a single point (one solution); (2) the lines are coincident (a single line in the plane) so an infinite number of solutions; (3) parallel lines (no solution).

Section 4.2

Preparing for Solving Systems of Linear Equations Using Substitution

P1. $3x - y = 2$
$-y = -3x + 2$
$y = 3x - 2$

P2. $2x + 5y = 8$
$2x = -5y + 8$
$x = -\frac{5}{2}y + 4$

P3. $3x - 2(5x + 1) = 12$
$3x - 10x - 2 = 12$
$-7x - 2 = 12$
$-7x = 14$
$x = -2$
The solution set is $\{-2\}$.

4.2 Quick Checks

1. When solving a system of equations by substitution, whenever possible solve one of the equations for the variable that has the coefficient of <u>1</u> or <u>–1</u>.

2. $\begin{cases} y = 3x - 2 & (1) \\ 2x - 3y = -8 & (2) \end{cases}$
Equation (1) is already solved for y. Substitute $3x - 2$ for y in equation (2).
$2x - 3(3x - 2) = -8$
$2x - 9x + 6 = -8$
$-7x + 6 = -8$
$-7x = -14$
$x = 2$
Let $x = 2$ in equation (1).
$y = 3(2) - 2$
$y = 6 - 2$
$y = 4$
Check (2, 4):
(1): $4 \stackrel{?}{=} 3(2) - 2$
$4 \stackrel{?}{=} 6 - 2$
$4 = 4$ True

(2): $2(2) - 3(4) \stackrel{?}{=} -8$
$4 - 12 \stackrel{?}{=} -8$
$-8 = -8$ True
The solution is (2, 4).

3. $\begin{cases} 2x + y = -1 & (1) \\ 4x + 3y = 3 & (2) \end{cases}$
Solve equation (1) for y.
$2x + y = -1$
$y = -2x - 1$
Substitute $-2x - 1$ for y in equation (2).
$4x + 3(-2x - 1) = 3$
$4x - 6x - 3 = 3$
$-2x - 3 = 3$
$-2x = 6$
$x = -3$
Let $x = -3$ in equation (1).
$2(-3) + y = -1$
$-6 + y = -1$
$y = 5$
Check (–3, 5):
(1): $2(-3) + 5 \stackrel{?}{=} -1$
$-6 + 5 \stackrel{?}{=} -1$
$-1 = -1$ True

(2): $4(-3) + 3(5) \stackrel{?}{=} 3$
$-12 + 15 \stackrel{?}{=} 3$
$3 = 3$ True
The solution is (–3, 5).

4. $\begin{cases} -4x + y = 1 & (1) \\ 8x - y = 5 & (2) \end{cases}$
Solve equation (1) for y.
$-4x + y = 1$
$y = 4x + 1$
Substitute $4x + 1$ for y in equation (2).
$8x - (4x + 1) = 5$
$8x - 4x - 1 = 5$
$4x - 1 = 5$
$4x = 6$
$x = \frac{3}{2}$

Let $x = \frac{3}{2}$ in equation (1).
$-4\left(\frac{3}{2}\right) + y = 1$
$-6 + y = 1$
$y = 7$
Check $\left(\frac{3}{2}, 7\right)$:

(1): $-4\left(\frac{3}{2}\right)+7 \stackrel{?}{=} 1$
$-6+7 \stackrel{?}{=} 1$
$1=1$ True

(2): $8\left(\frac{3}{2}\right)-7 \stackrel{?}{=} 5$
$12-7 \stackrel{?}{=} 5$
$5=5$ True

The solution is $\left(\frac{3}{2}, 7\right)$.

5. $\begin{cases} 3x+2y=-3 & (1) \\ -x+y=\frac{11}{6} & (2) \end{cases}$

Solve equation (2) for *y*.

$$-x+y=\frac{11}{6}$$
$$y=x+\frac{11}{6}$$

Substitute $x+\frac{11}{6}$ for *y* in equation (1).

$$3x+2\left(x+\frac{11}{6}\right)=-3$$
$$3x+2x+\frac{11}{3}=-3$$
$$5x+\frac{11}{3}=-\frac{9}{3}$$
$$5x=-\frac{20}{3}$$
$$x=-\frac{4}{3}$$

Let $x=-\frac{4}{3}$ in equation (2).

$$-\left(-\frac{4}{3}\right)+y=\frac{11}{6}$$
$$\frac{4}{3}+y=\frac{11}{6}$$
$$y=\frac{11}{6}-\frac{4}{3}$$
$$y=\frac{11}{6}-\frac{8}{6}=\frac{3}{6} \text{ or } \frac{1}{2}$$

Check $\left(-\frac{4}{3}, \frac{1}{2}\right)$:

(1): $3\left(-\frac{4}{3}\right)+2\left(\frac{1}{2}\right) \stackrel{?}{=} -3$
$-4+1 \stackrel{?}{=} -3$
$-3=-3$ True

(2): $-\left(-\frac{4}{3}\right)+\frac{1}{2} \stackrel{?}{=} \frac{11}{6}$
$\frac{8}{6}+\frac{3}{6} \stackrel{?}{=} \frac{11}{6}$
$\frac{11}{6}=\frac{11}{6}$ True

The solution is $\left(-\frac{4}{3}, \frac{1}{2}\right)$.

6. While solving a system of equations by substitution, the variable expression has been eliminated and a *false* statement such as $-3 = 0$ results. This means that the solution of the system is $\underline{\varnothing \text{ or } \{\ \}}$.

7. While solving a system of equations by substitution, the variable expression has been eliminated and a *true* statement such as $0 = 0$ results. This means that the system has infinitely many solutions.

8. False; the value of *x* found from equation (1) must be substituted into equation (2), not back into equation (1).

9. $\begin{cases} y=5x+2 & (1) \\ -10x+2y=4 & (2) \end{cases}$

Let $y = 5x + 2$ in equation (2).

$$-10x+2(5x+2)=4$$
$$-10x+10x+4=4$$
$$4=4$$

Since $4 = 4$ is a true statement, the system is consistent and the equations are dependent. The system has infinitely many solutions.

10. $\begin{cases} 3x-2y=0 & (1) \\ -9x+6y=5 & (2) \end{cases}$

Solve equation (1) for *x*.

$$3x-2y=0$$
$$3x=2y$$
$$x=\frac{2}{3}y$$

Let $x=\frac{2}{3}y$ in equation (2).

$$-9\left(\frac{2}{3}y\right)+6y=5$$
$$-6y+6y=5$$
$$0=5$$

The statement $0 = 5$ is false. The system is inconsistent and has no solution.

11. $\begin{cases} 2x-6y=2 & (1) \\ -3x+9y=4 & (2) \end{cases}$

Solve equation (1) for x.

$2x-6y=2$
$2x=6y+2$
$x=3y+1$

Let $x = 3y + 1$ in equation (2).

$-3(3y+1)+9y=-4$
$-9y-3+9y=-4$
$-3=-4$

The statement $-3 = -4$ is false. The system is inconsistent and has no solution.

12. We want the year when we can expect the participation rate of women in the workforce to be equal to that of men.
The variables are P, which represents the participation rate, and t, which represents the year. The system is

$\begin{cases} P=-0.08t+243.96 & (1) \\ P=0.24t-415.10 & (2) \end{cases}$

Substitute $-0.08t + 243.96$ for P in equation (2).

$-0.08t+243.96=0.24t-415.10$
$-0.08t=0.24t-659.06$
$-0.32t=-659.06$
$t\approx 2059.6$

We round this result to the nearest year, 2060.
Check:

(1): $P=-0.08(2060)+243.96$
$P=79.16$

(2): $P=0.24(2060)-415.10$
$P=79.3$

The answers differ slightly due to rounding, but the answer checks. The participation rate of men and women in the workforce will be the same in 2060.

4.2 Exercises

13. $\begin{cases} x+2y=2 \\ y=2x-9 \end{cases}$

Let $y = 2x - 9$ in the first equation.

$x+2(2x-9)=2$
$x+4x-18=2$
$5x=20$
$x=4$

$y = 2(4) - 9 = 8 - 9 = -1$

The solution is $(4, -1)$.

15. $\begin{cases} -2x+5y=7 \\ x=3y-4 \end{cases}$

Let $x = 3y - 4$ in the first equation.

$-2(3y-4)+5y=7$
$-6y+8+5y=7$
$-y=-1$
$y=1$

$x = 3(1) - 4 = 3 - 4 = -1$

The solution is $(-1, 1)$.

17. $\begin{cases} x+y=-7 \\ 2x-y=-2 \end{cases}$

Solve the first equation for y.

$y = -7 - x$

Let $y = -7 - x$ in the second equation.

$2x-(-7-x)=-2$
$2x+7+x=-2$
$3x=-9$
$x=-3$
$-3+y=-7$
$y=-4$

The solution is $(-3, -4)$.

19. $\begin{cases} y=\frac{1}{2}x-5 \\ y=-\frac{3}{4}x-10 \end{cases}$

Let $y=\frac{1}{2}x-5$ in the second equation.

$\frac{1}{2}x-5=-\frac{3}{4}x-10$
$5=-\frac{5}{4}x$
$\left(-\frac{4}{5}\right)5=x$
$-4=x$

$y=\frac{1}{2}(-4)-5=-2-5=-7$

The solution is $(-4, -7)$.

21. $\begin{cases} y=3x+4 \\ y=-\frac{1}{2}x+\frac{5}{3} \end{cases}$

Let $y = 3x + 4$ in the second equation.

$$3x+4=-\frac{1}{2}x+\frac{5}{3}$$
$$18x+24=-3x+10$$
$$21x=-14$$
$$x=-\frac{14}{21}=-\frac{2}{3}$$
$$y=-\frac{1}{2}\left(-\frac{2}{3}\right)+\frac{5}{3}=\frac{1}{3}+\frac{5}{3}=\frac{6}{3}=2$$

The solution is $\left(-\frac{2}{3}, 2\right)$.

23. $\begin{cases} x=-6y \\ x-3y=3 \end{cases}$

Let $x = -6y$ in the second equation.
$$-6y-3y=3$$
$$-9y=3$$
$$y=\frac{3}{-9}=-\frac{1}{3}$$
$$x=-6\left(-\frac{1}{3}\right)=2$$

The solution is $\left(2, -\frac{1}{3}\right)$.

25. $\begin{cases} 2x-3y=0 \\ 8x+6y=3 \end{cases}$

Solve the first equation for x.
$$2x=3y$$
$$x=\frac{3}{2}y$$

Let $x=\frac{3}{2}y$ in the second equation.
$$8\left(\frac{3}{2}y\right)+6y=3$$
$$12y+6y=3$$
$$18y=3$$
$$y=\frac{3}{18}=\frac{1}{6}$$
$$2x-3\left(\frac{1}{6}\right)=0$$
$$2x-\frac{1}{2}=0$$
$$2x=\frac{1}{2}$$
$$x=\frac{1}{4}$$

The solution is $\left(\frac{1}{4}, \frac{1}{6}\right)$.

27. $\begin{cases} x+3y=-12 \\ x+3y=6 \end{cases}$

Solve the first equation for x.
$x = -3y - 12$
Let $x = -3y - 12$ in the second equation.
$$-3y-12+3y=6$$
$$-12=6 \quad \text{False}$$

The system is inconsistent and has no solution.

29. $\begin{cases} y=4x-1 \\ 8x-2y=2 \end{cases}$

Let $y = 4x - 1$ in the second equation.
$$8x-2(4x-1)=2$$
$$8x-8x+2=2$$
$$2=2 \quad \text{True}$$

The system is dependent and has infinitely many solutions.

31. $\begin{cases} x+2y=6 \\ x=3-2y \end{cases}$

Let $x = 3 - 2y$ in the first equation.
$$3-2y+2y=6$$
$$3=6 \quad \text{False}$$

The system is inconsistent and has no solution.

33. $\begin{cases} 5x+6=2-y \\ y=-5x-4 \end{cases}$

Let $y = -5x - 4$ in the first equation.
$$5x+6=2-(-5x-4)$$
$$5x+6=2+5x+4$$
$$6=6 \quad \text{True}$$

The system is dependent and has infinitely many solutions.

35. $\begin{cases} x=2 \\ x-6y=4 \end{cases}$

Let $x = 2$ in the second equation.
$$2-6y=4$$
$$-6y=2$$
$$y=\frac{2}{-6}=-\frac{1}{3}$$

The solution is $\left(2, -\frac{1}{3}\right)$.

37. $\begin{cases} y=\frac{1}{2}x \\ x=2(y+1) \end{cases}$

Let $y=\frac{1}{2}x$ in the second equation.

$$x = 2\left(\frac{1}{2}x+1\right)$$
$$x = x+2$$
$$0 = 2 \quad \text{False}$$

The system is inconsistent and has no solution.

39. $\begin{cases} y = 2x-8 \\ x - \frac{1}{2}y = 4 \end{cases}$

Let $y = 2x - 8$ in the second equation.

$$x - \frac{1}{2}(2x-8) = 4$$
$$x - x + 4 = 4$$
$$4 = 4 \quad \text{True}$$

The system is dependent and has infinitely many solutions.

41. $\begin{cases} 3x+2y = 4 \\ 3x + y = \frac{9}{2} \end{cases}$

Solve the second equation for y.

$$y = -3x + \frac{9}{2}$$

Let $y = -3x + \frac{9}{2}$ in the first equation.

$$3x + 2\left(-3x+\frac{9}{2}\right) = 4$$
$$3x - 6x + 9 = 4$$
$$-3x + 9 = 4$$
$$-3x = -5$$
$$x = \frac{5}{3}$$

$$3\left(\frac{5}{3}\right) + y = \frac{9}{2}$$
$$5 + y = \frac{9}{2}$$
$$y = \frac{9}{2} - \frac{10}{2}$$
$$y = -\frac{1}{2}$$

The solution is $\left(\frac{5}{3}, -\frac{1}{2}\right)$.

43. $\begin{cases} x - 5y = 3 \\ -2x + 10y = 8 \end{cases}$

Solve the first equation for x.

$x = 5y + 3$

Let $x = 5y + 3$ in the second equation.

$$-2(5y+3) + 10y = 8$$
$$-10y - 6 + 10y = 8$$
$$-6 = 8 \quad \text{False}$$

The system is inconsistent and has no solution.

45. $\begin{cases} 3x - y = 1 \\ -6x + 2y = -2 \end{cases}$

Solve the first equation for y.

$$-y = -3x + 1$$
$$y = 3x - 1$$

Let $y = 3x - 1$ in the second equation.

$$-6x + 2(3x-1) = -2$$
$$-6x + 6x - 2 = -2$$
$$-2 = -2 \quad \text{True}$$

The system is dependent and has infinitely many solutions.

47. $\begin{cases} 4x + 8y = -9 \\ 2x + y = \frac{3}{4} \end{cases}$

Solve the second equation for y.

$$y = -2x + \frac{3}{4}$$

Let $y = -2x + \frac{3}{4}$ in the first equation.

$$4x + 8\left(-2x + \frac{3}{4}\right) = -9$$
$$4x - 16x + 6 = -9$$
$$-12x + 6 = -9$$
$$-12x = -15$$
$$x = \frac{-15}{-12} = \frac{5}{4}$$

$$2\left(\frac{5}{4}\right) + y = \frac{3}{4}$$
$$\frac{10}{4} + y = \frac{3}{4}$$
$$y = -\frac{7}{4}$$

The solution is $\left(\frac{5}{4}, -\frac{7}{4}\right)$.

49. $\begin{cases} \frac{3}{2}x - y = 1 \\ 3x - 2y = 2 \end{cases}$

Solve the first equation for y.

$$y = \frac{3}{2}x - 1$$

Let $y = \frac{3}{2}x - 1$ in the second equation.

$$3x - 2\left(\frac{3}{2}x - 1\right) = 2$$
$$3x - 3x + 2 = 2$$
$$2 = 2 \quad \text{True}$$

The system is dependent and has infinitely many solutions.

51. $\begin{cases} \frac{x}{2} + \frac{y}{3} = \frac{1}{12} \\ \frac{2x}{3} + \frac{y}{3} = -\frac{1}{3} \end{cases}$

Solve the first equation for y.

$$\frac{y}{3} = -\frac{x}{2} + \frac{1}{12}$$
$$y = 3\left(-\frac{x}{2} + \frac{1}{12}\right)$$

Let $y = 3\left(-\frac{x}{2} + \frac{1}{12}\right)$ in the second equation.

$$\frac{2}{3}x + \frac{1}{3}\left[3\left(-\frac{x}{2} + \frac{1}{12}\right)\right] = -\frac{1}{3}$$
$$\frac{2}{3}x + \left(-\frac{x}{2} + \frac{1}{12}\right) = -\frac{1}{3}$$
$$\frac{4}{6}x - \frac{3}{6}x + \frac{1}{12} = -\frac{1}{3}$$
$$\frac{1}{6}x + \frac{1}{12} = -\frac{4}{12}$$
$$\frac{1}{6}x = -\frac{5}{12}$$
$$x = 6\left(-\frac{5}{12}\right)$$
$$x = -\frac{5}{2}$$

$$\frac{1}{2}\left(-\frac{5}{2}\right) + \frac{1}{3}y = \frac{1}{12}$$
$$-\frac{5}{4} + \frac{1}{3}y = \frac{1}{12}$$
$$-15 + 4y = 1$$
$$4y = 16$$
$$y = 4$$

The solution is $\left(-\frac{5}{2}, 4\right)$.

53. $\begin{cases} 2l + 2w = 34 & (1) \\ l = w + 3 & (2) \end{cases}$

Let $l = w + 3$ in (1).

$$2(w + 3) + 2w = 34$$
$$2w + 6 + 2w = 34$$
$$4w = 28$$
$$w = 7$$

$l = 7 + 3 = 10$

The length is 10 feet and the width is 7 feet.

55. $\begin{cases} x = 2y & (1) \\ 0.025x + 0.09y = 560 & (2) \end{cases}$

Let $x = 2y$ in (2).

$$0.025(2y) + 0.09y = 560$$
$$0.05y + 0.09y = 560$$
$$0.14y = 560$$
$$y = \frac{560}{0.14} = 4000$$

$x = 2(4000) = 8000$

He should invest \$8000 in the money market fund and \$4000 in the international fund.

57. $y = 15{,}000 + 0.02x$
$y = 25{,}000 + 0.01x$

To find the sales (x) at which both salaries are equal, set (y) from option A equal to (y) from option B.

$$15{,}000 + 0.02x = 25{,}000 + 0.01x$$
$$0.01x = 10{,}000$$
$$x = \frac{10{,}000}{0.01}$$
$$x = 1{,}000{,}000$$

The options result in the same salary when annual sales are \$1,000,000.

59. $\begin{cases} x + y = 17 & (1) \\ x - y = 7 & (2) \end{cases}$

Solve (2) for x.

$x = y + 7$

Let $x = y + 7$ in (1).

$$y + 7 + y = 17$$
$$2y = 10$$
$$y = 5$$
$$x - 5 = 7$$
$$x = 12$$

The numbers are 5 and 12.

61. $\begin{cases} Ax + 3By = 2 \\ -3Ax + By = -11 \end{cases}$

Let $x = 3$ and $y = 1$ in the system.

$\begin{cases} 3A + 3B = 2 \\ -9A + B = -11 \end{cases}$

Solve the second equation for B.

$B = 9A - 11$

Let $B = 9A - 11$ in the first equation.

$$\begin{aligned} 3A+3(9A-11) &= 2 \\ 3A+27A-33 &= 2 \\ 30A-33 &= 2 \\ 30A &= 35 \\ A &= \frac{7}{6} \end{aligned}$$

Let $A=\frac{7}{6}$ in $-9A+B=-11$.

$$\begin{aligned} -9\left(\frac{7}{6}\right)+B &= -11 \\ -\frac{21}{2}+B &= -11 \\ B &= -\frac{22}{2}+\frac{21}{2} \\ B &= -\frac{1}{2} \end{aligned}$$

Therefore, $A=\frac{7}{6}$ and $B=-\frac{1}{2}$.

63. Answers may vary. One possibility:
$$\begin{cases} 2x-y=1 \\ -x+2y=7 \end{cases}$$

65. Answers may vary. One possibility:
$$\begin{cases} 2x-y=4 \\ 6x-3y=12 \end{cases}$$

67. A reasonable first step is to multiply the first equation by 6 to clear fractions and multiply the second equation by 4 to clear fractions. Then the system is $\begin{cases} 2x-y=4 \\ 6x+2y=-5 \end{cases}$.

Section 4.3

Preparing for Solving Systems of Linear Equations Using Elimination

P1. The additive inverse of 5 is -5 since $5+(-5)=0$.

P2. The additive inverse of -8 is $-(-8)=8$ since $-8+8=0$.

P3. $\frac{2}{3}(3x-9y)=\frac{2}{3}(3x)-\frac{2}{3}(9y)=2x-6y$

P4.
$$\begin{aligned} 2y-5y &= 12 \\ -3y &= 12 \\ y &= -4 \end{aligned}$$
The solution set is $\{-4\}$.

P5.
$$\begin{aligned} 4 &= 2\cdot 2 \\ 5 &= \quad\quad 5 \\ \hline & 2\cdot 2\cdot 5 \end{aligned}$$
The LCM is $2\cdot 2\cdot 5=20$.

4.3 Quick Checks

1. The basic idea in using the elimination method is to get the coefficients of one of the variables to be additive inverses, such as 3 and -3.

2.
$$\begin{cases} x-3y=2 & (1) \\ 2x+3y=-14 & (2) \end{cases}$$
Add the equations.
$$\begin{aligned} x-3y &= 2 \\ 2x+3y &= -14 \\ \hline 3x \quad\quad &= -12 \\ x &= -4 \end{aligned}$$
Let $x=-4$ in equation (1).
$$\begin{aligned} -4-3y &= 2 \\ -3y &= 6 \\ y &= -2 \end{aligned}$$
Check $(-4, -2)$:
$$\begin{aligned} (1)\text{: } -4-3(-2) &\stackrel{?}{=} 2 \\ -4+6 &\stackrel{?}{=} 2 \\ 2 &= 2 \quad \text{True} \end{aligned}$$

$$\begin{aligned} (2)\text{: } 2(-4)+3(-2) &\stackrel{?}{=} -14 \\ -8-6 &\stackrel{?}{=} -14 \\ -14 &= -14 \quad \text{True} \end{aligned}$$
The solution is $(-4, -2)$.

3.
$$\begin{cases} x-2y=2 & (1) \\ -2x+5y=-1 & (2) \end{cases}$$
Multiply (1) by 2. Then add.
$$\begin{aligned} 2x-4y &= 4 \\ -2x+5y &= -1 \\ \hline y &= 3 \end{aligned}$$
Let $y=3$ in equation (1).
$$\begin{aligned} x-2(3) &= 2 \\ x-6 &= 2 \\ x &= 8 \end{aligned}$$

Check (8, 3):

(1): $8-2(3) \stackrel{?}{=} 2$

$8-6 \stackrel{?}{=} 2$

$2=2$ True

(2): $-2(8)+5(3) \stackrel{?}{=} -1$

$-16+15 \stackrel{?}{=} -1$

$-1=-1$ True

The solution is (8, 3).

4. $\begin{cases} 5x+4y=10 & (1) \\ -2x+3y=-27 & (2) \end{cases}$

Multiply (1) by 2 and (2) by 5. Then add.

$\begin{cases} 10x+8y=20 \\ -10x+15y=-135 \end{cases}$

$23y=-115$

$y=-5$

Let $y=-5$ in equation (1).

$5x+4(-5)=10$

$5x-20=10$

$5x=30$

$x=6$

Check (6, –5):

(1): $5(6)+4(-5) \stackrel{?}{=} 10$

$30-20 \stackrel{?}{=} 10$

$10=10$ True

(2): $-2(6)+3(-5) \stackrel{?}{=} -27$

$-12-15 \stackrel{?}{=} -27$

$-27=-27$ True

The solution is (6, –5).

5. $\begin{cases} 4y=-6x+30 & (1) \\ 7x+10y=35 & (2) \end{cases}$

Start by writing equation (1) in standard form.

$\begin{cases} 6x+4y=30 & (1) \\ 7x+10y=35 & (2) \end{cases}$

Multiply (1) by 10 and (2) by –4. Then add.

$\begin{cases} 60x+40y=300 \\ -28x-40y=-140 \end{cases}$

$32x \quad =160$

$x=5$

Let $x=5$ in equation (1).

$4y=-6(5)+30$

$4y=-30+30$

$4y=0$

$y=0$

Check (5, 0):

(1): $4(0) \stackrel{?}{=} -6(5)+30$

$0 \stackrel{?}{=} -30+30$

$0=0$ True

(2): $7(5)+10(0) \stackrel{?}{=} 35$

$35+0 \stackrel{?}{=} 35$

$35=35$ True

The solution is (5, 0).

6. When using the elimination method to solve a system of equations, you add equation (1) and equation (2) resulting in the statement –50 = –50. This means that the equations are dependent have infinitely many solutions.

7. $\begin{cases} 2x-6y=10 & (1) \\ 5x-15y=4 & (2) \end{cases}$

Multiply (1) by 5 and (2) by –2. Then add.

$\begin{cases} 10x-30y=50 \\ -10x+30y=-8 \end{cases}$

$0=42$

The statement 0 = 42 is false. The system is inconsistent and has no solution.

8. $\begin{cases} -x+3y=2 & (1) \\ 3x-9y=-6 & (2) \end{cases}$

Multiply (1) by 3 and then add.

$\begin{cases} -3x+9y=6 \\ 3x-9y=-6 \end{cases}$

$0=0$

The statement 0 = 0 is true. The system is consistent but the equations are dependent. The system has infinitely many solutions.

9. $\begin{cases} -4x+8y=4 & (1) \\ 3x-6y=-3 & (2) \end{cases}$

Multiply (1) by 3 and (2) by 4. Then add.

$\begin{cases} -12x+24y=12 \\ 12x-24y=-12 \end{cases}$

$0=0$

The statement 0 = 0 is true. The system is consistent but the equations are dependent. The system has infinitely many solutions.

10. $\begin{cases} 5c+3s=15.50 & (1) \\ 3c+2s=9.75 & (2) \end{cases}$

Multiply (1) by 2 and (2) by –3. Then add.

$\begin{cases} 10c+6s=31.00 \\ -9c-6s=-29.25 \end{cases}$

$c \quad =1.75$

Let $c = 1.75$ in (2).

$$\begin{aligned} 3(1.75)+2s &= 9.75 \\ 5.25+2s &= 9.75 \\ 2s &= 4.50 \\ s &= 2.25 \end{aligned}$$

A cheeseburger costs \$1.75, and a medium shake costs \$2.25.

4.3 Exercises

11. $\begin{cases} 2x+y=3 & (1) \\ 5x-y=11 & (2) \end{cases}$

Add the equations, then solve for x.

$$\begin{cases} 2x+y=3 \\ 5x-y=11 \end{cases}$$
$$\begin{aligned} 7x \quad &= 14 \\ x &= 2 \end{aligned}$$

Let $x = 2$ in (1).

$$\begin{aligned} 2(2)+y &= 3 \\ 4+y &= 3 \\ y &= -1 \end{aligned}$$

The solution is (2, −1).

13. $\begin{cases} 3x-2y=10 & (1) \\ -3x+12y=30 & (2) \end{cases}$

Add the equations, then solve for y.

$$\begin{cases} 3x-2y=10 \\ -3x+12y=30 \end{cases}$$
$$\begin{aligned} 10y &= 40 \\ y &= 4 \end{aligned}$$

Let $y = 4$ in (1).

$$\begin{aligned} 3x-2(4) &= 10 \\ 3x-8 &= 10 \\ 3x &= 18 \\ x &= 6 \end{aligned}$$

The solution is (6, 4).

15. $\begin{cases} 2x+3y=-4 & (1) \\ -2x+y=6 & (2) \end{cases}$

Add the equations, then solve for y.

$$\begin{cases} 2x+3y=-4 \\ -2x+y=6 \end{cases}$$
$$\begin{aligned} 4y &= 2 \\ y &= \frac{1}{2} \end{aligned}$$

Let $y = \frac{1}{2}$ in (2).

$$\begin{aligned} -2x+\frac{1}{2} &= 6 \\ -2x &= \frac{11}{2} \\ x &= -\frac{11}{4} \end{aligned}$$

The solution is $\left(-\frac{11}{4}, \frac{1}{2}\right)$.

17. $\begin{cases} 6x-2y=0 & (1) \\ -9x-4y=21 & (2) \end{cases}$

Multiply (1) by −2, then add the equations and solve for x.

$$\begin{cases} -12x+4y=0 \\ -9x-4y=21 \end{cases}$$
$$\begin{aligned} -21x \quad &= 21 \\ x &= -1 \end{aligned}$$

Let $x = -1$ in (1).

$$\begin{aligned} 6(-1)-2y &= 0 \\ -6-2y &= 0 \\ -2y &= 6 \\ y &= -3 \end{aligned}$$

The solution is (−1, −3).

19. $\begin{cases} 2x+2y=1 & (1) \\ -2x-2y=1 & (2) \end{cases}$

Add the equations.

$$\begin{cases} 2x+2y=1 \\ -2x-2y=1 \end{cases}$$
$$0 = 2 \quad \text{False}$$

The system is inconsistent and has no solution.

21. $\begin{cases} 3x+y=-1 & (1) \\ 6x+2y=-2 & (2) \end{cases}$

Multiply (1) by −2, then add.

$$\begin{cases} -6x-2y=2 \\ 6x+2y=-2 \end{cases}$$
$$0 = 0 \quad \text{True}$$

The system is dependent and has infinitely many solutions.

23. $\begin{cases} 2x-3y=10 & (1) \\ -4x+6y=-20 & (2) \end{cases}$

Multiply (1) by 2 and add the equations.

$$\begin{cases} 4x-6y=20 \\ -4x+6y=-20 \end{cases}$$
$$0 = 0 \quad \text{True}$$

The system is dependent and has infinitely many solutions.

25. $\begin{cases} -4x+8y=1 & (1) \\ 3x-6y=1 & (2) \end{cases}$

Multiply (1) by 3 and (2) by 4 and add the equations.

$$\begin{cases} -12x+24y=3 \\ 12x-24y=4 \end{cases}$$

$$0=7 \quad \text{False}$$

The system is inconsistent and has no solution.

27. $\begin{cases} 2x+3y=14 & (1) \\ -3x+y=23 & (2) \end{cases}$

Multiply (2) by −3, then add.

$$\begin{cases} 2x+3y=14 \\ 9x-3y=-69 \end{cases}$$

$$11x = -55$$
$$x=-5$$

Let $x=-5$ in (2).

$$-3(-5)+y=23$$
$$15+y=23$$
$$y=8$$

The solution is (−5, 8).

29. $\begin{cases} 2x+4y=0 & (1) \\ 5x+2y=6 & (2) \end{cases}$

Multiply (2) by −2, then add.

$$\begin{cases} 2x+4y=0 \\ -10x-4y=-12 \end{cases}$$

$$-8x = -12$$
$$x=\frac{-12}{-8}=\frac{3}{2}$$

Let $x=\frac{3}{2}$ in (1).

$$2\left(\frac{3}{2}\right)+4y=0$$
$$3+4y=0$$
$$4y=-3$$
$$y=-\frac{3}{4}$$

The solution is $\left(\frac{3}{2}, -\frac{3}{4}\right)$.

31. $\begin{cases} x-3y=4 & (1) \\ -2x+6y=3 & (2) \end{cases}$

Multiply (1) by 2, then add.

$$\begin{cases} 2x-6y=8 \\ -2x+6y=3 \end{cases}$$

$$0=11 \quad \text{False}$$

The system is inconsistent and has no solution.

33. $\begin{cases} 2x+3y=-3 & (1) \\ 3x+5y=-9 & (2) \end{cases}$

Multiply (1) by −3 and (2) by 2, then add.

$$\begin{cases} -6x-9y=9 \\ 6x+10y=-18 \end{cases}$$

$$y=-9$$

Let $y=-9$ in (1).

$$2x+3(-9)=-3$$
$$2x-27=-3$$
$$2x=24$$
$$x=12$$

The solution is (12, −9).

35. $\begin{cases} 10y=4x-2 & (1) \\ 2x-5y=1 & (2) \end{cases}$

Rearrange (1).

$$\begin{cases} -4x+10y=-2 & (1) \\ 2x-5y=1 & (2) \end{cases}$$

Multiply (1) by $\frac{1}{2}$ and add.

$$\begin{cases} -2x+5y=-1 \\ 2x-5y=1 \end{cases}$$

$$0=0 \quad \text{True}$$

The system is dependent and has infinitely many solutions.

37. $\begin{cases} 4x+3y=0 & (1) \\ 3x-5y=2 & (2) \end{cases}$

Multiply (1) by 5 and (2) by 3, then add.

$$\begin{cases} 20x+15y=0 \\ 9x-15y=6 \end{cases}$$

$$29x = 6$$
$$x=\frac{6}{29}$$

Let $x=\frac{6}{29}$ in (1).

$4\left(\frac{6}{29}\right)+3y=0$

$\frac{24}{29}+3y=0$

$3y=-\frac{24}{29}$

$y=-\frac{8}{29}$

The solution is $\left(\frac{6}{29}, -\frac{8}{29}\right)$.

39. $\begin{cases} 4x-3y=-10 & (1) \\ -\frac{2}{3}x+y=\frac{11}{3} & (2) \end{cases}$

Multiply (2) by 3, then add.

$\begin{cases} 4x-3y=-10 \\ -2x+3y=11 \end{cases}$

$2x \quad =1$

$x=\frac{1}{2}$

Let $x=\frac{1}{2}$ in (1).

$4\left(\frac{1}{2}\right)-3y=-10$

$2-3y=-10$

$-3y=-12$

$y=4$

The solution is $\left(\frac{1}{2}, 4\right)$.

41. $\begin{cases} 1.5x+0.5y=-0.45 & (1) \\ -0.3x-0.4y=-0.54 & (2) \end{cases}$

Multiply (2) by 5, then add.

$\begin{cases} 1.5x+0.5y=-0.45 \\ -1.5x-2y=-2.7 \end{cases}$

$-1.5y=-3.15$

$y=\frac{-3.15}{-1.5}=2.1$

Let $y = 2.1$ in (1).

$1.5x+0.5(2.1)=-0.45$

$1.5x+1.05=-0.45$

$1.5x=-1.5$

$x=-1$

The solution is $(-1, 2.1)$.

43. $\begin{cases} \frac{1}{2}x+\frac{2}{3}y=-5 & (1) \\ \frac{5}{2}x+\frac{5}{6}y=-10 & (2) \end{cases}$

Multiply (1) by -5, then add.

$\begin{cases} -\frac{5x}{2}-\frac{10y}{3}=25 \\ \frac{5x}{2}+\frac{5y}{6}=-10 \end{cases}$

$-\frac{10y}{3}+\frac{5y}{6}=15$

$-\frac{20y}{6}+\frac{5y}{6}=15$

$-\frac{15y}{6}=15$

$y=\left(-\frac{6}{15}\right)15=-6$

Let $y = -6$ in (2).

$\frac{5x}{2}+\frac{5(-6)}{6}=-10$

$\frac{5x}{2}-5=-10$

$\frac{5}{2}x=-5$

$x=\left(\frac{2}{5}\right)(-5)=-2$

The solution is $(-2, -6)$.

45. $\begin{cases} 0.05x+0.10y=5.50 & (1) \\ x+y=80 & (2) \end{cases}$

Multiply (1) by 100 and (2) by -5, then add.

$\begin{cases} 5x+10y=550 \\ -5x-5y=-400 \end{cases}$

$5y=150$

$y=30$

Let $y = 30$ in (2).

$x+30=80$

$x=50$

The solution is $(50, 30)$.

47. $\begin{cases} x-y=-4 & (1) \\ 3x+y=8 & (2) \end{cases}$

$4x \quad =4$

$x=1$

Let $x = 1$ in (1).

$1-y=-4$

$-y=-5$

$y=5$

The solution is $(1, 5)$.

49. $\begin{cases} 3x-10y=-5 & (1) \\ 6x-8y=14 & (2) \end{cases}$

Multiply (1) by −2, then add.

$$\begin{cases} -6x+20y=10 \\ 6x-8y=14 \end{cases}$$
$$12y=24$$
$$y=2$$

Let $y = 2$ in (1).

$$3x-10(2)=-5$$
$$3x-20=-5$$
$$3x=15$$
$$x=5$$

The solution is (5, 2).

51. $\begin{cases} -x+3y=6 & (1) \\ 4x+5y=7 & (2) \end{cases}$

Solve (1) for x.

$$-x=-3y+6$$
$$x=3y-6$$

Let $x = 3y - 6$ in (2).

$$4(3y-6)+5y=7$$
$$12y-24+5y=7$$
$$17y=31$$
$$y=\frac{31}{17}$$

Let $y=\frac{31}{17}$ in (1).

$$-x+3\left(\frac{31}{17}\right)=6$$
$$-x+\frac{93}{17}=6$$
$$-x=\frac{102}{17}-\frac{93}{17}$$
$$-x=\frac{9}{17}$$
$$x=-\frac{9}{17}$$

The solution is $\left(-\frac{9}{17}, \frac{31}{17}\right)$.

53. $\begin{cases} 0.3x-0.7y=1.2 & (1) \\ 1.2x+2.1y=2 & (2) \end{cases}$

Multiply (1) by −4, then add.

$$\begin{cases} -1.2x+2.8y=-4.8 \\ 1.2x+2.1y=2 \end{cases}$$
$$4.9y=-2.8$$
$$y=-\frac{2.8}{4.9}=-\frac{4}{7}$$

Let $y=-\frac{4}{7}$ in (1).

$$0.3x-0.7\left(-\frac{4}{7}\right)=1.2$$
$$0.3x+0.4=1.2$$
$$0.3x=0.8$$
$$x=\frac{8}{3}$$

The solution is $\left(\frac{8}{3}, -\frac{4}{7}\right)$.

55. $\begin{cases} 3x-2y=6 & (1) \\ \frac{3}{2}x-y=3 & (2) \end{cases}$

Multiply (2) by −2 and then add.

$$\begin{cases} 3x-2y=6 \\ -3x+2y=-6 \end{cases}$$
$$0=0 \quad \text{True}$$

The system is dependent. There are infinitely many solutions.

57. $\begin{cases} y=-\frac{2}{3}x-\frac{7}{3} & (1) \\ y=\frac{3}{4}x-\frac{15}{4} & (2) \end{cases}$

Let $y=-\frac{2}{3}x-\frac{7}{3}$ in the second equation.

$$-\frac{2}{3}x-\frac{7}{3}=\frac{3}{4}x-\frac{15}{4}$$
$$12\left(-\frac{2}{3}x-\frac{7}{3}\right)=12\left(\frac{3}{4}x-\frac{15}{4}\right)$$
$$-8x-28=9x-45$$
$$-17x=-17$$
$$x=1$$

Let $x = 1$ in (1).

$$y=-\frac{2}{3}(1)-\frac{7}{3}=-\frac{2}{3}-\frac{7}{3}=-\frac{9}{3}=-3$$

The solution is (1, −3).

59. $\begin{cases} \frac{x}{2}+\frac{y}{4}=-2 & (1) \\ \frac{3x}{2}+\frac{y}{5}=-6 & (2) \end{cases}$

Multiply (1) by −8 and (2) by 10, then add.

$$\begin{cases} -4x-2y=16 \\ 15x+2y=-60 \end{cases}$$
$$11x=-44$$
$$x=-4$$

Let $x = -4$ in (1).

$$-\frac{4}{2}+\frac{y}{4}=-2$$
$$-2+\frac{y}{4}=-2$$
$$\frac{y}{4}=0$$
$$y=0$$

The solution is (–4, 0).

61. $\begin{cases} x-2y=-7 & (1) \\ 3x+4y=6 & (2) \end{cases}$

Multiply (1) by 2, then add.

$$\begin{cases} 2x-4y=-14 \\ 3x+4y=6 \end{cases}$$
$$5x = -8$$
$$x=-\frac{8}{5}$$

Multiply (1) by –3, then add.

$$\begin{cases} -3x+6y=21 \\ 3x+4y=6 \end{cases}$$
$$10y=27$$
$$y=\frac{27}{10}$$

The solution is $\left(-\frac{8}{5}, \frac{27}{10}\right)$.

63. $\begin{cases} 6x-5y=1 & (1) \\ 8x-2y=-22 & (2) \end{cases}$

Multiply (1) by –2 and (2) by 5, then add.

$$\begin{cases} -12x+10y=-2 \\ 40x-10y=-110 \end{cases}$$
$$28x = -112$$
$$x=-4$$

Let $x = -4$ in (1).

$$6(-4)-5y=1$$
$$-24-5y=1$$
$$-5y=25$$
$$y=-5$$

The solution is (–4, –5).

65. $\begin{cases} y=2x-4y \\ 4x+1=10y+3 \end{cases}$

$$\begin{cases} -2x+5y=0 & (1) \\ 4x-10y=2 & (2) \end{cases}$$

Multiply (1) by 2, then add.

$$\begin{cases} -4x+10y=0 \\ 4x-10y=2 \end{cases}$$
$$0=2$$

The system is inconsistent and has no solution.

67. $\begin{cases} \frac{x}{2}-y=1 \\ \frac{x}{5}+\frac{5y}{6}=\frac{14}{15} \end{cases}$

Multiply the first equation by 2 and the second equation by 30 to clear the fractions.

$$\begin{cases} x-2y=2 & (1) \\ 6x+25y=28 & (2) \end{cases}$$

Multiply (1) by –6, then add.

$$\begin{cases} -6x+12y=-12 \\ 6x+25y=28 \end{cases}$$
$$37y=16$$
$$y=\frac{16}{37}$$

Multiply (1) by 25 and (2) by 2, then add.

$$\begin{cases} 25x-50y=50 \\ 12x+50y=56 \end{cases}$$
$$37x = 106$$
$$x=\frac{106}{37}$$

The solution is $\left(\frac{106}{37}, \frac{16}{37}\right)$.

69. $\begin{cases} 2h+c=770 & (1) \\ 3h+2c=1260 & (2) \end{cases}$

Multiply (1) by –2, then add.

$$\begin{cases} -4h-2c=-1540 \\ 3h+2c=1260 \end{cases}$$
$$-h = -280$$
$$h=280$$

Let $h = 280$ in (1).

$$2(280)+c=770$$
$$560+c=770$$
$$c=210$$

A hamburger contains 280 calories and a medium Coke contains 210 calories.

71. $\begin{cases} 2t+3z=65 & (1) \\ 3t+4z=90 & (2) \end{cases}$

Multiply (1) by −3 and (2) by 2, then add.

$$\begin{cases} -6t-9z=-195 \\ 6t+8z=180 \end{cases}$$
$$-z=-15$$
$$z=15$$

Let $z = 15$ in (1).

$$2t+3(15)=65$$
$$2t+45=65$$
$$2t=20$$
$$t=10$$

It takes the farmer 10 hours to plant an acre of tomatoes and 15 hours to plant an acre of zucchini.
To plant 5 acres of tomatoes and 2 acres of zucchini would take
5(10) + 2(15) = 50 + 30 = 80 hours, so he won't get the crops in before the rain.

73. $\begin{cases} a+r=100 & (1) \\ 9a+11.50r=1000 & (2) \end{cases}$

Multiply (1) by −9, then add.

$$\begin{cases} -9a-9r=-900 \\ 9a+11.5r=1000 \end{cases}$$
$$2.5r=100$$
$$r=40$$

Let $r = 40$ in (1).

$$a+40=100$$
$$a=60$$

The blend will contain 60 pounds of Arabica beans and 40 pounds of Robusta beans.

75. $\begin{cases} A+B=90 \\ A=10+3B \end{cases}$

$\begin{cases} A+B=90 & (1) \\ A-3B=10 & (2) \end{cases}$

Multiply (1) by 3, then add.

$$\begin{cases} 3A+3B=270 \\ A-3B=10 \end{cases}$$
$$4A=280$$
$$A=70$$

Let $A = 70$ in (1).

$$70+B=90$$
$$B=20$$

The angles measure 70° and 20°.

77. $\begin{cases} ax+4y=1 & (1) \\ -2ax-y=3 & (2) \end{cases}$

Multiply (1) by 2, then add.

$$\begin{cases} 2ax+8y=2 \\ -2ax-3y=3 \end{cases}$$
$$5y=5$$
$$y=1$$

Let $y = 1$ in (1).

$$ax+4(1)=1$$
$$ax+4=1$$
$$ax=-3$$
$$x=-\frac{3}{a}$$

The solution is $\left(-\frac{3}{a}, 1\right)$.

79. $\begin{cases} -3x+2y=6a & (1) \\ x-2y=2b & (2) \end{cases}$

$$-2x=6a+2b$$
$$x=\frac{6a+2b}{-2}=-3a-b$$

Multiply (2) by 3, then add.

$$\begin{cases} -3x+2y=6a \\ 3x-6y=6b \end{cases}$$
$$-4y=6a+6b$$
$$y=\frac{6a+6b}{-4}=-\frac{3}{2}a-\frac{3}{2}b$$

The solution is $\left(-3a-b, -\frac{3}{2}a-\frac{3}{2}b\right)$.

81. It is easier to eliminate y. Multiply the second equation by 3, to form $6x + 3y = 15$. Then add the equations, obtaining $7x = 21$. So $x = 3$. Substitute $x = 3$ into either the first equation or the second equation and solve for y. $y = -1$.

83. Both of you will obtain the correct solution. Other strategies are: multiply equation (1) by one-third and adding; solving equation (2) for y and substituting this expression into equation (1); and graphing the equation and finding the point of intersection.

Putting the Concepts Together (Sections 4.1–4.3)

1. $\begin{cases} 4x+y=-20 & (1) \\ y=-\frac{1}{6}x+3 & (2) \end{cases}$

(a) $\left(3, \frac{5}{2}\right)$

$$\text{(1): } 4(3)+\frac{5}{2} \stackrel{?}{=} -20$$
$$12+\frac{5}{2} \stackrel{?}{=} -20$$
$$\frac{24}{2}+\frac{5}{2} \stackrel{?}{=} -20$$
$$\frac{29}{2}=-20 \text{ False}$$

$\left(3, \frac{5}{2}\right)$ is not a solution.

(b) (−6, 4)

$$\text{(1): } 4(-6)+4 \stackrel{?}{=} -20$$
$$-24+4 \stackrel{?}{=} -20$$
$$-20=-20 \text{ True}$$
$$\text{(2): } 4 \stackrel{?}{=} -\frac{1}{6}(-6)+3$$
$$4 \stackrel{?}{=} 1+3$$
$$4=4 \text{ True}$$

(−6, 4) is a solution.

(c) (−4, −4)

$$\text{(1): } 4(-4)+(-4) \stackrel{?}{=} -20$$
$$-16-4 \stackrel{?}{=} -20$$
$$-20=-20 \text{ True}$$
$$\text{(2): } -4 \stackrel{?}{=} -\frac{1}{6}(-4)+3$$
$$-4 \stackrel{?}{=} \frac{2}{3}+3$$
$$-4=\frac{11}{3} \text{ False}$$

(−4, −4) is not a solution.

2. (a) The lines coincide. Therefore, there are infinitely many solutions.

(b) The system is consistent.

(c) The system is dependent.

3. (a) Since the slopes are not equal, the lines intersect at one point. There is one solution.

(b) The system is consistent.

(c) The system is independent.

4. $\begin{cases} 4x+y=5 & (1) \\ -x+y=0 & (2) \end{cases}$

Solve (2) for *y*.
$y=x$
Let $y = x$ in (1).
$$4x+x=5$$
$$5x=5$$
$$x=1$$
Let $x = 1$ in (2).
$$-1+y=0$$
$$y=1$$
The solution is (1, 1).

5. $\begin{cases} y=-\frac{2}{5}x+1 & (1) \\ y=-x+4 & (2) \end{cases}$

Let $y = -x + 4$ in (1).
$$-x+4=-\frac{2}{5}x+1$$
$$-x+\frac{2}{5}x=1-4$$
$$-\frac{3}{5}x=-3$$
$$x=5$$
Let $x = 5$ in (2).
$y = -5 + 4$
$y = -1$
The solution is (5, −1).

6. $\begin{cases} x=2y+11 & (1) \\ 3x-y=8 & (2) \end{cases}$

Let $x = 2y + 11$ in (2).
$$3(2y+11)-y=8$$
$$6y+33-y=8$$
$$5y=-25$$
$$y=-5$$
Let $y = -5$ in (1).
$x = 2(-5) + 11$
$x = -10 + 11$
$x = 1$
The solution is (1, −5).

7. $\begin{cases} 4x+3y=-4 & (1) \\ x+5y=-1 & (2) \end{cases}$

Solve (2) for *x*.
$x = -5y - 1$
Let $x = -5y - 1$ in (1).

$4(-5y-1)+3y=-4$
$-20y-4+3y=-4$
$-17y=0$
$y=0$
Let $y = 0$ in (2).
$x+5(0)=-1$
$x=-1$
The solution is (–1, 0).

8. $\begin{cases} y=-2x-3 & (1) \\ y=\frac{1}{2}x+7 & (2) \end{cases}$
Let $y = -2x - 3$ in (2).
$-2x-3=\frac{1}{2}x+7$
$-2x-\frac{1}{2}x=7+3$
$-\frac{5}{2}x=10$
$x=-4$
Let $x = -4$ in (1).
$y = -2(-4) - 3$
$y = 8 - 3$
$y = 5$
The solution is (–4, 5).

9. $\begin{cases} -2x+4y=2 & (1) \\ 3x+5y=-14 & (2) \end{cases}$
Multiply (1) by 3 and (2) by 2. Then add.
$\begin{cases} -6x+12y=6 \\ 6x+10y=-28 \end{cases}$
$22y=-22$
$y=-1$
Let $y = -1$ in (1).
$-2x+4(-1)=2$
$-2x-4=2$
$-2x=6$
$x=-3$
The solution is (–3, –1).

10. $\begin{cases} -\frac{3}{4}x+\frac{2}{3}y=\frac{9}{4} & (1) \\ 3x-\frac{1}{2}y=-\frac{5}{2} & (2) \end{cases}$
Multiply (1) by 4 and add.
$\begin{cases} -3x+\frac{8}{3}y=9 \\ 3x-\frac{1}{2}y=-\frac{5}{2} \end{cases}$
$\frac{8}{3}y-\frac{1}{2}y=9-\frac{5}{2}$
$\frac{16}{6}y-\frac{3}{6}y=\frac{18}{2}-\frac{5}{2}$
$\frac{13}{6}y=\frac{13}{2}$
$y=3$
Let $y = 3$ in (2).
$3x-\frac{1}{2}(3)=-\frac{5}{2}$
$3x-\frac{3}{2}=-\frac{5}{2}$
$3x=-\frac{5}{2}+\frac{3}{2}$
$3x=-1$
$x=-\frac{1}{3}$
The solution is $\left(-\frac{1}{3}, 3\right)$.

11. $\begin{cases} 3(y+3)=1+4(3x-1) & (1) \\ x=\frac{y}{4}+1 & (2) \end{cases}$
Let $x=\frac{y}{4}+1$ in (1).
$3(y+3)=1+4\left[3\left(\frac{y}{4}+1\right)-1\right]$
$3y+9=1+4\left(\frac{3y}{4}+3-1\right)$
$3y+9=1+4\left(\frac{3y}{4}+2\right)$
$3y+9=1+3y+8$
$3y+9=3y+9$
$9=9$ True
The system is dependent. There are infinitely many solutions.

12. $\begin{cases} 0.4x-2.5y=-6.5 & (1) \\ x+y=5.5 & (2) \end{cases}$
Multiply (1) by 10 and (2) by –4. Then add.
$\begin{cases} 4x-25y=-65 \\ -4x-4y=-22 \end{cases}$
$-29y=-87$
$y=3$

Let $y = 3$ in (2).
$x + 3 = 5.5$
$x = 2.5$
The solution is (2.5, 3).

13. $\begin{cases} 2(y+1) = 3x+4 & (1) \\ x = \frac{2}{3}y - 3 & (2) \end{cases}$

Let $x = \frac{2}{3}y - 3$ in (1).

$2(y+1) = 3\left(\frac{2}{3}y - 3\right) + 4$
$2y + 2 = 2y - 9 + 4$
$2y + 2 = 2y - 5$
$2 = -5$ False
The system is inconsistent. There is no solution.

Section 4.4

Preparing for Solving Direct Translation, Geometry, and Uniform Motion Problems Using Systems of Linear Equations

P1. $r = 45$, $t = 3$
$d = rt$
$d = 45 \cdot 3$
$d = 135$
You will travel 135 miles.

P2. $P = 3600$, $r = 0.015$, $t = \frac{1}{12}$
$I = Prt$
$I = 3600 \cdot 0.015 \cdot \frac{1}{12}$
$I = 4.5$
The interest paid after 1 month is \$4.50.

4.4 Quick Checks

1. We are looking for two unknown numbers. Let x represent the first number and y represent the second number. The sum of the numbers is 104, so $x + y = 104$. The second number is 25 less than twice the first number, so $y = 2x - 25$. The system is
$\begin{cases} x + y = 104 & (1) \\ y = 2x - 25 & (2) \end{cases}$
Let $y = 2x - 25$ in equation (1).
$x + (2x - 25) = 104$
$3x - 25 = 104$
$3x = 129$
$x = 43$
Let $x = 43$ in equation (2).
$y = 2(43) - 25$
$y = 86 - 25$
$y = 61$
The sum of 43 and 61 is 104. Twice 43 minus 25 is 61. The numbers are 43 and 61.

2. We are looking for the length and the width of the yard. Let l represent the length and w represent the width of the yard. The perimeter is 400 yards, so $2l + 2w = 400$. The length is three times the width, so $l = 3w$. The system is
$\begin{cases} 2l + 2w = 400 & (1) \\ l = 3w & (2) \end{cases}$
Let $l = 3w$ in equation (1).
$2(3w) + 2w = 400$
$6w + 2w = 400$
$8w = 400$
$w = 50$
Let $w = 50$ in equation (2).
$l = 3(50) = 150$
With $l = 150$ and $w = 50$, the perimeter is $2(150) + 2(50) = 400$ yards. The length (150 yards) is three times the width (50 yards). The length of the yard is 150 yards and the width is 50 yards.

3. True

4. Supplementary angles are angles whose measures sum to $\underline{180°}$.

5. This is a complementary angle problem. We are looking for the measures of two angles whose sum is 90°. Let x represent the measure of the smaller angle and y represent the measure of the larger angle. Since the angles are complementary, we have $x + y = 90$. Since the larger angle measures 18° more than the smaller angle, we have
$y = x + 18$. The system is
$\begin{cases} x + y = 90 & (1) \\ y = x + 18 & (2) \end{cases}$
Let $y = x + 18$ in equation (1).
$x + (x + 18) = 90$
$2x + 18 = 90$
$2x = 72$
$x = 36$
Let $x = 36$ in equation (2).
$y = 36 + 18 = 54$
The sum of 36 and 54 is 90, and 54 is 18 more than 36. The angles measure 36° and 54°.

6. This is a supplementary angle problem. We are looking for the measures of two angles whose sum is 180°. Let x represent the measure of the smaller angle and y represent the measure of the larger angle. Since the angles are supplementary, $x + y = 180$. Since the larger angle measures 16° less than three times the measure of the smaller angle,
$y = 3x - 16$.
The system is

$$\begin{cases} x + y = 180 & (1) \\ y = 3x - 16 & (2) \end{cases}$$

Let $y = 3x - 16$ in equation (1).

$$\begin{aligned} x + (3x - 16) &= 180 \\ 4x - 16 &= 180 \\ 4x &= 196 \\ x &= 49 \end{aligned}$$

Let $x = 49$ in equation (2).
$y = 3(49) - 16 = 147 - 16 = 131$
The sum of 49 and 131 is 180, and 131 is 16 less than three times 49. The angles measure 49° and 131°.

7. True

8. This is a uniform motion problem. We want to determine the air speed of the plane and the effect of wind resistance. Let a represent the airspeed of the plane and w represent the impact of wind resistance.

	Distance	Rate	Time
East	1200	$a + w$	3
West	1200	$a - w$	4

The system is

$$\begin{cases} 3(a + w) = 1200 \\ 4(a - w) = 1200 \end{cases}$$

or

$$\begin{cases} 3a + 3w = 1200 & (1) \\ 4a - 4w = 1200 & (2) \end{cases}$$

Multiply (1) by 4 and (2) by 3. Then add.

$$\begin{cases} 12a + 12w = 4800 \\ 12a - 12w = 3600 \end{cases}$$

$$\begin{aligned} 24a &= 8400 \\ a &= 350 \end{aligned}$$

Let $a = 350$ in equation (1).

$$\begin{aligned} 3(350) + 3w &= 1200 \\ 1050 + 3w &= 1200 \\ 3w &= 150 \\ w &= 50 \end{aligned}$$

Flying west the groundspeed of the plane is $350 - 50 = 300$ miles per hour. Flying west, the plane flies 1200 miles in 4 hours for an average speed of 300 miles per hour. Flying east the groundspeed of the plane is $350 + 50 = 400$ miles per hour. Flying east, the plane flies 1200 miles in 3 hours for an average speed of 400 miles per hour. The airspeed of the plane is 350 miles per hour. The impact of wind resistance on the plane is 50 miles per hour.

4.4 Exercises

9. Two times the smaller number is written as $2x$. Twelve more than one-half the larger number is written as $12 + \frac{1}{2}y$. The equation is $12 + \frac{1}{2}y$.

11. Since the perimeter is 59 inches and the formula for perimeter is $P = 2l + 2w$, the equation is $2l + 2w = 59$.

13. The rate of the boat going upstream is $r - c$ and the time going upstream is 8 hours. The distance is 16 miles. Multiply the rate by the time and set equal to the distance. The equation is $8(r - c) = 16$.

15. Let x represent the first number and y represent the second number. The sum of x and y, $x + y$, is 82, while the difference of x and y, $x - y$, is 16.

$$\begin{cases} x + y = 82 \\ x - y = 16 \end{cases}$$

$$\begin{aligned} 2x &= 98 \\ x &= 49 \\ 49 + y &= 82 \\ y &= 33 \end{aligned}$$

The numbers are 33 and 49.

17. Let x represent the first number and y represent the second number. The sum of x and y, $x + y$, is 51, while twice the first subtracted from the second, $y - 2x$, is 9.

$$\begin{cases} x + y = 51 \\ y - 2x = 9 \end{cases}$$

$$\begin{cases} x + y = 51 & (1) \\ -2x + y = 9 & (2) \end{cases}$$

Multiply (1) by 2, then add.

$$\begin{cases} 2x+2y=102 \\ -2x+y=9 \end{cases}$$
$$3y=111$$
$$y=37$$
$$x+37=51$$
$$x=14$$
The numbers are 14 and 37.

19. Let t represent Thursday night's attendance, and let f represent Friday night's attendance. The total attendance for the two nights, $t+f$, was 77,000, while 7000 more than two-thirds of Friday's attendance, $7000+\frac{2}{3}f$, was Thursday's attendance, t.
$$\begin{cases} t+f=77{,}000 & (1) \\ 7000+\frac{2}{3}f=t & (2) \end{cases}$$
Let $t=7000+\frac{2}{3}f$ in (1).
$$7000+\frac{2}{3}f+f=77{,}000$$
$$\frac{5}{3}f=70{,}000$$
$$f=\frac{3}{5}(70{,}000)=42{,}000$$
$$t+42{,}000=77{,}000$$
$$t=35{,}000$$
Thursday night's attendance was 35,000; Friday night's attendance was 42,000.

21. Let s be the amount in stocks and b be the amount in bonds. The total amount of money, $s + b$, is \$21,000. Four times the amount in bonds equals three times the amount in stocks, or $4b = 3s$.
$$\begin{cases} s+b=21{,}000 & (1) \\ 4b=3s & (2) \end{cases}$$
Let $b=\frac{3s}{4}$ in (1).
$$s+\frac{3s}{4}=21{,}000$$
$$\frac{7s}{4}=21{,}000$$
$$s=12{,}000$$
$$12{,}000+b=21{,}000$$
$$b=9000$$
Therefore, \$12,000 should be invested in stocks and \$9000 in bonds.

23. Let l represent the length of the garden along the garage, and w represent the width. Since the length is 3 feet more than the width, $l = w + 3$. The sides that need to be fenced are w feet, l feet, and w feet, which total $2w + l$, or 30 feet.
$$\begin{cases} l=w+3 & (1) \\ 2w+l=30 & (2) \end{cases}$$
Let $l = w + 3$ in (2).
$$2w+w+3=30$$
$$3w=27$$
$$w=9$$
$$l=9+3=12$$
The length is 12 feet and the width is 9 feet.

25. Let l and w represent the length and width of the rectangle, respectively. The perimeter of 70 meters is $2l + 2w$. Since the width is 40% of the length, $w = 0.40l$.
$$\begin{cases} 2l+2w=70 & (1) \\ w=0.40l & (2) \end{cases}$$
Let $w = 0.40l$ in (1).
$$2l+2(0.40l)=70$$
$$2l+0.8l=70$$
$$2.8l=70$$
$$l=25$$
$$w=0.40(25)=10$$
The length is 25 meters and the width is 10 meters.

27. Let x and y represent the measures of the angles. Since the angles are complementary, the sum of their measures, $x + y$, is 90. Since one angle, say x, is 15° more than half its complement y, $x=\frac{1}{2}y+15$.
$$\begin{cases} x+y=90 & (1) \\ x=\frac{1}{2}y+15 & (2) \end{cases}$$
Let $x=\frac{1}{2}y+15$ in (1).
$$\frac{1}{2}y+15+y=90$$
$$\frac{3}{2}y=75$$
$$y=\frac{2}{3}(75)=2(25)=50$$
$$x+50=90$$
$$x=40$$
The angles measure 40° and 50°.

29. Let x and y represent the measures of the angles. Since the angles are supplementary, the sum of their measures, $x + y$, is 180°. Since one angle, say x, is 30° less than one-third of its supplement y, $x = \frac{1}{3}y - 30$.

$$\begin{cases} x + y = 180 & (1) \\ x = \frac{1}{3}y - 30 & (2) \end{cases}$$

Let $x = \frac{1}{3}y - 30$ in (1).

$$\frac{1}{3}y - 30 + y = 180$$
$$\frac{4}{3}y = 210$$
$$y = \frac{3}{4}(210) = 157.5$$
$$x + 157.5 = 180$$
$$x = 22.5$$

The angles measure 22.5° and 157.5°.

31. Let s be his paddling speed and c the speed of the current. Paddling with the current, his speed is $s + c$, or 4.3 mph. Paddling against the current, his speed is $s - c$, or 3.5 mph.

$$\begin{cases} s + c = 4.3 \\ s - c = 3.5 \end{cases}$$
$$2s = 7.8$$
$$s = 3.9$$
$$3.9 - c = 3.5$$
$$-c = -0.4$$
$$c = 0.4$$

The current is 0.4 mph and his still-water speed is 3.9 mph.

33. Let b be his biking speed and w the wind speed. Biking with the wind, his speed is $b + w$; biking against the wind, his speed is $b - w$. Biking for 6 hours against the wind is a distance of $6(b - w)$, while biking for 5 hours with the wind is a distance of $5(b + w)$. The distance is 60 miles.

$$\begin{cases} 6(b - w) = 60 & (1) \\ 5(b + w) = 60 & (2) \end{cases}$$

Divide (1) by 6 and (2) by 5, then add.

$$\begin{cases} b - w = 10 \\ b + w = 12 \end{cases}$$
$$2b = 22$$
$$b = 11$$
$$11 - w = 10$$
$$w = 1$$

The biking speed is 11 mph and the wind is 1 mph.

35. Let n be the speed of the northbound train and s the speed of the southbound train. Since the northbound train is going 12 mph slower, $n = s - 12$. After 4 hours, the northbound train will have gone $4n$ miles and the southbound train will have gone $4s$ miles. The total distance, $4n + 4s$, is 528 miles.

$$\begin{cases} n = s - 12 & (1) \\ 4n + 4s = 528 & (2) \end{cases}$$

Let $n = s - 12$ in (2).

$$4(s - 12) + 4s = 528$$
$$4s - 48 + 4s = 528$$
$$8s = 576$$
$$s = 72$$
$$n = 72 - 12 = 60$$

The northbound train is going 60 mph; the southbound train is going 72 mph.

37. Let v be the length of time that Vanessa rides, and r the length of time that Richie rides. Since Vanessa leaves the camp 30 minutes (0.5 hour) later, $r = v + 0.5$. In v hours, Vanessa goes $10v$ miles, while Richie goes $7r$ miles in r hours. The distance between them is 7 miles when $10v - 7r = 7$.

$$\begin{cases} r = v + 0.5 & (1) \\ 10v - 7r = 7 & (2) \end{cases}$$

Let $r = v + 0.5$ in (2).

$$10v - 7(v + 0.5) = 7$$
$$10v - 7v - 3.5 = 7$$
$$3v = 10.5$$
$$v = 3.5$$
$$r = 3.5 + 0.5 = 4.0$$

Richie has been riding for 4 hours when Vanessa is 7 miles ahead.

39. Let p be the speed of the Piper aircraft, and w the speed of the wind. Flying with the wind, the rate is $p + w$, and the distance traveled in 3 hours is $3(p + w)$, which is 600. Flying against the wind, the rate is $p - w$, and the distance traveled in 4 hours is $4(p - w)$, which is 600.

$$\begin{cases} 3(p + w) = 600 & (1) \\ 4(p - w) = 600 & (2) \end{cases}$$

Divide (1) by 3 and (2) by 4, then add.

$$\begin{cases} p+w=200 \\ p-w=150 \end{cases}$$
$$2p \quad =350$$
$$p=175$$
$$3(175+w)=600$$
$$175+w=200$$
$$w=25$$
The average wind speed was 25 mph and the speed of the Piper was 175 mph.

41. Let x be the digit in the tens place and y the digit in the ones place. Since the sum of the digits is 6, we have $x + y = 6$.
The original number equals $10x + y$. With the digits reversed, the value is $10y + x$. The difference between the new number and the original number is 18, so we have $(10y + x) - (10x + y) = 18$. Simplify this equation.
$$(10y+x)-(10x+y)=18$$
$$10y+x-10x-y=18$$
$$-9x+9y=18$$
$$-9(x-y)=18$$
$$x-y=-2$$
We have the following system. Add the equations.
$$\begin{cases} x+y=6 & (1) \\ x-y=-2 & (2) \end{cases}$$
$$2x \quad =4$$
$$x=2$$
Let $x = 2$ in equation (1).
$$2+y=6$$
$$y=4$$
The tens digit is 2 and the ones digit is 4. The number is 24.

Section 4.5

Preparing for Solving Mixture Problems Using Systems of Linear Equations

P1. $P = 1200,\ r = 0.14,\ t = \frac{1}{12}$
$$I = Prt$$
$$I = 1200 \cdot 0.14 \cdot \frac{1}{12}$$
$$I = 14$$
Roberta will be charged \$14 interest for one month. After one month, her balance will be \$1200 + \$14 = \$1214.

P2. $0.25x = 80$
$$\frac{0.25x}{0.25} = \frac{80}{0.25}$$
$$x = 320$$
The solution set is $\{320\}$.

4.5 Quick Checks

1. The general equation we use to solve mixture problems is
<u>number of units of the same kind</u> · <u>rate</u> = <u>amount</u>.

2. Let a represent the number of adult tickets and c represent the number of children's tickets.

	Number	Cost per person	
Adult	a	42.95	$42.95a$
Children	c	15.95	$15.95c$
Total	9		\$332.55

3. We want to know the number of dimes in the piggy bank. Let q represent the number of quarters and d represent the number of dimes.

	Number	Value	Total Value
Quarters	q	0.25	$0.25q$
Dimes	d	0.10	$0.10d$
Total	85		14.50

The system is
$$\begin{cases} q+d=85 & (1) \\ 0.25q+0.10d=14.50 & (2) \end{cases}$$
Multiply (1) by –0.10 and then add.
$$\begin{cases} -0.10q-0.10d=-8.50 \\ 0.25q+0.10d=14.50 \end{cases}$$
$$0.15q \quad =6.00$$
$$q=40$$
Let $q = 40$ in equation (1).
$$40+d=85$$
$$d=45$$
The total number of coins is 40 + 45 = 85. The value of the coins is
0.25(40) + 0.10(45) = 14.50. There are 45 dimes.

4. The simple interest formula states that interest = <u>principal</u> · <u>rate</u> · <u>time</u>.

5. *True or False:* When we solve a simple interest problem using the mixture model, we assume that $t = 1$. True

6. We need to determine how much should be placed in each investment. Let a represent the amount invested in Aa-bonds and b represent the amount invested in B-rated bonds.

	Principal	Rate	Interest
Aa-bonds	a	0.05	$0.05a$
B-rated	b	0.07	$0.07b$
Total	90,000		5500

The system is

$$\begin{cases} a+b=90{,}000 & (1) \\ 0.05a+0.07b=5500 & (2) \end{cases}$$

Multiply (1) by –0.05 and then add.

$$\begin{cases} -0.05a-0.05b=-4500 \\ 0.05a+0.07b=5500 \end{cases}$$

$$0.02b=1000$$
$$b=50{,}000$$

Let b = 50,000 in equation (1).

$$a+50{,}000=90{,}000$$
$$a=40{,}000$$

The sum of \$40,000 and \$50,000 is \$90,000, and the total interest is 0.05(40,000) + 0.07(50,000) = \$5500. Faye should invest \$40,000 in Aa-bonds and \$50,000 in B-rated bonds.

7. We need to know the number of pounds of each type of coffee that are required in the mix. Let b represent the pounds of Brazilian coffee and c represent the pounds of Colombian coffee.

	Price per pound	Number of pounds	Revenue
Brazilian	6	b	$6b$
Colombian	10	c	$10c$
Blend	9	20	9(20) = 180

The system is

$$\begin{cases} b+c=20 & (1) \\ 6b+10c=180 & (2) \end{cases}$$

Multiply (1) by –6 and then add.

$$\begin{cases} -6b-6c=-120 \\ 6b+10c=180 \end{cases}$$

$$4c=60$$
$$c=15$$

Let c = 15 in equation (1).

$$b+15=20$$
$$b=5$$

The total weight is 5 + 15 = 20 pounds. The total revenue is 6(5) + 10(15) = \$180. Mix 5 pounds of Brazilian coffee with 15 pounds of Colombian coffee.

8. We need to know how many gallons of each type of ice cream are required for the mixture. Let x represent the gallons of 5% butterfat ice cream and y represent the gallons of 15% butterfat ice cream.

	Gallons	Concentration	Amount of Alcohol
5% Butterfat	x	0.05	$0.05x$
15% Butterfat	y	0.15	$0.15y$
Total	200	0.09	$0.09(200) = 18$

The system is

$$\begin{cases} x + y = 200 & (1) \\ 0.05x + 0.15y = 18 & (2) \end{cases}$$

Multiply (1) by −0.05 and then add.

$$\begin{cases} -0.05x - 0.05y = -10 \\ 0.05x + 0.15y = 18 \end{cases}$$

$$0.10y = 8$$
$$y = 80$$

Let $y = 80$ in equation (1).

$$x + 80 = 200$$
$$x = 120$$

The total number of gallons is 120 + 80 = 200. The total butterfat is 0.05(120) + 0.15(80) = 18 gallons. Mix 120 gallons of 5% butterfat ice cream with 80 gallons of 15% butterfat ice cream.

4.5 Exercises

9. Let a be the number of adult tickets and s be the number of student tickets.

	number	cost per person	total value
adults' tickets	a	4	$4a$
students' tickets	s	1.5	$1.5s$
total	215		580

$$\begin{cases} a + s = 215 \\ 4a + 1.5s = 580 \end{cases}$$

11. Let s be the amount in savings and m be the amount in a money market.

	Principal	Rate	Interest
Savings account	s	0.05	$0.05s$
money market	m	0.03	$0.03m$
total	1600		50

$$\begin{cases} s + m = 1600 \\ 0.05s + 0.03m = 50 \end{cases}$$

13. Let m be the pounds of mild coffee and r be the pounds of robust coffee.

	number of pounds	price per pound	total value
mild coffee	m	7.50	$7.5m$
robust coffee	r	10	$10r$
total	12	8.75	8.75(12)

$$\begin{cases} m+r=12 \\ 7.5m+10r=8.75(12) \end{cases}$$

15.

	number	cost	total
adults	a	32	$32a$
children	c	24	$24c$
total	11		296

$$\begin{cases} a+c=11 \\ 32a+24c=296 \end{cases}$$

17.

	P	R	I
A account	A	0.10	$0.10A$
B account	B	0.07	$0.07B$
total	2250		195

$$\begin{cases} A+B=2250 \\ 0.01A+0.07B=195 \end{cases}$$

19.

	number bouquets	price per pound	total
Red	r	5.85	$5.85r$
Yellow	y	4.20	$4.20y$
total	$r+y$		128.85

$$\begin{cases} r=3+2y \\ 5.85r+4.20y=128.85 \end{cases}$$

21. Let s be the number of student tickets and n be the number of nonstudents.

	number	price	amount
students	s	8	$8s$
nonstudents	n	10	$10n$
total	390		3270

$$\begin{cases} s+n=390 & (1) \\ 8s+10n=3270 & (2) \end{cases}$$

Solve (1) for n.

$n = 390 - s$

Let $n = 390 - s$ in (2).

$$\begin{aligned} 8s+10(390-s) &= 3270 \\ 8s+3900-10s &= 3270 \\ -2s &= -630 \\ s &= 315 \end{aligned}$$

There were 315 student tickets sold.

23. Let b be the flats of bedding plants and h be the number of hanging baskets.

	number	price	amount
bedding	b	13	$13b$
hanging	h	18	$18h$
total			8800

$$\begin{cases} b=2h & (1) \\ 13b+18h=8800 & (2) \end{cases}$$

Solve (1) for h.

$$h=\frac{b}{2}$$

Let $h=\frac{b}{2}$ in (2).

$$\begin{aligned} 13b+18\left(\frac{b}{2}\right) &= 8800 \\ 13b+9b &= 8800 \\ 22b &= 8800 \\ b &= 400 \end{aligned}$$

They hope to sell 400 flats of bedding plants.

25. Let n be the number of nickels and d be the number of dimes.

	number	value	total amount
nickels	n	0.05	$0.05n$
dimes	d	0.10	$0.10d$
total	150		12

$$\begin{cases} n+d=150 & (1) \\ 0.05n+0.10d=12 & (2) \end{cases}$$

Solve (1) for n.

$n = 150 - d$

Let $n = 150 - d$ in (2).

$$\begin{aligned} 0.05(150-d)+0.10d &= 12 \\ 7.5-0.05d+0.10d &= 12 \\ 0.05d &= 4.5 \\ d &= 90 \end{aligned}$$

Let $d = 90$ in (1).

$$\begin{aligned} n+90 &= 150 \\ n &= 60 \end{aligned}$$

There are 60 nickels and 90 dimes in the jar.

27. Let f be the cost of each first-class stamp and p be the cost of each postcard stamp.

	number	price	amount
first-class	20	f	$20f$
postcard	10	p	$10p$
total			11.10

	number	price	amount
first-class	80	f	$80f$
postcard	50	p	$50p$
total			47.10

$$\begin{cases} 20f+10p=11.10 & (1) \\ 80f+50p=47.10 & (2) \end{cases}$$

Multiply (1) by −5, then add.

$$\begin{cases} -100f-50p=-55.50 \\ \;\;80f+50p=47.10 \end{cases}$$

$$\begin{aligned} -20f &= -8.4 \\ f &= 0.42 \end{aligned}$$

Let $f = 0.42$ in (1).

$$\begin{aligned} 20(0.37)+10p &= 9.30 \\ 7.4+10p &= 9.30 \\ 10p &= 1.9 \\ p &= 0.19 \end{aligned}$$

A first-class stamp was \$0.42 and a postcard stamp was \$0.19.

29. Let x be the amount invested at 5% and y be the amount invested at 8%.

	P	R	I
5% account	x	0.05	$0.05x$
8% account	y	0.08	$0.08y$
total	10,000		575

$$\begin{cases} x+y=10{,}000 & (1) \\ 0.05x+0.08y=575 & (2) \end{cases}$$

Solve (1) for x.

$x = 10{,}000 - y$

Let $x = 10{,}000 - y$ in (2).

$$\begin{aligned} 0.05(10{,}000-y)+0.08y &= 575 \\ 500-0.05y+0.08y &= 575 \\ 0.03y &= 75 \\ y &= 2500 \end{aligned}$$

Let $y = 2500$ in (1).

$$\begin{aligned} x+2500 &= 10{,}000 \\ x &= 7500 \end{aligned}$$

He should invest \$7500 in the 5% account and \$2500 in the 8% account.

31. Let r be the amount invested in the risky plan and s be the amount invested in the safer plan.

	P	R	I
risky	r	0.12	$0.12r$
safe	s	0.08	$0.08s$
total	5000		528

$$\begin{cases} r+s=5000 & (1) \\ 0.12r+0.08s=528 & (2) \end{cases}$$

Solve (1) for r.

$r = 5000 - s$

Let $r = 5000 - s$ in (2).

$$\begin{aligned} 0.12(5000-s)+0.08s &= 528 \\ 600-0.12s+0.08s &= 528 \\ -0.04s &= -72 \\ s &= 1800 \end{aligned}$$

Let $s = 1800$ in (1).

$r + 1800 = 5000$
$r = 3200$

She invested \$3200 in the risky plan and \$1800 in the safer plan.

33. Let a be the pounds of arbequina olives and g be the pounds of green olives.

	number pounds	cost	total amount
arbequina	a	9	$9a$
green	g	4	$4g$
total	5	6	5(6)

$$\begin{cases} a + g = 5 & (1) \\ 9a + 4g = 30 & (2) \end{cases}$$

Multiply (1) by –4, then add.

$$\begin{cases} -4a - 4g = -20 \\ 9a + 4g = 30 \end{cases}$$
$$5a = 10$$
$$a = 2$$

$2 + g = 5$
$g = 3$

2 pounds of arbequina and 3 pounds of green olives should be mixed.

35. Let x be the pounds of \$2.75 per pound coffee and y be the pounds of \$5 per pound coffee.

	number of pounds	price	total amount
\$2.75/lb	x	2.75	$2.75x$
\$5/lb	y	5	$5y$
total	100	3.90	3.90(100)

$$\begin{cases} x + y = 100 & (1) \\ 2.75x + 5y = 390 & (2) \end{cases}$$

Solve (1) for x.
$x = 100 - y$
Let $x = 100 - y$ in (2).
$2.75(100 - y) + 5y = 390$
$275 - 2.75y + 5y = 390$
$2.25y = 115$
$y = 51.1$
Let $y = 51.1$ in (1).
$x + 51.1 = 100$
$x = 48.9$

About 48.9 pounds of the \$2.75 per pound coffee should be blended with about 51.1 pounds of the \$5 per pound coffee.

37. Let r be the pounds of rye seed and b be the pounds of blue-grass seed.

	number of pounds	price	total amount
rye	r	4.20	$4.20r$
blue-grass	b	3.75	$3.75b$
total	180	3.95	3.95(180)

$$\begin{cases} r + b = 180 & (1) \\ 4.2r + 3.75b = 711 & (2) \end{cases}$$

Solve (1) for r.
$r = 180 - b$
Let $r = 180 - b$ in (2).
$4.2(180 - b) + 3.75b = 711$
$756 - 4.2b + 3.75b = 711$
$-0.45b = -45$
$b = 100$
Let $b = 100$ in (1).
$r + 100 = 180$
$r = 80$

The mixture has 80 pounds of rye seed and 100 pounds of blue-grass seed.

39. Let x be the ml of 30% saline solution and y be the ml of 60% saline solution.

	ml	concentration	amount
30%	x	0.3	$0.3x$
60%	y	0.6	$0.6y$
total	60	0.5	0.5(60)

$$\begin{cases} x + y = 60 & (1) \\ 0.3x + 0.6y = 30 & (2) \end{cases}$$

Solve (1) for x.
$x = 60 - y$
Let $x = 60 - y$ in (2).
$0.3(60 - y) + 0.6y = 30$
$18 - 0.3y + 0.6y = 30$
$0.3y = 12$
$y = 40$
Let $y = 40$ in (1).
$x + 40 = 60$
$x = 20$

She should add 20 ml of 30% saline solution to 40 ml of 60% saline solution.

41. Let x be the liters of 10% silver and y be the total liters of the 30% silver.

	liters	concentration	amount
10%	x	0.1	$0.1x$
50%	70	0.5	70(0.5)
total (30%)	y	0.3	$0.3y$

$$\begin{cases} x+70=y & (1) \\ 0.1x+35=0.3y & (2) \end{cases}$$

Let $y = x + 70$ in (2).

$$0.1x+35=0.3(x+70)$$
$$0.1x+35=0.3x+21$$
$$14=0.2x$$
$$70=x$$

So 70 liters of 10% silver should be added.

43. Let x be the gallons of 25% antifreeze that must remain. Let y be the gallons of water (or the amount of antifreeze drained).

	gallons	concentration	amount
25%	x	0.25	$0.25x$
water	y	0	0
total (15%)	3	0.15	0.15(3)

$$\begin{cases} x+y=3 & (1) \\ 0.25x=0.45 & (2) \end{cases}$$

Solve (2) for x.

$$x=\frac{0.45}{0.25}=1.8$$

Let $x = 1.8$ in (1).

$$1.8+y=3$$
$$y=1.2$$

Therefore, 1.8 gallons of 25% antifreeze should remain and 1.2 gallons should be drained and replaced with water.

45. Let x be the amount invested at 5% and y be the amount invested at a loss of 7.5%.

	P	R	I
5%	x	0.05	$0.05x$
7.5%	y	0.075	$0.075y$
total	10,000		25

$$\begin{cases} x+y=10{,}000 & (1) \\ 0.05x-0.075y=25 & (2) \end{cases}$$

Solve (1) for x.

$x = 10{,}000 - y$

Let $x = 10{,}000 - y$ in (2).

$$0.05(10{,}000-y)-0.075y=25$$
$$500-0.05y-0.075y=25$$
$$-0.125y=-475$$
$$y=3800$$

Let $y = 3800$ in (1).

$$x+3800=10{,}000$$
$$x=6200$$

He invested \$6200 at 5% and \$3800 at $7\frac{1}{2}\%$ loss.

47. The percentage ethanol in the final solution is greater than the percentage of ethanol in either of the two original solutions.

Section 4.6

Preparing for Systems of Linear Inequalities

P1. $3x-2\ge 7$

$$3x\ge 9$$
$$x\ge 3$$

The solution set is $\{x|x \ge 3\}$ or $[3, \infty)$.

P2. $4(x-1)<6x+4$

$$4x-4<6x+4$$
$$-2x-4<4$$
$$-2x<8$$
$$x>-4$$

The solution set is $\{x|x > -4\}$ or $(-4, \infty)$.

P3. $y > 2x - 5$

Graph the line $y = 2x - 5$ using a dashed line. Because (0, 0) satisfies the inequality, we shade the half-plane containing (0, 0).

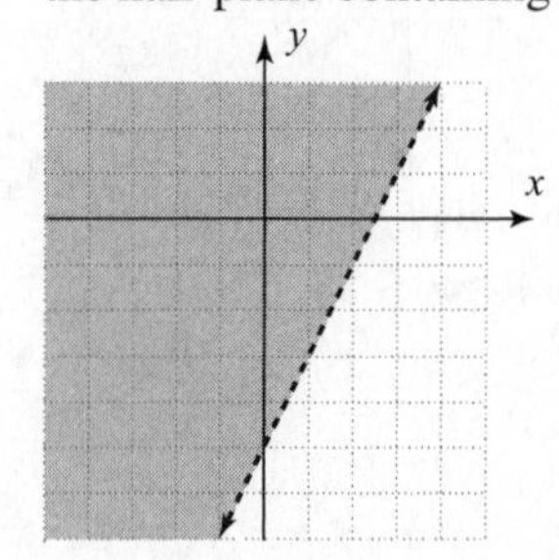

P4. $2x+3y\le 9$

$$3y\le -2x+9$$
$$y\le -\frac{2}{3}x+3$$

Graph $y=-\frac{2}{3}x+3$ using a solid line.

Because (0, 0) satisfies the inequality, we shade the half-plane containing (0, 0).

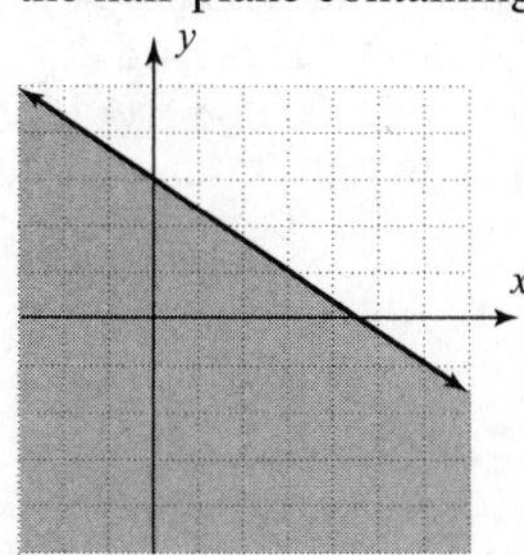

4.6 Quick Checks

1. An ordered pair is a solution of a system of linear inequalities if it makes each inequality in the system a true statement.

2. $\begin{cases} 4x+y\le 6 & (1) \\ 2x-5y<10 & (2) \end{cases}$

(a) (1, 2)

(1): $4(1)+2\le 6?$
$4+2\le 6?$
$6\le 6$ True

(2): $2(1)-5(2)<10?$
$2-10<10?$
$-8<10$ True

(1, 2) is a solution.

(b) (−1, −3)

(1): $4(-1)+(-3)\le 6?$
$-4-3\le 6?$
$-7\le 6$ True

(2): $2(-1)-5(-3)<10?$
$-2+15<10?$
$13<10$ False

(−1, −3) is not a solution.

3. When graphing linear inequalities, we use a dashed line when graphing strict inequalities (> or <) and a solid line when graphing nonstrict inequalities (≥ or ≤).

4. $\begin{cases} y\ge -3x+8 \\ y\ge 2x-7 \end{cases}$

Graph $y\ge -3x+8$ by graphing the line $y=-3x+8$ as a solid line because the inequality is nonstrict (≥). We choose the test point (0, 0). Since (0, 0) does not make the inequality true shade the half-plane not containing (0, 0).
Graph $y\ge 2x-7$ by graphing the line $y=2x-7$ as a solid line because the inequality is nonstrict (≥). We choose the test point (0, 0). Since (0, 0) makes the inequality true shade the half-plane containing (0, 0).
The overlapping shaded region is the solution of the system.

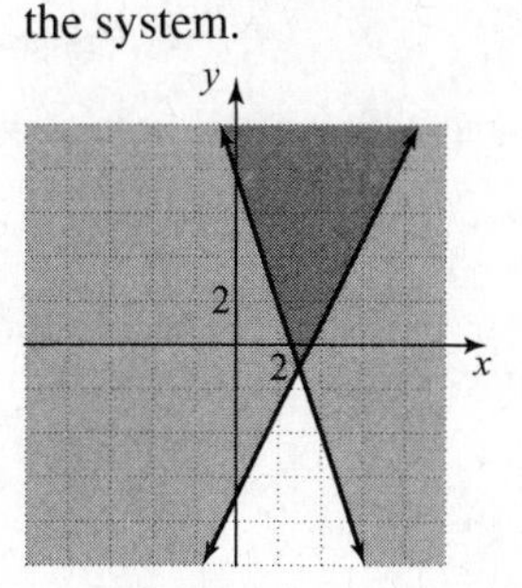

5. $\begin{cases} 4x+2y<-9 \\ x+3y<-1 \end{cases}$

Graph $4x+2y<-9\left(y<-2x-\frac{9}{2}\right)$. Use a dashed line because the inequality is strict (<).Graph $x+3y<-1\left(y<-\frac{1}{3}x-\frac{1}{3}\right)$. Use a dashed line because the inequality is strict (<).
The overlapping shaded region is the solution of the system.

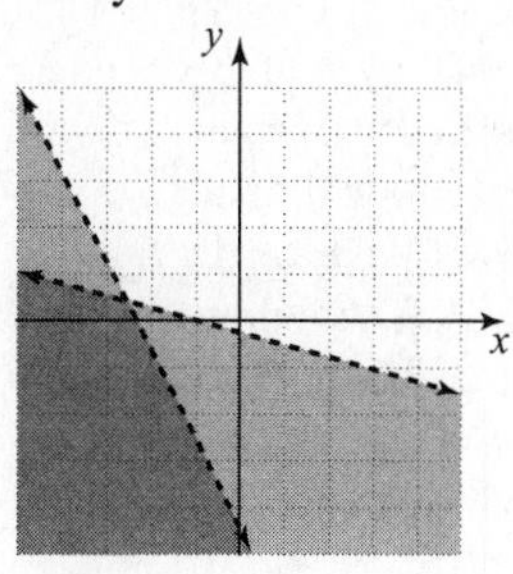

6. $\begin{cases} x+y\le 4 \\ -x+y>-4 \end{cases}$

Graph $x+y\le 4\,(y\le -x+4)$ Use a solid line because the inequality is nonstrict (≤). Graph $-x+y>-4\,(y>x-4)$ Use a dashed line because the inequality is strict (>). The overlapping shaded region is the solution of the system.

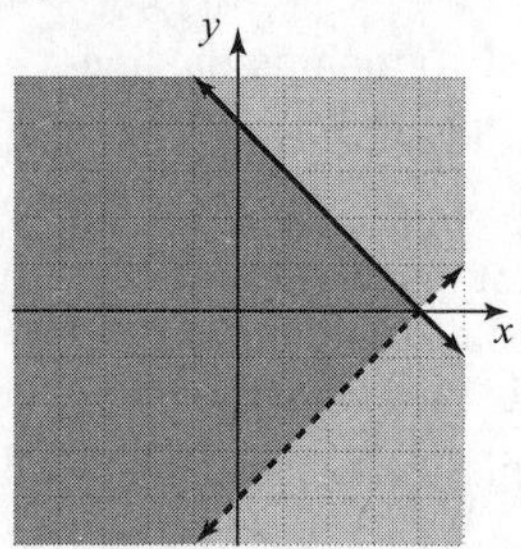

7. $\begin{cases} 3x+y>-5 \\ x+2y\le 0 \end{cases}$

Graph $3x+y>-5\,(y>-3x-5)$ Use a dashed line because the inequality is strict (>). Graph $x+2y\le 0\left(y\le -\frac{1}{2}x\right)$. Use a solid line because the inequality is nonstrict (≤). The overlapping shaded region is the solution of the system.

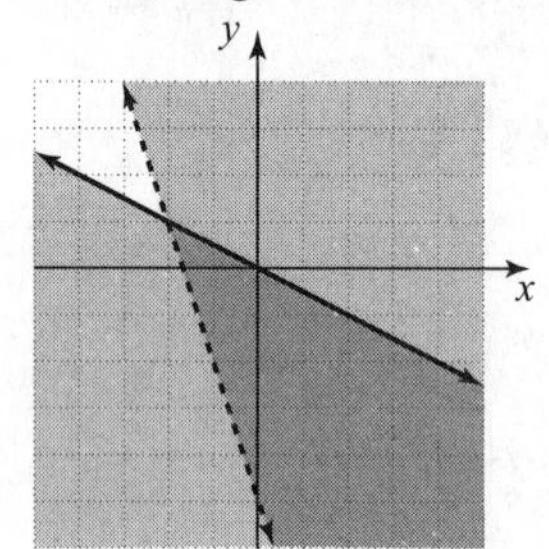

8. $\begin{cases} c+t\le 75{,}000 \\ c\le 50{,}000 \\ t\ge 25{,}000 \end{cases}$

(a) We draw a rectangular coordinate system with the horizontal axis labeled *c* and the vertical axis labeled *t*. Each axis is in thousands, so we draw the line $c+t=75$ and shade below. We draw the line $c=50$ and shade to the left. We draw the line $t=25$ and shade above.

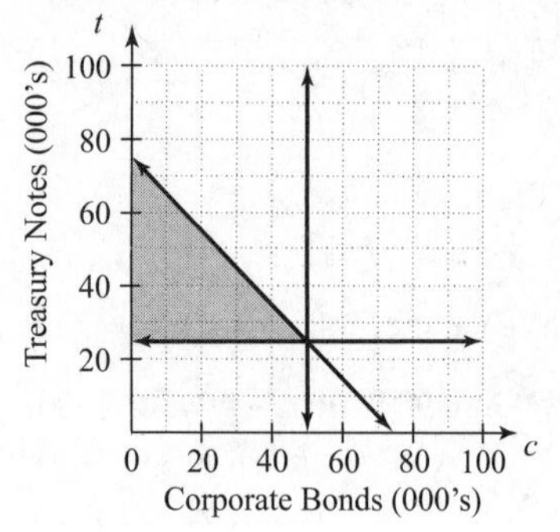

(b) Yes, Jack and Mary can invest $30,000 in corporate bonds and $35,000 in Treasury notes because these values lie within the shaded region. Put another way, $c = 30{,}000$ and $t = 35{,}000$ satisfies all three inequalities.

(c) No, Jack and Mary cannot invest $60,000 in corporate bonds and $15,000 in Treasury notes because these values do not lie within the shaded region. Put another way, $c = 60{,}000$ and $t = 15{,}000$ does not satisfy the inequality $c \le 50{,}000$.

4.6 Exercises

9. $\begin{cases} x\ge 5 & (1) \\ y<-\frac{1}{2}x+3 & (2) \end{cases}$

(a) (5, −2)
(1): $5\ge 5$ True
(2): $-2<-\frac{1}{2}(5)+3?$
$-2<-\frac{5}{2}+3?$
$-2<\frac{1}{2}$ True
(5, −2) is a solution.

(b) (10, −4)
(1): $10\ge 5$ True
(2): $-4<-\frac{1}{2}(10)+3?$
$-4<-5+3?$
$-4<-2$ True
(10, −4) is a solution.

(c) (8, −3)
(1): $8\ge 5$ True
(2): $-3<-\frac{1}{2}(8)+3?$
$-3<-4+3?$
$-3<-1$ True
(8, −3) is a solution.

11. $\begin{cases} 2x+y>-4 & (1) \\ x-y\le 1 & (2) \end{cases}$

(a) (−2, 1)
(1): $2(-2)+1>-4?$
$-4+1>-4?$
$-3>-4$ True

(2): $-2-1 \le 1$?
$-3 \le 1$ True
$(-2, 1)$ is a solution.

(b) $(-1, -2)$
(1): $2(-1)+(-2) > -4$?
$-2-2 > -4$?
$-4 > -4$ False
$(-1, -2)$ is not a solution.

(c) $(2, -3)$
(1): $2(2)+(-3) > -4$?
$4-3 > -4$?
$1 > -4$ True
(2): $2-(-3) \le 1$?
$2+3 \le 1$?
$5 \le 1$ False
$(2, -3)$ is not a solution.

13. $\begin{cases} x > 2 \\ y \le -1 \end{cases}$

Graph $x > 2$ with a dashed line and $y \le -1$ with a solid line. Graph on the same rectangular coordinate system. The overlapping shaded region represents the solution.

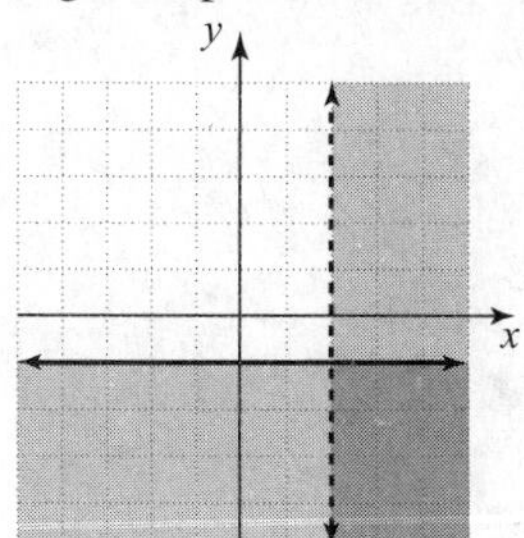

15. $\begin{cases} y > -2 \\ x > -3 \end{cases}$

Graph $y > -2$ with a dashed line and $x > -3$ with a dashed line. Graph on the same rectangular coordinate system. The overlapping shaded region represents the solution.

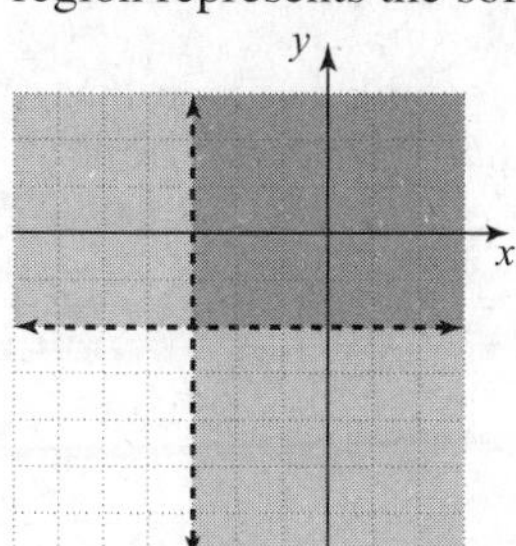

17. $\begin{cases} x+y < 3 \\ x-y > 5 \end{cases}$

Graph $x + y < 3$ $(y < -x + 3)$ with a dashed line and $x - y > 5$ $(y < x - 5)$ with a dashed line. Graph on the same rectangular coordinate system. The overlapping shaded region represents the solution.

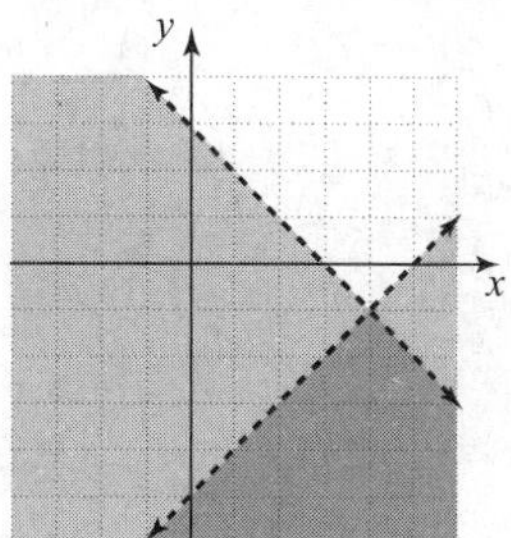

19. $\begin{cases} x+y > 3 \\ 2x-y > 4 \end{cases}$

Graph $x + y > 3$ $(y > -x + 3)$ with a dashed line and $2x - y > 4$ $(y < 2x - 4)$ with a dashed line. Graph on the same rectangular coordinate system. The overlapping shaded region represents the solution.

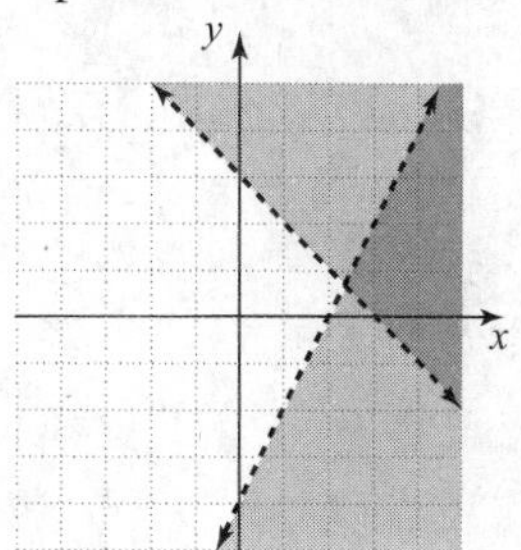

21. $\begin{cases} x < 2 \\ y < \frac{1}{2}x+3 \end{cases}$

Graph $x < 2$ with a dashed line and $y < \frac{1}{2}x+3$ with a dashed line. Graph on the same rectangular coordinate system. The overlapping shaded region represents the solution.

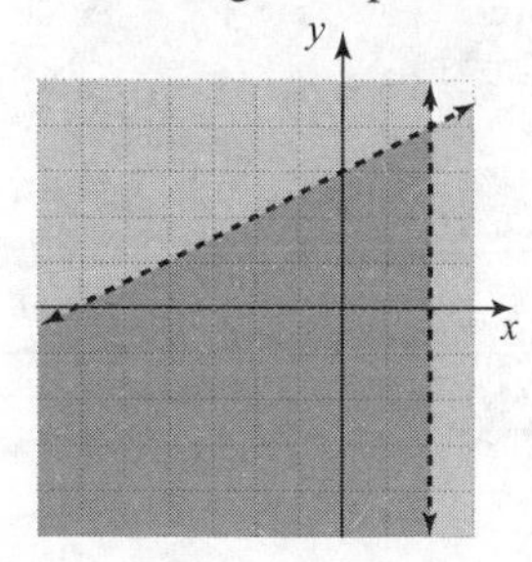

23. $\begin{cases} x \geq -2 \\ y < 2x+3 \end{cases}$

Graph $x \geq -2$ with a solid line and $y < 2x + 3$ with a dashed line. Graph on the same rectangular coordinate system. The overlapping shaded region represents the solution.

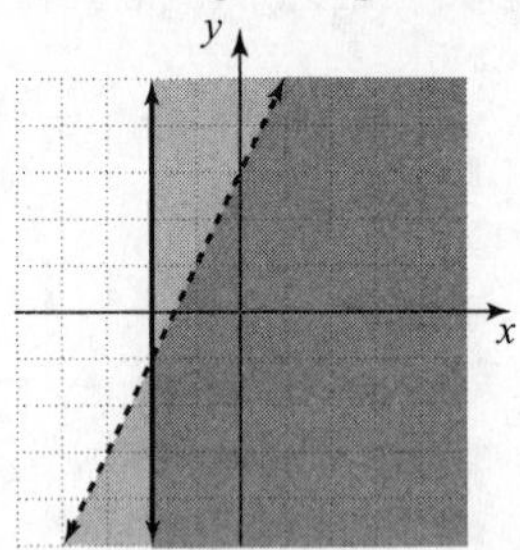

25. $\begin{cases} x > 0 \\ y \leq \frac{2}{5}x - 1 \end{cases}$

Graph $x > 0$ with a dashed line and $y \leq \frac{2}{5}x - 1$ with a solid line. Graph on the same rectangular coordinate system. The overlapping shaded region represents the solution.

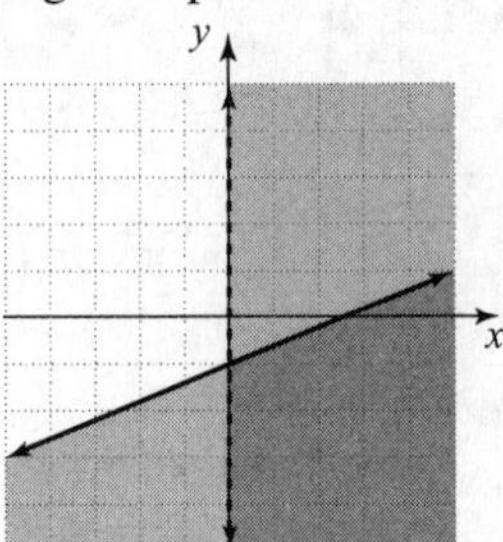

27. $\begin{cases} -y \leq x \\ 3x - y \geq -5 \end{cases}$

Graph $-y \leq x$ $(y \geq -x)$ with a solid line and $3x - y \geq -5$ $(y \leq 3x + 5)$ with a solid line. Graph on the same rectangular coordinate system. The overlapping shaded region represents the solution.

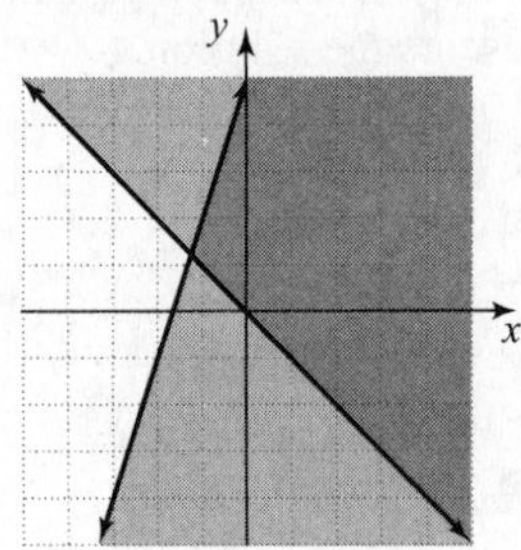

29. $\begin{cases} x + y \leq -2 \\ y \geq x + 3 \end{cases}$

Graph $x + y \leq -2$ $(y \leq -x - 2)$ with a solid line and $y \geq x + 3$ with a solid line. Graph on the same rectangular coordinate system. The overlapping shaded region represents the solution.

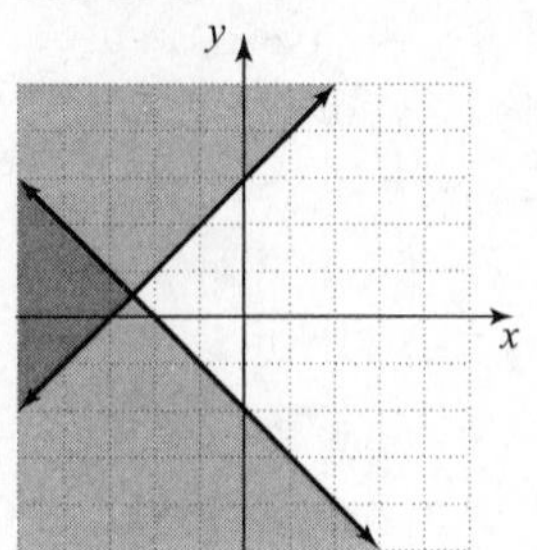

31. $\begin{cases} x + 3y \geq 0 \\ 2y < x + 1 \end{cases}$

Graph $x + 3y \geq 0$ $\left(y \geq -\frac{1}{3}x\right)$ with a solid line and $2y < x + 1$ $\left(y < \frac{1}{2}x + \frac{1}{2}\right)$ with a dashed line. Graph on the same rectangular coordinate system. The overlapping shaded region represents the solution.

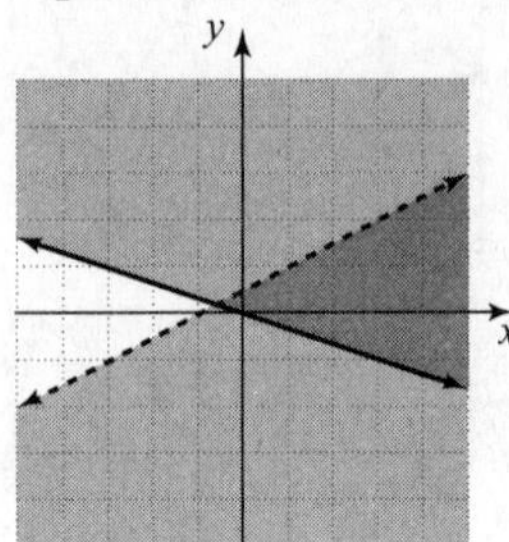

33. $\begin{cases} x + y \geq 0 \\ x < 2y + 4 \end{cases}$

Graph $x + y \geq 0$ $(y \geq -x)$ with a solid line and $x < 2y + 4$ $\left(y > \frac{1}{2}x - 2\right)$ with a dashed line.

Graph on the same rectangular coordinate system. The overlapping shaded region represents the solution.

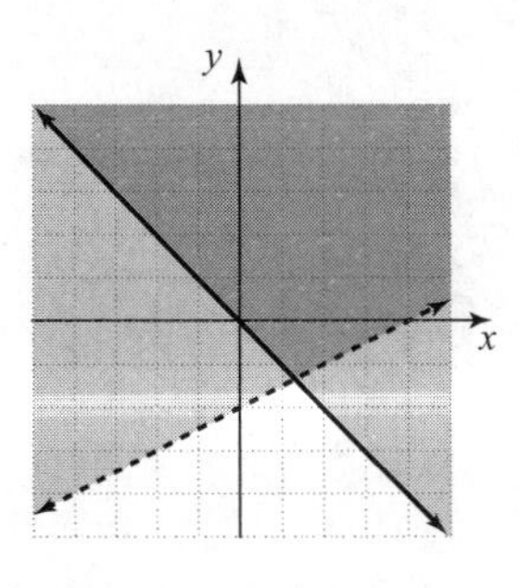

35. $\begin{cases} x+3y>6 \\ 2x-y\le 4 \end{cases}$

Graph $x + 3y > 6$ $\left(y > -\dfrac{1}{3}x+2\right)$ with a dashed line and $2x - y \le 4$ ($y \ge 2x - 4$) with a solid line. Graph on the same rectangular coordinate system. The overlapping shaded region represents the solution.

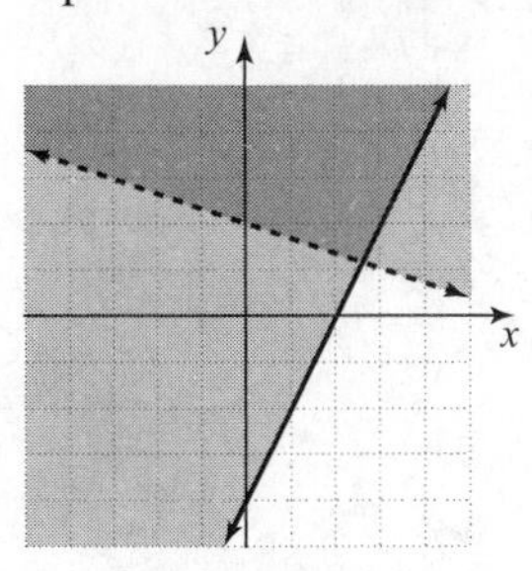

37. $\begin{cases} -y\le 3x-4 \\ 2x+3y\ge -3 \end{cases}$

Graph $-y \le 3x - 4$ ($y \ge -3x + 4$) with a solid line and $2x + 3y \ge -3$ $\left(y \ge -\dfrac{2}{3}x-1\right)$ with a solid line. Graph on the same rectangular coordinate system. The overlapping shaded region represents the solution.

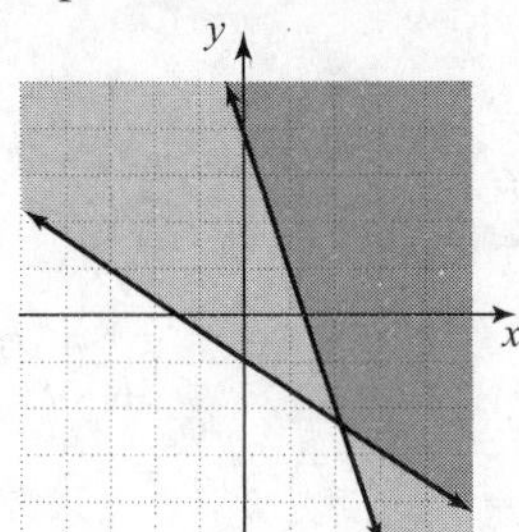

39. (a) $\begin{cases} f+h\le 30 \\ f\ge 2h \\ f\ge 0 \\ h\ge 0 \end{cases}$

Graph $f + h \le 30$ with a solid line and $f \ge 2h$ with a solid line.

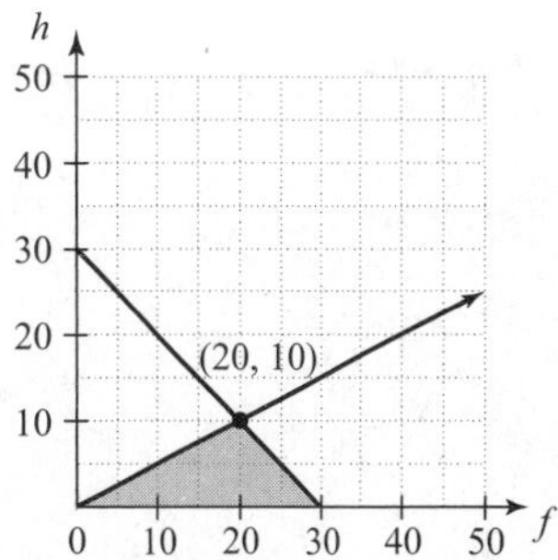

(b) $(f, h) = (18, 11)$
No, these values do not lie within the shaded region.

(c) $(f, h) = (8, 4)$
Yes, these values lie within the shaded region.

41. (a) $\begin{cases} 12c+18t\le 360 \\ t<2c-20 \\ c\le 24 \end{cases}$

Graph $12c + 18t \le 360$ and $c \le 24$ with solid lines and $t < 2c - 20$ with a dashed line.

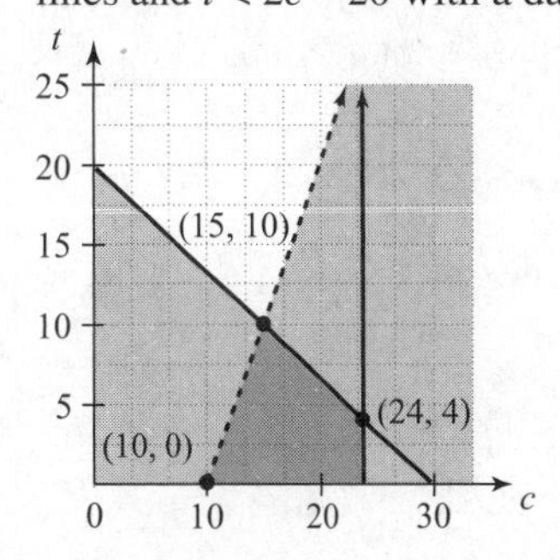

(b) $(c, t) = (17, 9)$
No, these values do not lie in the shaded region.

(c) $(c, t) = (12, 11)$
No, these values do not lie in the shaded region.

43. $\begin{cases} \dfrac{y}{2}-\dfrac{x}{6}\ge 1 & (1) \\ \dfrac{x}{3}-\dfrac{y}{1}\ge 1 & (2) \end{cases}$

(1): $\dfrac{y}{2}-\dfrac{x}{6}\ge 1$

$6\left(\dfrac{y}{2}-\dfrac{x}{6}\right)\ge 6(1)$

$3y-x\ge 6$

$3y\ge x+6$

$y\ge \dfrac{1}{3}x+2$

(2): $\dfrac{x}{3}-\dfrac{y}{1}\ge 1$

$\dfrac{x}{3}\ge \dfrac{y}{1}+1$

$\dfrac{x}{3}-1\ge y$

$y\le \dfrac{1}{3}x-1$

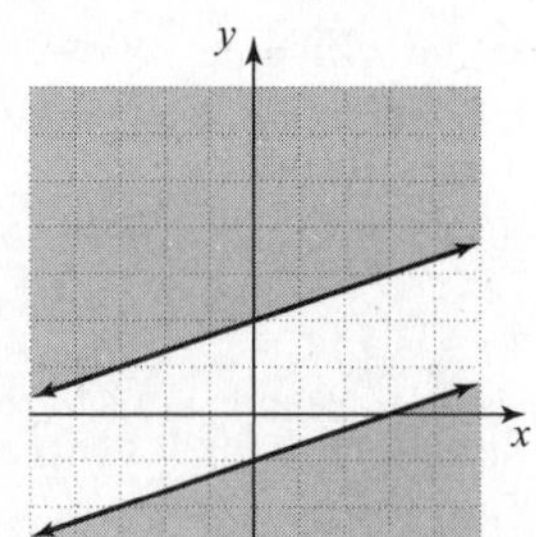

The regions do not overlap. There is no solution.

45. $\begin{cases} x<\dfrac{3}{2}y+\dfrac{9}{2} & (1) \\ -2x<-3(y+2) & (2) \end{cases}$

(1): $x<\dfrac{3}{2}y+\dfrac{9}{2}$

$2x<2\left(\dfrac{3}{2}y+\dfrac{9}{2}\right)$

$2x<3y+9$

$2x-9<3y$

$\dfrac{2}{3}x-3<y$ or $y>\dfrac{2}{3}x-3$

(2): $-2x<-3(y+2)$

$-2x<-3y-6$

$-2x+6<-3y$

$\dfrac{2}{3}x-2>y$ or $y<\dfrac{2}{3}x+2$

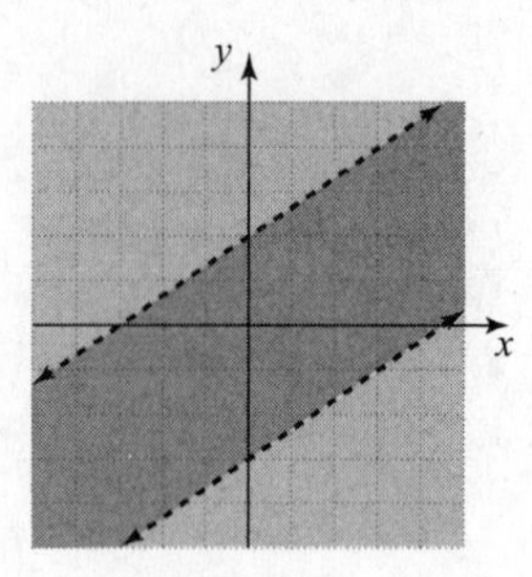

47. $\begin{cases} y\ge 0 \\ y\ge x \\ y\le \dfrac{1}{2}(x+6) \end{cases}$

49. Answers will vary. Graph each inequality in the system. The overlapping shaded region represents the solution of the system.

51. a. $\begin{cases} y\ge x \\ y\ge -2x \end{cases}$

b. $\begin{cases} y\le x \\ y\ge -2x \end{cases}$

c. $\begin{cases} y\ge x \\ y\le -2x \end{cases}$

d. $\begin{cases} y\le x \\ y\le -2x \end{cases}$

Chapter 4 Review

1. $\begin{cases} x+2y=6 & (1) \\ 3x-y=-10 & (2) \end{cases}$

(a) $(3, -1)$

(1): $3+2(-1)\stackrel{?}{=}6$

$3-2\stackrel{?}{=}6$

$1=6$ False

$(3, -1)$ is not a solution.

(b) $(-2, 4)$

(1): $-2+2(4)\stackrel{?}{=}6$

$-2+8\stackrel{?}{=}6$

$6=6$ True

(2): $3(-2)-4\stackrel{?}{=}-10$

$-6-4\stackrel{?}{=}-10$

$-10=-10$ True

$(-2, 4)$ is a solution.

(c) (4, 1)

(1): $4+2(1) \stackrel{?}{=} 6$

$4+2 \stackrel{?}{=} 6$

$6=6$ True

(2): $3(4)-1 \stackrel{?}{=} -10$

$12-1 \stackrel{?}{=} -10$

$11=-10$ False

(4, 1) is not a solution.

2. $\begin{cases} y=3x-5 & (1) \\ 3y=6x-5 & (2) \end{cases}$

(a) (2, 1)

(1): $1 \stackrel{?}{=} 3(2)-5$

$1 \stackrel{?}{=} 6-5$

$1=1$ True

(2): $3(1) \stackrel{?}{=} 6(2)-5$

$3 \stackrel{?}{=} 12-5$

$3=7$ False

(2, 1) is not a solution.

(b) $\left(0, -\frac{5}{3}\right)$

(1): $-\frac{5}{3} \stackrel{?}{=} 3(0)-5$

$-\frac{5}{3}=-5$ False

$\left(0, -\frac{5}{3}\right)$ is not a solution.

(c) $\left(\frac{10}{3}, 5\right)$

(1): $5 \stackrel{?}{=} 3\left(\frac{10}{3}\right)-5$

$5 \stackrel{?}{=} 10-5$

$5=5$ True

(2): $3(5) \stackrel{?}{=} 6\left(\frac{10}{3}\right)-5$

$15 \stackrel{?}{=} 20-5$

$15=15$ True

$\left(\frac{10}{3}, 5\right)$ is a solution.

3. $\begin{cases} 3x-4y=2 & (1) \\ 20y=15x-10 & (2) \end{cases}$

(a) $\left(\frac{1}{2}, -\frac{1}{8}\right)$

(1): $3\left(\frac{1}{2}\right)-4\left(-\frac{1}{8}\right) \stackrel{?}{=} 2$

$\frac{3}{2}+\frac{1}{2} \stackrel{?}{=} 2$

$2=2$ True

(2): $20\left(-\frac{1}{8}\right) \stackrel{?}{=} 15\left(\frac{1}{2}\right)-10$

$-\frac{5}{2} \stackrel{?}{=} \frac{15}{2}-\frac{20}{2}$

$-\frac{5}{2}=-\frac{5}{2}$ True

$\left(\frac{1}{2}, -\frac{1}{8}\right)$ is a solution.

(b) (6, 4)

(1): $3(6)-4(4) \stackrel{?}{=} 2$

$18-16 \stackrel{?}{=} 2$

$2=2$ True

(2): $20(4) \stackrel{?}{=} 15(6)-10$

$80 \stackrel{?}{=} 90-10$

$80=80$ True

(6, 4) is a solution.

(c) (0.4, –0.2)

(1): $3(0.4)-4(-0.2) \stackrel{?}{=} 2$

$1.2+0.8 \stackrel{?}{=} 2$

$2=2$ True

(2): $20(-0.2) \stackrel{?}{=} 15(0.4)-10$

$-4 \stackrel{?}{=} 6-10$

$-4=-4$ True

(0.4, –0.2) is a solution.

4. $\begin{cases} x=-4y+2 & (1) \\ 2x+8y=12 & (2) \end{cases}$

(a) (10, –1)

(1): $10 \stackrel{?}{=} -4(-1)+2$

$10 \stackrel{?}{=} 4+2$

$10=8$ False

(10, –1) is not a solution.

(b) $\left(-\frac{1}{2}, \frac{1}{2}\right)$

$$\begin{aligned}(1)\colon -\frac{1}{2} &\stackrel{?}{=} -4\left(\frac{1}{2}\right)+2\\ -\frac{1}{2} &\stackrel{?}{=} -2+2\\ -\frac{1}{2} &= 0 \quad \text{False}\end{aligned}$$

$\left(-\frac{1}{2}, \frac{1}{2}\right)$ is not a solution.

(c) (2, 1)

$$\begin{aligned}(1)\colon 2 &\stackrel{?}{=} -4(1)+2\\ 2 &\stackrel{?}{=} -4+2\\ 2 &= -2 \quad \text{False}\end{aligned}$$

(2, 1) is not a solution.

5. $\begin{cases} 2x-4y=8 & (1) \\ x+y=7 & (2) \end{cases}$

$$\begin{aligned}(1)\colon 2x-4y &= 8\\ -4y &= -2x+8\\ y &= \frac{1}{2}x-2\end{aligned}$$

$$\begin{aligned}(2)\colon x+y &= 7\\ y &= -x+7\end{aligned}$$

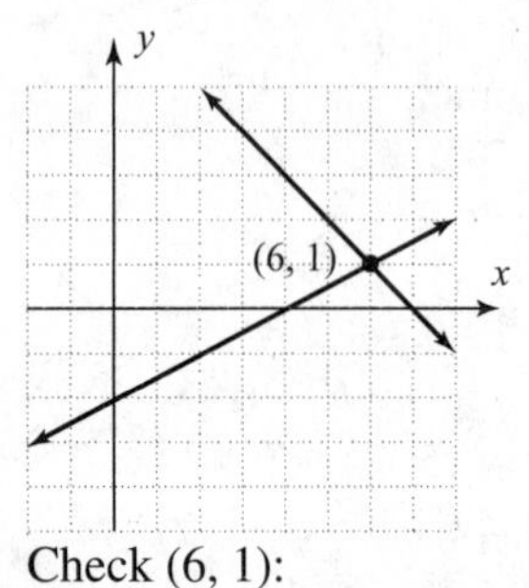

Check (6, 1):

$$\begin{aligned}(1)\colon 2(6)-4(1) &\stackrel{?}{=} 8\\ 12-4 &\stackrel{?}{=} 8\\ 8 &= 8 \quad \text{True}\end{aligned}$$

$$\begin{aligned}(2)\colon 6+1 &\stackrel{?}{=} 7\\ 7 &= 7 \quad \text{True}\end{aligned}$$

The solution is (6, 1).

6. $\begin{cases} x-y=-3 & (1) \\ 3x+2y=6 & (2) \end{cases}$

$$\begin{aligned}(1)\colon x-y &= -3\\ -y &= -x-3\\ y &= x+3\end{aligned}$$

$$\begin{aligned}(2)\colon 3x+2y &= 6\\ 2y &= -3x+6\\ y &= -\frac{3}{2}x+3\end{aligned}$$

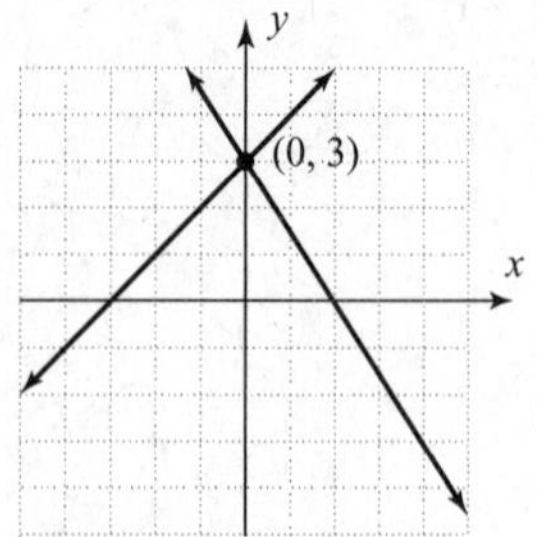

Check (0, 3):

$$\begin{aligned}(1)\colon 0-3 &\stackrel{?}{=} -3\\ -3 &= -3 \quad \text{True}\end{aligned}$$

$$\begin{aligned}(2)\colon 3(0)+2(3) &\stackrel{?}{=} 6\\ 6 &= 6 \quad \text{True}\end{aligned}$$

The solution is (0, 3).

7. $\begin{cases} y=-\frac{x}{2}+2 & (1) \\ y=x+8 & (2) \end{cases}$

The equations are already in slope-intercept form.

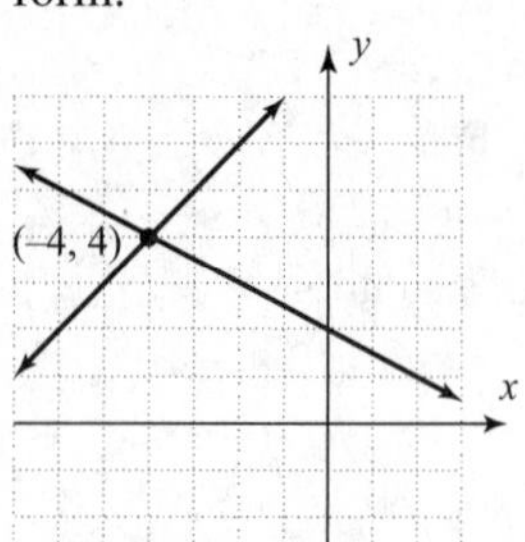

Check (−4, 4):

$$\begin{aligned}(1)\colon 4 &\stackrel{?}{=} -\frac{-4}{2}+2\\ 4 &\stackrel{?}{=} 2+2\\ 4 &= 4 \quad \text{True}\end{aligned}$$

$$\begin{aligned}(2)\colon 4 &\stackrel{?}{=} -4+8\\ 4 &= 4 \quad \text{True}\end{aligned}$$

The solution is (−4, 4).

8. $\begin{cases} y=-x-5 & (1) \\ y=\frac{3x}{4}+2 & (2) \end{cases}$

The equations are already in slope-intercept form.

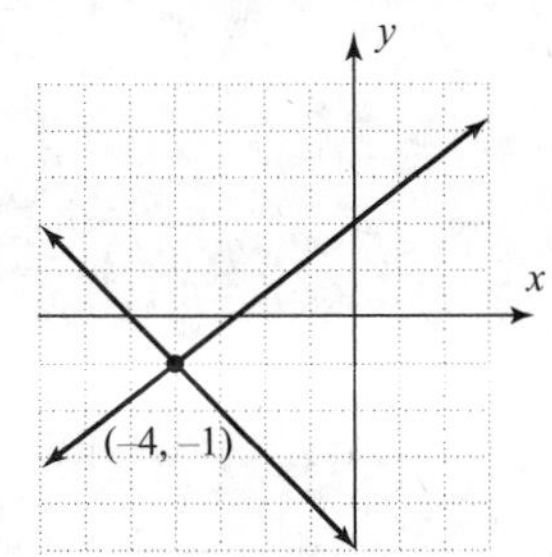

Check (−4, −1):

(1): $-1 \stackrel{?}{=} -(-4)-5$

$-1 \stackrel{?}{=} 4-5$

$-1=-1$ True

(2): $-1 \stackrel{?}{=} \frac{3(-4)}{4}+2$

$-1 \stackrel{?}{=} -3+2$

$-1=-1$ True

The solution is (−4, −1).

9. $\begin{cases} 4x-8=0 & (1) \\ 3y+9=0 & (2) \end{cases}$

(1): $4x-8=0$

$4x=8$

$x=2$

(2): $3y+9=0$

$3y=-9$

$y=-3$

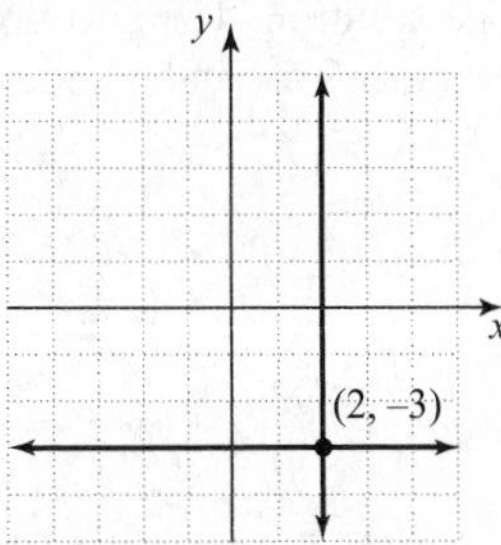

Check (2, −3):

(1): $4(2)-8 \stackrel{?}{=} 0$

$8-8 \stackrel{?}{=} 0$

$0=0$ True

(2): $3(-3)+9 \stackrel{?}{=} 0$

$-9+9 \stackrel{?}{=} 0$

$0=0$ True

The solution is (2, −3).

10. $\begin{cases} x=y & (1) \\ x+y=0 & (2) \end{cases}$

(1): $y=x$

(2): $x+y=0$

$y=-x$

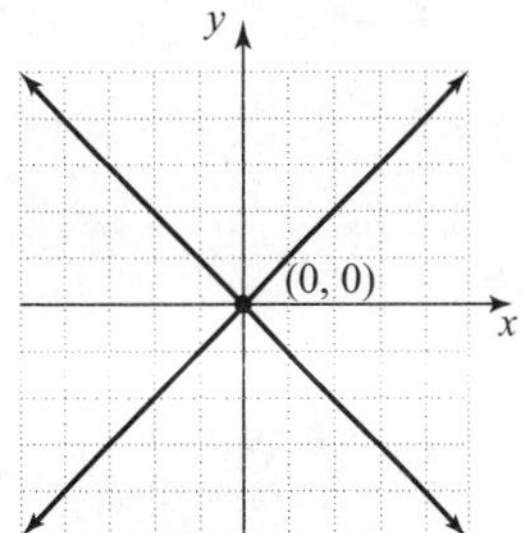

Check (0, 0):

(1): $0=0$ True

(2): $0+0 \stackrel{?}{=} 0$

$0=0$ True

The solution is (0, 0).

11. $\begin{cases} 0.6x+0.5y=2 & (1) \\ 10y=-12x+20 & (2) \end{cases}$

(1): $0.6x+0.5y=2$

$0.5y=-0.6x+2$

$y=-1.2x+4$

(2): $10y=-12x+20$

$y=-\frac{6}{5}x+2$

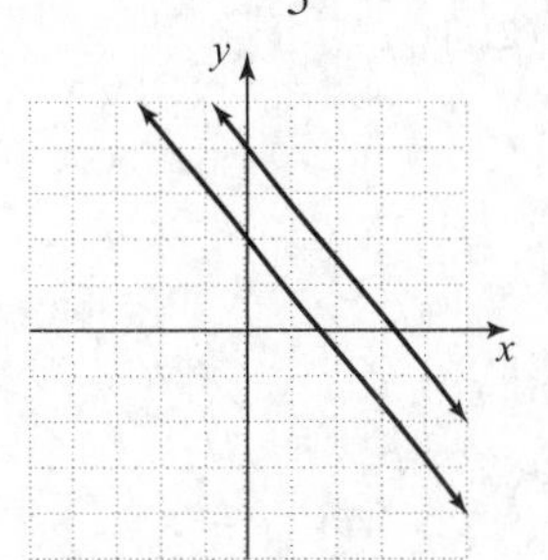

The lines are parallel, so there is no solution.

12. $\begin{cases} \frac{1}{4}x-\frac{1}{2}y=1 & (1) \\ 3x-6y=12 & (2) \end{cases}$

(1): $\frac{1}{4}x-\frac{1}{2}y=1$

$-\frac{1}{2}y=-\frac{1}{4}x+1$

$y=\frac{1}{2}x-2$

(2): $3x-6y=12$

$-6y=-3x+12$

$y=\frac{1}{2}x-2$

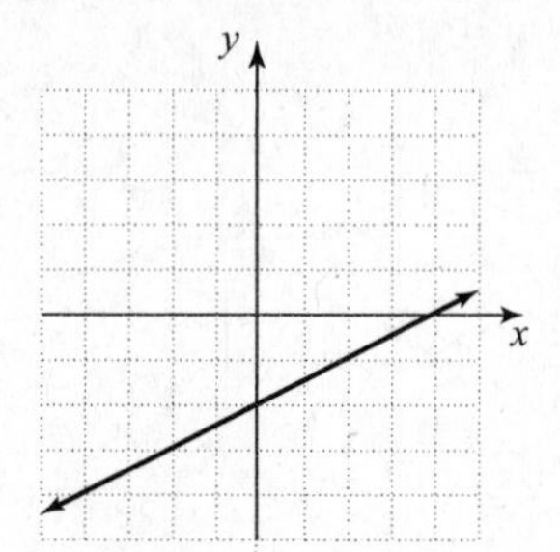

The lines coincide. There are infinitely many solutions.

13. $\begin{cases} 3x = y+4 & (1) \\ 3x - y = -4 & (2) \end{cases}$

(1): $3x = y+4$
$y = 3x-4$

(2): $3x - y = -4$
$y = 3x+4$

The lines have the same slope, but different y-intercepts, so the lines are parallel. The system has no solution. It is inconsistent.

14. $\begin{cases} 4y = 2x-8 & (1) \\ x-2y = -4 & (2) \end{cases}$

(1): $4y = 2x-8$
$y = \frac{1}{2}x-2$

(2): $x-2y = -4$
$-2y = -x-4$
$y = \frac{1}{2}x+2$

The lines have the same slope, but different y-intercepts, so the lines are parallel. The system has no solution. It is inconsistent.

15. $\begin{cases} -3x+3y = -3 & (1) \\ \frac{1}{2}x - \frac{1}{2}y = 0.5 & (2) \end{cases}$

(1): $-3x+3y = -3$
$3y = 3x-3$
$y = x-1$

(1): $\frac{1}{2}x - \frac{1}{2}y = 0.5$
$-\frac{1}{2}y = -\frac{1}{2}x+0.5$
$y = x-1$

The lines have the same slope and y-intercept, so the lines coincide. The system has infinitely many solutions. It is consistent and the equations are dependent.

16. $\begin{cases} x-2 = -\frac{2}{3}y - \frac{2}{3} & (1) \\ 3x = 4-2y \end{cases}$

(1): $x-2 = -\frac{2}{3}y - \frac{2}{3}$
$x - \frac{4}{3} = -\frac{2}{3}y$
$-\frac{3}{2}x+2 = y$ or $y = -\frac{3}{2}x+2$

(2): $3x = 4-2y$
$3x-4 = -2y$
$-\frac{3}{2}x+2 = y$ or $y = -\frac{3}{2}x+2$

The lines have the same slope and y-intercept, so the lines coincide. The system has infinitely many solutions. It is consistent and the equations are dependent.

17. $\begin{cases} 3-2x = y & (1) \\ \frac{y}{2} = x+1.5 & (2) \end{cases}$

(1): $3 - 2x = y$ or $y = -2x + 3$

(2): $\frac{y}{2} = x+1.5$
$y = 2x+3$

The lines have different slopes, so the lines intersect. There is one solution. The system is consistent and the equations are independent.

18. $\begin{cases} \frac{y}{2} = \frac{x}{4}+2 & (1) \\ \frac{x}{8}+\frac{y}{4} = -1 & (2) \end{cases}$

(1): $\frac{y}{2} = \frac{x}{4}+2$
$y = \frac{1}{2}x+4$

(2): $\frac{x}{8}+\frac{y}{4} = -1$
$\frac{y}{4} = -\frac{x}{8}-1$
$y = -\frac{1}{2}x-4$

The lines have different slopes, so the lines intersect. There is one solution. The system is consistent and the equations are independent.

19. (a) Let x be the number of fliers and y be the cost.
Printer A: $y = 0.10x + 70$
Printer B: $y = 0.04x + 100$

(b)

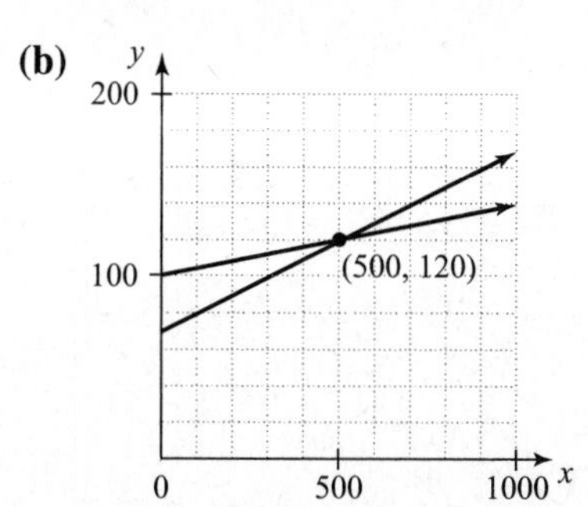

The cost is the same for 500 fliers.

(c) Since the graph for printer A is below that of printer B for fliers less than 500, she should choose Printer A.

20. (a) Let x be the square yards and y be the cost.
carpet: $y = 40x + 50$
tile: $y = 10x + 350$

(b)

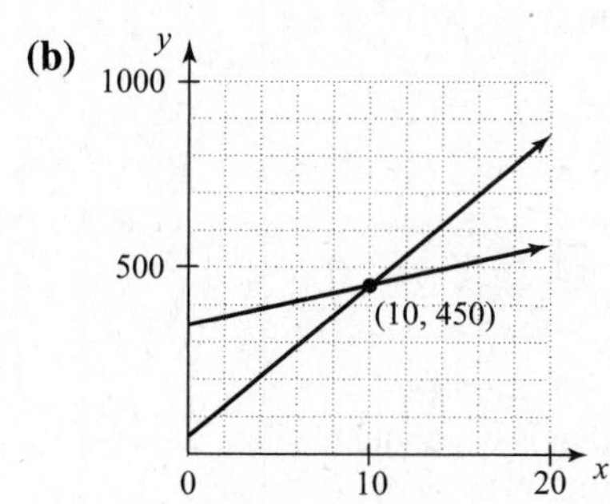

The cost is the same for 10 square yards.

(c) Since the graph for tile is below that for carpet for square yards greater than 10, he should choose tile.

21. $\begin{cases} x+4y=6 & (1) \\ y=2x-3 & (2) \end{cases}$

Let $y = 2x - 3$ in (1).

$$\begin{aligned} x+4(2x-3)&=6 \\ x+8x-12&=6 \\ 9x-12&=6 \\ 9x&=18 \\ x&=2 \end{aligned}$$

Let $x = 2$ in (2).
$y = 2(2) - 3$
$y = 4 - 3$
$y = 1$
The solution is (2, 1).

22. $\begin{cases} 7x-3y=10 & (1) \\ y=3x-4 & (2) \end{cases}$

Let $y = 3x - 4$ in (1).

$$\begin{aligned} 7x-3(3x-4)&=10 \\ 7x-9x+12&=10 \\ -2x+12&=10 \\ -2x&=-2 \\ x&=1 \end{aligned}$$

Let $x = 1$ in (2).
$y = 3(1) - 4$
$y = 3 - 4$
$y = -1$
The solution is (1, –1).

23. $\begin{cases} 2x+5y=4 & (1) \\ x=3-2y & (2) \end{cases}$

Let $x = 3 - 2y$ in (1).

$$\begin{aligned} 2(3-2y)+5y&=4 \\ 6-4y+5y&=4 \\ 6+y&=4 \\ y&=-2 \end{aligned}$$

Let $y = -2$ in (2).
$x = 3 - 2(-2)$
$x = 3 + 4$
$x = 7$
The solution is (7, –2).

24. $\begin{cases} 3x+y=10 & (1) \\ x=8+2y & (2) \end{cases}$

Let $x = 8 + 2y$ in (1).

$$\begin{aligned} 3(8+2y)+y&=10 \\ 24+6y+y&=10 \\ 24+7y&=10 \\ 7y&=-14 \\ y&=-2 \end{aligned}$$

Let $y = -2$ in (2).
$x = 8 + 2(-2)$
$x = 8 - 4$
$x = 4$
The solution is (4, –2).

25. $\begin{cases} y=\frac{2}{3}x-1 & (1) \\ y=\frac{1}{2}x+2 & (2) \end{cases}$

Let $y=\frac{2}{3}x-1$ in (2).

$$\frac{2}{3}x-1=\frac{1}{2}x+2$$
$$\frac{2}{3}x-\frac{1}{2}x=2+1$$
$$\frac{4}{6}x-\frac{3}{6}x=3$$
$$\frac{1}{6}x=3$$
$$x=18$$

Let $x = 18$ in (2).

$$y=\frac{1}{2}(18)+2$$
$$y=9+2$$
$$y=11$$

The solution is (18, 11).

26. $\begin{cases} y=-\frac{5}{6}x+3 & (1) \\ y=-\frac{4}{3}x & (2) \end{cases}$

Let $y=-\frac{4}{3}x$ in (1).

$$-\frac{4}{3}x=-\frac{5}{6}x+3$$
$$-\frac{4}{3}x+\frac{5}{6}x=3$$
$$-\frac{8}{6}x+\frac{5}{6}x=3$$
$$-\frac{1}{2}x=3$$
$$x=-6$$

Let $x = -6$ in (2).

$$y=-\frac{4}{3}(-6)$$
$$y=8$$

The solution is (–6, 8).

27. $\begin{cases} 2x-y=6 & (1) \\ 4x+3y=2 & (2) \end{cases}$

Solve (1) for y.

$$2x-y=6$$
$$-y=-2x+6$$
$$y=2x-6$$

Let $y = 2x - 6$ in (2).

$$4x+3(2x-6)=2$$
$$4x+6x-18=2$$
$$10x-18=2$$
$$10x=20$$
$$x=2$$

Let $x = 2$ in (1).

$$2(2)-y=6$$
$$4-y=6$$
$$-y=2$$
$$y=-2$$

The solution is (2, –2).

28. $\begin{cases} 5x+2y=13 & (1) \\ x+4y=-1 & (2) \end{cases}$

Solve (2) for x.

$$x+4y=-1$$
$$x=-1-4y$$

Let $x = -1 - 4y$ in (1).

$$5(-1-4y)+2y=13$$
$$-5-20y+2y=13$$
$$-5-18y=13$$
$$-18y=18$$
$$y=-1$$

Let $y = -1$ in (2).

$$x+4(-1)=-1$$
$$x-4=-1$$
$$x=3$$

The solution is (3, –1).

29. $\begin{cases} 6x+3y=12 & (1) \\ y=-2x+4 & (2) \end{cases}$

Let $y = -2x + 4$ in (1).

$$6x+3(-2x+4)=12$$
$$6x-6x+12=12$$
$$12=12 \quad \text{True}$$

The system is dependent. There are infinitely many solutions.

30. $\begin{cases} x=4y-2 & (1) \\ 8y-2x=4 & (2) \end{cases}$

Let $x = 4y - 2$ in (2).

$$8y-2(4y-2)=4$$
$$8y-8y+4=4$$
$$4=4 \quad \text{True}$$

The system is dependent. There are infinitely many solutions.

31. $\begin{cases} -6-2(3x-6y)=0 & (1) \\ 6-12(x-2y)=0 & (2) \end{cases}$

(1): $-6-6x+12y=0$

$$-6x+12y=6$$
$$x-2y=-1$$
$$x=2y-1$$

Let $x = 2y - 1$ in (2).

$$6-12[(2y-1)-2y]=0$$
$$6-12(2y-1-2y)=0$$
$$6-12(-1)=0$$
$$6+12=0$$
$$18=0 \quad \text{False}$$

The system is inconsistent. There is no solution.

32. $\begin{cases} 6-2(3y+4x)=0 & (1) \\ 9(y-1)+12x=0 & (2) \end{cases}$

(1): $6-2(3y+4x)=0$

$$6-6y-8x=0$$
$$-6y-8x=-6$$
$$-6y=8x-6$$
$$y=-\frac{4}{3}x+1$$

Let $y=-\frac{4}{3}x+1$ in (2).

$$9\left(-\frac{4}{3}x+1-1\right)+12x=0$$
$$9\left(-\frac{4}{3}x\right)+12x=0$$
$$-12x+12x=0$$
$$0=0$$

The system is dependent. There are infinitely many solutions.

33. $\begin{cases} \frac{1}{2}x-\frac{1}{4}y=\frac{1}{2} & (1) \\ \frac{1}{3}x-\frac{3}{4}y=-\frac{1}{4} & (2) \end{cases}$

Solve (1) for y.

$$\frac{1}{2}x-\frac{1}{4}y=\frac{1}{2}$$
$$-\frac{1}{4}y=-\frac{1}{2}x+\frac{1}{2}$$
$$y=2x-2$$

Let $y = 2x - 2$ in (2).

$$\frac{1}{3}x-\frac{3}{4}(2x-2)=-\frac{1}{4}$$
$$\frac{1}{3}x-\frac{3}{2}x+\frac{3}{2}=-\frac{1}{4}$$
$$\frac{2}{6}x-\frac{9}{6}x=-\frac{6}{4}-\frac{1}{4}$$
$$-\frac{7}{6}x=-\frac{7}{4}$$
$$x=\frac{3}{2}$$

Let $x=\frac{3}{2}$ in (1).

$$\frac{1}{2}\left(\frac{3}{2}\right)-\frac{1}{4}y=\frac{1}{2}$$
$$\frac{3}{4}-\frac{1}{4}y=\frac{1}{2}$$
$$-\frac{1}{4}y=-\frac{1}{4}$$
$$y=1$$

The solution is $\left(\frac{3}{2}, 1\right)$.

34. $\begin{cases} -\frac{5x}{4}+\frac{y}{6}=-\frac{7}{12} \\ \frac{3x}{2}-\frac{y}{10}=\frac{3}{5} \end{cases}$

Multiply the first equation by 12 and the second equation by 10 to clear fractions.

$\begin{cases} -15x+2y=-7 & (1) \\ 15x-y=6 & (2) \end{cases}$

Solve (2) for y.

$$15x-y=6$$
$$y=15x-6$$

Let $y = 15x - 6$ in (1).

$$-15x+2(15x-6)=-7$$
$$-15x+30x-12=-7$$
$$15x=5$$
$$x=\frac{1}{3}$$

Let $x=\frac{1}{3}$ in (2).

$$15\left(\frac{1}{3}\right)-y=6$$
$$5-y=6$$
$$-y=1$$
$$y=-1$$

The solution is $\left(\frac{1}{3}, -1\right)$.

35. Let w be the width and l be the length.

$\begin{cases} 2w+2l=650 & (1) \\ l=w+75 & (2) \end{cases}$

Let $l = w + 75$ in (1).

$$2w+2(w+75)=650$$
$$2w+2w+150=650$$
$$4w=500$$
$$w=125$$

Let $w = 125$ in (2).

$l = 125 + 75$

$l = 200$

The width is 125 meters and the length is 200 meters.

36. $\begin{cases} x+y=12 & (1) \\ x-2y=21 & (2) \end{cases}$

Solve (1) for y.

$y=12-x$

Let $y=12-x$ in (2).

$$\begin{aligned} x-2(12-x)&=21 \\ x-24+2x&=21 \\ -24+3x&=21 \\ 3x&=45 \\ x&=15 \end{aligned}$$

Let $x=15$ in (1).

$$\begin{aligned} 15+y&=12 \\ y&=-3 \end{aligned}$$

The smaller number is -3.

37. $\begin{cases} 4x-y=12 & (1) \\ 2x+y=-12 & (2) \end{cases}$

Add the equations.

$$\begin{array}{l} \begin{cases} 4x-y=12 \\ 2x+y=-12 \end{cases} \\ \hline 6x \quad\;\; =0 \\ \quad x=0 \end{array}$$

Let $x=0$ in (2).

$$\begin{aligned} 2(0)+y&=-12 \\ y&=-12 \end{aligned}$$

The solution is $(0, -12)$.

38. $\begin{cases} -2x+3y=27 & (1) \\ 2x-5y=-41 & (2) \end{cases}$

Add the equations.

$$\begin{array}{l} \begin{cases} -2x+3y=27 \\ \;\;\,2x-5y=-41 \end{cases} \\ \hline \qquad -2y=-14 \\ \qquad\quad y=7 \end{array}$$

Let $y=7$ in (2).

$$\begin{aligned} 2x-5(7)&=-41 \\ 2x-35&=-41 \\ 2x&=-6 \\ x&=-3 \end{aligned}$$

The solution is $(-3, 7)$.

39. $\begin{cases} -3x+4y=25 & (1) \\ x-5y=-23 & (2) \end{cases}$

Multiply (2) by 3 and add.

$$\begin{array}{l} \begin{cases} -3x+4y=25 \\ \;\;\,3x-15y=-69 \end{cases} \\ \hline \qquad -11y=-44 \\ \qquad\quad y=4 \end{array}$$

Let $y=4$ in (2).

$$\begin{aligned} x-5(4)&=-23 \\ x-20&=-23 \\ x&=-3 \end{aligned}$$

The solution is $(-3, 4)$.

40. $\begin{cases} 5x+8y=-15 & (1) \\ -2x+y=6 & (2) \end{cases}$

Multiply (2) by -8 and add.

$$\begin{array}{l} \begin{cases} 5x+8y=-15 \\ 16x-8y=-48 \end{cases} \\ \hline 21x \quad\;\; =-63 \\ \qquad x=-3 \end{array}$$

Let $x=-3$ in (2).

$$\begin{aligned} -2(-3)+y&=6 \\ 6+y&=6 \\ y&=0 \end{aligned}$$

The solution is $(-3, 0)$.

41. $\begin{cases} 4x-3y=-1 & (1) \\ 2x-\frac{3}{2}y=3 & (2) \end{cases}$

Multiply (2) by -2 and add.

$$\begin{array}{l} \begin{cases} 4x-3y=-1 \\ -4x+3y=-6 \end{cases} \\ \hline \qquad 0=-7 \end{array}$$

The system is inconsistent. There is no solution.

42. $\begin{cases} -2x+5y=3 & (1) \\ 4x-10y=-6 & (2) \end{cases}$

Multiply (1) by 2. Then add.

$$\begin{array}{l} \begin{cases} -4x+10y=6 \\ \;\;\,4x-10y=-6 \end{cases} \\ \hline \qquad 0=0 \;\; \text{True} \end{array}$$

The system is consistent and dependent. There are infinitely many solutions.

43. $\begin{cases} 1.3x-0.2y=-3 & (1) \\ -0.1x+0.5y=1.2 & (2) \end{cases}$

Multiply (1) by 5 and (2) by 2, then add.

$$\begin{cases} 6.5x - y = -15 \\ -0.2x + y = 2.4 \end{cases}$$
$$6.3x = -12.6$$
$$x = -2$$

Let $x = -2$ in (2).

$$-0.1(-2) + 0.5y = 1.2$$
$$0.2 + 0.5y = 1.2$$
$$0.5y = 1$$
$$y = 2$$

The solution is (−2, 2).

44. $\begin{cases} 2.5x + 0.5y = 6.25 & (1) \\ -0.5x - 1.2y = 1.5 & (2) \end{cases}$

Multiply (1) by 2 and (2) by 10, then add.

$$\begin{cases} 5x + y = 12.5 \\ -5x - 12y = 15 \end{cases}$$
$$-11y = 27.5$$
$$y = -\frac{27.5}{11} = -2.5$$

Let $y = -2.5$ in equation (1).

$$2.5x + 0.5(-2.5) = 6.25$$
$$2.5x - 1.25 = 6.25$$
$$2.5x = 7.5$$
$$x = 3$$

The solution is (3, −2.5).

45. $\begin{cases} 2x + y = -1 & (1) \\ -6x - 8y = 13 & (2) \end{cases}$

Solve (1) for y.

$$2x + y = -1$$
$$y = -2x - 1$$

Let $y = -2x - 1$ in (2).

$$-6x - 8(-2x - 1) = 13$$
$$-6x + 16x + 8 = 13$$
$$10x + 8 = 13$$
$$10x = 5$$
$$x = \frac{1}{2}$$

Let $x = \frac{1}{2}$ in (1).

$$2\left(\frac{1}{2}\right) + y = -1$$
$$1 + y = -1$$
$$y = -2$$

The solution is $\left(\frac{1}{2}, -2\right)$.

46. $\begin{cases} \frac{1}{6}x + y = \frac{3}{4} \\ \frac{2y - 8}{3} = 4x + 4 \end{cases}$

Multiply the first equation by 12 and rearrange.

$$12\left(\frac{1}{6}x + y\right) = 12\left(\frac{3}{4}\right)$$
$$2x + 12y = 9$$

Multiply the second equation by 3 and rearrange.

$$3\left(\frac{2y - 8}{3}\right) = 3(4x + 4)$$
$$2y - 8 = 12x + 12$$
$$-12x + 2y = 20$$
$$-6x + y = 10$$

$$\begin{cases} 2x + 12y = 9 & (1) \\ -6x + y = 10 & (2) \end{cases}$$

Multiply (1) by 3, then add.

$$\begin{cases} 6x + 36y = 27 \\ -6x + y = 10 \end{cases}$$
$$37y = 37$$
$$y = 1$$

Let $y = 1$ in (1).

$$2x + 12(1) = 9$$
$$2x = -3$$
$$x = -\frac{3}{2}$$

The solution is $\left(-\frac{3}{2}, 1\right)$.

47. $\begin{cases} y + 5 = \frac{2}{3}x + 3 \\ \frac{1}{3}x - \frac{1}{2}y = 1 \end{cases}$

Multiply the first equation by 3 and the second equation by 6.

$$\begin{cases} 3y + 15 = 2x + 9 \\ 2x - 3y = 6 \end{cases}$$

Rearrange the first equation, then add.

$$\begin{cases} -2x + 3y = -6 \\ 2x - 3y = 6 \end{cases}$$
$$0 = 0 \quad \text{True}$$

The system is dependent. There are infinitely many solutions.

48. $\begin{cases} \frac{1}{14} - \frac{x}{2} = -\frac{y}{7} \\ y = \frac{1}{2} + \frac{7x}{3} \end{cases}$

Multiply the first equation by 14 and the second equation by 6.

$\begin{cases} 1 - 7x = -2y & (1) \\ 6y = 3 + 14x & (2) \end{cases}$

Multiply (1) by 2 and rearrange both, then add.

$$\begin{array}{r} -14x + 4y = -2 \\ 14x - 6y = -3 \\ \hline -2y = -5 \\ y = \frac{5}{2} \end{array}$$

Let $y = \frac{5}{2}$ in (2) and solve for x.

$$\begin{aligned} 6\left(\frac{5}{2}\right) &= 3 + 14x \\ 15 &= 3 + 14x \\ 12 &= 14x \\ \frac{6}{7} &= x \end{aligned}$$

The solution is $\left(\frac{6}{7}, \frac{5}{2}\right)$.

49. $\begin{cases} -x + y = 7 & (1) \\ -3x + 4y = 8 & (2) \end{cases}$

Solve (1) for y.

$y = x + 7$

Let $y = x + 7$ in (2).

$$\begin{aligned} -3x + 4(x + 7) &= 8 \\ -3x + 4x + 28 &= 8 \\ x + 28 &= 8 \\ x &= -20 \end{aligned}$$

Let $x = -20$ in (1).

$$\begin{aligned} -(-20) + y &= 7 \\ 20 + y &= 7 \\ y &= -13 \end{aligned}$$

The solution is $(-20, -13)$.

50. $\begin{cases} 4x + y = 2 & (1) \\ 9y - 3x = 5 & (2) \end{cases}$

Solve (1) for y.

$y = 2 - 4x$

Let $y = 2 - 4x$ in (2).

$$\begin{aligned} 9(2 - 4x) - 3x &= 5 \\ 18 - 36x - 3x &= 5 \\ 18 - 39x &= 5 \\ -39x &= -13 \\ x &= \frac{1}{3} \end{aligned}$$

Let $x = \frac{1}{3}$ in (2).

$$\begin{aligned} 9y - 3\left(\frac{1}{3}\right) &= 5 \\ 9y - 1 &= 5 \\ 9y &= 6 \\ y &= \frac{2}{3} \end{aligned}$$

The solution is $\left(\frac{1}{3}, \frac{2}{3}\right)$.

51. $\begin{cases} 4x + 6 = 3y + 5 \\ 4(-2x - 4) = 6(-y - 3) \end{cases}$

Simplify and rearrange.

$\begin{cases} 4x - 3y = -1 & (1) \\ -8x + 6y = -2 & (2) \end{cases}$

Multiply (1) by 2 and add.

$$\begin{array}{r} 8x - 6y = -2 \\ -8x + 6y = -2 \\ \hline 0 = -4 \end{array} \quad \text{False}$$

The system is inconsistent. There is no solution.

52. $\begin{cases} 3y + 2x = 16x + 2 & (1) \\ x = -\frac{3}{14}y - \frac{1}{7} & (2) \end{cases}$

Rearrange and multiply the second equation by 14.

$$\begin{array}{r} -14x + 3y = 2 \\ 14x + 3y = -2 \\ \hline 6y = 0 \\ y = 0 \end{array}$$

Let $y = 0$ in (2).

$$\begin{aligned} x &= -\frac{3}{14}(0) - \frac{1}{7} \\ x &= -\frac{1}{7} \end{aligned}$$

The solution is $\left(-\frac{1}{7}, 0\right)$.

53. $\begin{cases} h + c = 4 & (1) \\ 1.50h + 7.75c = 16 & (2) \end{cases}$

Solve (1) for c.

$c = 4 - h$

Let $c = 4 - h$ in (2).

$$\begin{aligned} 1.5h + 7.75(4 - h) &= 4(4) \\ 1.5h + 31 - 7.75h &= 16 \\ -6.25h &= -15 \\ h &= 2.4 \end{aligned}$$

Let $h = 2.4$ in (1).

$$\begin{aligned} 2.4 + c &= 4 \\ c &= 1.6 \end{aligned}$$

Therefore, 2.4 pounds of cookies and 1.6 pounds of chocolates should be included.

54. $\begin{cases} t + b = 725 & (1) \\ t + 5b = 2025 & (2) \end{cases}$

Multiply (1) by -1, then add.

$$\begin{cases} -t - b = -725 \\ t + 5b = 2025 \end{cases}$$

$$\begin{aligned} 4b &= 1300 \\ b &= 325 \end{aligned}$$

Let $b = 325$ in (1).

$$\begin{aligned} t + 325 &= 725 \\ t &= 400 \end{aligned}$$

Therefore, 400 individual tickets and 325 block tickets were sold.

55. Let x be the first number and y be the second number. Their sum, $x + y$, is $\frac{17}{24}$. Their difference, $x - y$, is $\frac{1}{24}$.

$$\begin{cases} x + y = \frac{17}{24} & (1) \\ x - y = \frac{1}{24} & (2) \end{cases}$$

Add the equations.

$$\begin{cases} x + y = \frac{17}{24} \\ x - y = \frac{1}{24} \end{cases}$$

$$\begin{aligned} 2x &= \frac{18}{24} \\ x &= \frac{3}{8} \end{aligned}$$

Let $x = \frac{3}{8}$ in (1).

$$\begin{aligned} \frac{3}{8} + y &= \frac{17}{24} \\ y &= \frac{1}{3} \end{aligned}$$

The numbers are $\frac{3}{8}$ and $\frac{1}{3}$.

56. Let x be the smaller number. Then y is the larger number. Their sum, $x + y$, is 58. If twice the smaller is subtracted from the larger, $y - 2x$, the difference is -20.

$$\begin{cases} x + y = 58 & (1) \\ y - 2x = -20 & (2) \end{cases}$$

Solve (2) for y.

$y = 2x - 20$

Let $y = 2x - 20$ in (1).

$$\begin{aligned} x + 2x - 20 &= 58 \\ 3x &= 78 \\ x &= 26 \end{aligned}$$

The smaller number is 26.

57. Let s be the amount invested in stocks and b be the amount invested in bonds. The sum, $s + b$, is 50,000. The amount in stocks equals $4,000 less than twice the amount in bonds.

$$\begin{cases} s + b = 50{,}000 & (1) \\ s = 2b - 4000 & (2) \end{cases}$$

Let $s = 2b - 4000$ in (1).

$$\begin{aligned} 2b - 4000 + b &= 50{,}000 \\ 3b &= 54{,}000 \\ b &= 18{,}000 \end{aligned}$$

Let $b = 18{,}000$ in (2).

$s = 2(18{,}000) - 4000$

$s = 36{,}000 - 4000$

$s = 32{,}000$

He should invest $32,000 in stocks and $18,000 in bonds.

58. Let n be the number of notebooks sold and c be the number of scientific calculators sold. If a calculator sells for $7.50 and a notebook costs $\frac{1}{3}$ of that, notebooks sell for $2.50. The total number of items sold, $n + c$, is 24. The total amount, $2.5n + 7.5c$, is $110.

$$\begin{cases} n + c = 24 & (1) \\ 2.5n + 7.5c = 110 & (2) \end{cases}$$

Solve (1) for n.

$n = 24 - c$

Let $n = 24 - c$ in (2).

$$\begin{aligned} 2.5(24-c)+7.5c &= 110 \\ 60-2.5c+7.5c &= 110 \\ 5c &= 50 \\ c &= 10 \end{aligned}$$

Let $c = 10$ in (1).

$$\begin{aligned} n+10 &= 24 \\ n &= 14 \end{aligned}$$

They sold 14 notebooks and 10 calculators.

59. Let x be the measure of the angle and y be the measure of the other angle. Supplementary angles sum to 180, $x + y = 180$. One angle is 25° less than its supplement.

$$\begin{cases} x+y=180 & (1) \\ x=y-25 & (2) \end{cases}$$

Let $x = y - 25$ in (1).

$$\begin{aligned} y-25+y &= 180 \\ 2y &= 205 \\ y &= 102.5 \end{aligned}$$

Let $y = 102.5$ in (2).

$x = 102.5 - 25 = 77.5$

The angles measure 77.5° and 102.5°.

60. Let x be the measure of one angle and y be the measure of the other angle. Complementary angles sum to 90°, $x + y = 90$. One angle is 15° more than $\frac{1}{2}$ of its complement.

$$\begin{cases} x+y=90 & (1) \\ x=\frac{1}{2}y+15 & (2) \end{cases}$$

Let $x=\frac{1}{2}y+15$ in (1).

$$\begin{aligned} \frac{1}{2}y+15+y &= 90 \\ \frac{3}{2}y &= 75 \\ y &= 50 \end{aligned}$$

Let $y = 50$ in (2).

$$\begin{aligned} x &= \frac{1}{2}(50)+15 \\ x &= 25+15 \\ x &= 40 \end{aligned}$$

The angles measure 40° and 50°.

61. Let w be the width and l be the length. The three sides, $2w + l$, require 52 meters of fencing. The longest side is 8 meters less than twice the short side.

$$\begin{cases} 2w+l=52 & (1) \\ l=2w-8 & (2) \end{cases}$$

Let $l = 2w - 8$ in (1).

$$\begin{aligned} 2w+2w-8 &= 52 \\ 4w &= 60 \\ w &= 15 \end{aligned}$$

Let $w = 15$ in (2).

$l = 2(15) - 8$

$l = 30 - 8$

$l = 22$

The corral is 15 meters by 22 meters.

62. Let x be the measure of one angle and y be the measure of the other angle. The three angles sum to 180, $x + y + 30 = 180$. The first angle is 10° less than three times the second.

$$\begin{cases} x+y+30=180 & (1) \\ x=3y-10 & (2) \end{cases}$$

Let $x = 3y - 10$ in (1).

$$\begin{aligned} 3y-10+y+30 &= 180 \\ 4y &= 160 \\ y &= 40 \end{aligned}$$

Let $y = 40$ in (2).

$x = 3(40) - 10$

$x = 120 - 10$

$x = 110$

The angles measure 110° and 40°.

63. Let s be the speed of the plane in still air and w be the speed of the wind.

	d	r	t
with wind	2000	$s + w$	4
against wind	2000	$s - w$	5

$$\begin{cases} 4(s+w)=2000 & (1) \\ 5(s-w)=2000 & (2) \end{cases}$$

Divide (1) by 4 and (2) by 5, then add.

$$\begin{aligned} s+w &= 500 \\ s-w &= 400 \\ \hline 2s &= 900 \\ s &= 450 \end{aligned}$$

Let $s = 450$ in (1).

$$\begin{aligned} 4(450+w) &= 2000 \\ 1800+4w &= 2000 \\ 4w &= 200 \\ w &= 50 \end{aligned}$$

The plane's speed is 450 mph and the wind's speed is 50 mph.

64. Let c be the speed of the current and p be the paddling speed.

	d	r	t
upstream	10	$p-c$	2.5
downstream	10	$p+c$	1.25

$$\begin{cases} 10 = 2.5(p-c) & (1) \\ 10 = 1.25(p+c) & (2) \end{cases}$$

Divide (1) by 2.5 and (2) by 1.25, then add.

$$\begin{cases} 4 = p-c \\ 8 = p+c \end{cases}$$

$$12 = 2p$$
$$6 = p$$

Let $p = 6$ in (2).

$$10 = 1.25(6+c)$$
$$8 = 6+c$$
$$2 = c$$

The current is 2 mph and the paddling speed is 6 mph.

65. Let c be the speed of the cyclist and w be the speed of the wind.

	d	r	t
with wind	36	$c+w$	3
against wind	32	$c-w$	4

$$\begin{cases} 3(c+w) = 36 & (1) \\ 4(c-w) = 32 & (2) \end{cases}$$

Divide (1) by 3 and (2) by 4, then add.

$$\begin{cases} c+w = 12 \\ c-w = 8 \end{cases}$$

$$2c = 20$$
$$c = 10$$

Let $c = 10$ in (1).

$$3(10+w) = 36$$
$$10+w = 12$$
$$w = 2$$

The cyclist travels at 10 mph and the wind is 2 mph.

66. Let f be the speed of the faster jogger, and s be the speed of the slower jogger.

	d	r	$t = \frac{d}{r}$
faster	12	f	$\frac{12}{f}$
slower	9	s	$\frac{9}{s}$

The rate of the faster jogger is 2 mph more than that of the slower jogger. The times are the same.

$$\begin{cases} f = s+2 & (1) \\ \frac{12}{f} = \frac{9}{s} & (2) \end{cases}$$

Let $f = s + 2$ in (2).

$$\frac{12}{s+2} = \frac{9}{s}$$
$$12s = 9(s+2)$$
$$12s = 9s+18$$
$$3s = 18$$
$$s = 6$$

Let $s = 6$ in (1).

$f = 6 + 2 = 8$

The faster jogger's speed is 8 mph and the slower jogger's speed is 6 mph.

67. Let d be the number of dimes and n be the number of nickels.

	number of coins	· value of each coin	= total value
dimes	d	0.10	$0.10d$
nickels	n	0.05	$0.05n$
total	35		2.25

68. Let s be the amount invested in savings and m be the amount invested in a mutual fund.

	Principal	Rate	Interest
Savings account	s	0.065	$0.065s$
mutual fund	m	0.08	$0.08m$
total	15,000		2200

69. Let r be the number of regular tickets and m the number of matinee tickets.

	Number	Cost	Amount
Regular	r	\$8	$8r$
Matinee	m	\$5.50	$5.5m$
Total	1498		\$9929

$$\begin{cases} r+m=1498 & (1) \\ 8r+5.5m=9929 & (2) \end{cases}$$

Solve (1) for m.

$m = 1498 - r$

Let $m = 1498 - r = (2)$.

$$\begin{aligned} 8r+5.5(1498-r) &= 9929 \\ 8r+8239-5.5r &= 9929 \\ 2.5r+8239 &= 9929 \\ 2.5r &= 1690 \\ r &= 676 \end{aligned}$$

676 regular-priced tickets were sold.

70. Let d be the number of dimes and q be the number of quarters.

	number	value	amount
dimes	d	0.10	$0.10d$
quarters	q	0.25	$0.25q$
			1.70

$$\begin{cases} d=q+3 & (1) \\ 0.10d+0.25q=1.70 & (2) \end{cases}$$

Let $d = q + 3$ in (2).

$$\begin{aligned} 0.10(q+3)+0.25q &= 1.70 \\ 0.10q+0.3+0.25q &= 1.70 \\ 0.35q &= 1.40 \\ q &= 4 \end{aligned}$$

There are 4 quarters.

71. Let x be the amount invested at 5% and y be the amount invested at 9%.

	P	R	I
5%	x	0.05	$0.05x$
9%	y	0.09	$0.09y$
total			1430

$$\begin{cases} x=y-5000 & (1) \\ 0.05x+0.09y=1430 & (2) \end{cases}$$

Let $x = y - 5000$ in (2).

$$\begin{aligned} 0.05(y-5000)+0.09y &= 1430 \\ 0.05y-250+0.09y &= 1430 \\ 0.14y &= 1680 \\ y &= 12{,}000 \end{aligned}$$

Let $y = 12{,}000$ in (1).

$x = 12{,}000 - 5000 = 7000$

Carlos invested \$7000 at 5% and \$12,000 at 9%.

72. Let b be the amount invested in bonds and s be the amount invested in stocks.

	P	R	I
bonds	b	0.07	$0.07b$
stocks	s	0.08	$0.08s$
total	25,000		1900

$$\begin{cases} b+s=25{,}000 & (1) \\ 0.07b+0.08s=1900 & (2) \end{cases}$$

Solve (1) for b.

$b = 25{,}000 - s$

Let $b = 25{,}000 - s$ in (2).

$$\begin{aligned} 0.07(25{,}000-s)+0.08s &= 1900 \\ 1750-0.07s+0.08s &= 1900 \\ 0.01s &= 150 \\ s &= 15{,}000 \end{aligned}$$

Let $s = 15{,}000$ in (1).

$$\begin{aligned} b+15{,}000 &= 25{,}000 \\ b &= 10{,}000 \end{aligned}$$

Hilda invested \$10,000 in bonds and \$15,000 in stocks.

73. Let x be the quarts of 60% sugar solution and y be the quarts of 30% sugar solution.

	quarts	concentration	amount
60%	x	0.60	$0.60x$
30%	y	0.30	$0.30y$
total (51%)	10	0.51	0.51(10)

$$\begin{cases} x+y=10 & (1) \\ 0.60x+0.30y=5.1 & (2) \end{cases}$$

Solve (1) for x.

$x = 10 - y$

Let $x = 10 - y$ in (2).

$$0.60(10 - y) + 0.30y = 5.1$$
$$6 - 0.6y + 0.3y = 5.1$$
$$-0.3y = -0.9$$
$$y = 3$$

The baker should use 3 quarts of 30% sugar solution.

74. Let x be the pints of 25% peroxide solution and y be the pints of total solution (30% peroxide solution).

	pints	concentration	amount
25%	x	0.25	$0.25x$
60%	10	0.60	0.6(10)
total (30%)	y	0.30	$0.3y$

$$\begin{cases} x + 10 = y & (1) \\ 0.25x + 6 = 0.3y & (2) \end{cases}$$

Let $y = x + 10$ in (2).

$$0.25x + 6 = 0.3(x + 10)$$
$$0.25x + 6 = 0.3x + 3$$
$$3 = 0.05x$$
$$60 = x$$

Add 60 pints of 25% peroxide solution.

75. Let p be the pounds of peanuts and a be the pounds of almonds.

	pounds	value	amount
peanuts	p	4	$4p$
almonds	a	6.5	$6.5a$
total	5	5	5(5)

$$\begin{cases} p + a = 5 & (1) \\ 4p + 6.5a = 25 & (2) \end{cases}$$

Solve (1) for p.

$p = 5 - a$

Let $p = 5 - a$ in (2).

$$4(5 - a) + 6.5a = 25$$
$$20 - 4a + 6.5a = 25$$
$$2.5a = 5$$
$$a = 2$$

Let $a = 2$ in (1).

$$p + 2 = 5$$
$$p = 3$$

She should use 3 pounds of peanuts and 2 pounds of almonds.

76. Let x be the liters of 35% acid and y be the liters of 60% acid.

	liters	concentration	amount
35%	x	0.35	$0.35x$
60%	y	0.60	$0.60y$
total (55%)	20	0.55	0.55(20)

$$\begin{cases} x + y = 20 & (1) \\ 0.35x + 0.60y = 11 & (2) \end{cases}$$

Solve (1) for x.

$x = 20 - y$

Let $x = 20 - y$ in (2).

$$0.35(20 - y) + 0.60y = 11$$
$$7 - 0.35y + 0.60y = 11$$
$$0.25y = 4$$
$$y = 16$$

Let $y = 16$ in (1).

$$x + 16 = 20$$
$$x = 4$$

He should use 4 liters of 35% acid and 16 liters of 60% acid.

77. $$\begin{cases} x + y \le 2 & (1) \\ 3x - 2y > 6 & (2) \end{cases}$$

(a) $(-1, -5)$

(1): $-1 + (-5) \le 2?$

$-6 \le 2$ True

(2): $3(-1) - 2(-5) > 6?$

$-3 + 10 > 6?$

$7 > 6$ True

$(-1, -5)$ is a solution.

(b) $(3, -1)$

(1): $3 + (-1) \le 2?$

$2 \le 2$ True

(2): $3(3) - 2(-1) > 6?$

$9 + 2 > 6?$

$11 > 6$ True

$(3, -1)$ is a solution.

(c) $(4, 1)$

(1): $4 + 1 \le 2?$

$5 \le 2$ False

$(4, 1)$ is not a solution.

78. $\begin{cases} y \geq 3x+5 & (1) \\ y \geq -2x & (2) \end{cases}$

(a) $(-1, 0)$
(1): $0 \geq 3(-1)+5?$
$0 \geq -3+5?$
$0 \geq 2$ False
$(-1, 0)$ is not a solution.

(b) $(-2, 4)$
(1): $4 \geq 3(-2)+5?$
$4 \geq -6+5?$
$4 \geq -1$ True
(2): $4 \geq -2(-2)?$
$4 \geq 4$ True
$(-2, 4)$ is a solution.

(c) $(1, 8)$
(1): $8 \geq 3(1)+5?$
$8 \geq 3+5?$
$8 \geq 8$ True
(2): $8 \geq -2(1)?$
$8 \geq -2$ True
$(1, 8)$ is a solution.

79. $\begin{cases} x > 5 & (1) \\ y < -2 & (2) \end{cases}$

(a) $(10, -10)$
(1): $10 > 5$ True
(2): $-10 < -2$ True
$(10, -10)$ is a solution.

(b) $(5, -3)$
(1): $5 > 5$ False
$(5, -3)$ is not a solution.

(c) $(7, -2)$
(1): $7 > 5$ True
(2): $-2 < -2$ False
$(7, -2)$ is not a solution.

80. $\begin{cases} y > x & (1) \\ 2x - y \leq 3 & (2) \end{cases}$

(a) $(4, 3)$
(1): $3 > 4$ False
$(4, 3)$ is not a solution.

(b) $(-3, -2)$
(1): $-2 > -3$ True
(2): $2(-3)-(-2) \leq 3?$
$-6+2 \leq 3?$
$-4 \leq 3$ True
$(-3, -2)$ is a solution.

(c) $(-1, 4)$
(1): $4 > -1$ True
(2): $2(-1)-4 \leq 3?$
$-2-4 \leq 3?$
$-6 \leq 3$ True
$(-1, 4)$ is a solution.

81. $\begin{cases} x+2y < 6 & (1) \\ 4y-2x > 16 & (2) \end{cases}$

(a) $(-4, 2)$
(1): $-4+2(2) < 6?$
$-4+4 < 6?$
$0 < 6$ True
(2): $4(2)-2(-4) > 16?$
$8+8 > 16?$
$16 > 16$ False
$(-4, 2)$ is not a solution.

(b) $(-8, 6)$
(1): $-8+2(6) < 6?$
$-8+12 < 6?$
$4 < 6$ True
(2): $4(6)-2(-8) > 16?$
$24+16 > 16?$
$40 > 16$ True
$(-8, 6)$ is a solution.

(c) $(-2, -3)$
(1): $-2+2(-3) < 6?$
$-2-6 < 6?$
$-8 < 6$ True
(2): $4(-3)-2(-2) > 16?$
$-12+4 > 16?$
$-8 > 16$ False
$(-2, -3)$ is not a solution.

82. $\begin{cases} 2x - y \geq 3 & (1) \\ y \leq 2x + 1 & (2) \end{cases}$

(a) (1, 2)
(1): $2(1) - 2 \geq 3?$
$2 - 2 \geq 3?$
$0 \geq 3$ False
(1, 2) is not a solution.

(b) (3, −2)
(1): $2(3) - (-2) \geq 3?$
$6 + 2 \geq 3?$
$8 \geq 3$ True
(2): $-2 \leq 2(3) + 1?$
$-2 \leq 6 + 1?$
$-2 \leq 7$ True
(3, −2) is a solution.

(c) (−1, 4)
(1): $2(-1) - 4 \geq 3?$
$-2 - 43 \geq 3?$
$-6 \geq 3$ False
(−1, 4) is not a solution.

83. $\begin{cases} x > -2 \\ y > 1 \end{cases}$

Graph $x > -2$ and $y > 1$ with dashed lines. Graph on the same rectangular coordinate system. The overlapping shaded region represents the solution.

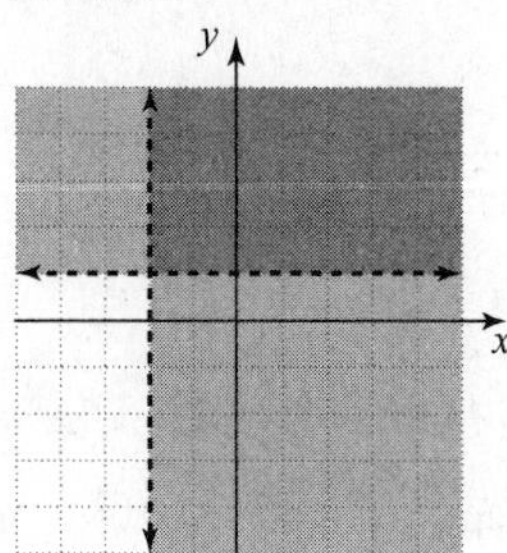

84. $\begin{cases} x \leq 3 \\ y > -1 \end{cases}$

Graph $x \leq 3$ with a solid line and $y > -1$ with a dashed line. Graph on the same rectangular coordinate system. The overlapping shaded region represents the solution.

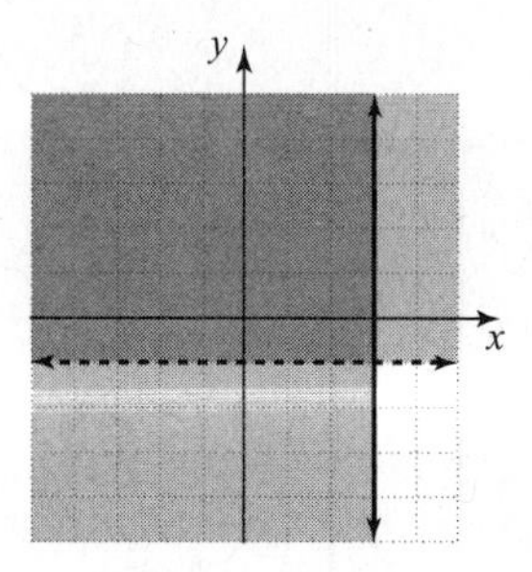

85. $\begin{cases} x + y \geq -2 \\ 2x - y \leq -4 \end{cases}$

Graph $x + y \geq -2$ ($y \geq -x - 2$) and $2x - y \leq -4$ ($y \geq 2x + 4$) with solid lines. Graph on the same rectangular coordinate system. The overlapping shaded region represents the solution.

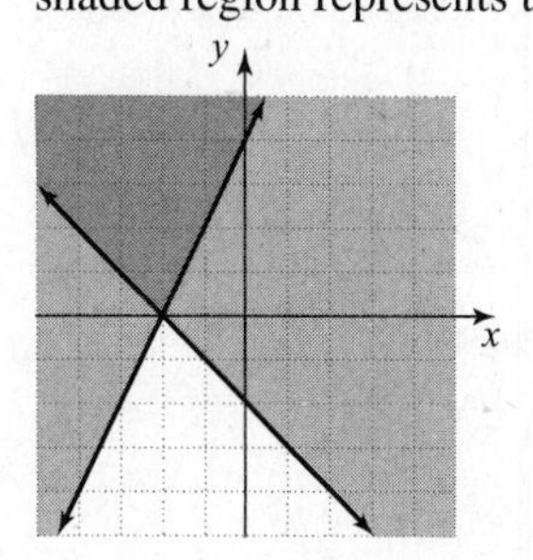

86. $\begin{cases} 3x + 2y < -6 \\ x - y < 2 \end{cases}$

Graph $3x + 2y < -6$ $\left(y < -\frac{3}{2}x - 3\right)$ and $x - y < 2$ ($y > x - 2$) with dashed lines. Graph on the same rectangular coordinate system. The overlapping shaded region represents the solution.

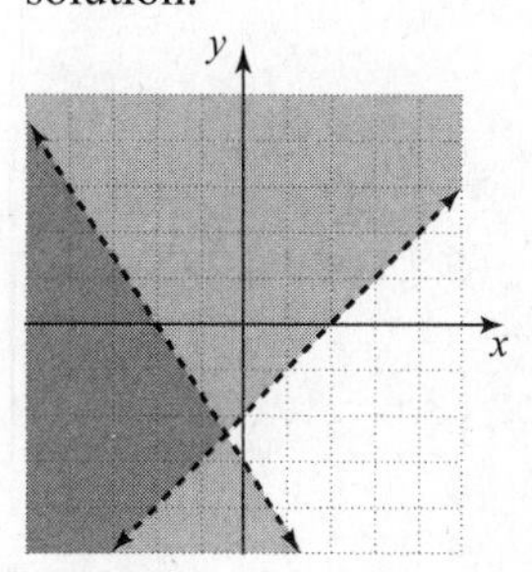

87. $\begin{cases} x > 0 \\ y \leq \frac{3}{4}x + 1 \end{cases}$

Graph $x > 0$ with a dashed line and $y \leq \frac{3}{4}x + 1$ with a solid line. Graph on the same rectangular coordinate system. The overlapping shaded region represents the solution.

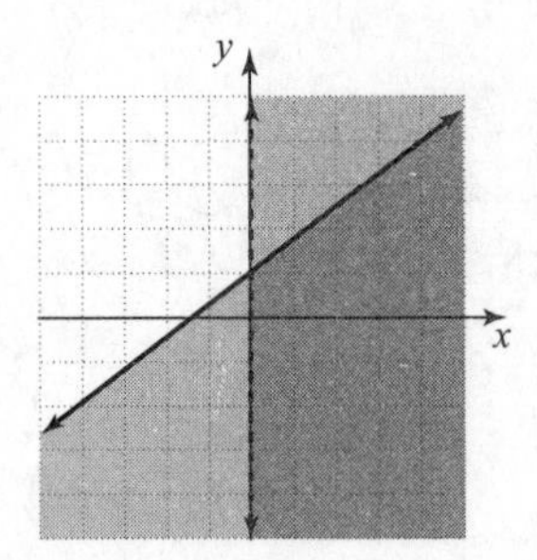

88. $\begin{cases} y \le 0 \\ y \le -\frac{1}{2}x - 3 \end{cases}$

Graph each with a solid line. Graph on the same rectangular coordinate system. The overlapping shaded region represents the solution.

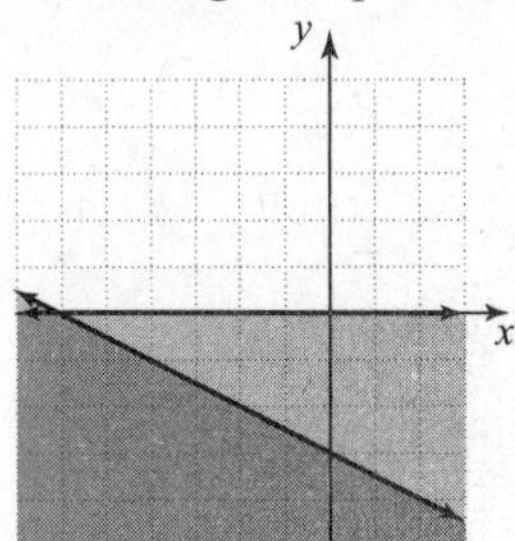

89. $\begin{cases} -y \ge x \\ 4x - 3y > -12 \end{cases}$

Graph $-y \ge x$ $(y \le -x)$ with a solid line and $4x - 3y > -12$ $\left(y < \frac{4}{3}x + 4\right)$ with a dashed line. Graph on the same rectangular coordinate system. The overlapping shaded region represents the solution.

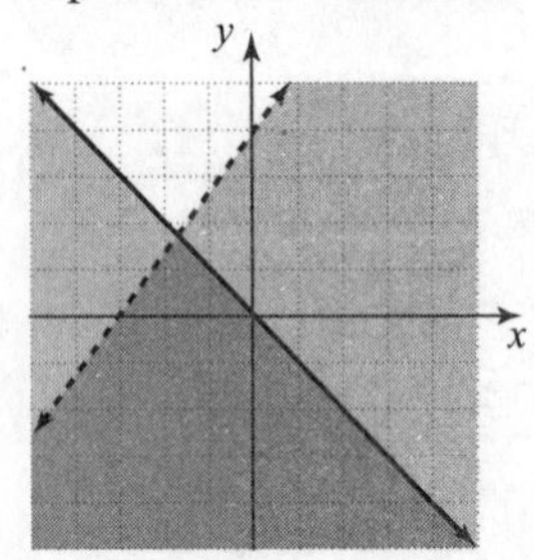

90. $\begin{cases} y \ge -x + 2 \\ x - y \ge -4 \end{cases}$

Graph $y \ge -x + 2$ with a solid line and $x - y \ge -4$ $(y \le x + 4)$ with a solid line. Graph on the same rectangular coordinate system. The overlapping shaded region represents the solution.

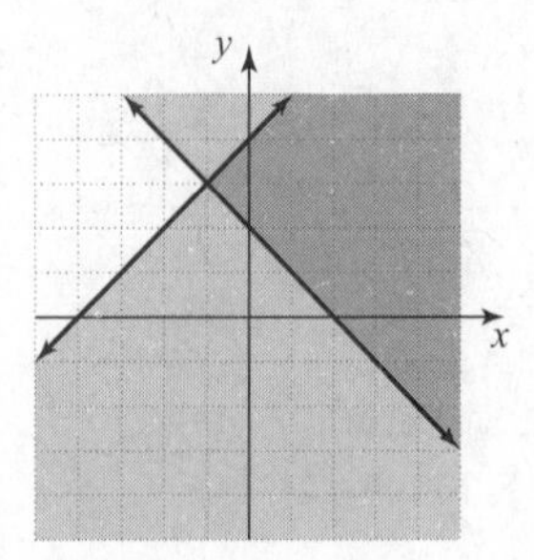

91. $\begin{cases} 2x + 3y \ge -3 \\ y > \frac{3}{2}x + 1 \end{cases}$

Graph $2x + 3y \ge -3$ $\left(y \ge -\frac{2}{3}x - 1\right)$ with a solid line and $y > \frac{3}{2}x + 1$ with a dashed line. Graph on the same rectangular coordinate system. The overlapping shaded region represents the solution.

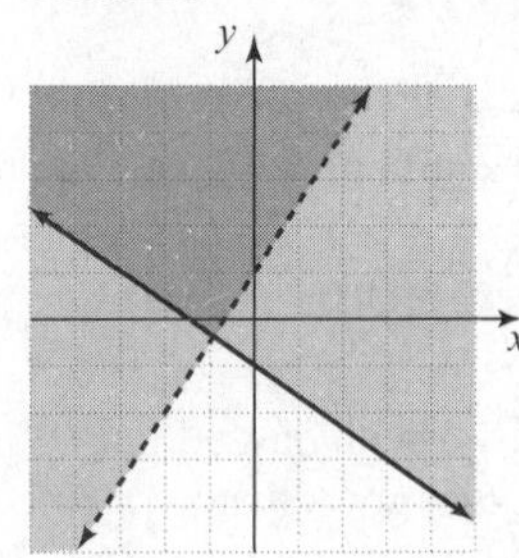

92. $\begin{cases} x + 4y \le -4 \\ y \ge \frac{1}{4}x + 3 \end{cases}$

Graph $x + 4y \le -4$ $\left(y \le -\frac{1}{4}x - 1\right)$ and $y \ge \frac{1}{4}x + 3$ with solid lines. Graph on the same rectangular coordinate system. The overlapping shaded region represents the solution.

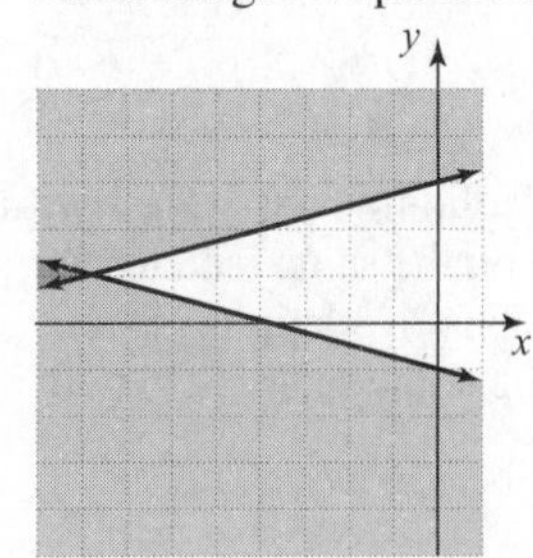

93. $\begin{cases} y > 2x - 5 \\ y - 2x \le 0 \end{cases}$

Graph $y > 2x - 5$ with a dashed line and $y - 2x \le 0$ ($y \le 2x$) with a solid line. Graph on the same rectangular coordinate system. The overlapping shaded region represents the solution.

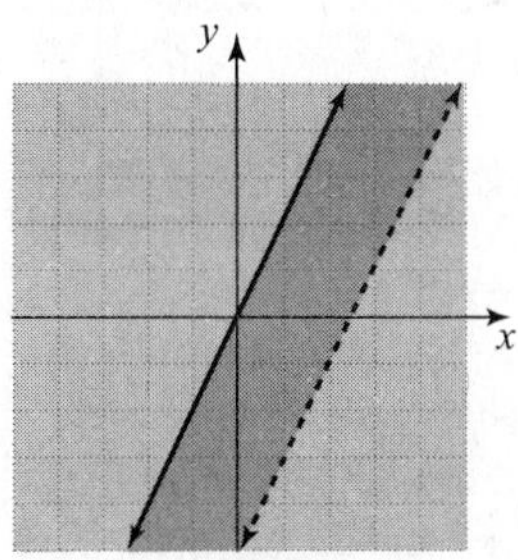

94. $\begin{cases} y < x + 2 \\ y > -\frac{5}{2}x + 2 \end{cases}$

Graph both with dashed lines. Graph on the same rectangular coordinate system. The overlapping shaded region represents the solution.

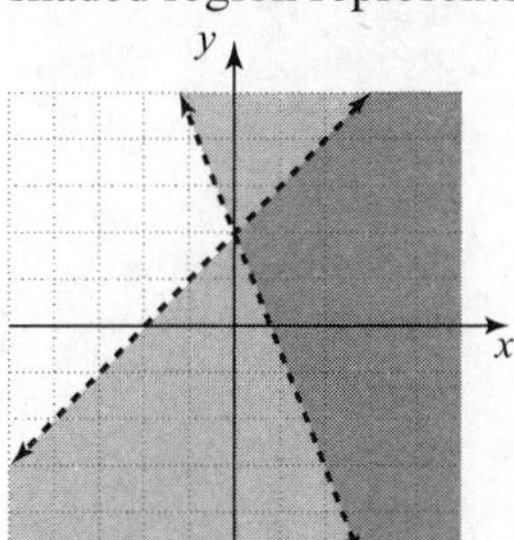

95. Let x be the pounds of fish and y be the pounds of carne asada.

$$\begin{cases} 8x + 5y \le 40 \\ x > 2y \\ x \ge 0 \\ y \ge 0 \end{cases}$$

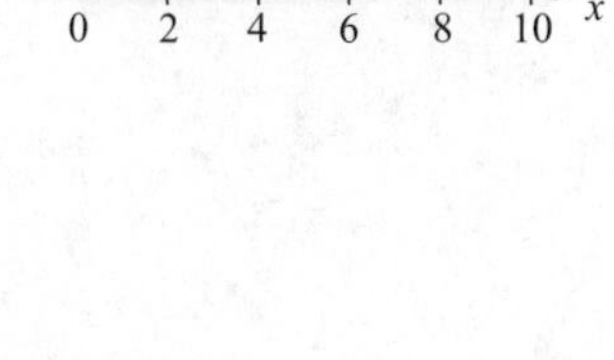

Chapter 4 Test

1. $\begin{cases} 3x - y = -5 & (1) \\ y = \frac{2}{3}x - 2 & (2) \end{cases}$

(a) $(-1, 2)$

(1): $3(-1) - 2 \stackrel{?}{=} -5$
$-3 - 2 \stackrel{?}{=} -5$
$-5 = -5$ True

(2): $2 \stackrel{?}{=} \frac{2}{3}(-1) - 2$
$2 \stackrel{?}{=} -\frac{2}{3} - 2$
$2 = -\frac{8}{3}$ False

$(-1, 2)$ is not a solution.

(b) $(-9, -8)$

(1): $3(-9) - (-8) \stackrel{?}{=} -5$
$-27 + 8 \stackrel{?}{=} -5$
$-19 = -5$ False

$(-9, -8)$ is not a solution.

(c) $(-3, -4)$

(1): $3(-3) - (-4) \stackrel{?}{=} -5$
$-9 + 4 \stackrel{?}{=} -5$
$-5 = -5$ True

(2): $-4 \stackrel{?}{=} \frac{2}{3}(-3) - 2$
$-4 \stackrel{?}{=} -2 - 2$
$-4 = -4$ True

$(-3, -4)$ is a solution.

2. $\begin{cases} 2x + y \ge 10 & (1) \\ y < \frac{x}{2} + 1 & (2) \end{cases}$

(a) $(5, 3)$

(1): $2(5) + 3 \ge 10?$
$10 + 3 \ge 10?$
$13 \ge 10$ True

(2): $3 < \frac{5}{2} + 1?$
$3 < \frac{7}{2}$ True

$(5, 3)$ is a solution.

(b) (−2, 14)

(1): $2(-2)+14 \ge 10?$

$-4+14 \ge 10?$

$10 \ge 10$ True

(2): $14 < \frac{-2}{2}+1?$

$14 < -1+1?$

$14 < 0$ False

(−2, 14) is not a solution.

(c) (−3, −1)

(1): $2(-3)+(-1) \ge 10?$

$-6-1 \ge 10?$

$-7 \ge 10$ False

(−3, −1) is not a solution.

3. (a) Since the slopes are negative reciprocals. The lines are perpendicular to each other. They intersect at one point, so there is one solution.

(b) The system is consistent.

(c) The equations are independent.

4. (a) Since the slopes are the same, but the *y*-intercepts are different, the lines are parallel. There is no solution.

(b) The system is inconsistent.

(c) The system is inconsistent.

5. $\begin{cases} 2x+3y=0 & (1) \\ x+4y=5 & (2) \end{cases}$

(1): $2x+3y=0$

$3y=-2x$

$y=-\frac{2}{3}x$

(2): $x+4y=5$

$4y=-x+5$

$y=-\frac{1}{4}x+\frac{5}{4}$

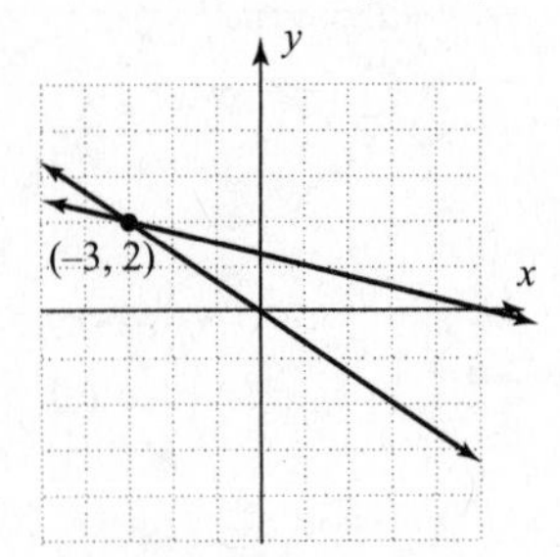

The solution is (−3, 2).

6. $\begin{cases} y=2x-6 \\ y=-\frac{1}{4}x+3 \end{cases}$

The equations are already written in slope-intercept form.

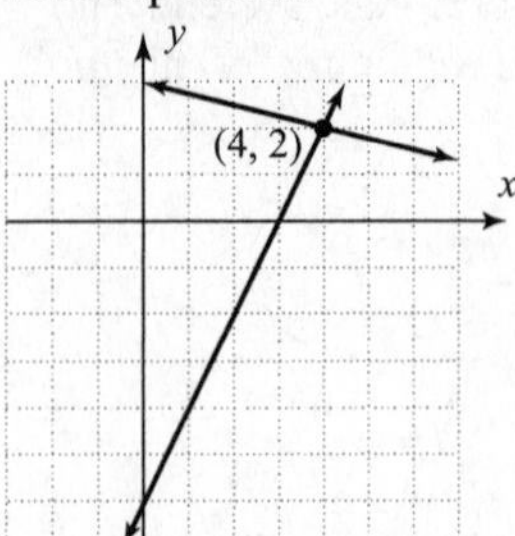

The solution is (4, 2).

7. $\begin{cases} 3x-y=3 & (1) \\ 4x+5y=-15 & (2) \end{cases}$

Solve (1) for *y*.

$y = 3x - 3$

Let $y = 3x - 3$ in (2).

$4x+5(3x-3)=-15$

$4x+15x-15=-15$

$4x+15x=0$

$19x=0$

$x=0$

Let $x = 0$ in (1).

$3(0)-y=3$

$-y=3$

$y=-3$

The solution is (0, −3).

8. $\begin{cases} y=\frac{2}{3}x+7 & (1) \\ y=\frac{1}{4}x+2 & (2) \end{cases}$

Let $y=\frac{2}{3}x+7$ in (2).

$\frac{2}{3}x+7=\frac{1}{4}x+2$

$\frac{2}{3}x-\frac{1}{4}x=2-7$

$\frac{5}{12}x=-5$

$x=-12$

Let $x = -12$ in (1).

$$y = \frac{2}{3}(-12) + 7$$
$$y = -8 + 7$$
$$y = -1$$

The solution is (−12, −1).

9. $\begin{cases} 3x + 2y = -3 & (1) \\ 5x - y = -18 & (2) \end{cases}$

Multiply (2) by 2, then add.

$$\begin{cases} 3x + 2y = -3 \\ 10x - 2y = -36 \end{cases}$$
$$13x = -39$$
$$x = -3$$

Let $x = -3$ in (2).

$$5(-3) - y = -18$$
$$-15 - y = -18$$
$$-y = -3$$
$$y = 3$$

The solution is (−3, 3).

10. $\begin{cases} 4x - 5y = 12 & (1) \\ 3x + 4y = -22 & (2) \end{cases}$

Multiply (1) by 4 and (2) by 5.

$$\begin{cases} 16x - 20y = 48 \\ 15x + 20y = -110 \end{cases}$$
$$31x = -62$$
$$x = -2$$

Let $x = -2$ in (2).

$$3(-2) + 4y = -22$$
$$-6 + 4y = -22$$
$$4y = -16$$
$$y = -4$$

The solution is (−2, −4).

11. $\begin{cases} \frac{2}{3}x - \frac{1}{6}y = \frac{25}{12} & (1) \\ -\frac{1}{4}x = \frac{3}{2}y & (2) \end{cases}$

Solve (2) for x.

$$-\frac{1}{4}x = \frac{3}{2}y$$
$$-4\left(-\frac{1}{4}x\right) = -4\left(\frac{3}{2}y\right)$$
$$x = -6y$$

Let $x = -6y$ in (1).

$$\frac{2}{3}(-6y) - \frac{1}{6}y = \frac{25}{12}$$
$$-4y - \frac{1}{6}y = \frac{25}{12}$$
$$-\frac{25}{6}y = \frac{25}{12}$$
$$y = -\frac{1}{2}$$

Let $y = -\frac{1}{2}$ in (2).

$$-\frac{1}{4}x = \frac{3}{2}\left(-\frac{1}{2}\right)$$
$$-\frac{1}{4}x = -\frac{3}{4}$$
$$x = 3$$

The solution is $\left(3, -\frac{1}{2}\right)$.

12. $\begin{cases} 0.4x - 2.5y = -6.5 & (1) \\ x + y = 5.5 & (2) \end{cases}$

Solve (2) for x.

$x = 5.5 - y$

Let $x = 5.5 - y$ in (1).

$$0.4(5.5 - y) - 2.5y = -6.5$$
$$2.2 - 0.4y - 2.5y = -6.5$$
$$-2.9y = -8.7$$
$$y = 3$$

Let $y = 3$ in (2).

$$x + 3 = 5.5$$
$$x = 2.5$$

The solution is (2.5, 3).

13. $\begin{cases} 4 + 3(y - 3) = -2(x + 1) & (1) \\ x + \frac{3y}{2} = \frac{3}{2} & (2) \end{cases}$

Simplify (1).

$$4 + 3(y - 3) = -2(x + 1)$$
$$4 + 3y - 9 = -2x - 2$$
$$2x + 3y = -2 + 9 - 4$$
$$2x + 3y = 3$$

Multiply (2) by −2, then add.

$$\begin{cases} 2x + 3y = 3 \\ -2x - 3y = -3 \end{cases}$$
$$0 = 0 \quad \text{True}$$

The system is dependent. There are infinitely many solutions.

14. Let p be the speed of the plane in still air and w be the speed of the wind.

	d	r	t
with wind	1575	$p+w$	7
against wind	1575	$p-w$	9

$$\begin{cases} 7(p+w)=1575 & (1) \\ 9(p-w)=1575 & (2) \end{cases}$$

Divide (1) by 7 and (2) by 9, then add.

$$\begin{cases} p+w=225 \\ p-w=175 \end{cases}$$
$$2p \quad =400$$
$$p=200$$

Let $p = 200$ in (1).

$$7(200+w)=1575$$
$$200+w=225$$
$$w=25$$

The plane's speed is 200 mph and the wind's speed is 25 mph.

15. Let v be the number of containers of vanilla ice cream and p be the number of containers of peach ice cream.

	containers	price	amount
vanilla	v	6	$6v$
peach	p	11.50	$11.5p$
total	11	8	8(11)

$$\begin{cases} v+p=11 & (1) \\ 6v+11.5p=88 & (2) \end{cases}$$

Solve (1) for v.

$v = 11 - p$

Let $v = 11 - p$ in (2).

$$6(11-p)+11.5p=88$$
$$66-6p+11.5p=88$$
$$5.5p=22$$
$$p=4$$

Let $p = 4$ in (1).

$$v+4=11$$
$$v=7$$

He should use 7 containers of vanilla and 4 containers of peach ice cream.

16. Let x be the measure of one angle and y be the measure of the other angle. Supplementary angles sum to 180°, $x + y = 180$. One angle is 30° less than three times its supplement, $x = 3y - 30$.

$$\begin{cases} x+y=180 & (1) \\ x=3y-30 & (2) \end{cases}$$

Let $x = 3y - 30$ in (1).

$$3y-30+y=180$$
$$4y=210$$
$$y=52.5$$

Let $y = 52.5$ in (1).

$$x+52.5=180$$
$$x=127.5$$

The angles are 52.5° and 127.5°.

17. Let b be the cost of a basketball and v be the cost of a volleyball. The total number of balls, $b + v$, is 40. The total cost, $25b + 15v$, is 750.

$$\begin{cases} b+v=40 & (1) \\ 25b+15v=750 & (2) \end{cases}$$

Solve (1) for b.

$b = 40 - v$

Let $b = 40 - v$ in (2).

$$25(40-v)+15v=750$$
$$1000-25v+15v=750$$
$$-10v=-250$$
$$v=25$$

Let $v = 25$ in (1).

$$b+25=40$$
$$b=15$$

The school will receive 15 basketballs and 25 volleyballs.

18. $\begin{cases} 4x-2y<8 \\ x+3y<6 \end{cases}$

Graph $4x - 2y < 8$ $(y > 2x - 4)$ and $x + 3y < 6$ $\left(y<-\frac{1}{3}x+2\right)$ with dashed lines.

Graph on the same rectangular coordinate system. The overlapping shaded region represents the solution.

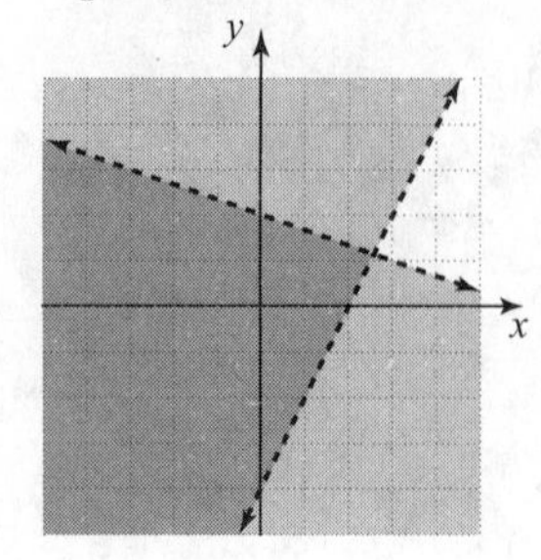

19. $\begin{cases} x \le -2 \\ -2x - 4y \le 8 \end{cases}$

Graph $x \le -2$ and $-2x - 4y \le 8$ $\left(y \ge -\frac{1}{2}x - 2 \right)$ with solid lines. Graph on the same rectangular coordinate system. The overlapping shaded region represents the solution.

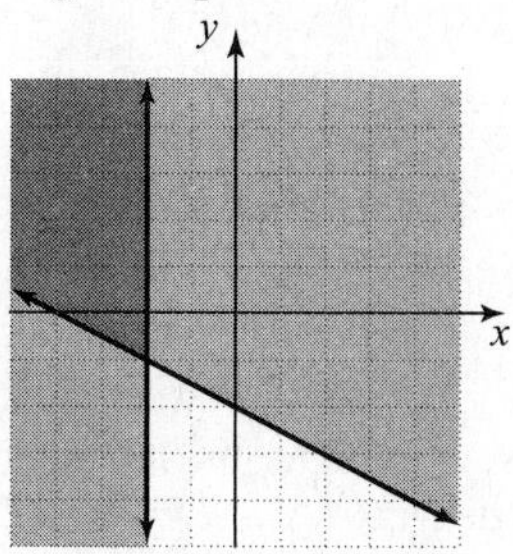

20. $\begin{cases} y \le \frac{2}{3}x + 4 \\ x + 4y < 0 \end{cases}$

Graph $y \le \frac{2}{3}x + 4$ with a solid line and $x + 4y < 0$ $\left(y < -\frac{1}{4}x \right)$ with a dashed line. Graph on the same rectangular coordinate system. The overlapping shaded region represents the solution.

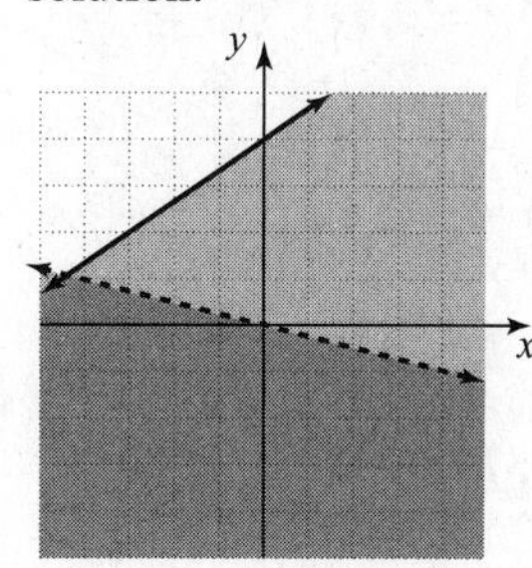

Chapter 5

Section 5.1

Preparing for Adding and Subtracting Polynomials

P1. The coefficient of $-4x^5$ is -4.

P2. $$\begin{aligned} -3x+2-2x-6x-7 &= -3x-2x-6x+2-7 \\ &= -5x-6x-5 \\ &= -11x-5 \end{aligned}$$

P3. $$\begin{aligned} -4(x-3) &= -4\cdot x-(-4)\cdot 3 \\ &= -4x-(-12) \\ &= -4x+12 \end{aligned}$$

P4. $5-3x$ for $x=-2$:
$5-3(-2)=5-(-6)=5+6=11$

5.1 Quick Checks

1. A monomial in one variable is the product of a number and a variable raised to a whole number power.

2. The coefficient of a monomial such as x^2 or z is 1.

3. $12x^6$ is a monomial. The coefficient is 12 and the degree is 6.

4. $3x^{-3}$ is not a monomial because the exponent of the variable x is -3 and -3 is not a whole number.

5. 10 is a monomial. The coefficient is 10 and the degree is 0.

6. $n^{1/3}$ is not a monomial because the exponent of the variable n is $\frac{1}{3}$ and $\frac{1}{3}$ is not a whole number.

7. The degree of a monomial in the form ax^my^n is $m+n$.

8. $3x^5y^2$ is a monomial in x and y of degree $5+2=7$. The coefficient is 3.

9. $4ab^{1/2}$ is not a monomial because the exponent on the variable b is $\frac{1}{2}$, and $\frac{1}{2}$ is not a whole number.

10. $-x^2y$ is a monomial in x and y of degree $2+1=3$. The coefficient is -1.

11. True

12. A polynomial is said to be written in standard form if it is written with the terms in descending order according to degree.

13. $-4x^3+2x^2-5x+3$ is a polynomial of degree 3.

14. $2m^{-1}+7$ is not a polynomial because the exponent on the first term, -1, is not a whole number.

15. $\frac{-1}{x^2+1}$ is not a polynomial because there is a variable expression in the denominator of the fraction.

16. $5p^3q-8pq^2+pq$ is a polynomial of degree $3+1=4$.

17. True or False: $4x^3+7x^3=11x^6$ False

18. $$\begin{aligned} &(9x^2-x+5)+(3x^2+4x-2) \\ &= 9x^2-x+5+3x^2+4x-2 \\ &= 9x^2+3x^2-x+4x+5-2 \\ &= (9+3)x^2+(-1+4)x+(5-2) \\ &= 12x^2+3x+3 \end{aligned}$$

19. $$\begin{array}{r} 4z^4-2z^3+z-5 \\ +\,(-2z^4+6z^3-z^2+4 \\ \hline 2z^4+4z^3-z^2+z-1 \end{array}$$

20. $$\begin{aligned}(7x^2y+x^2y^2-5xy^2)+(-2x^2y+5x^2y^2+4xy^2) &= 7x^2y+x^2y^2-5xy^2-2x^2y+5x^2y^2+4xy^2\\ &= 7x^2y-2x^2y+x^2y^2+5x^2y^2-5xy^2+4xy^2\\ &= (7-2)x^2y+(1+5)x^2y^2+(-5+4)xy^2\\ &= 5x^2y+6x^2y^2-xy^2\end{aligned}$$

21. $$\begin{array}{r} 7x^3-3x^2+2x+8\\ \underline{-(2x^3+12x^2-x+1)}\end{array}$$

Rewrite the subtraction as addition.

$$\begin{array}{r} 7x^3-3x^2+2x+8\\ \underline{+(-2x^3-12x^2+x-1)}\\ 5x^3-15x^2+3x+7\end{array}$$

22. $$\begin{aligned}(6y^3-3y^2+2y+4)-(-2y^3+5y+10) &= (6y^3-3y^2+2y+4)+(2y^3-5y-10)\\ &= 6y^3-3y^2+2y+4+2y^3-5y-10\\ &= 6y^3+2y^3-3y^2+2y-5y+4-10\\ &= 8y^3-3y^2-3y-6\end{aligned}$$

23. $$\begin{aligned}(9x^2y+6x^2y^2-3xy^2)-(-2x^2y+4x^2y^2+6xy^2) &= (9x^2y+6x^2y^2-3xy^2)+(2x^2y-4x^2y^2-6xy^2)\\ &= 9x^2y+6x^2y^2-3xy^2+2x^2y-4x^2y^2-6xy^2\\ &= 9x^2y+2x^2y+6x^2y^2-4x^2y^2-3xy^2-6xy^2\\ &= 11x^2y+2x^2y^2-9xy^2\end{aligned}$$

24. $$\begin{aligned}(4a^2b-3a^2b^3+2ab^2)-(3a^2b-4a^2b^3+5ab) &= (4a^2b-3a^2b^3+2ab^2)+(-3a^2b+4a^2b^3-5ab)\\ &= 4a^2b-3a^2b^3+2ab^2-3a^2b+4a^2b^3-5ab\\ &= a^2b+a^2b^3+2ab^2-5ab\end{aligned}$$

25. $$\begin{aligned}(3a^2-5ab+3b^2)+(4a^2-7b^2)-(8b^2-ab) &= 3a^2-5ab+3b^2+4a^2-7b^2-8b^2+ab\\ &= 3a^2+4a^2-5ab+ab+3b^2-7b^2-8b^2\\ &= 7a^2-4ab-12b^2\end{aligned}$$

26. **(a)** Let $x = 0$.

$$\begin{aligned}-2x^3+7x+1 &= -2(0)^3+7(0)+1\\ &= 0+0+1\\ &= 1\end{aligned}$$

(b) Let $x = 5$.

$$\begin{aligned}-2x^3+7x+1 &= -2(5)^3+7(5)+1\\ &= -250+35+1\\ &= -214\end{aligned}$$

(c) Let $x = -4$.

$$\begin{aligned}-2x^3+7x+1 &= -2(-4)^3+7(-4)+1\\ &= 128-28+1\\ &= 101\end{aligned}$$

27. **(a)** Let $m = -2$ and $n = -4$.

$$\begin{aligned}&-2m^2n+3mn-n^2\\&=-2(-2)^2(-4)+3(-2)(-4)-(-4)^2\\&=-2(4)(-4)+3(-2)(-4)-(16)\\&=32+24-16\\&=40\end{aligned}$$

(b) Let $m = 1$ and $n = 5$.

$$\begin{aligned}&-2m^2n+3mn-n^2\\&=-2(1)^2(5)+3(1)(5)-(5)^2\\&=-2(1)(5)+3(1)(5)-(25)\\&=-10+15-25\\&=-20\end{aligned}$$

28. $$\begin{aligned}40x-0.2x^2&=40(75)-0.2(75)^2\\&=40(75)-0.2(5625)\\&=3000-1125\\&=1875\end{aligned}$$

The revenue is \$1875.

5.1 Exercises

29. Yes, $\frac{1}{2}y^3$ is a monomial of degree 3 with coefficient $\frac{1}{2}$.

31. Yes, $\frac{x^2}{7}=\frac{1}{7}x^2$ is a monomial of degree 2 with coefficient $\frac{1}{7}$.

33. No, z^{-6} is not a monomial because the exponent of z is not a whole number.

35. Yes, $12mn^4$ is a monomial in m and n of degree $1 + 4 = 5$. The coefficient is 12.

37. No, $\frac{3}{n^2}$ is not a monomial because there is a variable in the denominator of the fraction.

39. Yes, 4 is a monomial of degree 0 with coefficient 4.

41. Yes, $6x^2-10$ is a binomial of degree 2.

43. No, $\frac{-20}{n}$ is not a polynomial because there is a variable in the denominator of a fraction.

45. No, $3y^{1/3}+2$ is not a polynomial because a variable has an exponent that is not a whole number.

47. Yes, $\frac{1}{8}$ is a monomial of degree 0.

49. Yes, $3t^2-\frac{1}{2}t^4+6t=-\frac{1}{2}t^4+3t^2+6t$ is a trinomial of degree 4.

51. No, $7x^{-1}+4$ is not a polynomial because a variable has an exponent that is not a whole number.

53. Yes, $5z^3-10z^2+z+12$ is a polynomial of degree 3.

55. Yes, $3x^2y^2+2xy^4+4=2xy^4+3x^2y^2+4$ is a trinomial of degree 5 (1 + 4).

57. $$\begin{aligned}(4x-3)+(3x-7)&=4x-3+3x-7\\&=4x+3x-3-7\\&=(4+3)x+(-3-7)\\&=7x-10\end{aligned}$$

59. $$\begin{aligned}&(-4m^2+2m-1)+(2m^2-2m+6)\\&=-4m^2+2m-1+2m^2-2m+6\\&=-4m^2+2m^2+2m-2m-1+6\\&=(-4+2)m^2+(2-2)m+(-1+6)\\&=-2m^2+0m+5\\&=-2m^2+5\end{aligned}$$

61. $$\begin{aligned}&(p-p^3+2)+(6-2p^2+p^3)\\&=p-p^3+2+6-2p^2+p^3\\&=-p^3+p^3-2p^2+p+2+6\\&=(-1+1)p^3-2p^2+p+(2+6)\\&=0p^3-2p^2+p+8\\&=-2p^2+p+8\end{aligned}$$

63. $$\begin{aligned}&(2y-10)+(-3y^2-4y+6)\\&=2y-10-3y^2-4y+6\\&=-3y^2+2y-4y-10+6\\&=-3y^2+(2-4)y+(-10+6)\\&=-3y^2-2y-4\end{aligned}$$

65. $\left(\frac{1}{2}p^2-\frac{2}{3}p+2\right)+\left(\frac{3}{4}p^2+\frac{5}{6}p-5\right)$
$=\frac{1}{2}p^2-\frac{2}{3}p+2+\frac{3}{4}p^2+\frac{5}{6}p-5$
$=\frac{1}{2}p^2+\frac{3}{4}p^2-\frac{2}{3}p+\frac{5}{6}p+2-5$
$=\left(\frac{1}{2}+\frac{3}{4}\right)p^2+\left(-\frac{2}{3}+\frac{5}{6}\right)p+(2-5)$
$=\frac{5}{4}p^2+\frac{1}{6}p-3$

67. $(5m^2-6mn+2n^2)+(m^2+2mn-3n^2)$
$=5m^2-6mn+2n^2+m^2+2mn-3n^2$
$=5m^2+m^2-6mn+2mn+2n^2-3n^2$
$=(5+1)m^2+(-6+2)mn+(2-3)n^2$
$=6m^2-4mn-n^2$

69.
$$\begin{array}{r} 4n^2-2n+1 \\ +\quad (-6n+4) \\ \hline 4n^2-8n+5 \end{array}$$

71. $(3x-10)-(4x+6)=(3x-10)+(-4x-6)$
$=3x-10-4x-6$
$=3x-4x-10-6$
$=-x-16$

73. $(12x^2-2x-4)-(-2x^2+x+1)$
$=(12x^2-2x-4)+(2x^2-x-1)$
$=12x^2-2x-4+2x^2-x-1$
$=12x^2+2x^2-2x-x-4-1$
$=14x^2-3x-5$

75. $(y^3-2y+1)-(-3y^3+y+5)$
$=(y^3-2y+1)+(3y^3-y-5)$
$=y^3-2y+1+3y^3-y-5$
$=y^3+3y^3-2y-y+1-5$
$=4y^3-3y-4$

77. $(3y^3-2y)-(2y+y^2+y^3)$
$=(3y^3-2y)+(-2y-y^2-y^3)$
$=3y^3-2y-2y-y^2-y^3$
$=3y^3-y^3-y^2-2y-2y$
$=2y^3-y^2-4y$

79. $\left(\frac{5}{3}q^2-\frac{5}{2}q+4\right)-\left(\frac{1}{9}q^2+\frac{3}{8}q+2\right)$
$=\left(\frac{5}{3}q^2-\frac{5}{2}q+4\right)+\left(-\frac{1}{9}q^2-\frac{3}{8}q-2\right)$
$=\frac{5}{3}q^2-\frac{5}{2}q+4-\frac{1}{9}q^2-\frac{3}{8}q-2$
$=\frac{5}{3}q^2-\frac{1}{9}q^2-\frac{5}{2}q-\frac{3}{8}q+4-2$
$=\frac{14}{9}q^2-\frac{23}{8}q+2$

81. $(-4m^2n^2-2mn+3)-(4m^2n^2+2mn+10)$
$=(-4m^2n^2-2mn+3)+(-4m^2n^2-2mn-10)$
$=-4m^2n^2-2mn+3-4m^2n^2-2mn-10$
$=-4m^2n^2-4m^2n^2-2mn-2mn+3-10$
$=-8m^2n^2-4mn-7$

83.
$$\begin{array}{r} 6x-3 \\ -\ (10x+2) \\ \hline \end{array} \quad\rightarrow\quad \begin{array}{r} 6x-3 \\ +\ (-10x-2) \\ \hline -4x-5 \end{array}$$

85. $(2x^2-3x+1)+(x^2+9)-(4x^2-2x-5)$
$=(2x^2-3x+1)+(x^2+9)+(-4x^2+2x+5)$
$=2x^2-3x+1+x^2+9-4x^2+2x+5$
$=2x^2+x^2-4x^2-3x+2x+1+9+5$
$=-x^2-x+15$

87. $(p^2+25)-(3p^2-p+4)-(-4p^2-9p+5)$
$=(p^2+25)+(-3p^2+p-4)+(4p^2+9p-5)$
$=p^2+25-3p^2+p-4+4p^2+9p-5$
$=p^2-3p^2+4p^2+p+9p+25-4-5$
$=2p^2+10p+16$

89. $(4y-3)+(y^3-8)-(2y^2-7y+3)$
$=(4y-3)+(y^3-8)+(-2y^2+7y-3)$
$=4y-3+y^3-8-2y^2+7y-3$
$=y^3-2y^2+4y+7y-3-8-3$
$=y^3-2y^2+11y-14$

91. $2x^2-x+3$

(a) $x=0$: $2(0)^2-0+3=2(0)-0+3=3$

(b) $x = 5$: $2(5)^2 - 5 + 3 = 2(25) - 5 + 3$
$= 50 - 5 + 3$
$= 48$

(c) $x = -2$: $2(-2)^2 - (-2) + 3 = 2(4) + 2 + 3$
$= 8 + 2 + 3$
$= 13$

93. $7 - x^2$

(a) $x = 3$: $7 - (3)^2 = 7 - 9 = -2$

(b) $x = -\frac{5}{2}$: $7 - \left(-\frac{5}{2}\right)^2 = 7 - \frac{25}{4} = \frac{3}{4}$

(c) $x = -1.5$: $7 - (-1.5)^2 = 7 - 2.25 = 4.75$

95. $-x^2y + 2xy^2 - 3$; $x = 2$, $y = -3$:
$-(2)^2(-3) + 2(2)(-3)^2 - 3$
$= -(4)(-3) + 2(2)(9) - 3$
$= 12 + 36 - 3$
$= 45$

97. $st + 2s^2t + 3st^2 - t^4$; $s = -2$, $t = 4$:
$(-2)(4) + 2(-2)^2(4) + 3(-2)(4)^2 - (4)^4$
$= (-2)(4) + 2(4)(4) + 3(-2)(16) - 256$
$= -8 + 32 - 96 - 256$
$= -328$

99. $(7t - 3) - 4t = 7t - 3 - 4t = 7t - 4t - 3 = 3t - 3$

101. $(5x^2 + x - 4) + (-2x^2 - 4x + 1)$
$= 5x^2 + x - 4 - 2x^2 - 4x + 1$
$= 5x^2 - 2x^2 + x - 4x - 4 + 1$
$= 3x^2 - 3x - 3$

103. $(2xy^2 - 3) + (7xy^2 + 4) = 2xy^2 - 3 + 7xy^2 + 4$
$= 2xy^2 + 7xy^2 - 3 + 4$
$= 9xy^2 + 1$

105. $(4 + 8y - 2y^2) - (3 - 7y - y^2)$
$= (4 + 8y - 2y^2) + (-3 + 7y + y^2)$
$= 4 + 8y - 2y^2 - 3 + 7y + y^2$
$= -2y^2 + y^2 + 8y + 7y + 4 - 3$
$= -y^2 + 15y + 1$

107. $\left(\frac{5}{6}q^2 - \frac{1}{3}\right) + \left(\frac{3}{2}q^2 + 2\right) = \frac{5}{6}q^2 - \frac{1}{3} + \frac{3}{2}q^2 + 2$
$= \frac{5}{6}q^2 + \frac{3}{2}q^2 - \frac{1}{3} + 2$
$= \frac{7}{3}q^2 + \frac{5}{3}$

109. $14d^2 - (2d - 10) - (d^2 - 3d)$
$= 14d^2 + (-2d + 10) + (-d^2 + 3d)$
$= 14d^2 - 2d + 10 - d^2 + 3d$
$= 14d^2 - d^2 - 2d + 3d + 10$
$= 13d^2 + d + 10$

111. $(4a^2 - 1) + (a^2 + 5a + 2) - (-a^2 + 4)$
$= (4a^2 - 1) + (a^2 + 5a + 2) + (a^2 - 4)$
$= 4a^2 - 1 + a^2 + 5a + 2 + a^2 - 4$
$= 4a^2 + a^2 + a^2 + 5a - 1 + 2 - 4$
$= 6a^2 + 5a - 3$

113. $(x^2 - 2xy - y^2) - (3x^2 + xy - y^2) + (xy + y^2)$
$= (x^2 - 2xy - y^2) + (-3x^2 - xy + y^2) + (xy + y^2)$
$= x^2 - 2xy - y^2 - 3x^2 - xy + y^2 + xy + y^2$
$= x^2 - 3x^2 - 2xy - xy + xy - y^2 + y^2 + y^2$
$= -2x^2 - 2xy + y^2$

115. $(3x + 10) + (-8x + 2) = 3x + 10 - 8x + 2$
$= 3x - 8x + 10 + 2$
$= -5x + 12$

117. $(-x^2 + 2x + 3) - (-4x^2 - 2x + 6)$
$= (-x^2 + 2x + 3) + (4x^2 + 2x - 6)$
$= -x^2 + 2x + 3 + 4x^2 + 2x - 6$
$= -x^2 + 4x^2 + 2x + 2x + 3 - 6$
$= 3x^2 + 4x - 3$

119. $(2x - 10) - (14x^2 - 2x + 3)$
$= (2x - 10) + (-14x^2 + 2x - 3)$
$= 2x - 10 - 14x^2 + 2x - 3$
$= -14x^2 + 2x + 2x - 10 - 3$
$= -14x^2 + 4x - 13$

121. $-(3x - 5) = -3x + 5$:
$3x - 5 + (-3x + 5) = 3x - 5 - 3x + 5$
$= 3x - 3x - 5 + 5$
$= 0$

123. $-16t^2 + 30;\ t = \frac{1}{2}$

$-16\left(\frac{1}{2}\right)^2 + 30 = -16\left(\frac{1}{4}\right) + 30 = -4 + 30 = 26$

The height of the ball is 26 feet.

125. (a) Let $x = 0$.

$-32.4x^2 + 905.6x + 438$
$= -32.4(0)^2 + 905.6(0) + 438$
$= 438$

Worldwide online social network advertising spending for 2006 is estimated to be \$438 million.

(b) Let $x = 2$.

$-32.4x^2 + 905.6x + 438$
$= -32.4(2)^2 + 905.6(2) + 438$
$= -32.4(4) + 905.6(2) + 438$
$= -129.6 + 1811.2 + 438$
$= 2119.6$

Worldwide online social network advertising spending for 2008 is estimated to be \$2119.6 million.

(c) Let $x = 5$.

$-32.4x^2 + 905.6x + 438$
$= -32.4(5)^2 + 905.6(5) + 438$
$= -32.4(25) + 905.6(5) + 438$
$= -810 + 4528 + 438$
$= 4156$

Worldwide online social network advertising spending for 2011 is estimated to be \$4156 million.

127. (a) $\text{profit} = (\text{revenue}) - (\text{cost})$
$= (-2x^2 + 120x) - (0.125x^2 + 15x)$
$= (-2x^2 + 120x) + (-0.125x^2 - 15x)$
$= -2x^2 + 120x - 0.125x^2 - 15x$
$= -2x^2 - 0.125x^2 + 120x - 15x$
$= -2.125x^2 + 105x$

(b) $-2.125x^2 + 105x;\ x = 20$

$-2.125(20)^2 + 105(20)$
$= -2.125(400) + 105(20)$
$= -850 + 2100$
$= 1250$

The profit is \$1250 if 20 calculators are produced and sold each day.

129. (a) For x lawns, Marissa's cost is $5x + 10x = 15x$, and her revenue is $25x$.

$\text{profit} = (\text{revenue}) - (\text{cost})$
$= (25x) - (15x)$
$= 10x$

(b) $10x;\ x = 50$

$10(50) = 500$

Her profit is \$500 per week if she fertilizes 50 lawns.

(c) Since her profit for one week is \$500, Marissa's profit for 30 weeks is 30(\$500) = \$15,000.

131. $(4x - 3) + (4x - 3) + (4x - 3) + (4x - 3)$
$= 4x + 4x + 4x + 4x - 3 - 3 - 3 - 3$
$= 16x - 12$

133. $(3x - 4) + (2x + 1) + (10) + (10) + (2x + 1)$
$= 3x + 2x + 2x - 4 + 1 + 10 + 10 + 1$
$= 7x + 18$

135. $(3x - 10) - (x - 5) = (3x - 10) + (-x + 5)$
$= 3x - 10 - x + 5$
$= 3x - x - 10 + 5$
$= 2x - 5$

137. Answers may vary.

139. To find the degree of a polynomial in one variable, use the largest exponent on any term in the polynomial. To find the degree of a polynomial in more than one variable, find the degree of each term in the polynomial. The degree of the term $ax^m y^n$ is the sum of the exponents, $m + n$. The highest degree of all of the terms is the degree of the polynomial.

Section 5.2

Preparing for Multiplying Monomials: The Product and Power Rules

P1. $4^3 = 4 \cdot 4 \cdot 4 = 16 \cdot 4 = 64$

P2. $3(4 - 5x) = 3 \cdot 4 - 3 \cdot 5x = 12 - 15x$

5.2 Quick Checks

1. The expression 12^3 is written in <u>exponential</u> form.

2. $a^m \cdot a^m = a^{\underline{m+m}}$

3. False; $3^4 \cdot 3^2 = 3^{4+2} = 3^6$.

4. $3^2 \cdot 3 = 3^{2+1} = 3^3 = 27$

5. $(-5)^2(-5)^3 = (-5)^{2+3} = (-5)^5 = -3125$

6. $c^6 \cdot c^2 = c^{6+2} = c^8$

7. $y^3 \cdot y \cdot y^5 = y^3 \cdot y^1 \cdot y^5 = y^{3+1+5} = y^9$

8. $a^4 \cdot a^5 \cdot b^6 = a^{4+5} \cdot b^6 = a^9b^6$

9. $(2^2)^4 = 2^{2 \cdot 4} = 2^8$

10. $[(-3)^3]^2 = (-3)^{3 \cdot 2} = (-3)^6$

11. $(b^2)^5 = b^{2 \cdot 5} = b^{10}$

12. False;

$$(ab^3)^2 \stackrel{?}{=} ab^6$$
$$(a)^2(b^3)^2 \stackrel{?}{=} ab^6$$
$$a^2b^6 \neq ab^6$$

13. $(2n)^3 = (2)^3(n)^3 = 8n^3$

14. $(-5x^4)^3 = (-5)^3(x^4)^3$

$$= -125x^{4 \cdot 3}$$
$$= -125x^{12}$$

15. $(-7a^3b)^2 = (-7)^2(a^3)^2(b)^2$

$$= 49a^{3 \cdot 2}b^2$$
$$= 49a^6b^2$$

16. $(2a^6)(4a^5) = 2 \cdot 4 \cdot a^6 \cdot a^5 = 8a^{6+5} = 8a^{11}$

17. $(3m^2n^4)(-6mn^5) = 3 \cdot (-6) \cdot m^2 \cdot m \cdot n^4 \cdot n^5$

$$= -18 \cdot m^{2+1} \cdot n^{4 \cdot 5}$$
$$= -18m^3n^9$$

18. $\left(\frac{8}{3}xy^2\right)\left(\frac{1}{2}x^2y\right)(-12xy)$

$$= \left(\frac{8}{3}\right)\left(\frac{1}{2}\right)\left(-\frac{12}{1}\right)x \cdot x^2 \cdot x \cdot y^2 \cdot y \cdot y$$
$$= \left(\frac{4}{3}\right)\left(-\frac{12}{1}\right)x^{1+2+1}y^{2+1+1}$$
$$= -16x^4y^4$$

5.2 Exercises

19. $4^2 \cdot 4^3 = 4^{2+3} = 4^5 = 1024$

21. $(-2)^3(-2)^4 = (-2)^{3+4} = (-2)^7 = -128$

23. $m^4 \cdot m^5 = m^{4+5} = m^9$

25. $b^9 \cdot b^{11} = b^{9+11} = b^{20}$

27. $x^7 \cdot x = x^7 \cdot x^1 = x^{7+1} = x^8$

29. $p \cdot p^2 \cdot p^6 = p^1 \cdot p^2 \cdot p^6 = p^{1+2+6} = p^9$

31. $(-n)^3(-n)^4 = (-n)^{3+4} = (-n)^7$

33. $(2^3)^2 = 2^{3 \cdot 2} = 2^6 = 64$

35. $[(-2)^2]^3 = (-2)^{2 \cdot 3} = (-2)^6 = 64$

37. $[(-m)^9]^2 = (-m)^{9 \cdot 2} = (-m)^{18}$

39. $(m^2)^7 = m^{2 \cdot 7} = m^{14}$

41. $[(-b)^4]^5 = (-b)^{4 \cdot 5} = (-b)^{20}$

43. $(3x^2)^3 = 3^3 \cdot (x^2)^3$

$$= 27 \cdot x^{2 \cdot 3}$$
$$= 27x^6$$

45. $(-5z^2)^2 = (-5)^2(z^2)^2 = (25)(z^{2 \cdot 2}) = 25z^4$

47. $\left(\frac{1}{2}a\right)^2 = \left(\frac{1}{2}\right)^2(a)^2 = \left(\frac{1}{4}\right)a^2 = \frac{1}{4}a^2$

49. $(-3p^7q^2)^4 = (-3)^4(p^7)^4(q^2)^4$
$= 81 \cdot p^{7\cdot 4} \cdot q^{2\cdot 4}$
$= 81p^{28}q^8$

51. $(-2m^2n)^3 = (-2)^3(m^2)^3 \cdot n^3$
$= -8 \cdot m^{2\cdot 3} \cdot n^3$
$= -8m^6n^3$

53. $(-5xy^2z^3)^2 = (-5)^2 \cdot x^2 \cdot (y^2)^2 \cdot (z^3)^2$
$= 25 \cdot x^2 \cdot y^{2\cdot 2} \cdot z^{3\cdot 2}$
$= 25x^2y^4z^6$

55. $(4x^2)(3x^3) = 4 \cdot 3 \cdot x^2 \cdot x^3$
$= 12 \cdot x^{2+3}$
$= 12x^5$

57. $(10a^3)(-4a^7) = 10 \cdot (-4) \cdot a^3 \cdot a^7$
$= -40 \cdot a^{3+7}$
$= -40a^{10}$

59. $(-m^3)(7m) = (-1 \cdot m^3) \cdot (7m)$
$= -1 \cdot 7 \cdot m^3 \cdot m$
$= -7 \cdot m^{3+1}$
$= -7m^4$

61. $\left(\frac{4}{5}x^4\right)\left(\frac{15}{2}x^3\right) = \frac{4}{5} \cdot \frac{15}{2} \cdot x^4 \cdot x^3$
$= 6 \cdot x^{4+3}$
$= 6x^7$

63. $(2x^2y^3)(3x^4y) = 2 \cdot 3 \cdot x^2 \cdot x^4 \cdot y^3 \cdot y$
$= 6 \cdot x^{2+4} \cdot y^{3+1}$
$= 6x^6y^4$

65. $\left(\frac{1}{4}mn^3\right)(-20mn) = \frac{1}{4} \cdot (-20) \cdot m \cdot m \cdot n^3 \cdot n$
$= -5 \cdot m^{1+1} \cdot n^{3+1}$
$= -5m^2n^4$

67. $(4x^2y)(-5xy^3)(2x^2y^2)$
$= 4 \cdot (-5) \cdot 2 \cdot x^2 \cdot x \cdot x^2 \cdot y \cdot y^3 \cdot y^2$
$= (-20) \cdot 2 \cdot x^{2+1+2} \cdot y^{1+3+2}$
$= -40x^5y^6$

69. $x^2 \cdot 4x = 4 \cdot x^2 \cdot x = 4 \cdot x^{2+1} = 4x^3$

71. $(-x)(5x^2) = (-1) \cdot (5) \cdot x \cdot x^2$
$= -5 \cdot x^{1+2}$
$= -5x^3$

73. $(b^3)^2 = b^{3\cdot 2} = b^6$

75. $(-3b)^4 = (-3)^4 \cdot b^4 = 81b^4$

77. $(-6p)^2\left(\frac{1}{4}p^3\right) = (-6)^2 \cdot p^2 \cdot \left(\frac{1}{4}p^3\right)$
$= 36p^2 \cdot \left(\frac{1}{4}p^3\right)$
$= 36 \cdot \frac{1}{4} \cdot p^2 \cdot p^3$
$= 9 \cdot p^{2+3}$
$= 9p^5$

79. $(-xy)(x^2y)(-3x)^3$
$= (-xy)(x^2y)(-3)^3(x)^3$
$= (-xy)(x^2y)(-27)(x^3)$
$= (-27) \cdot (-1) \cdot x \cdot x^2 \cdot x^3 \cdot y \cdot y$
$= (27) \cdot x^{1+2+3} \cdot y^{1+1}$
$= 27x^6y^2$

81. $\left(\frac{3}{5}\right)(5c)(-10d^2) = \left(\frac{3}{5}\right) \cdot (5) \cdot (-10) \cdot c \cdot d^2$
$= (3) \cdot (-10) \cdot c \cdot d^2$
$= -30cd^2$

83. $(5x)^2(x^2)^3 = 5^2 \cdot x^2 \cdot x^{2\cdot 3}$
$= 25 \cdot x^2 \cdot x^6$
$= 25 \cdot x^{2+6}$
$= 25x^8$

85. $(x^2y)^3(-2xy^3) = x^{2\cdot 3} \cdot y^3 \cdot (-2) \cdot x \cdot y^3$
$= x^6 \cdot y^3 \cdot (-2) \cdot x \cdot y^3$
$= -2 \cdot x^6 \cdot x \cdot y^3 \cdot y^3$
$= -2 \cdot x^{6+1}y^{3+3}$
$= -2x^7y^6$

87. $(-3x)^2(2x^4)^3 = (-3)^2 \cdot x^2 \cdot 2^3 \cdot (x^4)^3$
$= 9 \cdot x^2 \cdot 8 \cdot x^{4 \cdot 3}$
$= 9 \cdot x^2 \cdot 8 \cdot x^{12}$
$= 9 \cdot 8 \cdot x^2 \cdot x^{12}$
$= 72 \cdot x^{2+12}$
$= 72x^{14}$

89. $\left(\frac{4}{5}q\right)^2 (-5q)^2 \left(\frac{1}{4}q^2\right)$
$= \left(\frac{4}{5}\right)^2 \cdot q^2 \cdot (-5)^2 \cdot q^2 \cdot \frac{1}{4} \cdot q^2$
$= \frac{16}{25} \cdot q^2 \cdot 25 \cdot q^2 \cdot \frac{1}{4} \cdot q^2$
$= \frac{16}{25} \cdot 25 \cdot \frac{1}{4} \cdot q^2 \cdot q^2 \cdot q^2$
$= 4 \cdot q^{2+2+2}$
$= 4q^6$

91. $(3x)^2(-2x)^4\left(\frac{1}{3}x^4\right)^2$
$= 3^2 \cdot x^2 \cdot (-2)^4 \cdot x^4 \cdot \left(\frac{1}{3}\right)^2 \cdot (x^4)^2$
$= 9 \cdot x^2 \cdot 16 \cdot x^4 \cdot \frac{1}{9} \cdot x^{4 \cdot 2}$
$= 9 \cdot x^2 \cdot 16 \cdot x^4 \cdot \frac{1}{9} \cdot x^8$
$= 9 \cdot 16 \cdot \frac{1}{9} \cdot x^2 \cdot x^4 \cdot x^8$
$= 16 \cdot x^{2+4+8}$
$= 16x^{14}$

93. $-3(-5mn^3)^2\left(\frac{2}{5}m^5n\right)^3$
$= -3 \cdot (-5)^2 \cdot m^2 \cdot (n^3)^2 \cdot \left(\frac{2}{5}\right)^3 \cdot (m^5)^3 \cdot n^3$
$= -3 \cdot 25 \cdot m^2 \cdot n^{3 \cdot 2} \cdot \frac{8}{125} \cdot m^{5 \cdot 3} \cdot n^3$
$= -3 \cdot 25 \cdot m^2 \cdot n^6 \cdot \frac{8}{125} \cdot m^{15} \cdot n^3$
$= -3 \cdot 25 \cdot \frac{8}{125} \cdot m^2 \cdot m^{15} \cdot n^6 \cdot n^3$
$= -\frac{24}{5} \cdot m^{2+15} \cdot n^{6+3}$
$= -\frac{24}{5}m^{17}n^9$

95. volume $= (x^2)^3 = x^{2 \cdot 3} = x^6$

97. area $= (4x) \cdot (12x)$
$= 4 \cdot 12 \cdot x \cdot x$
$= 48 \cdot x^{1+1}$
$= 48x^2$

99. The product rule for exponential expressions is a "shortcut" for writing out each exponential expression in expanded form. For example, $x^2 \cdot x^3 = (x \cdot x) \cdot (x \cdot x \cdot x) = x^5$.

101. The product to a power rule generalizes the result of using a product as a factor several times. That is, $(a \cdot b)^n = (a \cdot b) \cdot (a \cdot b) \cdot (a \cdot b)...$
$= a \cdot a \cdot a \cdot ... \cdot b \cdot b \cdot b \cdot ...$
$= a^n \cdot b^n$.

103. The expression $4x^2 + 3x^2$ is a sum of terms: add the coefficients and retain the common base. The expression $(4x^2)(3x^2)$ is a product: multiply the constant terms and add the exponents on the variable bases.

Section 5.3

Preparing for Multiplying Polynomials

P1. $2(4x - 3) = 2 \cdot 4x - 2 \cdot 3 = 8x - 6$

P2. $(3x^2)(-5x^3) = 3 \cdot (-5) \cdot x^2 \cdot x^3$
$= -15x^{2+3}$
$= -15x^5$

P3. $(9a)^2 = 9a \cdot 9a = 9 \cdot 9 \cdot a \cdot a = 81a^2$

5.3 Quick Checks

1. $3x(x^2 - 2x + 4) = 3x(x^2) - 3x(2x) + 3x(4)$
$= 3x^3 - 6x^2 + 12x$

2. $-a^2b(2a^2b^2 - 4ab^2 + 3ab)$
$= -a^2b(2a^2b^2) - (-a^2b)(4ab^2) + (-a^2b)(3ab)$
$= -2a^4b^3 + 4a^3b^3 - 3a^3b^2$

3. $\left(5n^3 - \frac{15}{8}n^2 - \frac{10}{7}n\right)\frac{3}{5}n^2$
$= 5n^3\left(\frac{3}{5}n^2\right) - \frac{15}{8}n^2\left(\frac{3}{5}n^2\right) - \frac{10}{7}n\left(\frac{3}{5}n^2\right)$
$= 3n^5 - \frac{9}{8}n^4 - \frac{6}{7}n^3$

4. $(x+7)(x+2) = (x+7)x + (x+7)(2)$
$= x^2 + 7x + 2x + 14$
$= x^2 + 9x + 14$

5. $(2n+1)(n-3) = (2n+1)n + (2n+1)(-3)$
$= 2n^2 + n - 6n - 3$
$= 2n^2 - 5n - 3$

6. $(3p-2)(2p-1) = (3p-2)(2p) + (3p-2)(-1)$
$= 6p^2 - 4p - 3p + 2$
$= 6p^2 - 7p + 2$

7. $(x+3)(x+4) = (x)(x) + (x)(4) + (3)(x) + (3)(4)$
$= x^2 + 4x + 3x + 12$
$= x^2 + 7x + 12$

8. $(y+5)(y-3)$
$= (y)(y) + (y)(-3) + (5)(y) + (5)(-3)$
$= y^2 - 3 + 5y - 15$
$= y^2 + 2y - 15$

9. $(a-1)(a-5)$
$= (a)(a) + (a)(-5) + (-1)(a) + (-1)(-5)$
$= a^2 - 5a - a + 5$
$= a^2 - 6a + 5$

10. $(2x+3)(x+4)$
$= (2x)(x) + (2x)(4) + (3)(x) + (3)(4)$
$= 2x^2 + 8x + 3x + 12$
$= 2x^2 + 11x + 12$

11. $(3y+5)(2y-3)$
$= (3y)(2y) + (3y)(-3) + (5)(2y) + (5)(-3)$
$= 6y^2 - 9y + 10y - 15$
$= 6y^2 + y - 15$

12. $(2a-1)(3a-4) = 6a^2 - 8a - 3a + 4$
$= 6a^2 - 11a + 4$

13. $(5x+2y)(3x-4y)$
$= (5x)(3x) + (5x)(-4y) + (2y)(3x) + (2y)(-4y)$
$= 15x^2 - 20xy + 6xy - 8y^2$
$= 15x^2 - 14xy - 8y^2$

14. $(A+B)(A-B) = \underline{A^2 - B^2}$.

15. $(a-4)(a+4) = a^2 - 4^2 = a^2 - 16$

16. $(3w+7)(3w-7) = (3w)^2 - 7^2 = 9w^2 - 49$

17. $(x-2y^3)(x+2y^3) = x^2 - (2y^3)^2$
$= x^2 - 2^2(y^3)^2$
$= x^2 - 4y^6$

18. Let $x = 0$ and $y = 1$.
$(x-y)^2 \stackrel{?}{=} x^2 - y^2$
$(0-1)^2 \stackrel{?}{=} 0^2 - 1^2$
$(-1)^2 \stackrel{?}{=} -1^2$
$1 \neq -1$
The statement is false.

19. $x^2 + 2xy + y^2$ is referred to as a <u>perfect square trinomial</u>.

20. False; consider $(a+b)(a-b) = a^2 - b^2$.

21. $(z-9)^2 = z^2 - 2(z)(9) + 9^2$
$= z^2 - 18z + 81$

22. $(p+1)^2 = p^2 + 2(p)(1) + 1^2$
$= p^2 + 2p + 1$

23. $(4-a)^2 = 4^2 - 2(4)(a) + a^2 = 16 - 8a + a^2$

24. $(3z-4)^2 = (3z)^2 - 2(3z)(4) + 4^2$
$= 9z^2 - 24z + 16$

25. $(5p+1)^2 = (5p)^2 + 2(5p)(1) + (1)^2$
$= 25p^2 + 10p + 1$

26. $(2w+7y)^2 = (2w)^2 + 2(2w)(7y) + (7y)^2$
$= 4w^2 + 28wy + 49y^2$

27. False; the product $(x-y)(x^2+2xy+y^2)$ cannot be found using the FOIL method. The FOIL method can only be used with two binomials.

28.
$$\begin{aligned}(x-2)(x^2+2x+4) &= (x-2)x^2+(x-2)\cdot 2x+(x-2)\cdot 4\\ &= x(x^2)-2(x^2)+x(2x)-2(2x)+x(4)-2(4)\\ &= x^3-2x^2+2x^2-4x+4x-8\\ &= x^3-8\end{aligned}$$

29.
$$\begin{aligned}(3y-2)(y^2+2y+4) &= (3y-2)y^2+(3y-2)\cdot 2y+(3y-2)\cdot 4\\ &= 3y(y^2)-2(y^2)+3y(2y)-2(2y)+3y(4)-2(4)\\ &= 3y^3-2y^2+6y^2-4y+12y-8\\ &= 3y^3+4y^2+8y-8\end{aligned}$$

30.
$$\begin{array}{r} x^2+2x+4\\ \times\qquad\quad x-2\\ \hline -2x^2-4x-8\\ x^3+2x^2+4x\qquad\\ \hline x^3+0x^2+0x-8\end{array}$$

The product is x^3-8.

31.
$$\begin{array}{r} y^2+2y-5\\ \times\qquad\quad 3y-1\\ \hline -y^2-\ 2y+5\\ 3y^3+6y^2-15y\qquad\\ \hline 3y^3+5y^2-17y+5\end{array}$$

32.
$$\begin{aligned}-2a(4a-1)(3a+5) &= (-8a^2+2a)(3a+5)\\ &= -8a^2\cdot 3a+(-8a^2)\cdot 5+2a\cdot 3a+2a\cdot 5\\ &= -24a^3-40a^2+6a^2+10a\\ &= -24a^3-34a^2+10a\end{aligned}$$

33.
$$\begin{aligned}\frac{3}{2}(5x-2)(2x+8) &= \frac{3}{2}[5x\cdot 2x+5x\cdot 8+(-2)\cdot 2x+(-2)\cdot 8]\\ &= \frac{3}{2}(10x^2+40x-4x-16)\\ &= \frac{3}{2}(10x^2+36x-16)\\ &= \frac{3}{2}\cdot 10x^2+\frac{3}{2}\cdot 36x-\frac{3}{2}\cdot 16\\ &= 15x^2+54x-24\end{aligned}$$

34.
$$\begin{aligned}(x+2)(x-1)(x+3) &= (x^2-x+2x-2)(x+3)\\ &= (x^2+x-2)(x+3)\\ &= x^3+x^2-2x+3x^2+3x-6\\ &= x^3+4x^2+x-6\end{aligned}$$

5.3 Exercises

35. $2x(3x-5)=2x(3x)+2x(-5)=6x^2-10x$

37. $\frac{1}{2}n(4n-6)=\frac{1}{2}n(4n)+\frac{1}{2}n(-6)=2n^2-3n$

39. $3n^2(4n^2+2n-5)=3n^2(4n^2)+3n^2(2n)+3n^2(-5)$
$=12n^4+6n^3-15n^2$

41. $(4x^2y-3xy^2)x^2y=4x^2y(x^2y)+(-3xy^2)(x^2y)$
$=4x^4y^2-3x^3y^3$

43. $(x+5)(x+7)=(x+5)(x)+(x+5)(7)$
$=x\cdot x+5\cdot x+7\cdot x+5\cdot 7$
$=x^2+5x+7x+35$
$=x^2+12x+35$

45. $(y-5)(y+7)=(y-5)y+(y-5)7$
$=y\cdot y-5\cdot y+7\cdot y-5\cdot 7$
$=y^2-5y+7y-35$
$=y^2+2y-35$

47. $(3m-2y)(2m+5y)=(3m-2y)\cdot(2m)+(3m-2y)\cdot(5y)$
$=(3m)\cdot(2m)-(2y)\cdot(2m)+(3m)\cdot(5y)-(2y)\cdot(5y)$
$=6m^2-4my+15my-10y^2$
$=6m^2+11my-10y^2$

49. $(x+2)(x+3)=x^2+3x+2x+6$
$=x^2+5x+6$

51. $(q-6)(q-7)=q^2-7q-6q+42$
$=q^2-13q+42$

53. $(2x+3)(3x-1)=6x^2-2x+9x-3$
$=6x^2+7x-3$

55. $(x^2+3)(x^2+1)=x^4+x^2+3x^2+3$
$=x^4+4x^2+3$

57. $(7-x)(6-x)=42-7x-6x+x^2$
$=42-13x+x^2$

59. $(5u+6v)(2u+v)=10u^2+5uv+12uv+6v^2$
$=10u^2+17uv+6v^2$

61. $(2a-b)(5a+2b)=10a^2+4ab-5ab-2b^2$
$=10a^2-ab-2b^2$

63. $(x-3)(x+3)=(x)^2-(3)^2=x^2-9$

65. $(2z+5)(2z-5)=(2z)^2-(5)^2=4z^2-25$

67. $(4x^2+1)(4x^2-1)=(4x^2)^2-(1)^2=16x^4-1$

69. $(2x-3y)(2x+3y)=(2x)^2-(3y)^2=4x^2-9y^2$

71. $\left(x-\frac{1}{3}\right)\left(x+\frac{1}{3}\right)=(x)^2-\left(\frac{1}{3}\right)^2=x^2-\frac{1}{9}$

73. $(x-2)^2=(x)^2-2(x)(2)+(2)^2=x^2-4x+4$

75. $(5k-3)^2=(5k)^2-(2)(5k)(3)+(3)^2$
$=25k^2-30k+9$

77. $(x+2y)^2=(x)^2+2(x)(2y)+(2y)^2$
$=x^2+4xy+4y^2$

79. $(2a-3b)^2=(2a)^2-2(2a)(3b)+(3b)^2$
$=4a^2-12ab+9b^2$

81. $\left(x+\frac{1}{2}\right)^2=(x)^2+2(x)\left(\frac{1}{2}\right)+\left(\frac{1}{2}\right)^2=x^2+x+\frac{1}{4}$

83. $(x-2)(x^2+3x+1)=(x-2)x^2+(x-2)3x+(x-2)1$
$=x\cdot x^2+(-2)\cdot x^2+x\cdot 3x+(-2)\cdot 3x+x\cdot 1+(-2)\cdot 1$
$=x^3-2x^2+3x^2-6x+x-2$
$=x^3+x^2-5x-2$

85. $(2y^2-6y+1)(y-3)=(2y^2-6y+1)y+(2y^2-6y+1)(-3)$
$=2y^2\cdot y+(-6y)\cdot y+1\cdot y+2y^2\cdot(-3)+(-6y)\cdot(-3)+1\cdot(-3)$
$=2y^3-6y^2+y-6y^2+18y-3$
$=2y^3-12y^2+19y-3$

87. $(2x-3)(x^2-2x-1)=(2x-3)x^2+(2x-3)(-2x)+(2x-3)(-1)$
$=2x\cdot x^2+(-3)\cdot x^2+2x\cdot(-2x)+(-3)\cdot(-2x)+2x\cdot(-1)+(-3)\cdot(-1)$
$=2x^3-3x^2-4x^2+6x-2x+3$
$=2x^3-7x^2+4x+3$

89. $$\begin{aligned}2b(b-3)(b+4) &= (2b^2-6b)(b+4)\\ &= (2b^2-6b)b+(2b^2-6b)4\\ &= 2b^2\cdot b+(-6b)\cdot b+2b^2\cdot 4+(-6b)\cdot 4\\ &= 2b^3-6b^2+8b^2-24b\\ &= 2b^3+2b^2-24b\end{aligned}$$

91. $$\begin{aligned}-\frac{1}{2}x(2x+6)(x-3) &= (-x^2-3x)(x-3)\\ &= (-x^2-3x)x+(-x^2-3x)(-3)\\ &= (-x^2)\cdot x+(-3x)\cdot x+(-x^2)\cdot(-3)+(-3x)\cdot(-3)\\ &= -x^3-3x^2+3x^2+9x\\ &= -x^3+9x\end{aligned}$$

93. $$\begin{aligned}(5y^3-y^2+2)(2y^2+y+1) &= (5y^3-y^2+2)2y^2+(5y^3-y^2+2)y+(5y^3-y^2+2)1\\ &= 5y^3\cdot 2y^2+(-y^2)\cdot 2y^2+2\cdot 2y^2+5y^3\cdot y+(-y^2)\cdot y+2\cdot y+5y^3\cdot 1+(-y^2)\cdot 1+2\cdot 1\\ &= 10y^5-2y^4+4y^2+5y^4-y^3+2y+5y^3-y^2+2\\ &= 10y^5+3y^4+4y^3+3y^2+2y+2\end{aligned}$$

95. $$\begin{aligned}(b+1)(b-2)(b+3) &= (b^2-2b+b-2)(b+3)\\ &= (b^2-b-2)\cdot(b+3)\\ &= b^2\cdot(b+3)+(-b)\cdot(b+3)+(-2)\cdot(b+3)\\ &= b^3+3b^2-b^2-3b-2b-6\\ &= b^3+2b^2-5b-6\end{aligned}$$

97. $$\begin{aligned}4n(3-2n-n^2) &= 4n(3)+4n(-2n)+4n(-n^2)\\ &= 12n-8n^2-4n^3\end{aligned}$$

99. $$\begin{aligned}(2a-3b)(4a+b) &= 8a^2+2ab-12ab-3b^2\\ &= 8a^2-10ab-3b^2\end{aligned}$$

101. $$\begin{aligned}\left(\frac{1}{2}x+3\right)\left(\frac{1}{2}x-3\right) &= \left(\frac{1}{2}x\right)^2-(3)^2\\ &= \frac{1}{4}x^2-9\end{aligned}$$

103. $$\begin{aligned}(0.5b+3)^2 &= (0.5b)^2+(2)\cdot(0.5b)\cdot(3)+(3)^2\\ &= 0.25b^2+3b+9\end{aligned}$$

105. $$\begin{aligned}(x^2+1)(x^4-3) &= x^6-3x^2+x^4-3\\ &= x^6+x^4-3x^2-3\end{aligned}$$

107. $$\begin{aligned}(x+2)(2x^2-3x-1) &= x\cdot(2x^2-3x-1)+2(2x^2-3x-1)\\ &= 2x^3-3x^2-x+4x^2-6x-2\\ &= 2x^3+x^2-7x-2\end{aligned}$$

109. $-\frac{1}{2}x^2(10x^5-6x^4+12x^3)+(x^3)^2=-\frac{1}{2}x^2(10x^5)+\left(-\frac{1}{2}x^2\right)(-6x^4)+\left(-\frac{1}{2}x^2\right)(12x^3)+x^6$
$=-5x^7+3x^6-6x^5+x^6$
$=-5x^7+4x^6-6x^5$

111. $7x^2(x+3)-2x(x^2-1)=7x^2(x)+7x^2(3)+(-2x)(x^2)+(-2x)(-1)$
$=7x^3+21x^2-2x^3+2x$
$=5x^3+21x^2+2x$

113. $-3w(w-4)(w+3)=(-3w^2+12w)(w+3)$
$=(-3w^2)(w)+(-3w^2)(3)+(12w)(w)+(12w)(3)$
$=-3w^3-9w^2+12w^2+36w$
$=-3w^3+3w^2+36w$

115. $3a(a+4)^2=3a[(a)^2+2(a)(4)+(4)^2]=3a(a^2+8a+16)=3a(a^2)+3a(8a)+3a(16)=3a^3+24a^2+48a$

117. $(n+3)(n-3)+(n+3)^2=(n)^2-(3)^2+(n)^2+2(n)(3)+(3)^2=n^2-9+n^2+6n+9=2n^2+6n$

119. $(a+6b)^2-(a-6b)^2=(a)^2+2(a)(6b)+(6b)^2-[(a)^2-2(a)(6b)+(6b)^2]$
$=a^2+12ab+36b^2-(a^2-12ab+36b^2)$
$=a^2+12ab+36b^2-a^2+12ab-36b^2$
$=24ab$

121. $(x+1)^2-(2x+1)(x-1)=(x)^2+2(x)(1)+(1)^2-[(2x)(x)+(2x)(-1)+(1)(x)+(1)(-1)]$
$=x^2+2x+1-(2x^2-2x+x-1)$
$=x^2+2x+1-(2x^2-x-1)$
$=x^2+2x+1-2x^2+x+1$
$=-x^2+3x+2$

123. $(2x+1)^2=(2x)^2+2(2x)(1)+(1)^2=4x^2+4x+1$

125. $(x-1)^3=(x-1)^2(x-1)$
$=[(x)^2-2(x)(1)+(1)^2](x-1)$
$=(x^2-2x+1)(x-1)$
$=x^2(x-1)+(-2x)(x-1)+1(x-1)$
$=x^2(x)+x^2(-1)+(-2x)(x)+(-2x)(-1)+1(x)+1(-1)$
$=x^3-x^2-2x^2+2x+x-1$
$=x^3-3x^2+3x-1$

127. $(x+3)(2x-5)-(x-6)=(x)(2x)+(x)(-5)+(3)(2x)+3(-5)-x+6=2x^2-5x+6x-15-x+6=2x^2-9$

129. area $=(2x-3)(x+5)=(2x)(x)+(2x)(5)+(-3)(x)+(-3)(5)=2x^2+10x-3x-15=2x^2+7x-15$

131. $\text{area} = x(3x-1)-(x-1)^2$
$= x(3x)+x(-1)-[(x)^2-2(x)(1)+(1)^2]$
$= 3x^2-x-(x^2-2x+1)$
$= 3x^2-x-x^2+2x-1$
$= 2x^2+x-1$

133. $\text{Volume} = x(4x-3)(x+1)$
$= (4x^2-3x)(x+1)$
$= 4x^3+4x^2-3x^2-3x$
$= 4x^3+x^2-3x$

135. Find the area of the rectangle by adding the areas of the interior rectangles.
$\text{area} = (x)(x)+(x)(10)+(2)(x)+(2)(10)$
$= x^2+10x+2x+20$
$= x^2+12x+20$
Find the area of the rectangle by multiplying width and height.
$\text{area} = (x+10)(x+2)$
$= (x)(x)+(x)(2)+(10)(x)+(10)(2)$
$= x^2+2x+10x+20$
$= x^2+12x+20$

137. The second consecutive integer is $x + 1$; the third is $(x + 1) + 1 = x + 2$.
$(x+1)(x+2) = x(x)+x(2)+1(x)+1(2)$
$= x^2+2x+x+2$
$= x^2+3x+2$

139. The area of a triangle is one-half the product of the base and the altitude. In this exercise, the altitude is x and the base is $2x - 2$.
$\frac{1}{2}(2x-2)x \text{ ft}^2 = (x-1)x \text{ ft}^2 = (x^2-x) \text{ ft}^2$

141. The area of a circle with radius r is πr^2.
$\pi(x+2)^2$ square feet $= \pi[(x)^2+2(x)(2)+(2)^2]$ square feet
$= \pi[x^2+4x+4]$ square feet
$= \pi x^2+4\pi x+4\pi$ square feet

143. **(a)** Area of upper left quadrilateral: $a \cdot a = a^2$
Area of upper right quadrilateral: $a \cdot b = ab$
Area of lower left quadrilateral: $b \cdot a = ab$
Area of lower right quadrilateral: $b \cdot b = b^2$

(b) Area of entire region: $a^2+ab+ab+b^2 = a^2+2ab+b^2$

(c) Length of entire region: $a + b$
Width of entire region: $a + b$
Area of entire region: $(a+b)(a+b) = (a+b)^2$
Therefore, $(a+b)^2 = a^2 + 2ab + b^2$

145. $$\begin{aligned}(x-2)(x+2)^2 &= (x-2)[(x)^2 + 2(x)(2) + (2)^2] \\ &= (x-2)[x^2 + 4x + 4] \\ &= x(x^2 + 4x + 4) + (-2)(x^2 + 4x + 4) \\ &= x^3 + 4x^2 + 4x - 2x^2 - 8x - 8 \\ &= x^3 + 2x^2 - 4x - 8\end{aligned}$$

147. $$\begin{aligned}[3-(x+y)][3+(x+y)] &= (3)^2 - (x+y)^2 \\ &= 9 - [(x)^2 + 2(x)(y) + (y)^2] \\ &= 9 - (x^2 + 2xy + y^2) \\ &= 9 - x^2 - 2xy - y^2\end{aligned}$$

149. $$\begin{aligned}(x+2)^3 &= (x+2)^2(x+2) \\ &= [(x)^2 + 2(x)(2) + (2)^2](x+2) \\ &= (x^2 + 4x + 4)(x+2) \\ &= x^2(x+2) + (4x)(x+2) + 4(x+2) \\ &= x^2(x) + x^2(2) + 4x(x) + 4x(2) + 4(x) + 4(2) \\ &= x^3 + 2x^2 + 4x^2 + 8x + 4x + 8 \\ &= x^3 + 6x^2 + 12x + 8\end{aligned}$$

151. $$\begin{aligned}(z+3)^4 &= (z+3)^2(z+3)^2 \\ &= [(z)^2 + 2(z)(3) + (3)^2][(z)^2 + 2(z)(3) + (3)^2] \\ &= (z^2 + 6z + 9)(z^2 + 6z + 9) \\ &= z^2(z^2 + 6z + 9) + (6z)(z^2 + 6z + 9) + 9(z^2 + 6z + 9) \\ &= z^2(z^2) + z^2(6z) + z^2(9) + 6z(z^2) + 6z(6z) + 6z(9) + 9(z^2) + 9(6z) + 9(9) \\ &= z^4 + 6z^3 + 9z^2 + 6z^3 + 36z^2 + 54z + 9z^2 + 54z + 81 \\ &= z^4 + 12z^3 + 54z^2 + 108z + 81\end{aligned}$$

153. To square the binomial $(3a-5)^2$, use the pattern $A^2 - 2AB + B^2$. This means, square the first term, take the product of −2 times the first term times the second term, and square the second term. So $(3a-5)^2 = (3a)^2 - 2(3a)(5) + 5^2 = 9a^2 - 30a + 25$.

155. The FOIL method can be used to multiply only binomials.

Section 5.4

Preparing for Dividing Monomials: The Quotient Rule and Integer Exponents

P1. $(3a^2)^3 = 3^3(a^2)^3 = 27a^{2\cdot 3} = 27a^6$

P2. $\left(\frac{2}{3}\right)^2 = \frac{2}{3} \cdot \frac{2}{3} = \frac{2 \cdot 2}{3 \cdot 3} = \frac{4}{9}$

P3. (a) The reciprocal of $5 = \frac{5}{1}$ is $\frac{1}{5}$.

(b) The reciprocal of $-\frac{6}{7}$ is $-\frac{7}{6}$.

5.4 Quick Checks

1. $\frac{a^m}{a^n} = \underline{a^{m-n}}$ provided that $a \neq 0$.

2. False; $\frac{6^{10}}{6^4} \stackrel{?}{=} 1^6$

$6^{10-4} \stackrel{?}{=} 1^6$

$6^6 \stackrel{?}{=} 1^6$

$46,656 \neq 1$

3. $\frac{3^7}{3^5} = 3^{7-5} = 3^2 = 9$

4. $\frac{14c^6}{10c^5} = \frac{14}{10} \cdot \frac{c^6}{c^5} = \frac{7}{5} \cdot c^{6-5} = \frac{7}{5}c^1 = \frac{7}{5}c$ or $\frac{7c}{5}$

5. $\frac{-21w^4z^8}{14w^3z} = \frac{-21}{14} \cdot w^{4-3} \cdot z^{8-1}$

$= -\frac{3}{2}w^1z^7$

$= -\frac{3}{2}wz^7$ or $-\frac{3wz^7}{2}$

6. $\left(\frac{a}{b}\right)^n = \underline{\frac{a^n}{b^n}}$ provided that $\underline{b \neq 0}$.

7. $\left(\frac{p}{2}\right)^4 = \frac{p^4}{2^4} = \frac{p^4}{16}$

8. $\left(-\frac{2a^2}{b^4}\right)^3 = \left(\frac{-2a^2}{b^4}\right)^3$

$= \frac{(-2a^2)^3}{(b^4)^3}$

$= \frac{(-2)^3(a^2)^3}{b^{4\cdot 3}}$

$= \frac{-8a^{2\cdot 3}}{b^{12}}$

$= -\frac{8a^6}{b^{12}}$

9. (a) $10^0 = 1$

(b) $-10^0 = -1 \cdot 10^0 = -1$

(c) $(-10)^0 = 1$

10. (a) $(2b)^0 = 1$

(b) $2b^0 = 2 \cdot 1 = 2$

(c) $(-2b)^0 = 1$

11. (a) $2^{-4} = \frac{1}{2^4} = \frac{1}{16}$

(b) $(-2)^{-4} = \frac{1}{(-2)^4} = \frac{1}{16}$

(c) $-2^{-4} = -1 \cdot 2^{-4} = \frac{-1}{2^4} = \frac{-1}{16} = -\frac{1}{16}$

12. $4^{-1} - 2^{-3} = \frac{1}{4^1} - \frac{1}{2^3} = \frac{1}{4} - \frac{1}{8} = \frac{2}{8} - \frac{1}{8} = \frac{1}{8}$

13. (a) $y^{-8} = \frac{1}{y^8}$

(b) $(-y)^{-8} = \frac{1}{(-y)^8} = \frac{1}{(-1 \cdot y)^8} = \frac{1}{(-1)^8 y^8} = \frac{1}{y^8}$

(c) $-y^{-8} = -1 \cdot y^{-8} = -1 \cdot \frac{1}{y^8} = -\frac{1}{y^8}$

14. (a) $2m^{-5} = 2 \cdot \frac{1}{m^5} = \frac{2}{m^5}$

(b) $(2m)^{-5} = \frac{1}{(2m)^5} = \frac{1}{2^5 \cdot m^5} = \frac{1}{32m^5}$

(c) $(-2m)^{-5} = \frac{1}{(-2m)^5}$
$= \frac{1}{(-2)^5 \cdot m^5}$
$= \frac{1}{-32m^5}$
$= -\frac{1}{32m^5}$

15. $\frac{1}{3^{-2}} = 3^2 = 9$

16. $\frac{1}{-10^{-2}} = \frac{1}{-1 \cdot 10^{-2}} = \frac{10^2}{-1} = \frac{100}{-1} = -100$

17. $\frac{5}{2z^{-2}} = \frac{5z^2}{2}$

18. $\left(\frac{7}{8}\right)^{-1} = \left(\frac{8}{7}\right)^1 = \frac{8}{7}$

19. $\left(\frac{3a}{5}\right)^{-2} = \left(\frac{5}{3a}\right)^2 = \frac{5^2}{(3a)^2} = \frac{25}{3^2a^2} = \frac{25}{9a^2}$

20. $\left(-\frac{2}{3n^4}\right)^{-3} = \left(-\frac{3n^4}{2}\right)^3$
$= \frac{(-3n^4)^3}{2^3}$
$= \frac{(-3)^3(n^4)^3}{8}$
$= \frac{-27n^{4 \cdot 3}}{8}$
$= -\frac{27n^{12}}{8}$

21. $(-4a^{-3})(5a) = (-4 \cdot 5)(a^{-3} \cdot a)$
$= -20a^{-3+1}$
$= -20a^{-2}$
$= -20 \cdot \frac{1}{a^2}$
$= -\frac{20}{a^2}$

22. $\left(-\frac{2}{5}m^{-2}n^{-1}\right)\left(-\frac{15}{2}mn^0\right)$
$= \left[-\frac{2}{5} \cdot \left(-\frac{15}{2}\right)\right](m^{-2} \cdot m)(n^{-1} \cdot n^0)$
$= \frac{30}{10}m^{-2+1}n^{-1+0}$
$= 3m^{-1}n^{-1}$
$= 3 \cdot \frac{1}{m} \cdot \frac{1}{n}$
$= \frac{3}{mn}$

23. $-\frac{16a^4b^{-1}}{12ab^{-4}} = -\frac{16}{12} \cdot \frac{a^4}{a} \cdot \frac{b^{-1}}{b^{-4}}$
$= -\frac{4}{3} \cdot a^{4-1} \cdot b^{-1-(-4)}$
$= -\frac{4}{3} \cdot a^3 \cdot b^3$
$= -\frac{4a^3b^3}{3}$

24. $\frac{45x^{-2}y^{-2}}{35x^{-4}y} = \frac{45}{35} \cdot \frac{x^{-2}}{x^{-1}} \cdot \frac{y^{-2}}{y}$
$= \frac{9}{7} \cdot x^{-2-(-4)} \cdot y^{-2-1}$
$= \frac{9}{7} \cdot x^2 \cdot y^{-3}$
$= \frac{9x^2}{7y^3}$

25. $(3y^2z^{-3})^{-2} = \left(\frac{3y^2}{z^3}\right)^{-2}$
$= \left(\frac{z^3}{3y^2}\right)^2$
$= \frac{(z^3)^2}{(3y^2)^2}$
$= \frac{z^6}{9y^4}$

26. $\left(\frac{2wz^{-3}}{7w^{-1}}\right)^2 = \left(\frac{2\cdot w\cdot w}{7\cdot z^3}\right)^2$
$= \left(\frac{2w^2}{7z^3}\right)^2$
$= \frac{(2w^2)^2}{(7z^3)^2}$
$= \frac{4w^4}{49z^6}$

27. $\left(\frac{-6p^{-2}}{p}\right)(3p^8)^{-1} = \left(\frac{-6}{p\cdot p^2}\right)\left(\frac{1}{3p^8}\right)$
$= \left(\frac{-6}{p^3}\right)\left(\frac{1}{3p^8}\right)$
$= \frac{-6}{3}\cdot\frac{1}{p^3\cdot p^8}$
$= -2\cdot\frac{1}{p^{11}}$
$= -\frac{2}{p^{11}}$

28. $(-25k^5r^{-2})\left(\frac{2}{5}k^{-3}r\right)^2 = \left(-\frac{25k^5}{r^2}\right)\left(\frac{2r}{5k^3}\right)^2$
$= \left(-\frac{25k^5}{r^2}\right)\frac{(2r)^2}{(5k^3)^2}$
$= -\frac{25k^5}{r^2}\cdot\frac{4r^2}{25k^6}$
$= -4\cdot\frac{k^5}{k^6}\cdot\frac{r^2}{r^2}$
$= -4\cdot k^{5-6}\cdot 1$
$= -4\cdot k^{-1}$
$= -4\cdot\frac{1}{k}$
$= -\frac{4}{k}$

5.4 Exercises

29. $\frac{2^{23}}{2^{19}} = 2^{23-19} = 2^4 = 16$

31. $\frac{x^{15}}{x^6} = x^{15-6} = x^9$

33. $\frac{16y^4}{4y} = \frac{16}{4}\cdot\frac{y^4}{y^1} = 4\cdot y^{4-1} = 4y^3$

35. $\frac{-16m^{10}}{24m^3} = \frac{-16}{24}\cdot\frac{m^{10}}{m^3} = -\frac{2}{3}\cdot m^{10-3} = -\frac{2}{3}m^7$

37. $\frac{-12m^9n^3}{-6mn} = \frac{-12}{-6}\cdot\frac{m^9}{m^1}\cdot\frac{n^3}{n^1}$
$= 2m^{9-1}n^{3-1}$
$= 2m^8n^2$

39. $\left(\frac{3}{2}\right)^3 = \frac{3^3}{2^3} = \frac{27}{8}$

41. $\left(\frac{x^5}{3}\right)^3 = \frac{(x^5)^3}{3^3} = \frac{x^{5\cdot 3}}{27} = \frac{x^{15}}{27}$

43. $\left(-\frac{x^5}{y^7}\right)^4 = (-1)^4\frac{(x^5)^4}{(y^7)^4} = 1\cdot\frac{x^{5\cdot 4}}{y^{7\cdot 4}} = \frac{x^{20}}{y^{28}}$

45. $\left(\frac{7a^2b}{c^3}\right)^2 = \frac{(7a^2b)^2}{(c^3)^2}$
$= \frac{7^2(a^2)^2b^2}{(c^3)^2}$
$= \frac{49a^{2\cdot 2}b^2}{c^{3\cdot 2}}$
$= \frac{49a^4b^2}{c^6}$

47. $3^0 = 1$

49. $-\left(\frac{1}{2}\right)^0 = -1\left(\frac{1}{2}\right)^0 = -1(1) = -1$

51. $18\cdot 2^0 = 18(1) = 18$

53. $(-10)^0 = 1$

55. $(24ab)^0 = 1$

57. $24ab^0 = (24a)\cdot b^0 = 24a\cdot 1 = 24a$

59. $10^{-3} = \frac{1}{10^3} = \frac{1}{1000}$

61. $m^{-2} = \frac{1}{m^2}$

63. $-a^{-2} = -1\cdot a^{-2} = -1\cdot\frac{1}{a^2} = -\frac{1}{a^2}$

65. $-4y^{-3} = -4\cdot\frac{1}{y^3} = -\frac{4}{y^3}$

67. $2^{-1} + 3^{-2} = \frac{1}{2^1} + \frac{1}{3^2} = \frac{1}{2} + \frac{1}{9} = \frac{9}{18} + \frac{2}{18} = \frac{11}{18}$

69. $\left(\frac{2}{5}\right)^{-2} = \left(\frac{5}{2}\right)^2 = \frac{5^2}{2^2} = \frac{25}{4}$

71. $\left(\frac{3}{z^2}\right)^{-1} = \left(\frac{z^2}{3}\right)^1 = \frac{z^{2\cdot 1}}{3^1} = \frac{z^2}{3}$

73. $\left(-\frac{2n}{m^2}\right)^{-3} = \left(-\frac{m^2}{2n}\right)^3$
$= (-1)^3\frac{(m^2)^3}{(2n)^3}$
$= -1\cdot\frac{m^{2\cdot 3}}{2^3n^3}$
$= -\frac{m^6}{8n^3}$

75. $\frac{1}{4^{-2}} = 4^2 = 16$

77. $\frac{6}{x^{-4}} = 6\cdot x^4 = 6x^4$

79. $\frac{5}{2m^{-3}} = \frac{5m^3}{2}$

81. $\frac{5}{(2m)^{-3}} = 5\cdot(2m)^3$
$= 5\cdot 2^3\cdot m^3$
$= 5\cdot 8\cdot m^3$
$= 40m^3$

83. $\left(\frac{4}{3}y^{-2}z\right)\left(\frac{5}{8}y^{-2}z^4\right) = \left(\frac{4}{3}\right)\cdot\left(\frac{5}{8}\right)y^{-2-2}z^{1+4}$
$= \frac{20}{24}y^{-4}z^5$
$= \frac{5}{6}\cdot\frac{z^5}{y^4}$
$= \frac{5z^5}{6y^4}$

85. $\frac{-7m^7n^6}{3m^{-3}n^0} = -\frac{7}{3}m^{7+3}n^{6-0}$
$= -\frac{7}{3}m^{10}n^6$
$= -\frac{7m^{10}n^6}{3}$

87. $\frac{21y^2z^{-3}}{3y^{-2}z^{-1}} = \frac{21}{3}y^{2+2}z^{-3+1} = 7y^4z^{-2} = \frac{7y^4}{z^2}$

89. $(4x^2y^{-2})^{-2} = \left(\frac{4x^2}{y^2}\right)^{-2}$
$= \left(\frac{y^2}{4x^2}\right)^2$
$= \frac{(y^2)^2}{(4x^2)^2}$
$= \frac{y^4}{4^2(x^2)^2}$
$= \frac{y^4}{16x^4}$

91. $(3a^2b^{-1})\left(\frac{4a^{-1}b^2}{b^3}\right)^2 = (3a^2b^{-1})(4a^{-1}b^{2-3})^2$
$= (3a^2b^{-1})(4a^{-1}b^{-1})^2$
$= \left(3a^2 \cdot \frac{1}{b}\right)\left(4 \cdot \frac{1}{a} \cdot \frac{1}{b}\right)^2$
$= \left(\frac{3a^2}{6}\right)\left(\frac{4}{ab}\right)^2$
$= \left(\frac{3a^2}{b}\right)\left(\frac{4^2}{(ab)^2}\right)$
$= \left(\frac{3a^2}{b}\right)\left(\frac{16}{a^2b^2}\right)$
$= \frac{48a^2}{a^2b^3}$
$= \frac{48}{b^3}$

93. $2^5 \cdot 2^{-3} = 2^{5-3} = 2^2 = 4$

95. $2^{-7} \cdot 2^4 = 2^{-7+4} = 2^{-3} = \frac{1}{2^3} = \frac{1}{8}$

97. $\frac{3}{3^{-3}} = \frac{3^1}{3^{-3}} = 3^{1-(-3)} = 3^{1+3} = 3^4 = 81$

99. $\frac{x^6}{x^{15}} = x^{6-15} = x^{-9} = \frac{1}{x^9}$

101. $\frac{x^{10}}{x^{-3}} = x^{10-(-3)} = x^{10+3} = x^{13}$

103. $\frac{8x^2}{2x^5} = \frac{8}{2} \cdot \frac{x^2}{x^5} = 4 \cdot x^{2-5} = 4x^{-3} = \frac{4}{x^3}$

105. $\frac{-27xy^3z^4}{18x^4y^3z} = \frac{-27}{18} \cdot \frac{x}{x^4} \cdot \frac{y^3}{y^3} \cdot \frac{z^4}{z}$
$= -\frac{3}{2} \cdot x^{1-4} \cdot y^{3-3} \cdot z^{4-1}$
$= -\frac{3}{2} \cdot x^{-3} \cdot y^0 \cdot z^3$
$= -\frac{3}{2} \cdot \frac{1}{x^3} \cdot 1 \cdot z^3$
$= -\frac{3z^3}{2x^3}$

107. $(3x^2y^{-3})(12^{-1}x^{-5}y^{-6})$
$= 3 \cdot \frac{1}{12} \cdot x^2 \cdot x^{-5} \cdot y^{-3} \cdot y^{-6}$
$= \frac{1}{4} \cdot x^{2-5} \cdot y^{-3-6}$
$= \frac{1}{4} \cdot x^{-3} \cdot y^{-9}$
$= \frac{1}{4x^3y^9}$

109. $(-a^4)^{-3} = (-1)^{-3}(a^4)^{-3}$
$= -1 \cdot a^{4(-3)}$
$= -1 \cdot a^{-12}$
$= -\frac{1}{a^{12}}$

111. $(3m^{-2})^3 = 3^3(m^{-2})^3$
$= 27m^{-2 \cdot 3}$
$= 27m^{-6}$
$= \frac{27}{m^6}$

113. $(-3x^{-2}y^{-3})^{-2} = (-3)^{-2}(x^{-2})^{-2}(y^{-3})^{-2}$
$= \frac{1}{(-3)^2} \cdot x^{(-2)(-2)} \cdot y^{(-3)(-2)}$
$= \frac{1}{9} \cdot x^4 \cdot y^6$
$= \frac{x^4y^6}{9}$

115. $2p^{-4} \cdot p^{-3} \cdot p^0 = 2p^{-4-3+0} = 2p^{-7} = \frac{2}{p^7}$

117. $(-16a^3)(-3a^4)\left(\frac{1}{4}a^{-7}\right)$
$= (-16)(-3)\left(\frac{1}{4}\right)\cdot a^3\cdot a^4\cdot a^{-7}$
$= 12a^{3+4-7}$
$= 12a^{7-7}$
$= 12a^0$
$= 12(1)$
$= 12$

119. $\frac{8x^2\cdot x^7}{12x^{-3}\cdot x^4} = \frac{8}{12}\cdot\frac{x^{2+7}}{x^{-3+4}}$
$= \frac{2}{3}\cdot\frac{x^9}{x^1}$
$= \frac{2}{3}\cdot x^{9-1}$
$= \frac{2x^8}{3}$

121. $(2x^{-3}y^{-2})^4(3x^2y^{-3})^{-3}$
$= 2^4(x^{-3})^4(y^{-2})^4(3)^{-3}(x^2)^{-3}(y^{-3})^{-3}$
$= 16\cdot\frac{1}{3^3}\cdot x^{-3\cdot 4}\cdot x^{2(-3)}\cdot y^{-2(4)}\cdot y^{-3(-3)}$
$= \frac{16}{27}\cdot x^{-12}\cdot x^{-6}\cdot y^{-8}\cdot y^9$
$= \frac{16}{27}\cdot x^{-12-6}\cdot y^{-8+9}$
$= \frac{16}{27}x^{-18}y^1$
$= \frac{16y}{27x^{18}}$

123. $\left(\frac{y}{2z}\right)^{-3}\cdot\left(\frac{3y^2}{4z^3}\right)^2 = \left(\frac{2z}{y}\right)^3\cdot\left(\frac{3y^2}{4z^3}\right)^2$
$= \frac{(2z)^3}{y^3}\cdot\frac{(3y^2)^2}{(4z^3)^2}$
$= \frac{8z^3}{y^3}\cdot\frac{9y^4}{16z^6}$
$= 8\cdot\frac{9}{16}\cdot z^{3-6}y^{4-3}$
$= \frac{9}{2}z^{-3}y^1$
$= \frac{9}{2}\cdot\frac{1}{z^3}\cdot y$
$= \frac{9y}{2z^3}$

125. $(4x^2y)^3\cdot\left(\frac{2x}{3y}\right)^{-3} = (4x^2y)^3\cdot\left(\frac{3y}{2x}\right)^3$
$= 4^3(x^2)^3y^3\cdot\frac{3^3y^3}{2^3x^3}$
$= 64x^{2\cdot 3}y^3\cdot\frac{27y^3}{8x^3}$
$= 64x^6y^3\cdot\frac{27y^3}{8x^3}$
$= 64\cdot\frac{27}{8}\cdot x^{6-3}\cdot y^{3+3}$
$= 216x^3y^6$

127. $\frac{(5a^{-3}b^2)^2}{a^{-4}b^{-4}}(15a^{-3}b)^{-1}$
$= \frac{5^2(a^{-3})^2(b^2)^2}{a^{-4}b^{-4}}\cdot 15^{-1}(a^{-3})^{-1}b^{-1}$
$= \frac{25a^{-3\cdot 2}b^{2\cdot 2}}{a^{-4}b^{-4}}\cdot 15^{-1}a^{-3(-1)}b^{-1}$
$= \frac{25a^{-6}b^4}{a^{-4}b^{-4}}\cdot\frac{1}{15}a^3b^{-1}$
$= \frac{25}{15}a^{-6-(-4)+3}b^{4-(-4)-1}$
$= \frac{5}{3}a^1b^7$
$= \frac{5ab^7}{3}$

129. $\frac{4x^{-3}(2x^3)}{20(x^{-3})^2} = \frac{4\cdot 2\cdot x^{-3}\cdot x^3}{20x^{-3\cdot 2}}$
$= \frac{8x^{-3+3}}{20x^{-6}}$
$= \frac{2x^0}{5x^{-6}}$
$= \frac{2x^{0-(-6)}}{5}$
$= \frac{2x^6}{5}$

131. $x\cdot x\cdot 3x = 3\cdot x^1\cdot x^1\cdot x^1 = 3x^{1+1+1} = 3x^3$
The volume is $3x^3$ cubic meters.

133. The radius is one-half of the diameter, so $r = \frac{d}{2}$.

$V = \pi r^2 h = \pi\left(\frac{d}{2}\right)^2 h = \pi \cdot \frac{d^2}{2^2} \cdot h = \frac{\pi d^2 h}{4}$

135. The area of a circle is πr^2, so each button has area $\pi(3x)^2$. Since there are x buttons, $x \cdot \pi(3x)^2$ square units of material are needed.

$x \cdot \pi(3x)^2 = x \cdot \pi \cdot 3^2 \cdot x^2 = 9\pi x^{1+2} = 9\pi x^3$

$9\pi x^3$ square units of material are needed.

137. $\frac{x^{2n}}{x^{3n}} = x^{2n-3n} = x^{-n} = \frac{1}{x^n}$

139.
$$\begin{aligned}\left(\frac{x^n y^m}{x^{4n-1} y^{m+1}}\right)^{-2} &= \left(\frac{x^{4n-1} y^{m+1}}{x^n y^m}\right)^2 \\ &= (x^{4n-1-n} y^{m+1-m})^2 \\ &= (x^{3n-1} y^1)^2 \\ &= (x^{3n-1})^2 y^2 \\ &= x^{(3n-1)(2)} y^2 \\ &= x^{6n-2} y^2\end{aligned}$$

141.
$$\begin{aligned}(x^{2a} y^b z^{-c})^{3a} &= (x^{2a})^{3a} (y^b)^{3a} (z^{-c})^{3a} \\ &= x^{2a \cdot 3a} y^{b \cdot 3a} z^{-c \cdot 3a} \\ &= x^{6a^2} y^{3ab} z^{-3ac} \\ &= \frac{x^{6a^2} y^{3ab}}{z^{3ac}}\end{aligned}$$

143.
$$\begin{aligned}\frac{(3a^n)^2}{(2a^{4n})^3} &= \frac{3^2 (a^n)^2}{2^3 (a^{4n})^3} \\ &= \frac{9a^{n \cdot 2}}{8a^{4n \cdot 3}} \\ &= \frac{9a^{2n}}{8a^{12n}} \\ &= \frac{9}{8} a^{2n-12n} \\ &= \frac{9}{8} a^{-10n} \\ &= \frac{9}{8a^{10n}}\end{aligned}$$

145. The quotient $\frac{11^6}{11^4} \neq 1^2$ because
$$\begin{aligned}\frac{11^6}{11^4} &= \frac{11 \cdot 11 \cdot 11 \cdot 11 \cdot 11 \cdot 11}{11 \cdot 11 \cdot 11 \cdot 11} \\ &= \frac{\cancel{11} \cdot \cancel{11} \cdot \cancel{11} \cdot \cancel{11} \cdot 11 \cdot 11}{\cancel{11} \cdot \cancel{11} \cdot \cancel{11} \cdot \cancel{11}} \\ &= 11^2.\end{aligned}$$

147. When simplifying the expression $-12x^0$, the exponent 0 applies to the base x, so $-12x^0 = -12 \cdot 1 = -12$. However, $(-12x)^0 = 1$ because the entire expression is raised to the 0 power.

149. His answer is incorrect. To raise a factor to a power, multiply exponents. Your friend added exponents.

Putting the Concepts Together (Sections 5.1–5.4)

1. Yes, $6x^2y^4 - 8x^5 + 3$ is trinomial of degree 6 (2 + 4).

2. $-x^2 + 3x$

(a) $x = 0$: $-(0)^2 + 3(0) = 0 + 0 = 0$

(b) $x = -1$: $-(-1)^2 + 3(-1) = -1 - 3 = -4$

(c) $x = 2$: $-(2)^2 + 3(2) = -4 + 6 = 2$

3.
$$\begin{aligned}&(6x^4 - 2x^2 + 7) - (-2x^4 - 7 + 2x^2) \\ &= (6x^4 - 2x^2 + 7) + (2x^4 + 7 - 2x^2) \\ &= 6x^4 - 2x^2 + 7 + 2x^4 + 7 - 2x^2 \\ &= 8x^4 - 4x^2 + 14\end{aligned}$$

4.
$$\begin{aligned}&(2x^2y - xy + 3y^2) + (4xy - y^2 + 3x^2y) \\ &= 2x^2y - xy + 3y^2 + 4xy - y^2 + 3x^2y \\ &= 5x^2y + 3xy + 2y^2\end{aligned}$$

5.
$$\begin{aligned}&-2mn(3m^2n - mn^3) \\ &= -2mn(3m^2n) + (-2mn)(-mn^3) \\ &= -6m^3n^2 + 2m^2n^4\end{aligned}$$

6. $(5x+3)(x-4)=(5x)(x)+(5x)(-4)+(3)(x)+(3)(-4)$
$=5x^2-20x+3x-12$
$=5x^2-17x-12$

7. $(2x-3y)(4x-7y)=(2x)(4x)+(2x)(-7y)+(-3y)(4x)+(-3y)(-7y)$
$=8x^2-14xy-12xy+21y^2$
$=8x^2-26xy+21y^2$

8. $(5x+8)(5x-8)=(5x)^2-(8)^2=25x^2-64$

9. $(2x+3y)^2=(2x)^2+2(2x)(3y)+(3y)^2=4x^2+12xy+9y^2$

10. $2a(3a-4)(a+5)=2a[(3a)(a)+(3a)(5)+(-4)(a)+(-4)(5)]$
$=2a(3a^2+15a-4a-20)$
$=2a(3a^2+11a-20)$
$=2a(3a^2)+2a(11a)+2a(-20)$
$=6a^3+22a^2-40a$

11. $(4m+3)(4m^3-2m^2+4m-8)=4m(4m^3-2m^2+4m-8)+3(4m^3-2m^2+4m-8)$
$=4m(4m^3)+4m(-2m^2)+4m(4m)+4m(-8)+3(4m^3)+3(-2m^2)+3(4m)+3(-8)$
$=16m^4-8m^3+16m^2-32m+12m^3-6m^2+12m-24$
$=16m^4-8m^3+12m^3+16m^2-6m^2-32m+12m-24$
$=16m^4+4m^3+10m^2-20m-24$

12. $(-2x^7y^0z)(4xz^8)=-2\cdot 4\cdot x^7\cdot x^1\cdot y^0\cdot z^1\cdot z^8=-8\cdot x^{7+1}\cdot y^0\cdot z^{1+8}=-8\cdot x^8\cdot 1\cdot z^9=-8x^8z^9$

13. $(5m^3n^{-2})(-3m^{-4}n)=5\cdot(-3)\cdot m^3\cdot m^{-4}\cdot n^{-2}\cdot n^1=-15\cdot m^{3-4}\cdot n^{-2+1}=-15m^{-1}n^{-1}=-\dfrac{15}{mn}$

14. $\dfrac{-18a^8b^3}{6a^5b}=\dfrac{-18}{6}\cdot\dfrac{a^8}{a^5}\cdot\dfrac{b^3}{b}=-3\cdot a^{8-5}\cdot b^{3-1}=-3\cdot a^3\cdot b^2=-3a^3b^2$

15. $\dfrac{16x^7y}{32x^9y^3}=\dfrac{16}{32}\cdot\dfrac{x^7}{x^9}\cdot\dfrac{y}{y^3}=\dfrac{1}{2}\cdot x^{7-9}\cdot y^{1-3}=\dfrac{1}{2}\cdot x^{-2}\cdot y^{-2}=\dfrac{1}{2x^2y^2}$

16. $\dfrac{7}{2ab^{-2}}=\dfrac{7b^2}{2a}$

17. $\dfrac{q^{-6}rt^5}{qr^{-4}t^7}=q^{-6-1}r^{1+4}t^{5-7}=q^{-7}r^5t^{-2}=\dfrac{r^5}{q^7t^2}$

18. $\left(\dfrac{3}{2}r^2\right)^{-3}=\left(\dfrac{3}{2}\right)^{-3}\cdot(r^2)^{-3}=\left(\dfrac{2}{3}\right)^3\cdot r^{2(-3)}=\dfrac{8}{27}\cdot r^{-6}=\dfrac{8}{27r^6}$

19. $(4y^{-2}z^3)^{-2} = 4^{-2}(y^{-2})^{-2}(z^3)^{-2}$

$$= \frac{1}{4^2} \cdot y^{-2(-2)} \cdot z^{3(-2)}$$
$$= \frac{1}{16} \cdot y^4 \cdot z^{-6}$$
$$= \frac{y^4}{16z^6}$$

20. $\left(\frac{3x^3y^{-3}}{2x^{-1}y^0}\right)^{-4} \cdot (2x^{-4}y^{-2})^2$

$$= \left(\frac{3}{2}x^{3+1}y^{-3-0}\right)^{-4} \cdot (2x^{-4}y^{-2})^2$$
$$= \left(\frac{3}{2}x^4y^{-3}\right)^{-4} \cdot (2x^{-4}y^{-2})^2$$
$$= \left[\left(\frac{3}{2}\right)^{-4} x^{-4(4)}y^{-4(-3)}\right] \cdot (4x^{-8}y^{-4})$$
$$= \left[\left(\frac{81}{16}\right)^{-1} x^{-16}y^{12}\right] \cdot (4x^{-8}y^{-4})$$
$$= \left(\frac{16}{81}x^{-16}y^{12}\right) \cdot (4x^{-8}y^{-4})$$
$$= (4)\left(\frac{16}{81}\right)x^{-16-8}y^{12-4}$$
$$= \frac{64}{81}x^{-24}y^8$$
$$= \frac{64y^8}{81x^{24}}$$

Section 5.5

Preparing for Dividing Polynomials

P1. $\frac{24a^3}{6a} = \frac{24}{6} \cdot \frac{a^3}{a} = 4 \cdot a^{3-1} = 4a^2$

P2. $\frac{-49x^3}{21x^4} = \frac{-49}{21} \cdot \frac{x^3}{x^4}$

$$= \frac{-7}{3} \cdot x^{3-4}$$
$$= -\frac{7}{3} \cdot x^{-1}$$
$$= -\frac{7}{3} \cdot \frac{1}{x}$$
$$= -\frac{7}{3x}$$

P3. $3x(7x-2) = 3x \cdot 7x - 3x \cdot 2 = 21x^2 - 6x$

P4. The degree of $4x^2 - 2x + 1$ is 2.

5.5 Quick Checks

1. The first step to simplify $\frac{4x^4 + 8x^2}{2x}$ would be to rewrite $\frac{4x^4 + 8x^2}{2x}$ as $\underline{\frac{4x^4}{2x} + \frac{8x^2}{2x}}$.

2. $\frac{10n^4 - 20n^3 + 5n^2}{5n^2}$

$$= \frac{10n^4}{5n^2} - \frac{20n^3}{5n^2} + \frac{5n^2}{5n^2}$$
$$= \frac{10}{5}n^{4-2} - \frac{20}{5}n^{3-2} + \frac{5}{5}n^{2-2}$$
$$= 2n^2 - 4n^1 + 1n^0$$
$$= 2n^2 - 4n + 1$$

3. $\frac{12k^4 - 18k^2 + 5}{2k^2}$

$$= \frac{12k^4}{2k^2} - \frac{18k^2}{2k^2} + \frac{5}{2k^2}$$
$$= \frac{12}{2}k^{4-2} - \frac{18}{2}k^{2-2} + \frac{5}{2}k^{-2}$$
$$= 6k^2 - 9k^0 + \frac{5}{2}k^{-2}$$
$$= 6k^2 - 9 + \frac{5}{2k^2}$$

4. $\dfrac{x^4y^4+8x^2y^2-4xy}{4x^3y}$

$=\dfrac{x^4y^4}{4x^3y}+\dfrac{8x^2y^2}{4x^3y}-\dfrac{4xy}{4x^3y}$

$=\dfrac{1}{4}x^{4-3}y^{4-1}+\dfrac{8}{4}x^{2-3}y^{2-1}-\dfrac{4}{4}x^{1-3}y^{1-1}$

$=\dfrac{1}{4}x^1y^3+2x^{-1}y^1-1x^{-2}y^0$

$=\dfrac{xy^3}{4}+\dfrac{2y}{x}-\dfrac{1}{x^2}$

5. To begin a polynomial division problem, write the divisor and the dividend in <u>standard</u> form.

6. To check the result of long division, multiply the <u>quotient</u> and the divisor and add this result to the <u>remainder</u>. If correct, this result will be equal to the <u>dividend</u>.

7.
$$\begin{array}{r|l} & x-8 \\ \hline x+5 & x^2-3x-40 \\ & -(x^2+5x) \\ & \quad -8x-40 \\ & \quad -(-8x-40) \\ & \quad\quad 0 \end{array}$$

$\dfrac{x^2-3x-40}{x+5}=x-8$

8.
$$\begin{array}{r|l} & x-4 \\ \hline 2x+3 & 2x^2-5x-12 \\ & -(2x^2+3x) \\ & \quad -8x-12 \\ & \quad -(-8x-12) \\ & \quad\quad 0 \end{array}$$

$\dfrac{2x^2-5x-12}{2x+3}=x-4$

9.
$$\begin{array}{r|l} & 4x+5 \\ \hline x+3 & 4x^2+17x+21 \\ & -(4x^2+12x) \\ & \quad 5x+21 \\ & \quad -(5x+15) \\ & \quad\quad 6 \end{array}$$

$\dfrac{4x^2+17x+21}{x+3}=4x+5+\dfrac{6}{x+3}$

10. $\dfrac{x+1-3x^2+4x^3}{x+2}=\dfrac{4x^3-3x^2+x+1}{x+2}$

$$\begin{array}{r|l} & 4x^2-11x+23 \\ \hline x+2 & 4x^3-3x^2+\ x+\ 1 \\ & -(4x^3+8x^2) \\ & \quad -11x^2+\ x+\ 1 \\ & \quad -(-11x^2-22x) \\ & \quad\quad 23x+\ 1 \\ & \quad\quad -(23x+46) \\ & \quad\quad\quad -45 \end{array}$$

$\dfrac{x+1-3x^2+4x^3}{x+2}=4x^2-11x+23+\dfrac{-45}{x+2}$

11. $\dfrac{2x^3+3x^2+10}{2x-5}=\dfrac{2x^3+3x^2+0x+10}{2x-5}$

$$\begin{array}{r|l} & x^2+4x+10 \\ \hline 2x-5 & 2x^3+3x^2+\ 0x+10 \\ & -(2x^3-5x^2) \\ & \quad 8x^2+\ 0x+10 \\ & \quad -(8x^2-20x) \\ & \quad\quad 20x+10 \\ & \quad\quad -(20x-50) \\ & \quad\quad\quad 60 \end{array}$$

$\dfrac{2x^3+3x^2+10}{2x-5}=x^2+4x+10+\dfrac{60}{2x-5}$

12.
$$\begin{array}{r|l} & 4x-3 \\ \hline x^2+2 & 4x^3-3x^2+\ x+1 \\ & -(4x^3\qquad +8x) \\ & \quad -3x^2-7x+1 \\ & \quad -(-3x^2\qquad -6) \\ & \quad\quad -7x+7 \end{array}$$

$\dfrac{4x^3-3x^2+x+1}{x^2+2}=4x-3+\dfrac{-7x+7}{x^2+2}$

5.5 Exercises

13. $\dfrac{4x^2-2x}{2x}=\dfrac{4x^2}{2x}-\dfrac{2x}{2x}$

$=\dfrac{4}{2}x^{2-1}-\dfrac{2}{2}x^{1-1}$

$=2x^1-1\cdot x^0$

$=2x-1$

15. $\frac{9a^3+27a^2-3}{3a^2}=\frac{9a^3}{3a^2}+\frac{27a^2}{3a^2}-\frac{3}{3a^2}$
$=\frac{9}{3}a^{3-2}+\frac{27}{3}a^{2-2}-\frac{3}{3}a^{-2}$
$=3a^1+9a^0-1\cdot a^{-2}$
$=3a+9-\frac{1}{a^2}$

17. $\frac{5n^5-10n^3-25n}{25n}$
$=\frac{5n^5}{25n}-\frac{10n^3}{25n}-\frac{25n}{25n}$
$=\frac{5}{25}n^{5-1}-\frac{10}{25}n^{3-1}-\frac{25}{25}n^{1-1}$
$=\frac{1}{5}n^4-\frac{2}{5}n^2-1\cdot n^0$
$=\frac{n^4}{5}-\frac{2n^2}{5}-1$

19. $\frac{15r^5-27r^3}{9r^3}=\frac{15r^5}{9r^3}-\frac{27r^3}{9r^3}$
$=\frac{15}{9}r^{5-3}-\frac{27}{9}r^{3-3}$
$=\frac{5}{3}r^2-3r^0$
$=\frac{5r^2}{3}-3$

21. $\frac{3x^7-9x^6+27x^3}{-3x^5}$
$=\frac{3x^7}{-3x^5}-\frac{9x^6}{-3x^5}+\frac{27x^3}{-3x^5}$
$=\frac{3}{-3}x^{7-5}-\frac{9}{-3}x^{6-5}+\frac{27}{-3}x^{3-5}$
$=-1\cdot x^2-(-3)x^1+(-9)x^{-2}$
$=-x^2+3x-\frac{9}{x^2}$

23. $\frac{3z+4z^3-2z^2}{8z}=\frac{3z}{8z}+\frac{4z^3}{8z}-\frac{2z^2}{8z}$
$=\frac{3}{8}z^{1-1}+\frac{4}{8}z^{3-1}-\frac{2}{8}z^{2-1}$
$=\frac{3}{8}z^0+\frac{1}{2}z^2-\frac{1}{4}z^1$
$=\frac{3}{8}+\frac{z^2}{2}-\frac{z}{4}$

25. $\frac{14x-10y}{-2}=\frac{14x}{-2}-\frac{10y}{-2}$
$=\frac{14}{-2}x-\frac{10}{-2}y$
$=-7x-(-5)y$
$=-7x+5y$

27. $\frac{12y-30x}{-2x}=\frac{12y}{-2x}-\frac{30x}{-2x}$
$=\frac{12}{-2}\cdot\frac{y}{x}-\frac{30}{-2}x^{1-1}$
$=-6\cdot\frac{y}{x}-(-15)x^0$
$=-\frac{6y}{x}+15$

29. $\frac{25a^3b^2c+10a^2bc^3}{-5a^4b^2c}$
$=\frac{25a^3b^2c}{-5a^4b^2c}+\frac{10a^2bc^3}{-5a^4b^2c}$
$=\frac{25}{-5}a^{3-4}b^{2-2}c^{1-1}+\frac{10}{-5}a^{2-4}b^{1-2}c^{3-1}$
$=-5a^{-1}b^0c^0+(-2)a^{-2}b^{-1}c^2$
$=-\frac{5}{a}-\frac{2c^2}{a^2b}$

31.
$$\begin{array}{r|l} & x-7 \\ \hline x+3 & x^2-4x-21 \\ & \underline{-(x^2+3x)} \\ & -7x-21 \\ & \underline{-(-7x-21)} \\ & 0 \end{array}$$
$\frac{x^2-4x-21}{x+3}=x-7$

33.
$$\begin{array}{r|l} & x-5 \\ \hline x-4 & x^2-9x+20 \\ & \underline{-(x^2-4x)} \\ & -5x+20 \\ & \underline{-(-5x+20)} \\ & 0 \end{array}$$
$\frac{x^2-9x+20}{x-4}=x-5$

35.
$$\begin{array}{r}
x^2+6x-3 \\
x-2\overline{)\,x^3+4x^2-15x+6} \\
\underline{-(x^3-2x^2)} \\
6x^2-15x \\
\underline{-(6x^2-12x)} \\
-3x+6 \\
\underline{-(-3x+6)} \\
0
\end{array}$$

$$\frac{x^3+4x^2-15x+6}{x-2}=x^2+6x-3$$

37.
$$\begin{array}{r}
x^3-3x^2+6x-2 \\
x+2\overline{)\,x^4-x^3+0x^2+10x-4} \\
\underline{-(x^4+2x^3)} \\
-3x^3+0x^2 \\
\underline{-(-3x^3-6x^2)} \\
6x^2+10x \\
\underline{-(6x^2+12x)} \\
-2x-4 \\
\underline{-(-2x-4)} \\
0
\end{array}$$

$$\frac{x^4-x^3+10x-4}{x+2}=x^3-3x^2+6x-2$$

39.
$$\begin{array}{r}
x^2-2x+3 \\
x+1\overline{)\,x^3-x^2+x+8} \\
\underline{-(x^3+x^2)} \\
-2x^2+x \\
\underline{-(-2x^2-2x)} \\
3x+8 \\
\underline{-(3x+3)} \\
5
\end{array}$$

$$\frac{x^3-x^2+x+8}{x+1}=x^2-2x+3+\frac{5}{x+1}$$

41.
$$\begin{array}{r}
2x+3 \\
x-5\overline{)\,2x^2-7x-15} \\
\underline{-(2x^2-10x)} \\
3x-15 \\
\underline{-(3x-15)} \\
0
\end{array}$$

$$\frac{2x^2-7x-15}{x-5}=2x+3$$

43.
$$\begin{array}{r}
x^2+6x+7 \\
x-2\overline{)\,x^3+4x^2-5x+2} \\
\underline{-(x^3-2x^2)} \\
6x^2-5x \\
\underline{-(6x^2-12x)} \\
7x+2 \\
\underline{-(7x-14)} \\
16
\end{array}$$

$$\frac{x^3+4x^2-5x+2}{x-2}=x^2+6x+7+\frac{16}{x-2}$$

45.
$$\begin{array}{r}
2x^3+5x^2+9x-4 \\
x-4\overline{)\,2x^4-3x^3-11x^2-40x-1} \\
\underline{-(2x^4-8x^3)} \\
5x^3-11x^2 \\
\underline{-(5x^3-20x^2)} \\
9x^2-40x \\
\underline{-(9x^2-36x)} \\
-4x-1 \\
\underline{-(-4x+16)} \\
-17
\end{array}$$

$$\frac{2x^4-3x^3-11x^2-40x-1}{x-4}$$
$$=2x^3+5x^2+9x-4+\frac{-17}{x-4}$$

47.
$$\begin{array}{r}
x^2+4x-3 \\
2x-1\overline{)\,2x^3+7x^2-10x+5} \\
\underline{-(2x^3-x^2)} \\
8x^2-10x \\
\underline{-(8x^2-4x)} \\
-6x+5 \\
\underline{-(-6x+3)} \\
2
\end{array}$$

$$\frac{2x^3+7x^2-10x+5}{2x-1}=x^2+4x-3+\frac{2}{2x-1}$$

49. $\dfrac{-24+x^2+x}{5+x}=\dfrac{x^2+x-24}{x+5}$

$$\begin{array}{r|l} & x-4 \\ \hline x+5 & x^2+x-24 \\ & -(x^2+5x) \\ \hline & -4x-24 \\ & -(-4x-20) \\ \hline & -4 \end{array}$$

$\dfrac{-24+x^2+x}{5+x}=x-4+\dfrac{-4}{5+x}$

51. $\dfrac{4x^2+5}{1+2x}=\dfrac{4x^2+0x+5}{2x+1}$

$$\begin{array}{r|l} & 2x-1 \\ \hline 2x+1 & 4x^2+0x+5 \\ & -(4x^2+2x) \\ \hline & -2x+5 \\ & -(-2x-1) \\ \hline & 6 \end{array}$$

$\dfrac{4x^2+5}{1+2x}=2x-1+\dfrac{6}{1+2x}$

53. $\dfrac{x^3+2x^2-8}{x^2-2}=\dfrac{x^3+2x^2+0x-8}{x^2-2}$

$$\begin{array}{r|l} & x+2 \\ \hline x^2-2 & x^3+2x^2+0x-8 \\ & -(x^3 \quad -2x) \\ \hline & 2x^2+2x-8 \\ & -(2x^2 \quad -4) \\ \hline & 2x-4 \end{array}$$

$\dfrac{x^3+2x^2-8}{x^2-2}=x+2+\dfrac{2x-4}{x^2-2}$

55. $(a-5)(a+6)$
$=a(a)+a(6)+(-5)(a)+(-5)(6)$
$=a^2+6a-5a-30$
$=a^2+a-30$

57. $(2x-8)-(3x+x^2-2)$
$=(2x-8)+(-3x-x^2+2)$
$=2x-8-3x-x^2+2$
$=-x^2+2x-3x-8+2$
$=-x^2-x-6$

59. $(2ab+b^2-a^2)+(b^2-4ab+a^2)$
$=2ab+b^2-a^2+b^2-4ab+a^2$
$=-a^2+a^2+2ab-4ab+b^2+b^2$
$=-2ab+2b^2$

61. $\dfrac{4+7x^2-3x^4+6x^3}{2x^2}$
$=\dfrac{-3x^4+6x^3+7x^2+4}{2x^2}$
$=\dfrac{-3x^4}{2x^2}+\dfrac{6x^3}{2x^2}+\dfrac{7x^2}{2x^2}+\dfrac{4}{2x^2}$
$=-\dfrac{3}{2}x^{4-2}+\dfrac{6}{2}x^{3-2}+\dfrac{7}{2}x^{2-2}+\dfrac{4}{2}x^{-2}$
$=-\dfrac{3}{2}x^2+\dfrac{6}{2}x^1+\dfrac{7}{2}x^0+\dfrac{4}{2}x^{-2}$
$=\dfrac{-3x^2}{2}+3x+\dfrac{7}{2}+\dfrac{2}{x^2}$

63. $\dfrac{18x^3y^{-4}z^6}{27x^{-4}y^{-12}z^{-6}}$
$=\dfrac{18}{27}x^{3-(-4)}y^{-4-(-12)}z^{6-(-6)}$
$=\dfrac{2}{3}x^{3+4}y^{-4+12}z^{6+6}$
$=\dfrac{2}{3}x^7y^8z^{12}$

65.
$$\begin{array}{r|l} & 2x-6 \\ \hline 3x-5 & 6x^2-28x+30 \\ & -(6x^2-10x) \\ \hline & -18x+30 \\ & -(-18x+30) \\ \hline & 0 \end{array}$$

$\dfrac{6x^2-28x+30}{3x-5}=2x-6$

67. $(n-3)^2=(n)^2-2(n)(3)+(3)^2=n^2-6n+9$

69. $(x^3+x-4x^4)(-10x^2)$
$=x^3(-10x^2)+x(-10x^2)+(-4x^4)(-10x^2)$
$=-10x^5-10x^3+40x^6$

71. $(2pq-q^2)+(4pq-p^2-q^2)$
$= 2pq-q^2+4pq-p^2-q^2$
$= 2pq+4pq-q^2-q^2-p^2$
$= 6pq-2q^2-p^2$

73. $(x^2+x-1)(x+5)$
$= x^2(x+5)+x(x+5)+(-1)(x+5)$
$= x^3+5x^2+x^2+5x-x-5$
$= x^3+6x^2+4x-5$

75. $(7rs^2-2r^2s)-(2r^2s-8rs^2)$
$= (7rs^2-2r^2s)+(-2r^2s+8rs^2)$
$= 7rs^2-2r^2s-2r^2s+8rs^2$
$= 7rs^2+8rs^2-2r^2s-2r^2s$
$= 15rs^2-4r^2s$

77. $(x^4-2x^2+x)(-3x)$
$= (x^4)(-3x)+(-2x^2)(-3x)+(x)(-3x)$
$= -3x^5+6x^3-3x^2$

79. First perform the division.

$$\begin{array}{r|l} & 3x-1 \\ \hline 2x-3 & 6x^2-11x+3 \\ & -(6x^2-9x) \\ & \quad -2x+3 \\ & \quad -(-2x+3) \\ & \quad 0 \end{array}$$

Replace $\dfrac{3+6x^2-11x}{2x-3}$ with $3x-1$.

$$\frac{3+6x^2-11x}{2x-3}+(x+1) = (3x-1)+(x+1)$$
$$= 3x-1+x+1$$
$$= 4x$$

81. $\dfrac{3x^4-6x+12x^2}{-3x^3}$
$= \dfrac{3x^4}{-3x^3}-\dfrac{6x}{-3x^3}+\dfrac{12x^2}{-3x^3}$
$= \dfrac{3}{-3}x^{4-3}-\dfrac{6}{-3}x^{1-3}+\dfrac{12}{-3}x^{2-3}$
$= -1\cdot x^1-(-2)x^{-2}+(-4)x^{-1}$
$= -x+\dfrac{2}{x^2}-\dfrac{4}{x}$

83. $\dfrac{(x^2+3x-1)+(x-1)}{-2x^3}$
$= \dfrac{x^2+3x-1+x-1}{-2x^3}$
$= \dfrac{x^2+4x-2}{-2x^3}$
$= \dfrac{x^2}{-2x^3}+\dfrac{4x}{-2x^3}-\dfrac{2}{-2x^3}$
$= \dfrac{1}{-2}x^{2-3}+\dfrac{4}{-2}x^{1-3}-\dfrac{2}{-2}\cdot\dfrac{1}{x^3}$
$= -\dfrac{1}{2}x^{-1}+(-2)x^{-2}-(-1)\dfrac{1}{x^3}$
$= -\dfrac{1}{2x}-\dfrac{2}{x^2}+\dfrac{1}{x^3}$

85. The volume is the product of the lengths of the sides. Thus the length of the third side is the quotient of x^3-5x^2+6x and the product of $x-2$ and $x-3$.
$(x-2)(x-3)$
$= x(x)+x(-3)+(-2)(x)+(-2)(-3)$
$= x^2-3x-2x+6$
$= x^2-5x+6$

$$\begin{array}{r|l} & x \\ \hline x^2-5x+6 & x^3-5x^2+6x+0 \\ & -(x^3-5x^2+6x) \\ & \quad 0 \end{array}$$

Since $\dfrac{x^3-5x^2+6x}{(x-2)(x-3)} = \dfrac{x^3-5x^2+6x}{x^2-5x+6} = x$, the length of the third side is x feet.

87. The area is the product of the length and the width. Thus the width is the quotient of z^2+6z+9 and $z+3$.

$$\begin{array}{r|l} & z+3 \\ \hline z+3 & z^2+6z+9 \\ & -(z^2+3z) \\ & \quad 3z+9 \\ & \quad -(3z+9) \\ & \quad 0 \end{array}$$

Since $\dfrac{z^2+6z+9}{z+3} = z+3$, the width of the rectangle is $z+3$ inches.

89. The area is one-half of the product of the base and the height $\left(A = \frac{1}{2}bh\right)$, so twice the area is the product of the base and the height. Thus the base is the quotient of $2(6x^3 - 2x^2 - 8x)$ and $3x^2 - 4x$.

$$\begin{aligned} &2(6x^3 - 2x^2 - 8x) \\ &= 2(6x^3) + 2(-2x^2) + 2(-8x) \\ &= 12x^3 - 4x^2 - 16x \end{aligned}$$

$$\begin{array}{r} 4x+4 \\ 3x^2-4x \overline{\smash{)}\, 12x^3 - 4x^2 - 16x + 0} \\ \underline{-(12x^3 - 16x^2)} \\ 12x^2 - 16x \\ \underline{-(12x^2 - 16x)} \\ 0 \end{array}$$

Since $\frac{2(6x^3 - 2x^2 - 8x)}{3x^2 - 4x} = 4x + 4$, the length of the base is $4x + 4$ feet.

91. (a)
$$\begin{aligned} &\frac{0.004x^3 - 0.8x^2 + 180x + 5000}{x} \\ &= \frac{0.004x^3}{x} - \frac{0.8x^2}{x} + \frac{180x}{x} + \frac{5000}{x} \\ &= 0.004x^{3-1} - 0.8x^{2-1} + 180x^{1-1} + \frac{5000}{x} \\ &= 0.004x^2 - 0.8x + 180 + \frac{5000}{x} \end{aligned}$$

(b) When $x = 140$, the value of the expression is
$$\begin{aligned} &0.004(140)^2 - 0.8(140) + 180 + \frac{5000}{140} \\ &= 78.4 - 112 + 180 + \frac{250}{7} \\ &\approx 182.11 \end{aligned}$$
The average cost of manufacturing $x = 140$ computers in a day is \$182.11 per computer.

93.
$$\begin{array}{r} 3x-2 \\ 2x-3 \overline{\smash{)}\, 6x^2 - 13x + ?} \\ \underline{-(6x^2 - 9x)} \\ -4x + ? \\ \underline{-(-4x + 6)} \\ ? - 6 \end{array}$$

If the remainder is zero, then $? - 6 = 0$ or $? = 6$.

95. When the divisor is a monomial, divide each term in the numerator by the denominator. When dividing by a binomial, use long division. Answers may vary.

Section 5.6

Preparing for Applying Exponent Rules: Scientific Notation

P1. $$\begin{aligned} (3a^6)(4.5a^4) &= 3 \cdot 4.5 \cdot a^6 \cdot a^4 \\ &= 13.5 \cdot a^{6+4} \\ &= 13.5a^{10} \end{aligned}$$

P2. $$\begin{aligned} (7n^3)(2n^{-2}) &= 7 \cdot 2 \cdot n^3 \cdot n^{-2} \\ &= 14 \cdot n^{3+(-2)} \\ &= 14n^1 \\ &= 14n \end{aligned}$$

P3. $\frac{3.6b^9}{0.9b^{-2}} = \frac{3.6}{0.9} \cdot \frac{b^9}{b^{-2}} = 4 \cdot b^{9-(-2)} = 4b^{11}$

5.6 Quick Checks

1. A number written as 3.2×10^{-6} is said to be written in <u>scientific notation</u>.

2. When writing 47,000,000 in scientific notation, the power of 10 will be <u>positive</u> (positive or negative).

3. False; when a number is expressed in scientific notation, it is expressed as the product of a number x, $1 \le x < 10$, and a power of 10.

4. $432 = 4.32 \times 100 = 4.32 \times 10^2$

5. $10{,}302 = 1.0302 \times 10{,}000 = 1.0302 \times 10^4$

6. $5{,}432{,}000 = 5.432 \times 1{,}000{,}000 = 5.432 \times 10^6$

7. $0.093 = 9.3 \times 0.01 = 9.3 \times 10^{-2}$

8. $0.0000459 = 4.59 \times 0.00001 = 4.59 \times 10^{-5}$

9. $$\begin{aligned} 0.00000008 &= 8.0 \times 0.00000001 \\ &= 8.0 \times 10^{-8} \end{aligned}$$

10. To write 3.2×10^{-6} in decimal notation, move the decimal point in 3.2 six places to the <u>left</u>.

11. True

12. $3.1\times10^2 = 3.1\times100 = 310$

13. $9.01\times10^{-1} = 9.01\times0.1 = 0.901$

14. $1.7\times10^5 = 1.7\times100,000 = 170,000$

15. $7\times10^0 = 7\times1 = 7$

16. $8.9\times10^{-4} = 8.9\times0.0001 = 0.00089$

17. $(3\times10^4)\cdot(2\times10^3) = (3\cdot2)\times(10^4\cdot10^3)$
$= 6\times10^7$

18. $(2\times10^{-2})\cdot(4\times10^{-1}) = (2\cdot4)\times(10^{-2}\cdot10^{-1})$
$= 8\times10^{-3}$

19. $(5\times10^{-4})\cdot(3\times10^7) = (5\cdot3)\times(10^{-4}\cdot10^7)$
$= 15\times10^3$
$= (1.5\times10^1)\times10^3$
$= 1.5\times10^4$

20. $(8\times10^{-4})\cdot(3.5\times10^{-2})$
$= (8\cdot3.5)\times(10^{-4}\cdot10^{-2})$
$= 28\times10^{-6}$
$= (2.8\times10^1)\times10^{-6}$
$= 2.8\times10^{-5}$

21. $\dfrac{8\times10^6}{2\times10^1} = \dfrac{8}{2}\times\dfrac{10^6}{10^1} = 4\times10^5$

22. $\dfrac{2.8\times10^{-7}}{1.4\times10^{-3}} = \dfrac{2.8}{1.4}\times\dfrac{10^{-7}}{10^{-3}}$
$= 2\times10^{-7-(-3)}$
$= 2\times10^{-4}$

23. $\dfrac{3.6\times10^3}{7.2\times10^{-1}} = \dfrac{3.6}{7.2}\times\dfrac{10^3}{10^{-1}}$
$= 0.5\times10^{3-(-1)}$
$= (5\times10^{-1})\times10^4$
$= 5\times10^3$

24. $\dfrac{5\times10^{-2}}{8\times10^2} = \dfrac{5}{8}\times\dfrac{10^{-2}}{10^2}$
$= 0.625\times10^{-2-2}$
$= (6.25\times10^{-1})\times10^{-4}$
$= 6.25\times10^{-5}$

25. There are 31 days in August.
$31 = 3.1\times10^1$
$(3.1\times10^1)\cdot(8.72\times10^6)$
$= (3.1\cdot8.72)\times(10^1\cdot10^6)$
$= 27.032\times10^7$
$= (2.7032\times10^1)\times10^7$
$= 2.7032\times10^8$
$= 270,320,000$
Saudi Arabia produced
$2.7032\times10^8 = 270,320,000$ barrels of oil in August 2007.

26. There are 31 days in August: $31 = 3.1\times10^1$. In scientific notation, 12 million is written as 1.2×10^7.
$(3.1\times10^1)(1.2\times10^7) = (3.1\times1.2)\times10^{1+7}$
$= 3.72\times10^8$
$= 372,000,000$
Saudi Arabia is expected to produce 3.72×10^8 barrels or 372,000,000 barrels of oil in August 2009.

5.6 Exercises

27. $300,000 = 3.0\times100,000 = 3\times10^5$

29. $64,000,000 = 6.4\times10,000,000$
$= 6.4\times10^7$

31. $0.00051 = 5.1\times0.0001 = 5.1\times10^{-4}$

33. $0.000000001 = 1.0\times0.000000001$
$= 1\times10^{-9}$

35. $8,007,000,000 = 8.007\times1,000,000,000$
$= 8.007\times10^9$

37. $0.0000309 = 3.09\times0.00001 = 3.09\times10^{-5}$

39. $620 = 6.2\times100 = 6.2\times10^2$

41. $4 = 4.0 \times 1 = 4 \times 10^0$

43. $6,656,000,000 = 6.656 \times 10^9$

45. $70,510,000,000 = 7.051 \times 10^{10}$

47. $170,200,000 \text{ km} = 1.702 \times 10^8 \text{ km}$

49. $0.00003 = 3 \times 0.00001 = 3 \times 10^{-5}$

51. $0.00000025 \text{ m} = 2.5 \times 10^{-7} \text{ m}$

53. $0.00311 \text{ kg} = 3.11 \times 10^{-3} \text{ kg}$

55. $4.2 \times 10^5 = 4.2 \times 100,000 = 420,000$

57.
$$\begin{aligned} 1 \times 10^8 &= 1 \times 100,000,000 \\ &= 100,000,000 \end{aligned}$$

59. $3.9 \times 10^{-3} = 3.9 \times 0.001 = 0.0039$

61. $4 \times 10^{-1} = 4 \times 0.1 = 0.4$

63. $3.76 \times 10^3 = 3.76 \times 1000 = 3760$

65. $8.2 \times 10^{-3} = 8.2 \times 0.001 = 0.0082$

67. $6 \times 10^{-5} = 6 \times 0.00001 = 0.00006$

69. $7.05 \times 10^6 = 7.05 \times 1,000,000 = 7,050,000$

71.
$$\begin{aligned} 1 \times 10^{-15} &= 1 \times 0.000000000000001 \\ &= 0.000000000000001 \end{aligned}$$

73. $2.25 \times 10^{-3} = 2.25 \times 0.001 = 0.00225$

75. $5 \times 10^5 = 5 \times 100,000 = 500,000$

77.
$$\begin{aligned} (2 \times 10^6)(1.5 \times 10^3) &= (2 \cdot 1.5) \times (10^6 \cdot 10^3) \\ &= 3 \times 10^9 \end{aligned}$$

79.
$$\begin{aligned} (1.2 \times 10^0)(7 \times 10^{-3}) &= (1.2 \cdot 7) \times (10^0 \cdot 10^{-3}) \\ &= 8.4 \times 10^{-3} \end{aligned}$$

81. $\dfrac{9 \times 10^4}{3 \times 10^{-4}} = \dfrac{9}{3} \times \dfrac{10^4}{10^{-4}} = 3 \times 10^{4-(-4)} = 3 \times 10^8$

83.
$$\begin{aligned} \frac{2 \times 10^{-3}}{8 \times 10^{-5}} &= \frac{2}{8} \times \frac{10^{-3}}{10^{-5}} \\ &= \frac{1}{4} \times 10^{-3-(-5)} \\ &= 0.25 \times 10^2 \\ &= (2.5 \times 10^{-1}) \times 10^2 \\ &= 2.5 \times 10^1 \end{aligned}$$

85.
$$\begin{aligned} \frac{56,000}{0.00007} &= \frac{5.6 \times 10^4}{7 \times 10^{-5}} \\ &= \frac{5.6}{7} \times 10^{4-(-5)} \\ &= 0.8 \times 10^9 \\ &= (8 \times 10^{-1}) \times 10^9 \\ &= 8 \times 10^8 \end{aligned}$$

87.
$$\begin{aligned} &\frac{300,000 \times 15,000,000}{0.0005} \\ &= \frac{(3 \times 10^5) \cdot (1.5 \times 10^7)}{5 \times 10^{-4}} \\ &= \frac{(3 \cdot 1.5) \times (10^5 \cdot 10^7)}{5 \times 10^{-4}} \\ &= \frac{4.5 \times 10^{12}}{5 \times 10^{-4}} \\ &= \frac{4.5}{5} \times 10^{12-(-4)} \\ &= 0.9 \times 10^{16} \\ &= (9 \times 10^{-1}) \times 10^{16} \\ &= 9 \times 10^{15} \end{aligned}$$

89.
$$\begin{aligned} &(1.86 \times 10^5)(6.0 \times 10^1) \\ &= (1.86 \cdot 6.0) \times (10^5 \cdot 10^1) \\ &= 11.16 \times 10^6 \\ &= (1.116 \times 10^1) \times 10^6 \\ &= 1.116 \times 10^7 \end{aligned}$$

Light travels 1.116×10^7 miles in one minute.

91.
$$\begin{aligned} \frac{5.026 \times 10^{11} \text{ lb}}{3.0 \times 10^8 \text{ people}} &= \frac{5.026}{3} \times 10^{11-8} \left(\frac{\text{lb}}{\text{person}} \right) \\ &= \frac{5.026}{3} \times 10^3 \cdot \left(\frac{\text{lb}}{\text{person}} \right) \\ &\approx 1.68 \times 10^3 \text{ lb per person} \end{aligned}$$

93. **(a)** $1.55 \text{ billion} = 1.55\times10^9$

(b) $300,000,000 = 3\times10^8$

(c) $\frac{1.55\times10^9}{3\times10^8}\cdot\left(\frac{\text{gallons}}{\text{person}}\right)$
$= \frac{1.55}{3}\times10^{9-8}\cdot\left(\frac{\text{gallons}}{\text{person}}\right)$
$\approx 0.52\times10$ gallons per person
$= 5.2$ gallons per person

95. **(a)** $179.1 \text{ million} = 179.1\times10^6$
$= 1.791\times10^8$

(b) $44.8 \text{ million} = 44.8\times10^6 = 4.48\times10^7$

(c) $\frac{4.48\times10^7}{1.791\times10^8}\cdot 100\%$
$= \left(\frac{4.48}{1.791}\times10^{7-8}\right)\cdot 100\%$
$\approx (2.5\times10^{-1})\cdot 100\%$
$= 0.25\cdot 100\%$
$= 25\%$

97. $250\ \mu\text{m} = 250\cdot(1\times10^{-6})$ m
$= (2.5\times10^2)\cdot(1\times10^{-6})$ m
$= (2.5\cdot 1)\times(10^2\cdot 10^{-6})$ m
$= 2.5\times10^{-4}$ m

99. $800 \text{ pm} = 800\cdot(1\times10^{-12})$ m
$= (8\times10^2)\cdot(1\times10^{-12})$ m
$= (8\cdot 1)\times(10^2\cdot 10^{-12})$ m
$= 8\times10^{-10}$ m

101. $71.5 \text{ nm} = 71.5\cdot(1\times10^{-9})$ m
$= (7.15\times10^1)\cdot(1\times10^{-9})$ m
$= (7.15\cdot 1)\times(10^1\cdot 10^{-9})$ m
$= 7.15\times10^{-8}$ m

103. $21\ \mu\text{m} = 21\cdot(1\times10^{-6})$ m
$= (2.1\times10^1)\cdot(1\times10^{-6})$ m
$= (2.1\cdot 1)\times(10^1\cdot 10^{-6})$ m
$= 2.1\times10^{-5}$ m

$V = \frac{4}{3}\pi(21\ \mu\text{m})^3$
$= \frac{4}{3}\pi(2.1\times10^{-5}\text{ m})^3$
$= \frac{4}{3}\pi[2.1^3\times(10^{-5})^3\text{ m}^3]$
$= \frac{4}{3}\pi(9.261\times10^{-15})\text{ m}^3$
$\doteq \pi(12.348\times10^{-15})\text{ m}^3$
$= \pi[(1.2348\times10^1)\times10^{-15}]\text{ m}^3$
$= \pi(1.2348\times10^{-14})\text{ m}^3$
$= 1.2348\pi\times10^{-14}\text{ m}^3$

105. $6 \text{ nm} = 6\cdot(1\times10^{-9})$ m
$= (6\times10^0)\cdot(1\times10^{-9})$ m
$= (6\cdot 1)\times(10^0\cdot 10^{-9})$ m
$= 6\times10^{-9}$ m

$V = \frac{4}{3}\pi(6\text{ nm})^3$
$= \frac{4}{3}\pi(6\times10^{-9}\text{ m})^3$
$= \frac{4}{3}\pi[6^3\times(10^{-9})^3\text{ m}^3]$
$= \frac{4}{3}\pi(216\times10^{-27})\text{ m}^3$
$= \pi(288\times10^{-27})\text{ m}^3$
$= \pi[(2.88\times10^2)\times10^{-27}]\text{ m}^3$
$= \pi(2.88\times10^{-25})\text{ m}^3$
$= 2.88\pi\times10^{-25}\text{ m}^3$

107. To convert a number written in decimal notation to scientific notation, count the number of decimal places, N, that the decimal point must be moved to arrive at a number x such that $1 \le x < 10$. If the number is greater than or equal to 1, move the decimal point to the left that many places and write the number in the form $x\times10^N$. If the number is between 0 and 1, move the decimal point to the right that many places and write the number in the form $x\times10^{-N}$.

109. The number 34.5×10^4 is incorrect because the number 34.5 is not a number between 1 (inclusive) and 10. The correct answer is 3.45×10^5.

Chapter 5 Review

1. Yes, $4x^3$ is a monomial of degree 3 with coefficient 4.

2. No, $6x^{-3}$ is not a monomial because the exponent of x is not a whole number.

3. No, $m^{1/2}$ is not a monomial because the exponent of m is not a whole number.

4. Yes, mn^2 is monomial of degree 3 (1 + 2) with coefficient 1.

5. No, $4x^6 - 4x^{1/2}$ is not a polynomial because a variable has an exponent that is not a whole number.

6. No, $\frac{3}{x} - \frac{1}{x^2}$ is not a polynomial because there is a variable in the denominator of a fraction.

7. Yes, 6 is monomial of degree 0.

8. Yes, $3x^3 - 4xy^4$ is a binomial of degree 5 (1 + 4).

9. Yes, $-2x^5y - 7x^4y + 7$ is a trinomial of degree 6 (5 + 1).

10. Yes, $\frac{1}{2}x^3 + 2x^{10} - 5$ is trinomial of degree 10.

11. $(6x^2 - 2x + 1) + (3x^2 + 10x - 3)$
$= 6x^2 - 2x + 1 + 3x^2 + 10x - 3$
$= 6x^2 + 3x^2 - 2x + 10x + 1 - 3$
$= 9x^2 + 8x - 2$

12. $(-7m^3 - 2mn) + (8m^3 - 5m + 3mn)$
$= -7m^3 - 2mn + 8m^3 - 5m + 3mn$
$= -7m^3 + 8m^3 - 2mn + 3mn - 5m$
$= m^3 + mn - 5m$

13. $(4x^2y + 10x) - (5x^2y - 2x)$
$= (4x^2y + 10x) + (-5x^2y + 2x)$
$= 4x^2y + 10x - 5x^2y + 2x$
$= 4x^2y - 5x^2y + 10x + 2x$
$= -x^2y + 12x$

14. $(3y^2 - yz + 3z^2) - (10y^2 + 5yz - 6z^2)$
$= (3y^2 - yz + 3z^2) + (-10y^2 - 5yz + 6z^2)$
$= 3y^2 - yz + 3z^2 - 10y^2 - 5yz + 6z^2$
$= 3y^2 - 10y^2 - yz - 5yz + 3z^2 + 6z^2$
$= -7y^2 - 6yz + 9z^2$

15. $(-6x^2 + 5) + (4x^2 - 7) = -6x^2 + 5 + 4x^2 - 7$
$= -6x^2 + 4x^2 + 5 - 7$
$= -2x^2 - 2$

16. $(-18y + 10) - (20y^2 - 10y + 5)$
$= (-18y + 10) + (-20y^2 + 10y - 5)$
$= -18y + 10 - 20y^2 + 10y - 5$
$= -20y^2 - 18y + 10y + 10 - 5$
$= -20y^2 - 8y + 5$

17. $3x^2 - 5x$

(a) $x = 0$: $3(0)^2 - 5(0) = 0 - 0 = 0$

(b) $x = -1$: $3(-1)^2 - 5(-1) = 3 + 5 = 8$

(c) $x = 2$: $3(2)^2 - 5(2) = 12 - 10 = 2$

18. $-x^2 + 3$

(a) $x = 0$: $-(0)^2 + 3 = 0 + 3 = 3$

(b) $x = -1$: $-(-1)^2 + 3 = -1 + 3 = 2$

(c) $x = \frac{1}{2}$: $-\left(\frac{1}{2}\right)^2 + 3 = -\frac{1}{4} + 3 = -\frac{1}{4} + \frac{12}{4} = \frac{11}{4}$

19. $x^2y + 2xy^2$ for $x = -2$ and $y = 1$:
$(-2)^2(1) + 2(-2)(1)^2 = 4(1) + 2(-2)(1)$
$= 4 - 4$
$= 0$

20. $4a^2b^2 - 3ab + 2$ for $a = -1$ and $b = -3$:
$4(-1)^2(-3)^2 - 3(-1)(-3) + 2$
$= 4(1)(9) - 3(-1)(-3) + 2$
$= 36 - 9 + 2$
$= 29$

21. $6^2 \cdot 6^5 = 6^{2+5} = 6^7 = 279{,}936$

22. $\left(-\frac{1}{3}\right)^2\left(-\frac{1}{3}\right)^3 = \left(-\frac{1}{3}\right)^{2+3} = \left(-\frac{1}{3}\right)^5 = -\frac{1}{243}$

23. $(4^2)^6 = 4^{2\cdot 6} = 4^{12} = 16{,}777{,}216$

24. $[(-1)^4]^3 = (-1)^{4\cdot 3} = (-1)^{12} = 1$

25. $x^4 \cdot x^8 \cdot x = x^4 \cdot x^8 \cdot x^1 = x^{4+8+1} = x^{13}$

26. $m^4 \cdot m^2 = m^{4+2} = m^6$

27. $(r^3)^4 = r^{3\cdot 4} = r^{12}$

28. $(m^8)^3 = m^{8\cdot 3} = m^{24}$

29. $$\begin{aligned}(4x)^3(4x)^2 &= (4x)^{3+2}\\ &= (4x)^5\\ &= 4^5 \cdot x^5\\ &= 1024x^5\end{aligned}$$

30. $$\begin{aligned}(-2n)^3(-2n)^3 &= (-2n)^{3+3}\\ &= (-2n)^6\\ &= (-2)^6 \cdot n^6\\ &= 64n^6\end{aligned}$$

31. $$\begin{aligned}(-3x^2y)^4 &= (-3)^4 \cdot (x^2)^4 \cdot y^4\\ &= 81 \cdot x^{2\cdot 4} \cdot y^4\\ &= 81x^8y^4\end{aligned}$$

32. $$\begin{aligned}(2x^3y^4)^2 &= 2^2 \cdot (x^3)^2 \cdot (y^4)^2\\ &= 4 \cdot x^{3\cdot 2} \cdot y^{4\cdot 2}\\ &= 4x^6y^8\end{aligned}$$

33. $3x^2 \cdot 5x^4 = 3 \cdot 5 \cdot x^2 \cdot x^4 = 15x^{2+4} = 15x^6$

34. $-4a \cdot 9a^3 = -4 \cdot 9 \cdot a^1 \cdot a^3 = -36a^{1+3} = -36a^4$

35. $$\begin{aligned}-8y^4 \cdot (-2y) &= -8 \cdot (-2) \cdot y^4 \cdot y^1\\ &= 16y^{4+1}\\ &= 16y^5\end{aligned}$$

36. $$\begin{aligned}12p \cdot (-p^5) &= 12p^1 \cdot (-1 \cdot p^5)\\ &= 12 \cdot (-1) \cdot p^1 \cdot p^5\\ &= -12p^{1+5}\\ &= -12p^6\end{aligned}$$

37. $\frac{8}{3}w^3 \cdot \frac{9}{2}w = \frac{8}{3} \cdot \frac{9}{2} \cdot w^3 \cdot w^1 = 12w^{3+1} = 12w^4$

38. $$\begin{aligned}\frac{1}{3}z^2 \cdot \left(-\frac{9}{4}z\right) &= \frac{1}{3} \cdot \left(-\frac{9}{4}\right) \cdot z^2 \cdot z^1\\ &= -\frac{3}{4}z^{2+1}\\ &= -\frac{3}{4}z^3\end{aligned}$$

39. $$\begin{aligned}(3x^2)^3 \cdot (2x)^2 &= 3^3 \cdot (x^2)^3 \cdot 2^2 \cdot x^2\\ &= 27 \cdot x^{2\cdot 3} \cdot 4 \cdot x^2\\ &= 27 \cdot 4 \cdot x^6 \cdot x^2\\ &= 108x^{6+2}\\ &= 108x^8\end{aligned}$$

40. $$\begin{aligned}(-4a)^2 \cdot (5a^4) &= (-4)^2 \cdot a^2 \cdot 5 \cdot a^4\\ &= 16 \cdot 5 \cdot a^2 \cdot a^4\\ &= 80a^{2+4}\\ &= 80a^6\end{aligned}$$

41. $$\begin{aligned}&-2x^3(4x^2 - 3x + 1)\\ &= -2x^3(4x^2) + (-2x^3)(-3x) + (-2x^3)(1)\\ &= -8x^5 + 6x^4 - 2x^3\end{aligned}$$

42. $$\begin{aligned}&\frac{1}{2}x^4(4x^3 + 8x^2 - 2)\\ &= \frac{1}{2}x^4(4x^3) + \frac{1}{2}x^4(8x^2) + \frac{1}{2}x^4(-2)\\ &= 2x^7 + 4x^6 - x^4\end{aligned}$$

43. $$\begin{aligned}&(3x - 5)(2x + 1)\\ &= (3x - 5)(2x) + (3x - 5)(1)\\ &= 3x \cdot 2x + (-5) \cdot 2x + 3x \cdot 1 + (-5) \cdot 1\\ &= 6x^2 - 10x + 3x - 5\\ &= 6x^2 - 7x - 5\end{aligned}$$

44. $$\begin{aligned}(4x + 3)(x - 2) &= (4x + 3)(x) + (4x + 3)(-2)\\ &= 4x \cdot x + 3 \cdot x + 4x \cdot (-2) + 3 \cdot (-2)\\ &= 4x^2 + 3x - 8x - 6\\ &= 4x^2 - 5x - 6\end{aligned}$$

45. $(x+5)(x-8) = (x+5)x + (x+5)(-8)$
$= x \cdot x + 5 \cdot x + x \cdot (-8) + 5 \cdot (-8)$
$= x^2 + 5x - 8x - 40$
$= x^2 - 3x - 40$

46. $(w-1)(w+10) = (w-1)(w) + (w-1)(10)$
$= w \cdot w + (-1) \cdot w + w \cdot 10 + (-1) \cdot 10$
$= w^2 - w + 10w - 10$
$= w^2 + 9w - 10$

47. $(4m-3)(6m^2 - m + 1) = (4m-3)(6m^2) + (4m-3)(-m) + (4m-3)(1)$
$= 4m \cdot 6m^2 + (-3) \cdot 6m^2 + 4m \cdot (-m) + (-3) \cdot (-m) + 4m \cdot 1 + (-3) \cdot 1$
$= 24m^3 - 18m^2 - 4m^2 + 3m + 4m - 3$
$= 24m^3 - 22m^2 + 7m - 3$

48. $(2y+3)(4y^4 + 2y^2 - 3) = (2y+3)(4y^4) + (2y+3)(2y^2) + (2y+3)(-3)$
$= 2y \cdot 4y^4 + 3 \cdot 4y^4 + 2y \cdot 2y^2 + 3 \cdot 2y^2 + 2y \cdot (-3) + 3 \cdot (-3)$
$= 8y^5 + 12y^4 + 4y^3 + 6y^2 - 6y - 9$

49. $(x+5)(x+3) = (x)(x) + (x)(3) + (5)(x) + (5)(3) = x^2 + 3x + 5x + 15 = x^2 + 8x + 15$

50. $(2x-1)(x-8) = (2x)(x) + (2x)(-8) + (-1)(x) + (-1)(-8) = 2x^2 - 16x - x + 8 = 2x^2 - 17x + 8$

51. $(2m+7)(3m-2) = (2m)(3m) + (2m)(-2) + (7)(3m) + (7)(-2) = 6m^2 - 4m + 21m - 14 = 6m^2 + 17m - 14$

52. $(6m-4)(8m+1) = (6m)(8m) + (6m)(1) + (-4)(8m) + (-4)(1) = 48m^2 + 6m - 32m - 4 = 48m^2 - 26m - 4$

53. $(3x+2y)(7x-3y) = (3x)(7x) + (3x)(-3y) + (2y)(7x) + (2y)(-3y)$
$= 21x^2 - 9xy + 14xy - 6y^2$
$= 21x^2 + 5xy - 6y^2$

54. $(4x-y)(5x+3y) = (4x)(5x) + (4x)(3y) + (-y)(5x) + (-y)(3y)$
$= 20x^2 + 12xy - 5xy - 3y^2$
$= 20x^2 + 7xy - 3y^2$

55. $(x-4)(x+4) = (x)^2 - (4)^2 = x^2 - 16$

56. $(2x+5)(2x-5) = (2x)^2 - (5)^2 = 4x^2 - 25$

57. $(2x+3)^2 = (2x)^2 + 2(2x)(3) + (3)^2$
$= 4x^2 + 12x + 9$

58. $(7x-2)^2 = (7x)^2 - 2(7x)(2) + (2)^2$
$= 49x^2 - 28x + 4$

59. $(3x+4y)(3x-4y)=(3x)^2-(4y)^2$
$=9x^2-16y^2$

60. $(8m-6n)(8m+6n)=(8m)^2-(6n)^2$
$=64m^2-36n^2$

61. $(5x-2y)^2=(5x)^2-2(5x)(2y)+(2y)^2$
$=25x^2-20xy+4y^2$

62. $(2a+3b)^2=(2a)^2+2(2a)(3b)+(3b)^2$
$=4a^2+12ab+9b^2$

63. $(x-0.5)(x+0.5)=(x)^2-(0.5)^2=x^2-0.25$

64. $(r+0.25)(r-0.25)=(r)^2-(0.25)^2$
$=r^2-0.0625$

65. $\left(y+\frac{2}{3}\right)^2=(y)^2+2(y)\left(\frac{2}{3}\right)+\left(\frac{2}{3}\right)^2$
$=y^2+\frac{4}{3}y+\frac{4}{9}$

66. $\left(y-\frac{1}{2}\right)^2=(y)^2-2(y)\left(\frac{1}{2}\right)+\left(\frac{1}{2}\right)^2$
$=y^2-y+\frac{1}{4}$

67. $\frac{6^5}{6^3}=6^{5-3}=6^2=36$

68. $\frac{7}{7^4}=\frac{7^1}{7^4}=7^{1-4}=7^{-3}=\frac{1}{7^3}=\frac{1}{343}$

69. $\frac{x^{16}}{x^{12}}=x^{16-12}=x^4$

70. $\frac{x^3}{x^{11}}=x^{3-11}=x^{-8}=\frac{1}{x^8}$

71. $5^0=1$

72. $-5^0=-1\cdot 5^0=-1$

73. $m^0=1,\ m\neq 0$

74. $-m^0=-1\cdot m^0=-1,\ m\neq 0$

75. $\frac{25x^3y^7}{10xy^{10}}=\frac{25}{10}\cdot x^{3-1}\cdot y^{7-10}$
$=\frac{5}{2}\cdot x^2\cdot y^{-3}$
$=\frac{5x^2}{2y^3}$

76. $\frac{3x^4y^2}{9x^2y^{10}}=\frac{3}{9}\cdot x^{4-2}\cdot y^{2-10}$
$=\frac{1}{3}\cdot x^2\cdot y^{-8}$
$=\frac{x^2}{3y^8}$

77. $\left(\frac{x^3}{y^2}\right)^5=\frac{(x^3)^5}{(y^2)^5}=\frac{x^{3\cdot 5}}{y^{2\cdot 5}}=\frac{x^{15}}{y^{10}}$

78. $\left(\frac{7}{x^2}\right)^3=\frac{7^3}{(x^2)^3}=\frac{343}{x^{2\cdot 3}}=\frac{343}{x^6}$

79. $\left(\frac{2m^2n^2}{p^4}\right)^3=\frac{2^3(m^2)^3(n^2)^3}{(p^4)^3}$
$=\frac{8\cdot m^{2\cdot 3}\cdot n^{2\cdot 3}}{p^{4\cdot 3}}$
$=\frac{8m^6n^6}{p^{12}}$

80. $\left(\frac{3mn^2}{p^5}\right)^4=\frac{3^4m^4(n^2)^4}{(p^5)^4}$
$=\frac{81\cdot m^4\cdot n^{2\cdot 4}}{p^{5\cdot 4}}$
$=\frac{81m^4n^8}{p^{20}}$

81. $-5^{-2}=-1\cdot 5^{-2}=-1\cdot\frac{1}{5^2}=-\frac{1}{25}$

82. $\frac{1}{4^{-3}}=4^3=64$

83. $\left(\frac{2}{3}\right)^{-4} = \left(\frac{3}{2}\right)^{4} = \frac{3^4}{2^4} = \frac{81}{16}$

84. $\left(\frac{1}{3}\right)^{-3} = \left(\frac{3}{1}\right)^{3} = 3^3 = 27$

85. $2^{-2} + 3^{-1} = \frac{1}{2^2} + \frac{1}{3^1} = \frac{1}{4} + \frac{1}{3} = \frac{3}{12} + \frac{4}{12} = \frac{7}{12}$

86. $4^{-1} - 2^{-3} = \frac{1}{4^1} - \frac{1}{2^3} = \frac{1}{4} - \frac{1}{8} = \frac{2}{8} - \frac{1}{8} = \frac{1}{8}$

87.
$$\begin{aligned}\frac{16x^{-3}y^4}{24x^{-6}y^{-1}} &= \frac{16}{24}\cdot x^{-3-(-6)}\cdot y^{4-(-1)}\\ &= \frac{2}{3}\cdot x^3\cdot y^5\\ &= \frac{2x^3y^5}{3}\end{aligned}$$

88.
$$\begin{aligned}\frac{15x^0y^{-6}}{35xy^4} &= \frac{15}{35}\cdot x^{0-1}\cdot y^{-6-4}\\ &= \frac{3}{7}x^{-1}y^{-10}\\ &= \frac{3}{7xy^{10}}\end{aligned}$$

89.
$$\begin{aligned}&(2m^{-3}n)^{-4}(3m^{-4}n^2)^2\\ &= 2^{-4}\cdot(m^{-3})^{-4}\cdot n^{-4}\cdot 3^2\cdot(m^{-4})^2\cdot(n^2)^2\\ &= \frac{1}{2^4}\cdot m^{-3\cdot(-4)}\cdot n^{-4}\cdot 9\cdot m^{-4\cdot 2}\cdot n^{2\cdot 2}\\ &= \frac{1}{16}\cdot m^{12}\cdot n^{-4}\cdot 9\cdot m^{-8}\cdot n^4\\ &= \frac{9}{16}\cdot m^{12}\cdot m^{-8}\cdot n^{-4}\cdot n^4\\ &= \frac{9}{16}m^4n^0\\ &= \frac{9m^4}{16}\end{aligned}$$

90.
$$\begin{aligned}&(4m^{-6}n^0)^3(3m^{-6}n^3)^{-2}\\ &= 4^3\cdot(m^{-6})^3\cdot 1^3\cdot 3^{-2}\cdot(m^{-6})^{-2}\cdot(n^3)^{-2}\\ &= 64\cdot m^{-6\cdot 3}\cdot 1\cdot\frac{1}{3^2}\cdot m^{-6\cdot(-2)}\cdot n^{3\cdot(-2)}\\ &= 64\cdot m^{-18}\cdot 1\cdot\frac{1}{9}\cdot m^{12}\cdot n^{-6}\\ &= \frac{64}{9}\cdot m^{-18}\cdot m^{12}\cdot n^{-6}\\ &= \frac{64}{9}m^{-6}n^{-6}\\ &= \frac{64}{9m^6n^6}\end{aligned}$$

91.
$$\begin{aligned}&\left(\frac{3rs^{-1}}{4s^2}\right)^{-2}\cdot(2r^{-6}t^0)^{-1}\\ &= \left(\frac{3}{4}r\cdot s^{-1-2}\right)^{-2}\cdot(2r^{-6}\cdot 1)^{-1}\\ &= \left(\frac{3}{4}r\cdot s^{-3}\right)^{-2}\cdot(2r^{-6})^{-1}\\ &= \left(\frac{3}{4}\right)^{-2}\cdot r^{-2}\cdot(s^{-3})^{-2}\cdot 2^{-1}\cdot(r^{-6})^{-1}\\ &= \left(\frac{4}{3}\right)^{2}\cdot r^{-2}\cdot s^{-3\cdot(-2)}\cdot\frac{1}{2}\cdot r^{-6\cdot(-1)}\\ &= \frac{16}{9}\cdot r^{-2}\cdot s^6\cdot\frac{1}{2}\cdot r^6\\ &= \frac{16}{9}\cdot\frac{1}{2}\cdot r^{-2}\cdot r^6\cdot s^6\\ &= \frac{8}{9}\cdot r^4\cdot s^6\\ &= \frac{8r^4s^6}{9}\end{aligned}$$

92. $(6r^4s^{-3})^2 \cdot \left(\dfrac{3r^4s}{2r^{-2}s^{-2}}\right)^{-3}$

$= (6r^4s^{-3})^2 \cdot \left(\dfrac{3}{2}r^{4-(-2)}s^{1-(-2)}\right)^{-3}$

$= (6r^4s^{-3})^2 \cdot \left(\dfrac{3}{2}r^6s^3\right)^{-3}$

$= 6^2 \cdot (r^4)^2 \cdot (s^{-3})^2 \cdot \left(\dfrac{3}{2}\right)^{-3} \cdot (r^6)^{-3} \cdot (s^3)^{-3}$

$= 36 \cdot r^{4\cdot 2} \cdot s^{-3\cdot 2} \cdot \left(\dfrac{2}{3}\right)^3 \cdot r^{6\cdot(-3)} \cdot s^{3\cdot(-3)}$

$= 36 \cdot r^8 \cdot s^{-6} \cdot \dfrac{8}{27} \cdot r^{-18} \cdot s^{-9}$

$= 36 \cdot \dfrac{8}{27} \cdot r^8 \cdot r^{-18} \cdot s^{-6} \cdot s^{-9}$

$= \dfrac{32}{3} \cdot r^{-10} \cdot s^{-15}$

$= \dfrac{32}{3r^{10}s^{15}}$

93. $\dfrac{36x^7 - 24x^6 + 30x^2}{6x^2}$

$= \dfrac{36x^7}{6x^2} - \dfrac{24x^6}{6x^2} + \dfrac{30x^2}{6x^2}$

$= \dfrac{36}{6}x^{7-2} - \dfrac{24}{6}x^{6-2} + \dfrac{30}{6}x^{2-2}$

$= 6x^5 - 4x^4 + 5x^0$

$= 6x^5 - 4x^4 + 5$

94. $\dfrac{15x^5 + 25x^3 - 30x^2}{5x}$

$= \dfrac{15x^5}{5x} + \dfrac{25x^3}{5x} - \dfrac{30x^2}{5x}$

$= \dfrac{15}{5}x^{5-1} + \dfrac{25}{5}x^{3-1} - \dfrac{30}{5}x^{2-1}$

$= 3x^4 + 5x^2 - 6x^1$

$= 3x^4 + 5x^2 - 6x$

95. $\dfrac{16n^8 + 4n^5 - 10n}{4n^5}$

$= \dfrac{16n^8}{4n^5} + \dfrac{4n^5}{4n^5} - \dfrac{10n}{4n^5}$

$= \dfrac{16}{4}n^{8-5} + \dfrac{4}{4}n^{5-5} - \dfrac{10}{4}n^{1-5}$

$= 4n^3 + n^0 - \dfrac{5}{2}n^{-4}$

$= 4n^3 + 1 - \dfrac{5}{2n^4}$

96. $\dfrac{30n^6 - 20n^5 - 16n^3}{5n^5}$

$= \dfrac{30n^6}{5n^5} - \dfrac{20n^5}{5n^5} - \dfrac{16n^3}{5n^5}$

$= \dfrac{30}{5}n^{6-5} - \dfrac{20}{5}n^{5-5} - \dfrac{16}{5}n^{3-5}$

$= 6n^1 - 4n^0 - \dfrac{16}{5}n^{-2}$

$= 6n - 4 - \dfrac{16}{5n^2}$

97. $\dfrac{2p^8 + 4p^5 - 8p^3}{-16p^5}$

$= \dfrac{2p^8}{-16p^5} + \dfrac{4p^5}{-16p^5} - \dfrac{8p^3}{-16p^5}$

$= -\dfrac{2}{16}p^{8-5} - \dfrac{4}{16}p^{5-5} + \dfrac{8}{16}p^{3-5}$

$= -\dfrac{1}{8}p^3 - \dfrac{1}{4}p^0 + \dfrac{1}{2}p^{-2}$

$= -\dfrac{p^3}{8} - \dfrac{1}{4} + \dfrac{1}{2p^2}$

98. $\dfrac{3p^4 - 6p^2 + 9}{-6p^2}$

$= \dfrac{3p^4}{-6p^2} - \dfrac{6p^2}{-6p^2} + \dfrac{9}{-6p^2}$

$= -\dfrac{3}{6}p^{4-2} + \dfrac{6}{6}p^{2-2} - \dfrac{9}{6}p^{-2}$

$= -\dfrac{1}{2}p^2 + p^0 - \dfrac{3}{2}p^{-2}$

$= -\dfrac{p^2}{2} + 1 - \dfrac{3}{2p^2}$

99. $$\begin{array}{r} 4x-7 \\ 2x+3\overline{)8x^2-2x-21} \\ -(8x^2+12x) \\ \hline -14x-21 \\ -(-14x-21) \\ \hline 0 \end{array}$$

$$\frac{8x^2-2x-21}{2x+3}=4x-7$$

100. $$\begin{array}{r} x+6 \\ 3x-1\overline{)3x^2+17x-6} \\ -(3x^2-x) \\ \hline 18x-6 \\ -(18x-6) \\ \hline 0 \end{array}$$

$$\frac{3x^2+17x-6}{3x-1}=x+6$$

101. $$\frac{6x^2+x^3-2x+1}{x-1}=\frac{x^3+6x^2-2x+1}{x-1}$$

$$\begin{array}{r} x^2+7x+5 \\ x-1\overline{)x^3+6x^2-2x+1} \\ -(x^3-x^2) \\ \hline 7x^2-2x \\ -(7x^2-7x) \\ \hline 5x+1 \\ -(5x-5) \\ \hline 6 \end{array}$$

$$\frac{6x^2+x^3-2x+1}{x-1}=x^2+7x+5+\frac{6}{x-1}$$

102. $$\frac{-6x+2x^3-7x^2+8}{x-2}=\frac{2x^3-7x^2-6x+8}{x-2}$$

$$\begin{array}{r} 2x^2-3x-12 \\ x-2\overline{)2x^3-7x^2-6x+8} \\ -(2x^3-4x^2) \\ \hline -3x^2-6x \\ -(-3x^2+6x) \\ \hline -12x+8 \\ -(-12x+24) \\ \hline -16 \end{array}$$

$$\frac{-6x+2x^3-7x^2+8}{x-2}=2x^2-3x-12+\frac{-16}{x-2}$$

103. $$\frac{x^3+8}{x+2}=\frac{x^3+0x^2+0x+8}{x+2}$$

$$\begin{array}{r} x^2-2x+4 \\ x+2\overline{)x^3+0x^2+0x+8} \\ -(x^3+2x^2) \\ \hline -2x^2+0x \\ -(-2x^2-4x) \\ \hline 4x+8 \\ -(4x+8) \\ \hline 0 \end{array}$$

$$\frac{x^3+8}{x+2}=x^2-2x+4$$

104. $$\frac{3x^3+2x-7}{x-5}=\frac{3x^3+0x^2+2x-7}{x-5}$$

$$\begin{array}{r} 3x^2+15x+77 \\ x-5\overline{)3x^3+0x^2+2x-7} \\ -(3x^3-15x^2) \\ \hline 15x^2+2x \\ -(15x^2-75x) \\ \hline 77x-7 \\ -(77x-385) \\ \hline 378 \end{array}$$

$$\frac{3x^3+2x-7}{x-5}=3x^2+15x+77+\frac{378}{x-5}$$

105. $27,000,000=2.7\times 10,000,000=2.7\times 10^7$

106. $1,230,000,000=1.23\times 1,000,000,000$
$=1.23\times 10^9$

107. $0.00006=6\times 0.00001=6\times 10^{-5}$

108. $0.00000305=3.05\times 0.000001=3.05\times 10^{-6}$

109. $3=3\times 1=3\times 10^0$

110. $8=8\times 1=8\times 10^0$

111. $6\times 10^{-4}=6\times 0.0001=0.0006$

112. $1.25\times 10^{-3}=1.25\times 0.001=0.00125$

113. $6.13\times 10^5=6.13\times 100,000=613,000$

114. $8\times10^4 = 8\times10{,}000 = 80{,}000$

115. $3.7\times10^{-1} = 3.7\times0.1 = 0.37$

116. $5.4\times10^7 = 5.4\times10{,}000{,}000$
$= 54{,}000{,}000$

117. $(1.2\times10^{-5})(5\times10^8) = (1.2\cdot5)\times(10^{-5}\cdot10^8)$
$= 6\times10^3$

118. $(1.4\times10^{-10})(3\times10^2)$
$= (1.4\cdot3)\times(10^{-10}\cdot10^2)$
$= 4.2\times10^{-8}$

119. $\frac{2.4\times10^{-6}}{1.2\times10^{-8}} = \frac{2.4}{1.2}\times10^{-6-(-8)} = 2\times10^2$

120. $\frac{5\times10^6}{25\times10^{-3}} = \frac{5}{25}\times10^{6-(-3)}$
$= 0.2\times10^9$
$= (2\times10^{-1})\times10^9$
$= 2\times10^8$

121. $\frac{200{,}000\times4{,}000{,}000}{0.0002} = \frac{(2\times10^5)\cdot(4\times10^6)}{2\times10^{-4}}$
$= \frac{(2\cdot4)\times10^{5+6}}{2\times10^{-4}}$
$= \frac{8\times10^{11}}{2\times10^{-4}}$
$= \frac{8}{2}\times10^{11-(-4)}$
$= 4\times10^{15}$

122. $\frac{1{,}200{,}000}{0.003\times2{,}000{,}000} = \frac{1.2\times10^6}{(3\times10^{-3})\cdot(2\times10^6)}$
$= \frac{1.2\times10^6}{(3\cdot2)\times10^{-3+6}}$
$= \frac{1.2\times10^6}{6\times10^3}$
$= \frac{1.2}{6}\times10^{6-3}$
$= 0.2\times10^3$
$= (2\times10^{-1})\times10^3$
$= 2\times10^2$

Chapter 5 Test

1. Yes, $6x^5 - 2x^4$ is a binomial of degree 5.

2. $3x^2 - 2x + 5$

(a) $x = 0$: $3(0)^2 - 2(0) + 5 = 0 - 0 + 5 = 5$

(b) $x = -2$: $3(-2)^2 - 2(-2) + 5 = 12 + 4 + 5 = 21$

(c) $x = 3$: $3(3)^2 - 2(3) + 5 = 27 - 6 + 5 = 26$

3. $(3x^2y^2 - 2x + 3y) + (-4x - 6y + 4x^2y^2)$
$= 3x^2y^2 - 2x + 3y - 4x - 6y + 4x^2y^2$
$= 3x^2y^2 + 4x^2y^2 - 2x - 4x + 3y - 6y$
$= 7x^2y^2 - 6x - 3y$

4. $(8m^3 + 6m^2 - 4) - (5m^2 - 2m^3 + 2)$
$= (8m^3 + 6m^2 - 4) + (-5m^2 + 2m^3 - 2)$
$= 8m^3 + 6m^2 - 4 - 5m^2 + 2m^3 - 2$
$= 8m^3 + 2m^3 + 6m^2 - 5m^2 - 4 - 2$
$= 10m^3 + m^2 - 6$

5. $-3x^3(2x^2 - 6x + 5)$
$= -3x^3(2x^2) + (-3x^3)(-6x) + (-3x^3)(5)$
$= -6x^5 + 18x^4 - 15x^3$

6. $(x-5)(2x+7)$
$= (x)(2x) + (x)(7) + (-5)(2x) + (-5)(7)$
$= 2x^2 + 7x - 10x - 35$
$= 2x^2 - 3x - 35$

7. $(2x-7)^2 = (2x)^2 - 2(2x)(7) + (7)^2$
$= 4x^2 - 28x + 49$

8. $(4x-3y)(4x+3y) = (4x)^2 - (3y)^2$
$= 16x^2 - 9y^2$

9. $$\begin{aligned}(3x-1)(2x^2+x-8) &= (3x-1)(2x^2)+(3x-1)(x)+(3x-1)(-8)\\ &= 3x\cdot 2x^2+(-1)\cdot 2x^2+3x\cdot x+(-1)\cdot x+3x\cdot(-8)+(-1)\cdot(-8)\\ &= 6x^3-2x^2+3x^2-x-24x+8\\ &= 6x^3+x^2-25x+8\end{aligned}$$

10. $$\frac{6x^4-8x^3+9}{3x^3}=\frac{6x^4}{3x^3}-\frac{8x^3}{3x^3}+\frac{9}{3x^3}=\frac{6}{3}x^{4-3}-\frac{8}{3}x^{3-3}+\frac{9}{3}x^{-3}=2x^1-\frac{8}{3}x^0+3x^{-3}=2x-\frac{8}{3}+\frac{3}{x^3}$$

11. $$\frac{3x^3-2x^2+5}{x+3}=\frac{3x^3-2x^2+0x+5}{x+3}$$

$$\begin{array}{r} 3x^2-11x+33 \\ x+3\overline{)\,3x^3-2x^2+0x+5} \\ -(3x^3+9x^2) \\ \hline -11x^2+0x \\ -(-11x^2-33x) \\ \hline 33x+5 \\ -(33x+99) \\ \hline -94 \end{array}$$

$$\frac{3x^3-2x^2+5}{x+3}=3x^2-11x+33+\frac{-94}{x+3}$$

12. $$(4x^3y^2)(-3xy^4)=4\cdot x^3\cdot y^2\cdot(-3)\cdot x\cdot y^4=4\cdot(-3)\cdot x^3\cdot x\cdot y^2\cdot y^4=-12x^{3+1}y^{2+4}=-12x^4y^6$$

13. $$\frac{18m^5n}{27m^2n^6}=\frac{18}{27}\cdot m^{5-2}\cdot n^{1-6}=\frac{2}{3}m^3n^{-5}=\frac{2m^3}{n^5}$$

14. $$\left(\frac{m^{-2}n^0}{m^{-7}n^4}\right)^{-6}=(m^{-2-(-7)}n^{0-4})^{-6}=(m^5n^{-4})^{-6}=m^{5\cdot(-6)}n^{-4\cdot(-6)}=m^{-30}n^{24}=\frac{n^{24}}{m^{30}}$$

15. $$\begin{aligned}(4x^{-3}y)^{-2}(2x^4y^{-3})^4 &= 4^{-2}\cdot(x^{-3})^{-2}\cdot y^{-2}\cdot 2^4\cdot(x^4)^4\cdot(y^{-3})^4\\ &= \frac{1}{4^2}\cdot x^{-3\cdot(-2)}\cdot y^{-2}\cdot 16\cdot x^{4\cdot 4}\cdot y^{-3\cdot 4}\\ &= \frac{1}{16}\cdot x^6\cdot y^{-2}\cdot 16\cdot x^{16}\cdot y^{-12}\\ &= \frac{1}{16}\cdot 16\cdot x^6\cdot x^{16}\cdot y^{-2}\cdot y^{-12}\\ &= x^{22}y^{-14}\\ &= \frac{x^{22}}{y^{14}}\end{aligned}$$

16. $(2m^{-4}n^2)^{-1} \cdot \left(\dfrac{16m^0n^{-3}}{m^{-3}n^2}\right)$
$= (2m^{-4}n^2)^{-1} \cdot (16m^{0-(-3)}n^{-3-2})$
$= (2m^{-4}n^2)^{-1} \cdot (16m^3n^{-5})$
$= 2^{-1} \cdot (m^{-4})^{-1} \cdot (n^2)^{-1} \cdot 16 \cdot m^3 \cdot n^{-5}$
$= \dfrac{1}{2} \cdot m^{-4 \cdot (-1)} \cdot n^{2 \cdot (-1)} \cdot 16 \cdot m^3 \cdot n^{-5}$
$= \dfrac{1}{2} \cdot m^4 \cdot n^{-2} \cdot 16 \cdot m^3 \cdot n^{-5}$
$= \dfrac{1}{2} \cdot 16 \cdot m^4 \cdot m^3 \cdot n^{-2} \cdot n^{-5}$
$= 8m^7n^{-7}$
$= \dfrac{8m^7}{n^7}$

17. $0.000012 = 1.2 \times 0.00001 = 1.2 \times 10^{-5}$

18. $2.101 \times 10^5 = 2.101 \times 100{,}000 = 210{,}100$

19. $(2.1 \times 10^{-6}) \cdot (1.7 \times 10^{10})$
$= (2.1 \cdot 1.7) \times (10^{-6} \cdot 10^{10})$
$= 3.57 \times 10^4$

20. $\dfrac{3 \times 10^{-4}}{15 \times 10^2} = \dfrac{3}{15} \times 10^{-4-2}$
$= 0.2 \times 10^{-6}$
$= (2 \times 10^{-1}) \times 10^{-6}$
$= 2 \times 10^{-7}$

Cumulative Review Chapters 1–5

1. $-\dfrac{4}{2} = 2,\ \sqrt{25} = 5$

(a) $\{\sqrt{25}\}$

(b) $\{0, \sqrt{25}\}$

(c) $\left\{-6, -\dfrac{4}{2}, 0, \sqrt{25}\right\}$

(d) $\left\{-6, -\dfrac{4}{2}, 0, 1.4, \sqrt{25}\right\}$

(e) $\{\sqrt{7}\}$

(f) $\left\{-6, -\dfrac{4}{2}, 0, 1.4, \sqrt{7}, \sqrt{25}\right\}$

2. $-\dfrac{1}{2} + \dfrac{2}{3} \div 4 \cdot \dfrac{1}{3} = -\dfrac{1}{2} + \dfrac{2}{3} \cdot \dfrac{1}{4} \cdot \dfrac{1}{3}$
$= -\dfrac{1}{2} + \dfrac{1}{6} \cdot \dfrac{1}{3}$
$= -\dfrac{1}{2} + \dfrac{1}{18}$
$= -\dfrac{9}{18} + \dfrac{1}{18}$
$= -\dfrac{8}{18}$
$= -\dfrac{4}{9}$

3. $2 + 3[3 + 10(-1)] = 2 + 3[3 + (-10)]$
$= 2 + 3[-7]$
$= 2 + (-21)$
$= -19$

4. $6x^3 - (-2x^2 + 3x) + 3x^2$
$= 6x^3 + (2x^2 - 3x) + 3x^2$
$= 6x^3 + 2x^2 - 3x + 3x^2$
$= 6x^3 + 2x^2 + 3x^2 - 3x$
$= 6x^3 + 5x^2 - 3x$

5. $-4(6x - 1) + 2(3x + 2)$
$= -4(6x) + (-4)(-1) + 2(3x) + 2(2)$
$= -24x + 4 + 6x + 4$
$= -24x + 6x + 4 + 4$
$= -18x + 8$

6.
$$\begin{aligned} -2(3x-4)+6 &= 4x-6x+10 \\ -2(3x)+(-2)(-4)+6 &= -2x+10 \\ -6x+8+6 &= -2x+10 \\ -6x+14 &= -2x+10 \\ -6x+2x+14 &= -2x+2x+10 \\ -4x+14 &= 10 \\ -4x+14-14 &= 10-14 \\ -4x &= -4 \\ \frac{-4x}{-4} &= \frac{-4}{-4} \\ x &= 1 \end{aligned}$$
The solution set is $\{1\}$.

7. $4(x - 5) = 10 + 2x$

8. Let x be Kathy's earnings before the 3% raise. Then the amount of the raise is $0.03x$.

$$x + 0.3x = 659.20$$
$$1.03x = 659.20$$
$$\frac{1.03x}{1.03} = \frac{659.20}{1.03}$$
$$x = 640$$

Her earnings before the raise were $640 per month.

9. Let x be Amber's speed. Then $x + 12$ is Cheyenne's speed. In 3 hours, Amber drives $3x$ miles and Cheyenne drives $3(x + 12)$ miles.

$$3x + 3(x + 12) = 306$$
$$3x + 3x + 36 = 306$$
$$6x + 36 = 306$$
$$6x + 36 - 36 = 306 - 36$$
$$6x = 270$$
$$\frac{6x}{6} = \frac{270}{6}$$
$$x = 45$$

Amber's speed is 45 miles per hour.

10.

$$-5x + 2 > 17$$
$$-5x + 2 - 2 > 17 - 2$$
$$-5x > 15$$
$$\frac{-5x}{-5} < \frac{15}{-5}$$
$$x < -3$$

$\{x | x < -3\}$

$(-\infty, -3)$

−3 0

11. $\frac{x^2 - y^2}{z}$ when $x = 3$, $y = -2$, and $z = -10$:

$$\frac{(3)^2 - (-2)^2}{-10} = \frac{9-4}{-10} = \frac{5}{-10} = -\frac{1}{2}$$

12. Let $(x_1, y_1) = (-3, 8)$ and $(x_2, y_2) = (1, 2)$.

$$m = \frac{y_2 - y_1}{x_2 - x_1} = \frac{2-8}{1-(-3)} = \frac{-6}{4} = -\frac{3}{2}$$

13. $2x + 3y = 24$

Let $x = 0$.

$$2(0) + 3y = 24$$
$$3y = 24$$
$$y = 8$$

The y-intercept is 8 or (0, 8).

Let $y = 0$.

$$2x + 3(0) = 24$$
$$2x = 24$$
$$x = 12$$

The x-intercept is 12 or (12, 0).

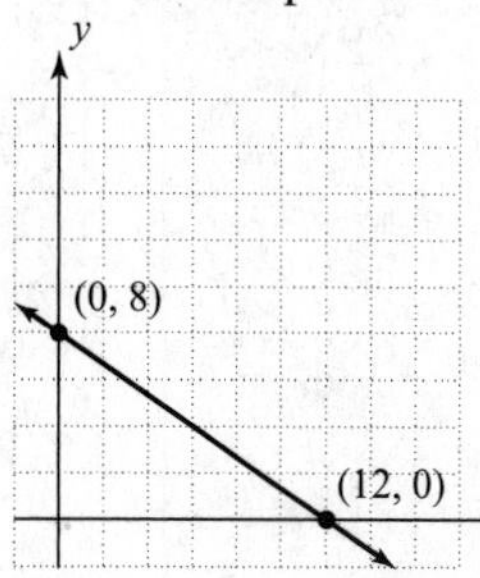

14. $y = -3x + 8$

The y-intercept is 8. Plot (0, 8). The slope is $-3 = \frac{-3}{1}$. Move 3 units down and 1 unit to the right to plot another point.

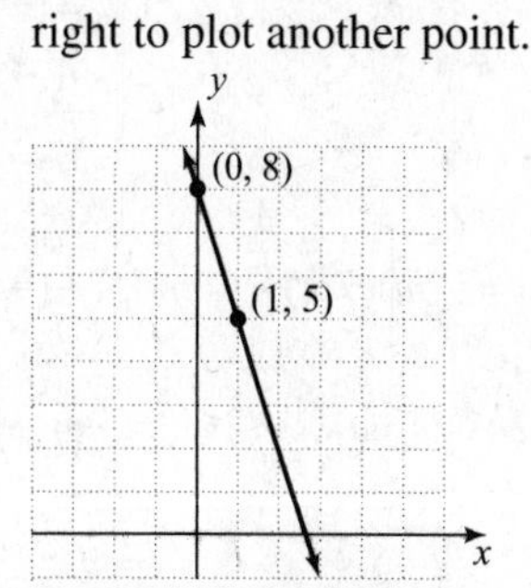

15. $\begin{cases} 2x - 3y = 27 & (1) \\ -4x + 2y = -27 & (2) \end{cases}$

Multiply (1) by 2. Then add.

$$\begin{cases} 4x - 6y = 54 \\ -4x + 2y = -27 \end{cases}$$
$$-4y = 27$$
$$y = \frac{27}{-4} = -\frac{27}{4}$$

Let $y = -\frac{27}{4}$ in (1).

$$2x-3\left(-\frac{27}{4}\right)=27$$
$$2x+\frac{81}{4}=27$$
$$2x=\frac{108}{4}-\frac{81}{4}$$
$$2x=\frac{27}{4}$$
$$x=\frac{27}{8}$$

The solution is $\left(\frac{27}{8},-\frac{27}{4}\right)$.

16. $\begin{cases} 3x-2y=8 & (1) \\ -6x+4y=8 & (2) \end{cases}$

Multiply (1) by 2. Then add.

$$\begin{cases} 6x-4y=16 \\ -6x+4y=8 \end{cases}$$
$$0=24 \quad \text{False}$$

The system is inconsistent. There is no solution.

17.
$$\begin{aligned}(4x^2+6x)-(-x+5x^2)+(6x^3-2x^2) &= (4x^2+6x)+(x-5x^2)+(6x^3-2x^2) \\ &= 4x^2+6x+x-5x^2+6x^3-2x^2 \\ &= 6x^3+4x^2-5x^2-2x^2+6x+x \\ &= 6x^3-3x^2+7x\end{aligned}$$

18. $(4m-3)(7m+2)=(4m)(7m)+(4m)(2)+(-3)(7m)+(-3)(2)=28m^2+8m-21m-6=28m^2-13m-6$

19. $(3m-2n)(3m+2n)=(3m)^2-(2n)^2=9m^2-4n^2$

20. $(7x+y)^2=(7x)^2+2(7x)(y)+(y)^2=49x^2+14xy+y^2$

21.
$$\begin{aligned}(2m+5)(2m^2-5m+3) &= (2m+5)(2m^2)+(2m+5)(-5m)+(2m+5)(3) \\ &= 2m(2m^2)+5(2m^2)+2m(-5m)+5(-5m)+2m(3)+5(3) \\ &= 4m^3+10m^2-10m^2-25m+6m+15 \\ &= 4m^3-19m+15\end{aligned}$$

22. $\frac{14xy^2+7x^2y}{7x^2y^2}=\frac{14xy^2}{7x^2y^2}+\frac{7x^2y}{7x^2y^2}=\frac{14}{7}x^{1-2}y^{2-2}+\frac{7}{7}x^{2-2}y^{1-2}=2x^{-1}y^0+1x^0y^{-1}=\frac{2}{x}+\frac{1}{y}$

23. $\dfrac{x^3+27}{x+3}=\dfrac{x^3+0x^2+0x+27}{x+3}$

$$\begin{array}{r} x^2-3x+9 \\ x+3\overline{)x^3+0x^2+0x+27} \\ -(x^3+3x^2) \\ \hline -3x^2+0x \\ -(-3x^2-9x) \\ \hline 9x+27 \\ -(9x+27) \\ \hline 0 \end{array}$$

$\dfrac{x^3+27}{x+3}=x^2-3x+9$

24. $(4m^0n^3)(-6n)=4\cdot m^0\cdot n^3\cdot(-6)\cdot n$
$=4\cdot(-6)\cdot m^0\cdot n^3\cdot n$
$=-24\cdot 1\cdot n^{3+1}$
$=-24n^4$

25. $\dfrac{25m^{-6}n^{-2}}{-10m^{-4}n^{-10}}=\dfrac{25}{-10}\cdot m^{-6-(-4)}\cdot n^{-2-(-10)}$
$=-\dfrac{5}{2}\cdot m^{-2}\cdot n^8$
$=-\dfrac{5n^8}{2m^2}$

26. $\left(\dfrac{2xy^4}{z^{-2}}\right)^{-6}=\left(\dfrac{z^{-2}}{2xy^4}\right)^6$
$=\dfrac{(z^{-2})^6}{2^6x^6(y^4)^6}$
$=\dfrac{z^{-2\cdot 6}}{64x^6y^{4\cdot 6}}$
$=\dfrac{z^{-12}}{64x^6y^{24}}$
$=\dfrac{1}{64x^6y^{24}z^{12}}$

27. $(x^4y^{-2})^{-4}\cdot\left(\dfrac{6x^{-4}y^3}{3y^{-8}}\right)^{-1}$
$=(x^4y^{-2})^{-4}\cdot\left(\dfrac{3y^{-8}}{6x^{-4}y^3}\right)$
$=(x^4)^{-4}(y^{-2})^{-4}\cdot\left(\dfrac{3}{6}\cdot x^4\cdot y^{-8-3}\right)$
$=x^{4\cdot(-4)}\cdot y^{-2\cdot(-4)}\cdot\dfrac{1}{2}\cdot x^4\cdot y^{-11}$
$=x^{-16}\cdot y^8\cdot\dfrac{1}{2}\cdot x^4\cdot y^{-11}$
$=\dfrac{1}{2}\cdot x^{-16}\cdot x^4\cdot y^8\cdot y^{-11}$
$=\dfrac{1}{2}x^{-12}y^{-3}$
$=\dfrac{1}{2x^{12}y^3}$

28. $0.0000605=6.05\times 0.00001$
$=6.05\times 10^{-5}$

29. $2.175\times 10^6=2.175\times 1{,}000{,}000$
$=2{,}175{,}000$

30. $(3.4\times 10^8)(2.1\times 10^{-3})$
$=(3.4\cdot 2.1)\times(10^8\cdot 10^{-3})$
$=7.14\times 10^5$

Chapter 6

Section 6.1

Preparing for Greatest Common Factor and Factoring by Grouping

P1. $48 = 2 \cdot 24$
$= 2 \cdot 2 \cdot 12$
$= 2 \cdot 2 \cdot 2 \cdot 6$
$= 2 \cdot 2 \cdot 2 \cdot 2 \cdot 3$
$= 2^4 \cdot 3$

P2. $2(5x - 3) = 2 \cdot 5x - 2 \cdot 3 = 10x - 6$

P3. $(2x+5)(x-3)$
$= 2x \cdot x + 2x \cdot (-3) + 5 \cdot x + 5 \cdot (-3)$
$= 2x^2 - 6x + 5x - 15$
$= 2x^2 - x - 15$

6.1 Quick Checks

1. The largest expression that divides evenly into a set of numbers is called the greatest common factor.

2. In the product $(3x-2)(x+4) = 3x^2 + 10x - 8$, the polynomials $(3x - 2)$ and $(x + 4)$ are called factors of the polynomial $3x^2 + 10x - 8$.

3. $32 = 8 \cdot 4 = 2 \cdot 2 \cdot 2 \cdot 2 \cdot 2$
$40 = 8 \cdot 5 = 2 \cdot 2 \cdot 2 \quad \cdot 5$
$\text{GCF} = 2 \cdot 2 \cdot 2 = 2^3 = 8$

4. $12 = 4 \cdot 3 = 2 \cdot 2 \cdot 3$
$45 = 3 \cdot 15 = \quad 3 \cdot 3 \cdot 5$
$\text{GCF} = 3$

5. $21 = 3 \cdot 7$
$35 = 5 \cdot 7$
$84 = 21 \cdot 4 = 3 \cdot 7 \cdot 2 \cdot 2$
$\text{GCF} = 7$

6. $14 = 2 \cdot 7$
$35 = 5 \cdot 7$
The GCF of the coefficients is 7.
The GCF of y^3 and y^2 is y^2.
The GCF of $14y^3$ and $35y^2$ is $7y^2$.

7. $6 = 2 \cdot 3$
$8 = 2 \cdot 2 \cdot 2$
$12 = 2 \cdot 2 \cdot 3$
. The GCF of the coefficients is 2.
The GCF of z^3, z^2, and z is z.
The GCF of $6z^3$, $8z^2$, and $12z$ is $2z$.

8. $4 = 2 \cdot 2$
$8 = 4 \cdot 2 = 2 \cdot 2 \cdot 2$
$24 = 4 \cdot 6 = 2 \cdot 2 \cdot 2 \cdot 3$
The GCF of the coefficients is $2 \cdot 2 = 2^2 = 4$.

The GCF of x^3, x^2, and x is x.
The GCF of y^5, y^3, and y^4 is y^3.
The GCF of $4x^3y^5$, $8x^2y^3$, and $24xy^4$ is $4xy^3$.

9. There is no common factor between the coefficients. Since each expression has $2x + 3$ as a factor, the GCF of $7(2x + 3)$ and $-4(2x + 3)$ is $2x + 3$.

10. The GCF between 9 and 12 is 3. The GCF between $(k + 8)(3k - 2)$ and $(k-1)(k+8)^2$ is $k + 8$. The GCF of the expressions $9(k + 8)(3k - 2)$ and $12(k-1)(k+8)^2$ is $3(k + 8)$.

11. To factor a polynomial means to write the polynomial as a product of two or more polynomials.

12. When we factor a polynomial using the GCF, we use the Distributive Property in reverse.

13. The GCF of 5 and 30 is 5.
The GCF of z^2 and z is z.
The GCF of $5z^2 + 30z$ is $5z$.
$5z^2 + 30z = 5z(z) + 5z(6) = 5z(z+6)$

14. The GCF of 12 and 12 is 12.
The GCF of p^2 and p is p.
The GCF of $12p^2 - 12p$ is $12p$.
$12p^2 - 12p = 12p(p) - 12p(1) = 12p(p-1)$

15. The GCF of 16, 12, and 4 is 4.
The GCF of y^3, y^2, and y is y.
The GCF of $16y^3 - 12y^2 + 4y$ is $4y$.
$16y^3 - 12y^2 + 4y$
$= 4y(4y^2) - 4y(3y) + 4y(1)$
$= 4y(4y^2 - 3y + 1)$

16. The GCF of 6, 18, and 22 is 2.
The GCF of m^4, m^3, and m^2 is m^2.
The GCF of n^2, n^4, and n^5 is n^2.
The GCF of $6m^4n^2 + 18m^3n^4 - 22m^2n^5$ is $2m^2n^2$.
$6m^4n^2 + 18m^3n^4 - 22m^2n^5$
$= 2m^2n^2(3m^2) + 2m^2n^2(9mn^2) - 2m^2n^2(11n^3)$
$= 2m^2n^2(3m^2 + 9mn^2 - 11n^3)$

17. Use $-4y$ as the GCF.
$-4y^2 + 8y = -4y(y) + (-4y)(-2) = -4y(y-2)$

18. Use $-3a$ as the GCF.
$-6a^3 + 12a^2 - 3a$
$= -3a(2a^2) + (-3a)(-4a) + (-3a)(1)$
$= -3a(2a^2 - 4a + 1)$

19. The GCF is $(2x + 1)$.
$(2x+1)(x-3) + (2x+1)(2x+7)$
$= (2x+1)(x-3+2x+7)$
$= (2x+1)(3x+4)$
$(2x+1)(3x+4) \neq (2x+1)^2(3x+4)$
False

20. The GCF is the binomial $a - 5$.
$2a(a-5) + 3(a-5) = (a-5)(2a+3)$

21. The GCF is the binomial $z + 5$.
$7z(z+5) - 4(z+5) = (z+5)(7z-4)$

22. $4x + 4y + bx + by = (4x + 4y) + (bx + by)$
$= 4(x + y) + b(x + y)$
$= (x + y)(4 + b)$

23. $6az - 2a - 9bz + 3b = (6az - 2a) + (-9bz + 3b)$
$= 2a(3z-1) + (-3b)(3z-1)$
$= (3z-1)(2a-3b)$

24. $8b + 4 - 10ab - 5a = (8b+4) + (-10ab - 5a)$
$= 4(2b+1) + (-5a)(2b+1)$
$= (2b+1)(4-5a)$

25. The GCF of $3z^3 + 12z^2 + 6z + 24$ is 3.
$3z^3 + 12z^2 + 6z + 24 = 3(z^3 + 4z^2 + 2z + 8)$
$= 3[(z^3 + 4z^2) + (2z+8)]$
$= 3[z^2(z+4) + 2(z+4)]$
$= 3(z+4)(z^2+2)$

26. The GCF of $2n^4 + 2n^3 - 4n^2 - 4n$ is $2n$.
$2n^4 + 2n^3 - 4n^2 - 4n$
$= 2n(n^3 + n^2 - 2n - 2)$
$= 2n[(n^3 + n^2) + (-2n - 2)]$
$= 2n[n^2(n+1) + (-2)(n+1)]$
$= 2n(n+1)(n^2 - 2)$

6.1 Exercises

27. $8 = 2 \cdot 2 \cdot 2$
$6 = 2 \cdot \quad 3$
The only common factor is 2. The GCF is 2.

29. $15 = 3 \cdot 5$
$14 = \quad 2 \cdot 7$
There are no common prime factors. The GCF is 1.

31. $12 = 2 \cdot 2 \cdot 3$
$28 = 2 \cdot 2 \cdot \quad 7$
$48 = 2 \cdot 2 \cdot 3 \cdot \quad 2 \cdot 2$
Because all three numbers contain two factors of 2, the GCF is $2 \cdot 2 = 4$.

33. $x^{10} = x \cdot x \cdot x \cdot x \cdot x \cdot x \cdot x \cdot x \cdot x \cdot x$
$x^2 = x \cdot x$
$x^8 = x \cdot x \cdot x \cdot x \cdot x \cdot x \cdot x \cdot x$
Each expression contains two factors of x, so the GCF is x^2.

35. The GCF of 7 and 14 is 7.
The GCF of x and x^3 is x.
The GCF is $7x$.

37. The GCF of 45 and 75 is 15.
The GCF of a^2 and a is a.
The GCF of b^3 and b^2 is b^2.
The GCF is $15ab^2$.

39. The GCF of 4, 6, and 8 is 2.
The GCF of a^2, a, and a^2 is a.
The GCF of b, b^2, and b^2 is b.
The GCF of c^3, c^2, and c^4 is c^2.
The GCF is $2abc^2$.

41. $3(x-1) = \quad 3 \cdot (x-1)$
$6(x+1) = 2 \cdot 3 \cdot (x+1)$
The GCF is 3.

43. $2(x-4)^2 = \quad 2 \cdot (x-4)^2$
$4(x-4)^3 = 2 \cdot 2 \cdot (x-4)^3$
The GCF of the variable expression is $(x-4)^2$. The GCF is $2(x-4)^2$.

45. $12(x+2)(x-3)^2 = 2 \cdot 2 \cdot 3 \cdot \quad (x+2)(x-3)^2$
$18(x-3)^2(x-2) = 2 \cdot \quad 3 \cdot 3 \cdot \quad (x-3)^2(x-2)$
The GCF of the variable expression is $(x-3)^2$. The GCF is $2 \cdot 3 \cdot (x-3)^2 = 6(x-3)^2$.

47. The GCF is 6.
$12x - 18 = 6(2x) - 6(3) = 6(2x - 3)$

49. The GCF is $-x$.
$-3x^2 + 12x = -3x(x) + (-3x)(-4) = -3x(x-4)$

51. The GCF is $5x^2y$.
$5x^2y - 15x^3y^2 = 5x^2y(1) - 5x^2y(3xy) = 5x^2y(1-3xy)$

53. The GCF is $3x$.
$3x^3 + 6x^2 - 3x = 3x(x^2) + 3x(2x) - 3x(1) = 3x(x^2 + 2x - 1)$

55. The GCF is $-5x$.
$-5x^3 + 10x^2 - 15x$
$= -5x(x^2) + (-5x)(-2x) + (-5x)(3)$
$= -5x(x^2 - 2x + 3)$

57. The GCF is 3.
$9m^5 - 18m^3 - 12m^2 + 81 = 3(3m^5) - 3(6m^3) - 3(4m^2) + 3(27) = 3(3m^5 - 6m^3 - 4m^2 + 27)$

59. The GCF is $-4z$.
$-12z^3 + 16z^2 - 8z$
$= -4z(3z^2) + (-4z)(-4z) + (-4z)(2)$
$= -4z(3z^2 - 4z + 2)$

61. The GCF is -5.
$10 - 5b - 15b^3 = -15b^3 - 5b + 10$
$= -5(3b^3) + (-5)b + (-5)(-2)$
$= -5(3b^3 + b - 2)$

63. The GCF is $15ab^2$.
$15a^2b^4 - 60ab^3 + 45a^3b^2$
$= 15ab^2(ab^2) - 15ab^2(4b) + 15ab^2(3a^2)$
$= 15ab^2(ab^2 - 4b + 3a^2)$

65. The GCF is $x - 3$.
$(x - 3)x - (x - 3)5 = (x - 3)(x - 5)$

67. The GCF is $x - 1$.
$x^2(x-1) + y^2(x-1) = (x-1)(x^2 + y^2)$

69. The GCF is $4x + 1$.
$x^2(4x+1) + 2x(4x+1) + 5(4x+1)$
$= (4x+1)(x^2 + 2x + 5)$

71. $xy + 3y + 4x + 12 = (xy + 3y) + (4x + 12)$
$= y(x+3) + 4(x+3)$
$= (x+3)(y+4)$

73. $yz + z - y - 1 = (yz + z) + (-y - 1)$
$= z(y+1) + (-1)(y+1)$
$= (y+1)(z-1)$

75. $x^3 - x^2 + 2x - 2 = (x^3 - x^2) + (2x - 2)$
$= x^2(x-1) + 2(x-1)$
$= (x-1)(x^2 + 2)$

77. $2t^3 - t^2 - 4t + 2 = (2t^3 - t^2) + (-4t + 2)$
$= t^2(2t-1) + (-2)(2t-1)$
$= (2t-1)(t^2 - 2)$

79. $2t^4 - t^3 - 6t + 3 = (2t^4 - t^3) + (-6t + 3)$
$= t^3(2t-1) + (-3)(2t-1)$
$= (2t-1)(t^3 - 3)$

81. $4y - 20 = 4(y) - 4(5) = 4(y - 5)$

83. $28m^3 + 7m^2 + 63m$
$= 7m(4m^2) + 7m(m) + 7m(9)$
$= 7m(4m^2 + m + 9)$

85. $12m^3n^2p - 18m^2n$
$= 6m^2n(2mnp) - 6m^2n(3)$
$= 6m^2n(2mnp - 3)$

87. $(2p-1)(p+3) + (7p+4)(p+3)$
$= (p+3)[(2p-1) + (7p+4)]$
$= (p+3)(9p+3)$
$= (p+3)[3(3p+1)]$
$= 3(p+3)(3p+1)$

89. $18ax - 9ay - 12bx + 6by$
$= 3(6ax - 3ay - 4bx + 2by)$
$= 3[(6ax - 3ay) + (-4bx + 2by)]$
$= 3[3a(2x - y) + (-2b)(2x - y)]$
$= 3(2x - y)(3a - 2b)$

91. $(x-2)(x-3) + (x-2)$
$= (x-2)[(x-3) + 1]$
$= (x-2)(x-2)$ or $(x-2)^2$

93. $15x^4 - 6x^3 + 30x^2 - 12x$
$= 3x(5x^3 - 2x^2 + 10x - 4)$
$= 3x[(5x^3 - 2x^2) + (10x - 4)]$
$= 3x[x^2(5x-2) + 2(5x-2)]$
$= 3x(5x-2)(x^2 + 2)$

95. $-3x^3 + 6x^2 - 9x$
$= -3x(x^2) + (-3x)(-2x) + (-3x)(3)$
$= -3x(x^2 - 2x + 3)$

97. $-12b + 16b^2 = 16b^2 - 12b$
$= 4b(4b) - 4b(3)$
$= 4b(4b - 3)$

99. $12xy + 9x - 8y - 6$
$= (12xy + 9x) + (-8y - 6)$
$= 3x(4y+3) + (-2)(4y+3)$
$= (4y+3)(3x-2)$

101. $\frac{1}{3}x^3 - \frac{2}{9}x^2 = \frac{1}{3}x^2(x) + \frac{1}{3}x^2\left(-\frac{2}{3}\right)$
$= \frac{1}{3}x^2\left(x - \frac{2}{3}\right)$

103. $-16t^2 + 150t = -2t(8t) + (-2t)(-75)$
$= -2t(8t - 75)$

105. $21{,}000 - 150p = 150(140) - 150(p)$
$= 150(140 - p)$

107. $8x^5 - 28x^3 = 4x^3(2x^2) - 4x^3(7)$
$= 4x^3(2x^2 - 7)$

109. $S = 2\pi r^2 + 8\pi r$
$= 2\pi r(r) + 2\pi r(4)$
$= 2\pi r(r + 4)$

111. $4x^2 - 8x = 4x(x) - 4x(2) = 4x(x - 2)$
The missing length is $x - 2$.

113. $8x^{3n} + 10x^n = 2x^n(4x^{2n}) + 2x^n(5)$
$= 2x^n(4x^{2n} + 5)$
The missing factor is $4x^{2n} + 5$.

115. $3 - 4x^{-1} + 2x^{-3}$
$= x^{-3}(3x^3) - x^{-3}(4x^2) + x^{-3}(2)$
$= x^{-3}(3x^3 - 4x^2 + 2)$
The missing factor is $3x^3 - 4x^2 + 2$.

117. $x^2 - 3x^{-1} + 2x^{-2}$
$= x^{-2}(x^4) - x^{-2}(3x) + x^{-2}(2)$
$= x^{-2}(x^4 - 3x + 2)$
The missing factor is $x^4 - 3x + 2$.

119. $\frac{6}{35}x^4 - \frac{1}{7}x^2 + \frac{2}{7}x$
$= \frac{2}{7}x\left(\frac{3}{5}x^3\right) - \frac{2}{7}x\left(\frac{1}{2}x\right) + \frac{2}{7}x(1)$
$= \frac{2}{7}x\left(\frac{3}{5}x^3 - \frac{1}{2}x + 1\right)$
The missing factor is $\frac{3}{5}x^3 - \frac{1}{2}x + 1$.

121. To find the greatest common factor, (1) determine the GCF of the coefficients of the variable expressions (the largest number that divides evenly into a set of numbers). (2) For each variable expression common to all the terms, determine the smallest exponent that the variable expression is raised to. (3) Find the product of these common factors. This is the GCF.
To factor the GCF from a polynomial, (1) identify the GCF of the terms that make up the polynomial. (2) Rewrite each term as the product of the GCF and the remaining factor. (3) Use the Distributive Property "in reverse" to factor out the GCF. (4) Use the Distributive Property to verify that the factorization is correct.

123. $3a(x+y) - 4b(x+y) \neq (3a-4b)(x+y)^2$ because when checking, the product $(3a-4b)(x+y)^2$ is equal to
$(3a-4b)(x+y)^2 = (3a-4b)(x^2+2xy+y^2)$, not
$3a(x + y) - 4b(x + y)$
$= 3ax + 3ay - 4bx - 4by$.

Section 6.2

Preparing for Factoring Trinomials of the Form $x^2 + bx + c$

P1. $2 \cdot 9 = 18$ and $2 + 9 = 11$, so the factors are 2 and 9.

P2. $4 \cdot (-6) = -24$ and $4 + (-6) = -2$, so the factors are 4 and -6.

P3. $4 \cdot (-3) = -12$ and $4 + (-3) = 1$, so the factors are 4 and -3.

P4. $(-7) \cdot (-5) = 35$ and $(-7) + (-5) = -12$, so the factors are -7 and -5.

P5. The coefficients of $3x^2 - x - 4 = 3x^2 + (-1)x + (-4)$ are 3, -1, and -4.

P6. $-5p(p+4) = -5p(p) + (-5p)(4)$
$= -5p^2 + (-20p)$
$= -5p^2 - 20p$

P7. $(z-1)(z+4) = z(z)+z(4)+(-1)(z)+(-1)(4)$
$= z^2 + 4z - z - 4$
$= z^2 + 3z - 4$

6.2 Quick Checks

1. A quadratic trinomial is a polynomial of the form $ax^2 + bx^2 + c,\ a \neq 0.$

2. When factoring $x^2 - 10x + 24,$ the signs of the factors of 24 must both be negative.

3. False; $4 + 4x + x^2 = x^2 + 4x + 4$
The leading coefficient is 1 not 4.

4. In $y^2 + 9y + 20,\ b = 9$ and $c = 20$.

Integers Whose Product is 20	1, 20	2, 10	4, 5
Sum	21	12	9

$4 \cdot 5 = 20 = c$ and $4 + 5 = 9 = b$ so $y^2 + 9y + 20 = (y+4)(y+5)$.
Check: $(y+4)(y+5) = y^2 + 5y + 4y + 4(5)$
$= y^2 + 9y + 20$

5. In $z^2 - 9z + 14,\ b = -9$ and $c = 14$.

Integers Whose Product is 14	−1, −14	−2, −7
Sum	−15	−9

$-2 \cdot (-7) = 14 = c$ and $-2 + (-7) = -9 = b$ so
$z^2 - 9z + 14 = (z+(-2))(z+(-7))$
$= (z-2)(z-7)$
Check: $(z-2)(z-7) = z^2 - 7z - 2z + (-2)(-7)$
$= z^2 - 9z + 14$

6. True

7. In $y^2 - 2y - 15,\ b = -2$ and $c = -15$.

Integers Whose Product is −15	−1, 15	−3, 5	1, −15	3, −5
Sum	14	2	−14	−2

$3 \cdot (-5) = -15 = c$ and $3 + (-5) = -2 = b$, so

$$y^2 - 2y - 15 = (y+3)(y+(-5))$$
$$= (y+3)(y-5)$$

Check: $(y+3)(y-5) = y^2 - 5y + 3y + 3(-5)$
$$= y^2 - 2y - 15$$

8. In $w^2 + w - 12$, $b = 1$ and $c = -12$.

Integers Whose Product is –12	–1, 12	–2, 6	–3, 4	1, –12	2, –6	3, –4
Sum	11	4	1	–11	–4	–1

$-3 \cdot 4 = -12 = c$ and $-3 + 4 = 1 = b$, so
$w^2 + w - 12 = (w+(-3))(w+4) = (w-3)(w+4)$

Check: $(w-3)(w+4) = w^2 + 4w - 3w + (-3)(4) = w^2 + w - 12$

9. In $z^2 - 5z + 8$, $b = -5$ and $c = 8$.

Integers Whose Product is 8	–1, –8	–2, –4
Sum	–9	–6

There are no factors of 8 whose sum is –5. Therefore $z^2 - 5z + 8$ is prime.

10. In $q^2 + 4q - 45$, $b = 4$ and $c = -45$.

Integers Whose Product is –45	–1, 45	–3, 15	–5, 9
Sum	44	12	4

$-5 \cdot 9 = -45 = c$ and $-5 + 9 = 4 = b$, so
$$q^2 + 4q - 45 = (q+(-5))(q+9)$$
$$= (q-5)(q+9)$$

Check: $(q-5)(q+9) = q^2 + 9q - 5q + (-5)(9)$
$$= q^2 + 4q - 45$$

11. In $x^2 + 9xy + 20y^2$, $b = 9$ and $c = 20$.

Integers Whose Product is 20	1, 20	2, 10	4, 5
Sum	21	12	9

$4 \cdot 5 = 20 = c$ and $4 + 5 = 9 = b$, so $x^2 + 9xy + 20y^2 = (x+4y)(x+5y)$

Check: $(x+4y)(x+5y) = x^2 + 5xy + 4xy + 20y^2$
$$= x^2 + 9xy + 20y^2$$

12. In $m^2 + mn - 42n^2$, $b = 1$ and $c = -42$.

Integers Whose Product is –42	–1, 42	–2, 21	–3, 14	–6, 7
Sum	41	19	11	1

$-6 \cdot 7 = -42 = c$ and $-6 + 7 = 1 = b$, so

$$m^2 + mn - 42n^2 = (m + (-6)n)(m + 7n) = (m - 6n)(m + 7n)$$

Check:

$$(m - 6n)(m + 7n) = m^2 + 7mn - 6mn - 42n^2 = m^2 + mn - 42n^2$$

13. **(a)** $-12 - x + x^2 = x^2 - x - 12$

(b) $9n + n^2 - 10 = n^2 + 9n - 10$

14. In $-56 + n^2 + n = n^2 + n - 56$, $b = 1$ and $c = -56$.

Integers Whose Product is –56	–1, 56	–2, 28	–4, 14	–7, 8
Sum	55	26	10	1

$-7 \cdot 8 = -56 = c$ and $-7 + 8 = 1 = b$, so

$$n^2 + n - 56 = (n - 7)(n + 8)$$

Check: $(n - 7)(n + 8) = n^2 + 8n - 7n + (-7)(8) = n^2 + n - 56$

15. In $y^2 + 35 - 12y = y^2 - 12y + 35$, $b = -12$ and $c = 35$.

Integers Whose Product is 35	–1, –35	–5, –7
Sum	–36	–12

$-5 \cdot (-7) = 35 = c$ and $-5 + (-7) = -12 = b$, so

$$y^2 - 12y + 35 = (y - 5)(y - 7)$$

Check: $(y - 5)(y - 7) = y^2 - 7y - 5y + (-5)(-7) = y^2 - 12y + 35$

16. $(x - 3)(3x + 6) = (x - 3)[3(x + 2)] = 3(x - 3)(x + 2)$

The polynomial was not factored completely. False

17. The GCF of $4m^2 - 16m - 84$ is 4.

$$4m^2 - 16m - 84 = 4(m^2 - 4m - 21)$$

In $m^2 - 4m - 21$, $b = -4$ and $c = -21$.

Integers Whose Product is –21	1, –21	3, –7
Sum	–20	–4

$3 \cdot (-7) = -21 = c$ and $3 + (-7) = -4 = b$, so

$$m^2 - 4m - 21 = (m + 3)(m - 7)$$

$$4m^2 - 16m - 84 = 4(m^2 - 4m - 21) = 4(m + 3)(m - 7)$$

Check: $4(m + 3)(m - 7) = 4(m^2 - 7m + 3m - 21) = 4(m^2 - 4m - 21) = 4m^2 - 16m - 84$

18. The GCF of $3z^3 + 12z^2 - 15z$ is $3z$.

$$3z^3 + 12z^2 - 15z = 3z(z^2 + 4z - 5)$$

In $z^2 + 4z - 5$, $b = 4$ and $c = -5$.

Integers Whose Product is –5	–1, 5	1, –5
Sum	4	–4

$-1 \cdot 5 = -5 = c$ and $-1 + 5 = 4 = b$, so

$$z^2 + 4z - 5 = (z - 1)(z + 5)$$

$$3z^3 + 12z^2 - 15z = 3z(z^2 + 4z - 5) = 3z(z - 1)(z + 5)$$

Check: $3z(z - 1)(z + 5) = 3z(z^2 + 5z - z - 5) = 3z(z^2 + 4z - 5) = 3z^3 + 12z^2 - 15z$

19. $-w^2-3w+10=-1(w^2+3w-10)$

In $w^2+3w-10$, $b=3$ and $c=-10$.

Integers Whose Product is –10	–1, 10	–2, 5
Sum	9	3

$-2\cdot 5=-10=c$ and $-2+5=3=b$, so

$w^2+3w-10=(w-2)(w+5)$.

$$\begin{aligned}-w^2-3w+10&=-1(w^2+3w-10)\\&=-1(w-2)(w+5)\\&=-(w-2)(w+5)\end{aligned}$$

20. The GCF of $-2a^2-8a+24$ is -2.

$-2a^2-8a+24=-2(a^2+4a-12)$

In $a^2+4a-24$, $b=4$ and $c=-12$.

Integers Whose Product is –12	–1, 12	–2, 6	–3, 4
Sum	11	4	1

$-2\cdot 6=-12=c$ and $-2+6=4=b$, so

$a^2+4a-12=(a-2)(a+6)$.

$$\begin{aligned}-2a^2-8a+24&=-2(a^2+4a-12)\\&=-2(a-2)(a+6)\end{aligned}$$

6.2 Exercises

21. $2\cdot 3=6$ and $2+3=5$

$x^2+5x+6=(x+2)(x+3)$

23. $3\cdot 6=18$ and $3+6=9$

$m^2+9m+18=(m+3)(m+6)$

25. $-3\cdot(-12)=36$ and $-3+(-12)=-15$

$x^2-15x+36=(x-3)(x-12)$

27. $-2\cdot(-6)=12$ and $-2+(-6)=-8$

$p^2-8p+12=(p-2)(p-6)$

29. $-4\cdot 3=-12$ and $-4+3=-1$

$x^2-x-12=(x-4)(x+3)$

31. $-4\cdot 5=-20$ and $-4+5=1$

$x^2+x-20=(x-4)(x+5)$

33. $-3\cdot 15=-45$ and $-3+15=12$

$z^2+12z-45=(z-3)(z+15)$

35. $-2\cdot(-3)=6$ and $-2+(-3)=-5$

$x^2-5xy+6y^2=(x-2y)(x-3y)$

37. $-2\cdot 3=-6$ and $-2+3=1$

$r^2+rs-6s^2=(r-2s)(r+3s)$

39. $-4\cdot 1=-4$ and $-4+1=-3$

$x^2-3xy-4y^2=(x-4y)(x+y)$

41. There are no factors of 1 whose sum is 7.

$z^2+7zy+y^2$ is prime.

43. $3x^2+3x-6=3(x^2+x-2)=3(x+2)(x-1)$

45. $$\begin{aligned}3n^3-24n^2+45n&=3n(n^2-8n+15)\\&=3n(n-3)(n-5)\end{aligned}$$

47. $$\begin{aligned}5x^2z-20xz-160z&=5z(x^2-4x-32)\\&=5z(x-8)(x+4)\end{aligned}$$

49. $$\begin{aligned}-3x^2+x^3-18x&=x^3-3x^2-18x\\&=x(x^2-3x-18)\\&=x(x-6)(x+3)\end{aligned}$$

51. $$\begin{aligned}&-2y^2+8y-8\\&=-2(y^2-4y+4)\\&=-2(y-2)(y-2)\text{ or }-2(y-2)^2\end{aligned}$$

53. $$\begin{aligned}4x^3-32x^2+x^4&=x^4+4x^3-32x^2\\&=x^2(x^2+4x-32)\\&=x^2(x+8)(x-4)\end{aligned}$$

55. $$\begin{aligned}2x^2+x^3-15x&=x^3+2x^2-15x\\&=x(x^2+2x-15)\\&=x(x+5)(x-3)\end{aligned}$$

57. $-7\cdot 4=-28$ and $-7+4=-3$

$x^2-3xy-28y^2=(x+4y)(x-7y)$

59. There are no factors of 6 whose product is 1.
x^2+x+6 is prime.

61. $-5 \cdot 4 = -20$ and $-5 + 4 = -1$
$k^2-k-20=(k-5)(k+4)$

63. $-5 \cdot 6 = -30$ and $-5 + 6 = 1$
$x^2+xy-30y^2=(x-5y)(x+6y)$

65. $-3 \cdot (-5) = 15$ and $-3 + (-5) = -8$
$s^2t^2-8st+15=(st-3)(st-5)$

67. $$\begin{aligned}-3p^3+3p^2+6p&=-3p(p^2-p-2)\\&=-3p(p-2)(p+1)\end{aligned}$$

69. $-x^2-x+6=-(x^2+x-6)=-(x+3)(x-2)$

71. $g^2-4g+21$ cannot be factored; it is prime.

73. $$\begin{aligned}n^4-30n^2-n^3&=n^4-n^3-30n^2\\&=n^2(n^2-n-30)\\&=n^2(n-6)(n+5)\end{aligned}$$

75. $m^2+2mn-15n^2=(m+5n)(m-3n)$

77. $35+12s+s^2=s^2+12s+35=(s+5)(s+7)$

79. $n^2-9n-45$ cannot be factored; it is prime.

81. $m^2n^2-8mn+12=(mn-6)(mn-2)$

83. $$\begin{aligned}-x^3+12x^2+28x&=-x(x^2-12x-28)\\&=-x(x-14)(x+2)\end{aligned}$$

85. $y^2-12y+36=(y-6)(y-6)$ or $(y-6)^2$

87. $$\begin{aligned}-36x+20x^2-2x^3&=-2x^3+20x^2-36x\\&=-2x(x^2-10x+18)\end{aligned}$$

89. $-21x^3y-14xy^2=-7xy(3x^2+2y)$

91. $$\begin{aligned}-16t^2+64t+80&=-16(t^2-4t-5)\\&=-16(t+1)(t-5)\end{aligned}$$

93. $x^2+9x+18=(x+6)(x+3)$
The binomials $x + 6$ and $x + 3$ represent the length and the width.

95. $$\begin{aligned}\frac{1}{2}x^2+x-\frac{15}{2}&=\frac{1}{2}(x^2+2x-15)\\&=\frac{1}{2}(x+5)(x-3)\end{aligned}$$
The binomials $x + 5$ and $x - 3$ represent the base and height.

97. If $b > 0$, then the constant 6 factors as (1)(6) or (2)(3). Thus, the coefficient of x could be $1 + 6 = 7$ or $2 + 3 = 5$. If $b < 0$, then the constant 6 factors as $(-1)(-6)$ or $(-2)(-3)$. Thus, the coefficient of x could be $-1 + (-6) = -7$ or $-2 + (-3) = -5$. So b could be -7, -5, 5, or 7.

99. If $b > 0$, then the constant -21 factors as $(-1)(21)$ or $(-3)(7)$. Thus, the coefficient of x could be $(-1) + 21 = 20$ or $(-3) + 7 = 4$. If $b < 0$, then the constant 21 factors as $(1)(-21)$ or $(3)(-7)$. Thus the coefficient of x could be $1 + (-21) = -20$ or $3 + (-7) = -4$. So, b could be -20, -4, 4, or 20.

101. Since $c > 0$, then c must have two negative factors that sum to -2.
$-1, -1$: $c = (-1)(-1) = 1$
Thus, c must be 1.

103. Since $c > 0$, then c must have two negative factors that sum to -7.
$-1, -6$: $c = (-1)(-6) = 6$
$-2, -5$: $c = (-2)(-5) = 10$
$-3, -4$: $c = (-3)(-4) = 12$
Thus, c could be 6, 10, or 12.

105. Your answer is marked correctly.
The product
$(1-x)(2-x)=2-x-2x+x^2=2-3x+x^2$.
The product
$(x-1)(x-2)=x^2-2x-x+2=x^2-3x+2$

Section 6.3

Preparing for Factoring Trinomials of the Form $ax^2 + bx + c, a \neq 1$

P1. $24=2\cdot 12=2\cdot 2\cdot 6=2\cdot 2\cdot 2\cdot 3=2^3\cdot 3$

P2. The coefficients of $5x^2 - 3x + 7 = 5x^2 + (-3)x + 7$ are 5, −3, and 7.

P3. $(2x+7)(3x-1)$
$= 2x \cdot 3x + 2x \cdot (-1) + 7 \cdot 3x + 7 \cdot (-1)$
$= 6x^2 - 2x + 21x - 7$
$= 6x^2 + 19x - 7$

6.3 Quick Checks

1. True

2. $3x^2 = 3x \cdot x$
$2 = 2 \cdot 1$ or $-2 \cdot (-1)$
Since the coefficient of *x*, 5, is positive, we do not use the factorization $-2 \cdot (-1)$.
$(3x+2)(x+1) = 3x^2 + 3x + 2x + 2$
$= 3x^2 + 5x + 2$
$(3x+1)(x+2) = 3x^2 + 6x + x + 2$
$= 3x^2 + 7x + 2$
$3x^2 + 5x + 2 = (3x+2)(x+1)$

3. $7y^2 = 7y \cdot y$
$3 = 3 \cdot 1$ or $-3 \cdot (-1)$
Since the coefficient of *y*, 22, is positive, we do not use the factorization $-3 \cdot (-1)$.
$(7y+3)(y+1) = 7y^2 + 7y + 3y + 3$
$= 7y^2 + 10y + 3$
$(7y+1)(y+3) = 7y^2 + 21y + y + 3$
$= 7y^2 + 22y + 3$
$7y^2 + 22y + 3 = (7y+1)(y+3)$

4. $3x^2 = 3x \cdot x$
$12 = 1 \cdot 12, 2 \cdot 6, 3 \cdot 4, -1 \cdot (-12), -2 \cdot (-6)$, or $-3 \cdot (-4)$
Since the coefficient of *x*, −13, is negative, we do not use the factorizations of 12 into positive factors.
$(3x-1)(x-12) = 3x^2 - 36x - x + 12$
$= 3x^2 - 37x + 12$
$(3x-12)(x-1) = 3x^2 - 3x - 12x + 12$
$= 3x^2 - 15x + 12$
$(3x-2)(x-6) = 3x^2 - 18x - 2x + 12$
$= 3x^2 - 20x + 12$
$(3x-6)(x-2) = 3x^2 - 6x - 6x + 12$
$= 3x^2 - 12x + 12$
$(3x-3)(x-4) = 3x^2 - 12x - 3x + 12$
$= 3x^2 - 15x + 12$
$(3x-4)(x-3) = 3x^2 - 9x - 4x + 12$
$= 3x^2 - 13x + 12$
$3x^2 - 13x + 12 = (3x-4)(x-3)$

5. $5p^2 = 5p \cdot p$
$4 = 1 \cdot 4, 2 \cdot 2, -1 \cdot (-4)$, or $-2 \cdot (-2)$
Since the coefficient of *p*, −21, is negative, we do not use the factorizations of 4 into positive factors.
$(5p-1)(p-4) = 5p^2 - 20p - p + 4$
$= 5p^2 - 21p + 4$
$(5p-4)(p-1) = 5p^2 - 5p - 4p + 4$
$= 5p^2 - 9p + 4$
$(5p-2)(p-2) = 5p^2 - 10p - 2p + 4$
$= 5p^2 - 12p + 4$
$5p^2 - 21p + 4 = (5p-1)(p-4)$

6. $2n^2 = 2n \cdot n$
$-9 = -1 \cdot 9, -3 \cdot 3$, or $1 \cdot (-9)$
$(2n-1)(n+9) = 2n^2 + 18n - n - 9$
$= 2n^2 + 17n - 9$
$(2n+9)(n-1) = 2n^2 - 2n + 9n - 9$
$= 2n^2 + 7n - 9$
$(2n-3)(n+3) = 2n^2 + 6n - 3n - 9$
$= 2n^2 + 3n - 9$
$(2n+3)(n-3) = 2n^2 - 6n + 3n - 9$
$= 2n^2 - 3n - 9$
$(2n+1)(n-9) = 2n^2 - 18n + n - 9$
$= 2n^2 - 17n - 9$
$(2n-9)(n+1) = 2n^2 + 2n - 9n - 9$
$= 2n^2 - 7n - 9$
$2n^2 - 17n - 9 = (2n+1)(n-9)$

7. $4w^2 = 4w \cdot w$ or $2w \cdot 2w$
$-6 = -1 \cdot 6, -2 \cdot 3, 1 \cdot (-6)$, or $2 \cdot (-3)$
$(4w-1)(w+6) = 4w^2 + 24w - w - 6$
$= 4w^2 + 23w - 6$
$(4w+6)(w-1) = 4w^2 - 4w + 6w - 6$
$= 4w^2 + 2w - 6$
$(4w-2)(w+3) = 4w^2 + 12w - 2w - 6$
$= 4w^2 + 10w - 6$
$(4w+3)(w-2) = 4w^2 - 8w + 3w - 6$
$= 4w^2 - 5w - 6$
$(4w+1)(w-6) = 4w^2 - 24w + w - 6$
$= 4w^2 - 23w - 6$
$(4w-6)(w+1) = 4w^2 + 4w - 6w - 6$
$= 4w^2 - 2w - 6$
$(4w+2)(w-3) = 4w^2 - 12w + 2w - 6$
$= 4w^2 - 10w - 6$
$(4w-3)(w+2) = 4w^2 + 8w - 3w - 6$
$= 4w^2 + 5w - 6$
$(2w-1)(2w+6) = 4w^2 + 12w - 2w - 6$
$= 4w^2 + 10w - 6$
$(2w-2)(2w+3) = 4w^2 + 6w - 4w - 6$
$= 4w^2 + 2w - 6$
$(2w+1)(2w-6) = 4w^2 - 12w + 2w - 6$
$= 4w^2 - 10w - 6$
$(2w+2)(2w-3) = 4w^2 - 6w + 4w - 6$
$= 4w^2 - 2w - 6$
$4w^2 - 5w - 6 = (4w+3)(w-2)$

8. When factoring $ax^2 + bx + c$ and b is a small number, choose factors of a and factors of c that are close to each other.

9. $12x^2 = 12x \cdot x$, $6x \cdot 2x$, or $4x \cdot 3x$
$6 = 6 \cdot 1, 3 \cdot 2, -6 \cdot (-1)$, or $-3 \cdot (-2)$
Since the coefficient of x, 17, is positive, we do not use the factorizations of 6 into negative factors. Since 17 is neither very large nor very small, try factors of 12 and 6 that are similar in size: $12x^2 = 4x \cdot 3x$ and $6 = 3 \cdot 2$.
$(4x+3)(3x+2) = 12x^2 + 8x + 9x + 6$
$= 12x^2 + 17x + 6$
$12x^2 + 17x + 6 = (4x+3)(3x+2)$

10. $12y^2 = 12y \cdot y$, $6y \cdot 2y$, or $4y \cdot 3y$
$-35 = -1 \cdot 35, -5 \cdot 7, 1 \cdot (-35)$, or $5 \cdot (-7)$
Since the coefficient of y, 32, is closer to $6 \cdot 5 = 30$ and $6 \cdot 7 = 42$ than to $12 \cdot 35 = 420$ or $12 \cdot 1 = 12$, try $12y^2 = 6y \cdot 2y$ and $-35 = -5 \cdot 7$.
$(6y-5)(2y+7) = 12y^2 + 42y - 10y - 35$
$= 12y^2 + 32y - 35$
$12y^2 + 32y - 35 = (6y-5)(2y+7)$

11. When factoring any polynomial, the first step is to look for a common factor.

12. The GCF of $8x^2 - 28x - 60$ is 4.
$8x^2 - 28x - 60 = 4(2x^2 - 7x - 15)$
$2x^2 = 2x \cdot x$
$-15 = -15 \cdot 1, 15 \cdot (-1), -5 \cdot 3$, or $5 \cdot (-3)$
Try $2x^2 = 2x \cdot x$
$-15 = -5 \cdot 3$
$(2x+3)(x-5) = 2x^2 - 10x + 3x - 15$
$= 2x^2 - 7x - 15$
$8x^2 - 28x - 60 = 4(2x+3)(x-5)$

13. The GCF of $90x^2 + 21xy - 6y^2$ is 3.
$90x^2 + 21xy - 6y^2 = 3(30x^2 + 7xy - 2y^2)$
$30x^2 = 30x \cdot x$, $15x \cdot 2x$, $10x \cdot 3x$, or $6x \cdot 5x$
$-2y^2 = -2y \cdot y$ or $2y \cdot (-y)$
Since the coefficient of xy, 7, is fairly small, try factors of $30x^2$ that are close together: $30x^2 = 6x \cdot 5x$.
$(6x-2y)(5x+y) = 30x^2 + 6xy - 10xy - 2y^2$
$= 30x^2 - 4xy - 2y^2$
$(6x+y)(5x-2y) = 30x^2 - 12xy + 5xy - 2y^2$
$= 30x^2 - 7xy - 2y^2$
To get the coefficient of xy to be 7, reverse the signs on the y-terms.
$(6x-y)(5x+2y) = 30x^2 + 12xy - 5xy - 2y^2$
$= 30x^2 + 7xy - 2y^2$
$90x^2 + 21xy - 6y^2 = 3(30x^2 + 7xy - 2y^2)$
$= 3(6x-y)(5x+2y)$

14. The GCF of $12x^2+22x+6$ is 2.

$12x^2+22x+6=2(6x^2+11x+3)$

$6x^2=6x\cdot x,\ 3x\cdot 2x$

$3=3\cdot 1$

$12x^2+22x+6=2(3x+1)(2x+3)$

Check

$$\begin{aligned}2(3x+1)(2x+3)&=2(6x^2+9x+2x+3)\\&=2(6x^2+11x+3)\end{aligned}$$

$12x^2+2x+6$ was not completely factored as $(4x+6)(3x+1)$. False

15. $-6y^2+23y+4=-1(6y^2-23y-4)$

$6y^2=6y\cdot y$ or $3y\cdot 2y$

$-4=-1\cdot 4, -2\cdot 2$, or $1\cdot(-4)$

Since the coefficient of y, -23, is close to $6\cdot(-4)$, try $6y^2=6y\cdot y$ and $-4=1\cdot(-4)$.

$$\begin{aligned}(6y+1)(y-4)&=6y^2-24y+y-4\\&=6y^2-23y-4\end{aligned}$$

$$\begin{aligned}-6y^2+23y+4&=-1(6y^2-23y-4)\\&=-1(6y+1)(y-4)\\&=-(6y+1)(y-4)\end{aligned}$$

16. The GCF of $-6x^2-3x+45$ is -3.

$-6x^2-3x+45=-3(2x^2+x-15)$

$2x^2=2x\cdot x$

$-15=-1\cdot 15, -3\cdot 5, 1\cdot(-15), 3\cdot(-5)$

Since the coefficient of x, 1, is small, try factors of -15 whose difference is small: $-15=-3\cdot 5$.

$$\begin{aligned}(2x-3)(x+5)&=2x^2+10x-3x-15\\&=2x^2+7x-15\end{aligned}$$

$$\begin{aligned}(2x+5)(x-3)&=2x^2-6x+5x-15\\&=2x^2-x-15\end{aligned}$$

To get the coefficient of x to be 1, reverse the signs on the constants.

$$\begin{aligned}(2x-5)(x+3)&=2x^2+6x-5x-15\\&=2x^2+x-15\end{aligned}$$

$$\begin{aligned}-6x^2-3x+45&=-3(2x^2+x-15)\\&=-3(2x-5)(x+3)\end{aligned}$$

17. When factoring $6x^2+x-1$ using grouping, $ac=\underline{-6}$ and $b=\underline{1}$.

18. False; since $ac=12$, we want to determine the integers whose product is 12 and whose sum is -13.

19. In $3x^2 - 2x - 8$, $a = 3$, $b = -2$, and $c = -8$.
$ac = 3 \cdot (-8) = -24$

Integers Whose Product is –24	1, –24	2, –12	3, –8	4, –6
Sum	–23	–10	–5	–2

$4 \cdot (-6) = -24 = ac$ and $4 + (-6) = -2 = b$, so we write $-2x$ as $4x - 6x$.

$$\begin{aligned} 3x^2 - 2x - 8 &= 3x^2 + 4x - 6x - 8 \\ &= (3x^2 + 4x) + (-6x - 8) \\ &= x(3x + 4) + (-2)(3x + 4) \\ &= (3x + 4)(x - 2) \end{aligned}$$

20. In $10z^2 + 21z + 9$, $a = 10$, $b = 21$, and $c = 9$.
$ac = 10 \cdot 9 = 90$

Integers Whose Product is 90	1, 90	2, 45	3, 30	5, 18	6, 15	9, 10
Sum	91	47	33	23	21	19

$6 \cdot 15 = 90 = ac$ and $6 + 15 = 21 = b$, so we write $21z$ as $6z + 15z$.

$$\begin{aligned} 10z^2 + 21z + 9 &= 10z^2 + 6z + 15z + 9 \\ &= (10z^2 + 6z) + (15z + 9) \\ &= 2z(5z + 3) + 3(5z + 3) \\ &= (5z + 3)(2z + 3) \end{aligned}$$

21. The GCF of $24x^2 + 6x - 9$ is 3.

$24x^2 + 6x - 9 = 3(8x^2 + 2x - 3)$

In $8x^2 + 2x - 3$, $a = 8$, $b = 2$, and $c = -3$.
$ac = 8 \cdot (-3) = -24$

Integers Whose Product is –24	–1, 24	–2, 12	–3, 8	–4, 6
Sum	23	10	5	2

$-4 \cdot 6 = -24 = ac$ and $-4 + 6 = 2 = b$, so we rewrite $2x$ as $-4x + 6x$.

$$\begin{aligned} 8x^2 + 2x - 3 &= 8x^2 - 4x + 6x - 3 \\ &= (8x^2 - 4x) + (6x - 3) \\ &= 4x(2x - 1) + 3(2x - 1) \\ &= (2x - 1)(4x + 3) \end{aligned}$$

$$\begin{aligned} 24x^2 + 6x - 9 &= 3(8x^2 + 2x - 3) \\ &= 3(2x - 1)(4x + 3) \end{aligned}$$

22. $-10n^2 + 17n - 3 = -1(10n^2 - 17n + 3)$

In $10n^2 - 17n + 3$, $a = 10$, $b = -17$, and $c = 3$.

$ac = 10 \cdot 3 = 30$

Integers Whose Product is 30	−1, −30	−2, −15	−3, −10	−5, −6
Sum	−31	−17	−13	−11

$-2 \cdot (-15) = 30 = ac$ and $-2 + (-15) = -17 = b$, so we rewrite $-17n$ as $-2n - 15n$.

$$\begin{aligned} 10n^2 - 17n + 3 &= 10n^2 - 2n - 15n + 3 \\ &= (10n^2 - 2n) + (-15n + 3) \\ &= 2n(5n - 1) + (-3)(5n - 1) \\ &= (5n - 1)(2n - 3) \end{aligned}$$

$$\begin{aligned} -10n^2 + 17n - 3 &= -1(10n^2 - 17n + 3) \\ &= -1(5n - 1)(2n - 3) \\ &= -(5n - 1)(2n - 3) \end{aligned}$$

6.3 Exercises

23. $2x^2 + 5x + 3$

$2x^2 = 2x \cdot x$

$3 = 3 \cdot 1$

Since the coefficient of x is positive, only the positive factors of 3 are listed.

$(2x + 3)(x + 1) = 2x^2 + 2x + 3x + 3 = 2x^2 + 5x + 3$

So, $2x^2 + 5x + 3 = (2x + 3)(x + 1)$.

25. $5n^2 + 7n + 2$

$5n^2 = 5n \cdot n$

$2 = 2 \cdot 1$

Since the coefficient of n is positive, only the positive factors of 2 are listed.

$(5n + 2)(n + 1) = 5n^2 + 5n + 2n + 2 = 5n^2 + 7n + 2$

So, $5n^2 + 7n + 2 = (5n + 2)(n + 1)$.

27. $5y^2 + 2y - 3$

$5y^2 = 5y \cdot y$

$-3 = -3 \cdot 1, -1 \cdot 3$

$(5y - 3)(y + 1) = 5y^2 + 5y - 3y - 3 = 5y^2 + 2y - 3$

So, $5y^2 + 2y - 3 = (5y - 3)(y + 1)$.

29. $-4p^2 + 11p + 3 = -(4p^2 - 11p - 3)$

$4p^2 = 4p \cdot p, 2p \cdot 2p$

$-3 = 3 \cdot (-1), (-3) \cdot 1$

$(4p+1)(p-3) = 4p^2 - 12p + p - 3$
$= 4p^2 - 11p - 3$
So, $-4p^2 + 11p + 3 = -(4p+1)(p-3)$.

31. $5w^2 + 13w - 6$
$5w^2 = 5w \cdot w$
$-6 = -6 \cdot 1, 6 \cdot (-1), -3 \cdot 2, 3 \cdot (-2)$
$(5w-2)(w+3) = 5w^2 + 15w - 2w - 6$
$= 5w^2 + 13w - 6$
So, $5w^2 + 13w - 6 = (5w-2)(w+3)$.

33. $7t^2 + 37t + 10$
$7t^2 = 7t \cdot t$
$10 = 10 \cdot 1, 5 \cdot 2$
Since the coefficient of t is positive, only the positive factors of 10 are listed.
$(7t+2)(t+5) = 7t^2 + 35t + 2t + 10$
$= 7t^2 + 37t + 10$
So, $7t^2 + 37t + 10 = (7t+2)(t+5)$.

35. $6n^2 - 17n + 10$
$6n^2 = 6n \cdot n, \ 3n \cdot 2n$
$10 = -10 \cdot (-1), -5 \cdot (-2)$
Since the coefficient of n is negative, only the negative factors of 10 are listed.
$(6n-5)(n-2) = 6n^2 - 12n - 5n + 10$
$= 6n^2 - 17n + 10$
So, $6n^2 - 17n + 10 = (6n-5)(n-2)$.

37. $2 - 11x + 5x^2 = 5x^2 - 11x + 2$
$5x^2 = 5x \cdot x$
$2 = -2 \cdot (-1)$
Since the coefficient of x is negative, only the negative factors of 2 are listed.
$(5x-1)(x-2) = 5x^2 - 10x - x + 2$
$= 5x^2 - 11x + 2$
So, $2 - 11x + 5x^2 = (5x-1)(x-2)$.

39. $2x^2 + 3xy + y^2$
$2x^2 = 2x \cdot x$
$y^2 = y \cdot y$
$(2x+y)(x+y) = 2x^2 + 2xy + xy + y^2$
$= 2x^2 + 3xy + y^2$
So, $2x^2 + 3xy + y^2 = (2x+y)(x+y)$.

41. $2m^2 - 3mn - 2n^2$

$2m^2 = 2m \cdot m$

$-2n^2 = 2n \cdot (-n),\ (-2n) \cdot n$

$(2m+n)(m-2n) = 2m^2 - 4mn + mn - 2n^2$

$= 2m^2 - 3mn - 2n^2$

So, $2m^2 - 3mn - 2n^2 = (2m+n)(m-2n)$.

43. $6x^2 + 2xy - 4y^2 = 2(3x^2 + xy - 2y^2)$

$3x^2 = 3x \cdot x$

$-2y^2 = -2y \cdot y,\ 2y \cdot (-y)$

$(3x-2y)(x+y) = 3x^2 + 3xy - 2xy - 2y^2$

$= 3x^2 + xy - 2y^2$

So, $6x^2 + 2xy - 4y^2 = 2(3x-2y)(x+y)$.

45. $-2x^2 - 7x + 15 = -(2x^2 + 7x - 15)$

$2x^2 = 2x \cdot x$

$-15 = 15 \cdot (-1),\ (-15) \cdot 1,\ 5 \cdot (-3),\ (-5) \cdot 3$

$(2x-3)(x+5) = 2x^2 + 10x - 3x - 15$

$= 2x^2 + 7x - 15$

So, $-2x^2 - 7x + 15 = -(2x-3)(x+5)$.

47. $2x^2 + 13x + 15,\ a = 2, b = 13, c = 15$

$ac = (2)(15) = 30$

Integers whose product is 30	30, 1	−30, −1	15, 2	−15, −2	10, 3	−10, −3	6, 5	−6, −5
Sum	31	−31	17	−17	13	−13	11	−11

The integers are 10 and 3.

$2x^2 + 13x + 15 = 2x^2 + 10x + 3x + 15$

$= (2x^2 + 10x) + (3x + 15)$

$= 2x(x+5) + 3(x+5)$

$= (2x+3)(x+5)$

49. $5w^2 + 13w - 6,\ a = 5, b = 13, c = -6$

$ac = (5)(-6) = -30$

Integers whose product is −30	−30, 1	30, −1	−15, 2	15, −2	−10, 3	10, −3	−6, 5	6, −5
Sum	−29	29	−13	13	−7	7	−1	1

The integers are 15 and −2.

$$\begin{aligned}5w^2+13w-6 &= 5w^2+15w-2w-6\\ &= (5w^2+15w)+(-2w-6)\\ &= 5w(w+3)+(-2)(w+3)\\ &= 5w(w+3)-2(w+3)\\ &= (5w-2)(w+3)\end{aligned}$$

51. $4w^2-8w-5,\ a=4,\ b=-8,\ c=-5$

$ac=(4)(-5)=-20$

Integers whose product is −20	−20, 1	20, −1	−10, 2	10, −2	−5, 4	5, −4
Sum	−19	19	−8	8	−1	1

The integers are −10 and 2.

$$\begin{aligned}4w^2-8w-5 &= 4w^2+2w-10w-5\\ &= (4w^2+2w)+(-10w-5)\\ &= 2w(2w+1)-5(2w+1)\\ &= (2w-5)(2w+1)\end{aligned}$$

53. $27z^2+3z-2,\ a=27,\ b=3,\ c=-2$

$ac=(27)(-2)=-54$

Integers whose product is −54	−54, 1	54, −1	−27, 2	27, −2	−18, 3	18, −3	−9, 6	9, −6
Sum	−53	53	−25	25	−15	15	−3	3

The integers are 9 and −6.

$$\begin{aligned}27z^2+3z-2 &= 27z^2+9z-6z-2\\ &= (27z^2+9z)+(-6z-2)\\ &= 9z(3z+1)-2(3z+1)\\ &= (9z-2)(3z+1)\end{aligned}$$

55. $6y^2-5y-6,\ a=6,\ b=-5,\ c=-6$

$ac=(6)(-6)=-36$

Integers whose product is −36	−36, 1	36, −1	−18, 2	18, −2	−12, 3	12, −3	−9, 4	9, −4	6, −6
Sum	−35	35	−16	16	−9	9	−5	5	0

The integers are –9 and 4.

$$\begin{aligned}6y^2-5y-6&=6y^2-9y+4y-6\\&=(6y^2-9y)+(4y-6)\\&=3y(2y-3)+2(2y-3)\\&=(3y+2)(2y-3)\end{aligned}$$

57. $4m^2+8m-5,\ a=4,\ b=8,\ c=-5$

$ac=(4)(-5)=-20$

Integers whose product is –20	–20, 1	20, –1	–10, 2	10, –2	–5, 4	5, –4
Sum	–19	19	–8	8	–1	1

The integers are 10 and –2.

$$\begin{aligned}4m^2+8m-5&=4m^2-2m+10m-5\\&=(4m^2-2m)+(10m-5)\\&=2m(2m-1)+5(2m-1)\\&=(2m+5)(2m-1)\end{aligned}$$

59. $12n^2+19n+5,\ a=12,\ b=19,\ c=5$

$ac=(12)(5)=60$

Integers whose product is 60	60, 1	30, 2	20, 3	15, 4	12, 5	10, 6
Sum	61	32	23	19	17	16

The integers are 15 and 4.

$$\begin{aligned}12n^2+19n+5&=12n^2+4n+15n+5\\&=(12n^2+4n)+(15n+5)\\&=4n(3n+1)+5(3n+1)\\&=(4n+5)(3n+1)\end{aligned}$$

61. $-5-9x+18x^2=18x^2-9x-5$

$a=18,\ b=-9,\ c=-5$

$ac=(18)(-5)=-90$

Integers whose product is –90	–90, 1	90, –1	–45, 2	45, –2	–30, 3	30, –3	–18, 5	18, –5	–15, 6	15, –6	–10, 9	10, –9
Sum	–89	89	–43	43	–27	27	–13	13	–9	9	–1	1

The integers are −15 and 6.

$$\begin{aligned}18x^2-9x-5&=18x^2+6x-15x-5\\&=(18x^2+6x)+(-15x-5)\\&=6x(3x+1)-5(3x+1)\\&=(6x-5)(3x+1)\end{aligned}$$

63. $12x^2+2xy-4y^2=2(6x^2+xy-2y^2)$

$a=6, b=1, c=-2$

$ac=(6)(-2)=-12$

Integers whose product is −12	−12, 1	12, −1	−6, 2	6, −2	−4, 3	4, −3
Sum	−11	11	−4	4	−1	1

The integers are 4 and −3.

$$\begin{aligned}6x^2+xy-2y^2&=6x^2-3xy+4xy-2y^2\\&=3x(2x-y)+2y(2x-y)\\&=(3x+2y)(2x-y)\end{aligned}$$

So, $12x^2+2xy-4y^2=2(3x+2y)(2x-y)$.

65. $8x^2+28x+12=4(2x^2+7x+3)$

$a=2, b=7, c=3$

$ac=(2)(3)=6$

Integers whose product is 6	6, 1	3, 2
Sum	7	5

The integers are 6 and 1.

$$\begin{aligned}2x^2+7x+3&=2x^2+6x+x+3\\&=2x(x+3)+(x+3)\\&=(2x+1)(x+3)\end{aligned}$$

So, $8x^2+28x+12=4(2x+1)(x+3)$.

67. $7x-5+6x^2=6x^2+7x-5$

$a=6, b=7, c=-5$

$ac=(6)(-5)=-30$

Integers whose product is −30	−30, 1	30, −1	−15, 2	15, −2	−10, 3	10, −3	−6, 5	6, −5
Sum	−29	29	−13	13	−7	7	−1	1

The integers are 10 and −3.

$$\begin{aligned}6x^2+7x-5 &= 6x^2-3x+10x-5\\ &= 3x(2x-1)+5(2x-1)\\ &= (3x+5)(2x-1)\end{aligned}$$

69. $-8p^2+6p+9=-(8p^2-6p-9)$

$a=8,\ b=-6,\ c=-9$

$ac=(8)(-9)=-72$

Integers whose product is −72	−72 1	72 −1	−36 2	36 −2	−24 3	24 −3	−18 4	18 −4	−12 6	12 −6	−9 8	9 −8
Sum	−71	71	−34	34	−21	21	−14	14	−6	6	−1	1

The integers are −12 and 6.

$$\begin{aligned}8p^2-6p-9 &= 8p^2-12p+6p-9\\ &= 4p(2p-3)+3(2p-3)\\ &= (4p+3)(2p-3)\end{aligned}$$

So, $-8p^2+6p+9=-(4p+3)(2p-3)$.

71. $15\cdot 4=60=-20\cdot(-3)$ and

$-20+(-3)=-23$

$$\begin{aligned}15x^2-23x+4 &= 15x^2-20x-3x+4\\ &= 5x(3x-4)-1(3x-4)\\ &= (3x-4)(5x-1)\end{aligned}$$

73. $$\begin{aligned}-13y+12-4y^2 &= -4y^2-13y+12\\ &= -(4y^2+13y-12)\\ &= -(4y^2+16y-3y-12)\\ &= -[4y(y+4)-3(y+4)]\\ &= -(y+4)(4y-3)\end{aligned}$$

75. $$\begin{aligned}10x^2-8xy-24y^2 &= 2(5x^2-4xy-12y^2)\\ &= 2(5x^2-10xy+6xy-12y^2)\\ &= 2[5x(x-2y)+6y(x-2y)]\\ &= 2(x-2y)(5x+6y)\end{aligned}$$

77. $$\begin{aligned}8-18x+9x^2 &= 9x^2-18x+8\\ &= 9x^2-12x-6x+8\\ &= 3x(3x-4)-2(3x-4)\\ &= (3x-4)(3x-2)\end{aligned}$$

79. $$\begin{aligned}-12x+9-24x^2 &= -24x^2-12x+9\\ &= -3(8x^2+4x-3)\end{aligned}$$

81. $4x^3y^2 - 8x^2y^3 - 4x^2y^2 = 4x^2y^2(x - 2y - 1)$

83. $4m^2 + 13mn + 3n^2 = 4m^2 + 12mn + mn + 3n^2$
$= 4m(m + 3n) + n(m + 3n)$
$= (m + 3n)(4m + n)$

85. $6x^2 - 17x - 12$ cannot be factored; it is prime.

87. $48xy + 24x^2 - 30y^2$
$= 24x^2 + 48xy - 30y^2$
$= 6(4x^2 + 8xy - 5y^2)$
$= 6(4x^2 + 10xy - 2xy - 5y^2)$
$= 6[2x(2x + 5y) - y(2x + 5y)]$
$= 6(2x + 5y)(2x - y)$

89. $18m^2 + 39mn - 24n^2$
$= 3(6m^2 + 13mn - 8n^2)$
$= 3(6m^2 + 16mn - 3mn - 8n^2)$
$= 3[2m(3m + 8n) - n(3m + 8n)]$
$= 3(3m + 8n)(2m - n)$

91. $-6x^3 + 10x^2 - 4x^4 = -4x^4 - 6x^3 + 10x^2$
$= -2x^2(2x^2 + 3x - 5)$
$= -2x^2(2x + 5)(x - 1)$

93. $30x + 22x^2 - 24x^3 = -24x^3 + 22x^2 + 30x$
$= -2x(12x^2 - 11x - 15)$
$= -2x(4x + 3)(3x - 5)$

95. $6x^2(x^2 + 1) - 25x(x^2 + 1) + 14(x^2 + 1)$
$= (x^2 + 1)(6x^2 - 25x + 14)$
$= (x^2 + 1)(6x^2 - 21x - 4x + 14)$
$= (x^2 + 1)[3x(2x - 7) - 2(2x - 7)]$
$= (x^2 + 1)(2x - 7)(3x - 2)$

97. $3x^2 + \frac{13}{2}x - 14$
$= \frac{1}{2}(6x^2 + 13x - 28)$
$= \frac{1}{2}(6x^2 + 21x - 8x - 28)$
$= \frac{1}{2}[(6x^2 + 21x) + (-8x - 28)]$
$= \frac{1}{2}[3x(2x + 7) - 4(2x + 7)]$
$= \frac{1}{2}(2x + 7)(3x - 4)$
The base and height are represented by $(2x + 7)$ meters and $(3x - 4)$ meters.

99. $6 = 3 \cdot 2$ and $-10 = 2 \cdot (-5)$, so the other factor is $2x - 5$.

101. $27z^4 + 42z^2 + 16 = (9z^2 + 8)(3z^2 + 2)$

103. $3x^{2n} + 19x^n + 6 = 3x^{2n} + x^n + 18x^n + 6$
$= x^n(3x^n + 1) + 6(3x^n + 1)$
$= (3x^n + 1)(x^n + 6)$

105. $3x^2 = 3x \cdot x$ and $-5 = -1 \cdot 5$ or $1 \cdot (-5)$.
$(3x + 5)(x - 1) = 3x^2 + 2x - 5$
$(3x - 5)(x + 1) = 3x^2 - 2x - 5$
$(3x + 1)(x - 5) = 3x^2 - 14x - 5$
$(3x - 1)(x + 5) = 3x^2 + 14x - 5$
Thus, b could be ± 2 or ± 14.

107. The trial and error method may be used when the value of a and/or the value of c are prime numbers, or numbers with few factors. The grouping method can be used when the product of a and c is not a very large number. Examples may vary.

Section 6.4

Preparing for Factoring Special Products

P1. $5^2 = 5 \cdot 5 = 25$

P2. $(-2)^3 = (-2) \cdot (-2) \cdot (-2) = -8$

P3. $(5p^2)^3 = 5^3(p^2)^3 = 125p^{2 \cdot 3} = 125p^6$

P4. $(3z + 2)^2 = (3z)^2 + 2(3z)(2) + 2^2 = 9z^2 + 12z + 4$

P5. $(4m+5)(4m-5)=(4m)^2-5^2=16m^2-25$

6.4 Quick Checks

1. The expression $A^2+2AB+B^2$ is called a <u>perfect square trinomial</u>.

2. $A^2-2AB+B^2=\underline{(A-B)^2}$

3. $x^2-12x+36=x^2-2\cdot x\cdot 6+6^2$
$=(x-6)^2$

4. $16x^2+40x+25=(4x)^2+2\cdot 4x\cdot 5+5^2$
$=(4x+5)^2$

5. $9a^2-60ab+100b^2$
$=(3a)^2-2\cdot 3a\cdot 10b+(10b)^2$
$=(3a-10b)^2$

6. $z^2-8z+16=z^2-2\cdot z\cdot 4+4^2$
$=(z-4)^2$

7. $4n^2+12n+9=(2n)^2+2\cdot 2n\cdot 3+3^2$
$=(2n+3)^2$

8. $35 = 1\cdot 35$ or $5\cdot 7$
$1+35=36$ and $5+7=12$, so none of the factors of 35 sum to 13. The polynomial is prime.

9. The first thing we look for when factoring a trinomial is the <u>greatest common factor</u>.

10. $4z^2+24z+36=4(z^2+6z+9)$
$=4(z^2+2\cdot z\cdot 3+3^2)$
$=4(z+3)^2$

11. $50a^3+80a^2+32a$
$=2a(25a^2+40a+16)$
$=2a[(5a)^2+2\cdot 5a\cdot 4+4^2]$
$=2a(5a+4)^2$

12. $4m^2-81n^2$ is called the <u>difference</u> of <u>two squares</u> and factors into two binomials.

13. $P^2-Q^2=\underline{(P-Q)(P+Q)}$

14. $z^2-25=z^2-5^2=(z-5)(z+5)$

15. $81m^2-16n^2=(9m)^2-(4n)^2$
$=(9m-4n)(9m+4n)$

16. $16a^2-\frac{4}{9}b^2=(4a)^2-\left(\frac{2}{3}b\right)^2$
$=\left(4a-\frac{2}{3}b\right)\left(4a+\frac{2}{3}b\right)$

17. True: x^2+9 is prime.

18. $100k^4-81w^2=(10k^2)^2-(9w)^2$
$=(10k^2-9w)(10k^2+9w)$

19. $x^4-16=(x^2)^2-4^2$
$=(x^2-4)(x^2+4)$
$=(x^2-2^2)(x^2+4)$
$=(x-2)(x+2)(x^2+4)$

20. False;
$4x^2-16y^2=4(x^2-4y^2)=4(x-2y)(x+2y)$

21. $147x^2-48=3(49x^2-16)$
$=3[(7x)^2-4^2]$
$=3(7x-4)(7x+4)$

22. $-27a^3b+75ab^3=-3ab(9a^2-25b^2)$
$=-3ab[(3a)^2-(5b)^2]$
$=-3ab(3a-5b)(3a+5b)$

23. The binomial $27x^3+64y^3$ is called the <u>sum</u> of <u>two cubes</u>.

24. False; $(x-1)(x^2-x+1)$ cannot be factored further.

25. $z^3+125=z^3+5^3$
$=(z+5)(z^2-z\cdot 5+5^2)$
$=(z+5)(z^2-5z+25)$

26. $8p^3-27q^6$
$=(2p)^3-(3q^2)^3$
$=(2p-3q^2)[(2p)^2+2p\cdot 3q^2+(3q^2)^2]$
$=(2p-3q^2)(4p^2+6pq^2+9q^4)$

27. $54a-16a^4 = 2a(27-8a^3)$
$= 2a[3^3-(2a)^3]$
$= 2a(3-2a)[3^2+3\cdot 2a+(2a)^2]$
$= 2a(3-2a)(9+6a+4a^2)$

28. $-375b^3+3 = -3(125b^3-1)$
$= -3[(5b)^3-1^3]$
$= -3(5b-1)[(5b)^2+5b\cdot 1+1^2]$
$= -3(5b-1)(25b^2+5b+1)$

6.4 Exercises

29. $x^2+10x+25 = x^2+2\cdot x\cdot 5+5^2 = (x+5)^2$

31. $4p^2-4p+1 = (2p)^2-2\cdot 2p\cdot 1+1^2$
$= (2p-1)^2$

33. $16x^2+24x+9 = (4x)^2+2\cdot 4x\cdot 3+3^2$
$= (4x+3)^2$

35. $x^2-4xy+4y^2 = x^2-2\cdot x\cdot 2y+(2y)^2$
$= (x-2y)^2$

37. $4z^2-12z+9 = (2z)^2-2\cdot 2z\cdot 3+3^2$
$= (2z-3)^2$

39. $16x^2-49y^2 = (4x)^2-(7y)^2$
$= (4x-7y)(4x+7y)$

41. $4x^2-25 = (2x)^2-5^2 = (2x-5)(2x+5)$

43. $100n^8-81p^4 = (10n^4)^2-(9p^2)^2$
$= (10n^4-9p^2)(10n^4+9p^2)$

45. $k^8-256 = (k^4)^2-16^2$
$= (k^4-16)(k^4+16)$
$= [(k^2)^2-4^2](k^4+16)$
$= (k^2-4)(k^2+4)(k^4+16)$
$= (k^2-2^2)(k^2+4)(k^4+16)$
$= (k-2)(k+2)(k^2+4)(k^2+16)$

47. $25p^4-49q^2 = (5p^2)^2-(7q)^2$
$= (5p^2-7q)(5p^2+7q)$

49. $27+x^3 = 3^3+x^3$
$= (3+x)(3^2-3\cdot x+x^2)$
$= (3+x)(9-3x+x^2)$

51. $8x^3-27y^3$
$= (2x)^3-(3y)^3$
$= (2x-3y)[(2x)^2+(2x)(3y)+(3y)^2]$
$= (2x-3y)(4x^2+6xy+9y^2)$

53. x^6-8y^3
$= (x^2)^3-(2y)^3$
$= (x^2-2y)[(x^2)^2+(x^2)(2y)+(2y)^2]$
$= (x^2-2y)(x^4+2x^2y+4y^2)$

55. $27c^3+64d^9$
$= (3c)^3+(4d^3)^3$
$= (3c+4d^3)[(3c)^2-(3c)(4d^3)+(4d^3)^2]$
$= (3c+4d^3)(9c^2-12cd^3+16d^6)$

57. $16x^3+250y^3$
$= 2(8x^3+125y^3)$
$= 2[(2x)^3+(5y)^3]$
$= 2(2x+5y)[(2x)^2-(2x)(5y)+(5y)^2]$
$= 2(2x+5y)(4x^2-10xy+25y^2)$

59. $16m^2+40mn+25n^2$
$= (4m)^2+2\cdot 4m\cdot 5n+(5n)^2$
$= (4m+5n)^2$

61. $18-12x+2x^2 = 2(9-6x+x^2)$
$= 2(3^2-2\cdot 3\cdot x+x^2)$
$= 2(3-x)^2$ or $2(x-3)^2$

63. $48a^3+72a^2+27a$
$= 3a(16a^2+24a+9)$
$= 3a[(4a)^2+2\cdot 4a\cdot 3+3^2]$
$= 3a(4a+3)^2$

65. $2x^3+10x^2+16x = 2x(x^2+5x+8)$

67. $x^4y^2-x^2y^4 = x^2y^2(x^2-y^2)$
$= x^2y^2(x-y)(x+y)$

69. $x^8 - 25y^{10} = (x^4)^2 - (5y^5)^2$
$= (x^4 - 5y^5)(x^4 + 5y^5)$

71. $2t^4 - 54t = 2t(t^3 - 27)$
$= 2t[(t)^3 - (3)^3]$
$= 2t(t-3)(t^2 + 3t + 9)$

73. $3s^7 + 24s = 3s(s^6 + 8)$
$= 3s[(s^2)^3 + (2)^3]$
$= 3s(s^2 + 2)[(s^2)^2 - 2s^2 + (2)^2]$
$= 3s(s^2 + 2)(s^4 - 2s^2 + 4)$

75. $2x^5 - 162x = 2x(x^4 - 81)$
$= 2x[(x^2)^2 - (9)^2]$
$= 2x(x^2 - 9)(x^2 + 9)$
$= 2x[(x)^2 - (3)^2](x^2 + 9)$
$= 2x(x-3)(x+3)(x^2 + 9)$

77. $x^3y - xy^3 = xy(x^2 - y^2) = xy(x-y)(x+y)$

79. $x^3y^3 + 1 = (xy)^3 + (1)^3$
$= (xy+1)[(xy)^2 - xy(1) + (1)^2]$
$= (xy+1)(x^2y^2 - xy + 1)$

81. $3n^2 + 14n + 36$ cannot be factored; it is prime.

83. $2x^2 - 8x + 8 = 2(x^2 - 4x + 4)$
$= 2(x-2)^2$

85. $9x^2 + y^2$ cannot be factored; it is prime.

87. $x^4y^3 + 216xy^3 = xy^3(x^3 + 216)$
$= xy^3[(x)^3 + (6)^3]$
$= xy^3(x+6)[(x)^2 - 6x + (6)^2]$
$= xy^3(x+y)(x^2 - 6x + 36)$

89. $48n^4 - 24n^3 + 3n^2 = 3n^2(16n^2 - 8n + 1)$
$= 3n^2(4n-1)^2$

91. $2x(x^2 - 4) + 5(x^2 - 4)$
$= (x^2 - 4)(2x + 5)$
$= (x^2 - 2^2)(2x + 5)$
$= (x-2)(x+2)(2x+5)$

93. $2y^3 + 5y^2 - 32y - 80$
$= (2y^3 + 5y^2) + (-32y - 80)$
$= y^2(2y+5) - 16(2y+5)$
$= (2y+5)(y^2 - 16)$
$= (2y+5)(y^2 - 4^2)$
$= (2y+5)(y-4)(y+4)$

95. $4x^2 + 20x + 25 = (2x)^2 + 2 \cdot 2x \cdot 5 + 5^2$
$= (2x+5)^2$
The length of one side is $2x + 5$.

97. $(x-2)^2 - (x+1)^2$
$= [(x-2) - (x+1)][(x-2) + (x+1)]$
$= -3(2x-1)$

99. $(x-y)^3 + y^3$
$= [(x-y) + y][(x-y)^2 - (x-y)y + y^2]$
$= x(x^2 - 2xy + y^2 - xy + y^2 + y^2)$
$= x(x^2 - 3xy + 3y^2)$

101. $(x+1)^2 - 9 = [(x+1) - 3][(x+1) + 3]$
$= (x-2)(x+4)$

103. $2a^2(x+1) - 17a(x+1) + 30(x+1)$
$= (x+1)(2a^2 - 17a + 30)$
$= (x+1)(2a-5)(a-6)$

105. $5(x+2)^2 - 7(x+2) - 6$
$= [5(x+2) + 3][(x+2) - 2]$
$= (5x+13)(x)$
$= x(5x+13)$

107. First identify the problem as a sum or a difference of cubes. Identify A and B and rewrite the problem as $A^3 + B^3$ or $A^3 - B^3$. The sum of cubes can be factored into the product of a binomial and a trinomial where the binomial factor is $A + B$ and the trinomial factor is A^2 minus the product AB plus B^2. The difference of cubes also can be factored into the product of a binomial and a trinomial, but here the binomial factor is $A - B$ and the trinomial factor is the sum of A^2, the product AB, and B^2.

109. The correct factorization of x^3+y^3 is $(x+y)(x^2-xy+y^2)$. The trinomial factor (x^2-xy+y^2) is not a perfect square trinomial. A perfect square trinomial is $(x-y)^2=(x^2-2xy+y^2)$.

Section 6.5

6.5 Quick Checks

1. The first step in any factoring problem is to look for the greatest common factor.

2. $2p^2+8p-90=2(p^2+4p-45)$
$=2(p-5)(p+9)$

Check: $2(p-5)(p+9)=2(p^2+9p-5p-45)$
$=2(p^2+4p-45)$
$=2p^2+8p-90$

3. $-45x^2+3xy+6y^2$
$=-3(15x^2-xy-2y^2)$
$=-3(15x^2+5xy-6xy-2y^2)$
$=-3[5x(3x+y)-2y(3x+y)]$
$=-3(3x+y)(5x-2y)$

Check: $-3(3x+y)(5x-2y)$
$=-3(15x^2-6xy+5xy-2y^2)$
$=-3(15x^2-xy-2y^2)$
$=-45x^2+3xy+6y^2$

4. $100x^2-81y^2=(10x)^2-(9y)^2$
$=(10x-9y)(10x+9y)$

Check: $(10x-9y)(10x+9y)$
$=100x^2+90xy-90xy-81y^2$
$=100x^2-81y^2$

5. $2ab^2-242a=2a(b^2-121)$
$=2a(b^2-11^2)$
$=2a(b-11)(b+11)$

Check:
$2a(b-11)(b+11)=2a(b^2+11b-11b-121)$
$=2a(b^2-121)$
$=2ab^2-242a$

6. $p^2-12pq+36q^2=p^2-2\cdot p\cdot 6q+(6q)^2$
$=(p-6q)^2$

Check: $(p-6q)^2=(p-6q)(p-6q)$
$=p^2-6pq-6pq+36q^2$
$=p^2-12pq+36q^2$

7. $75x^2+90x+27=3(25x^2+30x+9)$
$=3[(5x)^2+2\cdot 5x\cdot 3+3^2]$
$=3(5x+3)^2$

Check: $3(5x+3)^2=3(5x+3)(5x+3)$
$=3(25x^2+15x+15x+9)$
$=3(25x^2+30x+9)$
$=75x^2+90x+27$

8. $125y^3-64=(5y)^3-4^3$
$=(5y-4)[(5y)^2+5y\cdot 4+4^2]$
$=(5y-4)(25y^2+20y+16)$

Check:
$(5y-4)(25y^2+20y+16)$
$=125y^3+100y^2+80y-100y^2-80y-64$
$=125y^3-64$

9. $-24a^3+3b^3$
$=-3(8a^3-b^3)$
$=-3[(2a)^3-b^3]$
$=-3(2a-b)[(2a)^2+2a\cdot b+b^2]$
$=-3(2a-b)(4a^2+2ab+b^2)$

Check:
$-3(2a-b)(4a^2+2ab+b^2)$
$=-3(8a^3+4a^2b+2ab^2-4a^2b-2ab^2-b^3)$
$=-3(8a^3-b^3)$
$=-24a^3+3b^3$

10. False; $x^4-81=(x^2-9)(x^2+9)$
$=(x+3)(x-3)(x^2+9)$

11. $x^4-1=(x^2)^2-1^2$
$=(x^2-1)(x^2+1)$
$=(x^2-1^2)(x^2+1)$
$=(x-1)(x+1)(x^2+1)$

Check: $(x-1)(x+1)(x^2+1)$
$= (x^2+x-x-1)(x^2+1)$
$= (x^2-1)(x^2+1)$
$= x^4+x^2-x^2-1$
$= x^4-1$

12. $-36x^2y+16y = -4y(9x^2-4)$
$= -4y[(3x)^2-2^2]$
$= -4y(3x-2)(3x+2)$

Check: $-4y(3x-2)(3x+2)$
$= -4y(9x^2+6x-6x-4)$
$= -4y(9x^2-4)$
$= -36x^2y+16y$

13. When factoring a polynomial with four terms, try factoring by grouping.

14. True

15. $2x^3+3x^2+4x+6 = (2x^3+3x^2)+(4x+6)$
$= x^2(2x+3)+2(2x+3)$
$= (2x+3)(x^2+2)$

Check: $(2x+3)(x^2+2) = 2x^3+4x+3x^2+6$
$= 2x^3+3x^2+4x+6$

16. $6x^3-9x^2-6x-9$
$= 3(2x^3+3x^2-2x-3)$
$= 3[(2x^3+3x^2)+(-2x-3)]$
$= 3[x^2(2x+3)-1(2x+3)]$
$= 3(2x+3)(x^2-1)$
$= 3(2x+3)(x^2-1^2)$
$= 3(2x+3)(x-1)(x+1)$

Check: $3(2x+3)(x-1)(x+1)$
$= 3(2x+3)(x^2+x-x-1)$
$= 3(2x+3)(x^2-1)$
$= 3(2x^3-2x+3x^2-3)$
$= 3(2x^3+3x^2-2x-3)$
$= 6x^3+9x^2-6x-9$

17. $-3z^2+9z-21 = -3(z^2-3z+7)$
No further factorization is possible since z^2-3z+7 is prime.

18. $6xy^2+15x^3 = 3x(2y^2+5x^2)$
No further factorization is possible.

6.5 Exercises

19. $x^2-100 = x^2-10^2 = (x-10)(x+10)$

21. $t^2+t-6 = (t+3)(t-2)$

23. $x+y+2ax+2ay = (x+y)+2a(x+y)$
$= (x+y)(1+2a)$

25. $a^3-8 = a^3-2^3 = (a-2)(a^2+2a+4)$

27. $a^2-ab-6b^2 = (a-3b)(a+2b)$

29. $2x^2-5x-7 = (2x-7)(x+1)$

31. $2x^2-6xy-20y^2 = 2(x^2-3xy-10y^2)$
$= 2(x-5y)(x+2y)$

33. $9-a^2 = 3^2-a^2 = (3-a)(3+a)$

35. $u^2-14u+33 = (u-11)(u-3)$

37. $xy-ay-bx+ab = y(x-a)-b(x-a)$
$= (x-a)(y-b)$

39. $w^2+6w+8 = (w+2)(w+4)$

41. $36a^2-49b^4 = (6a)^2-(7b^2)^2$
$= (6a-7b^2)(6a+7b^2)$

43. $x^2+2xm-8m^2 = (x+4m)(x-2m)$

45. $6x^2y^2-13xy+6 = (3xy-2)(2xy-3)$

47. $x^3+x^2+x+1 = x^2(x+1)+(x+1)$
$= (x+1)(x^2+1)$

49. $12z^2+12z+18 = 6(2z^2+2z+3)$

51. $14c^2+19c-3 = (7c-1)(2c+3)$

53. $27m^3 + 64n^6$
$= (3m)^3 + (4n^2)^3$
$= (3m + 4n^2)(9m^2 - 12mn^2 + 16n^4)$

55. $2j^6 - 2j^2 = 2j^2(j^4 - 1)$
$= 2j^2[(j^2)^2 - 1^2]$
$= 2j^2(j^2 - 1)(j^2 + 1)$
$= 2j^2(j^2 - 1^2)(j^2 + 1)$
$= 2j^2(j - 1)(j + 1)(j^2 + 1)$

57. $8a^2 + 18ab - 5b^2 = (4a - b)(2a + 5b)$

59. $2a^3 + 6a = 2a(a^2 + 3)$

61. $12z^2 - 3 = 3(4z^2 - 1)$
$= 3[(2z)^2 - 1^2]$
$= 3(2z - 1)(2z + 1)$

63. $x^2 - x + 6$ cannot be factored; it is prime.

65. $16a^4 + 2ab^3 = 2a(8a^3 + b^3)$
$= 2a[(2a)^3 + b^3]$
$= 2a(2a + b)(4a^2 - 2ab + b^2)$

67. $p^2q^2 + 6pq - 7 = (pq + 7)(pq - 1)$

69. $s^2(s + 2) - 4(s + 2)$
$= (s + 2)(s^2 - 4)$
$= (s + 2)(s^2 - 2^2)$
$= (s + 2)(s - 2)(s + 2)$ or $(s + 2)^2(s - 2)$

71. $-12x^3 + 2x^2 + 2x = -2x(6x^2 - x - 1)$
$= -2x(3x + 1)(2x - 1)$

73. $10v^2 - 2 - v = 10v^2 - v - 2 = (5v + 2)(2v - 1)$

75. $4n^2 - n^4 + 3n^3 = -n^4 + 3n^3 + 4n^2$
$= -n^2(n^2 - 3n - 4)$
$= -n^2(n - 4)(n + 1)$

77. $-4a^3b + 2a^2b - 2ab = -2ab(2a^2 - a + 1)$

79. $12p - p^3 + p^2 = -p^3 + p^2 + 12p$
$= -p(p^2 - p - 12)$
$= -p(p - 4)(p + 3)$

81. $-32x^3 + 72xy^2 = -8x(4x^2 - 9y^2)$
$= -8x[(2x)^2 - (3y)^2]$
$= -8x(2x - 3y)(2x + 3y)$

83. $2n^3 - 10n^2 - 6n + 30$
$= 2(n^3 - 5n^2 - 3n + 15)$
$= 2[n^2(n - 5) - 3(n - 5)]$
$= 2(n - 5)(n^2 - 3)$

85. $16x^2 + 4x - 12 = 4(4x^2 + x - 3)$
$= 4(4x - 3)(x + 1)$

87. $14x^2 + 3x^4 + 8 = 3x^4 + 14x^2 + 8$
$= (3x^2 + 2)(x^2 + 4)$

89. $-2x^3y + x^2y^2 + 3xy^3 = -xy(2x^2 - xy - 3y^2)$
$= -xy(2x - 3y)(x + y)$

91. $(x^3 + 8x) - (8x^2 - 7x) = x^3 + 8x - 8x^2 + 7x$
$= x^3 - 8x^2 + 15x$
$= x(x^2 - 8x + 15)$
$= x(x - 5)(x - 3)$

93. $4x^3 - 10x^2 - 6x = 2x(2x^2 - 5x - 3)$
$= 2x(2x + 1)(x - 3)$

Possible dimensions for the box are $2x$, $2x + 1$, and $x - 3$.

95. $4m^2 - 4mn + n^2 - p^2$
$= (4m^2 - 4mn + n^2) - p^2$
$= (2m - n)^2 - p^2$
$= (2m - n + p)(2m - n - p)$

97. $x^2 - y^2 + 2yz - z^2$
$= x^2 - (y^2 - 2yz + z^2)$
$= x^2 - (y - z)^2$
$= [x + (y - z)][x - (y - z)]$
$= (x + y - z)(x - y + z)$

99. $x^2-2xy+y^2-6x+6y+8$
$=(x^2-2xy+y^2)+(-6x+6y)+8$
$=(x-y)^2-6(x-y)+8$
$=[(x-y)-4][(x-y)-2]$
$=(x-y-4)(x-y-2)$

101. $x^2+2xy+y^2-a^2+2ab-b^2$
$=(x^2+2xy+y^2)-(a^2-2ab+b^2)$
$=(x+y)^2-(a-b)^2$
$=[(x+y)-(a-b)][(x+y)+(a-b)]$
$=(x+y-a+b)(x+y+a-b)$

103. The student initially wrote the sum of the terms as a product of terms by enclosing the binomial $-3x-12$ in parentheses. Then in the second step, the student writes the factor $-3(x-4)$ and concludes that the factorization of $x^2+4x-3x-12$ is $(x-3)(x+4)$, even though both terms do not have the common factor of $(x+4)$. The student obtained the correct answer, but the first two steps are incorrect, the correct factorization is
$x^2+4x-3x-12=x(x+4)-3(x+4)$
$=(x+4)(x-3)$

Putting the Concepts Together (Sections 6.1–6.5)

1. The GCF of 10, 15, and 25 is 5.
The GCF of x^3, x^5, and x^2 is x^2.
The GCF of y^4, y, and y^7 is y.
The GCF is $5x^2y$.

2. $x^2-3x-4=(x-4)(x+1)$

3. $x^6-27=(x^2)^3-3^3$
$=(x^2-3)[(x^2)^2+x^2\cdot 3+3^2]$
$=(x^2-3)(x^4+3x^2+9)$

4. $6x(2x+1)+5z(2x+1)=(2x+1)(6x+5z)$

5. $x^2+5xy-6y^2=(x+6y)(x-y)$

6. $x^3+64=x^3+4^3=(x+4)(x^2-4x+16)$

7. $4x^2+49y^2$ cannot be factored; it is prime.

8. $3x^2+12xy-36y^2=3(x^2+4xy-12y^2)$
$=3(x+6y)(x-2y)$

9. $12z^5-44z^3-24z^2=4z^2(3z^3-11z-6)$

10. x^2+6x-5 cannot be factored; it is prime.

11. $4m^4+5m^3-6m^2$
$=m^2(4m^2+5m-6)$
$=m^2(4m^2+8m-3m-6)$
$=m^2[(4m^2+8m)+(-3m-6)]$
$=m^2[4m(m+2)-3(m+2)]$
$=m^2(m+2)(4m-3)$

12. $5p^2-17p+6=(5p-2)(p-3)$

13. $10m^2+25m-6m-15$
$=(10m^2+25m)+(-6m-15)$
$=5m(2m+5)-3(2m+5)$
$=(2m+5)(5m-3)$

14. $36m^2+6m-6=6(6m^2+m-1)$
$=6(3m-1)(2m+1)$

15. $4m^2-20m+25=(2m)^2-2\cdot 2m\cdot 5+5^2$
$=(2m-5)^2$

16. $5x^2-xy-4y^2=(5x+4y)(x-y)$

17. $S=2\pi rh+2\pi r^2$
$=2\pi r\cdot h+2\pi r\cdot r$
$=2\pi r(h+r)$

18. $h=48t-16t^2$
$=16t\cdot 3+16t\cdot(-t)$
$=16t(3-t)$

Section 6.6

Preparing for Solving Polynomial Equations by Factoring

P1. $x+5=0$
$x+5-5=0-5$
$x=-5$
The solution set is $\{-5\}$.

P2. $2(x-4)-10=0$
$$2x-8-10=0$$
$$2x-18=0$$
$$2x-18+18=0+18$$
$$2x=18$$
$$\frac{2x}{2}=\frac{18}{2}$$
$$x=9$$
The solution set is {9}.

P3. $2x^2+3x-4$

(a) $x = 2$: $2(2)^2+3\cdot 2-4=8+6-4=10$

(b) $x = -1$: $2(-1)^2+3(-1)-4=2-3-4=-5$

6.6 Quick Checks

1. The Zero-Product Property states that if $ab = 0$, then either $\underline{a = 0}$ or $\underline{b = 0}$.

2. $x(x + 3) = 0$
$$x=0 \quad \text{or} \quad x+3=0$$
$$x=-3$$
Check $x = 0$: $x(x+3)=0$
$$0(0+3) \stackrel{?}{=} 0$$
$$0(3) \stackrel{?}{=} 0$$
$$0=0 \quad \text{True}$$
Check $x = -3$: $x(x+3)=0$
$$-3(-3+3) \stackrel{?}{=} 0$$
$$-3(0) \stackrel{?}{=} 0$$
$$0=0 \quad \text{True}$$
The solution set is {0, −3}.

3. $(x - 2)(4x + 5) = 0$
$$x-2=0 \quad \text{or} \quad 4x+5=0$$
$$x=2 \quad \text{or} \quad 4x=-5$$
$$x=-\frac{5}{4}$$
Check $x = 2$: $(x-2)(4x+5)=0$
$$(2-2)(4\cdot 2+5) \stackrel{?}{=} 0$$
$$0(13) \stackrel{?}{=} 0$$
$$0=0 \quad \text{True}$$
Check $x=-\frac{5}{4}$:
$$(x-2)(4x+5)=0$$
$$\left(-\frac{5}{4}-2\right)\left(4\left(-\frac{5}{4}\right)+5\right) \stackrel{?}{=} 0$$
$$\left(-\frac{13}{4}\right)(-5+5) \stackrel{?}{=} 0$$
$$-\frac{13}{4}\cdot 0 \stackrel{?}{=} 0$$
$$0=0 \quad \text{True}$$
The solution set is $\left\{2, -\frac{5}{4}\right\}$.

4. A $\underline{\text{quadratic}}$ equation is an equation that can be written in the form $ax^2+bx+c=0$, where a, b, and c are real numbers and $a \neq 0$.

5. Quadratic equations are also known as $\underline{\text{second-degree}}$ equations.

6. False; the standard form of the equation $3x+x^2=6$ is $x^2+3x-6=0$.

7. $$p^2-6p+8=0$$
$$(p-2)(p-4)=0$$
$$p-2=0 \quad \text{or} \quad p-4=0$$
$$p=2 \quad \text{or} \quad p=4$$
Check $p = 2$: $p^2-6p+8=0$
$$2^2-6\cdot 2+8 \stackrel{?}{=} 0$$
$$4-12+8 \stackrel{?}{=} 0$$
$$0=0 \quad \text{True}$$
Check $p = 4$: $p^2-6p+8=0$
$$4^2-6\cdot 4+8 \stackrel{?}{=} 0$$
$$16-24+8 \stackrel{?}{=} 0$$
$$0=0 \quad \text{True}$$
The solution set is {2, 4}.

8. $$2t^2-5t=3$$
$$2t^2-5t-3=0$$
$$(2t+1)(t-3)=0$$
$$2t+1=0 \quad \text{or} \quad t-3=0$$
$$2t=-1 \quad \text{or} \quad t=3$$
$$t=-\frac{1}{2}$$

Check $t = -\frac{1}{2}$:
$$2t^2 - 5t = 3$$
$$2\left(-\frac{1}{2}\right)^2 - 5\left(-\frac{1}{2}\right) \stackrel{?}{=} 3$$
$$2\left(\frac{1}{4}\right) + \frac{5}{2} \stackrel{?}{=} 3$$
$$\frac{1}{2} + \frac{5}{2} \stackrel{?}{=} 3$$
$$3 = 3 \quad \text{True}$$

Check $t = 3$:
$$2t^2 - 5t = 3$$
$$2(3)^2 - 5 \cdot 3 \stackrel{?}{=} 3$$
$$18 - 15 \stackrel{?}{=} 3$$
$$3 = 3 \quad \text{True}$$

The solution set is $\left\{-\frac{1}{2}, 3\right\}$.

9.
$$2x^2 + 3x = 5$$
$$2x^2 + 3x - 5 = 0$$
$$(2x + 5)(x - 1) = 0$$
$$2x + 5 = 0 \quad \text{or} \quad x - 1 = 0$$
$$2x = -5 \quad \text{or} \quad x = 1$$
$$x = -\frac{5}{2}$$

Check $x = -\frac{5}{2}$:
$$2x^2 + 3x = 5$$
$$2\left(-\frac{5}{2}\right)^2 + 3\left(-\frac{5}{2}\right) \stackrel{?}{=} 5$$
$$2\left(\frac{25}{4}\right) - \frac{15}{2} \stackrel{?}{=} 5$$
$$\frac{25}{2} - \frac{15}{2} \stackrel{?}{=} 5$$
$$5 = 5 \quad \text{True}$$

Check $x = 1$:
$$2x^2 + 3x = 5$$
$$2(1)^2 + 3 \cdot 1 \stackrel{?}{=} 5$$
$$2 + 3 \stackrel{?}{=} 5$$
$$5 = 5 \quad \text{True}$$

The solution set is $\left\{-\frac{5}{2}, 1\right\}$.

10.
$$z^2 + 20 = -9z$$
$$z^2 + 9z + 20 = 0$$
$$(z + 5)(z + 4) = 0$$
$$z + 5 = 0 \quad \text{or} \quad z + 4 = 0$$
$$z = -5 \quad \text{or} \quad z = -4$$

Check $z = -5$:
$$z^2 + 20 = -9z$$
$$(-5)^2 + 20 \stackrel{?}{=} -9(-5)$$
$$25 + 20 \stackrel{?}{=} 45$$
$$45 = 45 \quad \text{True}$$

Check $z = -4$:
$$z^2 + 20 = -9z$$
$$(-4)^2 + 20 \stackrel{?}{=} -9(-4)$$
$$16 + 20 \stackrel{?}{=} 36$$
$$36 = 36 \quad \text{True}$$

The solution set is $\{-5, -4\}$.

11.
$$5k^2 + 3k - 1 = 3 - 5k$$
$$5k^2 + 3k - 1 + 5k - 3 = 3 - 5k + 5k - 3$$
$$5k^2 + 8k - 4 = 0$$
$$(5k - 2)(k + 2) = 0$$
$$5k - 2 = 0 \quad \text{or} \quad k + 2 = 0$$
$$5k = 2 \quad \text{or} \quad k = -2$$
$$k = \frac{2}{5}$$

Substitute $k = \frac{2}{5}$ and $k = -2$ into the original equation to check. The solution set is $\left\{-2, \frac{2}{5}\right\}$.

12.
$$3x^2 + 9x = 4 - 2x$$
$$3x^2 + 9x + 2x - 4 = 4 - 2x + 2x - 4$$
$$3x^2 + 11x - 4 = 0$$
$$(3x - 1)(x + 4) = 0$$
$$3x - 1 = 0 \quad \text{or} \quad x + 4 = 0$$
$$3x = 1 \quad \text{or} \quad x = -4$$
$$x = \frac{1}{3}$$

Substitute $x = \frac{1}{3}$ and $x = -4$ into the original equation to check. The solution set is $\left\{\frac{1}{3}, -4\right\}$.

13. False;
$$x(x - 3) = 4$$
$$x^2 - 3x = 4$$
$$x^2 - 3x - 4 = 4 - 4$$
$$x^2 - 3x - 4 = 0$$
$$(x - 4)(x + 1) = 0$$
$$x - 4 = 0 \quad \text{or} \quad x + 1 = 0$$
$$x = 4 \quad \text{or} \quad x = -1$$

The solution set is $\{-1, 4\}$.

14. $(x-3)(x+5)=9$
$x^2+2x-15=9$
$x^2+2x-24=0$
$(x+6)(x-4)=0$
$x+6=0$ or $x-4=0$
$x=-6$ or $x=4$
Substitute $x = -6$ and $x = 4$ into the original equation to check. The solution set is $\{-6, 4\}$.

15. $(x+3)(2x-1)=7x-3x^2$
$2x^2+5x-3=7x-3x^2$
$5x^2-2x-3=0$
$(5x+3)(x-1)=0$
$5x+3=0$ or $x-1=0$
$5x=-3$ or $x=1$
$x=-\frac{3}{5}$
Substitute $x=-\frac{3}{5}$ and $x = 1$ into the original equation to check. The solution set is $\left\{-\frac{3}{5}, 1\right\}$.

16. $9p^2+16=24p$
$9p^2-24p+16=0$
$(3p-4)^2=0$
$3p-4=0$ or $3p-4=0$
$3p=4$ or $3p=4$
$p=\frac{4}{3}$ or $p=\frac{4}{3}$
Substitute $p=\frac{4}{3}$ into the original equation to check. The solution set is $\left\{\frac{4}{3}\right\}$.

17. $4x^2+12x-72=0$
$4(x^2+3x-18)=0$
$4(x+6)(x-3)=0$
$4=0$ or $x+6=0$ or $x-3=0$
$x=-6$ or $x=3$
The statement $4 = 0$ is false. Substitute $x = -6$ and $x = 3$ into the original equation to check. The solution set is $\{-6, 3\}$.

18. $-2x^2+2x=-12$
$-2x^2+2x+12=0$
$-2(x^2-x-6)=0$
$-2(x+2)(x-3)=0$
$-2=0$ or $x+2=0$ or $x-3=0$
$x=-2$ or $x=3$
The statement $-2 = 0$ is false. Substitute $x = -2$ and $x = 3$ into the original equation to check. The solution set is $\{-2, 3\}$.

19. $-16t^2+80t=64$
$-16t^2+80t-64=0$
$-16(t^2-5t+4)=0$
$-16(t-1)(t-4)=0$
$-16=0$ or $t-1=0$ or $t-4=0$
$t=1$ or $t=4$
The statement $-16 = 0$ is false. The toy rocket is 64 feet from the ground after 1 second and 4 seconds.

20. $x^3-4x^2-12x=0$
$x(x^2-4x-12)=0$
$x(x+2)(x-6)=0$
$x=0$ or $x+2=0$ or $x-6=0$
$x=2$ $x=6$
The solution set is $\{0, 2, 6\}$.
True, x^3-4x^2-12x can be solved using the Zero-Product Property.

21. $(4x-5)(x^2-9)=0$
$(4x-5)(x-3)(x+3)=0$
$4x-5=0$ or $x-3=0$ or $x+3=0$
$4x=5$ or $x=3$ or $x=-3$
$x=\frac{5}{4}$
Substitute $x=\frac{5}{4}$, $x = 3$, and $x = -3$ into the original equation to check. The solution set is $\left\{\frac{5}{4}, 3, -3\right\}$.

22. $3x^3+9x^2+6x=0$
$3x(x^2+3x+2)=0$
$3x(x+1)(x+2)=0$
$3x=0$ or $x+1=0$ or $x+2=0$
$x=0$ or $x=-1$ or $x=-2$
Substitute $x = 0$, $x = -1$, and $x = -2$ into the original equation to check. The solution set is $\{0, -1, -2\}$.

6.6 Exercises

23. $2x(x+4)=0$
$2x=0$ or $x+4=0$
$x=0$ $\quad x=-4$
The solution set is $\{-4, 0\}$.

25. $(n+3)(n-9)=0$
$n+3=0$ or $n-9=0$
$n=-3$ $\quad n=9$
The solution set is $\{-3, 9\}$.

27. $(3p+1)(p-5)=0$
$3p+1=0$ or $p-5=0$
$3p=-1$ $\quad p=5$
$p=-\frac{1}{3}$
The solution set is $\left\{-\frac{1}{3}, 5\right\}$.

29. $(5y+3)(2-7y)=0$
$5y+3=0$ or $2-7y=0$
$5y=-3$ $\quad -7y=-2$
$y=-\frac{3}{5}$ $\quad y=\frac{2}{7}$
The solution set is $\left\{-\frac{3}{5}, \frac{2}{7}\right\}$.

31. The highest exponent is 1; the equation is linear.

33. The highest exponent is 2; the equation is quadratic.

35. $x^2-3x-4=0$
$(x-4)(x+1)=0$
$x-4=0$ or $x+1=0$
$x=4$ $\quad x=-1$
The solution set is $\{-1, 4\}$.

37. $n^2+9n+14=0$
$(n+7)(n+2)=0$
$n+7=0$ or $n+2=0$
$n=-7$ $\quad n=-2$
The solution set is $\{-7, -2\}$.

39. $4x^2+2x=0$
$2x(2x+1)=0$
$2x=0$ or $2x+1=0$
$x=0$ $\quad 2x=-1$
$x=-\frac{1}{2}$
The solution set is $\left\{-\frac{1}{2}, 0\right\}$.

41. $2x^2-3x-2=0$
$(2x+1)(x-2)=0$
$2x+1=0$ or $x-2=0$
$2x=-1$ $\quad x=2$
$x=-\frac{1}{2}$
The solution set is $\left\{-\frac{1}{2}, 2\right\}$.

43. $a^2-6a+9=0$
$(a-3)^2=0$
$a-3=0$ or $a-3=0$
$a=3$ $\quad a=3$
The solution set is $\{3\}$.

45. $6x^2=36x$
$6x^2-36x=0$
$6x(x-6)=0$
$6x=0$ or $x-6=0$
$x=0$ $\quad x=6$
The solution set is $\{0, 6\}$.

47. $n^2-n=6$
$n^2-n-6=0$
$(n-3)(n+2)=0$
$n-3=0$ or $n+2=0$
$n=3$ $\quad n=-2$
The solution set is $\{-2, 3\}$.

49. $b^2+18=11b$
$b^2-11b+18=0$
$(b-9)(b-2)=0$
$b-9=0$ or $b-2=0$
$b=9$ $\quad b=2$
The solution set is $\{2, 9\}$.

51.
$$1-5m=-4m^2$$
$$4m^2-5m+1=0$$
$$(4m-1)(m-1)=0$$
$$4m-1=0 \quad \text{or} \quad m-1=0$$
$$4m=1 \qquad m=1$$
$$m=\frac{1}{4}$$
The solution set is $\left\{\frac{1}{4}, 1\right\}$.

53.
$$n(n-2)=24$$
$$n^2-2n=24$$
$$n^2-2n-24=0$$
$$(n-6)(n+4)=0$$
$$n-6=0 \quad \text{or} \quad n+4=0$$
$$n=6 \qquad n=-4$$
The solution set is $\{-4, 6\}$.

55.
$$(x-2)(x-3)=56$$
$$x^2-5x+6=56$$
$$x^2-5x-50=0$$
$$(x-10)(x+5)=0$$
$$x-10=0 \quad \text{or} \quad x+5=0$$
$$x=10 \qquad x=-5$$
The solution set is $\{-5, 10\}$.

57.
$$(c+2)^2=9$$
$$c^2+4c+4=9$$
$$c^2+4c-5=0$$
$$(c+5)(c-1)=0$$
$$c+5=0 \quad \text{or} \quad c-1=0$$
$$c=-5 \qquad c=1$$
The solution set is $\{-5, 1\}$.

59.
$$2x^3+2x^2-12x=0$$
$$2x(x^2+x-6)=0$$
$$2x(x+3)(x-2)=0$$
$$2x=0 \quad \text{or} \quad x+3=0 \quad \text{or} \quad x-2=0$$
$$x=0 \qquad x=-3 \qquad x=2$$
The solution set is $\{-3, 0, 2\}$.

61.
$$y^3+3y^2-4y-12=0$$
$$y^2(y+3)-4(y+3)=0$$
$$(y+3)(y^2-4)=0$$
$$(y+3)(y-2)(y+2)=0$$
$$y+3=0 \quad \text{or} \quad y-2=0 \quad \text{or} \quad y+2=0$$
$$y=-3 \qquad y=2 \qquad y=-2$$
The solution set is $\{-3, -2, 2\}$.

63.
$$2x^3+3x^2=8x+12$$
$$2x^3+3x^2-8x-12=0$$
$$x^2(2x+3)-4(2x+3)=0$$
$$(2x+3)(x^2-4)=0$$
$$(2x+3)(x-2)(x+2)=0$$
$$2x+3=0 \quad \text{or} \quad x-2=0 \quad \text{or} \quad x+2=0$$
$$2x=-3 \qquad x=2 \qquad x=-2$$
$$x=-\frac{3}{2}$$
The solution set is $\left\{-\frac{3}{2}, 2, -2\right\}$.

65.
$$(5x+3)(x-4)=0$$
$$5x+3=0 \quad \text{or} \quad x-4=0$$
$$5x=-3 \qquad x=4$$
$$x=-\frac{3}{5}$$
The solution set is $\left\{-\frac{3}{5}, 4\right\}$.

67.
$$p^2-p-20=0$$
$$(p-5)(p+4)=0$$
$$p-5=0 \quad \text{or} \quad p+4=0$$
$$p=5 \qquad p=-4$$
The solution set is $\{-4, 5\}$.

69.
$$4w+3=2w-7$$
$$2w+3=-7$$
$$2w=-10$$
$$w=-5$$
The solution set is $\{-5\}$.

71. $4a^2 - 25a = 21$
$4a^2 - 25a - 21 = 0$
$(4a + 3)(a - 7) = 0$
$4a + 3 = 0$ or $a - 7 = 0$
$4a = -3$ $\quad a = 7$
$a = -\frac{3}{4}$

The solution set is $\left\{-\frac{3}{4}, 7\right\}$.

73. $2a(a + 1) = a^2 + 8$
$2a^2 + 2a = a^2 + 8$
$a^2 + 2a - 8 = 0$
$(a + 4)(a - 2) = 0$
$a + 4 = 0$ or $a - 2 = 0$
$a = -4$ $\quad a = 2$
The solution set is {−4, 2}.

75. $2x^3 + x^2 = 32x + 16$
$2x^3 + x^2 - 32x - 16 = 0$
$x^2(2x + 1) - 16(2x + 1) = 0$
$(2x + 1)(x^2 - 16) = 0$
$(2x + 1)(x - 4)(x + 4) = 0$
$2x + 1 = 0$ or $x - 4 = 0$ or $x + 4 = 0$
$2x = -1$ $\quad x = 4$ $\quad x = -4$
$x = -\frac{1}{2}$

The solution set is $\left\{-4, -\frac{1}{2}, 4\right\}$.

77. $4(b - 3) - 3b = 8$
$4b - 12 - 3b = 8$
$b - 12 = 8$
$b = 20$
The solution set is {20}.

79. $y^2 + 5y = 5(y + 20)$
$y^2 + 5y - 5(y + 20) = 0$
$y^2 + 5y - 5y - 100 = 0$
$y^2 - 100 = 0$
$(y + 10)(y - 10) = 10$
$y + 10 = 0$ or $y - 10 = 0$
$y = -10$ $\quad y = 10$
The solution set is {−10, 10}.

81. $(a + 3)(a - 5)(3a + 2) = 0$
$a + 3 = 0$ or $a - 5 = 0$ or $3a + 2 = 0$
$a = -3$ $\quad a = 5$ $\quad 3a = -2$
$a = -\frac{2}{3}$

The solution set is $\left\{-3, -\frac{2}{3}, 5\right\}$.

83. $(2k - 3)(2k^2 - 9k - 5) = 0$
$(2k - 3)(2k + 1)(k - 5) = 0$
$2k - 3 = 0$ or $2k + 1 = 0$ or $k - 5 = 0$
$2k = 3$ $\quad 2k = -1$ $\quad k = 5$
$k = \frac{3}{2}$ $\quad k = -\frac{1}{2}$

The solution set is $\left\{-\frac{1}{2}, \frac{3}{2}, 5\right\}$.

85. $(w - 3)^2 = 9 + 2w$
$w^2 - 6w + 9 = 9 + 2w$
$w^2 - 8w = 0$
$w(w - 8) = 0$
$w = 0$ or $w - 8 = 0$
$w = 8$
The solution set is {0, 8}.

87. $\frac{1}{2}x^2 + \frac{5}{4}x = 3$
$4\left(\frac{1}{2}x^2 + \frac{5}{4}x\right) = 4(3)$
$2x^2 + 5x = 12$
$2x^2 + 5x - 12 = 0$
$(2x - 3)(x + 4) = 0$
$2x - 3 = 0$ or $x + 4 = 0$
$2x = 3$ $\quad x = -4$
$x = \frac{3}{2}$

The solution set is $\left\{-4, \frac{3}{2}\right\}$.

89. $8x^2 + 44x = 24$

$8x^2 + 44x - 24 = 0$

$4(2x^2 + 11x - 6) = 0$

$4(2x - 1)(x + 6) = 0$

$4 = 0$ or $2x - 1 = 0$ or $x + 6 = 0$

False $\quad 2x = 1 \quad x = -6$

$x = \frac{1}{2}$

The solution set is $\left\{-6, \frac{1}{2}\right\}$.

91. $-16t^2 + 64t + 80 = 128$

$-16t^2 + 64t - 48 = 0$

$-16(t^2 - 4t + 3) = 0$

$-16(t - 3)(t - 1) = 0$

$-16 = 0$ or $t - 3 = 0$ or $t - 1 = 0$

false $\quad t = 3 \quad t = 1$

The ball is 128 feet from the ground after 1 second and after 3 seconds.

93. $-16t^2 + 40 = 24$

$-16t^2 + 16 = 0$

$-16(t^2 - 1) = 0$

$-16(t - 1)(t + 1) = 0$

$-16 = 0$ or $t - 1 = 0$ or $t + 1 = 0$

false $\quad t = 1 \quad t = -1$

The negative value has no meaning. The balloon is 24 feet from the ground after 1 second.

95. Let n and $n + 1$ be the integers.

$n(n + 1) = 12$

$n^2 + n - 12 = 0$

$(n + 4)(n - 3) = 0$

$n + 4 = 0$ or $n - 3 = 0$

$n = -4 \quad n = 3$

$n + 1 = -3 \quad n + 1 = 4$

The numbers are either –4 and –3 or 3 and 4.

97. Let n, $n + 2$, and $n + 4$ be the integers.

$n(n + 4) = 96$

$n^2 + 4n - 96 = 0$

$(n + 12)(n - 8) = 0$

$n + 12 = 0$ or $n - 8 = 0$

$n = -12 \quad n = 8$

$n + 2 = -10 \quad n + 2 = 10$

$n + 4 = -8 \quad n + 4 = 12$

The numbers are either –12, –10, and 8 or 8, 10, and 12.

99. Let n and $n + 2$ be the width and length of the rectangle.

$n(n + 2) = 255$

$n^2 + 2n - 255 = 0$

$(n + 17)(n - 15) = 0$

$n + 17 = 0$ or $n - 15 = 0$

$n = -17 \quad n = 15$

$n + 2 = 17$

The negative value has no meaning. The rectangle is 15 by 17.

101. $\frac{t^2 - t}{2} = 28$

$t^2 - t = 56$

$t^2 - t - 56 = 0$

$(t - 8)(t + 7) = 0$

$t - 8 = 0$ or $t + 7 = 0$

$t = 8 \quad t = -7$

The negative value has no meaning. The league has 8 teams.

103. Answers may vary. One solution is $(x - 3)(x + 5) = 0$; 2nd degree

105. Answers may vary. One solution is $(z - 6)^2 = 0$; 2nd degree

107. Answers may vary. One solution is $(x + 3)(x - 1)(x - 5) = 0$; 3rd degree

109. $x^2 - ax + bx - ab = 0$

$x(x - a) + b(x - a) = 0$

$(x + b)(x - a) = 0$

$x = -b, a$

111. $2x^3 - 4ax^2 = 0$

$2x^2(x - 2a) = 0$

$x = 0, 2a$

113. The student divided by the variable x instead of writing the quadratic equation in standard form, factoring, and using the Zero-Product Property. The correct solution is

$$15x^2 = 5x$$
$$15x^2 - 5x = 0$$
$$5x(3x - 1) = 0$$
$$5x = 0 \quad \text{or} \quad 3x - 1 = 0$$
$$x = 0 \quad \text{or} \quad 3x = 1$$
$$x = \frac{1}{3}$$

115. We write a quadratic equation in standard form, $ax^2 + bx + c = 0$, so that when the quadratic polynomial on the left side of the equation is factored, we can apply the Zero-Product Property, which says that if $a \cdot b = 0$, then $a = 0$ or $b = 0$.

Section 6.7

Preparing for Modeling and Solving Problems with Quadratic Equations

P1. $15^2 = 15 \cdot 15 = 225$

P2.
$$x^2 - 5x - 14 = 0$$
$$(x + 2)(x - 7) = 0$$
$$x + 2 = 0 \quad \text{or} \quad x - 7 = 0$$
$$x = -2 \qquad x = 7$$
The solution set is $\{-2, 7\}$.

6.7 Quick Checks

1. (a) Let $h = 384$ and solve for t.
$$-16t^2 + 160t = 384$$
$$-16t^2 + 160t - 384 = 0$$
$$-16(t^2 - 10t + 24) = 0$$
$$-16(t - 4)(t - 6) = 0$$
$$-16 = 0 \quad \text{or} \quad t - 4 = 0 \quad \text{or} \quad t - 6 = 0$$
$$t = 4 \quad \text{or} \quad t = 6$$
The equation $-16 = 0$ is false, so the solutions are $t = 4$ and $t = 6$. The height of the rocket will be 384 feet after 4 seconds and after 6 seconds.

(b) The rocket will strike the ground when $h = 0$, so we let $h = 0$ and solve for t.
$$-16t^2 + 160t = 0$$
$$-16t(t - 10) = 0$$
$$-16t = 0 \quad \text{or} \quad t - 10 = 0$$
$$t = 0 \quad \text{or} \quad t = 10$$
$t = 0$ indicates that the rocket was fired from the height of 0 feet (ground level). The rocket strikes the ground after 10 seconds.

2. Let w represent the width of the plot. Then the length is represented by $2w - 3$. We are given that the area is 104 square kilometers.
$$(\text{length})(\text{width}) = \text{area}$$
$$(2w - 3)(w) = 104$$
$$2w^2 - 3w = 104$$
$$2w^2 - 3w - 104 = 0$$
$$(2w + 13)(w - 8) = 0$$
$$2w + 13 = 0 \quad \text{or} \quad w - 8 = 0$$
$$2w = -13 \quad \text{or} \quad w = 8$$
$$w = -\frac{13}{2}$$
Since w represents the width of the plot, we discard the negative solution.
$w = 8$: $2w - 3 = 2(8) - 3 = 16 - 3 = 13$
The width of the plot is 8 kilometers and the length is 13 kilometers.

3. Let h represent the height of the triangle. Then the base is represented by $h + 4$. We are given that the area is 48 square yards.
$$\frac{1}{2}(\text{base})(\text{height}) = \text{area}$$
$$\frac{1}{2}(h + 4)(h) = 48$$
$$h^2 + 4h = 96$$
$$h^2 + 4h - 96 = 0$$
$$(h + 12)(h - 8) = 0$$
$$h + 12 = 0 \quad \text{or} \quad h - 8 = 0$$
$$h = -12 \quad \text{or} \quad h = 8$$
Since h represents the height of the triangle, we discard the negative solution.
$h = 8$: $h + 4 = 8 + 4 = 12$
The base is 12 yards and the height is 8 yards.

4. In the triangle pictured, the legs have lengths x and $x - 3$, while the hypotenuse has length 15.
$$a^2 + b^2 = c^2$$
$$x^2 + (x - 3)^2 = 15^2$$
$$x^2 + x^2 - 6x + 9 = 225$$
$$2x^2 - 6x + 9 = 225$$
$$2x^2 - 6x - 216 = 0$$
$$2(x^2 - 3x - 108) = 0$$
$$2(x + 9)(x - 12) = 0$$
$$2 = 0 \quad \text{or} \quad x + 9 = 0 \quad \text{or} \quad x - 12 = 0$$
$$x = -9 \quad \text{or} \quad x = 12$$
The equation $2 = 0$ is false and the solution $x = -9$ makes no sense, so $x = 12$.
$x = 12$: $x - 3 = 12 - 3 = 9$
The legs have lengths $x = 12$ and $x - 3 = 9$.

5. Let x = length of first leg in inches
$x + 17$ = length of second leg in inches
hypotenuse = 25 inches

$$x^2+(x+17)^2=25^2$$
$$x^2+x^2+34x+289=625$$
$$2x^2+34x-336=0$$
$$2(x^2+17x-168)=0$$
$$2(x+24)(x-7)=0$$

x cannot be negative, so the only solution for
$x = 7$ inches
$x + 17 = 24$ inches

6. Let x represent the height of the screen. Then $x + 2$ represents the width of the screen. We are given that the length of the diagonal is 10 inches.

$$a^2+b^2=c^2$$
$$x^2+(x+2)^2=10^2$$
$$x^2+x^2+4x+4=100$$
$$2x^2+4x+4=100$$
$$2x^2+4x-96=0$$
$$2(x^2+2x-48)=0$$
$$2(x+8)(x-6)=0$$
$$2=0 \quad \text{or} \quad x+8=0 \quad \text{or} \quad x-6=0$$
$$x=-8 \quad \text{or} \quad x=6$$

The equation $2 = 0$ is false and the solution $x = -8$ makes no sense, so $x = 6$.
$x = 6$: $x + 2 = 6 + 2 = 8$
The height of the television screen is 6 inches and the width is 8 inches.

6.7 Exercises

7.
$$x(2x+6)=56$$
$$2x^2+6x-56=0$$
$$2(x^2+3x-28)=0$$
$$2(x+7)(x-4)=0$$
$$2=0 \quad \text{or} \quad x+7=0 \quad \text{or} \quad x-4=0$$
$$\text{false} \qquad x=-7 \qquad x=4$$
$$2x+6=14$$

Discard $x = -7$. The sides are 4 and 14.

9.
$$x(x-7)=18$$
$$x^2-7x-18=0$$
$$(x-9)(x+2)=0$$
$$x=9 \quad \text{or} \quad x=-2$$
$$x-7=2$$

Discard $x = -2$. The sides are 2 and 9.

11.
$$\frac{1}{2}x(3x+2)=104$$
$$x(3x+2)=208$$
$$3x^2+2x-208=0$$
$$(3x+26)(x-8)=0$$
$$3x+26=0 \quad \text{or} \quad x=8$$
$$3x=-26 \qquad 3x+2=26$$
$$x=-\frac{26}{3}$$

Discard $x=-\frac{26}{3}$. The base is 26, and the height is 8.

13.
$$\frac{1}{2}(3x-6)(2x)=144$$
$$x(3x-6)=144$$
$$3x^2-6x-144=0$$
$$3(x^2-2x-48)=0$$
$$3(x-8)(x+6)=0$$
$$3=0 \quad \text{or} \quad x-8=0 \quad \text{or} \quad x+6=0$$
$$\text{false} \qquad x=8 \qquad x=-6$$

Discard $x = -6$. The base is $3x - 6 = 18$, and the height is $2x = 16$.

15.
$$(2x+1)(2x-1)=143$$
$$4x^2-1=143$$
$$4x^2-144=0$$
$$4(x^2-36)=0$$
$$4(x+6)(x-6)=0$$
$$4=0 \quad \text{or} \quad x+6=0 \quad \text{or} \quad x-6=0$$
$$\text{false} \qquad x=-6 \qquad x=6$$

Discard $x = -6$. The base is $2x + 1 = 13$ and the height is $2x - 1 = 11$.

17.
$$\frac{1}{2}(4)[(2x-10)+2x]=192$$
$$2(4x-10)=192$$
$$8x-20=192$$
$$8x=212$$
$$x=\frac{53}{2}$$

The base $B = 2x = 53$ and the base $b = 2x - 10 = 43$.

19. $x^2+(x+3)^2=15^2$

$x^2+x^2+6x+9=225$

$2x^2+6x-216=0$

$2(x^2+3x-108)=0$

$2(x+12)(x-9)=0$

$2=0$ or $x+12=0$ or $x-9=0$

false $\quad x=-12 \quad x=9$

Discard $x=-12$. The legs are 9 and $x+3=12$.

21. $(2x)^2+(2x-7)^2=13^2$

$4x^2+4x^2-28x+49=169$

$8x^2-28x-120=0$

$4(2x^2-7x-30)=0$

$4(2x+5)(x-6)=0$

$4=0$ or $2x+5=0$ or $x-6=0$

false $\quad 2x=-5 \quad x=6$

$x=-\frac{5}{2}$

Discard $x=-\frac{5}{2}$. The legs are $2x=12$ and $2x-7=5$.

23.

Time, in Sec.	0	0.5	1	1.5	2	2.5	3	3.5	4
Height, in Feet	256	252	240	220	192	156	112	60	0

25. $0=-16t^2+96t$

$0=-16t(t-6)$

$-16t=0$ or $t-6=0$

$t=0 \quad t=6$

Since $t=0$ corresponds to the time that the rocket is fired, the rocket will hit the ground after 6 seconds.

27. Let w = width of the room. Then $w+8$ = length of the room.

$w(w+8)=48$

$w^2+8w-48=0$

$(w+12)(w-4)=0$

$w+12=0$ or $w-4=0$

$w=-12 \quad w=4$

Discard $w=-12$. The width is 4 meters, and the length is $w+8=12$ meters.

29. Let b = base of the triangle. Then $3b$ = height of the sail.

$$\frac{1}{2}b(3b) = 54$$
$$3b^2 = 108$$
$$b^2 = 36$$
$$b^2 - 36 = 0$$
$$(b-6)(b+6) = 0$$
$$b-6=0 \quad \text{or} \quad b+6=0$$
$$b=6 \qquad\qquad b=-6$$

Discard $b = -6$. The base of the sail is 6 feet, and the height is $3b = 18$ feet.

31. Let h = height of the TV.

$$h^2 + 40^2 = 50^2$$
$$h^2 + 1600 = 2500$$
$$h^2 - 900 = 0$$
$$(h+30)(h-30) = 0$$
$$h+30=0 \quad \text{or} \quad h-30=0$$
$$h=-30 \qquad\qquad h=30$$

Discard $h = -30$. The height of the TV is 30 inches.

33. Let w = width of the rectangle. Then $2w + 1$ = length of the rectangle.

$$w(2w+1) = 300$$
$$2w^2 + w - 300 = 0$$
$$(2w+25)(w-12) = 0$$
$$2w+25=0 \quad \text{or} \quad w-12=0$$
$$2w=-25 \qquad\qquad w=12$$
$$w=-\frac{25}{2}$$

Discard $w=-\frac{25}{2}$. The width is 12 mm, and the length is $2w + 1 = 25$ mm.

35. Let s = a side of the square. Then $s - 6$ = width of the rectangle, and $2s + 5$ = length of the rectangle.

$$(2s+5)(s-6) = s^2$$
$$2s^2 - 7s - 30 = s^2$$
$$s^2 - 7s - 30 = 0$$
$$(s-10)(s+3) = 0$$
$$s-10=0 \quad \text{or} \quad s+3=0$$
$$s=10 \qquad\qquad s=-3$$

Discard $s = -3$. The width of the rectangle is $s - 6 = 4$ cm, and the length is $2s + 5 = 25$ cm.

37. (a) Let w = width of the TV screen. Then $w + 17$ = length of the TV screen. Including the casing, the width is $w + 3$ and the length is $w + 17 + 3 = w + 20$.

$$(w+3)^2 + (w+20)^2 = 53^2$$
$$w^2 + 6w + 9 + w^2 + 40w + 400 = 2809$$
$$2w^2 + 46w - 2400 = 0$$
$$2(w^2 + 23w - 1200) = 0$$
$$2(w+48)(w-25) = 0$$
$$2=0 \quad \text{or} \quad w+48=0 \quad \text{or} \quad w-25=0$$
$$\text{false} \qquad w=-48 \qquad w=25$$

Discard $w = -48$. The TV screen is 25 inches wide and $w + 17 = 42$ inches long.

(b) $25 + 3 = 28$; $42 + 3 = 45$
The opening should be 28 inches by 45 inches.

39. Let h = height of the pole. Then $h + 2$ is the distance along the ground.

$$h^2 + (h+2)^2 = 10^2$$
$$h^2 + h^2 + 4h + 4 = 100$$
$$2h^2 + 4h - 96 = 0$$
$$2(h^2 + 2h - 48) = 0$$
$$2(h-6)(h+8) = 0$$
$$2=0 \quad \text{or} \quad h-6=0 \quad \text{or} \quad h+8=0$$
$$\text{false} \qquad h=6 \qquad h=-8$$

Discard $h = -8$. The pole is 6 feet.

41. Let x = width of the garden. Then $28 - 2x$ = length of the garden.

$$x(28-2x) = 98$$
$$28x - 2x^2 - 98 = 0$$
$$-2(x^2 - 14x + 49) = 0$$
$$-2(x-7)(x-7) = 0$$
$$-2=0 \quad \text{or} \quad x-7=0$$
$$\text{false} \qquad x=7$$

The width is 7 feet and the length is $28 - 2x = 14$ feet.

43. If you have two positive numbers that satisfy the projectile motion word problem, then one answer is the height of the object one its upward path, and the other represents the height of the object on its downward path.

Chapter 6 Review

1. $24 = 2 \cdot 2 \cdot 2 \cdot 3$
$36 = 2 \cdot 2 \cdot \quad 3 \cdot 3$
The common factors are 2, 2, and 3. The GCF is $2 \cdot 2 \cdot 3 = 12$.

2. $27 = \quad 3 \cdot 3 \cdot 3$
$54 = 2 \cdot 3 \cdot 3 \cdot 3$
The common factors are 3, 3, and 3. The GCF is $3 \cdot 3 \cdot 3 = 27$.

3. $10 = 2 \cdot \quad 5$
$20 = 2 \cdot 2 \cdot \quad 5$
$30 = 2 \cdot \quad 3 \cdot 5$
The common factors are 2 and 5. The GCF is $2 \cdot 5 = 10$.

4. $8 = 2 \cdot 2 \cdot 2$
$16 = 2 \cdot 2 \cdot 2 \cdot 2$
$28 = 2 \cdot 2 \cdot \quad 7$
The common factors are 2 and 2. The GCF is $2 \cdot 2 = 4$.

5. $x^4 = x \cdot x \cdot x \cdot x$
$x^2 = x \cdot x$
$x^8 = x \cdot x \cdot x \cdot x \cdot x \cdot x \cdot x \cdot x$
Each expression contains two factors of x. The GCF is x^2.

6. $m^3 = m \cdot m \cdot m$
$m = m$
$m^5 = m \cdot m \cdot m \cdot m \cdot m$
The only common factor is m. The GCF is m.

7. The GCF of 30 and 45 is 15.
The GCF of a^2 and a is a.
The GCF of b^4 and b^2 is b^2.
The GCF is $15ab^2$.

8. The GCF of 18 and 24 is 6.
The GCF of x^4 and x^3 is x^3.
The GCF of y^2 and y^5 is y^2.
The GCF of z^3 and z is z.
The GCF is $6x^3y^2z$.

9. The GCF of 4 and 6 is 2.
The GCF of $(2a+1)^2$ and $(2a+1)^3$ is $(2a+1)^2$.
The GCF is $2(2a+1)^2$.

10. The GCF of 9 and 18 is 9.
The GCF of $(x - y)$ and $(x - y)$ is $(x - y)$.
The GCF is $9(x - y)$.

11. $-18a^3 - 24a^2 = -6a^2(3a) - 6a^2(4)$
$= -6a^2(3a+4)$

12. $-9x^2 + 12x = -3x(3x) - 3x(-4)$
$= -3x(3x-4)$

13. $15y^2z + 5y^7z + 20y^3z$
$= 5y^2z(3) + 5y^2z(y^5) + 5y^2z(4y)$
$= 5y^2z(3 + y^5 + 4y)$

14. $7x^3y - 21x^2y^2 + 14xy^3$
$= 7xy(x^2) + 7xy(-3xy) + 7xy(2y^2)$
$= 7xy(x^2 - 3xy + 2y^2)$

15. $x(5-y) + 2(5-y) = (5-y)x + (5-y)2$
$= (5-y)(x+2)$

16. $z(a+b) + y(a+b) = (a+b)z + (a+b)y$
$= (a+b)(z+y)$

17. $5m^2 + 2mn + 15mn + 6n^2$
$= (5m^2 + 2mn) + (15mn + 6n^2)$
$= m(5m+2n) + 3n(5m+2n)$
$= (5m+2n)(m+3n)$

18. $2xy + y^2 + 2x^2 + xy$
$= (2xy + y^2) + (2x^2 + xy)$
$= y(2x+y) + x(2x+y)$
$= (2x+y)(y+x)$

19. $8x + 16 - xy - 2y = (8x+16) + (-xy - 2y)$
$= 8(x+2) - y(x+2)$
$= (x+2)(8-y)$

20. $xy^2 + x - 3y^2 - 3 = (xy^2 + x) + (-3y^2 - 3)$
$= x(y^2+1) - 3(y^2+1)$
$= (y^2+1)(x-3)$

21. $3(2) = 6$ and $3 + 2 = 5$
$x^2 + 5x + 6 = (x+3)(x+2)$

22. $2(4) = 8$ and $2 + 4 = 6$
$x^2 + 6x + 8 = (x+2)(x+4)$

23. $x^2 - 21 - 4x = x^2 - 4x - 21 = (x-7)(x+3)$

24. $3x + x^2 - 10 = x^2 + 3x - 10 = (x+5)(x-2)$

25. $m^2 + m + 20$ cannot be factored; it is prime.

26. $m^2 - 6m - 5$ cannot be factored; it is prime.

27. $x^2 - 8xy + 15y^2 = (x-3y)(x-5y)$

28. $m^2 + 4mn - 5n^2 = (m+5n)(m-n)$

29. $-p^2 - 11p - 30 = -(p^2 + 11p + 30)$
$= -(p+6)(p+5)$

30. $-y^2 + 2y + 15 = -(y^2 - 2y - 15)$
$= -(y-5)(y+3)$

31. $3x^3 + 33x^2 + 36x = 3x(x^2 + 11x + 12)$

32. $4x^2 + 36x + 32 = 4(x^2 + 9x + 8)$
$= 4(x+1)(x+8)$

33. $2x^2 - 2xy - 84y^2 = 2(x^2 - xy - 42y^2)$
$= 2(x-7y)(x+6y)$

34. $4y^3 + 12y^2 - 40y = 4y(y^2 + 3y - 10)$
$= 4y(y+5)(y-2)$

35. $20(-6) = -120 = 5(-24)$ and $20 - 6 = 14$
$5y^2 + 14y - 24 = 5y^2 + 20y - 6y - 24$
$= 5y(y+4) - 6(y+4)$
$= (y+4)(5y-6)$

36. $-42(1) = -42 = 6(-7)$ and $-42 + 1 = -41$
$6y^2 - 41y - 7 = 6y^2 - 42y + y - 7$
$= 6y(y-7) + 1(y-7)$
$= (y-7)(6y+1)$

37. $-5x + 2x^2 + 3 = 2x^2 - 5x + 3$
$= (2x-3)(x-1)$

38. $23x + 6x^2 + 7 = 6x^2 + 23x + 7$
$21 \cdot 2 = 42 = 6 \cdot 7$ and $21 + 2 = 23$
$6x^2 + 23x + 7 = 6x^2 + 21x + 2x + 7$
$= 3x(2x+7) + 1(2x+7)$
$= (2x+7)(3x+1)$

39. $2x^2 - 7x - 6$ cannot be factored; it is prime.

40. $6 \cdot 12 = 72 = 8 \cdot 9$ and $6 + 12 = 18$
$8m^2 + 18m + 9 = 8m^2 + 6m + 12m + 9$
$= 2m(4m+3) + 3(4m+3)$
$= (4m+3)(2m+3)$

41. $9m^3 + 30m^2n + 21mn^2$
$= 3m(3m^2 + 10mn + 7n^2)$
$= 3m(3m^2 + 3mn + 7mn + 7n^2)$
$= 3m[3m(m+n) + 7n(m+n)]$
$= 3m(m+n)(3m+7n)$

42. $14m^2 + 16mn + 2n^2$
$= 2(7m^2 + 8mn + n^2)$
$= 2(7m^2 + mn + 7mn + n^2)$
$= 2[m(7m+n) + n(7m+n)]$
$= 2(7m+n)(m+n)$

43. $15x^3 + x^2 - 2x = x(15x^2 + x - 2)$
$= x(15x^2 + 6x - 5x - 2)$
$= x[3x(5x+2) - 1(5x+2)]$
$= x(5x+2)(3x-1)$

44. $6p^4 + p^3 - p^2 = p^2(6p^2 + p - 1)$
$= p^2(3p-1)(2p+1)$

45. $4x^2 - 12x + 9 = (2x)^2 - 2 \cdot 2x \cdot 3 + 3^2$
$= (2x-3)^2$

46. $x^2 - 10x + 25 = x^2 - 2 \cdot x \cdot 5 + 5^2 = (x-5)^2$

47. $x^2 + 6xy + 9y^2 = x^2 + 2 \cdot x \cdot 3y + (3y)^2$
$= (x+3y)^2$

48. $9x^2 + 24xy + 4y^2$ cannot be factored; it is prime.

49. $8m^2+8m+2=2(4m^2+4m+1)$
$=2[(2m)^2+2\cdot 2m\cdot 1+1^2]$
$=2(2m+1)^2$

50. $2m^2-24m+72=2(m^2-12m+36)$
$=2(m^2-2\cdot m\cdot 6+6^2)$
$=2(m-6)^2$

51. $4x^2-25y^2=(2x)^2-(5y)^2$
$=(2x-5y)(2x+5y)$

52. $49x^2-36y^2=(7x)^2-(6y)^2$
$=(7x-6y)(7x+6y)$

53. x^2+25 cannot be factored; it is prime.

54. x^2+100 cannot be factored; it is prime.

55. $x^4-81=(x^2)^2-9^2$
$=(x^2-9)(x^2+9)$
$=(x^2-3^2)(x^2+9)$
$=(x-3)(x+3)(x^2+9)$

56. $x^4-625=(x^2)^2-(25)^2$
$=(x^2-25)(x^2+25)$
$=(x^2-5^2)(x^2+25)$
$=(x-5)(x+5)(x^2+25)$

57. $m^3+27=m^3+3^3$
$=(m+3)(m^2-m\cdot 3+3^2)$
$=(m+3)(m^2-3m+9)$

58. $m^3+125=m^3+5^3$
$=(m+5)(m^2-m\cdot 5+5^2)$
$=(m+5)(m^2-5m+25)$

59. $27p^3-8=(3p)^3-2^3$
$=(3p-2)[(3p)^2+3p\cdot 2+2^2]$
$=(3p-2)(9p^2+6p+4)$

60. $64p^3-1=(4p)^3-1^3$
$=(4p-1)[(4p)^2+4p\cdot 1+1^2]$
$=(4p-1)(16p^2+4p+1)$

61. y^9+64z^6
$=(y^3)^3+(4z^2)^3$
$=(y^3+4z^2)[(y^3)^2-y^3\cdot 4z^2+(4z^2)^2]$
$=(y^3+4z^2)(y^6-4y^3z^2+16z^4)$

62. $8y^3+27z^6$
$=(2y)^3+(3z^2)^3$
$=(2y+3z^2)[(2y)^2-2y\cdot 3z^2+(3z^2)^2]$
$=(2y+3z^2)(4y^2-6yz^2+9z^4)$

63. $15a^3-6a^2b-25ab^2+10b^3$
$=3a^2(5a-2b)-5b^2(5a-2b)$
$=(5a-2b)(3a^2-5b^2)$

64. $12a^2-9ab+4ab-3b^2$
$=3a(4a-3b)+b(4a-3b)$
$=(4a-3b)(3a+b)$

65. $x^2-xy-48y^2$ cannot be factored; it is prime.

66. $x^2-10xy-24y^2=(x-12y)(x+2y)$

67. $x^3-x^2-42x=x(x^2-x-42)$
$=x(x-7)(x+6)$

68. $3x^6-30x^5+63x^4=3x^4(x^2-10x+21)$
$=3x^4(x-7)(x-3)$

69. $6x^2+11x+3=6x^2+2x+9x+3$
$=2x(3x+1)+3(3x+1)$
$=(3x+1)(2x+3)$

70. $10z^2+9z-9=10z^2+15z-6z-9$
$=5z(2z+3)-3(2z+3)$
$=(2z+3)(5z-3)$

71. $27x^3+8=(3x)^3+2^3$
$=(3x+2)[(3x)^2-(3x)\cdot 2+2^2]$
$=(3x+2)(9x^2-6x+4)$

72. $8z^3-1=(2z)^3-1^3$
$=(2z-1)[(2z)^2+2z\cdot 1+1^2]$
$=(2z-1)(4z^2+2z+1)$

73. $4z^2+18y-10=2(2y^2+9y-5)$
$=2(2y-1)(y+5)$

74. $5x^3y^2-8x^2y^2+3xy^2=xy^2(5x^2-8x+3)$
$=xy^2(5x-3)(x-1)$

75. $25k^2-81m^2=(5k)^2-(9m)^2$
$=(5k-9m)(5k+9m)$

76. $x^4-9=(x^2)^2-3^2=(x^2-3)(x^2+3)$

77. $16m^2+1$ cannot be factored; it is prime.

78. m^4+25 cannot be factored; it is prime.

79. $(x-4)(2x-3)=0$
$x-4=0$ or $2x-3=0$
$x=4$ $\quad 2x=3$
$x=\frac{3}{2}$
The solution set is $\left\{\frac{3}{2}, 4\right\}$.

80. $(2x+1)(x+7)=0$
$2x+1=0$ or $x+7=0$
$2x=-1$ $\quad x=-7$
$x=-\frac{1}{2}$
The solutions et is $\left\{-7, -\frac{1}{2}\right\}$.

81. $x^2-12x-45=0$
$(x-15)(x+3)=0$
$x-15=0$ or $x+3=0$
$x=15$ $\quad x=-3$
The solution set is $\{-3, 15\}$.

82. $x^2-7x+10=0$
$(x-2)(x-5)=0$
$x-2=0$ or $x-5=0$
$x=2$ $\quad x=5$
The solution set is $\{2, 5\}$.

83. $3x^2+6x=0$
$3x(x+2)=0$
$3x=0$ or $x+2=0$
$x=0$ $\quad x=-2$
The solution set is $\{0, -2\}$.

84. $4x^2+18x=0$
$2x(2x+9)=0$
$2x=0$ or $2x+9=0$
$x=0$ $\quad 2x=-9$
$x=-\frac{9}{2}$
The solution set is $\left\{-\frac{9}{2}, 0\right\}$.

85. $3x(x+1)=2x^2+5x+3$
$3x^2+3x=2x^2+5x+3$
$x^2-2x-3=0$
$(x-3)(x+1)=0$
$x-3=0$ or $x+1=0$
$x=3$ $\quad x=-1$
The solution set is $\{-1, 3\}$.

86. $2x^2+6x=(3x+1)(x+3)$
$2x^2+6x=3x^2+10x+3$
$-x^2-4x-3=0$
$-1(x^2+4x+3)=0$
$-1(x+3)(x+1)=0$
$-1=0$ or $x+3=0$ or $x+1=0$
false $\quad x=-3$ $\quad x=-1$
The solution set is $\{-3, -1\}$.

87. $5x^2+7x+16=3x^2-5x$
$2x^2+12x+16=0$
$2(x^2+6x+8)=0$
$2(x+4)(x+2)=0$
$x+4=0$ or $x+2=0$
$x=-4$ or $x=-2$
The solution set is $\{-4, -2\}$.

88. $8x^2-10x=-2x+6$
$8x^2-8x-6=0$
$2(4x^2-4x-3)=0$
$2(2x+1)(2x-3)=0$
$2=0$ or $2x+1=0$ or $2x-3=0$
false $\quad 2x=-1$ $\quad 2x=3$
$x=-\frac{1}{2}$ $\quad x=\frac{3}{2}$
The solution set is $\left\{-\frac{1}{2}, \frac{3}{2}\right\}$.

89.
$$x^3 = -11x^2 + 42x$$
$$x^3 + 11x^2 - 42x = 0$$
$$x(x^2 + 11x - 42) = 0$$
$$x(x+14)(x-3) = 0$$
$x = 0$ or $x + 14 = 0$ or $x - 3 = 0$
$x = -14$ $x = 3$
The solution set is $\{-14, 0, 3\}$.

90.
$$-3x^2 = -x^3 + 18x$$
$$x^3 - 3x^2 - 18x = 0$$
$$x(x^2 - 3x - 18) = 0$$
$$x(x-6)(x+3) = 0$$
$x = 0$ or $x - 6 = 0$ or $x + 3 = 0$
$x = 6$ $x = -3$
The solution set is $\{-3, 0, 6\}$.

91.
$$(3x-4)(x^2 - 9) = 0$$
$$(3x-4)(x+3)(x-3) = 0$$
$3x - 4 = 0$ or $x + 3 = 0$ or $x - 3 = 0$
$3x = 4$ $x = -3$ $x = 3$
$x = \frac{4}{3}$
The solution set is $\left\{-3, \frac{4}{3}, 3\right\}$.

92.
$$(2x+5)(x^2 + 4x + 4) = 0$$
$$(2x+5)(x+2)(x+2) = 0$$
$2x + 5 = 0$ or $x + 2 = 0$
$2x = -5$ $x = -2$
$x = -\frac{5}{2}$
The solution set is $\left\{-\frac{5}{2}, -2\right\}$.

93.
$$h = -16t^2 + 80t$$
$$0 = -16t^2 + 80t$$
$$0 = -16t(t-5)$$
$-16t = 0$ or $t - 5 = 0$
$t = 0$ $t = 5$
After 5 seconds, the water hits the ground.

94.
$$h = -16t^2 + 80t$$
$$96 = -16t^2 + 80t$$
$$16t^2 - 80t + 96 = 0$$
$$16(t^2 - 5t + 6) = 0$$
$$16(t-3)(t-2) = 0$$
$16 = 0$ or $t - 3 = 0$ or $t - 2 = 0$
false $t = 3$ $t = 2$
The water is 96 feet high after 2 seconds and 3 seconds.

95. Let w = width. Then $2w - 3$ = length.
$$w(2w-3) = 54$$
$$2w^2 - 3w - 54 = 0$$
$$(2w+9)(w-6) = 0$$
$2w + 9 = 0$ or $w - 6 = 0$
$2w = -9$ $w = 6$
$w = -\frac{9}{2}$
Discard $w = -\frac{9}{2}$. The width is 6 feet and the length is $2w - 3 = 9$ feet.

96. Let w = width. Then $2w - 1$ = length.
$$w(2w-1) = 15$$
$$2w^2 - w - 15 = 0$$
$$(2w+5)(w-3) = 0$$
$2w + 5 = 0$ or $w - 3 = 0$
$2w = -5$ $w = 3$
$w = -\frac{5}{2}$
Discard $w = -\frac{5}{2}$. The width is 3 yards and the length is $2w - 1 = 5$ yards.

97. Let l = length of longer leg. Then $l - 2$ = length of shorter leg.
$$l^2 + (l-2)^2 = 10^2$$
$$l^2 + l^2 - 4l + 4 = 100$$
$$2l^2 - 4l - 96 = 0$$
$$2(l^2 - 2l - 48) = 0$$
$$2(l-8)(l+6) = 0$$
$2 = 0$ or $l - 8 = 0$ or $l + 6 = 0$
false $l = 8$ $l = -6$
Discard $l = -6$. The longer leg is 8 feet and the shorter leg is $l - 2 = 6$ feet.

98. Let l = length of longer leg. Then $l - 14$ = length of shorter leg.

$$l^2 + (l-14)^2 = 26^2$$
$$l^2 + l^2 - 28l + 196 = 676$$
$$2l^2 - 28l - 480 = 0$$
$$2(l^2 - 14l - 240) = 0$$
$$2(l-24)(l+10) = 0$$

$2 = 0$ or $l - 24 = 0$ or $l + 10 = 0$
false $\quad l = 24 \quad l = -10$

Discard $l = -10$. The longer leg is 24 feet and the shorter leg is $l - 14 = 10$ feet.

Chapter 6 Test

1. The GCF of 16, 20, and 24 is 4.
The GCF of x^5, x^4, and x^6 is x^4.
The GCF of y^2, y^6, and y^8 is y^2.
The GCF is $4x^4y^2$.

2. $x^4 - 256 = (x^2)^2 - 16^2$
$= (x^2 - 16)(x^2 + 16)$
$= (x^2 - 4^2)(x^2 + 16)$
$= (x+4)(x-4)(x^2+16)$

3. $18x^3 - 9x^2 - 27x = 9x(2x^2 - x - 3)$
$= 9x(2x-3)(x+1)$

4. $xy - 7y - 4x + 28 = (xy - 7y) + (-4x + 28)$
$= y(x-7) - 4(x-7)$
$= (x-7)(y-4)$

5. $27x^3 + 125 = (3x)^3 + 5^3$
$= (3x+5)[(3x)^2 - (3x)5 + 5^2]$
$= (3x+5)(9x^2 - 15x + 25)$

6. $y^2 - 8y - 48 = (y-12)(y+4)$

7. $6m^2 - m - 5 = (6m+5)(m-1)$

8. $4x^2 + 25 = (2x)^2 + 5^2$
$4x^2 + 25$ cannot be factored; it is prime.

9. $4(x-5) + y(x-5) = (x-5)(4+y)$

10. $3x^2y - 15xy - 42y = 3y(x^2 - 5x - 14)$
$= 3y(x-7)(x+2)$

11. $x^2 + 4x + 12$ cannot be factored; it is prime.

12. $2x^6 - 54y^3 = 2(x^6 - 27y^3)$
$= 2[(x^2)^3 - (3y)^3]$
$= 2(x^3 - 3y)(x^4 + 3x^2y + 9y^2)$

13. $9x^3 + 39x^2 + 12x = 3x(3x^2 + 13x + 4)$
$= 3x(3x+1)(x+4)$

14. $6m^2 + 7m + 2 = (3m+2)(2m+1)$

15. $4m^2 - 6mn + 4 = 2(2m^2 - 3mn + 2)$

16. $25x^2 + 70xy + 49y^2$
$= (5x)^2 + 2(5x)(7y) + (7y)^2$
$= (5x+7y)^2$

17. $5x^2 = -16x - 3$
$5x^2 + 16x + 3 = 0$
$(5x+1)(x+3) = 0$
$5x + 1 = 0$ or $x + 3 = 0$
$5x = -1 \quad x = -3$
$x = -\frac{1}{5}$

The solution set is $\left\{-\frac{1}{5}, -3\right\}$.

18. $5x^3 - 20x^2 + 20x = 0$
$5x(x^2 - 4x + 4) = 0$
$5x(x-2)^2 = 0$
$5x = 0$ or $x - 2 = 0$
$x = 0 \quad x = 2$
The solution set is $\{0, 2\}$.

19. Let w = width. Then $3w - 8$ = length.
$w(3w-8) = 35$
$3w^2 - 8w - 35 = 0$
$(3w+7)(w-5) = 0$
$3w + 7 = 0$ or $w - 5 = 0$
$3w = -7 \quad w = 5$
$w = -\frac{7}{3}$

Discard $w = -\frac{7}{3}$. The width is 5 inches and the length is $3w - 8 = 7$ inches.

20. Let l = length of longer leg. Then
$l + 1$ = length of hypotenuse and
$l - 7$ = length of shorter leg.

$$l^2 + (l-7)^2 = (l+1)^2$$
$$l^2 + l^2 - 14l + 49 = l^2 + 2l + 1$$
$$l^2 - 16l + 48 = 0$$
$$(l-12)(l-4) = 0$$
$$l - 12 = 0 \quad \text{or} \quad l - 4 = 0$$
$$l = 12 \qquad\qquad l = 4$$

Discard $l = 4$ since $l - 7 = -3$. The legs are 12 inches and $l - 7 = 5$ inches and the hypotenuse is $l + 1 = 13$ inches.

Chapter 7

Section 7.1

Preparing for Simplifying Rational Expressions

P1. Substitute −1 for x and 7 for y.

$$\frac{3x+y}{2}=\frac{3(-1)+7}{2}=\frac{-3+7}{2}=\frac{4}{2}=2$$

P2. $2x^2+x-3=(2x+3)(x-1)$

P3.
$$\begin{aligned}3x^2-5x-2&=0\\(3x+1)(x-2)&=0\end{aligned}$$
$$3x+1=0 \quad \text{or} \quad x-2=0$$
$$3x=-1 \quad \text{or} \quad x=2$$
$$x=-\frac{1}{3} \quad \text{or} \quad x=2$$

The solution set is $\left\{-\frac{1}{3}, 2\right\}$.

P4. $\frac{21}{70}=\frac{3\cdot 7}{10\cdot 7}=\frac{3}{10}$

P5. $\frac{x^3y^4}{xy^2}=x^{3-1}y^{4-2}=x^2y^2$

7.1 Quick Checks

1. The quotient of two polynomials is called a rational expression.

2. (a)
$$\begin{aligned}\frac{3}{5x+1}&=\frac{3}{5(-5)+1}\\&=\frac{3}{-25+1}\\&=\frac{3}{-24}\\&=\frac{1\cdot 3}{-8\cdot 3}\\&=-\frac{1}{8}\end{aligned}$$

(b) $\frac{3}{5x+1}=\frac{3}{5(3)+1}=\frac{3}{15+1}=\frac{3}{16}$

3. (a)
$$\begin{aligned}\frac{x^2+6x+9}{x+1}&=\frac{(-5)^2+6(-5)+9}{-5+1}\\&=\frac{25-30+9}{-4}\\&=\frac{4}{-4}\\&=-1\end{aligned}$$

(b)
$$\begin{aligned}\frac{x^2+6x+9}{x+1}&=\frac{(3)^2+6(3)+9}{3+1}\\&=\frac{9+18+9}{4}\\&=\frac{36}{4}\\&=\frac{4\cdot 9}{4}\\&=9\end{aligned}$$

4. Substitute 2 for m, −1 for n, and 6 for p.

$$\frac{3m-5n}{p-4}=\frac{3(2)-5(-1)}{6-4}=\frac{6+5}{2}=\frac{11}{2}$$

5. Substitute 2 for y and 1 for z.

$$\frac{5y+2}{2y-z}=\frac{5(2)+2}{2(2)-1}=\frac{10+2}{4-1}=\frac{12}{3}=4$$

6. A rational expression is undefined for those values of the variable(s) that make the denominator zero.

7. We want to find all values of x in the rational expression $\frac{3}{x+7}$ that cause $x + 7$ to equal 0.

$$\begin{aligned}x+7&=0\\x&=-7\end{aligned}$$

So −7 causes the denominator, $x + 7$, to equal 0. Therefore, the expression $\frac{3}{x+7}$ is undefined for $x=-7$.

8. We want to find all values of n in the rational expression $\frac{-4}{3n+5}$ that cause $3n + 5$ to equal 0.

$$\begin{aligned}3n+5&=0\\3n&=-5\\n&=-\frac{5}{3}\end{aligned}$$

So $-\frac{5}{3}$ causes the denominator, $3n+5$, to equal 0. Therefore, the expression $\frac{-4}{3n+5}$ is undefined for $n=-\frac{5}{3}$.

9. We want to find all values of x in the rational expression $\frac{8x}{x^2+2x-3}$ that cause x^2+2x-3 to equal 0.

$$x^2+2x-3=0$$
$$(x+3)(x-1)=0$$
$$x+3=0 \quad \text{or} \quad x-1=0$$
$$x=-3 \quad \text{or} \quad x=1$$

Since -3 or 1 cause the denominator to equal 0, the rational expression $\frac{8x}{x^2+2x-3}$ is undefined for $x=-3$ or $x=1$.

10. To simplify a rational expression means to write the rational expression in the form $\frac{p}{q}$ where p and q are polynomials that have no common factors.

11. False; $\frac{2n+3}{2n+6}=\frac{2n+3}{2(n+3)}\neq\frac{3}{6}$ or $\frac{1}{2}$.

12. $\frac{3n^2+12n}{6n}=\frac{3n(n+4)}{3n(2)}=\frac{n+4}{2}$

13. $$\frac{2z^2+6z+4}{-4z-8}=\frac{2(z^2+3z+2)}{-4(z+2)}=\frac{2(z+1)(z+2)}{-2\cdot 2(z+2)}=-\frac{z+1}{2}$$

14. $\frac{a^2+3a-28}{2a^2-a-28}=\frac{(a+7)(a-4)}{(2a+7)(a-4)}=\frac{a+7}{2a+7}$

15. $$\frac{z^3+z^2-3z-3}{z^2-z-2}=\frac{z^2(z+1)-3(z+1)}{(z+1)(z-2)}=\frac{(z+1)(z^2-3)}{(z+1)(z-2)}=\frac{z^2-3}{z-2}$$

16. $\frac{4k^2+4k+1}{4k^2-1}=\frac{(2k+1)^2}{(2k+1)(2k-1)}=\frac{2k+1}{2k-1}$

17. False; $\frac{a+b}{a-b}=\frac{a+b}{-1(-a+b)}\neq -1.$

18. $\frac{7a-7b}{b-a}=\frac{7(a-b)}{b-a}=\frac{7(-1)(b-a)}{b-a}=-7$

19. $\frac{12-4x}{4x^2-13x+3}=\frac{-4(x-3)}{(x-3)(4x-1)}=\frac{-4}{4x-1}$

20. $$\frac{25z^2-1}{3-15z}=\frac{(5z+1)(5z-1)}{3(1-5z)}=\frac{(5z+1)(5z-1)}{3(-1)(5z-1)}=-\frac{5z+1}{3}$$

7.1 Exercises

21. $\frac{x}{x-5}$

(a) $x=10$: $\frac{10}{10-5}=\frac{10}{5}=2$

(b) $x=-5$: $\frac{-5}{-5-5}=\frac{-5}{-10}=\frac{1}{2}$

(c) $x=0$: $\frac{0}{0-5}=\frac{0}{-5}=0$

23. $\frac{2a-3}{a}$

(a) $a=3$: $\frac{2(3)-3}{3}=\frac{6-3}{3}=\frac{3}{3}=1$

(b) $a=-3$: $\frac{2(-3)-3}{-3}=\frac{-6-3}{-3}=\frac{-9}{-3}=3$

(c) $a=9$: $\frac{2(9)-3}{9}=\frac{18-3}{9}=\frac{15}{9}=\frac{5}{3}$

25. $\frac{x^2-2x}{x-4}$

(a) $x=3$: $\frac{3^2-2(3)}{3-4}=\frac{9-6}{-1}=\frac{3}{-1}=-3$

(b) $x=2$: $\frac{2^2-2(2)}{2-4}=\frac{4-4}{-2}=\frac{0}{-2}=0$

(c) $x=-3$: $\frac{(-3)^2-2(-3)}{-3-4}=\frac{9+6}{-7}=-\frac{15}{7}$

27. $\frac{x^2-y^2}{2x-y}$

(a) $x=2, y=2$: $\frac{2^2-2^2}{2(2)-2}=\frac{4-4}{4-2}=\frac{0}{2}=0$

(b) $x=3, y=4$: $\frac{3^2-4^2}{2(3)-4}=\frac{9-16}{6-4}=\frac{-7}{2}=-\frac{7}{2}$

(c) $x=1, y=-2$: $\frac{1^2-(-2)^2}{2(1)-(-2)}=\frac{1-4}{2+2}=\frac{-3}{4}=-\frac{3}{4}$

29. $\frac{2-4x}{3x}$ is undefined when $3x=0$.

$$3x=0$$
$$x=0$$

31. $\frac{3p}{p-5}$ is undefined when $p-5=0$.

$$p-5=0$$
$$p=5$$

33. $\frac{8}{3-2x}$ is undefined when $3-2x=0$.

$$3-2x=0$$
$$-2x=-3$$
$$x=\frac{3}{2}$$

35. $\frac{6z}{z^2-36}$ is undefined when $z^2-36=0$.

$$z^2-36=0$$
$$(z+6)(z-6)=0$$
$$z+6=0 \quad \text{or} \quad z-6=0$$
$$z=-6 \quad \text{or} \quad z=6$$

37. $\frac{x}{x^2-7x+10}$ is undefined when $x^2-7x+10=0$.

$$x^2-7x+10=0$$
$$(x-2)(x-5)=0$$
$$x-2=0 \quad \text{or} \quad x-5=0$$
$$x=2 \quad \text{or} \quad x=5$$

39. $\frac{12x+5}{x^3-x^2-6x}$ is undefined when $x^3-x^2-6x=0$.

$$x^3-x^2-6x=0$$
$$x(x+2)(x-3)=0$$
$$x=0 \quad \text{or} \quad x+2=0 \quad \text{or} \quad x-3=0$$
$$x=0 \quad \text{or} \quad x=-2 \quad \text{or} \quad x=3$$

41. $\frac{5x-10}{15}=\frac{5(x-2)}{5\cdot 3}=\frac{x-2}{3}$

43. $\frac{3z^2+6z}{3z}=\frac{3z(z+2)}{3z}=z+2$

45. $\frac{p-3}{p^2-p-6}=\frac{(p-3)\cdot 1}{(p-3)(p+2)}=\frac{1}{p+2}$

47. $\frac{2-x}{x-2}=\frac{-1(x-2)}{x-2}=-1$

49. $\frac{2k^2-14k}{7-k}=\frac{2k(k-7)}{-1\cdot(k-7)}=-2k$

51. $\frac{x^2-1}{x^2+5x+4}=\frac{(x-1)(x+1)}{(x+4)(x+1)}=\frac{x-1}{x+4}$

53. $\frac{6x+30}{x^2-25}=\frac{6(x+5)}{(x+5)(x-5)}=\frac{6}{x-5}$

55. $\frac{-3x-3y}{x+y}=\frac{-3(x+y)}{x+y}=-3$

57. $\frac{b^2-25}{4b+20}=\frac{(b-5)(b+5)}{4(b+5)}=\frac{b-5}{4}$

59. $\frac{x^2-2x-15}{x^2-8x+15}=\frac{(x+3)(x-5)}{(x-3)(x-5)}=\frac{x+3}{x-3}$

61. $\frac{x^3+x^2-12x}{x^3-x^2-20x}=\frac{x(x+4)(x-3)}{x(x+4)(x-5)}=\frac{x-3}{x-5}$

63. $\frac{x^2-y^2}{y-x}=\frac{(x+y)(x-y)}{y-x}$

$=\frac{-(x+y)(y-x)}{y-x}$

$=-(x+y)$

65. $\frac{x^3+4x^2+x+4}{x^2+5x+4}=\frac{x^2(x+4)+1\cdot(x+4)}{(x+1)(x+4)}$

$=\frac{(x+4)(x^2+1)}{(x+1)(x+4)}$

$=\frac{x^2+1}{x+1}$

67. $\frac{16-c^2}{(c-4)^2}=\frac{(4+c)(4-c)}{(c-4)(c-4)}$

$=\frac{-(4+c)(c-4)}{(c-4)(c-4)}$

$=\frac{-(4+c)}{c-4}$

69. $\frac{4x^2-20x+24}{6x^2-48x+90}=\frac{4(x-2)(x-3)}{6(x-3)(x-5)}$

$=\frac{2(x-3)\cdot 2(x-2)}{2(x-3)\cdot 3(x-5)}$

$=\frac{2(x-2)}{3(x-5)}$

71. $\frac{6+x-x^2}{x^2-4}=\frac{-1(x^2-x-6)}{x^2-4}$

$=\frac{-1(x-3)(x+2)}{(x-2)(x+2)}$

$=\frac{-(x-3)}{x-2}$

73. $\frac{2t^2-18}{t^4-81}=\frac{2(t^2-9)}{(t^2-9)(t^2+9)}=\frac{2}{t^2+9}$

75. $\frac{12w-3w^2}{w^3-5w^2+4w}=\frac{-3w(w-4)}{w(w-1)(w-4)}=\frac{-3}{w-1}$

77. (a) $C=\frac{50t}{t^2+25}$; find C when $t = 5$.

$\frac{50(5)}{5^2+25}=\frac{250}{25+25}=\frac{250}{50}=5$

The concentration is 5 mg/mL after 5 minutes.

(b) Find C when $t = 10$.

$C=\frac{50(10)}{10^2+25}=\frac{500}{100+25}=\frac{500}{125}=4$

The concentration is 4 mg/mL after 10 minutes.

79. $\text{BMI}=\frac{k}{m^2}$; find BMI when $k = 110$ and $m = 2$.

$\frac{110}{2^2}=\frac{110}{4}=27.5$

Yes, the person should consider looking for a weight-loss program.

81. $\bar{C}=\frac{0.2x^3-2.3x^2+14.3x+10.2}{x}$; find $\bar{C}$ when $x = 2$.

$\frac{0.2(2)^3-2.3(2)^2+14.3(2)+10.2}{2}$

$=\frac{1.6-9.2+28.6+10.2}{2}$

$=\frac{31.2}{2}$

$=15.6$

The average cost of producing 2 Cobalts is \$15,600 per car.

83. $\frac{c^8-1}{(c^4-1)(c^6+1)}=\frac{(c^4+1)(c^4-1)}{(c^4-1)(c^6+1)}=\frac{c^4+1}{c^6+1}$

85. $$\frac{x^4-x^2-12}{x^4+2x^3-9x-18}=\frac{(x^2-4)(x^2+3)}{x^3(x+2)-9(x+2)}$$
$$=\frac{(x+2)(x-2)(x^2+3)}{(x+2)(x^3-9)}$$
$$=\frac{(x-2)(x^2+3)}{x^3-9}$$

87. $$\frac{(t+2)^3(t^4-16)}{(t^3+8)(t+2)(t^2-4)}$$
$$=\frac{(t+2)^3(t^2+4)(t^2-4)}{(t+2)(t^2-2t+4)(t+2)(t+2)(t-2)}$$
$$=\frac{(t+2)^3(t^2+4)(t+2)(t-2)}{(t+2)(t^2-2t+4)(t+2)(t+2)(t-2)}$$
$$=\frac{(t+2)^3(t-2)\cdot(t^2+4)(t+2)}{(t+2)^3(t-2)\cdot(t^2-2t+4)}$$
$$=\frac{(t^2+4)(t+2)}{t^2-2t+4}$$

89. **(a)** The x^2s cannot be divided out because they are terms, not factors.

(b) When the factors of $x-2$ are divided out, the result is $\frac{1}{3}$, not 3.

91. A rational expression is undefined if the denominator is equal to zero. Division by zero is not allowed in the real number system. The expression $\frac{x^2-9}{3x-6}$ is undefined for $x = 2$ because when 2 is substituted for x, we get $\frac{2^2-9}{3(2)-6}=\frac{4-9}{6-6}=\frac{-5}{0}$, which is undefined.

93. $\frac{x-7}{7-x}=-1$ because $x-7$ and $7-x$ are opposites:
$$\frac{x-7}{7-x}=\frac{x-7}{-1(-7+x)}$$
$$=\frac{x-7}{-1(x-7)}$$
$$=\frac{\cancel{x-7}}{-1\cancel{(x-7)}}$$
$$=\frac{1}{-1}$$
$$=1.$$

The expression $\frac{x-7}{x+7}$ is in simplest form because $x-7$ and $x+7$ are not opposites and the terms cannot be divided out.

Section 7.2

Preparing for Multiplying and Dividing Rational Expressions

P1. $\frac{3}{14}\cdot\frac{28}{9}=\frac{3\cdot 28}{14\cdot 9}=\frac{3\cdot 2\cdot 14}{14\cdot 3\cdot 3}=\frac{2}{3}$

P2. The reciprocal of $\frac{5}{8}$ is $\frac{8}{5}$.

P3. $\frac{12}{25}\div\frac{12}{5}=\frac{12}{25}\cdot\frac{5}{12}=\frac{12\cdot 5}{25\cdot 12}=\frac{12\cdot 5}{5\cdot 5\cdot 12}=\frac{1}{5}$

P4. $$3x^3-27x=3x(x^2-9)$$
$$=3x(x+3)(x-3)$$

P5. $\frac{3x+12}{5x^2+20x}=\frac{3(x+4)}{5x(x+4)}=\frac{3}{5x}$

7.2 Quick Checks

1. To multiply two rational expressions, factor each numerator and denominator and then divide out any common factors.

2. False; $\frac{x+4}{24}\cdot\frac{24}{(x+4)(x-4)}=\frac{(x+4)\cdot 24\cdot 1}{24(x+4)(x-4)}$
$=\frac{\cancel{(x+4)}\cdot\cancel{24}\cdot 1}{\cancel{24}\cancel{(x+4)}(x-4)}$
$=\frac{1}{(x-4)}$
$\neq x-4$

3. $\frac{p^2-9}{5}\cdot\frac{25p}{2p-6}=\frac{(p+3)(p-3)}{5}\cdot\frac{5^2p}{2(p-3)}$
$=\frac{(p+3)(p-3)\cdot 5^2p}{5\cdot 2(p-3)}$
$=\frac{5p(p+3)}{2}$

4. $\frac{2x+8}{x^2+3x-4}\cdot\frac{7x-7}{6x+30}$
$=\frac{2(x+4)}{(x+4)(x-1)}\cdot\frac{7(x-1)}{2\cdot 3(x+5)}$
$=\frac{2(x+4)\cdot 7(x-1)}{(x+4)(x-1)\cdot 2\cdot 3(x+5)}$
$=\frac{7}{3(x+5)}$

5. False; $\frac{8}{x-7}\cdot\frac{7-x}{16x}=\frac{\overset{1}{\cancel{8}}}{-1(7-x)}\cdot\frac{7-6}{\underset{2}{\cancel{16}}x}$
$=-\frac{1}{7-x}\cdot\frac{7-x}{2x}$
$=-\frac{1}{\underset{1}{\cancel{(7-x)}}}\cdot\frac{\overset{1}{\cancel{(7-x)}}}{2x}$
$=-\frac{1}{2x}\neq\frac{1}{2x}$

6. $\frac{15a-3a^2}{7}\cdot\frac{3+2a}{2a^2-7a-15}$
$=\frac{3a(5-a)}{7}\cdot\frac{3+2a}{(2a+3)(a-5)}$
$=\frac{3a(-1)(a-5)(2a+3)}{7(2a+3)(a-5)}$
$=-\frac{3a}{7}$

7. $\frac{a^2+2ab+b^2}{3a+3b}\cdot\frac{a-b}{a^2-b^2}$
$=\frac{(a+b)^2}{3(a+b)}\cdot\frac{a-b}{(a+b)(a-b)}$
$=\frac{(a+b)^2(a-b)}{3(a+b)^2(a-b)}$
$=\frac{1}{3}$

8. $\frac{8}{9}\div\frac{2}{3}=\frac{8}{9}\cdot\frac{3}{2}=\frac{4\cdot 2\cdot 3}{3\cdot 3\cdot 2}=\frac{4\cdot\cancel{2}\cdot\cancel{3}}{\cancel{3}\cdot 3\cdot\cancel{2}}=\frac{4}{3}$

9. $\frac{12}{x^2-x}\div\frac{4x-2}{x^2-1}=\frac{12}{x^2-x}\cdot\frac{x^2-1}{4x-2}$
$=\frac{12}{x(x-1)}\cdot\frac{(x+1)(x-1)}{2(2x-1)}$
$=\frac{12(x+1)(x-1)}{2x(x-1)(2x-1)}$
$=\frac{6(x+1)}{x(2x-1)}$

10. $\frac{x^2-9}{x^2-16}\div\frac{x^2-x-12}{x^2+x-12}$
$=\frac{x^2-9}{x^2-16}\cdot\frac{x^2+x-12}{x^2-x-12}$
$=\frac{(x+3)(x-3)}{(x+4)(x-4)}\cdot\frac{(x+4)(x-3)}{(x+3)(x-4)}$
$=\frac{(x+3)(x+4)(x-3)^2}{(x+3)(x+4)(x-4)^2}$
$=\frac{(x-3)^2}{(x-4)^2}$

11. $\frac{q^2-6q-7}{q^2-25}\div(q-7)=\frac{q^2-6q-7}{q^2-25}\div\frac{q-7}{1}$
$=\frac{q^2-6q-7}{q^2-25}\cdot\frac{1}{q-7}$
$=\frac{(q+1)(q-7)}{(q+5)(q-5)}\cdot\frac{1}{q-7}$
$=\frac{(q+1)(q-7)}{(q+5)(q-5)(q-7)}$
$=\frac{q+1}{(q+5)(q-5)}$

12. $\dfrac{\frac{2}{x+1}}{\frac{8}{x^2-1}} = \dfrac{2}{x+1} \cdot \dfrac{x^2-1}{8}$
$= \dfrac{2}{x+1} \cdot \dfrac{(x+1)(x-1)}{2 \cdot 4}$
$= \dfrac{2 \cdot (x+1)(x-1)}{2 \cdot 4 \cdot (x+1)}$
$= \dfrac{x+1}{4}$

13. $\dfrac{\frac{x+3}{x^2-4}}{\frac{4x+12}{7x^2+14x}} = \dfrac{x+3}{x^2-4} \cdot \dfrac{7x^2+14x}{4x+12}$
$= \dfrac{x+3}{(x+2)(x-2)} \cdot \dfrac{7x(x+2)}{4(x+3)}$
$= \dfrac{(x+3) \cdot 7x(x+2)}{(x+2)(x-2) \cdot 4(x+3)}$
$= \dfrac{7x}{4(x-2)}$

14. $\dfrac{3m-6n}{5n} \div \dfrac{m^2-4n^2}{10mn}$
$= \dfrac{3m-6n}{5n} \cdot \dfrac{10mn}{m^2-4n^2}$
$= \dfrac{3(m-2n)}{5 \cdot n} \cdot \dfrac{5 \cdot 2 \cdot m \cdot n}{(m+2n)(m-2n)}$
$= \dfrac{3(m-2n) \cdot 5 \cdot 2 \cdot m \cdot n}{5 \cdot n \cdot (m+2n)(m-2n)}$
$= \dfrac{3 \cdot 2 \cdot m}{m+2n}$
$= \dfrac{6m}{m+2n}$

7.2 Exercises

15. $\dfrac{x+5}{7} \cdot \dfrac{14x}{x^2-25} = \dfrac{x+5}{7} \cdot \dfrac{7 \cdot 2x}{(x+5)(x-5)}$
$= \dfrac{7(x+5) \cdot 2x}{7(x+5) \cdot (x-5)}$
$= \dfrac{2x}{x-5}$

17. $\dfrac{x^2-x}{x^2-x-2} \cdot \dfrac{x-2}{x^2-1}$
$= \dfrac{x(x-1)}{(x+1)(x-2)} \cdot \dfrac{x-2}{(x+1)(x-1)}$
$= \dfrac{(x-1)(x-2) \cdot x}{(x-1)(x-2) \cdot (x+1)(x+1)}$
$= \dfrac{x}{(x+1)^2}$

19. $\dfrac{p^2-1}{2p-3} \cdot \dfrac{2p^2+p-6}{p^2+3p+2}$
$= \dfrac{(p+1)(p-1)}{2p-3} \cdot \dfrac{(2p-3)(p+2)}{(p+2)(p+1)}$
$= \dfrac{(p+1)(2p-3)(p+2) \cdot (p-1)}{(p+1)(2p-3)(p+2)}$
$= p-1$

21. $(x+1) \cdot \dfrac{x-6}{x^2-5x-6} = \dfrac{x+1}{1} \cdot \dfrac{x-6}{(x+1)(x-6)}$
$= \dfrac{(x+1)(x-6)}{(x+1)(x-6)}$
$= 1$

23. $\dfrac{x-4}{2x-8} \div \dfrac{3x}{2} = \dfrac{x-4}{2x-8} \cdot \dfrac{2}{3x}$
$= \dfrac{x-4}{2(x-4)} \cdot \dfrac{2}{3x}$
$= \dfrac{2(x-4) \cdot 1}{2(x-4) \cdot 3x}$
$= \dfrac{1}{3x}$

25. $\dfrac{m^2-16}{6m} \div \dfrac{m^2+8m+16}{12m}$
$= \dfrac{m^2-16}{6m} \cdot \dfrac{12m}{m^2+8m+16}$
$= \dfrac{(m+4)(m-4)}{6m} \cdot \dfrac{2 \cdot 6m}{(m+4)(m+4)}$
$= \dfrac{6m(m+4) \cdot 2(m-4)}{6m(m+4) \cdot (m+4)}$
$= \dfrac{2(m-4)}{m+4}$

27. $\frac{x^2-x}{x^2-1}\div\frac{x+2}{x^2+3x+2}$
$=\frac{x^2-x}{x^2-1}\cdot\frac{x^2+3x+2}{x+2}$
$=\frac{x(x-1)}{(x+1)(x-1)}\cdot\frac{(x+2)(x+1)}{x+2}$
$=\frac{(x-1)(x+2)(x+1)\cdot x}{(x-1)(x+2)(x+1)\cdot 1}$
$=\frac{x}{1}$
$=x$

29. $\frac{(x+2)^2}{x^2-4}\div\frac{x^2-x-6}{x^2-5x+6}$
$=\frac{(x+2)^2}{x^2-4}\cdot\frac{x^2-5x+6}{x^2-x-6}$
$=\frac{(x+2)(x+2)}{(x-2)(x+2)}\cdot\frac{(x-3)(x-2)}{(x-3)(x+2)}$
$=\frac{(x+2)(x+2)(x-3)(x-2)}{(x+2)(x+2)(x-3)(x-2)}$
$=1$

31. $\frac{14}{9}\cdot\frac{15}{7}=\frac{2\cdot 7\cdot 3\cdot 5}{3\cdot 3\cdot 7}=\frac{3\cdot 7\cdot 2\cdot 5}{3\cdot 7\cdot 3}=\frac{10}{3}$

33. $-\frac{20}{16}\div\left(-\frac{30}{24}\right)=-\frac{20}{16}\cdot\left(-\frac{24}{30}\right)$
$=\frac{-2\cdot 2\cdot 5}{2\cdot 2\cdot 2\cdot 2}\cdot\left(\frac{2\cdot 2\cdot 2\cdot 3}{-2\cdot 3\cdot 5}\right)$
$=\frac{-2\cdot 2\cdot 2\cdot 2\cdot 2\cdot 3\cdot 5}{-2\cdot 2\cdot 2\cdot 2\cdot 2\cdot 3\cdot 5}$
$=1$

35. $\frac{8}{11}\div(-2)=\frac{8}{11}\cdot\left(-\frac{1}{2}\right)$
$=\frac{2\cdot 4}{11}\cdot\left(-\frac{1}{2}\right)$
$=-\frac{2\cdot 4}{2\cdot 11}$
$=-\frac{4}{11}$

37. $\frac{3y}{y^2-y-6}\cdot\frac{4y+8}{9y^2}=\frac{3y}{(y+2)(y-3)}\cdot\frac{4(y+2)}{3y\cdot 3y}$
$=\frac{4}{3y(y-3)}$

39. $\frac{4a+8b}{a^2+2ab}\cdot\frac{a^2}{12}=\frac{4(a+2b)}{a(a+2b)}\cdot\frac{a\cdot a}{3\cdot 4}=\frac{a}{3}$

41. $\frac{3x^2-6x}{x^2-2x-8}\div\frac{x-2}{x+2}=\frac{3x^2-6x}{x^2-2x-8}\cdot\frac{x+2}{x-2}$
$=\frac{3x(x-2)}{(x-4)(x+2)}\cdot\frac{x+2}{x-2}$
$=\frac{(x-2)(x+2)\cdot 3x}{(x-2)(x+2)\cdot(x-4)}$
$=\frac{3x}{x-4}$

43. $\frac{(w-4)^2}{4-w^2}\div\frac{w^2-16}{w-2}$
$=\frac{(w-4)^2}{4-w^2}\cdot\frac{w-2}{w^2-16}$
$=\frac{(w-4)^2}{(2+w)(2-w)}\cdot\frac{-(2-w)}{(w+4)(w-4)}$
$=\frac{-(2-w)(w-4)^2}{(2-w)(2+w)(w+4)(w-4)}$
$=\frac{4-w}{(w+2)(w+4)}$

45. $\frac{3xy-2y^2-x^2}{x+y}\cdot\frac{x^2-y^2}{x^2-2xy}$
$=\frac{-(x-y)(x-2y)}{x+y}\cdot\frac{(x+y)(x-y)}{x(x-2y)}$
$=-\frac{(x-y)^2(x+y)(x-2y)}{x(x+y)(x-2y)}$
$=-\frac{(x-y)^2}{x}$

47. $\frac{\frac{2c-4}{8}}{\frac{2-c}{2}}=\frac{2c-4}{8}\cdot\frac{2}{2-c}$
$=\frac{2(c-2)}{2\cdot 2\cdot 2}\cdot\frac{2}{-(c-2)}$
$=-\frac{2\cdot 2(c-2)}{2\cdot 2(c-2)\cdot 2}$
$=-\frac{1}{2}$

49. $\frac{4n^2-9}{6n+18}\cdot\frac{9n^2-81}{2n^2+5n-12}$
$=\frac{(2n+3)(2n-3)}{2\cdot 3\cdot(n+3)}\cdot\frac{3\cdot 3\cdot(n+3)(n-3)}{(n+4)(2n-3)}$
$=\frac{(2n+3)(2n-3)\cdot 3\cdot 3\cdot(n+3)(n-3)}{2\cdot 3\cdot(n+3)(n+4)(2n-3)}$
$=\frac{3(2n+3)(n-3)}{2(n+4)}$

51. $\frac{x^2y}{2x^2-5xy+2y^2}\div\frac{(2xy^2)^2}{2x^2y-xy^2}$
$=\frac{x^2y}{2x^2-5xy+2y^2}\cdot\frac{2x^2y-xy^2}{(2xy^2)^2}$
$=\frac{x^2y}{(2x-y)(x-2y)}\cdot\frac{xy(2x-y)}{4x^2y^4}$
$=\frac{x^2y\cdot xy(2x-y)}{4x^2y^4(2x-y)(x-2y)}$
$=\frac{x}{4y^2(x-2y)}$

53. $\frac{2a^2+3ab-2b^2}{a^2-b^2}\cdot\frac{a^2-ab}{2a^3+4a^2b}$
$=\frac{(a+2b)(2a-b)}{(a+b)(a-b)}\cdot\frac{a(a-b)}{2a^2(a+2b)}$
$=\frac{a(a+2b)(a-b)\cdot(2a-b)}{a(a+2b)(a-b)\cdot 2a(a+b)}$
$=\frac{2a-b}{2a(a+b)}$

55. $\frac{3t^2-27}{t+2}\cdot\frac{t^2-4}{9t-27}$
$=\frac{3(t+3)(t-3)}{t+2}\cdot\frac{(t+2)(t-2)}{3\cdot 3(t-3)}$
$=\frac{3(t+2)(t-3)\cdot(t+3)(t-2)}{3(t+2)(t-3)\cdot 3}$
$=\frac{(t+3)(t-2)}{3}$

57. $\frac{(x+2)^2}{x^2-4}\div\frac{-x^2+x+6}{x^2-5x+6}$
$=\frac{(x+2)^2}{x^2-4}\cdot\frac{x^2-5x+6}{-x^2+x+6}$
$=\frac{(x+2)^2}{(x+2)(x-2)}\cdot\frac{(x-2)(x-3)}{-(x+2)(x-3)}$
$=-\frac{(x+2)^2(x-2)(x-3)}{(x+2)^2(x-2)(x-3)}$
$=-1$

59. $\frac{9-x^2}{x^2+5x+4}\div\frac{x^2-2x-3}{x^2+4x}$
$=\frac{9-x^2}{x^2+5x+4}\cdot\frac{x^2+4x}{x^2-2x-3}$
$=\frac{-(x+3)(x-3)}{(x+4)(x+1)}\cdot\frac{x(x+4)}{(x+1)(x-3)}$
$=-\frac{(x-3)(x+4)\cdot x(x+3)}{(x-3)(x+4)\cdot(x+1)(x+1)}$
$=-\frac{x(x+3)}{(x+1)^2}$

61. $\frac{\frac{a^2-b^2}{a^2+b^2}}{\frac{4a-4b}{2a^2+2b^2}}=\frac{a^2-b^2}{a^2+b^2}\cdot\frac{2a^2+2b^2}{4a-4b}$
$=\frac{(a+b)(a-b)}{a^2+b^2}\cdot\frac{2(a^2+b^2)}{2\cdot 2(a-b)}$
$=\frac{2(a-b)(a^2+b^2)\cdot(a+b)}{2(a-b)(a^2+b^2)\cdot 2}$
$=\frac{a+b}{2}$

63. $\frac{x^3-1}{x^4-1}\div\frac{3x^2+3x+3}{x^3+x^2+x+1}$
$=\frac{x^3-1}{x^4-1}\cdot\frac{x^3+x^2+x+1}{3x^2+3x+3}$
$=\frac{(x-1)(x^2+x+1)}{(x^2+1)(x+1)(x-1)}\cdot\frac{x^2(x+1)+1(x+1)}{3(x^2+x+1)}$
$=\frac{(x+1)(x-1)(x^2+1)(x^2+x+1)}{3(x+1)(x-1)(x^2+1)(x^2+x+1)}$
$=\frac{1}{3}$

65. $\frac{xy-ay+xb-ab}{xy+ay-xb-ab}\cdot\frac{2xy-2ay-2xb+2ab}{4b+4y}$
$=\frac{y(x-a)+b(x-a)}{y(x+a)-b(x+a)}\cdot\frac{2y(x-a)-2b(x-a)}{4(y+b)}$
$=\frac{(x-a)(y+b)}{(x+a)(y-b)}\cdot\frac{2(x-a)(y-b)}{2\cdot 2(y+b)}$
$=\frac{2(y+b)(y-b)\cdot(x-a)^2}{2(y+b)(y-b)\cdot 2(x+a)}$
$=\frac{(x-a)^2}{2(x+a)}$

67. $\frac{x^2+3xy+2y^2}{x^2-y^2}\cdot\frac{3x-3y}{9x^2+9xy-18y^2}$
$=\frac{(x+y)(x+2y)}{(x+y)(x-y)}\cdot\frac{3(x-y)}{3\cdot 3\cdot(x+2y)(x-y)}$
$=\frac{3(x-y)(x+y)(x+2y)}{3(x-y)(x+y)(x+2y)\cdot 3(x-y)}$
$=\frac{1}{3(x-y)}$

69. $\frac{\frac{x}{2y}}{\frac{(2xy)^2}{9xy^3}}=\frac{x}{2y}\cdot\frac{9xy^3}{(2xy)^2}$
$=\frac{x}{2y}\cdot\frac{9xy^3}{4x^2y^2}$
$=\frac{x^2y^3\cdot 9}{x^2y^3\cdot 8}$
$=\frac{9}{8}$

71. $\frac{\left(\frac{x-y}{x+y}\right)^2}{x^2-y^2}=\frac{(x-y)^2}{(x+y)^2}\cdot\frac{1}{x^2-y^2}$
$=\frac{(x-y)^2}{(x+y)^2}\cdot\frac{1}{(x+y)(x-y)}$
$=\frac{(x-y)\cdot(x-y)}{(x-y)\cdot(x+y)^2}$
$=\frac{x-y}{(x+y)^3}$

73. $\frac{\frac{6x}{x^2-4}}{\frac{3x-9}{2x+4}}=\frac{6x}{x^2-4}\cdot\frac{2x+4}{3x-9}$
$=\frac{2\cdot 3\cdot x}{(x+2)(x-2)}\cdot\frac{2(x+2)}{3(x-3)}$
$=\frac{3(x+2)\cdot 4x}{3(x+2)\cdot(x-2)(x-3)}$
$=\frac{4x}{(x-2)(x-3)}$

75. In a rectangle,
Area = (length)(width) $=\frac{3x+9}{27x^2}\cdot\frac{9x}{x+3}$
$=\frac{3(x+3)}{27x^2}\cdot\frac{9x}{x+3}$
$=\frac{27x(x+3)}{27x(x+3)\cdot x}$
$=\frac{1}{x}$

The area is $\frac{1}{x}$ square feet.

77. In a triangle,
Area $=\frac{1}{2}$(base)(height)
$=\frac{1}{2}\cdot\frac{1}{x^2-9}\cdot\frac{4x^2+20x+24}{1}$
$=\frac{1}{2}\cdot\frac{1}{(x+3)(x-3)}\cdot\frac{2\cdot 2(x+3)(x+2)}{1}$
$=\frac{2(x+3)\cdot 2(x+2)}{2(x+3)\cdot(x-3)}$
$=\frac{2(x+2)}{x-3}$

The area is $\frac{2(x+2)}{x-3}$ square inches.

79. $$\frac{x^2+x-12}{x^2-2x-35} \div \frac{x+4}{x^2+4x-5} \div \frac{12-4x}{x-7} = \frac{x^2+x-12}{x^2-2x-35} \cdot \frac{x^2+4x-5}{x+4} \cdot \frac{x-7}{12-4x}$$
$$= \frac{(x+4)(x-3)}{(x+5)(x-7)} \cdot \frac{(x+5)(x-1)}{x+4} \cdot \frac{x-7}{-4(x-3)}$$
$$= -\frac{(x+4)(x-3)(x+5)(x-7)\cdot(x-1)}{(x+4)(x-3)(x+5)(x-7)\cdot 4}$$
$$= \frac{1-x}{4}$$

81. $$\frac{a^2-2ab}{2b-3a} \div \frac{3a^2-4ab-4b^2}{16a^2b^2-36a^4} \div (6a) = \frac{a^2-2ab}{2b-3a} \cdot \frac{16a^2b^2-36a^4}{3a^2-4ab-4b^2} \cdot \frac{1}{6a}$$
$$= \frac{a(a-2b)}{2b-3a} \cdot \frac{4a^2(4b^2-9a^2)}{(3a+2b)(a-2b)} \cdot \frac{1}{6a}$$
$$= \frac{a(a-2b)}{2b-3a} \cdot \frac{2\cdot 2a^2\cdot(2b+3a)(2b-3a)}{(3a+2b)(a-2b)} \cdot \frac{1}{3\cdot 2\cdot a}$$
$$= \frac{2a(a-2b)(2b-3a)(2b+3a)\cdot 2a^2}{2a(a-2b)(2b-3a)(2b+3a)\cdot 3}$$
$$= \frac{2a^2}{3}$$

83. $$\frac{x^2+xy-3x-3y}{x^3+y^3} \cdot \frac{x^2+2xy+y^2}{x^2-x-6} = \frac{x(x+y)-3(x+y)}{x^3+y^3} \cdot \frac{x^2+2xy+y^2}{x^2-x-6}$$
$$= \frac{(x+y)(x-3)}{(x+y)(x^2-xy+y^2)} \cdot \frac{(x+y)^2}{(x+2)(x-3)}$$
$$= \frac{(x+y)(x-3)\cdot(x+y)^2}{(x+y)(x-3)\cdot(x^2-xy+y^2)(x+2)}$$
$$= \frac{(x+y)^2}{(x^2-xy+y^2)(x+2)}$$

85. $$\frac{p^3-27q^3}{9pq} \cdot \frac{(3p^2q)^3}{p^2-9q^2} \div \frac{1}{p^2+2pq-3q^2} = \frac{p^3-27q^3}{9pq} \cdot \frac{27p^6q^3}{p^2-9q^2} \cdot \frac{p^2+2pq-3q^2}{1}$$
$$= \frac{(p-3q)(p^2+3pq+9q^2)}{9pq} \cdot \frac{9pq\cdot 3p^5q^2}{(p+3q)(p-3q)} \cdot \frac{(p+3q)(p-q)}{1}$$
$$= \frac{9pq(p-3q)(p+3q)\cdot 3p^5q^2(p^2+3pq+9q^2)(p-q)}{9pq(p-3q)(p+3q)}$$
$$= 3p^5q^2(p^2+3pq+9q^2)(p-q)$$

87.
$$\frac{2x}{x^3-3x^2}\cdot\frac{x^2-x-6}{?}=\frac{1}{3x}$$
$$\frac{2x}{x\cdot x(x-3)}\cdot\frac{(x+2)(x-3)}{?}=\frac{1}{3x}$$
$$\frac{x(x-3)\cdot 2(x+2)}{x(x-3)\cdot x\cdot ?}=\frac{1}{3x}$$
$$\frac{2(x+2)}{x\cdot ?}=\frac{1}{3x}$$
$$?=3\cdot 2(x+2)$$
$$=6(x+2)$$
$$=6x+12$$

89. To multiply two rational expressions, (1) factor the polynomials in the numerators and denominators; (2) multiply the numerators and denominators using $\frac{a}{b}\cdot\frac{c}{d}=\frac{a\cdot c}{b\cdot d}$; (3) divide out common factors in the numerators and denominators. Leave the result in factored form.

91. The error is in incorrectly dividing out the factors $(n-6)$ and $(6-n)$.
$$\frac{n^2-2n-24}{6n-n^2}=\frac{(n-6)(n+4)}{n(6-n)}$$
$$=\frac{(n-6)(n+4)}{n(-1)(n-6)}$$
$$=\frac{n+4}{-n}$$
$$=-\frac{n+4}{n}$$

Section 7.3

Preparing for Adding and Subtracting Rational Expressions with a Common Denominator

P1. $\frac{12}{15}=\frac{2\cdot 2\cdot 3}{3\cdot 5}=\frac{2\cdot 2}{5}=\frac{4}{5}$

P2. (a) $\frac{7}{5}+\frac{2}{5}=\frac{7+2}{5}=\frac{9}{5}$

(b)
$$\frac{5}{6}+\frac{11}{6}=\frac{5+11}{6}$$
$$=\frac{16}{6}$$
$$=\frac{2\cdot 2\cdot 2\cdot 2}{2\cdot 3}$$
$$=\frac{2\cdot 2\cdot 2}{3}$$
$$=\frac{8}{3}$$

P3. (a) $\frac{7}{9}-\frac{5}{9}=\frac{7-5}{9}=\frac{2}{9}$

(b) $\frac{7}{8}-\frac{5}{8}=\frac{7-5}{8}=\frac{2}{8}=\frac{2}{2\cdot 2\cdot 2}=\frac{1}{2\cdot 2}=\frac{1}{4}$

P4. The additive inverse of 5 is -5 because $5+(-5)=0$.

P5. $-(x-2)=-1\cdot x+(-1)(-2)=-x+2=2-x$

7.3 Quick Checks

1. If $\frac{a}{c}$ and $\frac{b}{c}$, $c\neq 0$, are two rational expressions, then $\frac{a}{c}+\frac{b}{c}=\underline{\frac{a+b}{c}}$.

2. False; $\frac{2x}{a}+\frac{4x}{a}=\frac{2x+4x}{a}=\frac{6x}{a}\neq\frac{6x}{2a}$.

3. $\frac{1}{x-2}+\frac{3}{x-2}=\frac{1+3}{x-2}=\frac{4}{x-2}$

4. $\frac{2x+1}{x+1}+\frac{x^2}{x+1}=\frac{2x+1+x^2}{x+1}=\frac{(x+1)^2}{x+1}=x+1$

5. $\frac{9x}{6x-5}+\frac{2x-3}{6x-5}=\frac{9x+2x-3}{6x-5}=\frac{11x-3}{6x-5}$

6.
$$\frac{2x-2}{2x^2-7x-15}+\frac{5}{2x^2-7x-15}$$
$$=\frac{2x-2+5}{2x^2-7x-15}$$
$$=\frac{2x+3}{(2x+3)(x-5)}$$
$$=\frac{1}{x-5}$$

7. $\frac{8y}{2y-5}-\frac{6}{2y-5}=\frac{8y-6}{2y-5}=\frac{2(4y-3)}{2y-5}$

8. $$\begin{aligned}\frac{10+3z}{6z}-\frac{7}{6z}&=\frac{10+3z-7}{6z}\\&=\frac{3z+3}{6z}\\&=\frac{3(z+1)}{6z}\\&=\frac{z+1}{2z}\end{aligned}$$

9. $\frac{3x+1}{x-4}-\frac{x+3}{x-4}=\frac{3x+1-(x+3)}{x-4}$

10. $$\begin{aligned}\frac{2x^2-5x}{3x}-\frac{x^2-13x}{3x}&=\frac{2x^2-5x-(x^2-13x)}{3x}\\&=\frac{2x^2-5x-x^2+13x}{3x}\\&=\frac{x^2+8x}{3x}\\&=\frac{x(x+8)}{3x}\\&=\frac{x+8}{3}\end{aligned}$$

11. $$\begin{aligned}&\frac{3x^2+8x-1}{x^2-3x-28}-\frac{2x^2+2x-9}{x^2-3x-28}\\&=\frac{3x^2+8x-1-(2x^2+2x-9)}{x^2-3x-28}\\&=\frac{3x^2+8x-1-2x^2-2x+9}{x^2-3x-28}\\&=\frac{x^2+6x+8}{x^2-3x-28}\\&=\frac{(x+2)(x+4)}{(x-7)(x+4)}\\&=\frac{x+2}{x-7}\end{aligned}$$

12. True

13. $$\begin{aligned}\frac{3x}{x-5}+\frac{1}{5-x}&=\frac{3x}{x-5}+\frac{1}{-1(x-5)}\\&=\frac{3x}{x-5}+\frac{-1}{x-5}\\&=\frac{3x-1}{x-5}\end{aligned}$$

14. $$\begin{aligned}\frac{a^2+2a}{a-7}+\frac{a^2+14}{7-a}&=\frac{a^2+2a}{a-7}+\frac{a^2+14}{-1(a-7)}\\&=\frac{a^2+2a}{a-7}+\frac{-a^2-14}{a-7}\\&=\frac{a^2+2a-a^2-14}{a-7}\\&=\frac{2a-14}{a-7}\\&=\frac{2(a-7)}{a-7}\\&=2\end{aligned}$$

15. $$\begin{aligned}\frac{2n}{n^2-9}-\frac{6}{9-n^2}&=\frac{2n}{n^2-9}-\frac{6}{-1(n^2-9)}\\&=\frac{2n}{n^2-9}-\frac{-6}{n^2-9}\\&=\frac{2n-(-6)}{n^2-9}\\&=\frac{2n+6}{n^2-9}\\&=\frac{2(n+3)}{(n+3)(n-3)}\\&=\frac{2}{n-3}\end{aligned}$$

16. $$\begin{aligned}\frac{2k}{6k-6}-\frac{9+4k}{6-6k}&=\frac{2k}{6k-6}-\frac{9+4k}{-1(6k-6)}\\&=\frac{2k}{6k-6}-\frac{-9-4k}{6k-6}\\&=\frac{2k-(-9-4k)}{6k-6}\\&=\frac{2k+9+4k}{6k-6}\\&=\frac{6k+9}{6k-6}\\&=\frac{3(2k+3)}{2\cdot 3(k-1)}\\&=\frac{2k+3}{2(k-1)}\end{aligned}$$

7.3 Exercises

17. $\frac{3p}{8}+\frac{11p}{8}=\frac{3p+11p}{8}=\frac{14p}{8}=\frac{7p}{4}$

19. $\frac{n}{2}+\frac{3n}{2}=\frac{n+3n}{2}=\frac{4n}{2}=2n$

21. $\frac{4a-1}{3a}+\frac{2a-2}{3a}=\frac{4a-1+2a-2}{3a}$
$=\frac{6a-3}{3a}$
$=\frac{3(2a-1)}{3a}$
$=\frac{2a-1}{a}$

23. $\frac{8c-3}{c-1}+\frac{2c-1}{c-1}=\frac{8c-3+2c-1}{c-1}$
$=\frac{10c-4}{c-1}$
$=\frac{2(5c-2)}{c-1}$

25. $\frac{2x}{x+y}+\frac{2y}{x+y}=\frac{2x+2y}{x+y}=\frac{2(x+y)}{x+y}=2$

27. $\frac{14x}{x+2}+\frac{7x^2}{x+2}=\frac{14x+7x^2}{x+2}=\frac{7x(x+2)}{x+2}=7x$

29. $\frac{12a-1}{2a+6}+\frac{13-8a}{2a+6}=\frac{12a-1+13-8a}{2a+6}$
$=\frac{4a+12}{2a+6}$
$=\frac{4(a+3)}{2(a+3)}$
$=\frac{4}{2}$
$=2$

31. $\frac{a}{a^2-3a-10}+\frac{2}{a^2-3a-10}=\frac{a+2}{a^2-3a-10}$
$=\frac{a+2}{(a+2)(a-5)}$
$=\frac{1}{a-5}$

33. $\frac{x^2}{x^2-9}-\frac{3x}{x^2-9}=\frac{x^2-3x}{x^2-9}$
$=\frac{x(x-3)}{(x-3)(x+3)}$
$=\frac{x}{x+3}$

35. $\frac{x^2-x}{2x}-\frac{2x^2-x}{2x}=\frac{x^2-x-(2x^2-x)}{2x}$
$=\frac{x^2-x-2x^2+x}{2x}$
$=\frac{-x^2}{2x}$
$=-\frac{x}{2}$

37. $\frac{2c-3}{4c}-\frac{6c+9}{4c}=\frac{2c-3-(6c+9)}{4c}$
$=\frac{2c-3-6c-9}{4c}$
$=\frac{-4c-12}{4c}$
$=\frac{-4(c+3)}{4c}$
$=-\frac{c+3}{c}$

39. $\frac{x^2-6}{x-2}-\frac{x^2-3x}{x-2}=\frac{x^2-6-(x^2-3x)}{x-2}$
$=\frac{x^2-6-x^2+3x}{x-2}$
$=\frac{-6+3x}{x-2}$
$=\frac{3(x-2)}{x-2}$
$=3$

41. $\frac{2}{c^2-4}-\frac{c^2-2}{c^2-4}=\frac{2-(c^2-2)}{c^2-4}$
$=\frac{2-c^2+2}{c^2-4}$
$=\frac{4-c^2}{c^2-4}$
$=\frac{-(c^2-4)}{c^2-4}$
$=-1$

43. $$\frac{2x^2+5x}{2x+3}-\frac{4x+3}{2x+3}=\frac{2x^2+5x-(4x+3)}{2x+3}$$
$$=\frac{2x^2+5x-4x-3}{2x+3}$$
$$=\frac{2x^2+x-3}{2x+3}$$
$$=\frac{(2x+3)(x-1)}{2x+3}$$
$$=x-1$$

45. $$\frac{n}{n-3}+\frac{3}{3-n}=\frac{n}{n-3}+\frac{3}{-1(n-3)}$$
$$=\frac{n}{n-3}+\frac{-3}{n-3}$$
$$=\frac{n-3}{n-3}$$
$$=1$$

47. $$\frac{12q}{2p-2q}-\frac{8p}{2q-2p}=\frac{12q}{2p-2q}-\frac{8p}{-1(2p-2q)}$$
$$=\frac{12q}{2p-2q}-\frac{-8p}{2p-2q}$$
$$=\frac{12q-(-8p)}{2p-2q}$$
$$=\frac{12q+8p}{2p-2q}$$
$$=\frac{4(2p+3q)}{2(p-q)}$$
$$=\frac{2(2p+3q)}{p-q}$$

49. $$\frac{2p^2-1}{p-1}-\frac{2p+2}{1-p}=\frac{2p^2-1}{p-1}-\frac{2p+2}{-1(p-1)}$$
$$=\frac{2p^2-1}{p-1}-\frac{-(2p+2)}{p-1}$$
$$=\frac{2p^2-1+(2p+2)}{p-1}$$
$$=\frac{2p^2+2p+1}{p-1}$$

51. $$\frac{2a}{a-b}+\frac{2a-4b}{b-a}=\frac{2a}{a-b}+\frac{2a-4b}{-1(a-b)}$$
$$=\frac{2a}{a-b}+\frac{-(2a-4b)}{a-b}$$
$$=\frac{2a-(2a-4b)}{a-b}$$
$$=\frac{2a-2a+4b}{a-b}$$
$$=\frac{4b}{a-b}$$

53. $$\frac{3}{p^2-3}+\frac{p^2}{3-p^2}=\frac{3}{p^2-3}+\frac{p^2}{-1(p^2-3)}$$
$$=\frac{3}{p^2-3}+\frac{-p^2}{p^2-3}$$
$$=\frac{3-p^2}{p^2-3}$$
$$=\frac{-(p^2-3)}{p^2-3}$$
$$=-1$$

55. $$\frac{2x}{x^2-y^2}-\frac{2y}{y^2-x^2}=\frac{2x}{x^2-y^2}-\frac{2y}{-1(x^2-y^2)}$$
$$=\frac{2x}{x^2-y^2}-\frac{-2y}{x^2-y^2}$$
$$=\frac{2x-(-2y)}{x^2-y^2}$$
$$=\frac{2x+2y}{x^2-y^2}$$
$$=\frac{2(x+y)}{(x+y)(x-y)}$$
$$=\frac{2}{x-y}$$

57. $$\frac{4}{7x}-\frac{11}{7x}=\frac{4-11}{7x}=\frac{-7}{7x}=-\frac{1}{x}$$

59. $$\frac{5}{2x-5}-\frac{2x}{2x-5}=\frac{5-2x}{2x-5}=\frac{-(2x-5)}{2x-5}=-1$$

61. $$\frac{2n^3}{n-1}-\frac{2n^3}{n-1}=\frac{2n^3-2n^3}{n-1}=\frac{0}{n-1}=0$$

63. $\frac{n^2-3}{2n+3}+\frac{n^2+n}{2n+3}=\frac{n^2-3+n^2+n}{2n+3}$
$=\frac{2n^2+n-3}{2n+3}$
$=\frac{(2n+3)(n-1)}{2n+3}$
$=n-1$

65. $\frac{3v-1}{v^2-9}-\frac{8}{v^2-9}=\frac{3v-1-8}{v^2-9}$
$=\frac{3v-9}{v^2-9}$
$=\frac{3(v-3)}{(v+3)(v-3)}$
$=\frac{3}{v+3}$

67. $\frac{x^2}{x^2-1}+\frac{2x+1}{x^2-1}=\frac{x^2+2x+1}{x^2-1}$
$=\frac{(x+1)^2}{(x+1)(x-1)}$
$=\frac{x+1}{x-1}$

69. $\frac{2x^2+3}{x^2-3x}-\frac{x^2}{x^2-3x}=\frac{2x^2+3-x^2}{x^2-3x}$
$=\frac{x^2+3}{x^2-3x}$
$=\frac{x^2+3}{x(x-3)}$

71. $\frac{3x^2}{x+1}-\frac{x^2+2}{x+1}=\frac{3x^2-(x^2+2)}{x+1}$
$=\frac{3x^2-x^2-2}{x+1}$
$=\frac{2x^2-2}{x+1}$
$=\frac{2(x+1)(x-1)}{x+1}$
$=2(x-1)$

73. $\frac{3x-1}{x-y}+\frac{1-3y}{y-x}=\frac{3x-1}{x-y}+\frac{1-3y}{-1(x-y)}$
$=\frac{3x-1}{x-y}+\frac{-(1-3y)}{x-y}$
$=\frac{3x-1-(1-3y)}{x-y}$
$=\frac{3x-1-1+3y}{x-y}$
$=\frac{3x+3y-2}{x-y}$

75. $\frac{2p-3q}{3p^2-18q^2}-\frac{9q+p}{18q^2-3p^2}$
$=\frac{2p-3q}{3p^2-18q^2}-\frac{9q+p}{-1(3p^2-18q^2)}$
$=\frac{2p-3q}{3p^2-18q^2}-\frac{-(9q+p)}{3p^2-18q^2}$
$=\frac{2p-3q+9q+p}{3p^2-18q^2}$
$=\frac{3p+6q}{3p^2-18q^2}$
$=\frac{3(p+2q)}{3(p^2-6q^2)}$
$=\frac{p+2q}{p^2-6q^2}$

77. $\frac{7a^2}{a}+\frac{5a^2}{a}=\frac{7a^2+5a^2}{a}=\frac{12a^2}{a}=12a$

79. $\frac{2n}{n+1}-\frac{n+3}{n+1}=\frac{2n-(n+3)}{n+1}$
$=\frac{2n-n-3}{n+1}$
$=\frac{n-3}{n+1}$

81. $\frac{11}{2n}-\frac{3}{2n}=\frac{11-3}{2n}=\frac{8}{2n}=\frac{4}{n}$

83. Let $\frac{?}{x^2-9}$ be the unknown rational expression.

$$\frac{x^2}{x^2-9}-\frac{?}{x^2-9}=\frac{3x}{x^2-9}$$
$$\frac{x^2-?}{x^2-9}=\frac{3x}{x^2-9}$$
$$x^2-?=3x$$
$$x^2-?+?=3x+?$$
$$x^2=3x+?$$
$$x^2-3x=3x+?-3x$$
$$x^2-3x=?$$

The rational expression is

$$\frac{x^2-3x}{x^2-9}=\frac{x(x-3)}{(x+3)(x-3)}=\frac{x}{x+3}.$$

85. Let $\frac{?}{x+2}$ be the unknown rational expression.

Note that $1=\frac{x+2}{x+2}$.

$$\frac{-3x+4}{x+2}+\frac{?}{x+2}=1$$
$$\frac{-3x+4+?}{x+2}=\frac{x+2}{x+2}$$
$$-3x+4+?=x+2$$
$$-3x+4+?+3x=x+2+3x$$
$$4+?=4x+2$$
$$4+?-4=4x+2-4$$
$$?=4x-2$$

The rational expression is $\frac{4x-2}{x+2}$.

87. In a rectangle,
Perimeter = 2[(Length) + (Width)].

$$2\left(\frac{2x-3}{x}+\frac{4x+2}{x}\right)=2\left(\frac{2x-3+4x+2}{x}\right)$$
$$=2\left(\frac{6x-1}{x}\right)$$
$$=\frac{2}{1}\cdot\frac{6x-1}{x}$$
$$=\frac{2(6x-1)}{1\cdot x}$$
$$=\frac{2(6x-1)}{x}\text{ cm}$$

89. $\frac{7x-3y}{x^2-y^2}-\left[\frac{2x-3y}{x^2-y^2}-\frac{x+7y}{x^2-y^2}\right]$

$$=\frac{7x-3y-(2x-3y)+(x+7y)}{x^2-y^2}$$
$$=\frac{7x-3y-2x+3y+x+7y}{x^2-y^2}$$
$$=\frac{6x+7y}{x^2-y^2}$$

91. $\frac{x^2}{3x^2+5x-2}-\frac{x}{3x-1}\cdot\frac{2x-1}{x+2}$

$$=\frac{x^2}{3x^2+5x-2}-\frac{x(2x-1)}{(3x-1)(x+2)}$$
$$=\frac{x^2}{3x^2+5x-2}-\frac{2x^2-x}{3x^2+5x-2}$$
$$=\frac{x^2-(2x^2-x)}{3x^2+5x-2}$$
$$=\frac{x^2-2x^2+x}{3x^2+5x-2}$$
$$=\frac{-x^2+x}{3x^2+5x-2}$$
$$=\frac{-x(x-1)}{(3x-1)(x+2)}$$

93. $\frac{5x}{x-2}+\frac{2(x+3)}{x-2}-\frac{6(2x-1)}{x-2}$

$$=\frac{5x+2(x+3)-6(2x-1)}{x-2}$$
$$=\frac{5x+2x+6-12x+6}{x-2}$$
$$=\frac{-5x+12}{x-2}$$

95. $\frac{2n+1}{n-3}-\frac{?}{n-3}=\frac{6n+7}{n-3}$

$$\frac{2n+1-?}{n-3}=\frac{6n+7}{n-3}$$
$$2n+1-?=6n+7$$
$$2n+1-?-2n=6n+7-2n$$
$$1-?=4n+7$$
$$1-?-1=4n+7-1$$
$$-?=4n+6$$
$$-1(-?)=-1(4n+6)$$
$$?=-4n-6$$

97. The expression is incorrect because the subtraction symbol was not applied to both terms of the second numerator.

$$\frac{x-2}{x}-\frac{x+4}{x}=\frac{x-2-(x+4)}{x}$$
$$=\frac{x-2-x-4}{x}$$
$$=\frac{-6}{x}$$

99. The correct answer is (c). The answer (a) $\frac{7}{2x}$ is incorrect because the denominators were added; and the answer (b) $\frac{7}{x^2}$ is incorrect because the denominators were multiplied.

Section 7.4

Preparing for Finding the Least Common Denominator and Forming Equivalent Rational Expressions

P1. $\frac{5}{12}=\frac{5}{12}\cdot\frac{2}{2}=\frac{10}{24}$

P2. $15 = 3 \cdot 5$ and $25 = 5 \cdot 5$, so
LCD $= 3 \cdot 5 \cdot 5 = 75$.

7.4 Quick Checks

1. The least common denominator of two or more rational expressions is the smallest polynomial that is a multiple of each denominator in the rational expressions to be added or subtracted.

2. False; $4a^2b=2\cdot 2\cdot a^2\cdot b$
$4ab=2\cdot 2\cdot a\cdot b$
$\text{LCD}=2\cdot 2\cdot a^2\cdot b=4a^2b\neq 16a^2b^2$

3. $8x^2y=2^3\cdot x^2\cdot y$
$12xy^3=2^2\cdot 3\cdot x\cdot y^3$
$\text{LCD}=2^3\cdot 3\cdot x^2\cdot y^3=24x^2y^3$

4. $6a^3b^2=2\cdot 3\cdot a^3\cdot b^2$
$21ab^3=3\cdot 7\cdot a\cdot b^3$
$\text{LCD}=2\cdot 3\cdot 7\cdot a^3\cdot b^3=42a^3b^3$

5. True

6. $15z = 3 \cdot 5 \cdot z$
$5z^2+5z=5\cdot z\cdot (z+1)$
$\text{LCD} = 3 \cdot 5 \cdot z \cdot (z + 1) = 15z(z + 1)$

7. $x^2+4x-5=(x+5)(x-1)$
$x^2+10x+25=(x+5)^2$
$\text{LCD}=(x+5)^2(x-1)$

8. $21 - 3x = -3(x - 7)$
$x^2-49=(x-7)(x+7)$
$\text{LCD} = -3(x - 7)(x + 7)$

9. If we want to rewrite the rational expression $\frac{2x+1}{x-1}$ with a denominator of $(x - 1)(x + 3)$, we multiply the numerator and denominator of $\frac{2x+1}{x-1}$ by $\underline{x+3}$.

10. $4p^2-8p=4p(p-2)$
$16p^3(p-2)=4^2p^3(p-2)$
The "missing factor" is $4p^2$.
$$\frac{3}{4p^2-8p}=\frac{3}{4p(p-2)}\cdot\frac{4p^2}{4p^2}$$
$$=\frac{12p^2}{16p^3(p-2)}$$

11. $6a^2b=2\cdot 3\cdot a\cdot a\cdot b$
$20ab^3=2\cdot 2\cdot 5\cdot a\cdot b\cdot b\cdot b$
$\text{LCD}=2\cdot 2\cdot 3\cdot 5\cdot a\cdot a\cdot b\cdot b\cdot b$
$=60a^2b^3$
$$\frac{3}{6a^2b}=\frac{3}{6a^2b}\cdot\frac{10b^2}{10b^2}=\frac{30b^2}{60a^2b^3}$$
$$\frac{-5}{20ab^3}=\frac{-5}{20ab^3}\cdot\frac{3a}{3a}=\frac{-15a}{60a^2b^3}$$

12. $x^2-4x-5=(x+1)(x-5)$
$x^2-7x+10=(x-5)(x-2)$
$\text{LCD}=(x+1)(x-5)(x-2)$

$$\frac{5}{x^2-4x-5}=\frac{5}{(x+1)(x-5)}\cdot\frac{x-2}{x-2}=\frac{5(x-2)}{(x+1)(x-5)(x-2)}$$

$$\frac{-3}{x^2-7x+10}=\frac{-3}{(x-5)(x-2)}\cdot\frac{x+1}{x+1}=\frac{-3(x+1)}{(x+1)(x-5)(x-2)}$$

7.4 Exercises

13. $5x^2=5\cdot x^2$
$5x=5\cdot x$
$\text{LCD}=5x^2$

15. $12xy^2=2^2\cdot 3\cdot\quad x\cdot\ y^2$
$15x^3y=\quad 3\cdot 5\cdot x^3\cdot y$
$\text{LCD}=2^2\cdot 3\cdot 5\cdot x^3\cdot y^2=60x^3y^2$

17. $x=x$
$x+1=x+1$
$\text{LCD}=x(x+1)$

19. Note that $2=\frac{2}{1}$.
$2x+1=2x+1$
$1=1$
$\text{LCD}=1(2x+1)=2x+1$

21. $2b-6=2(b-3)$
$4b-12=4(b-3)=2^2(b-3)$
$\text{LCD}=2^2(b-3)=4(b-3)$

23. $p^2+p=p(p+1)$
$p^2-p-2=(p+1)(p-2)$
$\text{LCD}=p(p+1)(p-2)$

25. $r^2+4r+4=(r+2)^2$
$r^2-r-2=(r+1)(r-2)$
$\text{LCD}=(r+2)^2(r+1)(r-2)$

27. $x-4=x-4$
$4-x=-1(x-4)$
$\text{LCD}=-1(x-4)=-(x-4)$

29. $x^2-9=(x+3)(x-3)$
$6-2x=-2(x-3)$
$\text{LCD}=-2(x+3)(x-3)$

31. $(x-1)(x+2)=(x-1)(x+2)$
$(1-x)(2+x)=-1(x-1)(x+2)$
$\text{LCD}=-1(x-1)(x+2)=-(x-1)(x+2)$

33. $x=x$
$3x^3=3\cdot x^3=3x^2\cdot x$
The "missing factor" is $3x^2$.
$$\frac{4}{x}=\frac{4}{x}\cdot\frac{3x^2}{3x^2}=\frac{4\cdot 3x^2}{x\cdot 3x^2}=\frac{12x^2}{3x^3}$$

35. $a^2b^2c=a^2\cdot b^2\cdot c$
$a^2b^2c^2=a^2\cdot b^2\cdot c^2=c\cdot a^2b^2c$
The "missing factor" is c.
$$\frac{3+c}{a^2b^2c}=\frac{3+c}{a^2b^2c}\cdot\frac{c}{c}=\frac{(3+c)c}{a^2b^2c\cdot c}=\frac{3c+c^2}{a^2b^2c^2}$$

37. $x+4=x+4$
$x^2-16=(x+4)(x-4)$
The "missing factor" is $x-4$.
$$\frac{x-4}{x+4}=\frac{x-4}{x+4}\cdot\frac{x-4}{x-4}=\frac{(x-4)(x-4)}{(x+4)(x-4)}=\frac{(x-4)^2}{x^2-16}$$

39. $2n+2=2(n+1)$
$6n^2-6=2\cdot 3(n+1)(n-1)=3(n-1)\cdot 2(n+1)$
The "missing factor" is $3(n-1)$.
$$\frac{3n}{2n+2}=\frac{3n}{2(n+1)}\cdot\frac{3(n-1)}{3(n-1)}=\frac{3n\cdot 3(n-1)}{2(n+1)\cdot 3(n-1)}=\frac{9n^2-9n}{6n^2-6}$$

41. $4t=\frac{4t}{1}=\frac{4t}{1}\cdot\frac{t-1}{t-1}=\frac{4t(t-1)}{t-1}=\frac{4t^2-4t}{t-1}$

43. $3y = 3 \cdot y$
$9y^2 = 3^2 \cdot y^2$
$\text{LCD} = 3^2 \cdot y^2 = 9y^2$
$$\frac{2x}{3y} = \frac{2x}{3y} \cdot \frac{3y}{3y} = \frac{2x \cdot 3y}{3y \cdot 3y} = \frac{6xy}{9y^2}$$
$$\frac{4}{9y^2} = \frac{4}{9y^2} \cdot \frac{1}{1} = \frac{4}{9y^2}$$

45. $2a^3 = 2 \cdot a^3$
$4a = 2^2 \cdot a$
$\text{LCD} = 2^2 \cdot a^3 = 4a^3$
$$\frac{3a+1}{2a^3} = \frac{3a+1}{2a^3} \cdot \frac{2}{2} = \frac{(3a+1) \cdot 2}{2a^3 \cdot 2} = \frac{6a+2}{4a^3}$$
$$\frac{4a-1}{4a} = \frac{4a-1}{4a} \cdot \frac{a^2}{a^2} = \frac{(4a-1)a^2}{4a \cdot a^2} = \frac{4a^3 - a^2}{4a^3}$$

47. $m = m$
$m + 1 = m + 1$
$\text{LCD} = m(m + 1)$
$$\frac{2}{m} = \frac{2}{m} \cdot \frac{m+1}{m+1} = \frac{2(m+1)}{m(m+1)} = \frac{2m+2}{m(m+1)}$$
$$\frac{3}{m+1} = \frac{3}{m+1} \cdot \frac{m}{m} = \frac{3 \cdot m}{(m+1)m} = \frac{3m}{m(m+1)}$$

49. $4y = 2^2 \cdot y$
$8y - 4 = 2^2 \cdot (2y - 1)$
$\text{LCD} = 2^2 \cdot y \cdot (2y-1) = 4y(2y-1)$
$$\frac{y-2}{4y} = \frac{y-2}{4y} \cdot \frac{2y-1}{2y-1} = \frac{(y-2)(2y-1)}{4y(2y-1)} = \frac{2y^2 - 5y + 2}{4y(2y-1)}$$
$$\frac{y}{8y-4} = \frac{y}{4(2y-1)} \cdot \frac{y}{y} = \frac{y \cdot y}{4(2y-1) \cdot y} = \frac{y^2}{4y(2y-1)}$$

51. $x - 1 = x - 1$
$1 - x = -(x - 1)$
$\text{LCD} = -(x - 1)$
$$\frac{1}{x-1} = \frac{1}{x-1} \cdot \frac{-1}{-1} = \frac{1(-1)}{(x-1)(-1)} = \frac{-1}{-(x-1)}$$
$$\frac{2x}{1-x} = \frac{2x}{-(x-1)}$$

53. $x - 7 = x - 7$
$7 - x = -(x - 7)$
$\text{LCD} = -(x - 7)$
$$\frac{3x}{x-7} = \frac{3x}{x-7} \cdot \frac{-1}{-1} = \frac{3x(-1)}{(x-7)(-1)} = \frac{-3x}{-(x-7)}$$
$$\frac{-5}{7-x} = \frac{-5}{-(x-7)}$$

55. $x^2 - 4 = (x+2)(x-2)$
$x + 2 = x + 2$
$\text{LCD} = (x + 2)(x - 2)$
$$\frac{4x}{x^2-4} = \frac{4x}{(x+2)(x-2)}$$
$$\frac{2}{x+2} = \frac{2}{x+2} \cdot \frac{x-2}{x-2} = \frac{2(x-2)}{(x+2)(x-2)} = \frac{2x-4}{(x+2)(x-2)}$$

57. $x^2 - 9 = (x+3)(x-3)$
$x^2 - 2x - 3 = (x+1)(x-3)$
$\text{LCD} = (x + 3)(x - 3)(x + 1)$
$$\frac{x+1}{x^2-9} = \frac{x+1}{(x+3)(x-3)} \cdot \frac{x+1}{x+1} = \frac{(x+1)(x+1)}{(x+3)(x-3)(x+1)} = \frac{x^2+2x+1}{(x+3)(x-3)(x+1)}$$
$$\frac{x+2}{x^2-2x-3} = \frac{x+2}{(x+1)(x-3)} \cdot \frac{x+3}{x+3} = \frac{(x+2)(x+3)}{(x+1)(x-3)(x+3)} = \frac{x^2+5x+6}{(x+3)(x-3)(x+1)}$$

59. $x+4=x+4$

$2x^2+7x-4=(x+4)(2x-1)$

LCD = $(x+4)(2x-1)$

$$\frac{3}{x+4}=\frac{3}{x+4}\cdot\frac{2x-1}{2x-1}=\frac{3(2x-1)}{(x+4)(2x-1)}=\frac{6x-3}{(x+4)(2x-1)}$$

$$\frac{2x-1}{2x^2+7x-4}=\frac{2x-1}{(x+4)(2x-1)}$$

61. $\frac{1}{x}$ and $\frac{1}{3x}$

$x=x$

$3x=3\cdot x$

LCD = $3x$

$$\frac{1}{x}=\frac{1}{x}\cdot\frac{3}{3}=\frac{3}{3x}$$

$$\frac{1}{3x}=\frac{1}{3x}\cdot\frac{1}{1}=\frac{1}{3x}$$

63. $\frac{12}{r-4}$ and $\frac{12}{r+4}$

$r-4=r-4$

$r+4=r+4$

LCD = $(r-4)(r+4)$

$$\frac{12}{r-4}=\frac{12}{r-4}\cdot\frac{r+4}{r+4}=\frac{12(r+4)}{(r-4)(r+4)}=\frac{12r+48}{(r-4)(r+4)}$$

$$\frac{12}{r+4}=\frac{12}{r+4}\cdot\frac{r-4}{r-4}=\frac{12(r-4)}{(r+4)(r-4)}=\frac{12r-48}{(r-4)(r+4)}$$

65. $x^3+8=(x+2)(x^2-2x+4)$

$x^2-4=(x+2)(x-2)$

$x^3-8=(x-2)(x^2+2x+4)$

LCD = $(x+2)(x-2)(x^2-2x+4)(x^2+2x+4)$

67. $p^2-p=p(p-1)$

$p^3-p^2=p^2(p-1)$

$p^2-4p+3=(p-1)(p-3)$

LCD = $p^2(p-1)(p-3)$

69. $2a^4-2a^2b^2=2a^2(a^2-b^2)=2a^2(a+b)(a-b)$

$4ab^2+4b^3=2^2b^2(a+b)$

$a^3b-b^4=b(a^3-b^3)=b(a-b)(a^2+ab+b^2)$

LCD = $2^2a^2b^2(a+b)(a-b)(a^2+ab+b^2)=4a^2b^2(a+b)(a-b)(a^2+ab+b^2)$

71. To write equivalent rational expressions with a common denominator, (1) write each denominator in factored form; (2) determine the "missing factors" (3) Multiply the original expression by $1=\frac{\text{missing factors}}{\text{missing factors}}$; (3) find the product leaving the denominator in factored form.

Section 7.5

Preparing for Adding and Subtracting Rational Expressions with Unlike Denominators

P1. $15=\quad 3\cdot 5$

$24=2^3\cdot 3$

LCD = $2^3\cdot 3\cdot 5=120$

P2. $2x^2+3x+1=(2x+1)(x+1)$

P3. $\frac{3x+9}{6x}=\frac{3(x+3)}{3\cdot 2x}=\frac{x+3}{2x}$

P4. $$\frac{1-x^2}{2x-2}=\frac{(1-x)(1+x)}{2(x-1)}=\frac{-1(x-1)(1+x)}{2(x-1)}=\frac{-(1+x)}{2}$$

P5. $\frac{5}{2x}+\frac{7}{2x}=\frac{5+7}{2x}=\frac{12}{2x}=\frac{6}{x}$

7.5 Quick Checks

1. The least common denominator of $\frac{1}{8}$ and $\frac{5}{18}$ is $\underline{72}$.

2. $12 = 2^2 \cdot 3$
$18 = 2 \cdot 3^2$
$\text{LCD} = 2^2 \cdot 3^2 = 36$
$$\frac{5}{12}+\frac{5}{18}=\frac{5}{12}\cdot\frac{3}{3}+\frac{5}{18}\cdot\frac{2}{2}=\frac{15}{36}+\frac{10}{36}=\frac{25}{36}$$

3. $15 = 3 \cdot 5$
$10 = 2 \cdot 5$
$\text{LCD} = 2 \cdot 3 \cdot 5 = 30$
$$\frac{8}{15}-\frac{3}{10}=\frac{8}{15}\cdot\frac{2}{2}-\frac{3}{10}\cdot\frac{3}{3}=\frac{16}{30}-\frac{9}{30}=\frac{7}{30}$$

4. The first step in adding rational expressions with unlike denominators is to determine the <u>least common denominator</u>.

5. True

6. $8a^3b = 2^3 \cdot a^3 \cdot b$
$12ab^2 = 2^2 \cdot 3 \cdot a \cdot b^2$
$\text{LCD} = 2^3 \cdot 3 \cdot a^3 \cdot b^2 = 24a^3b^2$
$$\frac{1}{8a^3b}+\frac{5}{12ab^2}=\frac{1}{8a^3b}\cdot\frac{3b}{3b}+\frac{5}{12ab^2}\cdot\frac{2a^2}{2a^2}$$
$$=\frac{3b}{24a^3b^2}+\frac{10a^2}{24a^3b^2}$$
$$=\frac{10a^2+3b}{24a^3b^2}$$

7. $15xy^2 = 3 \cdot 5 \cdot x \cdot y^2$
$18x^3y = 2 \cdot 3^2 \cdot x^3 \cdot y$
$\text{LCD} = 2 \cdot 3^2 \cdot 5 \cdot x^3 \cdot y^2 = 90x^3y^2$
$$\frac{1}{15xy^2}+\frac{7}{18x^3y}=\frac{1}{15xy^2}\cdot\frac{6x^2}{6x^2}+\frac{7}{18x^3y}\cdot\frac{5y}{5y}$$
$$=\frac{6x^2}{90x^3y^2}+\frac{35y}{90x^3y^2}$$
$$=\frac{6x^2+35y}{90x^3y^2}$$

8. $$\frac{5}{x-4}+\frac{3}{x+2}$$
$$=\frac{5}{x-4}\cdot\frac{x+2}{x+2}+\frac{3}{x+2}\cdot\frac{x-4}{x-4}$$
$$=\frac{5x+10}{(x+2)(x-4)}+\frac{3x-12}{(x+2)(x-4)}$$
$$=\frac{5x+10+3x-12}{(x+2)(x-4)}$$
$$=\frac{8x-2}{(x+2)(x-4)}$$
$$=\frac{2(4x-1)}{(x+2)(x-4)}$$

9. $$\frac{-1}{n-3}+\frac{4}{n+1}=\frac{-1}{n-3}\cdot\frac{n+1}{n+1}+\frac{4}{n+1}\cdot\frac{n-3}{n-3}$$
$$=\frac{-n-1}{(n+1)(n-3)}+\frac{4n-12}{(n+1)(n-3)}$$
$$=\frac{-n-1+4n-12}{(n+1)(n-3)}$$
$$=\frac{3n-13}{(n+1)(n-3)}$$

10. $z^2+7z+10=(z+5)(z+2)$
$z+5=z+5$
$\text{LCD} = (z+5)(z+2)$
$$\frac{-z+1}{z^2+7z+10}+\frac{2}{z+5}$$
$$=\frac{-z+1}{(z+5)(z+2)}+\frac{2}{z+5}$$
$$=\frac{-z+1}{(z+5)(z+2)}+\frac{2}{z+5}\cdot\frac{z+2}{z+2}$$
$$=\frac{-z+1}{(z+5)(z+2)}+\frac{2z+4}{(z+5)(z+2)}$$
$$=\frac{-z+1+2z+4}{(z+5)(z+2)}$$
$$=\frac{z+5}{(z+5)(z+2)}$$
$$=\frac{1}{z+2}$$

11. $x^2+5x=x(x+5)$
$x^2-5x=x\quad\cdot(x-5)$
$\text{LCD}=x(x+5)(x-5)$

$$\frac{1}{x^2+5x}+\frac{1}{x^2-5x}$$
$$=\frac{1}{x(x+5)}+\frac{1}{x(x-5)}$$
$$=\frac{1}{x(x+5)}\cdot\frac{x-5}{x-5}+\frac{1}{x(x-5)}\cdot\frac{x+5}{x+5}$$
$$=\frac{x-5}{x(x+5)(x-5)}+\frac{x+5}{x(x+5)(x-5)}$$
$$=\frac{x-5+x+5}{x(x+5)(x-5)}$$
$$=\frac{2x}{x(x+5)(x-5)}$$
$$=\frac{2}{(x+5)(x-5)}$$

12. $5ab^2=\quad 5\cdot a\quad\cdot b^2$
$4a^2b^3=2^2\quad\cdot a^2\cdot b^3$
$\text{LCD}=2^2\cdot 5\cdot a^2\cdot b^3=20a^2b^3$

$$\frac{-4}{5ab^2}-\frac{3}{4a^2b^3}=\frac{-4}{5ab^2}\cdot\frac{4ab}{4ab}-\frac{3}{4a^2b^3}\cdot\frac{5}{5}$$
$$=\frac{-16ab}{20a^2b^3}-\frac{15}{20a^2b^3}$$
$$=\frac{-16ab-15}{20a^2b^3}$$

13. $\text{LCD}=x\cdot(x-4)=x(x-4)$

$$\frac{5}{x}-\frac{3}{x-4}=\frac{5}{x}\cdot\frac{x-4}{x-4}-\frac{3}{x-4}\cdot\frac{x}{x}$$
$$=\frac{5x-20}{x(x-4)}-\frac{3x}{x(x-4)}$$
$$=\frac{5x-20-3x}{x(x-4)}$$
$$=\frac{2x-20}{x(x-4)}$$
$$=\frac{2(x-10)}{x(x-4)}$$

14. $\text{LCD}=(x-5)(x+4)(x-1)$

$$\frac{3}{(x-5)(x+4)}-\frac{2}{(x-5)(x-1)}$$
$$=\frac{3}{(x-5)(x+4)}\cdot\frac{x-1}{x-1}-\frac{2}{(x-5)(x-1)}\cdot\frac{x+4}{x+4}$$
$$=\frac{3x-3}{(x-5)(x+4)(x-1)}-\frac{2x+8}{(x-5)(x+4)(x-1)}$$
$$=\frac{3x-3-2x-8}{(x-5)(x+4)(x-1)}$$
$$=\frac{x-11}{(x-5)(x+4)(x-1)}$$

15. $x^2-3x=x(x-3)$
$4x-12=4(x-3)$
$\text{LCD}=4x(x-3)$

$$\frac{x-2}{x^2-3x}+\frac{x+3}{4x-12}$$
$$=\frac{x-2}{x(x-3)}\cdot\frac{4}{4}+\frac{x+3}{4(x-3)}\cdot\frac{x}{x}$$
$$=\frac{4x-8}{4x(x-3)}+\frac{x^2+3x}{4x(x-3)}$$
$$=\frac{4x-8+x^2+3x}{4x(x-3)}$$
$$=\frac{x^2+7x-8}{4x(x-3)}$$
$$=\frac{(x+8)(x-1)}{4x(x-3)}$$

16. True

17.
$$\frac{7}{3p^2-3p}+\frac{5}{6-6p}=\frac{7}{3p^2-3p}+\frac{5}{-1(6p-6)}$$
$$=\frac{7}{3p^2-3p}+\frac{-5}{6p-6}$$

$3p^2-3p=3p(p-1)$
$6p-6=6(p-1)=2\cdot 3(p-1)$
$\text{LCD}=2\cdot 3p(p-1)=6p(p-1)$

$$\frac{7}{3p^2-3p}+\frac{-5}{6p-6}$$
$$=\frac{7}{3p^2-3p}\cdot\frac{2}{2}+\frac{-5}{6p-6}\cdot\frac{p}{p}$$
$$=\frac{14}{6p(p-1)}+\frac{-5p}{6p(p-1)}$$
$$=\frac{14-5p}{6p(p-1)}$$

18. $2-\frac{3}{x-1}=\frac{2}{1}-\frac{3}{x-1}$

LCD = $x-1$

$$\frac{2}{1}\cdot\frac{x-1}{x-1}-\frac{3}{x-1}=\frac{2(x-1)}{x-1}-\frac{3}{x-1}$$
$$=\frac{2x-2}{x-1}-\frac{3}{x-1}$$
$$=\frac{2x-2-3}{x-1}$$
$$=\frac{2x-5}{x-1}$$

19.
$$x+1=x+1$$
$$x^2-1=(x+1)(x-1)$$
$$x-1=\quad x-1$$
LCD = $(x+1)(x-1)$

$$\frac{1}{x+1}-\frac{2}{x^2-1}+\frac{3}{x-1}$$
$$=\frac{1}{x+1}\cdot\frac{x-1}{x-1}-\frac{2}{(x+1)(x-1)}+\frac{3}{x-1}\cdot\frac{x+1}{x+1}$$
$$=\frac{1(x-1)}{(x+1)(x-1)}-\frac{2}{(x+1)(x-1)}+\frac{3(x+1)}{(x+1)(x-1)}$$
$$=\frac{x-1-2+3x+3}{(x+1)(x-1)}$$
$$=\frac{4x}{(x+1)(x-1)}$$

20.
$$x=x$$
$$x-2=\quad x-2$$
$$x^2-2x=x(x-2)$$
LCD = $x(x-2)$

$$\frac{3}{x}-\left(\frac{1}{x-2}-\frac{6}{x^2-2x}\right)$$
$$=\frac{3}{x}-\frac{1}{x-2}+\frac{6}{x(x-2)}$$
$$=\frac{3}{x}\cdot\frac{x-2}{x-2}-\frac{1}{x-2}\cdot\frac{x}{x}+\frac{6}{x(x-2)}$$
$$=\frac{3(x-2)}{x(x-2)}-\frac{x}{x(x-2)}+\frac{6}{x(x-2)}$$
$$=\frac{3x-6-x+6}{x(x-2)}$$
$$=\frac{2x}{x(x-2)}$$
$$=\frac{2}{x-2}$$

7.5 Exercises

21. LCD = $3\cdot 2=6$

$$\frac{-4}{3}+\frac{1}{2}=\frac{-4}{3}\cdot\frac{2}{2}+\frac{1}{2}\cdot\frac{3}{3}$$
$$=\frac{-8}{6}+\frac{3}{6}$$
$$=\frac{-8+3}{6}$$
$$=\frac{-5}{6}$$

23. LCD = $3\cdot x=3x$

$$\frac{2}{3x}+\frac{1}{x}=\frac{2}{3x}+\frac{1}{x}\cdot\frac{3}{3}=\frac{2}{3x}+\frac{3}{3x}=\frac{2+3}{3x}=\frac{5}{3x}$$

25. LCD = $(2a-1)(2a+1)$

$$\frac{a}{2a-1}+\frac{3}{2a+1}$$
$$=\frac{a}{2a-1}\cdot\frac{2a+1}{2a+1}+\frac{3}{2a+1}\cdot\frac{2a-1}{2a-1}$$
$$=\frac{a(2a+1)}{(2a-1)(2a+1)}+\frac{3(2a-1)}{(2a-1)(2a+1)}$$
$$=\frac{2a^2+a+6a-3}{(2a-1)(2a+1)}$$
$$=\frac{2a^2+7a-3}{(2a-1)(2a+1)}$$

27. $x-4=x-4;\ 4-x=-(x-4)$

LCD = $-(x-4)$

$$\frac{3}{x-4}+\frac{5}{4-x}=\frac{3}{x-4}\cdot\frac{-1}{-1}+\frac{5}{-(x-4)}$$
$$=\frac{-3}{-(x-4)}+\frac{5}{-(x-4)}$$
$$=\frac{-3+5}{-(x-4)}$$
$$=\frac{2}{-(x-4)}$$
$$=\frac{-2}{x-4}$$

29. $5a-a^2=-a(a-5)$; $4a-20=4(a-5)$

LCD = $-4a(a-5)$

$$\frac{a+3}{5a-a^2}+\frac{2a+1}{4a-20}$$
$$=\frac{a+3}{-a(a-5)}\cdot\frac{4}{4}+\frac{2a+1}{4(a-5)}\cdot\frac{-a}{-a}$$
$$=\frac{4a+12}{-4a(a-5)}+\frac{-2a^2-a}{-4a(a-5)}$$
$$=\frac{4a+12-2a^2-a}{-4a(a-5)}$$
$$=\frac{-2a^2+3a+12}{-4a(a-5)}$$
$$=\frac{2a^2-3a-12}{4a(a-5)}$$

31. $x^2+2x=x(x+2)$; $x=x$

LCD = $x(x+2)$

$$\frac{2x+4}{x^2+2x}+\frac{3}{x}=\frac{2x+4}{x(x+2)}+\frac{3}{x}\cdot\frac{x+2}{x+2}$$
$$=\frac{2x+4}{x(x+2)}+\frac{3x+6}{x(x+2)}$$
$$=\frac{2x+4+3x+6}{x(x+2)}$$
$$=\frac{5x+10}{x(x+2)}$$
$$=\frac{5(x+2)}{x(x+2)}$$
$$=\frac{5}{x}$$

33. $15=3\cdot 5$; $25=5\cdot 5$

LCD = $3\cdot 5\cdot 5=75$

$$\frac{7}{15}-\frac{9}{25}=\frac{7}{15}\cdot\frac{5}{5}-\frac{9}{25}\cdot\frac{3}{3}$$
$$=\frac{35}{75}-\frac{27}{75}$$
$$=\frac{35-27}{75}$$
$$=\frac{8}{75}$$

35. LCD = m

$$m-\frac{16}{m}=\frac{m}{1}\cdot\frac{m}{m}-\frac{16}{m}$$
$$=\frac{m^2}{m}-\frac{16}{m}$$
$$=\frac{m^2-16}{m}$$
$$=\frac{(m-4)(m+4)}{m}$$

37. LCD = $(x-3)(x+3)$

$$\frac{x}{x-3}-\frac{x-2}{x+3}=\frac{x}{x-3}\cdot\frac{x+3}{x+3}-\frac{x-2}{x+3}\cdot\frac{x-3}{x-3}$$
$$=\frac{x(x+3)}{(x-3)(x+3)}-\frac{(x-2)(x-3)}{(x-3)(x+3)}$$
$$=\frac{x^2+3x-(x^2-5x+6)}{(x-3)(x+3)}$$
$$=\frac{x^2+3x-x^2+5x-6}{(x-3)(x+3)}$$
$$=\frac{8x-6}{(x-3)(x+3)}$$
$$=\frac{2(4x-3)}{(x-3)(x+3)}$$

39. $x+3=x+3$; $x^2+6x+9=(x+3)^2$

LCD = $(x+3)^2$

$$\frac{x+2}{x+3}-\frac{x^2-x}{x^2+6x+9}=\frac{x+2}{x+3}\cdot\frac{x+3}{x+3}-\frac{x^2-x}{(x+3)^2}$$
$$=\frac{x^2+5x+6}{(x+3)^2}-\frac{x^2-x}{(x+3)^2}$$
$$=\frac{x^2+5x+6-(x^2-x)}{(x+3)^2}$$
$$=\frac{x^2+5x+6-x^2+x}{(x+3)^2}$$
$$=\frac{6x+6}{(x+3)^2}$$
$$=\frac{6(x+1)}{(x+3)^2}$$

41. $2a + 6 = 2(a + 3)$; $a + 3 = a + 3$
LCD = $2(a + 3)$

$$\frac{6}{2a+6} - \frac{4a-1}{a+3} = \frac{6}{2(a+3)} - \frac{4a-1}{a+3}\cdot\frac{2}{2}$$
$$= \frac{6}{2(a+3)} - \frac{8a-2}{2(a+3)}$$
$$= \frac{6-(8a-2)}{2(a+3)}$$
$$= \frac{6-8a+2}{2(a+3)}$$
$$= \frac{-8a+8}{2(a+3)}$$
$$= \frac{-4\cdot 2(a-1)}{2(a+3)}$$
$$= \frac{-4(a-1)}{a+3}$$

43. $x^2 + x - 6 = (x+3)(x-2)$; $2 - x = -(x - 2)$
LCD = $-(x + 3)(x - 2)$

$$\frac{-3x-9}{x^2+x-6} - \frac{x+3}{2-x}$$
$$= \frac{-3x-9}{(x+3)(x-2)}\cdot\frac{-1}{-1} - \frac{x+3}{-(x-2)}\cdot\frac{x+3}{x+3}$$
$$= \frac{3x+9}{-(x+3)(x-2)} - \frac{x^2+6x+9}{-(x+3)(x-2)}$$
$$= \frac{3x+9-(x^2+6x+9)}{-(x+3)(x-2)}$$
$$= \frac{3x+9-x^2-6x-9}{-(x+3)(x-2)}$$
$$= \frac{-x^2-3x}{-(x+3)(x-2)}$$
$$= \frac{-x(x+3)}{-(x+3)(x-2)}$$
$$= \frac{x}{x-2}$$

45. LCD $= 3\cdot 4\cdot y^2 = 12y^2$

$$\frac{5}{3y^2} - \frac{3}{4y} = \frac{5}{3y^2}\cdot\frac{4}{4} - \frac{3}{4y}\cdot\frac{3y}{3y}$$
$$= \frac{20}{12y^2} - \frac{9y}{12y^2}$$
$$= \frac{20-9y}{12y^2}$$

47. LCD $= 2 \cdot 5x = 10x$

$$\frac{9}{5x} - \frac{6}{10x} = \frac{9}{5x}\cdot\frac{2}{2} - \frac{6}{10x}$$
$$= \frac{18}{10x} - \frac{6}{10x}$$
$$= \frac{18-6}{10x}$$
$$= \frac{12}{10x}$$
$$= \frac{2\cdot 6}{2\cdot 5x}$$
$$= \frac{6}{5x}$$

49. LCD $= 1(2x + 3) = 2x + 3$

$$\frac{2x}{2x+3} - 1 = \frac{2x}{2x+3} - \frac{1}{1}\cdot\frac{2x+3}{2x+3}$$
$$= \frac{2x}{2x+3} - \frac{2x+3}{2x+3}$$
$$= \frac{2x-(2x+3)}{2x+3}$$
$$= \frac{2x-2x-3}{2x+3}$$
$$= \frac{-3}{2x+3}$$

51. LCD $= (n - 2) \cdot n = n(n - 2)$

$$\frac{n}{n-2} + \frac{n+2}{n} = \frac{n}{n-2}\cdot\frac{n}{n} + \frac{n+2}{n}\cdot\frac{n-2}{n-2}$$
$$= \frac{n^2}{n(n-2)} + \frac{n^2-4}{n(n-2)}$$
$$= \frac{n^2+n^2-4}{n(n-2)}$$
$$= \frac{2n^2-4}{n(n-2)}$$
$$= \frac{2(n^2-2)}{n(n-2)}$$

53. LCD $= (x - 3) \cdot x = x(x - 3)$

$$\frac{2x}{x-3} - \frac{5}{x} = \frac{2x}{x-3}\cdot\frac{x}{x} - \frac{5}{x}\cdot\frac{x-3}{x-3}$$
$$= \frac{2x^2}{x(x-3)} - \frac{5x-15}{x(x-3)}$$
$$= \frac{2x^2-(5x-15)}{x(x-3)}$$
$$= \frac{2x^2-5x+15}{x(x-3)}$$

55. LCD $= a^2(a-1)$

$$\frac{a}{a^2}-\frac{1}{a-1}=\frac{a}{a^2}\cdot\frac{a-1}{a-1}-\frac{1}{a-1}\cdot\frac{a^2}{a^2}$$
$$=\frac{a^2-a}{a^2(a-1)}-\frac{a^2}{a^2(a-1)}$$
$$=\frac{a^2-a-a^2}{a^2(a-1)}$$
$$=\frac{-a}{a^2(a-1)}$$
$$=\frac{-1}{a(a-1)}$$

57. LCD = $1\cdot(2n+1) = 2n+1$

$$\frac{4}{2n+1}+1=\frac{4}{2n+1}+\frac{1}{1}\cdot\frac{2n+1}{2n+1}$$
$$=\frac{4}{2n+1}+\frac{2n+1}{2n+1}$$
$$=\frac{4+2n+1}{2n+1}$$
$$=\frac{2n+5}{2n+1}$$

59. $3x-6=3(x-2)$; $x-2=x-2$
LCD $= 3(x-2)$

$$\frac{-12}{3x-6}+\frac{4x-1}{x-2}=\frac{-12}{3(x-2)}+\frac{4x-1}{x-2}\cdot\frac{3}{3}$$
$$=\frac{-12}{3(x-2)}+\frac{12x-3}{3(x-2)}$$
$$=\frac{-12+12x-3}{3(x-2)}$$
$$=\frac{12x-15}{3(x-2)}$$
$$=\frac{3(4x-5)}{3(x-2)}$$
$$=\frac{4x-5}{x-2}$$

61. $x=x$; $x^2-9x=x(x-9)$
LCD $= x(x-9)$

$$\frac{3x-1}{x}-\frac{9}{x^2-9x}=\frac{3x-1}{x}\cdot\frac{x-9}{x-9}-\frac{9}{x(x-9)}$$
$$=\frac{3x^2-28x+9}{x(x-9)}-\frac{9}{x(x-9)}$$
$$=\frac{3x^2-28x+9-9}{x(x-9)}$$
$$=\frac{3x^2-28x}{x(x-9)}$$
$$=\frac{x(3x-28)}{x(x-9)}$$
$$=\frac{3x-28}{x-9}$$

63. $n^2-3n=n(n-3)$; $n^3-n^2=n^2(n-1)$

LCD $= n^2(n-3)(n-1)$

$$\frac{4}{n^2-3n}-\frac{3}{n^3-n^2}$$
$$=\frac{4}{n(n-3)}\cdot\frac{n(n-1)}{n(n-1)}-\frac{3}{n^2(n-1)}\cdot\frac{n-3}{n-3}$$
$$=\frac{4n^2-4n}{n^2(n-3)(n-1)}-\frac{3n-9}{n^2(n-1)(n-3)}$$
$$=\frac{4n^2-4n-(3n-9)}{n^2(n-3)(n-1)}$$
$$=\frac{4n^2-4n-3n+9}{n^2(n-3)(n-1)}$$
$$=\frac{4n^2-7n+9}{n^2(n-3)(n-1)}$$

65. $2a-a^2=-a(a-2)$; $2a^2-4a=2a(a-2)$
LCD $= -2a(a-2)$

$$\frac{5}{2a-a^2}-\frac{3}{2a^2-4a}$$
$$=\frac{5}{-a(a-2)}\cdot\frac{2}{2}-\frac{3}{2a(a-2)}\cdot\frac{-1}{-1}$$
$$=\frac{10}{-2a(a-2)}-\frac{-3}{-2a(a-2)}$$
$$=\frac{10-(-3)}{-2a(a-2)}$$
$$=\frac{13}{-2a(a-2)}$$
$$=\frac{-13}{2a(a-2)}$$

67. $n^2-4=(n+2)(n-2)$; $6-n^2-n=-n^2-n+6=-(n+3)(n-2)$

LCD $= -(n+2)(n-2)(n+3)$

$$\begin{aligned}\frac{2n+1}{n^2-4}+\frac{3n}{6-n^2-n} &= \frac{2n+1}{(n+2)(n-2)}\cdot\frac{-(n+3)}{-(n+3)}+\frac{3n}{-(n+3)(n-2)}\cdot\frac{n+2}{n+2}\\ &= \frac{-2n^2-7n-3}{-(n+2)(n-2)(n+3)}+\frac{3n^2+6n}{-(n+2)(n-2)(n+3)}\\ &= \frac{-2n^2-7n-3+3n^2+6n}{-(n+2)(n-2)(n+3)}\\ &= \frac{n^2-n-3}{-(n+2)(n-2)(n+3)}\end{aligned}$$

69. $x^2-1=(x+1)(x-1)$; $x^2-2x+1=(x-1)^2$

LCD $= (x+1)(x-1)^2$

$$\begin{aligned}\frac{x}{x^2-1}+\frac{x+1}{x^2-2x+1} &= \frac{x}{(x+1)(x-1)}\cdot\frac{x-1}{x-1}+\frac{x+1}{(x-1)^2}\cdot\frac{x+1}{x+1}\\ &= \frac{x^2-x}{(x+1)(x-1)^2}+\frac{x^2+2x+1}{(x+1)(x-1)^2}\\ &= \frac{x^2-x+x^2+2x+1}{(x+1)(x-1)^2}\\ &= \frac{2x^2+x+1}{(x+1)(x-1)^2}\end{aligned}$$

71. $n+3=n+3$; $n^2+n-6=(n+3)(n-2)$

LCD $= (n+3)(n-2)$

$$\begin{aligned}\frac{2n+1}{n+3}+\frac{7-2n^2}{n^2+n-6} &= \frac{2n+1}{n+3}\cdot\frac{n-2}{n-2}+\frac{7-2n^2}{(n+3)(n-2)}\\ &= \frac{2n^2-3n-2}{(n+3)(n-2)}+\frac{7-2n^2}{(n+3)(n-2)}\\ &= \frac{2n^2-3n-2+7-2n^2}{(n+3)(n-2)}\\ &= \frac{-3n+5}{(n+3)(n-2)}\end{aligned}$$

73. $x^2+x-6=(x+3)(x-2);\ 2-x=-(x-2)$

LCD = $-(x+3)(x-2)$

$$\begin{aligned}
\frac{-3x-9}{x^2+x-6}-\frac{x+3}{2-x} &= \frac{-3x-9}{(x+3)(x-2)}\cdot\frac{-1}{-1}-\frac{x+3}{-(x-2)}\cdot\frac{x+3}{x+3}\\
&= \frac{3x+9}{-(x+3)(x-2)}-\frac{x^2+6x+9}{-(x+3)(x-2)}\\
&= \frac{3x+9-(x^2+6x+9)}{-(x+3)(x-2)}\\
&= \frac{3x+9-x^2-6x-9}{-(x+3)(x-2)}\\
&= \frac{-x^2-3x}{-(x+3)(x-2)}\\
&= \frac{-x(x+3)}{-(x+3)(x-2)}\\
&= \frac{x}{x-2}
\end{aligned}$$

75. $x+2=x+2;\ x^2+2x=x(x+2);\ x=x$

LCD = $x(x+2)$

$$\begin{aligned}
\frac{4}{x+2}+\frac{-5x-2}{x^2+2x}-\frac{3-x}{x} &= \frac{4}{x+2}\cdot\frac{x}{x}+\frac{-5x-2}{x(x+2)}-\frac{3-x}{x}\cdot\frac{x+2}{x+2}\\
&= \frac{4x}{x(x+2)}+\frac{-5x-2}{x(x+2)}-\frac{-x^2+x+6}{x(x+2)}\\
&= \frac{4x-5x-2-(-x^2+x+6)}{x(x+2)}\\
&= \frac{x^2-2x-8}{x(x+2)}\\
&= \frac{(x+2)(x-4)}{x(x+2)}\\
&= \frac{x-4}{x}
\end{aligned}$$

77. $m + 2 = m + 2$; $m = m$; $m^2 - 4 = (m+2)(m-2)$

LCD = $m(m + 2)(m - 2)$

$$\frac{2}{m+2} - \frac{3}{m} + \frac{m+10}{m^2-4} = \frac{2}{m+2} \cdot \frac{m(m-2)}{m(m-2)} - \frac{3}{m} \cdot \frac{(m+2)(m-2)}{(m+2)(m-2)} + \frac{m+10}{(m+2)(m-2)} \cdot \frac{m}{m}$$

$$= \frac{2m^2 - 4m}{m(m+2)(m-2)} - \frac{3m^2 - 12}{m(m+2)(m-2)} + \frac{m^2 + 10m}{m(m+2)(m-2)}$$

$$= \frac{2m^2 - 4m - (3m^2 - 12) + m^2 + 10m}{m(m+2)(m-2)}$$

$$= \frac{2m^2 - 4m - 3m^2 + 12 + m^2 + 10m}{m(m+2)(m-2)}$$

$$= \frac{6m + 12}{m(m+2)(m-2)}$$

$$= \frac{6(m+2)}{m(m+2)(m-2)}$$

$$= \frac{6}{m(m-2)}$$

79. $$\frac{\frac{x-3}{x+2}}{\frac{x^2-9}{x^2+4}} = \frac{x-3}{x+2} \div \frac{x^2-9}{x^2+4} = \frac{x-3}{x+2} \cdot \frac{x^2+4}{x^2-9} = \frac{x-3}{x+2} \cdot \frac{x^2+4}{(x+3)(x-3)} = \frac{x^2+4}{(x+2)(x+3)}$$

81. $2x^2 + x - 3 = (2x+3)(x-1)$; $x^2 - 1 = (x+1)(x-1)$

LCD = $(2x + 3)(x + 1)(x - 1)$

$$\frac{4x+6}{2x^2+x-3} - \frac{x-1}{x^2-1} = \frac{4x+6}{(2x+3)(x-1)} \cdot \frac{x+1}{x+1} - \frac{x-1}{(x+1)(x-1)} \cdot \frac{2x+3}{2x+3}$$

$$= \frac{4x^2 + 10x + 6}{(2x+3)(x+1)(x-1)} - \frac{2x^2 + x - 3}{(2x+3)(x+1)(x-1)}$$

$$= \frac{4x^2 + 10x + 6 - (2x^2 + x - 3)}{(2x+3)(x+1)(x-1)}$$

$$= \frac{4x^2 + 10x + 6 - 2x^2 - x + 3}{(2x+3)(x+1)(x-1)}$$

$$= \frac{2x^2 + 9x + 9}{(2x+3)(x+1)(x-1)}$$

$$= \frac{(2x+3)(x+3)}{(2x+3)(x+1)(x-1)}$$

$$= \frac{x+3}{(x+1)(x-1)}$$

83. $x^2+x-6=(x+3)(x-2);\ x^2+5x+6=(x+3)(x+2)$

LCD = $(x + 3)(x - 2)(x + 2)$

$$\frac{x+2}{x^2+x-6}+\frac{x-3}{x^2+5x+6}=\frac{x+2}{(x+3)(x-2)}\cdot\frac{x+2}{x+2}+\frac{x-3}{(x+3)(x+2)}\cdot\frac{x-2}{x-2}$$
$$=\frac{(x+2)^2}{(x+3)(x-2)(x+2)}+\frac{(x-3)(x-2)}{(x+3)(x-2)(x+2)}$$
$$=\frac{x^2+4x+4}{(x+3)(x-2)(x+2)}+\frac{x^2-5x+6}{(x+3)(x-2)(x+2)}$$
$$=\frac{x^2+4x+4+x^2-5x+6}{(x+3)(x-2)(x+2)}$$
$$=\frac{2x^2-x+10}{(x+3)(x-2)(x+2)}$$

85. $\frac{2x-3}{x+6}\cdot\frac{x-1}{x-7}=\frac{(2x-3)(x-1)}{(x+6)(x-7)}=\frac{2x^2-5x+3}{(x+6)(x-7)}$

87. In a rectangle, Perimeter = [(Length) + (Width)]

$$2\left(\frac{2x+1}{4}+\frac{3x-7}{8}\right)=2\left(\frac{2x+1}{4}\cdot\frac{2}{2}+\frac{3x-7}{8}\right)$$
$$=2\left(\frac{4x+2}{8}+\frac{3x-7}{8}\right)$$
$$=2\left(\frac{4x+2+3x-7}{8}\right)$$
$$=2\left(\frac{7x-5}{8}\right)$$
$$=\frac{7x-5}{4}\text{ units}$$

89. $a=a;\ a-1=a-1;\ (a-1)^2=(a-1)^2$

LCD = $a(a-1)^2$

$$\frac{2}{a}-\left(\frac{2}{a-1}-\frac{3}{(a-1)^2}\right)=\frac{2}{a}\cdot\frac{(a-1)^2}{(a-1)^2}-\left(\frac{2}{a-1}\cdot\frac{a(a-1)}{a(a-1)}-\frac{3}{(a-1)^2}\cdot\frac{a}{a}\right)$$
$$=\frac{2a^2-4a+2}{a(a-1)^2}-\left(\frac{2a^2-2a}{a(a-1)^2}-\frac{3a}{a(a-1)^2}\right)$$
$$=\frac{2a^2-4a+2}{a(a-1)^2}-\left(\frac{2a^2-2a-3a}{a(a-1)^2}\right)$$
$$=\frac{2a^2-4a+2}{a(a-1)^2}-\frac{2a^2-5a}{a(a-1)^2}$$
$$=\frac{2a^2-4a+2-2a^2+5a}{a(a-1)^2}$$
$$=\frac{a+2}{a(a-1)^2}$$

91. $x = x;\ x^2 + x = x(x+1);\ x^3 - x^2 = x^2(x-1)$

$\text{LCD} = x^2(x+1)(x-1)$

$$\frac{1}{x} - \frac{2}{x^2+x} + \frac{3}{x^3-x^2} = \frac{1}{x}\cdot\frac{x(x+1)(x-1)}{x(x+1)(x-1)} - \frac{2}{x(x+1)}\cdot\frac{x(x-1)}{x(x-1)} + \frac{3}{x^2(x-1)}\cdot\frac{x+1}{x+1}$$

$$= \frac{x^3-x}{x^2(x+1)(x-1)} - \frac{2x^2-2x}{x^2(x+1)(x-1)} + \frac{3x+3}{x^2(x+1)(x-1)}$$

$$= \frac{x^3-x-(2x^2-2x)+3x+3}{x^2(x+1)(x-1)}$$

$$= \frac{x^3-x-2x^2+2x+3x+3}{x^2(x+1)(x-1)}$$

$$= \frac{x^3-2x^2+4x+3}{x^2(x+1)(x-1)}$$

93. $\frac{2a+b}{a-b}\cdot\frac{2}{a+b} - \frac{3a+3b}{a^2-b^2} = \frac{4a+2b}{(a-b)(a+b)} - \frac{3a+3b}{(a-b)(a+b)} = \frac{4a+2b-(3a+3b)}{(a-b)(a+b)} = \frac{a-b}{(a-b)(a+b)} = \frac{1}{a+b}$

95. $\text{LCD} = (x-4)(x+1)$

$$\frac{x-3}{x-4} + \frac{x+2}{x-4}\cdot\frac{4}{x+1} = \frac{x-3}{x-4}\cdot\frac{x+1}{x+1} + \frac{x+2}{x-4}\cdot\frac{4}{x+1}$$

$$= \frac{x^2-2x-3}{(x-4)(x+1)} + \frac{4x+8}{(x-4)(x+1)}$$

$$= \frac{x^2-2x-3+4x+8}{(x-4)(x+1)}$$

$$= \frac{x^2+2x+5}{(x-4)(x+1)}$$

97. $\text{LCD} = x-2$

$$\frac{x+2}{x-2} - \frac{x-2}{x+2} \div \frac{1}{x^2-4} = \frac{x+2}{x-2} - \frac{x-2}{x+2}\cdot\frac{(x+2)(x-2)}{1}$$

$$= \frac{x+2}{x-2} - \frac{(x-2)(x-2)}{1}$$

$$= \frac{x+2}{x-2} - \frac{(x^2-4x+4)}{1}\cdot\frac{x-2}{x-2}$$

$$= \frac{x+2}{x-2} - \frac{x^3-6x^2+12x-8}{x-2}$$

$$= \frac{x+2-x^3+6x^2-12x+8}{x-2}$$

$$= \frac{-x^3+6x^2-11x+10}{x-2}$$

99. The steps to add or subtract rational expressions with unlike denominators are:
(1) find the least common denominator; (2) rewrite each expression with the common denominator; (3) add or subtract the rational expressions from step 2; (4) simplify the result.

101. Since the fractions have a common denominator, subtract the second numerator from the first and write the result over the common denominator. Simplify the numerator and look for factors common to the numerator and denominator.

$$\frac{3x+1}{(2x+5)(x-1)}-\frac{x-4}{(2x+5)(x-1)}$$
$$=\frac{3x+1-(x-4)}{(2x+5)(x-1)}$$
$$=\frac{2x+5}{(2x+5)(x-1)}$$
$$=\frac{1}{x-1}$$

Section 7.6

Preparing for Complex Rational Expressions

P1. $6y^2-5y-6=(3y+2)(2y-3)$

P2. $\frac{x+3}{12}\div\frac{x^2-9}{15}=\frac{x+3}{12}\cdot\frac{15}{x^2-9}$
$$=\frac{15(x+3)}{12(x+3)(x-3)}$$
$$=\frac{15}{12(x-3)}$$
$$=\frac{3\cdot 5}{3\cdot 4(x-3)}$$
$$=\frac{5}{4(x-3)}$$

7.6 Quick Checks

1. An expression such as $\frac{\frac{x}{2}+\frac{5}{x}}{\frac{2x-1}{3}}$ is called a complex rational expression.

2. To simplify a complex rational expression means to write the rational expression in the form $\frac{p}{q}$, where p and q are polynomials that have no common factors.

3. $\frac{\frac{2}{k+3}}{\frac{4}{k^2+4k+3}}=\frac{2}{k+3}\cdot\frac{k^2+4k+3}{4}$
$$=\frac{2}{k+3}\cdot\frac{(k+3)(k+1)}{2\cdot 2}$$
$$=\frac{2(k+3)(k+1)}{2(k+3)\cdot 2}$$
$$=\frac{k+1}{2}$$

4. $\frac{\frac{1}{n+3}}{\frac{8n}{2n+6}}=\frac{1}{n+3}\cdot\frac{2n+6}{8n}$
$$=\frac{1}{n+3}\cdot\frac{2(n+3)}{2\cdot 4n}$$
$$=\frac{2(n+3)}{2(n+3)\cdot 4n}$$
$$=\frac{1}{4n}$$

5. False; $\frac{x-y}{\frac{1}{x}+\frac{2}{y}}=\frac{x-y}{\frac{1}{x}\cdot\frac{y}{y}+\frac{2}{y}\cdot\frac{x}{x}}$
$$=\frac{x-y}{\frac{y}{xy}+\frac{2x}{xy}}$$
$$=\frac{x-y}{\frac{y+2x}{xy}}$$
$$=\frac{x-y}{1}\cdot\frac{xy}{y+2x}$$
$$\neq\frac{x-y}{1}\cdot\left(\frac{x}{1}+\frac{y}{2}\right)$$

6. $\frac{\frac{1}{2}+\frac{2}{5}}{1-\frac{2}{5}}=\frac{\frac{1}{2}\cdot\frac{5}{5}+\frac{2}{5}\cdot\frac{2}{2}}{\frac{5}{5}-\frac{2}{5}}$
$$=\frac{\frac{5}{10}+\frac{4}{10}}{\frac{5-2}{5}}$$
$$=\frac{\frac{9}{10}}{\frac{3}{5}}$$
$$=\frac{9}{10}\cdot\frac{5}{3}$$
$$=\frac{3\cdot 3}{2\cdot 5}\cdot\frac{5}{3}$$
$$=\frac{3\cdot 5\cdot 3}{3\cdot 5\cdot 2}$$
$$=\frac{3}{2}$$

7. $$\frac{\frac{3}{y}-1}{\frac{9}{y}-y}=\frac{\frac{3}{y}-1\cdot\frac{y}{y}}{\frac{9}{y}-y\cdot\frac{y}{y}}=\frac{\frac{3}{y}-\frac{y}{y}}{\frac{9}{y}-\frac{y^2}{y}}=\frac{\frac{3-y}{y}}{\frac{9-y^2}{y}}=\frac{3-y}{y}\cdot\frac{y}{9-y^2}=\frac{3-y}{y}\cdot\frac{y}{(3-y)(3+y)}=\frac{y(3-y)}{y(3-y)(3+y)}=\frac{1}{3+y}$$

8. $$\frac{\frac{1}{3}+\frac{1}{x+5}}{\frac{x+8}{9}}=\frac{\frac{1}{3}\cdot\frac{x+5}{x+5}+\frac{1}{x+5}\cdot\frac{3}{3}}{\frac{x+8}{9}}=\frac{\frac{x+5}{3(x+5)}+\frac{3}{3(x+5)}}{\frac{x+8}{9}}=\frac{\frac{x+5+3}{3(x+5)}}{\frac{x+8}{3\cdot 3}}=\frac{x+8}{3(x+5)}\cdot\frac{3\cdot 3}{x+8}=\frac{3(x+8)\cdot 3}{3(x+8)(x+5)}=\frac{3}{x+5}$$

9. $$\frac{\frac{3}{2}+\frac{2}{3}}{\frac{1}{2}-\frac{1}{3}}=\frac{\frac{3}{2}+\frac{2}{3}}{\frac{1}{2}-\frac{1}{3}}\cdot\frac{6}{6}$$ LCD $= 2\cdot 3 = 6$
$$=\frac{\frac{3}{2}\cdot 6+\frac{2}{3}\cdot 6}{\frac{1}{2}\cdot 6-\frac{1}{3}\cdot 6}=\frac{9+4}{3-2}=\frac{13}{1}=13$$

10. LCD $= x\cdot y = xy$
$$\frac{\frac{1}{x}+\frac{2}{y}}{\frac{2}{x}-\frac{1}{y}}=\frac{\frac{1}{x}+\frac{2}{y}}{\frac{2}{x}-\frac{1}{y}}\cdot\frac{xy}{xy}=\frac{\frac{1}{x}\cdot xy+\frac{2}{y}\cdot xy}{\frac{2}{x}\cdot xy-\frac{1}{y}\cdot xy}=\frac{y+2x}{2y-x}$$

Exercises 7.6

11. $$\frac{1-\frac{3}{4}}{\frac{1}{8}+2}=\frac{\frac{4}{4}-\frac{3}{4}}{\frac{1}{8}+\frac{16}{8}}=\frac{\frac{1}{4}}{\frac{17}{8}}=\frac{1}{4}\cdot\frac{8}{17}=\frac{2}{17}$$

13. $$\frac{\frac{x^2}{12}-\frac{1}{3}}{\frac{x+2}{18}}=\frac{\frac{x^2}{12}-\frac{1}{3}\cdot\frac{4}{4}}{\frac{x+2}{18}}=\frac{\frac{x^2}{12}-\frac{4}{12}}{\frac{x+2}{18}}=\frac{\frac{x^2-4}{12}}{\frac{x+2}{18}}=\frac{x^2-4}{12}\cdot\frac{18}{x+2}=\frac{(x+2)(x-2)}{2\cdot 6}\cdot\frac{3\cdot 6}{x+2}=\frac{6(x+2)\cdot 3(x-2)}{6(x+2)\cdot 2}=\frac{3(x-2)}{2}$$

15. $$\frac{x+3}{\frac{x^2}{9}-1}=\frac{x+3}{\frac{x^2}{9}-1\cdot\frac{9}{9}}=\frac{x+3}{\frac{x^2-9}{9}}=\frac{x+3}{1}\cdot\frac{9}{x^2-9}=\frac{x+3}{1}\cdot\frac{9}{(x+3)(x-3)}=\frac{9(x+3)}{(x+3)(x-3)}=\frac{9}{x-3}$$

17. $\dfrac{\frac{m}{2}+n}{\frac{m}{n}} = \dfrac{\frac{m}{n}+n\cdot\frac{2}{2}}{\frac{m}{n}}$

$= \dfrac{\frac{m}{2}+\frac{2n}{2}}{\frac{m}{n}}$

$= \dfrac{\frac{m+2n}{2}}{\frac{m}{n}}$

$= \dfrac{m+2n}{2}\cdot\dfrac{n}{m}$

$= \dfrac{n(m+2n)}{2m}$

19. $\dfrac{\frac{5}{a}+\frac{4}{b^2}}{\frac{5b+4}{b^2}} = \dfrac{\frac{5}{a}\cdot\frac{b^2}{b^2}+\frac{4}{b^2}\cdot\frac{a}{a}}{\frac{5b+4}{b^2}}$

$= \dfrac{\frac{5b^2}{ab^2}+\frac{4a}{ab^2}}{\frac{5b+4}{b^2}}$

$= \dfrac{\frac{5b^2+4a}{ab^2}}{\frac{5b+4}{b^2}}$

$= \dfrac{5b^2+4a}{ab^2}\cdot\dfrac{b^2}{5b+4}$

$= \dfrac{b^2(5b^2+4a)}{b^2\cdot a(5b+4)}$

$= \dfrac{5b^2+4a}{a(5b+4)}$

21. $\dfrac{\frac{8}{y+3}-2}{y-\frac{4}{y+3}} = \dfrac{\frac{8}{y+3}-2\cdot\frac{y+3}{y+3}}{y\cdot\frac{y+3}{y+3}-\frac{4}{y+3}}$

$= \dfrac{\frac{8-2y-6}{y+3}}{\frac{y^2+3y-4}{y+3}}$

$= \dfrac{8-2y-6}{y+3}\cdot\dfrac{y+3}{y^2+3y-4}$

$= \dfrac{-2y+2}{y+3}\cdot\dfrac{y+3}{y^2+3y-4}$

$= \dfrac{-2(y-1)}{y+3}\cdot\dfrac{y+3}{(y+4)(y-1)}$

$= \dfrac{(y-1)(y+3)\cdot(-2)}{(y-1)(y+3)\cdot(y+4)}$

$= -\dfrac{2}{y+4}$

23. $\dfrac{\frac{1}{4}-\frac{6}{y}}{\frac{5}{6y}-y} = \dfrac{\frac{1}{4}\cdot\frac{y}{y}-\frac{6}{y}\cdot\frac{4}{4}}{\frac{5}{6y}-y\cdot\frac{6y}{6y}}$

$= \dfrac{\frac{y}{4y}-\frac{24}{4y}}{\frac{5}{6y}-\frac{6y^2}{y}}$

$= \dfrac{\frac{y-24}{4y}}{\frac{5-6y^2}{6y}}$

$= \dfrac{y-24}{4y}\cdot\dfrac{6y}{5-6y^2}$

$= \dfrac{y-24}{2\cdot 2y}\cdot\dfrac{3\cdot 2y}{5-6y^2}$

$= \dfrac{2y\cdot 3(y-24)}{2y\cdot 2(5-6y^2)}$

$= \dfrac{3(y-24)}{2(5-6y^2)}$

25. LCD = 12

$\dfrac{\frac{3}{2}-\frac{1}{4}}{\frac{5}{6}+\frac{1}{2}} = \dfrac{\frac{3}{2}-\frac{1}{4}}{\frac{5}{6}+\frac{1}{2}}\cdot\dfrac{12}{12} = \dfrac{\frac{3}{2}\cdot 12-\frac{1}{4}\cdot 12}{\frac{5}{6}\cdot 12+\frac{1}{2}\cdot 12} = \dfrac{18-3}{10+6} = \dfrac{15}{16}$

27. LCD = m^2

$\dfrac{\frac{3}{m}+\frac{2}{m^2}}{\frac{6}{m}+\frac{4}{m^2}} = \dfrac{\frac{3}{m}+\frac{2}{m^2}}{\frac{6}{m}+\frac{4}{m^2}}\cdot\dfrac{m^2}{m^2}$

$= \dfrac{3m+2}{6m+4}$

$= \dfrac{3m+2}{2(3m+2)}$

$= \dfrac{1}{2}$

29. LCD = b^2

$\dfrac{1-\frac{49}{b^2}}{1+\frac{7}{b}} = \dfrac{1-\frac{49}{b^2}}{1+\frac{7}{b}}\cdot\dfrac{b^2}{b^2}$

$= \dfrac{b^2-49}{b^2+7b}$

$= \dfrac{(b+7)(b-7)}{b(b+7)}$

$= \dfrac{b-7}{b}$

31. LCD = $x(x+4)$

$$\frac{1+\frac{5}{x}}{1+\frac{1}{x+4}} = \frac{1+\frac{5}{x}}{1+\frac{1}{x+4}} \cdot \frac{x(x+4)}{x(x+4)}$$
$$= \frac{x(x+4)+5(x+4)}{x(x+4)+x}$$
$$= \frac{x^2+4x+5x+20}{x^2+4x+x}$$
$$= \frac{x^2+9x+20}{x^2+5x}$$
$$= \frac{(x+4)(x+5)}{x(x+5)}$$
$$= \frac{x+4}{x}$$

33. LCD = x^2y^2

$$\frac{\frac{1}{x^2}-\frac{1}{y^2}}{x-y} = \frac{\frac{1}{x^2}-\frac{1}{y^2}}{x-y} \cdot \frac{x^2y^2}{x^2y^2}$$
$$= \frac{y^2-x^2}{x^2y^2(x-y)}$$
$$= \frac{-(x+y)(x-y)}{x^2y^2(x-y)}$$
$$= -\frac{x+y}{x^2y^2}$$

35. LCD = b

$$\frac{a+\frac{1}{b}}{a+\frac{2}{b}} = \frac{a+\frac{1}{b}}{a+\frac{2}{b}} \cdot \frac{b}{b} = \frac{ab+1}{ab+2}$$

37. LCD = $3n$

$$\frac{12}{\frac{4}{n}-\frac{2}{3n}} = \frac{12}{\frac{4}{n}-\frac{2}{3n}} \cdot \frac{3n}{3n} = \frac{36n}{12-2} = \frac{36n}{10} = \frac{18n}{5}$$

39. $$\frac{\frac{x}{x-y}+\frac{y}{x+y}}{\frac{xy}{x^2-y^2}} = \frac{(x+y)(x-y)\left(\frac{x}{x-y}+\frac{y}{x+y}\right)}{(x+y)(x-y)\left(\frac{xy}{(x+y)(x-y)}\right)}$$
$$= \frac{x(x+y)+y(x-y)}{xy}$$
$$= \frac{x^2+xy+xy-y^2}{xy}$$
$$= \frac{x^2+2xy-y^2}{xy}$$

41. $$\frac{\frac{2}{x+4}}{\frac{2}{x+4}-4} = \frac{(x+4)\left(\frac{2}{x+4}\right)}{(x+4)\left(\frac{2}{x+4}-4\right)}$$
$$= \frac{2}{2-4(x+4)}$$
$$= \frac{2}{2-4x-16}$$
$$= \frac{2}{-4x-14}$$
$$= \frac{2}{-2(2x+7)}$$
$$= -\frac{1}{2x+7}$$

43. $$\frac{\frac{b^2}{b^2-16}-\frac{b}{b+4}}{\frac{b}{b^2-16}-\frac{1}{b-4}}$$
$$= \frac{\frac{b^2}{(b+4)(b-4)}-\frac{b}{b+4}}{\frac{b}{(b+4)(b-4)}-\frac{1}{b-4}} \cdot \frac{(b+4)(b-4)}{(b+4)(b-4)}$$
$$= \frac{b^2-b(b-4)}{b-(b+4)}$$
$$= \frac{b^2-b^2+4b}{b-b-4}$$
$$= \frac{4b}{-4}$$
$$= -b$$

45. $$\frac{1-\frac{2}{x}-\frac{3}{x^2}}{1-\frac{9}{x^2}} = \frac{x^2\left(1-\frac{2}{x}-\frac{3}{x^2}\right)}{x^2\left(1-\frac{9}{x^2}\right)}$$
$$= \frac{x^2-2x-3}{x^2-9}$$
$$= \frac{(x+1)(x-3)}{(x+3)(x-3)}$$
$$= \frac{x+1}{x+3}$$

47. $$\frac{1-\frac{a^2}{4b^2}}{1+\frac{a}{2b}} = \frac{4b^2\left(1-\frac{a^2}{4b^2}\right)}{4b^2\left(1+\frac{a}{2b}\right)}$$
$$= \frac{4b^2-a^2}{4b^2+2ab}$$
$$= \frac{(2b+a)(2b-a)}{2b(2b+a)}$$
$$= \frac{2b-a}{2b}$$

49. $\dfrac{\frac{1}{y+z}+\frac{1}{y-z}}{\frac{1}{y^2-z^2}}=\dfrac{\frac{1}{y+z}+\frac{1}{y-z}}{\frac{1}{(y+z)(y-z)}}$

$=\dfrac{(y+z)(y-z)\left(\frac{1}{y+z}+\frac{1}{y-z}\right)}{(y+z)(y-z)\frac{1}{(y+z)(y-z)}}$

$=\dfrac{y-z+y+z}{1}$

$=\dfrac{2y}{1}$

$=2y$

51. $\dfrac{\frac{3}{n-2}+1}{5+\frac{1}{n-2}}=\dfrac{(n-2)\left(\frac{3}{n-2}+1\right)}{(n-2)\left(5+\frac{1}{n-2}\right)}$

$=\dfrac{3+n-2}{5(n-2)+1}$

$=\dfrac{1+n}{5n-10+1}$

$=\dfrac{n+1}{5n-9}$

53. $\dfrac{\frac{n}{6}+\frac{n+3}{2}+\frac{2n-1}{8}}{3}=\dfrac{24\left(\frac{n}{6}+\frac{n+3}{2}+\frac{2n-1}{8}\right)}{24(3)}$

$=\dfrac{4n+12(n+3)+3(2n-1)}{72}$

$=\dfrac{4n+12n+36+6n-3}{72}$

$=\dfrac{22n+33}{72}$

$=\dfrac{11(2n+3)}{72}$

55. $\dfrac{\frac{2x+3}{x^2}-\frac{3}{x}}{\frac{x^2-9}{x^5}}=\dfrac{x^5\left(\frac{2x+3}{x^2}-\frac{3}{x}\right)}{x^5\left(\frac{x^2-9}{x^5}\right)}$

$=\dfrac{x^3(2x+3)-3x^4}{x^2-9}$

$=\dfrac{2x^4+3x^3-3x^4}{x^2-9}$

$=\dfrac{-x^4+3x^3}{x^2-9}$

$=\dfrac{-x^3(x-3)}{(x+3)(x-3)}$

$=-\dfrac{x^3}{x+3}$

The length is $-\dfrac{x^3}{x+3}$ feet.

57. (a) $R=\dfrac{1}{\frac{1}{R_1}+\frac{1}{R_2}}$

$=\dfrac{R_1R_2(1)}{R_1R_2\left(\frac{1}{R_1}+\frac{1}{R_2}\right)}$

$=\dfrac{R_1R_2}{R_2+R_1}$

(b) $R_1=6$ ohms; $R_2=10$ ohms

$R=\dfrac{6(10)}{10+6}=\dfrac{60}{16}=\dfrac{15}{4}$ ohms

59. $\dfrac{x^{-1}-3}{x^{-2}-9}=\dfrac{\frac{1}{x}-3}{\frac{1}{x^2}-9}$

$=\dfrac{x^2\left(\frac{1}{x}-3\right)}{x^2\left(\frac{1}{x^2}-9\right)}$

$=\dfrac{x-3x^2}{1-9x^2}$

$=\dfrac{x(1-3x)}{(1+3x)(1-3x)}$

$=\dfrac{x}{1+3x}$

61. $\frac{2x^{-1}+5}{4x^{-2}-25}=\frac{\frac{2}{x}+5}{\frac{4}{x^2}-25}$

$=\frac{x^2\left(\frac{2}{x}+5\right)}{x^2\left(\frac{4}{x^2}-25\right)}$

$=\frac{2x+5x^2}{4-25x^2}$

$=\frac{x(2+5x)}{(2+5x)(2-5x)}$

$=\frac{x}{2-5x}$

63. $1+\frac{1}{1+\frac{1}{x}}=1+\frac{1}{\frac{x+1}{x}}$

$=1+\frac{x}{x+1}$

$=\frac{x+1+x}{x+1}$

$=\frac{2x+1}{x+1}$

65. $\frac{1}{1-\frac{1}{2-\frac{1}{3-x}}}=\frac{1}{1-\frac{1}{\frac{2(3-x)-1}{3-x}}}$

$=\frac{1}{1-\frac{1}{\frac{5-2x}{3-x}}}$

$=\frac{1}{1-\frac{3-x}{5-2x}}$

$=\frac{1}{\frac{5-2x-(3-x)}{5-2x}}$

$=\frac{1}{\frac{2-x}{5-2x}}$

$=\frac{5-2x}{2-x}$

67. The expression $\frac{\frac{2}{x+1}}{\frac{1}{x-5}}$ is not in simplified form, since the numerator and denominator are rational expressions, not polynomials. Simplified form is the form $\frac{p}{q}$ where p and q are polynomials that have no common factors.

69. Answers may vary.

Putting the Concepts Together (Sections 7.1–7.6)

1. $\frac{a^3-b^3}{a+b}=\frac{2^3-(-3)^3}{2+(-3)}$

$=\frac{8-(-27)}{2-3}$

$=\frac{8+27}{-1}$

$=-35$

2. (a) $\frac{-3a}{a-6}$ is undefined when $a-6=0$, so the expression is undefined for $a=6$.

(b) $\frac{y+2}{y^2+4y}$ is undefined when $y^2+4y=0$.

$y^2+4y=0$

$y(y+4)=0$

$y=0$ or $y+4=0$

$y=0$ or $y=-4$

The expression is undefined for $y=0$ or $y=-4$.

3. (a) $\frac{ax+ay-4bx-4by}{2x+2y}$

$=\frac{a(x+y)-4b(x+y)}{2(x+y)}$

$=\frac{(x+y)(a-4b)}{2(x+y)}$

$=\frac{a-4b}{2}$

(b) $\frac{x^2+x-2}{1-x^2}=\frac{(x+2)(x-1)}{(1+x)(1-x)}$

$=\frac{(x+2)(x-1)}{(x+1)(-1)(x-1)}$

$=\frac{-(x+2)}{x+1}$

4. $2a+4b=2(a+2b)$

$4a+8b=4(a+2b)=2^2(a+2b)$

$8a-32b=8(a-4b)=2^3(a-4b)$

$\text{LCD}=2^3(a+2b)(a-4b)$

$=8(a+2b)(a-4b)$

5. $\frac{7}{3x^2-x}=\frac{7}{x(3x-1)}\cdot\frac{5x}{5x}=\frac{35x}{5x^2(3x-1)}$

6. $$\frac{y^2-y}{3y}\cdot\frac{6y^2}{1-y^2}=\frac{y(y-1)}{3y}\cdot\frac{3y(2y)}{-1(y+1)(y-1)}$$
$$=\frac{3y(y-1)\cdot 2y^2}{3y(y-1)\cdot(-1)(y+1)}$$
$$=\frac{-2y^2}{y+1}$$

7. $$\frac{m^2+m-2}{m^3-6m^2}\cdot\frac{2m^2-14m+12}{m+2}$$
$$=\frac{(m+2)(m-1)}{m^2(m-6)}\cdot\frac{2(m-6)(m-1)}{m+2}$$
$$=\frac{(m+2)(m-6)\cdot 2(m-1)^2}{(m+2)(m-6)\cdot m^2}$$
$$=\frac{2(m-1)^2}{m^2}$$

8. $$\frac{4y+12}{5y-5}\div\frac{2y^2-18}{y^2-2y+1}$$
$$=\frac{4y+12}{5y-5}\cdot\frac{y^2-2y+1}{2y^2-18}$$
$$=\frac{2^2(y+3)}{5(y-1)}\cdot\frac{(y-1)^2}{2(y+3)(y-3)}$$
$$=\frac{2(y-1)(y+3)\cdot 2(y-1)}{2(y-1)(y+3)\cdot 5(y-3)}$$
$$=\frac{2(y-1)}{5(y-3)}$$

9. $$\frac{4x^2+x}{x+3}+\frac{12x+3}{x+3}=\frac{4x^2+x+12x+3}{x+3}$$
$$=\frac{4x^2+13x+3}{x+3}$$
$$=\frac{(4x+1)(x+3)}{x+3}$$
$$=4x+1$$

10. $$\frac{x^2}{x^3+1}-\frac{x-1}{x^3+1}=\frac{x^2-(x-1)}{(x+1)(x^2-x+1)}$$
$$=\frac{x^2-x+1}{(x+1)(x^2-x+1)}$$
$$=\frac{1}{x+1}$$

11. $$\frac{-8}{2x-1}-\frac{9}{1-2x}=\frac{-8}{2x-1}-\frac{9}{-1(2x-1)}$$
$$=\frac{-8}{2x-1}-\frac{-9}{2x-1}$$
$$=\frac{-8-(-9)}{2x-1}$$
$$=\frac{1}{2x-1}$$

12. $$\frac{4}{m+3}+\frac{3}{3m+2}$$
$$=\frac{4}{m+3}\cdot\frac{3m+2}{3m+2}+\frac{3}{3m+2}\cdot\frac{m+3}{m+3}$$
$$=\frac{4(3m+2)}{(m+3)(3m+2)}+\frac{3(m+3)}{(m+3)(3m+2)}$$
$$=\frac{12m+8+3m+9}{(m+3)(3m+2)}$$
$$=\frac{15m+17}{(m+3)(3m+2)}$$

13. $$\frac{3m}{m^2+7m+10}-\frac{2m}{m^2+6m+8}$$
$$=\frac{3m}{(m+5)(m+2)}-\frac{2m}{(m+4)(m+2)}$$
$$=\frac{3m}{(m+5)(m+2)}\cdot\frac{m+4}{m+4}-\frac{2m}{(m+4)(m+2)}\cdot\frac{m+5}{m+5}$$
$$=\frac{3m(m+4)-2m(m+5)}{(m+5)(m+4)(m+2)}$$
$$=\frac{3m^2+12m-2m^2-10m}{(m+5)(m+4)(m+2)}$$
$$=\frac{m^2+2m}{(m+5)(m+4)(m+2)}$$
$$=\frac{m(m+2)}{(m+5)(m+4)(m+2)}$$
$$=\frac{m}{(m+5)(m+4)}$$

14. $\dfrac{\frac{-1}{m+1}-1}{m-\frac{2}{m+1}}=\dfrac{\frac{-1}{m+1}-1}{m-\frac{2}{m+1}}\cdot\dfrac{m+1}{m+1}$

$=\dfrac{-1-(m+1)}{m(m+1)-2}$

$=\dfrac{-1-m-1}{m^2+m-2}$

$=\dfrac{-m-2}{(m+2)(m-1)}$

$=\dfrac{-(m+2)}{(m+2)(m-1)}$

$=-\dfrac{1}{m-1}$

15. $\dfrac{\frac{4}{a}+\frac{1}{6}}{\frac{3}{a^2}+\frac{1}{2}}=\dfrac{\frac{4}{a}+\frac{1}{6}}{\frac{3}{a^2}+\frac{1}{2}}\cdot\dfrac{6a^2}{6a^2}$

$=\dfrac{24a+a^2}{18+3a^2}$

$=\dfrac{a(24+a)}{3(6+a^2)}$

Section 7.7

Preparing for Rational Equations

P1. $3k-2(k+1)=6$
$3k-2k-2=6$
$k-2=6$
$k-2+2=6+2$
$k=8$
The solution set is $\{8\}$.

P2. $3p^2-7p-6=(3p+2)(p-3)$

P3. $8z^2-10z-3=0$
$(4z+1)(2z-3)=0$
$4z+1=0$ or $2z-3=0$
$4z=-1$ or $2z=3$
$z=-\dfrac{1}{4}$ or $z=\dfrac{3}{2}$
The solution set is $\left\{-\dfrac{1}{4}, \dfrac{3}{2}\right\}$.

P4. $\dfrac{x+4}{x^2-2x-24}$ is undefined when
$x^2-2x-24=0.$
$x^2-2x-24=0$
$(x+4)(x-6)=0$
$x+4=0$ or $x-6=0$
$x=-4$ or $x=6$
The expression is undefined for $x=-4$ or $x=6$.

P5. $4x-2y=10$
$-4x+4x-2y=-4x+10$
$-2y=-4x+10$
$\dfrac{-2y}{-2}=\dfrac{-4x+10}{-2}$
$y=2x-5$

Quick Checks 7.7

1. A rational equation is an equation that contains a rational expression.

2. Undefined value: $z=0$
LCD = $2z$
$\dfrac{5}{2}+\dfrac{1}{z}=4$
$2z\left(\dfrac{5}{2}+\dfrac{1}{z}\right)=2z(4)$
$5z+2=8z$
$2=3z$
$\dfrac{2}{3}=z$
Check: $\dfrac{5}{2}+\dfrac{1}{\frac{2}{3}}\stackrel{?}{=}4$
$\dfrac{5}{2}+\dfrac{3}{2}\stackrel{?}{=}4$
$\dfrac{5+3}{2}\stackrel{?}{=}4$
$\dfrac{8}{2}\stackrel{?}{=}4$
$4=4$ True
The solution set is $\left\{\dfrac{2}{3}\right\}$.

3. Undefined values: $x = -4$ and $x = 3$
LCD = $(x + 4)(x - 3)$

$$\frac{8}{x+4} = \frac{12}{x-3}$$
$$(x+4)(x-3)\left(\frac{8}{x+4}\right) = (x+4)(x-3)\left(\frac{12}{x-3}\right)$$
$$8(x-3) = 12(x+4)$$
$$8x-24 = 12x+48$$
$$-72 = 4x$$
$$-18 = x$$

Check: $\frac{8}{-18+4} \stackrel{?}{=} \frac{12}{-18-3}$
$$\frac{8}{-14} \stackrel{?}{=} \frac{12}{-21}$$
$$-\frac{4}{7} = -\frac{4}{7} \quad \text{True}$$

The solution set is $\{-18\}$.

4. Undefined value: $b = 0$
LCD = $6b$

$$\frac{4}{3b} + \frac{1}{6b} = \frac{7}{2b} + \frac{1}{3}$$
$$6b\left(\frac{4}{3b} + \frac{1}{6b}\right) = 6b\left(\frac{7}{2b} + \frac{1}{3}\right)$$
$$8+1 = 21+2b$$
$$9 = 21+2b$$
$$-12 = 2b$$
$$-6 = b$$

Check: $\frac{4}{3(-6)} + \frac{1}{6(-6)} \stackrel{?}{=} \frac{7}{2(-6)} + \frac{1}{3}$
$$-\frac{4}{18} - \frac{1}{36} \stackrel{?}{=} -\frac{7}{12} + \frac{1}{3}$$
$$-\frac{8}{36} - \frac{1}{36} \stackrel{?}{=} -\frac{7}{12} + \frac{4}{12}$$
$$-\frac{9}{36} \stackrel{?}{=} -\frac{3}{12}$$
$$-\frac{1}{4} = -\frac{1}{4} \quad \text{True}$$

The solution set is $\{-6\}$.

5. $x - 3 = 0$ when $x = 3$. $2x - 6 = 2(x - 3) = 0$ when $x = 3$.
The equation is undefined for $x = 3$.
LCD = $2 \cdot (x - 3) = 2x - 6 = 2(x - 3)$

$$\frac{3}{2} + \frac{5}{x-3} = \frac{x+9}{2x-6}$$
$$2(x-3)\left(\frac{3}{2} + \frac{5}{x-3}\right) = 2(x-3)\left(\frac{x+9}{2x-6}\right)$$
$$3(x-3) + 2(5) = (2x-6)\left(\frac{x+9}{2x-6}\right)$$
$$3x-9+10 = x+9$$
$$3x+1 = x+9$$
$$2x = 8$$
$$x = 4$$

Check: $\frac{3}{2} + \frac{5}{4-3} \stackrel{?}{=} \frac{4+9}{2(4)-6}$
$$\frac{3}{2} + \frac{5}{1} \stackrel{?}{=} \frac{13}{8-6}$$
$$\frac{3}{2} + \frac{10}{2} \stackrel{?}{=} \frac{13}{2}$$
$$\frac{13}{2} = \frac{13}{2} \quad \text{True}$$

The solution set is $\{4\}$.

6. Undefined values: $x = 3$ and $x = -3$
LCD = $(x + 3)(x - 3)$

$$\frac{4}{x-3} - \frac{3}{x+3} = 1$$
$$(x+3)(x-3)\left(\frac{4}{x-3} - \frac{3}{x+3}\right) = (x+3)(x-3)(1)$$
$$4(x+3) - 3(x-3) = x^2 - 9$$
$$4x+12-3x+9 = x^2 - 9$$
$$x+21 = x^2 - 9$$
$$0 = x^2 - x - 30$$
$$0 = (x+5)(x-6)$$

$x+5 = 0$ or $x-6 = 0$
$x = -5$ or $x = 6$

Check: $\frac{4}{-5-3} - \frac{3}{-5+3} \stackrel{?}{=} 1$
$$\frac{4}{-8} - \frac{3}{-2} \stackrel{?}{=} 1$$
$$-\frac{1}{2} + \frac{3}{2} \stackrel{?}{=} 1$$
$$\frac{2}{2} \stackrel{?}{=} 1$$
$$1 = 1 \quad \text{True}$$

$$\frac{4}{6-3}-\frac{3}{6+3} \stackrel{?}{=} 1$$
$$\frac{4}{3}-\frac{3}{9} \stackrel{?}{=} 1$$
$$\frac{12}{9}-\frac{3}{9} \stackrel{?}{=} 1$$
$$\frac{9}{9} \stackrel{?}{=} 1$$
$$1=1 \quad \text{True}$$

The solution set is $\{-5, 6\}$.

7. Solutions obtained through the solving process that do not satisfy the original equation are called <u>extraneous solutions</u>.

8. $z-4=0$ when $z=4$. $z-2=0$ when $z=2$.
$z^2-6z+8=(z-4)(z-2)=0$ when $z=4$ or $z=2$.
The equation is undefined for $z=4$ and $z=2$.
LCD = $(z-4)(z-2)$

$$\frac{5}{z-4}+\frac{3}{z-2}=\frac{z^2-z-2}{z^2-6z+8}$$
$$(z-4)(z-2)\left(\frac{5}{z-4}+\frac{3}{z-2}\right)=(z-4)(z-2)\left(\frac{z^2-z-2}{(z-4)(z-2)}\right)$$
$$5(z-2)+3(z-4)=z^2-z-2$$
$$5z-10+3z-12=z^2-z-2$$
$$8z-22=z^2-z-2$$
$$0=z^2-9z+20$$
$$0=(z-5)(z-4)$$

$z-5=0$ or $z-4=0$
$z=5$ or $z=4$
$z=4$ is extraneous, so the solution set is $\{5\}$.

9. True

10. $y+1=0$ when $y=-1$. $y-1=0$ when $y=1$. $y^2-1=(y+1)(y-1)=0$ when $y=1$ or $y=-1$.
The equation is undefined for $y=1$ and $y=-1$.
LCD = $(y+1)(y-1)$

$$\frac{4}{y+1}=\frac{7}{y-1}-\frac{8}{y^2-1}$$
$$(y+1)(y-1)\left(\frac{4}{y+1}\right)=(y+1)(y-1)\left(\frac{7}{y-1}-\frac{8}{(y+1)(y-1)}\right)$$
$$4(y-1)=7(y+1)-8$$
$$4y-4=7y+7-8$$
$$4y-4=7y-1$$
$$-3=3y$$
$$-1=y$$

Since y cannot equal -1, we reject $y=-1$ as a solution. Therefore, the equation has no solution. The solution set is $\{\ \}$ or $\varnothing$.

11.
$$C = \frac{50t}{t^2+25}$$
$$4 = \frac{50t}{t^2+25}$$
$$(t^2+25)(4) = (t^2+25)\left(\frac{50t}{t^2+25}\right)$$
$$4t^2+100 = 50t$$
$$4t^2-50t+100 = 0$$
$$2t^2-25t+50 = 0$$
$$(2t-5)(t-10) = 0$$
$$2t-5=0 \quad \text{or} \quad t-10=0$$
$$2t=5 \quad \text{or} \quad t=10$$
$$t=\frac{5}{2} \quad \text{or} \quad t=10$$

The concentration of the drug will be 4 milligrams per liter at $\frac{5}{2}$ hours and 10 hours.

12. $x \neq 0$
LCD $= x \cdot 1 = x$
$$R = \frac{4g}{x}$$
$$x(R) = x\left(\frac{4g}{x}\right)$$
$$xR = 4g$$
$$x = \frac{4g}{R}$$

13. $r \neq 1$
LCD $= 1 \cdot (1-r) = 1-r$
$$S = \frac{a}{1-r}$$
$$(1-r)(S) = (1-r)\left(\frac{a}{1-r}\right)$$
$$S(1-r) = a$$
$$1-r = \frac{a}{S}$$
$$-r = -1+\frac{a}{S}$$
$$r = 1-\frac{a}{S}$$

14. $f \neq 0, p \neq 0, q \neq 0$
LCD $= fpq$
$$\frac{1}{f} = \frac{1}{p}+\frac{1}{q}$$
$$fpq\left(\frac{1}{f}\right) = fpq\left(\frac{1}{p}+\frac{1}{q}\right)$$
$$pq = fq+fp$$
$$pq-fp = fq$$
$$p(q-f) = fq$$
$$p = \frac{fq}{q-f}$$

Exercises 7.7

15. Undefined value: $y = 0$.
$$\frac{5}{3y}-\frac{1}{2} = \frac{5}{6y}-\frac{1}{12}$$
$$12y\left(\frac{5}{3y}-\frac{1}{2}\right) = 12y\left(\frac{5}{6y}-\frac{1}{12}\right)$$
$$4(5)-6y = 2(5)-y$$
$$20-6y = 10-y$$
$$20 = 10+5y$$
$$10 = 5y$$
$$2 = y$$
The solution set is $\{2\}$.

17. Undefined value: $x = 0$
$$\frac{6}{x}+\frac{2}{3} = \frac{4}{2x}-\frac{14}{3}$$
$$6x\left(\frac{6}{x}+\frac{2}{3}\right) = 6x\left(\frac{4}{2x}-\frac{14}{3}\right)$$
$$6(6)+2x(2) = 3(4)-2x(14)$$
$$36+4x = 12-28x$$
$$36+32x = 12$$
$$32x = -24$$
$$x = -\frac{24}{32} = -\frac{3}{4}$$
The solution set is $\left\{-\frac{3}{4}\right\}$.

19. Undefined values: $x = 1$ and $x = -1$.
$$\frac{4}{x-1} = \frac{3}{x+1}$$
$$(x-1)(x+1)\left(\frac{4}{x-1}\right) = (x-1)(x+1)\left(\frac{3}{x+1}\right)$$
$$4(x+1) = 3(x-1)$$
$$4x+4 = 3x-3$$
$$x+4 = -3$$
$$x = -7$$
The solution set is $\{-7\}$.

21. Undefined value: $x = -2$.

$$\frac{2}{x+2}+2=\frac{7}{x+2}$$
$$(x+2)\left(\frac{2}{x+2}+2\right)=(x+2)\left(\frac{7}{x+2}\right)$$
$$2+2(x+2)=7$$
$$2+2x+4=7$$
$$2x+6=7$$
$$2x=1$$
$$x=\frac{1}{2}$$

The solution set is $\left\{\frac{1}{2}\right\}$.

23. Undefined value: $r = 0$.

$$\frac{r-4}{3r}+\frac{2}{5r}=\frac{1}{5}$$
$$15r\left(\frac{r-4}{3r}+\frac{2}{5r}\right)=15r\left(\frac{1}{5}\right)$$
$$5(r-4)+3(2)=3r$$
$$5r-20+6=3r$$
$$5r-14=3r$$
$$-14=-2r$$
$$7=r$$

The solution set is $\{7\}$.

25. Undefined values: $a = 1$ and $a = -1$.

$$\frac{2}{a-1}+\frac{3}{a+1}=\frac{-6}{a^2-1}$$
$$(a-1)(a+1)\left(\frac{2}{a-1}+\frac{3}{a+1}\right)=(a-1)(a+1)\left(\frac{-6}{(a-1)(a+1)}\right)$$
$$2(a+1)+3(a-1)=-6$$
$$2a+2+3a-3=-6$$
$$5a-1=-6$$
$$5a=-5$$
$$a=-1$$

$a = -1$ is an extraneous solution. The solution set is $\{\ \}$ or $\varnothing$.

27. Undefined values: $x = 4$ and $x = -4$.

$$\frac{1}{4-x}+\frac{2}{x^2-16}=\frac{1}{x-4}$$

$$(x-4)(x+4)\left(\frac{1}{-(x-4)}+\frac{2}{(x-4)(x+4)}\right)=(x-4)(x+4)\left(\frac{1}{x-4}\right)$$

$$\begin{aligned}-(x+4)+2&=x+4\\ -x-4+2&=x+4\\ -x-2&=x+4\\ -2&=2x+4\\ -6&=2x\\ -3&=x\end{aligned}$$

The solution set is $\{-3\}$.

29. $2t - 2 = 2(t - 1)$; $3t - 3 = 3(t - 1)$
Undefined value: $t = 1$.

$$\frac{3}{2t-2}-\frac{2t}{3t-3}=-4$$

$$2\cdot 3(t-1)\left(\frac{3}{2(t-1)}-\frac{2t}{3(t-1)}\right)=2\cdot 3(t-1)(-4)$$

$$\begin{aligned}3(3)-2(2t)&=-24(t-1)\\ 9-4t&=-24t+24\\ 9+20t&=24\\ 20t&=15\\ t&=\frac{3}{4}\end{aligned}$$

The solution set is $\left\{\frac{3}{4}\right\}$.

31. $10a - 20 = 10(a - 2)$
Undefined value: $a = 2$.

$$\frac{2}{5}+\frac{3-2a}{10a-20}=\frac{2a+1}{a-2}$$

$$10(a-2)\left(\frac{2}{5}+\frac{3-2a}{10(a-2)}\right)=10(a-2)\left(\frac{2a+1}{a-2}\right)$$

$$\begin{aligned}2\cdot 2(a-2)+3-2a&=10(2a+1)\\ 4a-8+3-2a&=20a+10\\ 2a-5&=20a+10\\ -5&=18a+10\\ -15&=18a\\ -\frac{15}{18}&=a\\ -\frac{5}{6}&=a\end{aligned}$$

The solution set is $\left\{-\frac{5}{6}\right\}$.

33. $j^2-1=(j+1)(j-1)$; $j^2-5j+4=(j-1)(j-4)$

Undefined values: $j=-1$, $j=1$, and $j=4$.

$$\frac{6}{j^2-1}-\frac{4j}{j^2-5j+4}=-\frac{4}{j-1}$$

$$(j+1)(j-1)(j-4)\left(\frac{6}{(j+1)(j-1)}-\frac{4j}{(j-1)(j-4)}\right)=(j+1)(j-1)(j-4)\left(-\frac{4}{j-1}\right)$$

$$6(j-4)-4j(j+1)=(-4j-4)(j-4)$$

$$6j-24-4j^2-4j=-4j^2+16j-4j+16$$

$$-4j^2+2j-24=-4j^2+12j+16$$

$$-10j=40$$

$$j=-4$$

The solution set is $\{-4\}$.

35.

$$\frac{x}{x-2}=\frac{3}{x+8}$$

$$(x-2)(x+8)\left(\frac{x}{x-2}\right)=(x-2)(x+8)\left(\frac{3}{x+8}\right)$$

$$x(x+8)=3(x-2)$$

$$x^2+8x=3x-6$$

$$x^2+5x+6=0$$

$$(x+2)(x+3)=0$$

$$x+2=0 \quad \text{or} \quad x+3=0$$

$$x=-2 \quad \text{or} \quad x=-3$$

Neither value is extraneous. The solution set is $\{-3, -2\}$.

37.

$$\frac{x}{x+3}=\frac{6}{x-3}+1$$

$$(x+3)(x-3)\left(\frac{x}{x+3}\right)=(x+3)(x-3)\left(\frac{6}{x-3}+1\right)$$

$$x(x-3)=6(x+3)+(x+3)(x-3)$$

$$x^2-3x=6x+18+x^2-9$$

$$x^2-3x=x^2+6x+9$$

$$-3x=6x+9$$

$$0=9x+9$$

$$-9=9x$$

$$-1=x$$

The value is not extraneous. The solution set is $\{-1\}$.

39.

$$x = \frac{2-x}{6x}$$
$$6x(x) = 6x\left(\frac{2-x}{6x}\right)$$
$$6x^2 = 2 - x$$
$$6x^2 + x - 2 = 0$$
$$(3x+2)(2x-1) = 0$$
$$3x+2=0 \quad \text{or} \quad 2x-1=0$$
$$x = -\frac{2}{3} \quad \text{or} \quad x = \frac{1}{2}$$

Neither value is extraneous. The solution set is $\left\{-\frac{2}{3}, \frac{1}{2}\right\}$.

41.

$$\frac{2x+3}{x-1} = \frac{x-2}{x+1} + \frac{6x}{x^2-1}$$
$$(x-1)(x+1)\left(\frac{2x+3}{x-1}\right) = (x-1)(x+1)\left(\frac{x-2}{x+1} + \frac{6x}{(x-1)(x+1)}\right)$$
$$(x+1)(2x+3) = (x-1)(x-2) + 6x$$
$$2x^2 + 5x + 3 = x^2 - 3x + 2 + 6x$$
$$2x^2 + 5x + 3 = x^2 + 3x + 2$$
$$x^2 + 2x + 1 = 0$$
$$(x+1)^2 = 0$$
$$x+1 = 0$$
$$x = -1$$

Since $x = -1$ causes a denominator to be zero, it is an extraneous solution. The solution set is { } or ∅.

43.

$$\frac{2x}{x+3} - \frac{2x^2+2}{x^2-9} = \frac{-6}{x-3} + 1$$
$$(x+3)(x-3)\left(\frac{2x}{x+3} - \frac{2x^2+2}{x^2-9}\right) = (x+3)(x-3)\left(\frac{-6}{x-3} + 1\right)$$
$$2x(x-3) - (2x^2+2) = -6(x+3) + (x+3)(x-3)$$
$$2x^2 - 6x - 2x^2 - 2 = -6x - 18 + x^2 - 9$$
$$-6x - 2 = x^2 - 6x - 27$$
$$0 = x^2 - 25$$
$$0 = (x+5)(x-5)$$
$$x+5=0 \quad \text{or} \quad x-5=0$$
$$x=-5 \quad \text{or} \quad x=5$$

Neither value is extraneous. The solution set is $\{-5, 5\}$.

45.
$$\begin{aligned}\frac{1}{n-3}&=\frac{3n-1}{9-n^2}\\ -(n-3)(n+3)\left(\frac{1}{n-3}\right)&=-(n-3)(n+3)\left(\frac{3n-1}{-(n-3)(n+3)}\right)\\ -1(n+3)&=3n-1\\ -n-3&=3n-1\\ -2&=4n\\ -\frac{1}{2}&=n\end{aligned}$$

The value is not extraneous. The solution set is $\left\{-\frac{1}{2}\right\}$.

47.
$$\begin{aligned}x&=\frac{2}{y}\\ y(x)&=y\left(\frac{2}{y}\right)\\ xy&=2\\ \frac{xy}{x}&=\frac{2}{x}\\ y&=\frac{2}{x}\end{aligned}$$

49.
$$\begin{aligned}I&=\frac{E}{R}\\ R(I)&=R\left(\frac{E}{R}\right)\\ IR&=E\\ \frac{IR}{I}&=\frac{E}{I}\\ R&=\frac{E}{I}\end{aligned}$$

51.
$$\begin{aligned}h&=\frac{2A}{B+b}\\ (B+b)(h)&=(B+b)\left(\frac{2A}{B+b}\right)\\ Bh+bh&=2A\\ bh&=2A-Bh\\ \frac{bh}{h}&=\frac{2A-Bh}{h}\\ b&=\frac{2A-Bh}{h}\text{ or }b=\frac{2A}{h}-B\end{aligned}$$

53.
$$\frac{x}{3+y} = z$$
$$(3+y)\left(\frac{x}{3+y}\right) = (3+y)z$$
$$x = 3z + yz$$
$$x - 3z = yz$$
$$\frac{x-3z}{z} = \frac{yz}{z}$$
$$y = \frac{x-3z}{z} \text{ or } y = \frac{x}{z} - 3$$

55.
$$\frac{1}{R} = \frac{1}{S} + \frac{1}{T}$$
$$RST\left(\frac{1}{R}\right) = RST\left(\frac{1}{S} + \frac{1}{T}\right)$$
$$ST = RT + RS$$
$$ST - RS = RT$$
$$S(T-R) = RT$$
$$\frac{S(T-R)}{T-R} = \frac{RT}{T-R}$$
$$S = \frac{RT}{T-R}$$

57.
$$m = \frac{n}{y} - \frac{p}{ay}$$
$$ay(m) = ay\left(\frac{n}{y} - \frac{p}{ay}\right)$$
$$amy = an - p$$
$$\frac{amy}{am} = \frac{an-p}{am}$$
$$y = \frac{an-p}{am}$$

59.
$$A = \frac{xy}{x+y}$$
$$(x+y)A = (x+y)\left(\frac{xy}{x+y}\right)$$
$$Ax + Ay = xy$$
$$Ay = xy - Ax$$
$$Ay = x(y-A)$$
$$\frac{Ay}{y-A} = \frac{x(y-A)}{y-A}$$
$$x = \frac{Ay}{y-A}$$

61.
$$\frac{2}{x} - \frac{1}{y} = \frac{6}{z}$$
$$xyz\left(\frac{2}{x} - \frac{1}{y}\right) = xyz\left(\frac{6}{z}\right)$$
$$2yz - xz = 6xy$$
$$2yz - 6xy = xz$$
$$y(2z-6x) = xz$$
$$\frac{y(2z-6x)}{2z-6x} = \frac{xz}{2z-6x}$$
$$y = \frac{xz}{2z-6x}$$

63.
$$y = \frac{x}{x-c}$$
$$(x-c)y = (x-c)\left(\frac{x}{x-c}\right)$$
$$xy - cy = x$$
$$-cy = x - xy$$
$$-cy = x(1-y)$$
$$\frac{-cy}{1-y} = \frac{x(1-y)}{1-y}$$
$$\frac{-cy}{1-y} = x$$
$$x = \frac{cy}{y-1}$$

65.
$$\frac{1}{x} + \frac{3}{x+5} = \frac{1}{x}\cdot\frac{x+5}{x+5} + \frac{3}{x+5}\cdot\frac{x}{x}$$
$$= \frac{x+5}{x(x+5)} + \frac{3x}{x(x+5)}$$
$$= \frac{x+5+3x}{x(x+5)}$$
$$= \frac{4x+5}{x(x+5)}$$

67.
$$x - \frac{6}{x} = 1$$
$$x\left(x - \frac{6}{x}\right) = x(1)$$
$$x^2 - 6 = x$$
$$x^2 - x - 6 = 0$$
$$(x+2)(x-3) = 0$$
$$x+2 = 0 \quad \text{or} \quad x-3 = 0$$
$$x = -2 \quad \text{or} \quad x = 3$$

Neither value is extraneous. The solution set is $\{-2, 3\}$.

69. $\frac{3}{x-1}\cdot\frac{x^2-1}{6}+3=\frac{3}{x-1}\cdot\frac{(x+1)(x-1)}{2\cdot 3}+3$

$$=\frac{x+1}{2}+3$$
$$=\frac{x+1}{2}+\frac{6}{2}$$
$$=\frac{x+1+6}{2}$$
$$=\frac{x+7}{2}$$

71.
$$2b-\frac{5}{3}=\frac{1}{3b}$$
$$3b\left(2b-\frac{5}{3}\right)=3b\left(\frac{1}{3b}\right)$$
$$6b^2-5b=1$$
$$6b^2-5b-1=0$$
$$(6b+1)(b-1)=0$$
$$6b+1=0 \quad \text{or} \quad b-1=0$$
$$b=-\frac{1}{6} \quad \text{or} \quad b=1$$

Neither value is extraneous. The solution set is $\left\{-\frac{1}{6}, 1\right\}$.

73.
$$\frac{5x}{2x-3}=\frac{3x}{x-1}-\frac{5}{2x^2-5x+3}$$
$$\frac{5x}{2x-3}=\frac{3x}{x-1}-\frac{5}{(2x-3)(x-1)}$$
$$(2x-3)(x-1)\left(\frac{5x}{2x-3}\right)=(2x-3)(x-1)\left(\frac{3x}{x-1}-\frac{5}{(2x-3)(x-1)}\right)$$
$$5x(x-1)=3x(2x-3)-5$$
$$5x^2-5x=6x^2-9x-5$$
$$0=x^2-4x-5$$
$$0=(x-5)(x+1)$$
$$x-5=0 \quad \text{or} \quad x+1=0$$
$$x=5 \quad\quad x=-1$$

Neither value is extraneous. The solution set is $\{-1, 5\}$.

75. $\frac{x^2}{x^2-4}-\frac{1}{x}=\frac{x^2}{x^2-4}\cdot\frac{x}{x}-\frac{1}{x}\cdot\frac{x^2-4}{x^2-4}$

$$=\frac{x^3}{x(x^2-4)}-\frac{x^2-4}{x(x^2-4)}$$
$$=\frac{x^3-(x^2-4)}{x(x+2)(x-2)}$$
$$=\frac{x^3-x^2+4}{x(x+2)(x-2)}$$

77. $$\frac{x^2-9}{x+1} \div \frac{3x^2+9x}{x^2-1} = \frac{x^2-9}{x+1} \cdot \frac{x^2-1}{3x^2+9x}$$
$$= \frac{(x+3)(x-3)}{x+1} \cdot \frac{(x+1)(x-1)}{3x(x+3)}$$
$$= \frac{(x+3)(x+1) \cdot (x-3)(x-1)}{(x+3)(x+1) \cdot 3x}$$
$$= \frac{(x-3)(x-1)}{3x}$$

79. $$\frac{\frac{1}{a}+3}{\frac{3a+1}{5}} = \frac{\frac{1}{a}+\frac{3a}{a}}{\frac{3a+1}{5}}$$
$$= \frac{\frac{1+3a}{a}}{\frac{3a+1}{5}}$$
$$= \frac{1+3a}{a} \cdot \frac{5}{3a+1}$$
$$= \frac{(3a+1) \cdot 5}{(3a+1) \cdot a}$$

81. $2a - 2 = 2(a - 1)$; $3a + 3 = 3(a + 1)$; $12a^2 - 12 = 2 \cdot 2 \cdot 3(a+1)(a-1)$

$$\frac{a}{2a-2} - \frac{2}{3a+3} = \frac{5a^2-2a+9}{12a^2-12}$$
$$12(a+1)(a-1)\left(\frac{a}{2(a-1)} - \frac{2}{3(a+1)}\right) = 12(a+1)(a-1)\left(\frac{5a^2-2a+9}{12(a+1)(a-1)}\right)$$
$$6a(a+1) - 8(a-1) = 5a^2 - 2a + 9$$
$$6a^2 + 6a - 8a + 8 = 5a^2 - 2a + 9$$
$$6a^2 - 2a + 8 = 5a^2 - 2a + 9$$
$$a^2 - 1 = 0$$
$$(a+1)(a-1) = 0$$

$a+1=0$ or $a-1=0$

$a=-1$ or $a=1$

Since both $a = -1$ and $a = 1$ cause denominators to be zero, they are extraneous solutions. The solution set is { } or $\varnothing$.

83. $\left(\frac{3}{b+c}+\frac{5}{b-c}\right)\div\left(\frac{4b+c}{b^2-c^2}\right)=\left(\frac{3}{b+c}\cdot\frac{b-c}{b-c}+\frac{5}{b-c}\cdot\frac{b+c}{b+c}\right)\div\left(\frac{4b+c}{(b+c)(b-c)}\right)$

$$=\left(\frac{3b-3c}{(b+c)(b-c)}+\frac{5b+5c}{(b+c)(b-c)}\right)\div\left(\frac{4b+c}{(b+c)(b-c)}\right)$$

$$=\left(\frac{3b-3c+5b+5c}{(b+c)(b-c)}\right)\div\left(\frac{4b+c}{(b+c)(b-c)}\right)$$

$$=\left(\frac{8b+2c}{(b+c)(b-c)}\right)\div\left(\frac{4b+c}{(b+c)(b-c)}\right)$$

$$=\frac{8b+2c}{(b+c)(b-c)}\cdot\frac{(b+c)(b-c)}{4b+c}$$

$$=\frac{2(4b+c)}{(b+c)(b-c)}\cdot\frac{(b+c)(b-c)}{4b+c}$$

$$=\frac{(b+c)(b-c)(4b+c)\cdot 2}{(b+c)(b-c)(4b+c)}$$

$$=2$$

85. $C=\frac{40t}{t^2+9}$; find t when $C = 4$.

$$4=\frac{40t}{t^2+9}$$

$$(t^2+9)(4)=(t^2+9)\left(\frac{40t}{t^2+9}\right)$$

$$4t^2+36=40t$$

$$4t^2-40t+36=0$$

$$t^2-10t+9=0$$

$$(t-9)(t-1)=0$$

$$t-9=0 \quad \text{or} \quad t-1=0$$

$$t=9 \quad \text{or} \quad t=1$$

The concentration is 4 milligrams per liter after 1 hour and 9 hours.

87. $C=\frac{25x}{100-x}$; find x when $C = 100$.

$$100=\frac{25x}{100-x}$$

$$(100-x)\cdot 100=(100-x)\cdot\frac{25x}{100-x}$$

$$10{,}000-100x=25x$$

$$10{,}000=125x$$

$$80=x$$

If the government budgets \$100 million, then 80% of the pollutants can be removed.

89. If the solution set is {2}, the value of k can be found by letting $x = 2$ and solving for k.

$$\frac{4x+3}{k}=\frac{x-1}{3}$$
$$\frac{4(2)+3}{k}=\frac{2-1}{3}$$
$$\frac{11}{k}=\frac{1}{3}$$
$$3k\left(\frac{11}{k}\right)=3k\left(\frac{1}{3}\right)$$
$$33=k$$

91. The error occurred when the student incorrectly multiplied $-2[(x-1)(x+1)]$ on the right side of the equation. The correct solution is

$$\frac{2}{x-1}-\frac{4}{x^2-1}=-2$$
$$\frac{2}{x-1}-\frac{4}{(x-1)(x+1)}=-2$$
$$[(x-1)(x+1)]\left(\frac{2}{x-1}-\frac{4}{(x-1)(x+1)}\right)=-2[(x-1)(x+1)]$$
$$2(x+1)-4=-2(x-1)(x+1)$$
$$2x+2-4=-2(x^2-1)$$
$$2x-2=-2x^2+2$$
$$2x^2+2x-4=0$$
$$x^2+x-2=0$$
$$(x+2)(x-1)=0$$

$x+2=0$ or $x-1=0$

$x=-2$ $\quad x=1$

The value of $x = 1$ is extraneous. The solution set is {−2}.

93. To *simplify* a rational expression means to add, subtract, multiply, or divide the rational expressions and express the result in a form in which there are no common factors in the numerator or denominator. To *solve* an equation means to find the value(s) of the variable that satisfy the equation.

Section 7.8

Preparing for Models Involving Rational Equations

P1.
$$\frac{150}{r}=\frac{250}{r+20}$$
$$r(r+20)\left(\frac{150}{r}\right)=r(r+20)\left(\frac{250}{r+20}\right)$$
$$150(r+20)=250r$$
$$150r+3000=250r$$
$$3000=100r$$
$$r=30$$

The solution set is {30}.

7.8 Quick Checks

1. A proportion is an equation of the form $\frac{a}{b}=\frac{c}{d}$, where $b \neq 0$ and $d \neq 0$.

2. $\frac{2p+1}{4} = \frac{p}{8}$
$8(2p+1) = 4p$
$16p+8 = 4p$
$12p = -8$
$p = -\frac{8}{12}$
$p = -\frac{2}{3}$
The solution set is $\left\{-\frac{2}{3}\right\}$.

3. $\frac{6}{x^2} = \frac{2}{x}$
$6x = 2x^2$
$0 = 2x^2 - 6x$
$0 = 2x(x-3)$
$2x = 0$ or $x-3 = 0$
$x = 0$ or $x = 3$
$x = 0$ is extraneous since it makes the denominators zero. The solution set is $\{3\}$.

4. Let x be the monthly payment.
$\frac{1000}{16.67} = \frac{14,000}{x}$
$1000x = 16.67(14,000)$
$\frac{1000x}{1000} = \frac{16.67(14,000)}{1000}$
$x = 16.67(14)$
$x = 233.38$
Clem's monthly payment is \$233.38.

5. Let x be the number of miles between the two cities.
$\frac{\frac{1}{4}}{15} = \frac{\frac{7}{2}}{x}$
$\frac{1}{4}x = \frac{7}{2}(15)$
$x = 4\left(\frac{7}{2}\right)(15)$
$x = 14(15)$
$x = 210$
The cities are 210 miles apart.

6. In geometry, two figures are <u>similar</u> if their corresponding angle measures are equal and their corresponding sides are proportional.

7. $\frac{XY}{XZ} = \frac{MN}{MP}$
$\frac{a}{10} = \frac{12}{15}$
$15a = 12(10)$
$15a = 120$
$\frac{15a}{15} = \frac{120}{15}$
$a = 8$
The length of side XY is 8 units.

8. Let h be the height of the tree.
$\frac{6}{2.5} = \frac{h}{25}$
$2.5h = 25(6)$
$2.5h = 150$
$\frac{2.5h}{2.5} = \frac{150}{2.5}$
$h = 60$
The tree is 60 feet tall.

9. Let t be the number of hours working together. Molly shovels $\frac{1}{3}$ of the driveway in 1 hour. The neighbor shovels $\frac{1}{6}$ of the driveway in 1 hour. Together they shovel $\frac{1}{t}$ of the driveway in 1 hour.
$\frac{1}{3} + \frac{1}{6} = \frac{1}{t}$
$6t\left(\frac{1}{3} + \frac{1}{6}\right) = 6t\left(\frac{1}{t}\right)$
$2t + t = 6$
$3t = 6$
$t = 2$
Molly and her neighbor would take 2 hours to shovel the driveway together.

10. Let t be the number of hours it would take Michael to seal the driveway alone. Leon seals $\frac{1}{5}$ of the driveway in 1 hour. Michael seals $\frac{1}{t}$ of the driveway in 1 hour. Together they seal $\frac{1}{2}$ of the driveway in 1 hour.

$$\frac{1}{5}+\frac{1}{t}=\frac{1}{2}$$
$$10t\left(\frac{1}{5}+\frac{1}{t}\right)=10t\left(\frac{1}{2}\right)$$
$$2t+10=5t$$
$$10=3t$$
$$t=\frac{10}{3}$$

It would take Michael $3\frac{1}{3}$ hours or 3 hours, 20 minutes to seal the driveway alone.

11. Let s be the speed of the wind. The speed into the wind is $120 - s$. The speed with the wind is $120 + s$. Since $d = rt$, then $t=\frac{d}{r}$. The times are the same for each trip. The plane flies 700 miles at $120 + s$ miles per hour and 500 miles at $120 - s$ miles per hour.

$$\frac{500}{120-s}=\frac{700}{120+s}$$
$$500(120+s)=700(120-s)$$
$$5(120+s)=7(120-s)$$
$$600+5s=840-7s$$
$$12s=240$$
$$s=20$$

The speed of the wind is 20 mph.

12. Let Sue's running speed be s. Sue's walking speed is $\frac{s}{12}$. Since $d = rt$, then $t=\frac{d}{r}$.

time running + time walking = total time

$$\frac{18}{s}+\frac{1}{\frac{s}{12}}=5$$
$$\frac{18}{s}+\frac{12}{s}=5$$
$$\frac{18+12}{s}=5$$
$$\frac{30}{s}=5$$
$$30=5s$$
$$s=6$$

Sue's running speed was 6 mph.

7.8 Exercises

13.
$$\frac{9}{x}=\frac{3}{4}$$
$$9(4)=3x$$
$$36=3x$$
$$12=x$$

The solution set is $\{12\}$.

15.
$$\frac{4}{7}=\frac{2x}{9}$$
$$4(9)=7(2x)$$
$$36=14x$$
$$\frac{36}{14}=x$$
$$\frac{18}{7}=x$$

The solution set is $\left\{\frac{18}{7}\right\}$.

17.
$$\frac{6}{5}=\frac{x+2}{15}$$
$$15\left(\frac{6}{5}\right)=15\left(\frac{x+2}{15}\right)$$
$$3(6)=x+2$$
$$18=x+2$$
$$16=x$$

The solution set is $\{16\}$.

19.
$$\frac{b}{b+6}=\frac{4}{9}$$
$$9(b)=4(b+6)$$
$$9b=4b+24$$
$$5b=24$$
$$b=\frac{24}{5}$$

The solution set is $\left\{\frac{24}{5}\right\}$.

21. $\dfrac{y}{y-10}=\dfrac{2}{27}$

$27(y)=2(y-10)$
$27y=2y-20$
$25y=-20$
$y=\dfrac{-20}{25}$
$y=-\dfrac{4}{5}$

The solution set is $\left\{-\dfrac{4}{5}\right\}$.

23. $\dfrac{p+2}{4}=\dfrac{2p+4}{5}$

$5(p+2)=4(2p+4)$
$5p+10=8p+16$
$10=3p+16$
$-6=3p$
$-2=p$

The solution set is $\{-2\}$.

25. $\dfrac{2z-1}{z}=\dfrac{3}{5}$

$5(2z-1)=3(z)$
$10z-5=3z$
$10z=3z+5$
$7z=5$
$z=\dfrac{5}{7}$

The solution set is $\left\{\dfrac{5}{7}\right\}$.

27. $\dfrac{2}{v^2-v}=\dfrac{1}{3-v}$

$2(3-v)=1(v^2-v)$
$6-2v=v^2-v$
$0=v^2+v-6$
$0=(v+3)(v-2)$
$v+3=0$ or $v-2=0$
$v=-3$ or $v=2$

The solution set is $\{-3, 2\}$.

29. $\dfrac{10-x}{4x}=\dfrac{1}{x-1}$

$(10-x)(x-1)=4x(1)$
$-x^2+11x-10=4x$
$0=x^2-7x+10$
$0=(x-2)(x-5)$
$x-2=0$ or $x-5=0$
$x=2$ or $x=5$

The solution set is $\{2, 5\}$.

31. $\dfrac{2p-3}{p^2+12p+6}=\dfrac{1}{p+4}$

$(2p-3)(p+4)=(p^2+12p+6)(1)$
$2p^2+5p-12=p^2+12p+6$
$p^2-7p-18=0$
$(p+2)(p-9)=0$
$p+2=0$ or $p-9=0$
$p=-2$ or $p=9$

The solution set is $\{-2, 9\}$.

33. $\dfrac{AB}{AC}=\dfrac{XY}{XZ}$

$\dfrac{6}{8}=\dfrac{9}{XZ}$
$6(XZ)=9(8)$
$6(XZ)=72$
$XZ=12$

35. $\dfrac{XY}{ZY}=\dfrac{AB}{BC}$

$\dfrac{n}{2n-1}=\dfrac{3}{5}$
$5(n)=3(2n-1)$
$5n=6n-3$
$-n=-3$
$n=3$
$ZY=2n-1=2(3)-1=5$

37. $\dfrac{AB}{BC}=\dfrac{EF}{FG}$

$\dfrac{x-2}{2x+3}=\dfrac{4}{9}$
$9(x-2)=4(2x+3)$
$9x-18=8x+12$
$x-18=12$
$x=30$
$AB=x-2=30-2=28$

39. hours Natalie worked $=2x$

41. days David painted = $t - 3$

43. Let x be the number of hours working together. Christina paints $\frac{1}{5}$ chair in one hour, Victoria paints $\frac{1}{3}$ chair in one hour, and together they paint $\frac{1}{t}$ chair in one hour.

$$\frac{1}{5}+\frac{1}{3}=\frac{1}{t}$$

45. Let b be the number of hours for Pipe B alone to fill the tank. Pipe A takes $(b + 4)$ hours. Pipe A fills $\frac{1}{b+4}$ tank in one hour, Pipe B fills $\frac{1}{b}$ tank in one hour, and together they fill $\frac{1}{7}$ tank in one hour.

$$\frac{1}{b+4}+\frac{1}{b}=\frac{1}{7}$$

47. Joe's rate upstream = $r - 2$

49. rate of the plan against the wind = $r - 48$

51. Let c be the speed of the current. Bob's speed upstream is $14 - c$, and his speed downstream is $14 + c$. Since $d = rt$, then $t=\frac{d}{r}$. The times are the same for 4 miles upstream and 7 miles downstream.

$$\frac{4}{14-c}=\frac{7}{14+c}$$

53. Let r be the speed of the boat in still water. Assen's speed downstream is $r + 3$, and his speed upstream is $r - 3$. Since $d = rt$, then $t=\frac{d}{r}$. The times are the same for 12 miles downstream and 8 miles upstream.

$$\frac{12}{r+3}=\frac{8}{r-3}$$

55. Let x be the distance between the two towns.

$$\frac{\frac{1}{4}}{10}=\frac{3\frac{5}{8}}{x}$$
$$\frac{1}{4}x=10\left(3\frac{5}{8}\right)$$
$$\frac{x}{4}=10\left(\frac{29}{8}\right)$$
$$\frac{x}{4}=\frac{145}{4}$$
$$x=145$$

The distance between the towns is 145 miles.

57. Let f be the flour for 5 loaves.

$$\frac{5}{3}=\frac{f}{5}$$
$$5(5)=3f$$
$$25=3f$$
$$\frac{25}{3}=f$$

It takes $\frac{25}{3}=8\frac{1}{3}$ lb of flour for 5 loaves.

59. Let r be the number of rubles.

$$\frac{5}{143.25}=\frac{1300}{r}$$
$$5r=186,225$$
$$r=37,245$$

She can purchase 37,245 rubles.

61. Let t be the height of the tree.

$$\frac{6}{2.5}=\frac{t}{10}$$
$$60=2.5t$$
$$24=t$$

The tree is 24 feet tall.

63. Let t be the number of hours working together. Josh cleans $\frac{1}{3}$ building in one hour, Ken cleans $\frac{1}{5}$ building one hour, and they clean $\frac{1}{t}$ building in one hour working together.

$$\frac{1}{3}+\frac{1}{5}=\frac{1}{t}$$
$$15t\left(\frac{1}{3}+\frac{1}{5}\right)=15t\left(\frac{1}{t}\right)$$
$$5t+3t=15$$
$$8t=15$$
$$t=\frac{15}{8}$$

It takes them $\frac{15}{8}=1.875$ hours working together.

65. Let t be the time working together. Dyanne collects $\frac{1}{8}$ of the balls in one minute, Makini collects $\frac{1}{6}$ of the balls in one minute, and they collect $\frac{1}{t}$ of the balls in one minute working together.

$$\frac{1}{8}+\frac{1}{6}=\frac{1}{t}$$
$$24t\left(\frac{1}{8}+\frac{1}{6}\right)=24t\left(\frac{1}{t}\right)$$
$$3t+4t=24$$
$$7t=24$$
$$t=\frac{24}{7}$$

It takes them $\frac{24}{7}\approx 3.4$ minutes together.

67. Since the part that is left is equal to what each of them has already completed, José can paint $\frac{1}{9}$ of the remaining part in one hour, and Joaquín can paint $\frac{1}{12}$ in one hour. Let t be the time it takes them working together. Then they paint $\frac{1}{t}$ of the remaining part in one hour.

$$\frac{1}{9}+\frac{1}{12}=\frac{1}{t}$$
$$36t\left(\frac{1}{9}+\frac{1}{12}\right)=36t\left(\frac{1}{t}\right)$$
$$4t+3t=36$$
$$7t=36$$
$$t=\frac{36}{7}$$

It takes them $\frac{36}{7}\approx 5.1$ hours to finish the job.

69. Let p be the hours it takes the experienced plumber. Then it takes the apprentice $2p$ hours. They complete $\frac{1}{5}$ of the job in one hour working together, the plumber completes $\frac{1}{p}$ of the job in one hour, and the apprentice completes $\frac{1}{2p}$ of the job in one hour.

$$\frac{1}{p}+\frac{1}{2p}=\frac{1}{5}$$
$$10p\left(\frac{1}{p}+\frac{1}{2p}\right)=10p\left(\frac{1}{5}\right)$$
$$10+5=2p$$
$$15=2p$$

It would take the apprentice $2p = 15$ hours to do the job alone.

71. Let t be the time to fill the pool. Then $\frac{1}{6}$ of the pool will be filled in one hour, while $\frac{1}{8}$ of the pool will be drained in one hour.

$$\frac{1}{6}-\frac{1}{8}=\frac{1}{t}$$
$$24t\left(\frac{1}{6}-\frac{1}{8}\right)=24t\left(\frac{1}{t}\right)$$
$$4t-3t=24$$
$$t=24$$

It will take 24 hours to fill the pool.

73. Let b be the speed of the boat in still water. The boat's speed upstream is $b-2$ and its speed downstream is $b+2$. Since $d=rt$, then $t=\frac{d}{r}$. The times are the same for 12 km downstream and 4 km upstream.

$$\frac{12}{b+2}=\frac{4}{b-2}$$
$$12(b-2)=4(b+2)$$
$$12b-24=4b+8$$
$$8b-24=8$$
$$8b=32$$
$$b=4$$

The speed of the boat in still water is 4 km per hr.

75. Let b be his running speed, so $8b$ is his biking speed. Since $d = rt$, then $t=\frac{d}{r}$. Tony's time running is $\frac{15}{b}$ and his time biking is $\frac{60}{8b}=\frac{15}{2b}$. The total time is 5 hours.

$$\frac{15}{b}+\frac{15}{2b}=5$$
$$2b\left(\frac{15}{b}+\frac{15}{2b}\right)=2b(5)$$
$$30+15=10b$$
$$45=10b$$
$$4.5=b$$

The time he spent biking is

$$\frac{15}{2b}=\frac{15}{2(4.5)}=\frac{15}{9}=1\frac{2}{3}\text{ hours}$$

or 1 hr, 40 min.

77. Let r be the average rate.

	d	r	$t=\frac{d}{r}$
slow	60	$r-20$	$\frac{60}{r-20}$
fast	90	r	$\frac{90}{r}$

The trip took a total of 3 hours.

$$\frac{60}{r-20}+\frac{90}{r}=3$$
$$r(r-20)\left(\frac{60}{r-20}+\frac{90}{r}\right)=3r(r-20)$$
$$60r+90(r-20)=3r^2-60r$$
$$60r+90r-1800=3r^2-60r$$
$$0=3r^2-210r+1800$$
$$0=r^2-70r+600$$
$$0=(r-60)(r-10)$$
$$r-60=0 \quad\text{or}\quad r-10=0$$
$$r=60 \quad\text{or}\quad r=10$$

$r = 10$ is extraneous since the slow speed would be negative ($r - 20 = 10 - 20 = -10$). The average rate during the last 60 miles was $r - 20 = 60 - 20 = 40$ mph.

79. Let r be the average speed going to work. Then $r - 7$ is the average speed going home. Since $d = rt$, then $t=\frac{d}{r}$. The time for 16 miles at $(r - 7)$ mph is the same as the time for 20 miles at r mph.

$$\frac{20}{r}=\frac{16}{r-7}$$
$$20(r-7)=16r$$
$$20r-140=16r$$
$$4r=140$$
$$r=35$$

The average speed going to work is 35 miles per hour.

81.
$$\frac{XA}{YA}=\frac{BA}{CA}$$
$$\frac{4}{x}=\frac{12}{18}$$
$$18(4)=12x$$
$$72=12x$$
$$6=x$$

83. Note that $AB = AX + XB$.

$$\frac{XA}{XY}=\frac{AB}{BC}$$
$$\frac{XA}{XY}=\frac{AX+XB}{BC}$$
$$\frac{5}{7}=\frac{5+9}{x}$$
$$\frac{5}{7}=\frac{14}{x}$$
$$5x=7(14)$$
$$5x=98$$
$$x=\frac{98}{5}$$

85. Note that $AB = AX + XB$ and $AC = AY + YC$.

$$\frac{AX}{AY} = \frac{AB}{AC}$$
$$\frac{AX}{AY} = \frac{AX + XB}{AY + YC}$$
$$\frac{6}{5} = \frac{6+x}{5+12}$$
$$\frac{6}{5} = \frac{6+x}{17}$$
$$17(6) = 5(6+x)$$
$$102 = 30 + 5x$$
$$72 = 5x$$
$$\frac{72}{5} = x$$

87. The time component is not correctly placed. The times in the table should be $\frac{10}{r+1}$ and $\frac{10}{r-1}$. The equation will incorporate the 5 hours: time traveled downstream + time traveled upstream = 5 hours.

89. One equation to solve the problem is $\frac{1}{8}+\frac{1}{6}=\frac{1}{t}$ where t is the number of hours required to complete the job together. Answers may vary.

Section 7.9

Preparing for Variation

P1. $10 = 2k$
$$\frac{10}{2} = \frac{2k}{2}$$
$$5 = k$$
The solution set is $\{5\}$.

P2. $3 = \frac{k}{5}$
$$5 \cdot 3 = 5 \cdot \frac{k}{5}$$
$$15 = k$$
The solution set is $\{15\}$.

7.9 Quick Checks

1. The statement t varies directly with s is given by the equation $t = ks$.

2. In the equation $y = kx$, k is called the constant of proportionality and represents the rate of change between the variables x and y.

3. $y = kx$ and $y = 15$ when $x = 5$.
$$15 = k \cdot 5$$
$$3 = k$$
Thus, $y = 3x$.

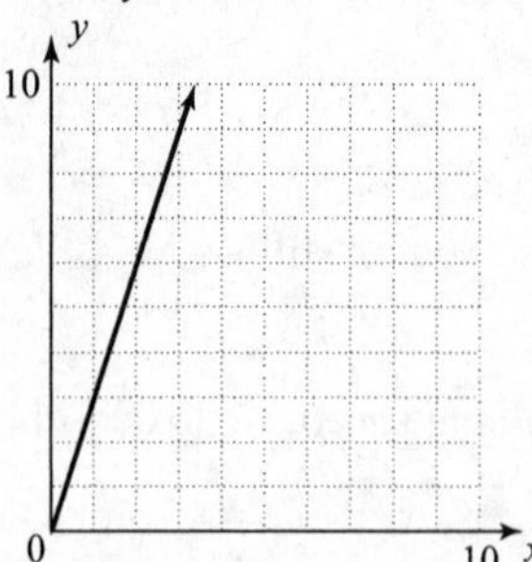

4. **(a)** $q = kw$ and $q = 10$ when $w = 40$.
$$10 = k \cdot 40$$
$$\frac{10}{40} = k$$
$$\frac{1}{4} = k$$
Thus, $q = \frac{1}{4}w$.

(b) $q = \frac{1}{4}w$; $w = 60$
$$q = \frac{1}{4} \cdot 60 = 15$$
When $w = 60$, $q = 15$.

(c)

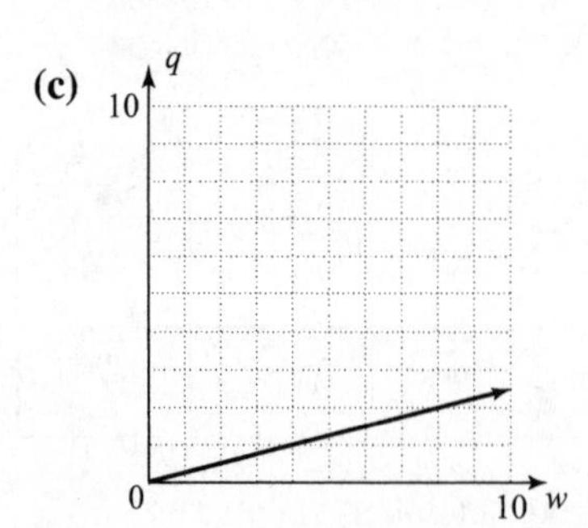

5. We want to know the cost of pumping 6.8 gallons of gasoline. $C = kg$ and $C = 34$ when $g = 8$.
$$34 = k \cdot 8$$
$$\frac{34}{8} = k$$
$$4.25 = k$$
Thus, $C = 4.25g$.
$C = 4.25g$; $g = 6.8$
$C = 4.25 \cdot 6.8 = 28.90$
The cost of pumping 6.8 gallons is \$28.90.

6. The statement f varies inversely with d is given by the equation $\underline{f = \frac{k}{d}}$.

7. False; the reimbursement is proportional to the miles you drive.

8. (a) $y = \frac{k}{x}$ and $y = 6$ when $x = 2$.

$$6 = \frac{k}{2}$$
$$12 = k$$

Thus, $y = \frac{12}{x}$.

(b) $y = \frac{12}{x}$; $x = 4$

$$y = \frac{12}{4} = 3$$

When $x = 4$, $y = 3$.

9. (a) $p = \frac{k}{q}$ and $p = 10$ when $q = 4$.

$$10 = \frac{k}{4}$$
$$40 = k$$

Thus, $p = \frac{40}{q}$.

(b) $p = \frac{40}{q}$; $q = 8$

$$p = \frac{40}{8} = 5$$

When $q = 8$, $p = 5$.

10. We want to find the rate of vibration, V, when the length, l, is 24 inches.

$V = \frac{k}{l}$ and $V = 20$ when $l = 30$.

$$20 = \frac{k}{30}$$
$$600 = k$$

Thus, $V = \frac{600}{l}$.

$$V = \frac{600}{l};\ l = 24$$

$$V = \frac{600}{24} = 25$$

A string that is 24 inches long vibrates 25 times per second.

7.9 Exercises

11.
$$y = kx$$
$$3 = k(6)$$
$$\frac{1}{2} = k$$
$$y = \frac{1}{2}x$$

13.
$$y = kx$$
$$24 = k(12)$$
$$2 = k$$
$$y = 2x$$

15.
$$y = kx$$
$$-6 = k(8)$$
$$-\frac{3}{4} = k$$
$$y = -\frac{3}{4}x$$

17. (a)
$$p = kg$$
$$12 = k(36)$$
$$\frac{1}{3} = k$$
$$p = \frac{1}{3}g$$

(b) $g = 9$: $p = \frac{1}{3}(9) = 3$

19. (a)
$$y = kx$$
$$-9 = k(12)$$
$$-\frac{3}{4} = k$$
$$y = -\frac{3}{4}x$$

(b) $y = \frac{5}{4}$:
$$\frac{5}{4} = -\frac{3}{4}x$$
$$\frac{5}{4}\left(-\frac{4}{3}\right) = x$$
$$-\frac{5}{3} = x$$

21. $y = \frac{k}{x}$

$3 = \frac{k}{2}$

$6 = k$

$y = \frac{6}{x}$

23. $y = \frac{k}{x}$

$3 = \frac{k}{4}$

$12 = k$

$y = \frac{12}{x}$

25. $y = \frac{k}{x}$

$\frac{4}{7} = \frac{k}{\frac{1}{2}}$

$\frac{4}{7}\left(\frac{1}{2}\right) = k$

$\frac{2}{7} = k$

$y = \frac{\frac{2}{7}}{x} = \frac{2}{7x}$

27. (a) $e = \frac{k}{n}$

$4 = \frac{k}{2}$

$8 = k$

$e = \frac{8}{n}$

(b) $n = 16$: $e = \frac{8}{16} = \frac{1}{2}$

29. (a) $b = \frac{k}{a}$

$\frac{3}{2} = \frac{k}{\frac{1}{2}}$

$\frac{3}{2}\left(\frac{1}{2}\right) = k$

$\frac{3}{4} = k$

$b = \frac{\frac{3}{4}}{a} = \frac{3}{4a}$

(b) $b = \frac{9}{10}$: $\frac{9}{10} = \frac{3}{4a}$

$36a = 30$

$a = \frac{30}{36} = \frac{5}{6}$

31. $y = \frac{2x}{3} = \frac{2}{3}x$ represents direct variation with constant of proportionality $\frac{2}{3}$.

33. $y = \frac{1}{2}$ does not involve x so it is not an equation representing variation.

35. $xy = 9$

$y = \frac{9}{x}$

This represents inverse variation with constant of proportionality 9.

37. $6x = 3y$

$2x = y$

$y = 2x$

This represents direct variation with constant of proportionality 2.

39. $x = \frac{k}{y}$

$6 = \frac{k}{3}$

$18 = k$

$x = \frac{18}{y}$

$y = 4$: $x = \frac{18}{4} = \frac{9}{2}$

41. $J = kP$

$80 = k(120)$

$\frac{2}{3} = k$

$J = \frac{2}{3}P$

$J = 50$: $50 = \frac{2}{3}P$

$50\left(\frac{3}{2}\right) = P$

$75 = P$

43. $s = kt$

$21 = k(18)$

$\frac{7}{6} = k$

$s = \frac{7}{6}t$

$s = 49$: $49 = \frac{7}{6}t$

$49\left(\frac{6}{7}\right) = t$

$42 = t$

45. $b = \frac{k}{a}$

$6 = \frac{k}{4}$

$24 = k$

$b = \frac{24}{a}$

$a = \frac{4}{3}$: $b = \frac{24}{\frac{4}{3}} = 24\left(\frac{3}{4}\right) = 18$

47. Let r = representatives and p = population. Then $r = kp$.

$18 = k(11.375)$

$\frac{18}{11.375} = k$

$r = \frac{18}{11.375}P$

$p = 6.356$: $r = \frac{18}{11.375}(6.356) \approx 10$

Massachusetts has 10 representatives.

49. Let c = cost and b = board feet. Then $c = kb$.

$717.50 = k(70)$

$\frac{717.5}{70} = k$

$10.25 = k$

$c = 10.25b$

$c = 1230$: $1230 = 10.25b$

$\frac{1230}{10.25} = b$

$120 = b$

Elizabeth bought 120 board feet.

51. $P = \frac{k}{V}$

$5 = \frac{k}{2.5}$

$12.5 = k$

$P = \frac{12.5}{V}$

$V = 2$: $P = \frac{12.5}{2} = 6.25$

The pressure will be 6.25 atmospheres.

53. $D = \frac{k}{p}$

$180 = \frac{k}{2.5}$

$450 = k$

$D = \frac{450}{p}$

$p = 3$: $D = \frac{450}{3} = 150$

The demand will be 150 bags of candy.

55. $m = krs$

$12 = k(0.5)(4)$

$12 = 2k$

$6 = k$

$m = 6rs$

$r = \frac{5}{3}$, $s = 9$: $m = 6\left(\frac{5}{3}\right)(9) = 90$

57. $p = kqr^2$

$162 = k(9)(6)^2$

$162 = k(9)(36)$

$162 = 324k$

$\frac{1}{2} = k$

$p = \frac{1}{2}qr^2$

$p = 300, r = 5$: $300 = \frac{1}{2}q(5)^2$

$300 = \frac{25}{2}q$

$24 = q$

59. **(a)** $e=\frac{k}{\lambda}$

$k=(6.63\times10^{-34})(3\times10^{8})$
$=19.89\times10^{-34+8}$
$=1.989\times10^{1}\times10^{-26}$
$=1.989\times10^{-25}$

$e=\frac{1.989\times10^{-25}}{\lambda}$

(b) $\lambda=500\times10^{-9}$:

$e=\frac{1.989\times10^{-25}}{500\times10^{-9}}$
$=\frac{1.989}{500}\times10^{-25-(-9)}$
$=0.003978\times10^{-25+9}$
$=3.978\times10^{-3}\times10^{-16}$
$=3.978\times10^{-19}$

The energy is 3.978×10^{-19} joules.

61. Answers will vary.

Chapter 7 Review

1. $\frac{2x}{x+3}$

(a) $x=1$: $\frac{2(1)}{1+3}=\frac{2}{4}=\frac{1}{2}$

(b) $x=-2$: $\frac{2(-2)}{-2+3}=\frac{-4}{1}=-4$

(c) $x=3$: $\frac{2(3)}{3+3}=\frac{6}{6}=1$

2. $\frac{3x}{x-4}$

(a) $x=0$: $\frac{3(0)}{0-4}=\frac{0}{-4}=0$

(b) $x=1$: $\frac{3(1)}{1-4}=\frac{3}{-3}=-1$

(c) $x=-2$: $\frac{3(-2)}{-2-4}=\frac{-6}{-6}=1$

3. $\frac{x^2+2xy+y^2}{x-y}$

(a) $x=1, y=-1$:
$\frac{1^2+2(1)(-1)+(-1)^2}{1-(-1)}=\frac{1-2+1}{2}=\frac{0}{2}=0$

(b) $x=2, y=1$: $\frac{2^2+2(2)(1)+1^2}{2-1}=\frac{4+4+1}{1}=9$

(c) $x=-3, y=-4$:
$\frac{(-3)^2+2(-3)(-4)+(-4)^2}{-3-(-4)}=\frac{9+24+16}{-3+4}$
$=\frac{49}{1}$
$=49$

4. $\frac{2x^2-x-3}{x+z}$

(a) $x=1, z=1$:
$\frac{2(1)^2-1-3}{1+1}=\frac{2-1-3}{2}=\frac{-2}{2}=-1$

(b) $x=1, z=2$:
$\frac{2(1)^2-1-3}{1+2}=\frac{2-1-3}{3}=\frac{-2}{3}=-\frac{2}{3}$

(c) $x=5, z=-4$: $\frac{2(5)^2-5-3}{5+(-4)}=\frac{50-5-3}{1}=42$

5. $\frac{3x}{3x-7}$ is undefined when $3x-7=0$.

$3x-7=0$
$3x=7$
$x=\frac{7}{3}$

6. $\frac{x+1}{4x-2}$ is undefined when $4x-2=0$.

$4x-2=0$
$4x=2$
$x=\frac{2}{4}=\frac{1}{2}$

7. $\dfrac{5}{x^2+25}$ is undefined when $x^2+25=0$.

$x^2+25=0$

$x^2=-25$

x^2 is always nonnegative. There are no values for which $\dfrac{5}{x^2+25}$ is undefined.

8. $\dfrac{17}{4x^2+49}$ is undefined when $4x^2+49=0$.

$4x^2+49=0$

$4x^2=-49$

$x^2=-\dfrac{49}{4}$

x^2 is always nonnegative. There are no values for which $\dfrac{17}{4x^2+49}$ is undefined.

9. $\dfrac{5x+2}{x^2+12x+20}$ is undefined when $x^2+12x+20=0$.

$x^2+12x+20=0$

$(x+10)(x+2)=0$

$x+10=0$ or $x+2=0$

$x=-10$ or $x=-2$

10. $\dfrac{3x-1}{x^2-3x-4}$ is undefined when $x^2-3x-4=0$.

$x^2-3x-4=0$

$(x+1)(x-4)=0$

$x+1=0$ or $x-4=0$

$x=-1$ or $x=4$

11. $\dfrac{5y^2+10y}{25y}=\dfrac{5y(y+2)}{5y\cdot 5}=\dfrac{y+2}{5}$

12. $\dfrac{2x^3-8x^2}{10x}=\dfrac{2x\cdot x(x-4)}{2x\cdot 5}=\dfrac{x(x-4)}{5}$

13. $\dfrac{3k-21}{k^2-5k-14}=\dfrac{3(k-7)}{(k+2)(k-7)}=\dfrac{3}{k+2}$

14. $\dfrac{-2x-2}{x^2-2x-3}=\dfrac{-2(x+1)}{(x-3)(x+1)}=-\dfrac{2}{x-3}$

15. $\dfrac{x^2+8x+15}{2x^2+5x-3}=\dfrac{(x+5)(x+3)}{(2x-1)(x+3)}=\dfrac{x+5}{2x-1}$

16. $\dfrac{3x^2+5x-2}{2x^3+16}=\dfrac{(3x-1)(x+2)}{2(x+2)(x^2-2x+4)}$

$=\dfrac{3x-1}{2(x^2-2x+4)}$

17. $\dfrac{12m^4n^3}{7}\cdot\dfrac{21}{18m^2n^5}$

$=\dfrac{2\cdot 6\cdot m^2\cdot m^2\cdot n^3}{7}\cdot\dfrac{7\cdot 3}{3\cdot 6\cdot m^2\cdot n^3\cdot n^2}$

$=\dfrac{3\cdot 6\cdot 7\cdot m^2\cdot n^3\cdot 2\cdot m^2}{3\cdot 6\cdot 7\cdot m^2\cdot n^3\cdot n^2}$

$=\dfrac{2m^2}{n^2}$

18. $\dfrac{3x^2y^4}{4}\cdot\dfrac{2}{9x^3y}=\dfrac{3\cdot x^2\cdot y\cdot y^3}{2\cdot 2}\cdot\dfrac{2}{3\cdot 3\cdot x^2\cdot x\cdot y}$

$=\dfrac{2\cdot 3\cdot x^2\cdot y\cdot y^3}{2\cdot 3\cdot x^2\cdot y\cdot 2\cdot 3\cdot x}$

$=\dfrac{y^3}{6x}$

19. $\dfrac{10m^2n^4}{9m^3n}\div\dfrac{15mn^6}{21m^2n}$

$=\dfrac{10m^2n^4}{9m^3n}\cdot\dfrac{21m^2n}{15mn^6}$

$=\dfrac{2\cdot 5\cdot m^2\cdot n^4}{3\cdot 3\cdot m^2\cdot m\cdot n}\cdot\dfrac{3\cdot 7\cdot m\cdot m\cdot n}{3\cdot 5\cdot m\cdot n^4\cdot n^2}$

$=\dfrac{3\cdot 5\cdot m^2\cdot m\cdot m\cdot n^4\cdot n\cdot 2\cdot 7}{3\cdot 5\cdot m^2\cdot m\cdot m\cdot n^4\cdot n\cdot 3\cdot 3\cdot n^2}$

$=\dfrac{14}{9n^2}$

20. $\frac{5ab^3}{3b^4} \div \frac{10a^2b^8}{6b^2}$

$= \frac{5ab^3}{3b^4} \cdot \frac{6b^2}{10a^2b^8}$

$= \frac{5 \cdot a \cdot b^3}{3 \cdot b^2 \cdot b^2} \cdot \frac{2 \cdot 3 \cdot b^2}{2 \cdot 5 \cdot a \cdot a \cdot b^3 \cdot b^5}$

$= \frac{3 \cdot 2 \cdot 5 \cdot a \cdot b^3 \cdot b^2 \cdot 1}{3 \cdot 2 \cdot 5 \cdot a \cdot b^3 \cdot b^2 \cdot a \cdot b^2 \cdot b^5}$

$= \frac{1}{ab^7}$

21. $\frac{5x-15}{x^2-x-12} \cdot \frac{x^2-6x+8}{3-x}$

$= \frac{5(x-3)}{(x-4)(x+3)} \cdot \frac{(x-4)(x-2)}{-1(x-3)}$

$= \frac{(x-3)(x-4) \cdot 5(x-2)}{(x-3)(x-4) \cdot (-1)(x+3)}$

$= -\frac{5(x-2)}{x+3}$

22. $\frac{4x-24}{x^2-18x+81} \cdot \frac{x^2-9}{6-x}$

$= \frac{4(x-6)}{(x-9)(x-9)} \cdot \frac{(x-3)(x+3)}{-1(x-6)}$

$= \frac{(x-6) \cdot 4(x-3)(x+3)}{(x-6) \cdot (-1)(x-9)(x-9)}$

$= -\frac{4(x-3)(x+3)}{(x-9)(x-9)}$

23. $\frac{\frac{x^2-4}{x^2-8x+15}}{\frac{12x+24}{3x-15}} = \frac{x^2-4}{x^2-8x+15} \cdot \frac{3x-15}{12x+24}$

$= \frac{(x-2)(x+2)}{(x-3)(x-5)} \cdot \frac{3(x-5)}{12(x+2)}$

$= \frac{3(x+2)(x-5) \cdot (x-2)}{3(x+2)(x-5) \cdot 4(x-3)}$

$= \frac{x-2}{4(x-3)}$

24. $\frac{\frac{5x^3+10x^2}{3x}}{\frac{2x+4}{18x^2}} = \frac{5x^3+10x^2}{3x} \cdot \frac{18x^2}{2x+4}$

$= \frac{5x^2(x+2)}{3x} \cdot \frac{2 \cdot 3 \cdot 3 \cdot x \cdot x}{2(x+2)}$

$= \frac{2 \cdot 3 \cdot x \cdot (x+2) \cdot 3 \cdot 5 \cdot x \cdot x^2}{2 \cdot 3 \cdot x \cdot (x+2) \cdot 1}$

$= 15x^3$

25. $\frac{3x^2+14x-5}{x^2+x-30} \cdot \frac{x^2-2x-15}{3x^2+8x-3}$

$= \frac{(3x-1)(x+5)}{(x+6)(x-5)} \cdot \frac{(x-5)(x+3)}{(3x-1)(x+3)}$

$= \frac{(3x-1)(x-5)(x+3) \cdot (x+5)}{(3x-1)(x-5)(x+3) \cdot (x+6)}$

$= \frac{x+5}{x+6}$

26. $\frac{y^2-5y-14}{y^2-2y-35} \cdot \frac{y^2+6y+5}{y^2-y-6}$

$= \frac{(y-7)(y+2)}{(y-7)(y+5)} \cdot \frac{(y+5)(y+1)}{(y-3)(y+2)}$

$= \frac{(y-7)(y+2)(y+5) \cdot (y+1)}{(y-7)(y+2)(y+5) \cdot (y-3)}$

$= \frac{y+1}{y-3}$

27. $\frac{x^2-9x}{x^2+3x+2} \div \frac{x^2-81}{x^2+2x}$

$= \frac{x^2-9x}{x^2+3x+2} \cdot \frac{x^2+2x}{x^2-81}$

$= \frac{x(x-9)}{(x+2)(x+1)} \cdot \frac{x(x+2)}{(x-9)(x+9)}$

$= \frac{x(x-9)(x+2) \cdot x}{(x-9)(x+2) \cdot (x+1)(x+9)}$

$= \frac{x^2}{(x+1)(x+9)}$

28. $\frac{y^2-9}{2y^2-y-15} \div \frac{3y^2+10y+3}{2y^2+y-10}$
$= \frac{y^2-9}{2y^2-y-15} \cdot \frac{2y^2+y-10}{3y^2+10y+3}$
$= \frac{(y-3)(y+3)}{(2y+5)(y-3)} \cdot \frac{(2y+5)(y-2)}{(3y+1)(y+3)}$
$= \frac{(y-3)(y+3)(2y+5) \cdot (y-2)}{(y-3)(y+3)(2y+5) \cdot (3y+1)}$
$= \frac{y-2}{3y+1}$

29. $\frac{4}{x-3} + \frac{5}{x-3} = \frac{4+5}{x-3} = \frac{9}{x-3}$

30. $\frac{3}{x+4} + \frac{7}{x+4} = \frac{3+7}{x+4} = \frac{10}{x+4}$

31. $\frac{m^2}{m+3} + \frac{3m}{m+3} = \frac{m^2+3m}{m+3} = \frac{m(m+3)}{m+3} = m$

32. $\frac{1}{6m} + \frac{5}{6m} = \frac{1+5}{6m} = \frac{6}{6m} = \frac{1}{m}$

33. $\frac{-m+1}{m^2-4} + \frac{2m+1}{m^2-4} = \frac{-m+1+2m+1}{m^2-4}$
$= \frac{m+2}{(m+2)(m-2)}$

34. $\frac{2m^2}{m+1} + \frac{5m+3}{m+1} = \frac{2m^2+5m+3}{m+1}$
$= \frac{(2m+3)(m+1)}{m+1}$
$= 2m+3$

35. $\frac{11}{15m} - \frac{6}{15m} = \frac{11-6}{15m} = \frac{5}{15m} = \frac{1}{3m}$

36. $\frac{15b}{2b^2} - \frac{11b}{2b^2} = \frac{15b-11b}{2b^2} = \frac{4b}{2b^2} = \frac{2b \cdot 2}{2b \cdot b} = \frac{2}{b}$

37. $\frac{2y^2}{y-7} - \frac{y^2+49}{y-7} = \frac{2y^2-(y^2+49)}{y-7}$
$= \frac{2y^2-y^2-49}{y-7}$
$= \frac{y^2-49}{y-7}$
$= \frac{(y-7)(y+7)}{y-7}$
$= y+7$

38. $\frac{6m+5}{m^2-36} - \frac{5m-1}{m^2-36} = \frac{6m+5-(5m-1)}{m^2-36}$
$= \frac{6m+5-5m+1}{m^2-36}$
$= \frac{m+6}{(m+6)(m-6)}$
$= \frac{1}{m-6}$

39. $\frac{3}{x-y} + \frac{10}{y-x} = \frac{3}{x-y} + \frac{10}{-1(x-y)}$
$= \frac{3}{x-y} + \frac{-10}{x-y}$
$= \frac{3-10}{x-y}$
$= -\frac{7}{x-y}$

40. $\frac{7}{a-b} + \frac{3}{b-a} = \frac{7}{a-b} + \frac{3}{-1(a-b)}$
$= \frac{7}{a-b} + \frac{-3}{a-b}$
$= \frac{7-3}{a-b}$
$= \frac{4}{a-b}$

41. $\frac{2x}{x^2-25} - \frac{x-5}{25-x^2} = \frac{2x}{x^2-25} - \frac{x-5}{-1(x^2-25)}$
$= \frac{2x}{x^2-25} - \frac{-(x-5)}{x^2-25}$
$= \frac{2x+x-5}{x^2-25}$
$= \frac{3x-5}{x^2-25}$

42. $\frac{x+5}{2x-6}-\frac{x+3}{6-2x}=\frac{x+5}{2x-6}-\frac{x+3}{-(2x-6)}$
$=\frac{x+5}{2x-6}-\frac{-(x+3)}{2x-6}$
$=\frac{x+5+x+3}{2x-6}$
$=\frac{2x+8}{2x-6}$
$=\frac{2(x+4)}{2(x-3)}$
$=\frac{x+4}{x-3}$

43. $4x^2y^7=2^2\cdot x^2\cdot y^7$
$6x^4y=2\cdot 3\cdot x^4\cdot y$
$\text{LCD}=2^2\cdot 3\cdot x^4\cdot y^7=12x^4y^7$

44. $20a^3bc^4=2^2\cdot 5\cdot a^3\cdot b\cdot c^4$
$30ab^4c^7=2\cdot 3\cdot 5\cdot a\cdot b^4\cdot c^7$
$\text{LCD}=2^2\cdot 3\cdot 5\cdot a^3\cdot b^4\cdot c^7=60a^3b^4c^7$

45. $4a=2^2\cdot a$
$8a+16=2^3\cdot\ (a+2)$
$\text{LCD}=2^3\cdot a\cdot(a+2)=8a(a+2)$

46. $5a=5\cdot a$
$a^2+6a=\ a\cdot(a+6)$
$\text{LCD}=5a\cdot(a+6)=5a(a+6)$

47. $4x-12=2^2\cdot(x-3)$
$x^2-2x-3=\ (x-3)\cdot(x+1)$
$\text{LCD}=2^2\cdot(x-3)\cdot(x+1)=4(x-3)(x+1)$

48. $x^2-7x=x\cdot\ (x-7)$
$x^2-49=\ (x-7)\cdot\ (x+7)$
$\text{LCD}=x(x-7)(x+7)$

49. $x^4y^7=xy^6\cdot x^3y$
The "missing factor" is xy^6.
$\frac{6}{x^3y}=\frac{6}{x^3y}\cdot\frac{xy^6}{xy^6}=\frac{6xy^6}{x^4y^7}$

50. $a^7b^5=a^4b^3\cdot a^3b^2$
The "missing factor" is a^4b^3.
$\frac{11}{a^3b^2}=\frac{11}{a^3b^2}\cdot\frac{a^4b^3}{a^4b^3}=\frac{11a^4b^3}{a^7b^5}$

51. $x^2-4=(x+2)(x-2)$
The "missing factor" is $x+2$.
$\frac{x-1}{x-2}=\frac{x-1}{x-2}\cdot\frac{x+2}{x+2}=\frac{(x-1)(x+2)}{(x-2)(x+2)}$

52. $m^2+5m-14=(m-2)(m+7)$
The "missing factor" is $m-2$.
$\frac{m+2}{m+7}=\frac{m+2}{m+7}\cdot\frac{m-2}{m-2}=\frac{(m+2)(m-2)}{(m+7)(m-2)}$

53. $6x^3=2\cdot 3\cdot x^3$
$8x^5=2^3\cdot\ x^5$
$\text{LCD}=2^3\cdot 3\cdot x^5=24x^5$
$\frac{5y}{6x^3}=\frac{5y}{6x^3}\cdot\frac{4x^2}{4x^2}=\frac{5y\cdot 4x^2}{6x^3\cdot 4x^2}=\frac{20x^2y}{24x^5}$
$\frac{7}{8x^5}=\frac{7}{8x^5}\cdot\frac{3}{3}=\frac{7\cdot 3}{8x^5\cdot 3}=\frac{21}{24x^5}$

54. $5a^3=\ 5\cdot a^3$
$10a=2\cdot 5\cdot a$
$\text{LCD}=2\cdot 5\cdot a^3=10a^3$
$\frac{6}{5a^3}=\frac{6}{5a^3}\cdot\frac{2}{2}=\frac{6\cdot 2}{5a^3\cdot 2}=\frac{12}{10a^3}$
$\frac{11b}{10a}=\frac{11b}{10a}\cdot\frac{a^2}{a^2}=\frac{11b\cdot a^2}{10a\cdot a^2}=\frac{11a^2b}{10a^3}$

55. $x-2=x-2$
$2-x=-(x-2)$
$\text{LCD}=-(x-2)$
$\frac{4x}{x-2}=\frac{4x}{x-2}\cdot\frac{-1}{-1}=\frac{-4x}{-(x-2)}$
$\frac{6}{2-x}=\frac{6}{-(x-2)}$

56. $m-5=m-5$
$5-m=-(m-5)$
$\text{LCD}=-(m-5)$
$$\frac{3}{m-5}=\frac{3}{m-5}\cdot\frac{-1}{-1}=\frac{-3}{-(m-5)}$$
$$\frac{-2m}{5-m}=\frac{-2m}{-(m-5)}$$

57. $m^2+5m-14=(m+7)(m-2)$
$m^2+9m+14=(m+7)(m+2)$
$\text{LCD}=(m+7)(m-2)(m+2)$
$$\frac{2}{m^2+5m-14}=\frac{2}{(m+7)(m-2)}\cdot\frac{m+2}{m+2}=\frac{2m+4}{(m+7)(m-2)(m+2)}$$
$$\frac{m+1}{m^2+9m+14}=\frac{m+1}{(m+7)(m+2)}\cdot\frac{m-2}{m-2}=\frac{m^2-m-2}{(m+7)(m-2)(m+2)}$$

58. $n^2-5n=n(n-5)$
$n^2-25=(n-5)(n+5)$
$\text{LCD}=n(n-5)(n+5)$
$$\frac{n-2}{n^2-5n}=\frac{n-2}{n(n-5)}\cdot\frac{n+5}{n+5}=\frac{n^2+3n-10}{n(n-5)(n+5)}$$
$$\frac{n}{n^2-25}=\frac{n}{(n-5)(n+5)}\cdot\frac{n}{n}=\frac{n^2}{n(n-5)(n+5)}$$

59. $xy^3=x\cdot y^3;\ x^2z=x^2\cdot z$
$\text{LCD}=x^2y^3z$
$$\frac{4}{xy^3}+\frac{8y}{x^2z}=\frac{4}{xy^3}\cdot\frac{xz}{xz}+\frac{8y}{x^2z}\cdot\frac{y^3}{y^3}$$
$$=\frac{4xz}{x^2y^3z}+\frac{8y^4}{x^2y^3z}$$
$$=\frac{4xz+8y^4}{x^2y^3z}$$

60. $2x^3y=2\cdot x^3\cdot y;\ 10xy^3=2\cdot 5\cdot x\cdot y^3$
$\text{LCD}=2\cdot 5\cdot x^3\cdot y^3=10x^3y^3$
$$\frac{x}{2x^3y}+\frac{y}{10xy^3}=\frac{x}{2x^3y}\cdot\frac{5y^2}{5y^2}+\frac{y}{10xy^3}\cdot\frac{x^2}{x^2}$$
$$=\frac{5xy^2}{10x^3y^3}+\frac{x^2y}{10x^3y^3}$$
$$=\frac{5xy^2+x^2y}{10x^3y^3}$$

61. $\text{LCD}=(x+7)(x-7)$
$$\frac{x}{x+7}+\frac{2}{x-7}$$
$$=\frac{x}{x+7}\cdot\frac{x-7}{x-7}+\frac{2}{x-7}\cdot\frac{x+7}{x+7}$$
$$=\frac{x^2-7x}{(x+7)(x-7)}+\frac{2x+14}{(x+7)(x-7)}$$
$$=\frac{x^2-7x+2x+14}{(x+7)(x-7)}$$
$$=\frac{x^2-5x+14}{(x+7)(x-7)}$$

62. $\text{LCD}=(2x+3)(2x-3)$
$$\frac{4}{2x+3}+\frac{x+1}{2x-3}$$
$$=\frac{4}{2x+3}\cdot\frac{2x-3}{2x-3}+\frac{x+1}{2x-3}\cdot\frac{2x+3}{2x+3}$$
$$=\frac{8x-12}{(2x+3)(2x-3)}+\frac{2x^2+5x+3}{(2x+3)(2x-3)}$$
$$=\frac{8x-12+2x^2+5x+3}{(2x+3)(2x-3)}$$
$$=\frac{2x^2+13x-9}{(2x+3)(2x-3)}$$

63. $\text{LCD}=x(x-2)$
$$\frac{x+5}{x}-\frac{x+7}{x-2}=\frac{x+5}{x}\cdot\frac{x-2}{x-2}-\frac{x+7}{x-2}\cdot\frac{x}{x}$$
$$=\frac{x^2+3x-10}{x(x-2)}-\frac{x^2+7x}{x(x-2)}$$
$$=\frac{x^2+3x-10-x^2-7x}{x(x-2)}$$
$$=\frac{-4x-10}{x(x-2)}$$

64. LCD = $x(x + 5)$

$$\frac{3x}{x+5}-\frac{x+1}{x}=\frac{3x}{x+5}\cdot\frac{x}{x}-\frac{x+1}{x}\cdot\frac{x+5}{x+5}$$
$$=\frac{3x^2}{x(x+5)}-\frac{x^2+6x+5}{x(x+5)}$$
$$=\frac{3x^2-(x^2+6x+5)}{x(x+5)}$$
$$=\frac{3x^2-x^2-6x-5}{x(x+5)}$$
$$=\frac{2x^2-6x-5}{x(x+5)}$$

65. $4x + 1 = 4x + 1$; $4x^2+9x+2=(4x+1)(x+2)$

LCD = $(4x + 1)(x + 2)$

$$\frac{3x-4}{4x+1}+\frac{3x+6}{4x^2+9x+2}$$
$$=\frac{3x-4}{4x+1}\cdot\frac{x+2}{x+2}+\frac{3x+6}{(4x+1)(x+2)}$$
$$=\frac{3x^2+2x-8}{(4x+1)(x+2)}+\frac{3x+6}{(4x+1)(x+2)}$$
$$=\frac{3x^2+2x-8+3x+6}{(4x+1)(x+2)}$$
$$=\frac{3x^2+5x-2}{(4x+1)(x+2)}$$
$$=\frac{(3x-1)(x+2)}{(4x+1)(x+2)}$$
$$=\frac{3x-1}{4x+1}$$

66. $3x^2+x-4=(3x+4)(x-1)$;

$3x^2-2x-8=(3x+4)(x-2)$

LCD = $(3x + 4)(x - 1)(x - 2)$

$$\frac{7}{3x^2+x-4}+\frac{9x+2}{3x^2-2x-8}$$
$$=\frac{7}{(3x+4)(x-1)}\cdot\frac{x-2}{x-2}+\frac{9x+2}{(3x+4)(x-2)}\cdot\frac{x-1}{x-1}$$
$$=\frac{7x-14}{(3x+4)(x-1)(x-2)}+\frac{9x^2-7x-2}{(3x+4)(x-1)(x-2)}$$
$$=\frac{7x-14+9x^2-7x-2}{(3x+4)(x-1)(x-2)}$$
$$=\frac{9x^2-16}{(3x+4)(x-1)(x-2)}$$
$$=\frac{(3x+4)(3x-4)}{(3x+4)(x-1)(x-2)}$$
$$=\frac{3x-4}{(x-1)(x-2)}$$

67. $m^2-9=(m+3)(m-3)$; $m + 3 = m + 3$

LCD = $(m + 3)(m - 3)$

$$\frac{m}{m^2-9}-\frac{4m-12}{m+3}$$
$$=\frac{m}{(m+3)(m-3)}-\frac{4m-12}{m+3}\cdot\frac{m-3}{m-3}$$
$$=\frac{m}{(m+3)(m-3)}-\frac{4m^2-24m+36}{(m+3)(m-3)}$$
$$=\frac{m-4m^2+24m-36}{(m+3)(m-3)}$$
$$=\frac{-4m^2+25m-36}{(m+3)(m-3)}$$
$$=\frac{-(4m-9)(m-4)}{(m+3)(m-3)}$$

68. $m - 5 = m - 5$; $m^2-3m-10=(m-5)(m+2)$

LCD = $(m - 5)(m + 2)$

$$\frac{2m+1}{m-5}-\frac{4}{m^2-3m-10}$$
$$=\frac{2m+1}{m-5}\cdot\frac{m+2}{m+2}-\frac{4}{(m-5)(m+2)}$$
$$=\frac{2m^2+5m+2}{(m-5)(m+2)}-\frac{4}{(m-5)(m+2)}$$
$$=\frac{2m^2+5m+2-4}{(m-5)(m+2)}$$
$$=\frac{2m^2+5m-2}{(m-5)(m+2)}$$

69. $m-2=m-2;\ 2-m=-(m-2)$
LCD $=-(m-2)$

$$\frac{3}{m-2}-\frac{1}{2-m}=\frac{3}{m-2}\cdot\frac{-1}{-1}-\frac{1}{-(m-2)}$$
$$=\frac{-3}{-(m-2)}-\frac{1}{-(m-2)}$$
$$=\frac{-3-1}{-(m-2)}$$
$$=\frac{-4}{-(m-2)}$$
$$=\frac{4}{m-2}$$

70. $m-n=m-n;\ n-m=-(m-n)$
LCD $=-(m-n)$

$$\frac{m}{m-n}-\frac{n}{n-m}=\frac{m}{m-n}\cdot\frac{-1}{-1}-\frac{n}{-(m-n)}$$
$$=\frac{-m}{-(m-n)}-\frac{n}{-(m-n)}$$
$$=\frac{-m-n}{-(m-n)}$$
$$=\frac{-(m+n)}{-(m-n)}$$
$$=\frac{m+n}{m-n}$$

71. LCD $=x+3$

$$4+\frac{x}{x+3}=\frac{4}{1}\cdot\frac{x+3}{x+3}+\frac{x}{x+3}$$
$$=\frac{4x+12}{x+3}+\frac{x}{x+3}$$
$$=\frac{4x+12+x}{x+3}$$
$$=\frac{5x+12}{x+3}$$

72. LCD $=x+2$

$$7-\frac{1}{x+2}=\frac{7}{1}\cdot\frac{x+2}{x+2}-\frac{1}{x+2}$$
$$=\frac{7x+14}{x+2}-\frac{1}{x+2}$$
$$=\frac{7x+14-1}{x+2}$$
$$=\frac{7x+13}{x+2}$$

73. $$\frac{\frac{1}{2}-\frac{2}{3}}{\frac{4}{9}+\frac{5}{6}}=\frac{\frac{3}{6}-\frac{4}{6}}{\frac{8}{18}+\frac{15}{18}}=\frac{-\frac{1}{6}}{\frac{23}{18}}=-\frac{1}{6}\cdot\frac{18}{23}=-\frac{3}{23}$$

74. $$\frac{\frac{1}{4}+\frac{1}{2}}{\frac{5}{8}-\frac{1}{6}}=\frac{\frac{1}{4}+\frac{2}{4}}{\frac{15}{24}-\frac{4}{24}}=\frac{\frac{3}{4}}{\frac{11}{24}}=\frac{3}{4}\cdot\frac{24}{11}=\frac{18}{11}$$

75. $$\frac{\frac{1}{5}-\frac{1}{m}}{\frac{1}{10}+\frac{1}{m^2}}=\frac{10m^2\left(\frac{1}{5}-\frac{1}{m}\right)}{10m^2\left(\frac{1}{10}+\frac{1}{m^2}\right)}=\frac{2m^2-10m}{m^2+10}$$

76. $$\frac{\frac{1}{m^2}+\frac{2}{3}}{\frac{1}{m}-\frac{5}{6}}=\frac{6m^2\left(\frac{1}{m^2}+\frac{2}{3}\right)}{6m^2\left(\frac{1}{m}-\frac{5}{6}\right)}=\frac{6+4m^2}{6m-5m^2}$$

77. $$\frac{\frac{x}{4}-\frac{1}{2}}{\frac{3x}{2}-3}=\frac{\frac{x}{4}-\frac{1}{2}}{\frac{3x}{2}-3}\cdot\frac{4}{4}=\frac{x-2}{6x-12}=\frac{x-2}{6(x-2)}=\frac{1}{6}$$

78. $$\frac{\frac{7x}{3}+7}{\frac{3x+9}{8}}=\frac{\frac{7x+21}{3}}{\frac{3x+9}{8}}$$
$$=\frac{7x+21}{3}\cdot\frac{8}{3x+9}$$
$$=\frac{7(x+3)\cdot 8}{3\cdot 3(x+3)}$$
$$=\frac{56}{9}$$

79. $$\frac{\frac{8}{y+4}+2}{\frac{12}{y+4}-2}=\frac{(y+4)\left(\frac{8}{y+4}+2\right)}{(y+4)\left(\frac{12}{y+4}-2\right)}$$
$$=\frac{8+2y+8}{12-2y-8}$$
$$=\frac{2y+16}{-2y+4}$$
$$=\frac{2(y+8)}{2(-y+2)}$$
$$=\frac{y+8}{-y+2}$$

80. $$\frac{\frac{25}{y+5}+5}{\frac{3}{y+5}-5}=\frac{(y+5)\left(\frac{25}{y+5}+5\right)}{(y+5)\left(\frac{3}{y+5}-5\right)}$$
$$=\frac{25+5y+25}{3-5y-25}$$
$$=\frac{5y+50}{-5y-22}$$
$$=\frac{5(y+10)}{-5y-22}$$

81. The rational equation is undefined when $x + 5 = 0$ or $x = -5$ and when $x = 0$.

82. The rational equation is undefined when

$$\begin{aligned} x^2 - 36 &= 0 \quad \text{or} \quad x+6=0 \\ (x+6)(x-6) &= 0 \qquad\qquad x=-6 \\ x+6=0 \quad &\text{or} \quad x-6=0 \\ x=-6 \quad &\text{or} \quad x=6 \end{aligned}$$

83.

$$\begin{aligned} \frac{2}{x}-\frac{3}{4} &= \frac{5}{x} \\ 4x\left(\frac{2}{x}-\frac{3}{4}\right) &= 4x\left(\frac{5}{x}\right) \\ 8-3x &= 20 \\ -3x &= 12 \\ x &= -4 \end{aligned}$$

The value is not extraneous. The solution set is $\{-4\}$.

84.

$$\begin{aligned} \frac{4}{x}+\frac{3}{4} &= \frac{2}{3x}+\frac{23}{4} \\ 12x\left(\frac{4}{x}+\frac{3}{4}\right) &= 12x\left(\frac{2}{3x}+\frac{23}{4}\right) \\ 48+9x &= 8+69x \\ 40 &= 60x \\ x &= \frac{40}{60} \\ x &= \frac{2}{3} \end{aligned}$$

The value is not extraneous. The solution set is $\left\{\frac{2}{3}\right\}$.

85.

$$\begin{aligned} \frac{4}{m}-\frac{3}{2m} &= \frac{1}{2} \\ 2m\left(\frac{4}{m}-\frac{3}{2m}\right) &= 2m\left(\frac{1}{2}\right) \\ 8-3 &= m \\ 5 &= m \end{aligned}$$

The value is not extraneous. The solution set is $\{5\}$.

86.

$$\begin{aligned} \frac{3}{m}+\frac{5}{3m} &= 1 \\ 3m\left(\frac{3}{m}+\frac{5}{3m}\right) &= 3m(1) \\ 9+5 &= 3m \\ 14 &= 3m \\ \frac{14}{3} &= m \end{aligned}$$

The value is not extraneous. The solution set is $\left\{\frac{14}{3}\right\}$.

87.

$$\begin{aligned} \frac{m+4}{m-3} &= \frac{m+10}{m+2} \\ (m-3)(m+2)\left(\frac{m+4}{m-3}\right) &= (m-3)(m+2)\left(\frac{m+10}{m+2}\right) \\ (m+2)(m+4) &= (m-3)(m+10) \\ m^2+6m+8 &= m^2+7m-30 \\ 6m+8 &= 7m-30 \\ 8 &= m-30 \\ 38 &= m \end{aligned}$$

The value is not extraneous. The solution set is $\{38\}$.

88.

$$\begin{aligned} \frac{m-6}{m+5} &= \frac{m-3}{m+1} \\ (m+5)(m+1)\left(\frac{m-6}{m+5}\right) &= (m+5)(m+1)\left(\frac{m-3}{m+1}\right) \\ (m+1)(m-6) &= (m+5)(m-3) \\ m^2-5m-6 &= m^2+2m-15 \\ -5m-6 &= 2m-15 \\ -7m &= -9 \\ m &= \frac{9}{7} \end{aligned}$$

The value is not extraneous. The solution set is $\left\{\frac{9}{7}\right\}$.

89.

$$\begin{aligned} \frac{2x}{x-1}-5 &= \frac{2}{x-1} \\ (x-1)\left(\frac{2x}{x-1}-5\right) &= (x-1)\left(\frac{2}{x-1}\right) \\ 2x-5x+5 &= 2 \\ -3x &= -3 \\ x &= 1 \end{aligned}$$

The value is extraneous. The solution set is $\{\ \}$ or $\varnothing$.

90.
$$\frac{2x}{x-2}-3=\frac{4}{x-2}$$
$$(x-2)\left(\frac{2x}{x-2}-3\right)=(x-2)\left(\frac{4}{x-2}\right)$$
$$2x-3x+6=4$$
$$-x=-2$$
$$x=2$$
The value is extraneous. The solution set is { } or ∅.

91.
$$\frac{1}{x+3}+\frac{1}{x-3}=\frac{-5}{x^2-9}$$
$$(x+3)(x-3)\left(\frac{1}{x+3}+\frac{1}{x-3}\right)=(x+3)(x-3)\left(\frac{-5}{(x+3)(x-3)}\right)$$
$$x-3+x+3=-5$$
$$2x=-5$$
$$x=-\frac{5}{2}$$
The value is not extraneous. The solution set is $\left\{-\frac{5}{2}\right\}$.

92.
$$\frac{3}{x-5}-\frac{11}{x^2-25}=\frac{4}{x+5}$$
$$(x-5)(x+5)\left(\frac{3}{x-5}-\frac{11}{x^2-25}\right)=(x-5)(x+5)\left(\frac{4}{x+5}\right)$$
$$3(x+5)-11=4(x-5)$$
$$3x+15-11=4x-20$$
$$3x+4=4x-20$$
$$4=x-20$$
$$24=x$$
The value is not extraneous. The solution set is {24}.

93.
$$y=\frac{4}{k}$$
$$yk=\frac{4}{k}\cdot k$$
$$yk=4$$
$$\frac{yk}{y}=\frac{4}{y}$$
$$k=\frac{4}{y}$$

94. $6 = \frac{x}{y}$

$$6y = \frac{x}{y} \cdot y$$
$$6y = x$$
$$\frac{6y}{6} = \frac{x}{6}$$
$$y = \frac{x}{6}$$

95. $\frac{1}{x} + \frac{1}{y} = \frac{1}{z}$

$$xyz\left(\frac{1}{x} + \frac{1}{y}\right) = xyz\left(\frac{1}{z}\right)$$
$$yz + xz = xy$$
$$yz - xy = -xz$$
$$y(z - x) = -xz$$
$$\frac{y(z-x)}{z-x} = \frac{-xz}{z-x}$$
$$y = \frac{-xz}{z-x}$$
$$y = \frac{xz}{x-z}$$

96. $\frac{1}{x} + \frac{1}{y} = \frac{1}{z}$

$$xyz\left(\frac{1}{x} + \frac{1}{y}\right) = xyz\left(\frac{1}{z}\right)$$
$$yz + xz = xy$$
$$z(y + x) = xy$$
$$\frac{z(y+x)}{y+x} = \frac{xy}{y+x}$$
$$z = \frac{xy}{x+y}$$

97. $\frac{6}{4y+5} = \frac{2}{7}$

$$6(7) = 2(4y+5)$$
$$42 = 8y + 10$$
$$32 = 8y$$
$$4 = y$$

The solution set is $\{4\}$.

98. $\frac{2}{y-3} = \frac{5}{y}$

$$2y = 5(y-3)$$
$$2y = 5y - 15$$
$$-3y = -15$$
$$y = 5$$

The solution set is $\{5\}$.

99. $\frac{y+1}{8} = \frac{1}{4}$

$$4(y+1) = 8 \cdot 1$$
$$4y + 4 = 8$$
$$4y = 4$$
$$y = 1$$

The solution set is $\{1\}$.

100. $\frac{6y+7}{10} = \frac{2y+9}{6}$

$$6(6y+7) = 10(2y+9)$$
$$36y + 42 = 20y + 90$$
$$16y + 42 = 90$$
$$16y = 48$$
$$y = 3$$

The solution set is $\{3\}$.

101. $\frac{2}{3} = \frac{10}{x}$

$$2x = 3 \cdot 10$$
$$2x = 30$$
$$x = 15$$

102. $\frac{14}{7} = \frac{7}{x}$

$$14x = 7 \cdot 7$$
$$14x = 49$$
$$x = \frac{49}{14}$$
$$x = 3.5$$

103. Let x be the number of tanks needed for 30 bags.

$$\frac{4}{15} = \frac{x}{30}$$
$$4 \cdot 30 = 15x$$
$$120 = 15x$$
$$8 = x$$

Therefore, 8 tanks of water are needed for 30 bags of cement.

104. Let x be the cost of 7 small pizzas.

$$\frac{4}{15}=\frac{7}{x}$$
$$4x=15\cdot 7$$
$$4x=105$$
$$x=26.25$$

The cost of 7 small pizzas is \$26.25.

105. Let t be the time working together. Lucille can complete $\frac{1}{3}$ of the job in one hour. Teresa can complete $\frac{1}{2}$ of the job in one hour.

$$\frac{1}{3}+\frac{1}{2}=\frac{1}{t}$$
$$6t\left(\frac{1}{3}+\frac{1}{2}\right)=6t\left(\frac{1}{t}\right)$$
$$2t+3t=6$$
$$5t=6$$
$$t=\frac{6}{5}$$

It will take them $\frac{6}{5}=1.2$ hours working together.

106. Let t be the time working together. Fred can carpet $\frac{1}{3}$ of the room in an hour and Barney can carpet $\frac{1}{5}$ of the room in one hour.

$$\frac{1}{3}+\frac{1}{5}=\frac{1}{t}$$
$$15t\left(\frac{1}{3}+\frac{1}{5}\right)=15t\left(\frac{1}{t}\right)$$
$$5t+3t=15$$
$$8t=15$$
$$t=\frac{15}{8}$$

It will take them $\frac{15}{8}\approx 1.9$ hours working together.

107. Let t be the time it takes Adrienne to wash the dishes. Then $t + 5$ is the time it takes Jake to wash the dishes. Adrienne can complete $\frac{1}{t}$ of the job in an hour, while Jake can complete $\frac{1}{t+5}$ of the job in an hour.

$$\frac{1}{t}+\frac{1}{t+5}=\frac{1}{6}$$
$$6t(t+5)\left(\frac{1}{t}+\frac{1}{t+5}\right)=6t(t+5)\cdot\frac{1}{6}$$
$$6(t+5)+6t=t(t+5)$$
$$6t+30+6t=t^2+5t$$
$$12t+30=t^2+5t$$
$$0=t^2-7t-30$$
$$0=(t-10)(t+3)$$
$$t-10=0 \quad\text{or}\quad t+3=0$$
$$t=10 \quad\text{or}\quad t=-3$$

Disregard the negative value. Jake can wash the dishes is $t + 5 = 10 + 5 = 15$ minutes.

108. Let t be the time it took Ben to paint the room alone. Donovan can complete $\frac{1}{7}$ of the room in one hour whereas Ben can complete $\frac{1}{t}$ of the room in one hour. It takes 4 hours working together.

$$\frac{1}{7}+\frac{1}{t}=\frac{1}{4}$$
$$28t\left(\frac{1}{7}+\frac{1}{t}\right)=28t\left(\frac{1}{4}\right)$$
$$4t+28=7t$$
$$28=3t$$
$$\frac{28}{3}=t$$

It will take Ben $\frac{28}{3}\approx 9.3$ hours working alone.

109. Let p be Paul's speed.

	d	r	$t=\frac{d}{r}$
upstream	15	$p-2$	$\frac{15}{p-2}$
downstream	27	$p+2$	$\frac{27}{p+2}$

The times are the same.

$$\frac{15}{p-2}=\frac{27}{p+2}$$
$$15(p+2)=27(p-2)$$
$$15p+30=27p-54$$
$$30=12p-54$$
$$84=12p$$
$$7=p$$

Paul travels $p + 2 = 7 + 2 = 9$ mph when traveling downstream.

110. Let r be the rate of the ship.

	d	r	$t=\frac{d}{r}$
with current	275	$r + 10$	$\frac{275}{r+10}$
against current	175	$r - 10$	$\frac{175}{r-10}$

The times are the same.

$$\frac{275}{r+10}=\frac{175}{r-10}$$
$$275(r-10)=175(r+10)$$
$$275r-2750=175r+1750$$
$$100r=4500$$
$$r=45$$

The ship cruises at $r + 10 = 45 + 10 = 55$ mph.

111. Let r be the speed of the train.

	d	r	$t=\frac{d}{r}$
train	135	r	$t=\frac{135}{r}$
plane	855	$3r$	$t=\frac{855}{3}$

The total time was 6 hours.

$$\frac{135}{r}+\frac{855}{3r}=6$$
$$3r\left(\frac{135}{r}+\frac{855}{3r}\right)=3r(6)$$
$$3(135)+855=18r$$
$$405+855=18r$$
$$1260=18r$$
$$70=r$$

The train's speed was 70 mph.

112. Let r be Tamika's speed in the city.

	d	r	$t=\frac{d}{r}$
city	90	r	$\frac{90}{r}$
highway	130	$r + 20$	$\frac{130}{r+20}$

Her total time was 4 hours.

$$\frac{90}{r}+\frac{130}{r+20}=4$$
$$r(r+20)\left(\frac{90}{r}+\frac{130}{r+20}\right)=r(r+20)\cdot 4$$
$$90(r+20)+130r=4r(r+20)$$
$$90r+1800+130r=4r^2+80r$$
$$0=4r^2-140r-1800$$
$$0=4(r-45)(r+10)$$
$$r-45=0 \quad \text{or} \quad r+10=0$$
$$r=45 \quad \text{or} \quad r=-10$$

Disregard the negative rate. Tamika's speed in the city was 45 mph.

113. $y=kx$
$$3=k(12)$$
$$\frac{1}{4}=k$$
$$y=\frac{1}{4}x$$

114. $y=kx$
$$18=k(3)$$
$$6=k$$
$$y=6x$$

115. $f=kg$
$$12=k(18)$$
$$\frac{2}{3}=k$$
$$f=\frac{2}{3}g$$

Let $g = 24$.
$$f=\frac{2}{3}(24)$$
$$f=16$$

116. $p = kq$
$25 = k(20)$
$\frac{5}{4} = k$
$p = \frac{5}{4}q$
Let $q = 88$.
$p = \frac{5}{4}(88)$
$p = 110$

117. Let i be income and n be the number of customers.
$i = kn$
$290 = k(20)$
$\frac{29}{2} = k$
$i = \frac{29}{2}n$
Let $n = 35$.
$i = \frac{29}{2}(35)$
$i = 507.5$
The income will be $507.50.

118. Let m be the map distance and a be the actual distance.
$m = ka$
$3 = k(60)$
$\frac{1}{20} = k$
$m = \frac{1}{20}a$
Let $m = 8$.
$8 = \frac{1}{20}a$
$160 = a$
The actual distance is 160 miles.

119. $y = \frac{k}{x}$
$6 = \frac{k}{8}$
$48 = k$
$y = \frac{48}{x}$

120. $y = \frac{k}{x}$
$9 = \frac{k}{6}$
$54 = k$
$y = \frac{54}{x}$

121. $r = \frac{k}{s}$
$18 = \frac{k}{2}$
$36 = k$
$r = \frac{36}{s}$
Let $s = 4$.
$r = \frac{36}{4}$
$r = 9$

122. $s = \frac{k}{u}$
$3 = \frac{k}{2}$
$6 = k$
$s = \frac{6}{u}$
Let $u = 24$.
$s = \frac{6}{24}$
$s = \frac{1}{4}$

123. Let t be the time and s be the speed.
$t = \frac{k}{s}$
$6.5 = \frac{k}{60}$
$390 = k$
$t = \frac{390}{s}$
Let $t = 5$,
$5 = \frac{390}{s}$
$s = \frac{390}{5}$
$s = 78$
A speed of 78 mph is necessary.

124. Let v be volume and p be pressure.

$$v=\frac{k}{p}$$
$$8=\frac{k}{100}$$
$$800=k$$
$$v=\frac{800}{p}$$

Let $p = 32$.

$$v=\frac{800}{32}$$
$$v=25$$

The volume is 25 liters.

Chapter 7 Test

1. $\dfrac{3x-2y^2}{6z}$; $x=2, y=-3, z=-1$

$$\frac{3(2)-2(-3)^2}{6(-1)}=\frac{3(2)-2(9)}{6(-1)}=\frac{6-18}{-6}=\frac{-12}{-6}=2$$

2. $\dfrac{x+5}{x^2-3x-10}$ is undefined when $x^2-3x-10=0$.

$$x^2-3x-10=0$$
$$(x-5)(x+2)=0$$
$$x-5=0 \quad \text{or} \quad x+2=0$$
$$x=5 \quad \text{or} \quad x=-2$$

3. $$\frac{x^2-4x-21}{14-2x}=\frac{(x-7)(x+3)}{-2(x-7)}=\frac{x+3}{-2}=-\frac{x+3}{2}$$

4. $$\frac{\frac{35x^6}{9x^4}}{\frac{25x^5}{18x}}=\frac{35x^6}{9x^4}\cdot\frac{18x}{25x^5}=\frac{35x^6\cdot 18x}{9x^4\cdot 25x^5}=\frac{5\cdot 9\cdot x^5\cdot x\cdot x\cdot 2\cdot 7}{5\cdot 9\cdot x^5\cdot x\cdot x\cdot 5\cdot x^2}=\frac{14}{5x^2}$$

5. $$\frac{5x-15}{3x+9}\cdot\frac{5x+15}{3x-9}=\frac{5(x-3)}{3(x+3)}\cdot\frac{5(x+3)}{3(x-3)}=\frac{5(x-3)\cdot 5(x+3)}{3\cdot(x+3)\cdot 3(x-3)}=\frac{(x+3)\cdot 5\cdot 5\cdot(x-3)}{(x+3)\cdot 3\cdot 3\cdot(x-3)}=\frac{25}{9}$$

6. $$\frac{\frac{2x^2-5xy-12y^2}{x^2+xy-20y^2}}{\frac{4x^2-9y^2}{x^2+4xy-5y^2}}$$
$$=\frac{2x^2-5xy-12y^2}{x^2+xy-20y^2}\cdot\frac{x^2+4xy-5y^2}{4x^2-9y^2}$$
$$=\frac{(2x+3y)(x-4y)\cdot(x+5y)(x-y)}{(x+5y)(x-4y)\cdot(2x-3y)(2x+3y)}$$
$$=\frac{x-y}{2x-3y}$$

7. $$\frac{y^2}{y+3}+\frac{3y}{y+3}=\frac{y^2+3y}{y+3}=\frac{y(y+3)}{y+3}=y$$

8. $$\frac{x^2}{x^2-9}-\frac{8x-15}{x^2-9}=\frac{x^2-(8x-15)}{x^2-9}=\frac{x^2-8x+15}{x^2-9}=\frac{(x-5)(x-3)}{(x+3)(x-3)}=\frac{x-5}{x+3}$$

9. $$\begin{aligned}\frac{6}{y-z}+\frac{7}{z-y}&=\frac{6}{y-z}+\frac{7}{-1\cdot(y-z)}\\&=\frac{6}{y-z}+\frac{-7}{y-z}\\&=\frac{6-7}{y-z}\\&=\frac{-1}{y-z}\text{ or }\frac{1}{z-y}\end{aligned}$$

10. LCD = $(x-2)(2x+1)$

$$\begin{aligned}&\frac{x}{x-2}+\frac{3}{2x+1}\\&=\frac{x}{x-2}\cdot\frac{2x+1}{2x+1}+\frac{3}{2x+1}\cdot\frac{x-2}{x-2}\\&=\frac{2x^2+x}{(x-2)(2x+1)}+\frac{3x-6}{(2x+1)(x-2)}\\&=\frac{2x^2+x+3x-6}{(x-2)(2x+1)}\\&=\frac{2x^2+4x-6}{(x-2)(2x+1)}\\&=\frac{2(x+3)(x-1)}{(x-2)(2x+1)}\end{aligned}$$

11. $x^2+5x+6=(x+2)(x+3)$

$x^2+2x-3=(x-1)(x+3)$

LCD = $(x+2)(x+3)(x-1)$

$$\begin{aligned}&\frac{2x}{x^2+5x+6}-\frac{x+1}{x^2+2x-3}\\&=\frac{2x}{(x+2)(x+3)}\cdot\frac{x-1}{x-1}-\frac{x+1}{(x-1)(x+3)}\cdot\frac{x+2}{x+2}\\&=\frac{2x^2-2x}{(x+2)(x+3)(x-1)}-\frac{x^2+3x+2}{(x+2)(x+3)(x-1)}\\&=\frac{2x^2-2x-(x^2+3x+2)}{(x+2)(x+3)(x-1)}\\&=\frac{2x^2-2x-x^2-3x-2}{(x+2)(x+3)(x-1)}\\&=\frac{x^2-5x-2}{(x+2)(x+3)(x-1)}\end{aligned}$$

12. $$\begin{aligned}\frac{\frac{1}{9}-\frac{1}{y^2}}{\frac{1}{3}+\frac{1}{y}}&=\frac{9y^2\left(\frac{1}{9}-\frac{1}{y^2}\right)}{9y^2\left(\frac{1}{3}+\frac{1}{y}\right)}\\&=\frac{y^2-9}{3y^2+9y}\\&=\frac{(y+3)(y-3)}{3y(y+3)}\\&=\frac{y-3}{3y}\end{aligned}$$

13. Undefined value: $m = 0$.

$$\begin{aligned}\frac{m}{5}+\frac{5}{m}&=\frac{m+3}{4}\\\frac{m}{5}\cdot\frac{m}{m}+\frac{5}{m}\cdot\frac{5}{5}&=\frac{m+3}{4}\\\frac{m^2}{5m}+\frac{25}{5m}&=\frac{m+3}{4}\\\frac{m^2+25}{5m}&=\frac{m+3}{4}\\4(m^2+25)&=5m(m+3)\\4m^2+100&=5m^2+15m\\0&=m+15m-100\\0&=(m+20)(m-5)\end{aligned}$$

$$\begin{aligned}m+20=0\quad&\text{or}\quad m-5=0\\m=-20\quad&\text{or}\quad m=5\end{aligned}$$

Neither value is extraneous. The solution set is $\{-20, 5\}$.

14. Undefined values: $x = -3$ and $x = 6$.

$$\begin{aligned}\frac{4}{x+3}+\frac{5}{x-6}&=\frac{4x+1}{x^2-3x-18}\\\frac{4}{x+3}\cdot\frac{x-6}{x-6}+\frac{5}{x-6}\cdot\frac{x+3}{x+3}&=\frac{4x+1}{(x-6)(x+3)}\\\frac{4x-24}{(x+3)(x-6)}+\frac{5x+15}{(x-6)(x+3)}&=\frac{4x+1}{(x-6)(x+3)}\\\frac{4x-24+5x+15}{(x-6)(x+3)}&=\frac{4x+1}{(x-6)(x+3)}\\\frac{9x-9}{(x-6)(x+3)}&=\frac{4x+1}{(x-6)(x+3)}\\9x-9&=4x+1\\5x&=10\\x&=2\end{aligned}$$

The value is not extraneous. The solution set is $\{2\}$.

15. $$\frac{1}{x}+\frac{1}{y}=\frac{1}{z}$$
$$xyz\left(\frac{1}{x}+\frac{1}{y}\right)=xyz\left(\frac{1}{z}\right)$$
$$yz+xz=xy$$
$$yz-xy=-xz$$
$$y(z-x)=-xz$$
$$y=\frac{-xz}{z-x}$$
$$y=\frac{xz}{x-z}$$

16. Undefined values: $y=-1$ and $y=2$.
$$\frac{2}{y+1}=\frac{1}{y-2}$$
$$2(y-2)=1(y+1)$$
$$2y-4=y+1$$
$$2y-y=1+4$$
$$y=5$$
The solution set is {5}.

17. $$\frac{5}{12}=\frac{2}{x}$$
$$5x=2(12)$$
$$5x=24$$
$$x=\frac{24}{5}$$

18. Let x be the cost of 12 hair barrettes.
$$\frac{8}{22}=\frac{12}{x}$$
$$8x=22(12)$$
$$8x=264$$
$$x=33$$
The cost is $33.

19. Let j be the time for Juan to wash the car alone. Then j + 18 is the time for Frank to wash the car alone. Juan can wash $\frac{1}{j}$ of the car in one minute, whereas Frank can wash $\frac{1}{j+18}$ of the car in one minute. Together, it takes them 12 minutes.
$$\frac{1}{j}+\frac{1}{j+18}=\frac{1}{12}$$
$$12j(j+18)\left(\frac{1}{j}+\frac{1}{j+18}\right)=12j(j+18)\frac{1}{12}$$
$$12(j+18)+12j=j(j+18)$$
$$12j+216+12j=j^2+18j$$
$$0=j^2-6j-216$$
$$0=(j-18)(j+12)$$
$$j-18=0 \quad \text{or} \quad j+12=0$$
$$j=18 \quad \text{or} \quad j=-12$$
Disregard the negative time. It takes Frank j + 18 = 18 + 18 = 36 minutes to wash the car alone.

20. Let r be the rate at which they hiked.

	d	r	$t=\frac{d}{r}$
nature path	7	$r+3$	$t=\frac{7}{r+3}$
mountainside	12	r	$t=\frac{12}{r}$

The total time was 4 hours.
$$\frac{7}{r+3}+\frac{12}{r}=4$$
$$r(r+3)\left(\frac{7}{r+3}+\frac{12}{r}\right)=r(r+3)\cdot 4$$
$$7r+12(r+3)=4r(r+3)$$
$$7r+12r+36=4r^2+12r$$
$$19r+36=4r^2+12r$$
$$0=4r^2-7r-36$$
$$0=(4r+9)(r-4)$$
$$4r+9=0 \quad \text{or} \quad r-4=0$$
$$r=-\frac{9}{4} \quad \text{or} \quad r=4$$
Disregard the negative rate. The tourists hiked at 4 mph.

21. $$m=kn$$
$$12=k(8)$$
$$\frac{12}{8}=k$$
$$k=\frac{3}{2}$$
$$m=\frac{3}{2}n$$

Let $n = 20$.

$$m = \frac{3}{2} \cdot 20$$
$$m = 30$$

Cumulative Review Chapters 1–7

1.
$$\begin{aligned} -6^2 + 4(-5+2)^3 &= -36 + 4(-3)^3 \\ &= -36 + 4(-27) \\ &= -36 + (-108) \\ &= -144 \end{aligned}$$

2.
$$\begin{aligned} 3(4x-2) - (3x+5) &= 3 \cdot 4x - 3 \cdot 2 - 3x - 5 \\ &= 12x - 6 - 3x - 5 \\ &= 12x - 3x - 6 - 5 \\ &= 9x - 11 \end{aligned}$$

3.
$$\begin{aligned} -3(x-5) + 2x &= 5x - 4 \\ -3x + 15 + 2x &= 5x - 4 \\ -x + 15 &= 5x - 4 \\ x - x + 15 &= x + 5x - 4 \\ 15 &= 6x - 4 \\ 15 + 4 &= 6x - 4 + 4 \\ 19 &= 6x \\ \frac{19}{6} &= \frac{6x}{6} \\ \frac{19}{6} &= x \end{aligned}$$

The solution set is $\left\{\frac{19}{6}\right\}$.

4.
$$\begin{aligned} 3(2x-1) + 5 &= 6x + 2 \\ 6x - 3 + 5 &= 6x + 2 \\ 6x + 2 &= 6x + 2 \\ -6x + 6x + 2 &= -6x + 6x + 2 \\ 2 &= 2 \end{aligned}$$

This is a true statement. It is an identity. The solution set is the set of all real numbers.

5.
$$\begin{aligned} 0.25x + 0.10(x-3) &= 0.05(22) \\ 100[0.25x + 0.10(x-3)] &= 100[0.05(22)] \\ 25x + 10(x-3) &= 5(22) \\ 25x + 10x - 30 &= 110 \\ 35x - 30 &= 110 \\ 35x - 30 + 30 &= 110 + 30 \\ 35x &= 140 \\ \frac{35x}{35} &= \frac{140}{35} \\ x &= 4 \end{aligned}$$

The solution set is $\{4\}$.

6. Let w be the width of the poster.
Then $2w - 3$ is the length of the poster.
Perimeter = 2(length + width)

$$\begin{aligned} 24 &= 2(2w - 3 + w) \\ 24 &= 2(3w - 3) \\ 24 &= 6w - 6 \\ 30 &= 6w \\ 5 &= w \end{aligned}$$

The length of the poster is
$2w - 3 = 2(5) - 3 = 10 - 3 = 7$ feet.

7. Let n be the first even integer. Then $n + 2$ and $n + 4$ are the second and third consecutive even integers, respectively. Their sum is 138.

$$\begin{aligned} n + (n+2) + (n+4) &= 138 \\ n + n + 2 + n + 4 &= 138 \\ 3n + 6 &= 138 \\ 3n &= 132 \\ n &= 44 \\ n + 2 &= 46 \\ n + 4 &= 48 \end{aligned}$$

The three integers are 44, 46, and 48.

8.
$$\begin{aligned} 2(x-3) - 5 &\le 3(x+2) - 18 \\ 2x - 6 - 5 &\le 3x + 6 - 18 \\ 2x - 11 &\le 3x - 12 \\ -3x + 2x - 11 &\le -3x + 3x - 12 \\ -x - 11 &\le -12 \\ -x - 11 + 11 &\le -12 + 11 \\ -x &\le -1 \\ x &\ge 1 \end{aligned}$$

0 1

9.
$$\begin{aligned} (3x - 2y)^2 &= (3x)^2 - 2(3x)(2y) + (2y)^2 \\ &= 9x^2 - 12xy + 4y^2 \end{aligned}$$

10.
$$\begin{aligned} \left(\frac{2a^5b}{4ab^{-2}}\right)^{-4} &= \frac{2^{-4}a^{-20}b^{-4}}{4^{-4}a^{-4}b^{8}} \\ &= \frac{4^4}{2^4} \cdot a^{-20-(-4)}b^{-4-8} \\ &= \frac{256}{16}a^{-16}b^{-12} \\ &= \frac{16}{a^{16}b^{12}} \end{aligned}$$

11. $(3x^0y^{-4}z^3)^3 = (3\cdot 1\cdot y^{-4}\cdot z^3)^3$
$$= \left(\frac{3z^3}{y^4}\right)^3 = \frac{3^3 z^{3\cdot 3}}{y^{4\cdot 3}} = \frac{27z^9}{y^{12}}$$

12. $8a^2b+34ab-84b = 2b(4a^2+17a-42) = 2b(4a-7)(a+6)$

13.
$$6x^3-31x^2=-5x$$
$$6x^3-31x^2+5x=0$$
$$x(6x^2-31x+5)=0$$
$$x(6x-1)(x-5)=0$$
$$x=0 \quad\text{or}\quad 6x-1=0 \quad\text{or}\quad x-5=0$$
$$x=\frac{1}{6} \quad\text{or}\quad x=5$$

The solution set is $\left\{0, \frac{1}{6}, 5\right\}$.

14. Let w be the width of the rectangle. Then $2w + 2$ is the length. The area is 40 square centimeters.
Area = (length)(width)
$$40=(2w+2)(w)$$
$$40=2w^2+2w$$
$$0=2w^2+2w-40$$
$$0=2(w^2+w-20)$$
$$0=2(w+5)(w-4)$$
$$2=0 \quad\text{or}\quad w+5=0 \quad\text{or}\quad w-4=0$$
$$\text{false} \quad\text{or}\quad w=-5 \quad\text{or}\quad w=4$$

Disregard a negative width. The length is $2w + 2 = 2(4) + 2 = 8 + 2 = 10$ centimeters.

15. $\frac{x+5}{x^2+25}$; $x=-5$
$$\frac{-5+5}{(-5)^2+25}=\frac{0}{25+25}=\frac{0}{50}=0$$

16.
$$\frac{3x+3}{5x-5x^2}\cdot\frac{2x^2+x-3}{4x^2-9} = \frac{3(x+1)}{-5x(x-1)}\cdot\frac{(2x+3)(x-1)}{(2x-3)(2x+3)} = -\frac{3(x+1)}{5x(2x-3)}$$

17.
$$\frac{x^2-x-2}{10}\div\frac{2x+4}{5} = \frac{x^2-x-2}{10}\cdot\frac{5}{2x+4} = \frac{(x-2)(x+1)}{5\cdot 2}\cdot\frac{5}{2(x+2)} = \frac{5\cdot(x-2)(x+1)}{5\cdot 4(x+2)} = \frac{(x-2)(x+1)}{4(x+2)}$$

18.
$$\frac{2x+3}{x^2-x-30}-\frac{x-2}{x^2-x-30}=\frac{2x+3-(x-2)}{x^2-x-30} = \frac{2x+3-x+2}{x^2-x-30} = \frac{x+5}{(x-6)(x+5)} = \frac{1}{x-6}$$

19.
$$\frac{15}{2x-4}+\frac{x}{x^2-4} = \frac{15}{2(x-2)}+\frac{x}{(x+2)(x-2)} = \frac{15}{2(x-2)}\cdot\frac{x+2}{x+2}+\frac{x}{(x+2)(x-2)}\cdot\frac{2}{2}$$
$$= \frac{15x+30}{2(x-2)(x+2)}+\frac{2x}{2(x-2)(x+2)} = \frac{15x+30+2x}{2(x-2)(x+2)} = \frac{17x+30}{2(x-2)(x+2)}$$

20.
$$\frac{\frac{2}{x^2}-\frac{3}{5x}}{\frac{4}{x}+\frac{1}{4x}} = \frac{20x^2\left(\frac{2}{x^2}-\frac{3}{5x}\right)}{20x^2\left(\frac{4}{x}+\frac{1}{4x}\right)} = \frac{40-12x}{80x+5x} = \frac{4(10-3x)}{85x}$$

21. LCD = $(x-4)(x+4)$
Undefined values: $x = 4$ and $x = -4$.

$$\frac{3}{x-4} = \frac{5x+4}{x^2-16} - \frac{4}{x+4}$$
$$\frac{3}{x-4} = \frac{5x+4}{(x+4)(x-4)} - \frac{4}{x+4} \cdot \frac{x-4}{x-4}$$
$$\frac{3}{x-4} = \frac{5x+4}{(x+4)(x-4)} - \frac{4x-16}{(x+4)(x-4)}$$
$$\frac{3}{x-4} = \frac{5x+4-4x+16}{(x+4)(x-4)}$$
$$\frac{3}{x-4} = \frac{x+20}{(x+4)(x-4)}$$
$$3(x+4)(x-4) = (x-4)(x+20)$$
$$3x^2 - 48 = x^2 + 16x - 80$$
$$2x^2 - 16x + 32 = 0$$
$$2(x-4)(x-4) = 0$$
$$2 = 0 \quad \text{or} \quad x-4 = 0$$
$$\text{false} \quad \text{or} \quad x = 4$$

Since $x = 4$ is undefined, there is no solution. The solution set is { } or ∅.

22. $$\frac{x-5}{3} = \frac{x+2}{2}$$
$$2(x-5) = 3(x+2)$$
$$2x - 10 = 3x + 6$$
$$-10 = x + 6$$
$$-16 = x$$

The solution set is {−16}.

23. Let x represent the number of inches for 125 miles.

$$\frac{4}{50} = \frac{x}{125}$$
$$4(125) = 50x$$
$$500 = 50x$$
$$10 = x$$

There are 10 inches for 125 miles.

24. Let s be the time for Sharona working alone. Then $s + 9$ is the time for Trent working alone.

Sharona can do $\frac{1}{s}$ of the job in an hour, whereas Trent can do $\frac{1}{s+9}$ of the job in an hour. It takes then 6 hours working together.

$$\frac{1}{s} + \frac{1}{s+9} = \frac{1}{6}$$
$$6s(s+9)\left(\frac{1}{s} + \frac{1}{s+9}\right) = 6s(s+9)\left(\frac{1}{6}\right)$$
$$6(s+9) + 6s = s(s+9)$$
$$6s + 54 + 6s = s^2 + 9s$$
$$0 = s^2 - 3s - 54$$
$$0 = (s+6)(s-9)$$
$$s + 6 = 0 \quad \text{or} \quad s - 9 = 0$$
$$s = -6 \quad \text{or} \quad s = 9$$

Disregard the negative time. It takes Sharona 9 hours working alone.

25. Let r be the rate he walks.

	d	r	$t = \frac{d}{r}$
jogged	35	$r + 4$	$\frac{35}{r+4}$
walked	6	r	$\frac{6}{r}$

It took him 7 hours total.

$$\frac{35}{r+4} + \frac{6}{r} = 7$$
$$r(r+4)\left(\frac{35}{r+4} + \frac{6}{r}\right) = r(r+4) \cdot 7$$
$$35r + 6(r+4) = 7r(r+4)$$
$$35r + 6r + 24 = 7r^2 + 28r$$
$$0 = 7r^2 - 13r - 24$$
$$0 = (7r+8)(r-3)$$
$$7r + 8 = 0 \quad \text{or} \quad r - 3 = 0$$
$$r = -\frac{8}{7} \quad \text{or} \quad r = 3$$

Disregard the negative rate. Francisco jogs at $r + 4 = 3 + 4 = 7$ mph.

26. $(x_1, y_1) = (-1, 3)$; $(x_2, y_2) = (5, 11)$

$$m = \frac{y_2 - y_1}{x_2 - x_1} = \frac{11-3}{5-(-1)} = \frac{8}{6} = \frac{4}{3}$$

27. $m = -\frac{3}{2}$; $(x_1, y_1) = (4, -1)$

$$y - y_1 = m(x - x_1)$$
$$y - (-1) = -\frac{3}{2}(x - 4)$$
$$y + 1 = -\frac{3}{2}x + 6$$
$$y = -\frac{3}{2}x + 5 \text{ or } 3x + 2y = 10$$

28. $2x - 3y = -12$

y-intercept: Let $x = 0$.

$$2(0) - 3y = -12$$
$$-3y = -12$$
$$y = 4$$

(0, 4)

x-intercept: Let $y = 0$.

$$2x - 3(0) = -12$$
$$2x = -12$$
$$x = -6$$

(−6, 0)

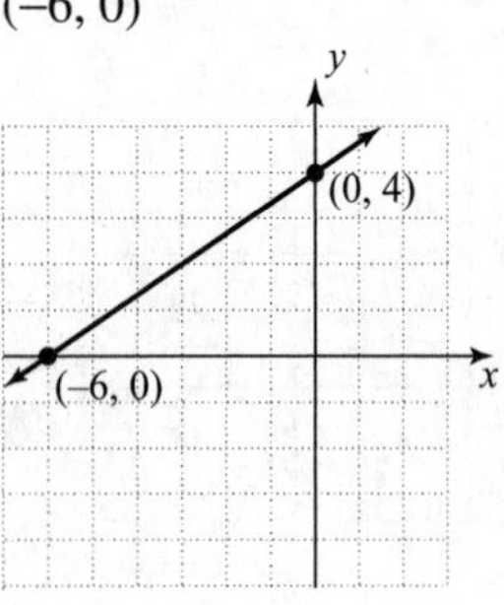

29. $3x + 5y = 15$

$$5y = -3x + 15$$
$$y = -\frac{3}{5}x + 3$$

$$m_{\text{parallel}} = -\frac{3}{5}$$

$$m_{\text{perpendicular}} = \frac{5}{3}$$

30. $\begin{cases} 2x + 3y = 1 & (1) \\ -3x + 2y = 18 & (2) \end{cases}$

Multiply equation (1) by 3 and equation (2) by 2.

$$\begin{cases} 6x + 9y = 3 \\ -6x + 4y = 36 \end{cases}$$
$$13y = 39$$
$$y = 3$$

Let $y = 3$ in equation (1).

$$2x + 3(3) = 1$$
$$2x + 9 = 1$$
$$2x = -8$$
$$x = -4$$

The solution is (−4, 3).

Chapter 8

Section 8.1

Preparing for Introduction to Square Roots

P1. –4, 0, and 13 are integers.

P2. –4, $\frac{5}{3}$, 0, 6.95, and 13 are rational numbers.

P3. $\sqrt{2}$ and π are irrational numbers.

P4. **(a)** $\left(\frac{3}{2}\right)^2 = \frac{3}{2}\cdot\frac{3}{2} = \frac{9}{4}$

(b) $(0.4)^2 = 0.4 \cdot 0.4 = 0.16$

8.1 Quick Checks

1. For any real numbers a and b, b is a square root of a if $\underline{b^2} = \underline{a}$.

2. 8 is a square root of 64 because $8^2 = 64$.
   –8 is a square root of 64 because $(-8)^2 = 64$.

3. $\frac{5}{7}$ is a square root of $\frac{25}{49}$ because $\left(\frac{5}{7}\right)^2 = \frac{25}{49}$.
   $-\frac{5}{7}$ is a square root of $\frac{25}{49}$ because $\left(-\frac{5}{7}\right)^2 = \frac{25}{49}$.

4. 0.6 is a square root of 0.36 because $0.6^2 = 0.36$.
   –0.6 is a square root of 0.36 because $(-0.6)^2 = 0.36$.

5. The number represented by $\sqrt{a}$ is called the <u>principal</u> <u>square</u> <u>root</u> of a.

6. In the expression $\sqrt{4x}$, the $4x$ is called the <u>radicand</u>.

7. $\sqrt{100} = 10$ because $10^2 = 100$.

8. $-\sqrt{9} = -3$ because $3^2 = 9$.

9. $\sqrt{\frac{25}{49}} = \frac{5}{7}$ because $\left(\frac{5}{7}\right)^2 = \frac{25}{49}$.

10. $\sqrt{0.36} = 0.6$ because $0.6^2 = 0.36$.

11. $2\sqrt{36} = 2 \cdot 6 = 12$

12. $\sqrt{25+144} = \sqrt{169} = 13$

13. $\sqrt{25} + \sqrt{144} = 5 + 12 = 17$

14. $\sqrt{25 - 4 \cdot 3 \cdot (-2)} = \sqrt{25+24} = \sqrt{49} = 7$

15. $\sqrt{25} = 5$ and $\sqrt{36} = 6$, so $\sqrt{35}$ is between 5 and 6.
    $\sqrt{35} \approx 5.92$

16. $-\sqrt{6} \approx -2.45$

17. True

18. $\sqrt{49}$ is a rational number because $7^2 = 49$, so $\sqrt{49} = 7$.

19. $\sqrt{71}$ is irrational because 71 is not a perfect square.
    $\sqrt{71} \approx 8.43$

20. $\sqrt{-25}$ is not a real number since there is no real number whose square is negative.

21. $-\sqrt{16}$ is a rational number because $4^2 = 16$, so $-\sqrt{16} = -4$.

22. For any real number a, $\sqrt{a^2} = \underline{|a|}$.

23. $\sqrt{b^2} = |b|$ for any real number b.

24. $\sqrt{(4p)^2} = |4p| = 4|p|$ for any real number p.

25. $\sqrt{(w+3)^2} = |w+3|$ for any real number w.

26. $\sqrt{(5b)^2} = 5b$ because $b \geq 0$.

27. $\sqrt{(z-6)^2} = z-6$ because $z - 6 \geq 0$.

28. $\sqrt{(2h-5)^2} = 2h-5$ because $2h - 5 \geq 0$.

8.1 Exercises

29. The square roots of 1 are the numbers that can be squared to get 1. Since $1^2 = 1$ and $(-1)^2 = 1$, the square roots of 1 are 1 and -1.

31. The square roots of $\frac{1}{9}$ are $\frac{1}{3}$ and $-\frac{1}{3}$ since $\left(\frac{1}{3}\right)^2 = \frac{1}{9}$ and $\left(-\frac{1}{3}\right)^2 = \frac{1}{9}$.

33. The square roots of 36 are 6 and -6 since $6^2 = 36$ and $(-6)^2 = 36$.

35. The square roots of 0.25 are 0.5 and -0.5 since $(0.5)^2 = 0.25$ and $(-0.5)^2 = 0.25$.

37. $\sqrt{144} = 12$ since $12^2 = 144$.

39. $-\sqrt{9} = -1 \cdot \sqrt{9} = -1 \cdot 3 = -3$

41. $\sqrt{225} = 15$ since $15^2 = 225$.

43. $\sqrt{\frac{1}{121}} = \frac{1}{11}$ since $\left(\frac{1}{11}\right)^2 = \frac{1}{121}$.

45. $\sqrt{0.04} = 0.2$ since $(0.2)^2 = 0.04$.

47. $-6\sqrt{9} = -6 \cdot 3 = -18$

49. $\sqrt{\frac{16}{81}} = \frac{4}{9}$ since $\left(\frac{4}{9}\right)^2 = \frac{16}{81}$.

51. $5\sqrt{49} = 5 \cdot 7 = 35$

53. $\sqrt{36+64} = \sqrt{100} = 10$

55. $\sqrt{36} + \sqrt{64} = 6 + 8 = 14$

57. $\sqrt{4-(4)(1)(-15)} = \sqrt{4+60} = \sqrt{64} = 8$

59. $\sqrt{49-4(15)(-2)} = \sqrt{49+120} = \sqrt{169} = 13$

61. $\sqrt{8} \approx 2.828$

63. $\sqrt{30} \approx 5.48$

65. $\sqrt{57} \approx 7.5$

67. $\sqrt{-4}$ is not a real number since the square of any real number is nonnegative.

69. $\sqrt{400} = 20$ is rational.

71. $\sqrt{\frac{1}{4}} = \frac{1}{2}$ is rational.

73. $\sqrt{54} \approx 7.35$ is irrational.

75. $\sqrt{50} \approx 7.07$ is irrational.

77. $\sqrt{d^2} = d$ since $d \geq 0$.

79. $\sqrt{(x-9)^2} = |x-9|$

81. $\sqrt{(2m-n)^2} = 2m-n$ since $2m - n \geq 0$.

83. $\sqrt{3} \approx 1.73$

85. $\sqrt{0.4} \approx 0.63$

87. $\sqrt{-2}$ is not a real number.

89. $3\sqrt{4} = 3 \cdot 2 = 6$

91. $3\sqrt{\frac{25}{9}} - \sqrt{169} = 3 \cdot \frac{5}{3} - 13 = 5 - 13 = -8$

93. $\sqrt{3-19} = \sqrt{-16}$ is not a real number.

95. $\sqrt{7^2 - (4)(-2)(-3)} = \sqrt{49-24} = \sqrt{25} = 5$

97.
$$\begin{aligned}\sqrt{(11-8)^2 + (11-5)^2} &= \sqrt{3^2 + 6^2} \\ &= \sqrt{9+36} \\ &= \sqrt{45} \\ &\approx 6.71\end{aligned}$$

99. $s = \sqrt{A} = \sqrt{625} = 25$
The length of the side of the square is 25 feet.

101. $s = \sqrt{A} = \sqrt{256} = 16$
The length of the side of the square is 16 kilometers.

103. $r = \sqrt{\frac{A}{\pi}} = \sqrt{\frac{49\pi}{\pi}} = \sqrt{49} = 7$
The radius is 7 meters.

105. $r = \sqrt{\frac{A}{\pi}} = \sqrt{\frac{196\pi}{\pi}} = \sqrt{196} = 14$
The radius is 14 inches.

107. $s = \sqrt{\frac{3V}{h}} = \sqrt{\frac{3(7,700,000)}{146}} \approx 398$
The length of the side of the pyramid is approximately 398 meters.

109. $\sqrt{\sqrt{16}} = \sqrt{4} = 2$

111. $\sqrt{5 \cdot \sqrt{25}} = \sqrt{5 \cdot 5} = \sqrt{25} = 5$

113. The number $-\sqrt{9}$ is a real number because it denotes the opposite, or negative, of the principal square root of 9, namely 3. The number $\sqrt{-9}$ is not a real number because there is no real number whose square is -9.

Section 8.2

Preparing for Simplifying Square Roots

P1. $2^1 = 2, 2^2 = 4, 2^3 = 8, 2^4 = 16, 2^5 = 32,$
$2^6 = 64$

P2. $3^1 = 3, 3^2 = 9, 3^3 = 27, 3^4 = 81$

P3. $\sqrt{49} = 7$ since $7^2 = 49$.

8.2 Quick Checks

1. If $\sqrt{a}$ and $\sqrt{b}$ are real numbers, then $\sqrt{ab} = \underline{\sqrt{a} \cdot \sqrt{b}}.$

2. To simplify $\sqrt{125}$ write $\sqrt{\underline{25} \cdot 5}$.

3. $\sqrt{20} = \sqrt{4 \cdot 5} = \sqrt{4} \cdot \sqrt{5} = 2\sqrt{5}$

4. $\sqrt{24} = \sqrt{4 \cdot 6} = \sqrt{4} \cdot \sqrt{6} = 2\sqrt{6}$

5. $\sqrt{72} = \sqrt{36 \cdot 2} = \sqrt{36} \cdot \sqrt{2} = 6\sqrt{2}$

6.
$$\begin{aligned}\frac{6+\sqrt{45}}{3} &= \frac{6+\sqrt{9 \cdot 5}}{3} \\ &= \frac{6+\sqrt{9} \cdot \sqrt{5}}{3} \\ &= \frac{6+3\sqrt{5}}{3} \\ &= \frac{3\left(2+\sqrt{5}\right)}{3} \\ &= 2+\sqrt{5}\end{aligned}$$

7.
$$\begin{aligned}\frac{-2+\sqrt{32}}{4} &= \frac{-2+\sqrt{16 \cdot 2}}{4} \\ &= \frac{-2+\sqrt{16} \cdot \sqrt{2}}{4} \\ &= \frac{-2+4\sqrt{2}}{4} \\ &= \frac{2\left(-1+2\sqrt{2}\right)}{2 \cdot 2} \\ &= \frac{-1+2\sqrt{2}}{2}\end{aligned}$$

8. False; write the radicand as a product of factors, one of which is a perfect square.

9. $\sqrt{36y^8} = \sqrt{36} \cdot \sqrt{y^8} = \sqrt{36} \cdot \sqrt{(y^4)^2} = 6y^4$

10.
$$\begin{aligned}\sqrt{45x^{12}} &= \sqrt{9 \cdot 5 \cdot x^{12}} \\ &= \sqrt{9x^{12} \cdot 5} \\ &= \sqrt{9x^{12}} \cdot \sqrt{5} \\ &= 3x^6\sqrt{5}\end{aligned}$$

11.
$$\begin{aligned}\sqrt{28a^2b^{10}} &= \sqrt{7 \cdot 4 \cdot a^2b^{10}} \\ &= \sqrt{4a^2b^{10} \cdot 7} \\ &= \sqrt{4a^2b^{10}} \cdot \sqrt{7} \\ &= 2ab^5\sqrt{7}\end{aligned}$$

12.
$$\begin{aligned}\sqrt{25y^{11}} &= \sqrt{25y^{10} \cdot y} \\ &= \sqrt{25y^{10}} \cdot \sqrt{y} \\ &= 5y^5\sqrt{y}\end{aligned}$$

13. $\sqrt{121a^5} = \sqrt{121a^4 \cdot a}$
$= \sqrt{121a^4} \cdot \sqrt{a}$
$= 11a^2\sqrt{a}$

14. $\sqrt{32ab^9} = \sqrt{16b^8 \cdot 2ab}$
$= \sqrt{16b^8} \cdot \sqrt{2ab}$
$= 4b^4\sqrt{2ab}$

15. $\sqrt{12x^7y^{13}} = \sqrt{4x^6y^{12} \cdot 3xy}$
$= \sqrt{4x^6y^{12}} \cdot \sqrt{3xy}$
$= 2x^3y^6\sqrt{3xy}$

16. If $\sqrt{a}$ and $\sqrt{b}$ are nonnegative real numbers and $b \neq 0$, then $\sqrt{\frac{a}{b}} = \frac{\sqrt{a}}{\sqrt{b}}$.

17. $\sqrt{\frac{49a^3}{100}} = \frac{\sqrt{49a^3}}{\sqrt{100}}$
$= \frac{\sqrt{49a^2 \cdot a}}{\sqrt{100}}$
$= \frac{\sqrt{49a^2} \cdot \sqrt{a}}{\sqrt{100}}$
$= \frac{7a\sqrt{a}}{10}$

18. $\sqrt{\frac{11}{16m^2}} = \frac{\sqrt{11}}{\sqrt{16m^2}} = \frac{\sqrt{11}}{4m}$

19. $\sqrt{\frac{125}{5}} = \sqrt{25} = 5$

20. $\sqrt{\frac{144b^5}{16}} = \sqrt{9b^5}$
$= \sqrt{9b^4 \cdot b}$
$= \sqrt{9b^4} \cdot \sqrt{b}$
$= 3b^2\sqrt{b}$

8.2 Exercises

21. $\sqrt{8} = \sqrt{4 \cdot 2} = \sqrt{4} \cdot \sqrt{2} = 2\sqrt{2}$

23. $\sqrt{40} = \sqrt{4 \cdot 10} = \sqrt{4} \cdot \sqrt{10} = 2\sqrt{10}$

25. $\sqrt{18} = \sqrt{9 \cdot 2} = \sqrt{9} \cdot \sqrt{2} = 3\sqrt{2}$

27. $\sqrt{33}$ cannot be simplified further.

29. $\sqrt{45} = \sqrt{9 \cdot 5} = \sqrt{9} \cdot \sqrt{5} = 3\sqrt{5}$

31. $\sqrt{125} = \sqrt{25 \cdot 5} = \sqrt{25} \cdot \sqrt{5} = 5\sqrt{5}$

33. $\sqrt{42}$ cannot be simplified further.

35. $\sqrt{98} = \sqrt{49 \cdot 2} = \sqrt{49} \cdot \sqrt{2} = 7\sqrt{2}$

37. $\sqrt{270} = \sqrt{9 \cdot 30} = \sqrt{9} \cdot \sqrt{30} = 3\sqrt{30}$

39. $\frac{4+\sqrt{36}}{2} = \frac{4+6}{2} = \frac{10}{2} = 5$

41. $\frac{9+\sqrt{18}}{3} = \frac{9+\sqrt{9 \cdot 2}}{3}$
$= \frac{9+\sqrt{9} \cdot \sqrt{2}}{3}$
$= \frac{9+3\sqrt{2}}{3}$
$= \frac{3(3+\sqrt{2})}{3}$
$= 3+\sqrt{2}$

43. $\sqrt{x^{10}} = \sqrt{(x^5)^2} = x^5$

45. $\sqrt{n^{144}} = \sqrt{(n^{72})^2} = n^{72}$

47. $\sqrt{9y^4} = \sqrt{9} \cdot \sqrt{y^4} = 3y^2$

49. $\sqrt{24z^{14}} = \sqrt{4z^{14} \cdot 6} = \sqrt{4z^{14}} \cdot \sqrt{6} = 2z^7\sqrt{6}$

51. $\sqrt{49x^7} = \sqrt{49x^6 \cdot x} = \sqrt{49x^6} \cdot \sqrt{x} = 7x^3\sqrt{x}$

53. $\sqrt{48b^5} = \sqrt{16b^4 \cdot 3b}$
$= \sqrt{16b^4} \cdot \sqrt{3b}$
$= 4b^2\sqrt{3b}$

55. $\sqrt{18m^3n^4} = \sqrt{9m^2n^4 \cdot 2m}$
$= \sqrt{9m^2n^4} \cdot \sqrt{2m}$
$= 3mn^2\sqrt{2m}$

57. $\sqrt{75c^4d^2} = \sqrt{25c^4d^2 \cdot 3}$
$= \sqrt{25c^4d^2} \cdot \sqrt{3}$
$= 5c^2d\sqrt{3}$

59. $\sqrt{\frac{25}{36}} = \frac{\sqrt{25}}{\sqrt{36}} = \frac{5}{6}$

61. $\sqrt{\frac{11}{9}} = \frac{\sqrt{11}}{\sqrt{9}} = \frac{\sqrt{11}}{3}$

63. $\sqrt{\frac{y^{16}}{121}} = \frac{\sqrt{y^{16}}}{\sqrt{121}} = \frac{y^8}{11}$

65. $\sqrt{\frac{3}{a^4}} = \frac{\sqrt{3}}{\sqrt{a^4}} = \frac{\sqrt{3}}{a^2}$

67. $\sqrt{\frac{x^5}{y^8}} = \frac{\sqrt{x^5}}{\sqrt{y^8}} = \frac{\sqrt{x^4 \cdot x}}{\sqrt{y^8}} = \frac{\sqrt{x^4} \cdot \sqrt{x}}{\sqrt{y^8}} = \frac{x^2\sqrt{x}}{y^4}$

69. $\sqrt{\frac{20x^2y^5}{z^4}} = \frac{\sqrt{20x^2y^5}}{\sqrt{z^4}}$
$= \frac{\sqrt{4x^2y^4 \cdot 5y}}{\sqrt{z^4}}$
$= \frac{\sqrt{4x^2y^4} \cdot \sqrt{5y}}{\sqrt{z^4}}$
$= \frac{2xy^2\sqrt{5y}}{z^2}$

71. $\sqrt{\frac{105x^2}{7x^6}} = \sqrt{\frac{7x^2 \cdot 15}{7x^2 \cdot x^4}} = \sqrt{\frac{15}{x^4}} = \frac{\sqrt{15}}{\sqrt{x^4}} = \frac{\sqrt{15}}{x^2}$

73. $\sqrt{\frac{100a^3}{16a}} = \sqrt{\frac{100a^2}{16}} = \frac{\sqrt{100a^2}}{\sqrt{16}} = \frac{10a}{4} = \frac{5a}{2}$

75. $\sqrt{\frac{64n^9}{9n^4}} = \frac{\sqrt{64n^9}}{\sqrt{9n^4}}$
$= \frac{\sqrt{64n^8 \cdot n}}{\sqrt{9n^4}}$
$= \frac{\sqrt{64n^8} \cdot \sqrt{n}}{\sqrt{9n^4}}$
$= \frac{8n^4\sqrt{n}}{3n^2}$
$= \frac{8n^2\sqrt{n}}{3}$

77. $\sqrt{63} = \sqrt{9 \cdot 7} = \sqrt{9} \cdot \sqrt{7} = 3\sqrt{7}$

79. $\sqrt{\frac{12}{25}} = \frac{\sqrt{12}}{\sqrt{25}} = \frac{\sqrt{4 \cdot 3}}{\sqrt{25}} = \frac{\sqrt{4} \cdot \sqrt{3}}{\sqrt{25}} = \frac{2\sqrt{3}}{5}$

81. $\sqrt{144a^6} = \sqrt{144} \cdot \sqrt{a^6} = 12a^3$

83. $\sqrt{\frac{10x^5}{90}} = \sqrt{\frac{10 \cdot x^5}{10 \cdot 9}}$
$= \sqrt{\frac{x^5}{9}}$
$= \frac{\sqrt{x^5}}{\sqrt{9}}$
$= \frac{\sqrt{x^4 \cdot x}}{\sqrt{9}}$
$= \frac{\sqrt{x^4} \cdot \sqrt{x}}{\sqrt{9}}$
$= \frac{x^2\sqrt{x}}{3}$

85. $\sqrt{12a^4b^2c^6} = \sqrt{4a^4b^2c^6 \cdot 3}$
$= \sqrt{4a^4b^2c^6} \cdot \sqrt{3}$
$= 2a^2bc^3\sqrt{3}$

87. $\sqrt{15pq^5r^4} = \sqrt{q^4r^4 \cdot 15pq}$
$= \sqrt{q^4r^4} \cdot \sqrt{15pq}$
$= q^2r^2\sqrt{15pq}$

89. $\sqrt{\frac{4x^7}{y^{12}}} = \frac{\sqrt{4x^7}}{\sqrt{y^{12}}}$
$= \frac{\sqrt{4x^6 \cdot x}}{\sqrt{y^{12}}}$
$= \frac{\sqrt{4x^6} \cdot \sqrt{x}}{\sqrt{y^{12}}}$
$= \frac{2x^3\sqrt{x}}{y^6}$

91. $\sqrt{27a^{25}} = \sqrt{9a^{24} \cdot 3a}$
$= \sqrt{9a^{24}} \cdot \sqrt{3a}$
$= 3a^{12}\sqrt{3a}$

93. $\frac{-4-\sqrt{162}}{6} = \frac{-4-\sqrt{81 \cdot 2}}{6} = \frac{-4-9\sqrt{2}}{6}$

95. $\frac{7-\sqrt{98}}{14} = \frac{7-\sqrt{49 \cdot 2}}{14}$
$= \frac{7-7\sqrt{2}}{14}$
$= \frac{7(1-\sqrt{2})}{7 \cdot 2}$
$= \frac{1-\sqrt{2}}{2}$

97. $c = \sqrt{a^2+b^2}$
$x = \sqrt{90^2+90^2}$
$= \sqrt{90^2 \cdot 2}$
$= \sqrt{90^2} \cdot \sqrt{2}$
$= 90\sqrt{2}$

99. $c = \sqrt{a^2+b^2}$
$x = \sqrt{14^2+7^2}$
$= \sqrt{196+49}$
$= \sqrt{245}$
$= \sqrt{49 \cdot 5}$
$= \sqrt{49} \cdot \sqrt{5}$
$= 7\sqrt{5}$

101. $c = \sqrt{a^2+b^2}$
$x = \sqrt{6^2+10^2}$
$= \sqrt{36+100}$
$= \sqrt{136}$
$= \sqrt{4 \cdot 34}$
$= \sqrt{4} \cdot \sqrt{34}$
$= 2\sqrt{34}$

103. $c = \sqrt{a^2+b^2}$
$x = \sqrt{6^2+16^2}$
$= \sqrt{36+256}$
$= \sqrt{292}$
$= \sqrt{4 \cdot 73}$
$= \sqrt{4} \cdot \sqrt{73}$
$= 2\sqrt{73}$

105. **(a)** $f = \frac{1}{2\pi}\sqrt{\frac{k}{m}}$
$= \frac{1}{2\pi}\sqrt{\frac{120}{5}}$
$= \frac{1}{2\pi}\sqrt{24}$
$= \frac{1}{2\pi}\sqrt{4 \cdot 6}$
$= \frac{1}{2\pi}2\sqrt{6}$
$= \frac{\sqrt{6}}{\pi}$

(b) $f = \frac{\sqrt{6}}{\pi} \approx 0.78$

107. Since $\sqrt{x^{24}} = \sqrt{(x^{12})^2}$, the missing expression is x^{24}.

109. Since $\sqrt{16s^{18n}} = \sqrt{4^2 \cdot (s^{9n})^2}$, the missing expression is $16s^{18n}$.

111. When simplifying $\sqrt{\frac{x^4}{y}}$, it must be stated that $y > 0$ because if $y = 0$, the denominator would be 0; division by zero is undefined in the set of real numbers. If $y < 0$ the radicand is negative, but the principal square root of a real number cannot be negative.

113. The expression $\sqrt{\frac{315}{5}} = \frac{3\sqrt{35}}{\sqrt{5}}$ is not completely simplified because $\frac{3\sqrt{35}}{\sqrt{5}} = 3\sqrt{\frac{35}{5}} = 3\sqrt{7}$.

Another strategy to simplify $\sqrt{\frac{315}{5}}$ is

$\sqrt{\frac{315}{5}} = \sqrt{63} = \sqrt{9 \cdot 7} = 3\sqrt{7}$.

Section 8.3

Preparing for Adding and Subtracting Square Roots

P1. **(a)** $3a$ and $5a$ are like terms.

(b) $4m$ and $-6n$ are not like terms.

P2. $(3x^2 - 2x + 4) + (x^2 - 4x - 1)$
$= 3x^2 - 2x + 4 + x^2 - 4x - 1$
$= (3x^2 + x^2) + (-2x - 4x) + (4 - 1)$
$= 4x^2 - 6x + 3$

P3. $(3 - 5k + k^2) - (6 - 3k - 4k^2)$
$= 3 - 5k + k^2 - 6 + 3k + 4k^2$
$= (k^2 + 4k^2) + (-5k + 3k) + (3 - 6)$
$= 5k^2 - 2k - 3$

P4. $3x(x - 7) = 3x \cdot x + 3x \cdot (-7) = 3x^2 - 21x$

8.3 Quick Checks

1. Square root expressions are like square roots if each square root has the same <u>radicand</u>.

2. $6\sqrt{3} + 8\sqrt{3} = (6 + 8)\sqrt{3} = 14\sqrt{3}$

3. $7\sqrt{11} - 8\sqrt{11} + 2\sqrt{10} = (7 - 8)\sqrt{11} + 2\sqrt{10}$
$= -\sqrt{11} + 2\sqrt{10}$

4. $3\sqrt{13} + 4\sqrt{6} - 11\sqrt{13} - 3\sqrt{6}$
$= (4 - 3)\sqrt{6} + (3 - 11)\sqrt{13}$
$= \sqrt{6} - 8\sqrt{13}$

5. $4\sqrt{5x} - 5\sqrt{5x} + 13\sqrt{5x} = (4 - 5 + 13)\sqrt{5x}$
$= 12\sqrt{5x}$

6. False; $\sqrt{5} + \sqrt{6} \neq \sqrt{5 + 6}$

7. False; the radicals simplify to like radicals:
$\sqrt{8} = \sqrt{4 \cdot 2} = 2\sqrt{2}$
$\sqrt{18} = \sqrt{9 \cdot 2} = 3\sqrt{2}$

8. $\sqrt{50} + \sqrt{32} = \sqrt{25 \cdot 2} + \sqrt{16 \cdot 2}$
$= 5\sqrt{2} + 4\sqrt{2}$
$= 9\sqrt{2}$

9. $\sqrt{48} - \sqrt{108} = \sqrt{16 \cdot 3} - \sqrt{36 \cdot 3}$
$= 4\sqrt{3} - 6\sqrt{3}$
$= -2\sqrt{3}$

10. $\sqrt{45} + \sqrt{20} - \frac{1}{2}\sqrt{72}$
$= \sqrt{9 \cdot 5} + \sqrt{4 \cdot 5} - \frac{1}{2}\sqrt{36 \cdot 2}$
$= 3\sqrt{5} + 2\sqrt{5} - \frac{1}{2} \cdot 6\sqrt{2}$
$= 5\sqrt{5} - 3\sqrt{2}$

11. $\sqrt{48z} + \sqrt{108z} - \sqrt{12z}$
$= \sqrt{16 \cdot 3z} + \sqrt{36 \cdot 3z} - \sqrt{4 \cdot 3z}$
$= 4\sqrt{3z} + 6\sqrt{3z} - 2\sqrt{3z}$
$= 8\sqrt{3z}$

12. $\frac{1}{3}\sqrt{108a^3} - 2a\sqrt{75a}$
$= \frac{1}{3}\sqrt{36a^2 \cdot 3a} - 2a\sqrt{25 \cdot 3a}$
$= \frac{1}{3} \cdot 6a\sqrt{3a} - 2a \cdot 5\sqrt{3a}$
$= 2a\sqrt{3a} - 10a\sqrt{3a}$
$= -8a\sqrt{3a}$

8.3 Exercises

13. $\sqrt{3} + 5\sqrt{3} = (1 + 5)\sqrt{3} = 6\sqrt{3}$

15. $a\sqrt{2a} + a\sqrt{2a} = (a + a)\sqrt{2a} = 2a\sqrt{2a}$

17. $\sqrt{15} + \sqrt{7}$ is completely simplified.

19. $5\sqrt{19} - 5\sqrt{19} = (5 - 5)\sqrt{19} = 0\sqrt{19} = 0$

21. $4\sqrt{2} - 5\sqrt{2} = (4 - 5)\sqrt{2} = (-1)\sqrt{2} = -\sqrt{2}$

23. $-3\sqrt{6} - 6\sqrt{11}$ is completely simplified.

25. $-\sqrt{22}-\left(-3\sqrt{22}\right)=-\sqrt{22}+3\sqrt{22}$
$=(-1+3)\sqrt{22}$
$=2\sqrt{22}$

27. $\sqrt{12}-\sqrt{27}=\sqrt{4\cdot 3}-\sqrt{9\cdot 3}$
$=2\sqrt{3}-3\sqrt{3}$
$=(2-3)\sqrt{3}$
$=-\sqrt{3}$

29. $-\sqrt{75}+6\sqrt{12}=-\sqrt{25\cdot 3}+6\sqrt{4\cdot 3}$
$=-5\sqrt{3}+6\cdot 2\sqrt{3}$
$=-5\sqrt{3}+12\sqrt{3}$
$=(-5+12)\sqrt{3}$
$=7\sqrt{3}$

31. $7\sqrt{5xy}+3\sqrt{45xy}=7\sqrt{5xy}+3\sqrt{9\cdot 5xy}$
$=7\sqrt{5xy}+3\cdot 3\sqrt{5xy}$
$=7\sqrt{5xy}+9\sqrt{5xy}$
$=(7+9)\sqrt{5xy}$
$=16\sqrt{5xy}$

33. $2\sqrt{3}-\frac{1}{2}\sqrt{300}=2\sqrt{3}-\frac{1}{2}\sqrt{100\cdot 3}$
$=2\sqrt{3}-\frac{1}{2}\cdot 10\sqrt{3}$
$=2\sqrt{3}-5\sqrt{3}$
$=(2-5)\sqrt{3}$
$=-3\sqrt{3}$

35. $6\sqrt{60}+3\sqrt{135}=6\sqrt{4\cdot 15}+3\sqrt{9\cdot 15}$
$=6\cdot 2\sqrt{15}+3\cdot 3\sqrt{15}$
$=12\sqrt{15}+9\sqrt{15}$
$=(12+9)\sqrt{15}$
$=21\sqrt{15}$

37. $4\sqrt{20}+\sqrt{45}-2\sqrt{24}$
$=4\sqrt{4\cdot 5}+\sqrt{9\cdot 5}-2\sqrt{4\cdot 6}$
$=4\cdot 2\sqrt{5}+3\cdot\sqrt{5}-2\cdot 2\sqrt{6}$
$=8\sqrt{5}+3\sqrt{5}-4\sqrt{6}$
$=(8+3)\sqrt{5}-4\sqrt{6}$
$=11\sqrt{5}-4\sqrt{6}$

39. $\sqrt{54x}-\sqrt{96x}=\sqrt{9\cdot 6x}-\sqrt{16\cdot 6x}$
$=3\sqrt{6x}-4\sqrt{6x}$
$=(3-4)\sqrt{6x}$
$=-\sqrt{6x}$

41. $\frac{1}{3}\sqrt{108a^3}-2a\sqrt{75a}$
$=\frac{1}{3}\sqrt{36a^2\cdot 3a}-2a\sqrt{25\cdot 3a}$
$=\frac{1}{3}\cdot 6a\sqrt{3a}-2a\cdot 5\sqrt{3a}$
$=2a\sqrt{3a}-10a\sqrt{3a}$
$=(2a-10a)\sqrt{3a}$
$=-8a\sqrt{3a}$

43. $\frac{2}{5}\sqrt{20}-\frac{4}{3}\sqrt{45}=\frac{2}{5}\sqrt{4\cdot 5}-\frac{4}{3}\sqrt{9\cdot 5}$
$=\frac{2}{5}\cdot 2\sqrt{5}-\frac{4}{3}\cdot 3\sqrt{5}$
$=\frac{4}{5}\sqrt{5}-4\sqrt{5}$
$=\left(\frac{4}{5}-4\right)\sqrt{5}$
$=\left(\frac{4}{5}-\frac{20}{5}\right)\sqrt{5}$
$=-\frac{16}{5}\sqrt{5}$

45. $15\sqrt{14}-8\sqrt{14}=(15-8)\sqrt{14}=7\sqrt{14}$

47. $3\sqrt{2}-3\sqrt{2}=(3-3)\sqrt{2}=0\sqrt{2}=0$

49. $\sqrt{y}-6\sqrt{y}=(1-6)\sqrt{y}=-5\sqrt{y}$

51. $15\sqrt{6}-14\sqrt{6}+9\sqrt{5}=(15-14)\sqrt{6}+9\sqrt{5}$
$=\sqrt{6}+9\sqrt{5}$

53. $\sqrt{2x}+8\sqrt{2x}-5\sqrt{2}=(1+8)\sqrt{2x}-5\sqrt{2}$
$=9\sqrt{2x}-5\sqrt{2}$

55. $2n\sqrt{5n}-4n\sqrt{5n}=(2-4)n\sqrt{5n}=-2n\sqrt{5n}$

57. $3\sqrt{20}-3\sqrt{125}+4\sqrt{45}$
$=3\sqrt{4\cdot 5}-3\sqrt{25\cdot 5}+4\sqrt{9\cdot 5}$
$=3\cdot 2\sqrt{5}-3\cdot 5\sqrt{5}+4\cdot 3\sqrt{5}$
$=6\sqrt{5}-15\sqrt{5}+12\sqrt{5}$
$=(6-15+12)\sqrt{5}$
$=3\sqrt{5}$

59. $2\sqrt{75}+6\sqrt{12}-\sqrt{48}$
$=2\sqrt{25\cdot 3}+6\sqrt{4\cdot 3}-\sqrt{16\cdot 3}$
$=2\cdot 5\sqrt{3}+6\cdot 2\sqrt{3}-4\sqrt{3}$
$=10\sqrt{3}+12\sqrt{3}-4\sqrt{3}$
$=(10+12-4)\sqrt{3}$
$=18\sqrt{3}$

61. $5\sqrt{72}-4\sqrt{243}-\sqrt{288}$
$=5\sqrt{36\cdot 2}-4\sqrt{81\cdot 3}-\sqrt{144\cdot 2}$
$=5\cdot 6\sqrt{2}-4\cdot 9\sqrt{3}-12\sqrt{2}$
$=30\sqrt{2}-36\sqrt{3}-12\sqrt{2}$
$=(30-12)\sqrt{2}-36\sqrt{3}$
$=18\sqrt{2}-36\sqrt{3}$

63. $5\sqrt{32y^4}-4\sqrt{50y^4}$
$=5\sqrt{16y^4\cdot 2}-4\sqrt{25y^4\cdot 2}$
$=5\cdot 4y^2\sqrt{2}-4\cdot 5y^2\sqrt{2}$
$=20y^2\sqrt{2}-20y^2\sqrt{2}$
$=(20-20)y^2\sqrt{2}$
$=0\sqrt{2}$
$=0$

65. $\sqrt{45x^2y}+\sqrt{20x^2y}-\sqrt{125x^2}$
$=\sqrt{9x^2\cdot 5y}+\sqrt{4x^2\cdot 5y}-\sqrt{25x^2\cdot 5}$
$=3x\sqrt{5y}+2x\sqrt{5y}-5x\sqrt{5}$
$=(3+2)x\sqrt{5y}-5x\sqrt{5}$
$=5x\sqrt{5y}-5x\sqrt{5}$

67. $2a^2\sqrt{40a^5}-a\sqrt{90a^7}+3\sqrt{12a^3}$
$=2a^2\sqrt{4a^4\cdot 10a}-a\sqrt{9a^6\cdot 10a}+3\sqrt{4a^2\cdot 3a}$
$=2a^2\cdot 2a^2\sqrt{10a}-a\cdot 3a^3\sqrt{10a}+3\cdot 2a\sqrt{3a}$
$=4a^4\sqrt{10a}-3a^4\sqrt{10a}+6a\sqrt{3a}$
$=(4-3)a^4\sqrt{10a}+6a\sqrt{3a}$
$=a^4\sqrt{10a}+6a\sqrt{3a}$

69. $3+2\sqrt{8}+4+3\sqrt{18}$
$=3+4+2\sqrt{8}+3\sqrt{18}$
$=7+2\sqrt{4\cdot 2}+3\sqrt{9\cdot 2}$
$=7+2\cdot 2\sqrt{2}+3\cdot 3\sqrt{2}$
$=7+4\sqrt{2}+9\sqrt{2}$
$=7+(4+9)\sqrt{2}$
$=7+13\sqrt{2}$

71. $-1-2\sqrt{27}+4\sqrt{48}-\sqrt{75}$
$=-1-2\sqrt{9\cdot 3}+4\sqrt{16\cdot 3}-\sqrt{25\cdot 3}$
$=-1-2\cdot 3\sqrt{3}+4\cdot 4\sqrt{3}-5\sqrt{3}$
$=-1-6\sqrt{3}+16\sqrt{3}-5\sqrt{3}$
$=-1+(-6+16-5)\sqrt{3}$
$=-1+5\sqrt{3}$

73. $\frac{1}{2}\sqrt{3}+\frac{2}{3}\sqrt{3}=\left(\frac{1}{2}+\frac{2}{3}\right)\sqrt{3}$
$=\left(\frac{3}{6}+\frac{4}{6}\right)\sqrt{3}$
$=\frac{7}{6}\sqrt{3}$

75. $\sqrt{\frac{3}{25}}+\sqrt{\frac{3}{16}}=\frac{\sqrt{3}}{\sqrt{25}}+\frac{\sqrt{3}}{\sqrt{16}}$
$=\frac{\sqrt{3}}{5}+\frac{\sqrt{3}}{4}$
$=\left(\frac{1}{5}+\frac{1}{4}\right)\sqrt{3}$
$=\left(\frac{4}{20}+\frac{5}{20}\right)\sqrt{3}$
$=\frac{9}{20}\sqrt{3}$

77. $4\sqrt{\frac{15}{9}}-2\sqrt{\frac{15}{16}}=4\frac{\sqrt{15}}{\sqrt{9}}-2\frac{\sqrt{15}}{\sqrt{16}}$
$=4\frac{\sqrt{15}}{3}-2\frac{\sqrt{15}}{4}$
$=\left(\frac{4}{3}-\frac{1}{2}\right)\sqrt{15}$
$=\left(\frac{8}{6}-\frac{3}{6}\right)\sqrt{15}$
$=\frac{5}{6}\sqrt{15}$

79. $\frac{2}{3}\sqrt{\frac{18}{3}}+\frac{3}{2}\sqrt{\frac{120}{5}}=\frac{2}{3}\sqrt{6}+\frac{3}{2}\sqrt{24}$
$=\frac{2}{3}\sqrt{6}+\frac{3}{2}\sqrt{4\cdot 6}$
$=\frac{2}{3}\sqrt{6}+\frac{3}{2}\cdot 2\sqrt{6}$
$=\frac{2}{3}\sqrt{6}+3\sqrt{6}$
$=\left(\frac{2}{3}+3\right)\sqrt{6}$
$=\left(\frac{2}{3}+\frac{9}{3}\right)\sqrt{6}$
$=\frac{11}{3}\sqrt{6}$

81. Perimeter of a square is 4 times the length of a side.
$P=4\cdot 3\sqrt{12}=12\sqrt{4\cdot 3}=12\cdot 2\sqrt{3}=24\sqrt{3}$
The perimeter is $24\sqrt{3}$ cm.

83. Perimeter of a rectangle is twice the sum of the lengths of the sides.
$P=2\left(3\sqrt{18}+4\sqrt{48}\right)$
$=2\left(3\sqrt{9\cdot 2}+4\sqrt{16\cdot 3}\right)$
$=2\left(3\cdot 3\sqrt{2}+4\cdot 4\sqrt{3}\right)$
$=2\left(9\sqrt{2}+16\sqrt{3}\right)$
$=18\sqrt{2}+32\sqrt{3}$
The perimeter is $\left(18\sqrt{2}+32\sqrt{3}\right)$ inches.

85. The metal bar should be twice the length of the hypotenuse of a right triangle with legs of length 6 feet and 10 feet.
$h=\sqrt{6^2+10^2}$
$=\sqrt{36+100}$
$=\sqrt{136}$
$=\sqrt{4\cdot 34}$
$=2\sqrt{34}$
Alex should buy a piece of metal $2\cdot 2\sqrt{34}=4\sqrt{34}\approx 23.3$ feet long.

87. $P=\left(8+3\sqrt{3}\right)+\left(1+\sqrt{12}\right)+\left(4+\sqrt{27}\right)$
$=8+3\sqrt{3}+1+\sqrt{4\cdot 3}+4+\sqrt{9\cdot 3}$
$=8+3\sqrt{3}+1+2\sqrt{3}+4+3\sqrt{3}$
$=8+1+4+3\sqrt{3}+2\sqrt{3}+3\sqrt{3}$
$=13+(3+2+3)\sqrt{3}$
$=13+8\sqrt{3}$

89. In the expression $6-2\sqrt{x}-10+5\sqrt{x}$, the terms 6 and -10 are like terms, and $-2\sqrt{x}$ and $5\sqrt{x}$ are like terms. The expression $6-2\sqrt{x}-10+5\sqrt{x}$ has not been correctly simplified. The correct answer is
$6-2\sqrt{x}-10+5\sqrt{x}=6-10-2\sqrt{x}+5\sqrt{x}$
$=-4+3\sqrt{x}.$

91. When multiplying square roots, we multiply the radicands: $\sqrt{a\cdot b}=\sqrt{a}\cdot\sqrt{b}$ is the Product Rule for square roots. However, when adding square roots, the square symbol acts as a grouping symbol. The operation under the radical sign must be done first, and then the radicand may be simplified, if possible. So $\sqrt{a+b}\neq\sqrt{a}+\sqrt{b}$.

Section 8.4

Preparing for Multiplying Expressions with Square Roots

P1. $(3x)^2=3x^2\cdot 3x^2=9x^2$

P2. $3(2x^2-4x+1)=3\cdot 2x^2-3\cdot 4x+3\cdot 1$
$=6x^2-12x+3$

P3. $(2a+3)(3a-4)$
$=2a\cdot 3a+2a\cdot(-4)+3\cdot 3a+3\cdot(-4)$
$=6a^2-8a+9a-12$
$=6a^2+a-12$

P4. $(5b+2)^2=(5b)^2+2(5b)(2)+2^2$
$=25b^2+20b+4$

P5. $(3y-4)(3y+4)=(3y)^2-4^2=9y^2-16$

8.4 Quick Checks

1. True

2. $\sqrt{3}\cdot\sqrt{5}=\sqrt{3\cdot 5}=\sqrt{15}$

3. $\sqrt{2z}\cdot\sqrt{7}=\sqrt{2z\cdot 7}=\sqrt{14z}$

4. $\sqrt{15b}\cdot\sqrt{5b}=\sqrt{15b\cdot 5b}$
$=\sqrt{75b^2}$
$=\sqrt{25b^2\cdot 3}$
$=5b\sqrt{3}$

5. $\sqrt{24k}\cdot\sqrt{3k^3}=\sqrt{24k\cdot 3k^3}$
$=\sqrt{72k^4}$
$=\sqrt{36k^4\cdot 2}$
$=6k^2\sqrt{2}$

6. $\sqrt{2z^3}\cdot\sqrt{10z^4}=\sqrt{2z^3\cdot 10z^4}$
$=\sqrt{20z^7}$
$=\sqrt{4z^6\cdot 5z}$
$=2z^3\sqrt{5z}$

7. If a is a positive number, then $\left(\sqrt{a}\right)^2=\underline{a}$ and $\left(-\sqrt{a}\right)^2=\underline{a}$.

8. $\left(\sqrt{11}\right)^2=11$

9. $\left(-\sqrt{31}\right)^2=31$

10. $\left(7\sqrt{3}\right)\left(\sqrt{3}\right)=7\cdot\sqrt{3}\cdot\sqrt{3}$
$=7\cdot\left(\sqrt{3}\right)^2$
$=7\cdot 3$
$=21$

11. $\left(2\sqrt{5}\right)\left(-3\sqrt{10}\right)=2\cdot(-3)\cdot\sqrt{5}\cdot\sqrt{10}$
$=-6\cdot\sqrt{50}$
$=-6\cdot\sqrt{25\cdot 2}$
$=-6\cdot 5\cdot\sqrt{2}$
$=-30\sqrt{2}$

12. $\left(-2\sqrt{11}\right)^2=\left(-2\sqrt{11}\right)\cdot\left(-2\sqrt{11}\right)$
$=(-2)^2\cdot\left(\sqrt{11}\right)^2$
$=4\cdot 11$
$=44$

13. $-3\left(5-\sqrt{7}\right)=-3\cdot 5-3\left(-\sqrt{7}\right)=-15+3\sqrt{7}$

14. $\sqrt{5}\left(3-\sqrt{5}\right)=\sqrt{5}\cdot 3-\sqrt{5}\cdot\sqrt{5}$
$=3\sqrt{5}-\left(\sqrt{5}\right)^2$
$=3\sqrt{5}-5$

15. $\sqrt{2}\left(-4+\sqrt{10}\right)=\sqrt{2}(-4)+\sqrt{2}\cdot\sqrt{10}$
$=-4\sqrt{2}+\sqrt{2\cdot 10}$
$=-4\sqrt{2}+\sqrt{20}$
$=-4\sqrt{2}+\sqrt{4\cdot 5}$
$=-4\sqrt{2}+2\sqrt{5}$

16. $-\sqrt{8}\left(\sqrt{2}-\sqrt{6}\right)=-\sqrt{8}\cdot\sqrt{2}-\sqrt{8}\left(-\sqrt{6}\right)$
$=-\sqrt{16}+\sqrt{48}$
$=-4+\sqrt{16\cdot 3}$
$=-4+4\sqrt{3}$

17. $\left(3+\sqrt{7}\right)\left(4+\sqrt{7}\right)$
$=3\cdot 4+3\sqrt{7}+4\sqrt{7}+\sqrt{7}\cdot\sqrt{7}$
$=12+7\sqrt{7}+7$
$=19+7\sqrt{7}$

18. $\left(7-\sqrt{5n}\right)\left(6+\sqrt{5n}\right)$
$=7\cdot 6+7\sqrt{5n}-6\sqrt{5n}-\sqrt{5n}\cdot\sqrt{5n}$
$=42+\sqrt{5n}-5n$

19. $\left(4-2\sqrt{3}\right)\left(3+5\sqrt{3}\right)$
$=4\cdot 3+4\cdot 5\sqrt{3}+\left(-2\sqrt{3}\right)\cdot 3+\left(-2\sqrt{3}\right)\left(5\sqrt{3}\right)$
$=12+20\sqrt{3}-6\sqrt{3}-10\cdot 3$
$=12+14\sqrt{3}-30$
$=-18+14\sqrt{3}$

20. $\left(2-3\sqrt{7}\right)\left(5+2\sqrt{7}\right)$
$=2\cdot 5+2\cdot 2\sqrt{7}+\left(-3\sqrt{7}\right)\cdot 5+\left(-3\sqrt{7}\right)\left(2\sqrt{7}\right)$
$=10+4\sqrt{7}-15\sqrt{7}-6\cdot 7$
$=10-11\sqrt{7}-42$
$=-32-11\sqrt{7}$

21. False; $\left(\sqrt{a}+\sqrt{b}\right)^2 = \left(\sqrt{a}\right)^2 + 2\cdot\sqrt{a}\cdot\sqrt{b} + \left(\sqrt{b}\right)^2$
$= a + 2\sqrt{ab} + b$
while $\left(\sqrt{a}\right)^2 + \left(\sqrt{b}\right)^2 = a + b.$

22. $\left(8+\sqrt{3}\right)^2 = 8^2 + 2\cdot 8\cdot\sqrt{3} + \left(\sqrt{3}\right)^2$
$= 64 + 16\sqrt{3} + 3$
$= 67 + 16\sqrt{3}$

23. $\left(2-\sqrt{7z}\right)^2 = 2^2 - 2\cdot 2\cdot\sqrt{7z} + \left(\sqrt{7z}\right)^2$
$= 4 - 4\sqrt{7z} + 7z$

24. $\left(-2+3\sqrt{5}\right)^2 = (-2)^2 - 2\cdot 2\cdot 3\sqrt{5} + \left(3\sqrt{5}\right)^2$
$= 4 - 12\sqrt{5} + 9\cdot 5$
$= 4 - 12\sqrt{5} + 45$
$= 49 - 12\sqrt{5}$

25. Square root expressions such as $-3+\sqrt{11}$ and $-3-\sqrt{11}$ are called <u>conjugates</u>.

26. $\left(6+\sqrt{3}\right)\left(6-\sqrt{3}\right) = 6^2 - \left(\sqrt{3}\right)^2 = 36 - 3 = 33$

27. $\left(\sqrt{2}-7\right)\left(\sqrt{2}+7\right) = \left(\sqrt{2}\right)^2 - 7^2$
$= 2 - 49$
$= -47$

28. $\left(1+4\sqrt{5n}\right)\left(1-4\sqrt{5n}\right) = 1^2 - \left(4\sqrt{5n}\right)^2$
$= 1 - 16\cdot 5n$
$= 1 - 80n$

8.4 Exercises

29. $\sqrt{3}\cdot\sqrt{3} = \left(\sqrt{3}\right)^2 = 3$

31. $\sqrt{7x}\cdot\sqrt{7x} = \left(\sqrt{7x}\right)^2 = 7x$

33. $\sqrt{3}\cdot\sqrt{13} = \sqrt{3\cdot 13} = \sqrt{39}$

35. $\sqrt{2}\cdot\sqrt{7a} = \sqrt{2\cdot 7a} = \sqrt{14a}$

37. $-\sqrt{10}\cdot\sqrt{6} = -\sqrt{10\cdot 6}$
$= -\sqrt{60}$
$= -\sqrt{4\cdot 15}$
$= -2\sqrt{15}$

39. $\sqrt{24}\cdot\sqrt{72} = \sqrt{4\cdot 6}\cdot\sqrt{36\cdot 2}$
$= 2\sqrt{6}\cdot 6\sqrt{2}$
$= 2\cdot 6\sqrt{6\cdot 2}$
$= 12\sqrt{12}$
$= 12\sqrt{4\cdot 3}$
$= 12\cdot 2\sqrt{3}$
$= 24\sqrt{3}$

41. $-\sqrt{3x}\cdot\sqrt{27x^3} = -\sqrt{3x\cdot 27x^3}$
$= -\sqrt{81x^4}$
$= -\sqrt{(9x^2)^2}$
$= -9x^2$

43. $3\sqrt{12}\cdot 4\sqrt{18} = 3\cdot 4\sqrt{12\cdot 18}$
$= 12\sqrt{4\cdot 3\cdot 9\cdot 2}$
$= 12\cdot 2\cdot 3\sqrt{3\cdot 2}$
$= 72\sqrt{6}$

45. $5\sqrt{12}\cdot 9\sqrt{18} = 5\sqrt{4\cdot 3}\cdot 9\sqrt{9\cdot 2}$
$= 5\cdot 2\sqrt{3}\cdot 9\cdot 3\sqrt{2}$
$= 5\cdot 2\cdot 9\cdot 3\sqrt{3\cdot 2}$
$= 270\sqrt{6}$

47. $-\sqrt{45}\cdot 3\sqrt{30} = -\sqrt{9\cdot 5}\cdot 3\sqrt{30}$
$= -3\sqrt{5}\cdot 3\sqrt{30}$
$= -3\cdot 3\sqrt{5\cdot 30}$
$= -9\sqrt{5\cdot 5\cdot 6}$
$= -9\cdot 5\sqrt{6}$
$= -45\sqrt{6}$

49. $2\sqrt{6x}\cdot 4\sqrt{8x^4} = 2\cdot 4\sqrt{6x\cdot 8x^4}$
$= 8\sqrt{48x^5}$
$= 8\sqrt{16x^4\cdot 3x}$
$= 8\cdot 4x^2\sqrt{3x}$
$= 32x^2\sqrt{3x}$

51. $\left(\sqrt{37}\right)^2 = 37$

53. $\left(-\sqrt{7x}\right)^2 = (-1)^2\left(\sqrt{7x}\right)^2 = 7x$

55. $4\left(6+2\sqrt{5}\right) = 4\cdot 6 + 4\cdot 2\sqrt{5} = 24 + 8\sqrt{5}$

57. $$\begin{aligned}\sqrt{6}\left(2\sqrt{6}-3\right) &= 2\cdot\sqrt{6}\cdot\sqrt{6} - 3\cdot\sqrt{6}\\ &= 2\left(\sqrt{6}\right)^2 - 3\sqrt{6}\\ &= 2\cdot 6 - 3\sqrt{6}\\ &= 12 - 3\sqrt{6}\end{aligned}$$

59. $$\begin{aligned}\sqrt{3}\left(7\sqrt{15}+4\right) &= 7\cdot\sqrt{3}\cdot\sqrt{15} + 4\cdot\sqrt{3}\\ &= 7\sqrt{3\cdot 15} + 4\sqrt{3}\\ &= 7\sqrt{45} + 4\sqrt{3}\\ &= 7\sqrt{9\cdot 5} + 4\sqrt{3}\\ &= 7\cdot 3\sqrt{5} + 4\sqrt{3}\\ &= 21\sqrt{5} + 4\sqrt{3}\end{aligned}$$

61. $$\begin{aligned}\sqrt{a}\left(\sqrt{2a}+2\right) &= \sqrt{a}\cdot\sqrt{2a} + 2\cdot\sqrt{a}\\ &= \sqrt{a\cdot 2a} + 2\sqrt{a}\\ &= \sqrt{a^2\cdot 2} + 2\sqrt{a}\\ &= a\sqrt{2} + 2\sqrt{a}\end{aligned}$$

63. $$\begin{aligned}\left(6+\sqrt{3}\right)\left(5+\sqrt{3}\right) &= 6\cdot 5 + 6\cdot\sqrt{3} + 5\cdot\sqrt{3} + \sqrt{3}\cdot\sqrt{3}\\ &= 30 + (6+5)\sqrt{3} + \left(\sqrt{3}\right)^2\\ &= 30 + 11\sqrt{3} + 3\\ &= 33 + 11\sqrt{3}\end{aligned}$$

65. $$\begin{aligned}\left(2+\sqrt{5}\right)\left(3-\sqrt{5}\right) &= 2\cdot 3 - 2\cdot\sqrt{5} + 3\cdot\sqrt{5} - \sqrt{5}\cdot\sqrt{5}\\ &= 6 + (-2+3)\sqrt{5} - \left(\sqrt{5}\right)^2\\ &= 6 + \sqrt{5} - 5\\ &= 1 + \sqrt{5}\end{aligned}$$

67. $$\begin{aligned}\left(5-\sqrt{3}\right)\left(4-\sqrt{3}\right) &= 5\cdot 4 - 5\cdot\sqrt{3} - 4\cdot\sqrt{3} + \left(-\sqrt{3}\right)\left(-\sqrt{3}\right)\\ &= 20 - (5+4)\sqrt{3} + \left(-\sqrt{3}\right)^2\\ &= 20 - 9\sqrt{3} + 3\\ &= 23 - 9\sqrt{3}\end{aligned}$$

69. $\left(3-2\sqrt{x}\right)\left(1-4\sqrt{x}\right)=3\cdot 1-3\cdot 4\sqrt{x}-1\cdot 2\sqrt{x}+\left(-2\sqrt{x}\right)\left(-4\sqrt{x}\right)$
$=3-12\sqrt{x}-2\sqrt{x}+(-2)(-4)\left(\sqrt{x}\right)^2$
$=3+(-12-2)\sqrt{x}+8x$
$=3-14\sqrt{x}+8x$

71. $\left(5+2\sqrt{3}\right)\left(4-3\sqrt{2}\right)=5\cdot 4-5\cdot 3\sqrt{2}+4\cdot 2\sqrt{3}-2\sqrt{3}\cdot 3\sqrt{2}$
$=20-15\sqrt{2}+8\sqrt{3}-2\cdot 3\cdot\sqrt{3}\cdot\sqrt{2}$
$=20-15\sqrt{2}+8\sqrt{3}-6\sqrt{3\cdot 2}$
$=20-15\sqrt{2}+8\sqrt{3}-6\sqrt{6}$

73. $\left(2\sqrt{3x}+\sqrt{x}\right)\left(5\sqrt{x}-2\sqrt{3x}\right)=2\sqrt{3x}\cdot 5\sqrt{x}-2\sqrt{3x}\cdot 2\sqrt{3x}+\sqrt{x}\cdot 5\sqrt{x}-\sqrt{x}\cdot 2\sqrt{3x}$
$=2\cdot 5\cdot\sqrt{3x}\cdot\sqrt{x}-2\cdot 2\cdot\left(\sqrt{3x}\right)^2+5\cdot\left(\sqrt{x}\right)^2-2\cdot\sqrt{x}\cdot\sqrt{3x}$
$=10\sqrt{3x\cdot x}-4\cdot 3x+5\cdot x-2\sqrt{x\cdot 3x}$
$=10\sqrt{3x^2}-12x+5x-2\sqrt{3x^2}$
$=10x\sqrt{3}+(-12+5)x-2x\sqrt{3}$
$=(10-2)x\sqrt{3}+(-7)x$
$=8x\sqrt{3}-7x$

75. $\left(4+\sqrt{3}\right)^2=4^2+2\cdot 4\cdot\sqrt{3}+\left(\sqrt{3}\right)^2=16+8\sqrt{3}+3=19+8\sqrt{3}$

77. $\left(3\sqrt{x}-2\right)^2=\left(3\sqrt{x}\right)^2-2\cdot 3\sqrt{x}\cdot 2+2^2=3^2\left(\sqrt{x}\right)^2-12\sqrt{x}+4=9x-12\sqrt{x}+4$

79. $\left(3+2\sqrt{2}\right)\left(3-2\sqrt{2}\right)=3^2-\left(2\sqrt{2}\right)^2=9-2^2\left(\sqrt{2}\right)^2=9-4\cdot 2=9-8=1$

81. $\left(5+2\sqrt{2}\right)^2=5^2+2\cdot 5\cdot 2\sqrt{2}+\left(2\sqrt{2}\right)^2$
$=25+20\sqrt{2}+2^2\left(\sqrt{2}\right)^2$
$=25+20\sqrt{2}+4\cdot 2$
$=25+20\sqrt{2}+8$
$=33+20\sqrt{2}$

83. $\left(4-9\sqrt{5}\right)\left(4+9\sqrt{5}\right)=4^2-\left(9\sqrt{5}\right)^2$
$=16-9^2\left(\sqrt{5}\right)^2$
$=16-81\cdot 5$
$=16-405$
$=-389$

85. $\left(3\sqrt{5}-\sqrt{11}\right)\left(3\sqrt{5}+\sqrt{11}\right)=\left(3\sqrt{5}\right)^2-\left(\sqrt{11}\right)^2$
$$=3^2\left(\sqrt{5}\right)^2-11$$
$$=9\cdot 5-11$$
$$=45-11$$
$$=34$$

87. $\sqrt{15m^5}\cdot\sqrt{3m}=\sqrt{15m^5\cdot 3m}=\sqrt{45m^6}=\sqrt{9m^6\cdot 5}=3m^3\sqrt{5}$

89. $3\sqrt{2}-4\sqrt{8}=3\sqrt{2}-4\sqrt{4\cdot 2}=3\sqrt{2}-4\cdot 2\sqrt{2}=3\sqrt{2}-8\sqrt{2}=(3-8)\sqrt{2}=-5\sqrt{2}$

91. $3\left(2\sqrt{2n}+\sqrt{8n}\right)=3\left(2\sqrt{2n}+\sqrt{4\cdot 2n}\right)=3\left(2\sqrt{2n}+2\sqrt{2n}\right)=3\left(4\sqrt{2n}\right)=3\cdot 4\sqrt{2n}=12\sqrt{2n}$

93. $\left(4\sqrt{2}\right)^2=4^2\left(\sqrt{2}\right)^2=16\cdot 2=32$

95. $\left(\sqrt{a}+\sqrt{b}\right)^2-\sqrt{ab}=\left(\sqrt{a}\right)^2+2\cdot\sqrt{a}\cdot\sqrt{b}+\left(\sqrt{b}\right)^2-\sqrt{ab}$
$$=a+2\sqrt{ab}+b-\sqrt{ab}$$
$$=a+\sqrt{ab}+b$$

97. $\left(3\sqrt{5}-2\sqrt{6}\right)\left(5\sqrt{5}+4\sqrt{6}\right)=3\sqrt{5}\cdot 5\sqrt{5}+3\sqrt{5}\cdot 4\sqrt{6}-2\sqrt{6}\cdot 5\sqrt{5}-2\sqrt{6}\cdot 4\sqrt{6}$
$$=3\cdot 5\left(\sqrt{5}\right)^2+3\cdot 4\cdot\sqrt{5}\cdot\sqrt{6}-2\cdot 5\cdot\sqrt{6}\cdot\sqrt{5}-2\cdot 4\left(\sqrt{6}\right)^2$$
$$=15\cdot 5+12\sqrt{5\cdot 6}-10\sqrt{6\cdot 5}-8\cdot 6$$
$$=75+12\sqrt{30}-10\sqrt{30}-48$$
$$=75-48+(12-10)\sqrt{30}$$
$$=27+2\sqrt{30}$$

99. $\sqrt{m}\left(\sqrt{2m}-\sqrt{m}\right)=\sqrt{m}\cdot\sqrt{2m}-\sqrt{m}\cdot\sqrt{m}$
$$=\sqrt{m\cdot 2m}-\sqrt{m\cdot m}$$
$$=\sqrt{m^2\cdot 2}-\sqrt{m^2}$$
$$=m\sqrt{2}-m$$

101. $\left(3\sqrt{s}-2\right)^2+12\sqrt{s}=\left(3\sqrt{s}\right)^2-2\cdot 3\sqrt{s}\cdot 2+2^2+12\sqrt{s}$
$$=3^2\left(\sqrt{s}\right)^2-12\sqrt{s}+4+12\sqrt{s}$$
$$=9s+4$$

103. $-3\sqrt{5}\cdot 4\sqrt{20}=-3\cdot 4\sqrt{5}\cdot\sqrt{20}$
$$=-12\sqrt{100}$$
$$=-12\cdot 10$$
$$=-120$$

105. $\sqrt{12a}\cdot\sqrt{3a^4}+4a^2\sqrt{a}=\sqrt{12a\cdot 3a^4}+4a^2\sqrt{a}$
$=\sqrt{12\cdot 3\cdot a\cdot a^4}+4a^2\sqrt{a}$
$=\sqrt{36\cdot a^5}+4a^2\sqrt{a}$
$=\sqrt{36a^4\cdot a}+4a^2\sqrt{a}$
$=6a^2\sqrt{a}+4a^2\sqrt{a}$
$=10a^2\sqrt{a}$

107. $\frac{\sqrt{27}}{2}-3\sqrt{12}=\frac{\sqrt{9\cdot 3}}{2}-3\sqrt{4\cdot 3}$
$=\frac{3\sqrt{3}}{2}-3\cdot 2\sqrt{3}$
$=\frac{3}{2}\sqrt{3}-6\sqrt{3}$
$=\left(\frac{3}{2}-6\right)\sqrt{3}$
$=\left(\frac{3}{2}-\frac{12}{2}\right)\sqrt{3}$
$=-\frac{9}{2}\sqrt{3}$

109. $(3-\sqrt{5})(3+\sqrt{5})=3^2-(\sqrt{5})^2=9-5=4$

111. $\sqrt{3}(\sqrt{3}+\sqrt{12})=\sqrt{3}\cdot\sqrt{3}+\sqrt{3}\cdot\sqrt{12}$
$=\sqrt{3\cdot 3}+\sqrt{3\cdot 12}$
$=\sqrt{9}+\sqrt{36}$
$=3+6$
$=9$

113. $A=\frac{1}{2}h(B+b)$
$=\frac{1}{2}\cdot 3\sqrt{6}\left(3\sqrt{32}+10\sqrt{18}\right)$
$=\frac{3}{2}\sqrt{6}\left(3\sqrt{16\cdot 2}+10\sqrt{9\cdot 2}\right)$
$=\frac{3}{2}\sqrt{6}\left(3\cdot 4\sqrt{2}+10\cdot 3\sqrt{2}\right)$
$=\frac{3}{2}\sqrt{6}\left(12\sqrt{2}+30\sqrt{2}\right)$
$=\frac{3}{2}\sqrt{6}\left(42\sqrt{2}\right)$
$=\frac{3}{2}\cdot 42\cdot\sqrt{6}\cdot\sqrt{2}$
$=63\sqrt{12}$
$=63\sqrt{4\cdot 3}$
$=63\cdot 2\sqrt{3}$
$=126\sqrt{3}$

The area is $126\sqrt{3}$ square units.

115. $A=\frac{1}{2}h(B+b)$
$=\frac{1}{2}\cdot 2\sqrt{14}\left(4\sqrt{128}+6\sqrt{32}\right)$
$=\sqrt{14}\left(4\sqrt{64\cdot 2}+6\sqrt{16\cdot 2}\right)$
$=\sqrt{14}\left(4\cdot 8\sqrt{2}+6\cdot 4\sqrt{2}\right)$
$=\sqrt{14}\left(32\sqrt{2}+24\sqrt{2}\right)$
$=\sqrt{14}\left(56\sqrt{2}\right)$
$=56\sqrt{14\cdot 2}$
$=56\sqrt{28}$
$=56\sqrt{4\cdot 7}$
$=56\cdot 2\sqrt{7}$
$=112\sqrt{7}$

The area is $112\sqrt{7}$ square units.

117. $A = \frac{1}{2}h(B+b)$
$= \frac{1}{2} \cdot 4\left(\sqrt{27} + 6 + 3\sqrt{12}\right)$
$= 2\left(\sqrt{9 \cdot 3} + 6 + 3\sqrt{4 \cdot 3}\right)$
$= 2\left(3\sqrt{3} + 6 + 3 \cdot 2\sqrt{3}\right)$
$= 2\left(3\sqrt{3} + 6 + 6\sqrt{3}\right)$
$= 2\left(6 + 9\sqrt{3}\right)$
$= 12 + 18\sqrt{3}$
The area is $\left(12 + 18\sqrt{3}\right)$ square units.

119. $A = \pi r^2 = \pi\left(7\sqrt{3}\right)^2$
$= \pi \cdot 7^2 \left(\sqrt{3}\right)^2$
$= \pi \cdot 49 \cdot 3$
$= 147\pi$
The area is 147π square meters.

121. The radius is half the length of the diameter, or $\frac{3}{2}\sqrt{56}$ inches.
$A = \pi r^2 = \pi\left(\frac{3}{2}\sqrt{56}\right)^2$
$= \pi\left(\frac{3}{2}\right)^2 \left(\sqrt{56}\right)^2$
$= \pi\left(\frac{9}{4}\right)(56)$
$= \pi\left(\frac{9}{4}\right)(4 \cdot 14)$
$= 126\pi$
The area is 126π square inches.

123. $s = \frac{1}{2}(11+12+17) = \frac{1}{2}(40) = 20$
$A = \sqrt{s(s-a)(s-b)(s-c)}$
$= \sqrt{20(20-11)(20-12)(20-17)}$
$= \sqrt{20 \cdot 9 \cdot 8 \cdot 3}$
$= \sqrt{4 \cdot 5 \cdot 9 \cdot 4 \cdot 2 \cdot 3}$
$= 2 \cdot 2 \cdot 3\sqrt{5 \cdot 2 \cdot 3}$
$= 12\sqrt{30}$
The area is $12\sqrt{30}$ square units.

125. $s = \frac{1}{2}(15+25+30) = \frac{1}{2}(70) = 35$
$A = \sqrt{s(s-a)(s-b)(s-c)}$
$= \sqrt{35(35-15)(35-25)(35-30)}$
$= \sqrt{35 \cdot 20 \cdot 10 \cdot 5}$
$= \sqrt{5 \cdot 7 \cdot 4 \cdot 5 \cdot 2 \cdot 5 \cdot 5}$
$= \sqrt{4 \cdot 5^4 \cdot 7 \cdot 2}$
$= 2 \cdot 5^2\sqrt{7 \cdot 2}$
$= 2 \cdot 25\sqrt{14}$
$= 50\sqrt{14}$
The area is $50\sqrt{14}$ square units.

127. No; $\left(\sqrt{5}+\sqrt{7}\right)^2 \neq \left(\sqrt{5}\right)^2 + \left(\sqrt{7}\right)^2$. When squaring a binomial, we use the pattern $A^2 + 2AB + B^2$. So
$\left(\sqrt{5}+\sqrt{7}\right)^2 = \left(\sqrt{5}\right)^2 + 2\left(\sqrt{5}\right)\left(\sqrt{7}\right) + \left(\sqrt{7}\right)^2$
$= 5 + 2\sqrt{35} + 7$
$= 12 + 2\sqrt{35}$

129. Answers will vary. The answer is correct but it is not in simplest form. The correct answer is $\sqrt{24x^5} = \sqrt{4x^4 \cdot 6x} = 2x^2\sqrt{6x}.$

Section 8.5

Preparing for Dividing Expressions with Square Roots

P1. $\frac{55x^8}{11x} = \frac{5 \cdot 11 \cdot x^7 \cdot x}{11 \cdot x} = 5x^7$

P2. $\sqrt{\frac{45x^7}{5x^3}} = \sqrt{\frac{5x^3 \cdot 9x^4}{5x^3}} = \sqrt{9x^4} = 3x^2$

8.5 Quick Checks

1. $\frac{\sqrt{a}}{\sqrt{b}} = \underline{\sqrt{\frac{a}{b}}}$

2. $\frac{\sqrt{108}}{\sqrt{12}} = \sqrt{\frac{108}{12}} = \sqrt{\frac{12 \cdot 9}{12}} = \sqrt{9} = 3$

3. $\frac{\sqrt{35y}}{\sqrt{5y}} = \sqrt{\frac{35y}{5y}} = \sqrt{\frac{5y \cdot 7}{5y}} = \sqrt{7}$

4. $\frac{\sqrt{32}}{\sqrt{50}} = \sqrt{\frac{32}{50}} = \sqrt{\frac{2 \cdot 16}{2 \cdot 25}} = \sqrt{\frac{16}{25}} = \frac{4}{5}$

5. $\frac{\sqrt{30}-\sqrt{75}}{\sqrt{5}} = \frac{\sqrt{30}}{\sqrt{5}} - \frac{\sqrt{75}}{\sqrt{5}}$

$= \sqrt{\frac{30}{5}} - \sqrt{\frac{75}{5}}$

$= \sqrt{\frac{5 \cdot 6}{5}} - \sqrt{\frac{5 \cdot 15}{5}}$

$= \sqrt{6} - \sqrt{15}$

6. To rationalize the denominator means to rewrite a quotient in an equivalent form which has a rational number in the denominator.

7. True

8. $\frac{1}{\sqrt{2}} = \frac{1}{\sqrt{2}} \cdot \frac{\sqrt{2}}{\sqrt{2}} = \frac{\sqrt{2}}{\sqrt{4}} = \frac{\sqrt{2}}{2}$

9. $-\frac{\sqrt{2}}{5\sqrt{3}} = -\frac{\sqrt{2}}{5\sqrt{3}} \cdot \frac{\sqrt{3}}{\sqrt{3}}$

$= -\frac{\sqrt{6}}{5\sqrt{9}}$

$= -\frac{\sqrt{6}}{5 \cdot 3}$

$= -\frac{\sqrt{6}}{15}$

10. $\frac{5}{\sqrt{20}} = \frac{5}{\sqrt{20}} \cdot \frac{\sqrt{5}}{\sqrt{5}} = \frac{5\sqrt{5}}{\sqrt{100}} = \frac{5\sqrt{5}}{10} = \frac{\sqrt{5}}{2}$

11. False; the conjugate of $-5-2\sqrt{7}$ is $-5+2\sqrt{7}$.

12. $\frac{2}{2+\sqrt{3}} = \frac{2}{2+\sqrt{3}} \cdot \frac{2-\sqrt{3}}{2-\sqrt{3}}$

$= \frac{2(2-\sqrt{3})}{(2+\sqrt{3})(2-\sqrt{3})}$

$= \frac{2(2-\sqrt{3})}{(2)^2-(\sqrt{3})^2}$

$= \frac{2(2-\sqrt{3})}{4-3}$

$= \frac{2(2-\sqrt{3})}{1}$

$= 4-2\sqrt{3}$

13. $\frac{12}{3-\sqrt{5}} = \frac{12}{3-\sqrt{5}} \cdot \frac{3+\sqrt{5}}{3+\sqrt{5}}$

$= \frac{12(3+\sqrt{5})}{(3-\sqrt{5})(3+\sqrt{5})}$

$= \frac{12(3+\sqrt{5})}{(3)^2-(\sqrt{5})^2}$

$= \frac{12(3+\sqrt{5})}{9-5}$

$= \frac{4 \cdot 3(3+\sqrt{5})}{4}$

$= 3(3+\sqrt{5})$

$= 9+3\sqrt{5}$

14. $\frac{\sqrt{3}}{4+\sqrt{7}} = \frac{\sqrt{3}}{4+\sqrt{7}} \cdot \frac{4-\sqrt{7}}{4-\sqrt{7}}$

$= \frac{\sqrt{3}(4-\sqrt{7})}{(4)^2-(\sqrt{7})^2}$

$= \frac{4\sqrt{3}-\sqrt{21}}{16-7}$

$= \frac{4\sqrt{3}-\sqrt{21}}{9}$

15. $\frac{\sqrt{3}}{3-\sqrt{x}} = \frac{\sqrt{3}}{3-\sqrt{x}} \cdot \frac{3+\sqrt{x}}{3+\sqrt{x}}$
$= \frac{\sqrt{3}\left(3+\sqrt{x}\right)}{(3)^2 - \left(\sqrt{x}\right)^2}$
$= \frac{3\sqrt{3}+\sqrt{3x}}{9-x}$

16. $\frac{-\sqrt{2}}{\sqrt{3}+\sqrt{2}} = \frac{-\sqrt{2}}{\sqrt{3}+\sqrt{2}} \cdot \frac{\sqrt{3}-\sqrt{2}}{\sqrt{3}-\sqrt{2}}$
$= \frac{-\sqrt{2}\left(\sqrt{3}-\sqrt{2}\right)}{\left(\sqrt{3}\right)^2 - \left(\sqrt{2}\right)^2}$
$= \frac{-\sqrt{6}+\sqrt{4}}{3-2}$
$= \frac{-\sqrt{6}+2}{1}$
$= 2-\sqrt{6}$

8.5 Exercises

17. $\frac{\sqrt{96}}{\sqrt{6}} = \sqrt{\frac{96}{6}} = \sqrt{16} = 4$

19. $-\frac{\sqrt{5}}{\sqrt{125}} = -\sqrt{\frac{5}{125}} = -\sqrt{\frac{1}{25}} = -\frac{1}{5}$

21. $\frac{\sqrt{54}}{\sqrt{3}} = \sqrt{\frac{54}{3}} = \sqrt{18} = \sqrt{9 \cdot 2} = 3\sqrt{2}$

23. $\frac{\sqrt{196}}{\sqrt{7}} = \sqrt{\frac{196}{7}} = \sqrt{28} = \sqrt{4 \cdot 7} = 2\sqrt{7}$

25. $\frac{-\sqrt{75y^2}}{\sqrt{3y}} = -\sqrt{\frac{75y^2}{3y}} = -\sqrt{25y} = -5\sqrt{y}$

27. $\sqrt{\frac{128a^2b^5}{2a^2b^5}} = \sqrt{64} = 8$

29. $\frac{3}{\sqrt{5}} = \frac{3}{\sqrt{5}} \cdot \frac{\sqrt{5}}{\sqrt{5}} = \frac{3\sqrt{5}}{\left(\sqrt{5}\right)^2} = \frac{3\sqrt{5}}{5}$

31. $\frac{20}{\sqrt{6}} = \frac{20}{\sqrt{6}} \cdot \frac{\sqrt{6}}{\sqrt{6}} = \frac{20\sqrt{6}}{\left(\sqrt{6}\right)^2} = \frac{20\sqrt{6}}{6} = \frac{10\sqrt{6}}{3}$

33. $-\frac{16}{3\sqrt{2}} = -\frac{16}{3\sqrt{2}} \cdot \frac{\sqrt{2}}{\sqrt{2}}$
$= -\frac{16\sqrt{2}}{3\left(\sqrt{2}\right)^2}$
$= -\frac{16\sqrt{2}}{3 \cdot 2}$
$= -\frac{8\sqrt{2}}{3}$

35. $\frac{2}{-\sqrt{x}} = -\frac{2}{\sqrt{x}} \cdot \frac{\sqrt{x}}{\sqrt{x}} = -\frac{2\sqrt{x}}{\left(\sqrt{x}\right)^2} = -\frac{2\sqrt{x}}{x}$

37. $\frac{9y}{\sqrt{2y}} = \frac{9y}{\sqrt{2y}} \cdot \frac{\sqrt{2y}}{\sqrt{2y}}$
$= \frac{9y\sqrt{2y}}{\left(\sqrt{2y}\right)^2}$
$= \frac{9y\sqrt{2y}}{2y}$
$= \frac{9\sqrt{2y}}{2}$

39. $\frac{6}{\sqrt{48}} = \frac{6}{\sqrt{48}} \cdot \frac{\sqrt{3}}{\sqrt{3}} = \frac{6\sqrt{3}}{\sqrt{144}} = \frac{6\sqrt{3}}{12} = \frac{\sqrt{3}}{2}$

41. $\frac{6n}{\sqrt{8n^3}} = \frac{6n}{\sqrt{8n^3}} \cdot \frac{\sqrt{2n}}{\sqrt{2n}}$
$= \frac{6n\sqrt{2n}}{\sqrt{16n^4}}$
$= \frac{6n\sqrt{2n}}{4n^2}$
$= \frac{3\sqrt{2n}}{2n}$

43. $\frac{4\sqrt{2}}{3\sqrt{6}} = \frac{4\sqrt{2}}{3\sqrt{6}} \cdot \frac{\sqrt{6}}{\sqrt{6}}$
$= \frac{4\sqrt{12}}{3(\sqrt{6})^2}$
$= \frac{4\sqrt{4 \cdot 3}}{3 \cdot 6}$
$= \frac{4 \cdot 2\sqrt{3}}{18}$
$= \frac{4\sqrt{3}}{9}$

45. $\frac{2\sqrt{a}}{7\sqrt{6b}} = \frac{2\sqrt{a}}{7\sqrt{6b}} \cdot \frac{\sqrt{6b}}{\sqrt{6b}}$
$= \frac{2\sqrt{6ab}}{7(\sqrt{6b})^2}$
$= \frac{2\sqrt{6ab}}{7 \cdot 6b}$
$= \frac{2\sqrt{6ab}}{42b}$
$= \frac{\sqrt{6ab}}{21b}$

47. $-\frac{7\sqrt{5}}{2\sqrt{18}} = -\frac{7\sqrt{5}}{2\sqrt{18}} \cdot \frac{\sqrt{2}}{\sqrt{2}}$
$= -\frac{7\sqrt{10}}{2\sqrt{36}}$
$= -\frac{7\sqrt{10}}{2 \cdot 6}$
$= -\frac{7\sqrt{10}}{12}$

49. The conjugate of $2+\sqrt{3}$ is $2-\sqrt{3}$.
$(2+\sqrt{3})(2-\sqrt{3}) = 2^2 - (\sqrt{3})^2 = 4 - 3 = 1$

51. The conjugate of $\sqrt{6}-1$ is $\sqrt{6}+1$.
$(\sqrt{6}-1)(\sqrt{6}+1) = (\sqrt{6})^2 - 1^2 = 6 - 1 = 5$

53. The conjugate of $-1-3\sqrt{14}$ is $-1+3\sqrt{14}$.
$(-1-3\sqrt{14})(-1+3\sqrt{14}) = (-1)^2 - (3\sqrt{14})^2$
$= 1 - 3^2(\sqrt{14})^2$
$= 1 - 9 \cdot 14$
$= 1 - 126$
$= -125$

55. The conjugate of $\sqrt{p}+3\sqrt{q}$ is $\sqrt{p}-3\sqrt{q}$.
$(\sqrt{p}+3\sqrt{q})(\sqrt{p}-3\sqrt{q}) = (\sqrt{p})^2 - (3\sqrt{q})^2$
$= p - 3^2(\sqrt{q})^2$
$= p - 9q$

57. $\frac{1}{3+\sqrt{5}} = \frac{1}{3+\sqrt{5}} \cdot \frac{3-\sqrt{5}}{3-\sqrt{5}}$
$= \frac{3-\sqrt{5}}{3^2-(\sqrt{5})^2}$
$= \frac{3-\sqrt{5}}{9-5}$
$= \frac{3-\sqrt{5}}{4}$

59. $\frac{9}{9-\sqrt{3}} = \frac{9}{9-\sqrt{3}} \cdot \frac{9+\sqrt{3}}{9+\sqrt{3}}$
$= \frac{9(9+\sqrt{3})}{9^2-(\sqrt{3})^2}$
$= \frac{9(9+\sqrt{3})}{81-3}$
$= \frac{9(9+\sqrt{3})}{78}$
$= \frac{3(9+\sqrt{3})}{26}$
$= \frac{27+3\sqrt{3}}{26}$

61. $\frac{4}{\sqrt{7}+6}=\frac{4}{\sqrt{7}+6}\cdot\frac{\sqrt{7}-6}{\sqrt{7}-6}$
$=\frac{4(\sqrt{7}-6)}{(\sqrt{7})^2-6^2}$
$=\frac{4(\sqrt{7}-6)}{7-36}$
$=\frac{4(\sqrt{7}-6)}{-29}$
$=-\frac{4\sqrt{7}-24}{29}$
$=\frac{24-4\sqrt{7}}{29}$

63. $\frac{3}{\sqrt{y}-6}=\frac{3}{\sqrt{y}-6}\cdot\frac{\sqrt{y}+6}{\sqrt{y}+6}$
$=\frac{3(\sqrt{y}+6)}{(\sqrt{y})^2-6^2}$
$=\frac{3\sqrt{y}+18}{y-36}$

65. $\frac{16}{-2\sqrt{2}+3}=\frac{16}{3-2\sqrt{2}}\cdot\frac{3+2\sqrt{2}}{3+2\sqrt{2}}$
$=\frac{16(3+2\sqrt{2})}{3^2-(2\sqrt{2})^2}$
$=\frac{48+32\sqrt{2}}{9-8}$
$=48+32\sqrt{2}$

67. $\frac{\sqrt{2}}{6-\sqrt{21}}=\frac{\sqrt{2}}{6-\sqrt{21}}\cdot\frac{6+\sqrt{21}}{6+\sqrt{21}}$
$=\frac{\sqrt{2}(6+\sqrt{21})}{6^2-(\sqrt{21})^2}$
$=\frac{6\sqrt{2}+\sqrt{2}\cdot\sqrt{21}}{36-21}$
$=\frac{6\sqrt{2}+\sqrt{42}}{15}$

69. $\frac{n}{\sqrt{n}+1}=\frac{n}{\sqrt{n}+1}\cdot\frac{\sqrt{n}-1}{\sqrt{n}-1}$
$=\frac{n(\sqrt{n}-1)}{(\sqrt{n})^2-1^2}$
$=\frac{n(\sqrt{n}-1)}{n-1}$
$=\frac{n\sqrt{n}-n}{n-1}$

71. $\frac{\sqrt{x}}{2+3\sqrt{x}}=\frac{\sqrt{x}}{2+3\sqrt{x}}\cdot\frac{2-3\sqrt{x}}{2-3\sqrt{x}}$
$=\frac{\sqrt{x}(2-3\sqrt{x})}{2^2-(3\sqrt{x})^2}$
$=\frac{2\sqrt{x}-3(\sqrt{x})^2}{4-3^2(\sqrt{x})^2}$
$=\frac{2\sqrt{x}-3x}{4-9x}$

73. $\frac{\sqrt{5}}{\sqrt{11}-\sqrt{6}}=\frac{\sqrt{5}}{\sqrt{11}-\sqrt{6}}\cdot\frac{\sqrt{11}+\sqrt{6}}{\sqrt{11}+\sqrt{6}}$
$=\frac{\sqrt{5}(\sqrt{11}+\sqrt{6})}{(\sqrt{11})^2-(\sqrt{6})^2}$
$=\frac{\sqrt{5}\cdot\sqrt{11}+\sqrt{5}\cdot\sqrt{6}}{11-6}$
$=\frac{\sqrt{55}+\sqrt{30}}{5}$

75. $\frac{\sqrt{2}}{\sqrt{6}-\sqrt{2}} = \frac{\sqrt{2}}{\sqrt{6}-\sqrt{2}} \cdot \frac{\sqrt{6}+\sqrt{2}}{\sqrt{6}+\sqrt{2}}$
$= \frac{\sqrt{2}\left(\sqrt{6}+\sqrt{2}\right)}{\left(\sqrt{6}\right)^2 - \left(\sqrt{2}\right)^2}$
$= \frac{\sqrt{2}\cdot\sqrt{6}+\left(\sqrt{2}\right)^2}{6-2}$
$= \frac{\sqrt{12}+2}{4}$
$= \frac{\sqrt{4\cdot 3}+2}{4}$
$= \frac{2\sqrt{3}+2}{4}$
$= \frac{2\left(\sqrt{3}+1\right)}{4}$
$= \frac{\sqrt{3}+1}{2}$

77. $\frac{\sqrt{6}-\sqrt{12}}{\sqrt{2}} = \frac{\sqrt{6}}{\sqrt{2}} - \frac{\sqrt{12}}{\sqrt{2}}$
$= \sqrt{\frac{6}{2}} - \sqrt{\frac{12}{2}}$
$= \sqrt{3} - \sqrt{6}$

79. $\sqrt{\frac{1}{2}} = \frac{\sqrt{1}}{\sqrt{2}} = \frac{1}{\sqrt{2}} \cdot \frac{\sqrt{2}}{\sqrt{2}} = \frac{\sqrt{2}}{\left(\sqrt{2}\right)^2} = \frac{\sqrt{2}}{2}$

81. $\frac{\sqrt{18}-\sqrt{20}}{2\sqrt{10}} = \frac{\sqrt{18}}{2\sqrt{10}} - \frac{\sqrt{20}}{2\sqrt{10}}$
$= \frac{1}{2}\sqrt{\frac{18}{10}} - \frac{1}{2}\sqrt{\frac{20}{10}}$
$= \frac{1}{2}\sqrt{\frac{9}{5}} - \frac{1}{2}\sqrt{2}$
$= \frac{1}{2} \cdot \frac{\sqrt{9}}{\sqrt{5}} - \frac{\sqrt{2}}{2}$
$= \frac{1}{2} \cdot \frac{3}{\sqrt{5}} - \frac{\sqrt{2}}{2}$
$= \frac{3}{2\sqrt{5}} \cdot \frac{\sqrt{5}}{\sqrt{5}} - \frac{\sqrt{2}}{2}$
$= \frac{3\sqrt{5}}{2\left(\sqrt{5}\right)^2} - \frac{\sqrt{2}}{2}$
$= \frac{3\sqrt{5}}{10} - \frac{\sqrt{2}}{2}$

83. $\frac{2}{3}\sqrt{18} + \frac{5}{2}\sqrt{8} = \frac{2}{3}\sqrt{9\cdot 2} + \frac{5}{2}\sqrt{4\cdot 2}$
$= \frac{2}{3}\cdot 3\sqrt{2} + \frac{5}{2}\cdot 2\sqrt{2}$
$= 2\sqrt{2} + 5\sqrt{2}$
$= 7\sqrt{2}$

85. $-\sqrt{\frac{10}{15}} = -\sqrt{\frac{2}{3}}$
$= -\frac{\sqrt{2}}{\sqrt{3}} \cdot \frac{\sqrt{3}}{\sqrt{3}}$
$= -\frac{\sqrt{6}}{\left(\sqrt{3}\right)^2}$
$= -\frac{\sqrt{6}}{3}$

87. $\frac{5}{14\sqrt{7}} = \frac{5}{14\sqrt{7}} \cdot \frac{\sqrt{7}}{\sqrt{7}}$
$= \frac{5\sqrt{7}}{14\left(\sqrt{7}\right)^2}$
$= \frac{5\sqrt{7}}{14\cdot 7}$
$= \frac{5\sqrt{7}}{98}$

89. $\frac{4x}{\sqrt{2x}} = \frac{4x}{\sqrt{2x}} \cdot \frac{\sqrt{2x}}{\sqrt{2x}}$
$= \frac{4x\sqrt{2x}}{\left(\sqrt{2x}\right)^2}$
$= \frac{4x\sqrt{2x}}{2x}$
$= 2\sqrt{2x}$

91. $\sqrt{\frac{4}{3}} = \frac{\sqrt{4}}{\sqrt{3}} = \frac{2}{\sqrt{3}} = \frac{2}{\sqrt{3}} \cdot \frac{\sqrt{3}}{\sqrt{3}} = \frac{2\sqrt{3}}{\left(\sqrt{3}\right)^2} = \frac{2\sqrt{3}}{3}$

93. $-4\sqrt{8} + 3\sqrt{20} = -4\sqrt{4\cdot 2} + 3\sqrt{4\cdot 5}$
$= -4\cdot 2\sqrt{2} + 3\cdot 2\sqrt{5}$
$= -8\sqrt{2} + 6\sqrt{5}$

95. $\sqrt{\frac{3}{4}}\cdot\sqrt{\frac{6}{15}}=\sqrt{\frac{3}{4}\cdot\frac{6}{15}}$
$=\sqrt{\frac{18}{60}}$
$=\sqrt{\frac{3}{10}}$
$=\frac{\sqrt{3}}{\sqrt{10}}\cdot\frac{\sqrt{10}}{\sqrt{10}}$
$=\frac{\sqrt{30}}{\left(\sqrt{10}\right)^2}$
$=\frac{\sqrt{30}}{10}$

97. $\frac{2}{2+\sqrt{3}}=\frac{2}{2+\sqrt{3}}\cdot\frac{2-\sqrt{3}}{2-\sqrt{3}}$
$=\frac{2\left(2-\sqrt{3}\right)}{2^2-\left(\sqrt{3}\right)^2}$
$=\frac{4-2\sqrt{3}}{4-3}$
$=\frac{4-2\sqrt{3}}{1}$
$=4-2\sqrt{3}$

99. $\frac{1}{\sqrt{50}}=\frac{1}{\sqrt{50}}\cdot\frac{\sqrt{2}}{\sqrt{2}}=\frac{\sqrt{2}}{\sqrt{100}}=\frac{\sqrt{2}}{10}$

101. $\sqrt{\frac{5x}{y}}\cdot\sqrt{\frac{1}{35x^4}}=\sqrt{\frac{5x}{y}\cdot\frac{1}{35x^4}}$
$=\sqrt{\frac{5x}{35x^4y}}$
$=\sqrt{\frac{1}{7x^3y}}$
$=\frac{\sqrt{1}}{\sqrt{7x^3y}}\cdot\frac{\sqrt{7xy}}{\sqrt{7xy}}$
$=\frac{\sqrt{7xy}}{\sqrt{49x^4y^2}}$
$=\frac{\sqrt{7xy}}{7x^2y}$

103. $\left(3-\sqrt{2}\right)^2=3^2-2\cdot3\sqrt{2}+\left(\sqrt{2}\right)^2$
$=9-6\sqrt{2}+2$
$=11-6\sqrt{2}$

105. **(a)** $\frac{1}{2+\sqrt{3}}\approx0.2679$

(b) $\frac{1}{2+\sqrt{3}}=\frac{1}{2+\sqrt{3}}\cdot\frac{2-\sqrt{3}}{2-\sqrt{3}}$
$=\frac{2-\sqrt{3}}{2^2-\left(\sqrt{3}\right)^2}$
$=\frac{2-\sqrt{3}}{4-3}$
$=\frac{2-\sqrt{3}}{1}$
$=2-\sqrt{3}$

(c) $2-\sqrt{3}\approx0.2679$

(d) They are the same.

107. **(a)** $\frac{5}{2\sqrt{3}-1}\approx2.0291$

(b) $\frac{5}{2\sqrt{3}-1}=\frac{5}{2\sqrt{3}-1}\cdot\frac{2\sqrt{3}+1}{2\sqrt{3}+1}$
$=\frac{5\left(2\sqrt{3}+1\right)}{\left(2\sqrt{3}\right)^2-1^2}$
$=\frac{10\sqrt{3}+5}{4\cdot3-1}$
$=\frac{10\sqrt{3}+5}{11}$

(c) $\frac{10\sqrt{3}+5}{11}\approx2.0291$

(d) They are the same.

109. **(a)** $\frac{\sqrt{8}}{\sqrt{2}-3}\approx-1.7836$

(b) $\frac{\sqrt{8}}{\sqrt{2}-3}=\frac{\sqrt{4\cdot 2}}{\sqrt{2}-3}\cdot\frac{\sqrt{2}+3}{\sqrt{2}+3}$
$=\frac{2\sqrt{2}\left(\sqrt{2}+3\right)}{\left(\sqrt{2}\right)^2-3^2}$
$=\frac{2\sqrt{2}\cdot\sqrt{2}+3\cdot 2\sqrt{2}}{2-9}$
$=\frac{2\left(\sqrt{2}\right)^2+6\sqrt{2}}{-7}$
$=-\frac{4+6\sqrt{2}}{7}$

(c) $-\frac{4+6\sqrt{2}}{7}\approx -1.7836$

(d) They are the same.

111. (a) $\frac{\sqrt{6}}{3+5\sqrt{3}}\approx 0.2101$

(b) $\frac{\sqrt{6}}{3+5\sqrt{3}}=\frac{\sqrt{6}}{3+5\sqrt{3}}\cdot\frac{3-5\sqrt{3}}{3-5\sqrt{3}}$
$=\frac{\sqrt{6}\left(3-5\sqrt{3}\right)}{3^2-\left(5\sqrt{3}\right)^2}$
$=\frac{3\cdot\sqrt{6}-5\sqrt{3}\cdot\sqrt{6}}{9-5^2\left(\sqrt{3}\right)^2}$
$=\frac{3\sqrt{6}-5\sqrt{18}}{9-25\cdot 3}$
$=\frac{3\sqrt{6}-5\sqrt{9\cdot 2}}{9-75}$
$=\frac{3\sqrt{6}-5\cdot 3\sqrt{2}}{-66}$
$=\frac{3\sqrt{6}-15\sqrt{2}}{-66}$
$=\frac{-3\left(-\sqrt{6}+5\sqrt{2}\right)}{-3(22)}$
$=\frac{5\sqrt{2}-\sqrt{6}}{22}$

(c) $\frac{5\sqrt{2}-\sqrt{6}}{22}\approx 0.2101$

(d) They are the same.

113. $\frac{1+\sqrt{3}}{1-\sqrt{3}}=\frac{1+\sqrt{3}}{1-\sqrt{3}}\cdot\frac{1+\sqrt{3}}{1+\sqrt{3}}$
$=\frac{\left(1+\sqrt{3}\right)^2}{1^2-\left(\sqrt{3}\right)^2}$
$=\frac{1^2+2\cdot 1\cdot\sqrt{3}+\left(\sqrt{3}\right)^2}{1-3}$
$=\frac{1+2\sqrt{3}+3}{-2}$
$=\frac{4+2\sqrt{3}}{-2}$
$=\frac{-2\left(-2-\sqrt{3}\right)}{-2}$
$=-2-\sqrt{3}$

115. $\frac{\sqrt{6}-\sqrt{3}}{\sqrt{2}-\sqrt{3}}$
$=\frac{\sqrt{6}-\sqrt{3}}{\sqrt{2}-\sqrt{3}}\cdot\frac{\sqrt{2}+\sqrt{3}}{\sqrt{2}+\sqrt{3}}$
$=\frac{\sqrt{6}\cdot\sqrt{2}+\sqrt{6}\cdot\sqrt{3}-\sqrt{3}\cdot\sqrt{2}-\left(\sqrt{3}\right)^2}{\left(\sqrt{2}\right)^2-\left(\sqrt{3}\right)^2}$
$=\frac{\sqrt{12}+\sqrt{18}-\sqrt{6}-3}{2-3}$
$=\frac{\sqrt{4\cdot 3}+\sqrt{9\cdot 2}-\sqrt{6}-3}{-1}$
$=-1\left(2\sqrt{3}+3\sqrt{2}-\sqrt{6}-3\right)$
$=-2\sqrt{3}-3\sqrt{2}+\sqrt{6}+3$

117. $\frac{2\sqrt{x}+\sqrt{y}}{\sqrt{x}+4\sqrt{y}}$
$=\frac{2\sqrt{x}+\sqrt{y}}{\sqrt{x}+4\sqrt{y}}\cdot\frac{\sqrt{x}-4\sqrt{y}}{\sqrt{x}-4\sqrt{y}}$
$=\frac{2\left(\sqrt{x}\right)^2-2\sqrt{x}\cdot 4\sqrt{y}+\sqrt{y}\cdot\sqrt{x}-4\left(\sqrt{y}\right)^2}{\left(\sqrt{x}\right)^2-\left(4\sqrt{y}\right)^2}$
$=\frac{2x-8\sqrt{xy}+\sqrt{xy}-4y}{x-4^2\left(\sqrt{y}\right)^2}$
$=\frac{2x-7\sqrt{xy}-4y}{x-16y}$

119. x cannot be divided out in the expression $\frac{3\sqrt{xy}}{x}$ because the x in the numerator is part of the radicand and the x in the denominator is not. The quotient rule states $\frac{\sqrt{a}}{\sqrt{b}} = \sqrt{\frac{a}{b}}$.

121. Neither student is correct. The correct approach is

$$\begin{aligned}\frac{3}{\sqrt{2}+\sqrt{3}} &= \frac{3}{\sqrt{2}+\sqrt{3}}\cdot\frac{\sqrt{2}-\sqrt{3}}{\sqrt{2}-\sqrt{3}}\\ &= \frac{3\left(\sqrt{2}-\sqrt{3}\right)}{\left(\sqrt{2}\right)^2-\left(\sqrt{3}\right)^2}\\ &= \frac{3\left(\sqrt{2}-\sqrt{3}\right)}{2-3}\\ &= \frac{3\left(\sqrt{2}-\sqrt{3}\right)}{-1}\\ &= -3\sqrt{2}-3\sqrt{3}\end{aligned}$$

Putting the Concepts Together (Sections 8.1–8.5)

1. (a) $\sqrt{98} \approx 9.90$ is irrational.

(b) $\sqrt{361} = 19$ is rational.

2. $-2\sqrt{18} = -2\sqrt{9\cdot 2} = -2\cdot 3\sqrt{2} = -6\sqrt{2}$

3. $\sqrt{32x^3y^6} = \sqrt{16x^2y^6\cdot 2x} = 4xy^3\sqrt{2x}$

4. $\sqrt{\frac{100x^3}{16x^5}} = \sqrt{\frac{4x^3\cdot 25}{4x^3\cdot 4x^2}} = \sqrt{\frac{25}{4x^2}} = \frac{5}{2x}$

5. $\sqrt{121-4(6)(3)} = \sqrt{121-72} = \sqrt{49} = 7$

6.
$$\begin{aligned}\frac{\sqrt{72m^6}}{\sqrt{8}} &= \sqrt{\frac{72m^6}{8}}\\ &= \sqrt{\frac{8\cdot 9m^6}{8}}\\ &= \sqrt{9m^6}\\ &= 3m^3\end{aligned}$$

7.
$$\begin{aligned}5\sqrt{3}-\sqrt{12} &= 5\sqrt{3}-\sqrt{4\cdot 3}\\ &= 5\sqrt{3}-2\sqrt{3}\\ &= (5-2)\sqrt{3}\\ &= 3\sqrt{3}\end{aligned}$$

8.
$$\begin{aligned}3\sqrt{50}-6\sqrt{8} &= 3\sqrt{25\cdot 2}-6\sqrt{4\cdot 2}\\ &= 3\cdot 5\sqrt{2}-6\cdot 2\sqrt{2}\\ &= 15\sqrt{2}-12\sqrt{2}\\ &= (15-12)\sqrt{2}\\ &= 3\sqrt{2}\end{aligned}$$

9.
$$\begin{aligned}\sqrt{12}\cdot\sqrt{18} &= \sqrt{12\cdot 18}\\ &= \sqrt{216}\\ &= \sqrt{36\cdot 6}\\ &= 6\sqrt{6}\end{aligned}$$

10.
$$\begin{aligned}\left(2\sqrt{7}\right)\left(-3\sqrt{7}\right) &= 2\cdot(-3)\left(\sqrt{7}\right)^2\\ &= -6\cdot 7\\ &= -42\end{aligned}$$

11.
$$\begin{aligned}6\sqrt{5}\left(2\sqrt{10}+3\right) &= 6\sqrt{5}\cdot 2\sqrt{10}+6\sqrt{5}\cdot 3\\ &= 12\sqrt{50}+18\sqrt{5}\\ &= 12\sqrt{25\cdot 2}+18\sqrt{5}\\ &= 12\cdot 5\sqrt{2}+18\sqrt{5}\\ &= 60\sqrt{2}+18\sqrt{5}\end{aligned}$$

12.
$$\begin{aligned}&\left(6-\sqrt{3}\right)\left(2-4\sqrt{3}\right)\\ &= 6\cdot 2-6\cdot 4\sqrt{3}-\sqrt{3}\cdot 2+4\cdot\left(\sqrt{3}\right)^2\\ &= 12-24\sqrt{3}-2\sqrt{3}+4\cdot 3\\ &= 12-26\sqrt{3}+12\\ &= 24-26\sqrt{3}\end{aligned}$$

13.
$$\begin{aligned}\left(3-2\sqrt{5}\right)\left(3+2\sqrt{5}\right) &= 3^2-\left(2\sqrt{5}\right)^2\\ &= 9-2^2\left(\sqrt{5}\right)^2\\ &= 9-4\cdot 5\\ &= 9-20\\ &= -11\end{aligned}$$

14.
$$\begin{aligned}\left(\sqrt{7}-2\right)^2 &= \left(\sqrt{7}\right)^2-2\cdot\sqrt{7}\cdot 2+2^2\\ &= 7-4\sqrt{7}+4\\ &= 11-4\sqrt{7}\end{aligned}$$

15. $\frac{-4+\sqrt{28}}{2}=\frac{-4+\sqrt{4\cdot 7}}{2}$
$=\frac{-4+2\sqrt{7}}{2}$
$=\frac{2\left(-2+\sqrt{7}\right)}{2}$
$=-2+\sqrt{7}$

16. $\frac{\sqrt{90x^2}}{\sqrt{5x^6}}=\sqrt{\frac{90x^2}{5x^6}}$
$=\sqrt{\frac{5x^2\cdot 18}{5x^2\cdot x^4}}$
$=\sqrt{\frac{18}{x^4}}$
$=\frac{\sqrt{9\cdot 2}}{\sqrt{x^4}}$
$=\frac{3\sqrt{2}}{x^2}$

17. $\frac{8}{\sqrt{24}}=\frac{8}{\sqrt{4\cdot 6}}$
$=\frac{8}{2\sqrt{6}}$
$=\frac{4}{\sqrt{6}}\cdot\frac{\sqrt{6}}{\sqrt{6}}$
$=\frac{4\sqrt{6}}{6}$
$=\frac{2\sqrt{6}}{3}$

18. $\frac{12}{\sqrt{3}+3}=\frac{12}{\sqrt{3}+3}\cdot\frac{\sqrt{3}-3}{\sqrt{3}-3}$
$=\frac{12\left(\sqrt{3}-3\right)}{\left(\sqrt{3}\right)^2-3^2}$
$=\frac{12\left(\sqrt{3}-3\right)}{3-9}$
$=\frac{12\left(\sqrt{3}-3\right)}{-6}$
$=-2\left(\sqrt{3}-3\right)$
$=-2\sqrt{3}+6$

Section 8.6

Preparing for Solving Equations Containing Square Roots

P1. $4x-5=0$
$4x-5+5=0+5$
$4x=5$
$\frac{4x}{4}=\frac{5}{4}$
$x=\frac{5}{4}$
The solution set is $\left\{\frac{5}{4}\right\}$.

P2. $2p^2+4p-16=0$
$2(p^2+2p-8)=0$
$p^2+2p-8=0$
$(p+4)(p-2)=0$
$p+4=0$ or $p-2=0$
$p=-4$ or $p=2$
The solution set is $\{-4, 2\}$.

P3. $(2x-3)^2=(2x)^2-2\cdot 2x\cdot 3+3^2$
$=4x^2-12x+9$

P4. $\left(\sqrt{x}+2\right)^2=\left(\sqrt{x}\right)^2+2\cdot\sqrt{x}\cdot 2+2^2$
$=x+4\sqrt{x}+4$

8.6 Quick Checks

1. An equation such as $\sqrt{x+3}=5$ is called a <u>radical</u> <u>equation</u>.

2. $\sqrt{6x+4}=8;\ x=10$
$\sqrt{6(10)+4}\stackrel{?}{=}8$
$\sqrt{60+4}\stackrel{?}{=}8$
$\sqrt{64}\stackrel{?}{=}8$
$8=8$ True
$x = 10$ is a solution of $\sqrt{6x+4}=8$.

3. $\sqrt{n+2}=n;\ n=-1$
$\sqrt{-1+2}\stackrel{?}{=}-1$
$\sqrt{1}\stackrel{?}{=}-1$
$1=-1$ False
$n=-1$ is not a solution of $\sqrt{n+2}=n$.

4. $\sqrt{2y+11}=3;\ y=-1$

$$\sqrt{2(-1)+11}\stackrel{?}{=}3$$
$$\sqrt{-2+11}\stackrel{?}{=}3$$
$$\sqrt{9}\stackrel{?}{=}3$$
$$3=3\quad \text{True}$$

$y=-1$ is a solution of $\sqrt{2y+11}=3$.

5. True

6.
$$\sqrt{y-5}=3$$
$$\left(\sqrt{y-5}\right)^2=3^2$$
$$y-5=9$$
$$y-5+5=9+5$$
$$y=14$$

Check: $\sqrt{14-5}\stackrel{?}{=}3$
$$\sqrt{9}\stackrel{?}{=}3$$
$$3=3\quad \text{True}$$

The solution set is $\{14\}$.

7.
$$\sqrt{x+17}+4=7$$
$$\sqrt{x+17}+4-4=7-4$$
$$\sqrt{x+17}=3$$
$$\left(\sqrt{x+17}\right)^2=3^2$$
$$x+17=9$$
$$x+17-17=9-17$$
$$x=-8$$

Check: $\sqrt{-8+17}+4\stackrel{?}{=}7$
$$\sqrt{9}+4\stackrel{?}{=}7$$
$$3+4\stackrel{?}{=}7$$
$$7=7\quad \text{True}$$

The solution set is $\{7\}$.

8.
$$2\sqrt{p}=12$$
$$\frac{2\sqrt{p}}{2}=\frac{12}{2}$$
$$\sqrt{p}=6$$
$$\left(\sqrt{p}\right)^2=6^2$$
$$p=36$$

Check: $2\sqrt{36}\stackrel{?}{=}12$
$$2\cdot 6\stackrel{?}{=}12$$
$$12=12\quad \text{True}$$

The solution set is $\{36\}$.

9.
$$4\sqrt{t-3}-2=10$$
$$4\sqrt{t-3}-2+2=10+2$$
$$4\sqrt{t-3}=12$$
$$\frac{4\sqrt{t-3}}{4}=\frac{12}{4}$$
$$\sqrt{t-3}=3$$
$$\left(\sqrt{t-3}\right)^2=3^2$$
$$t-3=9$$
$$t-3+3=9+3$$
$$t=12$$

Check: $4\sqrt{12-3}-2\stackrel{?}{=}10$
$$4\sqrt{9}-2\stackrel{?}{=}10$$
$$4\cdot 3-2\stackrel{?}{=}10$$
$$12-12\stackrel{?}{=}10$$
$$10=10\quad \text{True}$$

The solution set is $\{12\}$.

10. Apparent solutions that do not satisfy the original equation are called extraneous solutions.

11.
$$\sqrt{2a+3}=a$$
$$\left(\sqrt{2a+3}\right)^2=a^2$$
$$2a+3=a^2$$
$$2a-2a+3-3=a^2-2a-3$$
$$0=a^2-2a-3$$
$$0=(a-3)(a+1)$$
$$a-3=0 \quad \text{or} \quad a+1=0$$
$$a=3 \quad \text{or} \quad a=-1$$

Check $x=3$: $\sqrt{2(3)+3}\stackrel{?}{=}3$
$$\sqrt{6+3}\stackrel{?}{=}3$$
$$\sqrt{9}\stackrel{?}{=}3$$
$$3=3\quad \text{True}$$

Check $x=-1$:
$$\sqrt{2(-1)+3}\stackrel{?}{=}-1$$
$$\sqrt{-2+3}\stackrel{?}{=}-1$$
$$\sqrt{1}\stackrel{?}{=}-1$$
$$1=-1\quad \text{False}$$

The solution set is $\{3\}$.

12. $\sqrt{2p+8} = p$

$\left(\sqrt{2p+8}\right)^2 = p^2$

$2p+8 = p^2$

$2p-2p+8-8 = p^2-2p-8$

$0 = p^2-2p-8$

$0 = (p-4)(p+2)$

$p-4=0$ or $p+2=0$

$p=4$ or $p=-2$

Check $p=4$: $\sqrt{2(4)+8} \stackrel{?}{=} 4$

$\sqrt{8+8} \stackrel{?}{=} 4$

$\sqrt{16} \stackrel{?}{=} 4$

$4=4$ True

Check $p=-2$: $\sqrt{2(-2)+8} \stackrel{?}{=} -2$

$\sqrt{-4+8} \stackrel{?}{=} -2$

$\sqrt{4} \stackrel{?}{=} -2$

$2=-2$ False

The solution set is $\{4\}$.

13. False; the first step is to isolate the radical.

14. $\sqrt{m+20} = m+8$

$\left(\sqrt{m+20}\right)^2 = (m+8)^2$

$m+20 = m^2+16m+64$

$0 = m^2+15m+44$

$0 = (m+11)(m+4)$

$m+11=0$ or $m+4=0$

$m=-11$ or $m=-4$

Check: $m=-11$: $\sqrt{-11+20} \stackrel{?}{=} -11+8$

$\sqrt{9} \stackrel{?}{=} -3$

$3=-3$ False

Check $m=-4$: $\sqrt{-4+20} \stackrel{?}{=} -4+8$

$\sqrt{16} \stackrel{?}{=} 4$

$4=4$ True

The solution set is $\{-4\}$.

15. $\sqrt{17-2x}+1 = x$

$\sqrt{17-2x}+1-1 = x-1$

$\sqrt{17-2x} = x-1$

$\left(\sqrt{17-2x}\right)^2 = (x-1)^2$

$17-2x = x^2-2x+1$

$17-17-2x+2x = x^2-2x+1-17+2x$

$0 = x^2-16$

$0 = (x-4)(x+4)$

$x-4=0$ or $x+4=0$

$x=4$ or $x=-4$

Check $x=4$: $\sqrt{17-2(4)}+1 \stackrel{?}{=} 4$

$\sqrt{17-8}+1 \stackrel{?}{=} 4$

$\sqrt{9}+1 \stackrel{?}{=} 4$

$3+1 \stackrel{?}{=} 4$

$4=4$ True

Check $x=-4$: $\sqrt{17-2(-4)}+1 \stackrel{?}{=} -4$

$\sqrt{17+8}+1 \stackrel{?}{=} -4$

$\sqrt{25}+1 \stackrel{?}{=} -4$

$5+1 \stackrel{?}{=} -4$

$6=-4$ False

The solution set is $\{4\}$.

16. $\sqrt{3b-2}+8 = 5$

$\sqrt{3b-2}+8-8 = 5-8$

$\sqrt{3b-2} = -3$

$\left(\sqrt{3b-2}\right)^2 = (-3)^2$

$3b-2 = 9$

$3b-2+2 = 9+2$

$3b = 11$

$b = \frac{11}{3}$

Check: $\sqrt{3\left(\frac{11}{3}\right)-2}+8 \stackrel{?}{=} 5$

$\sqrt{11-2}+8 \stackrel{?}{=} 5$

$\sqrt{9}+8 \stackrel{?}{=} 5$

$3+8 \stackrel{?}{=} 5$

$11=5$ False

The solution set is { } or $\varnothing$.

17. $\sqrt{3x+1} = \sqrt{2x+7}$

$$\left(\sqrt{3x+1}\right)^2 = \left(\sqrt{2x+7}\right)^2$$
$$3x+1 = 2x+7$$
$$3x+1-1 = 2x+7-1$$
$$3x = 2x+6$$
$$3x-2x = 2x-2x+6$$
$$x = 6$$

Check: $\sqrt{3(6)+1} \stackrel{?}{=} \sqrt{2(6)+7}$

$$\sqrt{18+1} \stackrel{?}{=} \sqrt{12+7}$$
$$\sqrt{19} = \sqrt{19} \quad \text{True}$$

The solution set is $\{6\}$.

18. $\sqrt{2w^2-3w-4} = \sqrt{w^2+6w+6}$

$$\left(\sqrt{2w^2-3w-4}\right)^2 = \left(\sqrt{w^2+6w+6}\right)^2$$
$$2w^2-3w-4 = w^2+6w+6$$
$$2w^2-3w-4-w^2-6w-6 = w^2-w^2+6w-6w+6-6$$
$$w^2-9w-10 = 0$$
$$(w-10)(w+1) = 0$$
$$w-10 = 0 \quad \text{or} \quad w+1 = 0$$
$$w = 10 \quad \text{or} \quad w = -1$$

Check $w = 10$: $\sqrt{2(10)^2-3(10)-4} \stackrel{?}{=} \sqrt{10^2+6(10)+6}$

$$\sqrt{2\cdot 100-30-4} \stackrel{?}{=} \sqrt{100+60+6}$$
$$\sqrt{200-34} \stackrel{?}{=} \sqrt{166}$$
$$\sqrt{166} = \sqrt{166} \quad \text{True}$$

Check $w = -1$: $\sqrt{2(-1)^2-3(-1)-4} \stackrel{?}{=} \sqrt{(-1)^2+6(-1)+6}$

$$\sqrt{2\cdot 1+3-4} \stackrel{?}{=} \sqrt{1-6+6}$$
$$\sqrt{2+3-4} \stackrel{?}{=} \sqrt{1}$$
$$\sqrt{1} = \sqrt{1} \quad \text{True}$$

The solution set is $\{-1, 10\}$.

19. $2\sqrt{k+5} - \sqrt{8k+4} = 0$

$$2\sqrt{k+5} - \sqrt{8k+4} + \sqrt{8k+4} = \sqrt{8k+4}$$
$$2\sqrt{k+5} = \sqrt{8k+4}$$
$$\left(2\sqrt{k+5}\right)^2 = \left(\sqrt{8k+4}\right)^2$$
$$4(k+5) = 8k+4$$
$$4k+20 = 8k+4$$
$$4k-4k+20 = 8k+4-4k$$
$$20 = 4k+4$$
$$20-4 = 4k+4-4$$
$$16 = 4k$$
$$\frac{16}{4} = \frac{4k}{4}$$
$$4 = k$$

Check: $2\sqrt{4+5}-\sqrt{8(4)+4} \stackrel{?}{=} 0$
$2\sqrt{9}-\sqrt{32+4} \stackrel{?}{=} 0$
$2\cdot 3-\sqrt{36} \stackrel{?}{=} 0$
$6-6 \stackrel{?}{=} 0$
$0=0$ True
The solution set is {4}.

20. False; the first step is $\left(\sqrt{x}+3\right)^2=\left(\sqrt{2x+1}\right)^2$.

21.
$$\sqrt{x}=\sqrt{x+9}-1$$
$$\sqrt{x}+1=\sqrt{x+9}-1+1$$
$$\sqrt{x}+1=\sqrt{x+9}$$
$$\left(\sqrt{x}+1\right)^2=\left(\sqrt{x+9}\right)^2$$
$$x+2\sqrt{x}+1=x+9$$
$$x+2\sqrt{x}+1-x-1=x-x+9-1$$
$$2\sqrt{x}=8$$
$$\frac{2\sqrt{x}}{2}=\frac{8}{2}$$
$$\sqrt{x}=4$$
$$\left(\sqrt{x}\right)^2=4^2$$
$$x=16$$
Check: $\sqrt{16} \stackrel{?}{=} \sqrt{16+9}-1$
$4 \stackrel{?}{=} \sqrt{25}-1$
$4 \stackrel{?}{=} 5-1$
$4=4$ True
The solution set is {16}.

22. $N=10\sqrt{O}$
$70=10\sqrt{O}$
$7=\sqrt{O}$
$7^2=\left(\sqrt{O}\right)^2$
$49=O$
The original score is 49.

8.6 Exercises

23. $\sqrt{6x+1}=5;\ x=4$
$\sqrt{6\cdot 4+1} \stackrel{?}{=} 5$
$\sqrt{24+1} \stackrel{?}{=} 5$
$\sqrt{25} \stackrel{?}{=} 5$
$5=5$ True
$x=4$ is a solution of $\sqrt{6x+1}=5$.

25. $\sqrt{5n-6}+6=0;\ n=-6$
$\sqrt{5(-6)-6}+6 \stackrel{?}{=} 0$
$\sqrt{-30-6}+6 \stackrel{?}{=} 0$
$\sqrt{-36}+6 \stackrel{?}{=} 0$
Since $\sqrt{-36}$ is not a real number, $n=-6$ is not a solution to $\sqrt{5n-6}+6=0$.

27. $\sqrt{p+6}=p+4;\ p=-2$
$\sqrt{-2+6} \stackrel{?}{=} -2+4$
$\sqrt{4} \stackrel{?}{=} 2$
$2=2$ True
$p=-2$ is a solution of $\sqrt{p+6}=p+4$.

29. $\sqrt{2y+3}=\sqrt{y+5}-1;\ y=-1$
$\sqrt{2(-1)+3} \stackrel{?}{=} \sqrt{-1+5}-1$
$\sqrt{-2+3} \stackrel{?}{=} \sqrt{4}-1$
$\sqrt{1} \stackrel{?}{=} 2-1$
$1=1$ True
$y=-1$ is a solution to $\sqrt{2y+3}=\sqrt{y+5}-1$.

31.
$$\sqrt{a}=4$$
$$\left(\sqrt{a}\right)^2=4^2$$
$$a=16$$
Check: $\sqrt{16} \stackrel{?}{=} 4$
$4=4$ True
The solution set is {16}.

33.
$$\sqrt{5-x}=3$$
$$\left(\sqrt{5-x}\right)^2=3^2$$
$$5-x=9$$
$$5=9+x$$
$$-4=x$$
Check: $\sqrt{5-(-4)} \stackrel{?}{=} 3$
$\sqrt{5+4} \stackrel{?}{=} 3$
$\sqrt{9} \stackrel{?}{=} 3$
$3=3$ True
The solution set is {−4}.

35. $5=\sqrt{4x-3}$
$5^2=\left(\sqrt{4x-3}\right)^2$
$25=4x-3$
$28=4x$
$7=x$

Check: $5 \stackrel{?}{=} \sqrt{4(7)-3}$
$5 \stackrel{?}{=} \sqrt{28-3}$
$5 \stackrel{?}{=} \sqrt{25}$
$5 = 5$ True
The solution set is {7}.

37. $3\sqrt{p} = 6$
$\sqrt{p} = 2$
$\left(\sqrt{p}\right)^2 = 2^2$
$p = 4$
Check: $3\sqrt{4} \stackrel{?}{=} 6$
$3 \cdot 2 \stackrel{?}{=} 6$
$6 = 6$ True
The solution set is {4}.

39. $2\sqrt{x} + 2 = 8$
$2\sqrt{x} = 6$
$\sqrt{x} = 3$
$\left(\sqrt{x}\right)^2 = 3^2$
$x = 9$
Check: $2\sqrt{9} + 2 \stackrel{?}{=} 8$
$2 \cdot 3 + 2 \stackrel{?}{=} 8$
$6 + 2 \stackrel{?}{=} 8$
$8 = 8$ True
The solution set is {9}.

41. $3 = 6 + \sqrt{n}$
$-3 = \sqrt{n}$
Since the principal square root of a real number cannot be negative, the equation has no solution. The solution set is { } or ∅.

43. $\sqrt{2y+11} - 4 = -1$
$\sqrt{2y+11} = 3$
$\left(\sqrt{2y+11}\right)^2 = 3^2$
$2y + 11 = 9$
$2y = -2$
$y = -1$
Check: $\sqrt{2(-1)+11} - 4 \stackrel{?}{=} -1$
$\sqrt{-2+11} - 4 \stackrel{?}{=} -1$
$\sqrt{9} - 4 \stackrel{?}{=} -1$
$3 - 4 \stackrel{?}{=} -1$
$-1 = -1$ True
The solution set is {−1}.

45. $\sqrt{9z-18} = z$
$\left(\sqrt{9z-18}\right)^2 = z^2$
$9z - 18 = z^2$
$0 = z^2 - 9z + 18$
$0 = (z-6)(z-3)$
$0 = z - 6$ or $0 = z - 3$
$6 = z$ or $3 = z$
Check $z = 6$: $\sqrt{9(6)-18} \stackrel{?}{=} 6$
$\sqrt{54-18} \stackrel{?}{=} 6$
$\sqrt{36} \stackrel{?}{=} 6$
$6 = 6$ True
Check: $z = 3$: $\sqrt{9(3)-18} \stackrel{?}{=} 3$
$\sqrt{27-18} \stackrel{?}{=} 3$
$\sqrt{9} \stackrel{?}{=} 3$
$3 = 3$ True
The solution set is {3, 6}.

47. $y = \sqrt{4-3y}$
$y^2 = \left(\sqrt{4-3y}\right)^2$
$y^2 = 4 - 3y$
$y^2 + 3y - 4 = 0$
$(y+4)(y-1) = 0$
$y + 4 = 0$ or $y - 1 = 0$
$y = -4$ or $y = 1$
Check $y = -4$: $-4 \stackrel{?}{=} \sqrt{4-3(-4)}$
Since the principal square root of a real number cannot be negative, the statement is false and $y = -4$ is not a solution of the equation.
Check $y = 1$: $1 \stackrel{?}{=} \sqrt{4-3(1)}$
$1 \stackrel{?}{=} \sqrt{4-3}$
$1 \stackrel{?}{=} \sqrt{1}$
$1 = 1$ True
The solution set is {1}. ($y = -4$ is an extraneous solution.)

49. $n + 1 = \sqrt{n^2 - 3}$
$(n+1)^2 = \left(\sqrt{n^2-3}\right)^2$
$n^2 + 2n + 1 = n^2 - 3$
$2n + 1 = -3$
$2n = -4$
$n = -2$

Check: $-2+1 \stackrel{?}{=} \sqrt{(-2)^2-3}$
$-1 \stackrel{?}{=} \sqrt{4-3}$
$-1 \stackrel{?}{=} \sqrt{1}$
$-1 = 1$ False
$n=-2$ is an extraneous solution. The equation has no solution. The solution set is { } or ∅.

51. $\sqrt{3x-11} = x-5$
$\left(\sqrt{3x-11}\right)^2 = (x-5)^2$
$3x-11 = x^2-10x+25$
$0 = x^2-13x+36$
$0 = (x-4)(x-9)$
$x-4=0$ or $x-9=0$
$x=4$ or $x=9$
Check $x=4$: $\sqrt{3(4)-11} \stackrel{?}{=} 4-5$
$\sqrt{12-11} \stackrel{?}{=} -1$
$\sqrt{1} \stackrel{?}{=} -1$
$1=-1$ False
Check $x=9$: $\sqrt{3(9)-11} \stackrel{?}{=} 9-5$
$\sqrt{27-11} \stackrel{?}{=} 4$
$\sqrt{16} \stackrel{?}{=} 4$
$4=4$ True
The solution set is {9}. ($x=4$ is an extraneous solution.)

53. $\sqrt{31x+10}-8 = x$
$\sqrt{31x+10} = x+8$
$\left(\sqrt{31x+10}\right)^2 = (x+8)^2$
$31x+10 = x^2+16x+64$
$0 = x^2-15x+54$
$0 = (x-6)(x-9)$
$x-6=0$ or $x-9=0$
$x=6$ or $x=9$
Check $x=6$: $\sqrt{31(6)+10}-8 \stackrel{?}{=} 6$
$\sqrt{186+10}-8 \stackrel{?}{=} 6$
$\sqrt{196}-8 \stackrel{?}{=} 6$
$14-8 \stackrel{?}{=} 6$
$6=6$ True
Check $x=9$: $\sqrt{31(9)+10}-8 \stackrel{?}{=} 9$
$\sqrt{279+10}-8 \stackrel{?}{=} 9$
$\sqrt{289}-8 \stackrel{?}{=} 9$
$17-8 \stackrel{?}{=} 9$
$9=9$ True
The solution set is {6, 9}.

55. $6+\sqrt{x}-x = 0$
$\sqrt{x} = x-6$
$\left(\sqrt{x}\right)^2 = (x-6)^2$
$x = x^2-12x+36$
$0 = x^2-13x+36$
$0 = (x-4)(x-9)$
$x-4=0$ or $x-9=0$
$x=4$ or $x=9$
Check $x=4$: $6+\sqrt{4}-4 \stackrel{?}{=} 0$
$6+2-4 \stackrel{?}{=} 0$
$4=0$ False
Check $x=9$: $6+\sqrt{9}-9 \stackrel{?}{=} 0$
$6+3-9 \stackrel{?}{=} 0$
$0=0$ True
The solution set is {9}. ($x=4$ is an extraneous solution.)

57. $\sqrt{3x-4} = \sqrt{x+8}$
$\left(\sqrt{3x-4}\right)^2 = \left(\sqrt{x+8}\right)^2$
$3x-4 = x+8$
$2x-4 = 8$
$2x = 12$
$x = 6$
Check: $\sqrt{3(6)-4} \stackrel{?}{=} \sqrt{6+8}$
$\sqrt{14} = \sqrt{14}$ True
The solution set is {6}.

59. $\sqrt{x^2+3x} = \sqrt{x^2+6x-3}$
$\left(\sqrt{x^2+3x}\right)^2 = \left(\sqrt{x^2+6x-3}\right)^2$
$x^2+3x = x^2+6x-3$
$3x = 6x-3$
$-3x = -3$
$x = 1$
Check: $\sqrt{1^2+3(1)} \stackrel{?}{=} \sqrt{1^2+6(1)-3}$
$\sqrt{1+3} \stackrel{?}{=} \sqrt{1+6-3}$
$\sqrt{4} = \sqrt{4}$ True
The solution set is {1}.

61. $2\sqrt{a+5}-\sqrt{7a+20}=0$

$$2\sqrt{a+5}=\sqrt{7a+20}$$
$$\left(2\sqrt{a+5}\right)^2=\left(\sqrt{7a+20}\right)^2$$
$$4(a+5)=7a+20$$
$$4a+20=7a+20$$
$$4a=7a$$
$$0=-3a$$
$$0=a$$

Check: $2\sqrt{0+5}-\sqrt{7(0)+20}\stackrel{?}{=}0$

$$2\sqrt{5}-\sqrt{20}\stackrel{?}{=}0$$
$$2\sqrt{5}-\sqrt{4\cdot 5}\stackrel{?}{=}0$$
$$2\sqrt{5}-2\sqrt{5}\stackrel{?}{=}0$$
$$0=0 \quad \text{True}$$

The solution set is $\{0\}$.

63. $\sqrt{2x^2-10}-\sqrt{x^2-3x}=0$

$$\sqrt{2x^2-10}=\sqrt{x^2-3x}$$
$$\left(\sqrt{2x^2-10}\right)^2=\left(\sqrt{x^2-3x}\right)^2$$
$$2x^2-10=x^2-3x$$
$$x^2+3x-10=0$$
$$(x+5)(x-2)=0$$

$x+5=0$ or $x-2=0$

$x=-5$ or $x=2$

Check $x=-5$:

$$\sqrt{2(-5)^2-10}-\sqrt{(-5)^2-3(-5)}\stackrel{?}{=}0$$
$$\sqrt{2(25)-10}-\sqrt{25+15}\stackrel{?}{=}0$$
$$\sqrt{50-10}-\sqrt{40}\stackrel{?}{=}0$$
$$\sqrt{40}-\sqrt{40}\stackrel{?}{=}0$$
$$0=0 \quad \text{True}$$

Check $x=2$: $\sqrt{2(2)^2-10}-\sqrt{2^2-3(2)}\stackrel{?}{=}0$

$$\sqrt{2(4)-10}-\sqrt{4-6}\stackrel{?}{=}0$$
$$\sqrt{8-10}-\sqrt{-2}\stackrel{?}{=}0$$

Since the square root of a negative number is not a real number, $x=2$ is an extraneous solution. The solution set is $\{-5\}$.

65. $2\sqrt{n}=\sqrt{n^2-5}$

$$\left(2\sqrt{n}\right)^2=\left(\sqrt{n^2-5}\right)^2$$
$$4n=n^2-5$$
$$0=n^2-4n-5$$
$$0=(n-5)(n+1)$$

$0=n-5$ or $0=n+1$

$5=n$ or $-1=n$

Check $n=5$: $2\sqrt{5}\stackrel{?}{=}\sqrt{5^2-5}$

$$2\sqrt{5}\stackrel{?}{=}\sqrt{25-5}$$
$$2\sqrt{5}\stackrel{?}{=}\sqrt{20}$$
$$2\sqrt{5}=2\sqrt{5} \quad \text{True}$$

Check $n=-1$: $2\sqrt{-1}\stackrel{?}{=}\sqrt{(-1)^2-5}$

Since the square root of a negative number is not a real number, $n=-1$ is an extraneous solution. The solution set is $\{5\}$.

67. $\sqrt{2x}+2=\sqrt{2x-12}$

$$\left(\sqrt{2x}+2\right)^2=\left(\sqrt{2x-12}\right)^2$$
$$\left(\sqrt{2x}\right)^2+2\cdot\sqrt{2x}\cdot 2+2^2=2x-12$$
$$2x+4\sqrt{2x}+4=2x-12$$
$$4\sqrt{2x}=-16$$
$$\sqrt{2x}=-4$$

Since the principal square root of a real number cannot be negative, the equation has no solution. The solution set is $\{\ \}$ or $\varnothing$.

69. $\sqrt{z+10}=2+\sqrt{5z+6}$

$$\left(\sqrt{z+10}\right)^2=\left(2+\sqrt{5z+6}\right)^2$$
$$z+10=2^2+2\cdot 2\cdot\sqrt{5z+6}+\left(\sqrt{5z+6}\right)^2$$
$$z+10=4+4\sqrt{5z+6}+5z+6$$
$$z+10=5z+10+4\sqrt{5z+6}$$
$$-4z=4\sqrt{5z+6}$$
$$-z=\sqrt{5z+6}$$
$$(-z)^2=\left(\sqrt{5z+6}\right)^2$$
$$z^2=5z+6$$
$$z^2-5z-6=0$$
$$(z+1)(z-6)=0$$

$z+1=0$ or $z-6=0$

$z=-1$ or $z=6$

Check $z=-1$: $\sqrt{-1+10} \stackrel{?}{=} 2+\sqrt{5(-1)+6}$

$\sqrt{9} \stackrel{?}{=} 2+\sqrt{-5+6}$

$3 \stackrel{?}{=} 2+\sqrt{1}$

$3 \stackrel{?}{=} 2+1$

$3=3$ True

Check $z=6$: $\sqrt{6+10} \stackrel{?}{=} 2+\sqrt{5(6)+6}$

$\sqrt{16} \stackrel{?}{=} 2+\sqrt{30+6}$

$4 \stackrel{?}{=} 2+\sqrt{36}$

$4 \stackrel{?}{=} 2+6$

$4=8$ False

The solution set is $\{-1\}$.

71. $2\sqrt{t+9}=\sqrt{4t+3}+3$

$$\left(2\sqrt{t+9}\right)^2=\left(\sqrt{4t+3}+3\right)^2$$
$$4(t+9)=\left(\sqrt{4t+3}\right)^2+2\cdot 3\cdot\sqrt{4t+3}+3^2$$
$$4t+36=4t+3+6\sqrt{4t+3}+9$$
$$4t+36=4t+12+6\sqrt{4t+3}$$
$$24=6\sqrt{4t+3}$$
$$4=\sqrt{4t+3}$$
$$4^2=\left(\sqrt{4t+3}\right)^2$$
$$16=4t+3$$
$$13=4t$$
$$\frac{13}{4}=t$$

Check: $2\sqrt{\frac{13}{4}+9} \stackrel{?}{=} \sqrt{4\left(\frac{13}{4}\right)+3}+3$

$2\sqrt{\frac{13}{4}+\frac{36}{4}} \stackrel{?}{=} \sqrt{13+3}+3$

$2\sqrt{\frac{49}{4}} \stackrel{?}{=} \sqrt{16}+3$

$2\left(\frac{7}{2}\right) \stackrel{?}{=} 4+3$

$7=7$ True

The solution set is $\left\{\frac{13}{4}\right\}$.

73. $\sqrt{x+2}+2=\sqrt{2x+3}$

$$\left(\sqrt{x+2}+2\right)^2=\left(\sqrt{2x+3}\right)^2$$
$$\left(\sqrt{x+2}\right)^2+4\sqrt{x+2}+2^2=2x+3$$
$$x+2+4\sqrt{x+2}+4=2x+3$$
$$4\sqrt{x+2}=x-3$$
$$\left(4\sqrt{x+2}\right)^2=(x-3)^2$$
$$16(x+2)=x^2-6x+9$$
$$16x+32=x^2-6x+9$$
$$0=x^2-22x-23$$
$$0=(x-23)(x+1)$$

$0=x-23$ or $0=x+1$

$23=x$ or $-1=x$

Check $x=23$: $\sqrt{23+2}+2 \stackrel{?}{=} \sqrt{2(23)+3}$

$\sqrt{25}+2 \stackrel{?}{=} \sqrt{46+3}$

$5+2 \stackrel{?}{=} \sqrt{49}$

$7=7$ True

Check $x=-1$: $\sqrt{-1+2}+2 \stackrel{?}{=} \sqrt{2(-1)+3}$

$\sqrt{1}+2 \stackrel{?}{=} \sqrt{1}$

$1+2 \stackrel{?}{=} 1$

$3=1$ False

The solution set is $\{23\}$. ($x=-1$ is an extraneous solution.)

75. $\sqrt{3x+4}-\sqrt{4x+1}=-1$

$$\sqrt{3x+4}=\sqrt{4x+1}-1$$
$$\left(\sqrt{3x+4}\right)^2=\left(\sqrt{4x+1}-1\right)^2$$
$$3x+4=\left(\sqrt{4x+1}\right)^2-2\sqrt{4x+1}+1^2$$
$$3x+4=4x+1-2\sqrt{4x+1}+1$$
$$2\sqrt{4x+1}=x-2$$
$$\left(2\sqrt{4x+1}\right)^2=(x-2)^2$$
$$4(4x+1)=x^2-4x+4$$
$$16x+4=x^2-4x+4$$
$$0=x^2-20x$$
$$0=x(x-20)$$

$0=x$ or $0=x-20$

$20=x$

Check $x=0$: $\sqrt{3(0)+4}-\sqrt{4(0)+1} \stackrel{?}{=} -1$

$\sqrt{4}-\sqrt{1} \stackrel{?}{=} -1$

$2-1 \stackrel{?}{=} -1$

$1=-1$ False

Check $x = 20$:

$$\sqrt{3(20)+4} - \sqrt{4(20)+1} \stackrel{?}{=} -1$$
$$\sqrt{64} - \sqrt{81} \stackrel{?}{=} -1$$
$$8 - 9 \stackrel{?}{=} -1$$
$$-1 = -1 \quad \text{True}$$

The solution set is $\{20\}$.

77. $-2 = \sqrt{5x+6} - 3$

$$1 = \sqrt{5x+6}$$
$$1^2 = \left(\sqrt{5x+6}\right)^2$$
$$1 = 5x+6$$
$$-5 = 5x$$
$$-1 = x$$

Check: $-2 \stackrel{?}{=} \sqrt{5(-1)+6} - 3$

$$-2 \stackrel{?}{=} \sqrt{1} - 3$$
$$-2 \stackrel{?}{=} 1 - 3$$
$$-2 = -2 \quad \text{True}$$

The solution set is $\{-1\}$.

79. $2\sqrt{x} - 8 = 0$

$$2\sqrt{x} = 8$$
$$\sqrt{x} = 4$$
$$\left(\sqrt{x}\right)^2 = 4^2$$
$$x = 16$$

Check: $2\sqrt{16} - 8 \stackrel{?}{=} 0$

$$2(4) - 8 \stackrel{?}{=} 0$$
$$8 - 8 \stackrel{?}{=} 0$$
$$0 = 0 \quad \text{True}$$

The solution set is $\{16\}$.

81. $3x - 2 = \sqrt{9x^2 - 20}$

$$(3x-2)^2 = \left(\sqrt{9x^2-20}\right)^2$$
$$9x^2 - 12x + 4 = 9x^2 - 20$$
$$-12x + 4 = -20$$
$$-12x = -24$$
$$x = 2$$

Check: $3(2) - 2 \stackrel{?}{=} \sqrt{9(2)^2 - 20}$

$$6 - 2 \stackrel{?}{=} \sqrt{9 \cdot 4 - 20}$$
$$4 \stackrel{?}{=} \sqrt{36 - 20}$$
$$4 \stackrel{?}{=} \sqrt{16}$$
$$4 = 4 \quad \text{True}$$

The solution set is $\{2\}$.

83. $3 = \sqrt{w+12} - 2w$

$$2w + 3 = \sqrt{w+12}$$
$$(2w+3)^2 = \left(\sqrt{w+12}\right)^2$$
$$4w^2 + 12w + 9 = w + 12$$
$$4w^2 + 11w - 3 = 0$$
$$(w+3)(4w-1) = 0$$
$$w + 3 = 0 \quad \text{or} \quad 4w - 1 = 0$$
$$w = -3 \quad \text{or} \quad w = \frac{1}{4}$$

Check $w = -3$: $3 \stackrel{?}{=} \sqrt{-3+12} - 2(-3)$

$$3 \stackrel{?}{=} \sqrt{9} + 6$$
$$3 \stackrel{?}{=} 3 + 6$$
$$3 = 9 \quad \text{False}$$

Check $w = \frac{1}{4}$: $3 \stackrel{?}{=} \sqrt{\frac{1}{4} + 12} - 2\left(\frac{1}{4}\right)$

$$3 \stackrel{?}{=} \sqrt{\frac{49}{4}} - \frac{1}{2}$$
$$3 \stackrel{?}{=} \frac{7}{2} - \frac{1}{2}$$
$$3 \stackrel{?}{=} \frac{6}{2}$$
$$3 = 3 \quad \text{True}$$

The solution set is $\left\{\frac{1}{4}\right\}$. ($w = -3$ is an extraneous solution.)

85. $x(x-5) = 24$

$$x^2 - 5x = 24$$
$$x^2 - 5x - 24 = 0$$
$$(x-8)(x+3) = 0$$
$$x - 8 = 0 \quad \text{or} \quad x + 3 = 0$$
$$x = 8 \quad \text{or} \quad x = -3$$

Check $x = 8$: $8(8-5) \stackrel{?}{=} 24$

$$8(3) \stackrel{?}{=} 24$$
$$24 = 24 \quad \text{True}$$

Check $x = -3$: $-3(-3-5) \stackrel{?}{=} 24$

$$-3(-8) \stackrel{?}{=} 24$$
$$24 = 24 \quad \text{True}$$

The solution set is $\{-3, 8\}$.

87. $\sqrt{a^2-3a+5}=\sqrt{2a^2-6a-23}$

$$\left(\sqrt{a^2-3a+5}\right)^2=\left(\sqrt{2a^2-6a-23}\right)^2$$
$$a^2-3a+5=2a^2-6a-23$$
$$0=a^2-3a-28$$
$$0=(a+4)(a-7)$$
$$a+4=0 \quad \text{or} \quad a-7=0$$
$$a=-4 \quad \text{or} \quad a=7$$

Check $a=-4$:
$$\sqrt{(-4)^2-3(-4)+5}\stackrel{?}{=}\sqrt{2(-4)^2-6(-4)-23}$$
$$\sqrt{16+12+5}\stackrel{?}{=}\sqrt{32+24-23}$$
$$\sqrt{33}=\sqrt{33} \quad \text{True}$$

Check $a=7$:
$$\sqrt{7^2-3(7)+5}\stackrel{?}{=}\sqrt{2(7)^2-6(7)-23}$$
$$\sqrt{49-21+5}\stackrel{?}{=}\sqrt{98-42-23}$$
$$\sqrt{33}=\sqrt{33} \quad \text{True}$$

The solution set is $\{-4, 7\}$.

89. $\sqrt{2y+1}-2=1$
$$\sqrt{2y+1}=3$$
$$\left(\sqrt{2y+1}\right)^2=3^2$$
$$2y+1=9$$
$$2y=8$$
$$y=4$$

Check: $\sqrt{2(4)+1}-2\stackrel{?}{=}1$
$$\sqrt{8+1}-2\stackrel{?}{=}1$$
$$\sqrt{9}-2\stackrel{?}{=}1$$
$$3-2\stackrel{?}{=}1$$
$$1=1 \quad \text{True}$$

The solution set is $\{4\}$.

91. $x-\frac{2}{x}=1$
$$\left(x-\frac{2}{x}\right)x=1(x)$$
$$x^2-2=x$$
$$x^2-x-2=0$$
$$(x-2)(x+1)=0$$
$$x-2=0 \quad \text{or} \quad x+1=0$$
$$x=2 \quad \text{or} \quad x=-1$$

Check $x=2$: $2-\frac{2}{2}\stackrel{?}{=}1$
$$2-1\stackrel{?}{=}1$$
$$1=1 \quad \text{True}$$

Check $x=-1$: $-1-\frac{2}{-1}\stackrel{?}{=}1$
$$-1+2\stackrel{?}{=}1$$
$$1=1 \quad \text{True}$$

93. $\sqrt{p+4}+1=\sqrt{p-5}$
$$\left(\sqrt{p+4}+1\right)^2=\left(\sqrt{p-5}\right)^2$$
$$\left(\sqrt{p+4}\right)^2+2\cdot1\cdot\sqrt{p+4}+1^2=p-5$$
$$p+4+2\sqrt{p+4}+1=p-5$$
$$p+5+2\sqrt{p+4}=p-5$$
$$2\sqrt{p+4}=-10$$

Since the principal square root of a real number cannot be negative, the equation has no solution. The solution set is { } or $\varnothing$.

95. $2x+3(x-4)=7x-18$
$$2x+3x-12=7x-18$$
$$5x-12=7x-18$$
$$-2x-12=-18$$
$$-2x=-6$$
$$x=3$$

Check: $2(3)+3(3-4)\stackrel{?}{=}7(3)-18$
$$6+3(-1)\stackrel{?}{=}21-18$$
$$6-3\stackrel{?}{=}3$$
$$3=3 \quad \text{True}$$

The solution set is $\{3\}$.

97. Notice that if $x>0$, $-x<0$ and the equation has no solution because the principal square root of a real number cannot be negative. If $x\le 0$, then $11x-18<0$ and the equation has no solution because the square root of a negative number is not a real number.
Thus, the equation has no solution. The solution set is { } or $\varnothing$.

99. (a) $t=\sqrt{\frac{h}{16}}$
$$t=\sqrt{\frac{1024}{16}}=\sqrt{64}=8$$

It will take 8 seconds to hit the ground.

(b) $t = \sqrt{\frac{h}{16}}$

$$t = \sqrt{\frac{300}{16}}$$
$$= \frac{\sqrt{300}}{\sqrt{16}}$$
$$= \frac{\sqrt{100 \cdot 3}}{4}$$
$$= \frac{10\sqrt{3}}{4}$$
$$= \frac{5\sqrt{3}}{2}$$

It will take $\frac{5\sqrt{3}}{2}$ seconds to hit the ground.

101. Here $S = 25$.

$$S = \sqrt{22.5d}$$
$$25 = \sqrt{22.5d}$$
$$25^2 = \left(\sqrt{22.5d}\right)^2$$
$$625 = 22.5d$$
$$\frac{625}{22.5} = d$$
$$d \approx 27.78$$

The car will skid about 28 feet.

103. **(a)** Here $S = 25$.

$$S = \sqrt{15d}$$
$$25 = \sqrt{15d}$$
$$25^2 = \left(\sqrt{15d}\right)^2$$
$$625 = 15d$$
$$\frac{625}{15} = d$$
$$d \approx 41.67$$

The car will skid about 42 feet.

(b) Yes; Mark will hit the rabbit.

105. Let x be the unknown number.

$$\sqrt{2(x-3)} = 4$$
$$\left(\sqrt{2(x-3)}\right)^2 = 4^2$$
$$2(x-3) = 16$$
$$x - 3 = 8$$
$$x = 11$$

Check: $\sqrt{2(11-3)} \overset{?}{=} 4$

$$\sqrt{2(8)} \overset{?}{=} 4$$
$$\sqrt{16} \overset{?}{=} 4$$
$$4 = 4 \quad \text{True}$$

The number is 11.

107. Let n be the unknown number.

$$3\sqrt{n+1} = 2\sqrt{2n+3}$$
$$\left(3\sqrt{n+1}\right)^2 = \left(2\sqrt{2n+3}\right)^2$$
$$3^2(n+1) = 2^2(2n+3)$$
$$9(n+1) = 4(2n+3)$$
$$9n + 9 = 8n + 12$$
$$n = 3$$

Check: $3\sqrt{3+1} \overset{?}{=} 2\sqrt{2 \cdot 3 + 3}$

$$3\sqrt{4} \overset{?}{=} 2\sqrt{9}$$
$$3 \cdot 2 \overset{?}{=} 2 \cdot 3$$
$$6 = 6 \quad \text{True}$$

The number is 3.

109.

$$\sqrt{\frac{A}{\pi}} = r$$
$$\left(\sqrt{\frac{A}{\pi}}\right)^2 = r^2$$
$$\frac{A}{\pi} = r^2$$
$$A = \pi r^2$$

111.

$$S = \sqrt{\frac{3V}{h}}$$
$$S^2 = \left(\sqrt{\frac{3V}{h}}\right)^2$$
$$S^2 = \frac{3V}{h}$$
$$hS^2 = 3V$$
$$h = \frac{dV}{S^2}$$

113.

$$b = \sqrt{2a+b}$$
$$b^2 = \left(\sqrt{2a+b}\right)^2$$
$$b^2 = 2a + b$$
$$b^2 - b = 2a$$
$$\frac{b^2 - b}{2} = a$$

115. $2\sqrt{m-2n} = \sqrt{4n-2m}$

$$\left(2\sqrt{m-2n}\right)^2 = \left(\sqrt{4n-2m}\right)^2$$
$$4(m-2n) = 4n-2m$$
$$4m-8n = 4n-2m$$
$$4m = 12n-2m$$
$$6m = 12n$$
$$\frac{6m}{12} = n$$
$$\frac{1}{2}m = n$$

117. The student's work is incorrect. The student did not square the first step correctly to obtain the second step. Using the property $\left(\sqrt{a}\right)^2 = a$ for $a > 0$ is correct, but the student incorrectly squared the right side of the equation. The student should have used the pattern $A^2 + 2AB + B^2$ to simplify the right side of the equation, and then proceed to solve for x.

Section 8.7

Preparing for Higher Roots and Rational Exponents

P1. **(a)** $2^4 = 2\cdot 2\cdot 2\cdot 2 = 16$

(b) $(-4)^3 = (-4)\cdot(-4)\cdot(-4) = -64$

P2. $x^3 \cdot x^5 = x^{3+5} = x^8$

P3. $\frac{y^5}{y} = \frac{y^5}{y^1} = y^{5-1} = y^4$

P4. $(b^3)^4 = b^{3\cdot 4} = b^{12}$

8.7 Quick Checks

1. In the expression $\sqrt[n]{a}$, a is called the radicand and n is called the index.

2. Because $(-5)^3 = -125$, $\sqrt[3]{-125} = \underline{-5}$.

3. True

4. $\sqrt[5]{32} = 2$ because $2^5 = 32$.

5. $\sqrt[3]{-27} = -3$ because $(-3)^3 = -27$.

6. $\sqrt[4]{-16}$ is not a real number. There is no real number b such that $b^4 = -16$.

7. $\sqrt[6]{64} = 2$ because $2^6 = 64$.

8. $\sqrt[6]{7^6} = 7$

9. $\sqrt[4]{w^4} = w$ for $w \ge 0$.

10. $\sqrt[5]{(12p)^5} = 12p$

11. If $\sqrt[n]{a}$ and $\sqrt[n]{b}$ are real numbers and $n \ge 2$ is an integer, then $\sqrt[n]{ab} = \underline{\sqrt[n]{a}\cdot\sqrt[n]{b}}$.

12. False; $\sqrt[4]{32} = \sqrt[4]{16\cdot 4} = \sqrt[4]{2^4}\cdot\sqrt[4]{2} = 2\sqrt[4]{2}$.

13. $$\sqrt[3]{-32} = \sqrt[3]{-8\cdot 4}$$
$$= \sqrt[3]{-8}\cdot\sqrt[3]{4}$$
$$= \sqrt[3]{(-2)^3}\cdot\sqrt[3]{4}$$
$$= -2\sqrt[3]{4}$$

14. $$\sqrt[3]{250n^3} = \sqrt[3]{125n^3\cdot 2}$$
$$= \sqrt[3]{125n^3}\cdot\sqrt[3]{2}$$
$$= \sqrt[3]{(5n)^3}\cdot\sqrt[3]{2}$$
$$= 5n\sqrt[3]{2}$$

15. $\sqrt[4]{81b^{12}} = \sqrt[4]{(3b^3)^4} = 3b^3$

16. $\frac{\sqrt[4]{324}}{\sqrt[4]{4}} = \sqrt[4]{\frac{324}{4}} = \sqrt[4]{81} = \sqrt[4]{3^4} = 3$

17. $$\frac{\sqrt[3]{512m^{10}}}{\sqrt[3]{4m}} = \sqrt[3]{\frac{512m^{10}}{4m}}$$
$$= \sqrt[3]{128m^9}$$
$$= \sqrt[3]{64m^9\cdot 2}$$
$$= \sqrt[3]{64m^9}\cdot\sqrt[3]{2}$$
$$= \sqrt[3]{(4m^3)^3}\cdot\sqrt[3]{2}$$
$$= 4m^3\sqrt[3]{2}$$

18. If a is a nonnegative real number and $n \ge 2$ is an integer, then $a^{1/n} = \sqrt[n]{a}$.

19. **(a)** $49^{1/2} = \sqrt{49} = 7$

(b) $(-64)^{1/3} = \sqrt[3]{-64} = -4$

(c) $-100^{1/2} = -\sqrt{100} = -10$

(d) $(-100)^{1/2} = \sqrt{-100}$ is not a real number.

20. **(a)** $p^{1/5} = \sqrt[5]{p}$

(b) $(4n)^{1/3} = \sqrt[3]{4n}$

(c) $4n^{1/3} = 4 \cdot n^{1/3} = 4 \cdot \sqrt[3]{n} = 4\sqrt[3]{n}$

21. **(a)** $\sqrt{b} = b^{1/2}$

(b) $5\sqrt[3]{p} = 5 \cdot \sqrt[3]{p} = 5 \cdot p^{1/3} = 5p^{1/3}$

(c) $\sqrt[4]{10p} = (10p)^{1/4}$

22. False; it is best to begin by taking the cube root of 8.

23. If a is a real number and $\frac{m}{n}$ is a rational number in lowest terms with $n \geq 2$ then $a^{m/n} = \underline{\sqrt[n]{a^m}}$ or $\underline{\left(\sqrt[n]{a}\right)^m}$.

24. **(a)** $25^{3/2} = \left(\sqrt{25}\right)^3 = 5^3 = 125$

(b) $64^{2/3} = \left(\sqrt[3]{64}\right)^2 = 4^2 = 16$

(c) $-16^{3/4} = -\left(\sqrt[4]{16}\right)^3 = -(2^3) = -8$

(d) $(-8)^{4/3} = \left(\sqrt[3]{-8}\right)^4 = (-2)^4 = 16$

(e) $(-49)^{5/2} = \left(\sqrt{-49}\right)^5$ is not a real number.

25. **(a)** $\sqrt[4]{d^3} = d^{3/4}$

(b) $\sqrt[5]{y^8} = y^{8/5}$

(c) $\left(\sqrt[4]{11a^2b}\right)^3 = (11a^2b)^{3/4}$

26. **(a)** $z^{4/5} = \sqrt[5]{z^4} = \left(\sqrt[5]{z}\right)^4$

(b) $2n^{2/3} = 2\sqrt[3]{n^2} = 2\left(\sqrt[3]{n}\right)^2$

(c) $(2n)^{2/3} = \sqrt[3]{(2n)^2} = \left(\sqrt[3]{2n}\right)^2$

27. **(a)** $36^{-1/2} = \frac{1}{36^{1/2}} = \frac{1}{\sqrt{36}} = \frac{1}{6}$

(b) $8^{-4/3} = \frac{1}{8^{4/3}} = \frac{1}{\left(\sqrt[3]{8}\right)^4} = \frac{1}{2^4} = \frac{1}{16}$

(c)
$$\begin{aligned} -25^{-3/2} &= -1 \cdot 25^{-3/2} \\ &= \frac{-1}{25^{3/2}} \\ &= \frac{-1}{\left(\sqrt{25}\right)^3} \\ &= \frac{-1}{5^3} \\ &= -\frac{1}{125} \end{aligned}$$

28. **(a)** $7^{-5/12} \cdot 7^{7/12} = 7^{-5/12+7/12} = 7^{2/12} = 7^{1/6}$

(b) $\frac{11^{6/5}}{11^{3/5}} = 11^{6/5-3/5} = 11^{3/5}$

(c) $(2^{3/8})^{16/3} = 2^{(3/8)\cdot(16/3)} = 2^{16/8} = 2^2 = 4$

(d) $\frac{a^{3/2} \cdot a^{5/2}}{a^{-3/2}} = a^{3/2+5/2-(-3/2)} = a^{11/2}$

8.7 Exercises

29. The cube root of 8 is $\sqrt[3]{8}$ or 2.

31. The fourth root of 16 is $\sqrt[4]{16}$ or 2, since $2^4 = 16$.

33. The fourth root of 81 is $\sqrt[4]{81}$ or 3.

35. The fifth root of −32 is $\sqrt[5]{-32}$ or −2.

37. $\sqrt[3]{27} = 3$ since $3^3 = 27$.

39. $\sqrt[4]{-1}$ is not a real number since the index, 4, is even and the radicand, −1, is negative.

41. $\sqrt[3]{-125} = -5$ since $(-5)^3 = -125$.

43. $\sqrt[4]{625} = 5$ since $5^4 = 625$.

45. $\sqrt[6]{0} = 0$ since $0^6 = 0$.

47. $\sqrt[5]{243} = 3$ since $3^5 = 243$.

49. $\sqrt[6]{5^6} = 5$

51. $\sqrt[4]{n^4} = n$

53. $\sqrt[5]{r^5} = r$

55. $\sqrt[7]{(3a)^7} = 3a$

57. $\sqrt[3]{16} = \sqrt[3]{8 \cdot 2} = \sqrt[3]{2^3 \cdot 2} = 2\sqrt[3]{2}$

59. $\sqrt[4]{-64}$ is not a real number.

61. $\sqrt[3]{-40} = \sqrt[3]{-8 \cdot 5} = \sqrt[3]{(-2)^3 \cdot 5} = -2\sqrt[3]{5}$

63. $\dfrac{\sqrt[4]{243}}{\sqrt[4]{3}} = \sqrt[4]{\dfrac{243}{3}} = \sqrt[4]{81} = \sqrt[4]{3^4} = 3$

65.
$$\begin{aligned} -\sqrt[4]{32n^8} &= -\sqrt[4]{16n^8 \cdot 2} \\ &= -\sqrt[4]{2^4 (n^2)^4 \cdot 2} \\ &= -2n^2\sqrt[4]{2} \end{aligned}$$

67.
$$\begin{aligned} -\sqrt[3]{-128} &= -\sqrt[3]{-64 \cdot 2} \\ &= -\sqrt[3]{(-4)^3 \cdot 2} \\ &= -(-4)\sqrt[3]{2} \\ &= 4\sqrt[3]{2} \end{aligned}$$

69. $\sqrt[3]{81x^3} = \sqrt[3]{27x^3 \cdot 3} = \sqrt[3]{3^3 x^3 \cdot 3} = 3x\sqrt[3]{3}$

71. $\dfrac{\sqrt[3]{56b^5}}{\sqrt[3]{7b^2}} = \sqrt[3]{\dfrac{56b^5}{7b^2}} = \sqrt[3]{8b^3} = \sqrt[3]{(2b)^3} = 2b$

73.
$$\begin{aligned} \frac{\sqrt[3]{-108z^8}}{\sqrt[3]{2z^2}} &= \sqrt[3]{\frac{-108z^8}{2z^2}} \\ &= \sqrt[3]{-54z^6} \\ &= \sqrt[3]{-27z^6 \cdot 2} \\ &= \sqrt[3]{(-3z^2) \cdot 2} \\ &= -3z^2\sqrt[3]{2} \end{aligned}$$

75. $-16^{1/4} = -(2^4)^{1/4} = -2^{4/4} = -2^1 = -2$

77. $-8^{1/3} = -(2^3)^{1/3} = -2^{3/3} = -2^1 = -2$

79. $(-216)^{1/3} = [(-6)^3]^{1/3} = (-6)^{3/3} = -6^1 = -6$

81. $(-36)^{1/2} = \sqrt{-36}$ is not a real number.

83. $(2a)^{1/2} = \sqrt{2a}$

85. $7z^{1/4} = 7\sqrt[4]{z}$

87. $(-27r)^{1/3} = \sqrt[3]{-27r} = \sqrt[3]{(-3)^3 \cdot r} = -3\sqrt[3]{r}$

89. $\sqrt{c} = c^{1/2}$

91. $\sqrt{2y} = (2y)^{1/2}$

93. $7\sqrt{a} = 7a^{1/2}$

95. $64^{3/2} = \left(\sqrt{64}\right)^3 = 8^3 = 512$

97. $81^{-1/2} = \dfrac{1}{81^{1/2}} = \dfrac{1}{\sqrt{81}} = \dfrac{1}{9}$

99. $64^{2/3} = \left(\sqrt[3]{64}\right)^2 = 4^2 = 16$

101. $-9^{-3/2} = -\frac{1}{9^{3/2}}$
$= -\frac{1}{\left(\sqrt{9}\right)^3}$
$= -\frac{1}{3^3}$
$= -\frac{1}{27}$

103. $(-81)^{3/4} = \left(\sqrt[4]{-81}\right)^3$ is not a real number.

105. $49^{-3/2} = \frac{1}{49^{3/2}}$
$= \frac{1}{\left(\sqrt{49}\right)^3}$
$= \frac{1}{7^3}$
$= \frac{1}{343}$

107. $u^{2/3} = (u^2)^{1/3} = \sqrt[3]{u^2} = \left(\sqrt[3]{u}\right)^2$

109. $4x^{2/3} = 4(x^2)^{1/3} = 4\sqrt[3]{x^2} = 4\left(\sqrt[3]{x}\right)^2$

111. $(5n)^{2/3} = \sqrt[3]{(5n)^2} = \left(\sqrt[3]{5n}\right)^2$

113. $\sqrt[4]{p^5} = (p^5)^{1/4} = p^{5/4}$

115. $5\sqrt[4]{t^3} = 5 \cdot \sqrt[4]{t^3} = 5 \cdot t^{3/4} = 5t^{3/4}$

117. $\sqrt[3]{(5a)^2} = (5a)^{2/3}$

119. $4^{2/3} \cdot 4^{4/3} = 4^{\frac{2}{3}+\frac{4}{3}} = 4^{6/3} = 4^2 = 16$

121. $\frac{6^{3/2}}{6^{7/2}} = 6^{\frac{3}{2}-\frac{7}{2}} = 6^{-4/2} = 6^{-2} = \frac{1}{6^2} = \frac{1}{36}$

123. $(x^9)^{2/3} = x^{9 \cdot \frac{2}{3}} = x^{18/3} = x^6$

125. $(8n^{1/2})^{2/3} = 8^{2/3}(n^{1/2})^{2/3}$
$= (2^3)^{2/3}(n^{1/2})^{2/3}$
$= 2^{3 \cdot \frac{2}{3}} n^{\frac{1}{2} \cdot \frac{2}{3}}$
$= 2^{6/3} \cdot n^{2/6}$
$= 2^2 \cdot n^{1/3}$
$= 4n^{1/3}$ or $4\sqrt[3]{n}$

127. $\frac{x^{-1/4} \cdot x^{3/4}}{x^{1/4}} = x^{-\frac{1}{4}+\frac{3}{4}-\frac{1}{4}} = x^{1/4}$ or $\sqrt[4]{x}$

129. $\frac{(2x)^{-3/2}}{(2x)^{-1/2}} = (2x)^{-\frac{3}{2}-\left(-\frac{1}{2}\right)}$
$= (2x)^{-\frac{3}{2}+\frac{1}{2}}$
$= (2x)^{-2/2}$
$= (2x)^{-1}$
$= \frac{1}{2x}$

131. $-25^{3/2} = -(5^2)^{3/2} = -5^{6/2} = -5^3 = -125$

133. $\frac{6^{5/2}}{6^{7/2}} = 6^{\frac{5}{2}-\frac{7}{2}} = 6^{-\frac{2}{2}} = 6^{-1} = \frac{1}{6}$

135. $-125^{-2/3} = -\frac{1}{125^{2/3}}$
$= -\frac{1}{(125^{1/3})^2}$
$= -\frac{1}{\left(\sqrt[3]{125}\right)^2}$
$= -\frac{1}{5^2}$
$= -\frac{1}{25}$

137. $4^{1/2} + 25^{3/2} = 4^{1/2} + (25^{1/2})^3$
$= \sqrt{4} + \left(\sqrt{25}\right)^3$
$= 2 + 5^3$
$= 2 + 125$
$= 127$

139. $\left(\sqrt[4]{25}\right)^2 = (25^{1/4})^2$
$= 25^{2/4}$
$= 25^{1/2}$
$= \sqrt{25}$
$= 5$

141. $(25^{3/4} \cdot 4^{-3/4})^2 = 25^{2 \cdot \frac{3}{4}} \cdot 4^{2 \cdot \left(-\frac{3}{4}\right)}$
$= 25^{3/2} \cdot 4^{-3/2}$
$= (25^{1/2})^3 \cdot \dfrac{1}{(4^{1/2})^3}$
$= \dfrac{\left(\sqrt{25}\right)^3}{\left(\sqrt{4}\right)^3}$
$= \dfrac{5^3}{2^3}$
$= \dfrac{125}{8}$

143. $\dfrac{y^{-1/3} \cdot y^{-1/3}}{y^{4/3}} = y^{-\frac{1}{3}+\left(-\frac{1}{3}\right)-\frac{4}{3}}$
$= y^{-6/3}$
$= y^{-2}$
$= \dfrac{1}{y^2}$

145. $2x^{-3/2}$ for $x = 4$:
$2(4)^{-3/2} = \dfrac{2}{4^{3/2}} = \dfrac{2}{\left(\sqrt{4}\right)^3} = \dfrac{2}{2^3} = \dfrac{2}{8} = \dfrac{1}{4}$

147. $(27x)^{1/3}$ for $x = 8$:
$(27 \cdot 8)^{1/3} = (3^3 \cdot 2^3)^{1/3}$
$= (3^3)^{1/3} \cdot (2^3)^{1/3}$
$= 3 \cdot 2$
$= 6$

149. $x^{1/2} \cdot x^{-3/4} = x^{\frac{1}{2}+\left(-\frac{3}{4}\right)}$
$= x^{\frac{2}{4}-\frac{3}{4}}$
$= x^{-1/4}$
$= \dfrac{1}{x^{1/4}}$
$= \dfrac{1}{\sqrt[4]{x}}$

151. $\dfrac{(-8a^4)^{1/3}}{a^{5/6}} = \dfrac{(-8)^{1/3}a^{4/3}}{a^{5/6}}$
$= \sqrt[3]{-8}a^{\frac{4}{3}-\frac{5}{6}}$
$= -2a^{\frac{8}{6}-\frac{5}{6}}$
$= -2a^{3/6}$
$= -2a^{1/2}$
$= -2\sqrt{a}$

153. $\dfrac{(3x^{2/3})^4}{x^{5/6}} = \dfrac{3^4 x^{\frac{2}{3} \cdot 4}}{x^{5/6}}$
$= \dfrac{81x^{8/3}}{x^{5/6}}$
$= 81x^{\frac{8}{3}-\frac{5}{6}}$
$= 81x^{\frac{16}{6}-\frac{5}{6}}$
$= 81x^{11/6}$
$= 81\sqrt[6]{x^{11}}$
$= 81\sqrt[6]{x^6 \cdot x^5}$
$= 81x\sqrt[6]{x^5}$

155. The expression $a^{1/n}$ is undefined when a is negative and n is even because an even root of a negative number is not defined. For example, $\sqrt{-16}$ is undefined because the principal square root of a negative real number is not a real number.

Chapter 8 Review

1. The square roots of 4 are the numbers that can be squared to get 4. Since $2^2 = 4$ and $(-2)^2 = 4$, the square roots of 4 are 2 and –2.

2. The square roots of 81 are the numbers that can be squared to get 81. Since $9^2 = 81$ and $(-9)^2 = 81$, the square roots of 81 are 9 and –9.

3. $-\sqrt{1} = -1 \cdot \sqrt{1} = -1 \cdot 1 = -1$

4. $-\sqrt{25} = -1 \cdot \sqrt{25} = -1 \cdot 5 = -5$

5. $\sqrt{0.16} = 0.4$ since $(0.4)^2 = 0.16$.

6. $\sqrt{0.04} = 0.2$ since $(0.2)^2 = 0.04$.

7. $\frac{3}{2}\sqrt{\frac{25}{36}} = \frac{3}{2} \cdot \frac{5}{6} = \frac{3 \cdot 5}{2 \cdot 6} = \frac{3 \cdot 5}{3 \cdot 2 \cdot 2} = \frac{5}{4}$

8. $\frac{4}{3}\sqrt{\frac{81}{4}} = \frac{4}{3} \cdot \frac{9}{2} = \frac{4 \cdot 9}{3 \cdot 2} = \frac{2 \cdot 2 \cdot 3 \cdot 3}{3 \cdot 2} = 2 \cdot 3 = 6$

9. $\sqrt{25-9} = \sqrt{16} = 4$

10. $\sqrt{169-25} = \sqrt{144} = 12$

11. $\sqrt{9^2-(4)(5)(-18)} = \sqrt{81+360} = \sqrt{441} = 21$

12. $\sqrt{13^2-(4)(-3)(-4)} = \sqrt{169-48}$
$= \sqrt{121}$
$= 11$

13. $-\sqrt{9} = -1 \cdot \sqrt{9} = -1 \cdot 3 = -3$ is rational.

14. $-\frac{1}{2}\sqrt{48} \approx -3.46$ is irrational.

15. $\sqrt{14} \approx 3.74$ is irrational.

16. $\sqrt{-2}$ is not a real number.

17. $\sqrt{(4x-9)^2} = |4x-9|$

18. $\sqrt{(16m-25)^2} = |16m-25|$

19. $\sqrt{28} = \sqrt{4 \cdot 7} = \sqrt{4} \cdot \sqrt{7} = 2\sqrt{7}$

20. $\sqrt{45} = \sqrt{9 \cdot 5} = \sqrt{9} \cdot \sqrt{5} = 3\sqrt{5}$

21. $\sqrt{200} = \sqrt{100 \cdot 2} = \sqrt{100} \cdot \sqrt{2} = 10\sqrt{2}$

22. $\sqrt{150} = \sqrt{25 \cdot 6} = \sqrt{25} \cdot \sqrt{6} = 5\sqrt{6}$

23. $\frac{-2+\sqrt{8}}{2} = \frac{-2+\sqrt{4 \cdot 2}}{2}$
$= \frac{-2+\sqrt{4} \cdot \sqrt{2}}{2}$
$= \frac{-2+2\sqrt{2}}{2}$
$= \frac{2\left(-1+\sqrt{2}\right)}{2}$
$= -1+\sqrt{2}$

24. $\frac{-3+\sqrt{27}}{6} = \frac{-3+\sqrt{9 \cdot 3}}{6}$
$= \frac{-3+\sqrt{9} \cdot \sqrt{3}}{6}$
$= \frac{-3+3\sqrt{3}}{6}$
$= \frac{3\left(-1+\sqrt{3}\right)}{3 \cdot 2}$
$= \frac{-1+\sqrt{3}}{2}$

25. $\sqrt{a^{36}} = \sqrt{(a^{18})^2} = a^{18}$

26. $\sqrt{x^{16}} = \sqrt{(x^8)^2} = x^8$

27. $\sqrt{16x^{10}} = \sqrt{16} \cdot \sqrt{x^{10}} = 4x^5$

28. $\sqrt{49a^{12}} = \sqrt{49} \cdot \sqrt{a^{12}} = 7a^6$

29. $\sqrt{18n^9} = \sqrt{9n^8 \cdot 2n} = \sqrt{9n^8} \cdot \sqrt{2n} = 3n^4\sqrt{2n}$

30. $\sqrt{8y^{25}} = \sqrt{4y^{24} \cdot 2y}$
$= \sqrt{4y^{24}} \cdot \sqrt{2y}$
$= 2y^{12}\sqrt{2y}$

31. $\sqrt{\frac{27}{3x^8}} = \sqrt{\frac{3 \cdot 9}{3x^8}} = \sqrt{\frac{9}{x^8}} = \frac{\sqrt{9}}{\sqrt{x^8}} = \frac{3}{x^4}$

32. $\sqrt{\frac{8}{2x^4}} = \sqrt{\frac{2 \cdot 4}{2 \cdot x^4}} = \sqrt{\frac{4}{x^4}} = \frac{\sqrt{4}}{\sqrt{x^4}} = \frac{2}{x^2}$

33. $\sqrt{\frac{81y^5}{25y}} = \sqrt{\frac{81y^4}{25}} = \frac{\sqrt{81y^4}}{\sqrt{25}} = \frac{9y^2}{5}$

34. $\sqrt{\frac{36n}{121n^9}} = \sqrt{\frac{36}{121n^8}} = \frac{\sqrt{36}}{\sqrt{121n^8}} = \frac{6}{11n^4}$

35. $\sqrt{7} - 3\sqrt{7} = (1-3)\sqrt{7} = -2\sqrt{7}$

36. $4\sqrt{11} - \sqrt{11} = (4-1)\sqrt{11} = 3\sqrt{11}$

37. $4\sqrt{x} - 3\sqrt{x} = (4-3)\sqrt{x} = \sqrt{x}$

38. $5\sqrt{n}-6\sqrt{n}=(5-6)\sqrt{n}=-\sqrt{n}$

39. $20a\sqrt{ab}+5a\sqrt{ba}=20a\sqrt{ab}+5a\sqrt{ab}$
$=(20a+5a)\sqrt{ab}$
$=25a\sqrt{ab}$

40. $2x\sqrt{3xy}+x\sqrt{3yx}=2x\sqrt{3xy}+x\sqrt{3xy}$
$=(2x+x)\sqrt{3xy}$
$=3x\sqrt{3xy}$

41. $-2\sqrt{12}+3\sqrt{27}=-2\sqrt{4\cdot 3}+3\sqrt{9\cdot 3}$
$=-2\sqrt{4}\cdot\sqrt{3}+3\sqrt{9}\cdot\sqrt{3}$
$=-2\cdot 2\sqrt{3}+3\cdot 3\sqrt{3}$
$=-4\sqrt{3}+9\sqrt{3}$
$=(-4+9)\sqrt{3}$
$=5\sqrt{3}$

42. $-4\sqrt{18}+5\sqrt{32}=-4\sqrt{9\cdot 2}+5\sqrt{16\cdot 2}$
$=-4\cdot\sqrt{9}\cdot\sqrt{2}+5\cdot\sqrt{16}\cdot\sqrt{2}$
$=-4\cdot 3\sqrt{2}+5\cdot 4\sqrt{2}$
$=-12\sqrt{2}+20\sqrt{2}$
$=(-12+20)\sqrt{2}$
$=8\sqrt{2}$

43. $4+2\sqrt{20}-\sqrt{45}=4+2\sqrt{4\cdot 5}-\sqrt{9\cdot 5}$
$=4+2\sqrt{4}\cdot\sqrt{5}-\sqrt{9}\cdot\sqrt{5}$
$=4+2\cdot 2\sqrt{5}-3\sqrt{5}$
$=4+4\sqrt{5}-3\sqrt{5}$
$=4+(4-3)\sqrt{5}$
$=4+\sqrt{5}$

44. $6-2\sqrt{27}+\sqrt{48}=6-2\sqrt{9\cdot 3}+\sqrt{16\cdot 3}$
$=6-2\cdot\sqrt{9}\cdot\sqrt{3}+\sqrt{16}\cdot\sqrt{3}$
$=6-2\cdot 3\sqrt{3}+4\sqrt{3}$
$=6-6\sqrt{3}+4\sqrt{3}$
$=6+(-6+4)\sqrt{3}$
$=6-2\sqrt{3}$

45. $2n\sqrt{8n^3}+5\sqrt{18n^5}$
$=2n\sqrt{4n^2\cdot 2n}+5\sqrt{9n^4\cdot 2n}$
$=2n\sqrt{4n^2}\cdot\sqrt{2n}+5\sqrt{9n^4}\cdot\sqrt{2n}$
$=2n\cdot 2n\sqrt{2n}+5\cdot 3n^2\sqrt{2n}$
$=4n^2\sqrt{2n}+15n^2\sqrt{2n}$
$=(4n^2+15n^2)\sqrt{2n}$
$=19n^2\sqrt{2n}$

46. $3\sqrt{56a^5}+a^2\sqrt{126a}$
$=3\sqrt{4a^4\cdot 14a}+a^2\sqrt{19\cdot 14a}$
$=3\sqrt{4a^4}\cdot\sqrt{14a}+a^2\cdot\sqrt{9}\cdot\sqrt{14a}$
$=3\cdot 2a^2\sqrt{14a}+a^2\cdot 3\sqrt{14a}$
$=(6a^2+3a^2)\sqrt{14a}$
$=9a^2\sqrt{14a}$

47. $\frac{3}{4}\sqrt{32}+\frac{2}{3}\sqrt{27}-\frac{1}{2}\sqrt{8}$
$=\frac{3}{4}\sqrt{16\cdot 2}+\frac{2}{3}\sqrt{9\cdot 3}-\frac{1}{2}\sqrt{4\cdot 2}$
$=\frac{3}{4}\sqrt{16}\cdot\sqrt{2}+\frac{2}{3}\cdot\sqrt{9}\cdot\sqrt{3}-\frac{1}{2}\cdot\sqrt{4}\cdot\sqrt{2}$
$=\frac{3}{4}\cdot 4\cdot\sqrt{2}+\frac{2}{3}\cdot 3\cdot\sqrt{3}-\frac{1}{2}\cdot 2\sqrt{2}$
$=3\sqrt{2}+2\sqrt{3}-\sqrt{2}$
$=(3-1)\sqrt{2}+2\sqrt{3}$
$=2\sqrt{2}+2\sqrt{3}$

48. $\frac{5}{2}\sqrt{24}+\frac{2}{9}\sqrt{27}-\frac{1}{6}\sqrt{54}$
$=\frac{5}{2}\sqrt{4\cdot 6}+\frac{2}{9}\cdot\sqrt{9\cdot 3}-\frac{1}{6}\cdot\sqrt{9\cdot 6}$
$=\frac{5}{2}\cdot\sqrt{4}\cdot\sqrt{6}+\frac{2}{9}\cdot\sqrt{9}\cdot\sqrt{3}-\frac{1}{6}\cdot\sqrt{9}\cdot\sqrt{6}$
$=\frac{5}{2}\cdot 2\sqrt{6}+\frac{2}{9}\cdot 3\sqrt{3}-\frac{1}{6}\cdot 3\sqrt{6}$
$=5\sqrt{6}+\frac{2}{3}\sqrt{3}-\frac{1}{2}\sqrt{6}$
$=\left(5-\frac{1}{2}\right)\sqrt{6}+\frac{2}{3}\sqrt{3}$
$=\frac{9}{2}\sqrt{6}+\frac{2}{3}\sqrt{3}$

49. $3\sqrt{\frac{3}{16}}-\frac{1}{2}\sqrt{\frac{12}{25}}=3\frac{\sqrt{3}}{\sqrt{16}}-\frac{1}{2}\frac{\sqrt{12}}{\sqrt{25}}$
$=3\frac{\sqrt{3}}{4}-\frac{1}{2}\frac{\sqrt{12}}{5}$
$=\frac{3}{4}\sqrt{3}-\frac{1}{2}\frac{\sqrt{4\cdot 3}}{5}$
$=\frac{3}{4}\sqrt{3}-\frac{1}{2}\cdot\frac{2}{5}\cdot\sqrt{3}$
$=\frac{3}{4}\sqrt{3}-\frac{1}{5}\sqrt{3}$
$=\left(\frac{3}{4}-\frac{1}{5}\right)\sqrt{3}$
$=\left(\frac{15}{20}-\frac{4}{20}\right)\sqrt{3}$
$=\frac{11}{20}\sqrt{3}$

50. $2\sqrt{\frac{2}{25}}-3\sqrt{\frac{8}{9}}=2\frac{\sqrt{2}}{\sqrt{25}}-3\frac{\sqrt{8}}{\sqrt{9}}$
$=2\cdot\frac{\sqrt{2}}{5}-3\frac{\sqrt{8}}{3}$
$=\frac{2}{5}\sqrt{2}-\sqrt{8}$
$=\frac{2}{5}\sqrt{2}-\sqrt{4\cdot 2}$
$=\frac{2}{5}\sqrt{2}-2\sqrt{2}$
$=\left(\frac{2}{5}-2\right)\sqrt{2}$
$=\left(\frac{2}{5}-\frac{10}{5}\right)\sqrt{2}$
$=-\frac{8}{5}\sqrt{2}$

51. $\sqrt{12}\cdot\sqrt{8}=\sqrt{12\cdot 8}$
$=\sqrt{96}$
$=\sqrt{16\cdot 6}$
$=\sqrt{16}\cdot\sqrt{6}$
$=4\sqrt{6}$

52. $\sqrt{24}\cdot\sqrt{10}=\sqrt{24\cdot 10}$
$=\sqrt{240}$
$=\sqrt{16\cdot 15}$
$=\sqrt{16}\cdot\sqrt{15}$
$=4\sqrt{15}$

53. $\sqrt{14x^3}\cdot\sqrt{21x^2}=\sqrt{14x^3\cdot 21x^2}$
$=\sqrt{294x^5}$
$=\sqrt{49x^4\cdot 6x}$
$=7x^2\sqrt{6x}$

54. $\sqrt{6y}\cdot\sqrt{30y^4}=\sqrt{6y\cdot 30y^4}$
$=\sqrt{180y^5}$
$=\sqrt{36y^4\cdot 5y}$
$=6y^2\sqrt{5y}$

55. $-4\sqrt{20}\cdot 8\sqrt{8}=-4\cdot 8\sqrt{20\cdot 8}$
$=-32\sqrt{160}$
$=-32\sqrt{16\cdot 10}$
$=-32\cdot 4\sqrt{10}$
$=-128\sqrt{10}$

56. $-2\sqrt{10}\cdot 5\sqrt{40}=-2\cdot 5\sqrt{10\cdot 40}$
$=-10\sqrt{400}$
$=-10\cdot 20$
$=-200$

57. $\left(2\sqrt{2a}\right)^2=2^2\left(\sqrt{2a}\right)^2=4\cdot 2a=8a$

58. $\left(3\sqrt{5n}\right)^2=3^2\left(\sqrt{5n}\right)^2=9\cdot 5n=45n$

59. $4\sqrt{2}\cdot\left(-3\sqrt{2}\right)=4(-3)\sqrt{2\cdot 2}$
$=-12\sqrt{4}$
$=-12\cdot 2$
$=-24$

60. $7\sqrt{11}\cdot\left(-2\sqrt{11}\right)=7\cdot(-2)\left(\sqrt{11}\right)^2$
$=-14\cdot 11$
$=-154$

61. $\sqrt{3}\left(\sqrt{3a}+\sqrt{15a}\right)=\sqrt{3}\cdot\sqrt{3a}+\sqrt{3}\cdot\sqrt{15a}$
$=\sqrt{9a}+\sqrt{45a}$
$=\sqrt{9\cdot a}+\sqrt{9\cdot 5a}$
$=3\sqrt{a}+3\sqrt{5a}$

62. $\sqrt{x}\left(\sqrt{4x}+\sqrt{8x}\right)=\sqrt{x}\cdot\sqrt{4x}+\sqrt{x}\cdot\sqrt{8x}$
$=\sqrt{4x^2}+\sqrt{8x^2}$
$=\sqrt{4x^2}+\sqrt{4x^2\cdot 2}$
$=2x+2x\sqrt{2}$

63. $\left(3+2\sqrt{6}\right)\left(2+4\sqrt{3}\right)$
$=3\cdot 2+3\cdot 4\sqrt{3}+2\cdot 2\sqrt{6}+2\cdot 4\cdot\sqrt{6}\cdot\sqrt{3}$
$=6+12\sqrt{3}+4\sqrt{6}+8\sqrt{18}$
$=6+12\sqrt{3}+4\sqrt{6}+8\sqrt{9\cdot 2}$
$=6+12\sqrt{3}+4\sqrt{6}+8\cdot 3\sqrt{2}$
$=6+12\sqrt{3}+4\sqrt{6}+24\sqrt{2}$

64. $\left(5-3\sqrt{3}\right)\left(3+2\sqrt{6}\right)$
$=5\cdot 3+5\cdot 2\sqrt{6}-3\cdot 3\sqrt{3}-3\cdot 2\cdot\sqrt{3}\cdot\sqrt{6}$
$=15+10\sqrt{6}-9\sqrt{3}-6\sqrt{18}$
$=15+10\sqrt{6}-9\sqrt{3}-6\sqrt{9\cdot 2}$
$=15+10\sqrt{6}-9\sqrt{3}-6\cdot 3\sqrt{2}$
$=15+10\sqrt{6}-9\sqrt{3}-18\sqrt{2}$

65. $\left(4-2\sqrt{2y}\right)\left(\sqrt{y}+3\right)$
$=4\sqrt{y}+4\cdot 3-2\cdot\sqrt{2y}\cdot\sqrt{y}-2\cdot 3\sqrt{2y}$
$=4\sqrt{y}+12-2\sqrt{2y^2}-6\sqrt{2y}$
$=4\sqrt{y}+12-2y\sqrt{2}-6\sqrt{2y}$

66. $\left(1+3\sqrt{3x}\right)\left(\sqrt{x}+9\right)$
$=1\cdot\sqrt{x}+1\cdot 9+3\cdot\sqrt{3x}\cdot\sqrt{x}+3\cdot 9\cdot\sqrt{3x}$
$=\sqrt{x}+9+3\sqrt{3x^2}+27\sqrt{3x}$
$=\sqrt{x}+9+3x\sqrt{3}+27\sqrt{3x}$

67. $\left(3-\sqrt{2}\right)^2=3^2-2\cdot 3\cdot\sqrt{2}+\left(\sqrt{2}\right)^2$
$=9-6\sqrt{2}+2$
$=11-6\sqrt{2}$

68. $\left(4-\sqrt{5}\right)^2=4^2-2\cdot 4\cdot\sqrt{5}+\left(\sqrt{5}\right)^2$
$=16-8\sqrt{5}+5$
$=21-8\sqrt{5}$

69. $\left(\sqrt{2x}+3\sqrt{x}\right)^2$
$=\left(\sqrt{2x}\right)^2+2\sqrt{2x}\cdot 3\sqrt{x}+\left(3\sqrt{x}\right)^2$
$=2x+6\sqrt{2x^2}+3^2\cdot\left(\sqrt{x}\right)^2$
$=2x+6\sqrt{x^2\cdot 2}+9x$
$=11x+6x\sqrt{2}$

70. $\left(\sqrt{n}+2\sqrt{3n}\right)^2$
$=\left(\sqrt{n}\right)^2+2\sqrt{n}\cdot 2\sqrt{3n}+\left(2\sqrt{3n}\right)^2$
$=n+4\sqrt{3n^2}+2^2\left(\sqrt{3n}\right)^2$
$=n+4\sqrt{n^2\cdot 3}+4\cdot 3n$
$=n+4n\sqrt{3}+12n$
$=13n+4n\sqrt{3}$

71. $\left(3+5\sqrt{x}\right)\left(3-5\sqrt{x}\right)=3^2-\left(5\sqrt{x}\right)^2$
$=9-5^2\left(\sqrt{x}\right)^2$
$=9-25x$

72. $\left(2\sqrt{x}-4\right)\left(2\sqrt{x}+4\right)=\left(2\sqrt{x}\right)^2-4^2$
$=2^2\left(\sqrt{x}\right)^2-16$
$=4x-16$

73. $\dfrac{\sqrt{108}}{\sqrt{2}}=\sqrt{\dfrac{108}{2}}=\sqrt{54}=\sqrt{9\cdot 6}=3\sqrt{6}$

74. $\dfrac{\sqrt{135}}{\sqrt{3}}=\sqrt{\dfrac{135}{3}}=\sqrt{45}=\sqrt{9\cdot 5}=3\sqrt{5}$

75. $\sqrt{\dfrac{6x}{4x^3}}=\sqrt{\dfrac{3}{2x^2}}$
$=\dfrac{\sqrt{3}}{\sqrt{2x^2}}$
$=\dfrac{\sqrt{3}}{x\sqrt{2}}$
$=\dfrac{\sqrt{3}}{x\sqrt{2}}\cdot\dfrac{\sqrt{2}}{\sqrt{2}}$
$=\dfrac{\sqrt{6}}{2x}$

76. $\sqrt{\dfrac{15x^4}{36x^2}} = \dfrac{\sqrt{15x^4}}{\sqrt{36x^2}} = \dfrac{x^2\sqrt{15}}{6x} = \dfrac{x\sqrt{15}}{6}$

77. $$\begin{aligned}\frac{\sqrt{10}+\sqrt{12}}{\sqrt{2}} &= \frac{\sqrt{10}}{\sqrt{2}} + \frac{\sqrt{12}}{\sqrt{2}}\\ &= \sqrt{\frac{10}{2}} + \sqrt{\frac{12}{2}}\\ &= \sqrt{5}+\sqrt{6}\end{aligned}$$

78. $$\begin{aligned}\frac{\sqrt{30}-\sqrt{15}}{\sqrt{5}} &= \frac{\sqrt{30}}{\sqrt{5}} - \frac{\sqrt{15}}{\sqrt{5}}\\ &= \sqrt{\frac{30}{5}} - \sqrt{\frac{15}{5}}\\ &= \sqrt{6}-\sqrt{3}\end{aligned}$$

79. $\dfrac{14}{\sqrt{7}} = \dfrac{14}{\sqrt{7}} \cdot \dfrac{\sqrt{7}}{\sqrt{7}} = \dfrac{14\sqrt{7}}{7} = 2\sqrt{7}$

80. $\dfrac{25}{\sqrt{5}} = \dfrac{25}{\sqrt{5}} \cdot \dfrac{\sqrt{5}}{\sqrt{5}} = \dfrac{25\sqrt{5}}{5} = 5\sqrt{5}$

81. $\dfrac{3}{2\sqrt{6}} = \dfrac{3}{2\sqrt{6}} \cdot \dfrac{\sqrt{6}}{\sqrt{6}} = \dfrac{3\sqrt{6}}{2 \cdot 6} = \dfrac{\sqrt{6}}{4}$

82. $\dfrac{2}{5\sqrt{10}} = \dfrac{2}{5\sqrt{10}} \cdot \dfrac{\sqrt{10}}{\sqrt{10}} = \dfrac{2\sqrt{10}}{5 \cdot 10} = \dfrac{\sqrt{10}}{25}$

83. The conjugate of $-2+3\sqrt{5}$ is $-2-3\sqrt{5}$.
$$\begin{aligned}\left(-2+3\sqrt{5}\right)\left(-2-3\sqrt{5}\right) &= (-2)^2 - \left(3\sqrt{5}\right)^2\\ &= 4-9\cdot 5\\ &= 4-45\\ &= -41\end{aligned}$$

84. The conjugate of $-3-2\sqrt{7}$ is $-3+2\sqrt{7}$.
$$\begin{aligned}\left(-3-2\sqrt{7}\right)\left(-3+2\sqrt{7}\right) &= (-3)^2 - \left(2\sqrt{7}\right)^2\\ &= 9-4\cdot 7\\ &= 9-28\\ &= -19\end{aligned}$$

85. $$\begin{aligned}\frac{4}{2-\sqrt{2}} &= \frac{4}{2-\sqrt{2}} \cdot \frac{2+\sqrt{2}}{2+\sqrt{2}}\\ &= \frac{4\left(2+\sqrt{2}\right)}{2^2-\left(\sqrt{2}\right)^2}\\ &= \frac{4\left(2+\sqrt{2}\right)}{4-2}\\ &= \frac{4\left(2+\sqrt{2}\right)}{2}\\ &= 2\left(2+\sqrt{2}\right)\\ &= 4+2\sqrt{2}\end{aligned}$$

86. $$\begin{aligned}\frac{3}{6+\sqrt{3}} &= \frac{3}{6+\sqrt{3}} \cdot \frac{6-\sqrt{3}}{6-\sqrt{3}}\\ &= \frac{3\left(6-\sqrt{3}\right)}{6^2-\left(\sqrt{3}\right)^2}\\ &= \frac{3\left(6-\sqrt{3}\right)}{36-3}\\ &= \frac{3\left(6-\sqrt{3}\right)}{33}\\ &= \frac{6-\sqrt{3}}{11}\end{aligned}$$

87. $$\begin{aligned}\frac{\sqrt{12}}{\sqrt{3}+\sqrt{6}} &= \frac{\sqrt{12}}{\sqrt{3}+\sqrt{6}} \cdot \frac{\sqrt{3}-\sqrt{6}}{\sqrt{3}-\sqrt{6}}\\ &= \frac{\sqrt{12}\left(\sqrt{3}-\sqrt{6}\right)}{\left(\sqrt{3}\right)^2-\left(\sqrt{6}\right)^2}\\ &= \frac{\sqrt{36}-\sqrt{72}}{3-6}\\ &= \frac{6-\sqrt{36\cdot 2}}{-3}\\ &= \frac{6-6\sqrt{2}}{-3}\\ &= \frac{-3\left(2\sqrt{2}-2\right)}{-3}\\ &= 2\sqrt{2}-2\end{aligned}$$

88. $\dfrac{\sqrt{8}}{\sqrt{6}-\sqrt{2}}=\dfrac{\sqrt{8}}{\sqrt{6}-\sqrt{2}}\cdot\dfrac{\sqrt{6}+\sqrt{2}}{\sqrt{6}+\sqrt{2}}$

$=\dfrac{\sqrt{8}\left(\sqrt{6}+\sqrt{2}\right)}{\left(\sqrt{6}\right)^2-\left(\sqrt{2}\right)^2}$

$=\dfrac{\sqrt{48}+\sqrt{16}}{6-2}$

$=\dfrac{\sqrt{16\cdot 3}+\sqrt{16}}{4}$

$=\dfrac{4\sqrt{3}+4}{4}$

$=\dfrac{4\left(\sqrt{3}+1\right)}{4}$

$=\sqrt{3}+1$

89. $\sqrt{3x+1}=5;\ x=8$

$\sqrt{3(8)+1}\stackrel{?}{=}5$

$\sqrt{24+1}\stackrel{?}{=}5$

$\sqrt{25}\stackrel{?}{=}5$

$5=5$ True

Thus, $x=8$ is a solution of $\sqrt{3x+1}=5$.

90. $3\sqrt{x}-9=-3;\ x=4$

$3\sqrt{4}-9\stackrel{?}{=}-3$

$3\cdot 2-9\stackrel{?}{=}-3$

$6-9\stackrel{?}{=}-3$

$-3=-3$ True

Thus, $x=4$ is a solution of $3\sqrt{x}-9=-3$.

91. $\sqrt{3-x}=4$

$\left(\sqrt{3-x}\right)^2=4^2$

$3-x=16$

$3-16=x$

$-13=x$

Check: $\sqrt{3-(-13)}\stackrel{?}{=}4$

$\sqrt{3+13}\stackrel{?}{=}4$

$\sqrt{16}\stackrel{?}{=}4$

$4=4$ True

The solution set is $\{-13\}$.

92. $5=\sqrt{2x-1}$

$5^2=\left(\sqrt{2x-1}\right)^2$

$25=2x-1$

$26=2x$

$13=x$

Check: $5\stackrel{?}{=}\sqrt{2\cdot 13-1}$

$5\stackrel{?}{=}\sqrt{26-1}$

$5\stackrel{?}{=}\sqrt{25}$

$5=5$ True

The solution set is $\{13\}$.

93. $3\sqrt{2x}+4=10$

$3\sqrt{2x}=6$

$\sqrt{2x}=2$

$\left(\sqrt{2x}\right)^2=2^2$

$2x=4$

$x=2$

Check: $3\sqrt{2(2)}+4\stackrel{?}{=}10$

$3\sqrt{4}+4\stackrel{?}{=}10$

$3\cdot 2+4\stackrel{?}{=}10$

$6+4\stackrel{?}{=}10$

$10=10$ True

The solution set is $\{2\}$.

94. $5\sqrt{x}-3=12$

$5\sqrt{x}=15$

$\sqrt{x}=3$

$\left(\sqrt{x}\right)^2=3^2$

$x=9$

Check: $5\sqrt{9}-3\stackrel{?}{=}12$

$5\cdot 3-3\stackrel{?}{=}12$

$15-3\stackrel{?}{=}12$

$12=12$ True

The solution set is $\{9\}$.

95. $x=\sqrt{8x-15}$

$x^2=\left(\sqrt{8x-15}\right)^2$

$x^2=8x-15$

$x^2-8x+15=0$

$(x-5)(x-3)=0$

$x-5=0$ or $x-3=0$

$x=5$ or $x=3$

Check $x = 5$: $5 \stackrel{?}{=} \sqrt{8\cdot5-15}$
$5 \stackrel{?}{=} \sqrt{40-15}$
$5 \stackrel{?}{=} \sqrt{25}$
$5 = 5$ True

Check $x = 3$: $3 \stackrel{?}{=} \sqrt{8\cdot3-15}$
$3 \stackrel{?}{=} \sqrt{24-15}$
$3 \stackrel{?}{=} \sqrt{9}$
$3 = 3$ True

The solution set is $\{3, 5\}$.

96. $\sqrt{3x+18} = x$

$$\left(\sqrt{3x+18}\right)^2 = x^2$$
$$3x+18 = x^2$$
$$0 = x^2 - 3x - 18$$
$$0 = (x-6)(x+3)$$

$0 = x-6$ or $0 = x+3$
$6 = x$ or $-3 = x$

Check $x = 6$: $\sqrt{3\cdot6+18} \stackrel{?}{=} 6$
$\sqrt{18+18} \stackrel{?}{=} 6$
$\sqrt{36} \stackrel{?}{=} 6$
$6 = 6$ True

Check $x = -3$: $\sqrt{3(-3)+18} \stackrel{?}{=} -3$
$\sqrt{-9+18} \stackrel{?}{=} -3$
$\sqrt{9} \stackrel{?}{=} -3$
$3 = -3$ False

The solution set is $\{6\}$. ($x = -3$ is an extraneous solution.)

97. $n+2 = \sqrt{n^2+5}$

$$(n+2)^2 = \left(\sqrt{n^2+5}\right)^2$$
$$n^2+4n+4 = n^2+5$$
$$4n+4 = 5$$
$$4n = 1$$
$$n = \frac{1}{4}$$

Check: $\frac{1}{4}+2 \stackrel{?}{=} \sqrt{\left(\frac{1}{4}\right)^2+5}$
$\frac{1}{4}+\frac{8}{4} \stackrel{?}{=} \sqrt{\frac{1}{16}+\frac{80}{16}}$
$\frac{9}{4} \stackrel{?}{=} \sqrt{\frac{81}{16}}$
$\frac{9}{4} = \frac{9}{4}$ True

The solution set is $\left\{\frac{1}{4}\right\}$.

98. $n+3 = \sqrt{n^2-9}$

$$(n+3)^2 = \left(\sqrt{n^2-9}\right)^2$$
$$n^2+6n+9 = n^2-9$$
$$6n+9 = -9$$
$$6n = -18$$
$$n = -3$$

Check: $-3+3 \stackrel{?}{=} \sqrt{(-3)^2-9}$
$0 \stackrel{?}{=} \sqrt{9-9}$
$0 \stackrel{?}{=} \sqrt{0}$
$0 = 0$ True

The solution set is $\{-3\}$.

99. $\sqrt{2x+7} = \sqrt{5x-2}$

$$\left(\sqrt{2x+7}\right)^2 = \left(\sqrt{5x-2}\right)^2$$
$$2x+7 = 5x-2$$
$$9 = 3x$$
$$3 = x$$

Check: $\sqrt{2\cdot3+7} \stackrel{?}{=} \sqrt{5\cdot3-2}$
$\sqrt{6+7} \stackrel{?}{=} \sqrt{15-2}$
$\sqrt{13} = \sqrt{13}$ True

The solution set is $\{3\}$.

100. $3\sqrt{a} - \sqrt{4a+10} = 0$

$$3\sqrt{a} = \sqrt{4a+10}$$
$$\left(3\sqrt{a}\right)^2 = \left(\sqrt{4a+10}\right)^2$$
$$9a = 4a+10$$
$$5a = 10$$
$$a = 2$$

Check: $3\sqrt{2}-\sqrt{4(2)+10} \stackrel{?}{=} 0$
$3\sqrt{2}-\sqrt{8+10} \stackrel{?}{=} 0$
$3\sqrt{2}-\sqrt{18} \stackrel{?}{=} 0$
$3\sqrt{2}-\sqrt{9\cdot 2} \stackrel{?}{=} 0$
$3\sqrt{2}-3\sqrt{2} \stackrel{?}{=} 0$
$0=0$ True

The solution set is {2}.

101. $\sqrt{2x-8}=3+\sqrt{2x+1}$
$\left(\sqrt{2x-8}\right)^2=\left(3+\sqrt{2x+1}\right)^2$
$2x-8=9+6\sqrt{2x+1}+2x+1$
$-18=6\sqrt{2x+1}$
$-3=\sqrt{2x+1}$

Since the square root of a real number is never negative, there is no solution. The solution set is { } or $\varnothing$.

102. $\sqrt{x-6}=2-\sqrt{x+10}$
$\left(\sqrt{x-6}\right)^2=\left(2-\sqrt{x+10}\right)^2$
$x-6=4-4\sqrt{x+10}+x+10$
$-20=-4\sqrt{x+10}$
$5=\sqrt{x+10}$
$5^2=\left(\sqrt{x+10}\right)^2$
$25=x+10$
$15=x$

Check: $\sqrt{15-6} \stackrel{?}{=} 2-\sqrt{15+10}$
$\sqrt{9} \stackrel{?}{=} 2-\sqrt{25}$
$3 \stackrel{?}{=} 2-5$
$3=-3$ False

There is no solution. The solution set is $\varnothing$.

103. $r=\sqrt{\dfrac{A}{\pi}}$
$3\sqrt{2}=\sqrt{\dfrac{A}{\pi}}$
$\left(3\sqrt{2}\right)^2=\left(\sqrt{\dfrac{A}{\pi}}\right)^2$
$9\cdot 2=\dfrac{A}{\pi}$
$18\pi=A$

The area is 18π square centimeters.

104. $r=\sqrt{\dfrac{3V}{\pi h}}$
$r^2=\left(\sqrt{\dfrac{3V}{\pi h}}\right)^2$
$r^2=\dfrac{3V}{\pi h}$
$hr^2=\dfrac{3V}{\pi}$
$h=\dfrac{3V}{\pi r^2}$
$h=\dfrac{3(98\pi)}{\pi(7)^2}$
$h=6$

The height is 6 inches.

105. $\sqrt[3]{-8}=-2$ since $(-2)^3=-8$.

106. $\sqrt[4]{-16}$ is not a real number.

107. $\sqrt[4]{48x^4}=(48x^4)^{1/4}$
$=48^{1/4}\cdot x^{4\cdot\frac{1}{4}}$
$=\sqrt[4]{48}\cdot x^1$
$=\sqrt[4]{16\cdot 3}\cdot x$
$=\sqrt[4]{2^4\cdot 3}\cdot x$
$=2x\sqrt[4]{3}$

108. $\sqrt[3]{625x^3}=(625x^3)^{1/3}$
$=625^{1/3}\cdot(x^3)^{1/3}$
$=\sqrt[3]{625}\cdot x^1$
$=\sqrt[3]{125\cdot 5}\cdot x$
$=x\sqrt[3]{5^3\cdot 5}$
$=5x\sqrt[3]{5}$

109. $\dfrac{\sqrt[3]{48n^6}}{\sqrt[3]{2n^2}}=\sqrt[3]{\dfrac{48n^6}{2n^2}}$
$=\sqrt[3]{24n^4}$
$=\sqrt[3]{8n^3\cdot 3n}$
$=\sqrt[3]{(2n)^3\cdot 3n}$
$=2n\sqrt[3]{3n}$

110. $\dfrac{\sqrt[4]{324z^7}}{\sqrt[4]{2z}} = \sqrt[4]{\dfrac{324z^7}{2z}}$
$= \sqrt[4]{162z^6}$
$= \sqrt[4]{81z^4 \cdot 2z^2}$
$= \sqrt[4]{(3z)^4 \cdot 2z^2}$
$= 3z\sqrt[4]{2z^2}$

111. $-16^{1/4} = -\sqrt[4]{16} = -\sqrt[4]{2^4} = -2$

112. $(-16)^{1/4}$ is not a real number.

113. $\sqrt[4]{x^3} = (x^3)^{1/4} = x^{3/4}$

114. $\sqrt[3]{z} = z^{1/3}$

115. $2x^{3/2} = 2(x^{1/2})^3$
$= 2\left(\sqrt{x}\right)^3$
$= 2\sqrt{x^3}$
$= 2x\sqrt{x}$

116. $(3v)^{1/3} = \sqrt[3]{3v}$

117. $27^{2/3} = \left(27^{1/3}\right)^2$
$= \left(\sqrt[3]{27}\right)^2$
$= \left(\sqrt[3]{3^3}\right)^2$
$= 3^2$
$= 9$

118. $16^{5/4} = (16^{1/4})^5$
$= \left(\sqrt[4]{16}\right)^5$
$= \left(\sqrt[4]{2^4}\right)^5$
$= 2^5$
$= 32$

119. $-16^{-3/2} = -\dfrac{1}{16^{3/2}}$
$= -\dfrac{1}{(16^{1/2})^3}$
$= -\dfrac{1}{\left(\sqrt{16}\right)^3}$
$= -\dfrac{1}{4^3}$
$= -\dfrac{1}{64}$

120. $32^{-3/5} = \dfrac{1}{32^{3/5}}$
$= \dfrac{1}{(32^{1/5})^3}$
$= \dfrac{1}{\left(\sqrt[5]{32}\right)^3}$
$= \dfrac{1}{\left(\sqrt[5]{2^5}\right)^3}$
$= \dfrac{1}{2^3}$
$= \dfrac{1}{8}$

121. $2^{2/3} \cdot 2^{4/3} = 2^{\frac{2}{3}+\frac{4}{3}} = 2^{6/3} = 2^2 = 4$

122. $6^{1/2} \cdot 6^{3/2} = 6^{\frac{1}{2}+\frac{3}{2}} = 6^{4/2} = 6^2 = 36$

123. $\dfrac{x^{3/4} \cdot x^{-1/4}}{x^{5/4}} = x^{\frac{3}{4}-\frac{1}{4}-\frac{5}{4}} = x^{-3/4} = \dfrac{1}{x^{3/4}}$

124. $(27a^{-1/2})^{-2/3} = (27)^{-2/3}(a^{-1/2})^{-2/3}$
$= \dfrac{1}{27^{2/3}} \cdot a^{1/3}$
$= \dfrac{1}{\left(\sqrt[3]{27}\right)^2} \cdot a^{1/3}$
$= \dfrac{1}{3^2} \cdot a^{1/3}$
$= \dfrac{a^{1/3}}{9}$

Chapter 8 Test

1. $-3\sqrt{12} = -3\sqrt{4\cdot 3}$
$= -3\cdot\sqrt{4}\cdot\sqrt{3}$
$= -3\cdot 2\cdot\sqrt{3}$
$= -6\sqrt{3}$

2. $\sqrt{50x^2y^3} = \sqrt{25x^2y^2\cdot 2y}$
$= \sqrt{(5xy)^2\cdot 2y}$
$= 5xy\sqrt{2y}$

3. $\sqrt{\dfrac{16x^4}{9x^2}} = \dfrac{\sqrt{16x^4}}{\sqrt{9x^2}} = \dfrac{\sqrt{(4x^2)^2}}{\sqrt{(3x)^2}} = \dfrac{4x^2}{3x} = \dfrac{4x}{3}$

4. $\sqrt{25-4(3)(-2)} = \sqrt{25+24} = \sqrt{49} = 7$

5. $3\sqrt{2}-\sqrt{8} = 3\sqrt{2}-\sqrt{4\cdot 2}$
$= 3\sqrt{2}-2\sqrt{2}$
$= (3-2)\sqrt{2}$
$= 1\sqrt{2}$
$= \sqrt{2}$

6. $3\sqrt{27}+2\sqrt{48} = 3\sqrt{9\cdot 3}+2\sqrt{16\cdot 3}$
$= 3\sqrt{9}\cdot\sqrt{3}+2\sqrt{16}\cdot\sqrt{3}$
$= 3\cdot 3\sqrt{3}+2\cdot 4\sqrt{3}$
$= 9\sqrt{3}+8\sqrt{3}$
$= (9+8)\sqrt{3}$
$= 17\sqrt{3}$

7. $\sqrt{45}\cdot\sqrt{15} = \sqrt{45\cdot 15}$
$= \sqrt{675}$
$= \sqrt{225\cdot 3}$
$= 15\sqrt{3}$

8. $2\sqrt{2}\left(3\sqrt{6}-2\right) = 2\sqrt{2}\cdot 3\sqrt{6}-2\cdot 2\sqrt{2}$
$= 2\cdot 3\sqrt{2\cdot 6}-4\sqrt{2}$
$= 6\sqrt{12}-4\sqrt{2}$
$= 6\sqrt{4\cdot 3}-4\sqrt{2}$
$= 6\cdot 2\sqrt{3}-4\sqrt{2}$
$= 12\sqrt{3}-4\sqrt{2}$

9. $\left(4-\sqrt{3}\right)^2 = 4^2-2\cdot 4\cdot\sqrt{3}+\left(\sqrt{3}\right)^2$
$= 16-8\sqrt{3}+3$
$= 19-8\sqrt{3}$

10. $\dfrac{\sqrt{36x^4}}{\sqrt{3x}} = \sqrt{\dfrac{36x^4}{3x}}$
$= \sqrt{12x^3}$
$= \sqrt{4x^2\cdot 3x}$
$= \sqrt{4x^2}\cdot\sqrt{3x}$
$= 2x\sqrt{3x}$

11. $\left(2-3\sqrt{2}\right)\left(4+\sqrt{2}\right)$
$= 2\cdot 4+2\sqrt{2}-3\cdot 4\sqrt{2}-3\left(\sqrt{2}\right)^2$
$= 8+2\sqrt{2}-12\sqrt{2}-3\cdot 2$
$= (8-6)+(2-12)\sqrt{2}$
$= 2-10\sqrt{2}$

12. $\dfrac{-3+\sqrt{162}}{6} = \dfrac{-3+\sqrt{81\cdot 2}}{6}$
$= \dfrac{-3+9\sqrt{2}}{6}$
$= \dfrac{3\left(-1+3\sqrt{2}\right)}{3\cdot 2}$
$= \dfrac{-1+3\sqrt{2}}{2}$

13. $\dfrac{12}{\sqrt{20}} = \dfrac{12}{\sqrt{20}}\cdot\dfrac{\sqrt{5}}{\sqrt{5}}$
$= \dfrac{12\sqrt{5}}{\sqrt{100}}$
$= \dfrac{12\sqrt{5}}{10}$
$= \dfrac{6\sqrt{5}}{5}$

14. $\frac{6}{\sqrt{2}+2} = \frac{6}{\sqrt{2}+2} \cdot \frac{\sqrt{2}-2}{\sqrt{2}-2}$
$= \frac{6(\sqrt{2}-2)}{(\sqrt{2})^2 - 2^2}$
$= \frac{6(\sqrt{2}-2)}{2-4}$
$= \frac{6(\sqrt{2}-2)}{-2}$
$= -3(\sqrt{2}-2)$
$= -3\sqrt{2}+6$

15. $3\sqrt{2x-1} = 9$
$\sqrt{2x-1} = 3$
$(\sqrt{2x-1})^2 = 3^2$
$2x-1 = 9$
$2x = 10$
$x = 5$

Check: $3\sqrt{2(5)-1} \stackrel{?}{=} 9$
$3\sqrt{10-1} \stackrel{?}{=} 9$
$3\sqrt{9} \stackrel{?}{=} 9$
$3 \cdot 3 \stackrel{?}{=} 9$
$9 = 9$ True

The solution set is $\{5\}$.

16. $2\sqrt{x+11} - 3 = x$
$2\sqrt{x+11} = x+3$
$(2\sqrt{x+11})^2 = (x+3)^2$
$4(x+11) = x^2 + 6x + 9$
$4x + 44 = x^2 + 6x + 9$
$0 = x^2 + 2x - 35$
$0 = (x+7)(x-5)$
$0 = x+7$ or $0 = x-5$
$-7 = x$ or $5 = x$

Check $x = -7$: $2\sqrt{-7+11} - 3 \stackrel{?}{=} -7$
$2\sqrt{4} - 3 \stackrel{?}{=} -7$
$2 \cdot 2 - 3 \stackrel{?}{=} -7$
$4 - 3 \stackrel{?}{=} -7$
$1 = -7$ False

Check $x = 5$: $2\sqrt{5+11} - 3 \stackrel{?}{=} 5$
$2\sqrt{16} - 3 \stackrel{?}{=} 5$
$2 \cdot 4 - 3 \stackrel{?}{=} 5$
$8 - 3 \stackrel{?}{=} 5$
$5 = 5$ True

The solution set is $\{5\}$. ($x = -7$ is an extraneous solution.)

17. $\sqrt{x^2 - 3x} = \sqrt{x+21}$
$(\sqrt{x^2-3x})^2 = (\sqrt{x+21})^2$
$x^2 - 3x = x + 21$
$x^2 - 4x - 21 = 0$
$(x-7)(x+3) = 0$
$x - 7 = 0$ or $x + 3 = 0$
$x = 7$ or $x = -3$

Check $x = 7$: $\sqrt{7^2 - 3(7)} \stackrel{?}{=} \sqrt{7+21}$
$\sqrt{49-21} \stackrel{?}{=} \sqrt{28}$
$\sqrt{28} = \sqrt{28}$ True

Check $x = -3$: $\sqrt{(-3)^2 - 3(-3)} \stackrel{?}{=} \sqrt{-3+21}$
$\sqrt{9+9} \stackrel{?}{=} \sqrt{18}$
$\sqrt{18} = \sqrt{18}$ True

The solution set is $\{7, -3\}$.

18. $\sqrt[3]{108x^5} = \sqrt[3]{27x^3 \cdot 4x^2}$
$= \sqrt[3]{27x^3} \cdot \sqrt[3]{4x^2}$
$= 3x\sqrt[3]{4x^2}$

19. $27^{-2/3} = \frac{1}{27^{2/3}}$
$= \frac{1}{(27^{1/3})^2}$
$= \frac{1}{(\sqrt[3]{27})^2}$
$= \frac{1}{3^2}$
$= \frac{1}{9}$

20. $c = \sqrt{a^2 + b^2}$
$c = \sqrt{3^2 + 6^2}$
$c = \sqrt{9 + 36}$
$c = \sqrt{45}$
$c = \sqrt{9 \cdot 5}$
$c = 3\sqrt{5}$
$c \approx 6.7$
The hypotenuse measures $3\sqrt{5} \approx 6.7$ feet.

Chapter 9

Section 9.1

Preparing for Solving Quadratic Equations Using the Square Root Property

P1. **(a)** $\sqrt{25} = 5$ because $5^2 = 25$.

(b) $\sqrt{48} = \sqrt{16 \cdot 3} = \sqrt{16} \cdot \sqrt{3} = 4\sqrt{3}$

P2.
$$\frac{3-\sqrt{72}}{6} = \frac{3-\sqrt{36 \cdot 2}}{6} = \frac{3-6\sqrt{2}}{6} = \frac{3\left(1-2\sqrt{2}\right)}{6} = \frac{1-2\sqrt{2}}{2} \text{ or } \frac{1}{2}-\sqrt{2}$$

P3. $z^2 - 8z + 16 = z^2 - 2 \cdot 4 \cdot z + 4^2 = (z-4)^2$

P4. $\sqrt{20} = \sqrt{4 \cdot 5} = 2\sqrt{5}$

9.1 Quick Checks

1. If $x^2 = p$, then $x = \underline{\sqrt{p}}$ or $x = -\underline{\sqrt{p}}$.

2. False; the solution set of $n^2 = 49$, is $\{-7, 7\}$.

3.
$$s^2 = 36$$
$$s = \pm\sqrt{36}$$
$$s = \pm 6$$
The solution set is $\{-6, 6\}$.

4.
$$p^2 - 20 = 0$$
$$p^2 = 20$$
$$p = \pm\sqrt{20}$$
$$p = \pm\sqrt{4 \cdot 5}$$
$$p = \pm 2\sqrt{5}$$
The solution set is $\left\{-2\sqrt{5}, 2\sqrt{5}\right\}$.

5.
$$3p^2 = 75$$
$$\frac{3p^2}{3} = \frac{75}{3}$$
$$p^2 = 25$$
$$p = \pm\sqrt{25}$$
$$p = \pm 5$$
The solution set is $\{-5, 5\}$.

6.
$$\frac{1}{2}z^2 + 1 = 10$$
$$\frac{1}{2}z^2 + 1 - 1 = 10 - 1$$
$$\frac{1}{2}z^2 = 9$$
$$2 \cdot \frac{1}{2}z^2 = 2 \cdot 9$$
$$z^2 = 9 \cdot 2$$
$$z = \pm\sqrt{9 \cdot 2}$$
$$z = \pm 3\sqrt{2}$$
The solution set is $\left\{-3\sqrt{2}, 3\sqrt{2}\right\}$.

7.
$$x^2 = \frac{1}{2}$$
$$x = \pm\sqrt{\frac{1}{2}}$$
$$x = \pm\frac{\sqrt{1}}{\sqrt{2}}$$
$$x = \pm\frac{1}{\sqrt{2}}$$
$$x = \pm\frac{1}{\sqrt{2}} \cdot \frac{\sqrt{2}}{\sqrt{2}}$$
$$x = \pm\frac{\sqrt{2}}{2}$$
The solution set is $\left\{-\frac{\sqrt{2}}{2}, \frac{\sqrt{2}}{2}\right\}$.

8. $5x^2 - 16 = 0$

$$5x^2 = 16$$
$$x^2 = \frac{16}{5}$$
$$x = \pm\sqrt{\frac{16}{5}}$$
$$x = \pm\frac{\sqrt{16}}{\sqrt{5}}$$
$$x = \pm\frac{4}{\sqrt{5}}$$
$$x = \pm\frac{4}{\sqrt{5}} \cdot \frac{\sqrt{5}}{\sqrt{5}}$$
$$x = \pm\frac{4\sqrt{5}}{5}$$

The solution set is $\left\{-\frac{4\sqrt{5}}{5}, \frac{4\sqrt{5}}{5}\right\}$.

9. When using the Square Root Property to solve an equation in the form $(ax+b)^2 = p$ where p is a positive real number, you will have <u>two</u> unique solutions.

10. $(y-3)^2 = 121$

$$y - 3 = \pm\sqrt{121}$$
$$y - 3 = \pm 11$$
$$y = \pm 11 + 3$$
$$y = 14 \text{ or } -8$$

Check $y = 14$: $(14-3)^2 \stackrel{?}{=} 121$

$$11^2 \stackrel{?}{=} 121$$
$$121 = 121 \quad \text{True}$$

Check $y = -8$: $(-8-3)^2 \stackrel{?}{=} 121$

$$(-11)^2 \stackrel{?}{=} 121$$
$$121 = 121 \quad \text{True}$$

The solution set is $\{-8, 14\}$.

11. $(5k+1)^2 - 2 = 26$

$$(5k+1)^2 = 28$$
$$5k + 1 = \pm\sqrt{28}$$
$$5k + 1 = \pm\sqrt{4 \cdot 7}$$
$$5k + 1 = \pm 2\sqrt{7}$$
$$5k = -1 \pm 2\sqrt{7}$$
$$k = \frac{-1 \pm 2\sqrt{7}}{5}$$

Check $k = \frac{-1+2\sqrt{7}}{5}$:

$$\left[5\left(\frac{-1+2\sqrt{7}}{5}\right)+1\right]^2 - 2 \stackrel{?}{=} 26$$
$$\left(-1+2\sqrt{7}+1\right)^2 - 2 \stackrel{?}{=} 26$$
$$\left(2\sqrt{7}\right)^2 - 2 \stackrel{?}{=} 26$$
$$4 \cdot 7 - 2 \stackrel{?}{=} 26$$
$$28 - 2 \stackrel{?}{=} 26$$
$$26 = 26 \quad \text{True}$$

Check $k = \frac{-1-2\sqrt{7}}{5}$:

$$\left[5\left(\frac{-1-2\sqrt{7}}{5}\right)+1\right]^2 - 2 \stackrel{?}{=} 26$$
$$\left(-1-2\sqrt{7}+1\right)^2 - 2 \stackrel{?}{=} 26$$
$$\left(-2\sqrt{7}\right)^2 - 2 \stackrel{?}{=} 26$$
$$4 \cdot 7 - 2 \stackrel{?}{=} 26$$
$$28 - 2 \stackrel{?}{=} 26$$
$$26 = 26 \quad \text{True}$$

The solution set is $\left\{\frac{-1-2\sqrt{7}}{5}, \frac{-1+2\sqrt{7}}{5}\right\}$.

12. $y^2 + 8y + 16 = 9$

$$(y+4)^2 = 9$$
$$y + 4 = \pm\sqrt{9}$$
$$y + 4 = \pm 3$$
$$y = \pm 3 - 4$$
$$y = -1 \text{ or } -7$$

Solution set is $\{-7, -1\}$.

13. $4n^2 + 16n - 20 = -4$

$$n^2 + 4n - 5 = -1$$
$$n^2 + 4n - 5 + 9 = -1 + 9$$
$$n^2 + 4n + 4 = 8$$
$$(n+2)^2 = 8$$
$$n + 2 = \pm\sqrt{8}$$
$$n + 2 = \pm 2\sqrt{2}$$
$$n = -2 \pm 2\sqrt{2}$$

The solution set is $\left\{-2-2\sqrt{2}, -2+2\sqrt{2}\right\}$.

14. $(n+3)^2 = -4$ has no real solution because there is no real number whose square is negative. The solution set is $\varnothing$.

15. $(2x+5)^2 + 8 = 7$

$(2x+5)^2 = -1$

The square root of -1 is not a real number. The solution set is $\varnothing$.

16. False; in a right triangle, the square of the length of the hypotenuse is equal to the sum of the squares of the lengths of the two legs.

17. $a^2 + b^2 = c^2$

$5^2 + 12^2 = c^2$

$25 + 144 = c^2$

$169 = c^2$

$\pm\sqrt{169} = c$

$\pm 13 = c$

The length of the hypotenuse is 13.

18. $a^2 + b^2 = c^2$

$5^2 + 15^2 = c^2$

$25 + 225 = c^2$

$250 = c^2$

$\pm\sqrt{250} = c$

$\pm\sqrt{25 \cdot 10} = c$

$\pm 5\sqrt{10} = c$

The length of the hypotenuse is $5\sqrt{10} \approx 15.8$.

19. $a^2 + b^2 = c^2$

$6^2 + 6^2 = c^2$

$36 + 36 = c^2$

$72 = c^2$

$\pm\sqrt{72} = c$

$\pm\sqrt{36 \cdot 2} = c$

$\pm 6\sqrt{2} = c$

The length of the hypotenuse is $6\sqrt{2} \approx 8.5$.

20. $a^2 + b^2 = c^2$ where $a = 25$, $b = 8$, and c is the length of a ladder that will reach the gutter.

$a^2 + b^2 = c^2$

$25^2 + 8^2 = c^2$

$625 + 64 = c^2$

$689 = c^2$

$\pm\sqrt{689} = c$

$c = \sqrt{689} \approx 26.2$ feet < 30 feet, so Stefan can use the 30-foot ladder to clean his gutters.

9.1 Exercises

21. $x^2 = 144$

$x = \pm\sqrt{144}$

$x = \pm 12$

The solution set is $\{-12, 12\}$.

23. $u^2 = 0$

$u = \pm\sqrt{0}$

$u = 0$

The solution set is $\{0\}$.

25. $12 = x^2$

$\pm\sqrt{12} = x$

$\pm\sqrt{4 \cdot 3} = x$

$\pm 2\sqrt{3} = x$

The solution set is $\left\{-2\sqrt{3}, 2\sqrt{3}\right\}$.

27. $s^2 = \frac{4}{9}$

$s = \pm\sqrt{\frac{4}{9}}$

$s = \pm\frac{2}{3}$

The solution set is $\left\{-\frac{2}{3}, \frac{2}{3}\right\}$.

29. $x^2 = \frac{4}{3}$
$x = \pm\sqrt{\frac{4}{3}}$
$x = \pm\frac{2}{\sqrt{3}}$
$x = \pm\frac{2}{\sqrt{3}} \cdot \frac{\sqrt{3}}{\sqrt{3}}$
$x = \pm\frac{2\sqrt{3}}{3}$
The solution set is $\left\{-\frac{2\sqrt{3}}{3}, \frac{2\sqrt{3}}{3}\right\}$.

31. $\frac{1}{2}r^2 = 16$
$r^2 = 32$
$r = \pm\sqrt{32}$
$r = \pm\sqrt{16 \cdot 2}$
$r = \pm 4\sqrt{2}$
The solution set is $\left\{-4\sqrt{2}, 4\sqrt{2}\right\}$.

33. $x^2 - 169 = 0$
$x^2 = 169$
$x = \pm\sqrt{169}$
$x = \pm 13$
The solution set is $\{-13, 13\}$.

35. $x^2 - 50 = 0$
$x^2 = 50$
$x = \pm\sqrt{50}$
$x = \pm\sqrt{25 \cdot 2}$
$x = \pm 5\sqrt{2}$
The solution set is $\left\{-5\sqrt{2}, 5\sqrt{2}\right\}$.

37. $p^2 + 16 = 0$
$p^2 = -16$
The square root of -16 is not a real number. The solution set is $\varnothing$.

39. $27x^2 = 3$
$x^2 = \frac{3}{27}$
$x^2 = \frac{1}{9}$
$x = \pm\sqrt{\frac{1}{9}}$
$x = \pm\frac{1}{3}$
The solution set is $\left\{-\frac{1}{3}, \frac{1}{3}\right\}$.

41. $\frac{1}{16}n^2 - 4 = 0$
$\frac{1}{16}n^2 = 4$
$n^2 = 64$
$n = \pm\sqrt{64}$
$n = \pm 8$
The solution set is $\{-8, 8\}$.

43. $65 = 2n^2 - 7$
$72 = 2n^2$
$36 = n^2$
$\pm\sqrt{36} = n$
$\pm 6 = n$
The solution set is $\{-6, 6\}$.

45. $3x^2 + 20 = 45$
$3x^2 = 25$
$x^2 = \frac{25}{3}$
$x = \pm\sqrt{\frac{25}{3}}$
$x = \pm\frac{5}{\sqrt{3}}$
$x = \pm\frac{5}{\sqrt{3}} \cdot \frac{\sqrt{3}}{\sqrt{3}}$
$x = \pm\frac{5\sqrt{3}}{3}$
The solution set is $\left\{-\frac{5\sqrt{3}}{3}, \frac{5\sqrt{3}}{3}\right\}$.

47. $2k^2 + 12 = 10$

$2k^2 = -2$

$k^2 = -1$

The square root of -1 is not a real number. The solution set is $\varnothing$.

49. $2x^2 - 15 = 49$

$2x^2 = 64$

$x^2 = 32$

$x = \pm\sqrt{32}$

$x = \pm\sqrt{16 \cdot 2}$

$x = \pm 4\sqrt{2}$

The solution set is $\{-4\sqrt{2}, 4\sqrt{2}\}$.

51. $4 = (x+4)^2$

$\pm\sqrt{4} = x+4$

$\pm 2 = x+4$

$-4 \pm 2 = x$

$x = -4 - 2 = -6$ or $x = -4 + 2 = -2$

The solution set is $\{-6, -2\}$.

53. $27(x-1)^2 = 3$

$(x-1)^2 = \frac{3}{27}$

$x - 1 = \pm\sqrt{\frac{1}{9}}$

$x - 1 = \pm\frac{1}{3}$

$x = 1 \pm \frac{1}{3}$

$x = 1 - \frac{1}{3} = \frac{2}{3}$ or $x = 1 + \frac{1}{3} = \frac{4}{3}$

The solution set is $\left\{\frac{2}{3}, \frac{4}{3}\right\}$.

55. $\frac{(n+5)^2}{3} = 2$

$(n+5)^2 = 6$

$n + 5 = \pm\sqrt{6}$

$n = -5 \pm \sqrt{6}$

The solution set is $\{-5-\sqrt{6}, -5+\sqrt{6}\}$.

57. $\frac{2}{3}(x-6)^2 = \frac{16}{3}$

$(x-6)^2 = \frac{16}{3} \cdot \frac{3}{2}$

$(x-6)^2 = 8$

$x - 6 = \pm\sqrt{8}$

$x - 6 = \pm 2\sqrt{2}$

$x = 6 \pm 2\sqrt{2}$

The solution set is $\{6-2\sqrt{2}, 6+2\sqrt{2}\}$.

59. $\left(p+\frac{1}{3}\right)^2 = \frac{16}{9}$

$p + \frac{1}{3} = \pm\sqrt{\frac{16}{9}}$

$p + \frac{1}{3} = \pm\frac{4}{3}$

$p = -\frac{1}{3} \pm \frac{4}{3}$

$p = -\frac{1}{3} - \frac{4}{3} = -\frac{5}{3}$ or $p = -\frac{1}{3} + \frac{4}{3} = \frac{3}{3} = 1$

The solution set is $\left\{-\frac{5}{3}, 1\right\}$.

61. $\left(x-\frac{5}{2}\right)^2 = \frac{15}{4}$

$x - \frac{5}{2} = \pm\sqrt{\frac{15}{4}}$

$x - \frac{5}{2} = \pm\frac{\sqrt{15}}{2}$

$x = \frac{5}{2} \pm \frac{\sqrt{15}}{2} = \frac{5 \pm \sqrt{15}}{2}$

The solution set is $\left\{\frac{5-\sqrt{15}}{2}, \frac{5+\sqrt{15}}{2}\right\}$.

63. $(x-1)^2 - 7 = 9$

$(x-1)^2 = 16$

$x - 1 = \pm\sqrt{16}$

$x - 1 = \pm 4$

$x = 1 \pm 4$

$x = 1 - 4 = -3$ or $x = 1 + 4 = 5$

The solution set is $\{-3, 5\}$.

65. $(8k+3)^2-5=-4$

$$(8k+3)^2=1$$
$$8k+3=\pm\sqrt{1}$$
$$8k+3=\pm1$$
$$8k=-3\pm1$$
$$k=\frac{-3\pm1}{8}$$

$k=\frac{-3-1}{8}=\frac{-4}{8}=-\frac{1}{2}$ or $k=\frac{-3+1}{8}=\frac{-2}{8}=-\frac{1}{4}$

The solution set is $\left\{-\frac{1}{2}, -\frac{1}{4}\right\}$.

67. $30=(9x+2)^2-24$

$$54=(9x+2)^2$$
$$\pm\sqrt{54}=9x+2$$
$$\pm\sqrt{9\cdot6}=9x+2$$
$$\pm3\sqrt{6}=9x+2$$
$$-2\pm3\sqrt{6}=9x$$
$$\frac{-2\pm3\sqrt{6}}{9}=x$$

The solution set is $\left\{\frac{-2-3\sqrt{6}}{9}, \frac{-2+3\sqrt{6}}{9}\right\}$.

69. $(2w+8)^2-12=-24$

$$(2w+8)^2=-12$$

The square root of -12 is not a real number. The solution set is $\varnothing$.

71. $x^2+8x+16=25$

$$(x+4)^2=25$$
$$x+4=\pm\sqrt{25}$$
$$x+4=\pm5$$
$$x=-4\pm5$$

$x=-4-5=-9$ or $x=-4+5=1$

The solution set is $\{-9, 1\}$.

73. $49=4w^2-4w+1$

$$49=(2w-1)^2$$
$$\pm\sqrt{49}=2w-1$$
$$\pm7=2w-1$$
$$1\pm7=2w$$
$$\frac{1\pm7}{2}=w$$

$w=\frac{1-7}{2}=\frac{-6}{2}=-3$ or $w=\frac{1+7}{2}=\frac{8}{2}=4$

The solution set is $\{-3, 4\}$.

75. $a^2+b^2=c^2$

Here, $a=6$ and $b=8$.

$$6^2+8^2=c^2$$
$$36+64=c^2$$
$$100=c^2$$
$$\pm\sqrt{100}=c$$
$$\pm10=c$$

The length of the hypotenuse is 10.

77. $a^2+b^2=c^2$

Here, $a=3$ and $b=3$.

$$3^2+3^2=c^2$$
$$9+9=c^2$$
$$18=c^2$$
$$\pm\sqrt{18}=c$$
$$\pm3\sqrt{2}=c$$

The hypotenuse is $3\sqrt{2}\approx4.24$.

79. $a^2+b^2=c^2$

Here, $b=4$ and $c=12$.

$$a^2+4^2=12^2$$
$$a^2+16=144$$
$$a^2=128$$
$$a=\pm\sqrt{128}$$
$$a=\pm8\sqrt{2}$$

The length of leg a is $8\sqrt{2}\approx11.31$.

81. $x^2-13x+36=0$

$$(x-4)(x-9)=0$$

$x-4=0$ or $x-9=0$

$x=4$ or $x=9$

The solution set is $\{4, 9\}$.

83. $2m^2 = 1 - m$

$2m^2 + m - 1 = 0$

$(m+1)(2m-1) = 0$

$m+1=0 \quad \text{or} \quad 2m-1=0$

$m=-1 \quad \text{or} \quad m=\frac{1}{2}$

The solution set is $\left\{-1, \frac{1}{2}\right\}$.

85. $2n^2 = 16$

$n^2 = 8$

$n = \pm\sqrt{8}$

$n = \pm 2\sqrt{2}$

The solution set is $\left\{-2\sqrt{2}, 2\sqrt{2}\right\}$.

87. $3x^2 - 9 = 36$

$3x^2 = 45$

$x^2 = 15$

$x = \pm\sqrt{15}$

The solution set is $\left\{-\sqrt{15}, \sqrt{15}\right\}$.

89. $0 = 15x^2 - 9$

$9 = 15x^2$

$\frac{9}{15} = x^2$

$\frac{3}{5} = x^2$

$\pm\sqrt{\frac{3}{5}} = x$

$\pm\frac{\sqrt{3}}{\sqrt{5}} \cdot \frac{\sqrt{5}}{\sqrt{5}} = x$

$\pm\frac{\sqrt{15}}{5} = x$

The solution set is $\left\{-\frac{\sqrt{15}}{5}, \frac{\sqrt{15}}{5}\right\}$.

91. $2 = r^2 + 6$

$-4 = r^2$

The square root of -4 is not a real number. The solution set is $\varnothing$.

93. $12 = x^2 + x$

$0 = x^2 + x - 12$

$0 = (x+4)(x-3)$

$x+4=0 \quad \text{or} \quad x-3=0$

$x=-4 \quad \text{or} \quad x=3$

The solution set is $\{-4, 3\}$.

95. $d^2 - 27 = 0$

$d^2 = 27$

$d = \pm\sqrt{27}$

$d = \pm 3\sqrt{3}$

The solution set is $\left\{-3\sqrt{3}, 3\sqrt{3}\right\}$.

97. $x^2 - 10x + 25 = 0$

$(x-5)^2 = 0$

$x-5=0$

$x=5$

The solution set is $\{5\}$.

99. $2(x+2)^2 + 3 = 8$

$2(x+2)^2 = 5$

$(x+2)^2 = \frac{5}{2}$

$x+2 = \pm\sqrt{\frac{5}{2}}$

$x+2 = \pm\frac{\sqrt{5}}{\sqrt{2}} \cdot \frac{\sqrt{2}}{\sqrt{2}}$

$x+2 = \pm\frac{\sqrt{10}}{2}$

$x = -2 \pm \frac{\sqrt{10}}{2}$

$x = \frac{-4 \pm \sqrt{10}}{2}$

The solution set is $\left\{\frac{-4-\sqrt{10}}{2}, \frac{-4+\sqrt{10}}{2}\right\}$.

101. $3x^2+4=6$
$$3x^2=2$$
$$x^2=\frac{2}{3}$$
$$x=\pm\sqrt{\frac{2}{3}}$$
$$x=\pm\frac{\sqrt{6}}{3}$$
The solution set is $\left\{-\frac{\sqrt{6}}{3}, \frac{\sqrt{6}}{3}\right\}$.

103. $2x^2-5x-12=0$
$$(2x+3)(x-4)=0$$
$$2x+3=0 \quad \text{or} \quad x-4=0$$
$$x=-\frac{3}{2} \quad \text{or} \quad x=4$$
The solution set is $\left\{-\frac{3}{2}, 4\right\}$.

105. $\sqrt{x^2-15}=5$
$$\left(\sqrt{x^2-15}\right)^2=5^2$$
$$x^2-15=25$$
$$x^2=40$$
$$x=\pm\sqrt{40}$$
$$x=\pm 2\sqrt{10}$$
The solution set is $\{-2\sqrt{10}, 2\sqrt{10}\}$.

107. $\sqrt{q^2+13}+3=6$
$$\sqrt{q^2+13}=3$$
$$\left(\sqrt{q^2+13}\right)^2=3^2$$
$$q^2+13=9$$
$$q^2=-4$$
The square root of –4 is not a real number. The solution set is ∅.

109. $c^2=a^2+b^2$
$$c=\pm\sqrt{a^2+b^2}$$
$$c=\pm\sqrt{2^2+6^2}$$
$$c=\pm\sqrt{4+36}$$
$$c=\pm\sqrt{40}$$
$$c=\pm 2\sqrt{10}$$
The diagonal is $2\sqrt{10}$.

111. (a) $a^2+b^2=c^2$
Here $a = 7$ and $c = 13$.
$$7^2+b^2=13^2$$
$$49+b^2=169$$
$$b^2=120$$
$$b=\pm\sqrt{120}$$
$$b=\pm 2\sqrt{30}$$
The ladder will reach $2\sqrt{30}$ feet up the building.

(b) $2\sqrt{30}\approx 11.0$
The height is about 11.0 feet.

113. $c^2=a^2+b^2$
Here $a = 20$ and $b = 100$.
$$c^2=20^2+100^2$$
$$c^2=400+10{,}000$$
$$c^2=10{,}400$$
$$c=\pm\sqrt{10{,}400}$$
$$c=\pm\sqrt{400\cdot 26}$$
$$c=\pm 20\sqrt{26}$$
The ball is $20\sqrt{26}\approx 102$ yards from the center of the green.

115. (a) $a^2+b^2=c^2$
Here $a = 9.25$ and $b = 10.25$.
$$9.25^2+10.25^2=c^2$$
$$85.5625+105.0625=c^2$$
$$190.625=c^2$$
$$\pm\sqrt{190.625}=c$$
$$\pm 13.8067\approx c$$
The distance is about 13.8 feet.

(b) 1 foot = 12 inches
0.8 foot = 0.8(12) inches = 9.6 inches
Thus 13.8 feet is 13 feet and 9.6 inches.

117. $A = P(1+r)^2$

Here $A = 583.2$ and $P = 500$.

$$583.2 = 500(1+r)^2$$
$$\frac{583.2}{500} = (1+r)^2$$
$$1.1664 = (1+r)^2$$
$$\pm\sqrt{1.1664} = 1+r$$
$$\pm 1.08 = 1+r$$
$$-1 \pm 1.08 = r$$

$r = -1 - 1.08 = -2.08$ or

$r = -1 + 1.08 = 0.08$

$r = -2.08$ or $r = 0.08$

Since the interest rate must be positive, it is 0.08 or 8%.

119. We work backwards from the solutions.

$$x = \pm 4$$
$$x^2 = (\pm 4)^2$$
$$x^2 = 16$$
$$x^2 - 16 = 0$$

121. We work backwards from the solutions.

$$x = \pm\sqrt{6}$$
$$x^2 = \left(\pm\sqrt{6}\right)^2$$
$$x^2 = 6$$
$$x^2 - 6 = 0$$

123. We work backwards from the solutions.

$$x = \pm 3\sqrt{5}$$
$$x^2 = \left(\pm 3\sqrt{5}\right)^2$$
$$x^2 = 3^2\left(\sqrt{5}\right)^2$$
$$x^2 = 9 \cdot 5$$
$$x^2 = 45$$
$$x^2 - 45 = 0$$

125. The error is the third step where only the positive square root was taken.

$$(x-5)^2 - 4 = 0$$
$$(x-5)^2 = 4$$
$$x - 5 = \pm 2$$

$x - 5 = 2$ or $x - 5 = -2$

$x = 7$ $\quad$ $x = 3$

The solution set is $\{3, 7\}$.

Section 9.2

Preparing for Solving Quadratic Equations by Completing the Square

P1. $a^2 + 10a + 25 = a^2 + 2\cdot 5\cdot a + 5^2 = (a+5)^2$

P2. $(x+3)^2 = 16$

$$\sqrt{(x+3)^2} = \pm\sqrt{16}$$
$$x + 3 = \pm 4$$
$$x = -3 \pm 4$$

$x = -3 + 4 = 1$ or $x = -3 - 4 = -7$

The solution set is $\{-7, 1\}$.

P3. $\dfrac{3x^2 - 5x + 4}{3} = \dfrac{3x^2}{3} - \dfrac{5x}{3} + \dfrac{4}{3} = x^2 - \dfrac{5}{3}x + \dfrac{4}{3}$

9.2 Quick Checks

1. The polynomial $x^2 - 20x + \underline{100}$ is a perfect square trinomial and factors as $\underline{(x-10)^2}$.

2. In general $x^2 - bx + \left(\dfrac{b}{2}\right)^2 = \underline{\left(x - \dfrac{b}{2}\right)^2}$.

3. $p^2 + 6p$

$$\left(\frac{6}{2}\right)^2 = 3^2 = 9$$
$$p^2 + 6p + 9 = (p+3)^2$$

4. $b^2 - 18b$

$$\left(-\frac{18}{2}\right)^2 = (-9)^2 = 81$$
$$b^2 - 18b + 81 = (b-9)^2$$

5. $w^2 + 3w$

$$\left(\frac{3}{2}\right)^2 = \frac{9}{4}$$
$$w^2 + 3w + \frac{9}{4} = \left(w + \frac{3}{2}\right)^2$$

6. False; to complete the square, add $\left(\dfrac{18}{2}\right)^2 = 9^2 = 81$ to both sides.

7.
$$n^2+8n+7=0$$
$$n^2+8n=-7$$
$$n^2+8n+\left(\frac{8}{2}\right)^2=-7+\left(\frac{8}{2}\right)^2$$
$$n^2+8n+4^2=-7+4^2$$
$$(n+4)^2=-7+16$$
$$(n+4)^2=9$$
$$n+4=\pm3$$
$$n=-4\pm3$$
$n=-4+3=-1$ or $n=-4-3=-7$
The solution set is $\{-7, -1\}$.

8.
$$y^2+4y-32=0$$
$$y^2+4y=32$$
$$y^2+4y+\left(\frac{4}{2}\right)^2=32+\left(\frac{4}{2}\right)^2$$
$$y^2+4y+2^2=32+2^2$$
$$(y+2)^2=32+4$$
$$(y+2)^2=36$$
$$y+2=\pm\sqrt{36}$$
$$y+2=\pm6$$
$$y=-2\pm6$$
$y=-2+6=4$ or $y=-2-6=-8$
The solution set is $\{-8, 4\}$.

9.
$$z^2+8z=-9$$
$$z^2+8z+16=-9+16$$
$$(z+4)^2=7$$
$$z+4=\pm\sqrt{7}$$
$$z=-4\pm\sqrt{7}$$
Check $z=-4+\sqrt{7}$:
$$\left(-4+\sqrt{7}\right)^2+8\left(-4+\sqrt{7}\right)\stackrel{?}{=}-9$$
$$16-8\sqrt{7}+7-32+8\sqrt{7}\stackrel{?}{=}-9$$
$$-9=-9 \quad \text{True}$$
Check $z=-4-\sqrt{7}$:
$$\left(-4-\sqrt{7}\right)^2+8\left(-4-\sqrt{7}\right)\stackrel{?}{=}-9$$
$$16+8\sqrt{7}+7-32-8\sqrt{7}\stackrel{?}{=}-9$$
$$-9=-9 \quad \text{True}$$
The solution set is $\left\{-4-\sqrt{7}, -4+\sqrt{7}\right\}$.

10.
$$t^2-10t=-19$$
$$t^2-10t+25=-19+25$$
$$(t-5)^2=6$$
$$t-5=\pm\sqrt{6}$$
$$t=5\pm\sqrt{6}$$
Check $t=5+\sqrt{6}$:
$$\left(5+\sqrt{6}\right)^2-10\left(5+\sqrt{6}\right)\stackrel{?}{=}-19$$
$$25+10\sqrt{6}+6-50-10\sqrt{6}\stackrel{?}{=}-19$$
$$-19=-19 \quad \text{True}$$
Check $t=5-\sqrt{6}$:
$$\left(5-\sqrt{6}\right)^2-10\left(5-\sqrt{6}\right)\stackrel{?}{=}-19$$
$$25-10\sqrt{6}+6-50+10\sqrt{6}\stackrel{?}{=}-19$$
$$-19=-19 \quad \text{True}$$
The solution set is $\left\{5-\sqrt{6}, 5+\sqrt{6}\right\}$.

11. False; the factors on the left can only be set equal to the number on the right when the number on the right is 0.

12.
$$(n+5)(n+3)=24$$
$$n^2+8n+15=24$$
$$n^2+8n=9$$
$$n^2+8n+16=9+16$$
$$(n+4)^2=25$$
$$n+4=\pm\sqrt{25}$$
$$n+4=\pm5$$
$$n=-4\pm5$$
$n=-4+5=1$ or $n=-4-5=-9$
The solution set is $\{-9, 1\}$.

13. $(w+6)(w-1)=4$

$$w^2+5w-6=4$$
$$w^2+5w=10$$
$$w^2+5w+\frac{25}{4}=10+\frac{25}{4}$$
$$\left(w+\frac{5}{2}\right)^2=\frac{65}{4}$$
$$w+\frac{5}{2}=\pm\sqrt{\frac{65}{4}}$$
$$w+\frac{5}{2}=\pm\frac{\sqrt{65}}{2}$$
$$w=-\frac{5}{2}\pm\frac{\sqrt{65}}{2}$$
$$w=\frac{-5\pm\sqrt{65}}{2}$$

The solution set is $\left\{\frac{-5-\sqrt{65}}{2}, \frac{-5+\sqrt{65}}{2}\right\}$.

14. To solve $2x^2+4x=9$ by completing the square, the first step is to divide both sides of the equation by 2.

15. $2m^2=m+10$

$$2m^2-m-10=0$$
$$m^2-\frac{m}{2}-5=0$$
$$m^2-\frac{m}{2}=5$$
$$m^2-\frac{m}{2}+\frac{1}{16}=5+\frac{1}{16}$$
$$\left(m-\frac{1}{4}\right)^2=\frac{81}{16}$$
$$m-\frac{1}{4}=\pm\frac{9}{4}$$
$$m=\frac{1}{4}\pm\frac{9}{4}$$
$$m=\frac{1}{4}+\frac{9}{4}=\frac{10}{4}=\frac{5}{2} \text{ or } m=\frac{1}{4}-\frac{9}{4}=-\frac{8}{4}=-2$$

The solution set is $\left\{-2, \frac{5}{2}\right\}$.

16. $3y^2-4y-2=0$

$$y^2-\frac{4}{3}y-\frac{2}{3}=0$$
$$y^2-\frac{4}{3}y=\frac{2}{3}$$
$$y^2-\frac{4}{3}y+\frac{4}{9}=\frac{2}{3}+\frac{4}{9}$$
$$\left(y-\frac{2}{3}\right)^2=\frac{10}{9}$$
$$y-\frac{2}{3}=\pm\frac{\sqrt{10}}{3}$$
$$y=\frac{2\pm\sqrt{10}}{3}$$

The solution set is $\left\{\frac{2-\sqrt{10}}{3}, \frac{2+\sqrt{10}}{3}\right\}$.

17. $n(n-4)=-12$

$$n^2-4n=-12$$
$$n^2-4n+4=-12+4$$
$$(n-2)^2=-8$$

There is no real number whose square is –8, so there is no real solution to this equation. The solution set is ∅ or { }.

18. $-x^2+6x=7$

$$x^2-6x=-7$$
$$x^2-6x+\left(\frac{6}{2}\right)^2=-7+\left(\frac{6}{2}\right)^2$$
$$x^2-6x+9=-7+9$$
$$(x-3)^2=2$$
$$x-3=\pm\sqrt{2}$$
$$x=3\pm\sqrt{2}$$

The solution set is $\{3-\sqrt{2}, 3+\sqrt{2}\}$.

9.2 Exercises

19. x^2+16x

$$\left(\frac{16}{2}\right)^2=(8)^2=64$$
$$x^2+16x+64=(x+8)^2$$

21. $z^2 - 12z$

$\left(-\frac{12}{2}\right)^2 = (-6)^2 = 36$

$z^2 - 12z + 36 = (z-6)^2$

23. $k^2 + \frac{4}{3}k$

$\left(\frac{\frac{4}{3}}{2}\right)^2 = \left(\frac{4}{6}\right)^2 = \left(\frac{2}{3}\right)^2 = \frac{4}{9}$

$k^2 + \frac{4}{3}k + \frac{4}{9} = \left(k + \frac{2}{3}\right)^2$

25. $y^2 - 9y$

$\left(\frac{-9}{2}\right)^2 = \frac{81}{4}$

$y^2 - 9y + \frac{81}{4} = \left(y - \frac{9}{2}\right)^2$

27. $t^2 + t = t^2 + 1t$

$\left(\frac{1}{2}\right)^2 = \frac{1}{4}$

$t^2 + t + \frac{1}{4} = \left(t + \frac{1}{2}\right)^2$

29. $s^2 - \frac{s}{2}$

$\left[\frac{1}{2}\left(-\frac{1}{2}\right)\right]^2 = \left[-\frac{1}{4}\right]^2 = \frac{1}{16}$

$s^2 - \frac{s}{2} + \frac{1}{16} = \left(s - \frac{1}{4}\right)^2$

31.
$$x^2 - 4x = 21$$
$$x^2 - 4x + \left(-\frac{4}{2}\right)^2 = 21 + \left(-\frac{4}{2}\right)^2$$
$$x^2 - 4x + (-2)^2 = 21 + (-2)^2$$
$$x^2 - 4x + 4 = 21 + 4$$
$$(x-2)^2 = 25$$
$$x - 2 = \pm\sqrt{25}$$
$$x - 2 = \pm 5$$
$$x = 2 \pm 5$$

$x = 2 - 5 = -3$ or $x = 2 + 5 = 7$

The solution set is $\{-3, 7\}$.

33.
$$x^2 + 4x = 7$$
$$x^2 + 4x + \left(\frac{4}{2}\right)^2 = 7 + \left(\frac{4}{2}\right)^2$$
$$x^2 + 4x + 2^2 = 7 + 2^2$$
$$x^2 + 4x + 4 = 7 + 4$$
$$(x+2)^2 = 11$$
$$x + 2 = \pm\sqrt{11}$$
$$x = -2 \pm \sqrt{11}$$

The solution set is $\left\{-2-\sqrt{11}, -2+\sqrt{11}\right\}$.

35.
$$m^2 - 3m - 18 = 0$$
$$m^2 - 3m = 18$$
$$m^2 - 3m + \left(-\frac{3}{2}\right)^2 = 18 + \left(-\frac{3}{2}\right)^2$$
$$m^2 - 3m + \frac{9}{4} = 18 + \frac{9}{4}$$
$$\left(m - \frac{3}{2}\right)^2 = \frac{81}{4}$$
$$m - \frac{3}{2} = \pm\sqrt{\frac{81}{4}}$$
$$m - \frac{3}{2} = \pm\frac{9}{2}$$
$$m = \frac{3}{2} \pm \frac{9}{2}$$
$$m = \frac{3 \pm 9}{2}$$

$m = \frac{3-9}{2} = \frac{-6}{2} = -3$ or $m = \frac{3+9}{2} = \frac{12}{2} = 6$

The solution set is $\{-3, 6\}$.

37.
$$r^2 - 8r + 15 = 0$$
$$r^2 - 8r = -15$$
$$r^2 - 8r + \left(-\frac{8}{2}\right)^2 = -15 + \left(-\frac{8}{2}\right)^2$$
$$r^2 - 8r + (-4)^2 = -15 + (-4)^2$$
$$r^2 - 8r + 16 = -15 + 16$$
$$(r-4)^2 = 1$$
$$r - 4 = \pm\sqrt{1}$$
$$r - 4 = \pm 1$$
$$r = 4 \pm 1$$

$r = 4 - 1 = 3$ or $r = 4 + 1 = 5$

The solution set is $\{3, 5\}$.

39.

$$x^2+2x+24=0$$
$$x^2+2x=-24$$
$$x^2+2x+\left(\frac{2}{2}\right)^2=-24+\left(\frac{2}{2}\right)^2$$
$$x^2+2x+1^2=-24+1^2$$
$$x^2+2x+1=-24+1$$
$$(x+1)^2=-23$$

The square root of –23 is not a real number. The solution set is ∅.

41.

$$x^2=3x+4$$
$$x^2-3x=4$$
$$x^2-3x+\left(-\frac{3}{2}\right)^2=4+\left(-\frac{3}{2}\right)^2$$
$$x^2-3x+\frac{9}{4}=4+\frac{9}{4}$$
$$\left(x-\frac{3}{2}\right)^2=\frac{25}{4}$$
$$x-\frac{3}{2}=\pm\frac{5}{2}$$
$$x=\frac{3\pm5}{2}$$
$$x=\frac{3+5}{2}=\frac{8}{2}=4 \text{ or } x=\frac{3-5}{2}=\frac{-2}{2}=-1$$

The solution set is {–1, 4}.

43.

$$t^2=8t+12$$
$$t^2-8t=12$$
$$t^2-8t+\left(\frac{-8}{2}\right)^2=12+\left(\frac{-8}{2}\right)^2$$
$$t^2-8t+(-4)^2=12+(-4)^2$$
$$t^2-8t+16=12+16$$
$$(t-4)^2=28$$
$$t-4=\pm\sqrt{28}$$
$$t-4=\pm2\sqrt{7}$$
$$t=4\pm2\sqrt{7}$$

The solution set is $\{4-2\sqrt{7}, 4+2\sqrt{7}\}$.

45.

$$-10=x^2+7x$$
$$-10+\left(\frac{7}{2}\right)^2=x^2+7x+\left(\frac{7}{2}\right)^2$$
$$-10+\frac{49}{4}=x^2+7x+\frac{49}{4}$$
$$\frac{9}{4}=\left(x+\frac{7}{2}\right)^2$$
$$\pm\sqrt{\frac{9}{4}}=x+\frac{7}{2}$$
$$\pm\frac{3}{2}=x+\frac{7}{2}$$
$$-\frac{7}{2}\pm\frac{3}{2}=x$$
$$\frac{-7\pm3}{2}=x$$
$$x=\frac{-7-3}{2}=\frac{-10}{2}=-5 \text{ or } x=\frac{-7+3}{2}=\frac{-4}{2}=-2$$

The solution set is {–5, –2}.

47.

$$n^2+7n-50=0$$
$$n^2+7n=50$$
$$n^2+7n+\left(\frac{7}{2}\right)^2=50+\left(\frac{7}{2}\right)^2$$
$$n^2+7n+\frac{49}{4}=50+\frac{49}{4}$$
$$\left(n+\frac{7}{2}\right)^2=\frac{249}{4}$$
$$n+\frac{7}{2}=\pm\sqrt{\frac{249}{4}}$$
$$n+\frac{7}{2}=\pm\frac{\sqrt{249}}{2}$$
$$n=-\frac{7}{2}\pm\frac{\sqrt{249}}{2}$$
$$n=\frac{-7\pm\sqrt{249}}{2}$$

The solution set is $\left\{\frac{-7-\sqrt{249}}{2}, \frac{-7+\sqrt{249}}{2}\right\}$.

49.

$$10x = -16 - x^2$$
$$x^2 + 10x = -16$$
$$x^2 + 10x + \left(\frac{10}{2}\right)^2 = -16 + \left(\frac{10}{2}\right)^2$$
$$x^2 + 10x + 5^2 = -16 + 5^2$$
$$x^2 + 10x + 25 = -16 + 25$$
$$(x+5)^2 = 9$$
$$x + 5 = \pm\sqrt{9}$$
$$x + 5 = \pm 3$$
$$x = -5 \pm 3$$

$x = -5 - 3 = -8$ or $x = -5 + 3 = -2$

The solution set is $\{-8, -2\}$.

51.

$$3x^2 + 6x - 9 = 0$$
$$\frac{3x^2 + 6x - 9}{3} = \frac{0}{3}$$
$$x^2 + 2x - 3 = 0$$
$$x^2 + 2x = 3$$
$$x^2 + 2x + \left(\frac{2}{2}\right)^2 = 3 + \left(\frac{2}{2}\right)^2$$
$$x^2 + 2x + 1 = 3 + 1$$
$$(x+1)^2 = 4$$
$$x + 1 = \pm\sqrt{4}$$
$$x + 1 = \pm 2$$
$$x = -1 \pm 2$$

$x = -1 - 2 = -3$ or $x = -1 + 2 = 1$

The solution set is $\{-3, 1\}$.

53.

$$4x^2 - 4x - 3 = 0$$
$$4x^2 - 4x = 3$$
$$x^2 - x = \frac{3}{4}$$
$$x^2 - x + \left(\frac{-1}{2}\right)^2 = \frac{3}{4} + \left(\frac{-1}{2}\right)^2$$
$$x^2 - x + \frac{1}{4} = \frac{3}{4} + \frac{1}{4}$$
$$\left(x - \frac{1}{2}\right)^2 = 1$$
$$x - \frac{1}{2} = \pm\sqrt{1}$$
$$x - \frac{1}{2} = \pm 1$$
$$x = \frac{1}{2} \pm 1$$
$$x = \frac{1}{2} \pm \frac{2}{2}$$
$$x = \frac{1 \pm 2}{2}$$

$x = \frac{1-2}{2} = \frac{-1}{2} = -\frac{1}{2}$ or $x = \frac{1+2}{2} = \frac{3}{2}$

The solution set is $\left\{-\frac{1}{2}, \frac{3}{2}\right\}$.

55.

$$3y^2+4y=2$$
$$y^2+\frac{4}{3}y=\frac{2}{3}$$
$$y^2+\frac{4}{3}y+\left[\frac{1}{2}\left(\frac{4}{3}\right)\right]^2=\frac{2}{3}+\left[\frac{1}{2}\left(\frac{4}{3}\right)\right]^2$$
$$y^2+\frac{4}{3}y+\left[\frac{2}{3}\right]^2=\frac{2}{3}+\left[\frac{2}{3}\right]^2$$
$$y^2+\frac{4}{3}y+\frac{4}{9}=\frac{2}{3}+\frac{4}{9}$$
$$\left(y+\frac{2}{3}\right)^2=\frac{10}{9}$$
$$y+\frac{2}{3}=\pm\sqrt{\frac{10}{9}}$$
$$y+\frac{2}{3}=\pm\frac{\sqrt{10}}{3}$$
$$y=-\frac{2}{3}\pm\frac{\sqrt{10}}{3}$$
$$y=\frac{-2\pm\sqrt{10}}{3}$$

The solution set is $\left\{\frac{-2-\sqrt{10}}{3}, \frac{-2+\sqrt{10}}{3}\right\}$.

57.

$$4x^2=5-4x$$
$$4x^2+4x=5$$
$$x^2+x=\frac{5}{4}$$
$$x^2+x+\left(\frac{1}{2}\right)^2=\frac{5}{4}+\left(\frac{1}{2}\right)^2$$
$$x^2+x+\frac{1}{4}=\frac{5}{4}+\frac{1}{4}$$
$$\left(x+\frac{1}{2}\right)^2=\frac{6}{4}$$
$$x+\frac{1}{2}=\pm\sqrt{\frac{6}{4}}$$
$$x+\frac{1}{2}=\pm\frac{\sqrt{6}}{2}$$
$$x=-\frac{1}{2}\pm\frac{\sqrt{6}}{2}$$
$$x=\frac{-1\pm\sqrt{6}}{2}$$

The solution set is $\left\{\frac{-1-\sqrt{6}}{2}, \frac{-1+\sqrt{6}}{2}\right\}$.

59.

$$4m^2+5m+7=0$$
$$4m^2+5m=-7$$
$$m^2+\frac{5}{4}m=\frac{-7}{4}$$
$$m^2+\frac{5}{4}m+\left[\frac{1}{2}\left(\frac{5}{4}\right)\right]^2=-\frac{7}{4}+\left[\frac{1}{2}\left(\frac{5}{4}\right)\right]^2$$
$$m^2+\frac{5}{4}m+\left[\frac{5}{8}\right]^2=-\frac{7}{4}+\left[\frac{5}{8}\right]^2$$
$$m^2+\frac{5}{4}m+\frac{25}{64}=-\frac{7}{4}+\frac{25}{64}$$
$$\left(m+\frac{5}{8}\right)^2=-\frac{87}{64}$$

The square root of $-\frac{87}{64}$ is not a real number.

The solution set is $\varnothing$.

61.

$$4p^2-12p=-7$$
$$p^2-3p=\frac{-7}{4}$$
$$p^2-3p+\left(\frac{-3}{2}\right)^2=-\frac{7}{4}+\left(\frac{-3}{2}\right)^2$$
$$p^2-3p+\frac{9}{4}=-\frac{7}{4}+\frac{9}{4}$$
$$\left(p-\frac{3}{2}\right)^2=\frac{2}{4}$$
$$p-\frac{3}{2}=\pm\sqrt{\frac{2}{4}}$$
$$p-\frac{3}{2}=\pm\frac{\sqrt{2}}{2}$$
$$p=\frac{3}{2}\pm\frac{\sqrt{2}}{2}$$
$$p=\frac{3\pm\sqrt{2}}{2}$$

The solution set is $\left\{\frac{3-\sqrt{2}}{2}, \frac{3+\sqrt{2}}{2}\right\}$.

63.
$$\begin{aligned}4r^2-24r-1&=0\\4r^2-24r&=1\\r^2-6r&=\frac{1}{4}\\r^2-6r+\left(\frac{-6}{2}\right)^2&=\frac{1}{4}+\left(\frac{-6}{2}\right)^2\\r^2-6r+(-3)^2&=\frac{1}{4}+\frac{36}{4}\\r^2-6r+9&=\frac{37}{4}\\(r-3)^2&=\frac{37}{4}\\r-3&=\pm\sqrt{\frac{37}{4}}\\r-3&=\pm\frac{\sqrt{37}}{2}\\r&=3\pm\frac{\sqrt{37}}{2}\\r&=\frac{6}{2}\pm\frac{\sqrt{37}}{2}\\r&=\frac{6\pm\sqrt{37}}{2}\end{aligned}$$

The solution set is $\left\{\frac{6-\sqrt{37}}{2}, \frac{6+\sqrt{37}}{2}\right\}$.

65.
$$\begin{aligned}(x+2)(x-4)&=16\\x^2-4x+2x-8&=16\\x^2-2x-8&=16\\x^2-2x-24&=0\\(x+4)(x-6)&=0\end{aligned}$$
$x+4=0$ or $x-6=0$
$x=-4$ or $x=6$
The solution set is $\{-4, 6\}$.

67.
$$\begin{aligned}4x^2-4x+1&=0\\(2x-1)^2&=0\\2x-1&=0\\2x&=1\\x&=\frac{1}{2}\end{aligned}$$

The solution set is $\left\{\frac{1}{2}\right\}$.

69.
$$\begin{aligned}n^2+6n-3&=0\\n^2+6n&=3\\n^2+6n+\left(\frac{6}{2}\right)^2&=3+\left(\frac{6}{2}\right)^2\\n^2+6n+3^2&=3+3^2\\n^2+6n+9&=3+9\\(n+3)^2&=12\\n+3&=\pm\sqrt{12}\\n+3&=\pm2\sqrt{3}\\n&=-3\pm2\sqrt{3}\end{aligned}$$
The solution set is $\{-3-2\sqrt{3}, -3+2\sqrt{3}\}$.

71.
$$\begin{aligned}3z^2+12&=72\\3z^2&=60\\z^2&=20\\z&=\pm\sqrt{20}\\z&=\pm2\sqrt{5}\end{aligned}$$
The solution set is $\{-2\sqrt{5}, 2\sqrt{5}\}$.

73.
$$\begin{aligned}3t^2&=2-6t\\3t^2+6t&=2\\t^2+2t&=\frac{2}{3}\\t^2+2t+\left(\frac{2}{2}\right)^2&=\frac{2}{3}+\left(\frac{2}{2}\right)^2\\t^2+2t+1^2&=\frac{2}{3}+1^2\\t^2+2t+1&=\frac{2}{3}+1\\(t+1)^2&=\frac{5}{3}\\t+1&=\pm\sqrt{\frac{5}{3}}\\t+1&=\pm\frac{\sqrt{15}}{3}\\t&=-1\pm\frac{\sqrt{15}}{3}\\t&=-\frac{3}{3}\pm\frac{\sqrt{15}}{3}\\t&=\frac{-3\pm\sqrt{15}}{3}\end{aligned}$$

The solution set is $\left\{\frac{-3-\sqrt{15}}{3}, \frac{-3+\sqrt{15}}{3}\right\}$.

75.
$$5x = 18 - x^2$$
$$x^2 + 5x = 18$$
$$x^2 + 5x + \left(\frac{5}{2}\right)^2 = 18 + \left(\frac{5}{2}\right)^2$$
$$x^2 + 5x + \frac{25}{4} = 18 + \frac{25}{4}$$
$$\left(x + \frac{5}{2}\right)^2 = \frac{97}{4}$$
$$x + \frac{5}{2} = \pm\sqrt{\frac{97}{4}}$$
$$x + \frac{5}{2} = \pm\frac{\sqrt{97}}{2}$$
$$x = -\frac{5}{2} \pm \frac{\sqrt{97}}{2}$$
$$x = \frac{-5 \pm \sqrt{97}}{2}$$
The solution set is $\left\{\frac{-5-\sqrt{97}}{2}, \frac{-5+\sqrt{97}}{2}\right\}$.

77. $h^2 = 9$
$$h = \pm\sqrt{9}$$
$$h = \pm 3$$
The solution set is $\{-3, 3\}$.

79. $(x+2)^2 - 16 = 0$
$$(x+2)^2 = 16$$
$$x + 2 = \pm\sqrt{16}$$
$$x + 2 = \pm 4$$
$$x = -2 \pm 4$$
$x = -2 - 4 = -6$ or $x = -2 + 4 = 2$
The solution set is $\{-6, 2\}$.

81.
$$n^2 + 10n = 24$$
$$n^2 + 10n - 24 = 0$$
$$(n+12)(n-2) = 0$$
$n + 12 = 0$ or $n - 2 = 0$
$n = -12$ or $n = 2$
The solution set is $\{-12, 2\}$.

83. $y^2 - 18 = 0$
$$y^2 = 18$$
$$y = \pm\sqrt{18}$$
$$y = \pm 3\sqrt{2}$$
The solution set is $\left\{-3\sqrt{2}, 3\sqrt{2}\right\}$.

85. $2x^2 - 24 = 0$
$$2x^2 = 24$$
$$x^2 = 12$$
$$x = \pm\sqrt{12}$$
$$x = \pm 2\sqrt{3}$$
The solution set is $\left\{-2\sqrt{3}, 2\sqrt{3}\right\}$.

87. $12u + 2u^2 - 27 = 0$
$$2u^2 + 12u = 27$$
$$u^2 + 6u = \frac{27}{2}$$
$$u^2 + 6u + \left(\frac{6}{2}\right)^2 = \frac{27}{2} + \left(\frac{6}{2}\right)^2$$
$$u^2 + 6u + 3^2 = \frac{27}{2} + 3^2$$
$$u^2 + 6u + 9 = \frac{27}{2} + 9$$
$$(u+3)^2 = \frac{45}{2}$$
$$u + 3 = \pm\sqrt{\frac{45}{2}}$$
$$u + 3 = \pm\frac{\sqrt{90}}{2}$$
$$u + 3 = \pm\frac{3\sqrt{10}}{2}$$
$$u = -3 \pm \frac{3\sqrt{10}}{2}$$
$$u = \frac{-6}{2} \pm \frac{3\sqrt{10}}{2}$$
$$u = \frac{-6 \pm 3\sqrt{10}}{2}$$
The solution set is $\left\{\frac{-6-3\sqrt{10}}{2}, \frac{-6+3\sqrt{10}}{2}\right\}$.

89.
$$r(3r - 11) = -6$$
$$3r^2 - 11r = -6$$
$$3r^2 - 11r + 6 = 0$$
$$(3r-2)(r-3) = 0$$
$3r - 2 = 0$ or $r - 3 = 0$
$r = \frac{2}{3}$ or $r = 3$
The solution set is $\left\{\frac{2}{3}, 3\right\}$.

91.
$$(x+6)(x-4)=9$$
$$x^2-4x+6x-24=9$$
$$x^2+2x-24=9$$
$$x^2+2x=33$$
$$x^2+2x+\left(\frac{2}{2}\right)^2=33+\left(\frac{2}{2}\right)^2$$
$$x^2+2x+1^2=33+1^2$$
$$x^2+2x+1=34$$
$$(x+1)^2=34$$
$$x+1=\pm\sqrt{34}$$
$$x=-1\pm\sqrt{34}$$
The solution set is $\{-1-\sqrt{34}, -1+\sqrt{34}\}$.

93. $a^2+b^2=c^2$
Here $a = b$ and $c = 30$.
$$a^2+a^2=30^2$$
$$2a^2=900$$
$$a^2=450$$
$$a=\pm\sqrt{450}$$
$$a=\pm15\sqrt{2}$$
Since a length cannot be negative, the legs have length $15\sqrt{2}$.

95. $a^2+b^2=c^2$
Let a be the length of the shorter leg. Then $b=\sqrt{27}$ and $c = 2a$.
$$a^2+\left(\sqrt{27}\right)^2=(2a)^2$$
$$a^2+27=4a^2$$
$$27=3a^2$$
$$9=a^2$$
$$\pm\sqrt{9}=a$$
$$\pm3=a$$
$c = 2a = 2(\pm3) = \pm6$
Since lengths cannot be negative, the other two sides of the triangle have lengths 3 and 6.

97. Let n be the number.
$$n\left(\frac{1}{2}n-1\right)=12$$
$$\frac{1}{2}n^2-n=12$$
$$n^2-2n=24$$
$$n^2-2n-24=0$$
$$(n+4)(n-6)=0$$
$$n+4=0 \quad\text{or}\quad n-6=0$$
$$n=-4 \quad\text{or}\quad n=6$$
The number is –4 or 6.

99. Let n be the number.
$$\left(\frac{1}{3}n+1\right)^2=2n$$
$$\frac{1}{9}n^2+\frac{2}{3}n+1=2n$$
$$\frac{1}{9}n^2-\frac{4}{3}n+1=0$$
$$n^2-12n+9=0$$
$$n^2-12n=-9$$
$$n^2-12n+\left(\frac{-12}{2}\right)^2=-9+\left(\frac{-12}{2}\right)^2$$
$$n^2-12n+(-6)^2=-9+(-6)^2$$
$$n^2-12n+36=-9+36$$
$$(n-6)^2=27$$
$$n-6=\pm\sqrt{27}$$
$$n-6=\pm3\sqrt{3}$$
$$n=6\pm3\sqrt{3}$$
The number is $6-3\sqrt{3}$ or $6+3\sqrt{3}$.

101. **(a)** $x^2-9x+18=0$
The sum of the solutions is –(–9) = 9.

(b)
$$x^2-9x=-18$$
$$x^2-9x+\left(\frac{-9}{2}\right)^2=-18+\left(\frac{-9}{2}\right)^2$$
$$x^2-9x+\frac{81}{4}=-18+\frac{81}{4}$$
$$\left(x-\frac{9}{2}\right)^2=\frac{9}{4}$$
$$x-\frac{9}{2}=\pm\sqrt{\frac{9}{4}}$$
$$x-\frac{9}{2}=\pm\frac{3}{2}$$
$$x=\frac{9}{2}\pm\frac{3}{2}$$
$$x=\frac{9\pm3}{2}$$
$$x=\frac{9-3}{2}=\frac{6}{2}=3 \quad\text{or}\quad x=\frac{9+3}{2}=\frac{12}{2}=6$$
The solution set is $\{3, 6\}$.

(c) The sum of the solutions is 3 + 6 = 9, which agrees with the previous computation.

103. **(a)** $2x^2 - 4x - 1 = 0$

$$x^2 - 2x - \frac{1}{2} = 0$$

The sum of the solutions is $-(-2) = 2$.

(b)
$$x^2 - 2x = \frac{1}{2}$$
$$x^2 - 2x + \left(\frac{-2}{2}\right)^2 = \frac{1}{2} + \left(\frac{-2}{2}\right)^2$$
$$x^2 - 2x + (-1)^2 = \frac{1}{2} + (-1)^2$$
$$x^2 - 2x + 1 = \frac{1}{2} + 1$$
$$(x-1)^2 = \frac{3}{2}$$
$$x - 1 = \pm\sqrt{\frac{3}{2}}$$
$$x - 1 = \pm\frac{\sqrt{6}}{2}$$
$$x = 1 \pm \frac{\sqrt{6}}{2}$$
$$x = 1 - \frac{\sqrt{6}}{2} \text{ or } x = 1 + \frac{\sqrt{6}}{2}$$

The solution set is $\left\{1 - \frac{\sqrt{6}}{2}, 1 + \frac{\sqrt{6}}{2}\right\}$.

(c) The sum of the solutions is $1 - \frac{\sqrt{6}}{2} + 1 + \frac{\sqrt{6}}{2} = 2$, which agrees with the previous computation.

105. **(a)** $4x^2 + 4x - 3 = 0$

$$x^2 + x - \frac{3}{4} = 0$$

The product of the solutions is $-\frac{3}{4}$.

(b)
$$x^2 + x = \frac{3}{4}$$
$$x^2 + x + \left(\frac{1}{2}\right)^2 = \frac{3}{4} + \left(\frac{1}{2}\right)^2$$
$$x^2 + x + \frac{1}{4} = \frac{3}{4} + \frac{1}{4}$$
$$\left(x + \frac{1}{2}\right)^2 = \frac{4}{4}$$
$$\left(x + \frac{1}{2}\right)^2 = 1$$
$$x + \frac{1}{2} = \pm\sqrt{1}$$
$$x + \frac{1}{2} = \pm 1$$
$$x = -\frac{1}{2} \pm 1$$
$$x = -\frac{1}{2} - 1 = -\frac{3}{2} \text{ or } x = -\frac{1}{2} + 1 = \frac{1}{2}$$

The solution set is $\left\{-\frac{3}{2}, \frac{1}{2}\right\}$.

(c) The product of the solutions is $\left(-\frac{3}{2}\right)\left(\frac{1}{2}\right) = -\frac{3}{4}$, which agrees with the previous computation.

107. **(a)**
$$2x^2 = 4x + 14$$
$$2x^2 - 4x - 14 = 0$$
$$x^2 - 2x - 7 = 0$$

The product of the solutions is -7.

(b)
$$x^2 - 2x = 7$$
$$x^2 - 2x + \left(\frac{-2}{2}\right)^2 = 7 + \left(\frac{-2}{2}\right)^2$$
$$x^2 - 2x + (-1)^2 = 7 + (-1)^2$$
$$x^2 - 2x + 1 = 7 + 1$$
$$(x-1)^2 = 8$$
$$x - 1 = \pm\sqrt{8}$$
$$x - 1 = \pm 2\sqrt{2}$$
$$x = 1 \pm 2\sqrt{2}$$
$$x = 1 - 2\sqrt{2} \text{ or } x = 1 + 2\sqrt{2}$$

The solution set is $\{1 - 2\sqrt{2}, 1 + 2\sqrt{2}\}$.

(c) The product of the solutions is $\left(1-2\sqrt{2}\right)\left(1+2\sqrt{2}\right)=1^2-\left(2\sqrt{2}\right)^2=1-8=-7,$ which agrees with the previous computation.

109.
$$\begin{aligned} x^2+y^2-10x+6y+30&=0\\ (x^2-10x)+(y^2+6y)&=-30\\ (x^2-10x+25)+(y^2+6y+9)&=-30+25+9\\ (x-5)^2+(y+3)^2&=4\\ (x-5)^2+[y-(-3)]^2&=2^2 \end{aligned}$$
The center is $(h, k) = (5, -3)$ and the radius is $r = 2$.

111.
$$\begin{aligned} x^2+y^2-20x-16y+22&=0\\ (x^2-20x)+(y^2-16y)&=-22\\ (x^2-20x+100)+(y^2-16y+64)&=-22+100+64\\ (x-10)^2+(y-8)^2&=142\\ (x-10)^2+(y-8)^2&=\left(\sqrt{142}\right)^2 \end{aligned}$$
The center is $(h, k) = (10, 8)$ and the radius is $r=\sqrt{142}$.

113. To find the number that must be added to x^2-9x to make a perfect square trinomial, take half of the coefficient of x, which is -9, and square that value: $\left(\frac{1}{2}\cdot -9\right)^2=\left(-\frac{9}{2}\right)^2=\frac{81}{4}$. So $x^2-9x+\frac{81}{4}=\left(x-\frac{9}{2}\right)^2$.

Section 9.3

Preparing for Solving Quadratic Equations Using the Quadratic Formula

P1.
$$\begin{aligned} -4+\sqrt{16-4(2)(1)}&=-4+\sqrt{16-8}\\ &=-4+\sqrt{8}\\ &=-4+\sqrt{4\cdot 2}\\ &=-4+2\sqrt{2} \end{aligned}$$

P2.
$$\begin{aligned} \frac{6+\sqrt{6^2-4(3)(-2)}}{6}&=\frac{6+\sqrt{36+24}}{6}\\ &=\frac{6+\sqrt{60}}{6}\\ &=\frac{6+\sqrt{4\cdot 15}}{6}\\ &=\frac{6+2\sqrt{15}}{6}\\ &=\frac{2\left(3+\sqrt{15}\right)}{2\cdot 3}\\ &=\frac{3+\sqrt{15}}{3} \text{ or } 1+\frac{\sqrt{15}}{3} \end{aligned}$$

P3. (a)
$$4x - x^2 = 12$$
$$-x^2 + 4x - 12 = 0$$
$$x^2 - 4x + 12 = 0$$

(b)
$$(n+2)(n-3) = 6$$
$$n^2 - n - 6 = 6$$
$$n^2 - n - 12 = 0$$

P4.
$$\frac{3}{x} + 4 = \frac{15}{x}$$
$$x\left(\frac{3}{x} + 4\right) = \left(\frac{15}{x}\right)x$$
$$3 + 4x = 15$$
$$4x = 12$$
$$x = 3$$
The solution set is $\{3\}$.

9.3 Quick Checks

1. The solutions to the quadratic equation $ax^2 + bx + c = 0$, $a \neq 0$, are given by the quadratic formula $x = \frac{-b \pm \sqrt{b^2 - 4ac}}{2a}$.

2. (a)
$$x^2 + 5x - 3 = 0$$
$$a = 1, b = 5, c = -3$$

(b)
$$4x^2 = 3x - 1$$
$$4x^2 - 3x + 1 = 0$$
$$a = 4, b = -3, c = 1$$

3. $3x^2 + 7x + 4 = 0$
$a = 3$, $b = 7$, $c = 4$
$$x = \frac{-7 \pm \sqrt{7^2 - 4(3)(4)}}{2(3)}$$
$$= \frac{-7 \pm \sqrt{49 - 48}}{6}$$
$$= \frac{-7 \pm \sqrt{1}}{6}$$
$$= \frac{-7 \pm 1}{6}$$
$$x = \frac{-7+1}{6} = \frac{-6}{6} = -1 \text{ or } x = \frac{-7-1}{6} = \frac{-8}{6} = -\frac{4}{3}$$
The solution set is $\left\{-\frac{4}{3}, -1\right\}$.

4. $2x^2 + 13x + 6 = 0$
$a = 2$, $b = 13$, $c = 6$
$$x = \frac{-13 \pm \sqrt{13^2 - 4(2)(6)}}{2(2)}$$
$$= \frac{-13 \pm \sqrt{169 - 48}}{4}$$
$$= \frac{-13 \pm \sqrt{121}}{4}$$
$$= \frac{-13 \pm 11}{4}$$
$$x = \frac{-13+11}{4} = \frac{-2}{4} = -\frac{1}{2} \text{ or}$$
$$x = \frac{-13-11}{4} = \frac{-24}{4} = -6$$
The solution set is $\left\{-6, -\frac{1}{2}\right\}$.

5.
$$6x^2 + 6 = -13x$$
$$6x^2 + 13x + 6 = 0$$
$a = 6$, $b = 13$, $c = 6$
$$x = \frac{-13 \pm \sqrt{13^2 - 4(6)(6)}}{2(6)}$$
$$= \frac{-13 \pm \sqrt{169 - 144}}{12}$$
$$= \frac{-13 \pm \sqrt{25}}{12}$$
$$= \frac{-13 \pm 5}{12}$$
$$x = \frac{-13+5}{12} = \frac{-8}{12} = -\frac{2}{3} \text{ or}$$
$$x = \frac{-13-5}{12} = \frac{-18}{12} = -\frac{3}{2}$$
The solution set is $\left\{-\frac{3}{2}, -\frac{2}{3}\right\}$.

6. $8x^2+6x=5$

$8x^2+6x-5=0$

$a=8, b=6, c=-5$

$$x=\frac{-6\pm\sqrt{6^2-4(8)(-5)}}{2(8)}=\frac{-6\pm\sqrt{36+160}}{16}=\frac{-6\pm\sqrt{196}}{16}=\frac{-6\pm14}{16}$$

$x=\frac{-6+14}{16}=\frac{8}{16}=\frac{1}{2}$ or

$x=\frac{-6-14}{16}=\frac{-20}{16}=-\frac{5}{4}$

The solution set is $\left\{-\frac{5}{4}, \frac{1}{2}\right\}$.

7. False; to solve $2x^2+5x+3=0$ using the quadratic formula, we write

$$x=\frac{-5\pm\sqrt{5^2-4(2)(3)}}{2(2)}.$$

8. $y^2-2=-4y$

$y^2+4y-2=0$

$a=1, b=4, c=-2$

$$y=\frac{-4\pm\sqrt{4^2-4(1)(-2)}}{2(1)}=\frac{-4\pm\sqrt{16+8}}{2}=\frac{-4\pm\sqrt{24}}{2}=\frac{-4\pm2\sqrt{6}}{2}=\frac{2\left(-2\pm\sqrt{6}\right)}{2}=-2\pm\sqrt{6}$$

The solution set is $\left\{-2-\sqrt{6}, -2+\sqrt{6}\right\}$.

9. $16x^2-24x+7=0$

$a=16, b=-24, c=7$

$$x=\frac{-(-24)\pm\sqrt{(-24)^2-4(16)(7)}}{2(16)}=\frac{24\pm\sqrt{576-448}}{32}=\frac{24\pm\sqrt{128}}{32}=\frac{24\pm8\sqrt{2}}{32}=\frac{8\left(3\pm\sqrt{2}\right)}{8\cdot4}=\frac{3\pm\sqrt{2}}{4}$$

The solution set is $\left\{\frac{3-\sqrt{2}}{4}, \frac{3+\sqrt{2}}{4}\right\}$.

10. True

11. $x-\frac{3}{8}=\frac{1}{4}x^2$

$0=\frac{1}{4}x^2-x+\frac{3}{8}$

$0=2x^2-8x+3$

$a=2, b=-8, c=3$

$$x=\frac{-(-8)\pm\sqrt{(-8)^2-4(2)(3)}}{2(2)}=\frac{8\pm\sqrt{64-24}}{4}=\frac{8\pm\sqrt{40}}{4}=\frac{8\pm2\sqrt{10}}{4}=\frac{2\left(4\pm\sqrt{10}\right)}{2\cdot2}=\frac{4\pm\sqrt{10}}{2}$$

The solution set is $\left\{\frac{4-\sqrt{10}}{2}, \frac{4+\sqrt{10}}{2}\right\}$.

12. $16n^2 - 40n = -25$

$16n^2 - 40n + 25 = 0$

$a = 16,\ b = -40,\ c = 25$

$$n = \frac{-(-40) \pm \sqrt{(-40)^2 - 4(16)(25)}}{2(16)} = \frac{40 \pm \sqrt{1600 - 1600}}{32} = \frac{40}{32} = \frac{5}{4}$$

The solution set is $\left\{\frac{5}{4}\right\}$.

13. $4n^2 - 3n + 5 = 0$

$a = 4,\ b = -3,\ c = 5$

$$n = \frac{-(-3) \pm \sqrt{(-3)^2 - 4(4)(5)}}{2(4)} = \frac{3 \pm \sqrt{9 - 80}}{8} = \frac{3 \pm \sqrt{-71}}{8}$$

Because $\sqrt{-71}$ is not a real number, the equation has no real solution.

14. $R = -0.02x^2 + 24x$

$R = 4000$

$$4000 = -0.02x^2 + 24x$$

$$0.02x^2 - 24x + 4000 = 0$$

$$2x^2 - 2400x + 400{,}000 = 0$$

$$x^2 - 1200x + 200{,}000 = 0$$

$a = 1,\ b = -1200,\ c = 200{,}000$

$$x = \frac{-(-1200) \pm \sqrt{(-1200)^2 - 4(1)(200{,}000)}}{2(1)} = \frac{1200 \pm \sqrt{1{,}440{,}000 - 800{,}000}}{2} = \frac{1200 \pm \sqrt{640{,}000}}{2} = \frac{1200 \pm 800}{2} = 600 \pm 400$$

$x = 600 + 400 = 1000$ or

$x = 600 - 400 = 200$

Since $200 < 400$ and $1000 > 400$, the park must sell 1000 passes daily.

15. The discriminant of $ax^2 + bx + c = 0$ is given by the formula $\underline{b^2 - 4ac}$.

16. If the discriminant of a quadratic equation has the value of zero, the most efficient way of solving the quadratic equation is factoring.

17. If the value of the discriminant of a quadratic equation is a perfect square, the most efficient way of solving the quadratic equation is factoring.

18. If the value of the discriminant of a quadratic equation is a positive number that is not a perfect square, the most efficient way of solving the quadratic equation is the quadratic formula.

19. $9q^2 - 6q + 1 = 0$

$a = 9,\ b = -6,\ c = 1$

$$b^2 - 4ac = (-6)^2 - 4(9)(1) = 36 - 36 = 0$$

Solve by factoring.

$$9q^2 - 6q + 1 = 0$$

$$(3q - 1)^2 = 0$$

$$3q - 1 = 0$$

$$3q = 1$$

$$q = \frac{1}{3}$$

The solution set is $\left\{\frac{1}{3}\right\}$.

20. $3w^2 = 5w - 2$

$3w^2 - 5w + 2 = 0$

$a = 3,\ b = -5,\ c = 2$

$b^2 - 4ac = (-5)^2 - 4(3)(2) = 25 - 24 = 1$

Solve by factoring.

$$3w^2 - 5w + 2 = 0$$

$$(3w - 2)(w - 1) = 0$$

$3w - 2 = 0$ or $w - 1 = 0$

$3w = 2$ or $w = 1$

$w = \frac{2}{3}$ or $w = 1$

The solution set is $\left\{\frac{2}{3}, 1\right\}$.

21. $4z^2 - 2z = 1$
$4z^2 - 2z - 1 = 0$
$a = 4, b = -2, c = -1$
$b^2 - 4ac = (-2)^2 - 4(4)(-1) = 4 + 16 = 20$
Use the quadratic formula.
$$z = \frac{-(-2) \pm \sqrt{20}}{2(4)} = \frac{2 \pm 2\sqrt{5}}{8} = \frac{1 \pm \sqrt{5}}{4}$$
The solution set is $\left\{\frac{1-\sqrt{5}}{4}, \frac{1+\sqrt{5}}{4}\right\}$.

22. $4n^2 - 24 = 0$
$n^2 - 6 = 0$
$n^2 = 6$
$n = \pm\sqrt{6}$
The solution set is $\left\{-\sqrt{6}, \sqrt{6}\right\}$.

23. $2y^2 = 3y + 35$
$2y^2 - 3y - 35 = 0$
$a = 2, b = -3, c = -35$
$b^2 - 4ac = (-3)^2 - 4(2)(-35)$
$= 9 + 280$
$= 289$
$= 17^2$
Solve by factoring.
$2y^2 - 3y - 35 = 0$
$(2y + 7)(y - 5) = 0$
$2y + 7 = 0$ or $y - 5 = 0$
$y = -\frac{7}{2}$ or $y = 5$
The solution set is $\left\{-\frac{7}{2}, 5\right\}$.

24. $1 = 3q^2 + 4q$
$0 = 3q^2 + 4q - 1$
$a = 3, b = 4, c = -1$
$b^2 - 4ac = 4^2 - 4(3)(-1) = 16 + 12 = 28$
Use the quadratic formula.
$$q = \frac{-4 \pm \sqrt{28}}{2(3)} = \frac{-4 \pm 2\sqrt{7}}{6} = \frac{-2 \pm \sqrt{7}}{3}$$
The solution set is $\left\{\frac{-2-\sqrt{7}}{3}, \frac{-2+\sqrt{7}}{3}\right\}$.

9.3 Exercises

25. $x - 2x^2 = -4$
$-2x^2 + x + 4 = 0$
$a = -2, b = 1, c = 4$

27. $x^2 + 4 = 0$
$x^2 + 0x + 4 = 0$
$a = 1, b = 0, c = 4$

29. $2 + \frac{6}{5}x = \frac{3}{2}x^2$
$\frac{3}{2}x^2 - \frac{6}{5}x - 2 = 0$
$a = \frac{3}{2}, b = -\frac{6}{5}, c = -2$

31. $0.5x^2 = x - 3$
$0.5x^2 - x + 3 = 0$
$a = 0.5, b = -1, c = 3$

33. $x^2 + 5x - 24 = 0$
$a = 1, b = 5, c = -24$
$$x = \frac{-5 \pm \sqrt{5^2 - 4(1)(-24)}}{2(1)}$$
$$= \frac{-5 \pm \sqrt{25 + 96}}{2}$$
$$= \frac{-5 \pm \sqrt{121}}{2}$$
$$= \frac{-5 \pm 11}{2}$$
$x = \frac{-5-11}{2} = \frac{-16}{2} = -8$ or $x = \frac{-5+11}{2} = \frac{6}{2} = 3$
The solution set is $\{-8, 3\}$.

35. $2x^2 + 11x + 12 = 0$
$a = 2, b = 11, c = 12$
$$x = \frac{-11 \pm \sqrt{11^2 - 4(2)(12)}}{2(2)}$$
$$= \frac{-11 \pm \sqrt{121 - 96}}{4}$$
$$= \frac{-11 \pm \sqrt{25}}{4}$$
$$= \frac{-11 \pm 5}{4}$$
$x = \frac{-11-5}{4} = \frac{-16}{4} = -4$ or

$x = \frac{-11+5}{4} = \frac{-6}{4} = -\frac{3}{2}$

The solution set is $\left\{-4, -\frac{3}{2}\right\}$.

37. $x^2 - x - 5 = 0$

$a = 1, b = -1, c = -5$

$x = \frac{-(-1) \pm \sqrt{(-1)^2 - 4(1)(-5)}}{2(1)}$

$= \frac{1 \pm \sqrt{1+20}}{2}$

$= \frac{1 \pm \sqrt{21}}{2}$

The solution set is $\left\{\frac{1-\sqrt{21}}{2}, \frac{1+\sqrt{21}}{2}\right\}$.

39. $3r^2 + 4r - 1 = 0$

$a = 3, b = 4, c = -1$

$r = \frac{-4 \pm \sqrt{4^2 - 4(3)(-1)}}{2(3)}$

$= \frac{-4 \pm \sqrt{16+12}}{6}$

$= \frac{-4 \pm \sqrt{28}}{6}$

$= \frac{-4 \pm 2\sqrt{7}}{6}$

$= \frac{2\left(-2 \pm \sqrt{7}\right)}{2 \cdot 3}$

$= \frac{-2 \pm \sqrt{7}}{3}$

The solution set is $\left\{\frac{-2-\sqrt{7}}{3}, \frac{-2+\sqrt{7}}{3}\right\}$.

41. $2t^2 - 4t + \frac{3}{2} = 0$

$2\left(2t^2 - 4t + \frac{3}{2}\right) = 2 \cdot 0$

$4t^2 - 8t + 3 = 0$

$a = 4, b = -8, c = 3$

$t = \frac{-(-8) \pm \sqrt{(-8)^2 - 4(4)(3)}}{2(4)}$

$= \frac{8 \pm \sqrt{64-48}}{8}$

$= \frac{8 \pm \sqrt{16}}{8}$

$= \frac{8 \pm 4}{8}$

$t = \frac{8-4}{8} = \frac{4}{8} = \frac{1}{2}$ or $t = \frac{8+4}{8} = \frac{12}{8} = \frac{3}{2}$

The solution set is $\left\{\frac{1}{2}, \frac{3}{2}\right\}$.

43. $z^2 = -1 - 7z$

$z^2 + 7z + 1 = 0$

$a = 1, b = 7, c = 1$

$z = \frac{-7 \pm \sqrt{7^2 - 4(1)(1)}}{2(1)}$

$= \frac{-7 \pm \sqrt{49-4}}{2}$

$= \frac{-7 \pm \sqrt{45}}{2}$

$= \frac{-7 \pm 3\sqrt{5}}{2}$

The solution set is $\left\{\frac{-7-3\sqrt{5}}{2}, \frac{-7+3\sqrt{5}}{2}\right\}$.

45. $2x^2 + 1 = 5x$

$2x^2 - 5x + 1 = 0$

$a = 2, b = -5, c = 1$

$x = \frac{-(-5) \pm \sqrt{(-5)^2 - 4(2)(1)}}{2(2)}$

$= \frac{5 \pm \sqrt{25-8}}{4}$

$= \frac{5 \pm \sqrt{17}}{4}$

The solution set is $\left\{\frac{5-\sqrt{17}}{4}, \frac{5+\sqrt{17}}{4}\right\}$.

47. $x^2 - 7x = -12$

$x^2 - 7x + 12 = 0$

$a = 1, b = -7, c = 12$

$$x = \frac{-(-7) \pm \sqrt{(-7)^2 - 4(1)(12)}}{2(1)} = \frac{7 \pm \sqrt{49-48}}{2} = \frac{7 \pm \sqrt{1}}{2} = \frac{7 \pm 1}{2}$$

$$x = \frac{7-1}{2} = \frac{6}{2} = 3 \text{ or } x = \frac{7+1}{2} = \frac{8}{2} = 4$$

The solution set is $\{3, 4\}$.

49. $3z^2 - 4z = 4$

$3z^2 - 4z - 4 = 0$

$a = 3, b = -4, c = -4$

$$z = \frac{-(-4) \pm \sqrt{(-4)^2 - 4(3)(-4)}}{2(3)} = \frac{4 \pm \sqrt{16+48}}{6} = \frac{4 \pm \sqrt{64}}{6} = \frac{4 \pm 8}{6}$$

$$z = \frac{4-8}{6} = \frac{-4}{6} = -\frac{2}{3} \text{ or } z = \frac{4+8}{6} = \frac{12}{6} = 2$$

The solution set is $\left\{-\frac{2}{3}, 2\right\}$.

51. $\frac{3}{2}n + 1 + \frac{1}{4}n^2 = 0$

$$4\left(\frac{3}{2}n + 1 + \frac{1}{4}n^2\right) = 4 \cdot 0$$

$6n + 4 + n^2 = 0$

$n^2 + 6n + 4 = 0$

$a = 1, b = 6, c = 4$

$$n = \frac{-6 \pm \sqrt{6^2 - 4(1)(4)}}{2(1)} = \frac{-6 \pm \sqrt{36-16}}{2} = \frac{-6 \pm \sqrt{20}}{2} = \frac{-6 \pm 2\sqrt{5}}{2} = \frac{-6}{2} \pm \frac{2\sqrt{5}}{2} = -3 \pm \sqrt{5}$$

The solution set is $\left\{-3-\sqrt{5}, -3+\sqrt{5}\right\}$.

53. $2k^2 + 1 - 4k = 0$

$2k^2 - 4k + 1 = 0$

$a = 2, b = -4, c = 1$

$$k = \frac{-(-4) \pm \sqrt{(-4)^2 - 4(2)(1)}}{2(2)} = \frac{4 \pm \sqrt{16-8}}{4} = \frac{4 \pm \sqrt{8}}{4} = \frac{4 \pm 2\sqrt{2}}{4} = \frac{2\left(2 \pm \sqrt{2}\right)}{4} = \frac{2 \pm \sqrt{2}}{2}$$

The solution set is $\left\{\frac{2-\sqrt{2}}{2}, \frac{2+\sqrt{2}}{2}\right\}$.

55. $x(2x+5) - 1 = 0$

$2x^2 + 5x - 1 = 0$

$a = 2, b = 5, c = -1$

$$x = \frac{-5 \pm \sqrt{5^2 - 4(2)(-1)}}{2(2)} = \frac{-5 \pm \sqrt{25+8}}{4} = \frac{-5 \pm \sqrt{33}}{4}$$

The solution set is $\left\{\frac{-5-\sqrt{33}}{4}, \frac{-5+\sqrt{33}}{4}\right\}$.

57. $15m^2 - 8m = -1$

$15m^2 - 8m + 1 = 0$

$a = 15, b = -8, c = 1$

$$m = \frac{-(-8) \pm \sqrt{(-8)^2 - 4(15)(1)}}{2(15)} = \frac{8 \pm \sqrt{64 - 60}}{30} = \frac{8 \pm \sqrt{4}}{30} = \frac{8 \pm 2}{30}$$

$$m = \frac{8-2}{30} = \frac{6}{30} = \frac{1}{5} \text{ or } m = \frac{8+2}{30} = \frac{10}{30} = \frac{1}{3}$$

The solution set is $\left\{\frac{1}{5}, \frac{1}{3}\right\}$.

59. $4x^2 + 18x + 9 = 6x$

$4x^2 + 12x + 9 = 0$

$a = 4, b = 12, c = 9$

$$x = \frac{-12 \pm \sqrt{12^2 - 4(4)(9)}}{2(4)} = \frac{-12 \pm \sqrt{144 - 144}}{8} = \frac{-12 \pm \sqrt{0}}{8} = \frac{-12}{8} = -\frac{3}{2}$$

The solution set is $\left\{-\frac{3}{2}\right\}$.

61. $\frac{k^2}{3} - 6 = -k$

$3\left(\frac{k^2}{3} - 6\right) = 3(-k)$

$k^2 - 18 = -3k$

$k^2 + 3k - 18 = 0$

$a = 1, b = 3, c = -18$

$$k = \frac{-3 \pm \sqrt{3^2 - 4(1)(-18)}}{2(1)} = \frac{-3 \pm \sqrt{9 + 72}}{2} = \frac{-3 \pm \sqrt{81}}{2} = \frac{-3 \pm 9}{2}$$

$$k = \frac{-3-9}{2} = \frac{-12}{2} = -6 \text{ or } k = \frac{-3+9}{2} = \frac{6}{2} = 3$$

The solution set is $\{-6, 3\}$.

63. $4z = \frac{3}{2}z^2 + 1$

$2 \cdot (4z) = 2\left(\frac{3}{2}z^2 + 1\right)$

$8z = 3z^2 + 2$

$0 = 3z^2 - 8z + 2$

$a = 3, b = -8, c = 2$

$$z = \frac{-(-8) \pm \sqrt{(-8)^2 - 4(3)(2)}}{2(3)} = \frac{8 \pm \sqrt{64 - 24}}{6} = \frac{8 \pm \sqrt{40}}{6} = \frac{8 \pm 2\sqrt{10}}{6} = \frac{4 \pm \sqrt{10}}{3}$$

The solution set is $\left\{\frac{4 - \sqrt{10}}{3}, \frac{4 + \sqrt{10}}{3}\right\}$.

65. $x = \frac{x^2}{6} + \frac{3}{2}$

$6(x) = 6\left(\frac{x^2}{6} + \frac{3}{2}\right)$

$6x = x^2 + 9$

$0 = x^2 - 6x + 9$

$a = 1, b = -6, c = 9$

$$x = \frac{-(-6) \pm \sqrt{(-6)^2 - 4(1)(9)}}{2(1)}$$
$$= \frac{6 \pm \sqrt{36-36}}{2}$$
$$= \frac{6 \pm \sqrt{0}}{2}$$
$$= \frac{6}{2}$$
$$= 3$$

The solution set is $\{3\}$.

67. $(x+3)(x-2) = 9$

$$x^2 + x - 6 = 9$$
$$x^2 + x - 15 = 0$$

$a = 1, b = 1, c = -15$

$$x = \frac{-1 \pm \sqrt{1^2 - 4(1)(-15)}}{2(1)}$$
$$= \frac{-1 \pm \sqrt{1+60}}{2}$$
$$= \frac{-1 \pm \sqrt{61}}{2}$$

The solution set is $\left\{\frac{-1-\sqrt{61}}{2}, \frac{-1+\sqrt{61}}{2}\right\}$.

69. $t(1-2t) = 1$

$$t - 2t^2 = 1$$
$$-2t^2 + t - 1 = 0$$

$a = -2, b = 1, c = -1$

$$t = \frac{-1 \pm \sqrt{1^2 - 4(-2)(-1)}}{2(-2)}$$
$$= \frac{-1 \pm \sqrt{1-8}}{-4}$$
$$= \frac{-1 \pm \sqrt{-7}}{-4}$$

Since $\sqrt{-7}$ is not a real number, there is no real number solution. The solution set is $\varnothing$.

71. $3w^2 = 36$

$$3w^2 + 0w - 36 = 0$$

$a = 3, b = 0, c = -36$

$$w = \frac{-0 \pm \sqrt{0^2 - 4(3)(-36)}}{2(3)}$$
$$= \frac{0 \pm \sqrt{0+432}}{6}$$
$$= \frac{\pm\sqrt{432}}{6}$$
$$= \pm\frac{12\sqrt{3}}{6}$$
$$= \pm 2\sqrt{3}$$

The solution set is $\left\{-2\sqrt{3}, 2\sqrt{3}\right\}$.

73. $\frac{2}{9}w^2 - w - 2 = 0$

$$9\left(\frac{2}{9}w^2 - w - 2\right) = 9(0)$$
$$2w^2 - 9w - 18 = 0$$

$a = 2, b = -9, c = -18$

$$w = \frac{-(-9) \pm \sqrt{(-9)^2 - 4(2)(-18)}}{2(2)}$$
$$= \frac{9 \pm \sqrt{81+144}}{4}$$
$$= \frac{9 \pm \sqrt{225}}{4}$$
$$= \frac{9 \pm 15}{4}$$

$$w = \frac{9-15}{4} = \frac{-6}{4} = -\frac{3}{2} \text{ or } w = \frac{9+15}{4} = \frac{24}{4} = 6$$

The solution set is $\left\{-\frac{3}{2}, 6\right\}$.

75. $8x = 2x^2 - 2x$

$$0 = 2x^2 - 10x + 0$$

$a = 2, b = -10, c = 0$

$$x = \frac{-(-10) \pm \sqrt{(-10)^2 - 4(2)(0)}}{2(2)}$$
$$= \frac{10 \pm \sqrt{100+0}}{4}$$
$$= \frac{10 \pm \sqrt{100}}{4}$$
$$= \frac{10 \pm 10}{4}$$

$$x = \frac{10-10}{4} = \frac{0}{4} = 0 \text{ or } x = \frac{10+10}{4} = \frac{20}{4} = 5$$

The solution set is $\{0, 5\}$.

77. $2p^2 - p + 5 = 0$
$a = 2, b = -1, c = 5$
$b^2 - 4ac = (-1)^2 - 4(2)(5) = 1 - 40 = -39$
Since $-39 < 0$, the equation has no real solution.

79. $0 = 2x^2 + 3x - 2$
$a = 2, b = 3, c = -2$
$b^2 - 4ac = 3^2 - 4(2)(-2) = 9 + 16 = 25$
Since $25 = 5^2$ is a perfect square, use factoring to solve the equation.

81. $4x(x+3) = -9$
$4x^2 + 12x + 9 = 0$
$a = 4, b = 12, c = 9$
$b^2 - 4ac = 12^2 - 4(4)(9) = 144 - 144 = 0$
Since $b^2 - 4ac = 0$, use factoring to solve the equation.

83. $y = y^2 + 7y - 1$
$0 = y^2 + 6y - 1$
$a = 1, b = 6, c = -1$
$b^2 - 4ac = 6^2 - 4(1)(-1) = 36 + 4 = 40$
Since 40 is not a perfect square, use the quadratic formula to solve the equation.

85. $0 = -20z - 4z^2 - 25$
$4z^2 + 20z + 25 = 0$
$a = 4, b = 20, c = 25$
$b^2 - 4ac = 20^2 - 4(4)(25) = 400 - 400 = 0$
Since $b^2 - 4ac = 0$, use factoring to solve the equation.

87. $\frac{x^2}{2} - \frac{x}{3} + \frac{5}{6} = 0$
$6\left(\frac{x^2}{2} - \frac{x}{3} + \frac{5}{6}\right) = 6(0)$
$3x^2 - 2x + 5 = 0$
$a = 3, b = -2, c = 5$
$b^2 - 4ac = (-2)^2 - 4(3)(5) = 4 - 60 = -56$
Since $-56 < 0$, the equation has no real solution.

89. $x^2 - 2x + \frac{2}{3} = 0$
$3\left(x^2 - 2x + \frac{2}{3}\right) = 3(0)$
$3x^2 - 6x + 2 = 0$
$a = 3, b = -6, c = 2$
$$x = \frac{-(-6) \pm \sqrt{(-6)^2 - 4(3)(2)}}{2(3)} = \frac{6 \pm \sqrt{36 - 24}}{6} = \frac{6 \pm \sqrt{12}}{6} = \frac{6 \pm 2\sqrt{3}}{6} = \frac{2\left(3 \pm \sqrt{3}\right)}{6} = \frac{3 \pm \sqrt{3}}{3}$$
The solution set is $\left\{\frac{3 - \sqrt{3}}{3}, \frac{3 + \sqrt{3}}{3}\right\}$.

91. $(3p+1)^2 = 16$
$3p + 1 = \pm\sqrt{16}$
$3p + 1 = \pm 4$
$3p = -1 \pm 4$
$p = \frac{-1 \pm 4}{3}$
$p = \frac{-1 - 4}{3} = -\frac{5}{3}$ or $p = \frac{-1 + 4}{3} = \frac{3}{3} = 1$
The solution set is $\left\{-\frac{5}{3}, 1\right\}$.

93. $(y+1)(y+5) + 9 = 0$
$y^2 + 5y + y + 5 + 9 = 0$
$y^2 + 6y + 14 = 0$
$a = 1, b = 6, c = 14$
$$y = \frac{-6 \pm \sqrt{6^2 - 4(1)(14)}}{2(1)} = \frac{-6 \pm \sqrt{36 - 56}}{2} = \frac{-6 \pm \sqrt{-20}}{2}$$
Since $\sqrt{-20}$ is not a real number, the equation has no real solutions. The solution set is $\varnothing$.

95.
$$x+1=\frac{-1}{4x}$$
$$4x(x+1)=4x\left(\frac{-1}{4x}\right)$$
$$4x^2+4x=-1$$
$$4x^2+4x+1=0$$
$$(2x+1)^2=0$$
$$2x+1=0$$
$$x=-\frac{1}{2}$$
The solution set is $\left\{-\frac{1}{2}\right\}$.

97.
$$2u(u+5)=-10(u+1)$$
$$2u^2+10u=-10u-10$$
$$2u^2+20u+10=0$$
$$\frac{1}{2}(2u^2+20u+10)=\frac{1}{2}(0)$$
$$u^2+10u+5=0$$
$a=1$, $b=10$, $c=5$
$$u=\frac{-10\pm\sqrt{10^2-4(1)(5)}}{2(1)}$$
$$=\frac{-10\pm\sqrt{100-20}}{2}$$
$$=\frac{-10\pm\sqrt{80}}{2}$$
$$=\frac{-10}{2}\pm\frac{4\sqrt{5}}{2}$$
$$=-5\pm2\sqrt{5}$$
The solution set is $\left\{-5-2\sqrt{5}, -5+2\sqrt{5}\right\}$.

99.
$$2n^2=-12+6(n+2)$$
$$2n^2=-12+6n+12$$
$$2n^2=6n$$
$$2n^2-6n=0$$
$$2n(n-3)=0$$
$$2n=0 \quad \text{or} \quad n-3=0$$
$$n=0 \quad \text{or} \quad n=3$$
The solution set is $\{0, 3\}$.

101. $(2x+7)^2-9=0$
$$(2x+7)^2=9$$
$$2x+7=\pm\sqrt{9}$$
$$2x+7=\pm3$$
$$2x=-7\pm3$$
$$x=\frac{-7\pm3}{2}$$
$$x=\frac{-7-3}{2}=\frac{-10}{2}=-5 \text{ or } x=\frac{-7+3}{2}=\frac{-4}{2}=-2$$
The solution set is $\{-5, -2\}$.

103.
$$x+\frac{7}{2}-\frac{2}{x}=0$$
$$2x\left(x+\frac{7}{2}-\frac{2}{x}\right)=2x(0)$$
$$2x^2+7x-4=0$$
$$(x+4)(2x-1)=0$$
$$x+4=0 \quad \text{or} \quad 2x-1=0$$
$$x=-4 \quad \text{or} \quad x=\frac{1}{2}$$
The solution set is $\left\{-4, \frac{1}{2}\right\}$.

105. $w^2+10=2$
$$w^2=-8$$
$$w=\pm\sqrt{-8}$$
The square root of a negative number is not a real number. The solution set is $\varnothing$.

107. **(a)** $P=40+30x-0.01x^2$
When $x=50$,
$$P=40+30(50)-0.01(50)^2$$
$$=40+1500-25$$
$$=1515$$
The profit is \$1515.

(b) Find x when $P = 20{,}000$.

$$20{,}000 = 40 + 30x - 0.01x^2$$

$$0.01x^2 - 30x + 19{,}960 = 0$$

$a = 0.01$, $b = -30$, $c = 19{,}960$

$$x = \frac{-(-30) \pm \sqrt{(-30)^2 - 4(0.01)(19{,}960)}}{2(0.01)}$$

$$= \frac{30 \pm \sqrt{900 - 798.4}}{0.02}$$

$$= \frac{30 \pm \sqrt{101.6}}{0.02}$$

$$x = \frac{30 - \sqrt{101.6}}{0.02} \approx 996 \text{ or } x = \frac{30 + \sqrt{101.6}}{0.02} \approx 2004$$

The publisher would have to sell either 996 or 2004 subscriptions.

109. $N = 0.05x^2 - 1.5x + 43$

(a) When $x = 0$, $N = 0.05(0)^2 - 1.5(0) + 43 = 43$.
43 students chose engineering in 1990.

(b) When $x = 5$, $N = 0.05(5)^2 - 1.5(5) + 43 = 36.75$.
Approximately 37 students chose engineering 5 years later.

(c) For 2000, $x = 10$ and $N = 0.05(10)^2 - 1.5(10) + 43 = 33$.
33 students chose engineering in 2000.

(d) Find x when $N = 60$.

$$60 = 0.05x^2 - 1.5x + 43$$

$$0 = 0.05x^2 - 1.5x - 17$$

$a = 0.05$, $b = -1.5$, $c = -17$

$$x = \frac{-(-1.5) \pm \sqrt{(-1.5)^2 - 4(0.05)(-17)}}{2(0.05)} = \frac{1.5 \pm \sqrt{5.65}}{0.10}$$

$$x = \frac{1.5 - \sqrt{5.65}}{0.10} \approx -8.77 \text{ or } x = \frac{1.5 + \sqrt{5.65}}{0.10} \approx 38.77$$

Since x is the number of years after 1990, x cannot be negative. 60 students will chose engineering 39 years after 1990, or in 2029.

111. Let n be the number.

$$(n+3)^2 = 18$$

$$n + 3 = \pm\sqrt{18}$$

$$n + 3 = \pm 3\sqrt{2}$$

$$n = -3 \pm 3\sqrt{2}$$

The number is $-3 - 3\sqrt{2}$ or $-3 + 3\sqrt{2}$.

113. Let n be the number.

$$4(n+2)=n^2+3$$
$$4n+8=n^2+3$$
$$0=n^2-4n-5$$
$$0=(n-5)(n+1)$$
$$n-5=0 \quad \text{or} \quad n+1=0$$
$$n=5 \quad \text{or} \quad n=-1$$

The number is 5 or –1.

115. $x^2+3\sqrt{2}x-8=0$

$a=1,\ b=3\sqrt{2},\ c=-8$

$$x=\frac{-3\sqrt{2}\pm\sqrt{\left(3\sqrt{2}\right)^2-4(1)(-8)}}{2(1)}$$
$$=\frac{-3\sqrt{2}\pm\sqrt{18+32}}{2}$$
$$=\frac{-3\sqrt{2}\pm\sqrt{50}}{2}$$
$$=\frac{-3\sqrt{2}\pm5\sqrt{2}}{2}$$
$$x=\frac{-3\sqrt{2}-5\sqrt{2}}{2}=\frac{-8\sqrt{2}}{2}=-4\sqrt{2} \text{ or}$$
$$x=\frac{-3\sqrt{2}+5\sqrt{2}}{2}=\frac{2\sqrt{2}}{2}=\sqrt{2}$$

The solution set is $\left\{-4\sqrt{2},\ \sqrt{2}\right\}$.

117. $\sqrt{3}x^2-5x+\sqrt{12}=0$

$a=\sqrt{3},\ b=-5,\ c=\sqrt{12}$

$$x=\frac{-(-5)\pm\sqrt{(-5)^2-4\left(\sqrt{3}\right)\left(\sqrt{12}\right)}}{2\left(\sqrt{3}\right)}$$
$$=\frac{5\pm\sqrt{25-4\sqrt{36}}}{2\sqrt{3}}$$
$$=\frac{5\pm\sqrt{25-4(6)}}{2\sqrt{3}}$$
$$=\frac{5\pm\sqrt{25-24}}{2\sqrt{3}}$$
$$=\frac{5\pm\sqrt{1}}{2\sqrt{3}}$$
$$=\frac{5\pm1}{2\sqrt{3}}$$
$$=\frac{(5\pm1)\sqrt{3}}{2\sqrt{3}\left(\sqrt{3}\right)}$$
$$=\frac{5\sqrt{3}\pm\sqrt{3}}{2(3)}$$
$$=\frac{5\sqrt{3}\pm\sqrt{3}}{6}$$
$$x=\frac{5\sqrt{3}-\sqrt{3}}{6}=\frac{4\sqrt{3}}{6}=\frac{2\sqrt{3}}{3} \text{ or}$$
$$x=\frac{5\sqrt{3}+\sqrt{3}}{6}=\frac{6\sqrt{3}}{6}=\sqrt{3}$$

The solution set is $\left\{\frac{2\sqrt{3}}{3},\ \sqrt{3}\right\}$.

119. If the discriminant is a perfect square, or if the discriminant is equal to zero the quadratic equation is factorable.

Putting the Concepts Together (Sections 9.1–9.3)

1. (a) $w^2=\frac{5}{4}$

$$w=\pm\sqrt{\frac{5}{4}}$$
$$w=\pm\frac{\sqrt{5}}{2}$$

The solution set is $\left\{-\frac{\sqrt{5}}{2},\ \frac{\sqrt{5}}{2}\right\}$.

(b) $(y-2)^2 = 12$

$$y-2=\pm\sqrt{12}$$
$$y-2=\pm 2\sqrt{3}$$
$$y=2\pm 2\sqrt{3}$$

The solution set is $\{2-2\sqrt{3},\ 2+2\sqrt{3}\}$.

(c) $\left(y+\frac{3}{2}\right)^2=\frac{3}{4}$

$$y+\frac{3}{2}=\pm\sqrt{\frac{3}{4}}$$
$$y+\frac{3}{2}=\pm\frac{\sqrt{3}}{2}$$
$$y=-\frac{3}{2}\pm\frac{\sqrt{3}}{2}$$
$$y=\frac{-3\pm\sqrt{3}}{2}$$

The solution set is $\left\{\frac{-3-\sqrt{3}}{2}, \frac{-3+\sqrt{3}}{2}\right\}$.

2. (a) z^2-18z

$$\left(\frac{-18}{2}\right)^2=(-9)^2=81$$
$$z^2-18z+81=(z-9)^2$$

(b) y^2+9y

$$\left(\frac{9}{2}\right)^2=\frac{81}{4}$$
$$y^2+9y+\frac{81}{4}=\left(y+\frac{9}{2}\right)^2$$

(c) $m^2+\frac{4}{5}m$

$$\left[\frac{1}{2}\left(\frac{4}{5}\right)\right]^2=\left(\frac{2}{5}\right)^2=\frac{4}{25}$$
$$m^2+\frac{4}{5}m+\frac{4}{25}=\left(m+\frac{2}{5}\right)^2$$

3. (a) $x^2+12x+35=0$

$$x^2+12x=-35$$
$$x^2+12x+36=-35+36$$
$$(x+6)^2=1$$
$$x+6=\pm 1$$
$$x=-6\pm 1$$

$x=-6-1=-7$ or $x=-6+1=-5$

The solution set is $\{-7, -5\}$.

(b) $z^2-4z+9=0$

$$z^2-4z=-9$$
$$z^2-4z+4=-9+4$$
$$(z-2)^2=-5$$

There is no number whose square is –5, so the equation has no real solution.

(c) $m^2=8m+2$

$$m^2-8m=2$$
$$m^2-8m+16=2+16$$
$$(m-4)^2=18$$
$$m-4=\pm\sqrt{18}$$
$$m-4=\pm 3\sqrt{2}$$
$$m=4\pm 3\sqrt{2}$$

The solution set is $\{4-3\sqrt{2},\ 4+3\sqrt{2}\}$.

4. (a) $x^2+8x-4=0$

$a=1$, $b=8$, $c=-4$

$$x=\frac{-8\pm\sqrt{8^2-4(1)(-4)}}{2(1)}$$
$$=\frac{-8\pm\sqrt{64+16}}{2}$$
$$=\frac{-8\pm\sqrt{80}}{2}$$
$$=\frac{-8\pm 4\sqrt{5}}{2}$$
$$=-4\pm 2\sqrt{5}$$

The solution set is $\{-4-2\sqrt{5},\ -4+2\sqrt{5}\}$.

(b) $2y^2 = 5y + 12$

$0 = -2y^2 + 5y + 12$

$a = -2, b = 5, c = 12$

$$y = \frac{-5 \pm \sqrt{5^2 - 4(-2)(12)}}{2(-2)} = \frac{-5 \pm \sqrt{25+96}}{-4} = \frac{-5 \pm \sqrt{121}}{-4} = \frac{-5 \pm 11}{-4}$$

$$y = \frac{-5-11}{-4} = \frac{-16}{-4} = 4 \text{ or}$$

$$y = \frac{-5+11}{-4} = \frac{6}{-4} = -\frac{3}{2}$$

The solution set is $\left\{-\frac{3}{2}, 4\right\}$.

(c) $4p^2 + 36p + 81 = 0$

$a = 4, b = 36, c = 81$

$$p = \frac{-36 \pm \sqrt{36^2 - 4(4)(81)}}{2(4)} = \frac{-36 \pm \sqrt{1296-1296}}{8} = \frac{-36 \pm \sqrt{0}}{8} = \frac{-36}{8} = -\frac{9}{2}$$

The solution set is $\left\{-\frac{9}{2}\right\}$.

5. $(x+3)(x-4) = -6$

$x^2 - x - 12 = -6$

$x^2 - x - 6 = 0$

$(x-3)(x+2) = 0$

$x - 3 = 0$ or $x + 2 = 0$

$x = 3$ or $x = -2$

The solution set is $\{-2, 3\}$.

6. $n + 1 = \frac{5}{2n}$

$$2n(n+1) = 2n\left(\frac{5}{2n}\right)$$

$2n^2 + 2n = 5$

$2n^2 + 2n - 5 = 0$

$a = 2, b = 2, c = -5$

$$n = \frac{-2 \pm \sqrt{2^2 - 4(2)(-5)}}{2(2)} = \frac{-2 \pm \sqrt{4+40}}{4} = \frac{-2 \pm \sqrt{44}}{4} = \frac{-2 \pm 2\sqrt{11}}{4} = \frac{-1 \pm \sqrt{11}}{2}$$

The solution set is $\left\{\frac{-1-\sqrt{11}}{2}, \frac{-1+\sqrt{11}}{2}\right\}$.

7. $\frac{1}{2}z^2 + \frac{1}{8} = -z$

$$8\left(\frac{1}{2}z^2 + \frac{1}{8}\right) = 8(-z)$$

$4z^2 + 1 = -8z$

$4z^2 + 8z + 1 = 0$

$a = 4, b = 8, c = 1$

$$z = \frac{-8 \pm \sqrt{8^2 - 4(4)(1)}}{2(4)} = \frac{-8 \pm \sqrt{64-16}}{8} = \frac{-8 \pm \sqrt{48}}{8} = \frac{-8 \pm 4\sqrt{3}}{8} = \frac{-2 \pm \sqrt{3}}{2} \text{ or } -1 \pm \frac{\sqrt{3}}{2}$$

The solution set is $\left\{\frac{-2-\sqrt{3}}{2}, \frac{-2+\sqrt{3}}{2}\right\}$ or $\left\{-1-\frac{\sqrt{3}}{2}, -1+\frac{\sqrt{3}}{2}\right\}$.

8. $2x = -5 + 3x^2$
$0 = 3x^2 - 2x - 5$

(a) $a = 3$, $b = -2$, $c = -5$
$$b^2 - 4ac = (-2)^2 - 4(3)(-5) = 4 + 60 = 64$$

(b) Since $64 = 8^2$ is a perfect square, use factoring to solve the equation.

(c) Answers may vary.

9. (a) $a^2 + b^2 = c^2$
Here, $a = 5$ and $c = 20$.
$$5^2 + b^2 = 20^2$$
$$25 + b^2 = 400$$
$$b^2 = 375$$
$$b = \pm\sqrt{375}$$
$$b = \pm 5\sqrt{15}$$
Since distances are positive, the ladder will reach $5\sqrt{15}$ feet up the house.

(b) $5\sqrt{15} \approx 19.4$ feet

10. $R = -x^2 + 220x$

(a) $R = -100^2 + 220(100)$
$= -10{,}000 + 22{,}000$
$= 12{,}000$
The revenue from selling 100 calculus texts is \$12,000.

(b) $9600 = -x^2 + 220x$
$$x^2 - 220x + 9600 = 0$$
$$(x - 60)(x - 160) = 0$$
$x - 60 = 0$ or $x - 160 = 0$
$x = 60$ or $x = 160$
He bookstore must sell 60 or 160 texts.

Section 9.4

Preparing for Problem Solving Using Quadratic Equations

P1. $a^2 + b^2 = c^2$
Here, $b = 7$ and $c = 25$.
$$a^2 + 7^2 = 25^2$$
$$a^2 + 49 = 625$$
$$a^2 = 576$$
$$a = \pm\sqrt{576}$$
$$a = \pm 24$$
Since lengths are positive, $a = 24$.

9.4 Quick Checks

1. False; the width is $l - 5$.

2. Let b be the base of the triangle. Then the height is $b + 2$ and the area is
$\frac{1}{2}(\text{base})(\text{height}) = \frac{1}{2}b(b+2)$.
$$\frac{1}{2}b(b+2) = 40$$
$$b(b+2) = 80$$
$$b^2 + 2b = 80$$
$$b^2 + 2b - 80 = 0$$
$$(b+10)(b-8) = 0$$
$b + 10 = 0$ or $b - 8 = 0$
$b = -10$ or $b = 8$
Since a length cannot be negative, the base is 8 inches and the height is $8 + 2 = 10$ inches.

3. The Pythagorean Theorem tells the relationship between the lengths of the legs and the hypotenuse of a right triangle and is given by the formula $\underline{a^2 + b^2 = c^2}$.

4. Let w be the width of the rectangle. Then the length is $w + 4$.
$$a^2 + b^2 = c^2$$
$$w^2 + (w+4)^2 = 14^2$$
$$w^2 + w^2 + 8w + 16 = 196$$
$$2w^2 + 8w - 180 = 0$$
$$w^2 + 4w - 90 = 0$$
$a = 1$, $b = 4$, $c = -90$

$$w = \frac{-4 \pm \sqrt{4^2 - 4(1)(-90)}}{2(1)}$$
$$= \frac{-4 \pm \sqrt{16+360}}{2}$$
$$= \frac{-4 \pm \sqrt{376}}{2}$$
$$= \frac{-4 \pm 2\sqrt{94}}{2}$$
$$= -2 \pm \sqrt{94}$$
$$w \approx 7.7 \text{ or } -11.7$$

Since a width cannot be negative, the width is approximately 7.7 meters and the length is approximately 7.7 + 4 = 11.7 meters.

5. False; the length of a box must be a positive number and $2-\sqrt{7} \approx -0.65$.

6. Let w be the width of the bin. Then $w + 1$ is the length and the volume is (length)(width)(height) = $(w+1)w(4)$.

$$4(w+1)w = 50$$
$$4w^2 + 4w = 50$$
$$2w^2 + 2w = 25$$
$$2w^2 + 2w - 25 = 0$$
$$a = 2,\ b = 2,\ c = -25$$
$$w = \frac{-2 \pm \sqrt{2^2 - 4(2)(-25)}}{2(2)}$$
$$= \frac{-2 \pm \sqrt{4+200}}{4}$$
$$= \frac{-2 \pm \sqrt{204}}{4}$$
$$= \frac{-2 \pm 2\sqrt{51}}{4}$$
$$= \frac{-1 \pm \sqrt{51}}{2}$$
$$w \approx 3.1 \text{ or } -4.1$$

Since a width cannot be negative, the width is approximately 3.1 feet and the length is approximately 3.1 + 1 = 4.1 feet.

7. $I = -55a^2 + 5119a - 54,448$

Here, $I = 50,000$.

$$50,000 = -55a^2 + 5119a - 54,448$$
$$0 = -55a^2 + 5119a - 104,448$$
$$a = -55,\ b = 5119,\ c = -104,448$$
$$a = \frac{-5119 \pm \sqrt{5119^2 - 4(-55)(-104,448)}}{2(-55)}$$
$$= \frac{-5119 \pm \sqrt{3,225,601}}{-110}$$
$$a \approx 30 \text{ or } 63$$

The worker's average income I equals \$50,000 at ages 30 and 63.

8. $h = -16t^2 + 100t + 3$

Here, $h = 50$.

$$50 = -16t^2 + 100t + 3$$
$$0 = -16t^2 + 100t - 47$$
$$a = -16,\ b = 100,\ c = -47$$
$$t = \frac{-100 \pm \sqrt{100^2 - 4(-16)(-47)}}{2(-16)}$$
$$= \frac{-100 \pm \sqrt{6992}}{-32}$$
$$t \approx 0.51 \text{ or } 5.74$$

The cannon ball will be 50 feet above the ground at 0.51 seconds and 5.74 seconds.

9.4 Exercises

9. Let w be the width; then the length is $2w + 3$.

$$w(2w+3) = 90$$
$$2w^2 + 3w = 90$$
$$2w^2 + 3w - 90 = 0$$
$$(2w+15)(w-6) = 0$$
$$2w+15 = 0 \quad \text{or} \quad w-6 = 0$$
$$w = \frac{-15}{2} \quad \text{or} \quad w = 6$$

Since a length cannot be negative, the width is 6 meters and the length is 2(6) + 3 = 15 meters.

11. Let l be the length; then the width is $\frac{1}{2}l - 3$.

$$l\left(\frac{1}{2}l - 3\right) = 36$$
$$\frac{1}{2}l^2 - 3l = 36$$
$$l^2 - 6l = 72$$
$$l^2 - 6l - 72 = 0$$
$$(l+6)(l-12) = 0$$
$$l+6 = 0 \quad \text{or} \quad l-12 = 0$$
$$l = -6 \quad \text{or} \quad l = 12$$

Since a length cannot be negative, the length is 12 inches and the width is $\frac{1}{2}(12) - 3 = 3$ inches.

13. Let h be the height of the triangle. Then the base is $h + 3$ and the area is

$\frac{1}{2}(\text{base})(\text{height}) = \frac{1}{2}(h+3)(h).$

$$\frac{1}{2}(h+3)(h) = 20$$
$$\frac{1}{2}h^2 + \frac{3}{2}h = 20$$
$$h^2 + 3h = 40$$
$$h^2 + 3h - 40 = 0$$
$$(h+8)(h-5) = 0$$
$$h+8 = 0 \quad \text{or} \quad h-5 = 0$$
$$h = -8 \quad \text{or} \quad h = 5$$

Since a length cannot be negative, the height is 5 cm and the base is 5 + 3 = 8 cm.

15. Let b be the base of the triangle. Then the height is $\frac{1}{2}b+5$ and the area is

$\frac{1}{2}(\text{base})(\text{height}) = \frac{1}{2}(b)\left(\frac{1}{2}b+5\right).$

$$\frac{1}{2}(b)\left(\frac{1}{2}b+5\right) = 66$$
$$\frac{1}{4}b^2 + \frac{5}{2}b = 66$$
$$b^2 + 10b = 264$$
$$b^2 + 10b - 264 = 0$$
$$(b+22)(b-12) = 0$$
$$b+22 = 0 \quad \text{or} \quad b-12 = 0$$
$$b = -22 \quad \text{or} \quad b = 12$$

Since a height cannot be negative, the base is 12 feet and the height is $\frac{1}{2}(12)+5 = 11$ feet.

17. Let b be the longer leg. Then the other leg is $b - 3$.

$$b^2 + (b-3)^2 = \left(3\sqrt{2}\right)^2$$
$$b^2 + b^2 - 6b + 9 = 18$$
$$2b^2 - 6b - 9 = 0$$
$$a = 2, b = -6, c = -9$$

$$b = \frac{-(-6) \pm \sqrt{(-6)^2 - 4(2)(-9)}}{2(2)}$$
$$= \frac{6 \pm \sqrt{36+72}}{4}$$
$$= \frac{6 \pm \sqrt{108}}{4}$$
$$= \frac{6 \pm 6\sqrt{3}}{4}$$
$$= \frac{3 \pm 3\sqrt{3}}{2}$$

$$b = \frac{3-3\sqrt{3}}{2} \approx -1.1 \text{ or } b = \frac{3+3\sqrt{3}}{2} \approx 4.1$$

Since a length cannot be negative, the longer leg is $\frac{3+3\sqrt{3}}{2} \approx 4.1$ inches and the shorter leg is $\frac{3+3\sqrt{3}}{2} - 3 = \frac{-3+3\sqrt{3}}{2} \approx 1.1$ inches.

19. Let a be the length of one leg. Then the hypotenuse is $2a$.

$$a^2 + 9^2 = (2a)^2$$
$$a^2 + 81 = 4a^2$$
$$81 = 3a^2$$
$$27 = a^2$$
$$\pm\sqrt{27} = a$$
$$\pm 3\sqrt{3} = a$$

Since a length cannot be negative, the other leg is $3\sqrt{3} \approx 5.2$ meters and the hypotenuse is $2\left(3\sqrt{3}\right) = 6\sqrt{3} \approx 10.4$ meters.

21. Let x be the smaller integer.

$$x(x+1) = 272$$
$$x^2 + x = 272$$
$$x^2 + x - 272 = 0$$
$$(x+17)(x-16) = 0$$
$$x+17 = 0 \quad \text{or} \quad x-16 = 0$$
$$x = -17 \quad \text{or} \quad x = 16$$
$$x+1 = -16 \qquad x+1 = 17$$

The pairs are −17, −16 and 16, 17.

23. Let x be the smaller integer.

$$2x[(x+1)-10]=140$$
$$2x(x-9)=140$$
$$2x^2-18x-140=0$$
$$x^2-9x-70=0$$
$$(x+5)(x-14)=0$$
$$x+5=0 \quad \text{or} \quad x-14=0$$
$$x=-5 \quad \text{or} \quad x=14$$
$$x+1=-4 \qquad x+1=15$$

The positive integers are 14 and 15.

25. Let x be the number.

$$x^2+2x=23$$
$$x^2+2x-23=0$$
$$a=1, b=2, c=-23$$
$$x=\frac{-2\pm\sqrt{2^2-4(1)(-23)}}{2(1)}$$
$$=\frac{-2\pm\sqrt{4+92}}{2}$$
$$=\frac{-2\pm\sqrt{96}}{2}$$
$$=\frac{-2}{2}\pm\frac{4\sqrt{6}}{2}$$
$$=-1\pm2\sqrt{6}$$

The number is $-1-2\sqrt{6}$ or $-1+2\sqrt{6}$.

27. Let x be the number.

$$x^2=10x-13$$
$$x^2-10x+13=0$$
$$a=1, b=-10, c=13$$
$$x=\frac{-(-10)\pm\sqrt{(-10)^2-4(1)(13)}}{2(1)}$$
$$=\frac{10\pm\sqrt{100-52}}{2}$$
$$=\frac{10\pm\sqrt{48}}{2}$$
$$=\frac{10}{2}\pm\frac{4\sqrt{3}}{2}$$
$$=5\pm2\sqrt{3}$$

The number is $5-2\sqrt{3}$ or $5+2\sqrt{3}$.

29. $h=-16t^2+90t$

(a) When $t=0.5$, the height is

$$h=-16(0.5)^2+90(0.5)$$
$$=-16(0.25)+90(0.5)$$
$$=-4+45$$
$$=41$$

The rocket is 41 feet high 0.5 second after being fired.

(b) When $t=2$, the height is

$$h=-16(2)^2+90(2)$$
$$=-16(4)+90(2)$$
$$=-64+180$$
$$=116$$

The rocket is 116 feet high 2 seconds after being fired.

(c) Find t when $h=100$.

$$100=-16t^2+90t$$
$$16t^2-90t+100=0$$
$$8t^2-45t+50=0$$
$$a=8, b=-45, c=50$$
$$t=\frac{-(-45)\pm\sqrt{(-45)^2-4(8)(50)}}{2(8)}$$
$$=\frac{45\pm\sqrt{2025-1600}}{16}$$
$$=\frac{45\pm\sqrt{425}}{16}$$
$$=\frac{45\pm5\sqrt{17}}{16}$$
$$t=\frac{45-5\sqrt{17}}{16}\approx1.52 \quad \text{or}$$
$$t=\frac{45+5\sqrt{17}}{16}\approx4.10$$

The rocket is 100 feet off the ground 1.52 seconds and 4.10 seconds after being fired.

(d) Find t when $h=0$.

$$0=-16t^2+90t$$
$$0=-2t(8t-45)$$
$$-2t=0 \quad \text{or} \quad 8t-45=0$$
$$t=0 \quad \text{or} \quad t=\frac{45}{8}=5.625$$

It takes the rocket 5.625 seconds to return to Earth.

31. $P = \dfrac{4000t}{4t^2 + 90}$

(a) When $t = 1$, the population is
$$P = \frac{4000(1)}{4(1)^2 + 90} = \frac{4000}{4 + 90} = \frac{4000}{94} \approx 42.55$$
There are about 43 fruit flies after 1 hour.

(b) When $t = 2$, the population is
$$P = \frac{4000(2)}{4(2)^2 + 90} = \frac{8000}{16 + 90} = \frac{8000}{106} \approx 75.47$$
There are about 75 fruit flies after 2 hours.

(c) Find t when $P = 80$.
$$80 = \frac{4000t}{4t^2 + 90}$$
$$80(4t^2 + 90) = 4000t$$
$$320t^2 + 7200 = 4000t$$
$$320t^2 - 4000t + 7200 = 0$$
$$4t^2 - 50t + 90 = 0$$
$a = 4$, $b = -50$, $c = 90$
$$t = \frac{-(-50) \pm \sqrt{(-50)^2 - 4(4)(90)}}{2(4)} = \frac{50 \pm \sqrt{2500 - 1440}}{8} = \frac{50 \pm \sqrt{1060}}{8} = \frac{50 \pm 2\sqrt{265}}{8}$$
$t = \dfrac{50 - 2\sqrt{265}}{8} \approx 2.18$ or

$t = \dfrac{50 + 2\sqrt{265}}{8} \approx 10.32$

The population first reaches 80 flies after 2.18 hours.

33. $D = \dfrac{n(n-3)}{2}$

(a) A hexagon has $n = 6$ sides.
$$D = \frac{6(6-3)}{2} = \frac{6(3)}{2} = \frac{18}{2} = 9$$
A hexagon has 9 diagonals.

(b) An octagon has $n = 8$ sides.
$$D = \frac{8(8-3)}{2} = \frac{8(5)}{2} = \frac{40}{2} = 20$$
An hexagon has 9 diagonals.

(c) Find n when $D = 90$.
$$90 = \frac{n(n-3)}{2}$$
$$180 = n(n-3)$$
$$180 = n^2 - 3n$$
$$0 = n^2 - 3n - 180$$
$$0 = (n - 15)(n + 12)$$
$n - 15 = 0$ or $n + 12 = 0$

$n = 15$ or $n = -12$

The number of sides must be positive, so a polygon with 90 diagonals has 15 sides.

35. (a) The measure is $(180 - x)°$ since $x + (180 - x) = x + 180 - x = 180$.

(b) Let x be the degree measure of the angle.
$$\frac{1}{3}x[(180 - x) - 25] = 1250$$
$$x(180 - x - 25) = 3750$$
$$x(155 - x) = 3750$$
$$155x - x^2 = 3750$$
$$0 = x^2 - 155x + 3750$$
$$0 = (x - 30)(x - 125)$$
$x - 30 = 0$ or $x - 125 = 0$

$x = 30$ or $x = 125$

The angle has measure 30° or 125°.

37. $P = -0.1x^2 + 50x - 150$

(a) $x = 6$
$$P = -0.1(6)^2 + 50(6) - 150 = -0.1(36) + 300 - 150 = -3.6 + 300 - 150 = 146.4$$
Her monthly profit will be \$146.40 after 6 months.

(b) $x = 24$ after 2 years.

$$\begin{aligned} P &= -0.1(24)^2 + 50(24) - 150 \\ &= -0.1(576) + 1200 - 150 \\ &= -57.6 + 1200 - 150 \\ &= 992.4 \end{aligned}$$

Her monthly profit will be \$992.40 after 2 years.

(c) When she starts out, $x = 0$, $p = -150$, and she is not making a profit. Just before she begins to make a profit, she breaks even, which means that her profit is 0.

$0 = -0.1x^2 + 50x - 150$

$a = -0.1$, $b = 50$, $c = -150$

$$\begin{aligned} x &= \frac{-50 \pm \sqrt{50^2 - 4(-0.1)(-150)}}{2(-0.1)} \\ &= \frac{-50 \pm \sqrt{2500 - 60}}{-0.2} \\ &= \frac{-50 \pm \sqrt{2440}}{-0.2} \end{aligned}$$

$x = \dfrac{-50 - \sqrt{2440}}{-0.2} \approx 497$ or

$x = \dfrac{-50 + \sqrt{2440}}{-0.2} \approx 3.02$

She has to stay in business for approximately 4 months to earn a profit.

(d) Find x when $p = 500$.

$$500 = -0.1x^2 + 50x - 150$$

$0.1x^2 - 50x + 650 = 0$

$a = 0.1$, $b = -50$, $c = 650$

$$\begin{aligned} x &= \frac{-(-50) \pm \sqrt{(-50)^2 - 4(0.1)(650)}}{2(0.1)} \\ &= \frac{50 \pm \sqrt{2500 - 260}}{0.2} \\ &= \frac{50 \pm \sqrt{2240}}{0.2} \end{aligned}$$

$x = \dfrac{50 - \sqrt{2240}}{0.2} \approx 13.4$ or

$x = \dfrac{50 + \sqrt{2240}}{0.2} \approx 487$

She will be making \$500 per month after approximately 14 months.

39. If leg $= x = 1$, then the other leg $= x = 1$ and the hypotenuse $= x\sqrt{2} = 1 \cdot \sqrt{2} = \sqrt{2}$.

41. If leg $= x = \sqrt{6}$, then the other leg $= x = \sqrt{6}$ and the hypotenuse $= x\sqrt{2} = \sqrt{6} \cdot \sqrt{2} = \sqrt{12} = 2\sqrt{3}$.

43. If hypotenuse $= x\sqrt{2} = \sqrt{8}$, then both legs $= x = \dfrac{\sqrt{8}}{\sqrt{2}} = \sqrt{\dfrac{8}{2}} = \sqrt{4} = 2$.

45. If the short leg $= x = 5$, the hypotenuse $= 2x = 2 \cdot 5 = 10$, and the long leg $= x\sqrt{3} = 5\sqrt{3}$.

47. If the long leg $= 9$, then $x\sqrt{3} = 9$ or $x = \dfrac{9}{\sqrt{3}} = \dfrac{9\sqrt{3}}{3} = 3\sqrt{3}$. So, the short leg $= 3\sqrt{3}$ and the hypotenuse $= 2x = 2\left(3\sqrt{3}\right) = 6\sqrt{3}$.

Section 9.5

Preparing for the Complex Number System

P1. **(a)** 6 is a natural number.

(b) 6 and 0 are whole numbers.

(c) 6, -14, 0, and $-\dfrac{18}{3} = -6$ are integers.

(d) 6, $-\dfrac{5}{6}$, -14, 0, $1.\overline{65}$, and $-\dfrac{18}{3}$ are rational numbers.

(e) $\sqrt{3}$ is an irrational number.

(f) 6, $-\dfrac{5}{6}$, -14, 0, $\sqrt{3}$, $1.\overline{65}$, and $-\dfrac{18}{3}$ are real numbers.

P2. $2x(3x-4) = 2x \cdot 3x + 2x(-4) = 6x^2 - 8x$

P3.
$$\begin{aligned} &(m+2)(3m-2) \\ &= m \cdot 3m + m(-2) + 2 \cdot 3m + 2(-2) \\ &= 3m^2 - 2m + 6m - 4 \\ &= 3m^2 + 4m - 4 \end{aligned}$$

P4. $(6y+7)(6y-7) = (6y)^2 - 7^2 = 36y^2 - 49$

9.5 Quick Checks

1. The imaginary unit, whose symbol is i, is the number whose square is $\underline{-1}$.

2. If N is a positive real number, the principal square root of $-N$ denoted by $\sqrt{-N}$, is $\sqrt{-N} = \underline{\sqrt{N}i}$.

3. In the complex number $-3 + 4i$, -3 is called the $\underline{\text{real}}$ part and 4 is called the $\underline{\text{imaginary}}$ part.

4. True

5. $\sqrt{-16} = \sqrt{16\cdot(-1)} = \sqrt{16}\cdot\sqrt{-1} = 4i$

6. $\sqrt{-2} = \sqrt{2\cdot(-1)} = \sqrt{2}\cdot\sqrt{-1} = \sqrt{2}i$

7. $$\begin{aligned}\sqrt{-24} &= \sqrt{24\cdot(-1)}\\ &= \sqrt{24}\cdot\sqrt{-1}\\ &= \sqrt{4\cdot 6}\cdot\sqrt{-1}\\ &= 2\sqrt{6}i\end{aligned}$$

8. $$\begin{aligned}9+\sqrt{-144} &= 9+\sqrt{144\cdot(-1)}\\ &= 9+\sqrt{144}\cdot\sqrt{-1}\\ &= 9+12i\end{aligned}$$

9. $$\begin{aligned}-7-\sqrt{-20} &= -7-\sqrt{4\cdot 5\cdot(-1)}\\ &= -7-\sqrt{4}\cdot\sqrt{5}\cdot\sqrt{-1}\\ &= -7-2\sqrt{5}i\end{aligned}$$

10. $$\begin{aligned}\frac{8-\sqrt{-32}}{4} &= \frac{8-\sqrt{16\cdot 2\cdot(-1)}}{4}\\ &= \frac{8-\sqrt{16}\cdot\sqrt{2}\cdot\sqrt{-1}}{4}\\ &= \frac{8-4\sqrt{2}i}{4}\\ &= \frac{4\left(2-\sqrt{2}i\right)}{4}\\ &= 2-\sqrt{2}i\end{aligned}$$

11. $$\begin{aligned}(4-3i)+(-2+5i) &= [4+(-2)]+(-3+5)i\\ &= 2+2i\end{aligned}$$

12. $$\begin{aligned}(-3+7i)-(5-4i) &= (-3-5)+(7+4)i\\ &= -8+11i\end{aligned}$$

13. $$\begin{aligned}&\left(3+\sqrt{-12}\right)-\left(-2-\sqrt{-27}\right)\\ &= \left(3+2\sqrt{3}i\right)-\left(-2-3\sqrt{3}i\right)\\ &= (3+2)+\left(2\sqrt{3}+3\sqrt{3}\right)i\\ &= 5+5\sqrt{3}i\end{aligned}$$

14. $$\begin{aligned}4i(3-6i) &= 4i\cdot 3-4i\cdot 6i\\ &= 12i-24i^2\\ &= 12i-24(-1)\\ &= 24+12i\end{aligned}$$

15. $$\begin{aligned}(-2+4i)(3-i) &= -2\cdot 3-2(-i)+4i\cdot 3-4i\cdot i\\ &= -6+2i+12i-4i^2\\ &= -6+14i-4(-1)\\ &= -6+14i+4\\ &= -2+14i\end{aligned}$$

16. $\sqrt{-25}\cdot\sqrt{-4} = 5i\cdot 2i = 10i^2 = 10(-1) = -10$

17. $$\begin{aligned}&\left(2-\sqrt{-16}\right)\left(1-\sqrt{-4}\right)\\ &= (2-4i)(1-2i)\\ &= 2\cdot 1-2\cdot 2i-4i\cdot 1+4i\cdot 2i\\ &= 2-4i-4i+8i^2\\ &= 2-8i+8(-1)\\ &= 2-8i-8\\ &= -6-8i\end{aligned}$$

18. The complex conjugate of $-1 + 5i$ is $\underline{-1-5i}$.

19. $(2-7i)(2+7i) = 2^2+7^2 = 4+49 = 53$

20. $$\begin{aligned}(-4+9i)(-4-9i) &= (-4)^2+9^2\\ &= 16+81\\ &= 97\end{aligned}$$

21. $$\begin{aligned}\frac{26}{-2+3i} &= \frac{26}{-2+3i}\cdot\frac{-2-3i}{-2-3i}\\ &= \frac{26(-2-3i)}{(-2+3i)(-2-3i)}\\ &= \frac{-26(2+3i)}{(-2)^2+3^2}\\ &= \frac{13(-2)(2+3i)}{4+9}\\ &= \frac{13(-2)(2+3i)}{13}\\ &= -2(2+3i)\\ &= -4-6i\end{aligned}$$

22. $\frac{-12}{4-i}=\frac{-12}{4-i}\cdot\frac{4+i}{4+i}$
$=\frac{-12(4+i)}{(4-i)(4+i)}$
$=\frac{-12(4+i)}{4^2+1^2}$
$=\frac{-12(4+i)}{16+1}$
$=\frac{-48-12i}{17}$
$=-\frac{48}{17}-\frac{12}{17}i$

23. $\frac{8+3i}{3i}=\frac{8+3i}{3i}\cdot\frac{-3i}{-3i}$
$=\frac{(8+3i)(-3i)}{3i(-3i)}$
$=\frac{8(-3i)+3i(-3i)}{-9i^2}$
$=\frac{-24i-9i^2}{-9i^2}$
$=\frac{-24i-9(-1)}{-9(-1)}$
$=\frac{-24i+9}{9}$
$=1-\frac{24}{9}i$
$=1-\frac{8}{3}i$

24. $\frac{1+5i}{2i}=\frac{1+5i}{2i}\cdot\frac{-2i}{-2i}$
$=\frac{(1+5i)(-2i)}{2i(-2i)}$
$=\frac{1(-2i)+5i(-2i)}{-4i^2}$
$=\frac{-2i-10i^2}{-4i^2}$
$=\frac{-2i-10(-1)}{-4(-1)}$
$=\frac{-2i+10}{4}$
$=\frac{10}{4}-\frac{2}{4}i$
$=\frac{5}{2}-\frac{1}{2}i$

25. $2w^2+5=2w^2+0w+5=0$
$a=2,\ b=0,\ c=5$
$w=\frac{-0\pm\sqrt{0^2-4(2)(5)}}{2(2)}$
$=\frac{\pm\sqrt{-40}}{4}$
$=\frac{\pm 2\sqrt{-10}}{4}$
$=\frac{\pm\sqrt{10(-1)}}{2}$
$=\pm\frac{\sqrt{10}}{2}i$

The solution set is $\left\{-\frac{\sqrt{10}}{2}i,\ \frac{\sqrt{10}}{2}i\right\}$.

26. $n^2+6n=-17$
$n^2+6n+17=0$
$a=1,\ b=6,\ c=17$
$n=\frac{-6\pm\sqrt{6^2-4(1)(17)}}{2(1)}$
$=\frac{-6\pm\sqrt{36-68}}{2}$
$=\frac{-6\pm\sqrt{-32}}{2}$
$=\frac{-6\pm\sqrt{16\cdot 2\cdot(-1)}}{2}$
$=\frac{-6\pm 4\sqrt{2}i}{2}$
$=-3\pm 2\sqrt{2}i$

The solution set is $\{-3-2\sqrt{2}i,\ -3+2\sqrt{2}i\}$.

9.5 Exercises

27. $\sqrt{-625}=\sqrt{625\cdot(-1)}=\sqrt{625}\cdot\sqrt{-1}=25i$

29. $-\sqrt{-52}=-\sqrt{52\cdot(-1)}$
$=-\sqrt{52}\cdot\sqrt{-1}$
$=-\sqrt{4\cdot 13}\cdot\sqrt{-1}$
$=-2\sqrt{13}i$

31. $5+\sqrt{-4}=5+\sqrt{4}\cdot\sqrt{-1}=5+2i$

33. $12-\sqrt{-27}=12-\sqrt{27}\cdot\sqrt{-1}=12-3\sqrt{3}i$

35. $\frac{-9+\sqrt{-18}}{6}=\frac{-9}{6}+\frac{\sqrt{18}\cdot\sqrt{-1}}{6}$
$=-\frac{3}{2}+\frac{3\sqrt{2}i}{6}$
$=-\frac{3}{2}+\frac{\sqrt{2}}{2}i$

37. $\frac{2-\sqrt{-48}}{-4}=\frac{2}{-4}-\frac{\sqrt{48}\cdot\sqrt{-1}}{-4}$
$=-\frac{1}{2}-\frac{4\sqrt{3}i}{-4}$
$=-\frac{1}{2}-\left(-\sqrt{3}i\right)$
$=-\frac{1}{2}+\sqrt{3}i$

39. $(3-2i)+(-2+5i)=3-2i-2+5i$
$=3-2-2i+5i$
$=1+3i$

41. $(-4+6i)-(6+6i)=-4+6i+(-6-6i)$
$=-4+6i-6-6i$
$=-4-6+6i-6i$
$=-10$

43. $9i-(3-12i)=9i+(-3+12i)$
$=9i-3+12i$
$=-3+9i+12i$
$=-3+21i$

45. $\left(10-\sqrt{-4}\right)-\left(-3+\sqrt{-64}\right)$
$=(10-2i)-(-3+8i)$
$=10-2i+(3-8i)$
$=10-2i+3-8i$
$=10+3-2i-8i$
$=13-10i$

47. $\left(-6+\sqrt{-64}\right)+\left(-2-\sqrt{-81}\right)$
$=(-6+8i)+(-2-9i)$
$=-6+8i-2-9i$
$=-6-2+8i-9i$
$=-8-i$

49. $\left(11+\sqrt{-18}\right)+\left(1-\sqrt{-8}\right)$
$=\left(11+3\sqrt{2}i\right)+\left(1-2\sqrt{2}i\right)$
$=11+3\sqrt{2}i+1-2\sqrt{2}i$
$=11+1+3\sqrt{2}i-2\sqrt{2}i$
$=12+\sqrt{2}i$

51. $\left(-1-\sqrt{-40}\right)-\left(-1+\sqrt{-90}\right)$
$=\left(-1-2\sqrt{10}i\right)-\left(-1+3\sqrt{10}i\right)$
$=-1-2\sqrt{10}i+\left(1-3\sqrt{10}i\right)$
$=-1-2\sqrt{10}i+1-3\sqrt{10}i$
$=-1+1-2\sqrt{10}i-3\sqrt{10}i$
$=-5\sqrt{10}i$

53. $3i(4-5i)=3i(4)-3i(5i)$
$=12i-15i^2$
$=12i-15(-1)$
$=12i+15$
$=15+12i$

55. $(1-i)(-10i)=1(-10i)-i(-10i)$
$=-10i+10i^2$
$=-10i+10(-1)$
$=-10i-10$
$=-10-10i$

57. $(-2+7i)(-2+i)$
$=-2(-2)+(-2)i+7i(-2)+7i(i)$
$=4-2i-14i+7i^2$
$=4-16i+7(-1)$
$=4-16i-7$
$=-3-16i$

59. $(16i)^2=16^2i^2=256(-1)=-256$

61. $(6+12i)(-2-3i)$
$=6(-2)+6(-3i)+12i(-2)+12i(-3i)$
$=-12-18i-24i-36i^2$
$=-12-42i-36(-1)$
$=-12-42i+36$
$=24-42i$

63. $(4-3i)^2=4^2-2(4)(3i)+(3i)^2$
$=16-24i+9i^2$
$=16-24i+9(-1)$
$=16-24i-9$
$=7-24i$

65. $\sqrt{-25}\cdot\sqrt{-16}=5i\cdot4i=20i^2=20(-1)=-20$

67. $\sqrt{-6}\cdot\sqrt{-12} = \sqrt{6}i\cdot\sqrt{12}i$
$= \sqrt{6}\cdot\sqrt{12}\cdot i^2$
$= \sqrt{72}(-1)$
$= -\sqrt{36\cdot 2}$
$= -6\sqrt{2}$

69. $\left(-5-\sqrt{-1}\right)\left(1+\sqrt{-36}\right)$
$= (-5-i)(1+6i)$
$= -5(1)+(-5)(6i)+(-i)(1)+(-i)(6i)$
$= -5-30i-i-6i^2$
$= -5-31i-6(-1)$
$= -5-31i+6$
$= 1-31i$

71. $-3 + 2i$
conjugate: $-3 - 2i$
$(-3+2i)(-3-2i) = (-3)^2 + 2^2 = 9+4 = 13$

73. $6+\sqrt{-18} = 6+\sqrt{18}i = 6+3\sqrt{2}i$
conjugate: $6-3\sqrt{2}i$
$\left(6+3\sqrt{2}i\right)\left(6-3\sqrt{2}i\right) = 6^2+\left(3\sqrt{2}\right)^2$
$= 36+18$
$= 54$

75. $\dfrac{20}{1-2i} = \dfrac{20}{1-2i}\cdot\dfrac{1+2i}{1+2i}$
$= \dfrac{20(1+2i)}{1^2+2^2}$
$= \dfrac{20(1+2i)}{1+4}$
$= \dfrac{20(1+2i)}{5}$
$= 4(1+2i)$
$= 4+8i$

77. $\dfrac{-4+6i}{2i} = \dfrac{-4+6i}{2i}\cdot\dfrac{-2i}{-2i}$
$= \dfrac{(-4+6i)(-2i)}{(2i)(-2i)}$
$= \dfrac{8i-12i^2}{-4i^2}$
$= \dfrac{8i-12(-1)}{-4(-1)}$
$= \dfrac{8i+12}{4}$
$= \dfrac{8i}{4}+\dfrac{12}{4}$
$= 2i+3$
$= 3+2i$

79. $\dfrac{3}{1-i} = \dfrac{3}{1-i}\cdot\dfrac{1+i}{1+i}$
$= \dfrac{3(1+i)}{1^2+1^2}$
$= \dfrac{3+3i}{2}$
$= \dfrac{3}{2}+\dfrac{3}{2}i$

81. $\dfrac{-5i}{-2+4i} = \dfrac{-5i}{-2+4i}\cdot\dfrac{-2-4i}{-2-4i}$
$= \dfrac{-5i(-2-4i)}{(-2)^2+4^2}$
$= \dfrac{10i+20i^2}{4+16}$
$= \dfrac{10i+20(-1)}{20}$
$= \dfrac{10i-20}{20}$
$= \dfrac{10i}{20}-\dfrac{20}{20}$
$= \dfrac{1}{2}i-1$
$= -1+\dfrac{1}{2}i$

83. $\frac{8+27i}{-4i} = \frac{8+27i}{-4i} \cdot \frac{4i}{4i}$
$= \frac{(8+27i)(4i)}{(-4i)(4i)}$
$= \frac{32i+108i^2}{-16i^2}$
$= \frac{32i+108(-1)}{-16(-1)}$
$= \frac{32i-108}{16}$
$= \frac{32i}{16} - \frac{108}{16}$
$= 2i - \frac{27}{4}$
$= -\frac{27}{4} + 2i$

85. $\frac{-1}{3+2i} = \frac{-1}{3+2i} \cdot \frac{3-2i}{3-2i}$
$= \frac{-1(3-2i)}{3^2+2^2}$
$= \frac{-3+2i}{9+4}$
$= \frac{-3+2i}{13}$
$= -\frac{3}{13} + \frac{2}{13}i$

87. $2x^2 - 4x + 5 = 0$
$a = 2, b = -4, c = 5$
$x = \frac{-(-4) \pm \sqrt{(-4)^2 - 4(2)(5)}}{2(2)}$
$= \frac{4 \pm \sqrt{16-40}}{4}$
$= \frac{4 \pm \sqrt{-24}}{4}$
$= \frac{4 \pm 2\sqrt{6}i}{4}$
$= \frac{4}{4} \pm \frac{2\sqrt{6}}{4}i$
$= 1 \pm \frac{\sqrt{6}}{2}i$

The solution set is $\left\{1 - \frac{\sqrt{6}}{2}i, 1 + \frac{\sqrt{6}}{2}i\right\}$.

89. $(n+1)^2 + 12 = 0$
$(n+1)^2 = -12$
$n+1 = \pm\sqrt{-12}$
$n+1 = \pm 2\sqrt{3}i$
$n = -1 \pm 2\sqrt{3}i$

The solution set is $\left\{-1-2\sqrt{3}i, -1+2\sqrt{3}i\right\}$.

91. $(w-5)^2 + 6 = 2$
$(w-5)^2 = -4$
$w-5 = \pm\sqrt{-4}$
$w-5 = \pm 2i$
$w = 5 \pm 2i$

The solution set is $\{5 - 2i, 5 + 2i\}$.

93. $z^2 + 6z + 9 = 2z$
$z^2 + 4z + 9 = 0$
$a = 1, b = 4, c = 9$
$z = \frac{-4 \pm \sqrt{4^2 - 4(1)(9)}}{2(1)}$
$= \frac{-4 \pm \sqrt{16-36}}{2}$
$= \frac{-4 \pm \sqrt{-20}}{2}$
$= \frac{-4 \pm 2\sqrt{5}i}{2}$
$= -2 \pm \sqrt{5}i$

The solution set is $\left\{-2-\sqrt{5}i, -2+\sqrt{5}i\right\}$.

95. $2x^2 - 13x = 24$
$2x^2 - 13x - 24 = 0$
$(2x+3)(x-8) = 0$
$2x+3 = 0$ or $x-8 = 0$
$x = -\frac{3}{2}$ or $x = 8$

The solution set is $\left\{-\frac{3}{2}, 8\right\}$.

97. $2x^2 - 16 = 0$
$2x^2 = 16$
$x^2 = 8$
$x = \pm\sqrt{8}$
$x = \pm 2\sqrt{2}$

The solution set is $\left\{-2\sqrt{2}, 2\sqrt{2}\right\}$.

99. $3 = 2t - t^2$

$t^2 - 2t + 3 = 0$

$a = 1, b = -2, c = 3$

$$t = \frac{-(-2) \pm \sqrt{(-2)^2 - 4(1)(3)}}{2(1)} = \frac{2 \pm \sqrt{4-12}}{2} = \frac{2 \pm \sqrt{-8}}{2} = \frac{2 \pm 2\sqrt{2}i}{2} = \frac{2}{2} \pm \frac{2\sqrt{2}}{2}i = 1 \pm \sqrt{2}i$$

The solution set is $\{1-\sqrt{2}i, 1+\sqrt{2}i\}$.

101. $13 = k(2k-1)$

$13 = 2k^2 - k$

$0 = 2k^2 - k - 13$

$a = 2, b = -1, c = -13$

$$k = \frac{-(-1) \pm \sqrt{(-1)^2 - 4(2)(-13)}}{2(2)} = \frac{1 \pm \sqrt{1+104}}{4} = \frac{1 \pm \sqrt{105}}{4}$$

The solution set is $\left\{\frac{1-\sqrt{105}}{4}, \frac{1+\sqrt{105}}{4}\right\}$.

103. Let n be the number.

$(n+1)^2 = (2n)^2 + 6$

$n^2 + 2n + 1 = 4n^2 + 6$

$0 = 3n^2 - 2n + 5$

$a = 3, b = -2, c = 5$

$$n = \frac{-(-2) \pm \sqrt{(-2)^2 - 4(3)(5)}}{2(3)} = \frac{2 \pm \sqrt{4-60}}{6} = \frac{2 \pm \sqrt{-56}}{6} = \frac{2 \pm \sqrt{4 \cdot 14 \cdot (-1)}}{6} = \frac{2 \pm 2\sqrt{14}i}{6} = \frac{1}{3} \pm \frac{\sqrt{14}}{3}i$$

The number is $\frac{1}{3} - \frac{\sqrt{14}}{3}i$ or $\frac{1}{3} + \frac{\sqrt{14}}{3}i$.

105. Let x be the number.

$x^2 = 2x - 10$

$x^2 - 2x + 10 = 0$

$a = 1, b = -2, c = 10$

$$x = \frac{-(-2) \pm \sqrt{(-2)^2 - 4(1)(10)}}{2(1)} = \frac{2 \pm \sqrt{4-40}}{2} = \frac{2 \pm \sqrt{-36}}{2} = \frac{2 \pm 6i}{2} = \frac{2}{2} \pm \frac{6i}{2} = 1 \pm 3i$$

The number is $1 - 3i$ or $1 + 3i$.

107. The product of two complex conjugates is always a real number because the product of the conjugates is in the form $(a+bi)(a-bi) = a^2 - abi + abi - (bi)^2$. When simplified, the terms $-abi + abi = 0$, and $(bi)^2 = (b^2)(i)^2 = (b^2)(-1) = -b^2$. Thus, the product contains no imaginary numbers, only real numbers.

109. $x(10-x) = 40$

$10x - x^2 = 40$

$0 = x^2 - 10x + 40$

$a = 1, b = -10, c = 40$

$$x = \frac{-(-10) \pm \sqrt{(-10)^2 - 4(1)(40)}}{2(1)}$$
$$= \frac{10 \pm \sqrt{100-160}}{2}$$
$$= \frac{10 \pm \sqrt{-60}}{2}$$
$$= \frac{10 \pm 2\sqrt{15}i}{2}$$
$$= \frac{10}{2} \pm \frac{2\sqrt{15}i}{2}$$
$$= 5 \pm \sqrt{15}i$$

The answers are $5-\sqrt{15}i$ and $5+\sqrt{15}i$.

Chapter 9 Review

1. $x^2 - 25 = 0$
$$x^2 = 25$$
$$x = \pm\sqrt{25}$$
$$x = \pm 5$$
The solution set is $\{-5, 5\}$.

2. $0 = x^2 - 144$
$$144 = x^2$$
$$\pm\sqrt{144} = x$$
$$\pm 12 = x$$
The solution set is $\{-12, 12\}$.

3. $4p^2 = 9$
$$p^2 = \frac{9}{4}$$
$$p = \pm\sqrt{\frac{9}{4}}$$
$$p = \pm\frac{3}{2}$$
The solution set is $\left\{-\frac{3}{2}, \frac{3}{2}\right\}$.

4. $25t^2 = 4$
$$t^2 = \frac{4}{25}$$
$$t = \pm\sqrt{\frac{4}{25}}$$
$$t = \pm\frac{2}{5}$$
The solution set is $\left\{-\frac{2}{5}, \frac{2}{5}\right\}$.

5. $3x^2 - 10 = 26$
$$3x^2 = 36$$
$$x^2 = 12$$
$$x = \pm\sqrt{12}$$
$$x = \pm 2\sqrt{3}$$
The solution set is $\left\{-2\sqrt{3}, 2\sqrt{3}\right\}$.

6. $2x^2 + 5 = 41$
$$2x^2 = 36$$
$$x^2 = 18$$
$$x = \pm\sqrt{18}$$
$$x = \pm 3\sqrt{2}$$
The solution set is $\left\{-3\sqrt{2}, 3\sqrt{2}\right\}$.

7. $(3x+1)^2 = 16$
$$3x+1 = \pm\sqrt{16}$$
$$3x+1 = \pm 4$$
$$3x = -1 \pm 4$$
$$x = \frac{-1 \pm 4}{3}$$
$$x = \frac{-1-4}{3} = -\frac{5}{3} \text{ or } x = \frac{-1+4}{3} = \frac{3}{3} = 1$$
The solution set is $\left\{-\frac{5}{3}, 1\right\}$.

8. $(2x-5)^2 = 49$
$$2x-5 = \pm\sqrt{49}$$
$$2x-5 = \pm 7$$
$$2x = 5 \pm 7$$
$$x = \frac{5 \pm 7}{2}$$
$$x = \frac{5-7}{2} = \frac{-2}{2} = -1 \text{ or } x = \frac{5+7}{2} = \frac{12}{2} = 6$$
The solution set is $\{-1, 6\}$.

9.
$$\begin{aligned} 9 &= (2n+10)^2+1 \\ 8 &= (2n+10)^2 \\ \pm\sqrt{8} &= 2n+10 \\ \pm 2\sqrt{2} &= 2n+10 \\ -10\pm 2\sqrt{2} &= 2n \\ \frac{-10\pm 2\sqrt{2}}{2} &= n \\ -5\pm\sqrt{2} &= n \end{aligned}$$
The solution set is $\left\{-5-\sqrt{2}, -5+\sqrt{2}\right\}$.

10.
$$\begin{aligned} 25 &= (4d-8)^2-7 \\ 32 &= (4d-8)^2 \\ \pm\sqrt{32} &= 4d-8 \\ \pm 4\sqrt{2} &= 4d-8 \\ 8\pm 4\sqrt{2} &= 4d \\ \frac{8\pm 4\sqrt{2}}{4} &= d \\ 2\pm\sqrt{2} &= d \end{aligned}$$
The solution set is $\left\{2-\sqrt{2}, 2+\sqrt{2}\right\}$.

11.
$$\begin{aligned} (2x+7)^2+6 &= 2 \\ (2x+7)^2 &= -4 \\ 2x+7 &= \pm\sqrt{-4} \end{aligned}$$
The square root of –4 is not a real number. The solution set is ∅.

12.
$$\begin{aligned} (5x-1)^2-5 &= -30 \\ (5x-1)^2 &= -25 \\ 5x-1 &= \pm\sqrt{-25} \end{aligned}$$
The square root of –25 is not a real number. The solution set is ∅.

13.
$$\begin{aligned} x^2-4x+4 &= 24 \\ (x-2)^2 &= 24 \\ x-2 &= \pm\sqrt{24} \\ x-2 &= \pm 2\sqrt{6} \\ x &= 2\pm 2\sqrt{6} \end{aligned}$$
The solution set is $\left\{2-2\sqrt{6}, 2+2\sqrt{6}\right\}$.

14.
$$\begin{aligned} x^2+8x+16 &= 28 \\ (x+4)^2 &= 28 \\ x+4 &= \pm\sqrt{28} \\ x+4 &= \pm 2\sqrt{7} \\ x &= -4\pm 2\sqrt{7} \end{aligned}$$
The solution set is $\left\{-4-2\sqrt{7}, -4+2\sqrt{7}\right\}$.

15.
$$\begin{aligned} c^2 &= a^2+b^2 \\ c^2 &= 12^2+8^2 \\ c^2 &= 144+64 \\ c^2 &= 208 \\ c &= \pm\sqrt{208} \\ c &= \pm 4\sqrt{13} \\ c &= 4\sqrt{13} \approx 14.42 \end{aligned}$$

16.
$$\begin{aligned} c^2 &= a^2+b^2 \\ c^2 &= 6^2+12^2 \\ c^2 &= 36+144 \\ c^2 &= 180 \\ c &= \pm\sqrt{180} \\ c &= \pm 6\sqrt{5} \\ c &= 6\sqrt{5} \approx 13.42 \end{aligned}$$

17.
$$\begin{aligned} a^2+b^2 &= c^2 \\ a^2+8^2 &= 14^2 \\ a^2+64 &= 196 \\ a^2 &= 132 \\ a &= \pm\sqrt{132} \\ a &= \pm 2\sqrt{33} \\ a &= 2\sqrt{33} \approx 11.49 \end{aligned}$$

18.
$$\begin{aligned} a^2+b^2 &= c^2 \\ 4^2+b^2 &= 12^2 \\ 16+b^2 &= 144 \\ b^2 &= 128 \\ b &= \pm\sqrt{128} \\ b &= \pm 8\sqrt{2} \\ b &= 8\sqrt{2} \approx 11.31 \end{aligned}$$

19. (a) $a^2 + b^2 = c^2$
$a^2 + a^2 = 6^2$
$2a^2 = 36$
$a^2 = 18$
$a = \pm\sqrt{18}$
$a = \pm 3\sqrt{2}$
$a = 3\sqrt{2}$
The window is $3\sqrt{2}$ feet wide and $3\sqrt{2}$ feet tall.

(b) $3\sqrt{2} \approx 4.24$
The approximate dimensions are 4.24 feet by 4.24 feet.

20. (a) $a^2 + b^2 = c^2$
$a^2 + 5^2 = 15^2$
$a^2 + 25 = 225$
$a^2 = 200$
$a = \pm\sqrt{200}$
$a = \pm 10\sqrt{2}$
$a = 10\sqrt{2}$
The ladder reaches $10\sqrt{2}$ feet up the house.

(b) $10\sqrt{2} \approx 14.1$
The distance is approximately 14.1 feet.

21. $x^2 - 12x$
$\left(-\frac{12}{2}\right)^2 = (-6)^2 = 36$
$x^2 - 12x + 36 = (x-6)^2$

22. $x^2 + 10x$
$\left(\frac{10}{2}\right)^2 = 5^2 = 25$
$x^2 + 10x + 25 = (x+5)^2$

23. $y^2 + \frac{2}{3}y$
$\left(\frac{1}{2} \cdot \frac{2}{3}\right)^2 = \left(\frac{1}{3}\right)^2 = \frac{1}{9}$
$y^2 + \frac{2}{3}y + \frac{1}{9} = \left(y + \frac{1}{3}\right)^2$

24. $k^2 - \frac{4}{5}k$
$\left[\frac{1}{2} \cdot \left(-\frac{4}{5}\right)\right]^2 = \left(-\frac{2}{5}\right)^2 = \frac{4}{25}$
$k^2 - \frac{4}{5}k + \frac{4}{25} = \left(k - \frac{2}{5}\right)^2$

25. $n^2 - \frac{n}{2}$
$\left[\frac{1}{2} \cdot \left(-\frac{1}{2}\right)\right]^2 = \left(-\frac{1}{4}\right)^2 = \frac{1}{16}$
$n^2 - \frac{n}{2} + \frac{1}{16} = \left(n - \frac{1}{4}\right)^2$

26. $t^2 + \frac{t}{3}$
$\left(\frac{1}{2} \cdot \frac{1}{3}\right)^2 = \left(\frac{1}{6}\right)^2 = \frac{1}{36}$
$t^2 + \frac{t}{3} + \frac{1}{36} = \left(t + \frac{1}{6}\right)^2$

27. $x^2 - 6x = 7$
$x^2 - 6x + \left(-\frac{6}{2}\right)^2 = 7 + \left(-\frac{6}{2}\right)^2$
$x^2 - 6x + (-3)^2 = 7 + (-3)^2$
$x^2 - 6x + 9 = 7 + 9$
$(x-3)^2 = 16$
$x - 3 = \pm 4$
$x = 3 \pm 4$
$x = 3 - 4 = -1$ or $x = 3 + 4 = 7$
The solution set is $\{-1, 7\}$.

28. $x^2 - 12x = -11$
$x^2 - 12x + \left(-\frac{12}{2}\right)^2 = -11 + \left(-\frac{12}{2}\right)^2$
$x^2 - 12x + (-6)^2 = -11 + (-6)^2$
$x^2 - 12x + 36 = -11 + 36$
$(x-6)^2 = 25$
$x - 6 = \pm\sqrt{25}$
$x - 6 = \pm 5$
$x = 6 \pm 5$
$x = 6 - 5 = 1$ or $x = 6 + 5 = 11$
The solution set is $\{1, 11\}$.

29.

$$x^2 = 4-3x$$
$$x^2+3x = 4$$
$$x^2+3x+\left(\frac{3}{2}\right)^2 = 4+\left(\frac{3}{2}\right)^2$$
$$x^2+3x+\frac{9}{4} = 4+\frac{9}{4}$$
$$\left(x+\frac{3}{2}\right)^2 = \frac{25}{4}$$
$$x+\frac{3}{2} = \pm\sqrt{\frac{25}{4}}$$
$$x+\frac{3}{2} = \pm\frac{5}{2}$$
$$x = -\frac{3}{2}\pm\frac{5}{2}$$

$x = -\frac{3}{2}-\frac{5}{2} = -\frac{8}{2} = -4$ or $x = -\frac{3}{2}+\frac{5}{2} = \frac{2}{2} = 1$

The solution set is {−4, 1}.

30.

$$x^2 = -14-9x$$
$$x^2+9x = -14$$
$$x^2+9x+\left(\frac{9}{2}\right)^2 = -14+\left(\frac{9}{2}\right)^2$$
$$x^2+9x+\frac{81}{4} = -14+\frac{81}{4}$$
$$\left(x+\frac{9}{2}\right)^2 = \frac{25}{4}$$
$$x+\frac{9}{2} = \pm\sqrt{\frac{25}{4}}$$
$$x+\frac{9}{2} = \pm\frac{5}{2}$$
$$x = -\frac{9}{2}\pm\frac{5}{2}$$

$x = -\frac{9}{2}-\frac{5}{2} = -\frac{14}{2} = -7$ or

$x = -\frac{9}{2}+\frac{5}{2} = -\frac{4}{2} = -2$

The solution set is {−7, −2}.

31.

$$\frac{1}{4}x^2 - x - 4 = 0$$
$$4\left(\frac{1}{4}x^2 - x - 4\right) = 4(0)$$
$$x^2 - 4x - 16 = 0$$
$$x^2 - 4x = 16$$
$$x^2 - 4x + \left(-\frac{4}{2}\right)^2 = 16+\left(-\frac{4}{2}\right)^2$$
$$x^2 - 4x + 4 = 16 + 4$$
$$(x-2)^2 = 20$$
$$x-2 = \pm\sqrt{20}$$
$$x-2 = \pm 2\sqrt{5}$$
$$x = 2\pm 2\sqrt{5}$$

The solution set is $\left\{2-2\sqrt{5},\ 2+2\sqrt{5}\right\}$.

32.

$$\frac{1}{2}x^2 - x - \frac{7}{2} = 0$$
$$2\left(\frac{1}{2}x^2 - x - \frac{7}{2}\right) = 2\cdot 0$$
$$x^2 - 2x - 7 = 0$$
$$x^2 - 2x = 7$$
$$x^2 - 2x + \left(-\frac{2}{2}\right)^2 = 7+\left(-\frac{2}{2}\right)^2$$
$$x^2 - 2x + 1 = 7 + 1$$
$$(x-1)^2 = 8$$
$$x-1 = \pm\sqrt{8}$$
$$x-1 = \pm 2\sqrt{2}$$
$$x = 1\pm 2\sqrt{2}$$

The solution set is $\left\{1-2\sqrt{2},\ 1+2\sqrt{2}\right\}$.

33.
$$2z^2+z-10=0$$
$$\frac{1}{2}(2z^2+z-10)=\frac{1}{2}(0)$$
$$z^2+\frac{1}{2}z-5=0$$
$$z^2+\frac{1}{2}z=5$$
$$z^2+\frac{1}{2}z+\left(\frac{1}{2}\cdot\frac{1}{2}\right)^2=5+\left(\frac{1}{2}\cdot\frac{1}{2}\right)^2$$
$$z^2+\frac{1}{2}z+\frac{1}{16}=5+\frac{1}{16}$$
$$\left(z+\frac{1}{4}\right)^2=\frac{81}{16}$$
$$z+\frac{1}{4}=\pm\sqrt{\frac{81}{16}}$$
$$z+\frac{1}{4}=\pm\frac{9}{4}$$
$$z=-\frac{1}{4}\pm\frac{9}{4}$$
$$z=-\frac{1}{4}-\frac{9}{4}=-\frac{10}{4}=-\frac{5}{2} \text{ or } z=-\frac{1}{4}+\frac{9}{4}=\frac{8}{4}=2$$

The solution set is $\left\{-\frac{5}{2}, 2\right\}$.

34.
$$3n^2+11n+6=0$$
$$\frac{1}{3}(3n^2+11n+6)=\frac{1}{3}\cdot 0$$
$$n^2+\frac{11}{3}n+2=0$$
$$n^2+\frac{11}{3}n=-2$$
$$n^2+\frac{11}{3}n+\left(\frac{1}{2}\cdot\frac{11}{3}\right)^2=-2+\left(\frac{1}{2}\cdot\frac{11}{3}\right)^2$$
$$n^2+\frac{11}{3}n+\frac{121}{36}=-2+\frac{121}{36}$$
$$\left(n+\frac{11}{6}\right)^2=\frac{49}{36}$$
$$n+\frac{11}{6}=\pm\sqrt{\frac{49}{36}}$$
$$n+\frac{11}{6}=\pm\frac{7}{6}$$
$$n=-\frac{11}{6}\pm\frac{7}{6}$$
$$n=-\frac{11}{6}-\frac{7}{6}=-\frac{18}{6}=-3 \text{ or}$$
$$n=-\frac{11}{6}+\frac{7}{6}=-\frac{4}{6}=-\frac{2}{3}$$

The solution set is $\left\{-3, -\frac{2}{3}\right\}$.

35.
$$(x-5)(x+2)=1$$
$$x^2-3x-10=1$$
$$x^2-3x=11$$
$$x^2-3x+\left(-\frac{3}{2}\right)^2=11+\left(-\frac{3}{2}\right)^2$$
$$x^2-3x+\frac{9}{4}=11+\frac{9}{4}$$
$$\left(x-\frac{3}{2}\right)^2=\frac{53}{4}$$
$$x-\frac{3}{2}=\pm\sqrt{\frac{53}{4}}$$
$$x-\frac{3}{2}=\pm\frac{\sqrt{53}}{2}$$
$$x=\frac{3}{2}\pm\frac{\sqrt{53}}{2}$$
$$x=\frac{3\pm\sqrt{53}}{2}$$

The solution set is $\left\{\frac{3-\sqrt{53}}{2}, \frac{3+\sqrt{53}}{2}\right\}$.

36.
$$(x+3)(x-2)=2$$
$$x^2+x-6=2$$
$$x^2+x=8$$
$$x^2+x+\left(\frac{1}{2}\right)^2=8+\left(\frac{1}{2}\right)^2$$
$$x^2+x+\frac{1}{4}=8+\frac{1}{4}$$
$$\left(x+\frac{1}{2}\right)^2=\frac{33}{4}$$
$$x+\frac{1}{2}=\pm\sqrt{\frac{33}{4}}$$
$$x+\frac{1}{2}=\pm\frac{\sqrt{33}}{2}$$
$$x=-\frac{1}{2}\pm\frac{\sqrt{33}}{2}$$
$$x=\frac{-1\pm\sqrt{33}}{2}$$

The solution set is $\left\{\frac{-1-\sqrt{33}}{2}, \frac{-1+\sqrt{33}}{2}\right\}$.

37.
$$x(x-10)-15=5$$
$$x^2-10x-15=5$$
$$x^2-10x=20$$
$$x^2-10x+\left(-\frac{10}{2}\right)^2=20+\left(-\frac{10}{2}\right)^2$$
$$x^2-10x+25=20+25$$
$$(x-5)^2=45$$
$$x-5=\pm\sqrt{45}$$
$$x-5=\pm3\sqrt{5}$$
$$x=5\pm3\sqrt{5}$$
The solution set is $\{5-3\sqrt{5}, 5+3\sqrt{5}\}$.

38.
$$x(x-8)+2=10$$
$$x^2-8x+2=10$$
$$x^2-8x=8$$
$$x^2-8x+\left(-\frac{8}{2}\right)^2=8+\left(-\frac{8}{2}\right)^2$$
$$x^2-8x+16=8+16$$
$$(x-4)^2=24$$
$$x-4=\pm\sqrt{24}$$
$$x-4=\pm2\sqrt{6}$$
$$x=4\pm2\sqrt{6}$$
The solution set is $\{4-2\sqrt{6}, 4+2\sqrt{6}\}$.

39. Let n be the number.
$$n(n+2)=\frac{5}{4}$$
$$n^2+2n=\frac{5}{4}$$
$$n^2+2n+\left(\frac{2}{2}\right)^2=\frac{5}{4}+\left(\frac{2}{2}\right)^2$$
$$n^2+2n+1=\frac{5}{4}+1$$
$$(n+1)^2=\frac{9}{4}$$
$$n+1=\pm\sqrt{\frac{9}{4}}$$
$$n+1=\pm\frac{3}{2}$$
$$n=-1\pm\frac{3}{2}$$
$$n=-1-\frac{3}{2}=-\frac{5}{2} \text{ or } n=-1+\frac{3}{2}=\frac{1}{2}$$
The number is $-\frac{5}{2}$ or $\frac{1}{2}$.

40. Let n be the number.
$$n^2+6n=16$$
$$n^2+6n+\left(\frac{6}{2}\right)^2=16+\left(\frac{6}{2}\right)^2$$
$$n^2+6n+9=16+9$$
$$(n+3)^2=25$$
$$n+3=\pm\sqrt{25}$$
$$n+3=\pm5$$
$$n=-3\pm5$$
$n=-3-5=-8$ or $n=-3+5=2$
The number is -8 or 2.

41.
$$3=x-2x^2$$
$$2x^2-x+3=0$$
$a=2, b=-1, c=3$

42.
$$\frac{1}{2}x^2=\frac{3}{4}-x$$
$$\frac{1}{2}x^2+x-\frac{3}{4}=0$$
$a=\frac{1}{2}$, $b=1$, $c=-\frac{3}{4}$ or
$$\frac{1}{2}x^2+x-\frac{3}{4}=0$$
$$4\left(\frac{1}{2}x^2+x-\frac{3}{4}\right)=4(0)$$
$$2x^2+4x-3=0$$
$a=2, b=4, c=-3$

43. $2x^2-13x+15=0$
$a=2, b=-13, c=15$
$$x=\frac{-(-13)\pm\sqrt{(-13)^2-4(2)(15)}}{2(2)}$$
$$=\frac{13\pm\sqrt{169-120}}{4}$$
$$=\frac{13\pm\sqrt{49}}{4}$$
$$=\frac{13\pm7}{4}$$
$$x=\frac{13-7}{4}=\frac{6}{4}=\frac{3}{2} \text{ or } x=\frac{13+7}{4}=\frac{20}{4}=5$$
The solution set is $\left\{\frac{3}{2}, 5\right\}$.

44. $4x^2+8x+3=0$

$a=4, b=8, c=3$

$$x=\frac{-8\pm\sqrt{8^2-4(4)(3)}}{2(4)}=\frac{-8\pm\sqrt{64-48}}{8}=\frac{-8\pm\sqrt{16}}{8}=\frac{-8\pm4}{8}$$

$$x=\frac{-8-4}{8}=\frac{-12}{8}=-\frac{3}{2} \text{ or } x=\frac{-8+4}{8}=\frac{-4}{8}=-\frac{1}{2}$$

The solution set is $\left\{-\frac{3}{2}, -\frac{1}{2}\right\}$.

45. $x^2-12x=-5$

$x^2-12x+5=0$

$a=1, b=-12, c=5$

$$x=\frac{-(-12)\pm\sqrt{(-12)^2-4(1)(5)}}{2(1)}=\frac{12\pm\sqrt{144-20}}{2}=\frac{12\pm\sqrt{124}}{2}=\frac{12\pm2\sqrt{31}}{2}=6\pm\sqrt{31}$$

The solution set is $\{6-\sqrt{31}, 6+\sqrt{31}\}$.

46. $x^2-3x=9$

$x^2-3x-9=0$

$a=1, b=-3, c=-9$

$$x=\frac{-(-3)\pm\sqrt{(-3)^2-4(1)(-9)}}{2(1)}=\frac{3\pm\sqrt{9+36}}{2}=\frac{3\pm\sqrt{45}}{2}=\frac{3\pm3\sqrt{5}}{2}$$

The solution set is $\left\{\frac{3-3\sqrt{5}}{2}, \frac{3+3\sqrt{5}}{2}\right\}$.

47. $p(3p+5)-8=0$

$3p^2+5p-8=0$

$a=3, b=5, c=-8$

$$p=\frac{-5\pm\sqrt{5^2-4(3)(-8)}}{2(3)}=\frac{-5\pm\sqrt{25+96}}{6}=\frac{-5\pm\sqrt{121}}{6}=\frac{-5\pm11}{6}$$

$$p=\frac{-5-11}{6}=\frac{-16}{6}=-\frac{8}{3} \text{ or } p=\frac{-5+11}{6}=\frac{6}{6}=1$$

The solution set is $\left\{-\frac{8}{3}, 1\right\}$.

48. $v(2v+7)-4=0$

$2v^2+7v-4=0$

$a=2, b=7, c=-4$

$$v=\frac{-7\pm\sqrt{7^2-4(2)(-4)}}{2(2)}=\frac{-7\pm\sqrt{49+32}}{4}=\frac{-7\pm\sqrt{81}}{4}=\frac{-7\pm9}{4}$$

$$v=\frac{-7-9}{4}=-\frac{16}{4}=-4 \text{ or } v=\frac{-7+9}{4}=\frac{2}{4}=\frac{1}{2}$$

The solution set is $\left\{-4, \frac{1}{2}\right\}$.

49. $2x = 3 - 2x^2$

$2x^2 + 2x - 3 = 0$

$a = 2, b = 2, c = -3$

$$x = \frac{-2 \pm \sqrt{2^2 - 4(2)(-3)}}{2(2)} = \frac{-2 \pm \sqrt{4+24}}{4} = \frac{-2 \pm \sqrt{28}}{4} = \frac{-2 \pm 2\sqrt{7}}{4} = \frac{-1 \pm \sqrt{7}}{2}$$

The solution set is $\left\{\frac{-1-\sqrt{7}}{2}, \frac{-1+\sqrt{7}}{2}\right\}$.

50. $x^2 = 2 - 4x$

$x^2 + 4x - 2 = 0$

$a = 1, b = 4, c = -2$

$$x = \frac{-4 \pm \sqrt{4^2 - 4(1)(-2)}}{2(1)} = \frac{-4 \pm \sqrt{16+8}}{2} = \frac{-4 \pm \sqrt{24}}{2} = \frac{-4 \pm 2\sqrt{6}}{2} = -2 \pm \sqrt{6}$$

The solution set is $\left\{-2-\sqrt{6}, -2+\sqrt{6}\right\}$.

51. $\frac{z}{3} + \frac{1}{12z} = 1$

$12z\left(\frac{z}{3} + \frac{1}{12z}\right) = 12z(1)$

$4z^2 + 1 = 12z$

$4z^2 - 12z + 1 = 0$

$a = 4, b = -12, c = 1$

$$z = \frac{-(-12) \pm \sqrt{(-12)^2 - 4(4)(1)}}{2(4)} = \frac{12 \pm \sqrt{144-16}}{8} = \frac{12 \pm \sqrt{128}}{8} = \frac{12 \pm 8\sqrt{2}}{8} = \frac{3 \pm 2\sqrt{2}}{2}$$

The solution set is $\left\{\frac{3-2\sqrt{2}}{2}, \frac{3+2\sqrt{2}}{2}\right\}$.

52. $\frac{3n}{2} + \frac{1}{n} = 4$

$2n\left(\frac{3n}{2} + \frac{1}{n}\right) = 2n(4)$

$3n^2 + 2 = 8n$

$3n^2 - 8n + 2 = 0$

$a = 3, b = -8, c = 2$

$$n = \frac{-(-8) \pm \sqrt{(-8)^2 - 4(3)(2)}}{2(3)} = \frac{8 \pm \sqrt{64-24}}{6} = \frac{8 \pm \sqrt{40}}{6} = \frac{8 \pm 2\sqrt{10}}{6} = \frac{4 \pm \sqrt{10}}{3}$$

The solution set is $\left\{\frac{4-\sqrt{10}}{3}, \frac{4+\sqrt{10}}{3}\right\}$.

53. (a) $R = -\frac{1}{2}x^2 + 130x$

When $x = 150$,

$R = -\frac{1}{2}(150)^2 + 130(150) = 8250$

The revenue is \$8250.

(b) Find x when $R = 8000$.

$$8000 = -\frac{1}{2}x^2 + 130x$$
$$0 = -\frac{1}{2}x^2 + 130x - 8000$$
$$0 = x^2 - 260x + 16,000$$
$$0 = (x-100)(x-160)$$
$$x - 100 = 0 \quad \text{or} \quad x - 160 = 0$$
$$x = 100 \quad \text{or} \quad x = 160$$

The bookstore must sell 100 or 160 texts.

54. (a) $P = x^2 + 30x - 60$

When $x = 8$, $P = 8^2 + 30(8) - 60 = 244$.

Martin's profit is $244.

(b) Find x when $P = 615$.

$$615 = x^2 + 30x - 60$$
$$0 = x^2 + 30x - 675$$

$a = 1$, $b = 30$, $c = -675$

$$x = \frac{-30 \pm \sqrt{30^2 - 4(1)(-675)}}{2(1)}$$
$$= \frac{-30 \pm \sqrt{900 + 2700}}{2}$$
$$= \frac{-30 \pm \sqrt{3600}}{2}$$
$$= \frac{-30 \pm 60}{2}$$
$$x = \frac{-30 - 60}{2} = -45 \quad \text{or} \quad x = \frac{-30 + 60}{2} = 15$$

Martin must groom 15 dogs.

55. $5t^2 + 2 = 8t$

$5t^2 - 8t + 2 = 0$

$a = 5$, $b = -8$, $c = 2$

$b^2 - 4ac = (-8)^2 - 4(5)(2) = 64 - 40 = 24$

Since $24 > 0$, and 24 is not a perfect square, solve by using the quadratic formula.

$$t = \frac{-(-8) \pm \sqrt{24}}{2(5)} = \frac{8 \pm 2\sqrt{6}}{10} = \frac{4 \pm \sqrt{6}}{5}$$

The solution set is $\left\{\frac{4-\sqrt{6}}{5}, \frac{4+\sqrt{6}}{5}\right\}$.

56. $x^2 - 5x = 6$

$x^2 - 5x - 6 = 0$

$a = 1$, $b = -5$, $c = -6$

$b^2 - 4ac = (-5)^2 - 4(1)(-6) = 25 + 24 = 49$

Since 49 is a perfect square, solve by factoring.

$$x^2 - 5x - 6 = 0$$
$$(x-6)(x+1) = 0$$
$$x - 6 = 0 \quad \text{or} \quad x + 1 = 0$$
$$x = 6 \quad \text{or} \quad x = -1$$

The solution set is $\{-1, 6\}$.

57. $3x^2 + 7x = 10$

$3x^2 + 7x - 10 = 0$

$a = 3$, $b = 7$, $c = -10$

$b^2 - 4ac = 7^2 - 4(3)(-10) = 49 + 120 = 169$

Since 169 is a perfect square, solve by factoring.

$$3x^2 + 7x - 10 = 0$$
$$(3x+10)(x-1) = 0$$
$$3x + 10 = 0 \quad \text{or} \quad x - 1 = 0$$
$$x = -\frac{10}{3} \quad \text{or} \quad x = 1$$

The solution set is $\left\{-\frac{10}{3}, 1\right\}$.

58. $n^2 - 2 = -3n$

$n^2 + 3n - 2 = 0$

$a = 1$, $b = 3$, $c = -2$

$b^2 - 4ac = 3^2 - 4(1)(-2) = 9 + 8 = 17$

Since $17 > 0$ and 17 is not a perfect square, solve by using the quadratic formula.

$$n = \frac{-3 \pm \sqrt{3^2 - 4(1)(-2)}}{2(1)}$$
$$= \frac{-3 \pm \sqrt{9+8}}{2}$$
$$= \frac{-3 \pm \sqrt{17}}{2}$$

The solution set is $\left\{\frac{-3-\sqrt{17}}{2}, \frac{-3+\sqrt{17}}{2}\right\}$.

59. $2 + 3d^2 = d$

$3d^2 - d + 2 = 0$

$a = 3$, $b = -1$, $c = 2$

$b^2 - 4ac = (-1)^2 - 4(3)(2) = 1 - 24 = -23$

Since $-23 < 0$, there are two unequal solutions that are not real. The solution set is $\varnothing$.

60. $3(p^2+1)+5p=0$
$3p^2+3+5p=0$
$3p^2+5p+3=0$
$a=3, b=5, c=3$
$b^2-4ac=5^2-4(3)(3)=25-36=-11$
Since $-11<0$, there are two unequal solutions that are not real. The solution set is $\varnothing$.

61. Let n be the first even integer. Then $n+2$ is the next consecutive integer.
$n(n+2)=288$
$n^2+2n=288$
$n^2+2n-288=0$
$(n+18)(n-16)=0$
$n+18=0$ or $n-16=0$
$n=-18$ or $n=16$
$n+2=-16$ $n+2=18$
The pairs are -18, -16 and 16, 18.

62. Let x be the number. Then $\frac{1}{x}$ is its reciprocal.
$$x+\frac{1}{x}=\frac{13}{6}$$
$$6x\left(x+\frac{1}{x}\right)=6x\left(\frac{13}{6}\right)$$
$$6x^2+6=13x$$
$$6x^2-13x+6=0$$
$$(2x-3)(3x-2)=0$$
$2x-3=0$ or $3x-2=0$
$x=\frac{3}{2}$ or $x=\frac{2}{3}$
The number is $\frac{3}{2}$ and its reciprocal is $\frac{2}{3}$.

63. Let w be the width. Then the length is $2w-2$. Area $=lw$.
$112=(2w-2)w$
$112=2w^2-2w$
$0=2w^2-2w-112$
$0=2(w^2-w-56)$
$0=2(w-8)(w+7)$
$2=0$ or $w-8=0$ or $w+7=0$
False or $w=8$ or $w=-7$
Since length cannot be negative, the width is 8 meters and the length is $2(8)-2=14$ meters.

64. Let l be the length. Then the width is $\frac{1}{3}l+1$.
Area $=lw$.
$$60=l\left(\frac{1}{3}l+1\right)$$
$$60=\frac{1}{3}l^2+l$$
$$3(60)=3\left(\frac{1}{3}l^2+l\right)$$
$$180=l^2+3l$$
$$0=l^2+3l-180$$
$$0=(l+15)(l-12)$$
$l+15=0$ or $l-12=0$
$l=-15$ or $l=12$
Since length cannot be negative, the length is 12 yards and the width is $\frac{1}{3}(12)+1=5$ yards.

65. Let x be the length of the short leg. Then $2x$ is the length of the long leg.
$$a^2+b^2=c^2$$
$$x^2+(2x)^2=\left(4\sqrt{5}\right)^2$$
$$x^2+4x^2=80$$
$$5x^2=80$$
$$x^2=16$$
$$x=\pm\sqrt{16}$$
$$x=\pm 4$$
Since length cannot be negative, the legs are 4 feet and $2(4)=8$ feet.

66. Let x be the length of the leg. Then $2x+3$ is the length of the hypotenuse.
$$a^2+b^2=c^2$$
$$x^2+5^2=(2x+3)^2$$
$$x^2+25=4x^2+12x+9$$
$$0=3x^2+12x-16$$
$a=3, b=12, c=-16$

$$x=\frac{-12\pm\sqrt{12^2-4(3)(-16)}}{2(3)}$$
$$=\frac{-12\pm\sqrt{144+192}}{6}$$
$$=\frac{-12\pm\sqrt{336}}{6}$$
$$=\frac{-12\pm4\sqrt{21}}{6}$$
$$=\frac{-6\pm2\sqrt{21}}{3}$$

Since length cannot be negative, the leg is $\frac{-6+2\sqrt{21}}{3}$ centimeters and the hypotenuse is

$$2\left(\frac{-6+2\sqrt{21}}{3}\right)+3$$
$$=\frac{-12+4\sqrt{21}}{3}+\frac{9}{3}$$
$$=\frac{-3+4\sqrt{21}}{3}\text{ centimeters.}$$

67. Let h be the height. Then the base is $h + 5$. The area $=\frac{1}{2}$(base)(height).

$$42=\frac{1}{2}(h+5)(h)$$
$$84=h^2+5h$$
$$0=h^2+5h-84$$
$$0=(h+12)(h-7)$$
$$h+12=0 \quad\text{or}\quad h-7=0$$
$$h=-12 \quad\text{or}\quad h=7$$

Since length cannot be negative, the height is 7 meters and the base is 7 + 5 = 12 meters.

68. Let h be the height. Then the base is $h+\frac{3}{2}$. The area $=\frac{1}{2}$(base)(height).

$$\frac{9}{4}=\frac{1}{2}\left(h+\frac{3}{2}\right)(h)$$
$$\frac{9}{4}=\frac{1}{2}h^2+\frac{3}{4}h$$
$$4\left(\frac{9}{4}\right)=4\left(\frac{1}{2}h^2+\frac{3}{4}h\right)$$
$$9=2h^2+3h$$
$$0=2h^2+3h-9$$
$$0=(2h-3)(h+3)$$
$$2h-3=0 \quad\text{or}\quad h+3=0$$
$$h=\frac{3}{2} \quad\text{or}\quad h=-3$$

Since length cannot be negative, the height is $\frac{3}{2}$ inches and the base is $\frac{3}{2}+\frac{3}{2}=3$ inches.

69. Let l be the length. Then the base width is $\frac{1}{2}l-1$. The volume = lwh.

$$200=l\left(\frac{1}{2}l-1\right)(5)$$
$$200=\frac{5}{2}l^2-5l$$
$$2(200)=2\left(\frac{5}{2}l^2-5l\right)$$
$$400=5l^2-10l$$
$$0=5l^2-10l-400$$
$$0=5(l^2-2l-80)$$
$$0=5(l-10)(l+8)$$
$$0=5 \quad\text{or}\quad 0=l-10 \quad\text{or}\quad 0=l+8$$
$$\text{False} \quad\text{or}\quad 10=l \quad\text{or}\quad -8=l$$

Since length cannot be negative, the length is 10 inches and the width is $\frac{1}{2}(10)-1=4$ inches.

70. Let w be the width. Then $2w - 1$ is the length. The height is 6 inches = 0.5 foot. The volume = lwh.

$$60=(2w-1)(w)(0.5)$$
$$2(60)=2(2w-1)(w)(0.5)$$
$$120=(2w-1)w$$
$$120=2w^2-2$$
$$0=2w^2-w-120$$
$$0=(2w+15)(w-8)$$
$$2w+15=0 \quad\text{or}\quad w-8=0$$
$$w=-\frac{15}{2} \quad\text{or}\quad w=8$$

Since the length cannot be negative, the width is 8 feet and the length is 2(8) – 1 = 16 – 1 = 15 feet.

71. $h=-16t^2+80t+4$

$0=-16t^2+80t+4$

$0=-4(4t^2-20t-1)$

$0=4t^2-20t-1$

$a=4,\ b=-20,\ c=-1$

$$t=\frac{-(-20)\pm\sqrt{(-20)^2-4(4)(-1)}}{2(4)}=\frac{20\pm\sqrt{400+16}}{8}=\frac{20\pm\sqrt{416}}{8}=\frac{20\pm4\sqrt{26}}{8}=\frac{5\pm\sqrt{26}}{2}$$

$t=\frac{5-\sqrt{26}}{2}\approx-0.05$ or $t=\frac{5+\sqrt{26}}{2}=5.0$

It takes the rocket $\frac{5+\sqrt{26}}{2}\approx5.0$ seconds.

72. $h=-16t^2+120t$

$80=-16t^2+120t$

$0=-16t^2+120t-80$

$0=-8(2t^2-15t+10)$

$0=2t^2-15t+10$

$a=2,\ b=-15,\ c=10$

$$t=\frac{-(-15)\pm\sqrt{(-15)^2-4(2)(10)}}{2(2)}=\frac{15\pm\sqrt{225-80}}{4}=\frac{15\pm\sqrt{145}}{4}$$

$t=\frac{15-\sqrt{145}}{4}\approx0.7$ or $t=\frac{15+\sqrt{145}}{4}\approx6.8$

The rocket will be 80 feet off the ground after 0.7 seconds and 6.8 seconds.

73. $R-C=0$

$\left(-\frac{x^2}{5}+10x\right)-(4x+25)=0$

$-\frac{x^2}{5}+10x-4x-25=0$

$-\frac{x^2}{5}+6x-25=0$

$x^2-30x+125=0$

$(x-25)(x-5)=0$

$x-25=0$ or $x-5=0$

$x=25$ or $x=5$

The break-even point is 5 or 25 units.

74. $R-C=0$

$\left(-\frac{x^2}{10}+8x\right)-(3x-35)=0$

$-\frac{x^2}{10}+8x-3x-35=0$

$-\frac{x^2}{10}+5x-35=0$

$x^2-50x+350=0$

$a=1,\ b=-50,\ c=350$

$$x=\frac{-(-50)\pm\sqrt{(-50)^2-4(1)(350)}}{2(1)}=\frac{50\pm\sqrt{2500-1400}}{2}=\frac{50\pm\sqrt{1100}}{2}=\frac{50\pm10\sqrt{11}}{2}=25\pm5\sqrt{11}$$

$x=-5\left(-5+\sqrt{11}\right)\approx8$ units or

$x=5\left(5+\sqrt{11}\right)\approx42$ units.

75. $-\sqrt{-100}=-\sqrt{100}\cdot\sqrt{-1}=-10i$

76. $-\sqrt{-49}=-\sqrt{49}\cdot\sqrt{-1}=-7i$

77. $5+\sqrt{-20}=5+\sqrt{20}\cdot\sqrt{-1}=5+2\sqrt{5}i$

78. $4-\sqrt{-60}=4-\sqrt{60}\cdot\sqrt{-1}=4-2\sqrt{15}i$

79. $\frac{-3-\sqrt{-27}}{3}=\frac{-3-\sqrt{27}\cdot\sqrt{-1}}{3}$
$=\frac{-3-3\sqrt{3}i}{3}$
$=\frac{-3}{3}-\frac{3\sqrt{3}i}{3}$
$=-1-\sqrt{3}i$

80. $\frac{-5+\sqrt{-75}}{10}=\frac{-5+\sqrt{75}\cdot\sqrt{-1}}{10}$
$=\frac{-5+5\sqrt{3}i}{10}$
$=\frac{-5}{10}+\frac{5\sqrt{3}i}{10}$
$=-\frac{1}{2}+\frac{\sqrt{3}}{2}i$

81. $(4+10i)+(-3-2i)=4+10i-3-2i$
$=4-3+10i-2i$
$=1+8i$

82. $(-3+5i)+(3+6i)=-3+5i+3+6i$
$=-3+3+5i+6i$
$=0+11i$
$=11i$

83. $(-8+5i)-(8-5i)=-8+5i-8+5i$
$=-8-8+5i+5i$
$=-16+10i$

84. $(-13+i)-(13-12i)=-13+i-13+12i$
$=-13-13+i+12i$
$=-26+13i$

85. $\left(2+\sqrt{-9}\right)-\left(12-\sqrt{-16}\right)$
$=(2+3i)-(12-4i)$
$=2+3i-12+4i$
$=2-12+3i+4i$
$=-10+7i$

86. $\left(12-\sqrt{-25}\right)+\left(2+\sqrt{-1}\right)=(12-5i)+(2+i)$
$=12-5i+2+i$
$=12+2-5i+i$
$=14-4i$

87. $2i(6+5i)=2i(6)+2i(5i)$
$=12i+10i^2$
$=12i+10(-1)$
$=12i-10$
$=-10+12i$

88. $-3i(1-i)=-3i(1)-3i(-i)$
$=-3i+3i^2$
$=-3i+3(-1)$
$=-3i-3$
$=-3-3i$

89. $(4-5i)(3-2i)$
$=4(3)+4(-2i)-5i(3)-5i(-2i)$
$=12-8i-15i+10i^2$
$=12-23i+10(-1)$
$=12-23i-10$
$=2-23i$

90. $(12+i)(2-3i)$
$=12(2)+12(-3i)+i(2)+i(-3i)$
$=24-36i+2i-3i^2$
$=24-36i+2i-3(-1)$
$=24-34i+3$
$=27-34i$

91. $(2-3i)^2=2^2+2(2)(-3i)+(3i)^2$
$=4-12i+9i^2$
$=4-12i+9(-1)$
$=4-12i-9$
$=-5-12i$

92. $(3+5i)^2=3^2+2(3)(5i)+(5i)^2$
$=9+30i+25i^2$
$=9+30i+25(-1)$
$=9+30i-25$
$=-16+30i$

93. $3\sqrt{-12}\cdot2\sqrt{-6}=3\cdot2\sqrt{3}i\cdot2\sqrt{6}i$
$=3\cdot2\cdot2\cdot\sqrt{3}\cdot\sqrt{6}\cdot i^2$
$=12\sqrt{18}\cdot(-1)$
$=12\cdot3\sqrt{2}\cdot(-1)$
$=-36\sqrt{2}$

94. $4\sqrt{-10}\cdot 3\sqrt{-8} = 4\sqrt{10}i\cdot 3\cdot 2\sqrt{2}i$
$= 4\cdot 3\cdot 2\cdot\sqrt{10}\cdot\sqrt{2}\cdot i^2$
$= 24\cdot\sqrt{20}\cdot(-1)$
$= 24\cdot 2\sqrt{5}\cdot(-1)$
$= -48\sqrt{5}$

95. $5+\sqrt{-4} = 5+2i$
The conjugate is $5 - 2i$.
$(5+2i)(5-2i) = 5^2 + 2^2 = 25+4 = 29$

96. $3-\sqrt{-9} = 3-3i$
The conjugate is $3 + 3i$.
$(3-3i)(3+3i) = 3^2 + 3^2 = 9+9 = 18$

97. $\frac{9-3i}{9i} = \frac{9-3i}{9i}\cdot\frac{-9i}{-9i}$
$= \frac{(9-3i)(-9i)}{(9i)(-9i)}$
$= \frac{-81i+27i^2}{-81(i^2)}$
$= \frac{-81i+27(-1)}{-81(-1)}$
$= \frac{-81i-27}{81}$
$= \frac{-9i-3}{9}$
$= -\frac{3}{9}-\frac{9}{9}i$
$= -\frac{1}{3}-i$

98. $\frac{-6+4i}{2i} = \frac{-6+4i}{2i}\cdot\frac{-2i}{-2i}$
$= \frac{(-6+4i)(-2i)}{(2i)(-2i)}$
$= \frac{12i-8i^2}{-4i^2}$
$= \frac{12i-8(-1)}{-4(-1)}$
$= \frac{12i+8}{4}$
$= \frac{8}{4}+\frac{12}{4}i$
$= 2+3i$

99. $\frac{5}{3-\sqrt{-16}} = \frac{5}{3-4i}$
$= \frac{5}{3-4i}\cdot\frac{3+4i}{3+4i}$
$= \frac{5(3+4i)}{3^2+4^2}$
$= \frac{15+20i}{9+16}$
$= \frac{15+20i}{25}$
$= \frac{15}{25}+\frac{20}{25}i$
$= \frac{3}{5}+\frac{4}{5}i$

100. $\frac{6}{4+\sqrt{-4}} = \frac{6}{4+2i}$
$= \frac{6}{4+2i}\cdot\frac{4-2i}{4-2i}$
$= \frac{6(4-2i)}{4^2+2^2}$
$= \frac{24-12i}{16+4}$
$= \frac{24-12i}{20}$
$= \frac{24}{20}-\frac{12}{20}i$
$= \frac{6}{5}-\frac{3}{5}i$

101. $\frac{5i}{1-i} = \frac{5i}{1-i}\cdot\frac{1+i}{1+i}$
$= \frac{5i(1+i)}{1^2+1^2}$
$= \frac{5i+5i^2}{2}$
$= \frac{5i+5(-1)}{2}$
$= \frac{5i-5}{2}$
$= -\frac{5}{2}+\frac{5}{2}i$

102. $\frac{2i}{3+5i} = \frac{2i}{3+5i} \cdot \frac{3-5i}{3-5i}$
$= \frac{2i(3-5i)}{3^2+5^2}$
$= \frac{6i-10i^2}{9+25}$
$= \frac{6i-10(-1)}{34}$
$= \frac{6i+10}{34}$
$= \frac{10}{34} + \frac{6}{34}i$
$= \frac{5}{17} + \frac{3}{17}i$

103. $2x^2 + 2x + 5 = 0$
$a = 2, b = 2, c = 5$
$x = \frac{-2 \pm \sqrt{2^2 - 4(2)(5)}}{2(2)}$
$= \frac{-2 \pm \sqrt{4-40}}{4}$
$= \frac{-2 \pm \sqrt{-36}}{4}$
$= \frac{-2 \pm 6i}{4}$
$= \frac{2(-1 \pm 3i)}{2 \cdot 2}$
$= \frac{-1 \pm 3i}{2}$
$= -\frac{1}{2} \pm \frac{3}{2}i$
The solution set is $\left\{-\frac{1}{2} - \frac{3}{2}i, -\frac{1}{2} + \frac{3}{2}i\right\}$.

104. $2x^2 - 6x + 9 = 0$
$a = 2, b = -6, c = 9$
$x = \frac{-(-6) \pm \sqrt{(-6)^2 - 4(2)(9)}}{2(2)}$
$= \frac{6 \pm \sqrt{36-72}}{4}$
$= \frac{6 \pm \sqrt{-36}}{4}$
$= \frac{6 \pm 6i}{4}$
$= \frac{6}{4} \pm \frac{6}{4}i$
$= \frac{3}{2} \pm \frac{3}{2}i$
The solution set is $\left\{\frac{3}{2} - \frac{3}{2}i, \frac{3}{2} + \frac{3}{2}i\right\}$.

105. $(x-2)^2 + 12 = 4$
$(x-2)^2 = -8$
$x - 2 = \pm\sqrt{-8}$
$x - 2 = \pm 2\sqrt{2}i$
$x = 2 \pm 2\sqrt{2}i$
The solution set is $\{2 - 2\sqrt{2}i, 2 + 2\sqrt{2}i\}$.

106. $(x+3)^2 + 9 = -23$
$(x+3)^2 = -32$
$x + 3 = \pm\sqrt{-32}$
$x + 3 = \pm 4\sqrt{2}i$
$x = -3 \pm 4\sqrt{2}i$
The solution is $\{-3 - 4\sqrt{2}i, -3 + 4\sqrt{2}i\}$.

Chapter 9 Test

1. $4x^2 - 12 = 24$
$4x^2 = 36$
$x^2 = 9$
$x = \pm\sqrt{9}$
$x = \pm 3$
The solution set is $\{-3, 3\}$.

2. $x^2+6x-40=0$

$x^2+6x=40$

$x^2+6x+\left(\frac{6}{2}\right)^2=40+\left(\frac{6}{2}\right)^2$

$x^2+6x+3^2=40+3^2$

$x^2+6x+9=40+9$

$(x+3)^2=49$

$x+3=\pm\sqrt{49}$

$x+3=\pm 7$

$x=-3\pm 7$

$x=-3-7=-10$ or $x=-3+7=4$

The solution set is $\{-10, 4\}$.

3. $x^2+6x+1=0$

$a=1,\ b=6,\ c=1$

$x=\frac{-6\pm\sqrt{6^2-4(1)(1)}}{2(1)}$

$=\frac{-6\pm\sqrt{36-4}}{2}$

$=\frac{-6\pm\sqrt{32}}{2}$

$=\frac{-6\pm 4\sqrt{2}}{2}$

$=-3\pm 2\sqrt{2}$

The solution set is $\left\{-3-2\sqrt{2}, -3+2\sqrt{2}\right\}$.

4. $(x-2)(x+1)=4$

$x^2-x-2=4$

$x^2-x-6=0$

$(x-3)(x+2)=0$

$x-3=0$ or $x+2=0$

$x=3$ or $x=-2$

The solution set is $\{-2, 3\}$.

5. $2n=4+\frac{3}{n}$

$n(2n)=n\left(4+\frac{3}{n}\right)$

$2n^2=4n+3$

$2n^2-4n-3=0$

$a=2, b=-4, c=-3$

$n=\frac{-(-4)\pm\sqrt{(-4)^2-4(2)(-3)}}{2(2)}$

$=\frac{4\pm\sqrt{16+24}}{4}$

$=\frac{4\pm\sqrt{40}}{4}$

$=\frac{4\pm 2\sqrt{10}}{4}$

$=\frac{2\pm\sqrt{10}}{2}$

The solution set is $\left\{\frac{2-\sqrt{10}}{2}, \frac{2+\sqrt{10}}{2}\right\}$.

6. $\frac{1}{2}t^2=\frac{1}{8}+\frac{3}{4}t$

$8\left(\frac{1}{2}t^2\right)=8\left(\frac{1}{8}+\frac{3}{4}t\right)$

$4t^2=1+6t$

$4t^2-6t-1=0$

$a=4, b=-6, c=-1$

$t=\frac{-(-6)\pm\sqrt{(-6)^2-4(4)(-1)}}{2(4)}$

$=\frac{6\pm\sqrt{36+16}}{8}$

$=\frac{6\pm\sqrt{52}}{8}$

$=\frac{6\pm 2\sqrt{13}}{8}$

$=\frac{3\pm\sqrt{13}}{4}$

The solution set is $\left\{\frac{3-\sqrt{13}}{4}, \frac{3+\sqrt{13}}{4}\right\}$.

7. (a) $x=5-3x^2$

$3x^2+x-5=0$

$a=3, b=1, c=-5$

$b^2-4ac=1^2-4(3)(-5)=1+60=61$

(b) Quadratic equation

(c) $61>0$ and 61 is not a perfect square.

8. $\dfrac{-3-\sqrt{-81}}{6}=\dfrac{-3-9i}{6}$
$=\dfrac{3(-1-3i)}{3\cdot 2}$
$=\dfrac{-1-3i}{2}$
$=-\dfrac{1}{2}-\dfrac{3}{2}i$

9. $(-12+4i)+(-2-8i)=-12+4i-2-8i$
$=-12-2+4i-8i$
$=-14-4i$

10. $\left(3-\sqrt{-4}\right)-\left(2-\sqrt{-25}\right)=(3-2i)-(2-5i)$
$=3-2i-2+5i$
$=3-2-2i+5i$
$=1+3i$

11. $2\sqrt{-3}\cdot 3\sqrt{-12}=2\sqrt{3}i\cdot 3\cdot 2\sqrt{3}i$
$=2\cdot 3\cdot 2\cdot\sqrt{3}\cdot\sqrt{3}\cdot i^2$
$=12\cdot 3\cdot(-1)$
$=-36$

12. $(1-4i)(3-2i)$
$=1(3)+1(-2i)-4i(3)-4i(-2i)$
$=3-2i-12i+8i^2$
$=3-14i+8(-1)$
$=3-14i-8$
$=-5-14i$

13. $(4-6i)^2=4^2+2(4)(-6i)+(6i)^2$
$=16-48i+36i^2$
$=16-48i+36(-1)$
$=16-48i-36$
$=-20-48i$

14. $\dfrac{4+12i}{3i}=\dfrac{4+12i}{3i}\cdot\dfrac{-3i}{-3i}$
$=\dfrac{(4+12i)(-3i)}{(3i)(-3i)}$
$=\dfrac{-12i-36i^2}{-9i^2}$
$=\dfrac{-12i-36(-1)}{-9(-1)}$
$=\dfrac{-12i+36}{9}$
$=\dfrac{36}{9}-\dfrac{12}{9}i$
$=4-\dfrac{4}{3}i$

15. $\dfrac{6}{5+3i}=\dfrac{6}{5+3i}\cdot\dfrac{5-3i}{5-3i}$
$=\dfrac{6(5-3i)}{5^2+3^2}$
$=\dfrac{30-18i}{25+9}$
$=\dfrac{30-18i}{34}$
$=\dfrac{30}{34}-\dfrac{18}{34}i$
$=\dfrac{15}{17}-\dfrac{9}{17}i$

16. $x^2-4x+5=0$
$a=1,\ b=-4,\ c=5$
$x=\dfrac{-(-4)\pm\sqrt{(-4)^2-4(1)(5)}}{2(1)}$
$=\dfrac{4\pm\sqrt{16-20}}{2}$
$=\dfrac{4\pm\sqrt{-4}}{2}$
$=\dfrac{4\pm 2i}{2}$
$=2\pm i$
The solution set is $\{2-i,\ 2+i\}$.

17. $2p^2+8p+15=0$

$a = 2, b = 8, c = 15$

$$p=\frac{-8\pm\sqrt{8^2-4(2)(15)}}{2(2)}$$
$$=\frac{-8\pm\sqrt{64-120}}{4}$$
$$=\frac{-8\pm\sqrt{-56}}{4}$$
$$=\frac{-8\pm 2\sqrt{14}i}{4}$$
$$=-2\pm\frac{\sqrt{14}}{2}i$$

The solution set is $\left\{-2-\frac{\sqrt{14}}{2}i, -2+\frac{\sqrt{14}}{2}i\right\}$.

18. (a) Let w be the width. Then $w + 2$ is the length.

$$a^2+b^2=c^2$$
$$w^2+(w+2)^2=16^2$$
$$w^2+w^2+4w+4=256$$
$$2w^2+4w-252=0$$
$$w^2+2w-126=0$$

$a = 1, b = 2, c = -126$

$$w=\frac{-2\pm\sqrt{2^2-4(1)(-126)}}{2(1)}$$
$$=\frac{-2\pm\sqrt{4+504}}{2}$$
$$=\frac{-2\pm\sqrt{508}}{2}$$
$$=\frac{-2\pm 2\sqrt{127}}{2}$$
$$=-1\pm\sqrt{127}$$

$w=-1-\sqrt{127}$ or $w=-1+\sqrt{127}$

Since length cannot be negative, the width is $-1+\sqrt{127}$ feet and the length is $-1+\sqrt{127}+2=1+\sqrt{127}$ feet.

(b) width $=-1+\sqrt{127}\approx 10.3$ feet

length $=1+\sqrt{127}\approx 12.3$ feet

19. (a) $h=-16t^2+100t+40$

$$0=-16t^2+100t+40$$

$a = -16, b = 100, c = 40$

$$t=\frac{-100\pm\sqrt{100^2-4(-16)(40)}}{2(-16)}$$
$$=\frac{-100\pm\sqrt{10,000+2560}}{-32}$$
$$=\frac{-100\pm\sqrt{12,560}}{-32}$$
$$=\frac{-100\pm 4\sqrt{785}}{-32}$$
$$=\frac{-4\left(25\pm\sqrt{785}\right)}{-4(8)}$$
$$=\frac{25\pm\sqrt{785}}{8}$$

$t=\frac{25+\sqrt{785}}{8}$ (since time has to be positive)

It will hit the water line at $\frac{25+\sqrt{785}}{8}$ seconds.

(b) $\frac{25+\sqrt{785}}{8}\approx 6.6$ seconds

20. (a) Let w be the width. Then $2w - 100$ is the length. The area = lw.

$$200,000=(2w-100)(w)$$
$$200,000=2w^2-100w$$
$$0=2w^2-100w-200,000$$
$$0=w^2-50w-100,000$$

$a = 1, b = -50, c = -100,000$

$$w=\frac{-(-50)\pm\sqrt{(-50)^2-4(1)(-100,000)}}{2(1)}$$
$$=\frac{50\pm\sqrt{2500+400,000}}{2}$$
$$=\frac{50\pm\sqrt{402,500}}{2}$$
$$=\frac{50\pm 50\sqrt{161}}{2}$$
$$=25\pm 25\sqrt{161}$$

Sin ce length is positive, the width is $25+25\sqrt{161}$ yards and the length is

$$2\left(25+25\sqrt{161}\right)-100$$
$$=-50+50\sqrt{161}$$ yards.

(b) width $= 25+25\sqrt{161} \approx 342.21$ yards

length $= -50+50\sqrt{161} \approx 584.43$ yards

Cumulative Review Chapters 1–9

1. $\dfrac{-8+4(-2)^2}{-1-(-1)} = \dfrac{-8+4(4)}{-1+1} = \dfrac{-8+16}{0}$ is undefined.

2. $\dfrac{\frac{1}{2}\cdot\frac{8}{3}-\frac{3}{4}\cdot\frac{16}{9}}{\left(\frac{1}{2}\right)^2} = \dfrac{\frac{4}{3}-\frac{4}{3}}{\frac{1}{4}} = \dfrac{0}{\frac{1}{4}} = 0$

3. $\dfrac{4-7}{-3-(-9)} = \dfrac{-3}{-3+9} = \dfrac{-3}{6} = -\dfrac{1}{2}$

4.
$$\begin{aligned}
&(3x^2y^{-2})^3(-x^3y^2)^2\\
&= (3^3x^6y^{-6})[(-1)^2x^6y^4]\\
&= 27x^6y^{-6}\cdot 1x^6y^4\\
&= 27x^{6+6}y^{-6+4}\\
&= 27x^{12}y^{-2}\\
&= \frac{27x^{12}}{y^2}
\end{aligned}$$

5.
$$\begin{aligned}
\left(\frac{2x^2y}{6xy^{-2}}\right)^{-2} &= \left(\frac{x^{2-1}y^{1-(-2)}}{3}\right)^{-2}\\
&= \left(\frac{xy^3}{3}\right)^{-2}\\
&= \frac{x^{-2}y^{-6}}{3^{-2}}\\
&= \frac{3^2}{x^2y^6}\\
&= \frac{9}{x^2y^6}
\end{aligned}$$

6. $-2x^2-3x+1$

(a) For $x=-2$,
$$\begin{aligned}
-2(-2)^2-3(-2)+1 &= -2(4)-3(-2)+1\\
&= -8+6+1\\
&= -1
\end{aligned}$$

(b) The leading coefficient is -2.

(c) The degree of the polynomial is 2, the highest degree.

(d) It is a trinomial because it has three terms.

7. (a)
$$\begin{aligned}
3x-2y &= 12\\
-2y &= -3x+12\\
y &= \frac{-3x+12}{-2}\\
y &= \frac{-3x}{-2}+\frac{12}{-2}\\
y &= \frac{3}{2}x-6
\end{aligned}$$

(b) The coefficient of x is $\dfrac{3}{2}$.

(c) The constant is -6.

8.
$$\begin{aligned}
3-5(x+2) &= -3(x-7)\\
3-5x-10 &= -3x+21\\
-5x-7 &= -3x+21\\
3x-5x-7 &= 3x-3x+21\\
-2x-7 &= 21\\
-2x-7+7 &= 21+7\\
-2x &= 28\\
\frac{-2x}{-2} &= \frac{28}{-2}\\
x &= -14
\end{aligned}$$

The solution set is $\{-14\}$.

9.
$$\begin{aligned}
\frac{x-1}{4} &= \frac{2x+1}{5}\\
20\left(\frac{x-1}{4}\right) &= 20\left(\frac{2x+1}{5}\right)\\
5(x-1) &= 4(2x+1)\\
5x-5 &= 8x+4\\
-8x+5x-5 &= -8x+8x+4\\
-3x-5 &= 4\\
-3x-5+5 &= 4+5\\
-3x &= 9\\
\frac{-3x}{-3} &= \frac{9}{-3}\\
x &= -3
\end{aligned}$$

The solution set is $\{-3\}$.

10. $\frac{x^2}{3}+\frac{x}{6}=1$

$$6\left(\frac{x^2}{3}+\frac{x}{6}\right)=6(1)$$
$$2x^2+x=6$$
$$2x^2+x-6=0$$
$$(2x-3)(x+2)=0$$
$$2x-3=0 \quad \text{or} \quad x+2=0$$
$$x=\frac{3}{2} \quad \text{or} \quad x=-2$$

The solution set is $\left\{-2, \frac{3}{2}\right\}$.

11. $x(x-2)=3$

$$x^2-2x=3$$
$$x^2-2x-3=0$$
$$(x-3)(x+1)=0$$
$$x-3=0 \quad \text{or} \quad x+1=0$$
$$x=3 \quad \text{or} \quad x=-1$$

The solution set is $\{-1, 3\}$.

12.

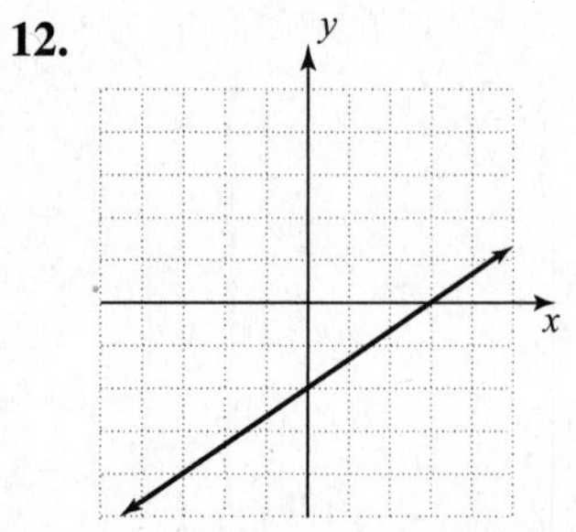

13. $(x_1, y_1)=(-2, 4); (x_2, y_2)=(4, -5)$

$$m=\frac{y_2-y_1}{x_2-x_1}=\frac{-5-4}{4-(-2)}=\frac{-9}{6}=-\frac{3}{2}$$
$$y-y_1=m(x-x_1)$$
$$y-4=-\frac{3}{2}[x-(-2)]$$
$$y-4=-\frac{3}{2}x-3$$
$$y=-\frac{3}{2}x+1$$

14. $\begin{cases}2x-3y=-12 & (1)\\ 3x+2y=-5 & (2)\end{cases}$

Multiply equation (1) by 2 and equation (2) by 3.

$$\begin{cases}4x-6y=-24\\ 9x+6y=-15\end{cases}$$
$$13x \quad =-39$$
$$x=-3$$

Let $x=-3$ in equation (1).

$$2(-3)-3y=-12$$
$$-6-3y=-12$$
$$-3y=-6$$
$$y=2$$

The solution is $(-3, 2)$.

15.

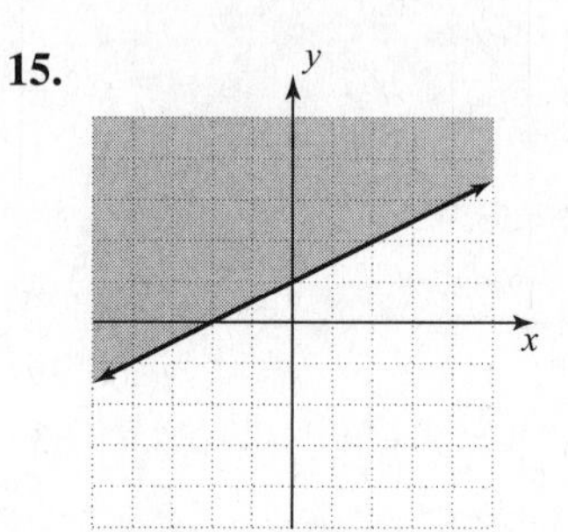

16. $x=ky$

$$2.5=k(16)$$
$$0.15625=k$$
$$x=0.15625y$$
$$15=0.15625y$$
$$96=y$$

17. $6h^2-11h-10=(2h-5)(3h+2)$

18. $3x^3+24=3(x^3+8)$

$$=3(x^3+2^3)$$
$$=3(x+2)(x^2-2x+4)$$

19. $xy+2x-3y-6=x(y+2)-3(y+2)$

$$=(y+2)(x-3)$$

20. $\dfrac{6x^3+5x^2-x+3}{2x+1}=3x^2+x-1+\dfrac{4}{2x+1}$

$$\begin{array}{r} 3x^2+x-1 \\ 2x+1\overline{)6x^3+5x^2-x+3} \\ \underline{6x^3+3x^2} \\ 2x^2-x \\ \underline{2x^2+x} \\ -2x+3 \\ \underline{-2x-1} \\ 4 \end{array}$$

21. $\dfrac{3x^2+6x-4}{2x}=\dfrac{3x^2}{2x}+\dfrac{6x}{2x}-\dfrac{4}{2x}=\dfrac{3}{2}x+3-\dfrac{2}{x}$

22. $-8^{2/3}=-\left(\sqrt[3]{8}\right)^2=-2^2=-4$

23. $9^{-3/2}=\dfrac{1}{9^{3/2}}=\dfrac{1}{\left(\sqrt{9}\right)^3}=\dfrac{1}{3^3}=\dfrac{1}{27}$

24. $34,000,000,000=3.4\times10^{10}$

25. $3.04\times10^{-4}=0.000304$

26. $\left(-3+\sqrt{2}\right)^2=(-3)^2=2(-3)\sqrt{2}+\left(\sqrt{2}\right)^2$
$=9-6\sqrt{2}+2$
$=11-6\sqrt{2}$

27.
$$\begin{aligned}\frac{4+3i}{6i}&=\frac{4+3i}{6i}\cdot\frac{-6i}{-6i}\\&=\frac{(4+3i)(-6i)}{(6i)(-6i)}\\&=\frac{-24i-18i^2}{-36i^2}\\&=\frac{-24i-18(-1)}{-36(-1)}\\&=\frac{-24i+18}{36}\\&=\frac{18}{36}-\frac{24}{36}i\\&=\frac{1}{2}-\frac{2}{3}i\end{aligned}$$

28. Let x be the number of \$20 seats sold; then $580-x$ is the number of \$50 seats sold.
$$\begin{aligned}20x+50(580-x)&=17,300\\20x+29,000-50x&=17,300\\-30x&=-11,700\\x&=\frac{-11,700}{-30}=390\end{aligned}$$
$580-x=580-390=190$
There were 390 general seating tickets and 190 reserved tickets sold.

29.
$$\begin{aligned}D&=\frac{k}{p}\\240&=\frac{k}{3.50}\\840&=k\\D&=\frac{840}{p}\\D&=\frac{840}{4}=210\end{aligned}$$
210 bags will be sold if the price is \$4 per bag.

30. $h=-2t^2+9t+6$

(a) When $t=4$,
$$\begin{aligned}h&=-2(4)^2+9(4)+6\\&=-2(16)+36+6\\&=-32+42\\&=10\end{aligned}$$
Misha is 10 feet high.

(b) When Misha lands, $h=0$.
$0=-2t^2+9t+6$
$a=-2$, $b=9$, $c=6$
$$\begin{aligned}t&=\frac{-9\pm\sqrt{9^2-4(-2)(6)}}{2(-2)}\\&=\frac{-9\pm\sqrt{81+48}}{-4}\\&=\frac{9\pm\sqrt{129}}{4}\end{aligned}$$
Since time is positive, it takes $\dfrac{9+\sqrt{129}}{4}$ seconds to reach the pit.

(c) $\dfrac{9+\sqrt{129}}{4}\approx5.09$ seconds

Chapter 10

Section 10.1

Preparing for Quadratic Equations in Two Variables

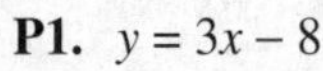

P1. $y = 3x - 8$

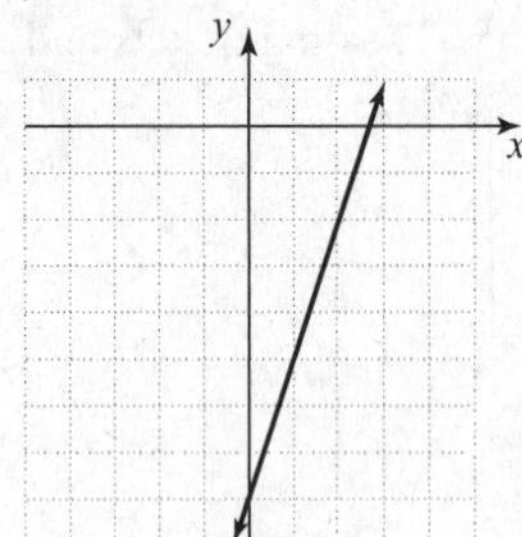

P2. $x = -1$

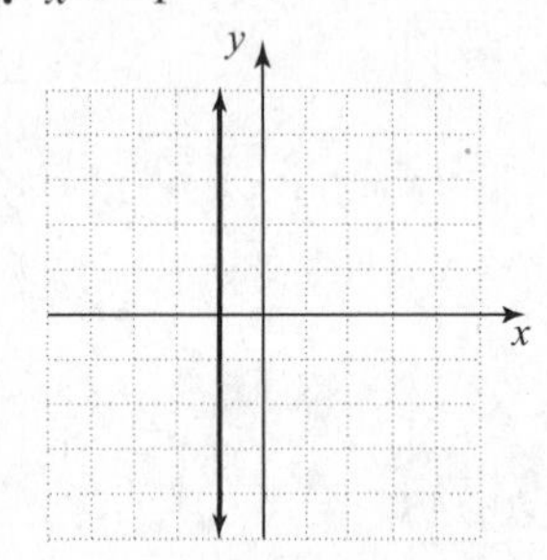

P3.
$$x^2 - 6x + 8 = 0$$
$$(x-2)(x-4) = 0$$
$$x-2=0 \quad \text{or} \quad x-4=0$$
$$x = 2 \quad \text{or} \quad x = 4$$
The solution set is {2, 4}.

P4. $2x^2 + x - 5 = 0$

$a = 2, b = 1, c = -5$
$$x = \frac{-1 \pm \sqrt{1^2 - 4(2)(-5)}}{2(2)}$$
$$= \frac{-1 \pm \sqrt{1+40}}{4}$$
$$= \frac{-1 \pm \sqrt{41}}{4}$$
The solution set is $\left\{\frac{-1-\sqrt{41}}{4}, \frac{-1+\sqrt{41}}{4}\right\}$.

P5. $3x - 5y = 30$

x-intercept, let $y = 0$:
$$3x - 5(0) = 30$$
$$3x = 30$$
$$x = 10$$
y-intercept, let $x = 0$:
$$3(0) - 5y = 30$$
$$-5y = 30$$
$$y = -6$$
The intercepts are (10, 0) and (0, −6).

10.1 Quick Checks

1. A <u>quadratic equation</u> is an equation in the form $y = ax^2 + bx + c$ where a, b, and c are real numbers, and $a \neq 0$.

2.

x	$y = x^2 - 3$	y
−3	$(-3)^2 - 3 = 6$	6
−2	$(-2)^2 - 3 = 1$	1
−1	$(-1)^2 - 3 = -2$	−2
0	$0^2 - 3 = -3$	−3
1	$1^2 - 3 = -2$	−2
2	$2^2 - 3 = 1$	1
3	$3^2 - 3 = 6$	6

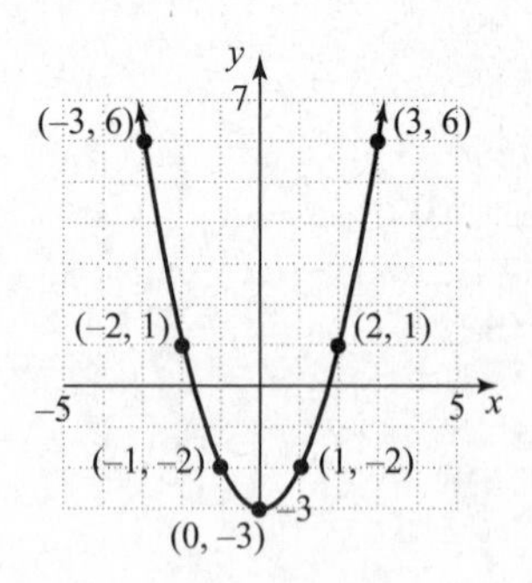

3.

x	$y = x^2 + 4x - 5$	y
−5	$(-5)^2 + 4(-5) - 5 = 0$	0
−4	$(-4)^2 + 4(-4) - 5 = -5$	−5
−3	$(-3)^2 + 4(-3) - 5 = -8$	−8
−2	$(-2)^2 + 4(-2) - 5 = -9$	−9
−1	$(-1)^2 + 4(-1) - 5 = -8$	−8
0	$0^2 + 4(0) - 5 = -5$	−5
1	$1^2 + 4(1) - 5 = 0$	0

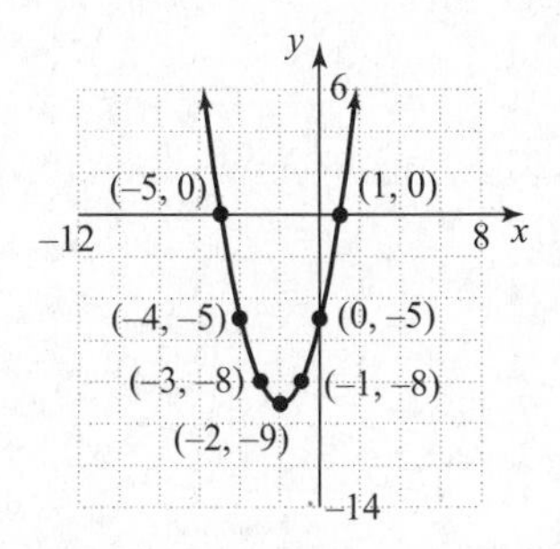

4.

x	$y = -x^2 + 8x - 15$	y
1	$-1^2 + 8(1) - 15 = -8$	−8
2	$-2^2 + 8(2) - 15 = -3$	−3
3	$-3^2 + 8(3) - 15 = 0$	0
4	$-4^2 + 8(4) - 15 = 1$	1
5	$-5^2 + 8(5) - 15 = 0$	0
6	$-6^2 + 8(6) - 15 = -3$	−3
7	$-7^2 + 8(7) - 15 = -8$	−8

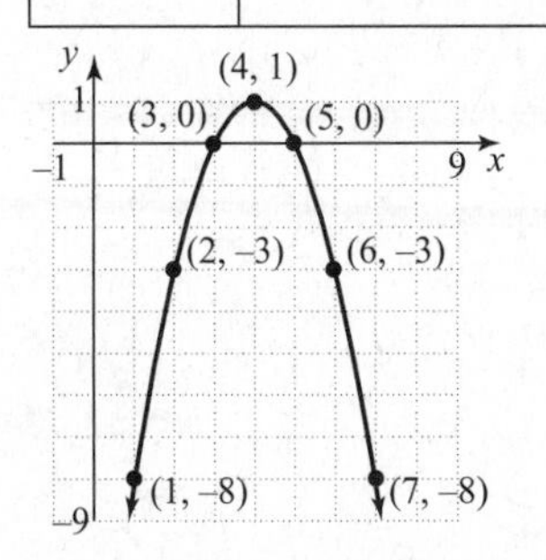

5. True

6. True

7. For a quadratic function $y = ax^2 + bx + c,\ a \neq 0,$ if $b^2 - 4ac = 0,$ the graph has <u>one</u> x-intercept.

8. $y = x^2 + 5x + 6$
$a = 1, b = 5, c = 6$
$b^2 - 4ac = 5^2 - 4(1)(6) = 25 - 24 = 1$
Since the discriminant > 0, there are two x-intercepts. To find the x-intercepts, let $y = 0$.
$$0 = x^2 + 5x + 6$$
$$0 = (x+2)(x+3)$$
$$x + 2 = 0 \quad \text{or} \quad x + 3 = 0$$
$$x = -2 \quad \text{or} \quad x = -3$$
The x-intercepts are −2 and −3. To find the y-intercept, let $x = 0$.
$$y = 0^2 + 5(0) + 6$$
$$y = 6$$
The y-intercept is 6.

9. $y = -3x^2 + 4x - 5$
$a = -3, b = 4, c = -5$
$b^2 - 4ac = 4^2 - 4(-3)(-5) = 16 - 60 = -44$
Since the discriminant < 0, there are no x-intercepts. To find the y-intercept, let $x = 0$.
$$y = -3(0)^2 + 4(0) - 5$$
$$y = -5$$
The y-intercept is −5.

10. $y = 4x^2 + 4x + 1$
$a = 4, b = 4, c = 1$
$b^2 - 4ac = 4^2 - 4(4)(1) = 16 - 16 = 0$
Since the discriminant = 0, there is one x-intercept. To find the x-intercept, let $y = 0$.

$$0 = 4x^2 + 4x + 1$$
$$0 = (2x+1)^2$$
$$0 = 2x + 1$$
$$-2x = 1$$
$$x = -\frac{1}{2}$$

The x-intercept is $-\frac{1}{2}$. To find the y-intercept, let $x = 0$.

$$y = 4(0)^2 + 4(0) + 1$$
$$y = 1$$

The y-intercept is 1.

11. When graphing a quadratic equation, if the leading coefficient is positive, the graph will open <u>up</u> and if the leading coefficient is negative, the graph will open <u>down</u>.

12. The lowest or highest point of the graph of $y = ax^2 + bx + c$ is called the <u>vertex</u>. Its x-coordinate can be determined using the formula $\underline{x = -\frac{b}{2a}}$.

13. True

14. $y = 3x^2 - 6x + 2$

$a = 3, b = -6, c = 2$

(a) Since $a > 0$, the parabola opens up.

(b) $x = -\frac{b}{2a} = -\frac{-6}{2(3)} = \frac{6}{6} = 1$

$y = 3(1)^2 - 6(1) + 2 = 3 - 6 + 2 = -1$

The vertex is $(1, -1)$.

(c) The axis of symmetry is $x = -\frac{b}{2a} = 1$.

15. $y = -2x^2 - 12x - 7$

$a = -2, b = -12, c = -7$

(a) Since $a < 0$, the parabola opens down.

(b) $x = -\frac{b}{2a} = -\frac{-12}{2(-2)} = \frac{12}{-4} = -3$

$$y = -2(-3)^2 - 12(-3) - 7$$
$$= -2 \cdot 9 + 36 - 7$$
$$= -18 + 29$$
$$= 11$$

The vertex is $(-3, 11)$.

(c) The axis of symmetry is $x = -\frac{b}{2a} = -3$.

16. $y = x^2 + 6x + 5$

$a = 1, b = 6, c = 5$

Since $a > 0$, the parabola opens up.

$x = -\frac{b}{2a} = -\frac{6}{2(1)} = -3$

$y = (-3)^2 + 6(-3) + 5 = 9 - 18 + 5 = -4$

The vertex is $(-3, -4)$.

The axis of symmetry is $x = -\frac{b}{2a} = -3$.

Let $x = 0$.

$$y = 0^2 + 6(0) + 5$$
$$y = 5$$

The y-intercept is 5.

$b^2 - 4ac = 6^2 - 4(1)(5) = 36 - 20 = 16$

Since the discriminant > 0, there are two x-intercepts. Let $y = 0$.

$$0 = x^2 + 6x + 5$$
$$0 = (x+5)(x+1)$$

$x + 5 = 0$ or $x + 1 = 0$

$x = -5$ or $x = -1$

The x-intercepts are -5 and -1.

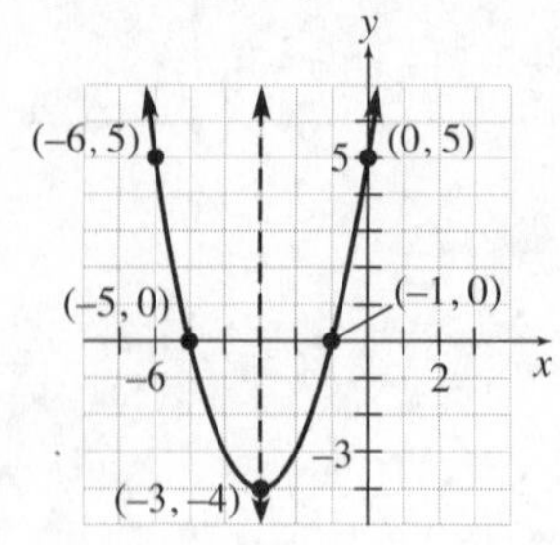

17. $y = -x^2 - 2x + 3$

$a = -1$, $b = -2$, $c = 3$

Since $a < 0$, the parabola opens down.

$$x = -\frac{b}{2a} = -\frac{-2}{2(-1)} = \frac{2}{-2} = -1$$

$$y = -(-1)^2 - 2(-1) + 3 = -1 + 2 + 3 = 4$$

The vertex is $(-1, 4)$.

The axis of symmetry is $x = -\dfrac{b}{2a} = -1$.

Let $x = 0$.

$$y = -0^2 - 2(0) + 3$$
$$y = 3$$

The y-intercept is 3.

$$b^2 - 4ac = (-2)^2 - 4(-1)(3) = 4 + 12 = 16$$

Since the discriminant is > 0, there are two x-intercepts. Let $y = 0$.

$$0 = -x^2 - 2x + 3$$
$$0 = x^2 + 2x - 3$$
$$0 = (x+3)(x-1)$$
$$x + 3 = 0 \quad \text{or} \quad x - 1 = 0$$
$$x = -3 \quad \text{or} \quad x = 1$$

The x-intercepts are -3 and 1.

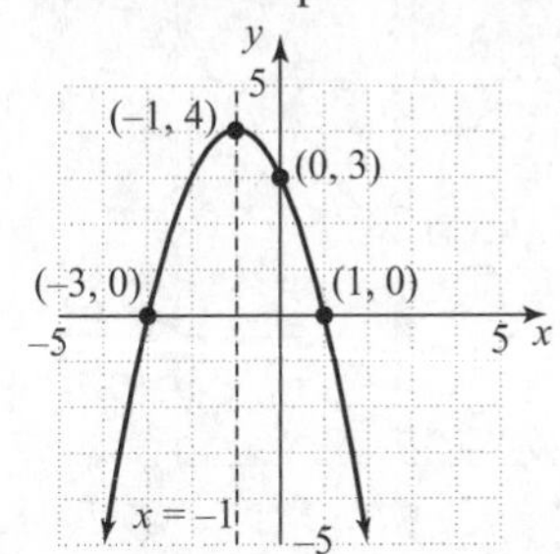

18. $y = 2x^2 + 8x - 1$

$a = 2$, $b = 8$, $c = -1$

Since $a > 0$, the parabola opens up.

$$x = -\frac{b}{2a} = -\frac{8}{2(2)} = -\frac{8}{4} = -2$$

$$\begin{aligned} y &= 2(-2)^2 + 8(-2) - 1 \\ &= 2 \cdot 4 - 16 - 1 \\ &= 8 - 17 \\ &= -9 \end{aligned}$$

The vertex is at $(-2, -9)$. The axis of symmetry is the line $x = -2$.

Let $x = 0$.

$$y = 2(0)^2 + 8(0) - 1 = -1$$

The y-intercept is -1.

Let $y = 0$.

$$0 = 2x^2 + 8x - 1$$

$$\begin{aligned} x &= \frac{-8 \pm \sqrt{8^2 - 4(2)(-1)}}{2(2)} \\ &= \frac{-8 \pm \sqrt{64 + 8}}{4} \\ &= \frac{-8 \pm \sqrt{72}}{4} \\ &= \frac{-8 \pm 6\sqrt{2}}{4} \\ &= -2 \pm \frac{3\sqrt{2}}{2} \end{aligned}$$

The x-intercepts are $-2 - \dfrac{3\sqrt{2}}{2} \approx -4.12$ and $-2 + \dfrac{3\sqrt{2}}{2} \approx 0.12$.

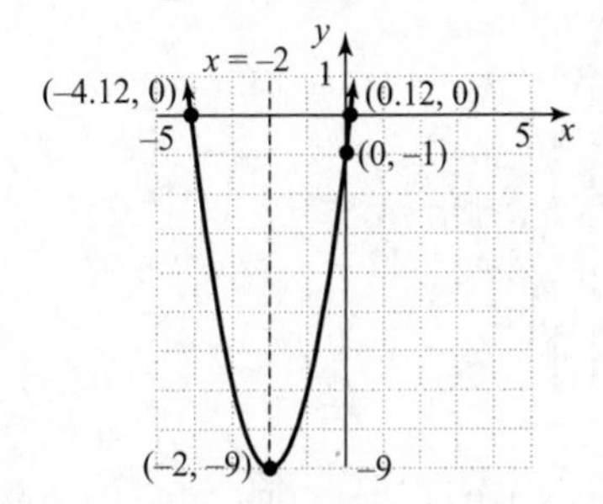

19. $y = -3x^2 - 12x + 5$

$a = -3$, $b = -12$, $c = 5$

Since $a < 0$, the parabola opens down.

$$x = -\frac{b}{2a} = -\frac{-12}{2(-3)} = \frac{12}{-6} = -2$$

$$\begin{aligned} y &= -3(-2)^2 - 12(-2) + 5 \\ &= -3 \cdot 4 + 24 + 5 \\ &= -12 + 29 \\ &= 17 \end{aligned}$$

The vertex is at $(-2, 17)$. The axis of symmetry is the line $x = -2$.

Let $x = 0$.

$$y = -3(0)^2 - 12(0) + 5 = 5$$

The y-intercept is 5.

Let $y = 0$.

$$0 = -3x^2 - 12x + 5$$

$$x = \frac{12 \pm \sqrt{(-12)^2 - 4(-3)(5)}}{2(-3)} = \frac{12 \pm \sqrt{144+60}}{-6} = \frac{12 \pm \sqrt{204}}{-6} = \frac{12 \pm 2\sqrt{51}}{-6} = -2 \pm \frac{\sqrt{51}}{3}$$

The x-intercepts are $-2 - \frac{\sqrt{51}}{3} \approx -4.38$ and $-2 + \frac{\sqrt{51}}{3} \approx 0.38$.

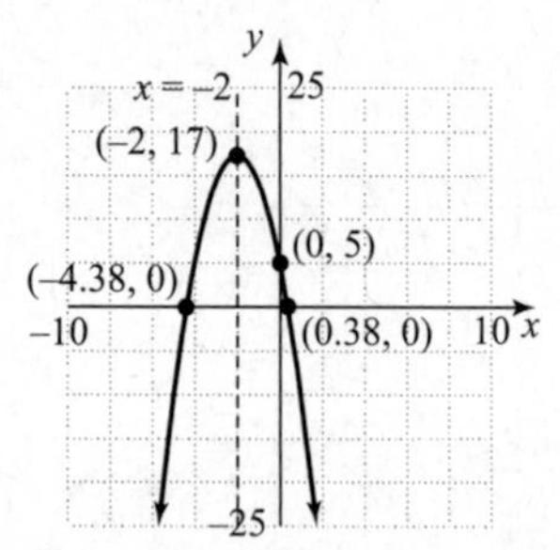

20. If the vertex is the highest point on the graph of a parabola, the y-coordinate of the vertex represents the maximum value of the equation.

21. True

22. $y = 2x^2 - 8x + 1$

$a = 2, b = -8, c = 1$

Since $a > 0$, the parabola opens up. Therefore, it has a minimum.

$$x = -\frac{b}{2a} = -\frac{-8}{2(2)} = \frac{8}{4} = 2$$

$$y = 2(2)^2 - 8(2) + 1 = 2 \cdot 4 - 16 + 1 = 8 - 15 = -7$$

The minimum value of the equation is −7.

23. $y = -x^2 + 10x + 8$

$a = -1, b = 10, c = 8$

Since $a < -1$, the parabola opens down. Therefore, it has a maximum.

$$x = -\frac{b}{2a} = -\frac{10}{2(-1)} = -\frac{10}{-2} = 5$$

$$y = -5^2 + 10(5) + 8 = -25 + 50 + 8 = 33$$

The maximum value of the equation is 33.

24. (a) $h = -16t^2 + 128t + 15$

$a = -16, b = 128, c = 15$

$$t = -\frac{b}{2a} = -\frac{128}{2(-16)} = -\frac{128}{-32} = 4$$

The time is 4 seconds.

(b) Let $t = 4$.

$$h = -16(4)^2 + 128(4) + 15 = -16 \cdot 16 + 512 + 15 = -256 + 527 = 271$$

10.1 Exercises

25.

x	$y = x^2$	y
−3	$(-3)^2 = 9$	9
−2	$(-2)^2 = 4$	4
−1	$(-1)^2 = 1$	1
0	$0^2 = 0$	0
1	$1^2 = 1$	1
2	$2^2 = 4$	4
3	$3^2 = 9$	9

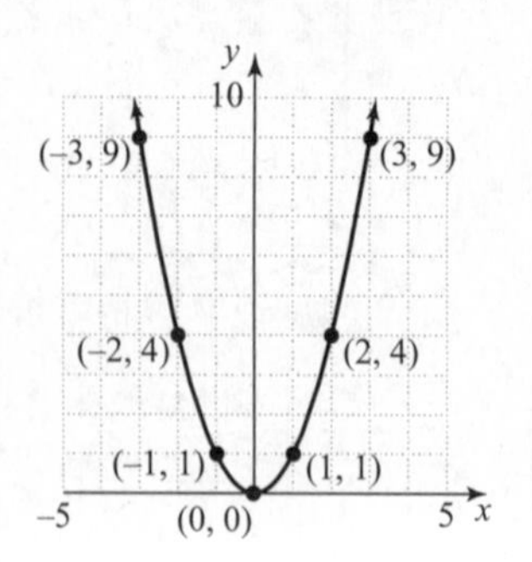

27.

x	$y=-x^2+4$	y
-3	$-(-3)^2+4=-5$	-5
-2	$-(-2)^2+4=0$	0
-1	$-(-1)^2+4=3$	3
0	$-0^2+4=4$	4
1	$-1^2+4=3$	3
2	$-2^2+4=0$	0
3	$-3^2+4=-5$	-5

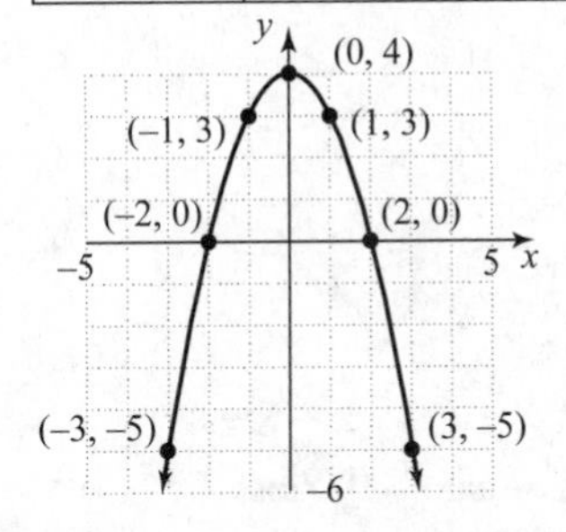

29.

x	$y=-x^2+3x+4$	y
-2	$-(-2)^2+3(-2)+4=-6$	-6
-1	$-(-1)^2+3(-1)+4=0$	0
0	$-0^2+3(0)+4=4$	4
1	$-1^2+3(1)+4=6$	6
2	$-2^2+3(2)+4=6$	6
3	$-3^2+3(3)+4=4$	4
4	$-4^2+3(4)+4=0$	0

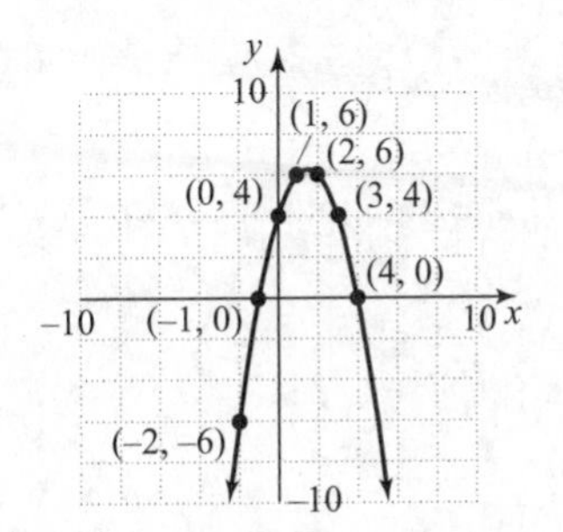

31.

x	$y=2x^2+5x-3$	y
-4	$2(-4)^2+5(-4)-3=9$	9
-3	$2(-3)^2+5(-3)-3=0$	0
-2	$2(-2)^2+5(-2)-3=-5$	-5
-1	$2(-1)^2+5(-1)-3=-6$	-6
0	$2(0)^2+5(0)-3=-3$	-3
1	$2(1)^2+5(1)-3=4$	4
2	$2(2)^2+5(2)-3=15$	15

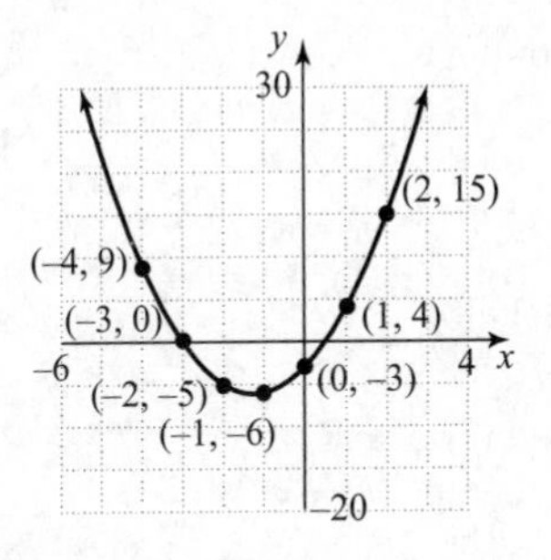

33. $y=-x^2+6x-8$

$a=-1$, $b=6$, $c=-8$

$b^2-4ac=6^2-4(-1)(-8)=36-32=4$

Since the discriminant > 0, there are two x-intercepts. To find the x-intercepts, let $y = 0$.

$$0=-x^2+6x-8$$
$$0=-1(x^2-6x+8)$$
$$0=-1(x-4)(x-2)$$
$$x-4=0 \quad \text{or} \quad x-2=0$$
$$x=4 \quad \text{or} \quad x=2$$

The x-intercepts are 2 and 4.
To find the y-intercept, let $x = 0$.

$$y=-0^2+6(0)-8$$
$$y=-8$$

The y-intercept is -8.

35. $y=6x^2-x-1$

$a=6$, $b=-1$, $c=-1$

$b^2-4ac=(-1)^2-4(6)(-1)=1+24=25$

Since the discriminant > 0, there are two x-intercepts. To find the x-intercepts, let $y = 0$.

$0 = 6x^2 - x - 1$
$0 = (3x+1)(2x-1)$
$3x+1=0 \quad \text{or} \quad 2x-1=0$
$x = -\frac{1}{3} \quad \text{or} \quad x = \frac{1}{2}$

The x-intercepts are $-\frac{1}{3}$ and $\frac{1}{2}$.
To find the y-intercept, let $x = 0$.
$y = 6(0)^2 - 0 - 1$
$y = -1$
The y-intercept is -1.

37. $y = x^2 + 3x$
$a = 1, b = 3, c = 0$
$b^2 - 4ac = 3^2 - 4(1)(0) = 9 - 0 = 9$
Since the discriminant > 0, there are two x-intercepts. To find the x-intercepts, let $y = 0$.
$0 = x^2 + 3x$
$0 = x(x+3)$
$x = 0 \quad \text{or} \quad x+3=0$
$x = -3$
The x-intercepts are -3 and 0.
To find the y-intercept, let $x = 0$.
$y = 0^2 + 3(0) = 0$
The y-intercept is 0.

39. $y = -x^2 - 5$
$a = -1, b = 0, c = -5$
$b^2 - 4ac = 0^2 - 4(-1)(-5) = 0 - 20 = -20$
Since the discriminant < 0, there are no x-intercepts. To find the y-intercept, let $x = 0$.
$y = -0^2 - 5$
$y = -5$
The y-intercept is -5.

41. $y = -2x^2 + x + 2$
$a = -2, b = 1, c = 2$
$b^2 - 4ac = 1^2 - 4(-2)(2) = 1 + 16 = 17$
Since the discriminant > 0, there are two x-intercepts. To find the x-intercepts, let $y = 0$.

$x = \frac{-b \pm \sqrt{b^2 - 4ac}}{2a}$
$x = \frac{-1 \pm \sqrt{17}}{2(-2)}$
$x = \frac{-1 \pm \sqrt{17}}{-4}$
$x = \frac{1 \pm \sqrt{17}}{4}$

The x-intercepts are $\frac{1-\sqrt{17}}{4} \approx -0.78$ and $\frac{1+\sqrt{17}}{4} \approx 1.28$.
To find the y-intercept, let $x = 0$.
$y = -2(0)^2 + 0 + 2$
$y = 2$
The y-intercept is 2.

43. $y = x^2 + 4x + 2$
$a = 1, b = 4, c = 2$
$b^2 - 4ac = 4^2 - 4(1)(2) = 16 - 8 = 8$
Since the discriminant > 0, there are two x-intercepts. To find the x-intercepts, let $y = 0$.
$0 = x^2 + 4x + 2$
$x = \frac{-b \pm \sqrt{b^2 - 4ac}}{2a}$
$= \frac{-4 \pm \sqrt{8}}{2(1)}$
$= \frac{-4 \pm 2\sqrt{2}}{2}$
$= -2 \pm \sqrt{2}$
The x-intercepts are $-2 - \sqrt{2} \approx -3.41$ and $-2 + \sqrt{2} \approx -0.59$.
To find the y-intercept, let $x = 0$.
$y = 0^2 + 4(0) + 2$
$y = 2$
The y-intercept is 2.

45. **(a)** $y = x^2 + 6$
$a = 1, b = 0, c = 6$
Since $a > 0$, the parabola opens up.

(b) $x = -\frac{b}{2a} = -\frac{0}{2(1)} = 0$
$y = 0^2 + 6 = 6$
The vertex is (0, 6).

(c) The axis of symmetry is $x = -\frac{b}{2a} = 0.$

47. (a) $y = -x^2 + 8x - 15$

$a = -1$, $b = 8$, $c = -15$

Since $a < 0$, the parabola opens down.

(b) $x = -\frac{b}{2a} = -\frac{8}{2(-1)} = 4$

$y = -4^2 + 8(4) - 15 = -16 + 32 - 15 = 1$

The vertex is (4, 1).

(c) The axis of symmetry is $x = -\frac{b}{2a} = 4.$

49. (a) $y = -2x^2 + 3x + 5$

$a = -2$, $b = 3$, $c = 5$

Since $a < 0$, the parabola opens down.

(b) $x = -\frac{b}{2a} = -\frac{3}{2(-2)} = \frac{3}{4}$

$$y = -2\left(\frac{3}{4}\right)^2 + 3\left(\frac{3}{4}\right) + 5$$
$$= -\frac{9}{8} + \frac{18}{8} + \frac{40}{8}$$
$$= \frac{49}{8}$$

The vertex is $\left(\frac{3}{4}, \frac{49}{8}\right)$.

(c) The axis of symmetry is $x = -\frac{b}{2a} = \frac{3}{4}.$

51. (a) $y = 3x^2 - 8x + 4$

$a = 3$, $b = -8$, $c = 4$

Since $a > 0$, the parabola opens up.

(b) $x = -\frac{b}{2a} = -\frac{-8}{2(3)} = \frac{8}{6} = \frac{4}{3}$

$$y = 3\left(\frac{4}{3}\right)^2 - 8\left(\frac{4}{3}\right) + 4$$
$$= \frac{16}{3} - \frac{32}{3} + \frac{12}{3}$$
$$= -\frac{4}{3}$$

The vertex is $\left(\frac{4}{3}, -\frac{4}{3}\right)$.

(c) The axis of symmetry is $x = -\frac{b}{2a} = \frac{4}{3}.$

53. $y = x^2 + 2x - 3$

$a = 1$, $b = 2$, $c = -3$

Since $a > 0$, the parabola opens up.

$b^2 - 4ac = 2^2 - 4(1)(-3) = 4 + 12 = 16$

Since the discriminant > 0, there are two x-intercepts. Let $y = 0$.

$$0 = x^2 + 2x - 3$$
$$0 = (x+3)(x-1)$$
$$x + 3 = 0 \quad \text{or} \quad x - 1 = 0$$
$$x = -3 \quad \text{or} \quad x = 1$$

The x-intercepts are −3 and 1.

Let $x = 0$.

$$y = 0^2 + 2(0) - 3$$
$$y = -3$$

The y-intercept is −3.

The axis of symmetry is $x = -\frac{b}{2a} = -\frac{2}{2(1)} = -1.$

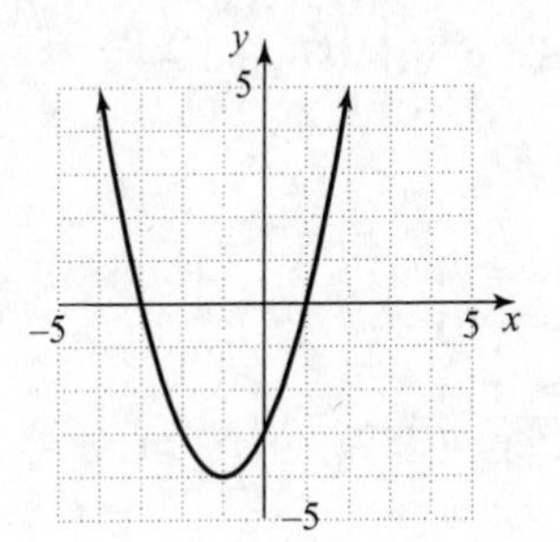

55. $y = -x^2 - 2x$

$a = -1$, $b = -2$, $c = 0$

Since $a < 0$, the parabola opens down.

$b^2 - 4ac = (-2)^2 - 4(-1)(0) = 4$

Since the discriminant > 0, there are two x-intercepts. Let $y = 0$.

$$0 = -x^2 - 2x$$
$$0 = -x(x+2)$$
$$-x = 0 \quad \text{or} \quad x + 2 = 0$$
$$x = 0 \quad \text{or} \quad x = -2$$

The x-intercepts are −2 and 0.

Let $x = 0$.

$y = -0^2 - 2(0) = 0$

The y-intercept is 0.

The axis of symmetry is

$$x = -\frac{b}{2a} = -\frac{-2}{2(-1)} = -1.$$

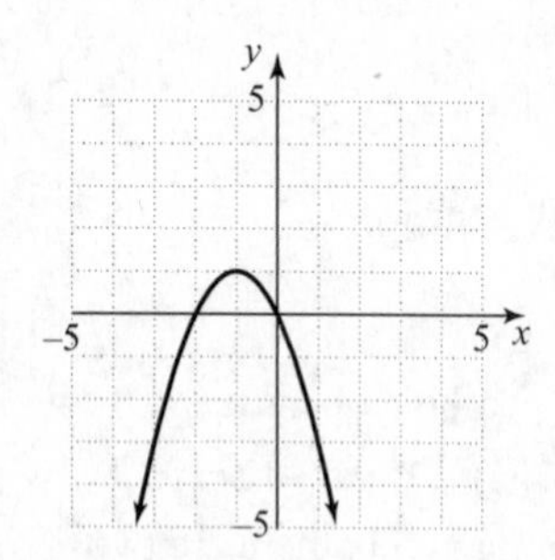

57. $y = -x^2 - 8x - 9$

$a = -1, b = -8, c = -9$

Since $a < 0$, the parabola opens down.

$b^2 - 4ac = (-8)^2 - 4(-1)(-9) = 64 - 36 = 28$

Since the discriminant > 0, there are two x-intercepts. Let $y = 0$.

$0 = -x^2 - 8x - 9$

$$x = \frac{-b \pm \sqrt{b^2 - 4ac}}{2a}$$

$$x = \frac{-(-8) \pm \sqrt{28}}{2(-1)} = \frac{8 \pm 2\sqrt{7}}{-2} = -4 \pm \sqrt{7}$$

The x-intercepts are $-4 - \sqrt{7}$ and $-4 + \sqrt{7}$.

Let $x = 0$.

$y = -0^2 - 8(0) - 9 = -9$

The y-intercept is -9.

The axis of symmetry is

$$x = -\frac{b}{2a} = -\frac{-8}{2(-1)} = -4.$$

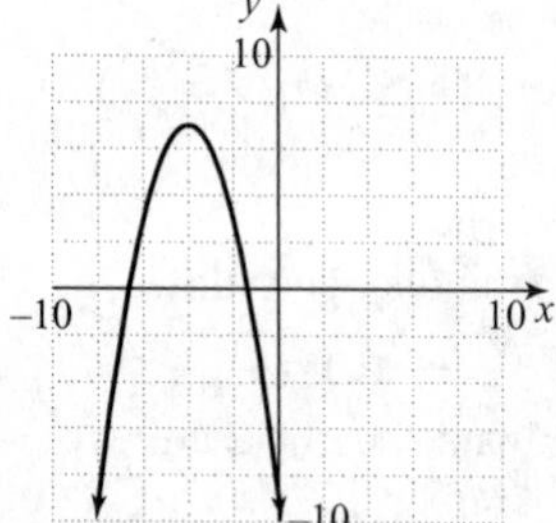

59. $y = x^2 - 9$

$a = 1, b = 0, c = -9$

Since $a > 0$, the parabola opens up.

$b^2 - 4ac = 0^2 - 4(1)(-9) = 0 + 36 = 36$

Since the discriminant > 0, there are two x-intercepts. Let $y = 0$.

$0 = x^2 - 9$

$0 = (x + 3)(x - 3)$

$x + 3 = 0$ or $x - 3 = 0$

$x = -3$ or $x = 3$

The x-intercepts are -3 and 3.

Let $x = 0$.

$y = 0^2 - 9 = -9$

The y-intercept is -9.

The axis of symmetry is $x = -\frac{b}{2a} = -\frac{0}{2(1)} = 0.$

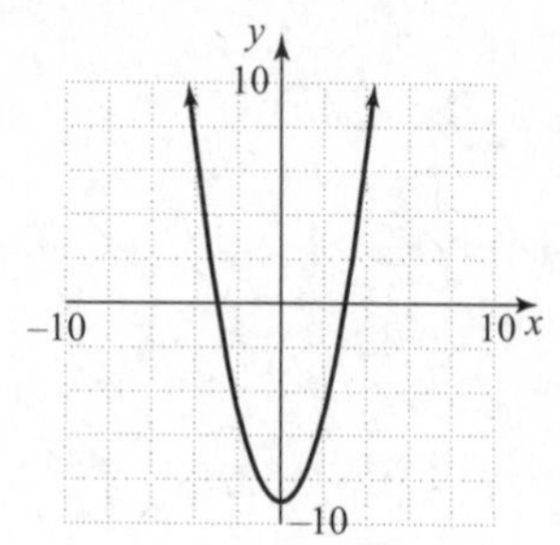

61. $y = x^2 - 7x + 10$

$a = 1, b = -7, c = 10$

Since $a > 0$, the parabola opens up.

$b^2 - 4ac = (-7)^2 - 4(1)(10) = 49 - 40 = 9$

Since the discriminant > 0, there are two x-intercepts. Let $y = 0$.

$0 = x^2 - 7x + 10$

$0 = (x - 5)(x - 2)$

$x - 5 = 0$ or $x - 2 = 0$

$x = 5$ or $x = 2$

The x-intercepts are 2 and 5.

Let $x = 0$.

$y = 0^2 - 7(0) + 10 = 10$

The y-intercept is 10.

The axis of symmetry is $x = -\frac{b}{2a} = -\frac{-7}{2(1)} = \frac{7}{2}.$

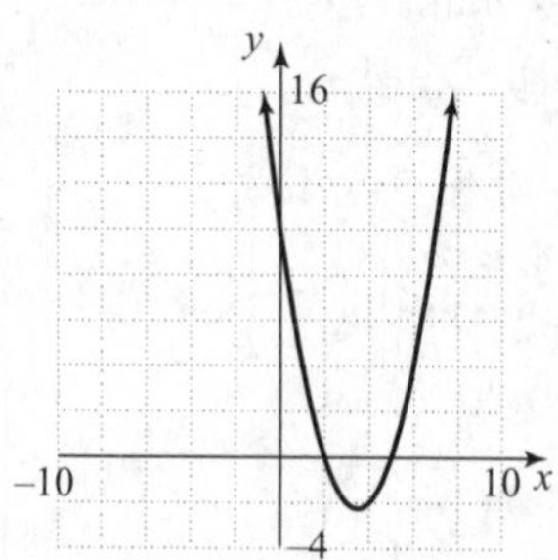

63. $y = x^2 - 6x + 9$

$a = 1, b = -6, c = 9$

Since $a > 0$, the parabola opens up.

$b^2 - 4ac = (-6)^2 - 4(1)(9) = 36 - 36 = 0$

Since the discriminant = 0, there is one x-intercept. Let $y = 0$.

$0 = x^2 - 6x + 9$
$0 = (x-3)(x-3)$
$x-3=0$ or $x-3=0$
$x=3$ or $x=3$
The x-intercept is 3. Let $x = 0$.
$y = 0^2 - 6(0) + 9 = 9$
The y-intercept is 9.
The axis of symmetry is
$x = -\frac{b}{2a} = -\frac{-6}{2(1)} = \frac{6}{2} = 3.$
$y = 3^2 - 6(3)9 = 9 - 18 + 9 = 0$
The vertex is (3, 0).

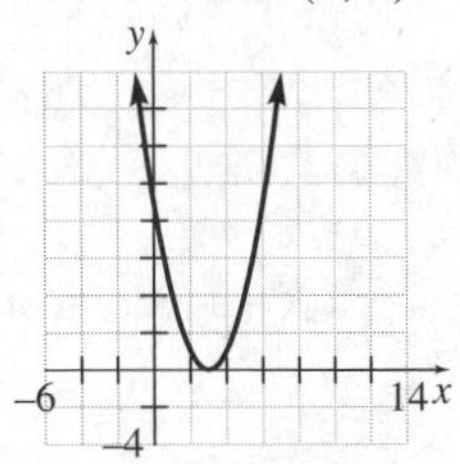

65. $y = -2x^2 - 4x + 5$
$a = -2, b = -4, c = 5$
Since $a < 0$, the parabola opens down.
$b^2 - 4ac = (-4)^2 - 4(-2)(5) = 16 + 40 = 56$
Since the discriminant > 0, there are two x-intercepts. Let $y = 0$.
$0 = -2x^2 - 4x + 5$
$x = \frac{-b \pm \sqrt{b^2 - 4ac}}{2a}$
$x = \frac{-(-4) \pm \sqrt{56}}{2(-2)} = \frac{4 \pm 2\sqrt{14}}{-4} = \frac{-2 \pm \sqrt{14}}{2}$
The x-intercepts are $\frac{-2-\sqrt{14}}{2}$ and $\frac{-2+\sqrt{14}}{2}$.
Let $x = 0$.
$y = -2(0)^2 - 4(0) + 5 = 5$
The y-intercept is 5.
The axis of symmetry is
$x = -\frac{b}{2a} = -\frac{-4}{2(-2)} = -1.$

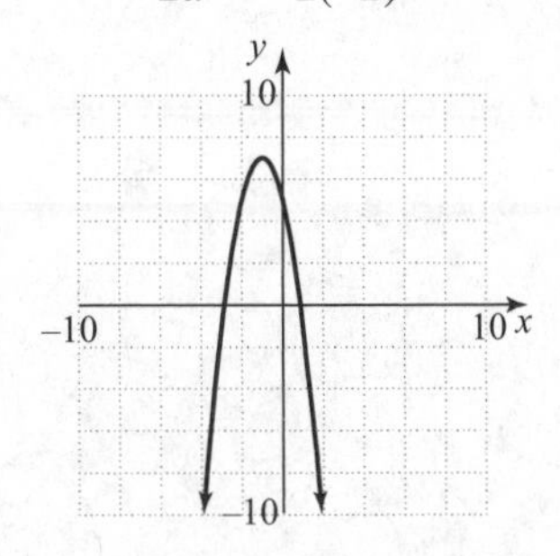

67. $y = x^2 - 2x - 5$
$a = 1, b = -2, c = -5$
Since $a > 0$, the parabola opens up.
$b^2 - 4ac = (-2)^2 - 4(1)(-5) = 4 + 20 = 24$
Since the discriminant > 0, there are two x-intercepts. Let $y = 0$.
$0 = x^2 - 2x - 5$
$x = \frac{-b \pm \sqrt{b^2 - 4ac}}{2a}$
$x = \frac{-(-2) \pm \sqrt{24}}{2(1)} = \frac{2 \pm 2\sqrt{6}}{2} = 1 \pm \sqrt{6}$
The x-intercepts are $1-\sqrt{6}$ are $1+\sqrt{6}$.
Let $x = 0$.
$y = 0^2 - 2(0) - 5 = -5$
The y-intercept is –5.
The axis of symmetry is $x = -\frac{b}{2a} = -\frac{(-2)}{2(1)} = 1.$

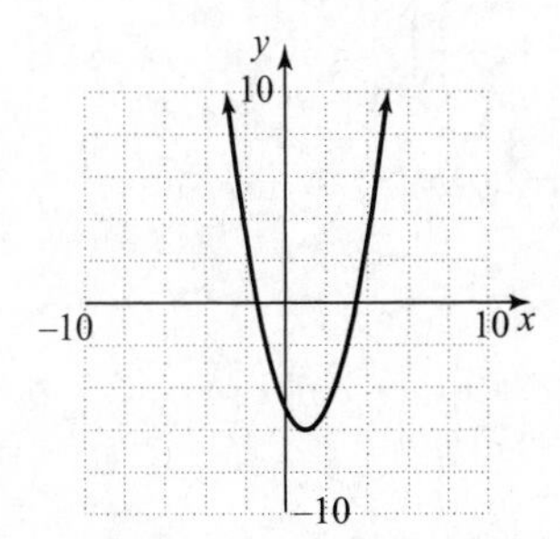

69. $y = x^2 - 6x + 5$
$a = 1, b = -6, c = 5$
Since $a > 0$, the parabola opens up. Therefore, it has a minimum.
$x = -\frac{b}{2a} = -\frac{-6}{2(1)} = 3$
$y = 3^2 - 6(3) + 5 = 9 - 18 + 5 = -4$
The minimum is –4.

71. $y = 2x^2 + 8x + 15$
$a = 2, b = 8, c = 15$
Since $a > 0$, the parabola opens up. Therefore, it has a minimum.
$x = -\frac{b}{2a} = -\frac{8}{2(2)} = -2$
$y = 2(-2)^2 + 8(-2) + 15 = 8 - 16 + 15 = 7$
The minimum is 7.

73. $y = -2x^2 + 6x + 3$

$a = -2$, $b = 6$, $c = 3$

Since $a < 0$, the parabola opens down. Therefore, it has a maximum.

$$x = -\frac{b}{2a} = -\frac{6}{2(-2)} = \frac{3}{2}$$

$$y = -2\left(\frac{3}{2}\right)^2 + 6\left(\frac{3}{2}\right) + 3 = -\frac{9}{2} + \frac{18}{2} + \frac{6}{2} = \frac{15}{2}$$

The maximum is $\frac{15}{2}$.

75. $y = -x^2 - 3x - 7$

$a = -1$, $b = -3$, $c = -7$

Since $a < 0$, the parabola opens down. Therefore, it has a maximum.

$$x = -\frac{b}{2a} = -\frac{-3}{2(-1)} = -\frac{3}{2}$$

$$y = -\left(-\frac{3}{2}\right)^2 - 3\left(-\frac{3}{2}\right) - 7$$
$$= -\frac{9}{4} + \frac{18}{4} - \frac{28}{4}$$
$$= -\frac{19}{4}$$

The maximum is $-\frac{19}{4}$.

77. **(a)** $m = -0.02s^2 + 2s - 25$

$a = -0.02$, $b = 2$, $c = -25$

$$s = -\frac{b}{2a} = -\frac{2}{2(-0.02)} = \frac{2}{0.04} = 50$$

The speed is 50 miles per hour.

(b) Let $x = 50$.

$$m = -0.02(50)^2 + 2(50) - 25$$
$$= -50 + 100 - 25$$
$$= 25$$

The maximum miles per gallon is 25.

79. **(a)** $A = -x^2 + 50x$

$a = -1$, $b = 50$, $c = 0$

$$x = -\frac{b}{2a} = -\frac{50}{2(-1)} = 25$$

The length that maximizes the area is 25 yards.

(b) Let $x = 25$.

$$A = -(25)^2 + 50(25)$$
$$= -625 + 1250$$
$$= 625$$

The maximum area is 625 square yards.

81. **(a)** $A = -2x^2 + 200x$

$a = -2$, $b = 200$, $c = 0$

$$x = -\frac{b}{2a} = -\frac{200}{2(-2)} = 50$$

The length that maximizes the area is 50 feet.

(b) Let $x = 50$.

$$A = -2(50)^2 + 200(50)$$
$$= -5000 + 10,000$$
$$= 5000$$

The maximum area is 5000 square feet.

83. $y = x^2 + 6x + 10$

$a = 1$, $b = 6$, $c = 10$

Since $a > 0$, the parabola opens up.

$$x = -\frac{b}{2a} = -\frac{6}{2(1)} = -3$$

$$y = (-3)^2 + 6(-3) + 10 = 9 - 18 + 10 = 1$$

The vertex is at $(-3, 1)$.

The axis of symmetry is $x = -3$.

Let $x = 0$.

$$y = 0^2 + 6(0) + 10 = 10$$

The y-intercept is 10.

$$b^2 - 4ac = 6^2 - 4(1)(10) = 36 - 40 = -4$$

Since the discriminant < 0, there are no x-intercepts.

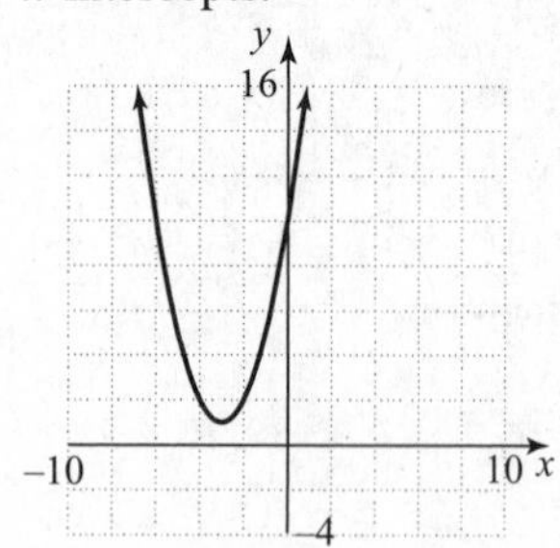

85. $y = -2x^2 + 6x - 7$
$a = -2, b = 6, c = -7$
Since $a < 0$, the parabola opens down.

$$x = -\frac{b}{2a} = -\frac{6}{2(-2)} = \frac{3}{2}$$

$$y = -2\left(\frac{3}{2}\right)^2 + 6\left(\frac{3}{2}\right) - 7$$
$$= -\frac{9}{2} + \frac{18}{2} - \frac{14}{2}$$
$$= -\frac{5}{2}$$

The vertex is $\left(\frac{3}{2}, -\frac{5}{2}\right)$.

The axis of symmetry is $x = \frac{3}{2}$.

Let $x = 0$.
$y = -2(0)^2 + 6(0) - 7 = -7$
The y-intercept is $y = -7$.
$b^2 - 4ac = 6^2 - 4(-2)(-7) = 36 - 56 = -20$
Since the discriminant < 0, there are no x-intercepts.

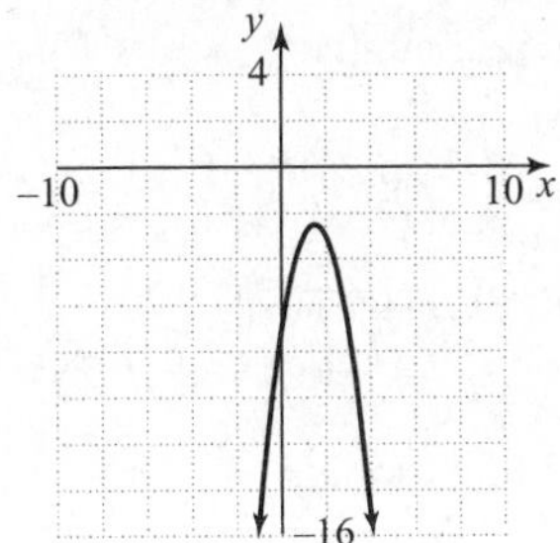

87. To find the number of x-intercepts the graph of a quadratic equation has, find the discriminant, $b^2 - 4ac$. If $b^2 - 4ac$ is positive, the parabola has two x-intercepts, which are found by letting $y = 0$ in the equation and solving for x. If $b^2 - 4ac = 0$, the vertex is the x-intercept. If $b^2 - 4ac$ is negative, there are no x-intercepts.

89. The axis of symmetry is found using the equation $x = -\frac{b}{2a}$. The axis of symmetry serves as a guide in graphing a parabola because every point that lies on the graph of a parabola contains a mirror-image point on the graph that lies on the opposite side of the axis of symmetry.

Section 10.2

Preparing for Relations

P1.

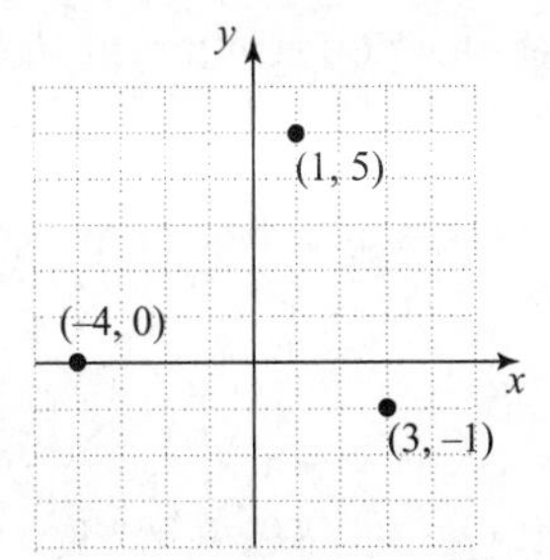

P2. $y = -2x + 6$

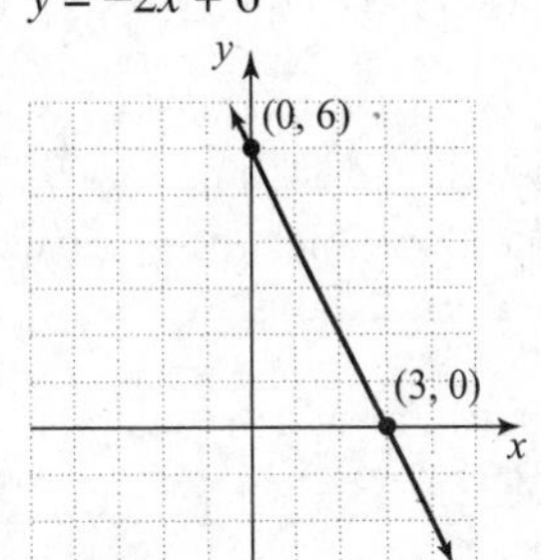

10.2 Quick Checks

1. When the elements in one set are associated with elements of a second set, we have a <u>relation</u>.

2. If a relation exists between the variables x and y, then we say that x <u>corresponds</u> to y, or that y <u>depends</u> on x.

3. {(Max, Nov. 8), (Alesia, Jan. 20), (Trent, Mar. 3), (Yolanda, Nov. 8), (Wanda, July 6), (Elvis, Jan. 8)}

4. {(1, 5), (2, 4), (3, 7), (8, 12)}

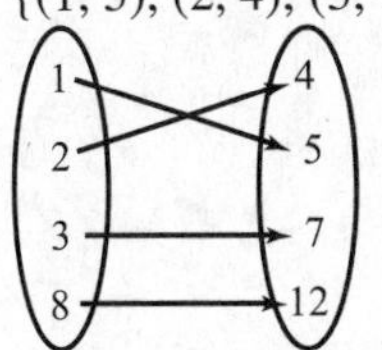

5. The <u>domain</u> of a relation is the set of all inputs to the relation. The <u>range</u> is the set of all outputs of the relation.

6. {(Max, Nov. 8), (Alesia, Jan. 20), (Trent, Mar. 3), (Yolanda, Nov. 8), (Wanda, July 6), (Elvis, Jan. 8)}
The domain is the set of all inputs:
{Max, Alesia, Trent, Yolanda, Wanda, Elvis}.
The range is the set of all outputs:
{Jan. 20, Mar. 3, July 6, Nov. 8, Jan. 8}.

7. The domain is the set of all inputs: {1, 2, 3, 8}. The range is the set of all outputs: {4, 5, 7}.

8. The ordered pairs of the points in the graph are (−3, 1), (−2, −1), (−1, 2), (2, −2), (2, 3), and (3, 4). The domain is the set of all x-coordinates: {−3, −2, −1, 2, 3}. The range is the set of all y-coordinates: {−2, −1, 1, 2, 3, 4}.

9. True

10. False; the domain of the graphed relation consists of all the x-coordinates for which the graph exists. The domain is $\{x | -2 \le x \le 2\}$.

11. Domain: x-coordinates for which the graph exists $(-2 \le x \le 4)$: $\{x | -2 \le x \le 4\}$.
Range: y-coordinates for which the graph exists $(-2 \le y \le 2)$: $\{y | -2 \le y \le 2\}$.

12. Domain: x-coordinates for which the graph exists (all real numbers): $\{x | x$ is a real number$\}$.
Range: y-coordinates for which the graph exists (all real numbers): $\{y | y$ is a real number$\}$.

13.

x	$y = 2x - 5$	y
0	$2(0) - 5 = -5$	−5
3	$2(3) - 5 = 1$	1

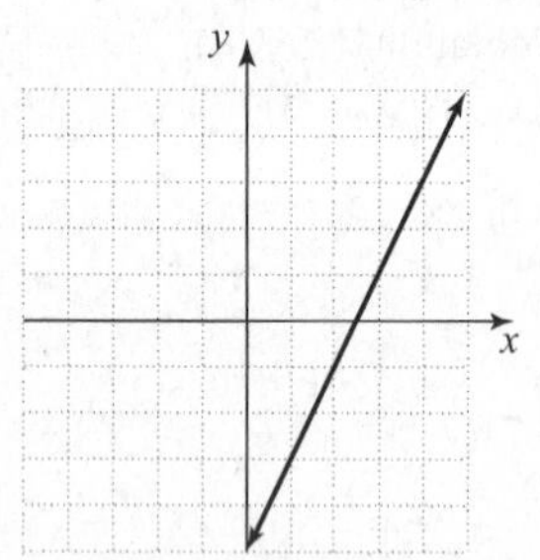

Domain: $\{x | x$ is a real number$\}$
Range: $\{y | y$ is a real number$\}$

14.

x	$y = x^2 - 3$	y
−3	$(-3)^2 - 3 = 6$	6
−2	$(-2)^2 - 3 = 1$	1
−1	$(-1)^2 - 3 = -2$	−2
0	$0^2 - 3 = -3$	−3
1	$1^2 - 3 = -2$	−2
2	$2^2 - 3 = 1$	1
3	$3^2 - 3 = 6$	6

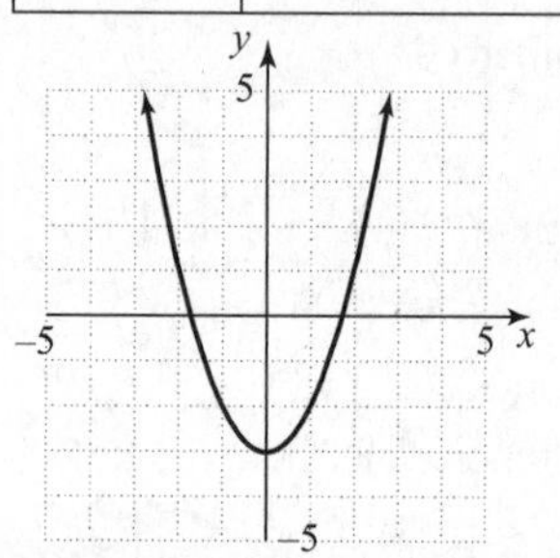

Domain: $\{x | x$ is a real number$\}$
Range: $\{y | y \ge -3\}$

10.2 Exercises

15. {(Physician, 120), (Teacher, 37), (Mathematician, 64), (Computer Engineer, 64), (Lawyer, 82)}
Domain is the set of all inputs.
{Physician, Teacher, Mathematician, Computer Engineer, Lawyer}
Range is the set of all outputs.
{120, 37, 64, 82}

17. {(Spain, 18), (Canada, 18), (Kenya, 14), (Ireland, 3), (Serbia, 3)}
Domain is the set of all inputs.
{Spain, Canada, Kenya, Ireland, Serbia}
Range is the set of all outputs.
{3, 8, 14}

19. {(−1, 3); (−4, 3); (0, 2); (1, 1)}

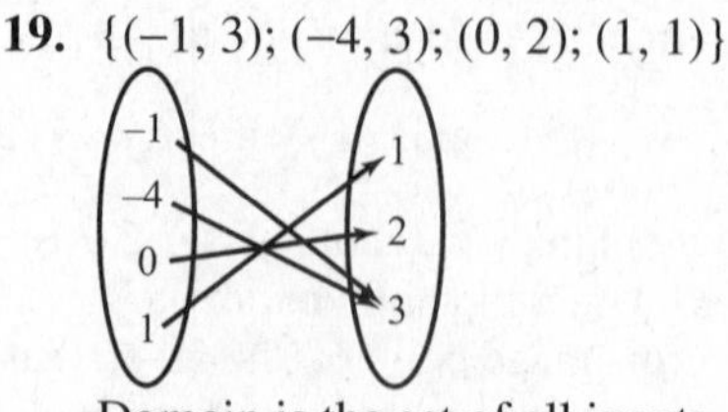

Domain is the set of all inputs.
{−1, −4, 0, 1}
Range is the set of all outputs.
{3, 2, 1}

21. $\{(2, -1); (2, -2); (3, -1); (-2, -2)\}$

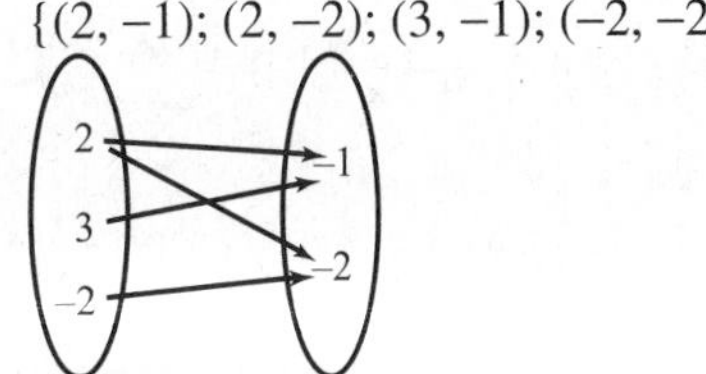

Domain is the set of all inputs.
$\{2, 3, -2\}$
Range is the set of all outputs.
$\{-1, -2\}$

23. $\{(a, z); (b, y); (c, x); (d, w)\}$

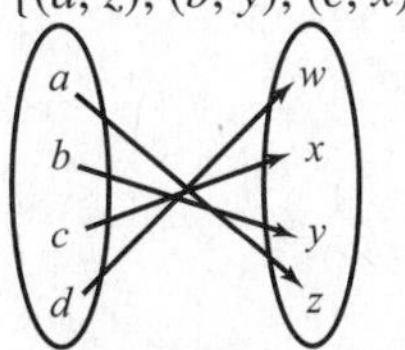

Domain is the set of all inputs.
$\{a, b, c, d\}$
Range is the set of all outputs.
$\{w, x, y, z\}$

25. The ordered pairs on the graph are (−3, 2), (−3, 0), (0, 0), (2, 4), (3, −2). The domain is the set of all x-coordinates: $\{-3, 0, 2, 3\}$.
The range is the set of all y-coordinates: $\{-2, 0, 2, 4\}$.

27. Domain: x-coordinates for which the graph exists (all real numbers): $\{x|x \text{ is a real number}\}$
Range: y-coordinates for which the graph exists (all real numbers):
$\{y|y \text{ is a real number}\}$

29. Domain: x-coordinate for which the graph exists ($x = 3$): $\{3\}$
Range: y-coordinates for which the graph exists (all real numbers):
$\{y|y \text{ is a real number}\}$

31. Domain: x-coordinates for which the graph exists (all real numbers):
$\{x|x \text{ is a real number}\}$
Range: y-coordinates for which the graph exists ($y \le 0$): $\{y|y \le 0\}$

33. Domain: x-coordinates for which the graph exists (all real numbers):
$\{x|x \text{ is a real number}\}$
Range: y-coordinates for which the graph exists ($y \ge -2$): $\{y|y \ge -2\}$

35. Domain: x-coordinates for which the graph exists ($x \ge -1$): $\{x|x \ge -1\}$
Range: y-coordinates for which the graph exists (all real numbers):
$\{y|y \text{ is a real number}\}$

37. Domain: x-coordinates for which the graph exists: $\{x|-1 \le x \le 5\}$
Range: y-coordinates for which the graph exists:
$\{y|-1 \le y \le 5\}$

39. Domain: x-coordinates for which the graph exists: $\{x|-4 \le x \le 4\}$
Range: y-coordinates for which the graph exists:
$\{y|-3 \le y \le 3\}$

41. $y = -2x + 1$

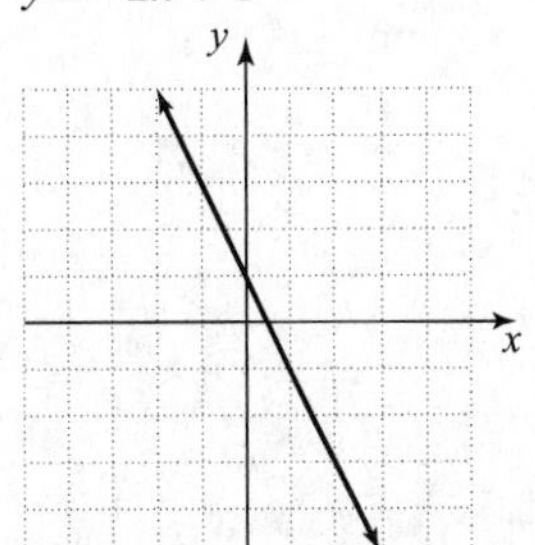

Domain: $\{x|x \text{ is a real number}\}$
Range: $\{y|y \text{ is a real number}\}$

43. $y = -\frac{2}{3}x - 1$

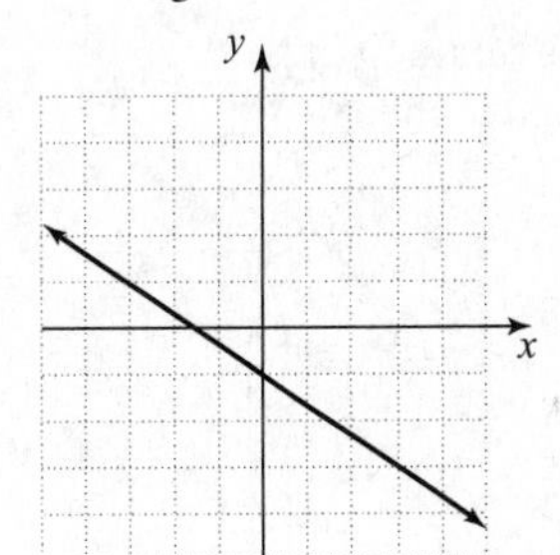

Domain: $\{x|x \text{ is a real number}\}$
Range: $\{y|y \text{ is a real number}\}$

45. $y = 4x - 2x^2$

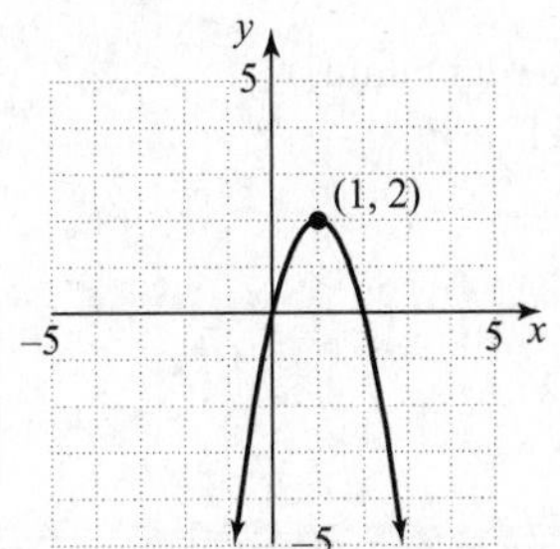

Domain: $\{x|x \text{ is a real number}\}$
Range: $\{y|y \le 2\}$

47. $y = x^2 - 7x + 10$

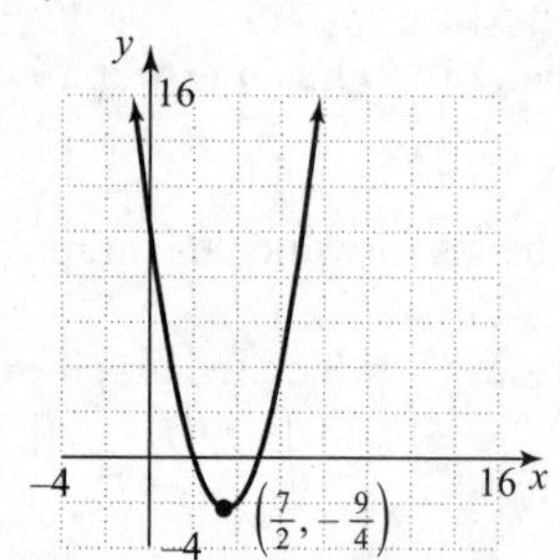

Domain: $\{x | x \text{ is a real number}\}$

Range: $\left\{y \middle| y \geq -\frac{9}{4}\right\}$

49. $y = 4x$

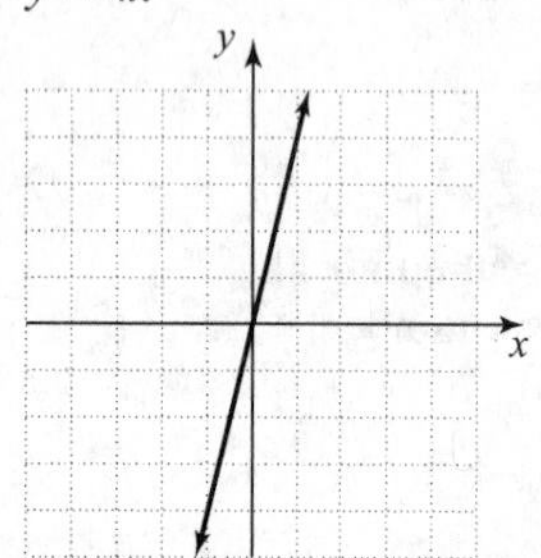

Domain: $\{x | x \text{ is a real number}\}$

Range: $\{y | y \text{ is a real number}\}$

51. $y = 2x^2 - 5$

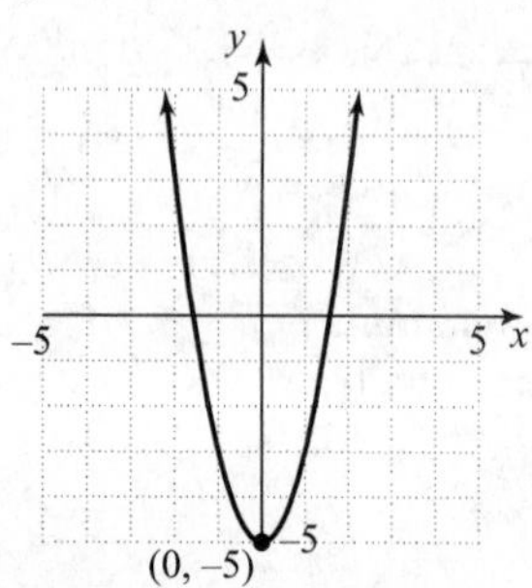

Domain: $\{x | x \text{ is a real number}\}$

Range: $\{y | y \geq -5\}$

53. $y = -x^2 + 4x + 2$

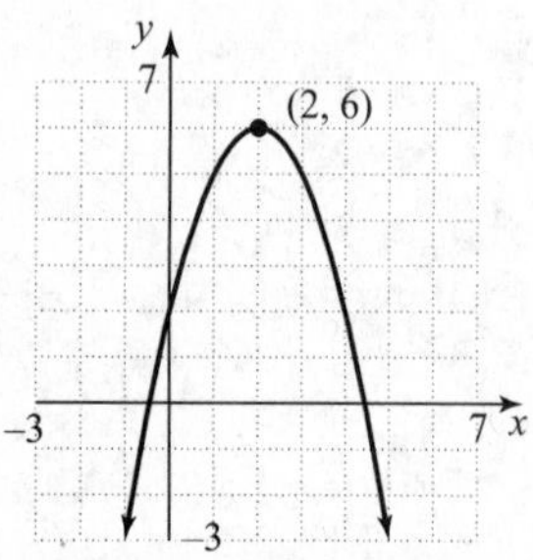

Domain: $\{x | x \text{ is a real number}\}$

Range: $\{y | y \leq 6\}$

55. $3x + 4y = 8 \left(\text{or } y = -\frac{3}{4}x + 2\right)$

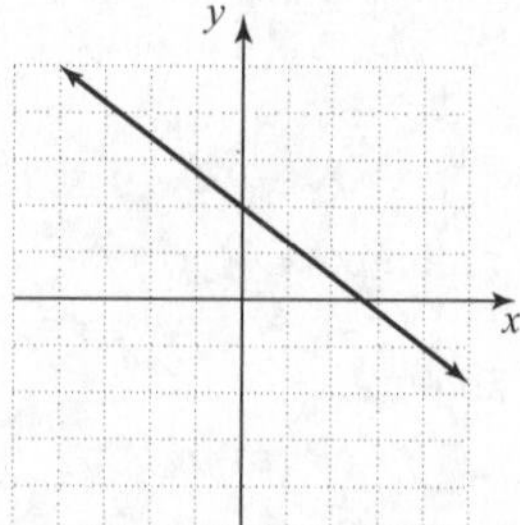

Domain: $\{x | x \text{ is a real number}\}$

Range: $\{y | y \text{ is a real number}\}$

57. **(a)** Domain: $\{x | 0 \leq t \leq 10\}$

Range: $\{y | 10 \leq p \leq 900\}$

(b) The peak is at (5, 900). So, the patient took the antibiotic on day 5.

(c) Find the number of days from (5, 90) to (10, 0). The number of days is $10 - 5 = 5$.

59. **(a)** Domain: $\{w | 0 < w \leq 5\}$

Range: $\{4.80, 5.20, 5.80, 6.45\}$

(b) Find the point on the graph where $w = \frac{10}{16} = 0.625$ lb. The point is (0.625, 4.80), so the cost is \$4.80.

(c) Find the point on the graph where $w = \frac{40}{16} = 2.5$. The point is (2.5, 5.20), so the cost is \$5.20.

61. Answers may vary. One possibility: $\{(1, 2), (3, 4), (5, 6)\}$

63. **(a)** No; there are an infinite number of rational and irrational numbers.

(b) Answers may vary. One possibility: $(1, 1), (2, 1), (3, 1), \left(\sqrt{2}, 0\right), \left(\sqrt{7}, 0\right), (\pi, 0)$

(c) No; answers may vary.

65. Answers may vary. One possibility: Any graph of the form $x = c$, where c is any real number. The graph will be a vertical line.

67. A relation is a correspondence between two related quantities. The domain of the relation is the set of all inputs into the relation and the range is the set of all outputs of the relation.

Putting the Concepts Together (Sections 10.1–10.2)

1. $y = x^2 - 2x - 3$
$a = 1, b = -2, c = -3$

(a) Since $a > 0$, the parabola opens up.

(b) $x = -\frac{b}{2a} = -\frac{-2}{2(1)} = 1$

$y = 1^2 - 2(1) - 3 = 1 - 2 - 3 = -4$
The vertex is $(1, -4)$.

(c) The axis of symmetry is
$x = -\frac{b}{2a} = -\frac{-2}{2(1)} = 1.$

(d) $b^2 - 4ac = (-2)^2 - 4(1)(-3)$
$= 4 + 12$
$= 16$
Since the discriminant > 0, there are two x-intercepts. Let $y = 0$.
$0 = x^2 - 2x - 3$
$0 = (x-3)(x+1)$
$x - 3 = 0$ or $x + 1 = 0$
$x = 3$ or $x = -1$
The x-intercepts are -1 and 3.
Let $x = 0$.
$y = 0^2 - 2(0) - 3 = -3$
The y-intercept is -3.

2. $y = x^2 - 6x$
$a = 1, b = -6, c = 0$
Since $a > 0$, the parabola opens up.
$x = -\frac{b}{2a} = -\frac{-6}{2(1)} = 3$

$y = 3^2 - 6(3) = 9 - 18 = -9$
The vertex is $(3, -9)$. The axis of symmetry is $x = 3$.
$b^2 - 4ac = (-6)^2 - 4(1)(0) = 36 - 0 = 36$
Since the discriminant > 0, there are two x-intercepts. Let $y = 0$.
$0 = x^2 - 6x$
$0 = x(x-6)$
$x = 0$ or $x - 6 = 0$
$x = 6$
The x-intercepts are 0 and 6.
Let $x = 0$.
$y = 0^2 - 6(0) = 0$
The y-intercept is 0.

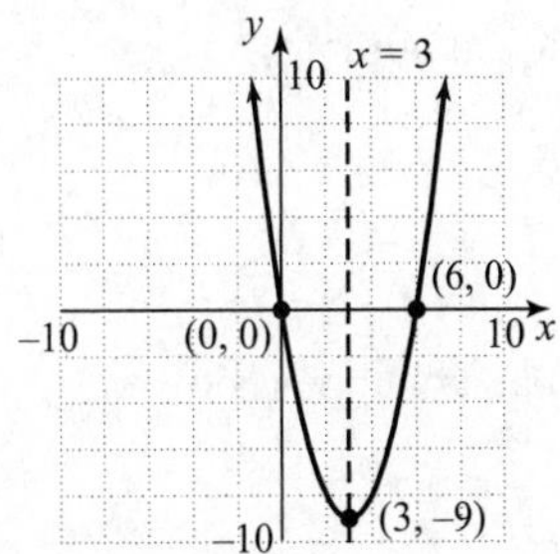

Domain: $\{x|x \text{ is a real number}\}$
Range: $\{y|y \geq -9\}$

3. $y = x^2 + 4x - 1$
$a = 1, b = 4, c = -1$
Since $a > 0$, the parabola opens up.
$x = -\frac{b}{2a} = -\frac{4}{2(1)} = -2$

$y = (-2)^2 + 4(-2) - 1 = 4 - 8 - 1 = -5$
The vertex is $(-2, -5)$. The axis of symmetry is $x = -2$.
$b^2 - 4ac = 4^2 - 4(1)(-1) = 16 + 4 = 20$
Since the discriminant > 0, there are two x-intercepts. Let $y = 0$.
$0 = x^2 + 4x - 1$

$$x = \frac{-b \pm \sqrt{b^2 - 4ac}}{2a} = \frac{-4 \pm \sqrt{20}}{2(1)} = \frac{-4 \pm 2\sqrt{5}}{2} = -2 \pm \sqrt{5}$$

The x-intercepts are $-2 - \sqrt{5}, -2 + \sqrt{5}$.
Let $x = 0$.
$y = 0^2 + 4(0) - 1 = -1$
The y-intercept is -1.

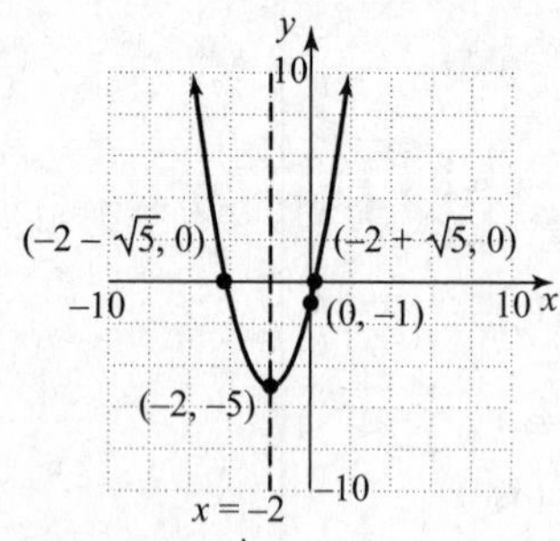

Domain: $\{x|x \text{ is a real number}\}$
Range: $\{y|y \geq -5\}$

4. $y = -x^2 + 9$
$a = -1, b = 0, c = 9$
Since $a < 0$, the parabola opens down.
$$x = -\frac{b}{2a} = -\frac{0}{2(-1)} = 0$$
$$y = -0^2 + 9 = 9$$
The vertex is (0, 9).
The axis of symmetry is $x = 0$.
$$b^2 - 4ac = 0^2 - 4(-1)(9) = 0 + 36 = 36$$
Since the discriminant > 0, there are two x-intercepts. Let $y = 0$.
$$0 = -x^2 + 9$$
$$0 = -(x^2 - 9)$$
$$0 = -(x-3)(x+3)$$
$$x - 3 = 0 \quad \text{or} \quad x + 3 = 0$$
$$x = 3 \quad \text{or} \quad x = -3$$
The x-intercepts are −3 and 3.
Let $x = 0$.
$$y = -0^2 + 9 = 9$$
The y-intercept is 9.

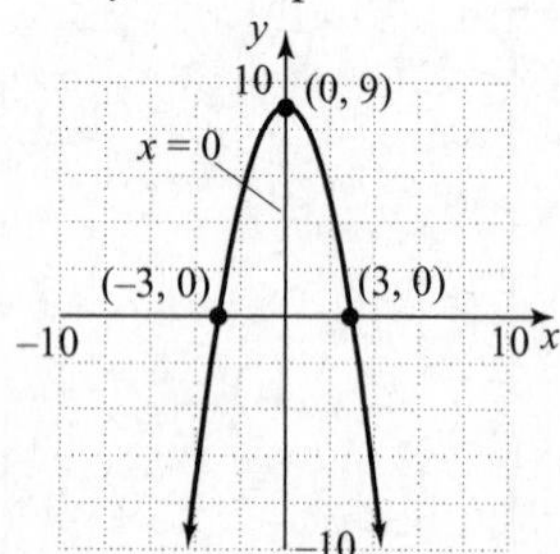

Domain: $\{x|x \text{ is a real number}\}$
Range: $\{y|y \leq 9\}$

5. $y = 2x^2 + 12x - 3$
$a = 2, b = 12, c = -3$

(a) Since $a > 0$, the parabola opens up. Therefore, it has a minimum.

(b) $x = -\frac{b}{2a} = -\frac{12}{2(2)} = -3$
$$y = 2(-3)^2 + 12(-3) - 3$$
$$= 18 - 36 - 3$$
$$= -21$$
The minimum value of the equation is −21.

6. (a) The domain is all the input values or $\{a, b, c\}$.

(b) The range is all the output values or $\{1, 2\}$.

7. (a) The domain is all the input values, or $\{1, 3\}$.

(b) The range is all the output values, or $\{-1, -3\}$.

8. (a) Domain: x-coordinates for which the graph exists: $\{x|x \geq -4\}$.

(b) Range: y-coordinates for which the graph exists: $\{y|y \text{ is a real number}\}$

9. (a) Domain: x-coordinates for which the graph exists: $\{x|1 \leq x \leq 7\}$

(b) Range: y-coordinates for which the graph exists: $\{y|-1 \leq y \leq 5\}$

Section 10.3

Preparing for An Introduction to Functions

P1. $2x^2 - 5x$

(a) For $x = 1$, we have $2(1)^2 - 5(1) = 2 - 5 = -3$

(b) For $x = 4$, we have
$$2(4)^2 - 5(4) = 32 - 20 = 12$$

(c) For $x = -3$, we have
$$2(-3)^2 - 5(-3) = 18 + 15 = 33$$

10.3 Quick Checks

1. A <u>function</u> is a relation in which each element in the domain of the relation corresponds to exactly one element in the range of the relation.

2. False; a relation in which an element in the domain corresponds to more than one element in the range is not a function.

3. The relation is a function because each element in the domain corresponds to exactly one element in the range. The domain of the function is {250, 300, 400, 500}. The range of the function is {\$41.20, \$43.04, \$55.39, \$64.03}.

4. The relation is a function because each element in the domain corresponds to exactly one element in the range. The domain of the function is {12, 15, 8, 7, 4}. The range of the function is {A, B, C, D}.

5. The relation is not a function because there is an element in the domain, 210, that corresponds to more than one element in the range.

6. The relation is a function because there are no ordered pairs with the same first coordinate but different second coordinates. The domain is {−3, −2, −1, 0, 1}. The range is {3, 2, 1, 0}.

7. The relation is not a function because there are two ordered pairs, (−3, 2) and (−3, 6), with the same first coordinate but different second coordinates.

8. The relation is a function because there are no ordered pairs with the same first coordinate but different second coordinates. The domain is {3, 4, 5, 6}. The range is {8, 10, 11}.

9. $y = -5x + 1$
Since there is only one output y that can result by performing these operations on any given input x, the equation is a function.

10. $y = \pm 6x$
Since there can be two outputs y that can result for an input x, for example (1, 6) and (1, −6), the equation is not that of a function.

11. $y = x^2 + 3x$
Since there is only one output y that can result by performing these operations on any given input x, the equation is a function.

12. $x^2 + y^2 = 4$
Since there can be two outputs y that can result for an input x, for example (2, 0) and (−2, 0), the equation is not that of a function.

13. True

14. The graph is a function because every vertical line intersects the graph in at most one point.

15. The graph is not a function because a vertical line intersects the graph in more than one point.

16. True

17. $f(x) = 4x + 1$
$f(4) = 4(4) + 1 = 16 + 1 = 17$

18. $f(x) = 4x + 1$
$f(-2) = 4(-2) + 1 = -8 + 1 = -7$

19. $g(x) = 2x^2 + 5x - 3$
$g(2) = 2(2)^2 + 5(2) - 3 = 8 + 10 - 3 = 15$

20. $g(x) = 2x^2 + 5x - 3$
$$\begin{aligned} g(-5) &= 2(-5)^2 + 5(-5) - 3 \\ &= 50 - 25 - 3 \\ &= 22 \end{aligned}$$

10.3 Exercises

21. The relation is not a function because there is an element in the domain, Folger, that corresponds to more than one element in the range.

23. The relation is a function because each element in the domain corresponds to exactly one element in the range. The domain is {Apples, Pears, Oranges, Grapes}. The range is {\$0.99, \$1.29, \$1.99}.

25. {(−1, −1), (0, 0), (1, 1), (2, 2)}
This relation is a function because there are no ordered pairs with the same first coordinate but different second coordinates. The domain is {−1, 0, 1, 2}. The range is {−1, 0, 1, 2}.

27. {(−1, 1), (0, 0), (1, −1), (−1, −2)}
This relation is not a function because there are two ordered pairs, (−1, 1) and (−1, −2), with the same first coordinate, but different second coordinates.

29. $\{(a, a), (b, a), (c, b), (d, b)\}$
This relation is a function because there are no ordered pairs with the same first coordinate but different second coordinates.
The domain is $\{a, b, c, d\}$.
The range is $\{a, b\}$.

31. $3x - y = 2$
Since there is only one output y that can result by performing these operations on any given input x, the equation is a function.

33. $x = 3$
All inputs are $x = 3$ for any output y. Because a single x corresponds to more than one y, the equation is not that of a function.

35. $y = -1$
Since there is only one output, $y = -1$, for any given input x, the equation represents a function.

37. $y = -x^2 + 2x - 3$
Since there is only one output y that can result by performing these operations on any given input x, the equation is a function.

39. $x = y^2$
Since there can be two outputs y that can result for an input x, for example (4, 2) and (4, −2), the equation does not represent a function.

41. $y = \pm x$
Since there can be two outputs y that can result for an input x, for example (3, 3) and (3, −3), the equation does not represent a function.

43. $x^2 + y^2 = 9$
Since there can be two outputs y that can result for an input x, for example (0, 3) and (0, −3), the equation does not represent a function.

45. The graph is not a function because a vertical line intersects the graph in more than one point.

47. The graph is a function because every vertical line intersects the graph in at most one point.

49. The graph is a function because every vertical line intersects the graph in at most one point.

51. The graph is not a function because a vertical line intersects the graph in more than one point.

53. The graph is a function because every vertical line intersects the graph in at most one point.

55. The graph is not a function because a vertical line intersects the graph in more than one point.

57. The graph is a function because every vertical line intersects the graph in at most one point.

59. $f(x) = -x - 3$

(a) $f(0) = -0 - 3 = -3$

(b) $f(3) = -3 - 3 = -6$

(c) $f(-2) = -(-2) - 3 = 2 - 3 = -1$

61. $f(x) = 4 - 2x$

(a) $f(0) = 4 - 2(0) = 4 - 0 = 4$

(b) $f(3) = 4 - 2(3) = 4 - 6 = -2$

(c) $f(-2) = 4 - 2(-2) = 4 + 4 = 8$

63. $f(x) = 3$

(a) $f(0) = 3$

(b) $f(3) = 3$

(c) $f(-2) = 3$

65. $f(x) = \frac{1}{2}x - 3$

(a) $f(0) = \frac{1}{2}(0) - 3 = 0 - 3 = -3$

(b) $f(3) = \frac{1}{2}(3) - 3 = \frac{3}{2} - \frac{6}{2} = -\frac{3}{2}$

(c) $f(-2) = \frac{1}{2}(-2) - 3 = -1 - 3 = -4$

67. $f(x) = 3 - 5x$

(a) $f(0) = 3 - 5(0) = 3 - 0 = 3$

(b) $f(3) = 3 - 5(3) = 3 - 15 = -12$

(c) $f(-2) = 3 - 5(-2) = 3 + 10 = 13$

69. $f(x) = x^2 - 3x$

(a) $f(0) = 0^2 - 3(0) = 0 - 0 = 0$

(b) $f(3)=3^2-3(3)=9-9=0$

(c) $f(-2)=(-2)^2-3(-2)=4+6=10$

71. $f(x)=-2x^2-5x+1$

(a) $f(0)=-2(0)^2-5(0)+1=0-0+1=1$

(b) $$\begin{aligned}f(3)&=-2(3)^2-5(3)+1\\&=-18-15+1\\&=-32\end{aligned}$$

(c) $$\begin{aligned}f(-2)&=-2(-2)^2-5(-2)+1\\&=-8+10+1\\&=3\end{aligned}$$

73. $f(x)=x^2-3x$

$f(-3)=(-3)^2-3(-3)=9+9=18$

75. $g(x)=-2x^2+x-5$

$g(-1)=-2(-1)^2+(-1)-5=-2-1-5=-8$

77. $a(x)=2+x-x^2$

$a(-1)=2+(-1)-(-1)^2=2-1-1=0$

79. $h(x)=\sqrt{x+2}$

$h(2)=\sqrt{2+2}=\sqrt{4}=2$

81. $h(x)=\dfrac{-2}{2x+3}$

$h\left(\dfrac{1}{2}\right)=\dfrac{-2}{2\left(\frac{1}{2}\right)+3}=\dfrac{-2}{1+3}=\dfrac{-2}{4}=-\dfrac{1}{2}$

83. $f(x)=\left|3x^2-10\right|$

$f(-1)=\left|3(-1)^2-10\right|=|3-10|=|-7|=7$

85. $g(x)=\sqrt{16-x^2}$

$$\begin{aligned}g(-2)&=\sqrt{16-(-2)^2}\\&=\sqrt{16-4}\\&=\sqrt{12}\\&=2\sqrt{3}\end{aligned}$$

87. $$\begin{aligned}f(x)&=2x+C\\f(3)&=2(3)+C\\11&=6+C\\5&=C\end{aligned}$$

89. $f(x)=-2x^2+5x+C$

$$\begin{aligned}f(-2)&=-2(-2)^2+5(-2)+C\\-15&=-8-10+C\\-15&=-18+C\\3&=C\end{aligned}$$

91. $f(x)=\dfrac{2x+5}{x-A}$

$$\begin{aligned}f(0)&=\frac{2(0)+5}{0-A}\\-1&=\frac{5}{-A}\\A&=5\end{aligned}$$

93. $G(h) = 12h$
$G(20) = 12(20) = 240$
Jackie's gross salary is \$240.

95. $h(x)=-0.00535x^2+0.353x+2.357$

(a) $$\begin{aligned}&h(30)\\&=-0.00535(30)^2+0.353(30)+2.357\\&=-4.815+10.59+2.357\\&=8.132\end{aligned}$$
In 1900 + 30 = 1930, there were 8.132 homicides per 100,000 people.

(b) $$\begin{aligned}&h(50)\\&=-0.00535(50)^2+0.353(50)+2.357\\&=-13.375+17.65+2.357\\&=6.632\end{aligned}$$
In 1900 + 50 = 1950, there were 6.632 homicides per 100,000 people.

97. $A(r)=\pi r^2$

$A(3)=\pi(3)^2=9\pi$

The area is 9π square inches.

99. $$\begin{aligned}f(x)&=2x+5\\9&=2x+5\\4&=2x\\2&=x\end{aligned}$$

101. $f(x) = 2 - \frac{1}{2}x$

$$1 = 2 - \frac{1}{2}x$$
$$-1 = -\frac{1}{2}x$$
$$2 = x$$

103. $f(x) = \frac{12}{x+2}$

$$-6 = \frac{12}{x+2}$$
$$x+2 = \frac{12}{-6}$$
$$x+2 = -2$$
$$x = -4$$

105. $f(x) = 1 - 2x$

(a) $f(3) = 1 - 2(3) = 1 - 6 = -5$

(b) $f(h) = 1 - 2h$

(c) $f(h+3) = 1 - 2(h+3)$
$$= 1 - 2h - 6$$
$$= -5 - 2h$$

(d) $f(h) + f(3) = (1-2h) + (-5)$
$$= 1 - 2h - 5$$
$$= -4 - 2h$$

(e) $f(h+3) - f(3) = (-5-2h) - (-5)$
$$= -5 - 2h + 5$$
$$= -2h$$

107. The four ways to represent a function are: maps, ordered pairs, equations, and graphs. Answers may vary.

Chapter 10 Review

1–6. Tables will vary.

1.

x	$y = -x^2$	y
−3	$-(-3)^2 = -9$	−9
−2	$-(-2)^2 = -4$	−4
−1	$-(-1)^2 = -1$	−1
0	$-0^2 = 0$	0
1	$-1^2 = -1$	−1
2	$-2^2 = -4$	−4
3	$-3^2 = -9$	−9

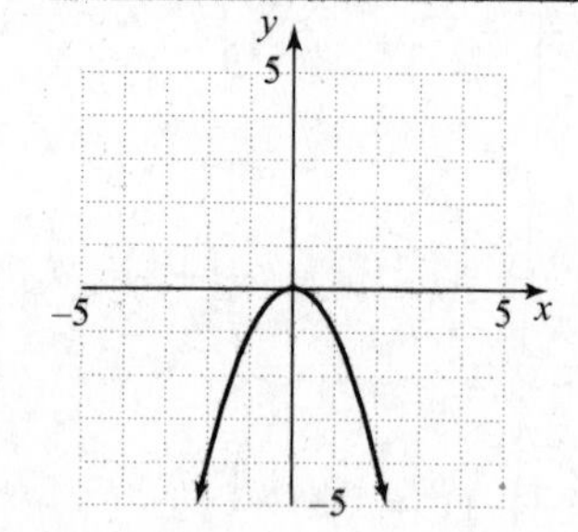

2.

x	$y = x^2$	y
−3	$(-3)^2 = 9$	9
−2	$(-2)^2 = 4$	4
−1	$(-1)^2 = 1$	1
0	$0^2 = 0$	0
1	$1^2 = 1$	1
2	$2^2 = 2$	2
3	$3^2 = 9$	9

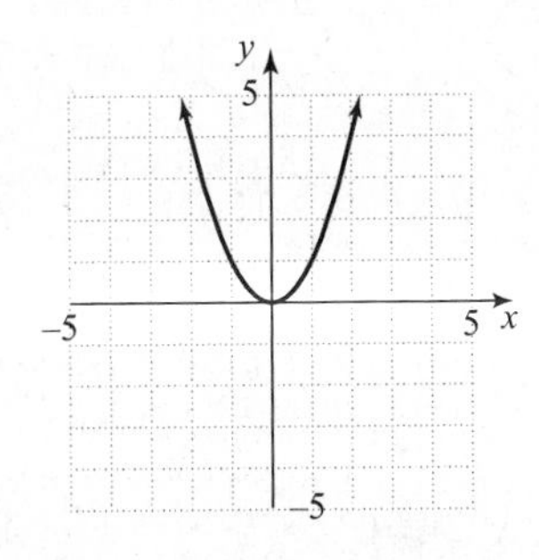

3.

x	$y = 2x^2 - 3$	y
−3	$2(-3)^2 - 3 = 15$	15
−2	$2(-2)^2 - 3 = 5$	5
−1	$2(-1)^2 - 3 = -1$	−1
0	$2(0)^2 - 3 = -3$	−3
1	$2(1)^2 - 3 = -1$	−1
2	$2(2)^2 - 3 = 5$	5
3	$2(3)^2 - 3 = 15$	15

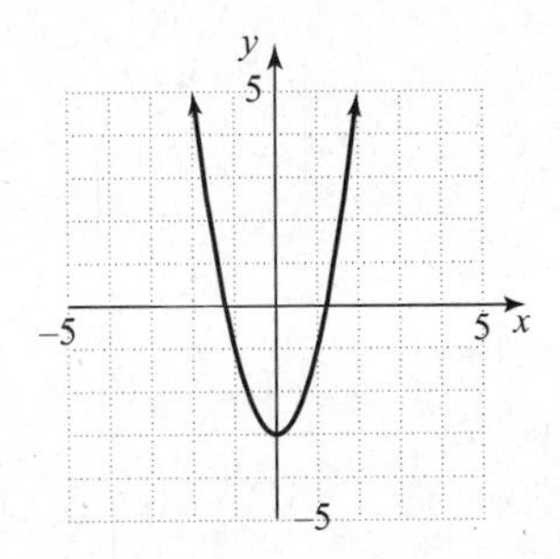

4.

x	$y = -\frac{1}{2}x^2 + 4$	y
−3	$y = -\frac{1}{2}(-3)^2 + 4$	$-\frac{1}{2}$
−2	$y = -\frac{1}{2}(-2)^2 + 4$	2
−1	$y = -\frac{1}{2}(-1)^2 + 4$	$3\frac{1}{2}$
0	$y = -\frac{1}{2}(0)^2 + 4$	4
1	$y = -\frac{1}{2}(1)^2 + 4$	$3\frac{1}{2}$
2	$y = -\frac{1}{2}(2)^2 + 4$	2
3	$y = -\frac{1}{2}(3)^2 + 4$	$-\frac{1}{2}$

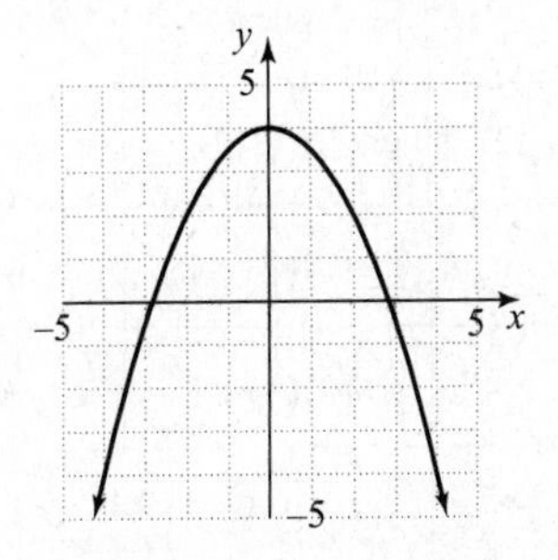

5.

x	$y = x^2 - 4x - 3$	y
−3	$y = (-3)^2 - 4(-3) - 3$	18
−2	$y = (-2)^2 - 4(-2) - 3$	9
−1	$y = (-1)^2 - 4(-1) - 3$	2
0	$y = (0)^2 - 4(0) - 3$	−3
1	$y = (1)^2 - 4(1) - 3$	−6
2	$y = (2)^2 - 4(2) - 3$	−7
3	$y = (3)^2 - 4(3) - 3$	−6

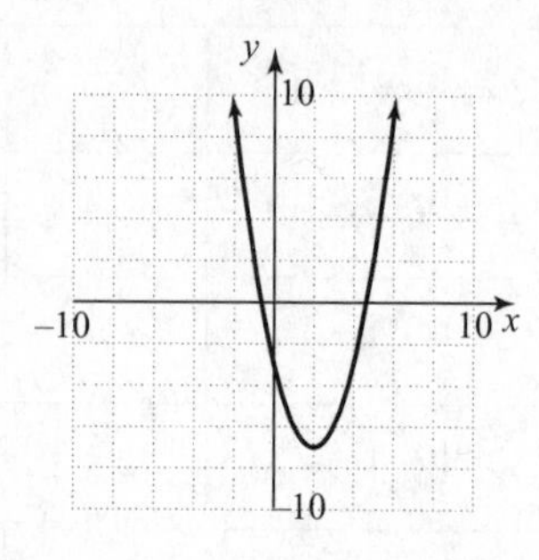

6.

x	$y=-2x^2+4x+1$	y
−3	$y=-2(-3)^2+4(-3)+1$	−29
−2	$y=-2(-2)^2+4(-2)+1$	−15
−1	$y=-2(-1)^2+4(-1)+1$	−5
0	$y=-2(0)^2+4(0)+1$	1
1	$y=-2(1)^2+4(1)+1$	3
2	$y=-2(2)^2+4(2)+1$	1
3	$y=-2(3)^2+4(3)+1$	−5

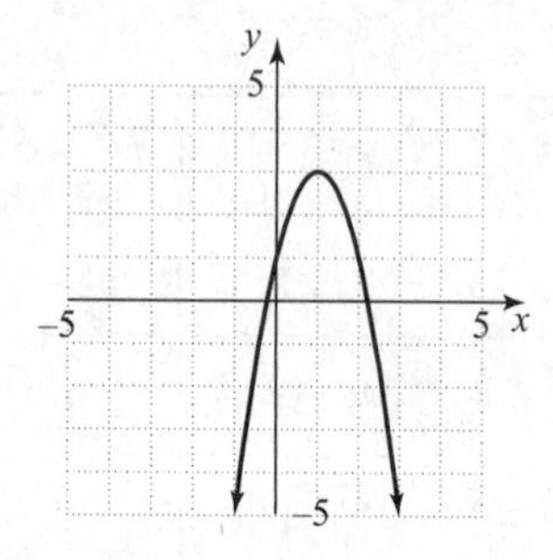

7. $y=x^2+x-6$

$a=1,\ b=1,\ c=-6$

$b^2-4ac=1^2-4(1)(-6)=1+24=25$

Since the discriminant > 0, there are two x-intercepts. Let $y=0$.

$0=x^2+x-6$

$0=(x+3)(x-2)$

$x+3=0$ or $x-2=0$

$x=-3$ or $x=2$

The x-intercepts are −3 and 2.
Let $x=0$.

$y=0^2+0-6=-6$

The y-intercept is −6.

8. $y=x^2+6x+8$

$a=1,\ b=6,\ c=8$

$b^2-4ac=6^2-4(1)(8)=36-32=4$

Since the discriminant > 0, there are two x-intercepts. Let $y=0$.

$0=x^2+6x+8$

$0=(x+2)(x+4)$

$x+2=0$ or $x+4=0$

$x=-2$ or $x=-4$

The x-intercepts are −2 and −4.
Let $x=0$.

$y=0^2+6(0)+8=8$

The y-intercept is 8.

9. $y=-4x^2+12x-9$

$a=-4,\ b=12,\ c=-9$

$b^2-4ac=12^2-4(-4)(-9)=144-144=0$

Since the discriminant = 0, there is one x-intercept. Let $y=0$.

$0=-4x^2+12x-9$

$0=-(4x^2-12x+9)$

$0=-(2x-3)^2$

$2x-3=0$

$x=\frac{3}{2}$

The x-intercept is $\frac{3}{2}$.

Let $x=0$.

$y=-4(0)^2+12(0)-9=-9$

The y-intercept is −9.

10. $y=-9x^2-6x-1$

$a=-9,\ b=-6,\ c=-1$

$b^2-4ac=(-6)^2-4(-9)(-1)=36-36=0$

Since the discriminant = 0, there is one x-intercept. Let $y=0$.

$0=-9x^2-6x-1$

$0=-(9x^2+6x+1)$

$0=-(3x+1)^2$

$3x+1=0$

$x=-\frac{1}{3}$

The x-intercept is $-\frac{1}{3}$.

Let $x=0$.

$y = -9(0)^2 - 6(0) - 1 = -1$

The y-intercept is -1.

11. $y = -4x^2 + 8x - 5$

$a = -4$, $b = 8$, $c = -5$

$b^2 - 4ac = 8^2 - 4(-4)(-5) = 64 - 80 = -16$

Since the discriminant < 0, there are no x-intercepts. Let $x = 0$.

$y = -4(0)^2 + 8(0) - 5 = -5$

The y-intercept is -5.

12. $y = -3x^2 + 2x - 1$

$a = -3$, $b = 2$, $c = -1$

$b^2 - 4ac = 2^2 - 4(-3)(-1) = 4 - 12 = -8$

Since the discriminant < 0, there are no x-intercepts. Let $x = 0$.

$y = -3(0)^2 + 2(0) - 1 = -1$

The y-intercept is -1.

13. $y = x^2 - 4x + 2$

$a = 1$, $b = -4$, $c = 2$

$b^2 - 4ac = (-4)^2 - 4(1)(2) = 16 - 8 = 8$

Since the discriminant > 0, there are two x-intercepts. Let $y = 0$.

$0 = x^2 - 4x + 2$

$$x = \frac{-b \pm \sqrt{b^2 - 4ac}}{2a}$$

$$x = \frac{-(-4) \pm \sqrt{8}}{2(1)} = \frac{4 \pm 2\sqrt{2}}{2} = 2 \pm \sqrt{2}$$

The x-intercepts are $2 - \sqrt{2}$ and $2 + \sqrt{2}$.

Let $x = 0$.

$y = 0^2 - 4(0) + 2 = 2$

The y-intercept is 2.

14. $y = -x^2 - 4x + 14$

$a = -1$, $b = -4$, $c = 14$

$b^2 - 4ac = (-4)^2 - 4(-1)(14) = 16 + 56 = 72$

Since the discriminant > 0, there are two x-intercepts. Let $y = 0$.

$0 = -x^2 - 4x + 14$

$$x = \frac{-b \pm \sqrt{b^2 - 4ac}}{2a}$$

$$x = \frac{-(-4) \pm \sqrt{72}}{2(-1)} = \frac{4 \pm 6\sqrt{2}}{-2} = -2 \pm 3\sqrt{2}$$

The x-intercepts are $-2 - 3\sqrt{2}$ and $-2 + 3\sqrt{2}$.

Let $x = 0$.

$y = -0^2 - 4(0) + 14 = 14$

The y-intercept is 14.

15. (a) $y = x^2 + 5$

$a = 1$, $b = 0$, $c = 5$

Since $a > 0$, the parabola opens up.

(b) $x = -\frac{b}{2a} = -\frac{0}{2(1)} = 0$

$y = 0^2 + 5 = 5$

The vertex is (0, 5).

(c) The axis of symmetry is $x = -\frac{b}{2a} = 0$.

16. (a) $y = -x^2 - 4$

$a = -1$, $b = 0$, $c = -4$

Since $a < 0$, the parabola opens down.

(b) $x = -\frac{b}{2a} = -\frac{0}{2(-1)} = 0$

$y = -0^2 - 4 = -4$

The vertex is (0, −4).

(c) The axis of symmetry is $x = -\frac{b}{2a} = 0$.

17. (a) $y = -2x^2 + 4x - 1$

$a = -2$, $b = 4$, $c = -1$

Since $a < 0$, the parabola opens down.

(b) $x = -\frac{b}{2a} = -\frac{4}{2(-2)} = 1$

$y = -2(1)^2 + 4(1) - 1 = -2 + 4 - 1 = 1$

The vertex is (1, 1).

(c) The axis of symmetry is $x = -\frac{b}{2a} = 1$.

18. (a) $y = 3x^2 - 12x + 4$

$a = 3$, $b = -12$, $c = 4$

Since $a > 0$, the parabola opens up.

(b) $x = -\frac{b}{2a} = -\frac{-12}{2(3)} = 2$

$y = 3(2)^2 - 12(2) + 4 = 12 - 24 + 4 = -8$

The vertex is $(2, -8)$.

(c) The axis of symmetry is $x = -\frac{b}{2a} = 2$.

19. (a) $y = \frac{3}{2}x^2 + 9x$

$a = \frac{3}{2}$, $b = 9$, $c = 0$

Since $a > 0$, the parabola opens up.

(b) $x = -\frac{b}{2a} = -\frac{9}{2\left(\frac{3}{2}\right)} = -3$

$y = \frac{3}{2}(-3)^2 + 9(-3) = \frac{27}{2} - \frac{54}{2} = -\frac{27}{2}$

The vertex is $\left(-3, -\frac{27}{2}\right)$.

(c) The axis of symmetry is $x = -\frac{b}{2a} = -3$.

20. (a) $y = -\frac{3}{2}x^2 - 6x$

$a = -\frac{3}{2}$, $b = -6$, $c = 0$

Since $a < 0$, the parabola opens down.

(b) $x = -\frac{b}{2a} = -\frac{-6}{2\left(-\frac{3}{2}\right)} = \frac{6}{-3} = -2$

$y = -\frac{3}{2}(-2)^2 - 6(-2) = -6 + 12 = 6$

The vertex is $(-2, 6)$.

(c) The axis of symmetry is $x = -\frac{b}{2a} = -2$.

21. (a) $y = -x^2 + 3x - 4$

$a = -1$, $b = 3$, $c = -4$

Since $a < 0$, the parabola opens down.

(b) $x = -\frac{b}{2a} = -\frac{3}{2(-1)} = \frac{3}{2}$

$$y = -\left(\frac{3}{2}\right)^2 + 3\left(\frac{3}{2}\right) - 4$$
$$= -\frac{9}{4} + \frac{18}{4} - \frac{16}{4}$$
$$= -\frac{7}{4}$$

The vertex is $\left(\frac{3}{2}, -\frac{7}{4}\right)$.

(c) The axis of symmetry is $x = -\frac{b}{2a} = \frac{3}{2}$.

22. (a) $y = x^2 + 5x + 5$

$a = 1$, $b = 5$, $c = 5$

Since $a > 1$, the parabola opens up.

(b) $x = -\frac{b}{2a} = -\frac{5}{2(1)} = -\frac{5}{2}$

$$y = \left(-\frac{5}{2}\right)^2 + 5\left(-\frac{5}{2}\right) + 5$$
$$= \frac{25}{4} - \frac{50}{4} + \frac{20}{4}$$
$$= -\frac{5}{4}$$

The vertex is $\left(-\frac{5}{2}, -\frac{5}{4}\right)$.

(c) The axis of symmetry is $x = -\frac{b}{2a} = -\frac{5}{2}$.

23. $y = x^2 + 2x - 3$

$a = 1$, $b = 2$, $c = -3$

Since $a > 0$, the parabola opens up.

$x = -\frac{b}{2a} = -\frac{2}{2(1)} = -1$

$y = (-1)^2 + 2(-1) - 3 = 1 - 2 - 3 = -4$

The vertex is $(-1, -4)$. The axis of symmetry is $x = -\frac{b}{2a} = -1$.

$b^2 - 4ac = 2^2 - 4(1)(-3) = 4 + 12 = 16$

Since the discriminant > 0, there are two x-intercepts. Let $y = 0$.

$0 = x^2 + 2x - 3$

$0 = (x + 3)(x - 1)$

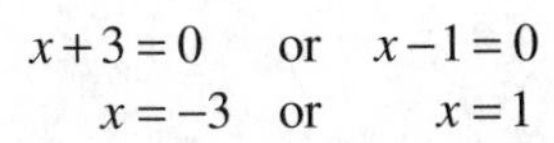

$x+3=0$ or $x-1=0$
$x=-3$ or $x=1$

The x-intercepts are -3 and 1.
Let $x = 0$.

$y = 0^2 + 2(0) - 3 = -3$

The y-intercept is -3.

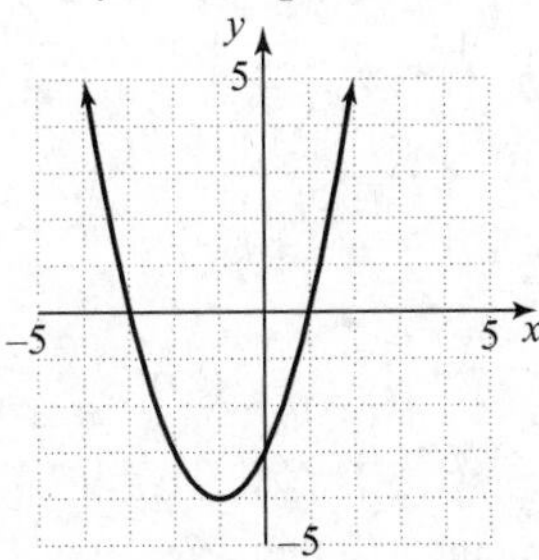

24. $y = -x^2 + 6x - 8$

$a = -1$, $b = 6$, $c = -8$
Since $a < 0$, the parabola opens down.

$x = -\frac{b}{2a} = -\frac{6}{2(-1)} = 3$

$y = -3^2 + 6(3) - 8 = -9 + 18 - 8 = 1$

The vertex is (3, 1). The axis of symmetry is

$x = -\frac{b}{2a} = 3.$

$b^2 - 4ac = 6^2 - 4(-1)(-8) = 36 - 32 = 4$

Since the discriminant > 0, there are two x-intercepts. Let $y = 0$.

$0 = -x^2 + 6x - 8$
$0 = -(x^2 - 6x + 8)$
$0 = -(x-4)(x-2)$

$x-4=0$ or $x-2=0$
$x=4$ or $x=2$

The x-intercepts are 2 and 4.
Let $y = 0$.

$y = -0^2 + 6(0) - 8 = -8$

The y-intercept is -8.

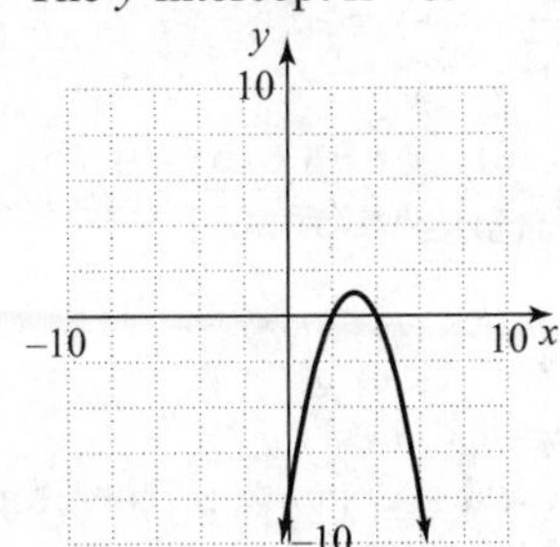

25. $y = -x^2 + 4$

$a = -1$, $b = 0$, $c = 4$
Since $a < 0$, the parabola opens down.

$x = -\frac{b}{2a} = -\frac{0}{2(-1)} = 0$

$y = -0^2 + 4 = 4$

The vertex is (0, 4). The axis of symmetry is

$x = -\frac{b}{2a} = 0.$

$b^2 - 4ac = 0^2 - 4(-1)(4) = 0 + 16 = 16$

Since the discriminant > 0, there are two x-intercepts. Let $y = 0$.

$0 = -x^2 + 4$
$0 = -(x^2 - 4)$
$0 = -(x+2)(x-2)$

$x+2=0$ or $x-2=0$
$x=-2$ or $x=2$

The x-intercepts are -2 or 2. Let $x = 0$.

$y = -0^2 + 4 = 4$

The y-intercept is 4.

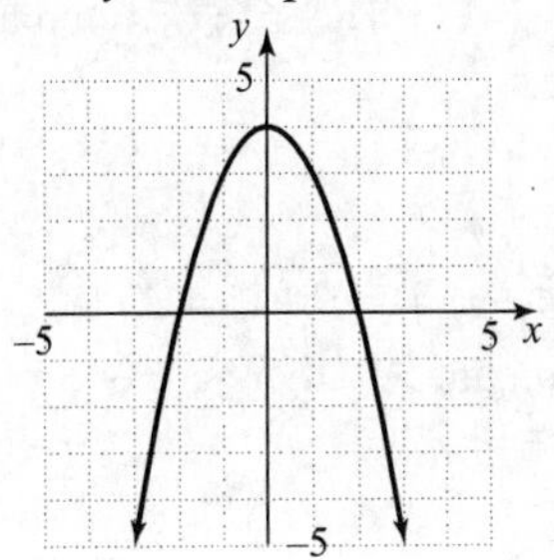

26. $y = x^2 - 9$

$a = 1$, $b = 0$, $c = -9$
Since $a > 0$, the parabola opens up.

$x = -\frac{b}{2a} = -\frac{0}{2(1)} = 0$

$y = 0^2 - 9 = -9$

The vertex is (0, -9). The axis of symmetry is

$x = -\frac{b}{2a} = 0.$

$b^2 - 4ac = 0^2 - 4(1)(-9) = 0 + 36 = 36$

Since the discriminant > 0, there are two x-intercepts. Let $y = 0$.

$0 = x^2 - 9$
$0 = (x+3)(x-3)$

$x+3=0$ or $x-3=0$
$x=-3$ or $x=3$

The x-intercepts are -3 and 3.
Let $x = 0$.

$y = 0^2 - 9 = -9$

The y-intercept is -9.

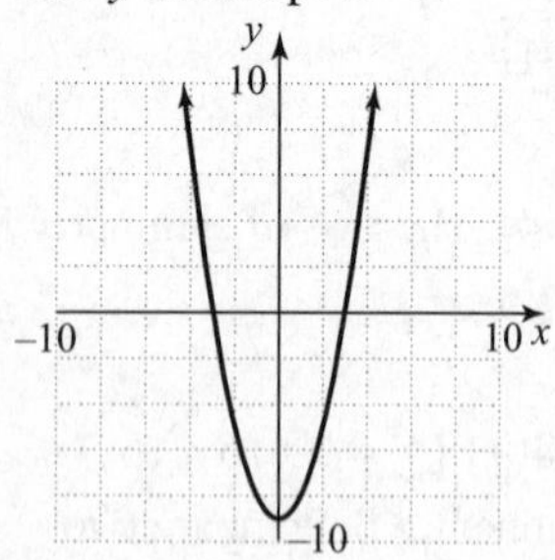

27. $y = -x^2 + 3x + 4$

$a = -1$, $b = 3$, $c = 4$
Since $a < 0$, the parabola opens down.

$x = -\frac{b}{2a} = -\frac{3}{2(-1)} = \frac{3}{2}$

$y = -\left(\frac{3}{2}\right)^2 + 3\left(\frac{3}{2}\right) + 4 = -\frac{9}{4} + \frac{18}{4} + \frac{16}{4} = \frac{25}{4}$

The vertex is $\left(\frac{3}{2}, \frac{25}{4}\right)$. The axis of symmetry is

$x = -\frac{b}{2a} = \frac{3}{2}$.

$b^2 - 4ac = 3^2 - 4(-1)(4) = 9 + 16 = 25$

Since the discriminant > 0, there are two x-intercepts. Let $y = 0$.

$0 = -x^2 + 3x + 4$
$0 = -(x^2 - 3x - 4)$
$0 = -(x - 4)(x + 1)$
$x - 4 = 0$ or $x + 1 = 0$
$x = 4$ or $x = -1$

The x-intercepts are -1 and 4.
Let $x = 0$.

$y = -0^2 + 3(0) + 4 = 4$

The y-intercept is 4.

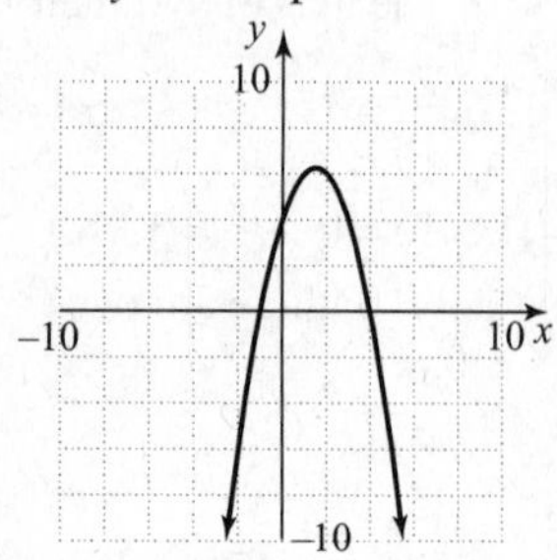

28. $y = x^2 - 5x + 4$

$a = 1$, $b = -5$, $c = 4$
Since $a > 0$, the parabola opens up.

$x = -\frac{b}{2a} = -\frac{-5}{2(1)} = \frac{5}{2}$

$y = \left(\frac{5}{2}\right)^2 - 5\left(\frac{5}{2}\right) + 4 = \frac{25}{4} - \frac{50}{4} + \frac{16}{4} = -\frac{9}{4}$

The vertex is $\left(\frac{5}{2}, -\frac{9}{4}\right)$. The axis of symmetry is

$x = -\frac{b}{2a} = \frac{5}{2}$.

$b^2 - 4ac = (-5)^2 - 4(1)(4) = 25 - 16 = 9$

Since the discriminant > 0, there are two x-intercepts. Let $y = 0$.

$0 = x^2 - 5x + 4$
$0 = (x - 4)(x - 1)$
$x - 4 = 0$ or $x - 1 = 0$
$x = 4$ or $x = 1$

The x-intercepts are 1 and 4.
Let $x = 0$.

$y = 0^2 - 5(0) + 4 = 4$

The y-intercept is 4.

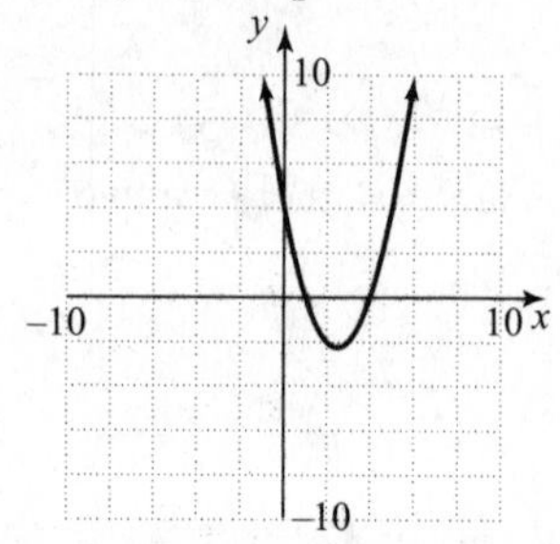

29. $y = -2x^2 - 8x + 7$

$a = -2$, $b = -8$, $c = 7$
Since $a < 0$, the parabola opens down. Therefore, it has a maximum.

$x = -\frac{b}{2a} = -\frac{-8}{2(-2)} = \frac{8}{-4} = -2$

$y = -2(-2)^2 - 8(-2) + 7 = -8 + 16 + 7 = 15$

The maximum is 15.

30. $y = 3x^2 + 12x - 4$

$a = 3$, $b = 12$, $c = -4$
Since $a > 0$, the parabola opens up. Therefore, it has a minimum.

$x = -\frac{b}{2a} = -\frac{12}{2(3)} = -2$

$y = 3(-2)^2 + 12(-2) - 4 = 12 - 24 - 4 = -16$

The minimum is −16.

31. $y = x^2 - 4x - 6$

$a = 1, b = -4, c = -6$

Since $a > 0$, the parabola opens up. Therefore, it has a minimum.

$x = -\frac{b}{2a} = -\frac{-4}{2(1)} = 2$

$y = 2^2 - 4(2) - 6 = 4 - 8 - 6 = -10$

The minimum is −10.

32. $y = -x^2 - 10x - 15$

$a = -1, b = -10, c = -15$

Since $a < 0$, the parabola opens down. Therefore, it has a maximum.

$x = -\frac{b}{2a} = -\frac{-10}{2(-1)} = -5$

$$\begin{aligned} y &= -(-5)^2 - 10(-5) - 15 \\ &= -25 + 50 - 15 \\ &= 10 \end{aligned}$$

The maximum is 10.

33. **(a)** $h = -0.00175x^2 + 0.325x + 5$

$a = -0.00175, b = 0.325, c = 5$

$x = -\frac{b}{2a} = -\frac{0.325}{2(-0.00175)} \approx 92.9$

The distance is about 92.9 feet.

(b) Let $x = 93$.

$h = -0.00175(92.9)^2 + 0.325(92.9) + 5$

$h \approx 20.1$

The maximum height is about 20.1 feet.

(c) When the ball lands, $h = 0$.

$0 = -0.00175x^2 + 0.325x + 5$

$$\begin{aligned} x &= \frac{-b \pm \sqrt{b^2 - 4ac}}{2a} \\ &= \frac{-0.325 \pm \sqrt{0.325^2 - 4(-0.00175)(5)}}{2(-0.00175)} \\ &\approx -14.3 \text{ or } 200 \end{aligned}$$

Sin ce $x > 0$ (the ball goes forward), the baseball lands 200 feet away.

34. **(a)** $A = -x^2 + 200x$

$a = -1, b = 200, c = 0$

$x = -\frac{b}{2a} = -\frac{200}{2(-1)} = 100$

The side length is 100 meters.

(b) Let $x = 100$.

$A = -100^2 + 200(100) = 10{,}000$

The maximum area is 10,000 square meters.

35. {(Alabama, 13.5), (California, 11.0), (Florida, 17.0), (Nebraska, 13.3), (New York, 13.2), (Massachusetts, 13.3), (South Dakota, 14.3)}

36. {(History, 400), (Math, 400), (Music, 200), (Psychology, 225)}

37. The domain is the set of inputs and the range is the set of outputs.

Domain: {Alabama, California, Florida, Nebraska, New York, Massachusetts, South Dakota}

Range: {14.3, 13.5, 13.3, 17.0, 11.0, 13.2}

38. The domain is the set of inputs and the range is the set of outputs.

Domain: {History, Math, Music, Psychology}

Range: {200, 225, 400}

39. $\{(-1, 3); (-1, 5); (0, -2); (0, -4)\}$

The domain is the set of inputs and the range is the set of outputs.

Domain: $\{-1, 0\}$

Range: $\{3, 5, -2, -4\}$

40. $\{(a, x^2); (b, x^3); (c, x^4); (d, x^5)\}$

The domain is the set of inputs and the range is the set of outputs.

Domain: $\{a, b, c, d\}$

Range: $\{x^2, x^3, x^4, x^5\}$

41. $\{(a_1, 1); (a_2, 4); (a_3, 9); (a_4, 16)\}$

The domain is the set of inputs and the range is the set of outputs.

Domain: $\{a_1, a_2, a_3, a_4\}$

Range: $\{1, 4, 9, 16\}$

42. $\{(1, -3); (3, -3); (5, 2); (7, -2)\}$

The domain is the set of inputs and the range is the set of outputs.

Domain: $\{1, 3, 5, 7\}$

Range: $\{-3, 2, -2\}$

43. The points on the graph: (–3, 0), (0, 3), (3, 3), (3, –3)
The domain is the set of all x-coordinates.
Domain: {–3, 0, 3}
The range is the set of all y-coordinates.
Range: {–3, 0, 3}

44. The points on the graph: (0, –4), (1, 0), (4, 0), (4, 4)
The domain is the set of all x-coordinates.
Domain: {0, 1, 4}
The range is the set of all y-coordinates.
Range: {–4, 0, 4}

45. Domain: x-coordinates for which the graph exists (all real numbers):
$\{x|x \text{ is a real number}\}$
Range: y-coordinates for which the graph exists (all real numbers):
$\{y|y \text{ is a real number}\}$

46. Domain: x-coordinates for which the graph exists (all real numbers):
$\{x|x \text{ is a real number}\}$
Range: y-coordinate for which the graph exists ($y = 4$): {4}

47. Domain: x-coordinate for which the graph exists ($x = 5$): {5}
Range: y-coordinates for which the graph exists (all real numbers):
$\{y|y \text{ is a real number}\}$

48. Domain: x-coordinates for which the graph exists (all real numbers):
$\{x|x \text{ is a real number}\}$
Range: y-coordinates for which the graph exists (all real numbers):
$\{y|y \text{ is a real number}\}$

49. Domain: x-coordinates for which the graph exists: $\{x|0 \le x \le 6\}$
Range: y-coordinates for which the graph exists:
$\{y|1 \le y \le 5\}$

50. Domain: x-coordinates for which the graph exists: $\{x|-4 \le x \le 0\}$
Range: y-coordinates for which the graph exists:
$\{y|-4 \le y \le 3\}$

51. $y = -\frac{3}{2}x + 4$

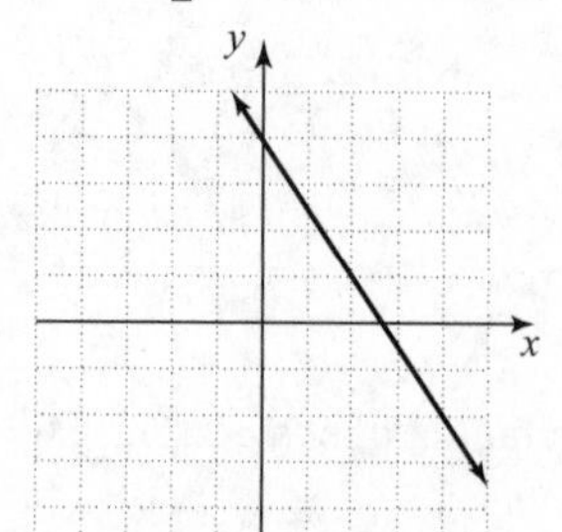

Domain: $\{x|x \text{ is a real number}\}$
Range: $\{y|y \text{ is a real number}\}$

52. $y = \frac{4}{3}x - 5$

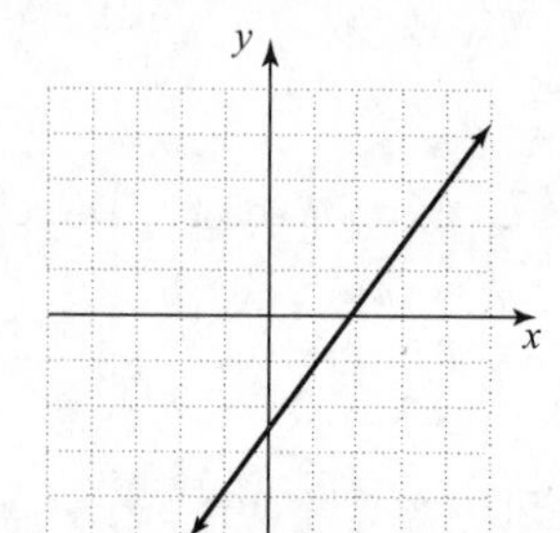

Domain: $\{x|x \text{ is a real number}\}$
Range: $\{y|y \text{ is a real number}\}$

53. $y = 8x - 4x^2$

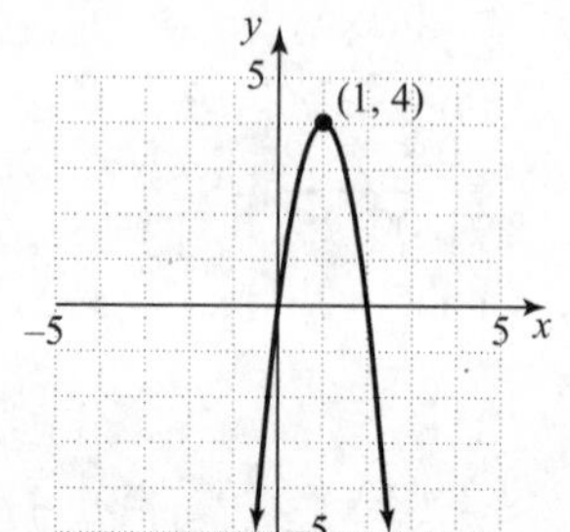

Domain: $\{x|x \text{ is a real number}\}$
Range: $\{y|y \le 4\}$

54. $y = x^2 - 4x + 3$

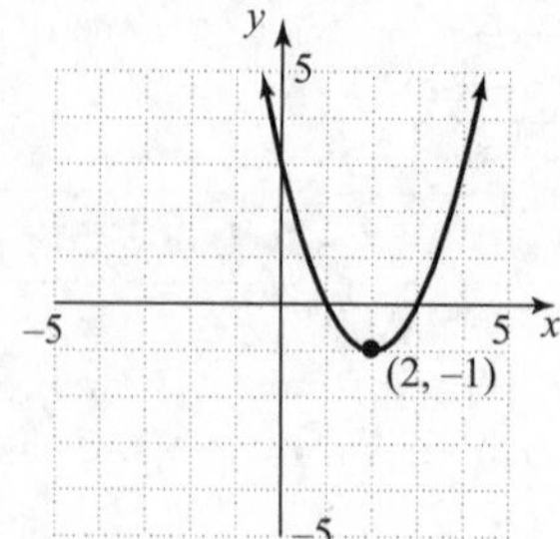

Domain: $\{x|x \text{ is a real number}\}$
Range: $\{y|y \ge -1\}$

55. The relation is not a function because there is an element in the domain, Monday (or Sunday), that corresponds to more than one element in the range.

56. The relation is a function because each element in the domain corresponds to exactly one element in the range.
The domain is {Alaska, Massachusetts, Minnesota, West Virginia}. The range is {16.9, 27.3, 30.7, 36.9}.

57. $\{(-4, 0), (-3, 0), (-2, 1), (-1, 1)\}$
This relation is a function because there are no ordered pairs with the same first coordinate but different second coordinates.
The domain is $\{-4, -3, -2, -1\}$. The range is $\{0, 1\}$.

58. $\{(1, 4), (3, 3), (1, 2), (-1, 1)\}$
The relation is not a function because there are two ordered pairs, (1, 4) and (1, 2), with the same first coordinate, but different second coordinates.

59. $\{(a, 9), (b, 8), (b, 7), (a, 6)\}$
The relation is not a function because there are ordered pairs, $(a, 9)$ and $(a, 6)$ or $(b, 8)$ and $(b, 7)$, with the same first coordinate, but different second coordinates.

60. $\{(5, z), (3, x), (1, x), (-1, z)\}$
This relation is a function because there are no ordered pairs with the same first coordinate but different second coordinates.
The domain is $\{5, 3, 1, -1\}$. The range is $\{x, z\}$.

61. $x = y + 2$
Since there is only one output y that can result by performing these operations on any given input x, the equation is a function.

62. $x = y^2$
Since there can be two outputs y that can result for an input x, for example (4, 2) and (4, −2), the equation does not represent a function.

63. $x = -3$
All inputs are $x = -3$ for any output y.
Because a single x corresponds to more than one y, the equation is not that of a function.

64. $y = 2x$
Since there is only one output y that can result by performing these operations on any given input x, the equation is a function.

65. $y = \pm\sqrt{x}$
Since there can be two outputs y that can result for an input x, for example (4, 2) and (4, −2), the equation does not represent a function.

66. $y = 7$
Since there is only one output, $y = 7$, for any given input x, the equation represents a function.

67. $y = x^2 + 4x$
Since there is only one output y that can result by performing these operations on any given input x, the equation is a function.

68. $x^2 + y^2 = 25$
Since there can be two outputs y that can result for an input x, for example (0, 5) and (0, −5), the equation does not represent a function.

69. $x^2 - y^2 = 16$
Since there can be two outputs y that can result for an input x, for example $\left(\sqrt{17}, 1\right)$ and $\left(\sqrt{17}, -1\right)$, the equation does not represent a function.

70. $x^2 = y - 1$
Since there is only one output y that can result by performing these operations on any given input x, the equation is a function.

71. The graph is not a function because a vertical line intersects the graph in more than one point.

72. The graph is a function because every vertical line intersects the graph in at most one point.

73. The graph is a function because every vertical line intersects the graph in at most one point.

74. The graph is not a function because a vertical line intersects the graph in more than one point.

75. The graph is a function because every vertical line intersects the graph in at most one point.

76. The graph is a function because every vertical line intersects the graph in at most one point.

77. The graph is not a function because a vertical line intersects the graph in more than one point.

78. The graph is not a function because a vertical line intersects the graph in more than one point.

79. $f(x) = -3x - 2$
$f(-1) = -3(-1) - 2 = 3 - 2 = 1$

80. $g(x) = 6 - 4x$
$g(-3) = 6 - 4(-3) = 6 + 12 = 18$

81. $h(x) = -3$
$h(2) = -3$

82. $F(x) = -10$
$F(-2) = -10$

83. $G(x) = x^2 - 3x + 4$

$G(-2) = (-2)^2 - 3(-2) + 4 = 4 + 6 + 4 = 14$

84. $H(x) = -x^2 + 6x - 1$

$H(4) = -4^2 + 6(4) - 1 = -16 + 24 - 1 = 7$

85. $A(x) = \sqrt{2x + x^2}$

$A(4) = \sqrt{2(4) + 4^2} = \sqrt{8 + 16} = \sqrt{24} = 2\sqrt{6}$

86. $B(x) = \left|\frac{3x+2}{2}\right|$

$$\begin{aligned} B(-4) &= \left|\frac{3(-4)+2}{2}\right| \\ &= \left|\frac{-12+2}{2}\right| \\ &= \left|\frac{-10}{2}\right| \\ &= |-5| \\ &= 5 \end{aligned}$$

87. **(a)** $C(m) = 0.10m + 175$

(b) $C(130) = 0.10(130) + 175 = 13 + 175 = 188$
The cost is $188 for driving 130 miles.

Chapter 10 Test

1. **(a)** $y = x^2 + 4x - 5$
$a = 1, b = 4, c = -5$
Since $a > 0$, the parabola opens up.

(b) $x = -\frac{b}{2a} = -\frac{4}{2(1)} = -2$

$y = (-2)^2 + 4(-2) - 5 = 4 - 8 - 5 = -9$
The vertex is $(-2, -9)$.

(c) The axis of symmetry is $x = -\frac{b}{2a} = -2$.

(d) $b^2 - 4ac = 4^2 - 4(1)(-5) = 16 + 20 = 36$
Since the discriminant > 0, there are two x-intercepts. Let $y = 0$.

$$\begin{aligned} 0 &= x^2 + 4x - 5 \\ 0 &= (x+5)(x-1) \end{aligned}$$

$x + 5 = 0$ or $x - 1 = 0$
$x = -5$ or $x = 1$

The x-intercepts are -5 and 1.
Let $x = 0$.
$y = 0^2 + 4(0) - 5 = -5$
The y-intercept is -5.

2. $y = -x^2 + 6x - 9$
$a = -1, b = 6, c = -9$
Since $a < 0$, the parabola opens down.
$x = -\frac{b}{2a} = -\frac{6}{2(-1)} = 3$

$y = -3^2 + 6(3) - 9 = -9 + 18 - 9 = 0$
The vertex is (3, 0). The axis of symmetry is $x = -\frac{b}{2a} = 3$.

$b^2 - 4ac = 6^2 - 4(-1)(-9) = 36 - 36 = 0$
Since the discriminant = 0, there is one x-intercept. Let $y = 0$.

$$\begin{aligned} 0 &= -x^2 + 6x - 9 \\ 0 &= -(x^2 - 6x + 9) \\ 0 &= -(x-3)^2 \\ x - 3 &= 0 \\ x &= 3 \end{aligned}$$

The x-intercept is 3.
Let $x = 0$.
$y = -0^2 + 6(0) - 9 = -9$
The y-intercept is -9.

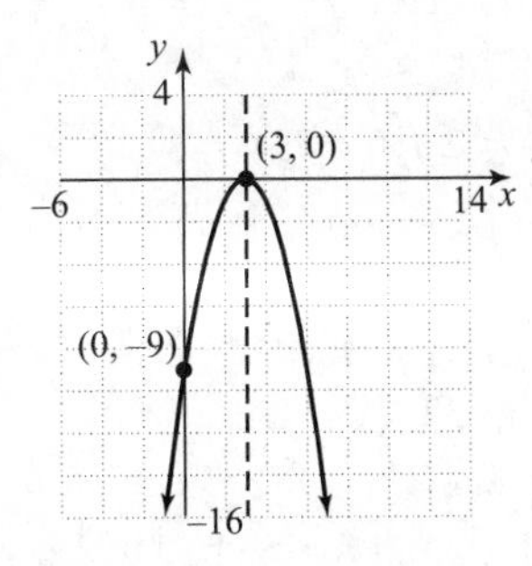

3. $y = x^2 - 4$

$a = 1, b = 0, c = -4$

Since $a > 0$, the parabola opens up.

$x = -\frac{b}{2a} = -\frac{0}{2(1)} = 0$

$y = 0^2 - 4 = -4$

The vertex is (0, −4). The axis of symmetry is

$x = -\frac{b}{2a} = 0.$

$b^2 - 4ac = 0^2 - 4(1)(-4) = 0 + 16 = 16$

Since the discriminant > 0, there are two x-intercepts. Let $y = 0$.

$0 = x^2 - 4$

$0 = (x+2)(x-2)$

$x+2=0$ or $x-2=0$

$x=-2$ or $x=2$

The x-intercepts are −2 and 2.

Let $x = 0$.

$y = 0^2 - 4$

$y = -4$

the y-intercept is −4.

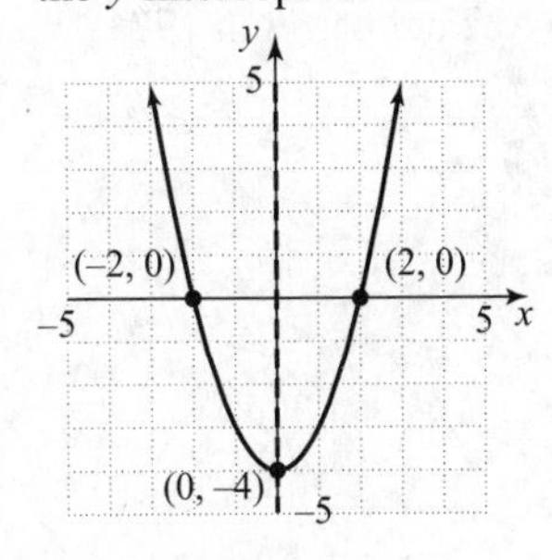

4. $y = x^2 - 4x - 3$

$a = 1, b = -4, c = -3$

Since $a > 0$, the parabola opens up.

$x = -\frac{b}{2a} = -\frac{-4}{2(1)} = 2$

$y = 2^2 - 4(2) - 3 = 4 - 8 - 3 = -7$

The vertex is (2, −7). The axis of symmetry is

$x = -\frac{b}{2a} = 2.$

$b^2 - 4ac = (-4)^2 - 4(1)(-3) = 16 + 12 = 28$

Since the discriminant > 0, there are two x-intercepts.

Let $y = 0$.

$0 = x^2 - 4x - 3$

$0 = (x-4)(x+1)$

$x-4=0$ or $x+1=0$

$x=4$ or $x=-1$

The x-intercepts are −1 and 4.

Let $x = 0$.

$y = 0^2 - 4(0) - 3 = -3$

The y-intercept is −3.

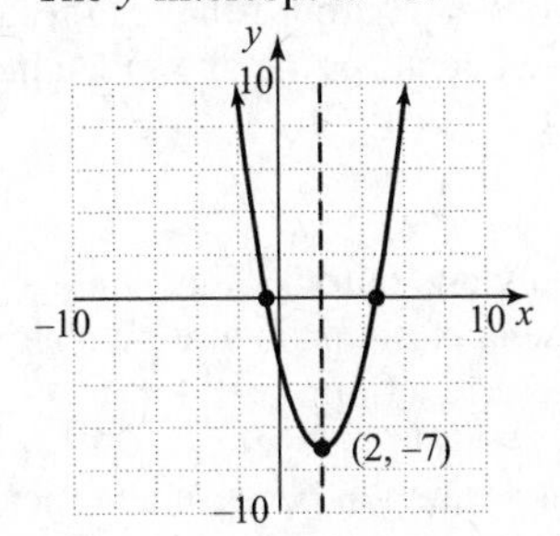

5. (a) $y = -2x^2 - 4x + 5$

$a = -2, b = -4, c = 5$

Since $a < 0$, the parabola opens down. Therefore, it has a maximum.

(b) $x = -\frac{b}{2a} = -\frac{-4}{2(-2)} = -1$

$y = -2(-1)^2 - 4(-1) + 5 = -2 + 4 + 5 = 7$

The maximum is 7.

6. {(1, −1), (1, −2), (1, −3)}

The domain is the set of inputs: {1}

The range is the set of outputs: {−1, −2, −3}.

7. {(−2, a), (−3, b), (−3, c)}

The domain is the set of inputs: {−2, −3}

The range is the set of outputs: {a, b, c}

8. The domain is the set of x-coordinates for which the graph exists:

$\{x | x \text{ is a real number}\}$

The range is the set of y-coordinates for which the graph exists:

$\{y | y \geq -2\}$

9. domain: x-coordinates for which the graph exists: $\{x|-5 \le x \le 3\}$
Range: y-coordinates for which the graph exists: $\{y|0 \le y \le 4\}$

10. $\{(-5, 10), (-5, 9), (-5, 8)\}$
This relation is not a function because all three ordered pairs have the same first coordinate and different second coordinates.

11. $\{(4, -2), (3, -2), (2, -2)\}$
This relation is a function because there are no ordered pairs with the same first coordinate but different second coordinates.

12. $x = \sqrt{y+2}$
Since there is only one output y that can result by performing these operations on any given input, x, the equation is a function.

13. $x = 4 + y$
Since there is only one output y that can result by performing these operations on any given input, x, the equation is a function.

14. The graph is not a function because a vertical line intersects the graph in more than one point.

15. The graph is a function because every vertical line intersects the graph in at most one point.

16. $f(x) = -x^2 + 3x - 1$

$f(-1) = -(-1)^2 + 3(-1) - 1 = -1 - 3 - 1 = -5$

17. $g(x) = \dfrac{-2}{3x+4}$

$g(2) = \dfrac{-2}{3(2)+4} = \dfrac{-2}{6+4} = \dfrac{-2}{10} = -\dfrac{1}{5}$

18. $h(x) = 2\sqrt{x+4}$

$h(12) = 2\sqrt{12+4} = 2\sqrt{16} = 2(4) = 8$

19. (a) $h(5) = -16t^2 + 32t$

$a = -16, b = 32, c = 0$

$t = -\dfrac{b}{2a} = -\dfrac{32}{2(-16)} = -\dfrac{32}{-32} = 1$

It will take 1 second for the ball to reach its maximum height.

(b) Let $t = 1$.

$h(1) = -16(1)^2 + 32(1) = -16 + 32 = 16$

The ball will reach 16 feet.

20. (a) $P(t) = -0.33t^2 + 6.5t + 3$

Let $t = 0$ for 1969.

$P(0) = -0.33(0)^2 + 6.5(0) + 3 = 3$

The price in 1969 was $3 per barrel.

(b) $a = -0.33, b = 6.5, c = 3$

$t = -\dfrac{b}{2a} = -\dfrac{6.5}{2(-0.33)} \approx 9.8$

It reached its maximum in $1969 + 9.8 = 1987.8$.

(c) Let $t = 9.8$.

$P(9.8) = -0.33(9.8)^2 + 6.5(9.8) + 3 \approx 35$

It reached $35 per barrel.

(d) $t = 1974 - 1969 = 5$

$P(5) = -0.33(5)^2 + 6.5(5) + 3 \approx 2.7$

The price is $27 per barrel in 1974.

Appendix B

B.1 Quick Checks

1. If two line segments have the same length, they are said to be <u>congruent</u>.

2. The amount of rotation from one ray to a second ray is called the <u>angle</u> between the rays.

3. An angle that measures 90° is called a <u>right</u> angle.

4. The measure of the angle is between 0° and 90°, so the angle is acute.

5. The measure of the angle is between 90° and 180°, so the angle is obtuse.

6. The measure of the angle is 180°, so the angle is straight.

7. The measure of the angle is 90°, so the angle is right.

8. False; two angles whose measures sum 180° are called supplementary.

9. Two angles are complementary if their sum is 90°. The measure of an angle that is complementary to an angle whose measure is 15° is $90°-15°=75°$.
   Two angles are supplementary if their sum is 180°. The measure of an angle that is supplementary to an angle whose measure is 15° is $180°-15°=165°$.

10. Two angles are complementary if their sum is 90°. The measure of an angle that is complementary to an angle whose measure is 60° is $90°-60°=30°$.
    Two angles are supplementary if their sum is 180°. The measure of an angle that is supplementary to an angle whose measure is 60° is $180°-60°=120°$.

11. <u>Parallel</u> lines are lines in the same plane that never meet.

12. False; while two intersecting lines do form four angles, the vertical angles have equal measures, so they are only supplementary when the lines meet at right angles.

13. True

14. $m\angle 1=180°-40°=140°$ because $\angle 1$ and the 40° angle are supplementary angles.
    $m\angle 2=40°$ because $\angle 2$ and the 40° angle are vertical angles.
    $m\angle 3=180°-40°=140°$ because $\angle 3$ and the 40° angle are supplementary angles.
    $m\angle 4=40°$ because $\angle 4$ and the 40° angle are corresponding angles.
    $m\angle 5=140°$ because $\angle 5$ and $\angle 3$ are alternate interior angles.
    $m\angle 6=40°$ because $\angle 6$ and $\angle 4$ are vertical angles.
    $m\angle 7=140°$ because $\angle 7$ and $\angle 5$ are vertical angles.

B.1 Exercises

15. The measure of the angle is between 0° and 90°, so the angle is acute.

17. The measure of the angle is 90°, so the angle is right.

19. The measure of the angle is 180°, so the angle is straight.

21. The measure of the angle is between 90° and 180°, so the angle is obtuse.

23. Two angles are complementary if their sum is 90°. The measure of an angle that is complementary to an angle whose measure is 32° is $90°-32°=58°$.

25. Two angles are complementary if their sum is 90°. The measure of an angle that is complementary to an angle whose measure is 73° is $90°-73°=17°$.

27. Two angles are supplementary if their sum is 180°. The measure of an angle that is supplementary to an angle whose measure is 67° is $180°-67°=113°$.

29. Two angles are supplementary if their sum is 180°. The measure of an angle that is supplementary to an angle whose measure is 8° is $180°-8°=172°$.

31. $m\angle 1 = 180° - 50° = 130°$ because $\angle 1$ and the 50° angle are supplementary angles.
$m\angle 2 = 50°$ because $\angle 2$ and the 50° angle are vertical angles.
$m\angle 3 = 180° - 50° = 130°$ because $\angle 3$ and the 50° angle are supplementary angles.
$m\angle 4 = 50°$ because $\angle 4$ and the 50° angle are corresponding angles.
$m\angle 5 = 130°$ because $\angle 5$ and $\angle 3$ are alternate interior angles.
$m\angle 6 = 50°$ because $\angle 6$ and $\angle 4$ are vertical angles.
$m\angle 7 = 130°$ because $\angle 7$ and $\angle 5$ are vertical angles.

B.2 Quick Checks

1. A triangle in which two sides are congruent is called an <u>isosceles</u> triangle.

2. A <u>right</u> triangle is a triangle that contains a 90° angle.

3. The sum of the measures of the interior angles in a triangle is <u>180</u> degrees.

4. There are 180° in a triangle and the right angle measures 90° so $m\angle B = 180° - 90° - 20° = 70°$.

5. There are 180° in a triangle so $m\angle B = 180° - 120° - 18° = 42°$.

6. Two triangles are <u>congruent</u> if the corresponding angles have the same measure and the corresponding sides have the same length.

7. Two triangles are <u>similar</u> if corresponding angles of the triangles are equal and the lengths of the corresponding sides are in proportion.

8. Because the triangles are similar, the corresponding sides are proportional. That is, $\frac{3}{7} = \frac{6}{x}$.

$$\frac{3}{7} = \frac{6}{x}$$
$$7x \cdot \left(\frac{3}{7}\right) = 7x \cdot \left(\frac{6}{x}\right)$$
$$3x = 42$$
$$x = 14$$

The missing length is 14 units.

9. A <u>radius</u> of a circle is a line segment drawn from the center of the circle to any point on the circle.

10. True

11. The length of the radius is one-half the length of the diameter.

$$r = \frac{1}{2} \cdot d$$
$$r = \frac{1}{2} \cdot 15 \text{ inches}$$
$$r = \frac{15}{2} \text{ or } 7\frac{1}{2} \text{ inches}$$

The radius of the circle is $7\frac{1}{2}$ inches.

12. The length of the radius is one-half the length of the diameter.

$$r = \frac{1}{2} \cdot d$$
$$r = \frac{1}{2} \cdot 24 \text{ feet}$$
$$r = 12 \text{ feet}$$

The radius of the circle is 12 feet.

13. The length of the diameter is twice the length of the radius.

$$d = 2 \cdot r$$
$$d = 2 \cdot 3.6 \text{ yards}$$
$$d = 7.2 \text{ yards}$$

The diameter of the circle is 7.2 yards.

14. The length of the diameter is twice the length of the radius.

$$d = 2 \cdot r$$
$$d = 2 \cdot 9 \text{ centimeters}$$
$$d = 18 \text{ centimeters}$$

The diameter of the circle is 18 centimeters.

B.2 Exercises

15. There are 180° in a triangle so the measure of the missing angle is $180° - 85° - 40° = 55°$.

17. There are 180° in a triangle and the right angle measures 90° so the measure of the missing angle $180° - 90° - 42° = 48°$.

19. Because the triangles are similar, the corresponding sides are proportional. That is, $\frac{4}{8} = \frac{2}{x}$.

$$\frac{4}{8} = \frac{2}{x}$$
$$8x \cdot \left(\frac{4}{8}\right) = 8x \cdot \left(\frac{2}{x}\right)$$
$$4x = 16$$
$$x = 4$$

The missing length is 4 units.

21. Because the triangles are similar, the corresponding sides are proportional. That is, $\frac{20}{30} = \frac{45}{x}$.

$$\frac{20}{30} = \frac{45}{x}$$
$$30x \cdot \left(\frac{20}{30}\right) = 30x \cdot \left(\frac{45}{x}\right)$$
$$20x = 1350$$
$$x = 67.5$$

The missing length is 67.5 units.

23. The length of the diameter is twice the length of the radius.

$$d = 2 \cdot r$$
$$d = 2 \cdot 5 \text{ inches}$$
$$d = 10 \text{ inches}$$

The diameter of the circle is 10 inches.

25. The length of the diameter is twice the length of the radius.

$$d = 2 \cdot r$$
$$d = 2 \cdot 2.5 \text{ centimeters}$$
$$d = 5 \text{ centimeters}$$

The diameter of the circle is 5 centimeters.

27. The length of the radius is one-half the length of the diameter.

$$r = \frac{1}{2} \cdot d$$
$$r = \frac{1}{2} \cdot 14 \text{ centimeters}$$
$$r = 7 \text{ centimeters}$$

The radius of the circle is 7 centimeters.

29. The length of the radius is one-half the length of the diameter.

$$r = \frac{1}{2} \cdot d$$
$$r = \frac{1}{2} \cdot 11 \text{ yards}$$
$$r = \frac{11}{2} \text{ yards}$$

The radius of the circle is $\frac{11}{2}$ or 5.5 yards.

B.3 Quick Checks

1. The perimeter of a polygon is the distance around the polygon.

2. The area of a polygon is the amount of surface the polygon covers.

3.
$$P = 2l + 2w$$
$$P = 2 \cdot 8 \text{ ft} + 2 \cdot 3 \text{ ft}$$
$$= 16 \text{ ft} + 6 \text{ ft}$$
$$= 22 \text{ ft}$$
$$A = lw$$
$$A = 8 \text{ ft} \cdot 3 \text{ ft}$$
$$= 24 \text{ square ft}$$

The perimeter of the rectangle is 22 feet, and the area is 24 square feet.

4.
$$P = 2l + 2w$$
$$P = 2 \cdot 10 \text{ m} + 2 \cdot 3 \text{ m}$$
$$= 20 \text{ m} + 6 \text{ m}$$
$$= 26 \text{ m}$$
$$A = lw$$
$$A = 10 \text{ m} \cdot 3 \text{ m}$$
$$= 30 \text{ square m}$$

The perimeter of the rectangle is 26 meters, and the area is 30 square meters.

5. False; the area of a square is the square of the length of a side.

6.
$$P = 4 \cdot s$$
$$P = 4 \cdot 4 \text{ cm}$$
$$= 16 \text{ cm}$$
$$A = s^2$$
$$A = (4 \text{ cm})^2$$
$$= 16 \text{ square cm}$$

The perimeter of the square is 16 centimeters, and the area is 16 square centimeters.

7. $P = 4 \cdot s$
$P = 4 \cdot 1.5$ yd
$= 6$ yd
$A = s^2$
$A = (1.5 \text{ yd})^2$
$= 2.25$ square yd
The perimeter of the square is 6 yards, and the area is 2.25 square yards.

8. The perimeter is the distance around the figure.
Perimeter
$= 45 \text{ yd} + 20 \text{ yd} + 20 \text{ yd} + 10 \text{ yd} + 25 \text{ yd} + 10 \text{ yd}$
$= 130$ yd
Area = Area of rectangle + Area of rectangle
$= (45 \text{ yd})(10 \text{ yd}) + (20 \text{ yd})(10 \text{ yd})$
$= 450 \text{ yd}^2 + 200 \text{ yd}^2$
$= 650 \text{ yd}^2$
The perimeter of the figure is 130 yards, and the area is 650 square yards.

9. To find the area of a trapezoid, we use the formula $A = \frac{1}{2}h(b+B)$ where h is the height of the trapezoid and the bases have lengths b and B.

10. $P = 2l + 2w$
$P = 2 \cdot 10 \text{ m} + 2 \cdot 8 \text{ m}$
$= 20 \text{ m} + 16 \text{ m}$
$= 36$ m
$A = b \cdot h$
$A = 10 \text{ m} \cdot 7 \text{ m} = 70$ square m
The perimeter of the parallelogram is 36 meters, and the area is 70 square meters.

11. $P = 5 \text{ yd} + 9 \text{ yd} + 12 \text{ yd} + 7 \text{ yd} = 33$ yd
$A = \frac{1}{2}h(b+B)$
$A = \frac{1}{2} \cdot 6 \text{ yd} \cdot (12 \text{ yd} + 5 \text{ yd})$
$= \frac{1}{2} \cdot 6 \text{ yd} \cdot 17 \text{ yd}$
$= 51$ square yd
The perimeter of the trapezoid is 33 yards, and the area is 51 square yards.

12. True

13. $P = 6 \text{ mm} + 5 \text{ mm} + 8 \text{ mm} = 19$ mm
$A = \frac{1}{2}bh$
$A = \frac{1}{2} \cdot 8 \text{ mm} \cdot 3 \text{ mm} = 12$ square mm
The perimeter of the triangle is 19 millimeters, and the area is 12 square millimeters.

14. $P = 5 \text{ ft} + 12 \text{ ft} + 13 \text{ ft} = 30$ ft
$A = \frac{1}{2}bh$
$A = \frac{1}{2} \cdot 12 \text{ ft} \cdot 5 \text{ ft} = 30$ square ft
The perimeter of the triangle is 30 feet, and the area is 30 square feet.

15. The circumference of a circle is the distance around the circle.

16. False: the area of a circle is given by the formula $A = \pi r^2$.

17. We know the length of the radius, so we use the formula $C = 2\pi r$ to find the circumference.
$C = 2\pi r$
$= 2 \cdot \pi \cdot 4$ ft
$= 8\pi$ ft
≈ 25.13 ft
$A = \pi r^2$
$= \pi \cdot (4 \text{ ft})^2$
$= 16\pi$ square ft
≈ 50.27 square ft
The circumference of the circle is exactly 8π feet or approximately 25.13 feet. The area of the circle is exactly 16π square feet or approximately 50.27 square feet.

18. The length of the diameter is given, so we use the formula $C = \pi d$ to find the circumference.
$C = \pi d$
$= \pi \cdot 24$ cm
$= 24\pi$ cm
≈ 75.40 cm
Since the diameter is 24 centimeters, the radius is 12 centimeters.
$A = \pi r^2$
$= \pi \cdot (12 \text{ cm})^2$
$= 144\pi$ square cm
≈ 452.39 square cm
The circumference of the circle is exactly 24π centimeters or approximately

75.40 centimeters. The area of the circle is exactly 144π square centimeters or approximately 452.39 square centimeters.

B.3 Exercises

19. $P = 2l + 2w$

$P = 2 \cdot 10 \text{ ft} + 2 \cdot 4 \text{ ft}$
$= 20 \text{ ft} + 8 \text{ ft}$
$= 28 \text{ ft}$

$A = lw$

$A = 10 \text{ ft} \cdot 4 \text{ ft}$
$= 40 \text{ square ft}$

The perimeter of the rectangle is 28 feet, and the area is 40 square feet.

21. $P = 2l + 2w$

$P = 2 \cdot 15 \text{ m} + 2 \cdot 5 \text{ m}$
$= 30 \text{ m} + 10 \text{ m}$
$= 40 \text{ m}$

$A = lw$

$A = 15 \text{ m} \cdot 5 \text{ m}$
$= 75 \text{ square m}$

The perimeter of the rectangle is
40 meters, and the area is 75 square meters.

23. $P = 4 \cdot s$

$P = 4 \cdot 6 \text{ km} = 24 \text{ km}$

$A = s^2$

$A = (6 \text{ km})^2 = 36 \text{ square km}$

The perimeter of the square is 24 kilometers, and the area is 36 square kilometers.

25. The perimeter is the distance around the figure.

Perimeter $= 15 \text{ ft} + 6 \text{ ft} + 7 \text{ ft} + 14 \text{ ft} + 22 \text{ ft} + 8 \text{ ft}$
$= 72 \text{ ft}$

Area = Area of rectangle + Area of rectangle
$= (7 \text{ ft})(6 \text{ ft}) + (22 \text{ ft})(8 \text{ ft})$
$= 42 \text{ ft}^2 + 176 \text{ ft}^2$
$= 218 \text{ ft}^2$

The perimeter of the figure is 72 feet, and the area is 218 square feet.

27. The perimeter is the distance around the figure.

Perimeter $= 13 \text{ m} + 6 \text{ m} + 13 \text{ m} + 2 \text{ m} + 8 \text{ m} + 2 \text{ m} + 8 \text{ m} + 2 \text{ m}$
$= 54 \text{ m}$

Area = Area of rectangle + Area of rectangle + Area of rectangle
$= (13 \text{ m})(2 \text{ m}) + (5 \text{ m})(2 \text{ m}) + (13 \text{ m})(2 \text{ m})$
$= 26 \text{ m}^2 + 10 \text{ m}^2 + 26 \text{ m}^2$
$= 62 \text{ m}^2$

The perimeter of the figure is 54 meters, and the area is 62 square meters.

29. $P = 2l + 2w$
$P = 2 \cdot 9 \text{ ft} + 2 \cdot 6 \text{ ft}$
$= 18 \text{ ft} + 12 \text{ ft}$
$= 30 \text{ ft}$
$A = b \cdot h$
$A = 9 \text{ ft} \cdot 5 \text{ ft}$
$= 45 \text{ square ft}$
The perimeter of the parallelogram is 30 feet, and the area is 45 square feet.

31. $P = 2l + 2w$
$P = 2 \cdot 10 \text{ mm} + 2 \cdot 4 \text{ mm}$
$= 20 \text{ mm} + 8 \text{ mm}$
$= 28 \text{ mm}$
$A = b \cdot h$
$A = 4 \text{ mm} \cdot 9 \text{ mm} = 36 \text{ square mm}$
The perimeter of the parallelogram is 28 millimeters, and the area is 36 square millimeters.

33. $P = 8 \text{ in.} + 8 \text{ in.} + 8 \text{ in.} + 16 \text{ in.} = 40 \text{ in.}$
$A = \frac{1}{2}h(b + B)$
$A = \frac{1}{2} \cdot 7 \text{ in.} \cdot (16 \text{ in.} + 8 \text{ in.})$
$= \frac{1}{2} \cdot 7 \text{ in.} \cdot 24 \text{ in.}$
$= 84 \text{ square in.}$
The perimeter of the trapezoid is 40 inches, and the area is 84 square inches.

35. $P = 8 \text{ cm} + 8 \text{ cm} + 10 \text{ cm} + 19 \text{ cm}$
$= 45 \text{ cm}$
$A = \frac{1}{2}h(b + B)$
$A = \frac{1}{2} \cdot 7 \text{ cm} \cdot (19 \text{ m} + 8 \text{ cm})$
$= \frac{1}{2} \cdot 7 \text{ cm} \cdot 27 \text{ cm}$
$= 94.5 \text{ square cm}$
The perimeter of the trapezoid is 45 centimeters, and the area is 94.5 square centimeters.

37. $P = 8 \text{ m} + 12 \text{ m} + 12 \text{ m} = 32 \text{ m}$
$A = \frac{1}{2}bh$
$A = \frac{1}{2} \cdot 12 \text{ m} \cdot 7 \text{ m} = 42 \text{ square m}$
The perimeter of the triangle is 32 meters, and the area is 42 square meters.

39. $P = 11 \text{ ft} + 15 \text{ ft} + 6 \text{ ft} = 32 \text{ ft}$
$A = \frac{1}{2}bh$
$A = \frac{1}{2} \cdot 6 \text{ ft} \cdot 8 \text{ ft} = 24 \text{ square ft}$
The perimeter of the triangle is 32 feet, and the area is 24 square feet.

41. We know the length of the radius, so we use the formula $C = 2\pi r$ to find the circumference.
$C = 2\pi r$
$= 2 \cdot \pi \cdot 16 \text{ in.}$
$= 32\pi \text{ in.}$
$\approx 100.53 \text{ in.}$
$A = \pi r^2$
$= \pi \cdot (16 \text{ in.})^2$
$= 256\pi \text{ square in.}$
$\approx 804.25 \text{ square in.}$
The circumference of the circle is exactly 32π inches or approximately 100.53 inches. The area of the circle is exactly 256π square inches or approximately 804.25 square inches.

43. The length of the diameter is given, so we use the formula $C = \pi d$. Since the diameter is 20 centimeters, the radius is 10 centimeters.
$C = \pi d = \pi \cdot 20 \text{ cm} = 20\pi \text{ cm} \approx 62.83 \text{ cm}$
$A = \pi r^2$
$= \pi \cdot (10 \text{ cm})^2$
$= 100\pi \text{ square cm}$
$\approx 314.16 \text{ square cm}$
The circumference of the circle is exactly 20π centimeters or approximately 62.83 centimeters. The area of the circle is exactly 100π square centimeters or approximately 314.16 square centimeters.

45. Area = Area of Circle $= \pi r^2 = \pi \cdot 1^2 = \pi$
The area of the shaded region is π square units.

47. The distance a wheel travels in one revolution is equal to the circumference of the circle.
$C = \pi d = 20\pi$ inches
In 5 revolutions, the wheel travels 100π inches, which is about 314.16 inches or 26.18 feet.

B.4 Quick Checks

1. A <u>polyhedron</u> is a three-dimensional solid formed by connecting polygons.

2. The <u>surface</u> <u>area</u> of a polyhedron is the sum of the areas of the faces of the polyhedron.

3. False; volume is measured in cubic units.

4. True

5. The figure is a cube.

$$\begin{aligned} V &= s^3 \\ &= (5\text{ m})^3 \\ &= 125\text{ cubic m} \end{aligned}$$

$$\begin{aligned} S &= 6s^2 \\ &= 6(5\text{ m})^2 \\ &= 6(25)\text{ square m} \\ &= 150\text{ square m} \end{aligned}$$

The volume of the cube is 125 cubic meters, and the surface area is 150 square meters.

6. The figure is a sphere.

$$\begin{aligned} V &= \frac{4}{3}\pi r^3 \\ &= \frac{4}{3}\cdot\pi\cdot(4\text{ in.})^3 \\ &= \frac{4}{3}\cdot\pi\cdot(64)\text{ cubic in.} \\ &= \frac{256}{3}\pi\text{ cubic in.} \\ &\approx 268.08\text{ cubic in.} \end{aligned}$$

$$\begin{aligned} S &= 4\pi r^2 \\ &= 4\cdot\pi\cdot(4\text{ in.})^2 \\ &= 4\cdot\pi\cdot(16)\text{ square in.} \\ &= 64\pi\text{ square in.} \\ &\approx 201.06\text{ square in.} \end{aligned}$$

The volume of the sphere is exactly $\frac{256\pi}{3}$ cubic inches or approximately 268.08 cubic inches. The surface area is exactly 64π square inches or approximately 201.06 square inches.

B.4 Exercises

7.

$$\begin{aligned} V &= lwh \\ &= 10\text{ ft}\cdot 5\text{ ft}\cdot 12\text{ ft} \\ &= 600\text{ cubic ft} \end{aligned}$$

$$\begin{aligned} S &= 2lw + 2lh + 2wh \\ &= 2(10\text{ ft})(5\text{ ft}) + 2(10\text{ ft})(12\text{ ft}) + 2(5\text{ ft})(12\text{ ft}) \\ &= 460\text{ square ft} \end{aligned}$$

The volume is 600 cubic feet, and the surface area is 460 square feet.

9.

$$\begin{aligned} V &= \frac{4}{3}\pi r^3 \\ &= \frac{4}{3}\cdot\pi\cdot(6\text{ cm})^3 \\ &= \frac{4}{3}\cdot\pi\cdot(216)\text{ cubic cm} \\ &= 288\pi\text{ cubic cm} \\ &\approx 904.78\text{ cubic cm} \end{aligned}$$

$$\begin{aligned} S &= 4\pi r^2 \\ &= 4\cdot\pi\cdot(6\text{ cm})^2 \\ &= 4\cdot\pi\cdot(36)\text{ square cm} \\ &= 144\pi\text{ square cm} \\ &\approx 452.39\text{ square cm} \end{aligned}$$

The volume is exactly 288π cubic centimeters or approximately 904.78 cubic centimeters. The surface area is exactly 144π square centimeters or approximately 452.39 square centimeters.

11.

$$\begin{aligned} V &= \pi r^2 h \\ &= \pi(2\text{ in.})^2(8\text{ in.}) \\ &= 32\pi\text{ cubic in.} \\ &\approx 100.53\text{ cubic in.} \end{aligned}$$

$$\begin{aligned} S &= 2\pi r^2 + 2\pi rh \\ &= 2\pi(2\text{ in.})^2 + 2\pi(2\text{ in.})(8\text{ in.}) \\ &= (8\pi + 32\pi)\text{ square in.} \\ &= 40\pi\text{ square in.} \\ &\approx 125.66\text{ square in.} \end{aligned}$$

The volume is exactly 32π cubic inches or approximately 100.53 cubic inches. The surface area is exactly 40π square inches or approximately 125.66 square inches.

13.

$$\begin{aligned} V &= \frac{1}{3}\pi r^2 h \\ &= \frac{1}{3}\pi(10\text{ mm})^2(8\text{ mm}) \\ &= \frac{800}{3}\pi\text{ cubic mm} \\ &\approx 837.76\text{ cubic mm} \end{aligned}$$

The volume is exactly $\frac{800}{3}\pi$ cubic millimeters or approximately 837.76 cubic millimeters.

15. $V = \frac{1}{3}b^2h$
$= \frac{1}{3}(8 \text{ ft})^2(10 \text{ ft})$
$= \frac{640}{3} \text{ cubic ft}$
$S = b^2 + 2bs$
$= (8 \text{ ft})^2 + 2(8 \text{ ft})(12 \text{ ft})$
$= 64 + 192 \text{ square ft}$
$= 256 \text{ square ft}$

The volume is $\frac{640}{3}$ cubic feet, and the surface area is 256 square feet.

17. We are looking for the volume of the gutter. Note that 12 feet is 12(12 inches) = 144 inches.
$V = lwh$
$= (144 \text{ in.})(3 \text{ in.})(4 \text{ in.})$
$= 1728 \text{ cubic in.}$
The gutter will hold 1728 cubic inches of water.

19. $V = \pi r^2 h$
$= \pi(2 \text{ in.})^2(6 \text{ in.})$
$= 24\pi \text{ cubic in.}$
$\approx 75.40 \text{ cubic in.}$
$S = 2\pi r^2 + 2\pi rh$
$= 2\pi(2 \text{ in.})^2 + 2\pi(2 \text{ in.})(6 \text{ in.})$
$= (8\pi + 24\pi) \text{ square in.}$
$= 32\pi \text{ square in.}$
$\approx 100.53 \text{ square in.}$
The volume of the can is approximately 75.40 cubic inches. The surface area is approximately 100.53 square inches.

21. We are looking for the volume of the waffle cone.
$V = \frac{1}{3}\pi r^2 h$
$= \frac{1}{3}\pi(4 \text{ cm})^2(16 \text{ cm})$
$= \frac{256}{3}\pi \text{ cubic cm}$
$\approx 268.08 \text{ cubic cm}$
The cone will hold approximately 268.08 cubic centimeters of ice cream.